Algebra

Subsets of the Real Numbers

Natural numbers $= \{1, 2, 3, \ldots\}$

Whole numbers $= \{0, 1, 2, 3, \ldots\}$

Integers $= \{\ldots -3, -2, -1, 0, 1, 2, 3, \ldots\}$

Rational $= \left\{\dfrac{a}{b} \middle| a \text{ and } b \text{ are integers with } b \neq 0\right\}$

Irrational $= \{x \mid x \text{ is not rational}\}$

Properties of the Real Numbers

For all real numbers a, b, and c

$a + b$ and ab are real numbers.	Closure
$a + b = b + a; a \cdot b = b \cdot a$	Commutative
$(a + b) + c = a + (b + c);$ $(ab)c = a(bc)$	Associative
$a(b + c) = ab + ac;$ $a(b - c) = ab - ac$	Distributive
$a + 0 = a; 1 \cdot a = a$	Identity
$a + (-a) = 0; a \cdot \dfrac{1}{a} = 1 \quad (a \neq 0)$	Inverse
$a \cdot 0 = 0$	Multiplication property of 0

Absolute Value

$$|a| = \begin{cases} a & \text{for } a \geq 0 \\ -a & \text{for } a < 0 \end{cases}$$

$\sqrt{x^2} = |x|$ for any real x

$|x| = k \Leftrightarrow x = k \text{ or } x = -k \qquad (k > 0)$

$|x| < k \Leftrightarrow -k < x < k \qquad (k > 0)$

$|x| > k \Leftrightarrow x < -k \text{ or } x > k \qquad (k > 0)$

(The symbol $\Leftrightarrow$ means "if and only if.")

Interval Notation

$(a, b) = \{x \mid a < x < b\}$	$[a, b] = \{x \mid a \leq x \leq b\}$
$(a, b] = \{x \mid a < x \leq b\}$	$[a, b) = \{x \mid a \leq x < b\}$
$(-\infty, a) = \{x \mid x < a\}$	$(a, \infty) = \{x \mid x > a\}$
$(-\infty, a] = \{x \mid x \leq a\}$	$[a, \infty) = \{x \mid x \geq a\}$

Exponents

$a^n = a \cdot a \cdot \cdots \cdot a \ (n \text{ factors of } a)$

$a^0 = 1 \qquad\qquad a^{-n} = \dfrac{1}{a^n}$

$a^r a^s = a^{r+s} \qquad\qquad \dfrac{a^r}{a^s} = a^{r-s}$

$(a^r)^s = a^{rs} \qquad\qquad (ab)^r = a^r b^r$

$\left(\dfrac{a}{b}\right)^r = \dfrac{a^r}{b^r} \qquad\qquad \left(\dfrac{a}{b}\right)^{-r} = \left(\dfrac{b}{a}\right)^r$

Radicals

$a^{1/n} = \sqrt[n]{a} \qquad\qquad a^{m/n} = \left(\sqrt[n]{a}\right)^m = \sqrt[n]{a^m}$

$\sqrt[n]{ab} = \sqrt[n]{a} \cdot \sqrt[n]{b} \qquad \sqrt[n]{\dfrac{a}{b}} = \dfrac{\sqrt[n]{a}}{\sqrt[n]{b}}$

Factoring

$a^2 + 2ab + b^2 = (a + b)^2$

$a^2 - 2ab + b^2 = (a - b)^2$

$a^2 - b^2 = (a + b)(a - b)$

$a^3 - b^3 = (a - b)(a^2 + ab + b^2)$

$a^3 + b^3 = (a + b)(a^2 - ab + b^2)$

Rational Expressions

$\dfrac{ac}{bc} = \dfrac{a}{b} \qquad\qquad \dfrac{a}{b} + \dfrac{c}{d} = \dfrac{ad + bc}{bd}$

$\dfrac{a}{b} \cdot \dfrac{c}{d} = \dfrac{ac}{bd} \qquad\qquad \dfrac{a}{b} \div \dfrac{c}{d} = \dfrac{a}{b} \cdot \dfrac{d}{c}$

Quadratic Formula

The solutions to $ax^2 + bx + c = 0$ with $a \neq 0$ are

$$x = \dfrac{-b \pm \sqrt{b^2 - 4ac}}{2a}.$$

Distance Formula

The distance from (x_1, y_1) to (x_2, y_2), is

$$\sqrt{(x_2 - x_1)^2 + (y_2 - y_1)^2}.$$

Mark Dugopolski
Department of Mathematics
Southeastern Louisiana University
Hammond, LA 70402

Dear Student:

Congratulations! By using this book, you've already increased your chances of success in this course. As you continue thumbing through the next few pages, you will see examples of special features that have been created to help you succeed in this course. If you use this book the way I've intended, your study time will be more productive, and you will be more confident as you take your tests.

Students often ask me, "What use is there for algebra and trigonometry?" As you begin your study, you may be wondering the same thing. To address this question, I've included a wide range of applications that relate to people's everyday lives, applications that show how algebra and trigonometry can be used to solve important problems that we face in our world today. I've taken care with these applications to help you connect the algebraic and trigonometric concepts to something real. Please take a look at the *applications index* that follows the table of contents for a complete list of the diverse topics included. I'm confident you'll find applications relevant to you and your major field of study.

I wish you success with your study of algebra and trigonometry. I am always eager for student feedback, so if you have any comments about this text, please feel free to contact me. Good luck!

Sincerely,

Mark Dugopolski

Mark Dugopolski

College Algebra and Trigonometry

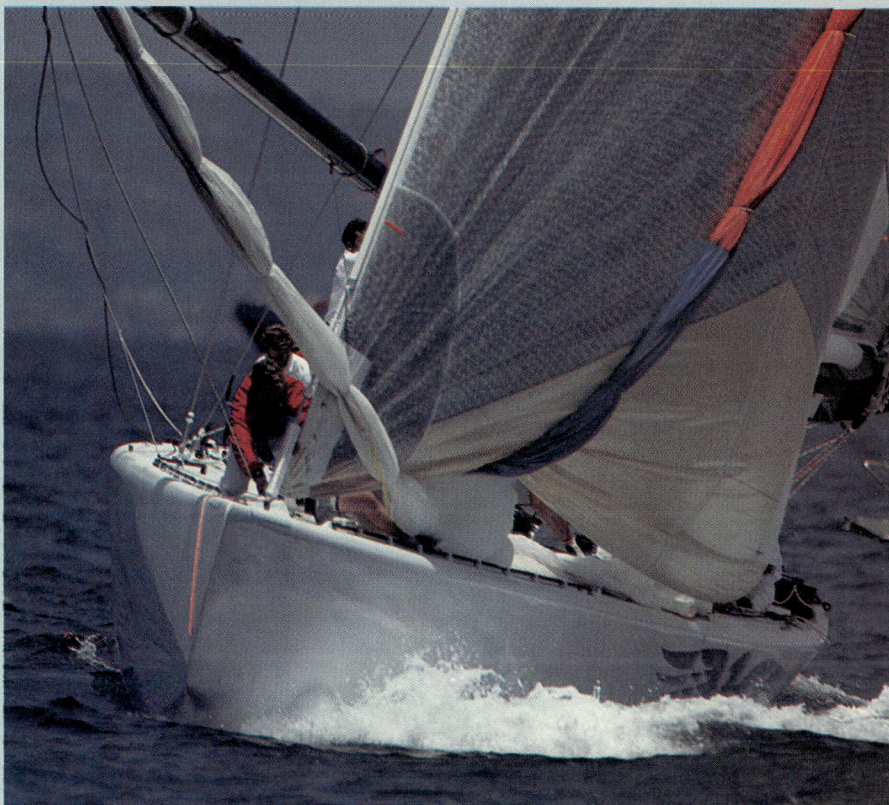

▼

Mark Dugopolski

Southeastern Louisiana University

Addison-Wesley Publishing Company

Reading, Massachusetts • Menlo Park, California • New York
Don Mills, Ontario • Wokingham, England • Amsterdam • Bonn
Sydney • Singapore • Tokyo • Madrid • San Juan • Milan • Paris

Sponsoring Editor	Bill Poole
Executive Editor, Development	Mary Clare McEwing
Managing Editor	Karen Guardino
Art and Design Director	Karen Rappaport
Development Editors	Bobbie Lewis, Barbara Willette
Art Development Supervisor	Meredith Nightingale
Art Development Editor	Claudia Simon
Production Supervisors	Karen Wernholm, Peggy McMahon
Prepress Buying Manager	Sarah McCracken
Technical Art Buyer	Joseph Vetere
Marketing Managers	Andy Fisher, Kate Derrick
Manufacturing Manager	Roy Logan
Text Design/Layout	Martha Podren
Cover Design	Peter Blaiwas
Technical Illustration	Tech Graphics
Reflective Illustration	Jim Bryant, Gary Torrisi, Howard Friedman
Composition	Progressive Information Technologies

About the cover   Between 1989 and 1992, The International America's Cup Committee (IACC) drafted rules setting standards for yachts competing in the America's Cup. The rules established design limits for yacht dimensions, including hull shape and length (L), sail area (S), and displacement (D), as determined by the inequality $(L + 1.25S^{1/2} - 9.8D^{1/3})/0.388 \le$ 42m. For the 1995 competition, the formula was revised to $(L + 1.25S^{1/2} - 9.8D^{1/3})/0.679 \le$ 24m, because the measure 24m (the approximate length of a hull) would be more relevant to lay persons working with the inequality. For the purpose of this text, and ease of use, both sides of the inequality were multiplied by 0.679 to give the result $L + 1.25S^{1/2} - 9.8D^{1/3} \le$ 16.296. (*Scientific American*, vol. 266, no. 3, May 1992, pp. 37–52)

Credits   Cover photograph: J. H. Peterson. Back cover: Computer-generated images © James A. Gretsky. Pages 88, 96: Graph reprinted courtesy of *The Boston Globe*. Pages 248–249: Data from "Technical Preparation of the Airplane *Spirit of St. Louis*," by Donald A. Hall. Technical Note No. 257, National Advisory Committee for Aeronautics, Washington, DC, July 1927. Pages 324–325: Based on "A New Look at Dinosaurs," by John H. Ostrom. *National Geographic*, August 1978, vol. 154, #2, pp. 152–185.

Library of Congress Cataloging-in-Publication Data

Reprinted with corrections, June 1996.
Dugopolski, Mark.
    College algebra and trigonometry / Mark Dugopolski.
        p.   cm.
    "Updated printing."
    Includes index.
    ISBN 0-201-88952-8
    1. Algebra.   2. Trigonometry.   I. Title.
  QA154.2.D83   1996   95-20696
  512′.13—dc20   CIP

4 5 6 7 8 9 10-DOC-97

To the

Instructor

The Vision for the Book

For a mathematics teacher, there's no greater satisfaction than seeing a student master a challenging concept or discover how mathematics relates to everyday life. It makes all of our hard work worthwhile. In fact, those goals are what guided me through the writing of this text. My vision has been to craft a clear, precise, student-oriented text that sets new standards for application, conceptualization, and visualization of college algebra and trigonometry.

Developing Applied Skills

As a teacher, you want your students to understand thoroughly the algebraic and trigonometric concepts they're studying, and to show them how real people use those concepts to solve real problems. You also want to be sure that the book they're using is mathematically precise. In writing this textbook, I've tried to achieve a balance between mathematical rigor and ease of understanding by blending simple but precise mathematical arguments and proofs with relevant, applied examples and exercises. A good example of this feature can be found in Chapter Three's *chapter opening application* that combines discussion of rain forest depletion with statistics from *Scientific American* to help students understand the composition of functions. I then weave this application into examples and exercises later in the chapter.

Another way I develop students' applied skills is through strong exercise sets that test every important concept. To help your students gain confidence with new ideas, I've organized the exercises so they proceed from the routine to the challenging. In addition, since the analysis of real data can bring algebra and trigonometry alive for students, I've based several examples and exercises on real-world situations with real data. Whenever possible, I've cited the actual sources of the data. For your convenience, I've also included an *Index of Applications* that lists all applied examples and exercises, and groups them according to subject matter or student major.

Enhancing Conceptual Understanding

It's a fact: Students in this course often have a difficult time grasping abstract concepts. They get discouraged easily and sometimes don't realize they're having difficulty until they're asked to test their skills. In this text, I help students avoid this scenario with features that get concepts across in manageable pieces. Throughout, you'll find features that spark discussion and help deepen students' understanding of concepts—such as nonthreatening *For Thought* questions. These true-and-false questions allow students to check their grasp of ideas.

Other tools I use to help students measure their grasp of new ideas include both *writing/discussion* and *cooperative learning* exercises. These exercises deepen understanding by getting students to express mathematical concepts in writing or to their classmates during small group or team discussions. Finally, I close each chapter with *Tying It All Together* exercises. These exercises move beyond the scope of traditional cumulative reviews by having students work problems that require them to integrate multiple concepts and skills from several chapters.

Developing Visualization Skills

The adage "a picture is worth a thousand words" rings true with this text. In teaching this course, I've found that students often have difficulty visualizing concepts. In response, I've included art that has been designed to enhance the underlying pedagogy and promote understanding. For example, in the figure shown at left I include *broad shift arrows* to help students visualize movement and change in curves. The shift arrows help to show that translations up and down give graphs of the same shape. I've adopted the convention of always drawing the most important curves in red in the graph art, with the highlighted elements appearing in blue. You'll also notice the use of boxed comments that provide additional information and help clarify the graphical representation. I believe that the use of full color in the art will also help students better understand the basic concepts.

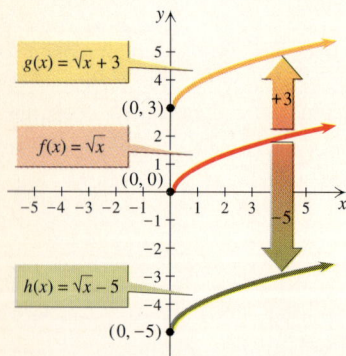

$g(x) = \sqrt{x} + 3$

$f(x) = \sqrt{x}$

$h(x) = \sqrt{x} - 5$

Graphic Calculator (Optional)

Visualization is further enhanced in this text through optional *graphing calculator* materials. Graphing calculator screens are incorporated throughout the text; optional graphing calculator exercises appear at the end of appropriate section exercise sets; and Appendix A, Graphing Calculator Lessons, provides chapter-by-chapter discussions, tips, examples, additional exercises, and even keystroke-level instructions. These materials make use of the power of the graphing calculator to help students solve applications not possible with traditional paper-and-pencil manipulation. Furthermore, many of the graphing calculator exercises extend algebraic ideas toward the calculus.

The graph of $y = 6x^5 + 24x^3$ has only one x-intercept at $(0, 0)$. See Appendix A for more examples.

Text Preview

The three themes of application, conceptualization, and visualization, along with many other features of the text, and the integrated supplements package are highlighted in the special preview that immediately follows this preface.

▼ Acknowledgments

I would like to express my appreciation to the reviewers whose comments and suggestions were invaluable in writing this book:

Victor A. Belfi, *Texas Christian University*
Daphne Bell, *Motlow State Community College*
Donald W. Bellairs, *Grossmont College*
Steven W. Blassberg, *West Valley College*
Robert R. Boltz, *Harrisburg Area Community College*
Jerry Burkhart, *Mankato State University*
Roger Carlson, *Moraine Valley Community College*
Floyd L. Downs, *Arizona State University*
Rebecca Gehrke-Griswold, *Jefferson Community College, SW*
Carolyn A. Goldberg, *Niagara County Community College*
Kenneth Grace, *Anoka-Ramsey Community College*
Mickey Levendusky Hoffman, *Pima Community College*
Carolyn T. Krause, *Delaware State College*
Claire Krukenberg, *Eastern Illinois University*
Anita D. Lesmeister, *University of North Dakota*
Duncan J. Melville, *St. Lawrence University*
Gael Mericle, *Mankato State University*

Ricardo Moena, *University of Cincinnati*
Robert E. Moyer, *Fort Valley State College*
E. James Peake, *Iowa State University*
James E. Peters II, *Weber State University*
Giles Wilson Maloof, *Boise State University*
Deborah J. Ritchie, *Moorpark College*
Kenneth Seydel, *Skyline College*
Gerald M. Smith, *Cayuga Community College*
Jan Vandever, *South Dakota State University*
Stewart Venit, *California State University*
Gerry Vidrine, *Louisiana State University*
Carol McRaney Walker, *Hinds Community College*
Christin Walker, *Utah Valley State College*
Lyndon C. Weberg, *University of Wisconsin*
S. K. Wyckoff, *Brigham Young University*
Mary F. Yorke, *Eastern Michigan University*

I would also like to thank those who took part in focus groups and who shared their ideas and advice: Kathleen J. Bavelas, Manchester Community-Technical College; Jane DeVoe, Northeastern University; Martha DiFazio, Salem State College; Fadia Harik, Wentworth Institute of Technology; Dwight Horan, Wentworth Institute of Technology; Harvey LeBoff, Massachusetts Bay Community College; and Edwin R. Sparks, Southern Connecticut State University.

I also want to express my thanks to Edgar Reyes, Southeastern Louisiana University, for working all exercises and writing the solutions manuals; James Morgan, Southeastern Louisiana University, for working all exercises, proofreading the book, and reviewing video scripts; Ian Gray, Georgia Tech student, for working all the exercises; Becky Muller, Southeastern Louisiana University, for writing the printed test bank; Bob Brown, Southeastern Louisiana University, for accuracy checking the printed test bank; Dale Nauta, Orange Coast College, Burnette Thompson, Texas Southern University, and Lori Vrionis for presenting the video lectures; and John DeValcourt, University of California-Santa Cruz, Zach Nightingale, and Eric Wurzbacher for researching applications.

My thanks go to all my friends at Addison-Wesley for their assistance, encouragement, and direction in this project. In particular, I thank Bill Poole, Karen Wernholm, Meredith Nightingale, Mary Clare McEwing, Bobbie Lewis, Claudia Simon, Kirsten Casteel, Barbara Willette, Kate Derrick, and Sylvia Herrera-Alaniz.

I cannot put into words how much I appreciate the love and encouragement of my wife, Cheryl, and my daughters, Sarah and Alisha. Thank you.

M. D.
Hammond, Louisiana

Text Preview

The following pages highlight the special features of the text and its integrated learning and teaching package.

Chapter opening applications offer real-world problem scenarios to show how relevant mathematics is to solving a wide variety of real problems.

Fast forward to the year 2200. . . A spacecraft streaks toward Earth on its return from the Outer Galaxy. As the astronauts approach the blue planet, they spot a band of vegetation encircling the equator. They recognize this emerald belt as tropical rainforest, but the ragged outline—like a scar on Earth's surface—is an ominous sign. "A little over two centuries ago," remarks the Science Officer, "that green line was practically unbroken. Now look at it. If only people had cared enough to do more

while they still had the chance!"

There may still be time to preserve the rainforests, a lush habitat for life that thrives in a warm, wet climate. Although covering less than an eighth of a percent of Earth's surface, rainforests are home to over half our animal and plant species. Destroying only a percentage of these millions of species could irrevocably change our planet's future. Rainforests are also a source of many common foods, such as chocolate, vanilla, pineapples, and cinnamon, as well as raw products

like latexes, resins, dyes, and lumber. In addition, they provide us with twenty to twenty-five percent of all medicines. (One or more rainforest plants may hold the cure for cancer.)

Why are we destroying this richest of resources? Why do we cut down or burn rainforests in Africa, Asia, and Latin America at a rate of some 25,000 square miles per year? This destruction is rooted in a complex combination of social, economic, and biological causes: Loggers looking for easy profits open up the rain-

Example 4 Composition of functions: rainforest application

Tropical rainforests are one of the earth's most ancient and complex ecosystems, yet each year deforestation claims an area the size of West Virginia. In "Deforestation in the Tropics" (*Scientific American,* April 1990) it is stated that loggers typically harvest only 10% of the area in a rainforest. However, the loggers destroy 50% of the area not harvested just to get at the valuable trees. In a forest of x square kilometers, the function $u = 0.9x$ gives the unharvested area when the valuable timber is removed. The function $w = 0.5u$ gives the amount of unharvested area that is destroyed and wasted. Write the wasted area w as a function of the original area x.

Solution

Since w is a function of u and u is a function of x, w is a function of x. Writing the composition of the two functions is simply a matter of replacing u by $0.9x$ in the formula for w:

$$w = 0.5u = 0.5(0.9x) = 0.45x$$

The composition function $w = 0.45x$ indicates that 45% of the area of a rainforest is wasted when the valuable timber is harvested.

Harvested area

Wasted area: $w = 0.5u$

Unharvested area: $u = 0.9x$

page 215

Mathematical concepts, such as composition of functions, are made lively by linking selected examples and exercises to chapter opening scenarios.

97. *Deforestation* In 1980 the world's tropical moist forest covered approximately 1080 million hectares (1 hectare = 10,000 m²). In 1992 the world's tropical moist forest covered approximately 1006 million hectares. What was the average rate of change of the area of tropical moist forest over those 12 years?

98. *Elimination of tropical moist forest* If the deforestation described in Exercise 97 continues at the same amount per year, then in which year will the tropical moist forest be totally eliminated?

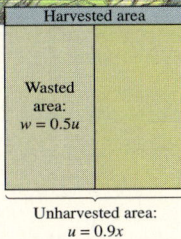

Figure for Exercises 97 and 98

page 184

The notations $C(x)$, $R(x)$, and $P(x)$ are used in business for functions that give the cost, revenue, and profit for the production of x units. The **marginal cost** function $MC(x)$ is the function that gives the change in cost if a production level of x is raised to $x + 1$. $MC(x)$ is equal to the difference quotient for $C(x)$ with $h = 1$. The definitions of **marginal revenue** $MR(x)$ and **marginal profit** $MP(x)$ are similar.

99. *Marginal cost* A dairy's weekly cost in dollars for producing x hundred gallons of milk can be modeled by the function $C(x) = 0.03x^2 + 40x + 6000$. Find $MC(x)$ and simplify it. What is the marginal cost when production is at 10,000 gallons per week?

Business—Marginal Cost

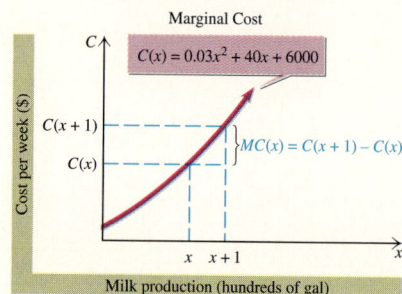

Marginal Cost

$C(x) = 0.03x^2 + 40x + 6000$

$MC(x) = C(x + 1) - C(x)$

Figure for Exercise 99

page 184

85. *Robot gap widens* The gap between industrial robot use in Japan and the United States is widening annually (*Forbes,* April 16, 1990). The equations $J = 30n + 100$ and $U = 4n + 20$ can be used to model the growth of industrial robot use in Japan and the United States. The letter n represents the number of years since 1985, while J and U represent the number of robots in Japan and the United States in thousands, respectively. Graph these two straight lines on the same coordinate system for n between 0 and 20, and see the robot gap for yourself.

page 172

Business—Modeling Production Growth

96. *Lowering cholesterol* According to the National Center for Health Statistics, cholesterol levels have been falling steadily since 1960 (*The Boston Globe,* June 16, 1993). The declining cholesterol levels can be modeled by the function

$$C(t) = \begin{cases} -0.35t + 220 & \text{for } 0 \le t \le 20 \\ -0.80t + 229 & \text{for } t > 20 \end{cases}$$

where $C(t)$ is the average number of milligrams of cholesterol per deciliter of blood in the adult population and t is the number of years since 1960. Graph this function. What was the average cholesterol level in 1980? Will the National Cholesterol Education Program reach its goal of getting average cholesterol under 200 by the year 2000?

Allied Health

page 197

The new International America's Cup Class rules, which took effect in 1989, establish the boundaries of a new class of longer, lighter, and faster yachts. In addition to satisfying 200 pages of other rules, the basic dimensions of any yacht competing for the silver trophy must satisfy the inequality

$$L + 1.25S^{1/2} - 9.8D^{1/3} \le 16.296$$

where L is the length in meters, S is the sail area in square meters, and D is the displacement in cubic meters. (*Scientific American,* May 1992)

123. *America's Cup inequality* Determine whether The New Zealand Challenge (considered a short, light boat), with a length of 20.95 m, a sail area of 277.3 m², and a displacement of 17.56 m³, satisfies the inequality.

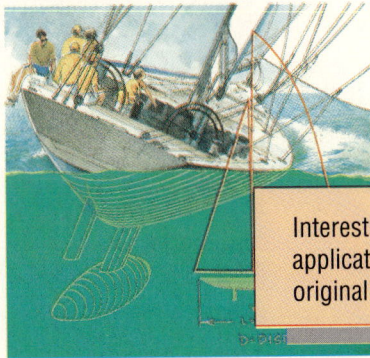

Interesting mathematical applications from original sources

Social Sciences— Demographics

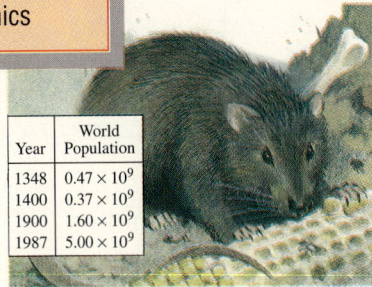

Year	World Population
1348	0.47×10^9
1400	0.37×10^9
1900	1.60×10^9
1987	5.00×10^9

104. *Black Death* Because of the Black Death, or plague, the only substantial period in recorded history when the earth's population was not increasing was from 1348 to 1400. During that period the world population decreased by about 100 million people. Use the exponential model $P = P_0 e^{rt}$ and the data from the accompanying table to find the annual growth rate r_1 for the period 1400 to 1900 and the annual growth rate r_2 for the period 1900 to 1987. Now, assume that the Black Death had not occurred and the growth rate r_1 was in effect for the years 1348 through 1900. Under these assumptions, predict the no-plague population P_1 in 1900. Start with the no-plague population P_1 in 1900 and use the growth rate r_2 to predict a no-plague population P_2 for the year 2000. How much larger is P_2 than the population predicted for the year 2000 in Exercise 103?

Foreshadowing Calculus

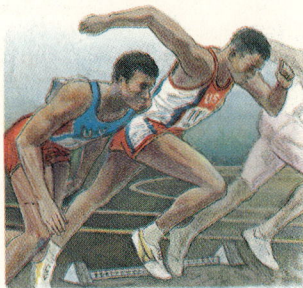

The Average Rate of Change of a Function

At the 1991 World Championship meet in Tokyo, Carl Lewis ran the fastest 100-meter race in the history of sports (*Runner's World,* December 1991). At 40 meters, Lewis's time was 4.77 seconds; at 100 meters, as he crossed the finish line, his time was 9.86 seconds. There is a function here pairing times to locations. Two ordered pairs of this function are (4.77, 40) and (9.86, 100). To calculate Lewis's average speed over the time interval from 4.77 seconds to 9.86 seconds, we find

$$\frac{\text{change in location}}{\text{change in time}} = \frac{100 - 40}{9.86 - 4.77} = \frac{60}{5.09} = 11.79 \text{ m/s.}$$

Lewis did not run at a constant 11.79 meters per second, but 11.79 meters per second was the average rate at which his location changed for the time between 4.77 seconds and 9.86 seconds.

4.77 sec   9.86 sec

Start   40 m   Finish
(0 m)   (100 m

For Thought

True or false? Explain.

1. The function $f(x) = (-2)^x$ is an exponential function.
2. The function $f(x) = 2^x$ is invertible.
3. If $2^x = -\dfrac{1}{8}$, then $x = -3$.
4. If $f(x) = 3^x$, then $f(0.5) = \sqrt{3}$.
5. If $f(x) = e^x$, then $f(0) = 1$.
6. If $f(x) = e^x$ and $f(t) = e^2$, then $t = 2$.
7. The x-axis is a horizontal asymptote for the graph of $y = e^x$.
8. The function $f(x) = (0.5)^x$ is increasing.
9. The functions $f(x) = 4^{x-1}$ and $g(x) = (0.25)^{1-x}$ have the same graph.
10. $2^{1.73} = \sqrt[100]{2^{173}}$

page 336

5.1 Exercises   Tape 9   Disk—

Let $f(x) = 3^x$, $g(x) = 2^{1-x}$, and $h(x) = (1/4)^x$. Find the following values.

1. $f(2)$ 2. $f(4)$ 3. $f(-2)$ 4. $f(-3)$
5. $g(2)$ 6. $g(1)$ 7. $g(-2)$ 8. $g(-3)$
9. $h(-1)$ 10. $h(-2)$ 11. $h(-1/2)$ 12. $h(3/2)$

Sketch the graph of each function by finding at least three ordered pairs on the graph. State the domain, the range, and whether the function is increasing or decreasing.

13. $f(x) = 5^x$ 14. $f(x) = 4^x$ 15. $y = 10^{-x}$
16. $y = e^{-x}$ 17. $f(x) = (1/4)^x$ 18. $f(x) = (0.2)^x$

Use transformations to help you graph each function.

19. $f(x) = 2^x - 3$ 20. $f(x) = 3^{-x} + 1$
21. $f(x) = 2^{x+3} - 5$ 22. $f(x) = 3^{1-x} - 4$

page 337

Either prove or disprove each statement. Use a graph only as a guide. Your proof should rely on algebraic calculations.

73. The points $(-1, 2)$, $(2, -1)$, $(3, 3)$, and $(-2, -2)$ are the vertices of a parallelogram.
74. The points $(-1, 1)$, $(-2, -5)$, $(2, -4)$, and $(3, 2)$ are the vertices of a parallelogram.
75. The points $(-5, -1)$, $(-3, -4)$, $(3, 0)$, and $(1, 3)$ are the vertices of a rectangle.
76. The points $(-5, -1)$, $(1, -4)$, $(4, 2)$, $(-1, 5)$ are the vertices of a square.

page 260

Graphing Calculator Exercises

These non-routine exercises help students visualize and take a fresh look at algebraic concepts.

Graphing Calculator Exercises

Find an approximate solution to each equation by graphing an appropriate function on a graphing calculator and reading from the graph. Round answers to two decimal places.

1. $\ln(x - 2) = 3.2$ **2.** $\log(x) = -\log(x + 2)$

3. $\log(x + 1) = -\ln(x + 2)$ **4.** $\log|x| = x^2 - 4$

Use your graphing calculator to help you sketch the graphs of each group of functions on the same coordinate plane.

5. $y_1 = \log(x)$, $y_2 = \log(10x)$, $y_3 = \log(100x)$

6. $y_1 = \log(x)$, $y_2 = \log(x/10)$, $y_3 = \log(x/100)$

7. $y_1 = \log(x)$, $y_2 = \log(x^3)$, $y_3 = \log(x^5)$

8. $y_1 = \log(x)$, $y_2 = \log(x^{1/3})$, $y_3 = \log(x^{1/5})$

Use the graphs drawn in Exercises 5–8 to answer the following questions.

9. How do the graphs of $y = \log(nx)$ and $y = \log(x/n)$ compare with the graph of $y = \log(x)$ for n a positive integral multiple of 10?

10. How do the graphs of $y = \log(x^n)$ and $y = \log(x^{1/n})$ compare with the graph of $y = \log(x)$ for n a positive odd integer?

11. Experiment with the graphs of $y = \log(x^n)$ and $y = \log(x^{1/n})$ for any nonzero integer n and write a summary of your findings.

page 349

95. *Cooperative learning* Work in a small group to find and simplify the difference quotient for the function $f(x) = x^2$. Evaluate the difference quotient for $h = 0$ and $x = -1, 0, 1, 2$, and 3, and draw the graph of $f(x) = x^2$. Do these values give any indication of how fast the curve is increasing or decreasing at the points with x-coordinates $-1, 0, 1, 2$, and 3? Use the difference quotient to define the concept of ''slope'' of a curve $y = f(x)$ at any point on the curve.

page 262

Writing/Discussion and Cooperative Learning Exercises

These exercises encourage students to think about and discuss important mathematical ideas.

For Writing/Discussion

138. *Egyptian area formula* Surrounded by thousands of square miles of arid desert, ancient Egypt was sustained by the green strips of land bordering the Nile. Villagers cooperated to control the annual flooding so that all might reap abundant harvests from a network of fields and orchards divided by irrigation canals. The Egyptians used the expression

$$\frac{(a + b)(c + d)}{4}$$

to approximate the area of a four-sided field, where a and b represent the lengths of two opposite sides and c and d represent the lengths of the other two opposite sides. Is the formula correct if the field is rectangular? Is it correct for every four-sided figure? Explain why the Egyptians used this formula.

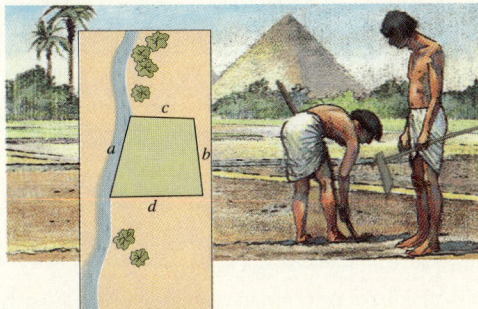

Figure for Exercise 138

page 50

The text offers numerous guides to help students focus their attention and assist them in learning.

Properties of Exponential Functions

The exponential function $f(x) = a^x$ has the following properties:

1. The function f is increasing for $a > 1$ and decreasing for $0 < a < 1$.
2. The y-intercept of the graph of f is $(0, 1)$.
3. The graph has the x-axis as a horizontal asymptote.
4. The range of f is $(0, \infty)$.
5. The function f is one-to-one.

Important procedures, formulas, definitions, properties, and theorems are boxed for quick reference and review.

page 329

Strategy: Problem Solving

1. Read the problem as many times as necessary to get an understanding of the problem.
2. If possible, draw a diagram to illustrate the problem.
3. Choose a variable, write down what it represents, and if possible, represent any other unknown quantities in terms of that variable.
4. Write an equation that models the situation. You may be able to use a known formula, or you may have to write an equation that models only that particular problem.
5. Solve the equation.
6. Check your answer by using it to solve the original problem (not just the equation).
7. Answer the question posed in the original problem.

Strategy boxes help students organize their thoughts, simplify techniques, and work efficiently to solve problems.

page 90

Strategy: Graphing Polynomial Functions

1. Check for symmetry of any type.
2. Find all real zeros of the polynomial function.
3. Determine the behavior at the corresponding x-intercepts.
4. Determine the behavior as $x \to \infty$ and as $x \to -\infty$.
5. Calculate several ordered pairs including the y-intercept to verify your suspicions about the shape of the graph.
6. Draw a smooth curve through the points to make the graph.

page 301

The art has been carefully designed to help students visualize key concepts.

$g(x) = 2\sqrt{x}$

(4, 4)

$\times 2$

(4, 2)

$f(x) = \sqrt{x}$

$\times \frac{1}{2}$

(4, 1)

$h(x) = \frac{1}{2}\sqrt{x}$

page 203

Stretching and Shrinking

$g(x) = x^2 + 2$

(−1, 3)

Translating

(−1, 1)

$f(x) = x^2$

$h(x) = x^2 - 3$

(−1, −2)

page 200

(−2, 4)

$f(x) = x^2$

$g(x) = -x^2$

(−2, −4)

Reflecting

page 199

Summary boxes recap important chapter concepts, and provide students with brief and handy overviews of important ideas.

Summary: Interval Notation for Unbounded Intervals

Set	Interval notation	Type	Graph
$\{x \mid x > a\}$	(a, ∞)	Open	a
$\{x \mid x < a\}$	$(-\infty, a)$	Open	a
$\{x \mid x \geq a\}$	$[a, \infty)$	Half open or half closed	a
$\{x \mid x \leq a\}$	$(-\infty, a]$	Half open or half closed	a
Real numbers	$(-\infty, \infty)$	Open	

page 134

The following excerpts show how all of the features function together to promote understanding over memorization, and to continually review and test all important ideas, such as finding inverse functions.

Detailed, annotated examples reinforce procedures.

Procedure: Finding $f^{-1}(x)$

To find the inverse of a one-to-one function given in f-notation:

1. **Replace $f(x)$ by y.**
2. **Interchange x and y.**
3. **Solve the equation for y.**
4. **Replace y by $f^{-1}(x)$.**
5. **Check that the domain of f is the range of f^{-1} and the range of f is the domain of f^{-1}.**

page 226

Example 6 Finding an inverse function

Find the inverse of the function $f(x) = \dfrac{2x + 1}{x - 3}$.

Solution

In Example 5(a) we showed that f is a one-to-one function. So we can find f^{-1} by interchanging x and y and solving for y:

$$y = \frac{2x + 1}{x - 3} \qquad \text{Replace } f(x) \text{ by } y.$$

$$x = \frac{2y + 1}{y - 3} \qquad \text{Interchange } x \text{ and } y.$$

$$x(y - 3) = 2y + 1 \qquad \text{Solve for } y.$$

$$xy - 3x = 2y + 1$$

$$xy - 2y = 3x + 1$$

$$y(x - 2) = 3x + 1$$

$$y = \frac{3x + 1}{x - 2}$$

page 227

Innovative visual images help clarify concepts.

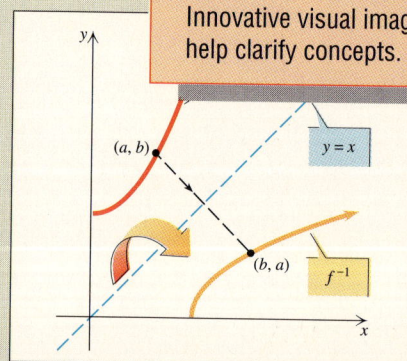

page 228

Text discussions encourage students to think about concepts.

Finding Inverse Functions Mentally

It is no surprise that the inverse of $f(x) = x^2$ for $x \geq 0$ is the function $f^{-1}(x) = \sqrt{x}$. For nonnegative numbers, taking the square root undoes what squaring does. It is also no surprise that the inverse of $f(x) = 3x$ is $f^{-1}(x) = x/3$ or that the inverse of $f(x) = x + 9$ is $f^{-1}(x) = x - 9$. If an invertible function involves a single opera-

Function	Inverse
$f(x) = 2x$	$f^{-1}(x) = x/2$
$f(x) = x - 5$	$f^{-1}(x) = x + 5$

page 229

Find the inverse of each function and graph both f and f^{-1} on the same coordinate plane.

83. $f(x) = 3x + 2$

84. $f(x) = -x - 8$

85. $f(x) = x^2 - 4$ for $x \geq 0$

86. $f(x) = 1 - x^2$ for $x \geq 0$

87. $f(x) = x^3$

88. $f(x) = -x^3$

89. $f(x) = \sqrt{x} - 3$

90. $f(x) = \sqrt{x - 3}$

page 233

For Writing/Discussion

97. Use a geometric argument to prove that if $a \neq b$, then (a, b) and (b, a) lie on a line perpendicular to the line $y = x$ and are equidistant from $y = x$.

98. Why is it difficult to find the inverse of a function such as $f(x) = (x - 3)/(x + 2)$ mentally?

page 233

Exercises, including writing and discussion, test student mastery of concepts.

Highlights

Each chapter ends with a Highlights section, a useful summary of the basic information students should have mastered in that chapter.

Section 3.6 Inverse Functions

1. If a function has no two ordered pairs with different first coordinates and the same second coordinate, then the function is a one-to-one function.

2. It is only one-to-one functions that are invertible.

3. The horizontal line test is a visual method of determining whether a function is one-to-one.

page 243

Chapter 3 Review Exercises

Determine whether each function is invertible. If the function is invertible, then find the inverse function, and state the domain and range of the inverse function.

87. $\{(\pi, 0), (\pi/2, 1), (2\pi/3, 0)\}$

88. $\{(-2, 1/3), (-3, 1/4), (-4, 1/5)\}$

Chapter review exercises following each chapter provide detailed review and prepare students for the Chapter Test.

page 245

Chapter 3 Test

Tying It All Together
Chapters 1–3

Tying It All Together requires integration of multiple concepts and skills.

◆ Supplements

For the Student

Printed Supplements

◆ **Student's Solutions Manual**
ISBN 0-201-62868-6
Edgar Reyes, Southeastern Louisiana University
- Complete solutions to all odd graphical and numerical exercises.
- Ask your bookstore about ordering.

◆ **Graphing Calculator Activities: Exploring Topics in Precalculus**
ISBN 0-201-54216-1
Charles Lund and Edwin E. Anderson
- Workbook format presents graphing calculator explorations of trigonometric functions, conic sections, polar equations, parametric equations, and explorations in data analysis.
- Ask your bookstore about ordering.

Media Supplements

◆ **InterAct Math Tutorial Software**
IBM and Macintosh
Demo IBM 5.25 ISBN 0-201-82507-4; Demo IBM 3.5 ISBN 0-201-82514-7
Demo Macintosh ISBN 0-201-82654-2
IBM 5.25 ISBN 0-201-62872-4; IBM 3.5 ISBN 0-201-62866-X
Macintosh ISBN 0-201-60969-X
- Keyed specifically to text.
- Provides students with a wealth of algorithm-based, randomly generated exercises for added practice.
- Interactive guided solutions involve students in solution process.
- Offers customized remediation for typical wrong answers.
- Offers sophisticated answer recognition capability.
- Can be ordered by mathematics instructors or departments.

◆ **Videotape Series**
Demonstration Video ISBN 0-201-60968-1; Videotape series ISBN 0-201-60007-2
- Keyed specifically to text.
- An engaging team of lecturers provide comprehensive coverage of each section and every topic.
- Utilizes worked-out examples, visual aids, and manipulatives to reinforce concepts.
- Emphasizes the relevance of material to the real world and relates mathematics to students' everyday lives.
- Can be ordered by mathematics instructors or departments.

◆ **Professor Wiessman's Software (IBM)**
- Offers randomly generated practice problems with step-by-step solutions.
- Provides on-disk record keeping.
- Students can receive a free sample by calling Professor Weissman at (718) 698-5219.

◆ **Precalculus Explorer**
IBM 5.25 ISBN 0-201-52909-2; IBM 3.5 ISBN 0-201-52908-4
- Sophisticated computer graphics software.
- Contains 10 programs that enliven the study and exploration of College Algebra and Trigonometry.
- Permits visualization while using an easy menu-driven system with on-line documentation, superior graphics capability, and fast operation.
- Ask your bookstore about ordering.

◆ **Master Grapher 3D Grapher**
Macintosh ISBN 0-201-50859-1
- Graphs functions, polar equations, parametric equations, conic sections, and functions of two variables.
- Allows students to focus on problem solving.
- Encourages generalizations based on geometric evidence.
- Ask your bookstore about ordering.

Supplements

For the Instructor

Printed Supplements

◆ **Instructor's Solutions Manual**
Edgar Reyes, Southeastern Louisiana University
ISBN 0-201-62861-9
- Complete solutions to all graphical and numerical exercises.
- Free to instructors with textbook adoption.

◆ **Instructor's Answer Book**
ISBN 0-201-60774-3
- Complete answers to all exercises, including For Thought questions.
- Free to instructors with textbook adoption.

◆ **Printed Test Bank**
ISBN 0-201-62871-6
- Contains six alternative forms of tests per chapter (two are multiple choice), and answer keys.
- Free to instructors with textbook adoption.

Media Supplements

◆ **OmniTest^3**
IBM and Macintosh
Demonstration Disk
Demo IBM 5.25 ISBN 0-201-51618-7
Demo IBM 3.5 ISBN 0-201-93376-4
Demo Macintosh ISBN 0-201-51619-5
IBM 5.25 ISBN 0-201-62869-4
IBM 3.5 ISBN 0-201-80914-1
Macintosh ISBN 0-201-62870-8
- New, easy to use Windows look-alike testing system.
- Allows the creation of up to 99 perfectly parallel forms of any test.
- New add and edit functions allow customization of items and level of difficulty.
- Allows for variable spacing between items.
- Compatible with SCANTRON and other OMR scanners.
- Contact your Addison-Wesley Sales Representative for more information.

◆ **InterAct Math Tutorial Software**
IBM and Macintosh
Demonstration Disk
Demo IBM 5.25 ISBN 0-201-82507-4
Demo IBM 3.5 ISBN 0-201-82514-7
Demo Macintosh ISBN 0-201-82654-2
IBM 5.25 ISBN 0-201-62872-4
IBM 3.5 ISBN 0-201-62866-X
Macintosh ISBN 0-201-60969-X
- Tracks students' performance.
- Provides scores that can be printed out.
- On-line testing function.
- Sophisticated Course Management System.
- Contact your Addison-Wesley Sales Representative for more information.

Learning Package

3.6 Exercises Tape 7 Disk—5.25″: 2 3.5″: 1 Macintosh: 1

For each function f, find f^{-1}, $f^{-1}(5)$, and $(f^{-1} \circ f)(2)$.

1. $f = \{(2, 1), (3, 5)\}$
2. $f = \{(-1, 5), (0, 0), (2, 6)\}$
3. $f = \{(-3, -3), (0, 5), (2, -7)\}$
4. $f = \{(3.2, 5), (2, 1.99)\}$

Determine whether each function is invertible. If it is invertible, find the inverse.

5. $\{(-1, 0), (0, 0), (1, 0)\}$
6. $\{(0, 0)\}$
7. $\{(3, 0), (2, 5), (4, 6), (7, 9)\}$
8. $\{(-1, -\sqrt{3}), (1, -\sqrt{3})\}$
9. $\{(1, 1), (2, 2), (4.5, 4.5)\}$
10. $\{(-3, 2), (3, 0), (6, 0)\}$
11. $\{(1, 1), (2, 4), (3, 9), (4, 16)\}$
12. $\{(3, 1/3), (4, 1/4), (5, 1/6)\}$
13. $\{(x, y) \mid y = x + 2\}$
14. $\{(x, y) \mid y = 3x - 1\}$
15. $\{(x, y) \mid y = 2x + 7\}$
16. $\{(x, y) \mid y = 1/x^2\}$
17. $\{(x, y) \mid y = |x|\}$
18. $\{(x, y) \mid y = \sqrt{x}\}$

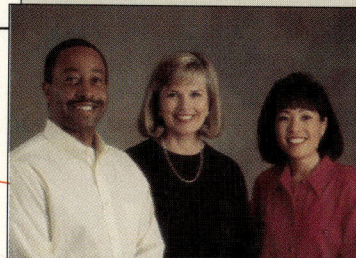

Textbook-Specific Videos

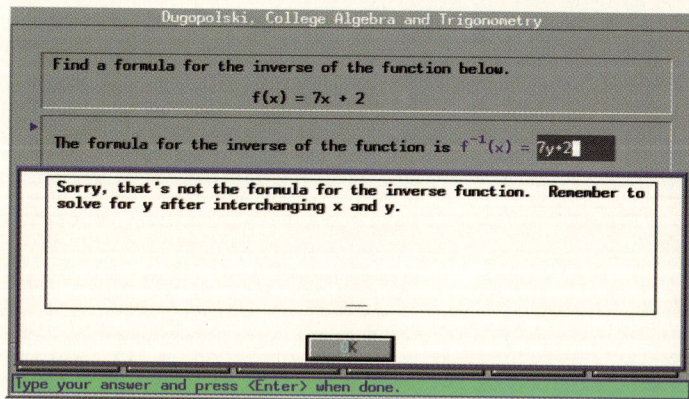

Dugopolski, College Algebra and Trigonometry

Find a formula for the inverse of the function below.

$$f(x) = 7x + 2$$

The formula for the inverse of the function is $f^{-1}(x) = \boxed{7y+2}$

Sorry, that's not the formula for the inverse function. Remember to solve for y after interchanging x and y.

K

Type your answer and press <Enter> when done.

InterAct Math Tutorial Software
• Keyed directly to the text

Teaching Package

23. $y = x^3 - x$

24. $y = \dfrac{1}{x}$

Determine whether each function is one-to-one.

25. $\{(2, 3), (3, 3), (4, 6)\}$

26. $\{(5, \pi), (6, \pi/2), (7, \pi/3)\}$

27. $\{(x, y) \mid y = |x + 1|\}$

28. $\{(x, y) \mid y = x^6\}$

29. $\left\{ (x, y) \mid y = \dfrac{x + 1}{x - 9} \right\}$

30. $\left\{ (x, y) \mid y = 2 + \dfrac{3 - 4x}{x + 2} \right\}$

31. $y = 3 + \dfrac{1 - x}{x - 5}$

32. $y = \dfrac{-3x}{x + 7}$

33. $y = x^2 + 3$

34. $y = 2x^2 - 4$

35. $y = x^2 - x$

36. $f(x) = 5 + \sqrt[3]{x + 9}$

Find the inverse of each function.

37. $f(x) = 3x - 7$

38. $f(x) = -2x + 5$

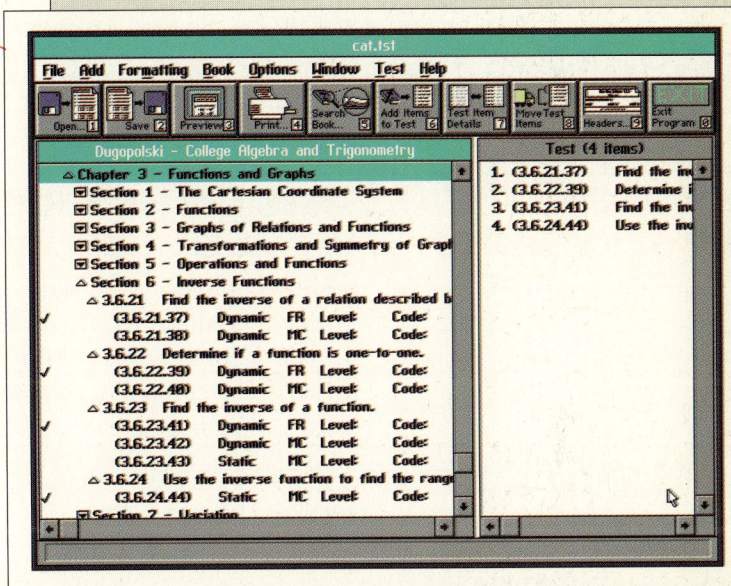

OmniTest³ (IBM and Macintosh)

cat.tst

File Add Formatting Book Options Window Test Help

Open... Save Preview Print... Search Book... Add Items to Test Test Item Details Move Test Items Headers Exit Program

Dugopolski – College Algebra and Trigonometry

Test (4 items)

△ Chapter 3 – Functions and Graphs
- ☑ Section 1 – The Cartesian Coordinate System
- ☑ Section 2 – Functions
- ☑ Section 3 – Graphs of Relations and Functions
- ☑ Section 4 – Transformations and Symmetry of Graphs
- ☑ Section 5 – Operations and Functions
- △ Section 6 – Inverse Functions
  - △ 3.6.21   Find the inverse of a relation described by
    - ✓ (3.6.21.37)   Dynamic   FR   Level: Code:
    - (3.6.21.38)   Dynamic   MC   Level: Code:
  - △ 3.6.22   Determine if a function is one-to-one.
    - (3.6.22.39)   Dynamic   FR   Level: Code:
    - (3.6.22.40)   Dynamic   MC   Level: Code:
  - △ 3.6.23   Find the inverse of a function.
    - ✓ (3.6.23.41)   Dynamic   FR   Level: Code:
    - (3.6.23.42)   Dynamic   MC   Level: Code:
    - (3.6.23.43)   Static   MC   Level: Code:
  - △ 3.6.24   Use the inverse function to find the range
    - ✓ (3.6.24.44)   Static   MC   Level: Code:
- ☑ Section 7 – Variation

1. (3.6.21.37)   Find the inv
2. (3.6.22.39)   Determine i
3. (3.6.23.41)   Find the inv
4. (3.6.24.44)   Use the inv

Contents

Index of Applications

"I will live forever." Believing that although the soul left the body at death it would return to it throughout eternity, the Fourth Dynasty pharaoh Khufu decreed, around the year 2575 B.C., that his subjects build a pyramid tomb at Giza. Upon his death, his body would be preserved there as a permanent shelter for his soul. To ensure that he would live his afterlife in the style to which he had become accustomed, he ordered his mummified remains to be surrounded by all the trappings of royalty.

To fulfill the pharaoh's quest for eternal life, over 4000 Egyptian free-citizens labored at Giza on a typical day. Although the work was back-breaking, many found toiling for the glory of their god-king exhilarating. Work gangs of 18 or more hauled 2 1/2-ton limestone blocks, secured only by palm-fiber ropes, over earthen ramps running up the pyramid's sides. In this way, 2.3 million blocks were dragged, fitted, and pushed into place. Ultimately, the Great Pyramid rose to a height of 481 feet and became known as one of the seven wonders of the ancient world.

How did the Egyptians—without using the wheel, draft animals, or block and tackle—erect an edifice so perfect that its four sides varied by no more than eight inches? In part they were able to do so because the monumental structure's design rested firmly on mathematical principles. In fact, we know from deciphering the

Rhind papyrus, part of a large mathematical text, that the study of algebra originated in Egypt about 4000 years ago.

In this first chapter, we will see how Egyptians used algebra to calculate the volume of pyramids, find the area of a field, and solve recreational word problems. Along the way, we hope that you, like the early Egyptians, will discover the beauty of mathematics and the importance of algebra as a tool in business, science, engineering, and many other fields of study. In preparation for our discussion of algebra, we begin by reviewing real numbers and their properties.

1

Basic

Concepts

of Algebra

1.1

Real Numbers and

Their Properties

In arithmetic we learn facts about the real numbers and how to perform operations with them. Since algebra is an extension of arithmetic, we begin our study of algebra with a discussion of the real numbers and their properties.

The Real Numbers

The most basic set of numbers is the set of **counting** or **natural numbers,** N,

$$\{1, 2, 3, \ldots\}.$$

The natural numbers together with zero form the set of **whole numbers,** W,

$$\{0, 1, 2, 3, \ldots\}.$$

Negative numbers are used to represent losses or debts. The whole numbers together with the negative counting numbers are referred to as the set of **integers,** J,

$$\{\ldots, -3, -2, -1, 0, 1, 2, 3, \ldots\}.$$

The integers can be pictured as points on a line, the **number line.** To draw a number line, draw a line and label any convenient point with the number 0. Now choose a convenient length, one **unit,** and use it to locate evenly spaced points corresponding to the positive integers to the right of zero and the negative integers to the left of zero as shown in Fig. 1.1.

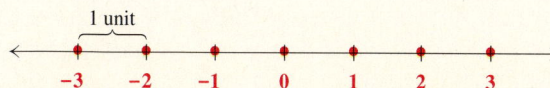

Figure 1.1

1

The numbers corresponding to the points on the line are called the **coordinates** of the points.

The integers and their ratios form the set of **rational numbers, Q.** The rational numbers also correspond to points on the number line. For example, the rational number 1/2 is found halfway between 0 and 1 on the number line. In set notation the set of rational numbers is written as

$$\left\{ \frac{a}{b} \,\middle|\, a \text{ and } b \text{ are integers with } b \neq 0 \right\}.$$

This notation is read ''The set of all numbers a/b such that a and b are integers with b not equal to zero.'' In our set notation we used letters to represent integers. A letter that is used to represent a number is called a **variable.**

There are infinitely many rational numbers located between each pair of consecutive integers, yet there are infinitely many points on the number line that do not correspond to rational numbers. The numbers that correspond to those points are called **irrational** numbers. In decimal notation, the rational numbers are the numbers that are repeating or terminating decimals, and the irrational numbers are the nonrepeating nonterminating decimals. For example, the number 0.595959 . . . is a rational number because the pair 59 repeats indefinitely. By contrast, notice that in the number 5.010010001 . . . , each group of zeros contains one more zero than the previous group. Because no group of digits repeats, 5.010010001 . . . is an irrational number.

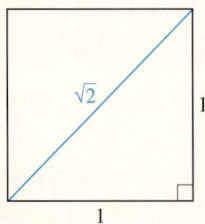

Figure 1.2

Numbers such as $\sqrt{2}$ or π are irrational also. We can visualize $\sqrt{2}$ as the length of the diagonal of a square whose sides are one unit in length. See Fig. 1.2. In any circle, the ratio of the circumference c to the diameter d is π ($\pi = c/d$). See Fig. 1.3. It is difficult to see that numbers like $\sqrt{2}$ and π are irrational because their decimal representations are not apparent. However, the irrationality of π was proven in 1767 by Johann Heinrich Lambert, and it can be shown that the square root of any positive integer that is not a perfect square is irrational. Since a calculator operates with a fixed number of decimal places, it gives us a *rational approximation* for an irrational number such as $\sqrt{2}$ or π.

The set of rational numbers, Q, together with the set of irrational numbers, I, is called the set of **real numbers, R.** The following are examples of real numbers:

$$-3, \quad -0.025, \quad \frac{1}{3}, \quad 0, \quad 0.595959\ldots, \quad \sqrt{2}, \quad \pi, \quad 5.010010001\ldots$$

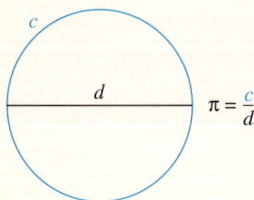

Figure 1.3

These numbers are **graphed** on a number line in Fig. 1.4.

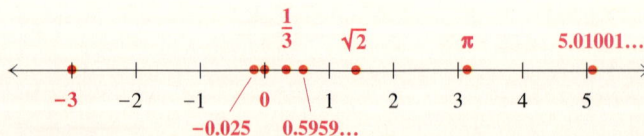

Figure 1.4

Since there is a one-to-one correspondence between the points of the number line and the real numbers, we often refer to a real number as a point. Figure 1.5 shows how the various subsets of the real numbers are related to one another.

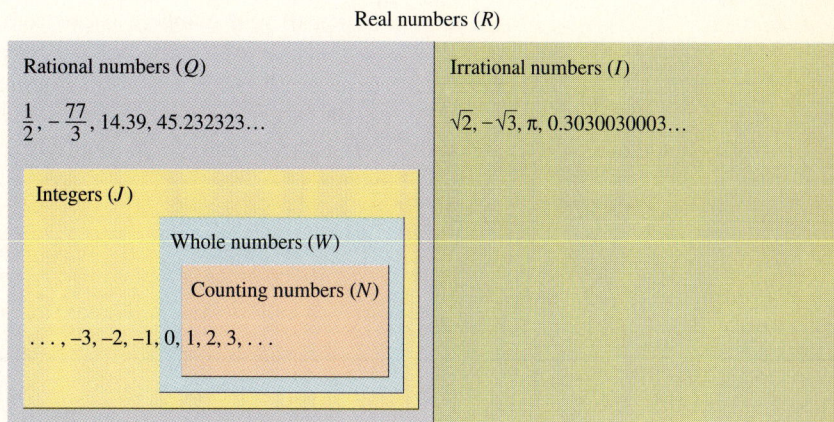

```
π
        3.141592654
√2
        1.414213562
■
```

A graphing calculator gives ten-digit rational approximations for the irrational numbers π and $\sqrt{2}$.

Real numbers (R)

Rational numbers (Q)	Irrational numbers (I)
$\frac{1}{2}, -\frac{77}{3}, 14.39, 45.232323\ldots$	$\sqrt{2}, -\sqrt{3}, \pi, 0.3030030003\ldots$

Integers (J)

Whole numbers (W)

Counting numbers (N)

$\ldots, -3, -2, -1, 0, 1, 2, 3, \ldots$

Figure 1.5

To indicate that a number is a member of a set, we write $a \in A$, which is read ''a is a member of set A.'' We write $a \notin A$ for ''a is not a member of set A.'' Set A is a subset of set B ($A \subseteq B$) means that every member of set A is also a member of set B, and A is not a subset of B ($A \not\subseteq B$) means that there is at least one member of A that is not a member of B.

Example 1   Classifying numbers and sets of numbers

Determine whether each statement is true or false and explain. See Fig. 1.5.

a) $0 \in R$ **b)** $\pi \in Q$ **c)** $R \subseteq Q$ **d)** $I \not\subseteq Q$ **e)** $\sqrt{5} \in Q$

Solution

a) True, because 0 is a member of the set of whole numbers, a subset of the set of real numbers.

b) False, because π is irrational.

c) False, because every irrational number is a member of R but not Q.

d) True, because the irrational numbers and the rational numbers have no numbers in common.

e) False, because the square root of any integer that is not a perfect square is irrational.

Properties of the Real Numbers

In arithmetic we can observe that $3 + 4 = 4 + 3$, $6 + 9 = 9 + 6$, etc. We get the same sum when we add two real numbers in either order. This property of addition of real numbers is the **commutative property.** Using variables, the commutative property of addition is stated as $a + b = b + a$ for any real numbers a and b. There is also a commutative property of multiplication, which is written as $a \cdot b = b \cdot a$ or $ab = ba$. There are many properties concerning the operations of addition and multiplication on the real numbers that are useful in algebra.

Properties of the Real Numbers

For any real numbers a, b, and c:

$a + b$ and ab are real numbers	**Closure property**
$a + b = b + a$ and $ab = ba$	**Commutative properties**
$a + (b + c) = (a + b) + c$ and $a(bc) = (ab)c$	**Associative properties**
$a(b + c) = ab + ac$	**Distributive property**
$0 + a = a$ and $1 \cdot a = a$ (Zero is the **additive identity,** and 1 is the **multiplicative identity.**)	**Identity properties**
$0 \cdot a = 0$	**Multiplication property of zero**
For each real number a, there is a unique real number $-a$ such that $a + (-a) = 0$. ($-a$ is the **additive inverse** of a.)	**Additive inverse property**
For each nonzero real number a, there is a unique real number $1/a$ such that $a \cdot 1/a = 1$. ($1/a$ is the **multiplicative inverse** or **reciprocal** of a.)	**Multiplicative inverse property**

The closure property indicates that the sum and product of any pair of real numbers is a real number. The commutative properties indicate that we can add or multiply in either order and get the same result. Since we can add or multiply only a pair of numbers, the associative properties indicate two different ways to obtain the result when adding or multiplying three numbers. The operations within parentheses are performed first. Because of the commutative property, the distributive property can be used also in the form $(b + c)a = ab + ac$.

Note that the properties stated here involve only addition and multiplication, considered the basic operations of the real numbers. Subtraction and division are defined in terms of addition and multiplication. By definition $a - b = a + (-b)$ and $a \div b = a \cdot 1/b$ for $b \neq 0$. Note that $a - b$ is called the **difference** of a and b and $a \div b$ is called the **quotient** of a and b.

Example 2 Using the properties

Complete each statement using the property named.

a) $a7 = $ _____ , commutative

b) $2x + 4 = $ _____ , distributive

c) $8($ _____ $) = 1$, multiplicative inverse

d) $\dfrac{1}{3}(3x) = $ _____ , associative

Solution

a) $a7 = 7a$

b) $2x + 4 = 2(x + 2)$

c) $8\left(\dfrac{1}{8}\right) = 1$

d) $\dfrac{1}{3}(3x) = \left(\dfrac{1}{3} \cdot 3\right)x$

Opposites and Negatives

The symbol $-$ has three meanings. In $a - b$ it means the operation of subtraction. In -7 it indicates the number negative seven. In the expression $-b$ it means the opposite or additive inverse of b. We read $-b$ as "the opposite of b" because $-b$ is not necessarily a negative number. If b is positive, then $-b$ is negative, but if b is negative, then $-b$ is positive.

Using two "opposite" signs has a cancellation effect. For example, $-(-5) = 5$ and $-(-(-3)) = -3$. Note that the additive inverse of a number can be obtained by multiplying the number by -1. For example, $-1 \cdot 3 = -3$.

We know that $a + b = b + a$ for any real numbers a and b, but is $a - b = b - a$ for any real numbers a and b? In general, $a - b$ is not equal to $b - a$. For example, $7 - 3 = 4$ and $3 - 7 = -4$. So subtraction is not commutative. Since $a - b + b - a = 0$, we can conclude that $a - b$ and $b - a$ are opposites or additive inverses of each other. We summarize these properties of opposites as follows.

Properties of Opposites

For any real numbers a and b:

1. $-1 \cdot a = -a$ (-1 times a is the opposite of a.)
2. $-(-a) = a$ (The opposite of the opposite of a is a.)
3. $-(a - b) = b - a$.   (The opposite of $a - b$ is $b - a$.)

```
-(-2)
                                    2
-(2-3)
                                    1
```

The negative sign (-) is used to indicate opposite or negative. The subtraction sign (−) is used only for subtraction.

Example 3   Using properties of opposites

Use the properties of opposites to complete each equation.

a) $-(-\pi) = $ _____   **b)** $-1(-2) = $ _____   **c)** $-1(x - h) = $ _____

Solution

a) $-(-\pi) = \pi$ **b)** $-1(-2) = -(-2) = 2$

c) $-1(x - h) = -(x - h) = h - x$ ◆

Relations

Symbols such as $<$, $>$, $=$, $\leq$, and $\geq$ are called **relations** because they indicate how numbers are related. We can visualize these relations by using a number line. For example, $\sqrt{2}$ is located to the right of 0 in Fig. 1.4, so $\sqrt{2} > 0$. Since $\sqrt{2}$ is to the left of π in Fig. 1.4, $\sqrt{2} < \pi$. In fact, if a and b are any two real numbers, we say that a is less than b (written $a < b$) provided that a is to the left of b on the number line. We say that a is greater than b (written $a > b$) if a is to the right of b on the number line. We say $a = b$ if a and b correspond to the same point on the number line. The fact that there are only three possibilities for ordering a pair of real numbers is called the **trichotomy property.**

▼

Trichotomy Property

> For any two real numbers a and b, exactly one of the following is true: $a < b$, $a = b$, or $a > b$.

The trichotomy property is very natural to use. For example, if we know that $r = t$ is false, then we can conclude (using the trichotomy property) that either $r > t$ or $r < t$ is true. If we know that $w + 6 > z$ is false, then we can conclude that $w + 6 \leq z$ is true. The following four properties of equality are also very natural to use, and we often use them without even thinking about them.

▼

Properties of Equality

> For any real numbers a, b, and c:
>
> **1.** $a = a$ **Reflexive property**
>
> **2.** If $a = b$, then $b = a$.   **Symmetric property**
>
> **3.** If $a = b$ and $b = c$, then $a = c$.   **Transitive property**
>
> **4.** If $a = b$, then a and b may be substituted   **Substitution property**
> for one another in any expression involving a or b.

The reflexive, symmetric, and transitive properties hold for other relations.

Example 4 Properties of inequality

Determine whether the relation $<$ is reflexive, symmetric, and/or transitive.

Solution

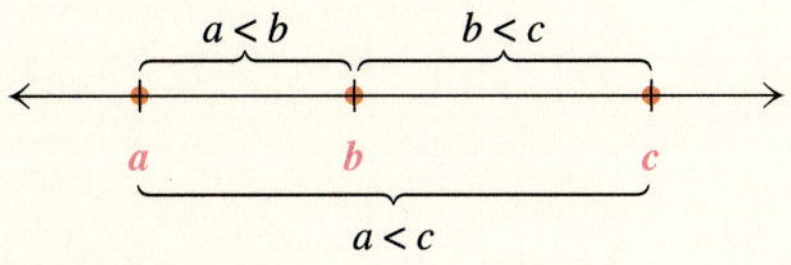

The relation $<$ is reflexive if $a < a$ is true for every real number a. Since $5 < 5$ is incorrect, the relation $<$ is not reflexive. The relation $<$ is symmetric if $a < b$ implies that $b < a$ for every pair of real numbers a and b. Since $7 < 8$ is correct but $8 < 7$ is incorrect, the relation $<$ is not symmetric. If a lies to the left of b and b lies to the left of c as shown in Fig. 1.6, then a lies to the left of c. In symbols, if $a < b$ and $b < c$, then $a < c$. So the relation $<$ is transitive. ◆

In Example 4 we proved that a general statement about all real numbers was false by giving a single example where the statement is incorrect, a **counterexample.**

Figure 1.6

Absolute Value

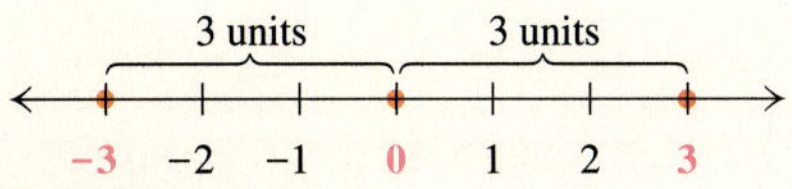

The **absolute value** of a (in symbols, $|a|$) can be thought of as the distance from a to 0 on a number line. Since both 3 and -3 are three units from 0 on a number line as shown in Fig. 1.7, $|3| = 3$ and $|-3| = 3$. A symbolic definition of absolute value is written as follows.

Figure 1.7

Definition: Absolute Value

For any real number a,

$$|a| = \begin{cases} a & \text{if } a \geq 0 \\ -a & \text{if } a < 0. \end{cases}$$

The symbolic definition of absolute value indicates that for $a \geq 0$ we use the equation $|a| = a$ (the absolute value of a is just a). For $a < 0$ we use the equation $|a| = -a$ (the absolute value of a is the opposite of a, a positive number).

Example 5 Using the definition of absolute value

Use the symbolic definition of absolute value to simplify each expression.

a) $|5.6|$ **b)** $|0|$ **c)** $|-3|$

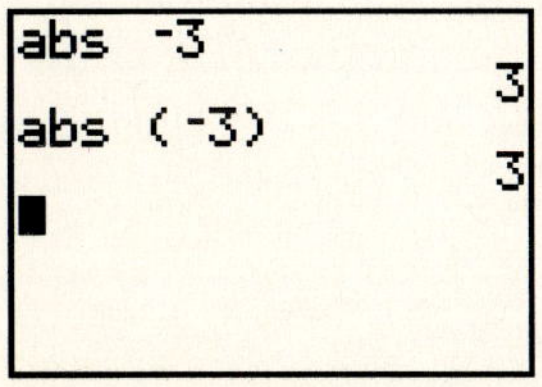

Solution

a) Since $5.6 \geq 0$, we use the equation $|a| = a$ to get $|5.6| = 5.6$.

b) Since $0 \geq 0$, we use the equation $|a| = a$ to get $|0| = 0$.

c) Since $-3 < 0$, we use the equation $|a| = -a$ to get $|-3| = -(-3) = 3$. ◆

The ABS key is used for absolute value. See Appendix A for more details.

The definition of absolute value guarantees that the absolute value of any number is nonnegative. The definition also implies that additive inverses (or opposites) have the same absolute value. So a and $-a$ have the same absolute value. Since $-(a - b) = b - a$, we know that $a - b$ and $b - a$ are additive inverses of each other. So $a - b$ and $b - a$ have the same absolute value. These properties of absolute value and two others are stated as follows.

Properties of Absolute Value

For any real numbers a and b:

1. $|a| \geq 0$ (The absolute value of any number is nonnegative.)

2. $|-a| = |a|$ (Additive inverses have the same absolute value.)

3. $|a - b| = |b - a|$ (Additive inverses have the same absolute value.)

4. $|a \cdot b| = |a| \cdot |b|$ (The absolute value of a product is the product of the absolute values.)

5. $\left| \dfrac{a}{b} \right| = \dfrac{|a|}{|b|}, b \neq 0$ (The absolute value of a quotient is the quotient of the absolute values.)

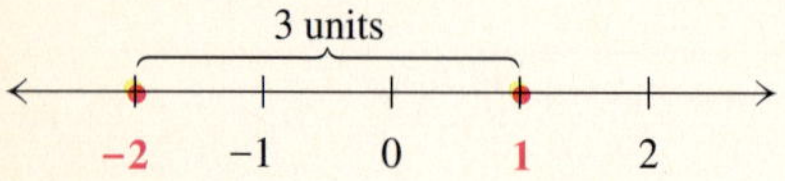

Absolute value is used in finding the distance between points on a number line. Since 9 lies four units to the right of 5, the distance between 5 and 9 is 4. In symbols, $d(5, 9) = 4$. We can obtain 4 by $9 - 5 = 4$ or $|5 - 9| = 4$. In general, $|a - b|$ gives the distance between a and b for any values of a and b. For example, the distance between -2 and 1 in Fig. 1.8 is three units and

$$d(-2, 1) = |-2 - 1| = |-3| = 3.$$

Figure 1.8

Distance Between Two Points on the Number Line

If a and b are any two points on the number line, then the distance between a and b is $|a - b|$. In symbols, $d(a, b) = |a - b|$.

Note that $d(a, 0) = |a - 0| = |a|$, which is consistent with the definition of absolute value of a as the distance between a and 0 on the number line. The property $|a - b| = |b - a|$ indicates that the distance between $-a$ and b is the same as the distance between b and a.

Example 6   Distance between two points on a number line

Find the distance between -3 and 5 on the number line.

Solution

The points corresponding to -3 and 5 are shown on the number line in Fig. 1.9. The distance between these points is found as follows:

$$d(-3, 5) = |-3 - 5| = |-8| = 8$$

Notice that $d(-3, 5) = d(5, -3)$:

$$d(5, -3) = |5 - (-3)| = |8| = 8$$

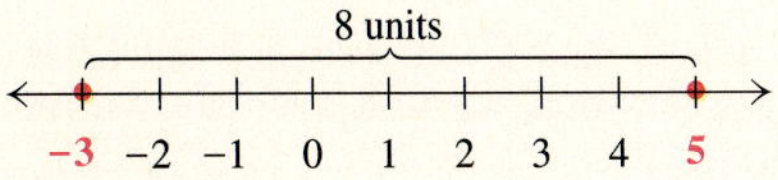

8 units

Figure 1.9

```
abs (-3-5)
              8
abs (5--3)
              8
```

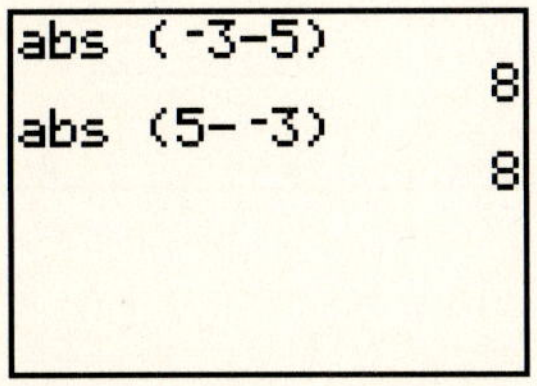

Parentheses must be used for the absolute value of a difference or sum.

Arithmetic Expressions

The result of writing numbers in a meaningful combination with the ordinary operations of arithmetic is called an **arithmetic expression** or simply an expression. Some expressions are

$$\frac{1 + 3}{2 - 5}, \qquad (36 + 8) + 2, \qquad 3 + 5 \cdot 6, \qquad \text{and} \qquad -1(7 - 9).$$

The **value** of an arithmetic expression is the real number obtained when all operations are performed. Symbols such as parentheses, brackets, braces, absolute value bars, and fraction bars are called **grouping symbols.** When no grouping symbols are used, we **evaluate** an expression (find its value) using the accepted order of operations.

The Order of Operations

1. Perform multiplication and division in order from left to right.
2. Perform addition and subtraction in order from left to right.

```
3-4*2
             -5
5*8/4*2
             20
```

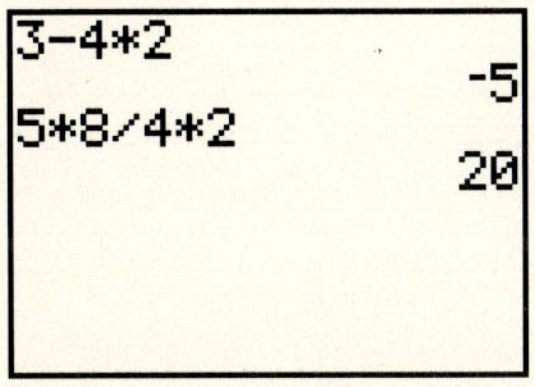

A graphing calculator follows the accepted order of operations. See Appendix A for more examples.

Example 7   Using the order of operations

Evaluate each expression.

a) $3 - 4 \cdot 2$ **b)** $5 \cdot 8 \div 4 \cdot 2$ **c)** $3 - 4 + 9 - 2$

Solution

a) In an expression that has multiplication and subtraction, multiplication is done first. So $3 - 4 \cdot 2 = 3 - 8 = -5$.

b) In an expression that has only multiplication and division, the operations are performed from left to right. So $5 \cdot 8 \div 4 \cdot 2 = 40 \div 4 \cdot 2 = 10 \cdot 2 = 20$.

c) In an expression that has only addition and subtraction, the operations are performed from left to right. So $3 - 4 + 9 - 2 = -1 + 9 - 2 = 8 - 2 = 6..$

The expression $8 \div 4 \cdot 2$ may be written on the display of a graphing calculator as $8/4 * 2$ or $8/4(2)$. Although we would give all three of these expressions the same value, some graphing calculators give the last two expressions different values. If you are using a graphing calculator or any other type, you should experiment with it so that you know the order of operations your calculator is using. When grouping symbols are used in expressions, we first evaluate within grouping symbols, using the order of operations.

Example 8   Using grouping symbols

Evaluate each expression.

a) $\dfrac{3 - 9}{-2 - (-5)}$ b) $2(3 - |5 - 2 \cdot 9|)$ c) $-17 \cdot 3 + (-17) \cdot 7$

Solution

a) Since the fraction bar acts as a grouping symbol, we evaluate the numerator and denominator before dividing:

$$\frac{3 - 9}{-2 - (-5)} = \frac{-6}{3} = -2$$

```
(3-9)/(-2-(-5))
              -2
■
```

The numerator and denominator must be enclosed in parentheses if the fraction is not in built-up form.

b) First evaluate within the innermost grouping symbols, the absolute value symbols:

$$
\begin{aligned}
2(3 - |5 - 2 \cdot 9|) &= 2(3 - |5 - 18|) &&\text{Multiply before subtracting.} \\
&= 2(3 - |-13|) &&\text{Subtract within absolute value.} \\
&= 2(3 - 13) &&\text{Evaluate absolute value.} \\
&= 2(-10) &&\text{Subtract within parentheses.} \\
&= -20
\end{aligned}
$$

```
2(3-abs (5-2*9))
             -20
■
```

The multiplication symbol is not necessary between a number and parentheses.

c) Using the order of operations, we would find each product before we add, but we can also use the distributive property to make the computation easier:

$$-17 \cdot 3 + (-17) \cdot 7 = -17(3 + 7) = -17(10) = -170 \quad \blacklozenge$$

Algebraic Expressions

When we write numbers and one or more variables in a meaningful combination with the ordinary operations of arithmetic, the result is called an **algebraic expression,** or simply an expression. Some examples are

$$7x, \qquad \frac{1}{x}, \qquad \frac{a^2 - b^2}{2}, \qquad \sqrt{x}, \qquad \text{and} \qquad 3x + 5x.$$

The **value of an algebraic expression** is the value of the arithmetic expression that is obtained when the variables are replaced by real numbers. Of course, we are not allowed to replace a variable by a number that will result in an expression that does not represent a real number. The **domain** of an algebraic expression is the set of real numbers that *are* allowed to be used for the variable. For example, the domain of $1/x$ is the set of nonzero real numbers. Zero is excluded because division by 0 is undefined. The domain of $\sqrt{x}$ is the set of nonnegative real numbers. The negative real numbers are excluded because the square root of a negative number is not a real number. Two algebraic expressions are **equivalent** if they have the same domain and if they have the same value for each x-value in the domain.

To **simplify** an expression means to find a simpler-looking equivalent expression. To simplify an expression such as $-1(7x)$ we use the associative property as follows:

$$-1(7x) = (-1 \cdot 7)x = -7x$$

Because the associative property is true for all real numbers, the equation $-1(7x) = -7x$ is true for any real number x. Therefore the expressions $-1(7x)$ and $-7x$ are equivalent.

An expression that is the product of a number and one or more variables raised to powers is called a **term.** Expressions such as $3x$, $2kab^3$, and πr^2 are terms. In the term $3x$, 3 and x are called **factors.** The **coefficient** of any variable part of a term is the product of the remaining factors in the term. For example, the coefficient of x in $3x$ is 3. The coefficient of ab^3 in $2kab^3$ is $2k$ and the coefficient of b^3 is $2ka$. The coefficient of x in the equation $y = mx + b$ is m. If two terms contain the same variables with the same exponents, then they are called **like terms.** It is the distributive property that allows us to **combine like terms:**

$$-4x + 7x = (-4 + 7)x \qquad \textcolor{blue}{\textbf{Distributive property}}$$
$$= 3x$$

Note that $-4x + 7x = 3x$ is true for any real number x. The expressions $-4x + 7x$ and $3x$ are equivalent.

The distributive property is used along with other properties to simplify expressions that have parentheses. We do not usually write down every step to simplify algebraic expressions, but we must be aware that we cannot take a step that is not justified by a property.

Example 9   Using properties to simplify an expression

Simplify each expression.

a) $-4x - (6 - 7x)$ \qquad\qquad **b)** $\dfrac{1}{2}x - \dfrac{3}{4}x$

c) $-6(x - 3) - 3(5 - 7x)$

Solution

a) $-4x - (6 - 7x) = -4x + [-(6 + (-7x))]$ Definition of subtraction

$\qquad = -4x + [-1(6 + (-7x))]$ First property of opposites

$\qquad = -4x + [(-6) + 7x]$ Distributive property

$\qquad = [-4x + 7x] + (-6)$ Commutative and associative properties

$\qquad = 3x - 6$ Combine like terms.

b) $\dfrac{1}{2}x - \dfrac{3}{4}x = \dfrac{2}{4}x - \dfrac{3}{4}x = -\dfrac{1}{4}x$ Write $\dfrac{1}{2}$ as $\dfrac{2}{4}$ to obtain a common denominator.

c) $-6(x - 3) - 3(5 - 7x) = -6x + 18 - 15 + 21x$

$\qquad = 15x + 3$ ◆

It is often necessary to replace the variables in an expression with numbers and evaluate the expression.

Example 10   Evaluating an expression

The expression

$$\frac{y_2 - y_1}{x_2 - x_1}$$

is used to find the slope of a line. (We will explore this concept in Section 4.1.) Find the value when $x_1 = -3$, $x_2 = 5$, $y_1 = 6$, and $y_2 = -1$.

Solution

Use the substitution property to replace the variables by the appropriate numbers:

$$\frac{y_2 - y_1}{x_2 - x_1} = \frac{-1 - 6}{5 - (-3)} = \frac{-7}{8} = -\frac{7}{8}$$ ◆

? For Thought

True or false? Explain.

1. Zero is the only number that is both rational and irrational.

2. Between any two distinct rational numbers there is another rational number.

3. Between any two distinct real numbers there is an irrational number.

4. Every real number has a multiplicative inverse.

5. If a is not less than and not equal to 3, then a is greater than 3.

6. If $a \le w$ and $w \le z$, then $a < z$.

7. For any real numbers a, b, and c, $a - (b - c) = (a - b) - c$.

8. If a and b are any two real numbers, then the distance between a and b on the number line is $a - b$.

9. If x is any real number, then $-3x + (-4x) = 7x$.

10. For any real numbers a and b, the opposite of $a + b$ is $a - b$.

1.1 Exercises   ⬛ Tape 1   💾 Disk—5.25": 1   3.5": 1   Macintosh: 1

Determine whether each statement is true or false. Explain. Sets are denoted by R for the real numbers, I for the irrational numbers, Q for the rational numbers, J for the integers, W for the whole numbers, and N for the natural numbers.

1. $\sqrt{2} \in R$ **2.** $\sqrt{3} \in Q$ **3.** $0 \notin I$

4. $-6 \notin J$ **5.** $J \subseteq R$ **6.** $I \subseteq Q$

7. $R \not\subseteq Q$ **8.** $N \not\subseteq W$

9. $\{\pi\} \subseteq I$ **10.** $\left\{\frac{1}{2}, -\frac{1}{2}\right\} \subseteq Q$

Determine which elements of the set
$\{-3.5, -\sqrt{2}, -1, 0, 1, \sqrt{3}, 3.14, \pi, 4.3535\ldots,$
$$5.090090009\ldots\}$$
are members of the following sets.

11. Real numbers   **12.** Rational numbers

13. Irrational numbers   **14.** Integers

15. Whole numbers   **16.** Natural numbers

Complete each statement using the property named.

17. $7 + x = $ _____, commutative

18. $5(4y) = $ _____, associative

19. $5(x + 3) = $ _____, distributive

20. $-3(x - 4) = $ _____, distributive

21. $\frac{1}{2}x + \frac{1}{2} = $ _____, distributive

22. $-5x + 10 = $ _____, distributive

23. $-13 + (4 + x) = $ _____, associative

24. $yx = $ _____, commutative

25. $0.125($_____$) = 1$, multiplicative inverse

26. $-(-3) + ($_____$) = 0$, additive inverse

Find the multiplicative inverse for each of the following real numbers.

27. $\frac{5}{7}$ **28.** -4.2 **29.** -1 **30.** 1

31. 0.6 **32.** $0.333333\ldots$

33. $0.666666\ldots$ **34.** 0.000001

Use the properties of opposites to complete each equation.

35. $-(-\sqrt{3}) = $ _____

36. $-1(-6.4) = $ _____

37. $-1(x^2 - y^2) = $ _____

38. $-(1 - a^2) = $ _____

Use the symbolic definition of absolute value to simplify each expression.

39. $|7.2|$ **40.** $|0/3|$

41. $|-\sqrt{5}|$ **42.** $|-3/4|$

Find the distance on the number line between each pair of numbers.

43. $8, 13.5$ **44.** $-7.5, 0$

45. $-5, 17$ **46.** $-3.2, -14.9$

Evaluate each expression, using the order of operations.

47. $4 \div 5 \cdot 6 \cdot 3$ **48.** $4 \cdot 3 - 7(-6)$

49. $3 - 4 + 5 - 7 \cdot 6$ **50.** $64 \cdot 2 \div 8 \div 4 \cdot 6$

51. $(52 + 39) + (-39)$ **52.** $(7 - 15) - 13$

53. $49 \cdot 6 + 49 \cdot 4$ **54.** $-12 \cdot 39 + 2 \cdot 39$

55. $26 \cdot \frac{1}{5} \div \frac{1}{2} \cdot 5$ **56.** $\frac{4}{3} \cdot 50(0.75)2$

57. $|4| - |-3|$ **58.** $-8 - |-9|$

59. $|3 - 4.6| - 5$ **60.** $5 - |4 - 2.3|$

61. $(12 - 1)(12 + 1)$ **62.** $-2 - 3(5 - 2 \cdot 8)$

63. $2 - 3|3 - 4 \cdot 6|$ **64.** $1 - (3 - |1 - 2 \cdot 3|)$

65. $(2 - 8 \div 2)(8 \div 2 + 3)$ **66.** $2(2 - |-3| + 1)$

67. $\dfrac{-2 - (-6)}{-5 - (-9)}$ **68.** $\dfrac{4 - (-3)}{-3 - (-1)}$

69. $\dfrac{\dfrac{2-8}{3}-5}{\dfrac{1}{4-2}+1}$

70. $\dfrac{\dfrac{-4-6}{-2}+\dfrac{-3}{6}}{\dfrac{-2}{3-5}+\dfrac{2-6}{2}}$

71. $(598+432)(-3\cdot9+27)$

72. $(23\cdot5+99)(-3\cdot17+51)$

73. $3.45-2\{4.6+2.3(-3.9)\}$

74. $6.7-\{9.8+3.2[7.6+4.1(5.8-9)]\}$

75. $45\div3\cdot5-3\cdot8\div(0.6)$

76. $6.7-9.8+3.2\cdot7.6+4.1\cdot5.8-9$

Use the properties of the real numbers to simplify each expression.

77. $-5x+3x$ **78.** $-5x-(-8x)$ **79.** $x-0.15x$

80. $x+3-0.9x$ **81.** $-3(2xy)$ **82.** $\dfrac{1}{2}(8wz)$

83. $-8w(3x)$ **84.** $\dfrac{1}{2}x(-3y)$

85. $(3-4x)+(x-9)$ **86.** $(9x-3)+(4-6x)$

87. $x-0.03(x+200)$ **88.** $y-0.9(y-3000)$

89. $3-4zy-(5-6zy)$ **90.** $4x-3z-(3x-5z)$

91. $-2(4-x)-3(3-3x)$ **92.** $5(4-2x)-2(x-5)$

93. $\dfrac{1}{3}x+\dfrac{1}{4}x$ **94.** $\dfrac{1}{3}y-\dfrac{1}{2}y$ **95.** $5x\left(\dfrac{3y}{20}\right)$

96. $-\dfrac{3}{4}z\left(\dfrac{7}{9}wt\right)$ **97.** $\dfrac{1}{2}(6-4x)$ **98.** $\dfrac{1}{4}(8x-4)$

99. $\dfrac{6x-2y}{2}$

100. $\dfrac{-9-6x}{-3}$

101. $-3\left(\dfrac{1}{2}-\dfrac{1}{4}y\right)+6\left(\dfrac{3}{4}-\dfrac{1}{3}y\right)$

102. $\dfrac{1}{3}\left(\dfrac{3}{5}x-\dfrac{1}{3}\right)-\dfrac{1}{2}\left(\dfrac{1}{3}x-\dfrac{2}{5}\right)$

103. $\dfrac{1}{2}\left(x-\dfrac{2}{5}x\right)-\dfrac{1}{3}\left(x+\dfrac{2}{3}x\right)-\dfrac{1}{4}\left(x-\dfrac{2}{3}x\right)$

104. $-\dfrac{1}{2}\left(x-\dfrac{2}{3}\right)-\dfrac{1}{3}\left(x-\dfrac{1}{3}x\right)$

Evaluate $\dfrac{y_2-y_1}{x_2-x_1}$ in each of the following cases.

105. $x_1=-2,\ x_2=3,\ y_1=5,\ y_2=-7$

106. $x_1=3,\ x_2=-4,\ y_1=6,\ y_2=-1$

107. $x_1=-8,\ x_2=5,\ y_1=3,\ y_2=3$

108. $x_1=3.9,\ x_2=4,\ y_1=5,\ y_2=8$

Solve each problem.

109. Find a rational number between $\dfrac{13}{25}$ and 0.5197.

110. Find an irrational number between 0.2316 and 0.2317.

111. Arrange the following numbers in order from smallest to largest: $\dfrac{1}{2},\ -\dfrac{1}{2},\dfrac{1}{3},\ -\dfrac{1}{3},\ 0,\ \dfrac{5}{12},$ and $-\dfrac{5}{12}$.

112. Use a calculator to help you arrange the following numbers in order from smallest to largest: $\dfrac{10}{3},\ \sqrt{10},\ \dfrac{22}{7},\ \pi,$ and $\dfrac{157}{50}$.

Volume of Stock Traded on New York Stock Exchange

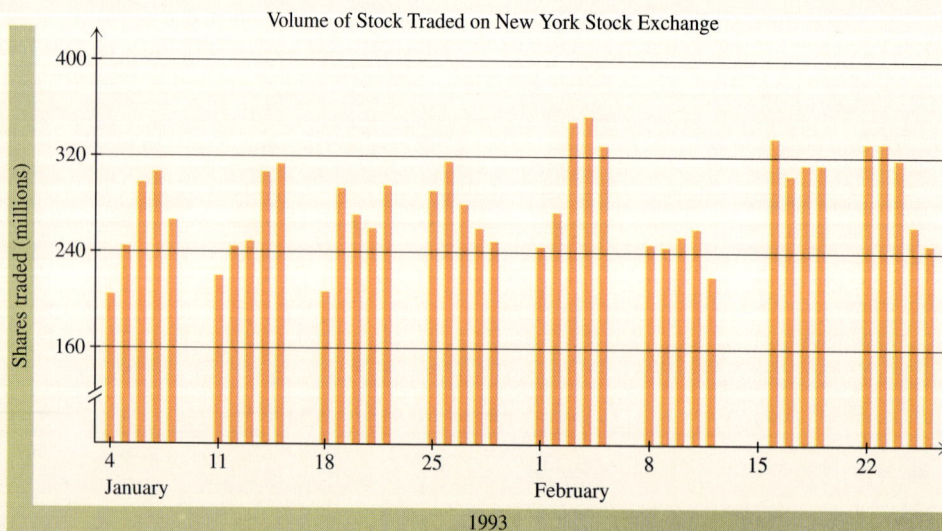

Figure for Exercises 113 and 114

113. *Volume of stock* The bar graph shows the volume of stock traded on the New York Stock Exchange for each trading day in January and February of 1993 *(New York Times)*. For how many days was the volume greater than 320 million shares?

114. *Maximum daily volume* Estimate the maximum daily volume of stock traded for the time period shown in the bar graph and give the date on which it occurred. For which week would you say that the volume was increasing all week?

115. *Estimating costs* The accompanying graph shows the estimated cost E and the actual cost A for repairs done on five cars at Bill Poole Ford. Bill does not want the actual cost to differ from the estimated cost by more than 10% of the estimated cost. For which cars is $|A - E|$ greater than 10% of E?

116. *Free wax* Bill Poole Ford does a free wax job on any car for which $|E - A| > 20$. Which of the cars listed in the accompanying graph get a free wax job?

Figure for Exercises 115 and 116

117. *Modern American trick* Think of a number between 1 and 10. Think of the product of your number and 9. Think of the sum of the digits in your answer. Think of the number that is 5 less than that sum. Think of the letter in the alphabet that corresponds to the number you are thinking about. (For example, if your number is seven, pick the seventh letter in the alphabet.) Think of a state that begins with the letter. Think of the second letter in the state. Think of a big animal that begins with that letter. Think of the color of that animal. Now check the answer in the back of this text to see what you are thinking about. Explain this trick.

118. *Ancient Egyptian trick* The following problem taken from the Rhind papyrus is believed to have originated in Egypt somewhere between 1849 and 1801 B.C.

Think of a number, and add 2/3 of this number to itself. From this sum, subtract 1/3 of its value and say what your answer is. From your answer subtract 1/10 of your answer. You will now have your original number. Try this procedure for yourself (with a multiple of 9 for simplicity). Verify that the following expression is what is described in the problem, then simplify the expression:

$$\left(x + \frac{2}{3}x\right) - \frac{1}{3}\left(x + \frac{2}{3}x\right)$$
$$- \frac{1}{10}\left\{\left(x + \frac{2}{3}x\right) - \frac{1}{3}\left(x + \frac{2}{3}x\right)\right\}$$

Figure for Exercise 118

For Writing/Discussion

119. *Number line* To make a number line, we arbitrarily select a length for one unit and a starting point 0. We use the one-unit length to locate points for the integers. Explain how the figure shown here locates the point corresponding to $\sqrt{2}$.

Figure for Exercise 119

120. *Locating points* Explain how you can use geometry to locate a point on a number line corresponding to $\sqrt{3}$.

121. *A relation* Determine whether the relation $>$ is reflexive, symmetric, and/or transitive and explain your answers.

122. *Another relation* Determine whether the relation $\geq$ is reflexive, symmetric, and/or transitive and explain your answers.

123. *Betazoid chocolates* On their fifth anniversary aboard the starship *Enterprise,* Counselor Troi presented Captain Picard with a box of Betazoid chocolates. The Captain promised to share the delicacies equally with First Officer Riker and Dr. Crusher; however, the number of sweets in the box was divisible by 2 and not by 3. Data was called to the captain's quarters to resolve the problem. Data instructed Riker to eat one chocolate and then count the rest. Riker discovered that the remainder was divisible by 3 and he took his third. Next, Data had Dr. Crusher eat one candy. Again, the remainder was divisible by 3, and Dr. Crusher took her share. Then Picard ate one sweet and found the remainder once more divisible by 3 and so took his third. Finally, Data observed that the number of remaining chocolates was divisible by 3 and gave each celebrant an equal share. What is the least number of chocolates that could have been in the box originally?

1.2

Integral Exponents

One of the most convenient notations of algebra is the notation of exponents. Exponents are used to indicate repeated multiplication, and to represent very large and very small numbers. In this section we will review the definitions and rules concerning exponents and simplify expressions that have integral exponents.

Positive Integral Exponents

We use positive integral exponents to indicate the number of times a factor occurs in a product. For example, $2 \cdot 2 \cdot 2 \cdot 2$ is written as 2^4 using exponents. We read 2^4 as "the fourth power of 2" or "2 to the fourth power." We have the following definition.

Definition: Positive Integral Exponents

For any positive integer n,
$$a^n = \underbrace{a \cdot a \cdot a \cdot \cdots \cdot a}_{n \text{ factors of } a}.$$

We call a the **base,** n the **exponent** or **power,** and a^n an **exponential expression.**

We read a^n as "a to the nth power." For a^1 we usually omit the exponent and just write a. We refer to expressions having the exponents 2 and 3 as squares and cubes. For example, 3^2 is read "3 squared," 2^3 is read "2 cubed," x^4 is read "x to the fourth," b^5 is read "b to the fifth," and so on.

Exponential expressions are given the first priority in the order of operations. For example, $5 + 2 \cdot 3^2 = 5 + 2 \cdot 9 = 5 + 18 = 23$. Operations within grouping symbols are always performed before other operations. For example, $2(3 + 1)^2 = 2 \cdot 4^2 = 2 \cdot 16 = 32$. In an expression such as -3^2 we square 3 first, then take the opposite. So $-3^2 = -9$ and $(-3)^2 = (-3)(-3) = 9$.

Example 1   Evaluating arithmetic expressions that have exponents

Evaluate each expression.

a) $(-5)^2 - 4(-1)(-3)$ **b)** $(3 - (-2))^2 + (-1 - 5)^2$ **c)** $-2^4 - (-10)^3$

Solution

```
(-5)²-4(-1)(-3)
             13
-2^4-(-10)^3
            984
```

The x^2 key can be used to square a number. For other powers, use the key labeled $\wedge$.

a) $(-5)^2 - 4(-1)(-3) = 25 - 12$ Powers and products before
$$= 13$$ subtraction

b) $(3 - (-2))^2 + (-1 - 5)^2 = 5^2 + (-6)^2$ Evaluate within parentheses first.
$$= 25 + 36$$
$$= 61$$

c) $-2^4 - (-10)^3 = -16 - (-1000)$ Evaluate powers first.
$$= -16 + 1000$$
$$= 984$$

Example 2   Evaluating algebraic expressions that have exponents

Evaluate $3xy^4 - z^3$ if $x = 2$, $y = -1$, and $z = -3$.

Solution

Replace x, y, and z by 2, -1, and -3, respectively, and then evaluate:

$$3 \cdot 2 \cdot (-1)^4 - (-3)^3 = 3 \cdot 2 \cdot 1 - (-27) = 6 + 27 = 33$$

Negative Integral Exponents

We use a negative sign in an exponent to represent multiplicative inverses or reciprocals. For example, 3^{-2} represents the reciprocal of 3^2. Because $3^2 = 9$, $3^{-2} = \frac{1}{9}$. Since $3^2 \cdot 3^{-2} = 1$, we have

$$3^{-2} = \frac{1}{3^2} \quad \text{and} \quad 3^2 = \frac{1}{3^{-2}}.$$

For negative exponents we do not allow the base to be zero because zero does not have a reciprocal. Negative exponents are used to represent reciprocals to make the rules for simplifying exponential expressions work smoothly.

Definition: Negative Integral Exponents

If a is a nonzero real number and n is a positive integer,

$$a^{-n} = \frac{1}{a^n}.$$

Example 3   Evaluating expressions that have negative exponents

Simplify each expression.

a) $4 \cdot 2^{-3}$ **b)** $3^{-1} \cdot 5^{-2} \cdot 10^2$ **c)** $\left(\dfrac{2}{3}\right)^{-3}$ **d)** $\dfrac{6^{-2}}{2^{-3}}$

Solution

a) $4 \cdot 2^{-3} = 4 \cdot \dfrac{1}{2^3} = 4 \cdot \dfrac{1}{8} = \dfrac{1}{2}$

b) $3^{-1} \cdot 5^{-2} \cdot 10^2 = \dfrac{1}{3} \cdot \dfrac{1}{5^2} \cdot 100 = \dfrac{1}{3} \cdot \dfrac{1}{25} \cdot 100 = \dfrac{100}{75} = \dfrac{4}{3}$

c) $\left(\dfrac{2}{3}\right)^{-3} = \dfrac{1}{\left(\dfrac{2}{3}\right)^3} = \dfrac{1}{\dfrac{2}{3} \cdot \dfrac{2}{3} \cdot \dfrac{2}{3}} = \dfrac{1}{\dfrac{8}{27}} = \dfrac{27}{8}$ Note that $\left(\dfrac{2}{3}\right)^{-3} = \left(\dfrac{3}{2}\right)^3$.

d) $\dfrac{6^{-2}}{2^{-3}} = \dfrac{\dfrac{1}{6^2}}{\dfrac{1}{2^3}} = \dfrac{1}{6^2} \cdot \dfrac{2^3}{1} = \dfrac{8}{36} = \dfrac{2}{9}$ Note that $\dfrac{6^{-2}}{2^{-3}} = \dfrac{2^3}{6^2}$.

Example 3(c) illustrates the fact that a fractional base can be inverted, if the sign of the exponent is changed. Example 3(d) illustrates the fact that a factor of the numerator or denominator can be moved from the numerator to the denominator or vice versa as long as we change the sign of the exponent. These rules follow from the definition of negative exponents.

```
(2/3)^-3
              3.375
(3/2)^3
              3.375
2/3^-3
                 54
```

To raise a fraction to a power, the fraction must be in parentheses. See Appendix A for more examples.

Rules for Negative Exponents and Fractions

If a and b are nonzero real numbers and m and n are integers, then

$$\left(\dfrac{a}{b}\right)^{-m} = \left(\dfrac{b}{a}\right)^{m} \quad \text{and} \quad \dfrac{a^{-m}}{b^{-n}} = \dfrac{b^n}{a^m}.$$

Using this rule, we could shorten Examples 3(c) and (d) as follows:

$$\left(\dfrac{2}{3}\right)^{-3} = \left(\dfrac{3}{2}\right)^3 = \dfrac{27}{8} \quad \text{and} \quad \dfrac{6^{-2}}{2^{-3}} = \dfrac{2^3}{6^2} = \dfrac{8}{36} = \dfrac{2}{9}.$$

Note that we cannot apply these rules when addition or subtraction is involved. You should verify that

$$\dfrac{1 + 3^{-2}}{2^{-3}} \neq \dfrac{1 + 2^3}{3^2} \quad \text{and} \quad \left(1 + \dfrac{2}{3}\right)^{-2} \neq \left(1 + \dfrac{3}{2}\right)^2.$$

Rules of Exponents

Consider the product $a^2 \cdot a^3$. Using the definition of exponents, we can simplify this product as follows:

$$a^2 \cdot a^3 = (a \cdot a)(a \cdot a \cdot a) = a^5$$

The definition of exponents can also be used to simplify the product $a^n \cdot a^m$ where n and m are any positive integers:

$$a^n \cdot a^m = \underbrace{\overbrace{a \cdot a \cdot \cdots \cdot a}^{n \text{ factors}} \cdot \overbrace{a \cdot a \cdot \cdots \cdot a}^{m \text{ factors}}}_{n + m \text{ factors}} = a^{n+m}$$

This equation indicates that the product of exponential expressions *with the same base* is obtained by adding the exponents. This fact is called the **product rule.**

Example 4   Using the product rule

Simplify each expression.

a) $(3x^8y^2)(-2xy^4)$ **b)** $2^3 \cdot 3^2$

Solution

a) Use the product rule to add the exponents for products of expressions with identical bases:

$$(3x^8y^2)(-2xy^4) = -6x^9y^6$$

b) Since the bases are different we cannot use the product rule, but we can simplify the expression using the definition of exponents:

$$2^3 \cdot 3^2 = 8 \cdot 9 = 72$$

◆

So far we have defined positive and negative integral exponents. The definition of zero as an exponent is given in the following box.

Definition: Zero Exponent

If a is a nonzero real number, then $a^0 = 1$.

This definition of zero exponent allows us to extend the product rule to any integral exponents. For example, using the definition of negative exponents, we get

$$2^{-3} \cdot 2^3 = \frac{1}{2^3} \cdot 2^3 = 1.$$

Adding exponents, we get $2^{-3} \cdot 2^3 = 2^{-3+3} = 2^0$. The answer is the same because 2^0 is defined to be 1. Note that the zero power of zero is not defined.

Using the definitions of positive, negative, and zero exponents, we can show that the product rule and several other rules hold for any integral exponents. We list these rules in the following box.

Rules for Integral Exponents

If a and b are nonzero real numbers and m and n are integers, then

1. $a^m a^n = a^{m+n}$ **Product rule**

2. $\dfrac{a^m}{a^n} = a^{m-n}$ **Quotient rule**

3. $(a^m)^n = a^{mn}$ **Power of a power rule**

4. $(ab)^n = a^n b^n$ **Power of a product rule**

5. $\left(\dfrac{a}{b}\right)^n = \dfrac{a^n}{b^n}$ **Power of a quotient rule**

```
2^(-3+3)
               1
2^-3+3
           3.125
■
```

To evaluate 2^{-3+3}, the expression $-3 + 3$ must be in parentheses.

The rules for integral exponents are used to simplify expressions.

Example 5   Simplifying expressions with integral exponents

Simplify each expression. Write your answer without negative exponents. Assume that all variables represent nonzero real numbers.

a) $(3x^2 y^3)(-4x^{-2} y^{-5})$ **b)** $\dfrac{-6a^5 b^{-1}}{2a^7 b^{-3}}$

Solution

a) $(3x^2 y^3)(-4x^{-2} y^{-5}) = -12x^{2+(-2)} y^{3+(-5)}$ Product rule

$= -12x^0 y^{-2}$ Simplify the exponents.

$= -\dfrac{12}{y^2}$ Definition of negative and zero exponents

b) $\dfrac{-6a^5 b^{-1}}{2a^7 b^{-3}} = -3a^{5-7} b^{-1-(-3)}$ Quotient rule

$= -3a^{-2} b^2$ Simplify the exponents.

$= -\dfrac{3b^2}{a^2}$ Definition of negative exponents

In the next example, we use the rules of exponents to simplify expressions that have variables in the exponents.

Example 6   Simplifying expressions with variable exponents

Simplify each expression. Assume that all bases are nonzero real numbers and all exponents are integers.

a) $(-3x^{a-5}y^{-3})^4$ **b)** $\left(\dfrac{3a^{2m-1}}{2a^{-3m}}\right)^{-3}$

Solution

a)
$$(-3x^{a-5}y^{-3})^4 = (-3)^4(x^{a-5})^4(y^{-3})^4 \qquad \text{Power of a product rule}$$
$$= 81x^{4a-20}y^{-12} \qquad \text{Power of a power rule}$$
$$= \frac{81x^{4a-20}}{y^{12}} \qquad \text{Definition of negative exponents}$$

b)
$$\left(\frac{3a^{2m-1}}{2a^{-3m}}\right)^{-3} = \frac{3^{-3}(a^{2m-1})^{-3}}{2^{-3}(a^{-3m})^{-3}} \qquad \text{Power of a quotient rule}$$
$$= \frac{2^3 a^{-6m+3}}{3^3 a^{9m}} \qquad \text{Power of a power rule and definition of negative exponents}$$
$$= \frac{8a^{-15m+3}}{27} \qquad \text{Quotient rule}$$

Scientific Notation

Archimedes (287–212 BC) was a brilliant Greek inventor and mathematician who studied at the Egyptian city of Alexandria, then the center of the scientific world. Archimedes used his knowledge of mathematics to calculate for King Gelon the number of grains of sand in the universe: 1 followed by 63 zeros. Of course, Archimedes' universe was different from our universe, and he performed his computations with Greek letter numerals, since the modern number system and scientific notation had not yet been invented. Although it is impossible to calculate the number of grains of sand in the universe, this story illustrates how long scientists have been interested in quantities ranging in size from the diameter of our galaxy to the diameter of an atom. Scientific notation offers a convenient way of expressing very large or very small numbers.

In scientific notation, a positive number is written as a product of a number between 1 and 10 and a power of 10. For example, 9.63×10^7 and 2.3×10^{-6} are numbers written in scientific notation. Conversion of numbers from scientific notation to standard notation is actually just multiplication.

Example 7   Scientific notation to standard notation

Convert each number to standard notation.

a) 9.63×10^7 **b)** 2.3×10^{-6}

Solution

a) $9.63 \times 10^7 = 9.63 \times 10,000,000$ Evaluate 10^7.

$= 96,300,000$ Move decimal point seven places to the right.

b) $2.3 \times 10^{-6} = 2.3 \times \dfrac{1}{1,000,000}$ Evaluate 10^{-6}.

$= 2.3 \times 0.000001$

$= 0.0000023$ Move decimal point six places to the left.

◆

Observe how the decimal point is relocated in Example 7. Converting from scientific notation to standard notation is simply a matter of moving the decimal point. To convert a number from scientific to standard notation, we move the decimal point the number of places indicated by the power of 10. Move the decimal point to the right for a positive power and to the left for a negative power. Note that in scientific notation a number greater than 10 is written with a positive power of 10 and a number less than 1 is written with a negative power of 10. Numbers between 1 and 10 are not written in scientific notation.

In the next example, we convert from standard notation to scientific notation by reversing the process used in Example 7.

Example 8   Standard notation to scientific notation

Convert each number to scientific notation.

a) 580,000,000,000   **b)** 0.0000683

Solution

a) Determine the power of 10 by counting the number of places that the decimal must move so that there is a single digit to the left of the decimal point (11 places). Since 580,000,000,000 is larger than 10, we use a positive power of 10:

$$580,000,000,000 = 5.8 \times 10^{11}$$

b) Determine the power of 10 by counting the number of places the decimal must move so that there is a single digit to the left of the decimal point (five places). Since 0.0000683 is smaller than 1, we use a negative power of 10:

$$0.0000683 = 6.83 \times 10^{-5}$$

◆

To convert a negative number to scientific notation, convert as you would a positive number and retain the negative sign:

$$-0.000000359 = -3.59 \times 10^{-7}$$

One advantage of scientific notation is that the rules of exponents can be used when performing certain computations involving scientific notation. Calculators can be used to perform computations with scientific notation, but it is good to practice some computation without a calculator.

Example 9   Using scientific notation in computations

Perform the indicated operations without a calculator. Write your answers in scientific notation.

a) $(4 \times 10^{13})(5 \times 10^{-9})$ **b)** $\dfrac{-1.2 \times 10^{-9}}{4 \times 10^{-7}}$ **c)** $\dfrac{(2{,}000{,}000{,}000)^3(0.00009)}{600{,}000{,}000}$

Solution

a)
$$(4 \times 10^{13})(5 \times 10^{-9}) = 20 \times 10^{13+(-9)} \qquad \text{Product rule for exponents}$$
$$= 20 \times 10^4 \qquad \text{Simplify the exponent.}$$
$$= 2 \times 10^1 \times 10^4 \qquad \text{Write 20 in scientific notation.}$$
$$= 2 \times 10^5 \qquad \text{Product rule for exponents}$$

b)
$$\frac{-1.2 \times 10^{-9}}{4 \times 10^{-7}} = \frac{-1.2}{4} \times 10^{-9-(-7)} \qquad \text{Quotient rule for exponents}$$
$$= -0.3 \times 10^{-2} \qquad \text{Simplify the exponent.}$$
$$= -3 \times 10^{-3} \qquad \text{Use } -0.3 = -3 \times 10^{-1}.$$

c) First convert each number to scientific notation, then use the rules of exponents to simplify:

$$\frac{(2{,}000{,}000{,}000)^3(0.00009)}{600{,}000{,}000} = \frac{(2 \times 10^9)^3(9 \times 10^{-5})}{6 \times 10^8}$$
$$= \frac{(8 \times 10^{27})(9 \times 10^{-5})}{6 \times 10^8}$$
$$= \frac{72 \times 10^{22}}{6 \times 10^8} = 12 \times 10^{14} = 1.2 \times 10^{15}$$

◆

In the next example we use scientific notation to perform the type of computation performed by Archimedes when he attempted to determine the number of grains of sand in the universe.

Example 10   The number of grains of sand in Archimedes' earth

If the radius of the earth is approximately 6.38×10^3 kilometers and the radius of a grain of sand is approximately 1×10^{-3} meters, then what number of grains of sand have a volume equal to the volume of the earth?

```
4/3*π(6.38E3)^3
   1.087803985E12
4/3*π(1E-6)^3
   4.1887902E-18
■
```

Use the EE or EXP key to write a number in scientific notation. Note that the 10 does not appear.

```
1.09E12/4.19E-18
   2.601431981E29
```

A number in scientific notation is treated as a single number, so parentheses are not needed in this quotient.

Solution

Since the volume of a sphere is given by $V = \frac{4}{3}\pi r^3$, the volume of the earth is

$$\frac{4}{3}\pi(6.38 \times 10^3)^3 \text{ km}^3 = 1.09 \times 10^{12} \text{ km}^3.$$

Since 1 km $= 10^3$ m, the radius of a grain of sand is 1×10^{-6} km and its volume is

$$\frac{4}{3}\pi(1 \times 10^{-6})^3 \text{ km}^3 = 4.19 \times 10^{-18} \text{ km}^3.$$

To get the number of grains of sand, divide the volume of the earth by the volume of a grain of sand:

$$\frac{1.09 \times 10^{12} \text{ km}^3}{4.19 \times 10^{-18} \text{ km}^3} = 2.60 \times 10^{29}$$

◆

? For Thought

True or false? Explain. Do not use a calculator.

1. $2^{-1} + 2^{-1} = 1$ **2.** $2^{100} = 4^{50}$ **3.** $9^8 \cdot 9^8 = 81^8$

4. $(0.25)^{-1} = 4$ **5.** $\dfrac{5^{10}}{5^{-12}} = 5^{-2}$ **6.** $2 \cdot 2 \cdot 2 \cdot 2^{-1} = \dfrac{1}{16}$

7. $-3^{-3} = -\dfrac{1}{27}$ **8.** $\left(\dfrac{3}{4}\right)^{-2} = \left(\dfrac{4}{3}\right)^2$ **9.** $10^{-4} = 0.00001$

10. $98.6 \times 10^8 = 9.86 \times 10^7$

1.2 Exercises   ▭ Tape 1   ▤ Disk—5.25″: 1   3.5″: 1   Macintosh: 1

Evaluate each expression.

1. 4^3 **2.** -3^4 **3.** -4^2 **4.** $5 \cdot 10^4$

5. $\left(\dfrac{1}{2}\right)^3$ **6.** $\left(-\dfrac{3}{4}\right)^4$

7. $7^2 - 2(-3)(-6)$ **8.** $(-3)^2 - 4(-2)(-5)$

9. $(-2 - 3)^2 + (4 - (-1))^2$ **10.** $(5 - (-2))^2 + (6 - 7)^2$

Use $a = -2$, $b = 3$, and $c = -4$ to evaluate each algebraic expression.

11. $a^2 - b^2$ **12.** $a^2 + b^2$

13. $(a - b)(a + b)$ **14.** $(a + b)^2$

15. $(a - b)(a^2 + ab + b^2)$ **16.** $(a + b)(a^2 - ab + b^2)$

17. $a^b + c^b$ **18.** $(a + c)^b$ **19.** $b^2 - 4ac$

20. $\dfrac{a^2 - c^2}{c - a}$ **21.** $\dfrac{a^b - c^b}{a^2 + ac + c^2}$ **22.** $4c(b + 1)^{a+3}$

Evaluate each expression.

23. 3^{-4} **24.** $\dfrac{1}{2^{-3}}$ **25.** $\dfrac{1}{5^{-2}}$

26. $-2 \cdot 10^{-3}$ **27.** $6^{-1} + 5^{-1}$ **28.** $2^0 + 2^{-1}$

29. $\dfrac{3^{-2}}{6^{-3}}$

30. $\dfrac{3^{-1}}{2^3}$

31. $\left(\dfrac{1}{2}\right)^{-3}$

32. $\left(-\dfrac{1}{10}\right)^{-4}$

33. $4 \cdot 4 \cdot 4 \cdot 4^{-1}$

34. $-1^{-1} \cdot (-2)^{-2}$

35. $5 \cdot 10^3 + 3 \cdot 10^2 + 6 \cdot 10^1$

36. $4 \cdot 10^{-2} + 3 \cdot 10^{-1} + 2 \cdot 10^0 + 7 \cdot 10^2 + 9 \cdot 10^3$

Simplify each expression.

37. $(-3x^2y^3)(2x^9y^8)$

38. $(-6a^7b^4)(3a^3b^5)$

39. $5^2 \cdot 3^2$

40. $x^2x^5 + x^3x^4$

Simplify each expression. Write answers without negative exponents. Assume that all variables represent nonzero real numbers.

41. $-1(2x^3)^2$

42. $(-3y^{-1})^{-1}$

43. $\left(\dfrac{-2x^2}{3}\right)^3$

44. $\left(\dfrac{-1}{2a}\right)^{-2}$

45. $\dfrac{6x^7}{2x^3}$

46. $\dfrac{-9x^2y}{3xy^2}$

47. $\left(\dfrac{y^2}{5}\right)^{-2}$

48. $\left(-\dfrac{y^2}{2a}\right)^4$

49. $\left(\dfrac{1}{2}x^{-4}y^3\right)\left(\dfrac{1}{3}x^4y^{-6}\right)$

50. $\left(\dfrac{1}{3}a^{-5}b\right)(a^4b^{-1})$

51. $\dfrac{-3m^{-1}n}{-6m^{-1}n^{-1}}$

52. $\dfrac{-p^{-1}q^{-1}}{-3pq^{-3}}$

53. $(2a^2)^3 + (-3a^3)^2$

54. $(b^{-4})^2 - (-b^{-2})^4$

55. $-3(3x^{-1}y^3)^{-2}$

56. $6(-2a^{-1}b^{-3})^{-1}$

57. $\left(\dfrac{-2x^4y^{-4}}{3x^{-1}y^{-2}}\right)^4$

58. $\left(\dfrac{-3a^3b^{-5}}{6a^{-2}b^{-2}}\right)^3$

59. $\left(\dfrac{-c^2d^{-3}}{c^{-1}d^{-4}}\right)^{-2}$

60. $\left(\dfrac{8x^8y^{-2}}{-4x^{-1}y^{-5}}\right)^{-3}$

Simplify each expression. Assume that all bases are nonzero real numbers and all exponents are integers.

61. $(x^{b-1})^3(x^{b-4})^{-2}$

62. $(a^2)^{m+2}(a^3)^{4m}$

63. $(-5a^{2t}b^{-3t})^3$

64. $(-2x^{-5v}y^3)^2$

65. $\dfrac{-9x^{3w}y^{9v}}{6x^{8w}y^{3v}}$

66. $\dfrac{6c^{9s}d^{4t}}{-9c^{3s}d^{8t}}$

67. $\left(\dfrac{a^{s+2}b^{t-3}}{a^{2s-3}b^{6-t}}\right)^3$

68. $\left(\dfrac{x^{2a-3}y^{-13}}{x^{-4a+1}y^{-9}}\right)^{-4}$

Convert each number given in standard notation to scientific notation and each number given in scientific notation to standard notation.

69. 4.3×10^4

70. -5.98×10^5

71. -3.56×10^{-5}

72. 9.333×10^{-9}

73. $5{,}000{,}000$

74. $-16{,}587{,}000$

75. -0.0000672

76. 0.000000981

77. 7×10^{-9}

78. -6×10^{-3}

79. $-20{,}000{,}000{,}000$

80. 0.00000000004

Perform the indicated operations without a calculator. Write your answers in scientific notation.

81. $(5 \times 10^8)(4 \times 10^7)$

82. $(5 \times 10^{-10})(6 \times 10^5)$

83. $\dfrac{8.2 \times 10^{-6}}{4.1 \times 10^{-3}}$

84. $\dfrac{9.3 \times 10^{12}}{3.1 \times 10^{-3}}$

85. $5(2 \times 10^{-10})^3$

86. $-2(2 \times 10^8)^{-4}$

87. $\dfrac{(2{,}000{,}000)^3(0.000005)}{(0.00002)^2}$

88. $\dfrac{(-6{,}000{,}000)^2(-0.000003)^{-3}}{(2000)^3(1{,}000{,}000)}$

Perform the indicated operations. Use a scientific calculator and give your answers in scientific notation.

89. $(4.32 \times 10^{-9})(-2.3 \times 10^4)$

90. $(-2.33 \times 10^{23})(3.98 \times 10^{-9})$

91. $(4{,}560{,}000{,}000{,}000)(0.00000000000345)$

92. $(88{,}000{,}000{,}000{,}000)^6$

93. $\dfrac{(5.63 \times 10^{-6})^3(3.5 \times 10^7)^{-4}}{\pi(8.9 \times 10^{-4})^2}$

94. $\dfrac{\pi(2.39 \times 10^{-12})^2}{(6.75 \times 10^{-8})^3}$

95. $(2.7 \times 10^9) + (3.6 \times 10^{-5})$

96. $(-6.3 \times 10^{12}) - (7.25 \times 10^{-4})$

Solve each problem. Use a scientific calculator.

97. *National debt* If the national debt is 1.3×10^{12} dollars and there are 2.53×10^8 people in the country, then how much does each person owe?

98. *Savings and loan bailout* It is estimated that the savings and loan bailout cost the United States $500 billion. If there are 2.53×10^8 people in the United States, then what is the cost per person?

99. *Energy from the sun* Merely an average star, our sun is a swirling mass of dense gases powered by thermonuclear reactions. The great solar furnace transforms 5 million tons of mass into energy every second. How many tons of mass will be transformed into energy during the sun's 10 billion-year lifetime?

At the sun's energy core, thermonuclear reactions, which convert hydrogen to helium, generate heat that radiates outward.

— Column of gas

Figure for Exercise 99

100. *Orbit of the earth* The earth orbits the sun in an approximately circular orbit with a radius of 1.495979×10^8 km. What is the area of the circle?

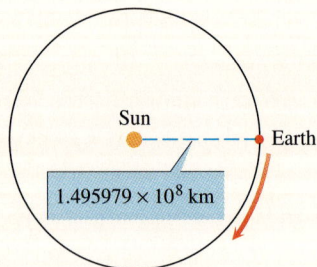

Sun
Earth
1.495979×10^8 km

Figure for Exercises 100 and 102

101. *Radius of the earth* The radius of the sun is 6.9599×10^5 km, which is 109.1 times the earth's radius. What is the radius of the earth?

102. *Distance to the sun* The distance from the sun to the earth is 1.495979×10^8 km. Use the fact that 1 km = 0.621 mi to find the distance in miles from the earth to the sun.

103. *Mass of the sun* The mass of the sun is 1.989×10^{30} kg and the mass of the earth is 5.976×10^{24} kg. How many times larger in mass is the sun than the earth?

104. *Speed of light* If the speed of light is 3×10^8 m/sec, then how long does it take light from the sun to reach the earth? (See Exercise 102.)

105. *Energy consumption in U.S.* According to the 1993 World Almanac, total world energy consumption for 1990 was 3.38×10^{17} Btu. Use the pie chart to determine the energy consumption for the United States.

1990 World Energy Consumption

United States 24% — 17% U.S.S.R.
9% China
All other countries 45% — 5% Japan

Figure for Exercise 105

106. *Toxic emissions* According to the EPA, the company responsible for the heaviest pollution in Louisiana in 1991 emitted 1.42×10^8 lb of toxic-chemical waste. In that same year, Louisiana was the worst state in emitting toxic-chemical waste (*USA Today,* May 26, 1993). Use the data from the accompanying graph to determine what percentage of Louisiana's toxic emissions was accounted for by that company.

Toxic waste emitted (millions of lbs)

458 410 215 171 136
La. Tex. Tenn. Ohio Ind.

Worst five states, 1991

Figure for Exercise 106

1.3

Rational Exponents

and Radicals

Raising a number to a power is reversed by finding the root of a number. We indicate roots by using rational exponents or radicals. Here we will review definitions and rules concerning rational exponents and radicals.

Roots

Since $2^4 = 16$ and $(-2)^4 = 16$, both 2 and -2 are fourth roots of 16. The nth root of a number is defined in terms of the nth power.

Definition: *n*th roots

> If n is a positive integer and $a^n = b$, then a is called an **nth root** of b. If $a^2 = b$, then a is a **square root** of b. If $a^3 = b$, then a is the **cube root** of b.

We also describe roots as even or odd, depending on whether the positive integer is even or odd. For example, if n is even (or odd) and a is an nth root of b, then a is called an **even** (or **odd**) **root** of b. Every positive real number has *two* real even roots, a positive root and a negative root. For example, both 5 and -5 are square roots of 25 because $5^2 = 25$ and $(-5)^2 = 25$. Moreover, every real number has exactly *one* real odd root. For example, because $2^3 = 8$ and 3 is odd, 2 is the one cube root of 8. Because $(-2)^3 = -8$ and 3 is odd, -2 is the only cube root of -8.

Finding an nth root is the reverse of finding an nth power, so we use the notation $a^{1/n}$ for the nth root of a. For example, since the positive square root of 25 is 5, we write $25^{1/2} = 5$.

Definition: Exponent 1/*n*

> If n is a positive even integer and a is positive, then $a^{1/n}$ denotes the **positive real nth root of a** and is called the **principal nth root of a.**
> If n is a positive odd integer and a is any real number, then $a^{1/n}$ denotes the real nth root of a.
> If n is a positive integer, then $0^{1/n} = 0$.

Example 1   Evaluating expressions involving exponent 1/n

Evaluate each expression.

a) $4^{1/2}$ **b)** $8^{1/3}$ **c)** $(-8)^{1/3}$ **d)** $(-4)^{1/2}$

Solution

a) The expression $4^{1/2}$ represents the positive real square root of 4. So $4^{1/2} = 2$.

b) $8^{1/3} = 2$ **c)** $(-8)^{1/3} = -2$

d) Since the definition of nth root does not include an even root of a negative number, $(-4)^{1/2}$ has not yet been defined. Even roots of negative numbers do exist in the complex number system, which we define in Chapter 2. So an even root of a negative number is not a real number.

Rational Exponents

We have defined $a^{1/n}$ as the nth root of a. We extend this definition to $a^{m/n}$, which is defined as the mth power of the nth root of a. A rational exponent indicates both a root and a power.

Definition: Rational Exponents

If m and n are positive integers, then

$$a^{m/n} = (a^{1/n})^m = (a^m)^{1/n}$$

provided that $a^{1/n}$ is a real number.

Of course a negative rational exponent indicates a reciprocal, just as a negative integral exponent does. Expressions such as $(-25)^{-3/2}$, $(-43)^{1/4}$, and $(-1)^{2/2}$ are not real numbers because each of them involves an even root of a negative number.

To evaluate an expression such as $8^{2/3}$, we must find the square of the cube root of 8 or the cube root of 8^2. In symbols,

$$8^{2/3} = (8^{1/3})^2 = 2^2 = 4 \qquad \text{or} \qquad 8^{2/3} = (8^2)^{1/3} = 64^{1/3} = 4.$$

To keep the numbers small we usually find the root and then the power.

We can evaluate $8^{-2/3}$ mentally as follows: The cube root of 8 is 2, 2 squared is 4, and the reciprocal of 4 is $\frac{1}{4}$. In written work it is usually best to eliminate the negative exponent first:

$$8^{-2/3} = \frac{1}{8^{2/3}} = \frac{1}{(8^{1/3})^2} = \frac{1}{2^2} = \frac{1}{4}$$

The three operations indicated by a negative rational exponent can be performed in any order. The simplest procedure for mental evaluation is summarized as follows.

Procedure: Evaluating $a^{-m/n}$

To evaluate $a^{-m/n}$ mentally,

1. find the nth root of a,
2. raise it to the m power,
3. find the reciprocal.

Rational exponents can be reduced to lowest terms. For example, we can evaluate $2^{6/2}$ by first reducing the exponent:

$$2^{6/2} = 2^3 = 8$$

Exponents can be reduced only on expressions that are real numbers. For example, $(-1)^{2/2} \neq (-1)^1$ because $(-1)^{2/2}$ is not a real number, while $(-1)^1$ is a real number.

```
16^(3/4)
            8
16^.75
            8
■
```

If the exponent is a fraction, it must be in parentheses. See Appendix A for more examples.

Example 2   Evaluating expressions with rational exponents

Evaluate each expression.

a) $16^{3/4}$ **b)** $(-8)^{2/3}$ **c)** $27^{-2/3}$ **d)** $(-27)^{-1/3}$ **e)** $100^{6/4}$

Solution

a) $16^{3/4} = (16^{1/4})^3 = 2^3 = 8$

b) $(-8)^{2/3} = ((-8)^{1/3})^2 = (-2)^2 = 4$

c) $27^{-2/3} = \dfrac{1}{27^{2/3}} = \dfrac{1}{(27^{1/3})^2} = \dfrac{1}{3^2} = \dfrac{1}{9}$

d) $(-27)^{-1/3} = \dfrac{1}{(-27)^{1/3}} = \dfrac{1}{-3} = -\dfrac{1}{3}$

e) $100^{6/4} = 100^{3/2} = 10^3 = 1000$

```
(-27)^(-1/3)
      -.3333333333
■
```

Your calculator might not evaluate certain powers of negative numbers. For example, try to find $(-27)^{-1/3}$ and $(-8)^{2/3}$.

Rules for Rational Exponents

The rules for integral exponents from Section 1.2 also hold for rational exponents.

▼

Rules for Rational Exponents

The following rules are valid for all real numbers a and b and rational numbers r and s, provided that all indicated powers are real and no denominator is zero.

1. $a^r a^s = a^{r+s}$ **2.** $\dfrac{a^r}{a^s} = a^{r-s}$ **3.** $(a^r)^s = a^{rs}$

4. $(ab)^r = a^r b^r$ **5.** $\left(\dfrac{a}{b}\right)^r = \dfrac{a^r}{b^r}$ **6.** $\left(\dfrac{a}{b}\right)^{-r} = \left(\dfrac{b}{a}\right)^r$ **7.** $\dfrac{a^{-r}}{b^{-s}} = \dfrac{b^s}{a^r}$

When variable expressions involve even roots, we must be careful with signs. For example, $(x^2)^{1/2} = x$ is not correct for all values of x, because $((-5)^2)^{1/2} = 25^{1/2} = 5$. However, using absolute value we can write

$$(x^2)^{1/2} = |x| \qquad \text{for every real number } x.$$

When finding an even root of an expression involving variables, remember that if n is even, $a^{1/n}$ is the *positive* nth root of a.

Example 3 Using absolute value with rational exponents

Simplify each expression, using absolute value when necessary. Assume that the variables can represent any real numbers.

a) $(64a^6)^{1/6}$ **b)** $(x^9)^{1/3}$ **c)** $(a^8)^{1/4}$ **d)** $(y^{12})^{1/4}$

Solution

a) For any nonnegative real number a, we have $(64a^6)^{1/6} = 2a$. If a is negative, $(64a^6)^{1/6}$ is positive and $2a$ is negative. So we write

$$(64a^6)^{1/6} = |2a| = 2|a| \qquad \text{for every real number } a.$$

b) For any nonnegative x, we have $(x^9)^{1/3} = x^{9/3} = x^3$. If x is negative, $(x^9)^{1/3}$ and x^3 are both negative. So we have

$$(x^9)^{1/3} = x^3 \qquad \text{for every real number } x.$$

c) For nonnegative a, we have $(a^8)^{1/4} = a^2$. Since $(a^8)^{1/4}$ and a^2 are both positive if a is negative, no absolute value sign is needed. So

$$(a^8)^{1/4} = a^2 \qquad \text{for every real number } a.$$

d) For nonnegative y, we have $(y^{12})^{1/4} = y^3$. If y is negative, $(y^{12})^{1/4}$ is positive but y^3 is negative. So

$$(y^{12})^{1/4} = |y^3| \qquad \text{for every real number } y. \qquad \blacklozenge$$

When simplifying expressions we will often assume that the variables represent positive real numbers so that we do not have to be concerned about undefined expressions or absolute value. In the following example we make that assumption as we use the rules of exponents to simplify expressions involving rational exponents.

Example 4 Simplifying expressions with rational exponents

Use the rules of exponents to simplify each expression. Assume that the variables represent positive real numbers. Write answers without negative exponents.

a) $x^{2/3}x^{4/3}$ **b)** $(x^4y^{1/2})^{1/4}$ **c)** $\left(\dfrac{a^{3/2}b^{2/3}}{a^2}\right)^3$

Solution

a) $x^{2/3}x^{4/3} = x^{6/3}$ Product rule
$= x^2$ Simplify the exponent.

b) $(x^4y^{1/2})^{1/4} = (x^4)^{1/4}(y^{1/2})^{1/4}$ Power of a product rule
$= xy^{1/8}$ Power of a power rule

c) $\left(\dfrac{a^{3/2}b^{2/3}}{a^2}\right)^3 = \dfrac{(a^{3/2})^3(b^{2/3})^3}{(a^2)^3}$ **Power of a quotient rule**

$\qquad\qquad = \dfrac{a^{9/2}b^2}{a^6}$ **Power of a power rule**

$\qquad\qquad = a^{-3/2}b^2$ **Quotient rule** $\left(\dfrac{9}{2} - 6 = -\dfrac{3}{2}\right)$

$\qquad\qquad = \dfrac{b^2}{a^{3/2}}$ **Definition of negative exponents**

Radical Notation

The exponent $1/n$ or the **radical sign** $\sqrt[n]{}$ are both used to indicate nth root.

Definition: Radical

> If n is a positive integer and a is a number for which $a^{1/n}$ is defined, then the expression $\sqrt[n]{a}$ is called a **radical,** and
> $$\sqrt[n]{a} = a^{1/n}.$$
> If $n = 2$, we write $\sqrt{a}$ rather than $\sqrt[2]{a}$.

The number a is called the **radicand** and n is the **index** of the radical. Expressions such as $\sqrt{-3}$, $\sqrt[4]{-81}$, and $\sqrt[6]{-1}$ do not represent real numbers because each is an even root of a negative number.

Example 5   Evaluating radicals

Evaluate each expression.

a) $\sqrt{49}$ **b)** $\sqrt[3]{-1000}$ **c)** $\sqrt[4]{\dfrac{16}{81}}$ **d)** $\sqrt[3]{125^2}$

Solution

a) The symbol $\sqrt{49}$ indicates the positive square root of 49. So $\sqrt{49} = 49^{1/2} = 7$. Writing $\sqrt{49} = \pm 7$ is incorrect.

b) $\sqrt[3]{-1000} = (-1000)^{1/3} = -10$ **Check that** $(-10)^3 = -1000.$

c) $\sqrt[4]{\dfrac{16}{81}} = \left(\dfrac{16}{81}\right)^{1/4} = \dfrac{2}{3}$ **Check that** $\left(\dfrac{2}{3}\right)^4 = \dfrac{16}{81}.$

d) $\sqrt[3]{125^2} = (125^2)^{1/3} = 125^{2/3} = 5^2 = 25$

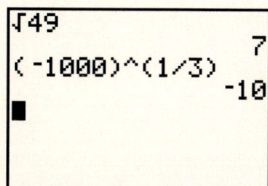

```
√49
              7
(-1000)^(1/3)
            -10
■
```

Use the radical symbol for square roots and fractional exponents for other roots.

Since $a^{1/n} = \sqrt[n]{a}$, expressions involving rational exponents can be written with radicals.

Rule for Converting $a^{m/n}$ to Radical Notation

If a is a real number and m and n are integers for which $\sqrt[n]{a}$ is real, then

$$a^{m/n} = \left(\sqrt[n]{a}\right)^m = \sqrt[n]{a^m}.$$

Example 6 Writing rational exponents as radicals

Write each expression in radical notation. Assume that all variables represent positive real numbers. Simplify the radicand if possible.

a) $2^{2/3}$ b) $(3x)^{3/4}$ c) $2(x^2 + 3)^{-1/2}$

Solution

a) $2^{2/3} = \sqrt[3]{2^2} = \sqrt[3]{4}$ b) $(3x)^{3/4} = \sqrt[4]{(3x)^3} = \sqrt[4]{27x^3}$

c) $2(x^2 + 3)^{-1/2} = 2 \cdot \dfrac{1}{(x^2 + 3)^{1/2}} = \dfrac{2}{\sqrt{x^2 + 3}}$

The Product and Quotient Rules for Radicals

Using radical notation, we can write the equation

$$(ab)^{1/n} = a^{1/n}b^{1/n} \qquad \text{as} \qquad \sqrt[n]{ab} = \sqrt[n]{a} \cdot \sqrt[n]{b}$$

and the equation

$$\left(\frac{a}{b}\right)^{1/n} = \frac{a^{1/n}}{b^{1/n}} \qquad \text{as} \qquad \sqrt[n]{\frac{a}{b}} = \frac{\sqrt[n]{a}}{\sqrt[n]{b}}.$$

These equations provide the basic rules for working with radicals.

Rules for Radicals

For any positive integer n and real numbers a and b ($b \neq 0$),

1. $\sqrt[n]{ab} = \sqrt[n]{a} \cdot \sqrt[n]{b}$ **Product rule for radicals**

2. $\sqrt[n]{\dfrac{a}{b}} = \dfrac{\sqrt[n]{a}}{\sqrt[n]{b}}$ **Quotient rule for radicals**

provided that all of the roots are real.

An expression that is the square of a term that is free of radicals is called a **perfect square**. For example, $9x^6$ is a perfect square because $9x^6 = (3x^3)^2$.

Likewise, $27y^{12}$ is a **perfect cube.** In general, an expression that is the nth power of an expression free of radicals is a **perfect nth power.** In the next example, the product and quotient rules for radicals are used to simplify radicals containing perfect squares, cubes, and so on.

Example 7   Using the product and quotient rules for radicals

Simplify each radical expression. Assume that all variables represent positive real numbers.

a) $\sqrt[3]{125a^6}$ b) $\sqrt{\dfrac{3}{16}}$ c) $\sqrt[5]{\dfrac{-32y^5}{x^{20}}}$

Solution

a) Both 125 and a^6 are perfect cubes. So use the product rule to simplify:
$$\sqrt[3]{125a^6} = \sqrt[3]{125} \cdot \sqrt[3]{a^6} = 5a^2 \qquad \text{Since } \sqrt[3]{a^6} = a^{6/3} = a^2$$

b) Since 16 is a perfect square, use the quotient rule to simplify the radical:
$$\sqrt{\frac{3}{16}} = \frac{\sqrt{3}}{\sqrt{16}} = \frac{\sqrt{3}}{4}$$

c) $\sqrt[5]{\dfrac{-32y^5}{x^{20}}} = \dfrac{\sqrt[5]{-32y^5}}{\sqrt[5]{x^{20}}} = \dfrac{-2y}{x^4} \qquad \text{Since } \sqrt[5]{x^{20}} = x^{20/5} = x^4$ ◆

To find the square root of a fraction, enclose the fraction in parentheses.

```
√(3/16)
        .4330127019
√3/4
        .4330127019
```

To find the square root of a fraction, enclose the fraction in parentheses.

Simplified Form and Rationalizing the Denominator

We have been simplifying radical expressions by just making them look simpler. However, a radical expression is in *simplified form* only if it satisfies the following three specific conditions. (You should check that the simplified expressions of Example 7 satisfy these conditions.)

▼

Definition: Simplified Form for Radicals of Index n

A radical of index n in **simplified form** has

1. *no* perfect nth powers as factors of the radicand,
2. *no* fractions inside the radical, and
3. *no* radicals in a denominator.

The product rule is used to remove the perfect nth powers that are factors of the radicand, and the quotient rule is used when fractions occur inside the radical. The process of removing radicals from a denominator is called **rationalizing the denominator.** Radicals can be removed from the numerator by using the same type of procedure.

Example 8   Simplified form of a radical expression

Write each radical expression in simplified form. Assume that all variables represent positive real numbers.

a) $\sqrt{20}$ **b)** $\sqrt{24x^8y^9}$ **c)** $\dfrac{9}{\sqrt{3}}$ **d)** $\sqrt[3]{\dfrac{3}{5a^4}}$

Solution

a) Since 4 is a factor of 20, $\sqrt{20}$ is not in its simplified form. Use the product rule for radicals to simplify it:

$$\sqrt{20} = \sqrt{4} \cdot \sqrt{5} = 2\sqrt{5}$$

b) Use the product rule to factor the radical, putting all perfect squares in the first factor:

$$\sqrt{24x^8y^9} = \sqrt{4x^8y^8} \cdot \sqrt{6y} \qquad \text{Product rule}$$
$$= 2x^4y^4\sqrt{6y} \qquad \text{Simplify the first radical.}$$

c) Since $\sqrt{3}$ appears in the denominator, we multiply the numerator and denominator by $\sqrt{3}$ to rationalize the denominator:

$$\frac{9}{\sqrt{3}} = \frac{9 \cdot \sqrt{3}}{\sqrt{3} \cdot \sqrt{3}} = \frac{9\sqrt{3}}{3} = 3\sqrt{3}$$

d) To rationalize this denominator, we must get a perfect cube in the denominator. The radicand $5a^4$ can be made into the perfect cube $125a^6$ by multiplying by $25a^2$:

$$\sqrt[3]{\frac{3}{5a^4}} = \frac{\sqrt[3]{3}}{\sqrt[3]{5a^4}} \qquad \text{Quotient rule for radicals}$$

$$= \frac{\sqrt[3]{3} \cdot \sqrt[3]{25a^2}}{\sqrt[3]{5a^4} \cdot \sqrt[3]{25a^2}} \qquad \text{Multiply numerator and denominator by } \sqrt[3]{25a^2}.$$

$$= \frac{\sqrt[3]{75a^2}}{\sqrt[3]{125a^6}} \qquad \text{Product rule for radicals}$$

$$= \frac{\sqrt[3]{75a^2}}{5a^2} \qquad \text{Since } (5a^2)^3 = 125a^6 \qquad \blacklozenge$$

Operations with Radical Expressions

We can use the distributive property to find the sum $2\sqrt{7} + 3\sqrt{7}$ as follows:

$$2\sqrt{7} + 3\sqrt{7} = (2 + 3)\sqrt{7} = 5\sqrt{7}$$

Because $2\sqrt{7}$ and $3\sqrt{7}$ are added in the same manner as like terms with the same variable, they are called **like terms.** We can subtract the like terms $3x\sqrt{5y}$ and $7x\sqrt{5y}$ as follows:

$$3x\sqrt{5y} - 7x\sqrt{5y} = (3 - 7)x\sqrt{5y} = -4x\sqrt{5y}$$

Note that sums such as $\sqrt{3} + \sqrt{5}$ or $\sqrt[3]{2y} + \sqrt{2y}$ cannot be written as a single radical because the terms are not like terms. The next two examples further illustrate basic operations with radicals.

```
√20+√5
         6.708203932
3√5
         6.708203932
```

Use a calculator to *support* the claim that $\sqrt{20} + \sqrt{5} = 3\sqrt{5}$. Note that the calculator does not prove it because the decimal numbers could be different in the unseen decimal places.

Example 9   Operations with radicals of the same index

Perform each operation and simplify each answer. Assume that each variable represents a positive real number.

a) $\sqrt{20} + \sqrt{5}$ **b)** $\sqrt[3]{24x} - \sqrt[3]{81x}$ **c)** $\sqrt[4]{4y^3} \cdot \sqrt[4]{12y^2}$ **d)** $\sqrt{40} \div \sqrt{5}$

Solution

a) $\sqrt{20} + \sqrt{5} = \sqrt{4} \cdot \sqrt{5} + \sqrt{5}$ Product rule for radicals

$\quad = 2\sqrt{5} + \sqrt{5} = 3\sqrt{5}$ Simplify. Add like terms.

b) $\sqrt[3]{24x} - \sqrt[3]{81x} = \sqrt[3]{8} \cdot \sqrt[3]{3x} - \sqrt[3]{27} \cdot \sqrt[3]{3x}$ Product rule for radicals

$\quad = 2\sqrt[3]{3x} - 3\sqrt[3]{3x} = -\sqrt[3]{3x}$ Simplify. Subtract like terms.

c) $\sqrt[4]{4y^3} \cdot \sqrt[4]{12y^2} = \sqrt[4]{48y^5}$ Product rule for radicals

$\quad = \sqrt[4]{16y^4} \cdot \sqrt[4]{3y} = 2y\sqrt[4]{3y}$ Factor out the perfect fourth powers. Simplify.

d) $\sqrt{40} \div \sqrt{5} = \sqrt{\dfrac{40}{5}} = \sqrt{8}$ Quotient rule for radicals Divide.

$\quad = \sqrt{4} \cdot \sqrt{2} = 2\sqrt{2}$ Product rule; simplify. ◆

Example 10   Combining radicals with different indices

Write each expression using a single radical symbol. Assume that each variable represents a positive real number.

a) $\sqrt[3]{2} \cdot \sqrt{3}$ **b)** $\sqrt[3]{y} \cdot \sqrt[4]{2y}$ **c)** $\sqrt{\sqrt[3]{2}}$

Solution

a) $\sqrt[3]{2} \cdot \sqrt{3} = 2^{1/3} \cdot 3^{1/2}$ Rewrite radicals as rational exponents.

$\quad = 2^{2/6} \cdot 3^{3/6}$ Write exponents with the least common denominator.

$\quad = \sqrt[6]{2^2 \cdot 3^3}$ Rewrite in radical notation using the product rule.

$\quad = \sqrt[6]{108}$ Simplify inside the radical.

b) $\sqrt[3]{y} \cdot \sqrt[4]{2y} = y^{1/3}(2y)^{1/4}$ Rewrite radicals as rational exponents.

$\quad = y^{4/12}(2y)^{3/12}$ Write exponents with the LCD.

$\quad = \sqrt[12]{y^4(2y)^3}$ Rewrite in radical notation using the product rule.

$\quad = \sqrt[12]{8y^7}$ Simplify inside the radical.

c) $\sqrt{\sqrt[3]{2}} = (2^{1/3})^{1/2} = 2^{1/6} = \sqrt[6]{2}$ ◆

In Example 10(c) we found that the square root of a cube root is a sixth root. In general, an mth root of an nth root is an mnth root.

▼

Theorem: mth Root of an nth Root

If m and n are positive integers for which all of the following roots are real, then

$$\sqrt[m]{\sqrt[n]{a}} = \sqrt[mn]{a}.$$

? **For Thought**

True or false? Explain. Do not use a calculator.

1. $8^{-1/3} = -2$

2. $16^{1/4} = 4^{1/2}$

3. $\sqrt{\dfrac{4}{6}} = \dfrac{2}{3}$

4. $(\sqrt{3})^3 = 3\sqrt{3}$

5. $(-1)^{2/2} = -1$

6. $\sqrt[3]{7^2} = 7^{3/2}$

7. $9^{1/2} = \sqrt{3}$

8. $\dfrac{1}{\sqrt{3}} = \dfrac{\sqrt{3}}{3}$

9. $\dfrac{2^{1/2}}{2^{1/3}} = \sqrt[6]{2}$

10. $\sqrt[3]{7^5} = 7\sqrt[3]{49}$

1.3 Exercises
Tape 1   Disk—5.25″: 1   3.5″: 1   Macintosh: 1

Evaluate each expression.

1. $-9^{1/2}$

2. $27^{1/3}$

3. $64^{1/2}$

4. $-144^{1/2}$

5. $(-64)^{1/3}$

6. $81^{1/4}$

7. $(-27)^{4/3}$

8. $125^{-2/3}$

9. $8^{-4/3}$

10. $4^{-3/2}$

Simplify each expression. Use absolute value when necessary.

11. $(x^6)^{1/6}$

12. $(x^{10})^{1/5}$

13. $(a^{15})^{1/5}$

14. $(y^2)^{1/2}$

15. $(a^8)^{1/4}$

16. $(z^{12})^{1/4}$

17. $(x^3y^6)^{1/3}$

18. $(16x^4y^8)^{1/4}$

19. $(a^2b^4)^{1/2}$

20. $(-x^3y^{12})^{1/3}$

Simplify each expression. Assume that all variables represent positive real numbers. Write your answers without negative exponents.

21. $(x^4y)^{1/2}$

22. $(a^{1/2}b^{1/3})^2$

23. $(2a^{1/2})(3a)$

24. $(-3y^{1/3})(-2y^{1/2})$

25. $\dfrac{6a^{1/2}}{2a^{1/3}}$

26. $\dfrac{-4y}{2y^{2/3}}$

27. $(a^2b^{1/2})(a^{1/3}b^{1/2})$

28. $(4^{3/4}a^2b^3)(4^{3/4}a^{-2}b^{-5})$

29. $(a^{4m}b^{6n})^{1/2}b^n$

30. $\dfrac{-3x^{1/n}y^{-1/m}}{6x^{2/n}y^{-7/m}}$

31. $\dfrac{(a^{2n-2})^{1/2}}{(a^{3-6n})^{2/3}}$

32. $(x^{2/n})^n(x^{n/6})^{1/n}$

33. $\left(\dfrac{x^6y^3}{z^9}\right)^{1/3}$

34. $\left(\dfrac{x^{1/2}y}{y^{1/2}}\right)^3$

35. $\left(\dfrac{a^{1/3}b^{-1}}{b^{3/2}}\right)^{-6}$

36. $\left(\dfrac{x^{-2/3}y^2}{x^{1/2}}\right)^{-3}$

Evaluate each radical expression.

37. $\sqrt{900}$

38. $\sqrt{400}$

39. $\sqrt[3]{-8}$

40. $\sqrt[3]{64}$

41. $\sqrt[5]{-32}$

42. $\sqrt[6]{64}$

43. $\sqrt[3]{-\dfrac{8}{1000}}$

44. $\sqrt[4]{\dfrac{1}{625}}$ **45.** $\sqrt[4]{16^3}$ **46.** $\sqrt[3]{8^5}$

Write each expression involving rational exponents in radical notation and each expression involving radicals in exponential notation.

47. $10^{2/3}$ **48.** $-2^{3/4}$ **49.** $3y^{-3/5}$

50. $a(b^4 + 1)^{-1/2}$ **51.** $\dfrac{1}{\sqrt{x}}$ **52.** $-4\sqrt{x^3}$

53. $\sqrt[5]{x^3}$ **54.** $\sqrt[3]{x^3 + y^3}$

Simplify each radical expression. Assume that all variables represent positive real numbers.

55. $\sqrt{16x^2}$ **56.** $\sqrt{121y^4}$

57. $\sqrt[3]{8y^9}$ **58.** $\sqrt[3]{125x^{18}}$

59. $\sqrt{\dfrac{xy}{100}}$ **60.** $\sqrt{\dfrac{t}{81}}$

61. $\sqrt[3]{\dfrac{-8a^3}{b^{15}}}$ **62.** $\sqrt[4]{\dfrac{16t^4}{y^8}}$

Write each radical expression in simplified form. Assume that all variables represent positive real numbers.

63. $\sqrt{28}$ **64.** $\sqrt{50}$ **65.** $\dfrac{1}{\sqrt{5}}$ **66.** $\dfrac{7}{\sqrt{7}}$

67. $\sqrt{\dfrac{x}{8}}$ **68.** $\sqrt{\dfrac{3y}{20}}$ **69.** $\sqrt[3]{40}$ **70.** $\sqrt[3]{54}$

71. $\sqrt[3]{-250x^4}$ **72.** $\sqrt[3]{-24a^5}$ **73.** $\sqrt[3]{\dfrac{1}{2}}$ **74.** $\sqrt[3]{\dfrac{3x}{25}}$

75. $\sqrt[4]{32x^4y^5}$ **76.** $\sqrt[4]{48a^{12}b^{21}}$

77. $\sqrt[4]{\dfrac{5}{2x}}$ **78.** $\sqrt[4]{\dfrac{3}{8x^2}}$

Perform the indicated operations and simplify your answer. Assume that all variables represent positive real numbers.

79. $3\sqrt{6} + 9 - 5\sqrt{6}$ **80.** $3\sqrt{2} + 8 - 5\sqrt{2}$

81. $\sqrt{8} + \sqrt{20} - \sqrt{12}$

82. $\sqrt{18} - \sqrt{50} + \sqrt{12} - \sqrt{75}$

83. $\left(-2\sqrt{3}\right)\left(5\sqrt{6}\right)$ **84.** $\left(-3\sqrt{2}\right)\left(-2\sqrt{3}\right)$

85. $\left(3\sqrt{5a}\right)\left(4\sqrt{5a}\right)$ **86.** $\left(-2\sqrt{6}\right)\left(3\sqrt{6}\right)$

87. $\left(-5\sqrt{3}\right)^2$ **88.** $\left(3\sqrt{5}\right)^2$

89. $\sqrt{18a} \div \sqrt{2a^4}$ **90.** $\sqrt{21x^7} \div \sqrt{3x^2}$

91. $5 \div \sqrt{x}$ **92.** $a \div \sqrt{b}$

93. $\sqrt{20x^3} + \sqrt{45x^3}$ **94.** $\sqrt[3]{16a^4} + \sqrt[3]{54a^4}$

Write each expression using a single radical sign. Assume that all variables represent positive real numbers. Simplify the radicand where possible.

95. $\sqrt[3]{3} \cdot \sqrt{2}$ **96.** $\sqrt{5} \cdot \sqrt[3]{4}$

97. $\sqrt[3]{3} \cdot \sqrt[4]{x}$ **98.** $\sqrt[3]{3} \cdot \sqrt[4]{4}$

99. $\sqrt{xy} \cdot \sqrt[3]{2xy}$ **100.** $\sqrt[3]{2a} \cdot \sqrt{2a}$

101. $\sqrt[3]{2ab^2} \cdot \sqrt[4]{3ab}$ **102.** $\sqrt[3]{3x^2y} \cdot \sqrt[5]{2xy^2}$

103. $(ab^2)^{1/3}a^{1/2}$ **104.** $(a^2b^3)^{1/4}(ab^2)^{1/3}$

105. $\sqrt[3]{\sqrt{7}}$ **106.** $\sqrt[3]{\sqrt[3]{2a}}$

107. $\sqrt[3]{\sqrt[5]{6}}$ **108.** $\sqrt[4]{\sqrt[4]{2}}$

109. $\sqrt[3]{\sqrt{2x}} \cdot \sqrt[4]{x}$ **110.** $\sqrt{x\sqrt[3]{x}} \cdot \sqrt[3]{x^2}$

The expression $\sqrt{b^2 - 4ac}$ will be used in the quadratic formula in Chapter 2. Evaluate this expression for each choice of a, b, and c.

111. $a = -2, b = -3, c = 5$

112. $a = 2, b = -2, c = -5$

113. $a = -1, b = -2, c = 5$

114. $a = \dfrac{1}{2}, b = 8, c = 2$

115. $a = -2.4, b = -2.3, c = 3.6$

116. $a = 1.9, b = -3.6, c = -4.7$

The expression $\sqrt{(x_1 - x_2)^2 + (y_1 - y_2)^2}$ will be used to find the distance between two points in Chapter 3. Evaluate the expression for each choice of x_1, x_2, y_1, and y_2.

117. $x_1 = 7, x_2 = 5, y_1 = 3, y_2 = 1$

118. $x_1 = 0, x_2 = 4, y_1 = -3, y_2 = 0$

119. $x_1 = -4, x_2 = 3, y_1 = -4, y_2 = 3$

120. $x_1 = -3, x_2 = -1, y_1 = 1, y_2 = -1$

121. $x_1 = -1.4, x_2 = 3.2, y_1 = -1.4, y_2 = 1.6$

122. $x_1 = -2.3, x_2 = 0, y_1 = -1.3, y_2 = 5.8$

The new International America's Cup Class rules, which took effect in 1989, establish the boundaries of a new class of longer, lighter, and faster yachts. In addition to satisfying 200 pages of other rules, the basic dimensions of any yacht competing for the silver trophy must satisfy the inequality

$$L + 1.25S^{1/2} - 9.8D^{1/3} \le 16.296$$

where L is the length in meters, S is the sail area in square meters, and D is the displacement in cubic meters. (*Scientific American,* May 1992)

123. *America's Cup inequality* Determine whether The New Zealand Challenge (considered a short, light boat), with a length of 20.95 m, a sail area of 277.3 m², and a displacement of 17.56 m³, satisfies the inequality.

Figure for Exercises 123 and 124

124. *Adjusting the displacement* The easiest way to make a yacht satisfy the inequality is to add weight to the boat to increase the displacement, without altering the length or sail area. Verify that the published size of The Challenge Australia (considered a long heavy boat), with a length of 21.87 m, a sail area of 311.78 m², and a displacement of 22.44 m³, does not satisfy the inequality. Is the inequality satisfied if the displacement is increased by 0.01 m³?

Solve each problem.

125. *Depreciation rate* The formula $r = 1 - (S/C)^{1/n}$ for $n \ge 1$ gives the annual depreciation rate for a piece of equipment that has a useful life of n years, a salvage value S (in dol-

lars), and an original cost C (in dollars). Find the annual depreciation rate for a computer that has a useful life of 5 years, a salvage value of $200, and an original cost of $5000.

Figure for Exercise 125

126. *Growth rate* The formula $r = (P/P_0)^{1/n} - 1$ gives the annual growth rate for a planet whose population grows from P_0 to P in n years. What annual growth rate would cause the population of the earth in the year 2000 to be double the 1990 population of 5.292 billion (*World Resources 1988–1989*)?

For Writing/Discussion

127. *The lost rule?* Is it true that the square root of a sum is equal to the sum of the square roots? Explain.

128. *Technicalities* If m and n are real numbers and $m^2 = n$, then m is a square root of n, but if $m^3 = n$ then m is *the* cube root of n. How do we know when to use "a" or "the"?

129. *Cooperative learning* On the planet Zoran, mathematicians developed algebra in the same manner as Earthlings did. However, Zoranians use exponent $\underline{n}$ for nth root and exponent $\bar{n}$ for reciprocal of the nth power. For example, on Zoran $8^{\underline{3}} = 2$ and $3^{\bar{2}} = 1/9$. Work in a small group to translate some of the expressions found in Exercises 21–36 into Zoranian and simplify them as a Zoranian would. State the Zoranian power of a power rule. What are the advantages of our system of exponents?

1.4

Polynomials

A polynomial is a type of algebraic expression that is as fundamental to algebra as the natural numbers are to arithmetic. In this section we will review some basic facts about polynomials.

Definitions

A polynomial is simply a single term or a finite sum of terms. Some examples of polynomials are

$$4, \qquad -\frac{1}{2}x^5, \qquad 2x^3 - 3x^2 + 5x - 9, \qquad \text{and} \qquad a^3 + 3a^2b + 3ab^2 + b^3.$$

A term was defined in Section 1.1 as the product of a number and one or more variables raised to powers. In a polynomial the powers of the variables must be whole numbers. We can give a formal definition of a polynomial in a single variable as follows.

Definition: Polynomial in *x*

If n is a nonnegative integer and $a_0, a_1, a_2, \ldots, a_n$ are real numbers, then

$$a_n x^n + a_{n-1}x^{n-1} + a_{n-2}x^{n-2} + \cdots + a_1 x + a_0$$

is a **polynomial** in a single variable x.

In algebra a single number is often referred to as a **constant.** The last term, a_0, is called the **constant term.** When the coefficient of a term is negative, we write a difference rather than a sum.

Example 1   Using the definition of polynomial

Determine whether each algebraic expression is a polynomial in x.

a) $x^3 - 4x + \sqrt{5}$ **b)** $4\sqrt{x} + 3$ **c)** $x^{-1} + 2$

Solution

The expression in (a) is a polynomial in x. The exponents on the variables in a polynomial must be nonnegative integers. So (b) and (c) are not polynomials. ◆

A polynomial with only one term is called a **monomial.** A polynomial with two terms is called a **binomial.** A polynomial with three terms is called a **trinomial.** We usually write the terms of a polynomial in a single variable so that the exponents are in descending order from left to right. When a polynomial is written in this manner, the coefficient of the first term is called the **leading coefficient.**

The **degree** of a polynomial in one variable is the highest power of the variable in the polynomial. The degree of $x^3 - 4x - \sqrt{5}$ is 3. A constant such as 5 is a polynomial with zero degree because $5 = 5x^0$. The number 0 is called the **zero polynomial.** It is a monomial without degree because $0 = 0x^n$ for any positive integer n. A first-degree polynomial is called a **linear polynomial,** a second-degree polynomial is called a **quadratic polynomial,** and a third-degree polynomial is called a **cubic polynomial.**

Example 2   Using the definitions of polynomial types

Find the degree and leading coefficient of each polynomial and determine whether the polynomial is a monomial, binomial, or trinomial.

a) $\dfrac{x^3}{2} - \dfrac{1}{8}$ **b)** $5x^2 + x - 9$ **c)** $3x$

Solution

Polynomial (a) is a third-degree binomial, (b) is a second-degree trinomial, and (c) is a first-degree monomial. The leading coefficients are $\frac{1}{2}$, 5, and 3, respectively. We can also describe (a) as a cubic polynomial, (b) as a quadratic polynomial, and (c) as a linear polynomial.

We will mainly study polynomials in one variable, but you will also see polynomials in more than one variable. The degree of a term in more than one variable is the sum of the powers of the variables. For example, the degree of $5x^3y^4z$ is 8. The degree of a polynomial in more than one variable is equal to the highest degree of any of its terms. For example, the degree of $3x^2y^2 + x^2y - 5y$ is 4.

Addition and Subtraction of Polynomials

Since the coefficients of a polynomial are real numbers and the variables represent real numbers, the properties of the real numbers are valid for polynomials. We add or subtract polynomials by adding or subtracting the like terms. The results are the same whether you arrange your work horizontally, as in Example 3, or vertically, as in Example 4.

Example 3   Adding and subtracting polynomials horizontally

Find each sum or difference.

a) $(3x^3 - x + 5) + (-8x^3 + 3x - 9)$ **b)** $(x^2 - 5x) - (3x^2 - 4x - 1)$

Solution

a) We use the commutative and associative properties of addition to rearrange the terms.

$$(3x^3 - x + 5) + (-8x^3 + 3x - 9) = (3x^3 - 8x^3) + (-x + 3x) + (5 - 9)$$
$$= -5x^3 + 2x - 4 \qquad \text{Combine like terms.}$$

b) The first step is to distribute the multiplication by -1 over the three terms of the second polynomial, changing the sign of every term.

$$(x^2 - 5x) - (3x^2 - 4x - 1) = x^2 - 5x - 3x^2 + 4x + 1 \qquad \text{Distributive property}$$
$$= -2x^2 - x + 1 \qquad \text{Combine like terms.}$$

Example 4   Adding and subtracting polynomials vertically

Find each sum or difference.

a) $(3x^3 - x + 5) + (-8x^3 + 3x - 9)$ **b)** $(x^2 - 5x) - (3x^2 - 4x - 1)$

Solution

a) Add: $\begin{array}{r} 3x^3 - x + 5 \\ -8x^3 + 3x - 9 \\ \hline -5x^3 + 2x - 4 \end{array}$ **b)** Subtract: $\begin{array}{r} x^2 - 5x \\ 3x^2 - 4x - 1 \\ \hline -2x^2 - x + 1 \end{array}$

Multiplication of Polynomials

We multiplied monomials when we learned the product rule. For example, $(3x^2)(4x^5) = 12x^7$. To multiply a monomial and a binomial, we use the distributive property. For example,

$$3x(2x + 5) = (3x)(2x) + (3x)(5) = 6x^2 + 15x.$$

Note that the commutative and associative properties are also used to get $(3x)(2x) = 6x^2$.

We multiply two binomials by using the distributive property twice and combining like terms. For example,

$$(3x + 1)(2x + 5) = 3x(2x + 5) + 1(2x + 5) \qquad \text{Distributive property}$$
$$= 6x^2 + 15x + 2x + 5 \qquad \text{Distributive property}$$
$$= 6x^2 + 17x + 5 \qquad \text{Combine like terms.}$$

When we multiply polynomials, we multiply each term of the first polynomial by every term of the second polynomial and then combine like terms. We can set up multiplication of polynomials vertically like multiplication of whole numbers:

Multiply:
$$2x + 5$$
$$\underline{3x + 1}$$
$$2x + 5$$ **Multiplying 1 and $2x + 5$ yields $2x + 5$.**
$$\underline{6x^2 + 15x}$$ **Multiplying $3x$ and $2x + 5$ yields $6x^2 + 15x$.**
$$6x^2 + 17x + 5$$ **Add.**

Example 5 Multiplying polynomials

Use the distributive property to find each product.

a) $-3x(2x - 3)$

b) $(x + 5)(2x - 3)$

c) $(x^2 - 3x + 4)(2x - 3)$

Solution

a) $-3x(2x - 3) = -6x^2 + 9x$ **Distributive property**

b)
$$
\begin{aligned}
(x + 5)(2x - 3) &= x(2x - 3) + 5(2x - 3) && \textbf{Distributive property}\\
&= 2x^2 - 3x + 10x - 15 && \textbf{Distributive property}\\
&= 2x^2 + 7x - 15 && \textbf{Combine like terms.}
\end{aligned}
$$

c)
$$
\begin{aligned}
(x^2 - 3x + 4)(2x - 3) &= x^2(2x - 3) - 3x(2x - 3) + 4(2x - 3)\\
&= 2x^3 - 3x^2 - 6x^2 + 9x + 8x - 12\\
&= 2x^3 - 9x^2 + 17x - 12
\end{aligned}
$$

Using FOIL

The product of two binomials (as in Example 5b) results in four terms. Such products can be found quickly if we memorize where those four terms come from. The word FOIL is used as a memory aid. Consider the following product:

$$(a + b)(c + d) = a(c + d) + b(c + d) = \overset{F}{ac} + \overset{O}{ad} + \overset{I}{bc} + \overset{L}{bd}.$$

The product of the two binomials consists of four terms:

the product of the *First* term of each (ac),

the product of the *Outer* terms (ad),

the product of the *Inner* terms (bc), and

the product of the *Last* term of each (bd).

Using FOIL to multiply two binomials is simply a way of speeding up the distributive property.

Example 6   Multiplying binomials using FOIL

Find each product using FOIL.

a) $(x + 4)(2x - 3)$ **b)** $(x^2 - 3)(2x - 3)$ **c)** $(a^2b + 5)(a^2b - 5)$ **d)** $(a + b)^2$

Solution

$$\overset{\text{F}\qquad\text{O}\qquad\text{I}\qquad\text{L}}{}$$

a) $(x + 4)(2x - 3) = \overbrace{2x^2} - \overbrace{3x} + \overbrace{8x} - \overbrace{12} = 2x^2 + 5x - 12$

b) $(x^2 - 3)(2x - 3) = 2x^3 - 3x^2 - 6x + 9$

c) $(a^2b + 5)(a^2b - 5) = a^4b^2 - 5a^2b + 5a^2b - 25 = a^4b^2 - 25$

d) $(a + b)^2 = (a + b)(a + b) = a^2 + ab + ab + b^2 = a^2 + 2ab + b^2$

Example 7   Multiplying radicals using FOIL

Use FOIL to find the product $\left(\sqrt{2} - 2\sqrt{3}\right)\left(\sqrt{2} + 4\sqrt{3}\right)$.

Solution

$$\left(\sqrt{2} - 2\sqrt{3}\right)\left(\sqrt{2} + 4\sqrt{3}\right) = \sqrt{2}\sqrt{2} + 4\sqrt{6} - 2\sqrt{6} - 2\sqrt{3} \cdot 4\sqrt{3}$$

$$= 2 + 2\sqrt{6} - 24 \qquad {\color{blue}2\sqrt{3} \cdot 4\sqrt{3} = 8 \cdot 3 = 24}$$

$$= -22 + 2\sqrt{6}$$

Special Products

The products found in Examples 6(c) and 6(d) are called **special products** because of the way that the result is simplified. We used FOIL to find those special products, but it is better to memorize the rules given here so that the products can be found quickly.

The Special Products

$(a + b)^2 = a^2 + 2ab + b^2$	**The square of a sum**
$(a - b)^2 = a^2 - 2ab + b^2$	**The square of a difference**
$(a + b)(a - b) = a^2 - b^2$	**The product of a sum and a difference**

It is helpful to memorize the verbal statements of the special products. For example, the first special product would be: "The square of a sum is equal to the square of the first term, plus twice the product of the first and last terms, plus the square of the last term." Or, for the third special product: "The product of a sum and a difference (of the same terms) is equal to the difference of their squares."

Example 8   Finding special products

Find each product by using the special product rules.

a) $(2x + 3)^2$ **b)** $(x^3 - 9)^2$ **c)** $(3x^m + 5)(3x^m - 5)$ **d)** $\left(3 - \sqrt{6}\right)\left(3 + \sqrt{6}\right)$

Solution

a) To find $(2x + 3)^2$ substitute $2x$ for a and 3 for b in $(a + b)^2 = a^2 + 2ab + b^2$:

$$(2x + 3)^2 = (2x)^2 + 2(2x)(3) + 3^2 = 4x^2 + 12x + 9$$

b) To find $(x^3 - 9)^2$ substitute x^3 for a and 9 for b in $(a - b)^2 = a^2 - 2ab + b^2$:

$$(x^3 - 9)^2 = (x^3)^2 - 2(x^3)(9) + 9^2 = x^6 - 18x^3 + 81$$

c) Substitute $3x^m$ for a and 5 for b in $(a + b)(a - b) = a^2 - b^2$:

$$(3x^m + 5)(3x^m - 5) = (3x^m)^2 - 5^2 = 9x^{2m} - 25$$

d) Substitute 3 for a and $\sqrt{6}$ for b in $(a - b)(a + b) = a^2 - b^2$:

$$\left(3 - \sqrt{6}\right)\left(3 + \sqrt{6}\right) = 3^2 - \left(\sqrt{6}\right)^2 = 9 - 6 = 3 \qquad \blacklozenge$$

In Example 8(d) the expressions $3 - \sqrt{6}$ and $3 + \sqrt{6}$ are called **conjugates.** Their product is a rational number. This fact is used to rationalize a denominator in the next example.

```
√3/(3-√6)
        3.14626437
√3+√2
        3.14626437
```

The approximate values on a calculator give you confidence in your exact answer.

Example 9   Using conjugates to rationalize a denominator

Simplify the expression

$$\frac{\sqrt{3}}{3 - \sqrt{6}}$$

Solution

Multiply the numerator and denominator by $3 + \sqrt{6}$, the conjugate of $3 - \sqrt{6}$:

$$\frac{\sqrt{3}}{3 - \sqrt{6}} = \frac{\sqrt{3}\left(3 + \sqrt{6}\right)}{\left(3 - \sqrt{6}\right)\left(3 + \sqrt{6}\right)} = \frac{3\sqrt{3} + \sqrt{18}}{3} = \frac{3\sqrt{3} + 3\sqrt{2}}{3} = \sqrt{3} + \sqrt{2}$$

$$\blacklozenge$$

Division of Polynomials

We use the quotient rule for exponents to divide monomials. For example,

$$8x^5 \div (2x^2) = \frac{8x^5}{2x^2} = 4x^3.$$

We check division by multiplication, and since $4x^3 \cdot 2x^2 = 8x^5$, we have found the correct quotient. We use the distributive property to divide a polynomial by a monomial. For example,

$$(6x^3 - 9x^2 + 12x) \div (3x) = \frac{6x^3}{3x} - \frac{9x^2}{3x} + \frac{12x}{3x} = 2x^2 - 3x + 4.$$

Since $3x(2x^2 - 3x + 4) = 6x^3 - 9x^2 + 12x$, we have found the correct quotient. You should check all division by multiplication.

To divide the trinomial $x^2 - 3x - 10$ by the binomial $x - 5$, we use a process that looks like long division of whole numbers:

$$
\begin{array}{r}
x + 2 \\
x - 5 \overline{)\, x^2 - 3x - 9\,} \\
\underline{x^2 - 5x} \\
2x - 9 \\
\underline{2x - 10} \\
1
\end{array}
$$

$x^2 \div x = x$
$x(x - 5) = x^2 - 5x$
Subtract: $-3x - (-5x) = 2x$. Bring down -9.
$2x \div x = 2$, $2(x - 5) = 2x - 10$
Subtract: $-9 - (-10) = 1$.

Check: $(x + 2)(x - 5) + 1 = x^2 - 3x - 10 + 1 = x^2 - 3x - 9$. For this division, 1 is the remainder, $x - 5$ is the divisor, $x + 2$ is the quotient, and $x^2 - 3x + 9$ is the dividend. These quantities are related as follows.

The Division Algorithm

> If the **dividend** $P(x)$ and the **divisor** $D(x)$ are polynomials such that $D(x) \neq 0$ and the degree of $P(x)$ is greater than or equal to the degree of $D(x)$, then there exist unique polynomials, the **quotient** $Q(x)$ and the **remainder** $R(x)$, such that
>
> $$P(x) = Q(x)D(x) + R(x),$$
>
> where $R(x) = 0$ or the degree of $R(x)$ is less than the degree of $D(x)$.

The division algorithm indicates that

$$\text{dividend} = (\text{quotient})(\text{divisor}) + \text{remainder}.$$

We can also express the relationship between these quantities as

$$\frac{\text{dividend}}{\text{divisor}} = \text{quotient} + \frac{\text{remainder}}{\text{divisor}}.$$

For the preceding division we can write

$$\frac{x^2 - 3x - 9}{x - 5} = x + 2 + \frac{1}{x - 5}.$$

Example 10   Dividing polynomials

Find the quotient when the first polynomial is divided by the second.

a) $-6x^7 \div (-3x^2)$ **b)** $(8x^3 - 4x) \div (-2x)$ **c)** $(x^3 - 8) \div (x - 2)$

Solution

a) $-6x^7 \div (-3x^2) = \dfrac{-6x^7}{-3x^2} = 2x^5$

Check by multiplying $2x^5$ and $-3x^2$ to get $-6x^7$.

b) Notice that $-2x$ is a factor of each term in $8x^3 - 4x$:

$$(8x^3 - 4x) \div (-2x) = \frac{8x^3 - 4x}{-2x} = \frac{8x^3}{-2x} - \frac{4x}{-2x} = -4x^2 + 2$$

Check by multiplying $-2x$ and $-4x^2 + 2$ to get $8x^3 - 4x$.

c) To keep the division organized, insert $0x^2$ and $0x$ for the missing x^2 and x terms.

$$
\begin{array}{r}
x^2 + 2x + 4 \\
x - 2 \overline{) x^3 + 0x^2 + 0x - 8} \\
\underline{x^3 - 2x^2} \qquad\qquad \\
2x^2 + 0x \qquad \\
\underline{2x^2 - 4x} \qquad \\
4x - 8 \\
\underline{4x - 8} \\
0
\end{array}
$$

$x^3 \div x = x^2$
$x^2(x - 2) = x^3 - 2x^2$
$0x^2 - (-2x^2) = 2x^2$
$2x(x - 2) = 2x^2 - 4x$
$4(x - 2) = 4x - 8$

The quotient is $x^2 + 2x + 4$.   ◆

The Value of a Polynomial

A polynomial such as $x^2 + 3x + 1$ is an expression. It expresses operations to be performed on x, and it has no numerical value unless we choose a number for x. If we choose $x = 2$, then the value of the polynomial is $2^2 + 3(2) + 1$ or 11. Polynomials are often used to calculate quantities such as profit or revenue. A profit polynomial might be named $P(x)$ (read ''P of x'') and a revenue polynomial might be named $R(x)$. For example, if we refer to the polynomial $x^2 + 3x + 1$ as $P(x)$, then we write $P(x) = x^2 + 3x + 1$. $P(2)$ stands for the value of the polynomial $P(x)$ when $x = 2$. So we write $P(2) = 11$.

```
Y1▊X²−5
Y2▊-X^3+5X−3
Y3=▊
Y4=
Y5=
Y6=
Y7=
Y8=
```

To evaluate polynomials, define the polynomials with the Y= key. See Appendix A for more examples.

Example 11   Evaluating a polynomial

Let $P(x) = x^2 - 5$ and $C(x) = -x^3 + 5x - 3$. Find the following.

a) $P(3)$ **b)** $P(-50)$ **c)** $C(10)$ **d)** $C(-20)$ **e)** $C(0)$

```
Y₁(3)
                4
Y₁(-50)
             2495
Y₂(10)
             -953
```

Use the Y-VARS menu to indicate which polynomials are to be evaluated.

Solution

a) $P(3) = 3^2 - 5 = 4$

b) $P(-50) = (-50)^2 - 5 = 2500 - 5 = 2495$

c) $C(10) = -10^3 + 5(10) - 3 = -1000 + 50 - 3 = -953$

d) $C(-20) = -(-20)^3 + 5(-20) - 3 = 8000 - 100 - 3 = 7897$

e) $C(0) = -0^3 + 5 \cdot 0 - 3 = -3$

◆

For Thought

True or false? Explain. Do not use a calculator.

1. The expression $x^{-2} + 3x^{-1} + 9$ is a trinomial.

2. The degree of the polynomial $x^3 + 5x^2 - 6x^2y^2$ is 3.

3. $(3 + 5)^2 = 3^2 + 5^2$

4. $(50 + 1)(50 - 1) = 2499$

5. If the side of a square is $x + 3$ ft, then $x^2 + 9$ ft^2 is its area.

6. $(2x - 1) - (3x + 5) = -x + 4$ for any real number x.

7. $(a + b)^3 = a^3 + b^3$ for any real numbers a and b.

8. The dividend times the quotient plus the remainder equals the divisor.

9. $\dfrac{x^2 + 5x + 7}{x + 2} = x + 3 + \dfrac{1}{x + 2}$ for any real number x.

10. If $P(x) = 3x^2 + 7$, $Q(x) = 5x^2 - 9$, and $S(x) = 8x^2 - 2$, then $P(17) + Q(17) = S(17)$.

1.4 Exercises

📼 **Tape 2**   💾 **Disk—5.25": 1   3.5": 1   Macintosh: 1**

Determine whether each algebraic expression is a polynomial in x. If it is a polynomial, find its degree and leading coefficient.

1. $x^3 - 4x^2 + \sqrt{5}$

2. $-x^7 - 6x^4$

3. $x^2 - 3x + 2 - \dfrac{1}{x}$

4. $x^{-1} + 5 + x^2$

5. 79

6. $-\dfrac{x}{\sqrt{2}}$

7. $0.07x^2 - 0.012x^5$

8. $\dfrac{x}{2} - \dfrac{x^2}{3} - \dfrac{x^3}{5} + \dfrac{1}{7}$

Find each sum or difference.

9. $(3x^2 - 4x) + (5x^2 + 7x - 1)$

10. $(-3x^2 - 4x + 2) + (5x^2 - 8x - 7)$

11. $(4x^2 - 3x) - (9x^2 - 4x + 3)$

12. $(x^2 + 2x + 4) - (x^2 + 4x + 4)$

13. $(4ax^3 - a^2x) - (5a^2x^3 - 3a^2x + 3)$

14. $(x^2y^2 - 3xy + 2x) - (6x^2y^2 + 4y - 6x)$

Find the sum or difference.

15. Add: $3x - 4$
            $\underline{-x + 3}$

16. Add: $-2x^2 - 5$
            $\underline{3x^2 - 6}$

17. Subtract: $\begin{array}{r} x^2 \quad\;\; - 8 \\ -2x^2 + 3x - 2 \\ \hline \end{array}$

18. Subtract: $\begin{array}{r} -2x^2 - 5x + 9 \\ 4x^2 - 7x \\ \hline \end{array}$

Use the distributive property to find each product.

19. $-3a^3(6a^2 - 5a + 2)$

20. $-2m(m^2 - 3m + 9)$

21. $(3b^2 - 5b + 2)(b - 3)$

22. $(-w^2 - 5w + 6)(w + 5)$

23. $(2x - 1)(4x^2 + 2x + 1)$

24. $(3x - 2)(9x^2 + 6x + 4)$

25. $(x + 5)(x^2 - 5x + 25)$

26. $(a + 3)(a^2 - 3a + 9)$

27. $(x - 4)(z + 3)$

28. $(a - 3)(b + c)$

29. $(a - b)(a^2 + ab + b^2)$

30. $(a + b)(a^2 - ab + b^2)$

Find each product using FOIL.

31. $(a + 9)(a - 2)$

32. $(z - 3)(z - 4)$

33. $(2y - 3)(y + 9)$

34. $(2y - 1)(3y + 4)$

35. $(2x - 9)(2x + 9)$

36. $(4x - 6y)(4x + 6y)$

37. $(2x + 5)^2$

38. $(5x - 3)^2$

39. $(2x^2 + 4)(3x^2 + 5)$

40. $(3x^3 - 2)(5x^3 + 6)$

Find each product using FOIL.

41. $\left(1 + \sqrt{2}\right)\left(3 + \sqrt{2}\right)$

42. $\left(5 + \sqrt{6}\right)\left(2 + \sqrt{6}\right)$

43. $\left(5 + \sqrt{2}\right)\left(4 - 3\sqrt{2}\right)$

44. $\left(2 - 3\sqrt{7}\right)\left(5 - \sqrt{7}\right)$

45. $\left(3\sqrt{2} + \sqrt{3}\right)\left(2\sqrt{2} - \sqrt{3}\right)$

46. $\left(2\sqrt{6} - \sqrt{3}\right)\left(4\sqrt{6} - \sqrt{3}\right)$

47. $\left(3\sqrt{5} - \sqrt{2}\right)\left(3\sqrt{5} + \sqrt{2}\right)$

48. $\left(2\sqrt{7} + \sqrt{3}\right)\left(2\sqrt{7} - \sqrt{3}\right)$

49. $\left(\sqrt{5} + \sqrt{3}\right)^2$

50. $\left(\sqrt{8} - \sqrt{3}\right)^2$

51. $\left(2 + \sqrt{x}\right)^2$

52. $\left(1 + \sqrt{x - 1}\right)^2$

Find each product using the special product rules.

53. $(3x + 5)^2$

54. $(x^3 - 2)^2$

55. $(x^n - 3)(x^n + 3)$

56. $(2z^b + 1)(2z^b - 1)$

57. $\left(\sqrt{2} - 5\right)\left(\sqrt{2} + 5\right)$

58. $\left(6 - \sqrt{3}\right)\left(6 + \sqrt{3}\right)$

59. $\left(3\sqrt{6} - 1\right)^2$

60. $\left(2\sqrt{5} + 2\right)^2$

61. $(3x^3 - 4)^2$

62. $(2x^2y^3 + 1)^2$

63. $\left(2 - \sqrt{x - 4}\right)^2$

64. $\left(\sqrt{x - 9} - 3\right)^2$

Simplify each expression by rationalizing the denominator.

65. $\dfrac{\sqrt{10}}{\sqrt{5} - 2}$

66. $\dfrac{\sqrt{3}}{\sqrt{2} - \sqrt{3}}$

67. $\dfrac{\sqrt{6}}{6 + \sqrt{3}}$

68. $\dfrac{\sqrt{2}}{\sqrt{8} + \sqrt{3}}$

Find each quotient.

69. $36x^6 \div (-4x^3)$

70. $-24y^{12} \div (3y^3)$

71. $(3x^2 - 6x) \div (-3x)$

72. $(6x^3 - 9x^2) \div (3x^2)$

73. $(x^2 + 6x + 9) \div (x + 3)$

74. $(x^2 - 3x - 54) \div (x - 9)$

75. $(a^3 - 1) \div (a - 1)$

76. $(b^6 + 8) \div (b^2 + 2)$

Find the quotient and remainder when the first polynomial is divided by the second.

77. $x^2 + 3x + 3, x - 2$

78. $3x^2 - x + 4, x + 2$

79. $2x^2 - 5, x + 3$

80. $-4x^2 + 1, x - 1$

81. $x^3 - 2x^2 - 2x - 3, x - 3$

82. $x^3 + 3x^2 - 3x + 4, x + 4$

83. $6x^2 - 7x + 2, 2x + 1$

84. $-3x^2 + 4x + 9, 2 - x$

85. $a^2b + 4ab + 3b, a + 3$

86. $3x^3y^2 - 4x^2y + x, xy - 1$

Use division to express each fraction in the form

$$\text{quotient} + \frac{\text{remainder}}{\text{divisor}}.$$

87. $\dfrac{x^2 - 2}{x - 1}$

88. $\dfrac{x^2 + 4x + 5}{x + 1}$

89. $\dfrac{2x^2 - 3x + 1}{x}$

90. $\dfrac{2x^2 + 1}{x^2}$

91. $\dfrac{x^2 + x + 1}{x}$

92. $\dfrac{-x^2 + x - 1}{x}$

93. $\dfrac{x^2 - 2x + 1}{x - 2}$

94. $\dfrac{x^2 - x - 9}{x - 3}$

Let $P(x) = x^2 - 3x + 2$ and $M(x) = -x^3 + 5x^2 - x + 2$.
Find the following.

95. $P(-2)$

96. $P(-1)$

97. $P(10)$

98. $P\left(\dfrac{1}{2}\right)$

99. $M(-3)$

100. $M(50)$

101. $M\left(\dfrac{1}{3}\right)$

102. $M\left(-\dfrac{1}{3}\right)$

Perform the indicated operations mentally. Write down only the answer.

103. $(x - 4)(x + 6)$

104. $(z^4 + 5)(z^4 - 4)$

105. $(2a^5 - 9)(a^5 + 3)$

106. $(3b^2 + 1)(2b^2 + 5)$

107. $(y - 3) - (2y + 6)$

108. $(a^2 + 3) - (a^2 - 6)$

109. $(7a - 9) + (3 - a)$

110. $(3m - 8) + (7 - m)$

111. $(w + 4)^2$

112. $(t - 2)^2$

113. $(4x - 9)(4x + 9)$

114. $(3x + 1)(3x - 1)$

115. $3y^2(y^3 - 3x)$

116. $a^4b(a^2b^2 + 1)$

117. $(6b^3 - 3b^2) \div (3b^2)$

118. $(3w - 8) \div (8 - 3w)$

119. $(x^3 + 8)(x^3 - 8)$

120. $(x^2 - 4)(x^2 + 4)$

121. $(3w^2 - 2n)^2$

122. $(7y^2 - 3x)^2$

123. $(\sqrt{5} - 2)(\sqrt{5} + 2)$

124. $(\sqrt{2} - \sqrt{3})(\sqrt{2} + \sqrt{3})$

125. $(1 + \sqrt{2})^2$

126. $(4 - \sqrt{3})^2$

Solve each problem.

127. *Area of a rectangle*   If the length of a field is $x + 3$ m and the width is $2x - 1$ m, then what trinomial represents the area of the field?

128. *Area of a triangle*   If the base of a triangular sign is $2x - 3$ cm and the height is $2x + 6$ cm, then what trinomial represents the area of the sign?

129. *Volume of a rectangular box*   If the volume of a box is $x^3 + 9x^2 + 26x + 24$ cm^3 and the height is $x + 4$ cm, then what polynomial represents the area of the bottom?

$x + 4$

Figure for Exercise 129

130. *Area of a triangle*   If the area of a triangle is $x^2 - x - 6$ m^2 and the base is $x - 3$ m, then what polynomial represents the height?

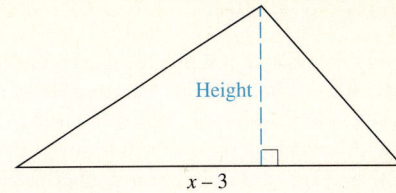

Height

$x - 3$

Figure for Exercise 130

131. *Available habitat*   A wild animal generally stays at least x mi from the edge of a forest. For a rectangular forest preserve that is 6 mi long and 4 mi wide, write a polynomial that represents the area of the available habitat for the wild animal.

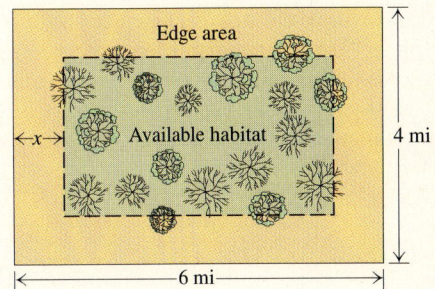

Edge area

$\leftarrow x \rightarrow$ Available habitat   4 mi

6 mi

Figure for Exercise 131

132. *Red fox*   What percent of the forest in Exercise 131 is available habitat for the red fox if $x = 0.5$ mi for this wild animal?

133. *Hazardous waste*   The polynomial $C(x) = 50x + 30$ is used to determine the cost in dollars of incinerating x tons of hazardous waste. Find the cost of incinerating 20 tons.

134. *Water pollution*   The polynomial $F(x) = 500 + 200x$ is used to determine the fine in dollars for violating water pollution standards, where x is the number of days that the violation persists after the citation. If a chemical plant persists in violating water pollution standards for 25 days after being cited, then how much is the fine?

135. *Profit for computer sales*   The polynomial $P(n) = -2n^2 + 800n$ gives the profit in dollars on the sale of n computers in a single order. Find the profit for the sale of 200 computers.

136. *Revenue for computer sales*   The polynomial $R(n) = 2400n$ gives the revenue in dollars on the sale of n computers. Find the revenue for a sale of 20 computers.

For Writing/Discussion

137. *Area of a rectangle*   Wilson thought that his lot was in the shape of a square. After he had it surveyed he discovered that it was rectangular in shape with the length 10 ft longer than he thought and the width 10 ft shorter than he thought. Does he have more or less area than he thought? How much? Explain.

138. *Egyptian area formula*   Surrounded by thousands of square miles of arid desert, ancient Egypt was sustained by the green strips of land bordering the Nile. Villagers cooperated to control the annual flooding so that all might reap abundant harvests from a network of fields and orchards divided by irrigation canals. The Egyptians used the expression

$$\frac{(a + b)(c + d)}{4}$$

to approximate the area of a four-sided field, where a and b represent the lengths of two opposite sides and c and d represent the lengths of the other two opposite sides. Is the formula correct if the field is rectangular? Is it correct for every four-sided figure? Explain why the Egyptians used this formula.

Figure for Exercise 138

139. *Cooperative learning*   Each student in your small group should write a polynomial $P(x)$ of a different degree. Use a calculator to evaluate each polynomial for $x = 5000$ and $x = -5000$. Repeat this procedure until you find a relationship between the sign of x, the sign of the leading coefficient, the degree of $P(x)$, and the value of $P(x)$.

1.5

Factoring

Polynomials

In Section 1.4 we studied multiplication of polynomials. In this section we will factor polynomials. **Factoring** ''reverses'' multiplication. By factoring, we can express a complicated polynomial as a product of several simpler expressions, which are often easier to study.

Factoring Out the Greatest Common Factor

To factor $6x^2 - 3x$, notice that $3x$ is a monomial that can be divided evenly into each term. We can use the distributive property to write

$$6x^2 - 3x = 3x(2x - 1).$$

We call this process **factoring out** $3x$. Both $3x$ and $2x - 1$ are **factors** of $6x^2 - 3x$. Since 3 is a factor of $6x^2$ and $3x$, 3 is a **common factor** of the terms of the polynomial. The **greatest common factor (GCF)** is a monomial that includes every number or variable that is a factor of all terms of the polynomial. The monomial $3x$ is the greatest common factor of $6x^2 - 3x$. Usually the common factor has a positive coefficient, but at times it is useful to factor out a common factor with a negative coefficient. For example, we could factor out $-3x$ from $6x^2 - 3x$:

$$6x^2 - 3x = -3x(-2x + 1)$$

Example 1 Factoring out the greatest common factor

Factor out the greatest common factor from each polynomial, first using the GCF with a positive coefficient and then using a negative coefficient.

a) $9x^4 - 6x^3 + 12x^2$ **b)** $x^2y + 10xy + 25y$ **c)** $-z - w$

Solution

a) $9x^4 - 6x^3 + 12x^2 = 3x^2(3x^2 - 2x + 4)$
$$= -3x^2(-3x^2 + 2x - 4)$$

b) $x^2y + 10xy + 25y = y(x^2 + 10x + 25)$
$$= -y(-x^2 - 10x - 25)$$

c) $-z - w = 1(-z - w)$
$$= -1(z + w)$$

Factoring by Grouping

Some four-term polynomials can be factored by **grouping** the terms in pairs and factoring out a common factor from each pair of terms. We show this technique in the next example. As we will see later, grouping can also be used to factor trinomials.

Example 2 Factoring four-term polynomials

Factor each polynomial by grouping.

a) $x^3 + x^2 + 3x + 3$ **b)** $aw + bc - bw - ac$

Solution

a) Factor the common factor x^2 out of the first two terms and the common factor 3 out of the last two terms:

$$x^3 + x^2 + 3x + 3 = x^2(x + 1) + 3(x + 1) \qquad \text{Factor out common factors.}$$
$$= (x + 1)(x^2 + 3) \qquad \text{Factor out the common factor } (x + 1).$$

b) We must first arrange the polynomial so that the first group of two terms has a common factor and the last group of two terms also has a common factor:

$$aw + bc - bw - ac = aw - bw - ac + bc \qquad \text{Rearrange.}$$
$$= w(a - b) - c(a - b) \qquad \text{Factor out common factors.}$$
$$= (w - c)(a - b) \qquad \text{Factor out } (a - b).$$

Factoring $ax^2 + bx + c$

Factoring by grouping was used in Example 2 to factor polynomials with four terms. Factoring by grouping can be used also to factor a trinomial that is the product of two binomials, because there are four terms in the product of two binomials (FOIL). If we examine the multiplication of two binomials using the distributive property, we can develop a strategy for factoring trinomials.

$$
\begin{aligned}
(x + 3)(x + 5) &= (x + 3)x + (x + 3)5 & &\text{Distributive property} \\
&= x^2 + 3x + 5x + 15 & &\text{Distributive property} \\
&= x^2 + 8x + 15 & &\text{Combine like terms.}
\end{aligned}
$$

To factor $x^2 + 8x + 15$ we simply reverse the steps in this multiplication. First write $8x$ as $3x + 5x$, and then factor by grouping. We could write $8x$ as $x + 7x$, $2x + 6x$, and so on, but we choose $3x + 5x$ because 3 and 5 have a product of 15. So to factor $x^2 + 8x + 15$ proceed as follows:

$$
\begin{aligned}
x^2 + 8x + 15 &= x^2 + 3x + 5x + 15 & &\text{Replace } 8x \text{ by } 3x + 5x. \\
&= (x + 3)x + (x + 3)5 & &\text{Factor out common factors.} \\
&= (x + 3)(x + 5) & &\text{Factor out } x + 3.
\end{aligned}
$$

The key to factoring $ax^2 + bx + c$ with $a = 1$ is to find two numbers that have a product equal to c and a sum equal to b, then proceed as above. After you have practiced this procedure, you should skip the middle two steps shown here and just write the answer.

Example 3   Factoring $ax^2 + bx + c$ with $a = 1$

Factor each trinomial.

a) $x^2 - 5x - 14$ **b)** $x^2 + 4x - 21$ **c)** $x^2 - 5x + 6$

Solution

a) Two numbers that have a product of -14 and a sum of -5 are -7 and 2.

$$
\begin{aligned}
x^2 - 5x - 14 &= x^2 - 7x + 2x - 14 & &\text{Replace } -5x \text{ by } -7x + 2x. \\
&= (x - 7)x + (x - 7)2 & &\text{Factor out common factors.} \\
&= (x - 7)(x + 2) & &\text{Factor out } (x - 7).
\end{aligned}
$$

Check by using FOIL.

b) Two numbers that have a product of -21 and a sum of 4 are 7 and -3.

$$
x^2 + 4x - 21 = (x + 7)(x - 3)
$$

Check by using FOIL.

c) Two numbers that have a product of 6 and a sum of -5 are -2 and -3.

$$
x^2 - 5x + 6 = (x - 2)(x - 3)
$$

Check by using FOIL.   ◆

Before trying to factor $ax^2 + bx + c$ with $a \neq 1$, consider the product of two linear factors:

$$(Mx + N)(Px + Q) = MPx^2 + (MQ + NP)x + NQ$$

Observe that the product of MQ and NP (from the coefficient of x) is $MNPQ$. The product of MP (the leading coefficient) and NQ (the constant term) is also $MNPQ$. So if the trinomial

$$ax^2 + bx + c$$

can be factored, there are two numbers that have a sum equal to b (the coefficient of x) and a product equal to ac (the product of the leading coefficient and the constant term). This fact is the key to the **ac-method** for factoring $ax^2 + bx + c$.

▼
Procedure: The *ac*-Method for Factoring

To factor $ax^2 + bx + c$ with $a \neq 1$:

1. **Find two numbers whose sum is b and whose product is ac.**

2. **Replace b by the sum of these two numbers.**

3. **Factor the resulting four-term polynomial by grouping.**

Example 4   Factoring $ax^2 + bx + c$ with $a \neq 1$

Factor each trinomial.

a) $2x^2 + 5x + 2$ **b)** $6x^2 - x - 12$ **c)** $15x^2 - 14x + 3$

Solution

a) Since $ac = 2 \cdot 2 = 4$ and $b = 5$, we need two numbers that have a product of 4 and a sum of 5. The numbers are 4 and 1.

$$\begin{aligned} 2x^2 + 5x + 2 &= 2x^2 + 4x + x + 2 && \text{\color{blue}{Replace $5x$ by $4x + x$.}} \\ &= (x + 2)2x + (x + 2)1 && \text{\color{blue}{Factor by grouping.}} \\ &= (x + 2)(2x + 1) && \text{\color{blue}{Check using FOIL.}} \end{aligned}$$

b) Since $ac = 6(-12) = -72$ and $b = -1$, we need two numbers that have a product of -72 and a sum of -1. The numbers are 8 and -9.

$$\begin{aligned} 6x^2 - x - 12 &= 6x^2 - 9x + 8x - 12 && \text{\color{blue}{Replace $-x$ by $-9x + 8x$.}} \\ &= (2x - 3)3x + (2x - 3)4 && \text{\color{blue}{Factor by grouping.}} \\ &= (2x - 3)(3x + 4) && \text{\color{blue}{Check using FOIL.}} \end{aligned}$$

c) Since $ac = 15 \cdot 3 = 45$ and $b = -14$, we need two numbers that have a product of 45 and a sum of -14. The numbers are -5 and -9.

$$\begin{aligned} 15x^2 - 14x + 3 &= 15x^2 - 5x - 9x + 3 && \text{\color{blue}{Replace $-14x$ by $-5x - 9x$.}} \\ &= (3x - 1)5x + (3x - 1)(-3) && \text{\color{blue}{Factor by grouping.}} \\ &= (3x - 1)(5x - 3) && \text{\color{blue}{Check using FOIL.}} \end{aligned}$$

Factoring the Special Products

In Section 1.4 we learned how to find the special products: the square of a sum, the square of a difference, and the product of a sum and a difference. The trinomial that results from squaring a sum or a difference is called a **perfect square trinomial.** We can write a factoring rule for each special product.

Factoring the Special Products

$a^2 - b^2 = (a + b)(a - b)$	**Difference of two squares**
$a^2 + 2ab + b^2 = (a + b)^2$	**Perfect square trinomial**
$a^2 - 2ab + b^2 = (a - b)^2$	**Perfect square trinomial**

Example 5   Factoring special products

Factor each polynomial.

a) $4x^2 - 1$ **b)** $x^2 - 6x + 9$ **c)** $9y^2 + 30y + 25$ **d)** $x^{2t} - 9$

Solution

a) $4x^2 - 1 = (2x)^2 - 1^2$ Recognize the difference of two squares.

$\qquad = (2x + 1)(2x - 1)$

b) $x^2 - 6x + 9 = x^2 - 2(3x) + 3^2$ Recognize the perfect square trinomial.

$\qquad = (x - 3)^2$

c) $9y^2 + 30y + 25 = (3y)^2 + 2(3y)(5) + 5^2$ Recognize the perfect square trinomial.

$\qquad = (3y + 5)^2$

d) $x^{2t} - 9 = (x^t - 3)(x^t + 3)$ Recognize the difference of two squares.

◆

Factoring the Difference and Sum of Two Cubes

The following formulas are used to factor the difference of two cubes and the sum of two cubes. You should verify these formulas using multiplication.

Factoring the Difference and Sum of Two Cubes

$a^3 - b^3 = (a - b)(a^2 + ab + b^2)$	**Difference of two cubes**
$a^3 + b^3 = (a + b)(a^2 - ab + b^2)$	**Sum of two cubes**

Example 6   Factoring differences and sums of cubes

Factor each polynomial.

a) $x^3 - 27$ **b)** $8w^6 + 125z^3$ **c)** $y^{3m} - 1$

Solution

a) Since $x^3 - 27 = x^3 - 3^3$, we use $a = x$ and $b = 3$ in the formula for factoring the difference of two cubes:

$$x^3 - 27 = (x - 3)(x^2 + 3x + 9)$$

b) Since $8w^6 + 125z^3 = (2w^2)^3 + (5z)^3$, we use $a = 2w^2$ and $b = 5z$ in the formula for factoring the sum of two cubes:

$$8w^6 + 125z^3 = (2w^2 + 5z)(4w^4 - 10w^2z + 25z^2)$$

c) Since $y^{3m} = (y^m)^3$, we can use the formula for the difference of two cubes with $a = y^m$ and $b = 1$:

$$y^{3m} - 1 = (y^m - 1)(y^{2m} + y^m + 1) \qquad \blacklozenge$$

Factoring by Substitution

When a polynomial involves a complicated expression, we can use two substitutions to help us factor. First we replace the complicated expression by a single variable and factor the simpler-looking polynomial. Then we replace the single variable by the complicated expression. This method is called **substitution.**

Example 7   Factoring higher-degree polynomials

Use substitution to factor each polynomial.

a) $w^4 - 6w^2 - 16$ **b)** $(a^2 - 3)^2 + 7(a^2 - 3) + 12$

Solution

a) Replace w^2 by x and w^4 by x^2.

$w^4 - 6w^2 - 16 = x^2 - 6x - 16$

$\qquad\qquad\qquad = (x - 8)(x + 2)$ Factor the trinomial.

$\qquad\qquad\qquad = (w^2 - 8)(w^2 + 2)$ Replace x by w^2.

b) Replace $a^2 - 3$ by b in the polynomial.

$(a^2 - 3)^2 + 7(a^2 - 3) + 12 = b^2 + 7b + 12$

$\qquad\qquad\qquad\qquad\qquad = (b + 3)(b + 4)$ Factor the trinomial.

$\qquad\qquad\qquad\qquad\qquad = (a^2 - 3 + 3)(a^2 - 3 + 4)$ Replace b by $a^2 - 3$.

$\qquad\qquad\qquad\qquad\qquad = a^2(a^2 + 1)$ Simplify. $\blacklozenge$

Factoring Completely

A polynomial is factored when it is written as a product. Unless noted otherwise, our discussion of factoring will continue to be limited to polynomials whose factors have integral coefficients. Polynomials that cannot be factored using integral coefficients are called **prime** or **irreducible over the integers.** For example, $a^2 + 1$, $b^2 + b + 1$, and $x + 5$ are prime polynomials because they cannot be expressed as a product (in a nontrivial manner). Any monomial is also considered a prime polynomial. A polynomial is said to be **factored completely** when it is written as a product of prime polynomials. For example, $3x^2(2x - 3)$ is factored completely because it is a product of a monomial and a binomial that is prime.

Example 8   Factoring completely

Factor each polynomial completely.

a) $2w^4 - 32$ **b)** $-6x^7 + 6x$

Solution

a) $2w^4 - 32 = 2(w^4 - 16)$ Factor out the greatest common factor.

$\qquad = 2(w^2 - 4)(w^2 + 4)$ Difference of two squares

$\qquad = 2(w - 2)(w + 2)(w^2 + 4)$ Difference of two squares

The polynomial is now factored completely because $w^2 + 4$ is prime.

b) $-6x^7 + 6x = -6x(x^6 - 1)$ Greatest common factor

$\qquad = -6x(x^3 - 1)(x^3 + 1)$ Difference of two squares

$\qquad = -6x(x - 1)(x^2 + x + 1)(x + 1)(x^2 - x + 1)$
    Difference of two cubes; sum of two cubes

The polynomial is factored completely because all of the factors are prime.

◆

If 7 is a factor of 147, then 7 is a divisor of 147. If we divide 147 by 7 to get 21 (with no remainder), then we have $147 = 7 \cdot 21$. By factoring 21, we factor 147 completely as $147 = 3 \cdot 7^2$. The same ideas hold for polynomials. If we know one factor of a polynomial, we can use division to find the other factor and then factor the polynomial completely.

Example 9   Using division to factor completely

Factor $x^3 - 7x + 6$ completely, given that $x - 2$ is a factor.

Solution

Use long division to find the other factor:

$$\begin{array}{r} x^2 + 2x - 3 \\ x - 2\overline{)x^3 + 0x^2 - 7x + 6} \\ \underline{x^3 - 2x^2} \\ 2x^2 - 7x \\ \underline{2x^2 - 4x} \\ -3x + 6 \\ \underline{-3x + 6} \\ 0 \end{array}$$

Because the remainder is 0, we have $x^3 - 7x + 6 = (x - 2)(x^2 + 2x - 3)$. Factoring completely gives

$$x^3 - 7x + 6 = (x - 2)(x + 3)(x - 1). \qquad \blacklozenge$$

? For Thought

Mark true if the polynomial is factored correctly and false otherwise. Explain.

1. $x^2 + 6x - 16 = x(x + 6) - 16$
2. $2x^4 - 5x^2 = -x^2(5 - 2x^2)$
3. $2x^4 - 5x^2 - 3 = (x^2 - 3)(2x^2 + 1)$
4. $a^2 - 1 = (a - 1)^2$
5. $a^3 - 1 = (a - 1)(a^2 + 1)$
6. $x^3 - y^3 = (x - y)(x^2 + 2xy + y^2)$
7. $8a^3 + 27b^3 = (2a + 3b)(4a^2 + 6ab + 9b^2)$
8. $x^2 + 8x - 12 = (x + 2)(x - 6)$
9. $(a + 3)6 - (a + 3)x = (a + 3)(6 - x)$
10. $a - b = -1(b - a)$

1.5 Exercises   Tape 2   Disk—5.25": 1   3.5": 1   Macintosh: 1

Factor out the greatest common factor from each polynomial, first using a positive coefficient on the GCF and then using a negative coefficient.

1. $6x^3 - 12x^2$
2. $12x^2 + 18x^3$
3. $4a - 8ab$
4. $3wm + 15wm^2$
5. $3x^6 - 6x^5 + 9x^2$
6. $12x^2y + 18xy - 30y$
7. $-ax^3 + 5ax^2 - 6ax$
8. $-sa^3 + sb^3 - sb$
9. $m - n$
10. $y - x$

Factor each polynomial by grouping.

11. $x^3 + 2x^2 + 5x + 10$

12. $2w^3 - 2w^2 + 3w - 3$

13. $y^3 - y^2 - 3y + 3$

14. $x^3 + x^2 - 7x - 7$

15. $ady - w + d - awy$

16. $xy + ab + by + ax$

17. $x^2y^2 + ab - ay^2 - bx^2$

18. $6yz - 3y - 10z + 5$

Factor each trinomial.

19. $x^2 + 10x + 16$

20. $x^2 + 7x + 12$

21. $x^2 - 4x - 12$

22. $y^2 + 3y - 18$

23. $m^2 - 12m + 20$

24. $n^2 - 8n + 7$

25. $t^2 + 5t - 84$

26. $s^2 - 6s - 27$

27. $2x^2 - 7x - 4$

28. $3x^2 + 5x - 2$

29. $8x^2 - 10x - 3$

30. $18x^2 - 15x + 2$

31. $6y^2 + 7y - 5$

32. $15x^2 - 14x - 8$

33. $12b^2 + 17b + 6$

34. $8h^2 + 22h + 9$

Factor each special product.

35. $t^2 - u^2$

36. $9t^2 - v^2$

37. $t^2 + 2t + 1$

38. $m^2 + 10m + 25$

39. $4w^2 - 4w + 1$

40. $9x^2 - 12xy + 4y^2$

41. $y^{4t} - 25$

42. $121w^{2t} - 1$

43. $9z^2x^2 + 24zx + 16$

44. $25t^2 - 20tw^3 + 4w^6$

45. $x^4 + 2x^2y + y^2$

46. $25t^2 - 80t + 64$

Factor each sum or difference of two cubes.

47. $t^3 - u^3$

48. $m^3 + n^3$

49. $a^3 - 8$

50. $b^3 + 1$

51. $27y^3 + 8$

52. $1 - 8a^6$

53. $27x^3y^6 - 8z^9$

54. $8t^3h^3 + n^9$

Use substitution to factor each polynomial.

55. $y^6 + 10y^3 + 25$

56. $y^8 + 8y^4 + 12$

57. $4a^4b^8 - 8a^2b^4 - 5$

58. $3c^2m^{14} - 22cm^7 + 7$

59. $(2a + 1)^2 + 2(2a + 1) - 24$

60. $2(w - 3)^2 + 3(w - 3) - 2$

61. $(b^2 + 2)^2 - 5(b^2 + 2) + 4$

62. $(t^3 + 5)^2 + 5(t^3 + 5) + 4$

Factor each polynomial completely.

63. $-3x^3 + 27x$

64. $a^4b^2 - 16b^2$

65. $16t^4 + 54w^3t$

66. $8a^6 - a^3b^3$

67. $a^3 + a^2 - 4a - 4$

68. $2b^3 + 3b^2 - 18b - 27$

69. $x^4 - 2x^3 - 8x + 16$

70. $a^4 - a^3 + a - 1$

71. $-36x^3 + 18x^2 + 4x$

72. $-6a^4 - a^3 + 15a^2$

73. $a^7 - a^6 - 64a + 64$

74. $a^5 - 4a^4 - 4a + 16$

75. $-6x^2 - x + 15$

76. $-6x^2 - 9x + 42$

77. $(a^2 + 2)^2 - 4(a^2 + 2) + 3$

78. $(z^3 + 5)^2 - 10(z^3 + 5) + 24$

79. $a^2 + 2ab + b^2 - 9$

80. $x^2 + 4xy + 4y^2 - 1$

For each pair of polynomials, determine whether the first polynomial is a factor of the second.

81. $13, 1261$

82. $7, 861$

83. $x + 3, x^3 + 4x^2 + 4x + 3$

84. $x - 2, x^3 - 3x^2 + 5x - 6$

85. $x - 1, 3x^3 + 5x^2 - 12x - 9$

86. $x + 1, 2x^3 + 4x^2 - 9x + 2$

In each pair of polynomials, the second polynomial is a factor of the first. Factor the first polynomial completely.

87. $x^3 + 4x^2 + x - 6, x - 1$

88. $x^3 - 13x - 12, x + 1$

89. $x^3 - x^2 - 4x - 6, x - 3$

90. $2x^3 + 7x^2 + 5x + 1, 2x + 1$

91. $x^4 + 5x^3 + 5x^2 - 5x - 6, x + 2$

92. $4x^4 + 4x^3 - 25x^2 - x + 6, x - 2$

Factor each polynomial. Assume that variables used as exponents represent positive integers.

93. $x^{2m} - 1$

94. $x^{10w} - y^{2z}$

95. $x^{2q} - 5x^q + 6$

96. $x^{2a+2} + 6x^{a+1} + 9$

97. $x^{3n} - 1$

98. $x^{12a} + 8$

99. $x^{2w+1} + x^{w+1} + 2x^w + 2$

100. $y^{3n} - 3y^{2n} - y^n + 3$

Polynomials can be factored using radicals. For example, $x^2 - 2 = (x - \sqrt{2})(x + \sqrt{2})$. Factor each of the following polynomials using radicals.

101. $x^2 - 5$

102. $3a^2 - 2b^2$

103. $a^3 - 2$

104. $x^3 + 5$

105. $x^4 - 2$

106. $9w^4 - 5$

107. $b^3 + \sqrt{2}$

108. $x^3 - \sqrt{3}$

Solve each problem.

109. *Volume of a rectangular box* If the volume of a box is $x^3 - 1$ ft^3 and the height is $x - 1$ ft, then what polynomial represents the area of the bottom?

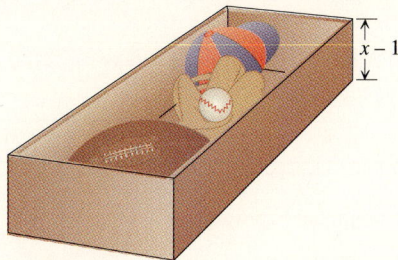

Figure for Exercise 109

110. *Area of a trapezoid* If the area of a trapezoid is $2x^2 + 5x + 2$ square meters and the two parallel sides are x meters and $x + 1$ meters, then what polynomial represents the height of the trapezoid?

Figure for Exercise 110

111. *Volume of an Egyptian pyramid* The Great Pyramid of Egypt was built with a square base 755 ft on each side and a height of 480 ft. The Egyptians knew that the formula $V = \frac{1}{3}ha^2$ gives the volume of a pyramid that has a base with area a^2 and a height h. Find the volume of the Great Pyramid.

112. *Pyramid in progress* The ancient Egyptians used the formula
$$V = \frac{H}{3}(a^2 + ab + b^2)$$
for the volume of a truncated pyramid with a square base of area a^2, a square top of area b^2, and height H as shown in the drawing. When the Great Pyramid was half as tall as its planned height (like a truncated pyramid), what volume of stone had been set in place? (See Exercise 111.) Was the pyramid half finished when it was half as high as planned?

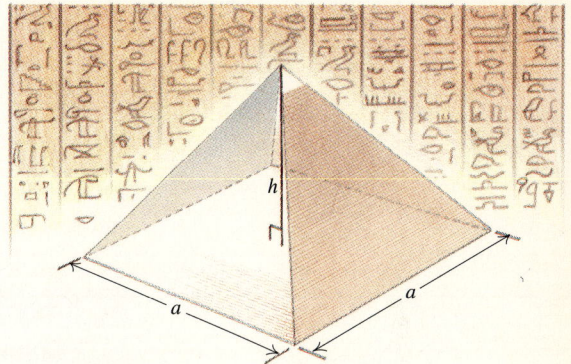

Figure for Exercise 111

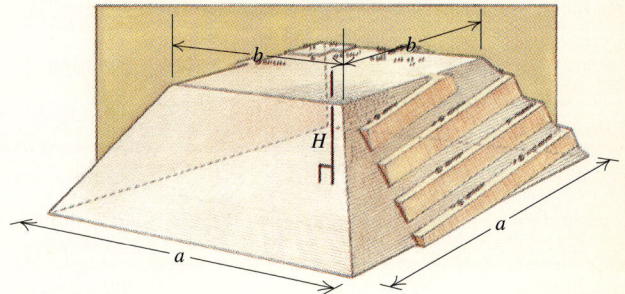

Figure for Exercise 112

For Writing/Discussion

113. *Volume of a truncated pyramid* Use the formula from Exercise 111 and the fact that $a^2 + ab + b^2$ is a factor of $a^3 - b^3$ to prove that the formula for the volume of a truncated pyramid given in Exercise 112 is correct.

114. *Another lost rule?* Is it true that the sum of two cubes factors as the cube of a sum? Explain.

115. *Cooperative learning* Work in a small group to write a summary of the techniques used for factoring polynomials. When given a polynomial to factor completely, what should you look for first, second, third, and so on? Evaluate your summary by using it as a guide to factor some polynomials selected from the exercises.

1.6

Rational Expressions

Rational expressions are to algebra what fractions are to arithmetic. In this section we will learn to reduce or build up rational expressions, and to perform basic operations with rational expressions.

Reducing

A **rational expression** is a ratio of two polynomials in which the denominator is not the zero polynomial. For example,

$$\frac{7}{2}, \quad \frac{2x - 1}{x + 3}, \quad 2a - 9, \quad \text{and} \quad \frac{2x + 4}{x^2 + 5x + 6}$$

are rational expressions. The **domain** of a rational expression is the set of all real numbers that can be used in place of the variable.

Example 1   Domain of a rational expression

Find the domain of each rational expression.

a) $\dfrac{2x - 1}{x + 3}$ b) $\dfrac{2x + 4}{(x + 2)(x + 3)}$ c) $\dfrac{1}{x^2 + 8}$

Solution

a) The domain is the set of all real numbers except those that cause $x + 3$ to have a value of 0. So -3 is excluded from the domain, because $x + 3$ has a value of 0 for $x = -3$. We write the domain in set notation as $\{x \mid x \neq -3\}$.

b) The domain is the set of all real numbers except -2 and -3, because replacing x by either of these numbers would cause the denominator to be 0. The domain is written in set notation as $\{x \mid x \neq -2 \text{ and } x \neq -3\}$.

c) The value of $x^2 + 8$ is positive for any real number x. So the domain is the set of all real numbers, R.   ◆

In arithmetic we learned that each rational number has infinitely many equivalent forms. This fact is due to the **basic principle of rational numbers.**

Basic Principle of Rational Numbers

If a, b, and c are integers with $b \neq 0$ and $c \neq 0$, then

$$\frac{ac}{bc} = \frac{a}{b}.$$

The basic principle holds true when a, b, c are real numbers as well as when they are integers, but we use it most often with integers, when we reduce fractions. For

```
3/6▶Frac
            1/2
408/1512▶Frac
            17/63
```

A fraction can be reduced using the fraction feature found in the MATH menu.

example, we reduce $\frac{3}{6}$ as follows:

$$\frac{3}{6} = \frac{1 \cdot 3}{2 \cdot 3} = \frac{1}{2}$$

Note that since 3 and 6 have a common factor of 3, we divide both the numerator and the denominator by 3 to reduce the fraction. We reduce rational expressions in the same manner: Factor the numerator and denominator completely, then *divide out* the common factors. A rational expression is in *lowest terms* when all common factors have been divided out.

Example 2   Reducing to lowest terms

Reduce each rational expression to lowest terms.

a) $\dfrac{2x + 4}{x^2 + 5x + 6}$ **b)** $\dfrac{b - a}{a^3 - b^3}$ **c)** $\dfrac{x^2 z^3}{x^5 z}$

Solution

a) $\dfrac{2x + 4}{x^2 + 5x + 6} = \dfrac{2(x + 2)}{(x + 2)(x + 3)}$ Factor the numerator and denominator.

$= \dfrac{2}{x + 3}$ Divide out the common factor $x + 2$.

b) $\dfrac{b - a}{a^3 - b^3} = \dfrac{-1(a - b)}{(a - b)(a^2 + ab + b^2)}$ Factor -1 out of $b - a$.

$= \dfrac{-1}{a^2 + ab + b^2}$

c) $\dfrac{x^2 z^3}{x^5 z} = \dfrac{(x^2 z)(z^2)}{(x^2 z)(x^3)} = \dfrac{z^2}{x^3}$ (The quotient rule yields the same result.)

◆

Be careful when reducing. The *only* way to reduce rational expressions is to factor and divide out the common *factors*. Identical terms that are not factors cannot be eliminated from a rational expression. For example,

$$\frac{x + 3}{3} \neq x$$

for all real numbers, because 3 is not a factor of the numerator.

Multiplication and Division

We multiply two rational numbers by multiplying their numerators and their denominators. For example,

$$\frac{2}{3} \cdot \frac{5}{7} = \frac{10}{21}.$$

Definition: Multiplication of Rational Numbers

If a/b and c/d are rational numbers, then

$$\frac{a}{b} \cdot \frac{c}{d} = \frac{ac}{bd}.$$

We multiply rational expressions in the same manner as rational numbers. Of course, any common factor can be divided out as we do when reducing rational expressions.

Example 3   Multiplying rational expressions

Multiply the rational expressions.

a) $\dfrac{2a - 2b}{6} \cdot \dfrac{9a}{a^2 - b^2}$ **b)** $\dfrac{x - 1}{x^2 + 4x + 4} \cdot \dfrac{x + 2}{x^2 + 2x - 3}$

Solution

a) $\dfrac{2a - 2b}{6} \cdot \dfrac{9a}{a^2 - b^2} = \dfrac{2(a - b)}{2 \cdot 3} \cdot \dfrac{3 \cdot 3a}{(a - b)(a + b)}$ Factor completely.

$= \dfrac{3a}{a + b}$ Divide out common factors.

b) $\dfrac{x - 1}{x^2 + 4x + 4} \cdot \dfrac{x + 2}{x^2 + 2x - 3} = \dfrac{x - 1}{(x + 2)^2} \cdot \dfrac{x + 2}{(x + 3)(x - 1)}$ Factor completely.

$= \dfrac{1}{x^2 + 5x + 6}$ Divide out common factors.

We divide rational numbers by multiplying by the reciprocal of the divisor, or *invert and multiply*. For example, $6 \div \frac{1}{2} = 6 \cdot 2 = 12$.

Definition: Division of Rational Numbers

If a/b and c/d are rational numbers with $c \neq 0$, then

$$\frac{a}{b} \div \frac{c}{d} = \frac{a}{b} \cdot \frac{d}{c}.$$

Rational expressions are divided in the same manner as rational numbers.

Example 4   Dividing rational expressions

Perform the indicated operations.

a) $\dfrac{9}{2x} \div \dfrac{3}{x}$

b) $\dfrac{4 - x^2}{6} \div \dfrac{x - 2}{2}$

Solution

a) $\dfrac{9}{2x} \div \dfrac{3}{x} = \dfrac{9}{2x} \cdot \dfrac{x}{3}$ Invert and multiply.

$= \dfrac{3 \cdot 3}{2x} \cdot \dfrac{x}{3}$ Factor completely.

$= \dfrac{3}{2}$ Divide out common factors.

b) $\dfrac{4 - x^2}{6} \div \dfrac{x - 2}{2} = \dfrac{-1(x - 2)(x + 2)}{2 \cdot 3} \cdot \dfrac{2}{x - 2}$ Invert and multiply.

$= \dfrac{-x - 2}{3}$ Divide out common factors.

Note that the division in Example 4(a) is valid only if $x \neq 0$. The division in Example 2(b) is valid only if $x \neq 2$ because $x - 2$ appears in the denominator after the rational expression is inverted.

Building Up the Denominator

The addition of fractions can be carried out only when their denominators are identical. To get a required denominator, we may **build up** the denominator of a fraction. To build up the denominator we use the basic principle of rational numbers in the reverse of the way we use it for reducing. We multiply the numerator and denominator of a fraction by the same nonzero number to get an equivalent fraction. For example, to get a fraction equivalent to $\frac{1}{3}$ with a denominator of 12, we multiply the numerator and denominator by 4:

$$\frac{1}{3} = \frac{1 \cdot 4}{3 \cdot 4} = \frac{4}{12}$$

To build up the denominator of a rational expression, we use the same procedure.

Example 5 Writing equivalent rational expressions

Convert the first rational expression into an equivalent one that has the indicated denominator.

a) $\dfrac{3}{2a}, \dfrac{?}{6ab}$ **b)** $\dfrac{x-1}{x+2}, \dfrac{?}{x^2+6x+8}$ **c)** $\dfrac{a}{3b-a}, \dfrac{?}{a^2-9b^2}$

Solution

a) Compare the two denominators. Since $6ab = 2a(3b)$, we multiply the numerator and denominator of the first expression by $3b$:

$$\frac{3}{2a} = \frac{3 \cdot 3b}{2a \cdot 3b} = \frac{9b}{6ab}$$

b) Factor the second denominator and compare it to the first. Since $x^2 + 6x + 8 = (x + 2)(x + 4)$, we multiply the numerator and denominator by $x + 4$:

$$\frac{x-1}{x+2} = \frac{(x-1)(x+4)}{(x+2)(x+4)} = \frac{x^2+3x-4}{x^2+6x+8}$$

c) Factor the second denominator as

$$a^2 - 9b^2 = (a - 3b)(a + 3b) = -1(3b - a)(a + 3b).$$

Since $3b - a$ is a factor of $a^2 - 9b^2$, we multiply the numerator and denominator by $-1(a + 3b)$:

$$\frac{a}{3b-a} = \frac{a(-1)(a+3b)}{(3b-a)(-1)(a+3b)} = \frac{-a^2-3ab}{a^2-9b^2}$$

Addition and Subtraction

Fractions can be added or subtracted only if their denominators are identical. For example, $\frac{1}{3} + \frac{1}{3} = \frac{2}{3}$ and $\frac{7}{12} - \frac{2}{12} = \frac{5}{12}$.

Definition: Addition and Subtraction of Rational Numbers

If a/b and c/b are rational numbers, then

$$\frac{a}{b} + \frac{c}{b} = \frac{a+c}{b} \quad \text{and} \quad \frac{a}{b} - \frac{c}{b} = \frac{a-c}{b}.$$

For fractions with different denominators, we build up one or both denominators to get denominators that are equal to the least common multiple (LCM) of the denominators. The **least common denominator (LCD)** is the smallest number that is a multiple of all of the denominators. Use the following steps to find the LCD.

> ▼
>
> ## Procedure: Finding the LCD
>
> 1. **Factor each denominator completely.**
> 2. **Write a product using each factor that appears in a denominator.**
> 3. **For each factor, use the highest power of that factor that occurs in the denominators.**

For example, to find the LCD for 10 and 12, we write $10 = 2 \cdot 5$ and $12 = 2^2 \cdot 3$. The LCD contains the factors 2, 3, and 5. Using the highest power of each, we get $2^2 \cdot 3 \cdot 5 = 60$ for the LCD. So, to add fractions with denominators of 10 and 12, we build up each fraction to a denominator of 60:

$$\frac{1}{10} + \frac{1}{12} = \frac{1 \cdot 6}{10 \cdot 6} + \frac{1 \cdot 5}{12 \cdot 5} = \frac{6}{60} + \frac{5}{60} = \frac{11}{60}$$

We use the same method to add or subtract rational expressions.

```
1/10+1/12▶Frac
          11/60
■
```

You can get fractional answers to operations with fractions using the fraction feature.

Example 6   Adding and subtracting rational expressions

Perform the indicated operations.

a) $\dfrac{x}{x - 1} + \dfrac{2x + 3}{x^2 - 1}$ **b)** $\dfrac{x}{x^2 + 6x + 9} - \dfrac{x - 3}{x^2 + 5x + 6}$

Solution

a) $\dfrac{x}{x - 1} + \dfrac{2x + 3}{x^2 - 1} = \dfrac{x}{x - 1} + \dfrac{2x + 3}{(x - 1)(x + 1)}$ Factor denominators completely.

$= \dfrac{x(x + 1)}{(x - 1)(x + 1)} + \dfrac{2x + 3}{(x - 1)(x + 1)}$ Build up using the LCD, $(x - 1)(x + 1)$.

$= \dfrac{x^2 + x + 2x + 3}{(x - 1)(x + 1)}$ Add the fractions.

$= \dfrac{x^2 + 3x + 3}{(x - 1)(x + 1)}$ Simplify the numerator.

b) $\dfrac{x}{x^2 + 6x + 9} - \dfrac{x - 3}{x^2 + 5x + 6} = \dfrac{x}{(x + 3)^2} - \dfrac{x - 3}{(x + 2)(x + 3)}$

$= \dfrac{x(x + 2)}{(x + 3)^2(x + 2)} - \dfrac{(x - 3)(x + 3)}{(x + 2)(x + 3)(x + 3)}$

$= \dfrac{x^2 + 2x}{(x + 3)^2(x + 2)} - \dfrac{x^2 - 9}{(x + 3)^2(x + 2)}$

$= \dfrac{2x + 9}{(x + 3)^2(x + 2)}$ ◆

Complex Fractions

A **complex fraction** is a fraction having rational expressions in the numerator, denominator, or both. Complex fractions can be simplified quickly by multiplying the numerator and denominator by the LCD of all of the denominators.

Example 7 Simplifying a complex fraction

Simplify each complex fraction.

a) $\dfrac{4 - \dfrac{3}{4x}}{\dfrac{1}{x^2} - \dfrac{1}{6}}$ **b)** $\dfrac{\dfrac{x+6}{x^2-9}}{\dfrac{x}{x-3} - \dfrac{x+4}{x+3}}$

Solution

a) The LCD for the denominators 6, $4x$, and x^2 is $12x^2$. Multiply the numerator and denominator of the complex fraction by $12x^2$.

$$\frac{4 - \dfrac{3}{4x}}{\dfrac{1}{x^2} - \dfrac{1}{6}} = \frac{\left(4 - \dfrac{3}{4x}\right)12x^2}{\left(\dfrac{1}{x^2} - \dfrac{1}{6}\right)12x^2} = \frac{48x^2 - 9x}{12 - 2x^2}$$

b) The LCD for the denominators $x^2 - 9$, $x - 3$, and $x + 3$ is $x^2 - 9$, because $x^2 - 9 = (x - 3)(x + 3)$. Multiply the numerator and denominator by $(x - 3)(x + 3)$, or $x^2 - 9$:

$$\frac{\dfrac{x+6}{x^2-9}}{\dfrac{x}{x-3} - \dfrac{x+4}{x+3}} = \frac{\left(\dfrac{x+6}{x^2-9}\right)(x-3)(x+3)}{\left(\dfrac{x}{x-3} - \dfrac{x+4}{x+3}\right)(x-3)(x+3)}$$

$$= \frac{x+6}{x(x+3) - (x+4)(x-3)}$$

$$= \frac{x+6}{x^2 + 3x - (x^2 + x - 12)}$$

$$= \frac{x+6}{2x+12} = \frac{x+6}{2(x+6)} = \frac{1}{2}$$

The fractions in a complex fraction can be written with negative exponents. To simplify complex fractions with negative exponents we still multiply the numerator and denominator by the LCD.

Example 8   A complex fraction with negative exponents

Simplify each complex fraction. Write answers with positive exponents only.

a) $\dfrac{a^{-1} + b^{-3}}{ab^{-2} + ba^{-4}}$ **b)** $pq + p^{-1}q^{-2}$

Solution

a) If we use the definition of negative exponent to rewrite each term, then the denominators would be a, b^3, b^2, and a^4. The LCD for these denominators is a^4b^3. Multiply the numerator and denominator by a^4b^3:

$$\frac{a^{-1} + b^{-3}}{ab^{-2} + ba^{-4}} = \frac{(a^{-1} + b^{-3})a^4b^3}{(ab^{-2} + ba^{-4})a^4b^3} = \frac{a^3b^3 + a^4}{a^5b + b^4}$$

Note that the exponents in a^4b^3 are just large enough to eliminate all negative exponents in the multiplication. Note also that we could have rewritten the complex fraction without negative exponents before multiplying by a^4b^3, but it is not necessary to do so.

b) Although this expression is not exactly a complex fraction, we can use the same technique to eliminate the negative exponents. Multiply the numerator and denominator by pq^2:

$$pq + p^{-1}q^{-2} = \frac{(pq + p^{-1}q^{-2})pq^2}{1 \cdot pq^2} = \frac{p^2q^3 + 1}{pq^2}.$$ ◆

? For Thought

True or false? Explain.

1. The rational expression $\dfrac{2x + 5}{2y}$ reduces to $\dfrac{x + 5}{y}$.

2. The rational expression $\dfrac{-3}{a - 5}$ is equivalent to $\dfrac{3}{5 - a}$.

3. The expressions $\dfrac{x(x + 2)}{x(x + 3)}$ and $\dfrac{x + 2}{x + 3}$ are equivalent.

4. The expression $\dfrac{a^2 - b^2}{a - b}$ reduced to lowest terms is $a - b$.

5. The LCD for the rational expressions $\dfrac{1}{x}$ and $\dfrac{1}{x + 1}$ is $x + 1$.

6. $\dfrac{2x - 1}{x - 3} + \dfrac{x + 5}{x - 3} = \dfrac{3x + 4}{x - 3}$ provided that $x \neq 3$.

7. $\dfrac{x}{2} = \dfrac{1}{2}x$ for all nonzero values of x.

8. $\dfrac{x}{3} - 1 = \dfrac{x - 3}{3}$ for any real number x.

9. If $x = 500$, then the approximate value of $\dfrac{2x + 1}{x - 3}$ is 2.

10. If $|x|$ is very large, then $\dfrac{5x + 1}{x}$ has an approximate value of 5.

1.6 Exercises Tape 2   Disk—5.25″: 1   3.5″: 1   Macintosh: 1

Find the domain of each rational expression.

1. $\dfrac{x - 3}{x + 2}$

2. $\dfrac{x^2 - 1}{x - 5}$

3. $\dfrac{x^2 - 9}{(x - 4)(x + 2)}$

4. $\dfrac{2x - 3}{(x + 1)(x - 3)}$

5. $\dfrac{x + 1}{x^2 - 9}$

6. $\dfrac{x + 2}{x^2 + 3x + 2}$

7. $\dfrac{3x^2 - 2x + 1}{x^2 + 3}$

8. $\dfrac{-2x^2 - 7}{3x^2 + 8}$

Reduce each rational expression to lowest terms.

9. $\dfrac{3x - 9}{x^2 - x - 6}$

10. $\dfrac{-2x - 4}{x^2 - 3x - 10}$

11. $\dfrac{10a - 8b}{12b - 15a}$

12. $\dfrac{a^2 - b^2}{b - a}$

13. $\dfrac{a^3b^6}{a^2b^3 - a^4b^2}$

14. $\dfrac{18u^6v^5 + 24u^3v^3}{42u^2v^5}$

15. $\dfrac{x^4y^5z^2}{x^7y^3z}$

16. $\dfrac{t^3u^7}{-t^8u^5}$

17. $\dfrac{a^3 - b^3}{a^2 - b^2}$

18. $\dfrac{a^3 + b^3}{a^2 + b^2}$

19. $\dfrac{ab + 3a - by - 3y}{a^2 - 2ay + y^2}$

20. $\dfrac{x^4 - 16}{x^4 + 8x^2 + 16}$

Find the products or quotients.

21. $\dfrac{2a}{3b^2} \cdot \dfrac{9b}{14a^2}$

22. $\dfrac{14w}{51y} \cdot \dfrac{3w}{7y}$

23. $\dfrac{12a}{7} \div \dfrac{2a^3}{49}$

24. $\dfrac{20x}{y^3} \div \dfrac{30}{y^5}$

25. $\dfrac{a^2 - 9}{3a - 6} \cdot \dfrac{a^2 - 4}{a^2 - a - 6}$

26. $\dfrac{6x^2 + x - 1}{6x + 3} \cdot \dfrac{15}{9x^2 - 1}$

27. $\dfrac{x^2 - y^2}{9} \div \dfrac{x^2 + 2xy + y^2}{18}$

28. $\dfrac{a^3 - b^3}{a^2 - 2ab + b^2} \div \dfrac{2a^2 + 2ab + 2b^2}{9a^2 - 9b^2}$

29. $\dfrac{x^2 - y^2}{-3xy} \cdot \dfrac{6x^2y^3}{2y - 2x}$

30. $\dfrac{a^2 - a - 2}{2} \cdot \dfrac{1}{4 - a^2}$

31. $\dfrac{wx - x}{x^2} \div \dfrac{1 - w^2}{2}$

32. $\dfrac{2 + x}{2} \div \dfrac{x^2 - 4}{4}$

In Exercises 33–40, convert the first rational expression into an equivalent one that has the indicated denominator.

33. $\dfrac{4}{3a}, \dfrac{?}{12a^2}$

34. $\dfrac{a + 2}{4a^2}, \dfrac{?}{20a^3b}$

35. $\dfrac{x-5}{x+3}, \dfrac{?}{x^2-9}$

36. $\dfrac{x+2}{x-8}, \dfrac{?}{16-2x}$

37. $\dfrac{x}{x+5}, \dfrac{?}{x^2+6x+5}$

38. $\dfrac{3a-b}{a+b}, \dfrac{?}{9b^2-9a^2}$

39. $\dfrac{t}{2t+2}, \dfrac{?}{2t^2+4t+2}$

40. $\dfrac{x-1}{2x+4}, \dfrac{?}{4x^2-16x-48}$

Find the least common denominator (LCD) for each given pair of rational expressions.

41. $\dfrac{1}{4ab^2}, \dfrac{7}{6a^2b^3}$

42. $\dfrac{3}{2x^2y}, \dfrac{a}{5xy}$

43. $\dfrac{-7a}{3a+3b}, \dfrac{5b}{2a+2b}$

44. $\dfrac{1}{3a-3b}, \dfrac{2}{a^2-b^2}$

45. $\dfrac{2x}{x^2+5x+6}, \dfrac{3x}{x^2-x-6}$

46. $\dfrac{x+7}{2x^2+7x-15}, \dfrac{x-5}{2x^2-5x+3}$

Perform the indicated operations.

47. $\dfrac{3}{2x}+\dfrac{1}{6}$

48. $\dfrac{-7}{3a^2b}+\dfrac{4}{6ab^2}$

49. $\dfrac{x+3}{x-1}-\dfrac{x+4}{x+1}$

50. $\dfrac{x+2}{x-3}-\dfrac{x^2+3x-2}{x^2-9}$

51. $3+\dfrac{1}{a}$

52. $-1-\dfrac{3}{c}$

53. $t-1-\dfrac{1}{t+1}$

54. $w+\dfrac{1}{w-1}$

55. $\dfrac{x}{x^2+3x+2}+\dfrac{x-1}{x^2+5x+6}$

56. $\dfrac{x-1}{x^2+x-6}-\dfrac{x-2}{x^2+4x+3}$

57. $\dfrac{1}{x-3}-\dfrac{5}{6-2x}$

58. $\dfrac{5}{4-x^2}-\dfrac{2x}{x-2}$

59. $\dfrac{y^2}{x^3-y^3}+\dfrac{x+y}{x^2+xy+y^2}$

60. $\dfrac{ab}{a^3+b^3}+\dfrac{a}{2a^2-2ab+2b^2}$

61. $\dfrac{x-2}{2x^2+7x-15}-\dfrac{x+1}{2x^2-5x+3}$

62. $\dfrac{x-1}{2x^2-5x-3}-\dfrac{x}{4x^2-1}$

63. $\dfrac{1}{x}+\dfrac{1}{x-1}-\dfrac{1}{x+1}$

64. $\dfrac{3}{x}-\dfrac{x-1}{x^2-9}+\dfrac{1}{x-3}$

Simplify each complex fraction.

65. $\dfrac{\dfrac{4}{a}-\dfrac{3}{b}}{\dfrac{1}{ab}+\dfrac{2}{b^2}}$

66. $\dfrac{\dfrac{2}{6xy}-\dfrac{1}{4x}}{\dfrac{1}{3y^2}+\dfrac{1}{2x}}$

67. $\dfrac{\dfrac{a}{ab^2}-\dfrac{b}{ab^3}}{\dfrac{3}{a^2}+\dfrac{a}{a^3b}}$

68. $\dfrac{\dfrac{1}{a^2b^3c}}{\dfrac{c}{ab^2}+\dfrac{a}{b^2c}}$

69. $\dfrac{a+\dfrac{4}{a+4}}{a-\dfrac{4a+4}{a+4}}$

70. $\dfrac{y-\dfrac{y+6}{y+2}}{y-\dfrac{4y+15}{y+2}}$

71. $\dfrac{\dfrac{t+2}{t-1}-\dfrac{t-3}{t}}{\dfrac{t+4}{t}+\dfrac{t-2}{t-1}}$

72. $\dfrac{\dfrac{3}{2+x}-\dfrac{4}{2-x}}{\dfrac{1}{x+2}-\dfrac{3}{x-2}}$

Simplify. Write answers with positive exponents only.

73. $\dfrac{x^{-1}+1}{x^{-1}-1}$

74. $\dfrac{x^{-2}-4}{x^{-1}-2}$

75. $a^2+a^{-1}b^{-3}$

76. $m+m^{-1}+m^{-2}$

77. $\dfrac{x^2-y^2}{x^{-1}-y^{-1}}$

78. $\dfrac{x^{-2}-y^{-2}}{(x-y)^2}$

79. $(m^{-1}-n^{-1})^{-2}$

80. $(a^{-1}+y^{-1})^{-1}$

Let

$$R(x) = \frac{3}{x + 6}, \; S(x) = \frac{2x - 5}{2x - 9}, \text{ and } T(x) = \frac{9x^2 - 1}{3x^2 - 2}.$$

Find the following.

81. $R(1)$ **82.** $R(-1)$ **83.** $R(500)$

84. $R(-1000)$ **85.** $S(2)$ **86.** $S(-2)$

87. $S(600)$ **88.** $S(-600)$ **89.** $T(-4)$

90. $T(7)$ **91.** $T(-400)$ **92.** $T(500)$

Solve each problem.

93. *Filing invoices* Gina can file all of the daily invoices in 4 hr and Bert can do the same job in 6 hr. If they work together, then what portion of the invoices can they file in 1 hr?

94. *Painting a house* Melanie can paint the entire house in x hours and Timothy can do the same job in y hours. Write a rational expression that represents the portion of the house that they can paint in 2 hr working together.

95. *Average speed* Barry Allen, alias the Flash, finishes a meal at the Golden Buddha restaurant only to find he's left his wallet at home. Not wanting to reveal his secret identity to his date, he excuses himself and slips outside. Dashing home at 250 mph, he snatches his wallet from his nightstand and races back to the restaurant at 300 mph. Discounting the time it took to find his wallet, what was his average speed for the trip? (Average speed is total distance divided by total time for the trip.)

96. *Average cost* Every day Denise spends the same amount on eggs for Denise's Diner. On Monday eggs were 50 cents per dozen, on Tuesday they were 60 cents per dozen, and on Wednesday they were 70 cents per dozen. What was her average cost for a dozen eggs for the three-day period? (The average cost is the total cost divided by the total number of dozens purchased.)

For Writing/Discussion

97. *Domain* Why is it important to know the domain of a rational expression?

98. *Cooperative learning* Each student in your small group should write down a rational expression in which the degree of the denominator is greater than or equal to the degree of the numerator. Evaluate your rational expression for several ''very large'' values of x, using a calculator. Discuss your results. Can you predict the approximate value by looking at the rational expression?

Highlights

Section 1.1   Real Numbers and Their Properties
1. Every real number is either rational (a ratio of integers) or irrational.
2. The properties of real numbers are the basic properties of algebra.
3. The absolute value of a real number a is a if $a \geq 0$ and $-a$ if $a < 0$.
4. The absolute value of a number indicates its distance from 0 on a number line.
5. The distance between a and b on a number line is $|a - b|$.

Section 1.2   Integral Exponents
1. A positive integral exponent indicates the number of times the base is used as a factor.

2. An expression with a negative exponent is the reciprocal of the expression with a positive exponent.

3. In scientific notation, a positive number less than 1 or greater than 10 is written as a product of a number between 1 and 10 and a power of 10.

Section 1.3   Rational Exponents and Radicals

1. The fractional exponent $1/n$ indicates nth root.

2. In a fractional exponent the numerator indicates the power and the denominator indicates the root.

3. An even root of a negative number is not a real number.

4. The same rules that apply to expressions with integral exponents apply to expressions with real numbers as exponents.

5. A radical expression of index n is in simplified form when it has no perfect nth powers as factors of the radicand, no fractions inside the radical, and no radicals in a denominator.

Section 1.4   Polynomials

1. If n is a nonnegative integer and a_0, a_1, a_2, ..., a_n are real numbers, then a polynomial in x is an expression of the form $a_n x^n + a_{n-1}x^{n-1} + \cdots + a_1 x + a_0$.

2. Polynomials can be added, subtracted, multiplied, and divided.

3. FOIL is a rule based on the distributive property that helps us find the product of two binomials quickly.

4. The value of a polynomial in x is the value of the expression obtained when x is replaced by a real number.

Section 1.5   Factoring Polynomials

1. To factor a polynomial means to write it as a product of two or more polynomials.

2. A perfect square trinomial is the square of a binomial.

3. Each special product rule is also a factoring rule.

4. All factoring can be checked by multiplication.

Section 1.6   Rational Expressions

1. A rational expression is a ratio of two polynomials, in which the denominator is not the zero polynomial.

2. The domain of a rational expression is the set of all real numbers for which the denominator does not have a value of zero.

3. Operations with rational expressions are done in the same manner as operations with fractions.

4. To simplify a complex fraction, multiply the numerator and denominator by the LCD of all denominators.

Chapter 1 Review Exercises

Determine whether each statement is true or false and explain your answer.

1. Every real number is a rational number.

2. Zero is neither rational nor irrational.

3. There are no negative integers.

4. Every repeating decimal number is a rational number.

5. The terminating decimal numbers are irrational numbers.

6. The number $\sqrt{289}$ is a rational number.

7. Zero is a natural number.

8. The multiplicative inverse of 8 is 0.125.

9. The reciprocal of 0.333 is 3.

10. The real number π is irrational.

11. The additive inverse of 0.5 is 0.

12. The distributive property is used in adding like terms.

Simplify each expression.

13. $-3x - 4(3 - 5x)$

14. $x - 0.02(x - 9)$

15. $\dfrac{x}{5} + \dfrac{x}{10}$

16. $\dfrac{1}{3}x - \dfrac{1}{8}x$

17. $\dfrac{3x - 6}{9}$

18. $\dfrac{1}{2}(4x - 6)$

19. $\dfrac{-7 - (-1)}{3 - (-5)}$

20. $\dfrac{6 - (3 - x)}{2 - (-1)}$

21. $|-3| - |-5|$

22. $|5 - (-2)|$

23. $|3 - 7|$

24. $|-3 - (-4)|$

25. $8 - 9 \cdot 2 \div 3 + 5$

26. $3 - 4(2 - 3 \cdot 5^2)$

27. $12 \div 4 \cdot 3 \div 6 + 3^3$

28. $8 \cdot 3^2 - 3\sqrt{3^2 + 4^2}$

Simplify each expression. Assume that all variables represent positive real numbers.

29. 5^4

30. 2^{-4}

31. $(-2)^2 - 4(-2)(5)$

32. $6^2 - 4(-1)(-3)$

33. $2^{-1} + 2^0$

34. $\dfrac{3^{-1}}{-3^2}$

35. $\dfrac{-3^{-1}2^3}{2^{-1}}$

36. $\dfrac{-1}{-1^{-1}}$

37. $8^{-2/3}$

38. $-16^{-3/4}$

39. $(125x^6)^{1/3}$

40. $\dfrac{1}{(27t^{12})^{-1/3}}$

41. $\sqrt{121}$

42. $\sqrt[3]{-1000}$

43. $\sqrt{28s^3}$

44. $\sqrt{75a^2b^9}$

45. $\sqrt[3]{-2000}$

46. $\sqrt[3]{56w^4}$

47. $\sqrt{\dfrac{5}{2a}}$

48. $\sqrt{\dfrac{1}{18z^3}}$

49. $\sqrt[3]{\dfrac{2}{5}}$

50. $\sqrt[3]{\dfrac{3}{4y}}$

51. $\sqrt{18n^3} + \sqrt{50n^3}$

52. $\sqrt[3]{24} - \sqrt[3]{81}$

53. $\dfrac{2\sqrt{3}}{\sqrt{3} - 1}$

54. $\dfrac{2}{\sqrt{6} - 2}$

55. $\dfrac{\sqrt{6}}{\sqrt{8} + \sqrt{18}}$

56. $\dfrac{\sqrt{15}}{\sqrt{75} + \sqrt{20}}$

Convert each number given in scientific notation to standard notation and each number given in standard notation to scientific notation.

57. 3.2×10^8

58. -4.543×10^9

59. -1.85×10^{-4}

60. 9.44×10^{-5}

61. 0.000056

62. -0.000341

63. $-2,340,000$

64. $88,300,000,000$

Perform the indicated operations. Write the answer in scientific notation.

65. $(5 \times 10^6)^3$

66. $\dfrac{(0.00000046)(3,000)}{2,300,000}$

67. $\dfrac{(800)^2(0.00001)^{-3}}{(2,000,000)^3(0.00002)}$

68. $\dfrac{(5.1 \times 10^8)(-2 \times 10^{-3})}{1.7 \times 10^{-6}}$

Perform the indicated operations.

69. $(3x^2 - x - 2) + (-x^2 + 2x - 5)$

70. $(4y^3 - y^2 + 5y - 9) - (y^3 - 6y^2 + 3y - 2)$

71. $(-4x^4 - 3x^3 + x) - (x^4 - 6x^3 - 2x)$

72. $(3y^4 - 4y^2 - 6) + (-y^4 - 8y + 7)$

73. $(3a^2 - 2a + 5)(a - 2)$ **74.** $(w - 5)(w^2 + 5w + 25)$

75. $(b - 3y)^2$ **76.** $(x - 1)^3$

77. $(t - 3)(3t + 2)$ **78.** $(5y - 9)(5y + 9)$

79. $-35y^5 \div (7y^2)$ **80.** $(3x^3 - 6x^2) \div (3x)$

81. $(3 + \sqrt{2})(3 - \sqrt{2})$ **82.** $(2\sqrt{3} - 1)(3\sqrt{3} + 2)$

83. $(2\sqrt{5} + \sqrt{3})^2$ **84.** $(2\sqrt{x} + 3)^2$

85. $(1 + \sqrt{2x - 1})^2$ **86.** $(3 + \sqrt{y - 4})^2$

Find the quotient and remainder when the first polynomial is divided by the second.

87. $x^3 + 2x^2 - 9x + 3, \ x - 2$

88. $x^3 - 6x^2 + 3x - 9, \ x + 3$

89. $6x^2 + x + 2, \ 2x - 1$

90. $12x^2 - x - 21, \ 3x - 4$

Use division to express each fraction in the form

$$\text{quotient} + \frac{\text{remainder}}{\text{divisor}}.$$

91. $\dfrac{x^2 - 3}{x + 2}$ **92.** $\dfrac{a^2 + 5a + 2}{a}$

93. $\dfrac{2x + 3}{x - 5}$ **94.** $\dfrac{-3x + 2}{5x - 4}$

Factor each polynomial completely.

95. $6x^3 - 6x$ **96.** $4u^2 - 9v^2$

97. $9h^2 + 24ht + 16t^2$ **98.** $b^2 + 6b - 16$

99. $t^3 + y^3$ **100.** $8a^3 - 27$

101. $x^3 + 3x^2 - 9x - 27$ **102.** $3x - by + bx - 3y$

103. $t^6 - 1$ **104.** $y^4 - 625$

105. $18x^2 - 9x - 20$ **106.** $(x - 1)^2 - (x - 1) - 2$

107. $a^3b + 3a^2b - 18ab$ **108.** $x^3y^4 + 4x^2y^2 - 12x$

109. $2x^3 + x^2y - y - 2x$ **110.** $12x^3 - 2x^2y - 24xy^2$

Perform the indicated operations with the rational expressions.

111. $\dfrac{x - 2}{x + 3} + \dfrac{x + 8}{x + 3}$ **112.** $\dfrac{3x - 5}{x^2 - 4} - \dfrac{3 - x}{x^2 - 4}$

113. $\dfrac{x - 1}{x - 2} - \dfrac{x + 3}{x + 4}$ **114.** $\dfrac{1 - x}{x} + \dfrac{3 - x}{x - 2}$

115. $\dfrac{x^2 - 9}{x + 3} \cdot \dfrac{1}{6 - 2x}$ **116.** $\dfrac{x^3 - 8}{6} \cdot \dfrac{3x + 6}{x^2 - 4}$

117. $\dfrac{a^3bc^8}{a^9b^3c} \cdot \dfrac{(ab^3c^5)^2}{a^4b^3}$ **118.** $\dfrac{(x^3z^2)^5}{xz^4} \cdot \left(\dfrac{xz^2}{x^3}\right)^2$

119. $\dfrac{1}{x^2 - 4} + \dfrac{3}{x - 2}$ **120.** $\dfrac{3}{2x - 4} + \dfrac{5}{3x - 6}$

121. $\dfrac{1}{6x} - \dfrac{7}{10x^2}$ **122.** $\dfrac{1}{3y^2} - \dfrac{2}{9y}$

123. $\dfrac{a^2 - 25}{a^2 - 4a - 5} \div \dfrac{2a + 10}{a^2 - 1}$

124. $\dfrac{y^4 - 16}{y^2 + y - 2} \div \dfrac{y^3 + 4y}{y^3 - y}$

125. $\dfrac{x^2 - 16}{x^2 + 5x + 4} \div \dfrac{8 - 2x}{x^3 + 1}$

126. $\dfrac{x^2 + ax + bx + ab}{x^2 + 2bx + b^2} \div \dfrac{x^2 + 2ax + a^2}{x^3 + b^3}$

127. $\dfrac{a - 2}{a^2 + 6a + 5} + \dfrac{2a + 1}{a^2 - 1}$

128. $\dfrac{y - 1}{y^2 - 2y - 24} - \dfrac{y - 3}{y^2 + 2y - 8}$

Simplify.

129. $\dfrac{\dfrac{5}{2x} - \dfrac{3}{4x}}{\dfrac{1}{2} - \dfrac{2}{x}}$ **130.** $\dfrac{\dfrac{4}{4 - y^2} - \dfrac{5}{y - 2}}{\dfrac{1}{2 - y} - \dfrac{3}{y + 2}}$

131. $\dfrac{\dfrac{1}{y^2 - 2} - 3}{\dfrac{5}{y^2 - 2} + 4}$ **132.** $\dfrac{\dfrac{1}{6a^2b^3}}{\dfrac{5}{8a^3b} - \dfrac{3}{10a^3b^4}}$

133. $\dfrac{a^{-2} - b^{-3}}{a^{-1}b^{-1}}$ **134.** $\dfrac{x^{-1} - y^{-1}}{x^{-3} - y^{-3}}$

135. $p^{-1} + pq^{-3}$ **136.** $a^{-1} + x^{-1}$

Given that

$$P(x) = x^3 - 3x^2 + x - 9 \text{ and } R(x) = \frac{3x - 1}{2x - 9},$$

find each of the following.

137. $P(2)$ **138.** $P(-1)$ **139.** $P(0)$

140. $P\left(\dfrac{1}{2}\right)$ **141.** $R(-1)$ **142.** $R(3)$

143. $R(50)$ **144.** $R(-40)$

Solve each problem.

145. *Weight of a hydrogen atom* One hydrogen atom weighs 1.7×10^{-24} g. Find the number of hydrogen atoms in 1 kg of hydrogen.

146. *Moon talk* Radio waves travel at 3.0×10^8 m/sec (the same as the speed of light) and the distance from the earth to the moon is 3.84×10^8 m. How many seconds did it take for a radio wave to travel from mission control in Houston to the astronauts on the surface of the moon and then back to Houston? (A good demonstration of the speed of light actually occurred when Houston controllers heard an echo to their words that traveled to the moon and back.)

147. *Distance* What is the distance between -2.35 and 8.77 on the number line?

148. *Absolute value* Write an expression involving absolute value that gives the distance between the points a and -5 on the number line.

149. *Mowing a lawn* Howard and Will get summer jobs doing yard work. If Howard can mow an entire lawn in 6 hr and Will can mow it in 4 hr, then what portion of it can they mow in 2 hr working together?

150. *Cost of watermelons* Write a polynomial that expresses the total cost of $x + 3$ watermelons at \$2 apiece and $x + 9$ watermelons at \$3 apiece.

3.84×10^8 m

Radio waves travel at 3.0×10^8 m/sec

Moon

Earth

Figure for Exercise 146

Chapter 1 Test

Determine which elements of the set

$$\{-\pi, -\sqrt{3}, -1.22, -1, 0, 2, \sqrt{5}, 10/3, 6.020020002\ldots\}$$

are members of the following sets.

  1. Real numbers   **2.** Rational numbers

  3. Irrational numbers   **4.** Nonnegative integers

Evaluate each expression.

  5. $|2 \cdot 3 - 5^2| - 6$ **6.** $\dfrac{1}{8^{-1/3}}$

  7. $-27^{-2/3}$ **8.** $\dfrac{(-3)^2 - 4(-3) + 9}{(-3 - 2)(5 - (-1))}$

Simplify each expression. Assume that all variables represent positive real numbers.

  9. $(-2x^4y^2)(-3xy^5)$ **10.** $(4x^4)^{1/2} + (-8x^6)^{1/3}$

  11. $\dfrac{(ab^2 + a^2b)^2}{a^2b^2}$ **12.** $\dfrac{(-2a^{-1}b^6)^3}{(8a^9b^{-12})^{-2/3}}$

Simplify each expression. Assume that all variables represent positive real numbers.

  13. $\sqrt{27} - \sqrt{8} + \sqrt{32}$ **14.** $\dfrac{\sqrt{8}}{\sqrt{6} - \sqrt{2}}$

  15. $\sqrt[3]{\dfrac{1}{4x^4}}$ **16.** $\sqrt{12x^3y^9z^0}$

Perform the indicated operations.

17. $(-x^3 - 5x) + (4x^3 + 3x^2 - 7x)$

18. $(-x^2 + 3x - 4) - (4x^2 - 6x + 9)$

19. $(x + 3)(x^2 - 2x - 1)$ **20.** $(8h^3 - 1) \div (2h - 1)$

21. $(x - 9y)(x + 3y)$ **22.** $(x^3 - 2x - 4) \div (x - 2)$

23. $(3x - 8)^2$ **24.** $(2t^4 - 1)(2t^4 + 1)$

Perform each operation.

25. $\dfrac{x^3 - 5x^2 + 6x}{2x^2 - 6x} \cdot \dfrac{4x^3 + 32}{2x^3 - 8x}$

26. $\dfrac{x + 5}{x^2 - 4x + 3} + \dfrac{x - 1}{x^2 + x - 12}$

27. $\dfrac{a - 1}{4a^2 - 9} - \dfrac{a - 2}{3 - 2a}$ **28.** $\dfrac{\dfrac{1}{2a^2b} - 2a}{\dfrac{1}{4ab^3} + \dfrac{1}{3b}}$

Factor completely.

29. $ax^2 - 11ax + 18a$ **30.** $m^5 - m$

31. $3x^2 + 14x - 5$ **32.** $bx^2 - 3bx + wx - 3w$

Solve each problem.

33. In 1989 (a record year), 2.9×10^6 people attended conventions in New York City and spent a total of \$1.1 billion. What amount was spent per person?

34. If one light year is equal to 6.3240×10^4 astronomical units and one astronomical unit is equal to 92.95582×10^6 mi, then what number of miles is equal to one light year?

35. The altitude in feet of an arrow t seconds after it is shot straight upward is calculated from the polynomial $A(t) = -16t^2 + 120t$. Find the altitude of the arrow 2 sec after it is shot upward.

Even infamous Heartbreak Hill couldn't break the winning spirit of Russian runner Olga Markova, as she focused mind and muscle on out-distancing her competitors. It was the 96th running of the Boston Marathon, the world's oldest and most presti-gious race of its kind, a 26.2-mile ordeal that one runner called "14 miles of fun, 8 miles of sweat, and 4 miles of hell!"

While Markova's rivals got off to a fast start, she shrewdly waited them out, breaking away at Mile 18 and never looking back until she burst through the finish line tape with a personal-best time of 2 hours, 23 minutes, 43 seconds—one of the best women's times ever in Boston.

Major sporting events like the Marathon have come a long way since the first Olympic Games, held in ancient Greece over 2500 years ago. In the 20th century, satellites beam the Games to a worldwide audience. And modern athletes and coaches often utilize both mathematics and computers to analyze the critical variables that help competitors increase aerobic capacity, reduce air resistance, or strengthen muscles in order to jump higher, run faster, or throw farther than ever before. Scientists are particularly fascinated with long-distance running, and the Boston race has been studied more than most because of the area's many research centers. The Marathon also boasts a top medical-response team that's always searching for new ways to interpret the effects of the grueling pace on muscles, bones, heart, and other organs.

In this chapter you'll encounter various examples of sports applications while studying two of the most important skills in algebra: solving equations and creating mathematical models. In Chapter 1 we reviewed fundamentals of algebra. Now you're ready to learn how to use these tools to solve problems that can be expressed in the form of equations and inequalities. By the time you reach the finish line, you should be able to solve both linear and quadratic equations, as well as a few other special types of equations. You'll also be able to create and use several types of mathematical models.

2 Equations and Inequalities

2.1

Linear Equations

One of our main goals in algebra is to develop techniques for solving a wide variety of equations. In this section we will solve linear equations and other similar equations.

Definitions

An **equation** is a statement (or sentence) indicating that two algebraic expressions are equal. The verb in an equation is the equality symbol. For example, $2x + 8 = 0$ is an equation. If we replace x by -4, we get $2 \cdot (-4) + 8 = 0$, a true statement. So we say that -4 is a **solution** or **root** to the equation or that -4 **satisfies** the equation. If we replace x by 3, we get $2 \cdot 3 + 8 = 0$, a false statement. Whether the equation $2x + 8 = 0$ is true or false depends on the value of x, and so it is called an **open sentence.** The equation is neither true nor false until we choose a value for x. The set of all solutions to an equation is called the **solution set** to the equation. To **solve** an equation means to find the solution set. The solution set for $2x + 8 = 0$ is $\{-4\}$. The equation $2x + 8 = 0$ is an example of a linear equation.

Definition: Linear Equation in One Variable

A **linear equation in one variable** is an equation of the form $ax + b = 0$, where a and b are real numbers, with $a \neq 0$.

Solving Equations

The equations $2x + 8 = 0$ and $2x = -8$ both have the solution set $\{-4\}$. Two equations with the same solution set are called **equivalent** equations. Adding the same real number to or subtracting the same real number from each side of an equation results in an equivalent equation. Multiplying or dividing each side of an equation by the same nonzero real number also results in an equivalent equation. These **properties of equality** are stated in symbols in the following box.

Properties of Equality

If A and B are algebraic expressions and C is a real number, then the following equations are equivalent to $A = B$:

$A + C = B + C$	**Addition property of equality**
$A - C = B - C$	**Subtraction property of equality**
$CA = CB \quad (C \neq 0)$	**Multiplication property of equality**
$\dfrac{A}{C} = \dfrac{B}{C} \quad (C \neq 0)$	**Division property of equality**

When C is a real number, the properties of equality allow us to perform the four basic operations of arithmetic on each side of an equation without changing the solution set. It is also correct to apply the properties of equality using an *algebraic expression* for C, because the value of an algebraic expression is a real number. However, if C involves a variable, we might obtain an equation that is not equivalent to the original equation. For example, if $1/x$ is subtracted from each side of the equation

$$x + \frac{1}{x} = 0 + \frac{1}{x}$$

we get $x = 0$. But replacing x by 0 in the original equation results in undefined expressions, so the equations are not equivalent. In an equation, if there is an expression whose domain excludes some real numbers, then we must check our solutions.

The strategy in solving equations is to use the properties of equality to simplify an equation until we get an equivalent equation whose solution set is obvious.

Example 1 Using the properties of equality

Solve each equation.

a) $3x - 4 = 8$ **b)** $\dfrac{1}{2}x - 6 = \dfrac{3}{4}x - 9$ **c)** $3(4x - 1) = 4 - 6(x - 3)$

Solution

a)
$$3x - 4 = 8$$
$$3x - 4 + 4 = 8 + 4 \qquad \text{Add 4 to each side.}$$
$$3x = 12 \qquad \text{Simplify.}$$
$$\frac{3x}{3} = \frac{12}{3} \qquad \text{Divide each side by 3.}$$
$$x = 4 \qquad \text{Simplify.}$$

Since the last equation is equivalent to the original, the solution set to the original equation is $\{4\}$. We can check by replacing x by 4 in $3x - 4 = 8$. Since $3 \cdot 4 - 4 = 8$ is correct, we are confident that the solution set is $\{4\}$.

b)
$$\frac{1}{2}x - 6 = \frac{3}{4}x - 9$$
$$4\left(\frac{1}{2}x - 6\right) = 4\left(\frac{3}{4}x - 9\right) \qquad \text{Multiply each side by 4, the LCD.}$$
$$2x - 24 = 3x - 36 \qquad \text{Distributive property}$$
$$2x - 24 - 3x = 3x - 36 - 3x \qquad \text{Subtract } 3x \text{ from each side.}$$
$$-x - 24 = -36 \qquad \text{Simplify.}$$
$$-x = -12 \qquad \text{Add 24 to each side.}$$
$$(-1)(-x) = (-1)(-12) \qquad \text{Multiply each side by } -1.$$
$$x = 12 \qquad \text{Simplify.}$$

Check 12 in the original equation. The solution set is $\{12\}$.

c) $3(4x - 1) = 4 - 6(x - 3)$
$$12x - 3 = 4 - 6x + 18 \qquad \text{Distributive property}$$
$$12x - 3 = 22 - 6x \qquad \text{Simplify.}$$
$$18x - 3 = 22 \qquad \text{Add } 6x \text{ to each side.}$$
$$18x = 25 \qquad \text{Add 3 to each side.}$$
$$x = \frac{25}{18} \qquad \text{Divide each side by 18.}$$

Check 25/18 in the original equation. The solution set is $\{\frac{25}{18}\}$.

Any linear equation, $ax + b = 0$, can be solved in two steps. Subtract b from each side and then divide each side by a ($a \neq 0$), to get $x = -b/a$. Although the equations of Example 1 are not exactly in the form $ax + b = 0$, they are often called linear equations because they are equivalent to linear equations. The next example of a linear equation involves radicals.

Example 2   An equation involving radicals

Solve $\sqrt{2}x - 4 = \sqrt{6}$. Express the answer in simplified radical form.

Solution

We proceed as in any other linear equation.

$$\sqrt{2}x - 4 = \sqrt{6}$$

$$\sqrt{2}x = 4 + \sqrt{6} \qquad \text{\color{blue}{Add 4 to each side.}}$$

$$x = \frac{4 + \sqrt{6}}{\sqrt{2}} \qquad \text{\color{blue}{Divide each side by $\sqrt{2}$.}}$$

$$x = \frac{(4 + \sqrt{6})\sqrt{2}}{\sqrt{2} \cdot \sqrt{2}} \qquad \text{\color{blue}{Rationalize the denominator.}}$$

$$x = \frac{4\sqrt{2} + 2\sqrt{3}}{2}$$

$$x = 2\sqrt{2} + \sqrt{3}$$

```
2√2+√3
            4.560477932
√2*Ans-4-√6
                      0
```

The ANS key contains the result of the last calculation and so it can be used to check. We can perform the check in one computation by subtracting the right side of the equation from the left side.

The exact solution is $2\sqrt{2} + \sqrt{3}$. Using a calculator and rounding to three decimal places, we get $x \approx 4.560$. The symbol $\approx$ means "is approximately equal to." Check 4.560 or $2\sqrt{2} + \sqrt{3}$ in the original equation. The solution set is $\{2\sqrt{2} + \sqrt{3}\}$. ◆

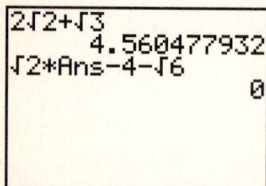

Identities, Conditional Equations, and Inconsistent Equations

An equation that is satisfied by every real number for which both sides are defined is an **identity.** Some examples of identities are

$$3x - 1 = 3x - 1, \qquad 2x + 5x = 7x, \qquad \text{and} \qquad \frac{x}{x} = 1.$$

The solution set to the first two identities is the set of all real numbers, R. Since $0/0$ is undefined, the solution set to $x/x = 1$ is the set of nonzero real numbers, $\{x | x \neq 0\}$.

A **conditional equation** is an equation that is satisfied by at least one real number but is not an identity. The equation $3x - 4 = 8$ is true only on condition that $x = 4$, and it is a conditional equation. The equations of Examples 1 and 2 are conditional equations.

An **inconsistent equation** is an equation that has no solution. Some inconsistent equations are

$$0 \cdot x + 1 = 2, \qquad x + 3 = x + 5, \qquad \text{and} \qquad 9x - 9x = 8.$$

Note that each of these inconsistent equations is equivalent to a false statement: $1 = 2, 3 = 5,$ and $0 = 8$, respectively.

Example 3   Classifying an equation

Determine whether the equation $3(x - 1) - 2x(4 - x) = (2x + 1)(x - 3)$ is an identity, an inconsistent equation, or a conditional equation.

Solution

$$3(x - 1) - 2x(4 - x) = (2x + 1)(x - 3)$$
$$3x - 3 - 8x + 2x^2 = 2x^2 - 5x - 3 \qquad \text{Simplify each side.}$$
$$2x^2 - 5x - 3 = 2x^2 - 5x - 3$$

Since the last equation is equivalent to the original and the last equation is an identity, the original equation is an identity. ◆

Equations Involving Rational Expressions

In Example 1(b) we solved an equation involving fractions. To simplify that equation, the first step was to multiply by the LCD of the fractions. The equations in the next example all involve rational expressions. Notice that multiplying each side of these equations by the LCD greatly simplifies the equations.

Example 4   Equations involving rational expressions

Solve each equation and identify each as an identity, an inconsistent equation, or a conditional equation.

a) $\dfrac{y}{y - 3} + 3 = \dfrac{3}{y - 3}$ **b)** $\dfrac{1}{x - 1} - \dfrac{1}{x + 1} = \dfrac{2}{x^2 - 1}$ **c)** $\dfrac{1}{2} + \dfrac{1}{x - 1} = 1$

Solution

a) Since $y - 3$ is the denominator in each rational expression, $y - 3$ is the LCD.

$$(y - 3)\left(\frac{y}{y - 3} + 3\right) = (y - 3)\frac{3}{y - 3} \qquad \text{Multiply each side by the LCD.}$$

$$(y - 3)\frac{y}{y - 3} + (y - 3)3 = 3 \qquad \text{Distributive property}$$

$$y + 3y - 9 = 3$$
$$4y - 9 = 3$$
$$4y = 12 \qquad \text{Add 9 to each side.}$$
$$y = 3 \qquad \text{Divide each side by 4.}$$

If we replace y by 3 in the original equation, then we get two undefined expressions. So 3 is not a solution to the original equation. The original equation has no solution. The equation is inconsistent.

b) Since $x^2 - 1 = (x - 1)(x + 1)$, the LCD is $(x - 1)(x + 1)$.

$$\frac{1}{x - 1} - \frac{1}{x + 1} = \frac{2}{x^2 - 1}$$

$$(x - 1)(x + 1)\left(\frac{1}{x - 1} - \frac{1}{x + 1}\right) = (x - 1)(x + 1)\frac{2}{x^2 - 1}$$ Multiply by the LCD.

$$(x - 1)(x + 1)\frac{1}{x - 1} - (x - 1)(x + 1)\frac{1}{x + 1} = 2$$ Distributive property

$$x + 1 - (x - 1) = 2$$

$$2 = 2$$

Since the last equation is an identity, the original equation is also an identity. The solution set is $\{x \mid x \neq 1 \text{ and } x \neq -1\}$, because 1 and -1 cannot be used for x in the original equation.

c)

$$\frac{1}{2} + \frac{1}{x - 1} = 1$$

$$2(x - 1)\left(\frac{1}{2} + \frac{1}{x - 1}\right) = 2(x - 1)1$$ Multiply by the LCD.

$$x - 1 + 2 = 2x - 2$$

$$x + 1 = 2x - 2$$

$$3 = x$$

Check 3 in the original equation. The solution set is $\{3\}$, and the equation is a conditional equation. ◆

In Example 4(a) the final equation had a root that did not satisfy the original equation, because the domain of the rational expression excluded the root. Such a root is called an **extraneous root.** If an equation has no solution, then its solution set is the **empty set** (the set with no members). The symbol $\emptyset$ is used to represent the empty set.

Example 5   Using a calculator in solving an equation

Solve

$$\frac{7}{2.4x} + \frac{3}{5.9} = \frac{1}{8.2}$$

with the aid of a calculator. Round the answer to three decimal places.

Solution

We could multiply each side by the LCD, but since we are using a calculator, we can subtract 3/5.9 from each side to isolate x.

$$\frac{7}{2.4x} + \frac{3}{5.9} = \frac{1}{8.2}$$

$$\frac{7}{2.4x} = \frac{1}{8.2} - \frac{3}{5.9}$$

$$\frac{7}{2.4x} \approx -0.38652335 \qquad \text{Use a calculator to simplify.}$$

$$\frac{7}{2.4} \approx -0.38652335x \qquad \text{Multiply each side by } x.$$

$$\frac{7}{2.4(-0.38652335)} \approx x \qquad \text{Divide each side by } -0.38652335.$$

$$x \approx -7.546 \qquad \text{Round to three decimal places.}$$

The solution set is $\{-7.546\}$. Since -0.38652335 is an approximate value, the sign $\approx$ (for "approximately equal to") is used instead of the equal sign. To get three-decimal-place accuracy in the final answer, use as many digits as your calculator allows until you get to the final computation.

The ANS key contains the result of the first computation.

Now the ANS key contains the solution to the equation and we can use it to check. Since -1×10^{-14} is very close to 0, this computation supports the solution -7.546.

▼

Strategy: Equation-Solving Overview

In solving equations, there are usually many different sequences of correct steps that lead to the correct solution. If the approach that you try first does not work, *try another approach,* but learn from your failures as well as your successes. Solving equations successfully takes patience and practice.

? **For Thought**

True or false? Explain.

1. The number 3 is in the solution set to $5(4 - x) = 2x - 1$.
2. The equation $3x - 1 = 8$ is equivalent to $3x - 2 = 7$.
3. The equation $x + \sqrt{x} = -2 + \sqrt{x}$ is equivalent to $x = -2$.
4. The solution set to $x - x = 7$ is the empty set.
5. The equation $12x = 0$ is an inconsistent equation.
6. The equation $x - 0.02x = 0.98x$ is an identity.

(continued on next page)

7. The equation $x + 3 + \dfrac{1}{x} = 5 + \dfrac{1}{x}$ is a linear equation.

8. The equations $\dfrac{x}{x-5} = \dfrac{5}{x-5}$ and $x = 5$ are equivalent.

9. To solve $-\dfrac{2}{3}x = \dfrac{3}{4}$, we should multiply each side by $-\dfrac{2}{3}$.

10. If a and b are real numbers, then $ax + b = 0$ has a solution.

2.1 Exercises

Tape 3   Disk—5.25″: 1   3.5″: 1   Macintosh: 1

Determine whether each given number is a solution to the equation following it.

1. $3, 2x - 4 = 9$

2. $-2, \dfrac{1}{x} - \dfrac{1}{2} = -1$

3. $-3, (x - 1)^2 = 16$

4. $4, \sqrt{3x + 4} = -4$

Solve each equation and check your answer.

5. $3x - 5 = 0$

6. $-2x + 3 = 0$

7. $-3x + 6 = 12$

8. $5x - 3 = -13$

9. $8x - 6 = 1 - 6x$

10. $4x - 3 = 6x - 1$

11. $7 + 3x = 4(x - 1)$

12. $-3(x - 5) = 4 - 2x$

13. $-\dfrac{3}{4}x = 18$

14. $\dfrac{2}{3}x = -9$

15. $\dfrac{x}{2} - 5 = -12 - \dfrac{2x}{3}$

16. $\dfrac{x}{4} - 3 = \dfrac{x}{2} + 3$

Find the exact solution to each equation. Use a calculator to check your answer.

17. $2x + 4 = \sqrt{8}$

18. $3x - 6 = \sqrt{18}$

19. $\sqrt{3}w - 5 = \sqrt{6}$

20. $\sqrt{2}z + 4 = \sqrt{10}$

21. $2\pi x - 3 = 4\pi x - 1$

22. $\pi r = 3 - 2\pi r$

23. $\sqrt[3]{2x - 1} = 5$

24. $\sqrt[3]{4x + 3} = 9$

Solve each equation. Identify each equation as an identity, an inconsistent equation, or a conditional equation.

25. $3(x - 6) = 3x - 18$

26. $2a + 3a = 6a$

27. $3(x - 6) = 3x + 18$

28. $2x + 3x = 5x$

29. $\dfrac{3x}{x} = 3$

30. $\dfrac{x(x + 2)}{x + 2} = x$

Solve each equation involving rational expressions. Identify each equation as an identity, an inconsistent equation, or a conditional equation.

31. $\dfrac{1}{w - 1} - \dfrac{1}{2w - 2} = \dfrac{1}{2w - 2}$

32. $\dfrac{1}{x - 3} - \dfrac{1}{x + 3} = \dfrac{6}{x^2 - 9}$

33. $\dfrac{1}{x} - \dfrac{1}{3x} = \dfrac{1}{2x} + \dfrac{1}{6x}$

34. $\dfrac{1}{5x} - \dfrac{1}{4x} + \dfrac{1}{3x} = -\dfrac{17}{60}$

35. $\dfrac{z + 2}{z - 3} = \dfrac{5}{-3}$

36. $\dfrac{2x - 3}{x - 4} = \dfrac{5}{x - 4}$

37. $\dfrac{1}{x} + \dfrac{1}{x - 3} = \dfrac{9}{x^2 - 3x}$

38. $\dfrac{4}{x - 1} - \dfrac{9}{x + 1} = \dfrac{3}{x^2 - 1}$

39. $4 + \dfrac{6}{y - 3} = \dfrac{2y}{y - 3}$

40. $\dfrac{x}{x + 6} - 3 = 1 - \dfrac{6}{x + 6}$

41. $\dfrac{t}{t + 3} + 4 = \dfrac{2}{t + 3}$

42. $\dfrac{3x}{x + 1} - 5 = \dfrac{x - 11}{x + 1}$

Use a calculator to help you solve each equation. Round each answer to three decimal places.

43. $0.27x - 3.9 = 0.48x + 0.29$

44. $0.06(x - 3.78) = 1.95$

45. $\sqrt{2}x - \sqrt{6} = \sqrt{3}$

46. $\sqrt[3]{2x - 3} = 0$

47. $2a + 1 = -\sqrt{17}$

48. $3c + 4 = \sqrt{38}$

49. $\dfrac{0.001}{y - 0.333} = 3$

50. $1 + \dfrac{0.001}{t - 1} = 0$

51. $\dfrac{x}{0.376} + \dfrac{x}{0.135} = 2$

52. $\dfrac{1}{x} + \dfrac{5}{6.72} = 10.379$

53. $(x + 3.25)^2 = (x - 4.1)^2$

54. $0.25(2x - 1.6)^2 = (x - 0.9)^2$

55. $(2.3 \times 10^6)x + 8.9 \times 10^5 = 1.63 \times 10^4$

56. $(-3.4 \times 10^{-9})x + 3.45 \times 10^{-8} = 1.63 \times 10^4$

Solve each equation.

57. $x - 0.05x = 190$ **58.** $x + 0.1x = 121$

59. $0.2x + 6 = 0.06x + 0.4$

60. $0.01x - 0.3 = 0.001x$

61. $0.1x - 0.05(x - 20) = 1.2$

62. $0.03x - 0.2 = 0.2(x + 0.03)$

63. $(x + 2)^2 = x^2 + 4$ **64.** $(x - 3)^2 = x^2 - 9$

65. $5 - 3(x - 6) = 2(x - 6)$

66. $(x + 4)x + 3 = x^2 + 7x + 12$

67. $\dfrac{x}{2} + 1 = \dfrac{1}{4}(x - 6)$ **68.** $-\dfrac{1}{6}(x + 3) = \dfrac{1}{4}(3 - x)$

69. $\dfrac{y - 3}{2} + \dfrac{y}{5} = 3 - \dfrac{y + 1}{6}$ **70.** $\dfrac{y - 3}{5} - \dfrac{y - 4}{2} = 5$

71. $\dfrac{a}{2} - \dfrac{2 - a}{8} = \dfrac{3a - 10}{2} - \dfrac{2a - 3}{3}$

72. $\dfrac{a}{3} - \dfrac{a - 3}{6} = \dfrac{a + 3}{2} + \dfrac{a - 5}{5}$

73. $\dfrac{3}{x - 2} + \dfrac{4}{x + 2} = \dfrac{7x - 2}{x^2 - 4}$

74. $\dfrac{2}{x - 1} - \dfrac{3}{x + 2} = \dfrac{8 - x}{x^2 + x - 2}$

75. $\dfrac{4}{x + 3} - \dfrac{3}{2 - x} = \dfrac{7x + 1}{x^2 + x - 6}$

76. $\dfrac{3}{x} - \dfrac{4}{1 - x} = \dfrac{7x - 3}{x^2 - x}$ **77.** $\dfrac{1}{4}(2x - 3)^2 = (x + 5)^2$

78. $(m + 3)^2 = (m - 9)^2$ **79.** $\dfrac{x - 2}{x - 3} = \dfrac{x - 3}{x - 4}$

80. $\dfrac{y - 1}{y + 4} = \dfrac{y + 1}{y - 2}$

81. $\dfrac{x + 5}{x + 2} = \dfrac{x - 1}{x - 3}$

82. $\dfrac{4 - x}{1 - x} = \dfrac{x - 3}{x - 7}$

83. $x - 3[x - 4(x + 3)] = 9(x - 1)$

84. $2x - 3[x - 4(x - 6)] = 1$

Determine the value of k that would make each equation equivalent to the equation $x + 3 = 7$.

85. $x + k = 5$ **86.** $kx + 3 = 11$

87. $5 - kx = 71$ **88.** $\dfrac{1}{x} + k = \dfrac{3}{4}$

In a study of oxygen uptake rate for marathon runners (Costill and Fox, *Medicine and Science in Sports, Vol. 1*) it was found that the power expended in kilocalories per minute is given by the formula $P = M(av - b)$, where M is the mass of the runner in kilograms and v is the speed in meters per minute. The constants a and b have values

$$a = 1.02 \times 10^{-3} \text{ kcal/kg m} \quad \text{and}$$
$$b = 2.62 \times 10^{-2} \text{ kcal/kg min.}$$

89. *Constant power* A 50-kg runner in training has a velocity of 480 m/min while carrying weights of 2 kg. If her power expenditure remains constant when the weights are removed, the decreased mass will result in increased velocity. Find her velocity without the weights by solving the equation $52(a480 - b) = 50(av - b)$.

2 kg

480 m/min

50 kg

Figure for Exercise 89

90. *Running velocity* An 80-kg runner is expending power at the rate of 38.7 kcal/min. Solve the equation $38.7 = 80(av - b)$ to find the velocity at which he is running.

Solve each problem.

91. *Production cost* An automobile manufacturer, who spent $500 million to develop a new line of cars, wants the cost of development and production to be $12,000 per vehicle. If the production costs are $10,000 per vehicle, then the cost per vehicle for development and production of x vehicles is

$(10,000x + 500,000,000)/x$ dollars. Solve the equation

$$\frac{10,000x + 500,000,000}{x} = 12,000$$

to find the number of vehicles that must be sold so that the cost of development and production is $12,000 per vehicle.

Figure for Exercise 91

92. *Comparing costs* The total cost for renting a copy machine for 60 months at $100 per month plus 5 cents per copy is $0.05x + 6000$ dollars, if x copies are made. If the same copy machine is purchased for $2000, then it costs 7 cents per copy for supplies and maintenance, for a total cost of $2000 + 0.07x$ dollars for x copies. Solve the equation

$$0.05x + 6000 = 2000 + 0.07x$$

to find the number of copies for which the five-year cost is

the same for renting or purchasing. If you plan to make 2000 copies per month for five years, is it cheaper to rent or purchase the copier? Assume that the copier has no value after five years.

Figure for Exercise 92

For Writing/Discussion

93. *Definitions* Without looking back in the text, write the definitions of linear equation, identity, and inconsistent equation. Use complete sentences.

94. *Cooperative learning* Each student in a small group should write a linear equation, an identity, an inconsistent equation, and an equation that has an extraneous root. The group should solve each equation and determine whether each equation is of the required type.

2.2

Applications

In Section 2.1 we solved equations of various types. In this section we will put those new techniques to work in solving problems.

Formulas

A **formula** is an equation involving two or more variables. What distinguishes a formula from an ordinary equation is its application. For example, the equation $D = RT$ is a formula giving the relationship between distance, rate, and time in uniform motion. Formulas are often used to **model** real-life situations. A list of formulas commonly used as mathematical models is given at the end of this text.

In general, a formula is a recipe for finding the value of one variable when the values of the other variables are known. For example, the formula $P = 2L + 2W$, which expresses the perimeter of a rectangle in terms of its length and width, is said to be **solved** for P. We can solve this formula for W as follows.

$$2L + 2W = P \qquad \text{Write the formula with } W \text{ on the left.}$$

$$2W = P - 2L \qquad \text{Subtract } 2L \text{ from each side.}$$

$$W = \frac{P - 2L}{2} \qquad \text{Divide each side by 2.}$$

When a formula is solved for a specified variable, that variable must not occur on both sides of the equal sign.

Example 1   Solving a formula for a specified variable

Solve the formula $S = P + Prt$ for P.

Solution

$$P + Prt = S \qquad \text{Write the formula with } P \text{ on the left.}$$

$$P(1 + rt) = S \qquad \text{Factor out } P.$$

$$P = \frac{S}{1 + rt} \qquad \text{Divide each side by } 1 + rt. \qquad \blacklozenge$$

In some situations we know the values of all variables except one. After we substitute values for those variables, the formula is an equation in one variable. We can then solve the equation to find the value of the remaining variable.

Example 2   Finding the value of a variable in a formula

The total resistance R (in ohms) in the parallel circuit shown in Fig. 2.1 is modeled by the formula

$$\frac{1}{R} = \frac{1}{R_1} + \frac{1}{R_2}.$$

The subscripts 1 and 2 indicate that R_1 and R_2 represent the resistance for two different receivers. If $R = 7$ ohms and $R_1 = 10$ ohms, then what is the value of R_2?

Solution

Substitute the values for R and R_1 and solve for R_2.

$$\frac{1}{7} = \frac{1}{10} + \frac{1}{R_2} \qquad \text{The LCD of 7, 10, and } R_2 \text{ is } 70R_2.$$

$$70R_2 \cdot \frac{1}{7} = 70R_2\left(\frac{1}{10} + \frac{1}{R_2}\right) \qquad \text{Multiply each side by } 70R_2.$$

$$10R_2 = 7R_2 + 70$$

$$3R_2 = 70$$

$$R_2 = \frac{70}{3}$$

Check the solution in the original formula. The resistance R_2 is 70/3 ohms. $\blacklozenge$

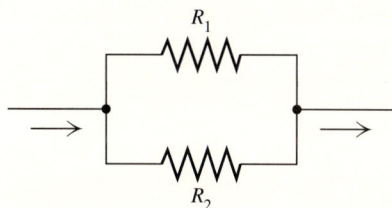

R_1

R_2

Figure 2.1

In the next example we have a formula that relates two variables that you might think would be unrelated.

Example 3 Injuries in the Boston Marathon

In the Boston Marathon there is a strong correlation between runner injuries and the temperature at the time of the race (*The Boston Globe,* April 20, 1992). See the bar graph in Fig. 2.2. When the Fahrenheit temperature at the time of the race is t, p% of the runners get injured, where $p = 0.27t - 8.46$. (You can find this equation from the data on the bar graph using the technique of linear regression described in your calculator manual.)

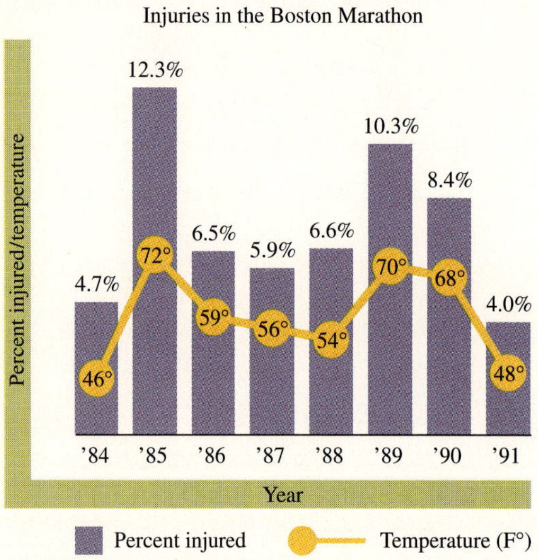

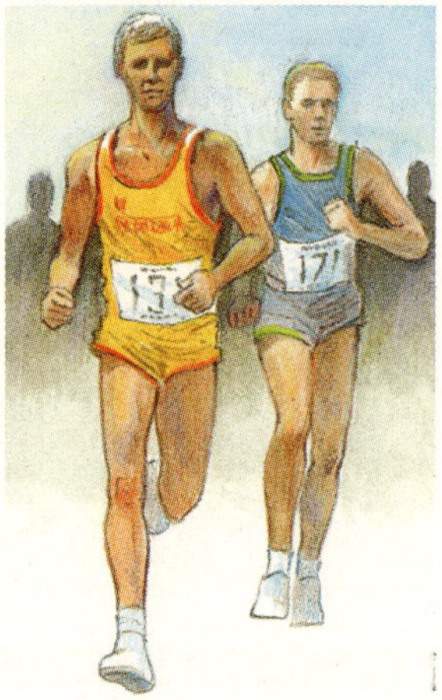

Injuries in the Boston Marathon

Figure 2.2

a) Estimate the percentage of runners that would be injured if the marathon was run on a day when the temperature was 82°F.

b) If 9% of the runners are injured, then estimate the temperature at the start of the race.

Solution

a) Use $t = 82$ in the formula $p = 0.27t - 8.46$:

$$p = 0.27(82) - 8.46 = 13.68$$

When the temperature is 82°F, we expect about 13.7% of the runners to be injured.

b) Now solve the formula for t:

$$0.27t - 8.46 = p$$
$$0.27t = p + 8.46$$
$$t = \frac{p + 8.46}{0.27}$$

Use $p = 9$ in this formula:

$$t = \frac{9 + 8.46}{0.27} \approx 65$$

If 9% of the runners are injured, then we would expect that the temperature was about 65°F. Note that we could have substituted $p = 9$ into the formula and then solved for t. ◆

Using a Calculator in Solving Equations

In Example 5 of Section 2.1 we used a calculator in solving an equation, but in this section we use a slightly different approach. We solve the equation for x using the technique of solving for a specified variable. All computations are performed at the end, to minimize round-off errors.

Example 4   Saving calculations until the end

Solve the equation for x, then use a calculator to find the value of x to three decimal places:

$$\frac{3.291}{x} + \frac{1}{2.35} = 4.76$$

Solution

$$2.35x\left(\frac{3.291}{x} + \frac{1}{2.35}\right) = 2.35x(4.76) \qquad \text{Multiply each side by the LCD.}$$

$$(2.35)(3.291) + x = (2.35)(4.76)x$$

$$x - (2.35)(4.76)x = -(2.35)(3.291) \qquad \text{Move all } x\text{-terms to the left side.}$$

$$x[1 - (2.35)(4.76)] = -(2.35)(3.291) \qquad \text{Factor out } x.$$

$$x = \frac{-2.35(3.291)}{1 - (2.35)(4.76)} \qquad \text{Divide each side by } 1 - (2.35)(4.76).$$

$$x \approx 0.759$$

Check 0.759 in the original equation. The solution set is $\{0.759\}$. ◆

```
-2.35*3.291/(1-2
.35*4.76)
        .7592627135
3.291/Ans+1/2.35
                4.76
■
```

Use your calculator to find the value of x and to check. See Appendix A for more examples.

Problem-Solving Techniques

Applied problems in mathematics often involve solving an equation. The equations for some problems come from known formulas, as in Example 2, while in other problems we must write an equation that models a particular problem situation.

   The best way to learn to solve problems is to study a few examples and then solve lots of problems. We will first look at an example of problem solving, then give a problem-solving strategy.

Example 5   Solving a problem involving sales tax

Jeannie Fung bought a new, fire-engine red Mazda RX-7 for a total cost of $18,966, including sales tax. If the sales tax rate is 9%, then what amount of tax did she pay?

Solution

There are two unknown quantities here, the price of the car and the amount of the tax. Represent the unknown quantities as follows:

$$x = \text{the price of the car}$$

$$0.09x = \text{amount of sales tax}$$

The price of the car plus the amount of sales tax is the total cost of the car. We can write this relationship as an equation and solve it for x.

$$x + 0.09x = 18,966$$

$$1.09x = 18,966$$

$$x = \frac{18,966}{1.09} = 17,400$$

$$0.09x = 1566$$

You can check this answer by adding $17,400 and $1566 to get $18,966, the total cost. The amount of tax that Jeannie Fung paid was $1566.   ◆

No two problems are exactly alike, but there are similarities. The following strategy will assist you in solving problems on your own.

▼

Strategy: Problem Solving

1. **Read the problem as many times as necessary to get an understanding of the problem.**
2. **If possible, draw a diagram to illustrate the problem.**
3. **Choose a variable, write down what it represents, and if possible, represent any other unknown quantities in terms of that variable.**
4. **Write an equation that models the situation. You may be able to use a known formula, or you may have to write an equation that models only that particular problem.**
5. **Solve the equation.**
6. **Check your answer by using it to solve the original problem (not just the equation).**
7. **Answer the question posed in the original problem.**

In the next example we have a geometric situation. Notice how we are following the strategy for problem solving.

Example 6   Solving a geometric problem

In 1974, Chinese workers found three pits that contained life-size sculptures of warriors, which were created to guard the tomb of an emperor (Nigel Hawkes, *Structures: The Way Things Are Built,* New York: Macmillan Co., 1990). The largest rectangular pit has a length that is 40 yards longer than three times the width. If its perimeter is 640 yards, what are the length and width?

Solution

First draw a diagram as shown in Fig. 2.3. Use the fact that the length is 40 yards longer than three times the width to represent the width and length as follows:

$$x = \text{the width in yards}$$
$$3x + 40 = \text{the length in yards}$$

The formula for the perimeter of a rectangle is $2L + 2W = P$. Replace W by x, L by $3x + 40$, and P by 640.

$$
\begin{aligned}
2L + 2W &= P &&\text{Perimeter formula} \\
2(3x + 40) + 2x &= 640 &&\text{Substitution} \\
6x + 80 + 2x &= 640 \\
8x + 80 &= 640 \\
8x &= 560 \\
x &= 70
\end{aligned}
$$

If $x = 70$, then $3x + 40 = 250$. Check that the dimensions of 70 yards and 250 yards really give a perimeter of 640 yards. We conclude that the length of the pit is 250 yards, and its width is 70 yards. ◆

The next problem is called a **uniform-motion** problem because it involves motion at a constant rate. Of course, people do not usually move at a constant rate, but their average speed can be assumed to be constant over some time interval. The problem also illustrates how a table can be used as an effective technique for organizing information.

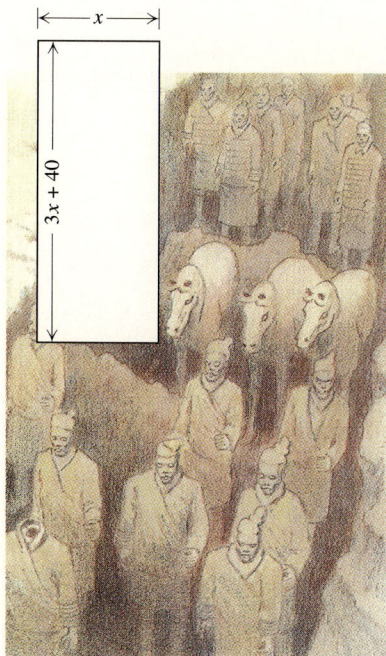

Figure 2.3

Example 7   Solving a uniform-motion problem

A group of hikers from Tulsa hiked down into the Grand Canyon in 3 hours 30 minutes. Coming back up on a trail that was 4 miles shorter, they hiked 2 mph slower and it took them 1 hour longer. What was their rate going down?

Solution

Let x represent the rate going down into the canyon. Next make a table like Table 2.1 to show the distance, rate, and time for both the trip down and the trip back up.

	Rate	Time	Distance
Down	x	3.5	$3.5x$
Up	$x - 2$	4.5	$4.5(x - 2)$

Table 2.1

Once we fill in any two entries in a row of the table, we can use $D = RT$ to obtain an expression for the third entry. Using the fact that the distance up was 4 miles shorter, we can write the following equation.

$$4.5(x - 2) = 3.5x - 4$$
$$4.5x - 9 = 3.5x - 4$$
$$x - 9 = -4$$
$$x = 5$$

After checking that 5 mph satisfies the conditions given in the problem, we conclude that the hikers traveled at 5 mph going down into the canyon. ◆

The next example involves mixing beverages with two different concentrations of orange juice. In other **mixture problems,** we may mix chemical solutions, candy, or even people.

Example 8 Solving a mixture problem

A beverage producer makes two products, Orange Drink, containing 10% orange juice, and Orange Delight, containing 50% orange juice. How many gallons of Orange Delight must be mixed with 300 gallons of Orange Drink to create a new product containing 40% orange juice?

Solution

Let x represent the number of gallons of Orange Delight shown in Fig. 2.4.

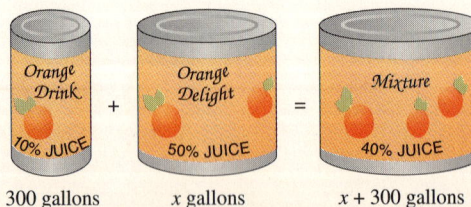

Figure 2.4

Orange Drink 10% JUICE 300 gallons + Orange Delight 50% JUICE x gallons = Mixture 40% JUICE $x + 300$ gallons

Table 2.2 shows three pertinent expressions for each product: the quantity of the product, the percentage of orange juice in the product, and the actual amount of orange juice in that quantity.

	Amount of product	Percent orange juice	Amount orange juice
Drink	300	10%	0.10(300)
Delight	x	50%	0.50x
Mixture	$x + 300$	40%	0.40($x + 300$)

Table 2.2

We write the following equation expressing the fact that the actual amount of orange juice in the mixture is the sum of the amounts of orange juice in the Orange Drink and in the Orange Delight.

$$0.40(x + 300) = 0.10(300) + 0.50x$$
$$0.4x + 120 = 30 + 0.5x$$
$$90 = 0.1x$$
$$900 = x$$

Mix 900 gallons of Orange Delight with the 300 gallons of Orange Drink to obtain the proper mixture. ◆

Work problems are problems in which people or machines are working together to accomplish a task. A typical situation might have two people painting a house at different rates. Suppose Joe and Frank are painting a house together for 2 hours and Joe paints at the rate of one-sixth of the house per hour while Frank paints at the rate of one-third of the house per hour. Note that

$$\left(\frac{1}{6} \text{ of house per hour}\right)(2 \text{ hr}) = \frac{1}{3} \text{ of house}$$

and

$$\left(\frac{1}{3} \text{ of house per hour}\right)(2 \text{ hr}) = \frac{2}{3} \text{ of house.}$$

The product of the rate and the time gives the fraction of the house completed by each person, and these fractions have a sum of 1 because the entire job is completed. Note how similar this situation is to a uniform-motion problem where $RT = D$.

Example 9 Solving a work problem

A gasoline-powered pump can drain Betty's entire swimming pool in 24 hours. An electric pump can drain the entire pool in 36 hours. After the electric pump was operating for 1 hour, the gasoline pump was started and both pumps were used until the pool was drained. How long did the gasoline pump operate?

Solution

Let x represent the number of hours of operation for the gasoline pump and $x + 1$ represent the number of hours of operation for the electric pump. The gasoline pump drains 1/24 of the pool per hour; the electric pump drains 1/36 of the pool per hour. Table 2.3 shows all of the pertinent quantities.

	Rate	Time	Work completed
Gasoline	$\dfrac{1}{24}$	x	$\dfrac{1}{24}x$
Electric	$\dfrac{1}{36}$	$x + 1$	$\dfrac{1}{36}(x + 1)$

Table 2.3

The following equation expresses the fact that the work completed by the pumps working together is the sum of the work completed by each pump alone.

$$\frac{1}{24}x + \frac{1}{36}(x + 1) = 1$$

$$72\left[\frac{1}{24}x + \frac{1}{36}(x + 1)\right] = 72 \cdot 1 \qquad \textcolor{blue}{\textbf{Multiply by the LCD 72.}}$$

$$3x + 2x + 2 = 72$$

$$5x = 70$$

$$x = 14$$

The gasoline pump operated for 14 hours before the pool was drained. ◆

```
14/24+(14+1)/36
               1
```

Check that 14 satisfies the equation.

? For Thought

True or false? Explain.

1. If we solve $P + Prt = S$ for P, we get $P = S - Prt$.
2. The perimeter of any rectangle is the product of its length and width.

3.   If n is an odd integer, then $n + 1$ and $n + 3$ represent odd integers.

4.   Solving $x - y = 1$ for y gives us $y = x - 1$.

5.   Two numbers that have a sum of -3 can be represented by x and $-3 - x$.

6.   If P is the number of professors and S is the number of students at the play, and there are twice as many professors as students, then $2P = S$.

7.   If you need $100,000 for your house, and the agent gets 9% of the selling price, then the agent gets $9,000, and the house sells for $109,000.

8.   If John can mow the lawn in x hours, then he mows the lawn at the rate of $1/x$ of the lawn per hour.

9.   If George hiked $3x$ miles and Anita hiked $4(x - 2)$ miles, and George hiked 5 more miles than Anita, then $3x + 5 = 4(x - 2)$.

10.   Two numbers that differ by 9 can be represented as 9 and $x + 9$.

2.2 Exercises   ▭ Tape 3   ▭ Disk—5.25″: 1   3.5″: 1   **Macintosh: 1**

Solve each formula for the specified variable.

1. $I = Prt$ for r (simple interest)

2. $D = RT$ for R (uniform motion)

3. $F = \dfrac{9}{5} C + 32$ for C (temperature)

4. $A = \dfrac{1}{2} bh$ for b (area of a triangle)

5. $C = 2\pi r$ for r (circumference of a circle)

6. $A = \dfrac{1}{2} h(b_1 + b_2)$ for h (area of a trapezoid)

7. $Ax + By = C$ for y (equation of a line)

8. $V = \pi r^2 h$ for h (volume of a cylinder)

9. $\dfrac{1}{R} = \dfrac{1}{R_1} + \dfrac{1}{R_2} + \dfrac{1}{R_3}$ for R_2 (resistance)

10. $S = \dfrac{a_1}{1 - r}$ for r (geometric series)

11. $a_n = a_1 + (n - 1)d$ for d (arithmetic sequence)

12. $S_n = \dfrac{n}{2} (a_1 + a_n)$ for a_1 (arithmetic series)

Use the appropriate formula to solve each problem.

13.   *Simple interest*   If $51.30 in interest is earned on a deposit of $950 in one year, then what is the simple interest rate?

14.   *Simple interest*   If you borrow $100 and pay back $105 at the end of one month, then what is the simple annual interest rate?

15.   *Uniform motion*   How long does it take an SR-71 Blackbird, one of the fastest U.S. jets, to make a surveillance run of 5570 mi if it travels at an average speed of Mach 3 (2228 mph)?

16.   *Circumference of a circle*   If the circumference of a circular sign is 72π in., then what is the radius?

17.   *Volume of a cylinder*   If the volume of a cylinder is 126π in^3 and the radius is 6 in., then what is the height?

18.   *Celsius temperature*   If the temperature at 1 P.M. on July 9 in Toronto was 30°C, then what was the temperature in degrees Fahrenheit?

19.   *Injuries in the Boston Marathon*   A study of the 4386 male runners in the 1985 Boston Marathon showed a high correlation between age and the percentage of injured runners for men aged 59 and below (*The Boston Globe*, April 20, 1992).

The bar graph shows the percentage of injured men age 59 and below grouped into five age categories. The percentage p of those injured is determined approximately from the age group A by the formula $p = -6.9A + 40.3$. Find the percentage of runners that are predicted to be injured in age group 4 according to the formula and compare the answer to the actual percentage injured.

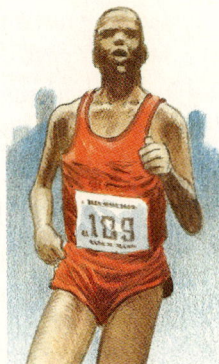

1: under 20   4: 40–49
2: 20–29   5: 50–59
3: 30–39

Figure for Exercise 19

20. *Injuries in the Boston Marathon* Use the formula of Exercise 19 to predict the percentage of runners injured in age group 6 (60-plus). In the 1985 race, 21% of the runners in the 60-plus age group were actually injured. Can you offer an explanation for the difference between the predicted and the actual figures?

Solve each equation for x, using your calculator only on the last step.

21. $3.458x - 2.347 = 4.782$

22. $5.76 - 3.49x = 2.388x - 1.42$

23. $\dfrac{3.33}{x} + \dfrac{2.391}{3.4} = 9.876$

24. $\dfrac{4.57}{x} - \dfrac{1}{4.59} = \dfrac{1}{3.6}$

25. $\dfrac{x - 3.45}{x + 4.98} = 7.53$

26. $\dfrac{x + 2.41}{x - 3.98} = \dfrac{x - 2.31}{x + 4.55}$

27. $\dfrac{1}{x - 3.44} - \dfrac{1}{2} = 2.95$

28. $\dfrac{1}{5.6 - x} + \dfrac{1}{3} = 4.55$

Solve each problem.

29. *Cost of a car* Jeff knows that his neighbor Sarah paid $28,728, including sales tax, for a new Buick Park Avenue. If the sales tax rate is 8%, then what is the cost of the car before the tax?

30. *Real estate commission* To be able to afford the house of their dreams, Dave and Leslie must clear $128,000 from the sale of their first house. If they must pay $780 in closing costs and 6% of the selling price for the sales commission, then what is the minimum selling price for which they will get $128,000?

31. *Investment income* Tara paid one-half of her game-show winnings to the government for taxes. She invested one-third of her winnings in Jeff's copy shop at 14% interest and one-sixth of her winnings in Kaiser's German Bakery at 12% interest. If she earned a total of $4000 on the investments in one year, then how much did she win on the game show?

32. *Construction penalties* Gonzales Construction contracted Kentwood High and Memorial Stadium for a total cost of $4.7 million. Because the construction was not completed on time, Gonzales paid 5% of the amount of the high school contract in penalties and 4% of the amount of the stadium contract in penalties. If the total penalty was $223,000, then what was the amount of each contract?

33. *Trimming a garage door* A carpenter used 30 ft of molding in three pieces to trim a garage door. If the long piece was 2 ft longer than twice the length of each shorter piece, then how long was each piece?

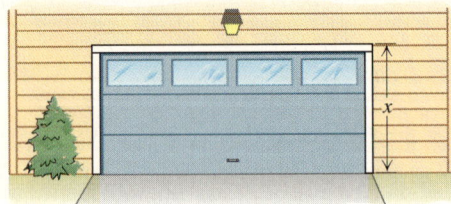

Figure for Exercise 33

34. *Increasing area of a field* Julia's soybean field is 3 m longer than it is wide. To increase her production, she plans to increase both the length and width by 2 m. If the new field is 46 m^2 larger than the old field, then what are the dimensions of the old field?

35. *Fencing a feed lot* Peter plans to fence off a square feed lot and then cross-fence to divide the feed lot into four smaller square feed lots. If he uses 480 ft of fencing, then how much area will be fenced in?

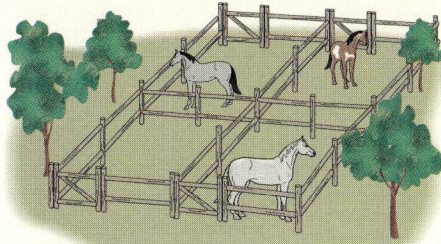

Figure for Exercise 35

36. *Fencing dog pens* Clint is constructing two adjacent rectangular dog pens. Each pen will be three times as long as it is wide, and the pens will share a common long side. If Clint has 65 ft of fencing, what are the dimensions of each pen?

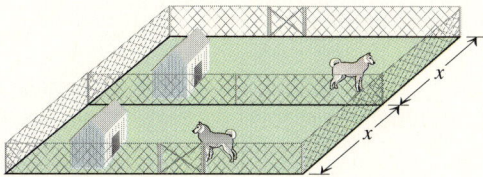

Figure for Exercise 36

37. *Racing speed* Bobby and Rick are in a 10-lap race on a one-mile oval track. Bobby, averaging 90 mph, has completed two laps just as Rick is getting his car onto the track. What speed does Rick have to average to be even with Bobby at the end of the tenth lap?

38. *Rowing a boat* Boudreaux rowed his pirogue from his camp on the bayou to his crab traps. Going down the bayou, he caught a falling tide that increased his normal speed by 2 mph, but coming back it decreased his normal speed by 2 mph. Going with the tide, the trip took only 10 min; going against the tide, the trip took 30 mins. How far is it from Boudreaux's camp to his crab traps?

39. *Average speed* Junior drove his rig on Interstate 10 from San Antonio to El Paso. At the halfway point he noticed that he had been averaging 80 mph, while his company requires his average speed to be 60 mph. What must be his speed for the last half of the trip so that he will average 60 mph for the trip?

40. *Average speed* Michelle drove her rig on Interstate 55 from St. Louis to Chicago. When she had traveled two-thirds of the way, she noticed that she had been averaging 40 mph, while her company requires that she average 60 mph. What must be her speed for the last one-third of the trip so that she will average 60 mph for the trip?

41. *Start-up capital* Norma decided to delay starting her catering business for one year and invest her start-up capital in two banks. The annual percentage rate (APR) was 5% at one bank and 6% at the other. After one year she made $5880. If the amount that she invested at 6% was $10,000 larger than the amount invested at 5%, then what was the amount of her start-up capital?

42. *Combining investments* Brent lent his brother Bob some money at 8% simple interest, and he lent his sister Betty half as much money at twice the interest rate. If Brent made a total of 24 cents in interest, then how much did he lend to each one?

43. *Percentage of minority workers* At the Northside assembly plant, 5% of the workers were classified as minority, while at the Southside assembly plant, 80% of the workers were classified as minority. When Northside and Southside were closed, all workers transferred to the new Eastside plant to make up its entire work force. If 50% of the 1500 employees at Eastside are minority, then how many employees did Northside and Southside have originally?

44. *Mixing alcohol solutions* A pharmacist needs to obtain a 70% alcohol solution. How many ounces of a 30% alcohol solution must be mixed with 40 ounces of an 80% alcohol solution to obtain a 70% alcohol solution?

45. *Harvesting wheat* With the old combine, Nikita's entire wheat crop can be harvested in 72 hr, but a new combine can do the same job in 48 hr. How many hours would it take to harvest the crop with both combines operating?

46. *Processing forms* Rita can process a batch of insurance claims in 4 hr working alone. Eduardo can process a batch of insurance claims in 2 hr working alone. How long would it take them to process a batch of claims if they worked together?

47. *Batman and Robin* Batman can clean up all of the crime in Gotham City in 8 hr working alone. Robin can do the same job alone in 12 hr. If Robin starts crime-fighting at 8 A.M. and Batman joins him at 10 A.M., then at what time will they have all of the crime cleaned up?

48. *Scraping barnacles* Della can scrape the barnacles from a 70-ft yacht in 10 hr using an electric barnacle scraper. Don can do the same job in 15 hr using a manual barnacle scraper. If Don starts scraping at noon and Della joins him at 3 P.M., then at what time will they finish the job?

Equations involving two variables *x* and *y* will be studied in later chapters. Solve each of the following equations for *y*.

49. $3x + 2y = -6$ **50.** $5x - 4y = 16$

51. $\dfrac{1}{2} x - \dfrac{1}{3} y = 3$

52. $x = \dfrac{5}{3} y + 10$

53. $y - 3 = \dfrac{3}{4}(x - 1)$

54. $y + 1 = -\dfrac{1}{2}(x - 3)$

55. $\dfrac{y - y_1}{x - x_1} = m$

56. $\dfrac{y - 3}{x + 5} = -\dfrac{2}{3}$

57. $y + xy = 3$

58. $xy + 3 = 2y - 9$

59. $xy = 2x + 3y + 1$

60. $-2xy = 3x - y + 4$

61. $\sqrt{2}y + x = -y$

62. $\sqrt{5}y - x = 2y$

63. $xy - x^2 = 1$

64. $(x - 2)(y - 3) = 1$

Solve each problem.

65. *Planning a race track*   If Mario plans to develop a circular race track one mile in circumference on a square plot of land, then what is the minimum number of acres that he needs? (One acre is equal to 43,560 ft².)

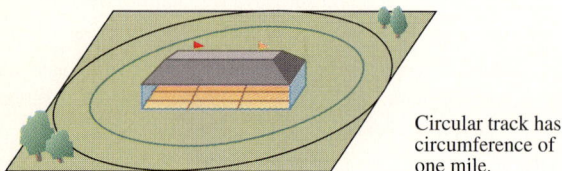

Figure for Exercise 65

Circular track has circumference of one mile.

66. *Volume of a can of Coke*   If a can of Coke contains 12 fluid ounces and the diameter of the can is 2.375 in., then what is the height of the can? (One fluid ounce equals approximately 1.8 in³.)

67. *Area of a lot*   Julio owns a four-sided lot that lies between two parallel streets. If his 90,000-ft² lot has 500 ft frontage on one street and 300 ft frontage on the other, then how far apart are the streets?

300 ft

500 ft

Figure for Exercise 67

68. *Width of a football field*   If the perimeter of a football field in the NFL including the end zones is 1040 ft and the field is 120 yd long, then what is the width of the field in feet?

69. *Depth of a swimming pool*   A circular swimming pool with a diameter of 30 ft and a horizontal bottom contains 22,000 gal of water. What is the depth of the water in the pool? (One cubic foot contains approximately 7.5 gal of water.)

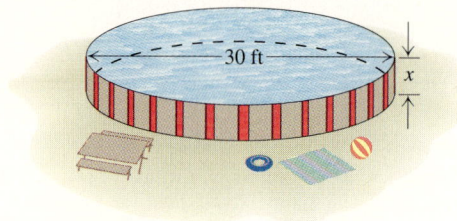

30 ft

x

Figure for Exercise 69

70. *Depth of a reflecting pool*   A rectangular reflecting pool with a horizontal bottom is 100 ft by 150 ft and contains 200,000 gal of water. How deep is the water in the pool?

71. *Olympic track*   To host the Summer Olympics, a city plans to build an eight-lane track. The track will consist of parallel 100-m straightaways with semicircular turns on either end as shown in the figure. The distance around the outside edge of the oval track is 514 m. If the track is built on a rectangular lot as shown in the drawing, then how many hectares (1 hectare = 10,000 m²) of land are needed in the rectangular lot?

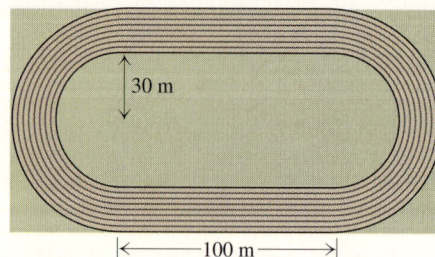

30 m

100 m

Figure for Exercises 71 and 72

72. *Green space*   If the inside radius of the turns is 30 m and grass is to be planted inside and outside the track of Exercise 71, then how many square meters of the rectangular lot will be planted in grass?

73. *Integers*   If the sum of three consecutive integers is 105, then what are the integers?

74. *Odd integers*   If the sum of three consecutive odd integers is 87, then what are the integers?

75. *Taxable income*   According to the 1992 federal income tax rate schedule, a single taxpayer with taxable income over $51,900 paid $11,743.50 plus 31% of the amount over $51,900 in federal income tax. If Lorinda paid $18,731.10 in federal income tax in 1992, then how much was her taxable income that year?

76. *Living comfortably*   Glen, a single taxpayer in 1992, figured that he needed $62,000 in after-federal-tax income to live comfortably. Use the information from Exercise 75 to find the amount of taxable income that would allow Glen to live comfortably in 1992.

77. *Diluting bleach*   How much pure water must be added to a 4-liter solution that is 5% sodium hypochlorite (the main ingredient in household bleach) to get a solution that is 3% sodium hypochlorite?

78. *Diluting antifreeze*   A mechanic is working on a car with a 20-quart radiator containing a 60% antifreeze solution. How much of the solution should he drain and replace with pure water to get a solution that is 50% antifreeze?

79. *Recording experience*   Kim has been in the recording business twice as long as Eric. In four years they will be able to advertise that their studio has 50 years of combined experience in recording. How much experience does Kim have now?

80. *Ratio of professors to students*   There were twice as many professors as students in attendance at the start of the play. At intermission, 14 professors departed and 5 students arrived, making the number of students 1 greater than the number of professors. How many professors were in attendance at the start?

81. *Mixing dried fruit*   The owner of a health-food store sells dried apples for $1.20 per quarter-pound, and dried apricots for $1.80 per quarter-pound. How many pounds of each must he mix together to get 20 lb of a mixture that sells for $1.68 per quarter-pound?

82. *Mixing breakfast cereal*   Raisins sell for $4.50/lb, and bran flakes sell for $2.80/lb. How many pounds of raisins should be mixed with 12 lb of bran flakes to get a mixture that sells for $3.14/lb?

83. *Coins in a vending machine*   Dana inserted eight coins, consisting of dimes and nickels, into a vending machine to purchase a Snickers bar for 55 cents. How many coins of each type did she use?

84. *Cost of a newspaper*   Ravi took eight coins from his pocket, which contained only dimes, nickels, and quarters, and bought the Sunday Edition of *The Daily Star* for 75 cents. If the number of nickels he used was one more than the number of dimes, then how many of each type of coin did he use?

85. *Robot gap*   Japan employs more robots than any other country (*Forbes,* April 16, 1990). The formulas $J = 30n + 100$ and $U = 4n + 20$ can be used to model the growth of industrial robot use in Japan and the United States. The letter n represents the number of years since 1985, while J and U represent the number of robots in Japan and the United States in thousands, respectively. If this trend continues, what is the first year in which Japan will have 600,000 more industrial robots than the United States?

Figure for Exercise 85

2.3

Complex Numbers

Our system of numbers developed as the need arose. Numbers were first used for counting. As society advanced, the rational numbers were formed to express fractional parts and ratios. Negative numbers were invented to express losses or debts. When it was discovered that the exact size of some very real objects could not be expressed with rational numbers, the irrational numbers were added to the system, forming the set of real numbers. Later still, there was a need for another expansion to the number system. In this section we study that expansion, the set of complex numbers.

Definitions

In Section 2.1 we solved equations of the form $ax + b = 0$ (linear equations). We learned that any linear equation has a real solution. In Section 2.4 we will study equations of the form $ax^2 + bx + c = 0$ (quadratic equations). For example, a simple quadratic equation such as $x^2 = 4$ has two solutions (2 and -2). However, $x^2 = -1$ has no real solution because the square of every real number is non-negative. So that equations such as $x^2 = -1$ will have solutions, imaginary numbers are defined. They are combined with the set of real numbers to form the set of complex numbers.

The imaginary numbers are based on the solution of the equation $x^2 = -1$. Since no real number solves this equation, a solution is called an *imaginary number*. The imaginary number i is defined to be a solution to this equation.

Definition: Imaginary Number *i*

The **imaginary number *i*** is defined as
$$i = \sqrt{-1} \quad \text{and} \quad i^2 = -1.$$

A complex number is formed as a real number plus a real multiple of i.

Definition: Complex Numbers

The set of **complex numbers** is the set of all numbers of the form $a + bi$, where a and b are real numbers.

In the complex number $a + bi$, a is called the **real part** and b is called the **imaginary part.** If $b \neq 0$, then $a + bi$ is called an **imaginary number.** Two complex numbers $a + bi$ and $c + di$ are **equal** if and only if $a = c$ and $b = d$.

The form $a + bi$ is called the **standard form** of a complex number, but for convenience we use a few variations of that form. If $a = 0$, then a is omitted and only the imaginary part is written. If $b = 0$, then only a is written and the complex number is a real number. If b is a radical, then i is usually written before b. For example, we write $2 + i\sqrt{3}$ rather than $2 + \sqrt{3}i$, which could be confused with

$2 + \sqrt{3}i$. If b is negative, we use the minus symbol to separate the real and imaginary parts. For example, instead of $3 + (-2)i$ we write $3 - 2i$. If both a and b are zero, the complex number $0 + 0i$ is the real number 0. For a complex number involving fractions, such as $\frac{1}{3} - \frac{2}{3}i$, we may write $(1 - 2i)/3$.

Example 1   Standard form of a complex number

Determine whether each complex number is real or imaginary and write it in the standard form $a + bi$.

a) $3i$ **b)** 87 **c)** $4 - 5i$ **d)** 0 **e)** $\dfrac{1 + \pi i}{2}$

Solution

a) The complex number $3i$ is imaginary, and $3i = 0 + 3i$.

b) The complex number 87 is a real number, and $87 = 87 + 0i$.

c) The complex number $4 - 5i$ is imaginary, and $4 - 5i = 4 + (-5)i$.

d) The complex number 0 is real, and $0 = 0 + 0i$.

e) The complex number $\dfrac{1 + \pi i}{2}$ is imaginary, and $\dfrac{1 + \pi i}{2} = \dfrac{1}{2} + \dfrac{\pi}{2}i$. ◆

Just as there are two types of real numbers (rational and irrational), there are two types of complex numbers, the real numbers and the imaginary numbers. The relationship between these sets of numbers is shown in Fig. 2.5.

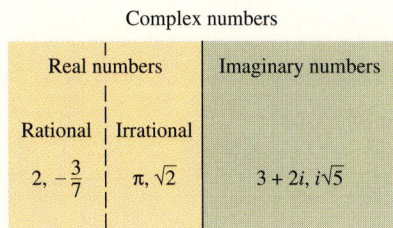

Addition, Subtraction, and Multiplication

Writing and thinking of complex numbers as binomials makes it easy to perform operations with complex numbers. The sum of $4 + 5i$ and $1 - 7i$ is found just as if we were adding binomials with i being the variable:

$$(4 + 5i) + (1 - 7i) = 5 - 2i$$

To find the sum, we add the real parts and add the imaginary parts. Subtraction is done similarly:

$$(9 - 2i) - (6 + 5i) = 3 - 7i$$

To multiply complex numbers, we use the FOIL method for multiplying binomials:

$$
\begin{aligned}
(2 + 3i)(5 - 4i) &= 10 - 8i + 15i - 12i^2 \\
&= 10 + 7i - 12(-1) \qquad \text{\textcolor{blue}{Replace } } i^2 \text{ \textcolor{blue}{by} } -1. \\
&= 22 + 7i
\end{aligned}
$$

In the box on the following page we give formal definitions of addition, subtraction, and multiplication of complex numbers.

Complex numbers

Real numbers		Imaginary numbers
Rational	Irrational	
$2, -\frac{3}{7}$	$\pi, \sqrt{2}$	$3 + 2i, i\sqrt{5}$

Figure 2.5

Definition: Addition, Subtraction, and Multiplication

If $a + bi$ and $c + di$ are complex numbers, we define their sum, difference, and product as follows.

$$(a + bi) + (c + di) = (a + c) + (b + d)i$$

$$(a + bi) - (c + di) = (a - c) + (b - d)i$$

$$(a + bi)(c + di) = (ac - bd) + (bc + ad)i$$

Note that it is not necessary to memorize these definitions to perform these operations with complex numbers. We get the same results by working with complex numbers as if they were binomials in which i is the variable, replacing i^2 by -1 wherever it occurs.

Example 2 Operations with complex numbers

Perform the indicated operations with the complex numbers.

a) $(-2 + 3i) + (-4 - 9i)$　**b)** $(-1 - 5i) - (3 - 2i)$　**c)** $2i(3 + i)$
d) $(3i)^2$　**e)** $(-3i)^2$　**f)** $(5 - 2i)(5 + 2i)$

Solution

a) $(-2 + 3i) + (-4 - 9i) = -6 - 6i$

b) $(-1 - 5i) - (3 - 2i) = -4 - 3i$

c) $2i(3 + i) = 6i + 2i^2 = 6i + 2(-1) = -2 + 6i$

d) $(3i)^2 = 3^2i^2 = 9(-1) = -9$

e) $(-3i)^2 = (-3)^2i^2 = 9(-1) = -9$

f) $(5 - 2i)(5 + 2i) = 25 - 4i^2 = 25 - 4(-1) = 29$　◆

Since $2^6 = 64$, 2^7 must be $2 \cdot 2^6$ or 128. Likewise, if we know any whole-number power of i, we can find the next higher power of i. Since $i^2 = -1$, we can find $i^3 = i \cdot i^2 = -i$. The first eight powers of i are listed here.

$$i^1 = i \qquad i^5 = i$$
$$i^2 = -1 \qquad i^6 = -1$$
$$i^3 = -i \qquad i^7 = -i$$
$$i^4 = 1 \qquad i^8 = 1$$

This list could be continued in this pattern, but any other whole-number power of i can be obtained from knowing the first four powers. We can simplify a power of i by using the fact that $i^4 = 1$ and $(i^4)^n = 1$ for any integer n.

Example 3   Simplifying a power of i

Simplify i^{83}.

Solution

Divide 83 by 4 and write $83 = 4 \cdot 20 + 3$. So

$$i^{83} = (i^4)^{20} \cdot i^3 = 1 \cdot i^3 = -i.$$

Division of Complex Numbers

The complex numbers $a + bi$ and $a - bi$ are called **complex conjugates** of each other. The symbol used for conjugate is a line segment over the complex number. We write $\overline{a + bi} = a - bi$ and $\overline{a - bi} = a + bi$.

Example 4   Complex conjugates

Find the product of the given complex number and its conjugate.

a) $3 - i$

b) $4 + 2i$

c) $-i$

Solution

a) The conjugate of $3 - i$ is $3 + i$, and $(3 - i)(3 + i) = 9 - i^2 = 10$.

b) The conjugate of $4 + 2i$ is $4 - 2i$, and $(4 + 2i)(4 - 2i) = 16 - 4i^2 = 20$.

c) The conjugate of $-i$ is i, and $-i \cdot i = -i^2 = 1$.

In general we have the following theorem about complex conjugates.

Theorem: Complex Conjugates

If a and b are real numbers, then the product of $a + bi$ and its conjugate $a - bi$ is the real number $a^2 + b^2$. In symbols,

$$(a + bi)(a - bi) = a^2 + b^2.$$

We use the theorem about complex conjugates to divide imaginary numbers, in a process that is similar to rationalizing a denominator.

Example 5   Dividing imaginary numbers

Write each quotient in the form $a + bi$.

a) $\dfrac{8 - i}{2 + i}$ b) $\dfrac{1}{5 - 4i}$ c) $\dfrac{3 - 2i}{i}$

Solution

a) Multiply the numerator and denominator by $2 - i$, the conjugate of $2 + i$:

$$\frac{8 - i}{2 + i} = \frac{(8 - i)(2 - i)}{(2 + i)(2 - i)} = \frac{16 - 10i + i^2}{4 - i^2} = \frac{15 - 10i}{5} = 3 - 2i$$

Check division using multiplication: $(3 - 2i)(2 + i) = 8 - i$.

b) $\dfrac{1}{5 - 4i} = \dfrac{1(5 + 4i)}{(5 - 4i)(5 + 4i)} = \dfrac{5 + 4i}{25 + 16} = \dfrac{5 + 4i}{41} = \dfrac{5}{41} + \dfrac{4}{41}i$

Check: $\left(\dfrac{5}{41} + \dfrac{4}{41}i\right)(5 - 4i) = \dfrac{25}{41} + \dfrac{20}{41}i - \dfrac{20}{41}i - \dfrac{16}{41}i^2$

$$= \frac{25}{41} + \frac{16}{41} = 1.$$

c) $\dfrac{3 - 2i}{i} = \dfrac{(3 - 2i)(-i)}{i(-i)} = \dfrac{-3i + 2i^2}{-i^2} = \dfrac{-2 - 3i}{1} = -2 - 3i$

Check: $(-2 - 3i)(i) = -3i^2 - 2i = 3 - 2i$.   ◆

Roots of Negative Numbers

In Examples 2(d) and 2(e), we saw that both $(3i)^2 = -9$ and $(-3i)^2 = -9$. This means that in the complex number system there are two square roots of -9, $3i$ and $-3i$. For any positive real number b, we have $(i\sqrt{b})^2 = -b$ and $(-i\sqrt{b})^2 = -b$. So there are two square roots of $-b$, $i\sqrt{b}$ and $-i\sqrt{b}$. We call $i\sqrt{b}$ the **principal square root** of $-b$ and make the following definition.

Definition: Square Root of a Negative Number

For any positive real number b, $\sqrt{-b} = i\sqrt{b}$.

In the real number system, $\sqrt{-2}$ and $\sqrt{-8}$ are undefined, but in the complex number system they are defined as $\sqrt{-2} = i\sqrt{2}$ and $\sqrt{-8} = i\sqrt{8}$. Even though we now have meaning for a symbol such as $\sqrt{-2}$, *all operations with complex numbers must be performed after converting to the $a + bi$ form.* If we perform operations with roots of negative numbers using properties of the real numbers, we can get contradictory results:

$$\sqrt{-2} \cdot \sqrt{-8} = \sqrt{(-2)(-8)} = \sqrt{16} = 4 \qquad \text{Incorrect.}$$

$$i\sqrt{2} \cdot i\sqrt{8} = i^2 \cdot \sqrt{16} = -4 \qquad \text{Correct.}$$

The product rule $\sqrt{a} \cdot \sqrt{b} = \sqrt{ab}$ is used *only* for nonnegative numbers a and b.

Example 6 Square roots of negative numbers

Write each expression in the form $a + bi$, where a and b are real numbers.

a) $\sqrt{-8} + \sqrt{-18}$ **b)** $\dfrac{-4 + \sqrt{-50}}{4}$ **c)** $\sqrt{-27}\left(\sqrt{9} - \sqrt{-2}\right)$

Solution

The first step in each case is to replace the square roots of negative numbers by expressions with i.

a) $\sqrt{-8} + \sqrt{-18} = i\sqrt{8} + i\sqrt{18} = 2i\sqrt{2} + 3i\sqrt{2} = 5i\sqrt{2}$

b) $\dfrac{-4 + \sqrt{-50}}{4} = \dfrac{-4 + i\sqrt{50}}{4} = \dfrac{-4 + 5i\sqrt{2}}{4} = -1 + \dfrac{5}{4}i\sqrt{2}$

c) $\sqrt{-27}\left(\sqrt{9} - \sqrt{-2}\right) = 3i\sqrt{3}\left(3 - i\sqrt{2}\right) = 9i\sqrt{3} - 3i^2\sqrt{6} = 3\sqrt{6} + 9i\sqrt{3}$

Complex Solutions to Equations

In Section 2.4 we will be solving equations that have complex number solutions. We can use the operations for complex numbers defined in this section to determine whether a given complex number satisfies an equation.

Example 7 Complex number solutions to equations

Determine whether the complex number $3 + i\sqrt{2}$ satisfies $x^2 - 6x + 11 = 0$.

Solution

Replace x by $3 + i\sqrt{2}$ in the polynomial $x^2 - 6x + 11$:

$$\left(3 + i\sqrt{2}\right)^2 - 6\left(3 + i\sqrt{2}\right) + 11 = 9 + 6i\sqrt{2} + \left(i\sqrt{2}\right)^2 - 18 - 6i\sqrt{2} + 11$$
$$= 9 + 6i\sqrt{2} - 2 - 18 - 6i\sqrt{2} + 11$$
$$= 0$$

Since the value of the polynomial is zero for $x = 3 + i\sqrt{2}$, the complex number $3 + i\sqrt{2}$ satisfies the equation $x^2 - 6x + 11 = 0$

For Thought

True or false? Explain.

1. The multiplicative inverse of i is $-i$.
2. The conjugate of i is $-i$.
3. The set of complex numbers is a subset of the set of real numbers.
4. $(\sqrt{3} - i\sqrt{2})(\sqrt{3} + i\sqrt{2}) = 5$
5. $\overline{(2 - 5i)}(2 + 5i) = 4 + 25$
6. $5 - \sqrt{-9} = 5 - 9i$
7. If $P(x) = x^2 + 9$, then $P(3i) = 0$.
8. The imaginary number $-3i$ is a root to the equation $x^2 + 9 = 0$.
9. $i^4 = 1$
10. $i^{18} = 1$

2.3 Exercises   Tape 3   Disk—5.25″: 1   3.5″: 1   Macintosh: 1

Determine whether each complex number is real or imaginary and write it in the standard form $a + bi$.

1. $6i$
2. $-3i + \sqrt{6}$
3. $\dfrac{1 + i}{3}$
4. -72
5. $\sqrt{7}$
6. $-i\sqrt{5}$
7. $\dfrac{\pi}{2}$
8. 0

Perform the indicated operations and write your answers in the form $a + bi$, where a and b are real numbers.

9. $(3 - 3i) + (4 + 5i)$
10. $(-3 + 2i) + (5 - 6i)$
11. $(1 - i) - (3 + 2i)$
12. $(6 - 7i) - (3 - 4i)$
13. $-6i(3 - 2i)$
14. $-3i(5 + 2i)$
15. $(2 - 3i)(4 + 6i)$
16. $(3 - i)(5 - 2i)$
17. $(5 - 2i)(5 + 2i)$
18. $(4 + 3i)(4 - 3i)$
19. $(\sqrt{3} - i)(\sqrt{3} + i)$
20. $(\sqrt{2} + i\sqrt{3})(\sqrt{2} - i\sqrt{3})$
21. $(3 + 4i)^2$
22. $(-6 - 2i)^2$
23. $(\sqrt{5} - 2i)^2$
24. $(\sqrt{6} + i\sqrt{3})^2$
25. i^{17}
26. i^{24}
27. i^{98}
28. i^{19}
29. i^{-4}
30. i^{-13}
31. i^{-1}
32. i^{-27}

Find the product of the given complex number and its conjugate.

33. $3 - 9i$
34. $4 + 3i$
35. $\dfrac{1}{2} + 2i$
36. $\dfrac{1}{3} - i$
37. i
38. $-i\sqrt{5}$
39. $3 - i\sqrt{3}$
40. $\dfrac{5}{2} + i\dfrac{\sqrt{2}}{2}$

Write each quotient in the form $a + bi$.

41. $\dfrac{1}{2 - i}$
42. $\dfrac{1}{5 + 2i}$
43. $\dfrac{-3i}{1 - i}$
44. $\dfrac{3i}{-2 + i}$
45. $\dfrac{-2 + 6i}{2}$
46. $\dfrac{-6 - 9i}{-3}$
47. $\dfrac{-3 + 3i}{i}$
48. $\dfrac{-2 - 4i}{-i}$
49. $\dfrac{1 - i}{3 + 2i}$
50. $\dfrac{4 + 2i}{2 - 3i}$
51. $\dfrac{\sqrt{2} - i\sqrt{3}}{\sqrt{3} + i\sqrt{2}}$
52. $\dfrac{\sqrt{3} - i\sqrt{2}}{\sqrt{3} + i\sqrt{2}}$

Write each expression in the form $a + bi$, where a and b are real numbers.

53. $\sqrt{-4} - \sqrt{-9}$
54. $\sqrt{-16} + \sqrt{-25}$

55. $\sqrt{-4} - \sqrt{16}$ **56.** $\sqrt{-3} \cdot \sqrt{-3}$

57. $(\sqrt{-6})^2$ **58.** $(\sqrt{-5})^3$

59. $\sqrt{-2} \cdot \sqrt{-50}$ **60.** $\dfrac{-6 + \sqrt{-3}}{3}$

61. $\dfrac{-2 + \sqrt{-20}}{2}$ **62.** $\dfrac{9 - \sqrt{-18}}{-6}$

63. $-3 + \sqrt{3^2 - 4(1)(5)}$ **64.** $1 - \sqrt{(-1)^2 - 4(1)(1)}$

65. $\sqrt{-8}(\sqrt{-2} + \sqrt{8})$ **66.** $\sqrt{-6}(\sqrt{2} - \sqrt{-3})$

67. $(\sqrt{-4} + \sqrt{8})(\sqrt{-1} - \sqrt{2})$

68. $(\sqrt{-54} - \sqrt{2})^2$

Evaluate the expression $\dfrac{-b + \sqrt{b^2 - 4ac}}{2a}$ for each choice of a, b, and c.

69. $a = 1, b = 2, c = 5$ **70.** $a = 5, b = -4, c = 1$

Evaluate the expression $\dfrac{-b - \sqrt{b^2 - 4ac}}{2a}$ for each choice of a, b, and c.

71. $a = 1, b = 6, c = 17$ **72.** $a = 1, b = -12, c = 84$

Let $P(x) = x^2 + 4x + 5$, $T(x) = 2x^2 + 1$, and $W(x) = x^2 - 6x + 14$. Find each of the following.

73. $P(-2 + i)$ **74.** $T\left(\dfrac{i\sqrt{2}}{2}\right)$ **75.** $P(-2 - i)$

76. $T\left(\dfrac{-i\sqrt{2}}{2}\right)$ **77.** $P(1 + i)$ **78.** $T(3 - i)$

79. $W(3 + i\sqrt{5})$ **80.** $W(3 - i\sqrt{5})$ **81.** $W(-1)$

82. $W(2 - i)$

Determine whether the given complex number satisfies the equation following it.

83. $2i, x^2 + 4 = 0$ **84.** $-4i, 2x^2 + 32 = 0$

85. $1 - i, x^2 + 2x - 2 = 0$ **86.** $1 + i, x^2 - 2x + 2 = 0$

87. $3 - 2i, x^2 - 6x + 13 = 0$

88. $2 - i, x^2 - 4x - 6 = 0$

89. $\dfrac{i\sqrt{3}}{3}, 3x^2 + 1 = 0$ **90.** $\dfrac{-i\sqrt{3}}{3}, 3x^2 - 1 = 0$

91. $2 + i\sqrt{3}, x^2 - 4x + 7 = 0$

92. $\sqrt{2} + i\sqrt{3}, x^2 - 2x\sqrt{2} + 5 = 0$

Find the product of each pair of binomials. Note that in each case the product is a quadratic polynomial with real coefficients.

93. $(x - i)(x + i)$ **94.** $(x - 3i)(x + 3i)$

95. $(x - (1 + i))(x - (1 - i))$

96. $(x - (2 - i))(x - (2 + i))$

97. $(x - (2 + 3i))(x - (2 - 3i))$

98. $(x - (1 - 2i))(x - (1 + 2i))$

99. $\left(x - \dfrac{1 + i}{2}\right)\left(x - \dfrac{1 - i}{2}\right)$

100. $\left(x - \dfrac{3 + i}{2}\right)\left(x - \dfrac{3 - i}{2}\right)$

Write each expression in the standard form for a complex number.

101. $\overline{2 - 3i} + \overline{2 - 3i}$ **102.** $\overline{4 + 5i} - (4 + 5i)$

103. $\overline{3 - i} \cdot \overline{4 + 5i}$ **104.** $\overline{2 - i} - \overline{3 + 2i}$

105. $\overline{(3 - i)(4 + 5i)}$ **106.** $\overline{(3 - 2i) + (-1 + 4i)}$

107. $\overline{3 - 2i} + \overline{-1 + 4i}$ **108.** $\overline{(2 - i) - (3 + 2i)}$

For Writing/Discussion

109. Explain in detail how to find i^n for any positive integer n.

110. Find a number $a + bi$ such that $a^2 + b^2$ is irrational.

111. Let $w = a + bi$ and $\overline{w} = a - bi$, where a and b are real numbers. Show that $w + \overline{w}$ is real and that $w - \overline{w}$ is imaginary. Write sentences (containing no mathematical symbols) stating these results.

112. Is it true that the product of a complex number and its conjugate is a real number? Explain.

113. Is it true that the conjugate of the sum of two complex numbers is equal to the sum of their conjugates? Explain.

114. Prove that the reciprocal of $a + bi$, where a and b are not both zero, is
$$\frac{a}{a^2 + b^2} - \frac{b}{a^2 + b^2} i.$$

115. Prove that the conjugate of the difference of two complex numbers is equal to the difference of their conjugates.

116. *Cooperative learning* Work in a small group to find the two square roots of 1 and two square roots of -1 in the complex number system. How many fourth roots of 1 are there in the complex number system and what are they? Explain how to find all of the fourth roots in the complex number system for any positive real number.

2.4

Quadratic Equations

One of our main goals is to solve polynomial equations. In Section 2.1 we learned to solve equations of the form $ax + b = 0$, the linear equations. Linear equations are first-degree polynomial equations. In this section we will solve second-degree polynomial equations, the quadratic equations.

Definition

Quadratic equations have a term that linear equations do not have, the x^2-term.

Definition: Quadratic Equation

A **quadratic equation** is an equation of the form

$$ax^2 + bx + c = 0,$$

where a, b, and c are real numbers with $a \neq 0$.

The condition that $a \neq 0$ in the definition ensures that the equation actually does have an x^2-term. There are several methods for solving quadratic equations. Which method is most appropriate depends on the type of quadratic equation that we are solving. We first consider a method for solving quadratic equations in which $b = 0$ and later consider methods for solving equations in which $b \neq 0$.

Solving $ax^2 + c = 0$

If $b = 0$ in the general form of the quadratic equation, then the equation is of the form $ax^2 + c = 0$. Equations of this form are solved by using the square root property.

The Square Root Property

For any real number k, the equation $x^2 = k$ is equivalent to $x = \pm\sqrt{k}$.

If $k > 0$, then $x^2 = k$ has two real solutions. If $k < 0$, then $x^2 = k$ has two imaginary solutions. If $k = 0$, then 0 is the only solution to $x^2 = k$.

Example 1   Using the square root property

Solve each equation.

a) $x^2 - 9 = 0$

b) $2x^2 - 1 = 0$

c) $(x - 3)^2 = -8$

Solution

a) Before using the square root property, isolate x^2.

$$x^2 - 9 = 0$$
$$x^2 = 9$$
$$x = \pm\sqrt{9} = \pm 3 \qquad \text{Square root property}$$

Check: $3^2 - 9 = 0$ and $(-3)^2 - 9 = 0$. The solution set is $\{-3, 3\}$.

b) $2x^2 - 1 = 0$

$$2x^2 = 1$$
$$x^2 = \frac{1}{2} \qquad \text{Isolate } x^2.$$
$$x = \pm\sqrt{\frac{1}{2}} = \pm\frac{\sqrt{2}}{2} \qquad \text{Square root property}$$

Check: $2\left(\pm\frac{\sqrt{2}}{2}\right)^2 - 1 = 2 \cdot \frac{1}{2} - 1 = 0.$

The solution set is $\left\{-\frac{\sqrt{2}}{2}, \frac{\sqrt{2}}{2}\right\}$.

c) $(x - 3)^2 = -8$

$$x - 3 = \pm\sqrt{-8} \qquad \text{Square root property}$$
$$x = 3 \pm 2i\sqrt{2}$$

Check in the original equation. The solution set is $\left\{3 - 2i\sqrt{2}, 3 + 2i\sqrt{2}\right\}$.

The key idea in solving $ax^2 + c = 0$ is to solve for x^2 and then apply the square root property. We now turn our attention to equations of the form $ax^2 + bx + c = 0$ in which the trinomial can be factored.

Solving Quadratic Equations by Factoring

Many second-degree polynomials can be factored as a product of first-degree binomials. When one side of an equation is a product and the other side is 0, we can write an equivalent equation by setting each factor equal to zero. This idea is called the **zero factor property.**

The Zero Factor Property

If A and B are algebraic expressions, then the equation $AB = 0$ is equivalent to the compound statement $A = 0$ or $B = 0$.

Example 2   Quadratic equations solved by factoring

Solve each equation by factoring.

a) $x^2 - x - 12 = 0$ **b)** $(x + 3)(x - 4) = 8$

Solution

a)
$$x^2 - x - 12 = 0$$
$$(x - 4)(x + 3) = 0 \qquad \text{Factor the left-hand side.}$$
$$x - 4 = 0 \quad \text{or} \quad x + 3 = 0 \qquad \text{Zero factor property}$$
$$x = 4 \quad \text{or} \qquad x = -3$$

Check: $(-3)^2 - (-3) - 12 = 0$ and $4^2 - 4 - 12 = 0$. The solution set is $\{-3, 4\}$.

b) We must first rewrite the equation with 0 on one side, because the zero factor property applies only when the factors have a product of 0.

$$(x + 3)(x - 4) = 8$$
$$x^2 - x - 12 = 8 \qquad \text{Multiply on the left-hand side.}$$
$$x^2 - x - 20 = 0 \qquad \text{Get 0 on the right-hand side.}$$
$$(x - 5)(x + 4) = 0 \qquad \text{Factor.}$$
$$x - 5 = 0 \quad \text{or} \quad x + 4 = 0 \qquad \text{Zero factor property}$$
$$x = 5 \quad \text{or} \qquad x = -4$$

Check in the original equation. The solution set is $\{-4, 5\}$.

Completing the Square

We cannot solve every quadratic equation by factoring because factoring methods are limited to integral coefficients. However, every quadratic equation can be written in the form of Example 1(c) and solved by using the square root property. This process is called **completing the square.**

Completing the square involves finding the third term of a perfect square trinomial when given the first two terms. For example, we can recognize $x^2 + 6x$ as the first two terms of the perfect square trinomial $x^2 + 6x + 9 = (x + 3)^2$. Note that one-half of 6 is 3 and 3^2 is 9. We can recognize $x^2 + 10x$ as the first two terms of the perfect square trinomial $x^2 + 10x + 25 = (x + 5)^2$. Note that one-half of 10 is 5, and 5^2 is 25. These examples suggest the following rule.

Rule for Finding the Last Term of $x^2 + bx + ?$

The last term of a perfect square trinomial (with $a = 1$) is the square of one-half of the coefficient of the middle term. In symbols, the perfect square trinomial whose first two terms are $x^2 + bx$ is

$$x^2 + bx + \left(\frac{b}{2}\right)^2.$$

A perfect square trinomial may have a leading coefficient that is not equal to 1. Since it is simpler to complete the square when $a = 1$, we rewrite any such polynomial so that $a = 1$ before completing the square.

Example 3   Solving a quadratic equation by completing the square

Solve each equation by completing the square.

a) $x^2 + 6x + 7 = 0$

b) $2x^2 - 3x - 4 = 0$

Solution

a) Since $x^2 + 6x + 7$ is not a perfect square trinomial, we must find a perfect square trinomial that has $x^2 + 6x$ as its first two terms. Since one-half of 6 is 3 and $3^2 = 9$, our goal is to get $x^2 + 6x + 9$ on the left-hand side:

$$x^2 + 6x + 7 = 0$$

$$x^2 + 6x = -7 \qquad \text{Subtract 7 from each side.}$$

$$x^2 + 6x + 9 = -7 + 9 \qquad \text{Add 9 to each side.}$$

$$(x + 3)^2 = 2 \qquad \text{Factor the left-hand side.}$$

$$x + 3 = \pm\sqrt{2} \qquad \text{Square root property}$$

$$x = -3 \pm \sqrt{2}$$

Check in the original equation. The solution set is $\left\{-3 - \sqrt{2}, -3 + \sqrt{2}\right\}$.

b) $2x^2 - 3x - 4 = 0$

$$x^2 - \frac{3}{2}x - 2 = 0 \qquad \text{Divide each side by 2 to get } a = 1.$$

$$x^2 - \frac{3}{2}x \phantom{+ \frac{9}{16}} = 2 \qquad \text{Add 2 to each side.}$$

$$x^2 - \frac{3}{2}x + \frac{9}{16} = 2 + \frac{9}{16} \qquad \frac{1}{2} \cdot \frac{3}{2} = \frac{3}{4} \text{ and } \left(\frac{3}{4}\right)^2 = \frac{9}{16}.$$

$$\left(x - \frac{3}{4}\right)^2 = \frac{41}{16} \qquad \text{Factor the left-hand side.}$$

$$x - \frac{3}{4} = \pm\frac{\sqrt{41}}{4} \qquad \text{Square root property}$$

$$x = \frac{3}{4} \pm \frac{\sqrt{41}}{4}$$

The solution set is $\left\{\dfrac{3 - \sqrt{41}}{4}, \dfrac{3 + \sqrt{41}}{4}\right\}$.

```
(3-√41)/4
        -.8507810594
2Ans²-3Ans-4
            1E-13
■
```

Check that $\dfrac{3 - \sqrt{41}}{4}$ satisfies the equation.

> **Strategy: Completing the Square**
>
> In completing the square we want an *equivalent equation* that has a perfect square trinomial on one side. To get an equivalent equation, the number that completes the square must be added to *both sides* of the equation.

The Quadratic Formula

The method of completing the square can be applied to any quadratic equation

$$ax^2 + bx + c = 0, \qquad \text{where } a \neq 0.$$

Assume for now that $a > 0$ and divide each side by a.

$$x^2 + \frac{b}{a}x + \frac{c}{a} = 0$$

$$x^2 + \frac{b}{a}x = -\frac{c}{a} \qquad \text{Subtract } \frac{c}{a} \text{ from each side.}$$

$$x^2 + \frac{b}{a}x + \frac{b^2}{4a^2} = -\frac{c}{a} + \frac{b^2}{4a^2} \qquad \frac{1}{2} \cdot \frac{b}{a} = \frac{b}{2a} \text{ and } \left(\frac{b}{2a}\right)^2 = \frac{b^2}{4a^2}.$$

Now factor the perfect square trinomial on the left-hand side. On the right-hand side get a common denominator and add.

$$\left(x + \frac{b}{2a}\right)^2 = \frac{b^2 - 4ac}{4a^2} \qquad \frac{c}{a} \cdot \frac{4a}{4a} = \frac{4ac}{4a^2}$$

$$x + \frac{b}{2a} = \pm\sqrt{\frac{b^2 - 4ac}{4a^2}} \qquad \text{Square root property}$$

$$x = -\frac{b}{2a} \pm \frac{\sqrt{b^2 - 4ac}}{2a} \qquad \text{Because } a > 0, \sqrt{4a^2} = 2a.$$

$$x = \frac{-b \pm \sqrt{b^2 - 4ac}}{2a}$$

We assumed that $a > 0$ so that $\sqrt{4a^2} = 2a$ would be correct. If a is negative, then $\sqrt{4a^2} = -2a$, and we get

$$x = -\frac{b}{2a} \pm \frac{\sqrt{b^2 - 4ac}}{-2a}.$$

However, the negative sign in $-2a$ can be deleted because of the $\pm$ symbol preceding it. For example, $5 \pm (-3)$ gives the same values as 5 ± 3. After deleting the negative sign on $-2a$, we get the same formula for the solution. It is called the **quadratic formula.** Its importance lies in its wide applicability. Any quadratic equation can be solved by using this formula.

The Quadratic Formula

The solution to $ax^2 + bx + c = 0$, with $a \neq 0$, is given by the formula

$$x = \frac{-b \pm \sqrt{b^2 - 4ac}}{2a}.$$

Example 4   Using the quadratic formula

Solve each equation using the quadratic formula.

a) $x^2 + 8x + 6 = 0$ **b)** $5x^2 - 4x + 1 = 0$

Solution

a) For $x^2 + 8x + 6 = 0$ we use $a = 1$, $b = 8$, and $c = 6$ in the formula:

$$x = \frac{-8 \pm \sqrt{8^2 - 4(1)(6)}}{2(1)} = \frac{-8 \pm \sqrt{40}}{2} = \frac{-8 \pm 2\sqrt{10}}{2} = -4 \pm \sqrt{10}$$

Check by evaluating either by hand or by using a calculator:

$$\left(-4 + \sqrt{10}\right)^2 + 8\left(-4 + \sqrt{10}\right) + 6 = 0 \qquad \text{and}$$
$$\left(-4 - \sqrt{10}\right)^2 + 8\left(-4 - \sqrt{10}\right) + 6 = 0$$

The solution set is $\left\{ -4 - \sqrt{10}, -4 + \sqrt{10} \right\}$.

```
-4+√10
        -.8377223398
Ans²+8Ans+6
             1E-13
■
```

Check that $-4 + \sqrt{10}$ satisfies the equation. See Appendix A for more examples.

b) For $5x^2 - 4x + 1 = 0$ we use $a = 5$, $b = -4$, and $c = 1$ in the formula:

$$x = \frac{-(-4) \pm \sqrt{(-4)^2 - 4(5)(1)}}{2(5)} = \frac{4 \pm \sqrt{-4}}{10} = \frac{4 \pm 2i}{10} = \frac{4}{10} \pm \frac{2}{10}i$$

$$= \frac{2}{5} \pm \frac{1}{5}i$$

You should check that these solutions are correct using operations with complex numbers. The solution set is $\{\frac{2}{5} - \frac{1}{5}i, \frac{2}{5} + \frac{1}{5}i\}$. Note that the solutions to this quadratic equation are complex conjugates.   ◆

To decide which of the four methods to use for solving a given quadratic equation, use the following strategy.

Strategy: Solving $ax^2 + bx + c = 0$

1. **If $b = 0$, solve $ax^2 + c = 0$ for x^2 and apply the square root property.**
2. **If $ax^2 + bx + c$ can be easily factored, then solve by factoring.**
3. **The quadratic formula or completing the square may be used on any quadratic equation, but the quadratic formula is usually easier to use.**

The Discriminant

Some quadratic equations have two real solutions, some have two imaginary solutions, and some have only one real solution. If we examine the quadratic formula, we can see what determines the number and type of solutions. If $b^2 - 4ac > 0$ in the formula, then $\sqrt{b^2 - 4ac}$ is a real number and we get two real solutions. If $b^2 - 4ac < 0$, then $\sqrt{b^2 - 4ac}$ is an imaginary number and we get two imaginary solutions. If $b^2 - 4ac = 0$ in the quadratic formula, then there will be only one real solution, $x = -b/(2a)$. Because of the $\pm$ sign in the formula, the imaginary solutions always occur in conjugate pairs. If a, b, and c are integers, and $b^2 - 4ac$ is a perfect square, then the solutions are rational numbers. Since $b^2 - 4ac$ determines the number and type of solutions, $b^2 - 4ac$ is called the **discriminant.** Table 2.4 summarizes this information.

Value of $b^2 - 4ac$	Type of solutions
Positive	Two real
Zero	One real
Negative	Two imaginary

Table 2.4   Number and type of solutions to a quadratic equation

Example 5   Using the discriminant

For each equation, state the value of the discriminant, the number of solutions, and whether the solutions are real or imaginary.

a) $x^2 + 8x + 6 = 0$ **b)** $5x^2 - 4x + 1 = 0$ **c)** $4x^2 + 12x + 9 = 0$

Solution

a) Find the value of $b^2 - 4ac$ using $a = 1$, $b = 8$, and $c = 6$:
$$b^2 - 4ac = 8^2 - 4(1)(6) = 40$$

The value of the discriminant is 40 and the equation has two real solutions.

b) Find the value of the discriminant for the equation $5x^2 - 4x + 1 = 0$:
$$b^2 - 4ac = (-4)^2 - 4(5)(1) = -4$$

Because the discriminant is negative, the equation has two imaginary solutions.

c) For $4x^2 + 12x + 9 = 0$, we have $b^2 - 4ac = 12^2 - 4(4)(9) = 0$. So the equation has one real solution. ◆

Applications

The problems that we solve in this section are very similar to those in Section 2.2. However, in this section the mathematical model of the situation results in a quadratic equation.

___

Example 6   A problem solved with a quadratic equation

It took Susan 30 minutes longer to drive 275 miles on I-70 west of Green River, Utah, than it took her to drive 300 miles east of Green River. Because of a sand storm and a full load of cantaloupes, she averaged 10 mph less while traveling west of Green River. What was her average speed for each part of the trip?

Solution

Let x represent Susan's average speed east of Green River and $x - 10$ represent her average speed west of Green River. We can organize all of the given information as in Table 2.5. Since $D = RT$, the time is determined by $T = D/R$.

	Distance	Rate	Time
East	300	x	$\dfrac{300}{x}$
West	275	$x - 10$	$\dfrac{275}{x - 10}$

Table 2.5

The following equation expresses the fact that her time west of Green River was $\frac{1}{2}$ hour greater than her time east of Green River.

$$\frac{275}{x - 10} = \frac{300}{x} + \frac{1}{2}$$

$$2x(x - 10) \cdot \frac{275}{x - 10} = 2x(x - 10)\left(\frac{300}{x} + \frac{1}{2}\right)$$

$$550x = 600(x - 10) + x(x - 10)$$

$$-x^2 - 40x + 6000 = 0$$

$$x^2 + 40x - 6000 = 0$$

$$(x + 100)(x - 60) = 0 \qquad \text{Factor.}$$

$$x = -100 \quad \text{or} \quad x = 60$$

The solution $x = -100$ is a solution to the equation, but not a solution to the problem. The other solution, $x = 60$, means that $x - 10 = 50$. Check that these two average speeds are a solution to the problem. Susan's average speed east of Green River was 60 mph, and her average speed west of Green River was 50 mph. ◆

The next example involves the formula for the height of a rising or falling object, $S = -16t^2 + v_0t + s_0$, where S is the height in feet, t is the time in seconds, v_0 is the initial velocity, and s_0 is the initial height. In this example, the solutions are irrational numbers. In this case it is usually best to give both exact and approximate answers to the problem.

Example 7 Using the quadratic formula in a problem

Michael Chang, top-ranked professional tennis player, skillfully uses the lob as both a defensive and an offensive shot. Chang knows that a well-placed lob can buy the time needed to prepare for the next shot. How long does it take for a tennis ball to hit the earth if it is hit straight upward with a velocity of 60 feet per second from a height of 5 feet?

Solution

We are looking for the value of t for which the height S is 0. Use initial velocity $v_0 = 60$ feet per second, initial height $s_0 = 5$ feet, and $S = 0$ in the formula $S = -16t^2 + v_0 t + s_0$:

$$0 = -16t^2 + 60t + 5$$

Use the quadratic formula to solve the equation:

$$t = \frac{-60 \pm \sqrt{60^2 - 4(-16)(5)}}{2(-16)} = \frac{-60 \pm \sqrt{3920}}{-32} = \frac{-60 \pm 28\sqrt{5}}{-32}$$

$$= \frac{15 \pm 7\sqrt{5}}{8}$$

Now $(15 - 7\sqrt{5})/8 \approx -0.082$ and $(15 + 7\sqrt{5})/8 \approx 3.83$. Since the time at which the ball returns to the earth must be positive, it will take exactly $(15 + 7\sqrt{5})/8$ seconds or approximately 3.83 seconds for the ball to hit the earth. ◆

Quadratic equations often arise from applications that involve the Pythagorean theorem from geometry: *A triangle is a right triangle if and only if the sum of the squares of the legs is equal to the square of the hypotenuse.*

Example 8 Using the Pythagorean theorem

In the house shown in Fig. 2.6, the ridge of the roof at point B is 6 feet above point C. If the distance from A to C is 18 feet, then what is the length of a rafter from A to B?

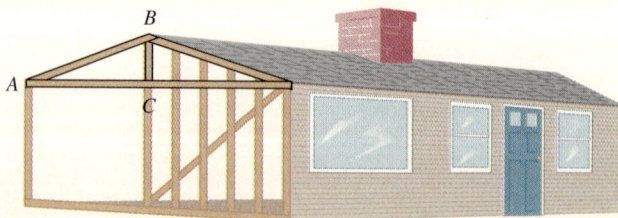

Figure 2.6

(margin)

$v_0 = 60$ ft/sec

S

5

3.83 t

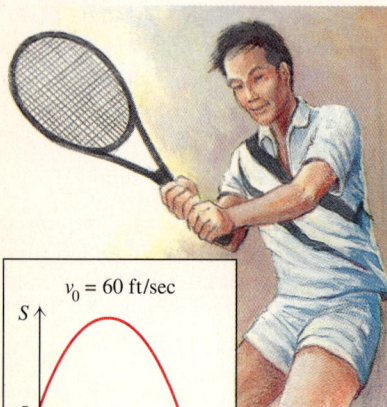

```
(15+7√5)/8
        3.83155948
-16Ans²+60Ans+5
            -1ε-11
```

Find the approximate answer and check.

Solution

Let x be the distance from A to B. Use the Pythagorean theorem to write the following equation.

$$x^2 = 18^2 + 6^2$$
$$x^2 = 360$$
$$x = \pm\sqrt{360} = \pm 6\sqrt{10}$$

Since x must be positive in this problem, $x = 6\sqrt{10}$ feet or $x \approx 18.97$ feet.

? For Thought

True or false? Explain.

1. The equation $(x - 3)^2 = 4$ is equivalent to $x - 3 = 2$.

2. Every quadratic equation can be solved by factoring.

3. The trinomial $x^2 + \dfrac{2}{3}x + \dfrac{4}{9}$ is a perfect square trinomial.

4. The equation $(x - 3)(2x + 5) = 0$ is equivalent to $x = 3$ or $x = \dfrac{5}{2}$.

5. All quadratic equations have two distinct solutions.

6. If m, n, and p are real numbers such that $mx^2 - nx + p = 0$ with $m \neq 0$, then $x = \dfrac{n \pm \sqrt{n^2 - 4mp}}{2m}$.

7. If $b = 0$, then $ax^2 + bx + c = 0$ cannot be solved by the quadratic formula.

8. All quadratic equations have at least one real solution.

9. The value of the discriminant is 0 for $4x^2 + 12x + 9 = 0$.

10. A quadratic equation with real coefficients can have one real and one imaginary solution.

2.4 Exercises   ▭ Tape 4   ▤ Disk—5.25″: 1   3.5″: 1   Macintosh: 1

Solve each equation by using the square root property.

1. $x^2 - 5 = 0$

2. $x^2 - 8 = 0$

3. $3x^2 + 2 = 0$

4. $2x^2 + 16 = 0$

5. $(x - 3)^2 = 9$

6. $(x + 1)^2 = \dfrac{9}{4}$

7. $\left(x - \dfrac{1}{2}\right)^2 = \dfrac{25}{4}$

8. $(3x - 1)^2 = \dfrac{1}{4}$

9. $(x + 2)^2 = -4$

10. $(x - 3)^2 = -20$

11. $\left(x - \dfrac{2}{3}\right)^2 = -\dfrac{4}{9}$

12. $\left(x + \dfrac{3}{2}\right)^2 = -\dfrac{1}{2}$

Solve each equation by factoring.

13. $x^2 - x - 20 = 0$

14. $x^2 + 2x - 8 = 0$

15. $a^2 + 3a = -2$ **16.** $b^2 - 4b = 12$

17. $2x^2 - 5x - 3 = 0$ **18.** $2x^2 - 5x + 2 = 0$

19. $6x^2 - 7x + 2 = 0$ **20.** $12x^2 - 17x + 6 = 0$

21. $(y - 3)(y + 4) = 30$ **22.** $(w - 1)(w - 2) = 6$

23. $(2z - 1)(z + 3) = 15$ **24.** $(2t - 3)(2t + 1) = 5$

Find the perfect square trinomial whose first two terms are given.

25. $x^2 - 12x$ **26.** $y^2 + 20y$ **27.** $r^2 + 3r$

28. $t^2 - 7t$ **29.** $w^2 + \dfrac{1}{2}w$ **30.** $p^2 - \dfrac{2}{3}p$

Solve each equation by completing the square.

31. $x^2 + 6x + 1 = 0$ **32.** $x^2 - 10x + 5 = 0$

33. $n^2 - 2n - 1 = 0$ **34.** $m^2 - 12m + 33 = 0$

35. $h^2 + 3h - 1 = 0$ **36.** $t^2 - 5t + 2 = 0$

37. $x^2 + 2x + 5 = 0$ **38.** $x^2 - 4x + 5 = 0$

39. $2x^2 + 5x = 12$ **40.** $3x^2 + x = 2$

41. $3x^2 + 2x + 1 = 0$ **42.** $5x^2 + 4x + 3 = 0$

Solve each equation using the quadratic formula.

43. $x^2 + 3x - 4 = 0$ **44.** $x^2 + 8x + 12 = 0$

45. $2x^2 - 5x - 3 = 0$ **46.** $2x^2 + 3x - 2 = 0$

47. $9x^2 + 6x + 1 = 0$ **48.** $16x^2 - 24x + 9 = 0$

49. $2x^2 - 3 = 0$ **50.** $-2x^2 + 5 = 0$

51. $x^2 - 4x + 5 = 0$ **52.** $x^2 - 6x + 13 = 0$

53. $9x^2 + 6x = 1$ **54.** $2x^2 + 3 = 6x$

Use a calculator and the quadratic formula to find all real and imaginary solutions to each equation. Round answers to two decimal places.

55. $3.2x^2 + 7.6x - 9 = 0$ **56.** $1.5x^2 - 6.3x = 10.1$

57. $3.25x^2 - 4.6x + 20 = 0$

58. $4.76x^2 + 6.12x + 55.3 = 0$

For each equation, state the value of the discriminant, the number of solutions, and whether the solutions are real or imaginary.

59. $9x^2 - 30x + 25 = 0$ **60.** $3x^2 - 7x + 3 = 0$

61. $5x^2 - 6x + 2 = 0$ **62.** $3x^2 + 5x + 5 = 0$

63. $7x^2 + 12x - 1 = 0$ **64.** $4x^2 + 28x + 49 = 0$

Solve each equation. Use the method of your choice.

65. $x^2 = \dfrac{4}{3}x - \dfrac{5}{9}$ **66.** $x^2 = \dfrac{2}{7}x - \dfrac{2}{49}$

67. $x^2 + \sqrt{2} = 0$ **68.** $\sqrt{2}x^2 - 1 = 0$

69. $12x^2 + x\sqrt{6} - 1 = 0$ **70.** $-10x^2 - x\sqrt{5} + 1 = 0$

71. $x(x + 6) = 72$ **72.** $x = \dfrac{96}{x + 4}$

73. $x = 1 + \dfrac{1}{x}$ **74.** $x = \dfrac{1}{x}$

75. $\dfrac{x - 12}{3 - x} = \dfrac{x + 4}{x + 7}$ **76.** $\dfrac{x - 9}{x - 2} = -\dfrac{x + 3}{x + 1}$

Use the methods for solving quadratic equations to solve each formula for the indicated variable.

77. $A = \pi r^2$ for r **78.** $S = 2\pi rh + 2\pi r^2$ for r

79. $x^2 + 2kx + 3 = 0$ for x **80.** $hy^2 - ky = p$ for y

81. $2y^2 + 4xy = x^2$ for y **82.** $\dfrac{\dfrac{1}{x + h} - \dfrac{1}{x}}{h} = 1$ for x

Find an exact solution to each problem. If the solution is irrational, then find an approximate solution also.

83. *Demand equation*   The demand equation for a certain product is $P = 40 - 0.001x$, where x is the number of units sold per week and P is the price in dollars at which each one is sold. The weekly revenue R is given by $R = xP$. What number of units sold produces a weekly revenue of $175{,}000$?

84. *Average cost*   The total cost in dollars of producing x items is given by $C = 0.02x^3 + 5x$. For what number of items is the average cost per item equal to 5.50?

85. *Height of a ball*   A juggler can toss a ball into the air with a velocity of 40 ft/sec from a height of 4 ft. How long will it take for the ball to return to the height of 4 ft?

86. *Height of a sky diver*   If a sky diver steps out of an airplane at 5000 ft, then how long does it take her to reach 4000 ft? What assumptions are you making to solve this problem?

87. *Diagonal of a football field*   A football field is 100 yd long from goal line to goal line and 160 ft wide. If a player ran diagonally across the field from one goal line to the other, then how far did he run?

88. *Dimensions of a flag*   If the perimeter of a rectangular flag is 34 in. and the diagonal is 13 in., then what are the length and width?

89. *Two unknown numbers* If the sum of two numbers is $12\sqrt{2}$ and their product is 64, then what are the numbers?

90. *Ages of siblings* Joe and Jan were born three years apart. If the product of their ages is 1404, then what are their ages?

91. *Bordering a flower bed* Juan, a landscaper in Albuquerque, is designing a spring display of tulips and daffodils. The city has requested a 6-ft by 8-ft area of tulips surrounded by a uniform border of daffodils in an overall area of 100 ft². How wide should the border be?

92. *Bracing a gate* The width of a rectangular gate is 2 ft more than its height. If a diagonal board of length 8 ft is used for bracing, then what are the dimensions of the gate?

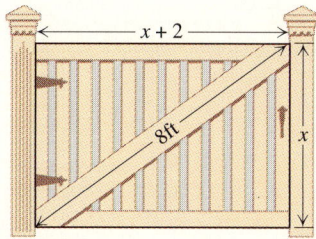

Figure for Exercise 92

93. *Speed of a tortoise* When a tortoise crosses a highway, his speed is 2 ft/hr faster than normal. If he can cross a 24-ft lane in 24 min less time than he can travel that same distance off the highway, then what is his normal speed?

94. *Speed of an electric car* An experimental electric-solar car completed a 1000-mi race in 35 hr. For the 600 mi traveled during daylight the car averaged 20 mph more than it did for the 400 mi traveled at night. What was the average speed of the car during the daytime?

95. *Painting a parking lot* Painting all of the parking lines at the Sunshine Mall can be done by using a new striper in two days less time than it took using the old striper. If Roberto uses the new striper and Curt uses the old striper, and together they complete the job in 3.5 days, then how long would it have taken Curt to do the job alone using the old striper?

96. *Making a dress* Rafael designed a sequined dress to be worn at the Academy Awards. His top seamstress, Maria, could sew on all the sequins in 10 hr less time than his next-best seamstress, Stephanie. To save time, he gave the job to both women and got all of the sequins attached in 17 hr. How long would it have taken Stephanie working alone?

97. *Percentage of white meat* The Kansas Fried Chicken store sells a Party Size bucket that weighs 10 lb more than the Big Family Size bucket. The Party Size bucket contains 8 lb of white meat, while the Big Family Size bucket contains 3 lb of white meat. If the percentage of white meat in the Party Size is 10 percentage points greater than the percentage of white meat in the Big Family Size, then how much does the Party Size bucket weigh?

98. *Mixing antifreeze in a radiator* Steve's car had a large radiator that contained an unknown amount of pure water. He added two quarts of antifreeze to the radiator. After testing, he decided that the percentage of antifreeze in the radiator was not large enough. Not knowing how to solve mixture problems, Steve decided to add one quart of water and another quart of antifreeze to the radiator to see what he would get. After testing he found that the last addition increased the percentage of antifreeze by three percentage points. How much water did the radiator contain originally?

99. *Initial velocity of a basketball player* Michael Jordan, former star guard of the Chicago Bulls, is famous for his high leaps and "hang time," especially while slam-dunking. If Jordan leaps for a dunk and reaches a peak height at the basket of 1.07 m, what is his upward velocity in meters per second at the moment his feet leave the floor? The formula $v_1^2 = v_0^2 + 2gS$ gives the relationship between final velocity v_1, initial velocity v_0, acceleration of gravity g, and his height S. Use $g = -9.8$ m/sec² and the fact that his final velocity is zero at his peak height.

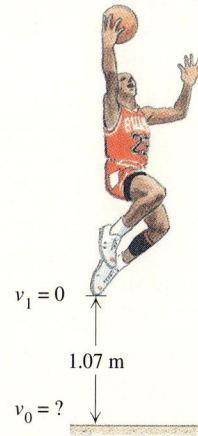

Figure for Exercise 99

100. *Hang time for a slam-dunk* Use the initial velocity for Michael Jordan determined in Exercise 99 and the formula $S = \frac{1}{2}gt^2 + v_0t$ to find the amount of time he is in the air.

101. *Games behind in baseball standings* Baseball fans keep up with their favorite teams through charts where the teams are ranked according to the percentage of games won. The chart usually has a column indicating the games behind (GB) for each team. If the win-loss record of the number one team is (A, B), then the games behind of another team whose win-loss record is (a, b) is calculated by the formula

$$GB = \frac{(A - a) + (b - B)}{2}.$$

Using the standings in Table 2.6, verify that Atlanta is 4 games behind Cincinnati and find how many games San Diego is behind Cincinnati.

Team	Won	Lost	Pct.	GB
Cincinnati	38	24	.613	—
Atlanta	35	29	.547	4
San Diego	34	31	.523	?

Table 2.6

102. *Misleading baseball standings* The formula for games behind can produce unusual results. In Table 2.7, Pittsburgh has a higher percentage of wins than Chicago, and so Pittsburgh is in first place. Use the formula for GB to find the number of games that Chicago is behind Pittsburgh. Is Chicago actually behind Pittsburgh in terms of the statistic GB?

Team	Won	Lost	Pct.	GB
Pittsburgh	18	13	.581	—
Chicago	22	16	.579	?

Table 2.7

103. *A new baseball statistic* Since the calculation of games behind (GB) in baseball standings can produce misleading results, another statistic to replace GB has been proposed. This new statistic, called the deficit D, is the number of games that two teams would have to play against each other to get equal percentages of wins, with the higher-ranked team losing all of the games. The statistic D gives a better indication of the distance from the leader than the statistic GB, but is not very popular with sports fans because a quadratic equation must be solved to find D. Find the value of D for Chicago in Table 2.8 that would give Pittsburgh and Chicago the same percentage of wins.

Team	Won	Lost
Pittsburgh	18	$13 + D$
Chicago	$22 + D$	16

Table 2.8

For Writing/Discussion

104. Write down a quadratic equation that has nonintegral coefficients and two real solutions. Use the quadratic formula to solve it and a calculator to check your solutions.

105. *Cooperative learning* Work in a small group to make up a problem involving uniform motion whose solution involves a quadratic equation. Give your problem to another group to solve.

106. *Cooperative learning* Find out more about Pythagoras and tell your class what you have learned.

2.5

More Equations

The techniques that we learned for solving quadratic equations in Section 2.4 can be applied to a wide variety of equations. The solution to most of the equations in this section will involve the square root property, factoring, or the quadratic formula.

Factoring Higher-Degree Equations

In Section 2.4 we used factoring to solve quadratic equations. Since we can also factor many higher-degree polynomials, we can solve many higher-degree equations by factoring. In Examples 1 and 2 we will use factoring to find all real and imaginary solutions to equations of degree higher than 2.

Example 1   Solving an equation by factoring

Solve $x^3 + 3x^2 + x + 3 = 0$.

Solution

Factor the polynomial on the left-hand side by grouping.

$$x^2(x + 3) + 1(x + 3) = 0 \qquad \text{Factor by grouping.}$$

$$(x^2 + 1)(x + 3) = 0 \qquad \text{Factor out } x + 3.$$

$$x^2 + 1 = 0 \quad \text{or} \quad x + 3 = 0 \qquad \text{Zero factor property}$$

$$x^2 = -1 \quad \text{or} \quad x = -3$$

$$x = \pm i \quad \text{or} \quad x = -3$$

Check these solutions in the original equation. The solution set is $\{-3, -i, i\}$.

◆

Example 2   Solving an equation by factoring

Solve $2x^5 = 16x^2$.

Solution

Write the equation with 0 on the right-hand side, then factor completely.

$$2x^5 - 16x^2 = 0$$

$$2x^2(x^3 - 8) = 0 \qquad \text{Factor out greatest common factor.}$$

$$2x^2(x - 2)(x^2 + 2x + 4) = 0 \qquad \text{Factor difference of two cubes.}$$

$$2x^2 = 0 \quad \text{or} \quad x - 2 = 0 \quad \text{or} \quad x^2 + 2x + 4 = 0$$

$$x = 0 \quad \text{or} \quad x = 2 \quad \text{or} \quad x = \frac{-2 \pm \sqrt{-12}}{2} = -1 \pm i\sqrt{3}$$

The solution set is $\left\{0, 2, -1 \pm i\sqrt{3}\right\}$.

◆

Note that in Example 2, if we had divided each side by x^2 as our first step, we would have lost the solution $x = 0$. *We do not usually divide each side of an equation by a variable expression.* Instead, bring all expressions to the same side and factor out the common factors.

Equations Involving Square Roots

When an equation involves a radical symbol, we raise each side to a power to eliminate the radical. For example, we square each side of $\sqrt{x} = 3$ to get $x = 9$, an equivalent equation. So the solution set is $\{9\}$. But squaring does not always lead to an equivalent equation. If we square both sides of $\sqrt{x} = -3$, we also get $x = 9$. The equation $x = 9$ is not equivalent to $\sqrt{x} = -3$, because 9 does not satisfy $\sqrt{x} = -3$. Since 9 appeared in our attempt to solve $\sqrt{x} = -3$, but does not solve it, 9 is an *extraneous root*. When we raise each side of an equation to a power, we cannot lose a solution, but we might gain one. So you must check for extraneous roots if you have squared each side of an equation.

Example 3   Squaring each side to solve an equation

Solve $\sqrt{x} + 2 = x$.

Solution

Isolate the radical before squaring each side.

$$\sqrt{x} = x - 2$$

$$\left(\sqrt{x}\right)^2 = (x - 2)^2 \qquad \text{Square each side.}$$

$$x = x^2 - 4x + 4 \qquad \text{Use the special product } (a - b)^2 = a^2 - 2ab + b^2.$$

$$0 = x^2 - 5x + 4 \qquad \text{Write in the form } ax^2 + bx + c = 0.$$

$$0 = (x - 4)(x - 1) \qquad \text{Factor the quadratic polynomial.}$$

$$x - 4 = 0 \quad \text{or} \quad x - 1 = 0 \qquad \text{Zero factor property}$$

$$x = 4 \quad \text{or} \quad x = 1$$

Checking $x = 4$, we get $\sqrt{4} + 2 = 4$, which is correct. Checking $x = 1$, we get $\sqrt{1} + 2 = 1$, which is incorrect. So 1 is an extraneous root and the solution set is $\{4\}$. ◆

The next example involves two radicals. In this example we will isolate the more complicated radical before squaring each side. But not all radicals are eliminated upon squaring each side. So we isolate the remaining radical and square each side again.

Example 4   Squaring each side twice

Solve $\sqrt{2x + 1} - \sqrt{x} = 1$.

Solution

First we write the equation so that the more complicated radical is isolated. Then we square each side. On the left side, when we square $\sqrt{2x + 1}$ we get $2x + 1$. On the right side, when we square $1 + \sqrt{x}$, we use the special product rule $(a + b)^2 = a^2 + 2ab + b^2$.

$$\sqrt{2x + 1} = 1 + \sqrt{x}$$

$$(\sqrt{2x + 1})^2 = (1 + \sqrt{x})^2 \qquad \text{Square each side.}$$

$$2x + 1 = 1 + 2\sqrt{x} + x$$

$$x = 2\sqrt{x} \qquad \text{All radicals are not eliminated by the first squaring.}$$

$$x^2 = (2\sqrt{x})^2 \qquad \text{Square each side a second time.}$$

$$x^2 = 4x$$

$$x^2 - 4x = 0$$

$$x(x - 4) = 0$$

$$x = 0 \qquad \text{or} \qquad x - 4 = 0$$

$$x = 0 \qquad \text{or} \qquad x = 4$$

Both 0 and 4 satisfy the original equation. So the solution set is $\{0, 4\}$.   ◆

Equations with Rational Exponents

To solve equations of the form $x^{m/n} = k$ in which m and n are positive integers and m/n is in lowest terms, we adapt the methods of Examples 3 and 4 of raising each side to a power. Cubing each side of $x^{2/3} = 4$, yields $(x^{2/3})^3 = 4^3$ or $x^2 = 64$. By the square root property, $x = \pm 8$. We can shorten this solution by raising each side of the equation to the power $3/2$ (the reciprocal of $2/3$) and inserting the $\pm$ symbol to obtain the two square roots.

$$x^{2/3} = 4$$

$$(x^{2/3})^{3/2} = \pm 4^{3/2} \qquad \text{Raise each side to the power 3/2 and insert } \pm.$$

$$x = \pm 8$$

The equation $x^{2/3} = 4$ has two solutions because the numerator of the exponent $2/3$ is an even number. An equation such as $x^{-3/2} = 1/8$ has only one solution because the numerator of the exponent $-3/2$ is odd. To solve $x^{-3/2} = 1/8$, raise each side to the power $-2/3$ (the reciprocal of $-3/2$).

$$x^{-3/2} = \frac{1}{8}$$

$$(x^{-3/2})^{-2/3} = \left(\frac{1}{8}\right)^{-2/3} \qquad \text{Raise each side to the power } -2/3.$$

$$x = 4$$

In the next example we solve two more equations of this type by raising each side to a fractional power. Note that we use the $\pm$ symbol only when the numerator of the original exponent is even.

Example 5 Equations with rational exponents

Solve each equation.

a) $x^{4/3} = 625$

b) $(y - 2)^{-5/2} = 32$

Solution

a) Raise each side of the equation to the power 3/4. Use the $\pm$ symbol because the numerator of 4/3 is even.

$$x^{4/3} = 625$$
$$(x^{4/3})^{3/4} = \pm 625^{3/4}$$
$$x = \pm 125$$

Check in the original equation. The solution set is $\{-125, 125\}$.

b) Raise each side to the power $-2/5$. Because the numerator in $-5/2$ is an odd number, there is only one solution.

$$(y - 2)^{-5/2} = 32$$
$$((y - 2)^{-5/2})^{-2/5} = 32^{-2/5} \qquad \text{Raise each side to the power } -2/5.$$
$$y - 2 = \frac{1}{4}$$
$$y = 2 + \frac{1}{4} = \frac{9}{4}$$

Check 9/4 in the original equation. The solution set is $\left\{\frac{9}{4}\right\}$. ◆

```
(9/4-2)^(-5/2)
              32
■
```

Check 9/4 in the original equation.

Equations of Quadratic Type

In some cases, an equation can be converted to a quadratic equation by substituting a single variable for a more complicated expression. Such equations are called **equations of quadratic type.** An equation of quadratic type has the form $au^2 + bu + c = 0$, where $a \neq 0$ and u is an algebraic expression.

In the next example, the expression x^2 in a fourth-degree equation is replaced by u, yielding a quadratic equation. After the quadratic equation is solved, u is replaced by x^2 so that we find values for x that satisfy the original fourth-degree equation.

Example 6   Solving a fourth-degree polynomial equation

Solve $x^4 - 14x^2 + 45 = 0$.

Solution

We let $u = x^2$ so that $u^2 = (x^2)^2 = x^4$.

$$(x^2)^2 - 14x^2 + 45 = 0$$

$$u^2 - 14u + 45 = 0 \qquad \text{Replace } x^2 \text{ by } u.$$

$$(u - 9)(u - 5) = 0$$

$$u - 9 = 0 \qquad \text{or} \qquad u - 5 = 0$$

$$u = 9 \qquad \text{or} \qquad u = 5$$

$$x^2 = 9 \qquad \text{or} \qquad x^2 = 5 \qquad \text{Replace } u \text{ by } x^2.$$

$$x = \pm 3 \qquad \text{or} \qquad x = \pm\sqrt{5}$$

Check in the original equation. The solution set is $\{-5, -3, 3, 5\}$. ◆

Note that the equation of Example 6 could be solved by factoring without doing substitution, because $x^4 - 14x^2 + 45 = (x^2 - 9)(x^2 - 5)$. Since the next example involves a more complicated algebraic expression, we use substitution to simplify it, although it too could be solved by factoring, without substitution.

Example 7   Another equation of quadratic type

Solve $(x^2 - x)^2 - 18(x^2 - x) + 72 = 0$.

Solution

If we let $u = x^2 - x$, then the equation becomes a quadratic equation.

$$(x^2 - x)^2 - 18(x^2 - x) + 72 = 0$$

$$u^2 - 18u + 72 = 0 \qquad \text{Replace } x^2 - x \text{ by } u.$$

$$(u - 6)(u - 12) = 0$$

$$u - 6 = 0 \qquad \text{or} \qquad u - 12 = 0$$

$$u = 6 \qquad \text{or} \qquad u = 12$$

$$x^2 - x = 6 \qquad \text{or} \qquad x^2 - x = 12 \qquad \text{Replace } u \text{ by } x^2 - x.$$

$$x^2 - x - 6 = 0 \qquad \text{or} \qquad x^2 - x - 12 = 0$$

$$(x - 3)(x + 2) = 0 \qquad \text{or} \qquad (x - 4)(x + 3) = 0$$

$$x = 3 \quad \text{or} \quad x = -2 \qquad \text{or} \qquad x = 4 \quad \text{or} \quad x = -3$$

Check in the original equation. The solution set is $\{-3, -2, 3, 4\}$. ◆

The next equations of quadratic type have rational exponents.

Example 8 Quadratic type and rational exponents

Find all real solutions to each equation.

a) $x^{2/3} - 9x^{1/3} + 8 = 0$

b) $(11x^2 - 18)^{1/4} = x$

Solution

a) If we let $u = x^{1/3}$, then $u^2 = (x^{1/3})^2 = x^{2/3}$.

$$u^2 - 9u + 8 = 0 \qquad \text{Replace } x^{2/3} \text{ by } u^2 \text{ and } x^{1/3} \text{ by } u.$$

$$(u - 8)(u - 1) = 0$$

$$u = 8 \qquad \text{or} \qquad u = 1$$

$$x^{1/3} = 8 \qquad \text{or} \qquad x^{1/3} = 1 \qquad \text{Replace } u \text{ by } x^{1/3}.$$

$$(x^{1/3})^3 = 8^3 \qquad \text{or} \qquad (x^{1/3})^3 = 1^3$$

$$x = 512 \qquad \text{or} \qquad x = 1$$

Check in the original equation. The solution set is $\{1, 512\}$.

b)
$$(11x^2 - 18)^{1/4} = x$$

$$((11x^2 - 18)^{1/4})^4 = x^4 \qquad \text{Raise each side to the power 4.}$$

$$11x^2 - 18 = x^4$$

$$x^4 - 11x^2 + 18 = 0$$

$$(x^2 - 9)(x^2 - 2) = 0$$

$$x^2 = 9 \qquad \text{or} \qquad x^2 = 2$$

$$x = \pm 3 \qquad \text{or} \qquad x = \pm\sqrt{2}$$

Since the exponent 1/4 means principal fourth root, the right-hand side of the equation cannot be negative. So -3 and $-\sqrt{2}$ are extraneous roots. Since 3 and $\sqrt{2}$ satisfy the original equation, the solution set is $\{\sqrt{2}, 3\}$. ◆

Equations Involving Absolute Value

To solve equations involving absolute value, remember that $|x| = x$ if $x \geq 0$, and $|x| = -x$ if $x < 0$. The absolute value of x is greater than or equal to 0 for any real number x. So an equation such as $|x| = -6$ has no solution. Since a number and its opposite have the same absolute value, $|x| = 4$ is equivalent to $x = 4$ or $x = -4$. The only number that has 0 absolute value is 0. These ideas are summarized as follows.

Summary: Basic Absolute Value Equations

Absolute value equation	Equivalent statement	Solution set
$\lvert x \rvert = k \ (k > 0)$	$x = k$ or $x = -k$	$\{k, -k\}$
$\lvert x \rvert = 0$	$x = 0$	$\{0\}$
$\lvert x \rvert = k \ (k < 0)$		$\varnothing$

Example 9   Equations involving absolute value

Solve each equation.

a) $\lvert x - 5 \rvert = 4$

b) $\lvert x^2 - 2x - 16 \rvert = 8$

Solution

a) First write an equivalent statement without using absolute value symbols.

$$\lvert x - 5 \rvert = 4$$

$$x - 5 = 4 \quad \text{or} \quad x - 5 = -4$$
$$x = 9 \quad \text{or} \quad x = 1$$

The solution set is $\{1, 9\}$.

b) $\lvert x^2 - 2x - 16 \rvert = 8$

$$x^2 - 2x - 16 = 8 \quad \text{or} \quad x^2 - 2x - 16 = -8$$
$$x^2 - 2x - 24 = 0 \quad \text{or} \quad x^2 - 2x - 8 = 0$$
$$(x - 6)(x + 4) = 0 \quad \text{or} \quad (x - 4)(x + 2) = 0$$
$$x = 6 \quad \text{or} \quad x = -4 \quad \text{or} \quad x = 4 \quad \text{or} \quad x = -2$$

The solution set is $\{-4, -2, 4, 6\}$.

Consider the absolute value equation $\lvert 2x - 5 \rvert = 3x$. Since $\lvert 2x - 5 \rvert$ is nonnegative for any real number x, the right-hand side, $3x$, cannot be negative. If we assume that $3x$ is nonnegative, then $\lvert 2x - 5 \rvert = 3x$ is equivalent to

$$2x - 5 = 3x \quad \text{or} \quad 2x - 5 = -3x,$$
$$-5 = x \quad \text{or} \quad 5x = 5,$$
$$x = -5 \quad \text{or} \quad x = 1.$$

The original equation is not satisfied for $x = -5$, because $3x$ is negative for $x = -5$. The only solution is 1.

Example 10 shows two more equations of this type.

Example 10   More equations involving absolute value

Solve each equation.

a) $|x^2 - 6| = 5x$

b) $|a - 1| = |2a - 3|$

Solution

a) Write the equivalent statement assuming that $5x$ is nonnegative:

$$x^2 - 6 = 5x \qquad \text{or} \qquad x^2 - 6 = -5x$$
$$x^2 - 5x - 6 = 0 \qquad \text{or} \qquad x^2 + 5x - 6 = 0$$
$$(x - 6)(x + 1) = 0 \qquad \text{or} \qquad (x + 6)(x - 1) = 0$$
$$x = 6 \quad \text{or} \quad x = -1 \qquad \text{or} \qquad x = -6 \quad \text{or} \quad x = 1$$

The expression $|x^2 - 6|$ is nonnegative for any real number x. But $5x$ is negative if $x = -1$ or if $x = -6$. So -1 and -6 are extraneous roots. They do not satisfy the original equation. The solution set is $\{1, 6\}$.

b) The equation $|a - 1| = |2a - 3|$ indicates that $a - 1$ and $2a - 3$ have the same absolute value. If two quantities have the same absolute value, they are either equal or opposites. Use this fact to write an equivalent statement without absolute value signs.

$$a - 1 = 2a - 3 \qquad \text{or} \qquad a - 1 = -(2a - 3)$$
$$a + 2 = 2a \qquad \text{or} \qquad a - 1 = -2a + 3$$
$$2 = a \qquad \text{or} \qquad a = \frac{4}{3}$$

Check that both 2 and $\frac{4}{3}$ satisfy the original absolute value equation. The solution set is $\left\{\frac{4}{3}, 2\right\}$.   ◆

Applications

The **break-even point** for a business is the point at which the cost of doing business is equal to the revenue generated by the business. The business is profitable when the revenue is greater than the cost.

Example 11   Break-even point for a bus tour

A tour operator uses the equation $C = 3x + \sqrt{50x + 9000}$ to find his cost in dollars for taking x people on a tour of San Francisco.

a) For what value of x is the cost $160?

b) If he charges $10 per person for the tour, then what is his break-even point?

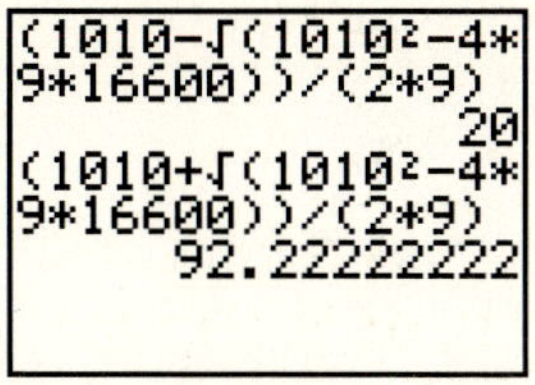

```
(1010-√(1010²-4*
9*16600))/(2*9)
                20
(1010+√(1010²-4*
9*16600))/(2*9)
        92.2222222
```

Use a calculator to evaluate the quadratic formula. See Appendix A for more examples.

Figure 2.7

Solution

a) Replace C by \$160 and solve the equation.

$$160 = 3x + \sqrt{50x + 9000}$$
$$160 - 3x = \sqrt{50x + 9000} \qquad \text{Isolate the radical.}$$
$$25{,}600 - 960x + 9x^2 = 50x + 9000 \qquad \text{Square each side.}$$
$$9x^2 - 1010x + 16{,}600 = 0$$
$$x = \frac{-(-1010) \pm \sqrt{1010^2 - 4(9)(16{,}600)}}{2(9)}$$
$$= 20 \quad \text{or} \quad 92.2$$

Check that 20 satisfies the original equation but 92.2 does not. The cost is \$160 when 20 people take the tour.

b) At \$10 per person, the revenue in dollars is given by $R = 10x$. When the revenue is equal to the cost as shown in Fig. 2.7, the operator breaks even.

$$10x = 3x + \sqrt{50x + 9000}$$
$$7x = \sqrt{50x + 9000}$$
$$49x^2 = 50x + 9000 \qquad \text{Square each side.}$$
$$49x^2 - 50x - 9000 = 0$$
$$x = \frac{50 \pm \sqrt{50^2 - 4(49)(-9000)}}{2(49)}$$
$$= -13.052 \quad \text{or} \quad 14.072$$

If 14.072 people took the tour, the operator would break even. Since the break-even point is not a whole number, the operator actually needs 15 people to make a profit. ◆

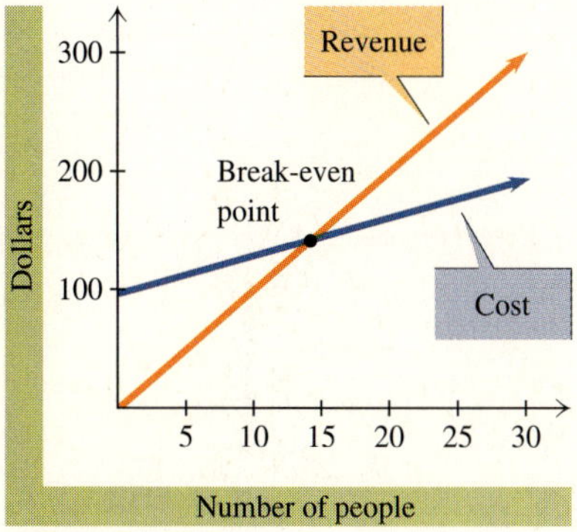

❓ For Thought

True or false? Explain.

1. Squaring each side of $\sqrt{x - 1} + \sqrt{x} = 6$ yields $x - 1 + x = 36$.

2. The equations $(2x - 1)^2 = 9$ and $2x - 1 = 3$ are equivalent.

3. The equations $x^{2/3} = 9$ and $x = 27$ have the same solution set.

4. To solve $2x^{1/4} - x^{1/2} + 3 = 0$, we let $u = x^{1/2}$ and $u^2 = x^{1/4}$.

5. If $(x - 1)^{-2/3} = 4$, then $x = 1 \pm 4^{-3/2}$.

6. No negative number satisfies $x^{-2/5} = 4$.

7. The solution set to $|2x + 10| = 3x$ is $\{-2, 10\}$.

8. No negative number satisfies $|x^2 - 3x + 2| = 7x$.

9. The equation $|2x + 1| = |x|$ is equivalent to $2x + 1 = x$ or $2x + 1 = -x$.

10. The equation $x^9 - 5x^3 + 6 = 0$ is an equation of quadratic type.

2.5 Exercises
Tape 4   Disk—5.25": 2   3.5": 1   **Macintosh: 1**

Find all real and imaginary solutions to each equation. Check your answers.

1. $x^3 + 3x^2 - 4x - 12 = 0$

2. $x^3 - x^2 - 5x + 5 = 0$

3. $2x^3 + 6x^2 - x - 3 = 0$

4. $3x^3 + 15x^2 - 2x - 10 = 0$

5. $a^3 + 5a = 15a^2$ **6.** $b^3 + 20b = 9b^2$

7. $3y^4 - 12y^2 = 0$ **8.** $5m^4 - 10m^3 + 5m^2 = 0$

9. $a^4 - 16 = 0$ **10.** $w^4 + 8w = 0$

Find all real solutions to each equation. Check your answers.

11. $\sqrt{x + 1} = x - 5$ **12.** $\sqrt{x - 1} = x - 7$

13. $\sqrt{x - 2} = x - 22$ **14.** $3 + \sqrt{x} = 1 + x$

15. $w = \dfrac{\sqrt{1 - 3w}}{2}$ **16.** $t = \dfrac{\sqrt{2 - 3t}}{3}$

17. $\dfrac{1}{z} = \dfrac{3}{\sqrt{4z + 1}}$ **18.** $\dfrac{1}{p} - \dfrac{2}{\sqrt{9p + 1}} = 0$

19. $\sqrt{x^2 - 2x - 15} = 3$ **20.** $\sqrt{3x^2 + 5x - 3} = x$

21. $\sqrt{x + 40} - \sqrt{x} = 4$ **22.** $\sqrt{x} + \sqrt{x - 36} = 2$

23. $\sqrt{n + 4} + \sqrt{n - 1} = 5$ **24.** $\sqrt{y + 10} - \sqrt{y - 2} = 2$

25. $\sqrt{2x + 5} + \sqrt{x + 6} = 9$ **26.** $\sqrt{3x - 2} - \sqrt{x - 2} = 2$

Find all real solutions to each equation. Check your answers.

27. $x^{2/3} = 2$ **28.** $x^{2/3} = \dfrac{1}{2}$ **29.** $w^{-4/3} = 16$

30. $w^{-3/2} = 27$ **31.** $t^{-1/2} = 7$ **32.** $t^{-1/2} = \dfrac{1}{2}$

33. $(s - 1)^{-1/2} = 2$ **34.** $(s - 2)^{-1/2} = \dfrac{1}{3}$

Find all real and imaginary solutions to each equation. Check your answers.

35. $x^4 - 12x^2 + 27 = 0$ **36.** $x^4 + 10 = 7x^2$

37. $\left(\dfrac{2c - 3}{5}\right)^2 + 2\left(\dfrac{2c - 3}{5}\right) = 8$

38. $\left(\dfrac{b - 5}{6}\right)^2 - \left(\dfrac{b - 5}{6}\right) - 6 = 0$

39. $\dfrac{1}{(5x - 1)^2} + \dfrac{1}{5x - 1} - 12 = 0$

40. $\dfrac{1}{(x - 3)^2} + \dfrac{2}{x - 3} - 24 = 0$

41. $(v^2 - 4v)^2 - 17(v^2 - 4v) + 60 = 0$

42. $(u^2 + 2u)^2 - 2(u^2 + 2u) - 3 = 0$

43. $x - 4\sqrt{x} + 3 = 0$ **44.** $2x + 3\sqrt{x} - 20 = 0$

45. $q - 7q^{1/2} + 12 = 0$ **46.** $h + 1 = 2h^{1/2}$

47. $x^{2/3} + 10 = 7x^{1/3}$ **48.** $x^{1/2} - 3x^{1/4} + 2 = 0$

Solve each absolute value equation.

49. $|3 - 2x| = 6$ **50.** $|5x - 1| = 9$

51. $|3x - 2| = 0$ **52.** $|6 - 3x| = 0$

53. $|w^2 - 4| = 3$ **54.** $|a^2 - 1| = 1$

55. $|v^2 - 3v| = 5v$ **56.** $|z^2 - 12| = z$

57. $|x^2 - x - 6| = 6$ **58.** $|2x^2 - x - 2| = 1$

59. $|x + 5| = |2x + 1|$ **60.** $|3x - 4| = |x|$

61. $3|x - 2| - 4 = -2$ **62.** $|x - 5| + 6 = 1$

Solve each equation. Find imaginary solutions when possible.

63. $\sqrt{16x + 1} - \sqrt{6x + 13} = -1$

64. $\sqrt{16x + 1} - \sqrt{6x + 13} = 1$

65. $v^6 - 64 = 0$ **66.** $t^4 - 1 = 0$

67. $(7x^2 - 12)^{1/4} = x$ **68.** $(10x^2 - 1)^{1/4} = 2x$

69. $|y^2 - 3y + 2| = -9$ **70.** $-3|z + 8| + 4 = 1$

71. $\sqrt[3]{2 + x - 2x^2} = x$ **72.** $\sqrt{48 + \sqrt{x}} - 4 = \sqrt[4]{x}$

73. $\left(\dfrac{x - 2}{3}\right)^2 - 2\left(\dfrac{x - 2}{3}\right) + 10 = 0$

74. $\dfrac{1}{(x + 1)^2} - \dfrac{2}{x + 1} + 2 = 0$

75. $(3u - 1)^{2/5} = 2$

76. $(2u + 1)^{2/3} = 3$

77. $x^2 - 11\sqrt{x^2 + 1} + 31 = 0$

78. $2x^2 - 3\sqrt{2x^2 - 3} - 1 = 0$

79. $|x^2 - 2x| = |3x - 6|$ **80.** $|x^2 + 5x| = |3 - x^2|$

81. $(3m + 1)^{-3/5} = -\dfrac{1}{8}$ **82.** $(1 - 2m)^{-5/3} = -\dfrac{1}{32}$

83. $|x^2 - 4| = x - 2$ **84.** $|x^2 + 7x| = x^2 - 4$

The International America's Cup Rules, which took effect in 1989, establish the boundaries of a new class of longer, lighter, and faster yachts. In addition to 200 pages of other rules, the basic dimensions are governed by an equation that balances length L in meters, sail area S in square meters, and displacement D in cubic meters. (*Scientific American*, May 1992)

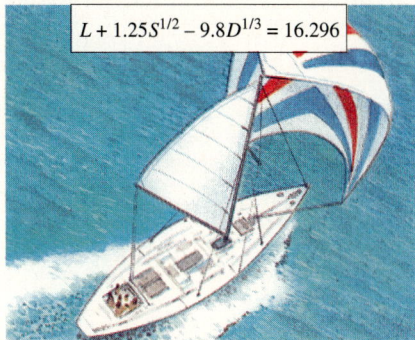

$$L + 1.25S^{1/2} - 9.8D^{1/3} = 16.296$$

Figure for Exercises 85 and 86

85. *Maximum sail area for a yacht*   The maximum sail area S for a boat with length 21.24 m and a displacement of 18.34 m³ is determined by the equation $L + 1.25S^{1/2} - 9.8D^{1/3} = 16.296$. Find the maximum sail area for this boat.

86. *Minimum displacement for a yacht*   The minimum displacement D for a boat with length 21.52 m and a sail area of 310.64 m² is determined by the equation $L + 1.25S^{1/2} - 9.8D^{1/3} = 16.296$. Find this boat's minimum displacement.

Solve each problem.

87. *Cost of baking bread*   The daily cost for baking x loaves of bread at Juanita's Bakery is given in dollars by $C = 0.5x + \sqrt{8x} + 5000$. Find the number of loaves for which the cost is $83.50.

88. *Break-even analysis*   If the bread in Exercise 87 sells for $1.10 per loaf, then what is the minimum number of loaves that Juanita must bake and sell to make a profit?

89. *Square roots*   Find two numbers that differ by 6 and whose square roots differ by 1.

90. *Right triangle*   One leg of a right triangle is 1 cm longer than the other leg. What is the length of the short leg if the total length of the hypotenuse and the short leg is 10 cm?

91. *Perimeter of a right triangle*   A sign in the shape of a right triangle has one leg that is 7 in. longer than the other leg. What is the length of the shorter leg if the perimeter is 30 in.?

92. *Right triangle inscribed in a semicircle*   One leg of a right triangle is 1 ft longer than the other leg. If the triangle is inscribed in a circle and the hypotenuse of the triangle is the diameter of the circle, then what is the length of the radius for which the length of the radius is equal to the length of the shortest side?

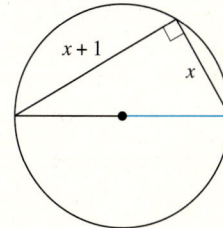

Figure for Exercise 92

93. *Area of a foundation*   The original plans for Jennifer's house called for a square foundation. After increasing one side by 30 ft and decreasing the other by 10 ft, the area of the rectangular foundation was 2100 ft². What was the area of the original square foundation?

94. *Shipping carton*   Heloise designed a cubic box for shipping paper. The height of the box was acceptable, but the paper would not fit into the box. After increasing the length by 0.5 in. and the width by 6 in. the area of the bottom was 119 in² and the paper would fit. What was the original volume of the cubic box?

Figure for Exercise 94

95. *Insulated carton*   Nina is designing a box for shipping frozen shrimp. The box is to have a square base and a height that is 2 in. greater than the width of the base. The box will be surrounded with a 1-in. thick layer of styrofoam. If the volume of the inside of the box must be equal to the volume of the styrofoam used, then what volume of shrimp can be shipped in the box?

Figure for Exercise 95

96. *Volume of a cubic box*   If the width of the base of a cubic container is increased by 3 m and the length of the base decreased by 1 m, then the volume of the new container is 6 m³. What is the height of the cubic container?

97. *Hiking time*   William, Nancy, and Edgar met at the lodge at 8 A.M., and William began hiking west at 4 mph. At 10 A.M., Nancy began hiking north at 5 mph and Edgar went east on a three-wheeler at 12 mph. At what time was the distance between Nancy and Edgar 14 mi greater than the distance between Nancy and William?

98. *Distance in three-dimensional space*   A balloonist was ready to launch her balloon at the intersection of two roads. A car passed the intersection heading north at 30 ft/sec. One second later a car passed the intersection heading east at 30 ft/sec, and at that same instant the balloon began rising at 50 ft/sec. How many seconds after the balloon began rising was the distance between the cars equal to the height of the balloon? How many seconds after the launch of the balloon was the distance between the balloon and the second car equal to the distance between the two cars?

99. *View from an airplane*   It is well known that you get a better view of the land if you rise up into the air. The formula $V = 1.22\sqrt{A}$ gives the approximate view V in miles at an altitude of A feet, where V is the distance from horizon to horizon. What is the altitude of an airplane from which the view is 200 mi?

100. *Boston molasses disaster*   In the city of Boston, during the afternoon of January 15, 1919, a cylindrical metal tank containing 25,850,000 kg of molasses ruptured. The sticky liquid poured into the streets in a 9-m-deep stream that knocked down buildings and killed pedestrians and horses (*The Boston Globe,* Sept. 19, 1991). If the diameter of the tank was equal to its height and the weight of molasses is 1600 kg/m³, then what was the height of the tank in meters?

Figure for Exercise 100

101. *Time swimming and running*   Lauren is competing in her town's cross-country competition. Early in the event, she must race from point A on the Greenbriar River to point B, which is 5 mi downstream and on the opposite bank. The Greenbriar is 1 mi wide. In planning her strategy, Lauren knows she can use any combination of running and swimming. She can run 10 mph and swim 8 mph. How long would it take if she ran 5 mi downstream and then swam across? Find the time it would take if she swam diagonally from A to B. Find x so that she could run x miles along the bank, swim diagonally to B, and complete the race in 36 min. (Ignore the current in the river.)

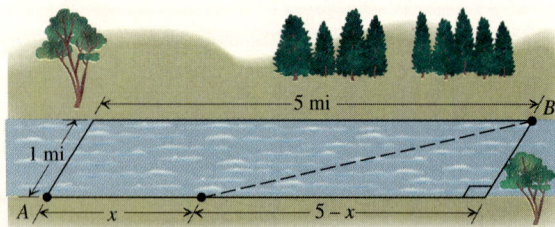

Figure for Exercise 101

102. *Installing a water pipe* Jonathan wants to run a water pipe
from his house to a water terminal, as shown in the diagram.
The terminal is 30 ft down the 10-ft-wide driveway and on
the other side. The labor charge is $3/ft alongside the drive-
way and $4/ft for underneath the driveway. The contractor
can do the job for $120 by going x feet down the driveway
and then cutting across on a diagonal as shown in the diagram. Find the value of x.

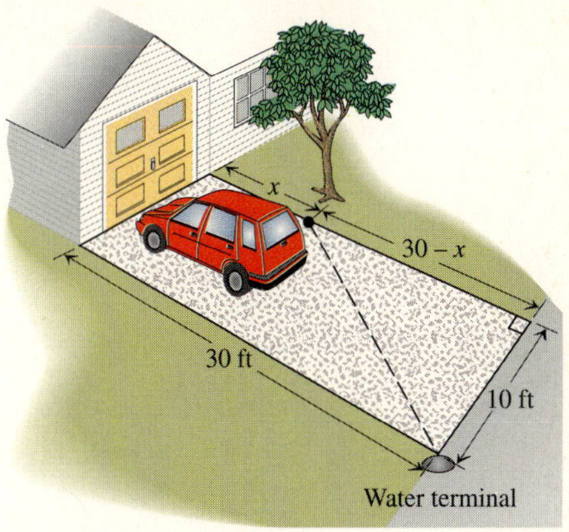

Figure for Exercise 102

2.6

Inequalities

An equation states that two algebraic expressions are equal, while an **inequality**
or **simple inequality** is a statement that two algebraic expressions are not equal in
a particular way. Inequalities are stated using less than ($<$), less than or equal to
($\leq$), greater than ($>$), or greater than or equal to ($\geq$). In this section we begin a
study of inequalities that continues through Section 2.7, where we study quadratic
and rational inequalities.

Interval Notation

The solution set to an inequality is the set of all real numbers for which the
inequality is true. The solution set to the inequality $x > 3$ is written $\{x \mid x > 3\}$
and consists of all real numbers to the right of 3 on the number line. This set is also
called the **interval** of numbers greater than 3, and it is written in **interval notation** as $(3, \infty)$. The graph of the interval $(3, \infty)$ is shown in Fig. 2.8. A parenthesis
is used next to the 3 to indicate that 3 is not in the interval or the solution set. The
infinity symbol (∞) is not used as a number, but only to indicate that there is no
bound on the numbers greater than 3.

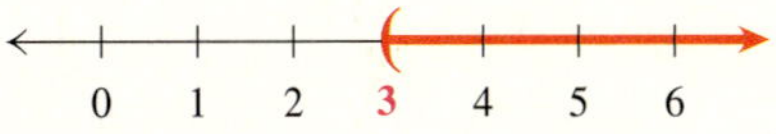

The solution set to $x \leq 4$ is written in set notation as $\{x \mid x \leq 4\}$ and in
interval notation as $(-\infty, 4]$. The symbol $-\infty$ means that all numbers to the left of
4 on the number line are in the set and the bracket means that 4 is in the set. The
graph of the interval $(-\infty, 4]$ is shown in Fig. 2.9. Intervals that use the infinity
symbol are **unbounded** intervals. The following summary lists the different types
of unbounded intervals used in interval notation and the graphs of those intervals
on a number line. If only parentheses are used, the interval is called **open.** If a
bracket is used only on one end, the interval is called **half open** or **half closed.**

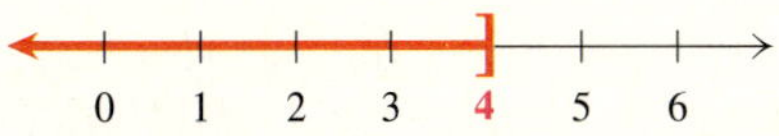

Figure 2.8

Figure 2.9

Summary: Interval Notation for Unbounded Intervals

Set	Interval notation	Type	Graph
$\{x \mid x > a\}$	(a, ∞)	Open	
$\{x \mid x < a\}$	$(-\infty, a)$	Open	
$\{x \mid x \geq a\}$	$[a, \infty)$	Half open or half closed	
$\{x \mid x \leq a\}$	$(-\infty, a]$	Half open or half closed	
Real numbers	$(-\infty, \infty)$	Open	

We use a parenthesis when an endpoint of an interval is not included in the solution set and a bracket when an endpoint is included. A bracket is never used next to ∞ because infinity is not a number. On the graphs above, the number lines are shaded, showing that the solutions include all real numbers in the given interval.

Example 1   Interval notation

Write an inequality whose solution set is the given interval.

a) $(-\infty, -9)$ **b)** $[0, \infty)$

Solution

a) The interval $(-\infty, -9)$ represents all real numbers less than -9. It is the solution set to $x < -9$.

b) The interval $[0, \infty)$ represents all real numbers greater than or equal to 0. It is the solution set to $x \geq 0$.

Linear Inequalities

Replacing the equal sign in the general linear equation $ax + b = 0$ by any of the symbols $<$, $\leq$, $>$, or $\geq$, gives a **linear inequality.** (Quadratic inequalities, discussed in Section 2.7, have an x^2-term.) Two inequalities are **equivalent** if they have the same solution set. To solve linear inequalities, we use techniques similar to those used in solving linear equations. It can be shown that the following properties of inequality are valid. They allow us to write equivalent inequalities. These properties are stated for $<$, but they also hold for $>$, $\leq$, and $\geq$.

Properties of Inequality

If A and B are algebraic expressions and C is a real number, then the inequality $A < B$ is equivalent to

1. $A \pm C < B \pm C$,
2. $CA < CB$ (for C positive), $CA > CB$ (for C negative),
3. $\dfrac{A}{C} < \dfrac{B}{C}$ (for C positive), $\dfrac{A}{C} > \dfrac{B}{C}$ (for C negative).

Solving linear inequalities is very similar to solving linear equations, but pay particular attention to the second and third properties. *When an inequality is multiplied or divided by a negative number, the direction of the inequality symbol is reversed.* The inequality is reversed because of the rules for multiplying or dividing signed numbers. For example, multiplying the inequality $-4 < 7$ on each side by -2 yields $(-2)(-4) > (-2)(7)$, or $8 > -14$.

Example 2   Solving a linear inequality

Solve $-3x - 5 < 4$. Write the solution set in interval notation and graph it.

Solution

Isolate the variable as is done in solving equations.

$$-3x - 5 < 4$$
$$-3x - 5 + 5 < 4 + 5 \qquad \text{Add 5 to each side.}$$
$$-3x < 9$$
$$x > -3 \qquad \text{Divide each side by } -3, \text{ reversing the inequality.}$$

Figure 2.10

The solution set is the interval $(-3, \infty)$ and its graph is shown in Fig. 2.10. Checking the solution to an inequality is generally not as simple as checking an equation, because usually there are infinitely many solutions. We can do a "partial check" by checking one number in $(-3, \infty)$ and one number not in $(-3, \infty)$. For example, $0 > -3$ and $-3(0) - 5 < 4$ is correct, while $-6 < -3$ and $-3(-6) - 5 < 4$ is incorrect.

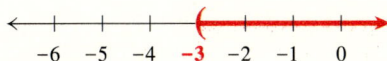

We can also perform operations on each side of an inequality using a variable expression. Addition or subtraction with variable expressions will give equivalent inequalities. However, we must always watch for undefined expressions. *Multiplication and division with a variable expression is usually avoided because we do not know whether the expression is positive or negative.*

Example 3   Solving a linear inequality

Solve $\frac{1}{2}x - 3 \geq \frac{1}{4}x + 2$ and graph the solution set.

Solution

Multiply each side by the LCD to eliminate the fractions.

$$\frac{1}{2}x - 3 \geq \frac{1}{4}x + 2$$

$$4\left(\frac{1}{2}x - 3\right) \geq 4\left(\frac{1}{4}x + 2\right) \qquad \text{Multiply each side by 4.}$$

$$2x - 12 \geq x + 8$$

$$x - 12 \geq 8$$

$$x \geq 20$$

Figure 2.11

The solution set is the interval $[20, \infty)$. See Fig. 2.11 for the graph.

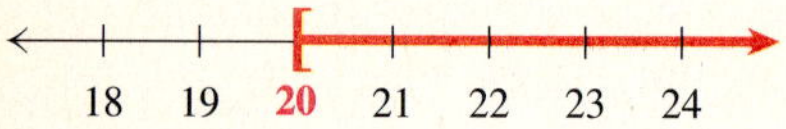

Compound Inequalities

A **compound inequality** is a sentence containing two simple inequalities connected with "and" or "or." The solution to a compound inequality can be an interval of real numbers that does not involve infinity, a **bounded** interval of real numbers. For example, the solution set to the compound inequality $x \geq 2$ and $x \leq 5$ is the set of real numbers between 2 and 5, inclusive. This inequality is also written as $2 \leq x \leq 5$. Its solution set is $\{x \mid 2 \leq x \leq 5\}$, which is written in interval notation as $[2, 5]$. Because $[2, 5]$ contains both of its endpoints, the interval is **closed.** The following summary lists the different types of bounded intervals used in interval notation and the graphs of those intervals on a number line.

Summary: Interval Notation for Bounded Intervals

Set	Interval notation	Type	Graph
$\{x \mid a < x < b\}$	(a, b)	Open	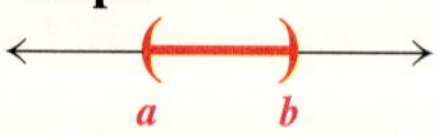
$\{x \mid a \leq x \leq b\}$	$[a, b]$	Closed	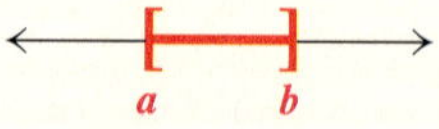
$\{x \mid a \leq x < b\}$	$[a, b)$	Half open or half closed	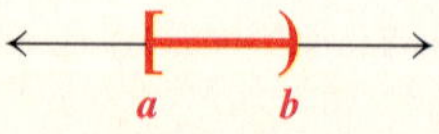
$\{x \mid a < x \leq b\}$	$(a, b]$	Half open or half closed	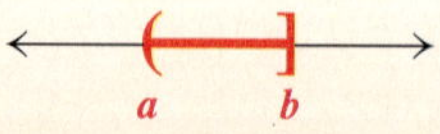

The notation $a < x < b$ is used only when x is actually between a and b, and a is less than b. We do not write $x > 5$ and $x < -3$ as $5 < x < -3$, nor do we write $a < x > b$ for $x > a$ and $x > b$.

The **intersection** of sets A and B is the set $A \cap B$, where $x \in A \cap B$ if and only if $x \in A$ and $x \in B$. The **union** of sets A and B is the set $A \cup B$, where $x \in A \cup B$ if and only if $x \in A$ or $x \in B$. In solving compound inequalities it is often necessary to find intersections and unions of intervals.

Figure 2.12

Example 4 Intersections and unions of intervals

Let $A = (1, 5)$, $B = [3, 7)$, and $C = (6, \infty)$. Write $A \cup B$, $A \cap B$, $A \cap C$, and $A \cup C$ in interval notation.

Solution

Figure 2.12 shows the intervals A, B, and C on the number line. Since $A \cup B$ consists of points that are either in A or in B, $A \cup B = (1, 7)$. Since $A \cap B$ consists of points that are in both A and B, $A \cap B = [3, 5)$. Since A and C have no points in common, $A \cap C = \varnothing$ and $A \cup C = (1, 5) \cup (6, \infty)$. ◆

The solution set to a compound inequality using the connector ''or'' is the union of the two solution sets, and the solution set to a compound inequality using ''and'' is the intersection of the two solution sets.

Example 5 Solving compound inequalities

Solve each compound inequality. Write the solution set using interval notation and graph it.

a) $2x - 3 > 5$ and $4 - x \le 3$ **b)** $4 - 3x < -2$ or $3(x - 2) \le -6$

c) $-4 \le 3x - 1 < 5$

Solution

a)
$$\begin{aligned} 2x - 3 &> 5 & \text{and} & & 4 - x &\le 3 \\ 2x &> 8 & \text{and} & & -x &\le -1 \\ x &> 4 & \text{and} & & x &\ge 1 \end{aligned}$$

Figure 2.13

The intersection of the intervals $(4, \infty)$ and $[1, \infty)$ is $(4, \infty)$, because only numbers larger than 4 belong to both intervals. The solution set to the compound inequality is the open interval $(4, \infty)$. Its graph is shown in Fig. 2.13.

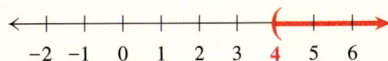

b)
$$\begin{aligned} 4 - 3x &< -2 & \text{or} & & 3(x - 2) &\le -6 \\ -3x &< -6 & \text{or} & & x - 2 &\le -2 \\ x &> 2 & \text{or} & & x &\le 0 \end{aligned}$$

Figure 2.14

The solution set is $(-\infty, 0] \cup (2, \infty)$, and its graph is shown in Fig. 2.14.

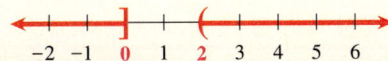

c) We could write $-4 \le 3x - 1 < 5$ as the compound inequality $-4 \le 3x - 1$ and $3x - 1 < 5$, and then solve each simple inequality. Since each is solved using the same sequence of steps, we can solve the original inequality without separating it:

$$-4 \le 3x - 1 < 5$$

$$-4 + 1 \le 3x - 1 + 1 < 5 + 1 \qquad \text{Add 1 to each part of the inequality.}$$

$$-3 \le 3x < 6$$

$$\frac{-3}{3} \le \frac{3x}{3} < \frac{6}{3} \qquad \text{Divide each part by 3.}$$

$$-1 \le x < 2$$

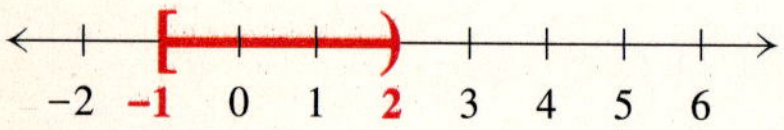

Figure 2.15

The solution set is the half-open interval $[-1, 2)$, graphed in Fig. 2.15. ◆

It is possible that all real numbers satisfy a compound inequality or no real numbers satisfy a compound inequality.

Example 6   Solving compound inequalities

Solve each compound inequality.

a) $3x - 9 \le 9$ or $4 - x \le 3$ b) $-\dfrac{2}{3}x < 4$ and $\dfrac{3}{4}x < -6$

Solution

a) Solve each simple inequality and find the union of their solution sets:

$$
\begin{array}{ccc}
3x - 9 \le 9 & \text{or} & 4 - x \le 3 \\
3x \le 18 & \text{or} & -x \le -1 \\
x \le 6 & \text{or} & x \ge 1
\end{array}
$$

The union of $(-\infty, 6]$ and $[1, \infty)$ is the set of all real numbers, $(-\infty, \infty)$.

b) Solve each simple inequality and find the intersection of their solution sets:

$$
\begin{array}{ccc}
-\dfrac{2}{3}x < 4 & \text{and} & \dfrac{3}{4}x < -6 \\[2mm]
\left(-\dfrac{3}{2}\right)\left(-\dfrac{2}{3}x\right) > \left(-\dfrac{3}{2}\right)4 & \text{and} & \left(\dfrac{4}{3}\right)\left(\dfrac{3}{4}x\right) < \left(\dfrac{4}{3}\right)(-6) \\[2mm]
x > -6 & \text{and} & x < -8
\end{array}
$$

Since $(-6, \infty) \cap (-\infty, -8) = \varnothing$, there is no solution to the compound inequality. ◆

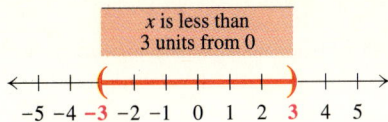

x is less than
3 units from 0

Figure 2.16

x is more than
5 units from 0

x is more than
5 units from 0

Figure 2.17

Absolute Value Inequalities

Recall that the absolute value of a number is the number's distance from 0 on the number line. The inequality $|x| < 3$ means that x is less than three units from 0. See Fig. 2.16. The real numbers that are less than three units from 0 are precisely the numbers that satisfy $-3 < x < 3$. So the solution set to $|x| < 3$ is the open interval $(-3, 3)$. The inequality $|x| > 5$ means that x is more than five units from 0 on the number line, which is equivalent to the compound inequality $x > 5$ or $x < -5$. See Fig. 2.17. So the solution to $|x| > 5$ is the union of two intervals, $(-\infty, -5) \cup (5, \infty)$. These ideas about absolute value inequalities are summarized as follows.

Summary: Basic Absolute Value Inequalities (for $k > 0$)

Absolute value inequality	Equivalent statement	Solution set in interval notation	Graph of solution set
$|x| > k$	$x > k$ or $x < -k$	$(-\infty, -k) \cup (k, \infty)$	
$|x| \geq k$	$x \geq k$ or $x \leq -k$	$(-\infty, -k] \cup [k, \infty)$	
$|x| < k$	$-k < x < k$	$(-k, k)$	
$|x| \leq k$	$-k \leq x \leq k$	$[-k, k]$	

Example 7   Absolute value inequalities

Solve each absolute value inequality and graph the solution set.

a) $|3x + 2| < 7$ **b)** $-2|4 - x| \leq -4$ **c)** $|7x - 9| \geq -3$

Solution

a)
$$|3x + 2| < 7$$
$$-7 < 3x + 2 < 7 \qquad \text{Write the equivalent compound inequality.}$$
$$-7 - 2 < 3x + 2 - 2 < 7 - 2 \qquad \text{Subtract 2 from each part.}$$
$$-9 < 3x < 5$$
$$-3 < x < \frac{5}{3} \qquad \text{Divide each part by 3.}$$

Figure 2.18

The solution set is the open interval $\left(-3, \frac{5}{3}\right)$. The graph is shown in Fig. 2.18.

b)
$$-2|4 - x| \le -4$$
$$|4 - x| \ge 2 \qquad \text{Divide each side by } -2, \text{ reversing the inequality.}$$
$$4 - x \ge 2 \quad \text{or} \quad 4 - x \le -2 \qquad \text{Write the equivalent compound inequality.}$$
$$-x \ge -2 \quad \text{or} \quad -x \le -6$$
$$x \le 2 \quad \text{or} \quad x \ge 6 \qquad \text{Multiply each side by } -1.$$

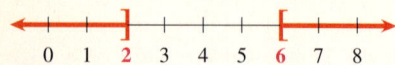

0 1 **2** 3 4 5 **6** 7 8

Figure 2.19

The solution set is $(-\infty, 2] \cup [6, \infty)$. Its graph is shown in Fig. 2.19. Check 8, 4, and 0 in the original inequality. If $x = 8$, we get $-2|4 - 8| \le -4$, which is true. If $x = 4$, we get $-2|4 - 4| \le -4$, which is false. If $x = 0$, we get $-2|4 - 0| \le -4$, which is true.

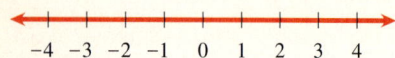

−4 −3 −2 −1 0 1 2 3 4

Figure 2.20

c) The expression $|7x - 9|$ has a nonnegative value for every real number x. So the inequality $|7x - 9| \ge -3$ is satisfied by every real number. The solution set is $(-\infty, \infty)$, and the graph is shown in Fig. 2.20. ◆

Applications

Inequalities occur in applications just as equations do. In fact, in real life, equality in anything is usually the exception. The solution to a problem involving an inequality is generally an interval of real numbers. In this case we often ask for the range of values that solve the problem.

Example 8   An application involving inequality

Remington scored 74 on his midterm exam in history. If he is to get a B, the average of his midterm and final exam must be between 80 and 89 inclusive. In what range must his final exam score lie for him to get a B in the course?

Solution

Let x represent Remington's final exam score. The average of his midterm and final must satisfy the following inequality.

$$80 \le \frac{74 + x}{2} \le 89$$
$$160 \le 74 + x \le 178$$
$$86 \le x \le 104$$

His final exam score must lie in the interval [86, 104] if he is to get a B. ◆

Example 9   Application of absolute value inequality

Michelle is writing a program for computer-controlled chemistry equipment. The computer must maintain a water temperature that does not deviate by more than six degrees from 78°C. Write an inequality for the acceptable temperature of the water and solve the inequality.

Solution

Let x represent the water temperature. The value of x might be larger or smaller than 78, as long as x and 78 differ by 6 units or less. The water temperature must satisfy the following absolute value inequality.

$$|x - 78| \leq 6$$
$$-6 \leq x - 78 \leq 6$$
$$72 \leq x \leq 84$$

The computer must keep the water temperature in the interval $[72°, 84°]$. ◆

? For Thought

True or false? Explain.

1. The inequality $-3 < x + 6$ is equivalent to $x + 6 > -3$.

2. The inequality $-2x < -6$ is equivalent to $\dfrac{-2x}{-2} < \dfrac{-6}{-2}$.

3. The smallest real number that satisfies $x > 12$ is 13.

4. The number -6 satisfies $|x - 6| > -1$.

5. $(-\infty, -3) \cap (-\infty, -2) = (-\infty, -2)$

6. $(5, \infty) \cap (-\infty, -3) = (-3, 5)$

7. All negative numbers satisfy $|x - 2| < 0$.

8. The compound inequality $x < -3$ or $x > 3$ is equivalent to $|x| < -3$.

9. The inequality $|x| + 2 < 5$ is equivalent to $-5 < x + 2 < 5$.

10. The fact that the difference between your age, y, and my age, m, is at most 5 years is written $|y - m| \leq 5$.

2.6 Exercises   ▭ Tape 4   ▱ Disk—5.25″: 2   3.5″: 1   Macintosh: 1

For each given interval, write an inequality whose solution is the interval, and for each given inequality, write the solution set in interval notation.

1. $(-\infty, 12)$ 2. $(-\infty, -3]$ 3. $x \geq -8$ 4. $x < 54$

5. $x < \pi/2$ 6. $x \geq \sqrt{3}$ 7. $[-7, \infty)$ 8. $(1.2, \infty)$

Solve each inequality. Write the solution set using interval notation and graph it.

9. $3x - 6 > 9$ 10. $2x + 1 < 6$

11. $7 - 5x \leq -3$ 12. $-1 - 4x \geq 7$

13. $\dfrac{1}{2}x - 4 < \dfrac{1}{3}x + 5$ 14. $\dfrac{1}{2} - x > \dfrac{x}{3} + \dfrac{1}{4}$

15. $\dfrac{7 - 3x}{2} \geq -3$ 16. $\dfrac{5 - 3x}{-7} \leq 0$

17. $-2(3x - 2) \geq 4 - x$

18. $-5x \leq 3(x - 9)$

19. $\dfrac{2(76) + x}{3} \geq 80$

20. $x + 0.09x \leq 872$

Write as a single interval.

21. $(-3, \infty) \cup (5, \infty)$ 22. $(-\infty, -2) \cap (-5, \infty)$

23. $(-\infty, -5) \cap (-2, \infty)$ 24. $(-\infty, -3) \cup (-7, \infty)$

25. $(-\infty, 4) \cup [4, 5]$ 26. $(4, 7) \cap (3, \infty)$

27. $(3, 5) \cup (-3, \infty)$ 28. $(-\infty, 0) \cup (-\infty, 6)$

29. $[3, 5] \cup [5, 7]$ 30. $(-3, \infty) \cap (2, \infty)$

31. $(3, 7) \cap [5, 7]$ 32. $(-\infty, 0) \cap [-4, 3]$

Determine whether the given number satisfies the inequality following it.

33. $-3, 4 - 2x \le 9$ 34. $4, 5 - 2x > -4x + 5$

35. $-5, 7 - x > 6$ or $2x + 3 > 9$

36. $-2, 3(x + 4) \le 6$ and $x < 1$

37. $8, 2|x - 9| \le 6$ 38. $1, -2 < \dfrac{5x - 9}{2} < 4$

Solve each compound inequality. Write the solution set using interval notation and graph it.

39. $5 > 8 - x$ and $1 + 0.5x < 4$

40. $\dfrac{3x + 4}{2} < -1$ and $5 - 2x < 5 - (4 + 3x)$

41. $\dfrac{2x - 5}{-2} < 2$ and $\dfrac{2x + 1}{3} > 0$

42. $1 - 2x \ge -2$ and $1 - x \le \dfrac{1}{2} - 2x$

43. $x - 5 > -3$ or $3 - x > 6$

44. $5 - x > 2$ or $5 < x - 7$

45. $1 - x < 7 + x$ or $4x + 3 > x$

46. $1 - 2(x + 3) > -15$ or $4x < x - 3$

47. $\dfrac{1}{2}(x + 1) > 3$ or $0 < 7 - x$

48. $\dfrac{1}{2}(x + 6) > 3$ or $4(x - 1) < 3x - 4$

49. $1 - \dfrac{3}{2}x < 4$ and $\dfrac{1}{4}x - 2 \le -3$

50. $\dfrac{3}{5}x - 1 > 2$ and $5 - \dfrac{2}{5}x \ge 3$

Solve each absolute value inequality. Write the solution set using interval notation and graph it.

51. $|3x - 1| < 2$ 52. $|4x - 3| \le 5$ 53. $|5 - 4x| \le 1$

54. $|6 - x| < 6$ 55. $3 < |2x - 1|$ 56. $5 \ge |4 - x|$

57. $|5 - 4x| < 0$ 58. $|3x - 7| \ge -5$

59. $|2x - 8| \le 0$ 60. $|x - 6| > 0$

61. $|4 - 5x| < -1$ 62. $|2 - 9x| \ge 0$

63. $|x - 2| + 6 > 9$ 64. $3|x - 1| + 2 < 8$

65. $\left| \dfrac{x - 3}{2} \right| > 1$ 66. $\left| \dfrac{9 - 4x}{2} \right| < 3$

Write an inequality of the form $|x - a| < k$ or of the form $|x - a| > k$ so that the inequality has the given solution set.

67. $(-\infty, 3) \cup (5, \infty)$ 68. $(-\infty, -4) \cup (5, \infty)$

69. $(4, 8)$ 70. $(-3, 8)$

71. $(-\infty, 4) \cup (4, \infty)$ 72. $(-\infty, 0) \cup (0, \infty)$

73. $(-3, 7)$ 74. $(3, 4)$

For each graph write an absolute value inequality that has the given solution set.

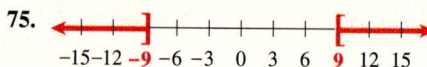

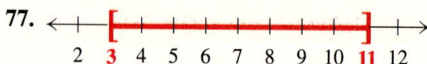

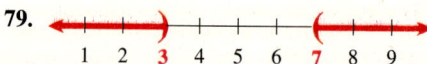

75.

76.

77.

78.

79.

80.

Find all values of x for which each expression is a real number.

81. $\sqrt{x - 2}$ 82. $\sqrt{3x - 1}$ 83. $\dfrac{1}{\sqrt{2 - x}}$

84. $\dfrac{5}{\sqrt{3 - 2x}}$ 85. $\sqrt{|x| - 3}$ 86. $\sqrt{5 - |x|}$

Solve each inequality. Write the solution set using interval notation and graph it.

87. $|x| > x$

88. $|x| > x + 4$

89. $|x - 6| < 2x$

90. $|x| \leq 1 - x$

Solve each problem.

91. *Price range for a car* Yolanda is shopping for a car in a city where the sales tax is 10% and the title and license fee is $300. If the maximum that she can spend is $8000, then she should look at cars in what price range?

92. *Price range of a hamburger* The price of a Coke at Jerry's Drive In is 10 cents more than the price of a hamburger. For 10 hamburgers and 5 Cokes, Whimpy is certain that he spent more than $9.14 but not more than $13.19 including tax at 8% and a 50-cent tip. In what price range is a hamburger?

93. *Final exam score* Lucky scored 65 on his Psychology 101 midterm. If the average of his midterm and final must be between 79 and 90 for a B, then for what range of scores on the final exam would Lucky get a B?

94. *Bringing up your average* Felix scored 52 and 64 on his first two tests in Sociology 212. What must he get on the third test to get an average for the three tests above 70?

95. *Weighted average with whole numbers* Ingrid scored 65 on her calculus midterm. If her final exam counts twice as much as her midterm exam, then for what range of scores on her final would she get an average between 79 and 90?

96. *Weighted average with fractions* Elizabeth scored 62, 76, and 80 on three equally weighted tests in French. If the final exam score counts for two-thirds of the grade and the tests count for one-third, then what range of scores on the final exam would give her a final average over 70?

97. *Difference in ages* In 1981, the British public speculated about the suitability of a marriage between Diana Spencer and the 32-year-old Prince of Wales, partly because there was a difference of more than 12 years in their ages. If you did not know Diana's age or that she was younger than Charles, then how would you describe this relationship with an absolute value inequality? What are the possibilities for Diana's age in 1981?

98. *Differences in prices* There is less than $800 difference between the price of a $14,200 Ford and a comparable Chevrolet. Express this as an absolute value inequality. What are the possibilities for the price of the Chevrolet?

99. *Stock exchange* (Continuation of Exercise 113, Section 1.1) The bar graph (repeated here) shows the daily volume of stock traded on the New York Stock Exchange during January and February of 1993 *(Wall Street Journal)*. If v is the volume of stock traded, then for how many days is the inequality $|v - 280,000,000| > 40,000,000$ true?

100. *Making a profit* A strawberry farmer paid $4200 for planting and fertilizing her strawberry crop. She must also pay $2.40 per flat (12 pints) for picking and packing the berries and $300 rent for space in a farmers' market where she sells the berries for $11 per flat. For what number of flats will her revenue exceed her costs?

Volume of Stock Traded on New York Stock Exchange

Figure for Exercise 99

2.7

More Inequalities

In this section we continue the study of inequalities with quadratic inequalities and rational inequalities.

Quadratic Inequalities

If the equal sign in the quadratic equation $ax^2 + bx + c = 0$ is replaced by $<, \leq, >$, or $\geq$, we have a **quadratic inequality.** The method that we use for solving a quadratic inequality is very different from the methods used in Section 2.6.

Consider the quadratic inequality $x^2 - x - 6 > 0$. Since the left-hand side can be factored, we can write the inequality as $(x - 3)(x + 2) > 0$. A number x satisfies this inequality if the product of $x - 3$ and $x + 2$ is positive. To analyze this situation, we make a **sign graph** showing where each factor is positive, negative, or zero. The value of the factor $x - 3$ is positive for $x > 3$, negative for $x < 3$, and zero for $x = 3$. So on the sign graph of Fig. 2.21 we put positive signs above the interval $(3, \infty)$, negative signs above the interval $(-\infty, 3)$, and 0 above 3.

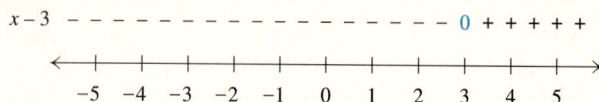

Figure 2.21

Now analyze the factor $x + 2$. The value of $x + 2$ is positive for $x > -2$, negative for $x < -2$, and zero for $x = -2$. So we add a line for the factor $x + 2$ to the sign graph of Fig. 2.21. In Fig. 2.22 we put positive signs above the interval $(-2, \infty)$, negative signs above $(-\infty, -2)$, and 0 above -2.

Figure 2.22

The product $(x - 3)(x + 2)$ is positive where both factors are positive or both factors are negative. From Fig. 2.22 we see that the solution set to the inequality $(x - 3)(x + 2) > 0$ is $(-\infty, -2) \cup (3, \infty)$.

Example 1   Solving quadratic inequalities

Solve each inequality. Write the solution set in interval notation and graph it.

a) $x^2 - x \leq 6$

b) $9 - x^2 < 0$

Solution

a) Write the inequality with 0 on the right-hand side before factoring.

$$x^2 - x \leq 6$$
$$x^2 - x - 6 \leq 0$$
$$(x - 3)(x + 2) \leq 0$$

The solution set to $(x - 3)(x + 2) \leq 0$ includes any value of x for which the product of $x - 3$ and $x + 2$ is negative or zero. Note that prior to this example we solved $(x - 3)(x + 2) > 0$. The signs of the factors $x - 3$ and $x + 2$ are already shown in Fig. 2.22. The product is negative where one factor is positive and the other is negative, between -2 and 3. The product is zero if $x = -2$ or $x = 3$. The solution set is $[-2, 3]$ and its graph is shown in Fig. 2.23. Note that the solution to $(x - 3)(x + 2) \leq 0$ includes -2 and 3, while the solution to $(x - 3)(x + 2) > 0$ did not include them.

b) If we factor the left-hand side of $9 - x^2 < 0$, we get $(3 - x)(3 + x) < 0$. The factor $3 - x$ is positive if $x < 3$, negative if $x > 3$, and 0 if $x = 3$. The factor $3 + x$ is positive if $x > -3$, negative if $x < -3$, and 0 if $x = -3$. This information is shown on the sign graph in Fig. 2.24.

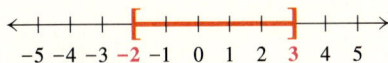

Figure 2.23

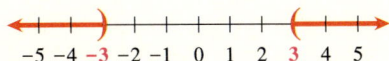

Figure 2.25

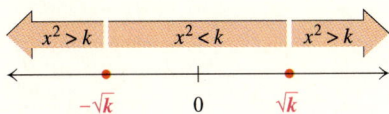

Figure 2.26

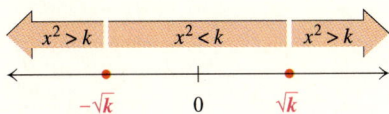

Figure 2.24

The solution set includes the values of x for which the product of the two factors is negative. Since the product is negative when the factors have opposite signs, the solution set to the inequality is $(-\infty, -3) \cup (3, \infty)$. The graph of the solution set is shown in Fig. 2.25. ◆

The inequality in Example 1(b) is of the type $x^2 > k$, where k is a positive real number. We can use the same steps as in Example 1(b) to show that the solution set to $x^2 > k$ is $\left(-\infty, -\sqrt{k}\right) \cup \left(\sqrt{k}, \infty\right)$. Likewise, the solution set to $x^2 < k$ is $\left(-\sqrt{k}, \sqrt{k}\right)$. See Fig. 2.26.

Rational Inequalities

Rational inequalities are inequalities that involve rational expressions. Some examples are

$$\frac{x + 3}{x - 2} \leq 2, \qquad \frac{x^2 - 4}{x - 3} \geq 0, \qquad \text{and} \qquad \frac{2}{x - 1} \geq \frac{1}{x + 1}.$$

When we multiply each side of an inequality by a real number, we must know whether that number is positive or negative. *So we generally do not multiply each side of a rational inequality by an expression that involves a variable.* As we did for quadratic inequalities, we can use sign graphs to solve rational inequalities.

Example 2 Solving rational inequalities

Solve the inequality. Write the solution set using interval notation and graph it.

$$\frac{x + 3}{x - 2} \le 2$$

Solution

To use the sign graph method, we must have a 0 on one side of the inequality.

$$\frac{x + 3}{x - 2} \le 2$$

$$\frac{x + 3}{x - 2} - 2 \le 0 \qquad \text{Subtract 2 from each side to get 0 on the right-hand side.}$$

$$\frac{x + 3}{x - 2} - \frac{2(x - 2)}{x - 2} \le 0 \qquad \text{Get a common denominator.}$$

$$\frac{x + 3 - 2(x - 2)}{x - 2} \le 0 \qquad \text{Subtract.}$$

$$\frac{7 - x}{x - 2} \le 0 \qquad \text{Get a single rational expression on the left.}$$

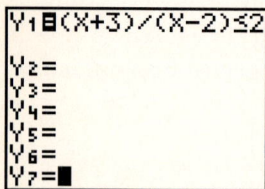

The denominator $x - 2$ is positive if $x > 2$, negative if $x < 2$, and 0 if $x = 2$. The numerator $7 - x$ is positive if $x < 7$, negative if $x > 7$, and 0 if $x = 7$. This information is shown on the sign graph in Fig. 2.27.

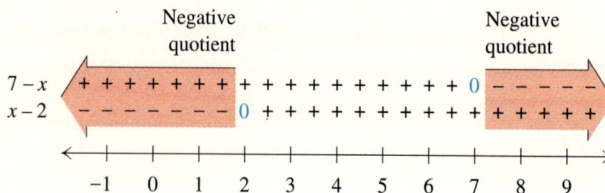

Figure 2.27

To discuss this sign graph, let

$$R(x) = \frac{7 - x}{x - 2}.$$

We want the values of x for which $R(x)$ is negative or zero. Since $R(x)$ is a quotient, $R(x)$ will be zero only if the numerator is zero (the denominator may not be zero), and $R(x)$ will be negative where $7 - x$ and $x - 2$ have opposite signs on the sign graph. The solution set to the inequality is $(-\infty, 2) \cup [7, \infty)$. Note that 2 is not in the solution set, because $R(2)$ is undefined; 7 is in the solution set because $R(7) = 0$. The graph of the solution set is shown in Fig. 2.28. ◆

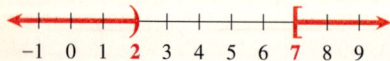

On a calculator, an inequality in x has value 1 when it is satisfied and value 0 when it is not satisfied. Setting y_1 equal to the inequality means that y_1 has value 0 or 1.

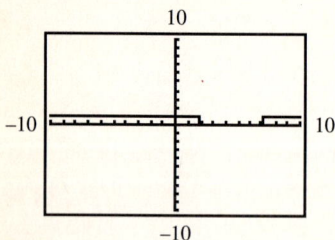

The graph of y_1 in the standard viewing window supports the solution $(-\infty, 2)$ $\cup [7, \infty)$. The horizontal line appears only above values for x that satisfy the inequality.

Figure 2.28

Example 3   Rational inequality with three factors

Solve the inequality. Write the solution set in interval notation and graph it.

$$\frac{2}{x-1} \geq \frac{1}{x+1}$$

Solution

Instead of multiplying by the LCD, we use subtraction to get both rational expressions on the same side of the inequality.

$$\frac{2}{x-1} - \frac{1}{x+1} \geq 0$$

$$\frac{2(x+1)}{(x-1)(x+1)} - \frac{1(x-1)}{(x+1)(x-1)} \geq 0 \qquad \text{\textcolor{blue}{Get a common denominator.}}$$

$$\frac{2x+2 - 1(x-1)}{(x-1)(x+1)} \geq 0 \qquad \text{\textcolor{blue}{Subtract.}}$$

$$\frac{x+3}{(x-1)(x+1)} \geq 0$$

Let

$$R(x) = \frac{x+3}{(x-1)(x+1)}.$$

The solution set includes any value of x for which $R(x)$ is positive or zero. The signs for each binomial in $R(x)$ are shown in Fig. 2.29.

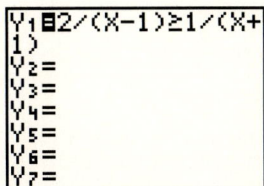

Inequality symbols are found in the TEST menu.

The horizontal line appears only for x in $[-3, -1) \cup (1, \infty)$. So the graph supports our solution to the inequality.

Figure 2.29

Since $R(x)$ involves only multiplication and division of the three binomials on the sign graph, $R(x)$ is positive if all three binomials are positive or if exactly one is positive. $R(x) = 0$ only if $x = -3$, because the denominator must not be 0. From the sign graph the solution set is $[-3, -1) \cup (1, \infty)$. The graph of the solution is shown in Fig. 2.30.

Figure 2.30

The Test Point Method

In Example 3 we worked with the rational expression

$$R(x) = \frac{x + 3}{(x - 1)(x + 1)}.$$

From Fig 2.29 we see that $R(x)$ is negative for x in $(-\infty, -3)$, positive for x in $(-3, -1)$, negative for x in $(-1, 1)$, and positive for x in $(1, \infty)$. The value of $R(x)$ changes sign only at a value of x for which the expression is zero or undefined. This idea is used in solving inequalities by the **test point method.** In this method, we determine all of the points at which the expression is zero or undefined. These points divide the line into intervals. We test a number from each interval in the expression to determine the sign of the expression for that interval. Then we write the solution set from the interval or intervals that satisfy the inequality.

Example 4 Solving an inequality with test points

Solve the inequality. State the solution set in interval notation and graph it.

$$\frac{x^2 - 6x + 8}{x^2 + 4x + 4} \geq 0$$

Solution

$$\frac{(x - 2)(x - 4)}{(x + 2)^2} \geq 0 \qquad \textbf{Factor the numerator and denominator.}$$

Let

$$R(x) = \frac{(x - 2)(x - 4)}{(x + 2)^2}.$$

We want to solve $R(x) \geq 0$. Now $R(x) = 0$ if $x = 2$ or $x = 4$. The denominator is 0 if $x = -2$, but -2 is not in the domain of the rational expression. So $R(x)$ is undefined for -2. Mark these points on a number line with a ''U'' for undefined or a zero as shown in Fig. 2.31. Then select one test point in each of the intervals determined by these points. We have selected $-3, 0, 3,$ and 6.

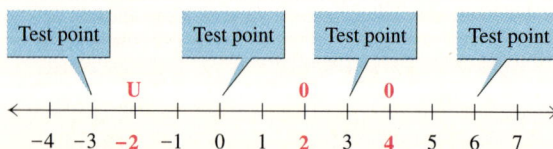

Figure 2.31

The value of $R(x)$ for each test point is listed in Table 2.9 on the next page. If $R(x)$ is positive for a test point, we conclude that $R(x)$ is positive on the corresponding interval. From the table, we see three intervals on which $R(x) > 0$. Since $R(x) = 0$ for $x = 2$ and 4, those points are included in the solution set. So the solution set to $R(x) \geq 0$ is $(-\infty, -2) \cup (-2, 2] \cup [4, \infty)$. Its graph is shown in Fig. 2.32.

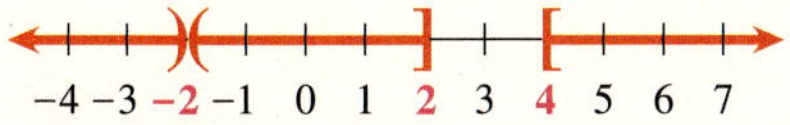

Figure 2.32

x	$R(x)$	Conclusion
-3	35	$R(x) > 0$ on $(-\infty, -2)$
0	2	$R(x) > 0$ on $(-2, 2)$
3	$-1/25$	$R(x) < 0$ on $(2, 4)$
6	1/8	$R(x) > 0$ on $(4, \infty)$

Table 2.9

In the next example we use the test point method to solve a quadratic inequality that involves a prime polynomial.

Example 5   Quadratic inequality with test points

Solve $x^2 < x + 1$. State the solution set in interval notation and graph it.

Solution

Write the inequality as $x^2 - x - 1 < 0$. Let $P(x) = x^2 - x - 1$. Since $P(x)$ is a prime polynomial, use the quadratic formula to solve $x^2 - x - 1 = 0$:

$$x = \frac{1 \pm \sqrt{(-1)^2 - 4(1)(-1)}}{2(1)} = \frac{1 \pm \sqrt{5}}{2}$$

Since $\frac{1 + \sqrt{5}}{2} \approx 1.6$ and $\frac{1 - \sqrt{5}}{2} \approx -0.6$, we mark these two points on a number line and choose a test point in each of the resulting three regions (Fig. 2.33). Table 2.10 lists the value of $P(x) = x^2 - x - 1$ at test points -2, 0, and 4.

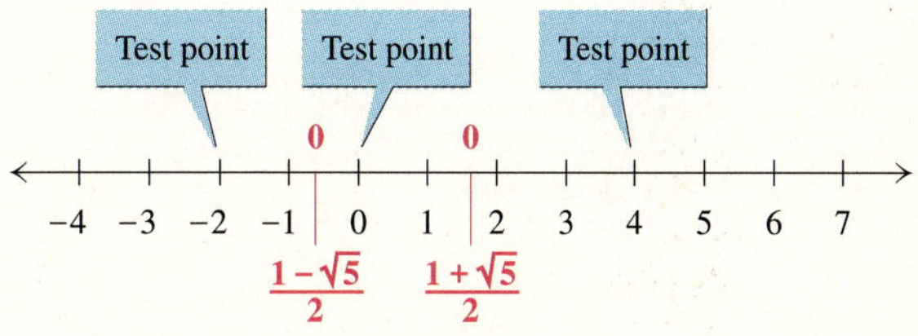

Figure 2.33

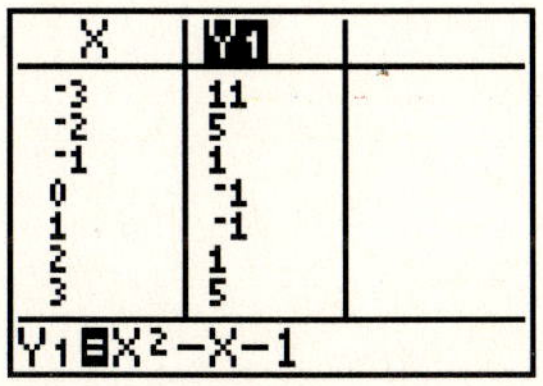

Some calculators can make tables. This table supports our conclusion that the values of $P(x)$ go from positive to negative to positive.

x	$P(x)$	Conclusion
-2	5	$P(x) > 0$ on $\left(-\infty, \dfrac{1 - \sqrt{5}}{2}\right)$
0	-1	$P(x) < 0$ on $\left(\dfrac{1 - \sqrt{5}}{2}, \dfrac{1 + \sqrt{5}}{2}\right)$
4	11	$P(x) > 0$ on $\left(\dfrac{1 + \sqrt{5}}{2}, \infty\right)$

Table 2.10

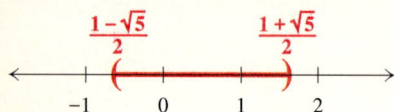

Figure 2.34

We conclude that the sign of $P(x)$ on each corresponding interval is the same as the sign of $P(x)$ at the test point. The original inequality is equivalent to $P(x) < 0$ and there is only one interval for which $P(x) < 0$. So the solution set is

$$\left(\frac{1 - \sqrt{5}}{2}, \frac{1 + \sqrt{5}}{2} \right)$$

and its graph is shown in Fig. 2.34.   ◆

The polynomial equation $P(x) = 0$ in Example 5 had two real solutions. If the equation $P(x) = 0$ has no real solutions, then there are no points on the real line where $P(x)$ changes sign. So $P(x) < 0$ will be satisfied by either all real numbers or no real numbers.

Applications

In Section 2.4 we introduced the equation that is used to model the motion of a projectile: $S = -16t^2 + v_0 t + s_0$, where S is the height in feet at time t in seconds, v_0 is the initial velocity in feet per second, and s_0 is the initial height in feet. In the next example, we write a quadratic inequality based on this model for the motion of a projectile.

Example 6   Application of a quadratic inequality

During the summer of 1992, Mark Lenzi of Virginia won a gold medal in the springboard diving competition at Barcelona. The complex maneuvers that he performs require split-second timing. If Mark springs into the air at 20 feet per second from a diving board 10 feet above the water, then for how many seconds is he at least 15 feet above the water?

Solution

Use an initial velocity of 20 feet per second and an initial height of 10 feet. The height of the diver at any time t is given by the formula $S = -16t^2 + 20t + 10$. See Fig. 2.35. We want the values of t that satisfy $-16t^2 + 20t + 10 \geq 15$, or $-16t^2 + 20t - 5 \geq 0$. Use the quadratic formula to solve $-16t^2 + 20t - 5 = 0$:

$$t = \frac{-20 \pm \sqrt{20^2 - 4(-16)(-5)}}{2(-16)} = \frac{-5 \pm \sqrt{5}}{-8} = \frac{5 \pm \sqrt{5}}{8} \approx 0.3, 0.9$$

Pick test points 0, 0.5, and 1. If $P(t) = -16t^2 + 20t - 5$, then

$$P(0) = -5, \qquad P(0.5) = 1, \qquad \text{and} \qquad P(1) = -1.$$

So $-16t^2 + 20t - 5 \geq 0$ is satisfied only for t in the interval

$$\left[\frac{5 - \sqrt{5}}{8}, \frac{5 + \sqrt{5}}{8} \right].$$

Figure 2.35

The length of this interval is the amount of time for which the diver is at least 15 feet above the water:

$$\frac{5 + \sqrt{5}}{8} - \frac{5 - \sqrt{5}}{8} = \frac{\sqrt{5}}{4} \approx 0.6$$

He is at least 15 feet above the water for exactly $\sqrt{5}/4$ seconds, or approximately 0.6 seconds. ◆

❓ For Thought

True or false? Explain.

1. The solution set to $x^2 > 9$ is the interval $(3, \infty)$.

2. The inequality $\dfrac{3}{x} > 2$ is equivalent to $3 > 2x$.

3. The inequality $x(2x - 1)(x - 3) < 0$ is equivalent to $x < 0$ or $2x - 1 < 0$ or $x - 3 < 0$.

4. We can solve any quadratic inequality.

5. If $\dfrac{-3}{a} > 0$, then $a < 0$.

6. The solution set to $(x - 2)(x + 5) > 0$ is $(-\infty, -5) \cup (2, \infty)$.

7. Any real number a satisfies $(a - 3)^2 > 0$.

8. The solution set to $\dfrac{x - 1}{(x + 3)^2} > 0$ is $(1, \infty)$.

9. Every real number satisfies $(x - (2 + i))(x - (2 - i)) > 0$.

10. The solution to $(x - 3)(x + 1)(x - 1) < 0$ consists of all values of x for which exactly one factor is negative or all three factors are negative.

2.7 Exercises   📼 Tape 5   💾 Disk—5.25": 2   3.5": 1   Macintosh: 1

Solve each inequality. State the solution set in interval notation and graph it.

1. $2x - 3 > 0$
2. $3x + 2 < 0$
3. $w + 1 < 0$
4. $w - 2 > 0$
5. $5 - x < 0$
6. $1 - x > 0$
7. $z + 3 < 0$
8. $z - 4 < 0$

11. $2x + 15 < x^2$
12. $5x - x^2 < 4$
13. $w^2 - 4w - 12 \geq 0$
14. $y^2 + 8y + 15 \leq 0$
15. $t^2 \leq 16$
16. $36 \leq h^2$
17. $a^2 + 6a + 9 \leq 0$
18. $c^2 + 4 \leq 4c$
19. $4z^2 - 12z + 9 > 0$
20. $9s^2 + 6s + 1 \geq 0$

Solve using a sign graph. State the solution set in interval notation and graph it.

9. $2x^2 - x < 3$
10. $3x^2 - 4x \leq 4$

Find the indicated value of each polynomial or rational expression in Exercises 21–28.

21. $P(3)$ if $P(x) = x^2 - 3x + 1$

22. $P(-2)$ if $P(x) = 2x^2 - 5x + 6$

23. $R(-1)$ if $R(x) = \dfrac{x-1}{x+3}$ **24.** $R(-3)$ if $R(x) = \dfrac{4-x}{2x-1}$

25. $P(2)$ if $P(x) = (x-1)(x-5)(x-6)$

26. $P(4)$ if $P(x) = (3-x)(x-2)(x+2)$

27. $R(-4)$ if $R(x) = \dfrac{(x-3)(5-x)}{(x+2)(x-1)}$

28. $R(3)$ if $R(x) = \dfrac{x^2+2x+7}{(x-5)(x-1)}$

Solve using a sign graph. State the solution set in interval notation and graph it.

29. $\dfrac{x-4}{x+2} \le 0$ **30.** $\dfrac{x+3}{x+5} \ge 0$ **31.** $\dfrac{x-7}{x+7} < 0$

32. $\dfrac{x+3}{x-1} > 0$ **33.** $\dfrac{1}{y} \le 0$ **34.** $\dfrac{1}{y^2} \le 0$

35. $\dfrac{w^2-w-6}{w-6} \ge 0$ **36.** $\dfrac{z-5}{z^2+2z-8} \le 0$

37. $\dfrac{3-t}{2t^2-9t-5} > 0$ **38.** $\dfrac{m^2-5m+6}{1-2m} < 0$

39. $\dfrac{q-2}{q+3} < 2$ **40.** $\dfrac{p+1}{2p-1} \ge 1$

41. $\dfrac{1}{x+2} > \dfrac{1}{x-3}$ **42.** $\dfrac{1}{x+1} > \dfrac{2}{x-1}$

43. $x < \dfrac{3x-8}{5-x}$ **44.** $\dfrac{2}{x+3} \ge \dfrac{1}{x-1}$

Solve using the test point method. State the solution set in interval notation and graph it.

45. $\dfrac{(x-3)(x+1)}{x-5} \ge 0$ **46.** $\dfrac{(x+2)(x-1)}{(x+4)^2} \le 0$

47. $\dfrac{x^2-25}{9-x^2} \le 0$ **48.** $\dfrac{x^2+2x+1}{x^2-2x-15} \ge 0$

49. $x^2 - 2x - 24 > 0$ **50.** $4x + 60 \le x^2$

51. $4x^2 - 9 < 0$ **52.** $2x^2 - 7 > 0$

53. $y^2 + 23 > 10y$ **54.** $y^2 + 3 \ge 6y$

55. $6t < t^2 + 10$ **56.** $8t < t^2 + 25$

57. $\dfrac{1}{w} > \dfrac{1}{w^2}$ **58.** $\dfrac{1}{w} > w^2$

59. $w > \dfrac{w-5}{w-3}$ **60.** $w < \dfrac{w-2}{w+1}$

Solve by the method of your choice. State the solution set in interval notation and graph it.

61. $(x-5)(3x+2)(x-1) < 0$

62. $(3-x)(x+4)(3x-2) \ge 0$

63. $(y-1)(y-3)(y-5)(y-7) > 0$

64. $y(y+2)(y-3)(y-6) < 0$

65. $y^3 - 9y \ge 0$

66. $2y^3 - 5y \le 0$

67. $t + 5t^2 > 4t + 20$ **68.** $t^3 - t^2 < 16t - 16$

69. $\dfrac{5z-12}{z-3} \le z$ **70.** $\dfrac{1}{z} \ge z$

71. $\dfrac{d-1}{d+2} > \dfrac{d+3}{d-4}$ **72.** $\dfrac{v+1}{v-2} \le \dfrac{v-1}{v+2}$

73. $\dfrac{u^2-3u+2}{u^2-7u+12} \le 0$ **74.** $\dfrac{4b^2-1}{b^2-5} \le 0$

75. $\dfrac{2t}{t-1} \ge \dfrac{t+2}{t-2}$ **76.** $\dfrac{2q+3}{q-1} \le \dfrac{q+2}{q-4}$

Find the values of w for which each expression represents a real number. Use interval notation.

77. $\sqrt{2w-1}$ **78.** $\sqrt{9-w}$

79. $\sqrt{9-w^2}$ **80.** $\sqrt{w^2-4}$

81. $\dfrac{1}{\sqrt{w^2-w-6}}$ **82.** $\dfrac{1}{\sqrt{w^2-9}}$

83. $\sqrt{\dfrac{w-1}{w+2}}$ **84.** $\sqrt{\dfrac{1-w}{w-2}}$

Solve each problem.

85. *Fencing a rectangular area* Hector has 100 ft of fencing to fence a rectangular area. What are the possible lengths for the north side that would give an area of less than or equal to 400 ft²?

86. *Profit from sales* Shelley's weekly profit in dollars for selling x bottles of skin toner is given by the formula $P(x) = -2x^2 + 50x + 20$. For what values of x is her profit at least $250?

87. *Height of a flare* A flare is fired straight into the air from a height of 6 ft above the water. If the flare leaves the barrel of the gun at 80 ft/sec, then for how long is the flare more than 100 ft above the water?

Figure for Exercise 87

Figure for Exercise 88

88. *Height of a baseball* Oakland A's first baseman, Mark McGwire, is a power hitter whose swing produces a lot of fly balls. If McGwire's bat connects with the ball at a height of 3 ft and sends it straight up with a velocity of 50 ft/sec, how long is the baseball more than 20 ft above the ground?

89. *Two numbers* If two numbers must have a sum of 12, then what are the possible values for the first number so that their product is at least 10?

90. *Two more numbers* If two numbers must have a difference of 6, then what are the possible values for the smaller number so that their product is at least 9?

91. *Inflation of car prices* The cost of a BMW in two years is given by the formula $C = P(1 + r)^2$, where P is the present cost and r is the annual inflation rate. According to a market analyst, today's $38,000 BMW will cost at least $43,000 in two years. For what possible values for r would this statement be true?

Figure for Exercise 91

92. *Painting a fence* Joe can paint a fence working alone in 40 hr. Joe plans to work with a helper who could do the job alone in x hours. For what values of x could Joe and the helper get the fence done in less than 30 hr?

93. *Velocity for a vertical leap* Boston Celtics player Dee Brown, at 6 ft and 160 lb, may have the highest vertical leap in the NBA. If Brown makes a vertical leap of 1.17 m, what is his initial velocity at the time his feet leave the floor? The formula $v_1^2 = v_0^2 + 2gS$ gives the relationship between terminal velocity v_1, initial velocity v_0, the acceleration of gravity g, and his height S. Use $g = -9.8$ m/sec² and the fact that his terminal velocity at his peak height is zero.

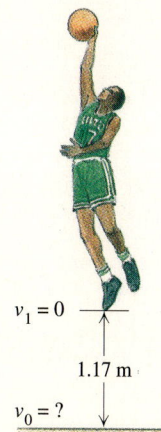

Figure for Exercise 93

94. *Time for a vertical leap* Use the initial velocity from Exercise 93 to find the amount of time that Dee Brown is more than 0.7 m above the court when he makes his vertical leap. Use the formula $S = -4.9t^2 + v_0t$, where S is height in meters, t is time in seconds, and v_0 is initial velocity in meters per second.

95. *Cleaning up the river*   Pollution in the Tangipahoa River has been blamed primarily on the area dairy farmers. The formula

$$C = \frac{400,000p}{100 - p}$$

is used to model the cost in dollars to area dairy farmers for removing $p\%$ of the coliform bacteria from the river. If less than \$1.2 million is spent, then what percentage of the coliform bacteria can be removed?

96. *Product awareness*   An advertising agency estimates that the cost in dollars of an advertising campaign that reaches $p\%$ of the consumers in Chicago is given by

$$C = \frac{100,000 + 500,000p}{100 - p}.$$

If less than \$752,500 is spent on the campaign, then what percentage of the consumers will be reached?

For Writing/Discussion

97. *General summary*   Write a general summary concerning the solution set to the quadratic inequality $ax^2 + bx + c > 0$, based on the value of the discriminant.

98. *Extended summary*   Extend the summary of Exercise 97 to include quadratic inequalities using $\geq$, $<$, and $\leq$.

99. *Cooperative learning*   Each student in your small group should write a linear inequality, an absolute value inequality, a quadratic inequality, and a rational inequality. The group should solve each inequality and determine whether it is the required type.

◆ Highlights

Section 2.1   Linear Equations

1. An equation is a statement indicating that two algebraic expressions are equal.
2. A linear equation is an equation of the form $ax + b = 0$, where a and b are real numbers, with $a \neq 0$.
3. To solve an equation means to find the set of all numbers for which the equation is true.
4. The properties of equality are used to simplify an equation until we get an equivalent equation whose solution set is obvious.
5. An equation may be an identity, an inconsistent equation, or a conditional equation.

Section 2.2   Applications

1. A formula is an equation in two or more variables that is usually of an applied nature.
2. The value of one variable can usually be found if values are known for all of the other variables in the formula.
3. The strategy for problem solving presented gives a step-by-step procedure for solving verbal problems.
4. Three main categories of applied problems studied here are uniform-motion problems, work problems, and mixture problems.

Section 2.3   Complex Numbers

1. Complex numbers are of the form $a + bi$, where a and b are real numbers and where $i = \sqrt{-1}$ and $i^2 = -1$.

2. Operations with complex numbers can be performed as if they are binomials with i being a variable, and using $i^2 = -1$ to simplify.

3. Division of complex numbers is performed by multiplying numerator and denominator by the complex conjugate of the denominator.

4. Expressions involving square roots of negative numbers must be converted to standard form using $\sqrt{-b} = i\sqrt{b}$ ($b > 0$) before we can perform any operations.

Section 2.4   Quadratic Equations

1. A quadratic equation is an equation of the form $ax^2 + bx + c = 0$, where a, b, and c are real numbers with $a \neq 0$.

2. A quadratic equation can have two distinct real solutions, one real solution, or two imaginary solutions that are conjugates.

3. The four methods presented for solving quadratic equations are factoring, the square root property, completing the square, and the quadratic formula.

Section 2.5   More Equations

1. Equations of higher degree can be solved by factoring.

2. Squaring each side of an equation can produce extraneous roots.

3. Equations of quadratic type are not quadratic equations, but they can be solved by the techniques used for quadratic equations.

4. To solve an equation involving absolute value, we first write an equivalent equation without absolute value.

Section 2.6   Inequalities

1. The inequality symbol is reversed if the inequality is multiplied or divided by a negative number.

2. We usually do not multiply or divide an inequality by a variable expression.

3. Interval notation is used to state the solution sets of most inequalities.

4. To solve an absolute value inequality, remember that the absolute value of x is the distance from x to 0 on the number line.

5. Most absolute value inequalities can be written as equivalent compound inequalities.

Section 2.7   More Inequalities

1. Polynomial and rational inequalities are solved by sign graphs and the rules for multiplying and dividing signed numbers. The test point method is also used for solving polynomial or rational inequalities.

2. The test point method is based on the fact that the value of a polynomial or rational expression can change sign only at a point on the number line where it has value 0 or is undefined.

Chapter 2 Review Exercises

Find all real and imaginary solutions to each equation.

1. $3x - 2 = 0$

2. $3x - 5 = 5(x + 7)$

3. $3x^2 - 2 = 0$

4. $x^4 - 16 = 0$

5. $a^3 - 8 = 0$

6. $p^3 + 1 = 0$

7. $\dfrac{1}{2}y - \dfrac{1}{3} = \dfrac{1}{4}y + \dfrac{1}{5}$

8. $\dfrac{1}{2} - \dfrac{w}{5} = \dfrac{w}{4} - \dfrac{1}{8}$

9. $\dfrac{2}{x} = \dfrac{3}{x - 1}$

10. $\dfrac{5}{x + 1} = \dfrac{2}{x - 3}$

11. $\dfrac{x + 1}{x - 1} = \dfrac{x + 2}{x - 3}$

12. $\dfrac{x + 3}{x - 8} = \dfrac{x + 7}{x - 4}$

13. $\dfrac{x + 2}{x + 4} = \dfrac{x - 2}{x - 4}$

14. $\dfrac{2x - 5}{x + 2} = \dfrac{x - 3}{x - 2}$

15. $\dfrac{1}{x} + \dfrac{1}{x - 1} = \dfrac{3}{2}$

16. $\dfrac{2}{x - 2} - \dfrac{3}{x + 2} = \dfrac{1}{2}$

17. $b^2 + 10 = 6b$

18. $4t^2 + 17 = 16t$

19. $s^2 - 4s + 1 = 0$

20. $3z^2 - 2z - 1 = 0$

21. $|3q - 4| = 2$

22. $|2v - 1| = 3v$

23. $|2h - 3| = |h|$

24. $|3y - 1| = -2$

25. $x^4 + 7x^2 = 18$

26. $2x^{-2} + 5x^{-1} = 12$

27. $\sqrt{x + 6} - \sqrt{x - 5} = 1$

28. $\sqrt{2x - 1} = \sqrt{x - 1} + 1$

29. $\left| \dfrac{3}{2x - 1} \right| = 4$

30. $\sqrt[3]{x^2} + \sqrt[3]{x} = 2$

31. $x^4 - 3x^2 - 4 = 0$

32. $(y - 1)^2 - (y - 1) = 2$

Solve each inequality. State the solution set using interval notation and graph the solution set.

33. $4x - 1 > 3x + 2$

34. $6(x - 3) < 5(x + 4)$

35. $5 - 2x > -3$

36. $7 - x > -6$

37. $\dfrac{1}{2}x - \dfrac{1}{3} > x + 2$

38. $0.06x + 1000 > x + 60$

39. $-2 < \dfrac{x - 3}{2} \le 5$

40. $-1 \le \dfrac{3 - 2x}{4} < 3$

41. $3 - 4x < 1$ and $5 + 3x < 8$

42. $-3x < 6$ and $2x + 1 > -1$

43. $-2x < 8$ or $3x > -3$

44. $1 - x < 6$ or $-5 + x < 1$

45. $|x - 3| > 2$

46. $|4 - x| \le 3$

47. $|2x - 7| \le 0$

48. $|6 - 5x| < 0$

49. $|7 - 3x| > -4$

50. $|4 - 3x| \ge 1$

Solve each inequality using a sign graph. State the solution set using interval notation.

51. $8x^2 + 1 < 6x$

52. $x^2 + 2x \le 63$

53. $(3 - x)(x + 5) \ge 0$

54. $-x^2 - 2x + 15 < 0$

55. $\dfrac{x - 3}{5 - x} \ge 0$

56. $\dfrac{4 - x}{x + 7} \le 0$

Solve each inequality using the test point method. State the solution set using interval notation.

57. $\dfrac{x + 10}{x + 2} < 5$

58. $\dfrac{x - 6}{2x + 1} \ge 1$

59. $\dfrac{12 - 7x}{x^2} > -1$

60. $x - \dfrac{2}{x} \le -1$

61. $x^2 \le 2x + 4$

62. $x^2 - x \le \dfrac{1}{2}$

Solve each inequality using the method of your choice. State the solution set using interval notation and graph it.

63. $\dfrac{x^2 - 3x + 2}{x^2 - 7x + 12} \ge 0$

64. $\dfrac{x^2 + 4x + 3}{x^2 - 2x - 15} \le 0$

65. $\dfrac{x + 3}{x + 3} \le 0$

66. $\dfrac{1}{(x - 2)^2} > 1$

67. $\dfrac{51}{(x - 1)^2} \ge 0$

68. $\dfrac{-1}{x} > 2$

Solve each equation for y.

69. $2x - 3y = 6$

70. $x(y - 2) = 1$

71. $xy = 1 + 3y$

72. $x^2y = 1 + 9y$

73. $ax + by = c$

74. $y^2 + my + n = 0$

75. $\dfrac{1}{y} = \dfrac{1}{x} + \dfrac{1}{2}$

76. $y^2 + by = ay + ab$

Write each expression in the form $a + bi$, where a and b are real numbers.

77. $(3 - 7i) + (-4 + 6i)$

78. $(-6 - 3i) - (3 - 2i)$

79. $(4 - 5i)^2$

80. $7 - i(2 - 3i)^2$

81. $(1 - 3i)(2 + 6i)$

82. $(0.3 + 2i)(0.3 - 2i)$

83. $(2 - 3i) \div i$

84. $(-2 + 4i) \div (-i)$

85. $(1 - i) \div (2 + i)$

86. $(3 + 6i) \div (4 - i)$

87. $\dfrac{1 + i}{2 - 3i}$

88. $\dfrac{3 - i}{4 - 3i}$

89. $\dfrac{6 + \sqrt{-8}}{2}$

90. $\dfrac{-2 - \sqrt{-18}}{2}$

91. $\dfrac{-6 + \sqrt{(-2)^2 - 4(-1)(-6)}}{-8}$

92. $\dfrac{-9 - \sqrt{(-9)^2 - 4(-3)(-9)}}{-6}$

93. $i^{34} + i^{19}$

94. $\sqrt{6} + \sqrt{-3}\sqrt{-2}$

Find all real solutions to each equation.

95. $(x - 1)^{2/3} = 4$

96. $(2x - 3)^{-1/2} = \dfrac{1}{2}$

97. $(x + 3)^{-3/4} = -8$

98. $\left(\dfrac{1}{x - 3}\right)^{-1/4} = \dfrac{1}{2}$

99. $\sqrt[3]{3x - 7} = \sqrt[3]{4 - x}$

100. $\sqrt[3]{x + 1} = \sqrt[6]{4x + 9}$

Find the values for x for which each expression represents a real number. Use interval notation.

101. $\sqrt{3x - 4}$

102. $\dfrac{1}{\sqrt{x - 1}}$

103. $\sqrt{x^2 - 25}$

104. $\sqrt{\dfrac{x^2 - x - 2}{4 - x}}$

Evaluate the discriminant for each equation, and use it to determine the type and number of solutions to the equation.

105. $x^2 + 2 = 4x$

106. $y^2 + 2 = 3y$

107. $4w^2 = 20w - 25$

108. $2x^2 - 3x + 10 = 0$

Solve each problem. Use an equation or an inequality as appropriate. For problems in which the answer involves an irrational number, find the exact answer and an approximate answer.

109. *Folding sheet metal* A square is to be cut from each corner of an 8-in. by 11-in. rectangular piece of copper and the sides are to be folded up to form a box as shown in the figure. If the area of the bottom is to be 50 in², what is the length of the side of the square to be cut from the corner?

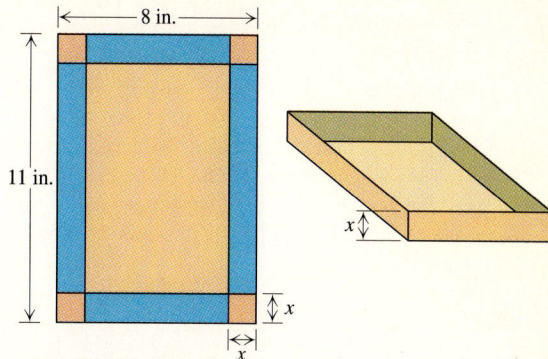

Figure for Exercise 109

110. *Peeling apples* Bart can peel a batch of 3000 lb of apples in 12 hr with the old machine, while Mona can do the same job in 8 hr using the new machine. If Bart starts peeling at 8 A.M. and Mona joins him at 9 A.M., then at what time will they finish the batch working together?

111. *Meeting between two cities* Lisa, an architect from Huntsville, has arranged a business luncheon with her client, Taro Numata from Norwood. They plan to meet at a restaurant off the 300-mi highway connecting their two cities. If they leave their offices simultaneously and arrive at the restaurant simultaneously, and Lisa averages 50 mph while Taro averages 60 mph, then how far from Huntsville is the restaurant?

112. *Driving speed* Lisa and Taro agree to meet at the construction site located 100 mi from Norwood on the 300-mi highway connecting Huntsville and Norwood. Lisa leaves Huntsville at noon, while Taro departs from Norwood one hour later. If they arrive at the construction site simultaneously and Lisa's driving speed averaged 10 mph faster than Taro's, then how fast did she drive?

113. *Fish population* A channel was dug to connect Homer and Mirror Lakes. Before the channel was dug, a biologist estimated that 20% of the fish in Homer Lake and 30% of the fish in Mirror Lake were bass. After the lakes were joined, the biologist made another estimate of the bass population in order to set fishing quotas. If she decided that 28% of the total fish population of 8000 were bass, then how many fish were in Homer Lake originally?

114. *Support for gambling* Eighteen pro-gambling representatives in the state house of representatives bring up a gambling bill every year. After redistricting, four new representatives are added to the house, causing the percentage of pro-gambling representatives to increase by 5 percentage points. If all four of the new representatives are pro-gambling and they still do not constitute a majority of the house, then how many representatives are there in the house after redistricting?

115. *Hiking distance* The distance between Marjorie and the dude ranch was $4\sqrt{34}$ mi straight across a rattlesnake-infested canyon. Instead of crossing the canyon, she hiked due north for a long time and then hiked due east for the shorter leg of the journey to the ranch. If she averaged 4 mph and it took her 8 hr to get to the ranch, then how far did she hike in a northerly direction?

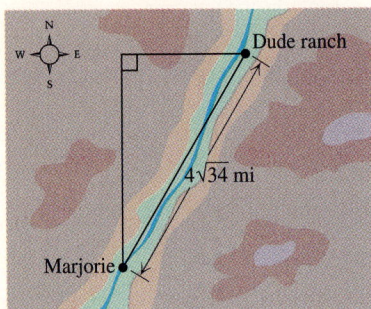

Figure for Exercise 115

116. *Price range for a haircut* A haircut at Joe's Barber Shop costs $50 less than a haircut at Renee's French Salon. In fact, you can get five haircuts at Joe's for less than the cost of one haircut at Renee's. What is the price range for a haircut at Joe's?

117. *Selling price of a home* Elena must get at least $120,000 for her house in order to break even with the original purchase price. If the real estate agent gets 6% of the selling price, then what should the selling price be?

118. *Dimensions of a picture frame* The length of a picture frame must be 2 in. longer than the width. If Reginald can use between 32 and 50 in. of frame molding for the frame, then what are the possibilities for the width?

119. *Saving gasoline* If Americans drive 10^{12} mi per year and the average gas mileage is raised from 27.5 mpg to 29.5 mpg, then how many gallons of gasoline are saved?

120. *Increasing gas mileage* If Americans continue to drive 10^{12} mi per year and the average gas mileage is raised from 29.5 to 31.5 mpg, then how many gallons of gasoline are saved? If the average mpg is presently 29.5, then what should it be increased to in order to achieve the same savings in gallons as the increase from 27.5 to 29.5 mpg?

121. *Thickness of concrete* Alfred used 40 yd^3 of concrete to pour a section of interstate highway that was 12 ft wide and 54 ft long. How thick was the concrete?

Figure for Exercise 121

122. *Marketing plans* A company has determined through market research that the demand equation for a new product is $P = 30 - \sqrt{0.002x + 3}$, where x is the number of units sold per month and P is the price in dollars of one unit. If the company sets the price of the product at $27 per unit, then how many units can it expect to sell per month?

Chapter 2 Test

Find all real and imaginary solutions to each equation.

1. $\dfrac{1}{2}x - \dfrac{1}{6} = \dfrac{1}{3}x$

2. $3x^2 - 2 = 0$

3. $x^3 - 27 = 0$

4. $x^2 + 14 = 9x$

Solve each inequality. State the solution set using interval notation and graph it.

5. $3 - 2x > 7$

6. $\dfrac{x}{2} > 3$ and $5 < x$

7. $|2x - 1| \le 3$

8. $2|x - 3| + 1 > 5$

Perform the indicated operations and write the answer in the form $a + bi$, where a and b are real.

9. $(4 - 3i)^2$

10. $\dfrac{2 - i}{3 + i}$

11. $i^6 - i^{35}$

12. $\sqrt{-8}(\sqrt{-2} + \sqrt{6})$

Solve each inequality using the method of your choice.

13. $x^2 - 2x < 8$

14. $\dfrac{x + 2}{x - 3} > -1$

15. $\dfrac{x + 3}{(4 - x)(x + 1)} \geq 0$

Find all real solutions to each equation.

16. $(x - 3)^{-2/3} = \dfrac{1}{3}$

17. $\sqrt{x} - \sqrt{x - 7} = 1$

18. $\dfrac{x - 1}{x + 3} = \dfrac{x + 2}{x - 6}$

Solve each problem.

19. Find the value of the discriminant for $x^2 - 5x + 9 = 0$.

20. Find the product of $7 - 2i$ and its conjugate.

21. Solve $5 - 2y = 4 + 3xy$ for y.

22. For what values of x is $\sqrt{x - 5}$ a real number?

23. Terry had a square patio. After expanding the length by 20 ft and the width by 10 ft the area was 999 ft². What was the original area?

24. How many gallons of 20% alcohol solution must be mixed with 10 gal of a 50% alcohol solution to obtain a 30% alcohol solution?

Tying It All Together
Chapters 1–2

Perform the indicated operations.

1. $3x + 4x$

2. $5x \cdot 6x$

3. $\dfrac{1}{x} + \dfrac{1}{2x}$

4. $(x + 3)^2$

5. $(2x - 1)(3x + 2)$

6. $\dfrac{(x + h)^2 - x^2}{h}$

7. $\dfrac{1}{x - 1} + \dfrac{1}{x + 1}$

8. $\left(x + \dfrac{3}{2}\right)^2$

Solve each equation.

9. $3x + 4x = 7x$

10. $5x \cdot 6x = 11x$

11. $\dfrac{1}{x} + \dfrac{1}{2x} = \dfrac{3}{2x}$

12. $(x + 3)^2 = x^2 + 9$

13. $6x^2 + x - 2 = 0$

14. $\dfrac{1}{x - 1} + \dfrac{1}{x + 1} = \dfrac{5}{8}$

15. $3x + 4x = 7x^2$

16. $3x + 4x = 7$

Given that $P(x) = x^3 - 3x^2 + 2$ and $T(x) = -x^2 + 3x - 4$, find the value of each polynomial for the indicated values of x.

17. $P(1)$

18. $P(-1)$

19. $P(-1/2)$

20. $P(-1/3)$

21. $T(-1)$

22. $T(2)$

23. $T(0)$

24. $T(0.5)$

GARY TORRISI

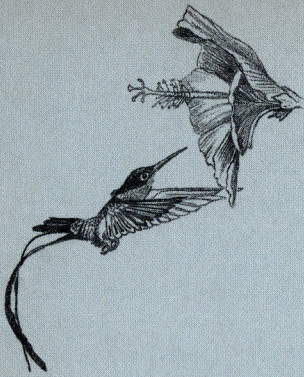

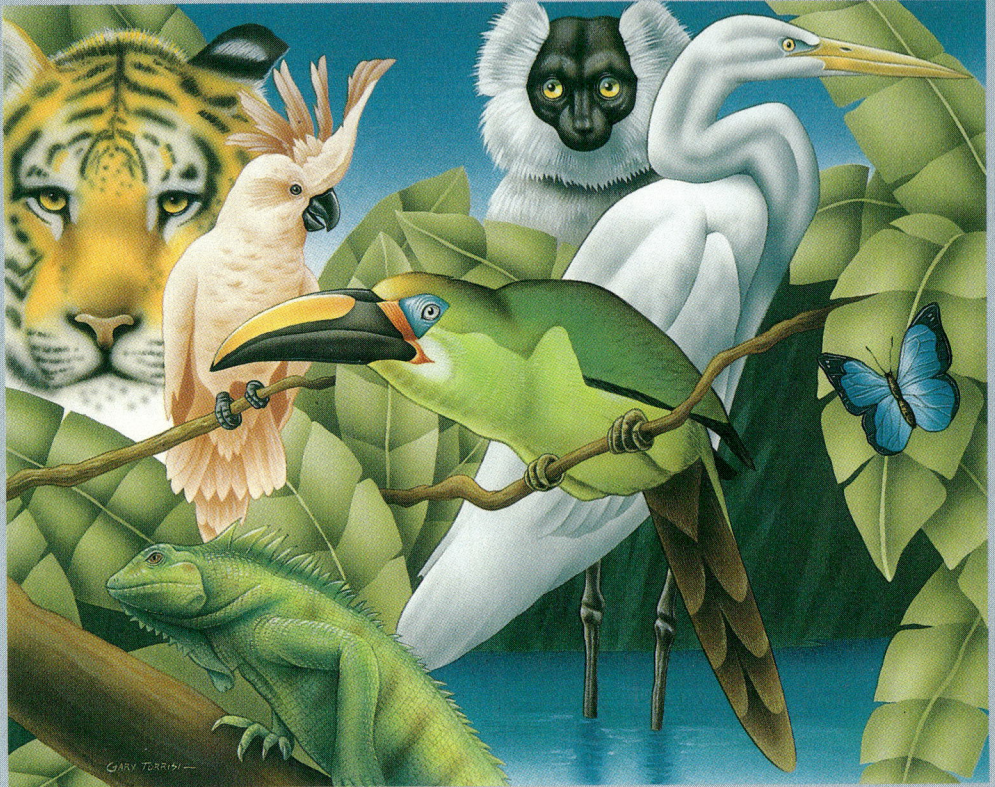

Fast forward to the year 2200. . . A spacecraft streaks toward Earth on its return from the Outer Galaxy. As the astronauts approach the blue planet, they spot a band of vegetation encircling the equator. They recognize this emerald belt as tropical rainforest, but the ragged outline—like a scar on Earth's surface—is an ominous sign. "A little over two centuries ago," remarks the Science Officer, "that green line was practically unbroken. Now look at it. If only people had cared enough to do more

while they still had the chance!"

There may still be time to preserve the rainforests, a lush habitat for life that thrives in a warm, wet climate. Although covering less than an eighth of a percent of Earth's surface, rainforests are home to over half our animal and plant species. Destroying only a percentage of these millions of species could irrevocably change our planet's future. Rainforests are also a source of many common foods, such as chocolate, vanilla, pineapples, and cinnamon, as well as raw products

like latexes, resins, dyes, and lumber. In addition, they provide us with twenty to twenty-five percent of all medicines. (One or more rainforest plants may hold the cure for cancer.)

Why are we destroying this richest of resources? Why do we cut down or burn rainforests in Africa, Asia, and Latin America at a rate of some 25,000 square miles per year? This destruction is rooted in a complex combination of social, economic, and biological causes: Loggers looking for easy profits open up the rainforests; poor migrant farmers follow, seeking fertile soil for their crops. And, as habitats vanish, animals disappear at the rate of a thousand species a year.

Depletion of the rainforests is only one of many hot ecological issues that include damage to the ozone layer, effects of radiation, and disposal of hazardous wastes. In assessing the impact humans have on the earth, scientists are looking for relationships between variables. In mathematics such relationships are called functions. This chapter discusses functions from both an algebraic and a geometric (graphing) point of view. We begin by looking at a two-dimensional coordinate system.

3.1

The Cartesian

Coordinate System

In Chapter 2 we studied equations and inequalities in one variable, whose solution sets are found on the number line. In this section we begin studying pairs of variables. For example, p might represent the price of gasoline and n the number of gallons that you consume in a month; x might be the number of toppings on a medium pizza and y the cost of that pizza; or h might be the height of a two-year-old child and w the weight. Since we cannot study relationships between pairs of variables on a one-dimensional number line, in this section we develop a two-dimensional coordinate system.

Ordered Pairs

Suppose that the cost of a medium pizza is $5 plus $2 per topping. Table 3.1 shows a chart in which x is the number of toppings and y is the cost. To indicate that $x = 3$ and $y = 11$ go together, we can use the **ordered pair** (3, 11). The ordered pair (2, 9) indicates that a two-topping pizza costs $9. The **first coordinate** of the ordered pair represents the number of toppings and the **second coordinate** represents the cost. The assignment of what the coordinates represent is arbitrary, but once made it is kept fixed. For example, in the present context the ordered pair (11, 3) indicates that an 11-topping pizza costs $3, and it is not the same as (3, 11). This is the reason they are called "ordered pairs." Recall that in Chapter 2 we used parentheses to indicate intervals of real numbers, and now we are using the same notation for ordered pairs. However, the meaning of this notation should always be clear from the discussion.

Toppings x	Cost y
0	$ 5
1	7
2	9
3	11
4	13

Table 3.1

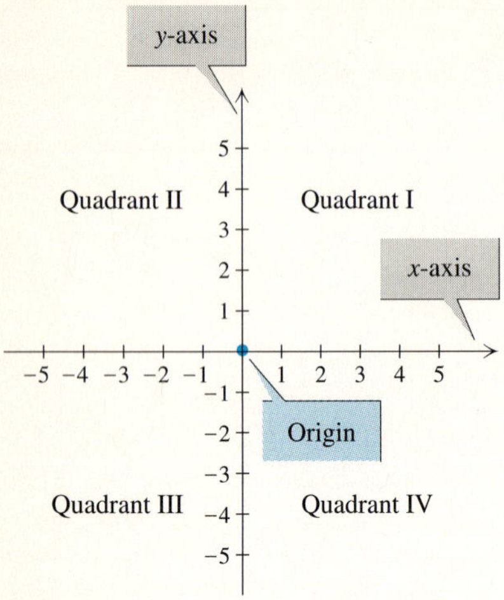

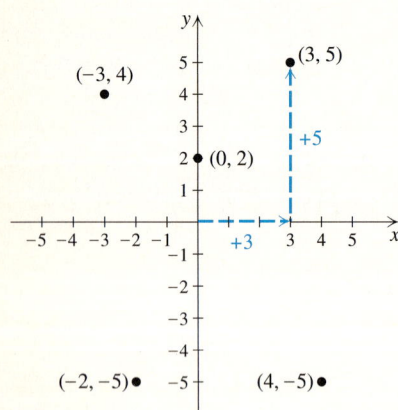

Figure 3.1

Figure 3.2

Just as sets of real numbers are pictured on a number line, sets of ordered pairs are pictured on a plane in the **rectangular coordinate system** or **Cartesian coordinate system,** named after the French mathematician René Descartes (1596–1650). The Cartesian coordinate system consists of two number lines drawn perpendicular to one another, intersecting at zero on each number line as in Fig. 3.1. The point at which they intersect is called the **origin.** On the horizontal number line, the **x-axis,** the positive numbers are to the right of the origin, and on the vertical number line, the **y-axis,** the positive numbers are above the origin. The two number lines divide the plane into four regions called **quadrants,** numbered as in Fig. 3.1. The quadrants do not include any points on the axes.

We call a plane with a rectangular coordinate system the **coordinate plane** or the **xy-plane.** Just as every real number corresponds to a point on the number line, every ordered pair of real numbers (a, b) corresponds to a point P in the xy-plane. For this reason, ordered pairs of numbers are often called **points.** The numbers a and b are called the **coordinates** of point P. Locating the point P that corresponds to (a, b) in the xy-plane is referred to as **plotting** or **graphing** the point, and P is called the **graph** of (a, b).

Example 1 Plotting points

Plot the points $(3, 5)$, $(4, -5)$, $(-3, 4)$, $(-2, -5)$, and $(0, 2)$ in the xy-plane.

Solution

The point $(3, 5)$ is located three units to the right of the origin and five units above the x-axis as shown in Fig. 3.2. The point $(4, -5)$ is located four units to the right of the origin and five units below the x-axis. The point $(-3, 4)$ is located three units to the left of the origin and four units above the x-axis. The point $(-2, -5)$ is located two units to the left of the origin and five units below the x-axis. The point $(0, 2)$ is on the y-axis because its first coordinate is zero. ◆

Note that for points in quadrant I, both coordinates are positive. In quadrant II the first coordinate is negative and the second is positive, while in quadrant III, both coordinates are negative. In quadrant IV the first coordinate is positive and the second is negative.

Graphing a Line

If a medium pizza costs $5 plus $2 per topping, then the equation $y = 2x + 5$ expresses the cost in terms of the number of toppings. If $x = 3$ and $y = 11$, then the equation becomes $11 = 2 \cdot 3 + 5$, a true statement. The ordered pair $(3, 11)$ **satisfies** the equation. If $x = 4$, then $y = 2 \cdot 4 + 5 = 13$. So $(4, 13)$ satisfies the equation. The following table shows five ordered pairs that satisfy $y = 2x + 5$.

x	0	1	2	3	4
$y = 2x + 5$	5	7	9	11	13

```
Y₁■2X+5
Y₂=
Y₃=
Y₄=
Y₅=
Y₆=
Y₇=
Y₈=
```

To graph $y = 2x + 5$, define the function on the calculator with the Y= key.

```
WINDOW FORMAT
 Xmin=-5
 Xmax=5
 Xscl=1
 Ymin=-6
 Ymax=14
 Yscl=1
```

Set the viewing window for $-5 \leq x \leq 5$ and $-6 \leq y \leq 14$ to get the desired view.

Press GRAPH. Because of the resolution, the line is not very "straight." See Appendix A for more examples.

The locations of these ordered pairs in a rectangular coordinate system are shown in Fig. 3.3.

Figure 3.3 shows the graph of $\{(0, 5), (1, 7), (2, 9), (3, 11), (4, 13)\}$. If we think of $y = 2x + 5$ as an equation without its application to the cost of a pizza, then there are infinitely many ordered pairs that satisfy $y = 2x + 5$, all lying along a line through the points of Fig. 3.3. For example, $(0.5, 6)$, $(-1, 3)$, and $(-2, 1)$ satisfy $y = 2x + 5$ and lie on the same line as the other points. The set of all ordered pairs that satisfy $y = 2x + 5$ is the **solution set** of the equation. The solution set is expressed in set notation as $\{(x, y) \mid y = 2x + 5\}$, and its graph is shown in Fig. 3.4. The graph of the solution set to an equation is called the **graph of the equation.**

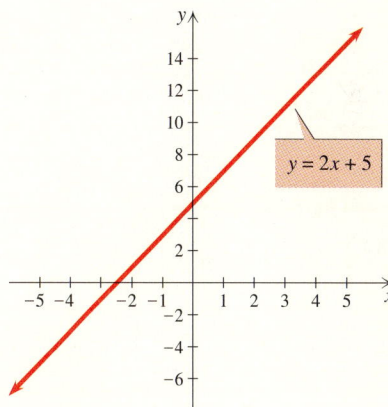

Figure 3.3

Figure 3.4

$y = 2x + 5$

In this chapter we will learn that graphs of equations come in many different shapes. It will save a great deal of time and effort in graphing if we can identify the type of graph an equation has before plotting any points. For example, the graph of any linear equation in two variables is a straight line, and that is why it is called a *linear equation.*

Definition: Linear Equation in Two Variables

A **linear equation** in two variables x and y is an equation of the form

$$Ax + By = C,$$

where A, B, and C are real numbers and A and B are not both zero.

The form $Ax + By = C$ is the **standard form** for a linear equation. So the graphs of the equations

$$x + y = 5, \qquad y = \frac{1}{2}x - 9, \qquad x = 4, \qquad \text{and} \qquad y = 5$$

are straight lines, because they are all variations of the standard form.

There is only one line containing any two distinct points in the xy-plane. So to graph a linear equation, we need locate only two points that satisfy the equation and draw a line through them. However, it is good practice to find three points as a check, because it is unlikely that all three points will be on the same line if an error has been made.

Example 2 Graphing a linear equation

Sketch the graph of each linear equation.

a) $x - 2y = 4$ **b)** $y = 40 - x$

Solution

a) It is easier to find ordered pairs that satisfy the equation if the equation is solved for y.

$$x - 2y = 4$$
$$-2y = -x + 4$$
$$y = \frac{1}{2}x - 2$$

Arbitrarily select $x = 0, 2,$ and 4 and find the corresponding y-values shown in the table.

x	0	2	4
$y = \dfrac{1}{2}x - 2$	-2	-1	0

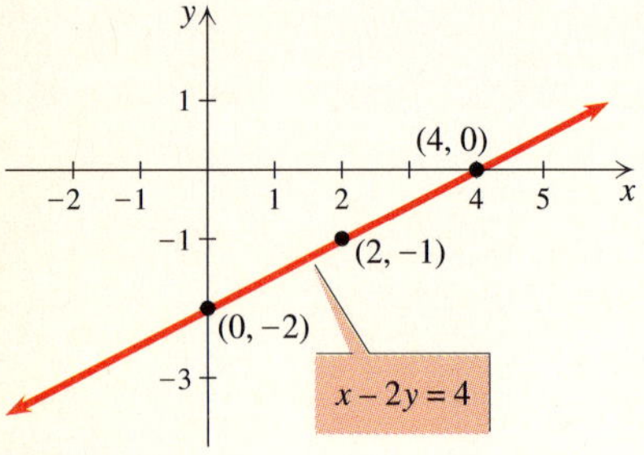

Plot $(0, -2)$, $(2, -1)$, and $(4, 0)$ and draw a line through them as in Fig. 3.5.

b) Arbitrarily select $x = 0, 10,$ and 20 and make a table of ordered pairs.

x	0	10	20
$y = 40 - x$	40	30	20

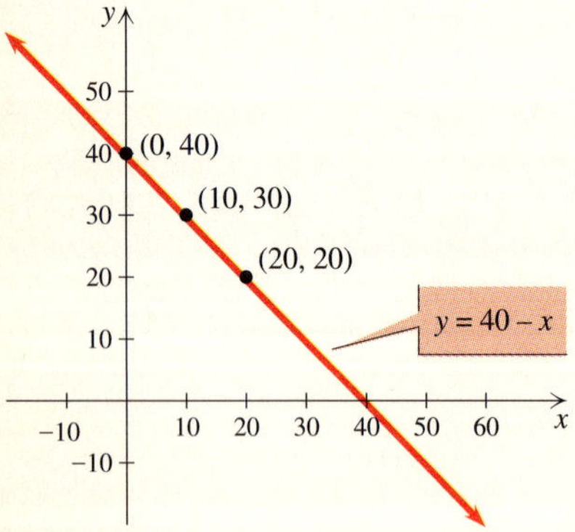

Label the axes to accommodate these ordered pairs. Plot $(0, 40)$, $(10, 30)$, and $(20, 20)$ and draw a line through them as shown in Fig. 3.6. ◆

Figure 3.5

Figure 3.6

Graphing Lines Using the Intercepts

In Fig. 3.5, the line crosses the x-axis at $(4, 0)$ and the y-axis at $(0, -2)$. A point at which a graph intersects the x-axis is an **x-intercept,** and a point at which a graph intersects the y-axis is a **y-intercept.** Since these points have 0 for one of the coordinates, they are usually easier to find than other points on the line. We often graph a linear equation by finding the intercepts.

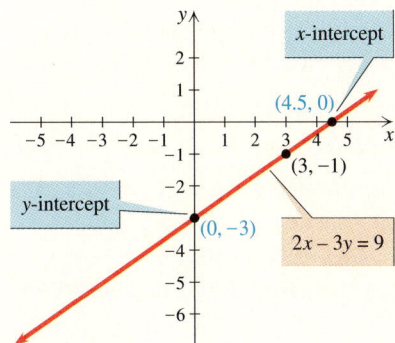

Example 3 Using the intercepts to graph a line

Graph the equation $2x - 3y = 9$ by finding the intercepts.

Solution

Since the y-coordinate of the x-intercept is 0, we replace y by 0 in the equation:

$$2x - 3(0) = 9$$
$$x = 4.5$$

Since the x-coordinate of the y-intercept is 0, we replace x by 0 in the equation:

$$2(0) - 3y = 9$$
$$y = -3$$

The x-intercept is $(4.5, 0)$, and the y-intercept is $(0, -3)$. Locate the intercepts and draw the line as shown in Fig. 3.7. To check, locate a point such as $(3, -1)$, which also satisfies the equation, and see if the line goes through it. ◆

Figure 3.7

Example 4 Graphing horizontal and vertical lines

Sketch the graph of each equation in the rectangular coordinate system.

a) $y = 3$

b) $x = 4$

Solution

a) The equation $y = 3$ is equivalent to $0 \cdot x + y = 3$. Because x is multiplied by 0, we can choose any value for x as long as we choose 3 for y. So ordered pairs such as $(-3, 3)$, $(-2, 3)$, and $(4, 3)$ satisfy the equation $y = 3$. The graph of $y = 3$ is the horizontal line shown in Fig. 3.8.

b) The equation $x = 4$ is equivalent to $x + 0 \cdot y = 4$. Because y is multiplied by 0, we can choose any value for y as long as we choose 4 for x. So ordered pairs such as $(4, -3)$, $(4, 2)$, and $(4, 5)$ satisfy the equation $x = 4$. The graph of $x = 4$ is the vertical line shown in Fig. 3.9. ◆

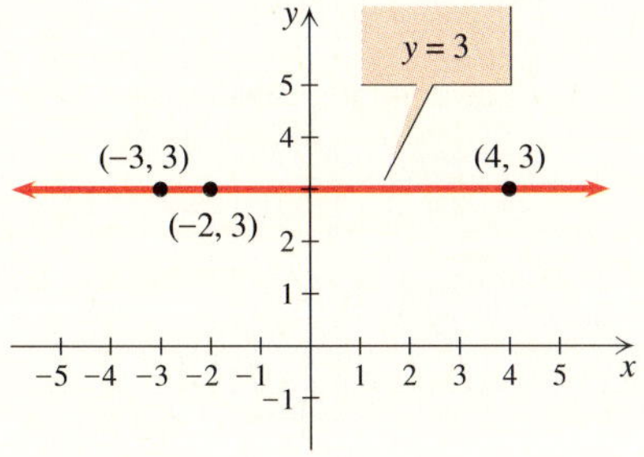

Figure 3.8

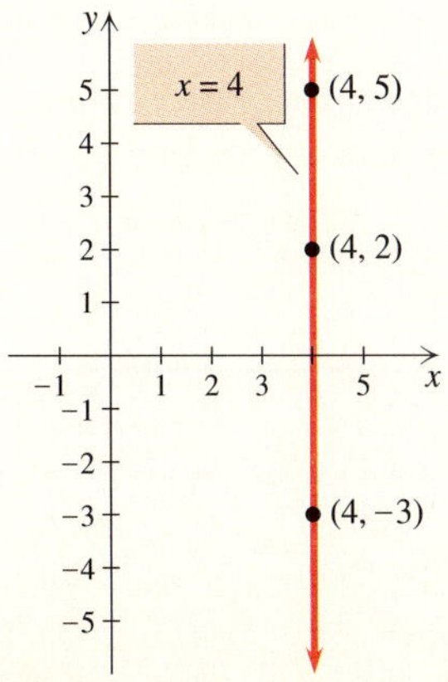

Figure 3.9

Distance Between Two Points

We have been studying sets of points that satisfy linear equations. Their graphs all have the same geometric shape, the straight line. We now turn our attention to some other geometric figures.

Consider the points $A(-1, 5)$ and $B(3, 2)$ as shown in Fig. 3.10. We use the notation $\overline{AB}$ for the line segment with endpoints A and B and AB for the length of that line segment. $\overline{AB}$ is the hypotenuse of the right triangle with a right angle at $C(3, 5)$ shown in Fig. 3.10. If A and C are thought of as lying on a horizontal number line, then we can use the formula for the distance between two points on a number line to get $AC = |-1 - 3| = 4$. Likewise $CB = |5 - 2| = 3$. The Pythagorean theorem says that a triangle is a right triangle if and only if the sum of the squares of the legs is equal to the hypotenuse squared. So by the Pythagorean theorem

$$AB = \sqrt{3^2 + 4^2} = \sqrt{25} = 5.$$

The Pythagorean theorem can also be used to find a general formula for the distance between two arbitrary points $A(x_1, y_1)$ and $B(x_2, y_2)$ shown in Fig. 3.11. BC depends only on the y-coordinates of points B and C, and $BC = |y_2 - y_1|$. Similarly, $AC = |x_2 - x_1|$. Using the Pythagorean theorem with d being the distance between A and B, we have

$$d^2 = |x_2 - x_1|^2 + |y_2 - y_1|^2.$$

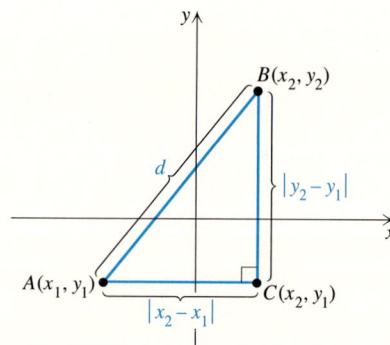

Figure 3.10   **Figure 3.11**

The absolute value symbols can be replaced by parentheses, because $|a - b|^2 = (a - b)^2$ for any values of a and b. Since the distance d must be a nonnegative number, we have the following result.

The Distance Formula

The distance d between the points (x_1, y_1) and (x_2, y_2) is given by the formula

$$d = \sqrt{(x_2 - x_1)^2 + (y_2 - y_1)^2}.$$

Example 5 Finding the distance between two points

Find the distance between $(5, -3)$ and $(-1, -6)$.

Solution

Let $(x_1, y_1) = (5, -3)$ and $(x_2, y_2) = (-1, -6)$. These points are shown on the graph in Fig. 3.12. Substitute these values into the distance formula:

$$d = \sqrt{(-1 - 5)^2 + (-6 - (-3))^2}$$
$$= \sqrt{(-6)^2 + (-3)^2}$$
$$= \sqrt{36 + 9} = \sqrt{45} = 3\sqrt{5}$$

The exact distance between the points is $3\sqrt{5}$.

Note that the distance between two points is the same regardless of which point is chosen as (x_1, y_1) or (x_2, y_2).

The distance formula can be used to establish properties of geometric figures drawn in the coordinate plane. The study of geometric figures in the coordinate plane is called **coordinate geometry** or **analytical geometry.** The main purpose of the Cartesian coordinate system is to help us visualize abstract concepts. In every problem involving coordinate geometry, you should first plot the points, draw the picture, and get a visual idea of what is going on.

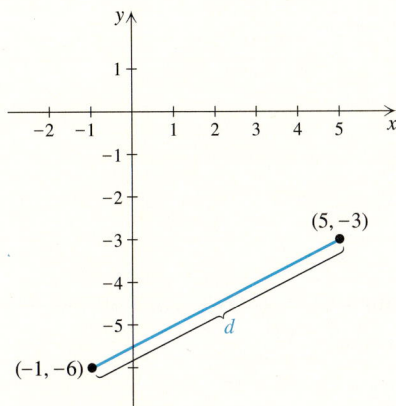

Figure 3.12

Example 6 Deciding whether a triangle is a right triangle

Determine whether the triangle with vertices $(-3, -5)$, $(3, -3)$, and $(0, 6)$ is a right triangle.

Solution

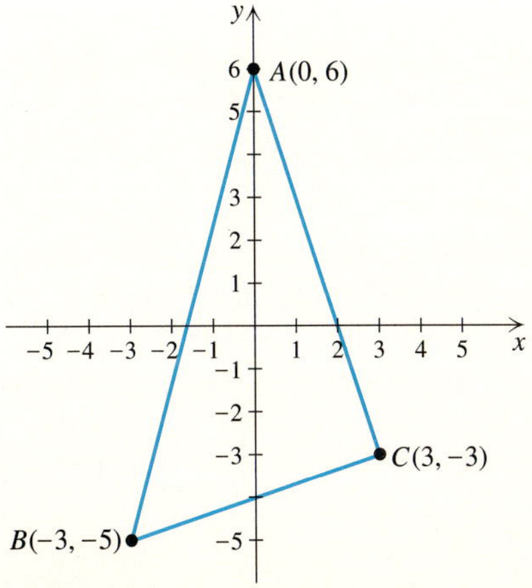

The Pythagorean theorem can be used to determine whether a triangle is a right triangle. Plot the three points and find the length of each side of the triangle of Fig. 3.13:

$$AC = \sqrt{(3 - 0)^2 + (-3 - 6)^2} = \sqrt{90}$$
$$BC = \sqrt{(3 - (-3))^2 + (-3 - (-5))^2} = \sqrt{40}$$
$$AB = \sqrt{(-3 - 0)^2 + (-5 - 6)^2} = \sqrt{130}$$

Since $\left(\sqrt{90}\right)^2 + \left(\sqrt{40}\right)^2 = \left(\sqrt{130}\right)^2$, we have $(AC)^2 + (BC)^2 = (AB)^2$. So by the Pythagorean theorem, the triangle with the given vertices is a right triangle and $\overline{AB}$ is the hypotenuse.

Figure 3.13

The Midpoint Formula

With geometric figures we are often interested in knowing whether one line segment bisects another. This question may be easy to decide if we know the location of the midpoint of a line segment. For the line segment with endpoints -3 and 9 on the number line shown in Fig. 3.14, the midpoint is 3. We find the midpoint just as we would average two test scores, $(-3 + 9)/2 = 3$.

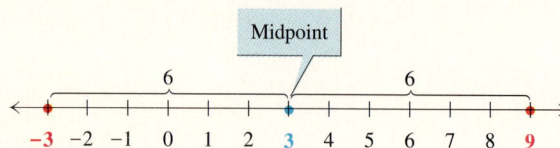

Midpoint

6   6

-3 -2 -1 0 1 2 **3** 4 5 6 7 8 **9**

Figure 3.14

In general, $(a + b)/2$ is the midpoint of the line segment with endpoints a and b shown in Fig. 3.15. (See Exercises 81 and 82.)

Midpoint

a $\dfrac{a + b}{2}$ b

Figure 3.15

The formula for the midpoint on the number line can be extended to a similar formula for the midpoint of a line segment in the xy-plane.

Theorem: The Midpoint Formula

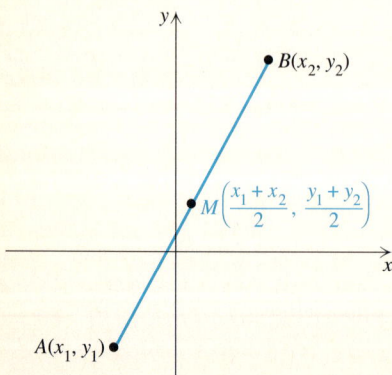

The midpoint of the line segment with endpoints (x_1, y_1) and (x_2, y_2) is

$$\left(\frac{x_1 + x_2}{2}, \frac{y_1 + y_2}{2} \right).$$

$B(x_2, y_2)$

$M\left(\dfrac{x_1 + x_2}{2}, \dfrac{y_1 + y_2}{2} \right)$

$A(x_1, y_1)$

Figure 3.16

Proof   To prove that M is the midpoint of $\overline{AB}$ as shown in Fig. 3.16, we will show that $AM = MB$ and that A, M, and B are colinear (on the same line). To prove that they are colinear, it is sufficient to show that $AM + MB = AB$. (If $AM + MB > AB$, then they form a triangle.) First we use the distance formula to find AM and MB:

$$AM = \sqrt{\left(\frac{x_1 + x_2}{2} - x_1 \right)^2 + \left(\frac{y_1 + y_2}{2} - y_1 \right)^2} = \sqrt{\left(\frac{x_2 - x_1}{2} \right)^2 + \left(\frac{y_2 - y_1}{2} \right)^2}$$

$$MB = \sqrt{\left(x_2 - \frac{x_1 + x_2}{2} \right)^2 + \left(y_2 - \frac{y_1 + y_2}{2} \right)^2} = \sqrt{\left(\frac{x_2 - x_1}{2} \right)^2 + \left(\frac{y_2 - y_1}{2} \right)^2}$$

Since $AM = MB$, we can conclude that M is equidistant from A and B. The expressions for AM and MB can be further simplified by factoring 2^2 out of each denominator:

$$AM = MB = \frac{1}{2}\sqrt{(x_2 - x_1)^2 + (y_2 - y_1)^2}$$

Since $AB = \sqrt{(x_2 - x_1)^2 + (y_2 - y_1)^2}$, we have $AM + MB = AB$. So $A, M,$ and B are colinear. ◆

Example 7   Using the midpoint formula

Prove that the diagonals of the parallelogram with vertices $(0, 0)$, $(1, 3)$, $(5, 0)$, and $(6, 3)$ bisect each other.

Solution

Figure 3.17

The parallelogram is shown in Fig. 3.17. The midpoint of the diagonal from $(0, 0)$ to $(6, 3)$ is

$$\left(\frac{0 + 6}{2}, \frac{0 + 3}{2}\right),$$

or $(3, 1.5)$. The midpoint of the diagonal from $(1, 3)$ to $(5, 0)$ is

$$\left(\frac{1 + 5}{2}, \frac{3 + 0}{2}\right),$$

or $(3, 1.5)$. Since the diagonals have the same midpoint, they bisect each other. ◆

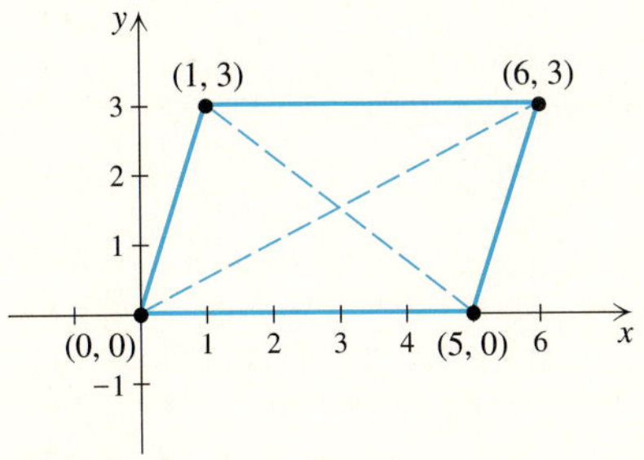

Graphing Calculators

Scientific calculators have proven to be a great tool for computing as well as helping in the process of understanding mathematics. For some time, computers with graphing utilities have been used as an aid in studying graphs and the relationships between variables, but access to a computer is sometimes a problem. Now that inexpensive graphing calculators are available, we have another tool that can help us to understand mathematics. In what follows we will refer to a graphing calculator, but the discussion applies equally to a computer with a graphing utility.

A typical graphing calculator can calculate and plot almost instantly many ordered pairs that satisfy an equation such as $y = 2x + 5$. But remember that the graph of $y = 2x + 5$ is an infinite set of points. The rectangular coordinate system is an infinite plane and the screen on a graphing calculator is just a small window for viewing that infinite plane. Consequently, the minimum and maximum values for x and y must be given to the calculator for it to show ordered pairs in a particular region called the **viewing rectangle.** Of course, if we look in the wrong place, we will not see any points of the graph. Figure 3.18 shows the graph of $y = 2x + 5$ for three different viewing rectangles.

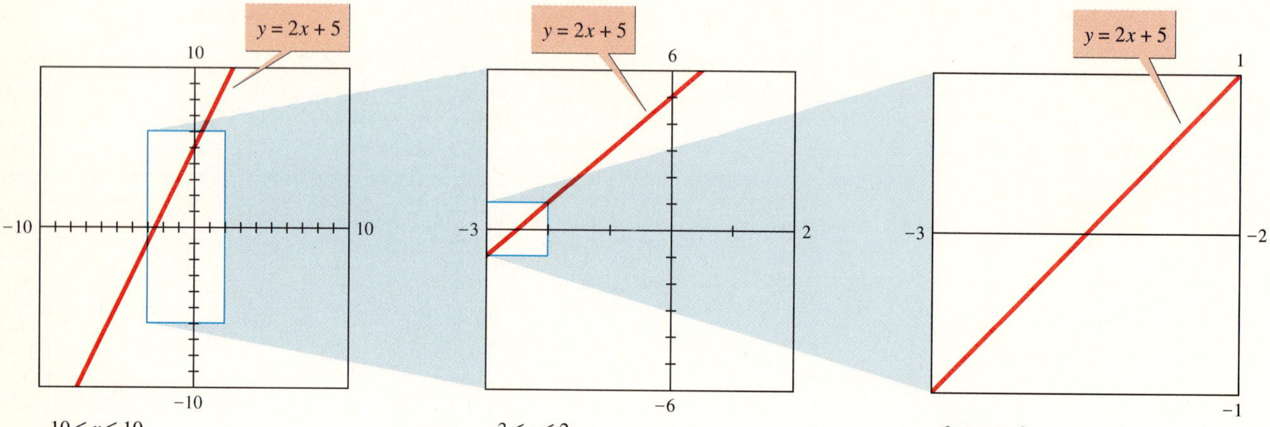

$-10 \leq x \leq 10$
$-10 \leq y \leq 10$

$-3 \leq x \leq 2$
$-6 \leq y \leq 6$

$-3 \leq x \leq -2$
$-1 \leq y \leq 1$

Figure 3.18

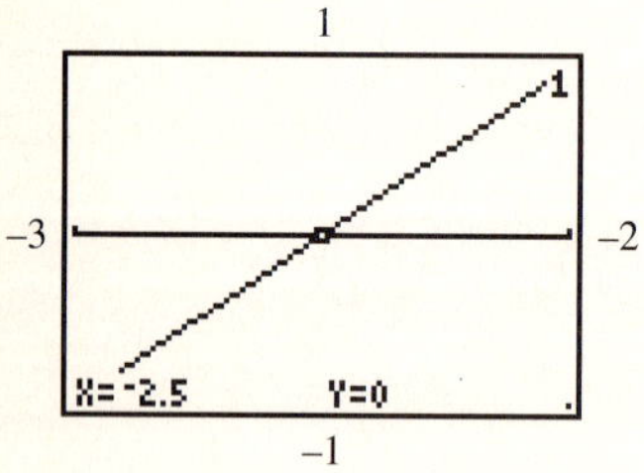

Use TRACE to see the coordinates of each point that is graphed for $y = 2x + 5$.

Starting in this section, **graphing calculator exercises** will be found at the end of many exercise sets. A typical exercise might ask you to graph a line and estimate the coordinates of the x-intercept. The **trace** feature of a graphing calculator gives the coordinates of each point that is graphed. You can use the trace feature to estimate the x-intercept. However, the accuracy of your estimate depends on the size of the viewing rectangle. The smaller the viewing rectangle, the more accurate your estimate will be. Consult your calculator manual for specific instructions on how to use its graphing capabilities.

For Thought

True or false? Explain.

1. The point $(2, -3)$ is in quadrant II.

2. The point $(4, 0)$ is in quadrant I.

3. The distance between (a, b) and (c, d) is $\sqrt{(a - b)^2 + (c - d)^2}$.

4. The equation $3x^2 + y = 5$ is a linear equation.

5. The equation $x = 5$ is a linear equation.

6. $\sqrt{7^2 + 9^2} = 7 + 9$

7. The origin lies midway between $(1, 3)$ and $(-1, -3)$.

8. The distance between $(3, -7)$ and $(3, 3)$ is 10.

9. The x-intercept for the graph of $3x - 2y = 7$ is $(7/3, 0)$.

10. For a and b on a number line, $\dfrac{a + b}{3}$ is a point between a and b.

3.1 Exercises

📼 **Tape 5**   💾 **Disk—5.25": 2**   **3.5": 1**   **Macintosh: 1**

In Exercises 1–10, for each point shown in the xy-plane, write the corresponding ordered pair and name the quadrant in which it lies or the axis on which it lies.

1. A **2.** B **3.** C **4.** D

5. E **6.** F **7.** G **8.** H

9. I **10.** J

Graph each set of ordered pairs.

11. $\{(-1, 0), (0, 1), (1, 2)\}$ **12.** $\{(-3, 3), (0, 0), (3, 3)\}$

13. $\{(x, y) \mid y = x\}$ **14.** $\{(x, y) \mid y = -x\}$

15. $\{(2, 2), (-2, 2), (-2, -2), (2, -2)\}$

16. $\{(-2, 4), (-1, 1), (0, 0), (1, 1), (2, 4)\}$

17. $\{(x, y) \mid y = -2x + 3\}$ **18.** $\{(x, y) \mid y = -3x + 5\}$

19. $\{(1980, 100), (1981, 200), (1982, 400)\}$

20. $\{(1, 300), (5, 400), (3, 100)\}$

21. $\{(x, y) \mid x + 0 \cdot y = 3\}$ **22.** $\{(x, y) \mid 0 \cdot x + y = -2\}$

Sketch the graph of each linear equation.

23. $y = 3x - 4$ **24.** $y = 5x - 5$

25. $3x - y = 6$ **26.** $5x - 2y = 10$

27. $x + y = 80$ **28.** $2x + y = -100$

29. $x + y = 0.5$ **30.** $x - 0.4y = 0.8$

31. $x = 3y - 9$ **32.** $x = 8 - 2y$

33. $\dfrac{1}{2}x - \dfrac{1}{3}y = 6$ **34.** $\dfrac{2}{3}y - \dfrac{1}{2}x = 4$

Graph each equation by finding the x- and y-intercepts.

35. $5x + 4y = 20$ **36.** $3x - 2y = -6$

37. $y = 3x - 2$ **38.** $y = -2x + 5$

39. $2x + 4y = 0.01$ **40.** $3x - 5y = 1.5$

41. $4y - 3x = 60$ **42.** $2y - 4x = 1$

43. $\dfrac{1}{2}x + \dfrac{1}{3}y = 20$ **44.** $\dfrac{1}{4}x - \dfrac{2}{3}y = 60$

45. $0.03x + 0.06y = 150$ **46.** $0.09x - 0.06y = 54$

Graph each equation in the rectangular coordinate system.

47. $x = 5$ **48.** $y = -2$ **49.** $y = 4$

50. $x = -3$ **51.** $x = -4$ **52.** $y = 5$

53. $y - 1 = 0$ **54.** $5 - x = 4$

For each pair of points find the distance between them and the midpoint of the line segment joining them.

55. $(1, 3), (4, 7)$ **56.** $(-3, -2), (9, 3)$

57. $(-1, -2), (1, 0)$ **58.** $(-1, 0), (1, 2)$

59. $\left(-1, 1\right), \left(-1 + 3\sqrt{3}, 4\right)$

60. $\left(1 + \sqrt{2}, -2\right), \left(1 - \sqrt{2}, 2\right)$

61. $\left(\sqrt{18}, \sqrt{12}\right), \left(\sqrt{8}, \sqrt{27}\right)$ **62.** $\left(\sqrt{50}, \sqrt{20}\right), \left(\sqrt{72}, \sqrt{45}\right)$

63. $(1.2, 4.8), (-3.8, -2.2)$ **64.** $(-2.3, 1.5), (4.7, -7.5)$

65. $(a, 0), (b, 0)$ **66.** $(a, 0), \left(\dfrac{a + b}{2}, 0\right)$

67. $(\pi, 0), (\pi/2, 1)$ **68.** $(0, 0), (\pi/2, 1)$

Solve each problem.

69. Determine whether the points $(-6, 3)$, $(-2, -2)$, and $(3, 1)$ are the vertices of a right triangle.

70. Determine whether the points $(-5, 3)$, $(-1, -2)$, and $(4, 2)$ are the vertices of a right triangle.

71. Determine whether the points $(-2, 2)$, $(-2, -3)$, and $(1, 1)$ are the vertices of an isosceles triangle.

72. Determine whether the points $(-1, -1)$, $(-1, 5)$, and $(4, 2)$ are the vertices of an equilateral triangle.

73. Find the value of k so that the distance between $(3, -2)$ and $(5, k)$ is 6.

74. Find all points on the x-axis that are eight units from $(2, 6)$.

75. Find (a, b) such that $(2, 5)$ is the midpoint of the line segment joining $(-1, 3)$ and (a, b).

76. Find a and b such that $(2, 5)$ is the midpoint of the line segment joining $(a, 3)$ and $(7, b)$.

77. Show that the points $(0, 0)$, $(2, -2)$, $(-2, -3)$, and $(-4, -1)$ are the vertices of a parallelogram.

78. Show that the points $(-4, -3)$, $(-2, 2)$, $(3, 4)$, and $(1, -1)$ are the vertices of a rhombus.

79. Show that the points $(-5, 1)$, $(-2, 4)$, $(2, 0)$, and $(-1, -3)$ are the vertices of a rectangle.

80. Show that the points $(-4, 3)$, $(1, 4)$, $(-3, -2)$, and $(2, -1)$ are the vertices of a square.

81. Prove that if a and b are real numbers such that $a < b$, then $a < (a + b)/2 < b$.

Figure for Exercises 81 and 82

82. Prove that the point $(a + b)/2$ is equidistant from a and b on the number line for any real numbers a and b. The proof of this statement together with the previous exercise proves that $(a + b)/2$ is the midpoint of the segment with endpoints a and b on a number line.

83. Show that the points $A(-4, -5)$, $B(1, 1)$, and $C(6, 7)$ are colinear. (Hint: Points A, B, and C lie on a straight line if $AB + BC = AC$.)

84. Sketch a rectangle with vertices $(0, 0)$, $(a, 0)$, (a, b), and $(0, b)$ for $a > 0$ and $b > 0$. Prove that the diagonals of this rectangle are equal in length.

85. *Robot gap widens* The gap between industrial robot use in Japan and the United States is widening annually (*Forbes*, April 16, 1990). The equations $J = 30n + 100$ and $U = 4n + 20$ can be used to model the growth of industrial robot use in Japan and the United States. The letter n represents the number of years since 1985, while J and U represent the number of robots in Japan and the United States in thousands, respectively. Graph these two straight lines on the same coordinate system for n between 0 and 20, and see the robot gap for yourself.

86. *Movie prices* The average price of a movie ticket in 1988 was \$4.11 and in 1992 it was \$5.05 (*USA Today*, July 6, 1993). Assuming that the average price of a movie ticket is growing in a linear manner, find the midpoint of the line segment joining the points $(1988, 4.11)$ and $(1992, 5.05)$, and interpret the result.

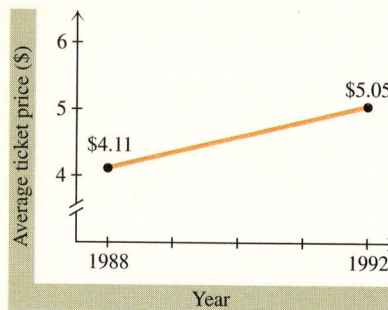

Figure for Exercise 86

For Writing/Discussion

87. *Finding points* Can you find two points such that their coordinates are integers and the distance between them is 10? $\sqrt{10}$? $\sqrt{19}$? Explain.

88. *Plotting points* Plot at least five points in the xy-plane that satisfy the inequality $y > 2x$. Give a verbal description of the solution set to $y > 2x$.

89. *Cooperative learning* Work in a small group to plot the points $(-1, 3)$ and $(4, 1)$ on graph paper. Assuming that these two points are adjacent vertices of a square, find the other two vertices. Now select your own pair of adjacent vertices and "complete the square." Next, start with the points (x_1, y_1) and (x_2, y_2) as adjacent vertices of a square and write expressions for the coordinates of the other two vertices.

90. *Cooperative learning* Repeat the previous exercise assuming that the first two points are opposite vertices of a square.

🖩 **Graphing Calculator Exercises**

Graph each of the following equations on a graphing calculator. The viewing window may be selected by trial and error, or by using our knowledge of x- and y-intercepts. Choose an appropriate viewing window so that you can see both the x- and y-intercepts, then draw the graph on paper and label the x- and y-axes.

1. $y = x - 20$

2. $y = 999x - 100$

3. $500x + y = 3000$

4. $200x - 300y = 1$

A graphing calculator can be used to find an approximate solution to an equation. For example, the x-coordinate of the x-intercept of the graph of $y = 2.2(x + 3.1) - 1.68$ gives the solution to the equation $2.2(x + 3.1) - 1.68 = 0$. With the graphing calculator

it is not necessary to first simplify the equation. Use a graphing calculator to estimate the solution to each equation to two decimal places. Then find the solution algebraically and compare it with your estimate.

5. $1.2x + 3.4 = 0$

6. $3.2x - 4.5 = 0$

7. $1.23x - 687 = 0$

8. $-2.46x + 1500 = 0$

9. $0.03x - 3497 = 0$

10. $0.09x + 2000 = 0$

11. $4.3 - 3.1(2.3x - 9.9) = 0$

12. $3.7(2.3x + 9.4) - 4.37(3.5x - 9.76) = 0$

3.2

Functions

In Section 3.1 we studied ordered pairs of numbers and graphed sets of ordered pairs. For some sets of ordered pairs we used an equation such as $y = 2x + 5$ to determine the second coordinate corresponding to a particular first coordinate. In this section we continue to study sets of ordered pairs, particularly those in which the second coordinate is determined uniquely by the first coordinate.

The Idea of a Function

Any set of ordered pairs is called a **relation.** Consider the relations defined by Tables 3.2 and 3.3. This data is from the 1990 Statistical Abstract of the United States. Table 3.2 shows the number of iron ore mines in operation in the United States for each year beginning in 1980 and the average price per ton at the mine. Table 3.3 shows the energy production of the United States for various years. Each table defines a relation, but there is an important difference between the tables. In Table 3.2 the second coordinate, price per ton, is not determined by the first coordinate, number of mines, because when 20 mines were in operation the price was either $43.61, $44.17, or $42.03. Energy production is determined by the year in Table 3.3, because each year corresponds to a single production level. For example, in 1988 the production was 66 quadrillion Btu of energy. Note that in Table 3.3 no first coordinate is paired with two different second coordinates. In this case we say that the second coordinate *is a function of* the first coordinate.

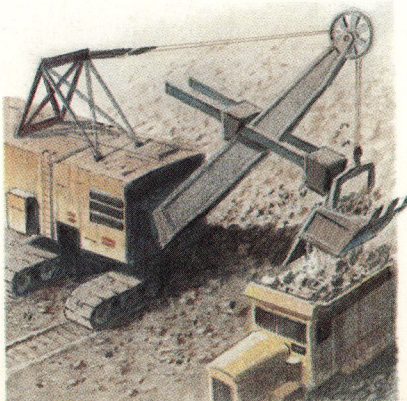

Number of mines	Price per ton
37	$36.56
32	40.39
27	41.72
20	43.61
20	44.17
20	42.03
19	35.63
19	31.88

Table 3.2 U.S. Iron Ore Production

Year	Quadrillion Btu
1980	65
1985	65
1986	64
1987	65
1988	66
1989	66

Table 3.3 U.S. Primary Energy Production

The Definition of Function

To keep the definition of function simple and precise, we state it in terms of ordered pairs.

Definition: Function

> A **function** is a set of ordered pairs in which no two ordered pairs have the same first coordinate and different second coordinates.

The values of the first coordinate of an ordered pair are thought of as values of a first or **independent variable,** while the values of the second coordinate are thought of as values of a second or **dependent variable.** A function is thought of as a rule that indicates how values of the independent variable are paired with values of the dependent variable. We say that the dependent variable **is a function of** the independent variable if and only if the set of ordered pairs determined by these variables is a function. Ordered pairs in the following example are given in set notation and as a table, but as we will see later, ordered pairs may also be given by a verbal description, a formula, or a graph.

Example 1 Using the definition of function

Determine whether each relation is a function.

a) $\{(1, 3), (2, 3), (4, 9)\}$

b) $\{(9, -3), (9, 3), (4, 2), (0, 0)\}$

c)

Quantity	Price each
1–5	$9.40
5–10	$8.75

Solution

a) This set of ordered pairs is a function because no two ordered pairs have the same first coordinate and different second coordinates.

b) This set of ordered pairs is not a function because both $(9, 3)$ and $(9, -3)$ are in the set and they have the same first coordinate and different second coordinates.

c) The quantity 5 corresponds to a price of $9.40 and also to a price of $8.75. Assuming that quantity is the first coordinate, the ordered pairs (5, $9.40) and (5, $8.75) both belong to this relation. If you are purchasing items whose price was determined from this table, you would certainly say that something is wrong, the table has a mistake in it, or the price you should pay is not clear. The price is not a function of the quantity purchased. ◆

Certain keys on a calculator are referred to as the **function keys.** These are the keys marked $\sqrt{x}$, x^2, $x!$, $1/x$, 10^x, e^x, $\sin(x)$, $\cos(x)$, $\tan(x)$, etc. If an appropriate number for x is on the display and one of these keys is pressed, then the corresponding y-coordinate is calculated and displayed. The ordered pairs certainly satisfy the definition of function because the calculator will not produce two different second coordinates corresponding to one first coordinate.

We have seen and used many functions as formulas. For example, the formula $c = \pi d$ defines a set of ordered pairs in which the first coordinate is the diameter of a circle and the second coordinate is the circumference. Since each diameter corresponds to a unique circumference, the circumference is a function of the diameter. If the set of ordered pairs satisfying an equation is a function, then we say that the equation is a function or the equation defines a function. Other well-known formulas such as $C = \frac{5}{9}(F - 32)$, $A = \pi r^2$, and $V = \frac{4}{3}\pi r^3$ are also functions, but, as we will see in the next example, not every equation defines a function. The variables in the next example and all others in this text represent real numbers unless indicated otherwise.

Example 2   Using the definition of function

Determine whether each equation defines y as a function of x.

a) $|y| = x$

b) $y = x^2 - 3x + 2$

c) $x^2 + y^2 = 1$

d) $3x - 4y = 8$

Solution

a) Since $|2| = 2$ and $|-2| = 2$, the ordered pairs $(2, 2)$ and $(2, -2)$ both satisfy the equation $|y| = x$. So this equation does *not* define y as a function of x.

b) Since y is determined by $y = x^2 - 3x + 2$, for every x-coordinate there is only one y-coordinate. So this equation *does* define y as a function of x.

c) Since the ordered pairs $(0, 1)$ and $(0, -1)$ both satisfy $x^2 + y^2 = 1$, this equation does *not* define y as a function of x. Note that $x^2 + y^2 = 1$ is equivalent to $y = \pm\sqrt{1 - x^2}$.

d) The equation $3x - 4y = 8$ is equivalent to $y = \frac{3}{4}x - 2$. Since each x determines only one y, this equation *does* define y as a function of x. ◆

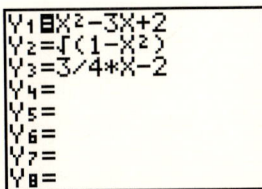

```
Y1▨X2-3X+2
Y2=√(1-X2)
Y3=3/4*X-2
Y4=
Y5=
Y6=
Y7=
Y8=
```

To turn a function "on" or "off," position the cursor on the equal sign and press ENTER. Here, y_1 is on and y_2 and y_3 are off.

In the next example we find a formula for a function relating two measurements in a geometric figure.

Example 3   Finding a formula for a function

Given that a square has diagonal of length d and side of length s, write the area A as a function of the length of the diagonal.

Solution

The area of any square is given by $A = s^2$. The diagonal is the hypotenuse of a right triangle as shown in Fig. 3.19. By the Pythagorean theorem, $d^2 = s^2 + s^2$, $d^2 = 2s^2$, or $s^2 = d^2/2$. Since $A = s^2$ and $s^2 = d^2/2$, we get the formula

$$A = \frac{d^2}{2}$$

expressing the area of the square as a function of the length of the diagonal. ◆

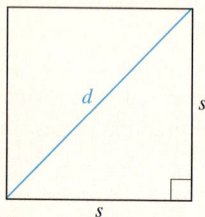

Figure 3.19

Domain and Range

A relation is a set of ordered pairs. The **domain** of a relation is the set of all first coordinates of the ordered pairs. The **range** of a relation is the set of all second coordinates of the ordered pairs. The relation

$$\{(19, 2.4), (27, 3.0), (19, 3.6), (22, 2.4), (36, 3.8)\}$$

shows the ages and grade point averages of five randomly selected students. The domain of this relation is the set of ages, $\{19, 22, 27, 36\}$. The range is the set of grade point averages, $\{2.4, 3.0, 3.6, 3.8\}$. This relation matches elements of the domain (ages) with elements of the range (grade point averages), as shown in Fig. 3.20. Is this relation a function?

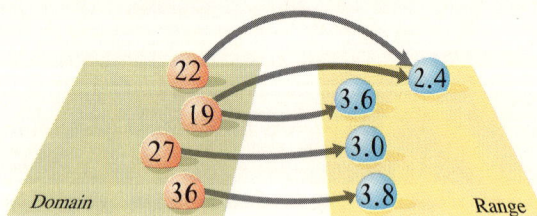

Figure 3.20

For some relations, all of the ordered pairs are listed, but for others, only an equation is given for determining the ordered pairs. When the domain of the relation is not stated, it is understood that the domain consists of only values of the independent variable that can be used in the expression defining the relation. When we use x and y for the variables, we always assume that x is the independent variable and y is the dependent variable.

Example 4   Determining domain and range

State the domain and range of each relation and whether the relation is a function.

a) $\{(-1, 1), (3, 9), (3, -9)\}$

b) $y = \sqrt{2x - 1}$

c) $x = |y|$

Solution

a) The domain is the set of first coordinates $\{-1, 3\}$, and the range is the set of second coordinates $\{1, 9, -9\}$. Note that an element of a set is not listed more than once. Since $(3, 9)$ and $(3, -9)$ are in the relation, the relation is not a function.

b) Since y is determined uniquely from x by the formula $y = \sqrt{2x - 1}$, y is a function of x. We are discussing only real numbers. So $\sqrt{2x - 1}$ is a real number provided that $2x - 1 \geq 0$, or $x \geq 1/2$. So the domain of the function is the interval $[1/2, \infty)$. If $2x - 1 \geq 0$ and $y = \sqrt{2x - 1}$, we have $y \geq 0$. So the range of the function is the interval $[0, \infty)$.

c) The expression $|y|$ is defined for any real number y. So the range is the interval of all real numbers, $(-\infty, \infty)$. Since $|y|$ is nonnegative, the values of x must be nonnegative. So the domain is $[0, \infty)$. Since ordered pairs such as $(2, 2)$ and $(2, -2)$ satisfy $x = |y|$, this equation does not give y as a function of x. ◆

X=.5106383   Y=.14586499

Look at the graph of $y = \sqrt{2x - 1}$ to verify that $x \geq 1/2$ and $y \geq 0$.

In Example 4(b) and (c) we found the domain and range by examining an equation defining a relation. In Section 3.3 we will use the graph of a relation to help us determine the domain and range. The range is usually easier to determine from the graph than from the equation.

The f-Notation

If y is a function of x, we may use the expression $f(x)$, rather than y, for the dependent variable. The notation $f(x)$ is read "the value of f at x" or simply "f of x." So y and $f(x)$ are both symbols for the value of the second coordinate when the first coordinate is x, and we may write $y = f(x)$. For example, we could write $f(x) = \sqrt{2x - 1}$ for the function $y = \sqrt{2x - 1}$ from Example 4(b). The notation $f(x)$ is called **function notation** or **f-notation.**

Example 5   Using f-notation

Find each of the following for the function $f(x) = \sqrt{2x - 1}$.

a) $f(5)$ b) x, if $f(x) = 0$ c) $f(-3)$ d) $f(a + 2)$

Solution

a) The expression $f(5)$ is the second coordinate when the first coordinate is 5. To find the value of $f(5)$ replace x by 5 in the formula $f(x) = \sqrt{2x - 1}$:

$$f(5) = \sqrt{2 \cdot 5 - 1} = \sqrt{9} = 3$$

b) To find a value of x so that the second coordinate is 0, we must solve $\sqrt{2x - 1} = 0$. The solution is $x = 1/2$.

c) The expression $f(-3)$ is undefined, because the domain of $f(x) = \sqrt{2x - 1}$ is the interval $[1/2, \infty)$.

d) The expression $f(a + 2)$ is the second coordinate when the first coordinate is $a + 2$. So replace x by $a + 2$ in the formula $f(x) = \sqrt{2x - 1}$:

$$f(a + 2) = \sqrt{2(a + 2) - 1} = \sqrt{2a + 3} \qquad \blacklozenge$$

In Example 5(d) the binomial $a + 2$ was used as the x-coordinate. In fact, any expression (whose values are in the domain of the function) can be used in place of x. The function notation such as $f(x) = 3x + 1$ provides a rule for finding the second coordinate: Multiply the first coordinate (whatever it is) by 3 and then add 1. The x in this notation is called a **dummy variable** because the letter used is unimportant. We could write $f(t) = 3t + 1$,

$$f(\text{first coordinate}) = 3(\text{first coordinate}) + 1,$$

or even $f(\ ) = 3(\ ) + 1$ to convey the same idea. It is often necessary to even use expressions involving x as the first coordinate. For example, if $f(x) = 3x + 1$, then to find $f(x - 4)$, replace x by $x - 4$:

$$f(x - 4) = 3(x - 4) + 1 = 3x - 11$$

Study the additional examples given in Table 3.4.

First coordinate	Find second coordinate using $f(x) = 3x + 1$	Ordered pair
4	$f(4) = 3(4) + 1 = 13$	$(4, 13)$
a	$f(a) = 3a + 1$	$(a, 3a + 1)$
$a + 2$	$f(a + 2) = 3(a + 2) + 1 = 3a + 7$	$(a + 2, 3a + 7)$
$x - 5$	$f(x - 5) = 3(x - 5) + 1 = 3x - 14$	$(x - 5, 3x - 14)$

Table 3.4

The f-notation is especially convenient for naming functions when more than one function is under discussion. For clarity, we name different functions with different letters. For example, let

$$f = \{(1, 4), (2, 5), (3, 6)\} \qquad \text{and} \qquad g = \{(1, 9), (5, 7)\}.$$

The expression $f(1)$ is the value of the second coordinate when the first coordinate is 1 in the function named f. So $f(1) = 4$, while $g(1) = 9$.

Example 6   Using f-notation

Given that $f(x) = x^2 - 2$ and $g(x) = 2x - 3$, find and simplify each of the following expressions.

a) $f(-3)$ **b)** $g(5)$ **c)** $f(a + 1)$

d) $g(x + h)$ **e)** $f(g(x))$

Solution

a) $f(-3) = (-3)^2 - 2 = 7$

b) $g(5) = 2 \cdot 5 - 3 = 7$

c) $f(a + 1) = (a + 1)^2 - 2$ Replace x by $a + 1$ in $f(x) = x^2 - 2$.
$$= a^2 + 2a + 1 - 2$$
$$= a^2 + 2a - 1$$

d) $g(x + h) = 2(x + h) - 3$ Replace x by $x + h$ in $g(x) = x - 3$.
$$= 2x + 2h - 3$$

e) $f(g(x)) = f(2x - 3)$ Replace $g(x)$ by $2x - 3$, because $g(x) = 2x - 3$.
$$= (2x - 3)^2 - 2$$ Use $2x - 3$ in place of x in $f(x) = x^2 - 2$.
$$= 4x^2 - 12x + 7$$ Simplify. ◆

If a function describes some real application, then a letter that fits the situation is usually used. For example, if watermelons are three dollars apiece, then the cost of x watermelons is given by the function $C(x) = 3x$. The cost of five watermelons is $C(5) = 3 \cdot 5 = \$15$. In trigonometry the abbreviations sin, cos, and tan are used rather than a single letter to name the trigonometric functions. The dependent variables are written as $\sin(x)$, $\cos(x)$, and $\tan(x)$.

The Average Rate of Change of a Function

At the 1991 World Championship meet in Tokyo, Carl Lewis ran the fastest 100-meter race in the history of sports (*Runner's World,* December 1991). At 40 meters, Lewis's time was 4.77 seconds; at 100 meters, as he crossed the finish line, his time was 9.86 seconds. There is a function here pairing times to locations. Two ordered pairs of this function are (4.77, 40) and (9.86, 100). To calculate Lewis's average speed over the time interval from 4.77 seconds to 9.86 seconds, we find

$$\frac{\text{change in location}}{\text{change in time}} = \frac{100 - 40}{9.86 - 4.77} = \frac{60}{5.09} = 11.79 \text{ m/s}.$$

Lewis did not run at a constant 11.79 meters per second, but 11.79 meters per second was the average rate at which his location changed for the time between 4.77 seconds and 9.86 seconds.

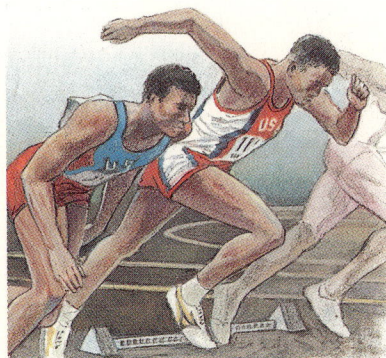

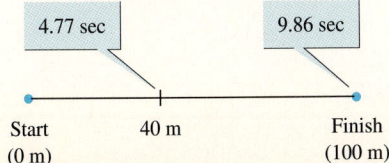

A calculator uses y_1 and y_2 for function names rather than f and g.

The notation used to evaluate a function on a calculator is similar to f-notation. See Appendix A for more examples.

4.77 sec

9.86 sec

Start
(0 m)

40 m

Finish
(100 m)

Definition: Average Rate of Change from x_1 to x_2

If (x_1, y_1) and (x_2, y_2) are two ordered pairs of a function, we define the **average rate of change** of the function as x varies from x_1 to x_2 as

$$\frac{y_2 - y_1}{x_2 - x_1}.$$

Example 7 Average rate of change of a function

A plumber charges $560 for installing 40 feet of sewer pipe and $780 for installing 60 feet of sewer pipe. Find the average rate of change of the cost as length varies from 40 to 60 feet.

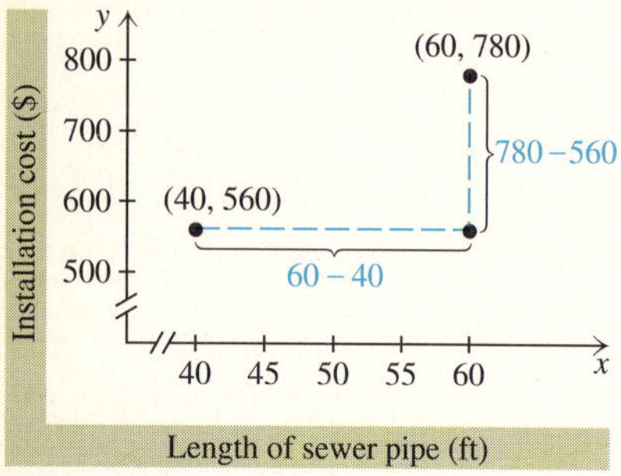

Solution

Figure 3.21 shows the ordered pairs (40, 560) and (60, 780). Note that the change in cost is the vertical distance between the points and the change in length is the horizontal distance. The average rate of change of the cost is the change in cost divided by the change in length.

$$\frac{\text{change in cost}}{\text{change in length}} = \frac{y_2 - y_1}{x_2 - x_1} = \frac{780 - 560}{60 - 40} = \frac{220}{20} = 11$$

The average rate of change of the cost is $11 per foot as the length varies from 40 to 60 feet. Note that the plumber does not charge $11 per foot for installing 40 feet of pipe, but his charge changes by $11 per foot for every foot between 40 and 60.

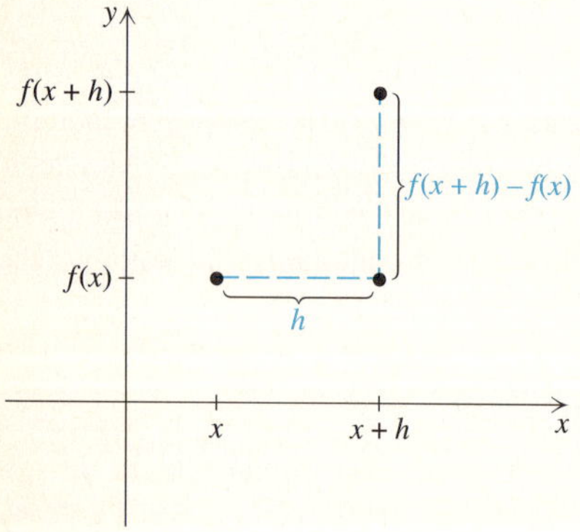

In calculus we are often interested in the average rate of change of a function f from $x_1 = x$ to $x_2 = x + h$, where $h \neq 0$. The function contains the points $(x, f(x))$ and $(x + h, f(x + h))$. See Fig. 3.22. In this case, $x_2 - x_1 = x + h - x = h$ and $y_2 - y_1 = f(x + h) - f(x)$. With this notation the average rate of change is written as

$$\frac{y_2 - y_1}{x_2 - x_1} = \frac{f(x + h) - f(x)}{h}.$$

Figure 3.21

Figure 3.22

Definition: Difference Quotient

The expression

$$\frac{f(x + h) - f(x)}{h}$$

for $h \neq 0$ is called the **difference quotient**.

In the difference quotient, h is usually thought of as a number very close to 0, so that the average rate of change can be found for a very short interval.

Example 8 Finding a difference quotient

Find and simplify the difference quotient for each of the following functions.

a) $j(x) = 3x + 2$ **b)** $f(x) = x^2 - 2x$ **c)** $g(x) = \sqrt{x}$

Solution

a)
$$\frac{j(x + h) - j(x)}{h} = \frac{3(x + h) + 2 - (3x + 2)}{h}$$

$$= \frac{3x + 3h + 2 - 3x - 2}{h}$$

$$= \frac{3h}{h} = 3$$

b)
$$\frac{f(x + h) - f(x)}{h} = \frac{[(x + h)^2 - 2(x + h)] - (x^2 - 2x)}{h}$$

$$= \frac{x^2 + 2xh + h^2 - 2x - 2h - x^2 + 2x}{h}$$

$$= \frac{2xh + h^2 - 2h}{h}$$

$$= 2x + h - 2$$

c)
$$\frac{g(x + h) - g(x)}{h} = \frac{\sqrt{x + h} - \sqrt{x}}{h}$$

$$= \frac{(\sqrt{x + h} - \sqrt{x})(\sqrt{x + h} + \sqrt{x})}{h(\sqrt{x + h} + \sqrt{x})} \qquad \text{Rationalize the numerator.}$$

$$= \frac{x + h - x}{h(\sqrt{x + h} + \sqrt{x})}$$

$$= \frac{1}{\sqrt{x + h} + \sqrt{x}}$$

Note that in Examples 8(b) and 8(c) the average rate of change of the function depends on the values of x and h, while in Example 8(a) the average rate of change of the function is constant. In Example 8(c) the expression does not look much different after rationalizing the numerator than it did before. However, we did remove h as a factor of the denominator, and in calculus it is often necessary to perform this step.

? For Thought

True or false? Explain.

1. Any set of ordered pairs is a function.

2. If $f = \{(1, 1), (2, 4), (3, 9)\}$, then $f(5) = 25$.

3. The domain of $f(x) = 1/x$ is $(-\infty, 0) \cup (0, \infty)$.

4. Each student's exam grade is a function of the student's IQ.

5. If $f(x) = x^2$, then $f(x + h) = x^2 + h$.

6. The domain of $g(x) = |x - 3|$ is $[3, \infty)$.

7. The range of $y = 8 - x^2$ is $(-\infty, 8]$.

8. The equation $x = y^2$ does not define y as a function of x.

9. If $f(t) = \dfrac{t - 2}{t + 2}$, then $f(0) = -1$.

10. The set $\left\{ \left(\frac{3}{8}, 8\right), \left(\frac{4}{7}, 7\right), (0.16, 6), (0.375, 5) \right\}$ is a function.

3.2 Exercises ▭ Tape 5   ▱ Disk—5.25″: 2   3.5″: 1   Macintosh: 1

Determine whether each relation is a function.

1. $\{(-1, -1), (2, 2), (3, 3)\}$

2. $\{(0.5, 7), (0, 7), (1, 7), (9, 7)\}$

3. $\{(25, 5), (25, -5), (0, 0)\}$

4. $\{(1, \pi), (30, \pi/2), (60, \pi/4)\}$

5.
x	y
3	6
4	9
3	12

6.
x	y
1	6.98
5	5.98
9	6.98

7. $\{(x, y) | y = (x + 2)^2\}$

8. $\{(x, y) | x = y^2 + 1\}$

9. $\{(x, y) | x = |2y|\}$

10. $\{(x, y) | 2x + y = 5\}$

11. $\{(x, y) | x = 6y\}$

12. $\{(x, y) | y^2 = x^4\}$

13. $\{(x, y) | x = y^2 \text{ and } y \geq 0\}$

14. $\{(x, y) | x + 0 \cdot y = 5\}$

Determine whether each equation defines y as a function of x.

15. $y = ax + b$

16. $y = ax^2 + bx + c$

17. $x = 3y - 9$

18. $x = y^2 - 6y + 9$

19. $x^2 = y^2$

20. $y^2 - x^2 = 9$

21. $y^3 = x$

22. $x = \sqrt{y}$

23. $y + 2 = |x|$

24. $\dfrac{x^2}{4} + \dfrac{y^2}{9} = 1$

Determine the domain and range of each relation.

25. $\left\{ (-3, 1), \left(\pi, \sqrt{2}\right), (-3, 6), (5, 6) \right\}$

26. $\{(1, 1/2), (2, 1/4), (3, 1/8), (4, 1/16)\}$

27. $y = |x| + 5$ 28. $y = x^2 + 8$ 29. $x = -y^2$

30. $x + 3 = |y|$ 31. $y = 5 + \sqrt{x}$ 32. $y = x^{1/3}$

33. $\{(x, y) | 2x + 3y = 6\}$ 34. $\{(x, y) | x = 5\}$

35. $\{(x, y) | 0 < x < 3 \text{ and } y = x + 2\}$

36. $\{(x, y) | y = x^2 \text{ and } |x| < 3\}$

Let $f(x) = 3x^2 - x$, $g(x) = 4x - 2$, and $k(x) = |x + 3|$. Find the following.

37. $f(2)$ 38. $f(-4)$ 39. $g(-1)$ 40. $g(-2)$

41. $k(5)$ 42. $k(-4)$ 43. $f(0.56)$ 44. $g(-0.5)$

45. $f(-1) + g(-1)$ 46. $f(3) \cdot k(3)$

47. $g(-5) - k(-5)$ 48. $\dfrac{f(1)}{g(1)}$

49. $f(a)$ **50.** $g(b)$ **51.** $f(a + 3)$

52. $g(b + 2)$ **53.** $f(x + h)$ **54.** $g(x + h)$

55. $f(x + h) - f(x)$ **56.** $g(x + h) - g(x)$

57. $f(x) + g(x)$ **58.** $f(x) - g(x)$ **59.** $f(x) \cdot g(x)$

60. $g\left(\dfrac{x + 2}{4}\right)$ **61.** x, if $f(x) = 0$ **62.** x, if $g(x) = 3$

63. a, if $k(a) = 4$ **64.** t, if $f(t) = 10$ **65.** $f(g(-1))$

66. $g(f(2))$ **67.** $f(k(-4))$ **68.** $k(g(-2))$

Find the difference quotient for each function and simplify it.

69. $f(x) = 3x + 5$ **70.** $f(x) = -2x + 3$

71. $g(x) = 3x^2 + 1$ **72.** $g(x) = -2x^2 - 4$

73. $y = -x^2 + x - 2$ **74.** $y = x^2 - x + 3$

75. $f(x) = \sqrt{x + 2}$ **76.** $f(x) = \sqrt{\dfrac{x}{2}}$

77. $g(x) = \dfrac{1}{x}$ **78.** $g(x) = 3 + \dfrac{2}{x - 1}$

79. $f(x) = ax + b$ **80.** $f(x) = ax^2 + bx + c$

Consider a square with side of length s, diagonal of length d, perimeter P, and area A.

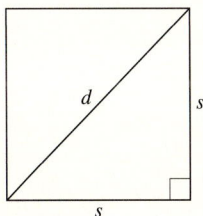

Figure for Exercises 81–88

81. Write A as a function of s.   **82.** Write s as a function of A.

83. Write s as a function of d.   **84.** Write d as a function of s.

85. Write P as a function of s.   **86.** Write s as a function of P.

87. Write A as a function of P.

88. Write d as a function of A.

89. *Aerobics* A woman in an aerobic dance class burns 353 calories per hour. Express the number of calories burned, C, as a function of the number of hours danced, n.

90. *Paper consumption* The average American uses 580 lb of paper annually. Express the total annual paper consumption of the United States, P, as a function of the size of the population, n (The Earth Works Group, 1989).

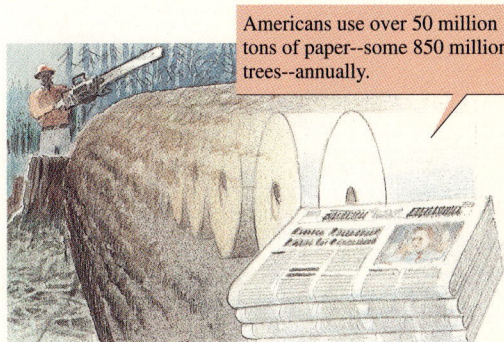

Americans use over 50 million tons of paper--some 850 million trees--annually.

Figure for Exercise 90

91. *Cost of window cleaning* A window cleaner charges $50 per visit plus $35 per hour. Express the total charge as a function of the number of hours worked, n.

92. *Cost of a sundae* At Doubledipski's ice cream parlor, the cost of a sundae is $2.50 plus $0.50 for each topping. Express the total cost of a sundae, C, as a function of the number of toppings, n.

93. *Cost of business cards* If Speedy Printers will print 100 business cards for $20 and 500 business cards for $65, then what is the average rate of change of the cost as the number of cards varies from 100 to 500? Interpret your result.

94. *Depreciation of a Mustang* If a new Mustang LX is valued at $16,000 and five years later it is valued at $4000, then what is the average rate of change of its value during those five years?

Figure for Exercise 94

95. *Cost of gravel* Wilson's Sand and Gravel will deliver 12 yd³ of gravel for $240, 30 yd³ for $528, and 60 yd³ for $948. What is the average rate of change of the cost as the number of cubic yards varies from 12 to 30? What is the average rate of change as cubic yards varies from 30 to 60?

Figure for Exercise 95

96. *Bungee jumping* Billy Joe McCallister jumped off the Tallahatchie Bridge, 70 ft above the water, with a bungee cord tied to his legs. If he was 6 ft above the water 2 sec after jumping, then what was the average rate of change of his altitude as the time varied from 0 to 2 sec?

97. *Deforestation* In 1980 the world's tropical moist forest covered approximately 1080 million hectares (1 hectare = 10,000 m²). In 1992 the world's tropical moist forest covered approximately 1006 million hectares. What was the average rate of change of the area of tropical moist forest over those 12 years?

98. *Elimination of tropical moist forest* If the deforestation described in Exercise 97 continues at the same amount per year, then in which year will the tropical moist forest be totally eliminated?

The notations $C(x)$, $R(x)$, and $P(x)$ are used in business for functions that give the cost, revenue, and profit for the production of x units. The **marginal cost** function $MC(x)$ is the function that gives the change in cost if a production level of x is raised to $x + 1$. $MC(x)$ is equal to the difference quotient for $C(x)$ with $h = 1$. The definitions of **marginal revenue** $MR(x)$ and **marginal profit** $MP(x)$ are similar.

99. *Marginal cost* A dairy's weekly cost in dollars for producing x hundred gallons of milk can be modeled by the function $C(x) = 0.03x^2 + 40x + 6000$. Find $MC(x)$ and simplify it. What is the marginal cost when production is at 10,000 gallons per week?

Figure for Exercise 99

100. *Marginal revenue* A manufacturer's weekly revenue in dollars for producing x items can be modeled by the function $R(x) = 200x - x^2$. Find $MR(x)$ and simplify it. What is the marginal revenue when production is at 30 units per week?

Figure for Exercises 97 and 98

For Writing/Discussion

101. *Find a function* Give an example of a function and an example of a relation that is not a function, from situations that you have encountered outside of this textbook.

102. *Cooperative learning* Work in a small group to consider the equation $y^n = x^m$ for any integers n and m. For which integers n and m does the equation define y as a function of x?

3.3

Graphs of Relations

and Functions

When we graph the set of ordered pairs that satisfy an equation, we are combining algebra with geometry. We saw in Section 3.1 that any equation of the form $Ax + By = C$ has a graph that is a straight line. In this section we will see that graphs of equations can take on many different geometric shapes and see how to determine whether a relation is a function by its graph.

The Circle

A **circle** is the set of all points in a plane that lie a fixed distance from a given point in the plane. The fixed distance is called the **radius,** and the given point is the **center.** The distance formula of Section 3.1 can be used to write an equation for the circle shown in Fig. 3.23 with center (h, k) and radius r for $r > 0$. A point (x, y) is on the circle if and only if it satisfies the equation

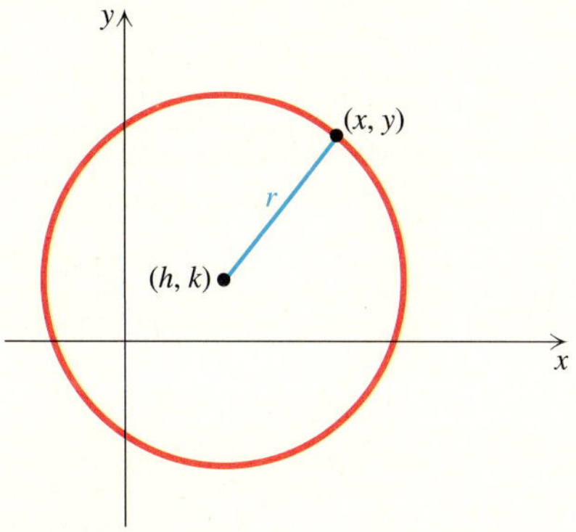

$$\sqrt{(x - h)^2 + (y - k)^2} = r.$$

Squaring both sides of this equation yields $(x - h)^2 + (y - k)^2 = r^2$. We have proved the following theorem.

Figure 3.23

Theorem: Equation for a Circle

The equation for a circle with center (h, k) and radius r for $r > 0$ is

$$(x - h)^2 + (y - k)^2 = r^2.$$

The form $(x - h)^2 + (y - k)^2 = r^2$ is called the **standard form** for the equation of a circle. If we have an equation in this form (or one that can be rewritten in this form), we can conclude that its graph is a circle centered at (h, k) with radius r. On the other hand, if we know the center and radius of a circle, we can write an equation for it in the standard form. Note that if a circle is centered at the origin, then its standard equation is $x^2 + y^2 = r^2$.

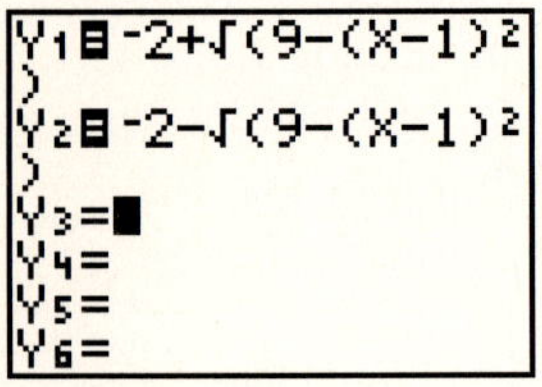

Example 1 Graphing a circle

Sketch the graph of the equation $(x - 1)^2 + (y + 2)^2 = 9$ and state the domain and range of the relation.

Solution

The equation is in the standard form of the equation of a circle. Its center is $(1, -2)$ and its radius is 3. The graph is shown in Fig. 3.24. To find the domain, observe from the graph that all x-coordinates on the graph are within three units of the x-coordinate of the center. So the domain is the closed interval $[-2, 4]$. Likewise, all y-coordinates are within three units of the y-coordinate of the center. So the range is the closed interval $[-5, 1]$.

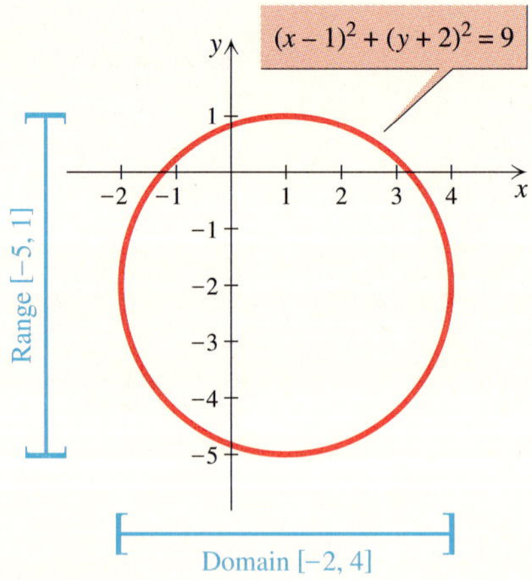

```
Y₁⊟-2+√(9-(X-1)²
>
Y₂⊟-2-√(9-(X-1)²
>
Y₃=■
Y₄=
Y₅=
Y₆=
```

To graph $(x - 1)^2 + (y + 2)^2 = 9$, solve the equation for y and graph two semicircles.

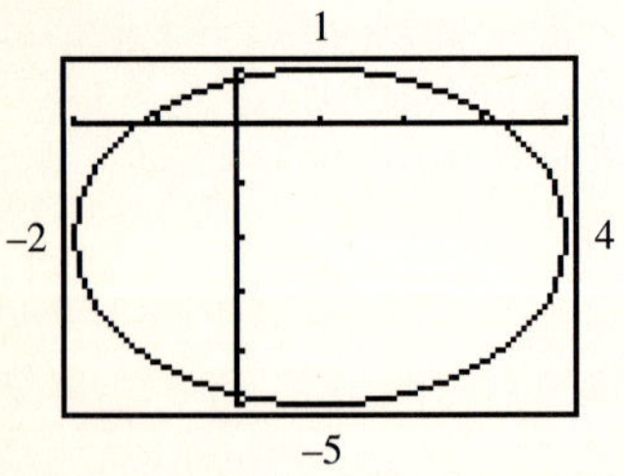

A circle looks more or less round depending on the viewing window.

Figure 3.24

Note that an equation such as $(x - 1)^2 + (y + 2)^2 = -9$ is not satisfied by any pair of real numbers, because the left-hand side is a nonnegative real number, while the right-hand side is negative. The equation $(x - 1)^2 + (y + 2)^2 = 0$ is satisfied only by $(1, -2)$. Since only one point satisfies $(x - 1)^2 + (y + 2)^2 = 0$, its graph is sometimes called a degenerate circle with radius zero. We will study circles again later in this text when we study the conic sections.

In Example 1 we were given an equation and we sketched its graph. In the next example, we are given a description of the circle and we are asked to find its equation.

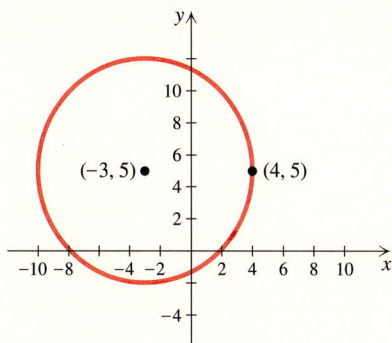

Figure 3.25

Example 2   Writing an equation of a circle

Write the standard equation for the circle with the center $(-3, 5)$ and passing through $(4, 5)$ as shown in Fig. 3.25.

Solution

Since the distance between the center $(-3, 5)$ and $(4, 5)$ is seven units, the radius of the circle is 7. Use $h = -3$, $k = 5$, and $r = 7$ in the standard equation of a circle:

$$(x - (-3))^2 + (y - 5)^2 = 7^2$$

So the equation of the circle is $(x + 3)^2 + (y - 5)^2 = 49$. ◆

If the equation of a circle is not given in the standard form, we can still recognize it by completing the square.

Example 3   Changing an equation of a circle to standard form

Graph the equation $x^2 + 6x + y^2 - 5y = -\frac{1}{4}$.

Solution

Complete the square for both x and y to get the standard form:

$$x^2 + 6x + 9 + y^2 - 5y + \frac{25}{4} = -\frac{1}{4} + 9 + \frac{25}{4} \qquad \left(\frac{1}{2} \cdot 6\right)^2 = 9,$$

$$\left(\frac{1}{2} \cdot 5\right)^2 = \frac{25}{4}$$

$$(x + 3)^2 + \left(y - \frac{5}{2}\right)^2 = 15 \qquad \text{\color{blue}Factor the trinomials on the left side.}$$

The graph is a circle with center $\left(-3, \frac{5}{2}\right)$ and radius $\sqrt{15}$. See Fig. 3.26. ◆

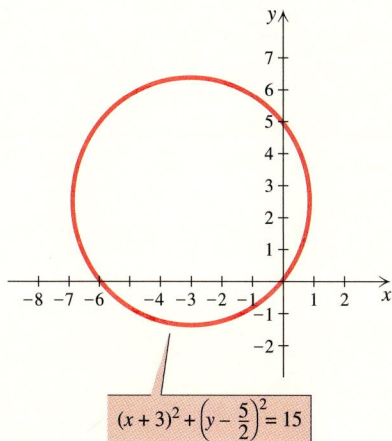

Figure 3.26

$$(x + 3)^2 + \left(y - \frac{5}{2}\right)^2 = 15$$

Graphing Equations

The circle and the line provide nice examples of how algebra and geometry are interrelated. When you see an equation that you recognize as the equation of a circle or a line, sketching a graph is easy to do. Other equations have graphs that are not such familiar shapes. Until we learn to recognize the kinds of graphs that other equations have, we graph other equations by calculating enough ordered pairs to determine the shape of the graph. When you graph equations, try to anticipate what the graph will look like, and after the graph is drawn, pause to reflect on the shape of the graph and the type of equation that produced it. If you have a graphing calculator, use it to help you graph the equations in the following example. Remember that a graphing calculator shows only finitely many points and a graph consists of infinitely many points. After looking at the display of a graphing calculator, you must still decide what the entire graph looks like.

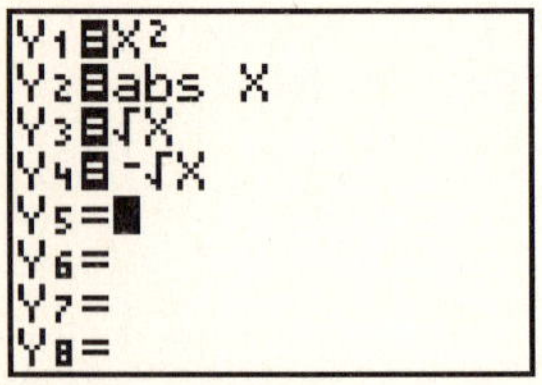

Example 4   Graphing by plotting ordered pairs

Graph each equation and state the domain and range of the relation.

a) $y = x^2$ **b)** $x = y^2$

c) $y = |x|$ **d)** $y = \sqrt{x}$

Solution

a) Make a table of ordered pairs that satisfy $y = x^2$:

x	0	1	-1	2	-2
$y = x^2$	0	1	1	4	4

These ordered pairs indicate a graph in the shape shown in Fig. 3.27. This curve is called a **parabola.** We will study parabolas in greater detail in Chapter 4. The domain is $(-\infty, \infty)$ because any real number can be used for x in $y = x^2$. Since all y-coordinates are nonnegative, the range is $[0, \infty)$.

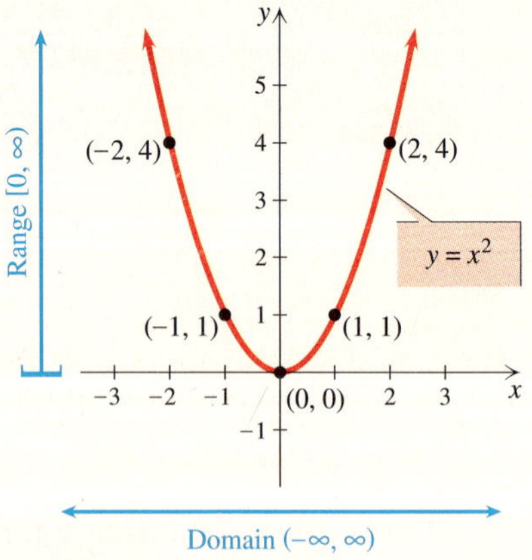

Define four functions on your calculator to see the graphs of Example 4.

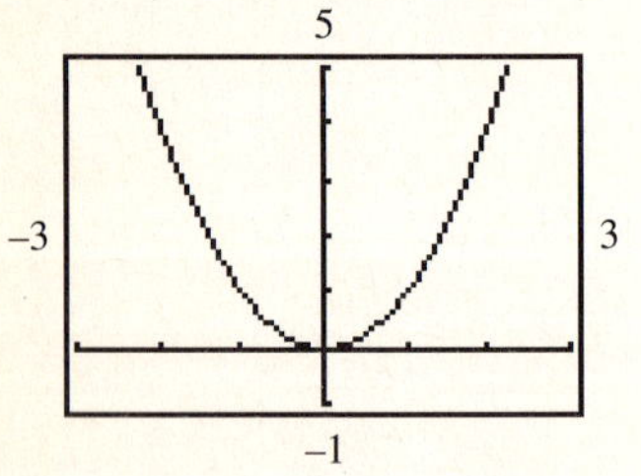

Press GRAPH with y_1 on and the rest of the functions off.

Figure 3.27

b) Make a table of ordered pairs that satisfy $x = y^2$. In this case choose y and calculate x.

$x = y^2$	0	1	1	4	4
y	0	1	-1	2	-2

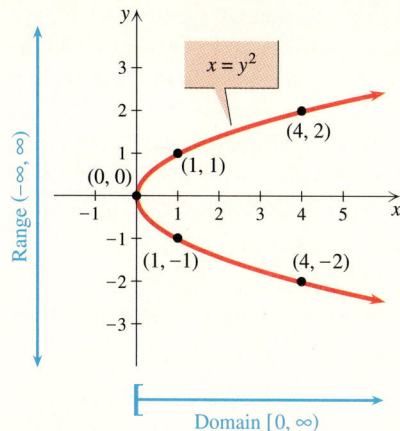

Figure 3.28

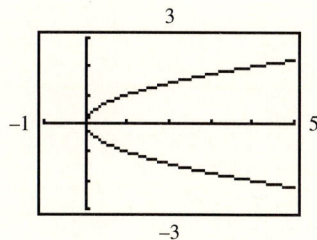

To graph $x = y^2$, graph $y_3 = \sqrt{x}$ and $y_4 = -\sqrt{x}$.

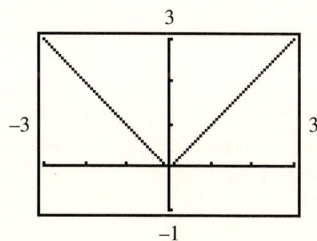

The graph of $y = |x|$ is v-shaped. See Appendix A for more examples.

The ordered pairs $(0, 0)$, $(1, 1)$, $(1, -1)$, $(4, 2)$, and $(4, -2)$ satisfy $x = y^2$. Note that these ordered pairs could be obtained from the ordered pairs that satisfy $y = x^2$ by interchanging the x- and y-coordinates. For this reason, the graph of $x = y^2$ in Fig. 3.28 has the same shape as the parabola in Fig. 3.27, and it is also a parabola. The domain of $x = y^2$ is $[0, \infty)$, and the range is $(-\infty, \infty)$.

c) Make a table of ordered pairs that satisfy $y = |x|$:

x	0	1	-1	2	-2	3	-3
$y = |x|$	0	1	1	2	2	3	3

Plotting these ordered pairs suggests the V-shaped graph of Fig. 3.29. The function $y = |x|$ is known as the **absolute value function**. The domain is $(-\infty, \infty)$, and the range is $[0, \infty)$.

d) Make a table of ordered pairs that satisfy $y = \sqrt{x}$:

x	0	1	4	9
$y = \sqrt{x}$	0	1	2	3

Plotting these ordered pairs suggests the graph shown in Fig. 3.30. The domain is $[0, \infty)$, and the range is $[0, \infty)$. Note that $x = y^2$ is equivalent to $y = \pm\sqrt{x}$. The graph of $y = \sqrt{x}$ is the same as the top half of the parabola of Fig. 3.28, and $y = -\sqrt{x}$ is the same as the bottom half.

Figure 3.29

Figure 3.30

The Vertical Line Test

The graphs shown in Figs. 3.27 through 3.30 are all graphs of relations. But which of these relations are functions? The graph of $x = y^2$ is not the graph of a function because the points $(4, 2)$ and $(4, -2)$ are both on the graph. The graph of the circle in Fig. 3.26 is not the graph of a function, because there are also ordered pairs on this graph with the same x-coordinate and different y-coordinates. Note that on these graphs at least one of the points is directly above another. If a vertical line can be drawn that crosses the graph at more than one point, then there are at least two ordered pairs with the same x-coordinate and different y-coordinates, and the graph is not the graph of a function. This criterion is referred to as the **vertical line test.** In general, the vertical line test can be used on any graph that has the independent variable on the horizontal axis, which is always the case in this text. The vertical line test is a convenient visual test for determining whether a graph is the graph of a function. However, it might not work for some graphs because of the inaccuracy of the graphs.

Example 5 Applying the vertical line test

Determine which of the graphs shown in Fig. 3.31 are graphs of functions.

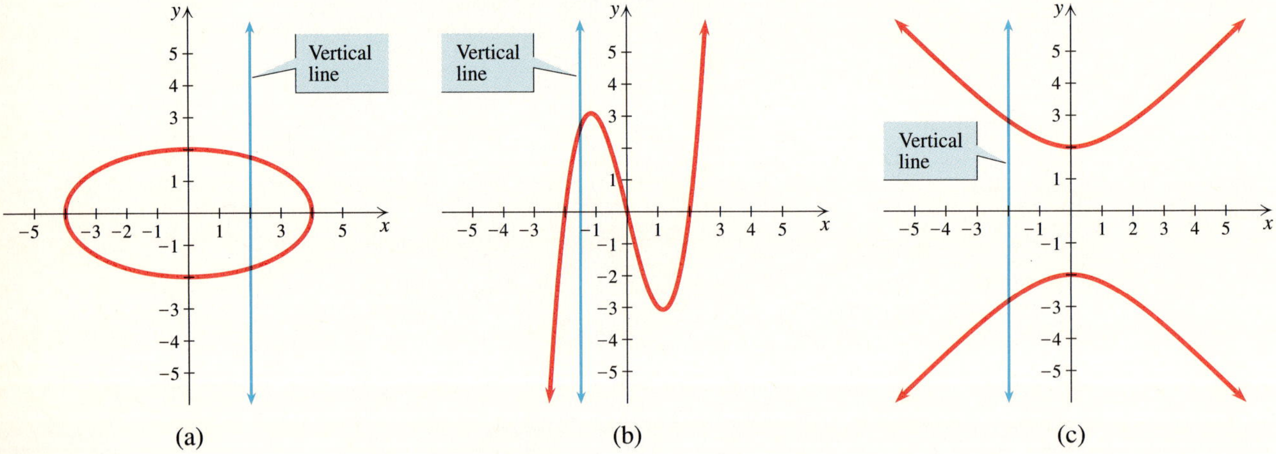

(a)   (b)   (c)

Figure 3.31

Solution

Neither (a) nor (c) is the graph of a function because we can draw a vertical line that crosses the graph more than once. The graph in Fig. 3.31(b) is the graph of a function because every vertical line crosses the graph at most once.

Semicircles

The graph of $x^2 + y^2 = r^2$ $(r > 0)$ is a circle centered at the origin of radius r. The circle does not pass the vertical line test, and a circle is not the graph of a function. We can find an equivalent equation by solving for y:

$$x^2 + y^2 = r^2$$

$$y^2 = r^2 - x^2$$

$$y = \pm \sqrt{r^2 - x^2}$$

The equation $y = \sqrt{r^2 - x^2}$ does define y as a function of x. Because y is nonnegative in this equation, the graph is the top semicircle in Fig. 3.32. The top semicir-

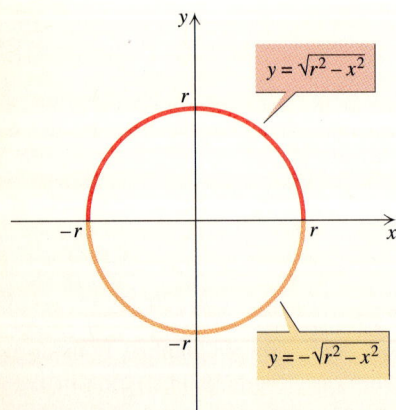

Figure 3.32

cle passes the vertical line test. Likewise, the equation $y = -\sqrt{r^2 - x^2}$ defines y as a function of x, and its graph is the bottom semicircle in Fig. 3.32.

Example 6   Graphing a semicircle

Sketch the graph of each function and state the domain and range of the function.

a) $y = -\sqrt{4 - x^2}$

b) $y = \sqrt{9 - x^2}$

Solution

a) Rewrite the equation in the standard form for a circle:

$$y = -\sqrt{4 - x^2}$$

$$y^2 = 4 - x^2 \qquad \text{Square each side.}$$

$$x^2 + y^2 = 4 \qquad \text{Standard form for the equation of a circle}$$

The graph of $x^2 + y^2 = 4$ is a circle of radius 2 centered at $(0, 0)$. Since y must be negative in $y = -\sqrt{4 - x^2}$, the graph of $y = -\sqrt{4 - x^2}$ is the semicircle shown in Fig. 3.33. We can see from the graph that the domain is $[-2, 2]$ and the range is $[-2, 0]$.

b) Rewrite the equation in the standard form for a circle:

$$y = \sqrt{9 - x^2}$$

$$y^2 = 9 - x^2 \qquad \text{Square each side.}$$

$$x^2 + y^2 = 9$$

The graph of $x^2 + y^2 = 9$ is a circle with center $(0, 0)$ and radius 3. But this equation is not equivalent to the original. The value of y in $y = \sqrt{9 - x^2}$ is nonnegative. So the graph of the original equation is the semicircle shown in Fig. 3.34. We can read the domain $[-3, 3]$ and the range $[0, 3]$ from the graph. ◆

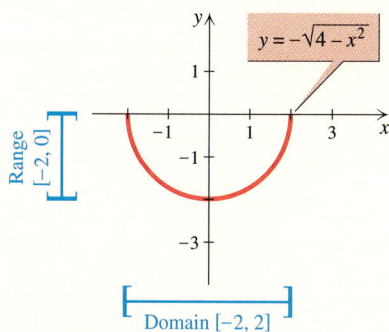

Figure 3.33

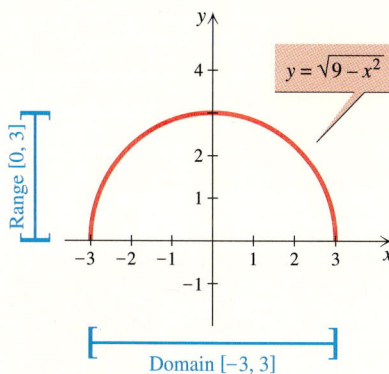

Figure 3.34

Piecewise Functions

For some functions, different formulas are used in different regions of the domain. Such functions are called **piecewise functions.** The simplest example of such a function is the absolute value function $f(x) = |x|$, which can be written as

$$f(x) = \begin{cases} x & \text{for} \quad x \geq 0 \\ -x & \text{for} \quad x < 0. \end{cases}$$

For $x \geq 0$ the equation $f(x) = x$ is used to obtain the second coordinate, and for $x < 0$ the equation $f(x) = -x$ is used. The graph of the absolute value function was drawn in Fig. 3.29.

Example 7 Graphing a piecewise function

Sketch the graph of each function and state the domain and range.

a) $f(x) = \begin{cases} 1 & \text{for} \quad x < 0 \\ -1 & \text{for} \quad x \geq 0 \end{cases}$

b) $f(x) = \begin{cases} x^2 - 4 & \text{for} \quad -2 \leq x \leq 2 \\ x - 2 & \text{for} \quad x > 2 \end{cases}$

Solution

a) For $x < 0$ the graph is the horizontal line $y = 1$. For $x \geq 0$ the graph is the horizontal line $y = -1$. Note that $(0, -1)$ is on the graph shown in Fig. 3.35, but $(0, 1)$ is not. The domain is the interval $(-\infty, \infty)$ and the range consists of only two numbers, -1 and 1. The range is not an interval. It is written in set notation as $\{-1, 1\}$.

b) Make a table of ordered pairs using $y = x^2 - 4$ for x between -2 and 2 and $y = x - 2$ for $x > 2$.

x	-2	-1	0	1	2
$y = x^2 - 4$	0	-3	-4	-3	0

x	2.1	3	4	5
$y = x - 2$	0.1	1	2	3

For x in the interval $[-2, 2]$ the graph is a portion of a parabola as shown in Fig. 3.36. For $x > 2$, the graph is a straight line through $(3, 1)$, $(4, 2)$, and $(5, 3)$. The domain is $[-2, \infty)$, and the range is $[-4, \infty)$.

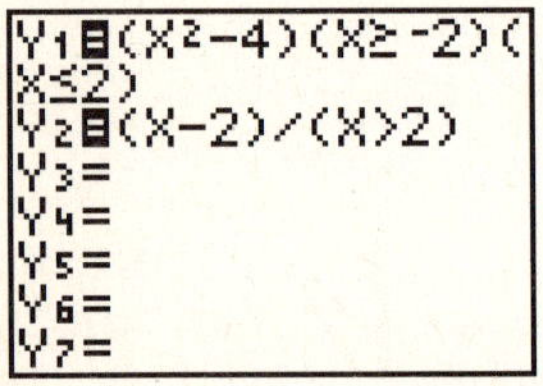

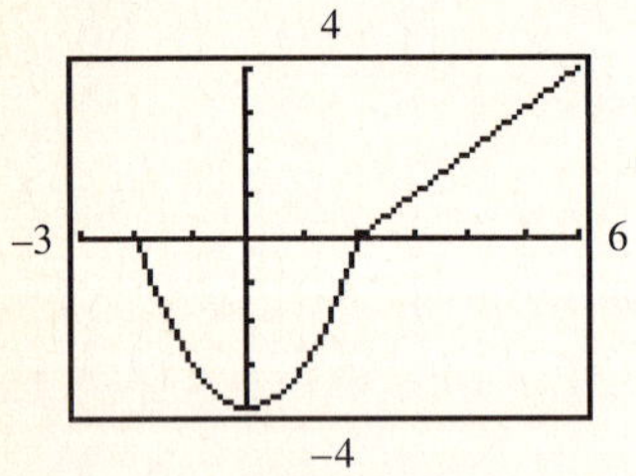

Using the TEST feature, you can define the piecewise function of Example 7(b) and see its graph.

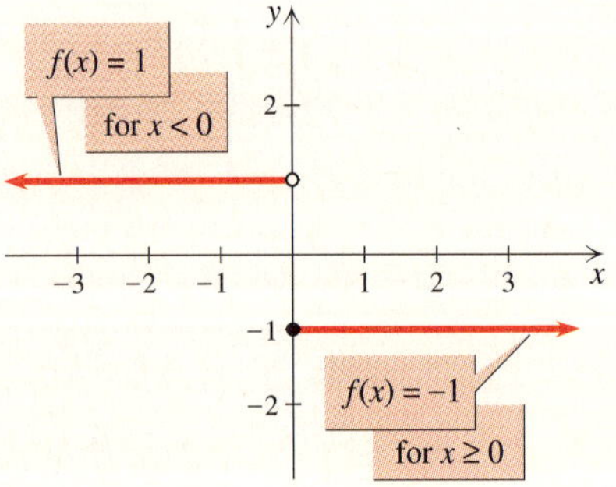

Figure 3.35

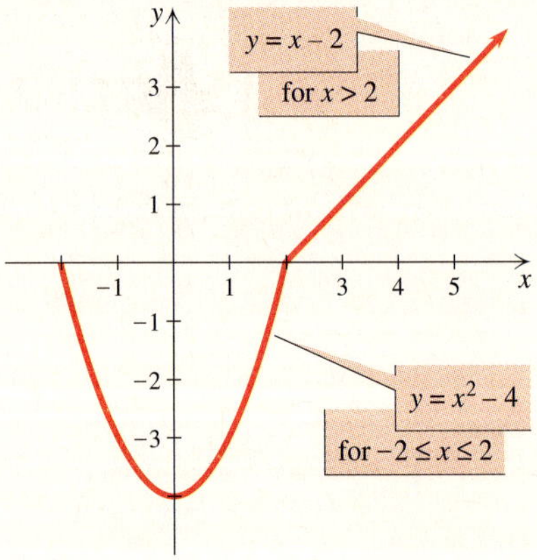

Figure 3.36

Piecewise functions are often found in shipping charges. For example, if the weight in pounds of an order is in the interval $(0, 1]$, the shipping and handling charge is \$3. If the weight is in the interval $(1, 2]$, the shipping and handling charge is \$4, and so on. The next example is a function that is similar to a shipping and handling charge. This function is referred to as the **greatest integer function** and is written $f(x) = [\![x]\!]$. The symbol $[\![x]\!]$ is defined to be the largest integer that is less than or equal to x. For example, $[\![5.01]\!] = 5$, because the greatest integer less than or equal to 5.01 is 5. Likewise, $[\![3.2]\!] = 3$, $[\![-2.2]\!] = -3$, and $[\![7]\!] = 7$.

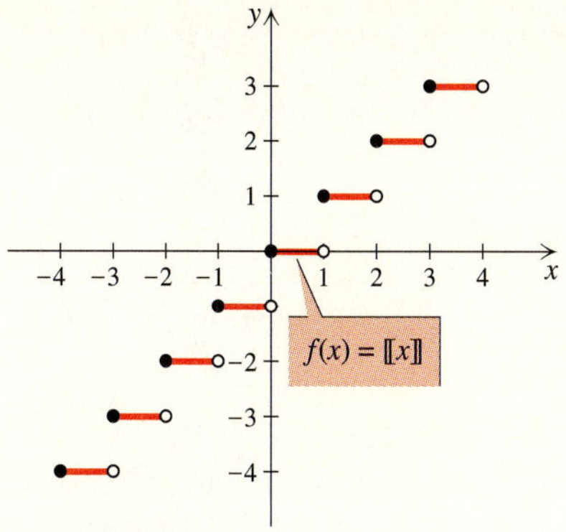

Figure 3.37

Example 8   Graphing the greatest integer function

Sketch the graph of $f(x) = [\![x]\!]$ and state the domain and range.

Solution

For any x in the interval $[0, 1)$ the greatest integer less than or equal to x is 0. For any x in $[1, 2)$ the greatest integer less than or equal to x is 1. For any x in $[-1, 0)$ the greatest integer less than or equal to x is -1. The definition of $[\![x]\!]$ causes the function to be constant between the integers and to "jump" at each integer. The graph of $f(x) = [\![x]\!]$ is shown in Fig. 3.37. The domain is $(-\infty, \infty)$, and the range is the set of integers.   ◆

In the next example we vary the form of the greatest integer function, but the graph is still similar to the graph in Fig. 3.37.

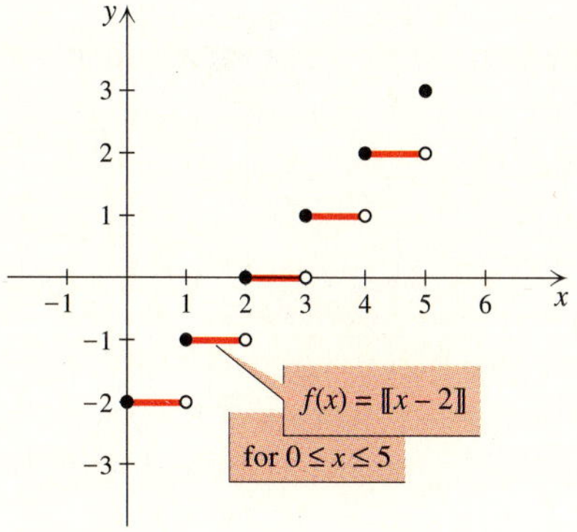

Figure 3.38

Example 9   A variation of the greatest integer function

Sketch the graph of $f(x) = [\![x - 2]\!]$ for $0 \leq x \leq 5$.

Solution

If $x = 0$, $f(0) = [\![-2]\!] = -2$. If $x = 0.5$, $f(0.5) = [\![-1.5]\!] = -2$. In fact, $f(x) = -2$ for any x in the interval $[0, 1)$. Similarly, $f(x) = -1$ for any x in the interval $[1, 2)$. This pattern continues with $f(x) = 2$ for any x in the interval $[4, 5)$, and $f(x) = 3$ for $x = 5$. The graph of $f(x) = [\![x - 2]\!]$ is shown in Fig. 3.38.   ◆

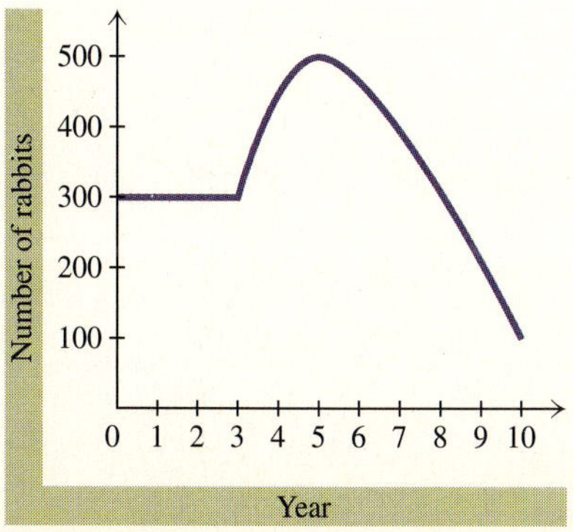

Increasing, Decreasing, and Constant

The graph of a function is a picture of the relationship between the independent variable and the dependent variable. Suppose that the population of rabbits in a forest preserve from 1980 through 1990 is as shown in Fig. 3.39, where 0 corresponds to 1980, 1 to 1981, and so on. Using the vertical line test, we can say that the number of rabbits is a function of time. From the graph we can see that the number of rabbits was constant from 1980 through 1983, increasing from 1983 through 1985, and decreasing from 1985 through 1990.

To better understand the ideas of increasing, decreasing, and constant, imagine a point moving from left to right along the graph of the function. In an interval where the function is increasing, the point will be rising; in an interval where the function is decreasing, the point will be falling; in an interval where the function is constant, the point will be moving horizontally. A formal definition of these ideas is given as follows.

Figure 3.39

Definitions: Increasing, Decreasing, and Constant

1. If $f(x_1) < f(x_2)$ for every x_1 and x_2 in (a, b) with $x_1 < x_2$, then f is **increasing** on (a, b).

2. If $f(x_1) > f(x_2)$ for every x_1 and x_2 in (a, b) with $x_1 < x_2$, then f is **decreasing** on (a, b).

3. If $f(x_1) = f(x_2)$ for every x_1 and x_2 in (a, b), then f is **constant** on (a, b).

In this text we will not work with the inequalities shown in the definition of increasing, decreasing, and constant. We will simply identify the intervals on which a function is increasing, decreasing, or constant, by inspecting a graph as we did in the case of the rabbit population. Note that we have defined these terms only for open intervals (a, b).

Example 10   Increasing, decreasing, or constant

Sketch the graph of each function and identify any open intervals on which the function is increasing, decreasing, or constant.

a) $f(x) = 4 - x^2$ **b)** $g(x) = \begin{cases} 0 & \text{for} \quad x \leq 0 \\ \sqrt{x} & \text{for} \quad 0 < x < 4 \\ 2 & \text{for} \quad x \geq 4 \end{cases}$

Solution

a) The graph of $f(x) = 4 - x^2$ includes the points $(-2, 0)$, $(-1, 3)$, $(0, 4)$, $(1, 3)$, and $(2, 0)$. The graph is shown in Fig. 3.40. The function is increasing on the interval $(-\infty, 0)$ and decreasing on $(0, \infty)$.

b) The graph of g is shown in Fig. 3.41. The function g is constant on the intervals $(-\infty, 0)$ and $(4, \infty)$ and increasing on $(0, 4)$.

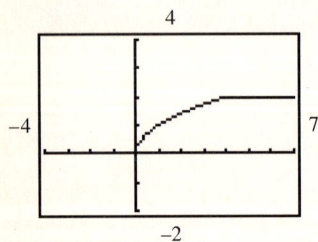

Using the TEST feature, you can define the function of Example 10(b) and see its graph. See Appendix A for more examples.

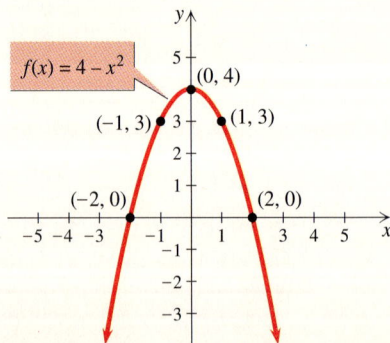

Figure 3.40

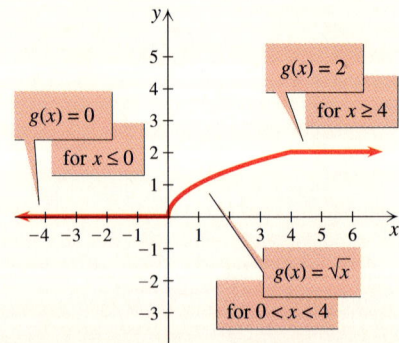

Figure 3.41

? For Thought

True or false? Explain.

1. The graph of $\sqrt{(x - 2)^2 + (y + 3)^2} = 9$ is a circle with center $(2, -3)$ and radius 9.

2. The graph of $(y - 2)^2 + (x - 3)^2 = 25$ is a circle with center $(2, 3)$ and radius 5.

3. The graph of $(x + 1)^2 + (y + 1)^2 = -16$ is a circle centered at $(-1, -1)$ with radius 4.

4. If $f(x) = [\![x + 3]\!]$, then $f(-4.5) = -2$.

5. The range of the function $f(x) = \dfrac{|x|}{x}$ is the interval $(-1, 1)$.

6. The range of $f(x) = [\![x - 1]\!]$ is the set of integers.

7. The only ordered pair that satisfies $(x - 5)^2 + (y + 6)^2 = 0$ is $(5, -6)$.

8. The domain of the function $y = \sqrt{4 - x^2}$ is the interval $[-2, 2]$.

9. The range of the function $y = \sqrt{16 - x^2}$ is the interval $[0, \infty)$.

10. The function $y = \sqrt{4 - x^2}$ is increasing on $(-2, 0)$ and decreasing on $(0, 2)$.

3.3 Exercises   Tape 6   Disk—5.25″: 2   3.5″: 1   Macintosh: 1

Determine the center and radius of each circle and sketch the graph.

1. $x^2 + y^2 = 16$ **2.** $x^2 + y^2 = 1$

3. $(x + 6)^2 + y^2 = 36$ **4.** $x^2 = 9 - (y - 3)^2$

5. $(x - 2)^2 = 8 - (y + 2)^2$ **6.** $(y + 2)^2 = 20 - (x - 4)^2$

Write the standard equation for each circle.

7. Center at $(0, 0)$ with radius $\sqrt{7}$

8. Center at $(0, 0)$ with radius $2\sqrt{3}$

9. Center at $(-2, 5)$ with radius $1/2$

10. Center at $(-1, -6)$ with radius $1/3$

11. Center at $(3, 5)$ and passing through the origin

12. Center at $(-3, 9)$ and passing through the origin

13. Center at $(5, -1)$ and passing through $(1, 3)$

14. Center at $(-2, -3)$ and passing through $(2, 5)$

Determine the center and radius of each circle and sketch the graph.

15. $x^2 + y^2 + 6y = 0$ **16.** $x^2 + y^2 = 4x$

17. $x^2 - 6x + y^2 - 8y = 0$

18. $x^2 + 10x + y^2 - 8y = -40$

19. $x^2 + y^2 = 4x + 3y$ **20.** $x^2 + y^2 = 5x - 6y$

21. $x^2 + y^2 = \dfrac{x}{2} - \dfrac{y}{3} - \dfrac{1}{16}$ **22.** $x^2 + y^2 = x - y + \dfrac{1}{2}$

Graph each equation by plotting ordered pairs of numbers. Determine the domain and range, and whether the relation is a function.

23. $x = \sqrt{y}$ **24.** $x = |y|$ **25.** $y = x^2 - 1$

26. $y = 1 + \sqrt{x}$ **27.** $y = 5$ **28.** $y = 2x$

29. $x = 2y$ **30.** $x - y = 0$ **31.** $x - y = 2$

32. $x = 3$ **33.** $y = 2|x|$ **34.** $y = 2x^2$

35. $x = y^2 + 1$ **36.** $x = 1 - y^2$ **37.** $y = |x - 1|$

38. $x = |y| + 1$ **39.** $x = |y + 2|$ **40.** $y = |x| - 3$

41. $y = \sqrt{1 - x^2}$ **42.** $y = -\sqrt{25 - x^2}$

43. $y^2 = 1 - x^2$ **44.** $y = x^3$ **45.** $x + y^2 = 0$

46. $x^2 + y^2 = 0$ **47.** $y = 1 - x^2$ **48.** $y^2 = -1 - x^2$

49. $xy = 1$ **50.** $xy = -1$ **51.** $y = -|x|$

52. $y = \dfrac{|x|}{x}$

Use the vertical line test to determine whether each graph in Exercises 53–58 is the graph of a function.

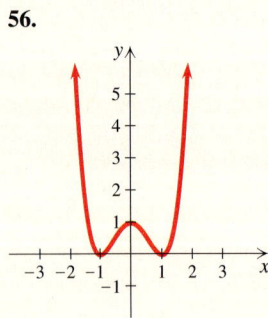

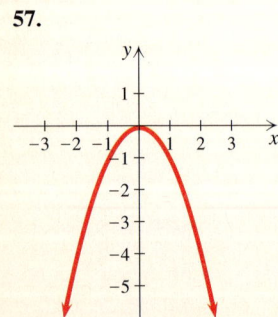

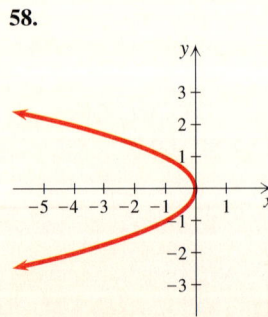

53.

54.

55.

56.

57.

58.

Sketch the graph of each function and state the domain and range.

59. $f(x) = \begin{cases} 2 & \text{for } x < -1 \\ -2 & \text{for } x \geq -1 \end{cases}$

60. $f(x) = \begin{cases} \sqrt{x + 2} & \text{for } -2 \leq x \leq 2 \\ 4 - x & \text{for } x > 2 \end{cases}$

61. $f(x) = \begin{cases} \sqrt{-x} & \text{for } x < 0 \\ \sqrt{x} & \text{for } x \geq 0 \end{cases}$

62. $f(x) = \begin{cases} x & \text{for } x < -1 \\ -x & \text{for } x \geq -1 \end{cases}$

63. $f(x) = \begin{cases} 4 - x^2 & \text{for } -2 \leq x \leq 2 \\ x - 2 & \text{for } x > 2 \end{cases}$

64. $f(x) = \begin{cases} 3 & \text{for } x < 0 \\ 3 + \sqrt{x} & \text{for } x \geq 0 \end{cases}$

65. $f(x) = [\![x + 1]\!]$ **66.** $f(x) = 2[\![x]\!]$

67. $f(x) = [\![x]\!] + 2$ for $0 \leq x < 4$

68. $f(x) = [\![x - 3]\!]$ for $0 < x \leq 5$

From the graph of each function in Exercises 69–74, state the domain, the range, and the intervals on which the function is increasing, decreasing, or constant.

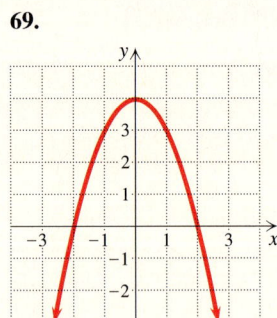

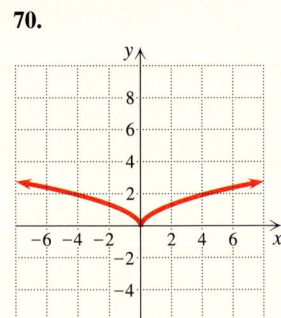

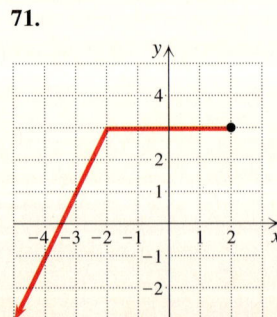

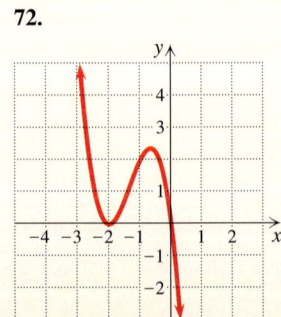

69.

70.

71.

72.

73.

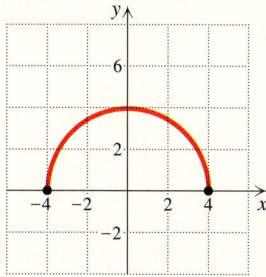

74.

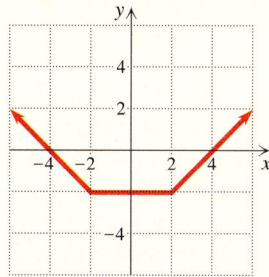

Sketch the graph of each function. Determine the domain and range. Identify any intervals on which f is increasing, decreasing, or constant.

75. $f(x) = 2x + 1$ **76.** $f(x) = x^2 - 3$ **77.** $f(x) = |x - 1|$

78. $f(x) = |x| + 1$ **79.** $f(x) = \dfrac{2x}{|x|}$ **80.** $f(x) = -3x$

81. $f(x) = -\sqrt{1 - x^2}$ **82.** $f(x) = \sqrt{9 - x^2}$

83. $f(x) = \begin{cases} x + 1 & \text{for} \quad x \neq 3 \\ x + 2 & \text{for} \quad x = 3 \end{cases}$

84. $f(x) = \begin{cases} \sqrt{-x} & \text{for} \quad x < 0 \\ -\sqrt{x} & \text{for} \quad x \geq 0 \end{cases}$

85. $f(x) = \begin{cases} x + 2 & \text{for} \quad x \leq -2 \\ \sqrt{4 - x^2} & \text{for} \quad -2 < x < 2 \\ -x + 2 & \text{for} \quad x \geq 2 \end{cases}$

86. $f(x) = \begin{cases} 8 + 2x & \text{for} \quad x \leq -2 \\ x^2 & \text{for} \quad -2 < x < 2 \\ 8 - 2x & \text{for} \quad x \geq 2 \end{cases}$

Solve each problem.

87. *Missing circle* Find the standard equation of the circle that has (4, 1) and (−6, 5) as endpoints of a diameter.

88. *Missing circle* Find the standard equation of the circle that has (−3, −2) and (4, 5) as endpoints of a diameter.

89. *Parking charges* A garage charges $4/hr up to 3 hr, with any fraction of an hour charged as a whole hour. Any time over 3 hr is charged at the all-day rate of $15. Use function notation to write the charge as a function of the number of hours x, where $0 < x \leq 8$, and graph this function.

90. *Delivering concrete* A concrete company charges $150 for delivering less than 3 yd³ of concrete. For 3 yd³ and more, the charge is $50/yd³. Use function notation to write the charge as a function of the number x of cubic yards delivered, where $0 < x \leq 10$, and graph this function.

91. *Water bill* The monthly water bill in Hammond is a function of the number of gallons used. The cost is $10.30 for 10,000 gal or less. Over 10,000 gal, the cost is $10.30 plus $1 for each 1000 gal over 10,000. On what interval is the cost constant? On what interval is the cost increasing?

92. *Gas mileage* The number of miles per gallon obtained with a new Firebird is a function of the speed at which it is driven. Is this function increasing or decreasing on its domain? Explain.

93. *Filing a tax return* An accountant determines the charge for filing a tax return by using the function $C = 50 + 40[\![t]\!]$ for $t > 0$, where C is in dollars and t is in hours. Sketch the graph of this function. For what values of t is the charge over $235?

94. *Shipping machinery* The cost in dollars of shipping a machine is given by the function $C = 200 + 37[\![w/100]\!]$ for $w > 0$, where w is the weight of the machine in pounds. For which values of w is the cost less than $862?

95. *Going to the movies* The average price of a movie ticket between 1988 and 1992 can be modeled by the function

$$P(t) = \begin{cases} 0.32t + 4.11 & \text{for} \quad 0 \leq t \leq 2 \\ 0.15t + 4.45 & \text{for} \quad t > 2 \end{cases}$$

where $P(t)$ is in dollars and t is the number of years since 1988 (*USA Today,* July 6, 1993). Graph this function. What was the average price in 1990? What will be the average price in 2000? What was the average rate of change of the price from 1988 to 1992?

96. *Lowering cholesterol* According to the National Center for Health Statistics, cholesterol levels have been falling steadily since 1960 (*The Boston Globe,* June 16, 1993). The declining cholesterol levels can be modeled by the function

$$C(t) = \begin{cases} -0.35t + 220 & \text{for} \quad 0 \leq t \leq 20 \\ -0.80t + 229 & \text{for} \quad t > 20 \end{cases}$$

where $C(t)$ is the average number of milligrams of cholesterol per deciliter of blood in the adult population and t is the number of years since 1960. Graph this function. What was the average cholesterol level in 1980? Will the National Cholesterol Education Program reach its goal of getting average cholesterol under 200 by the year 2000?

For Writing/Discussion

97. *Steps* Find an example of a function in real life whose graph has "steps" like that of the greatest integer function. Graph your function and find a formula for it if possible.

98. *Cooperative learning* Define a piecewise function using different formulas on the intervals $(-\infty, a]$, (a, b), and $[b, \infty)$ so that the graph does not "jump" at a or b. Give your function to a classmate to graph and check.

Graphing Calculator Exercises

Most graphing calculators can graph more than one function at a time. This feature allows us to graph relations that can be separated into two or more functions.

1. Graph the circle $(x - 2)^2 + (y + 1)^2 = 5$ on a graphing calculator by solving the equation for y and graphing the equations for the two semicircles.

2. Solve the equation $x = y^2$ for y and graph the two equations to get the complete graph of $x = y^2$.

The graphing calculator gives you the means to explore new ideas easily. In the next two exercises you can discover for yourself the effect on a curve of a slight change in its formula.

3. Graph the equations $y = \sqrt{x}$, $y = 2\sqrt{x}$, $y = \sqrt{x + 2}$, and $y = \sqrt{x} + 2$ at the same time on a graphing calculator. What can you conclude about the effect of the number 2 on the graph of $y = \sqrt{x}$?

4. Graph the equations $y = x^2$, $y = (x - 3)^2$, $y = x^2 - 3$, and $y = 3x^2$ at the same time on a graphing calculator. What can you conclude about the effect of the number 3 on the graph of $y = x^2$?

5. Graph the function $y = x^3 - 3x$ on a graphing calculator and determine the intervals for which the function is increasing and decreasing.

6. Graph the function $y = x^4 - 11x^2 + 18$ on a graphing calculator and determine the intervals for which the function is increasing and decreasing.

Most graphing calculators have the capability of graphing piecewise functions. Read your manual to learn the procedure for your calculator.

7. Graph the piecewise functions in Exercises 59–64 and 83–86 of this section, this time using your graphing calculator. Compare the results with your earlier graphs.

3.4

Transformations and

Symmetry of Graphs

In Section 3.3 we studied the graphs of relations and functions. In this section we continue that study by discussing how the graphs of some functions are transformed into the graphs of other functions. We also study the idea of symmetry of a graph.

Reflection

The graph of $f(x) = x^2$ goes through $(0, 0)$, $(1, 1)$, and $(2, 4)$. The graph of $g(x) = -x^2$ goes through $(0, 0)$, $(1, -1)$, and $(2, -4)$. Consider the graphs of $f(x) = x^2$ and $g(x) = -x^2$ shown in Fig. 3.42. Notice that the graph of g is a mirror image (or reflection) of the graph of f. For every ordered pair (x, y) on the graph of f, the ordered pair $(x, -y)$ is on the graph of g.

Definition: Reflection

The graph of $y = -f(x)$ is a **reflection** in the x-axis of the graph of $y = f(x)$.

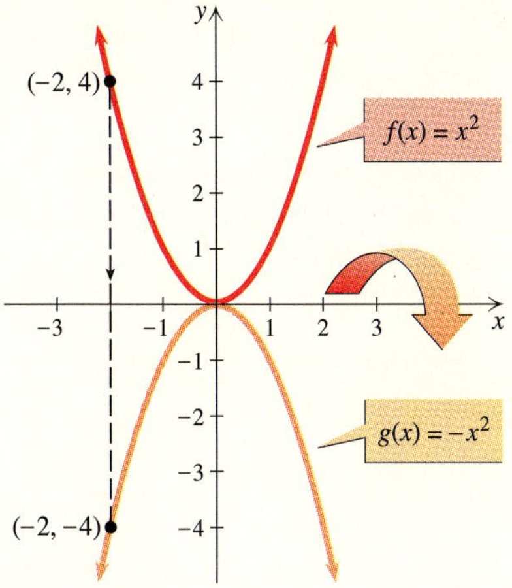

Figure 3.42

Knowing that a graph is a reflection of a familiar graph makes graphing easier. We reflect the familiar graph in the x-axis to obtain a new graph.

Example 1 Graphing using reflection

Sketch the graphs of each pair of functions on the same coordinate plane.

a) $f(x) = x^3$, $g(x) = -x^3$ **b)** $f(x) = |x|$, $g(x) = -|x|$

Solution

a) Make a table of ordered pairs for f as follows:

x	-2	-1	0	1	2
$f(x) = x^3$	-8	-1	0	1	8

Sketch the graph of f through these ordered pairs as shown in Fig. 3.43. Since $g(x) = -f(x)$, the graph of g can be obtained by reflecting the graph of f in the x-axis. Each point on the graph of f corresponds to a point on the graph of g with the opposite y-coordinate. For example, $(2, 8)$ on the graph of f corresponds to $(2, -8)$ on the graph of g. Both graphs are shown in Fig. 3.43.

b) The graph of f is the familiar V-shaped graph of the absolute value function as shown in Fig. 3.44. Since $g(x) = -f(x)$, the graph of g can be obtained by reflecting the graph of f in the x-axis. Each point on the graph of f corresponds to a point on the graph of g with the opposite y-coordinate. For example, $(2, 2)$ on f corresponds to $(2, -2)$ on g. Both graphs are shown in Fig. 3.44.

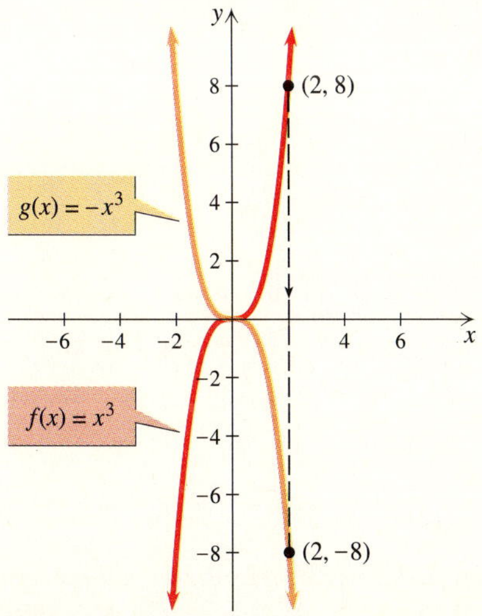

Figure 3.43

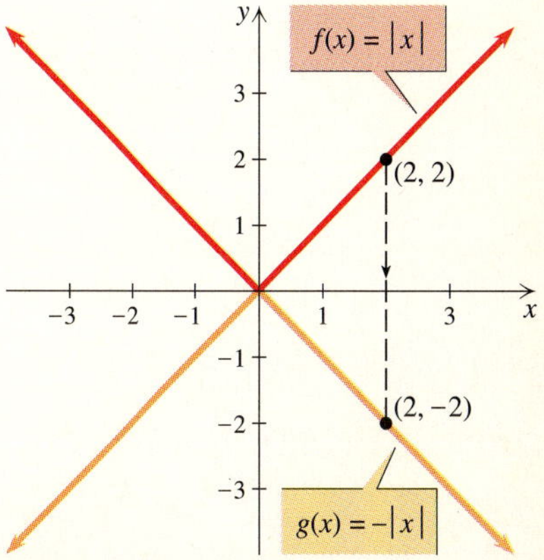

Figure 3.44

Translation

Consider the graphs of the functions $f(x) = \sqrt{x}$, $g(x) = \sqrt{x} + 3$, and $h(x) = \sqrt{x} - 5$ shown in Fig. 3.45. In the expression $\sqrt{x} + 3$, adding 3 is the last operation performed to obtain the y-coordinate. So every point on the graph of g is exactly three units above a corresponding point on the graph of f. This translating of points gives the graph of g the same shape as the graph of f. Likewise, every point on the graph of h is exactly five units below a corresponding point on the graph of f.

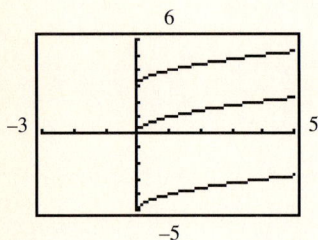

The graphs of $y_1 = \sqrt{x}$, $y_2 = \sqrt{x} + 3$, and $y_3 = \sqrt{x} - 5$ illustrate vertical translation.

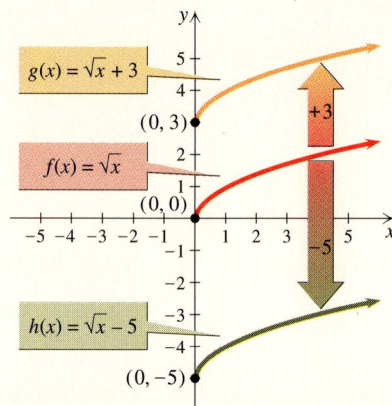

Figure 3.45

Definition: Translation Upward or Downward

If $k > 0$, then the graph of $y = f(x) + k$ is an **upward translation** of the graph of $y = f(x)$ and the graph of $y = f(x) - k$ is a **downward translation** of the graph of $y = f(x)$.

Knowing that the graph of one function is a translation of the graph of a familiar function makes graphing easier.

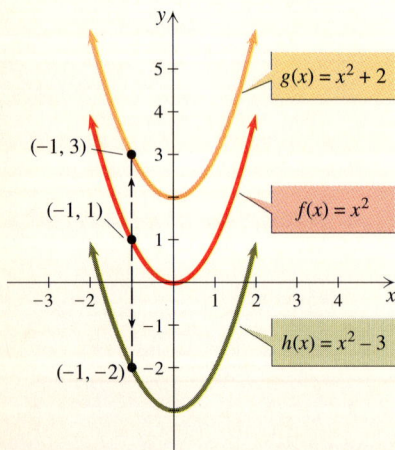

Figure 3.46

Example 2   Graphing using translation

Graph the functions $f(x) = x^2$, $g(x) = x^2 + 2$, and $h(x) = x^2 - 3$ on the same coordinate plane.

Solution

First sketch the familiar graph of $f(x) = x^2$ through $(-2, 4)$, $(-1, 1)$, $(0, 0)$, $(1, 1)$, and $(2, 4)$, as shown in Fig. 3.46. Since $g(x) = f(x) + 2$, the graph of g can be obtained by translating the graph of f upward two units. Since $h(x) = f(x) - 3$, the graph of h can be obtained by translating the graph of f downward three units.

For example, the point $(-1, 1)$ on the graph of f moves up to $(-1, 3)$ on the graph of g and down to $(-1, -2)$ on the graph of h. All three graphs are shown in Fig. 3.46.

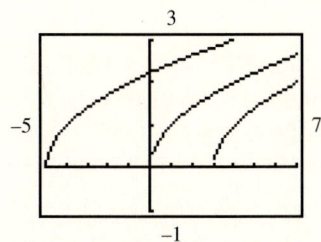

Consider the graphs of $f(x) = \sqrt{x}$, $g(x) = \sqrt{x - 3}$, and $h(x) = \sqrt{x + 5}$ in Fig. 3.47. In the expression $\sqrt{x - 3}$, subtracting 3 is the first operation to perform. So every point on the graph of g is exactly three units to the right of a corresponding point on the graph of f. (We must start with a larger value of x to get the same y-coordinate, because we first subtract 3.) Every point on the graph of h is exactly five units to the left of a corresponding point on the graph of f. Compare the graphs in Figs. 3.47 and 3.45. Notice that a slight difference in the formula makes a big difference in the location of the graph.

The graphs of $y_1 = \sqrt{x}$, $y_2 = \sqrt{x - 3}$, and $y_3 = \sqrt{x + 5}$ illustrate horizontal translation.

Figure 3.47

Definition: Translation to the Right or Left

If $h > 0$, then the graph of $y = f(x - h)$ is a **translation to the right** of the graph of $y = f(x)$ and the graph of $y = f(x + h)$ is a **translation to the left** of the graph of $y = f(x)$.

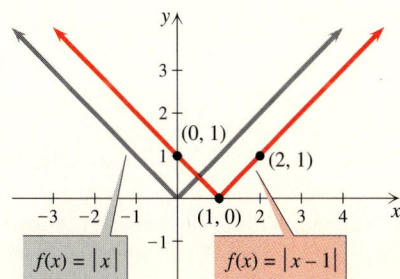

Example 3   Graphing using translation

Sketch the graph of each function.

a) $f(x) = |x - 1|$

b) $f(x) = (x + 3)^2$

Solution

a) The graph of $f(x) = |x - 1|$ is a translation one unit to the right of the familiar graph of $f(x) = |x|$. Calculate a few ordered pairs to get an accurate graph. The points $(0, 1)$, $(1, 0)$, and $(2, 1)$ are on the graph of $f(x) = |x - 1|$ shown in Fig. 3.48.

Figure 3.48

$f(x) = (x + 3)^2$

$f(x) = x^2$

$(-4, 1)$

$(-2, 1)$

$(-3, 0)$

Figure 3.49

b) The graph of $f(x) = (x + 3)^2$ is a translation three units to the left of the familiar graph of $f(x) = x^2$. Calculate a few ordered pairs to get an accurate graph. The points $(-3, 0), (-2, 1)$, and $(-4, 1)$ are on the graph shown in Fig. 3.49.

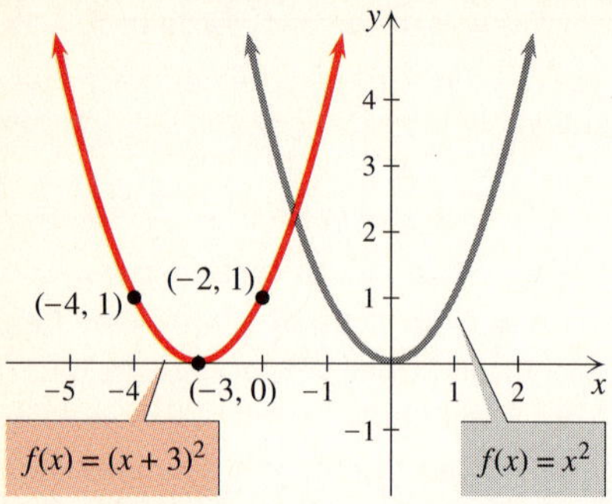

Nonrigid Transformation

Consider the graphs of the functions $f(x) = x^2$, $g(x) = 2x^2$, and $h(x) = \frac{1}{2}x^2$ shown in Fig. 3.50. For any given x-coordinate, the y-coordinate on g is twice as large as the corresponding y-coordinate on f, and the y-coordinate on h is one-half as large as the corresponding y-coordinate on f. Such transformations are referred to as stretching and shrinking transformations.

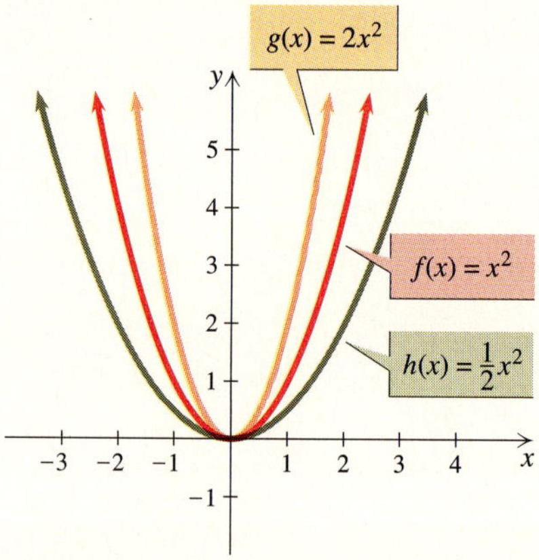

$g(x) = 2x^2$

$f(x) = x^2$

$h(x) = \frac{1}{2}x^2$

Figure 3.50

Definitions: Stretching and Shrinking

The graph of $y = af(x)$ is obtained from the graph of $y = f(x)$ by

1. **stretching** the graph of $y = f(x)$ when $a > 1$, or
2. **shrinking** the graph of $y = f(x)$ when $0 < a < 1$.

Note that the last operation to be performed in stretching or shrinking is multiplication by a. The function $f(x) = 2\sqrt{x}$ is obtained by stretching $f(x) = \sqrt{x}$ by a factor of 2. However, $f(x) = \sqrt{2x}$ is obtained from $f(x) = \sqrt{x}$ by stretching, not by a factor of 2, but by a factor of $\sqrt{2}$.

Reflecting, translating, stretching, and shrinking are all referred to as **transformations** because they transform one graph into another. Since reflection and translation do not change the shape of the graph, they are **rigid transformations.** Stretching and shrinking change the shape of a graph, so they are **nonrigid transformations.**

Example 4   Graphing using stretching and shrinking

Graph the functions $f(x) = \sqrt{x}$, $g(x) = 2\sqrt{x}$, and $h(x) = \frac{1}{2}\sqrt{x}$ on the same coordinate plane.

Solution

The graph of g is obtained by stretching the graph of f, and the graph of h is obtained by shrinking the graph of f. The graph of f includes the points $(0, 0)$, $(1, 1)$, and $(4, 2)$. The graph of g includes the points $(0, 0)$, $(1, 2)$, and $(4, 4)$. The graph of h includes the points $(0, 0)$, $(1, 0.5)$, and $(4, 1)$. The graphs are shown in Fig. 3.51.

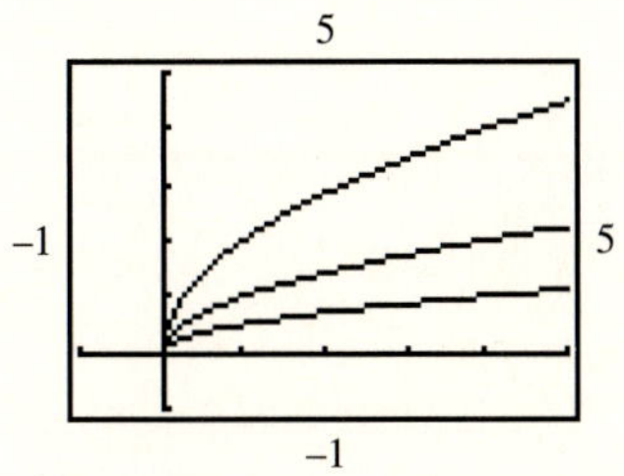

The graphs of $y_1 = \sqrt{x}$, $y_2 = 2\sqrt{x}$, and $y_3 = \frac{1}{2}\sqrt{x}$ illustrate stretching and shrinking.

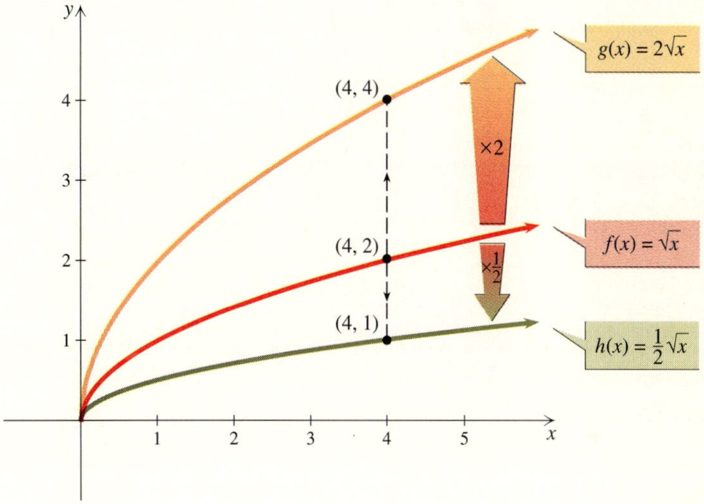

Figure 3.51

A function involving more than one transformation may be graphed using the following procedure.

Procedure: Multiple Transformations

Graph a function involving more than one transformation in the following order:

1. **Horizontal translation**
2. **Stretching or shrinking**
3. **Reflecting**
4. **Vertical translation**

The function in the next example involves all four of the above transformations.

Example 5 Graphing using several transformations

Graph the function $y = 4 - 2\sqrt{x + 1}$.

Solution

The graph of $y = \sqrt{x + 1}$ is a *horizontal* translation one unit to the left of the graph of $y = \sqrt{x}$. The graph of $y = 2\sqrt{x + 1}$ is obtained from $y = \sqrt{x + 1}$ by *stretching* it by a factor of 2. *Reflect* $y = 2\sqrt{x + 1}$ in the x-axis to obtain the graph of $y = -2\sqrt{x + 1}$. Finally, the graph of $y = 4 - 2\sqrt{x + 1}$ is a *vertical* translation of $y = -2\sqrt{x + 1}$, four units upward. All of these graphs are shown in Fig. 3.52.

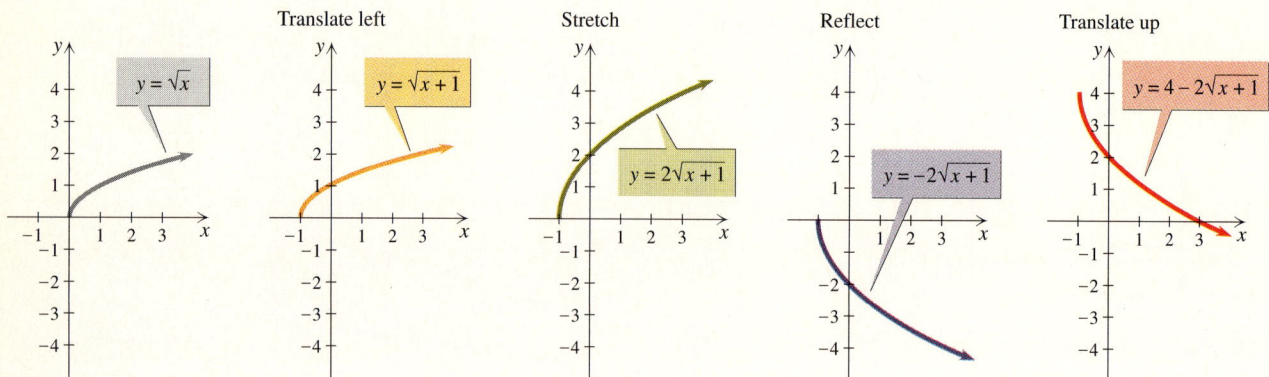

Figure 3.52

Linear Functions

A **linear function** is a function of the form $y = mx + b$, where m and b are real numbers. The simplest linear function is $y = x$, and the graphs of all others are obtained by transforming $y = x$. The function $y = x$ or $f(x) = x$ is called the **identity function.** We will study linear functions in more detail in Section 4.1.

Example 6 Graphing linear functions using transformations

Sketch the graphs of $y = x$, $y = 2x$, $y = -2x$, and $y = -2x - 3$.

Solution

The graph of $y = x$ is a line through $(0, 0)$, $(1, 1)$, and $(2, 2)$ as shown in Fig. 3.53. Stretch the graph of $y = x$ by a factor of 2 to get the graph of $y = 2x$ shown in Fig. 3.53. The graph of $y = -2x - 3$ is obtained by reflecting $y = 2x$ in the x-axis, and then translating downward three units as shown in Fig. 3.53.

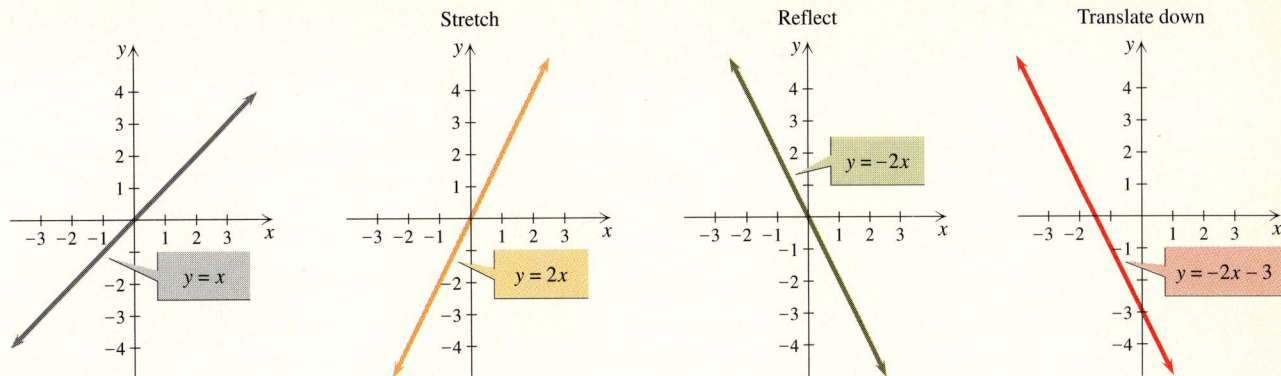

Stretch

Reflect

Translate down

$y = x$

$y = 2x$

$y = -2x$

$y = -2x - 3$

Figure 3.53

Symmetry

The graph of $g(x) = -x^2$ is a reflection in the x-axis of the graph of $f(x) = x^2$. If the paper were folded along the x-axis, the graphs would coincide. See Fig. 3.54. The symmetry that we call reflection occurs between two functions, but the graph of $f(x) = x^2$ has a symmetry within itself. Points such as $(2, 4)$ and $(-2, 4)$ are on the graph and are equidistant from the y-axis. Folding the paper along the y-axis brings all such pairs of points together. See Fig. 3.55. The reason for this symmetry about the y-axis is the fact that $f(x) = f(-x)$ for any value of x. We get the same y-coordinate whether we evaluate the function at a number or at its opposite.

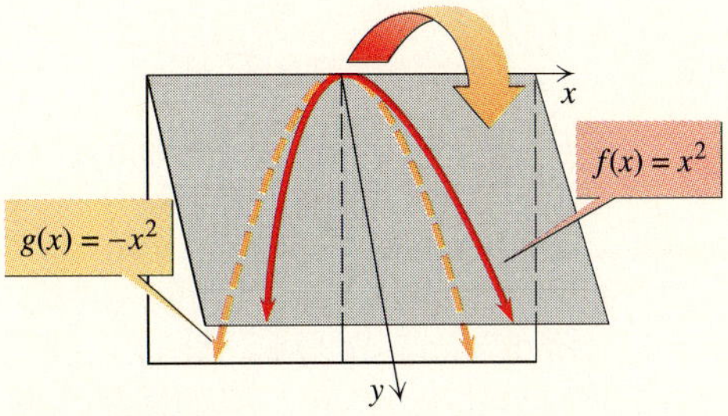

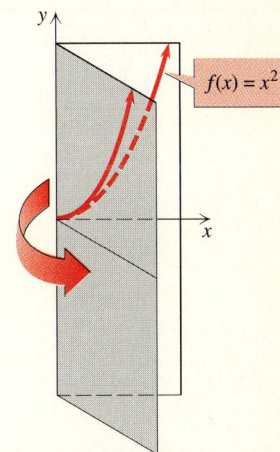

$f(x) = x^2$

$f(x) = x^2$

$g(x) = -x^2$

Figure 3.54

Figure 3.55

Definition: Symmetric about the y-Axis

If $f(x) = f(-x)$ for any value of x in the domain of the function f, then f is called an **even function** and its graph is **symmetric about the y-axis**.

The function $f(x) = x^2$ is an example of an even function and its graph is symmetric about the y-axis.

Consider the graph of $f(x) = x^3$ shown in Fig. 3.56. It is not symmetric about the y-axis like the graph of $f(x) = x^2$, but it has another kind of symmetry. On the graph of $f(x) = x^3$ we find pairs of points such as $(2, 8)$ and $(-2, -8)$. These points are equidistant from the origin and on opposite sides of the origin. So the symmetry of this graph is about the origin. In this case, $f(x)$ and $f(-x)$ are not equal, but $f(-x) = -f(x)$.

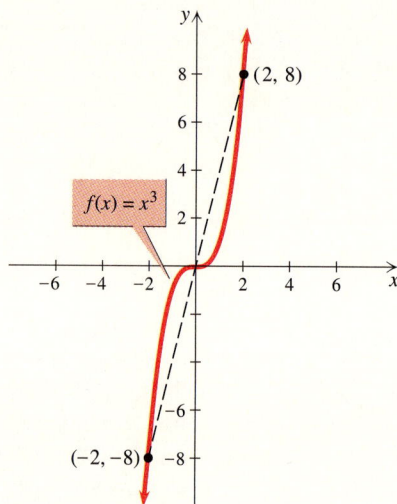

Figure 3.56

Definition: Symmetric about the Origin

If $f(-x) = -f(x)$ for any value of x in the domain of the function f, then f is called an **odd function** and its graph is **symmetric about the origin.**

We can look at a graph and see if it is symmetric about the y-axis or the origin, but it makes graphing easier and more accurate if we can identify symmetry *before* graphing. Using the definitions, we can determine whether a function is even, odd, or neither even nor odd from the formula defining the function. In this way we know the type of graph to expect before we start plotting points.

Example 7 Determining symmetry in a graph

Discuss the symmetry of the graph of each function.

a) $f(x) = 5x^3 - x$ **b)** $f(x) = |x| + 3$ **c)** $f(x) = x^2 - 3x + 6$

Solution

a) Replace x by $-x$ in the formula for $f(x)$ and simplify:

$$f(-x) = 5(-x)^3 - (-x) = -5x^3 + x$$

Is $f(-x)$ equal to $f(x)$ or the opposite of $f(x)$? Since $-f(x) = -5x^3 + x$, we have $f(-x) = -f(x)$. So f is an odd function and the graph is symmetric about the origin.

b) Since $|-x| = |x|$ for any x, we have $f(-x) = |-x| + 3 = |x| + 3$. Because $f(-x) = f(x)$, the function is even and the graph is symmetric about the y-axis.

c) In this case, $f(-x) = (-x)^2 - 3(-x) + 6 = x^2 + 3x + 6$. So $f(-x) \neq f(x)$, and $f(-x) \neq -f(x)$. This function is neither odd nor even and its graph has neither type of symmetry. ◆

Using Graphs to Solve Inequalities

In Chapter 2 we solved a variety of inequalities using various techniques. We can solve many inequalities just by examining a graph of a function. Consider the inequality $x^2 - 4 < 0$. The graph of $y = x^2 - 4$ is shown in Fig. 3.57. For any value of x for which $x^2 - 4 < 0$, the y-coordinate on the graph of $y = x^2 - 4$ is negative. The y-coordinate is negative precisely for x in the open interval $(-2, 2)$ as shown in Fig. 3.57. The solution set to $x^2 - 4 < 0$ is the interval $(-2, 2)$. The solution set to $x^2 - 4 \geq 0$ is also read from the graph of $y = x^2 - 4$ as $(-\infty, -2] \cup [2, \infty)$.

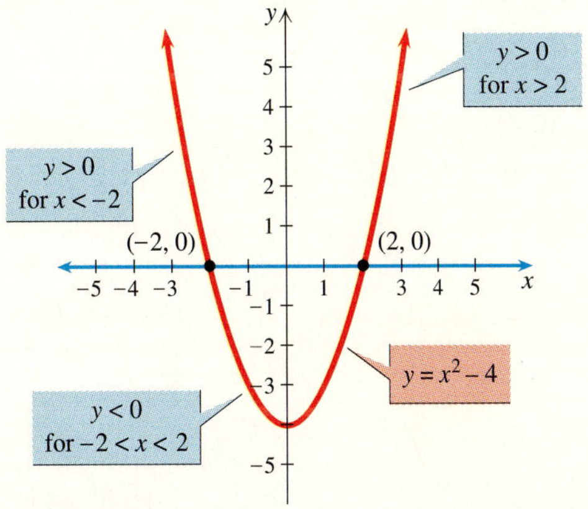

Figure 3.57

Since the x-intercepts are the points where the y-coordinate may change sign, they are critical for solving an inequality by inspecting a graph of a function.

Example 8   Using a graph to solve an inequality

Solve the inequality $(x - 1)^2 - 2 < 0$ by graphing.

Solution

The graph of $y = (x - 1)^2 - 2$ is obtained by translating the graph of $y = x^2$ one unit to the right and two units downward. See Fig. 3.58. To find the x-intercepts we solve $(x - 1)^2 - 2 = 0$:

$$(x - 1)^2 = 2$$

$$x - 1 = \pm\sqrt{2}$$

$$x = 1 \pm \sqrt{2}$$

The x-intercepts are $\left(1 - \sqrt{2}, 0\right)$ and $\left(1 + \sqrt{2}, 0\right)$. If the y-coordinate of a point on the graph is negative, then the x-coordinate satisfies $(x - 1)^2 - 2 < 0$. So the solution set to $(x - 1)^2 - 2 < 0$ is the open interval $\left(1 - \sqrt{2}, 1 + \sqrt{2}\right)$.

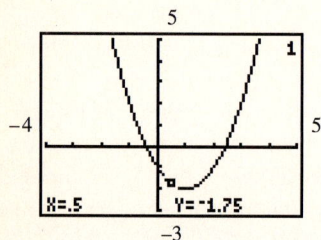

Use the TRACE feature on the graph of $y = (x - 1)^2 - 2$ to read the coordinates of the points. When $y < 0$, the x-coordinate satisfies the inequality.

Figure 3.58

Note that the solution set to $(x - 1)^2 - 2 \geq 0$ can also be obtained from the graph in Fig. 3.58. If the y-coordinate of a point on the graph is positive or zero, then the x-coordinate satisfies $(x - 1)^2 - 2 \geq 0$. So the solution set to $(x - 1)^2 - 2 \geq 0$ is $\left(-\infty, 1 - \sqrt{2}\right] \cup \left[1 + \sqrt{2}, \infty\right)$.

? For Thought

True or false? Explain.

1. The graph of $f(x) = (-x)^4$ is a reflection in the x-axis of the graph of $g(x) = x^4$.

2. The graph of $f(x) = x - 4$ lies four units to the right of the graph of $f(x) = x$.

3. The graph of $y = |x + 2| + 2$ is a translation two units to the right and two units upward of the graph of $y = |x|$.

4. The graph of $f(x) = -3$ is a reflection in the x-axis of the graph of $g(x) = 3$.

5. The functions $y = x^2 + 4x + 1$ and $y = (x + 2)^2 - 3$ have the same graph.

6. The graph of $y = -(x - 3)^2 - 4$ can be obtained by moving $y = x^2$ three units to the right and down four units, and then reflecting in the x-axis.

7. If $f(x) = -x^3 + 2x^2 - 3x + 5$, then $f(-x) = x^3 + 2x^2 + 3x + 5$.

8. The graphs of $f(x) = -\sqrt{x}$ and $g(x) = \sqrt{-x}$ are identical.

9. If $f(x) = x^3 - x$, then $f(-x) = -f(x)$.

10. The solution set to $|x| - 1 \leq 0$ is $[-1, 1]$.

3.4 Exercises Tape 6 Disk—5.25": 2 3.5": 1 Macintosh: 1

Sketch the graphs of each pair of functions on the same coordinate plane.

1. $f(x) = \sqrt{x}$, $g(x) = -\sqrt{x}$

2. $f(x) = x^2 + 1$, $g(x) = -x^2 - 1$

3. $y = x$, $y = -x$

4. $y = \sqrt{4 - x^2}$, $y = -\sqrt{4 - x^2}$

5. $f(x) = |x|$, $g(x) = |x| - 4$

6. $f(x) = \sqrt{x}$, $g(x) = \sqrt{x} + 3$

7. $f(x) = x$, $g(x) = x + 3$

8. $f(x) = x^2$, $g(x) = x^2 - 5$

9. $y = x^2$, $y = (x - 3)^2$

10. $y = |x|$, $y = |x + 2|$

11. $f(x) = x^3$, $g(x) = (x + 1)^3$

12. $f(x) = \sqrt{x}$, $g(x) = \sqrt{x - 3}$

13. $y = \sqrt{x}$, $y = 3\sqrt{x}$

14. $y = |x|$, $y = \frac{1}{3}|x|$

15. $y = x^2$, $y = \frac{1}{4}x^2$

16. $y = \sqrt{1 - x^2}$, $y = 4\sqrt{1 - x^2}$

Match each function in Exercises 17–24 with its graph (a)–(h).

17. $y = x^2$

18. $y = (x - 4)^2 + 2$

19. $y = (x + 4)^2 - 2$

20. $y = -2(x - 2)^2$

21. $y = -2(x + 2)^2$

22. $y = -\frac{1}{2}x^2 - 4$

23. $y = \frac{1}{2}(x + 4)^2 + 2$

24. $y = -2(x - 4)^2 - 2$

Use transformations to graph each function.

25. $y = (x - 1)^2 + 2$

26. $y = \sqrt{x + 5} - 4$

27. $y = |x - 1| + 3$

28. $y = |x + 3| - 4$

29. $y = 3x - 4$

30. $y = -4x + 2$

31. $y = \dfrac{1}{2}x - 2$

32. $y = -\dfrac{1}{2}x + 4$

33. $y = -\dfrac{1}{2}|x| + 4$

34. $y = 3|x| - 2$

35. $y = -\dfrac{1}{2}(|x| + 4)$

36. $y = 3(|x| - 2)$

37. $y = -\sqrt{x - 3} + 1$

38. $y = -(x + 2)^2 - 4$

39. $y = 2(x + 3)^2 - 4$

40. $y = 3(x + 1)^2 - 5$

41. $y = -2\sqrt{x + 3} + 2$

42. $y = -\dfrac{1}{2}\sqrt{x + 2} + 4$

43. $y = [\![x]\!] - 4$

44. $y = [\![x + 2]\!]$

Discuss the symmetry of the graph of each function and determine whether the function is even, odd, or neither.

45. $f(x) = x^4$

46. $f(x) = x^4 - 2x^2$

47. $f(x) = x^4 - x^3$

48. $f(x) = x^3 - x$

49. $f(x) = (x + 3)^2$

50. $f(x) = (x - 1)^2$

51. $f(x) = \sqrt{x}$

52. $f(x) = |x| - 9$

53. $f(x) = x$

54. $f(x) = -x$

55. $f(x) = 3x + 2$

56. $f(x) = x - 3$

57. $f(x) = x^3 - 5x + 1$

58. $f(x) = x^6 - x^4 + x^2$

59. $f(x) = |x - 2|$

60. $f(x) = (x^2 - 2)^3$

61. $f(x) = 1 + \dfrac{1}{x^2}$

62. $f(x) = \sqrt{9 - x^2}$

63. $f(x) = |x^2 - 3|$

64. $f(x) = \sqrt{x^2 + 3}$

Match each function with its graph (a)–(h).

65. $y = 2 + \sqrt{x}$

66. $y = \sqrt{2 + x}$

67. $y = \sqrt{x^2}$

68. $y = \sqrt{\dfrac{x}{2}}$

69. $y = \dfrac{1}{2}\sqrt{x}$

70. $y = 2 - \sqrt{x - 2}$

71. $y = -2\sqrt{x}$

72. $y = -\sqrt{-x}$

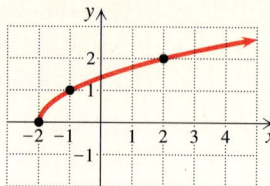

(a)

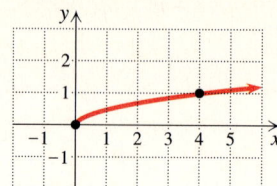

(b)

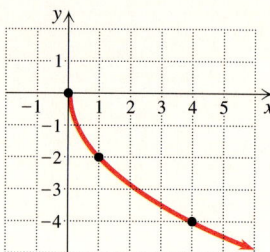

(c)

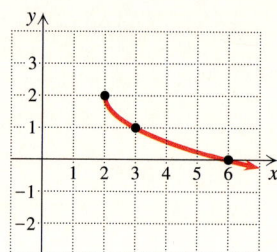

(d)

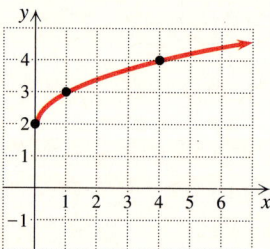

(e)

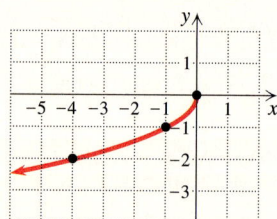

(f)

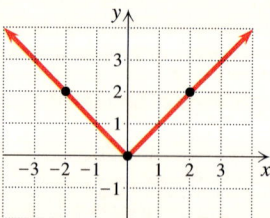

(g)

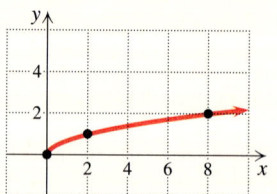

(h)

Solve each inequality by reading the corresponding graph.

73. $x^2 - 1 \geq 0$

$y = x^2 - 1$

$(-1, 0)$ $(1, 0)$

74. $2x^2 - 3 < 0$

$y = 2x^2 - 3$

$\left(-\frac{\sqrt{6}}{2}, 0\right)$ $\left(\frac{\sqrt{6}}{2}, 0\right)$

75. $|x - 2| - 3 > 0$

$y = |x - 2| - 3$

$(5, 0)$

$(-1, 0)$

76. $2 - |x + 1| \geq 0$

$y = 2 - |x + 1|$

$(-3, 0)$ $(1, 0)$

Solve each inequality by graphing an appropriate function. State the solution set using interval notation.

77. $(x - 1)^2 - 9 < 0$

78. $\left(x - \frac{1}{2}\right)^2 - \frac{9}{4} \geq 0$

79. $5 - \sqrt{x} \geq 0$

80. $\sqrt{x + 3} - 2 \geq 0$

81. $(x - 2)^2 > 3$

82. $(x - 1)^2 < 4$

83. $\sqrt{25 - x^2} > 0$

84. $\sqrt{4 - x^2} \geq 0$

Solve each problem.

85. *Across-the-board raise* Each teacher at C. F. Gauss Elementary School is given an across-the-board raise of $2000. Write a function that *transforms* each old salary x into a new salary $N(x)$.

86. *Cost-of-living raise* Each registered nurse at Blue Hills Memorial Hospital is first given a 5% cost-of-living raise and then a $3000 merit raise. Write a function that *transforms* each old salary x into a new salary $N(x)$. Does it make any difference in which order these raises are given? Explain.

87. *Unemployment versus inflation* The Phillips curve shows the relationship between the unemployment rate x and the inflation rate y. If the equation of the curve is $y = 1 - \sqrt{x}$ for a certain Third World country, then for what values of x is the inflation rate less than 50%?

Phillips Curve

$y = 1 - \sqrt{x}$

Figure for Exercise 87

88. *Production function* The production function shows the relationship between inputs and outputs. A manufacturer of custom windows produces y windows per week using x hours of labor per week, where $y = 1.75\sqrt{x}$. How many hours of labor are required to keep production at or above 28 windows per week?

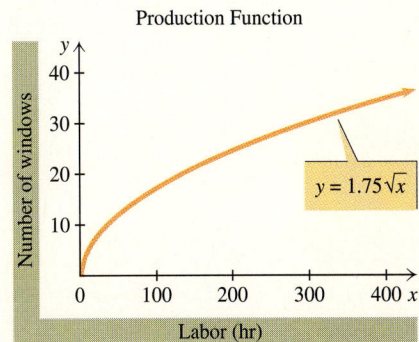

Production Function

$y = 1.75\sqrt{x}$

Figure for Exercise 88

Graphing Calculator Exercises

Many of the exercises in this text involve simple numbers so that we can concentrate on the ideas without getting lost in computation. When the numbers are not as nice, a graphing calculator can be a big help. However, the graphing calculator will usually give only approximate answers.

1. Graph the function $y = \sqrt{2} \cdot x^2 + \pi x - 0.03$ on a graphing calculator. Use the trace feature of the calculator to estimate the range of the function.

2. Graph $y = x^3 + 6x^2 + 12x + 8$ on a graphing calculator. This graph is a transformation of a graph that we have seen in this section. Use this fact to factor the polynomial $x^3 + 6x^2 + 12x + 8$.

Graph each pair of functions (without simplifying the second function) on the same screen and explain what each exercise illustrates.

3. $y = x^4 - x^2, y = (-x)^4 - (-x)^2$

4. $y = x^3 - x, y = (-x)^3 - (-x)$

5. $y = x^4 - x^2, y = (x + 1)^4 - (x + 1)^2$

6. $y = x^3 - x, y = (x - 2)^3 - (x - 2) + 3$

Find an approximate solution to each inequality by reading the graph of an appropriate function.

7. $\sqrt{3}x^2 + \pi x - 9 < 0$ 8. $x^3 - 5x^2 + 6x - 1 > 0$

3.5

Operations with

Functions

In Sections 3.3 and 3.4 we studied the graphs of functions to see the relationships between types of functions and their graphs. In this section we will study various ways in which two or more functions can be combined to make new functions. The emphasis here will be on formulas that define functions.

Basic Operations with Functions

A college student is hired to deliver new telephone books and collect the old ones for recycling. She is paid $4 per hour plus $0.30 for each old phone book she collects. Her salary for a 40-hour week is a function of the number of phone books collected. If x represents the number of phone books collected in one week, then the function $S(x) = 0.30x + 160$ gives her salary in dollars. However, she must use her own car for this job. She figures that her car expenses average $0.20 per phone book collected plus a fixed cost of $20 per week for insurance. We can write her expenses as a function of the number of phone books collected, $E(x) = 0.20x + 20$. Her profit for one week is her salary minus her expenses:

$$
\begin{aligned}
P(x) &= S(x) - E(x) \\
&= 0.30x + 160 - (0.20x + 20) \\
&= 0.10x + 140
\end{aligned}
$$

By subtracting, we get $P(x) = 0.10x + 140$. Her weekly profit is written as a function of the number of phone books collected. In this example we obtained a new function by subtracting two functions. In general, there are four basic arithmetic operations defined for functions.

Definition: Sum, Difference, Product, and Quotient Functions

For two functions f and g, the **sum, difference, product,** and **quotient functions,** functions $f + g, f - g, f \cdot g$, and f/g, respectively, are defined as follows:

$$(f + g)(x) = f(x) + g(x)$$

$$(f - g)(x) = f(x) - g(x)$$

$$(f \cdot g)(x) = f(x) \cdot g(x)$$

$$(f/g)(x) = f(x)/g(x) \qquad \text{provided that } g(x) \neq 0.$$

```
Y₁█3√X-2
Y₂█X²+5
Y₃=
Y₄=
Y₅=
Y₆=
Y₇=
Y₈=
```

Define the functions f and g on your calculator.

```
Y₁(4)+Y₂(4)
               25
Y₁(0)*Y₂(0)
              -10
■
```

Use y_1 and y_2 to find $(f + g)(4) = 25$ and $(f \cdot g)(0) = -10$.

Example 1   Evaluating functions

Let $f(x) = 3\sqrt{x} - 2$ and $g(x) = x^2 + 5$. Find and simplify each expression.

a) $(f + g)(4)$

b) $(f - g)(x)$

c) $(f \cdot g)(0)$

d) $\left(\dfrac{f}{g}\right)(9)$

Solution

a) $(f + g)(4) = f(4) + g(4) = 3\sqrt{4} - 2 + 4^2 + 5 = 25$

b) $(f - g)(x) = f(x) - g(x) = 3\sqrt{x} - 2 - (x^2 + 5) = 3\sqrt{x} - x^2 - 7$

c) $(f \cdot g)(0) = f(0) \cdot g(0) = \left(3\sqrt{0} - 2\right)(0^2 + 5) = (-2)(5) = -10$

d) $\left(\dfrac{f}{g}\right)(9) = \dfrac{f(9)}{g(9)} = \dfrac{3\sqrt{9} - 2}{9^2 + 5} = \dfrac{7}{86}$

Think of $f + g, f - g, f \cdot g$, and f/g as generic names for the sum, difference, product, and quotient of the functions f and g. If any of these functions has a particular meaning, as in the phone-book example, we can use a new letter to identify it. The domain of $f + g, f - g, f \cdot g$, or f/g is the intersection of the domain of f with the domain of g. Of course, we exclude from the domain of f/g any number for which $g(x) = 0$.

Example 2 The sum, product, and quotient functions

Let $f = \{(1, 3), (2, 8), (3, 6), (5, 9)\}$ and $g = \{(1, 6), (2, 11), (3, 0), (4, 1)\}$. Find $f + g, f \cdot g$, and f/g. State the domain of each function.

Solution

The domain of $f + g$ and $f \cdot g$ is $\{1, 2, 3\}$ because that is the intersection of the domains of f and g. The ordered pair $(1, 9)$ belongs to $f + g$ because

$$(f + g)(1) = f(1) + g(1) = 3 + 6 = 9.$$

The ordered pair $(2, 19)$ belongs to $f + g$ because $(f + g)(2) = 19$. The pair $(3, 6)$ also belongs to $f + g$. So

$$f + g = \{(1, 9), (2, 19), (3, 6)\}.$$

Since $(f \cdot g)(1) = f(1) \cdot g(1) = 3 \cdot 6 = 18$, the pair $(1, 18)$ belongs to $f \cdot g$. Likewise, $(2, 88)$ and $(3, 0)$ also belong to $f \cdot g$. So

$$f \cdot g = \{(1, 18), (2, 88), (3, 0)\}.$$

The domain of f/g is $\{1, 2\}$ because $g(3) = 0$. So

$$\frac{f}{g} = \left\{ \left(1, \frac{1}{2}\right), \left(2, \frac{8}{11}\right) \right\}.$$

$\blacklozenge$

In Example 2 the functions are given as sets of ordered pairs, and the results of performing operations with these functions are sets of ordered pairs. In the next example the sets of ordered pairs are defined by means of equations, so the result of performing operations with these functions will be new equations that determine the ordered pairs of the function.

Example 3 The sum, quotient, product, and difference functions

Let $f(x) = \sqrt{x}$, $g(x) = 3x + 1$, and $h(x) = x - 1$. Find each function and state its domain.

a) $f + g$ **b)** $\dfrac{g}{f}$ **c)** $g \cdot h$ **d)** $g - h$

Solution

a) Since the domain of f is $[0, \infty)$ and the domain of g is $(-\infty, \infty)$, the domain of $f + g$ is $[0, \infty)$. Since $(f + g)(x) = f(x) + g(x) = \sqrt{x} + 3x + 1$, the equation defining the function $f + g$ is

$$(f + g)(x) = \sqrt{x} + 3x + 1.$$

b) The number 0 is not in the domain of g/f because $f(0) = 0$. So the domain of g/f is $(0, \infty)$. The equation defining g/f is

$$\left(\frac{g}{f}\right)(x) = \frac{3x + 1}{\sqrt{x}}.$$

c) The domain of both g and h is $(-\infty, \infty)$. So the domain of $g \cdot h$ is $(-\infty, \infty)$. Since $(3x + 1)(x - 1) = 3x^2 - 2x - 1$, the equation defining the function $g \cdot h$ is

$$(g \cdot h)(x) = 3x^2 - 2x - 1.$$

d) The domain of both g and h is $(-\infty, \infty)$. So the domain of $g - h$ is $(-\infty, \infty)$. Since $(3x + 1) - (x - 1) = 2x + 2$, the equation defining $g - h$ is

$$(g - h)(x) = 2x + 2. \qquad \blacklozenge$$

Composition of Functions

The area of a circle is a function of the radius, $A = \pi r^2$. The radius is a function of the diameter, $r = d/2$. So the area is also a function of the diameter. The formula for A as a function of the diameter is obtained by substituting $d/2$ for r:

$$A = \pi \frac{d^2}{4}$$

This example illustrates the **composition of functions.**

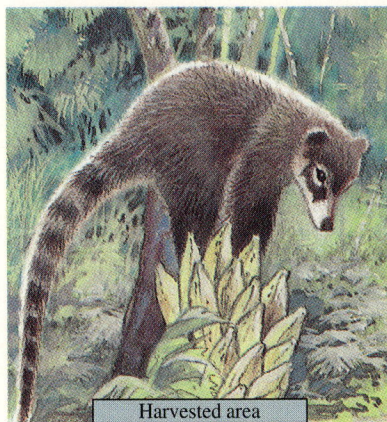

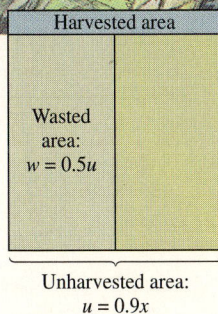

Example 4   Composition of functions: rainforest application

Tropical rainforests are one of the earth's most ancient and complex ecosystems, yet each year deforestation claims an area the size of West Virginia. In "Deforestation in the Tropics" (*Scientific American,* April 1990) it is stated that loggers typically harvest only 10% of the area in a rainforest. However, the loggers destroy 50% of the area not harvested just to get at the valuable trees. In a forest of x square kilometers, the function $u = 0.9x$ gives the unharvested area when the valuable timber is removed. The function $w = 0.5u$ gives the amount of unharvested area that is destroyed and wasted. Write the wasted area w as a function of the original area x.

Solution

Since w is a function of u and u is a function of x, w is a function of x. Writing the composition of the two functions is simply a matter of replacing u by $0.9x$ in the formula for w:

$$w = 0.5u = 0.5(0.9x) = 0.45x$$

The composition function $w = 0.45x$ indicates that 45% of the area of a rainforest is wasted when the valuable timber is harvested. $\qquad \blacklozenge$

Harvested area

Wasted
area:
$w = 0.5u$

Unharvested area:
$u = 0.9x$

Example 4 illustrates how composition is handled using functions defined by formulas. The salary of the phone-book collector mentioned at the start of this section was given in the f-notation as $S(x) = 0.30x + 160$, where x is the number of phone books collected. Now suppose that 20% of her salary is withheld for income taxes. The withholding function is $W(x) = 0.20x$, where x is her salary. The salary is a function of the number of phone books, and the withholding is a function of the salary, so the withholding is a function of the number of phone books, and we can write

$$W(S(x)) = W(0.30x + 160) = 0.20(0.30x + 160) = 0.06x + 32.$$

We have a new function pairing the number of phone books collected with the amount of withholding. We have found the composition of the functions W and S. The notation $W \circ S$ is used for the composition of W and S. In this case, $(W \circ S)(x) = 0.06x + 32$. For example, if the student collects 100 phone books, the amount of withholding is $(W \circ S)(100) = 0.06(100) + 32 = \$38$.

Definition: Composition of Functions

If f and g are two functions, the **composition** of f and g, written $f \circ g$, is defined by the equation

$$(f \circ g)(x) = f(g(x)),$$

provided that $g(x)$ is in the domain of f. The composition of g and f is written $g \circ f$.

Note that $f \circ g$ is not the same function as $f \cdot g$, the product of f and g.

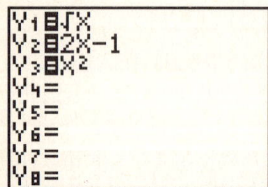

```
Y₁⊟√X
Y₂⊟2X−1
Y₃⊟X²
Y₄=
Y₅=
Y₆=
Y₇=
Y₈=
```

Use the Y= key to define the functions f, g, and h.

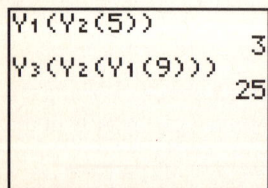

```
Y₁(Y₂(5))
                3
Y₃(Y₂(Y₁(9)))
               25
```

Use the Y-VARS key to find the compositions $(f \circ g)(5)$ and $(h \circ g \circ f)(9)$.

Example 5 Evaluating compositions defined by equations

Let $f(x) = \sqrt{x}$, $g(x) = 2x - 1$, and $h(x) = x^2$. Find the value of each expression.

a) $(f \circ g)(5)$ **b)** $(g \circ f)(5)$ **c)** $(h \circ g \circ f)(9)$

Solution

a) $(f \circ g)(5) = f(g(5))$ Definition of composition

$\qquad\qquad\quad = f(9)$ $g(5) = 2 \cdot 5 - 1 = 9$

$\qquad\qquad\quad = \sqrt{9}$

$\qquad\qquad\quad = 3$

b) $(g \circ f)(5) = g(f(5)) = g(\sqrt{5}) = 2\sqrt{5} - 1$

c) $(h \circ g \circ f)(9) = h(g(f(9))) = h(g(3)) = h(5) = 5^2 = 25$

Note that for the composition $f \circ g$ to be defined at x, $g(x)$ must be in the domain of f. So the domain of $f \circ g$ is the set of all values of x in the domain of g for which $g(x)$ is in the domain of f. The diagram shown in Fig. 3.59 will help you to understand the composition of functions. In Example 5 we found specific values of compositions. In the next example we will find the entire composition function.

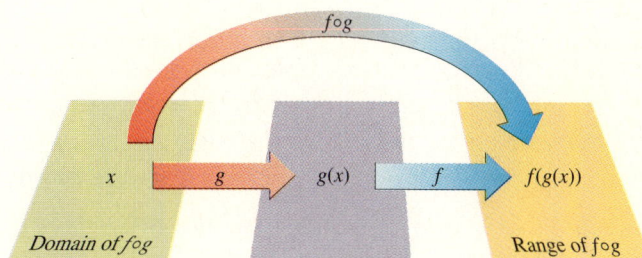

Figure 3.59

Example 6   Composition of functions defined by sets

Let $g = \{(1, 4), (2, 5), (3, 6)\}$ and $f = \{(3, 8), (4, 9), (5, 10)\}$. Find $f \circ g$.

Solution

Since $g(1) = 4$ and $f(4) = 9$, $(f \circ g)(1) = 9$. So the ordered pair $(1, 9)$ is in $f \circ g$. Since $g(2) = 5$ and $f(5) = 10$, $(f \circ g)(2) = 10$. So $(2, 10)$ is in $f \circ g$. Now $g(3) = 6$, but 6 is not in the domain of f. So there are only two ordered pairs in $f \circ g$:

$$f \circ g = \{(1, 9), (2, 10)\}$$

In Example 6, the domain of g is $\{1, 2, 3\}$ while the domain of $f \circ g$ is $\{1, 2\}$. To find the domain of $f \circ g$, we remove from the domain of g any number x such that $g(x)$ is not in the domain of f. In the next example we find compositions of functions defined by equations.

Example 7   Composition of functions defined by equations

Let $f(x) = \sqrt{x}$, $g(x) = 2x - 1$, and $h(x) = x^2$. Find each composition function and state its domain.

a) $f \circ g$ **b)** $g \circ f$ **c)** $h \circ g$

Solution

a) In the composition $f \circ g$, $g(x)$ must be in the domain of f. The domain of f is $[0, \infty)$. If $g(x)$ is in $[0, \infty)$, then $2x - 1 \geq 0$, or $x \geq \frac{1}{2}$. So the domain of $f \circ g$ is $[\frac{1}{2}, \infty)$. Since

$$(f \circ g)(x) = f(g(x)) = f(2x - 1) = \sqrt{2x - 1},$$

the function $f \circ g$ is defined by the equation $(f \circ g)(x) = \sqrt{2x - 1}$, for $x \geq \frac{1}{2}$.

b) Since the domain of g is $(-\infty, \infty)$, $f(x)$ is certainly in the domain of g. So the domain of $g \circ f$ is the same as the domain of f, $[0, \infty)$. Since

$$(g \circ f)(x) = g(f(x)) = g(\sqrt{x}) = 2\sqrt{x} - 1,$$

the function $g \circ f$ is defined by the equation $(g \circ f)(x) = 2\sqrt{x} - 1$. Note that $g \circ f$ is generally not equal to $f \circ g$, but in Section 3.6 we will study special types of functions for which they are equal.

c) Since the domain of h is $(-\infty, \infty)$, $g(x)$ is certainly in the domain of h. So the domain of $h \circ g$ is the same as the domain of g, $(-\infty, \infty)$. Since

$$(h \circ g)(x) = h(g(x)) = h(2x - 1) = (2x - 1)^2 = 4x^2 - 4x + 1,$$

the function $h \circ g$ is defined by $(h \circ g)(x) = 4x^2 - 4x + 1$. ◆

When studying rates of change of functions in calculus, we often need to look at a function that has several operations as a composition of simpler functions. We can also better understand the transformations of graphs in Section 3.4 when we think of a complicated function as a composition of simpler functions. For example, to find the second coordinate of an ordered pair for the function $H(x) = (x + 3)^2$, we start with x, add 3 to x, then square the result. These two operations can be accomplished by composition, using $f(x) = x + 3$ followed by $g(x) = x^2$:

$$(g \circ f)(x) = g(f(x)) = g(x + 3) = (x + 3)^2$$

So the function H is the same as the composition of g and f, $H = g \circ f$. Notice that $(f \circ g)(x) = x^2 + 3$ and it is not the same as $H(x)$.

Example 8 Writing a function as a composition

Let $f(x) = \sqrt{x}$, $g(x) = x - 3$, and $h(x) = 2x$. Write each given function as a composition of appropriate functions chosen from f, g, and h.

a) $F(x) = \sqrt{x - 3}$ **b)** $G(x) = x - 6$ **c)** $H(x) = 2\sqrt{x} - 3$

Solution

a) The function F consists of subtracting 3 from x and finding the square root of that result. These two operations can be accomplished by composition, using g followed by f. So $F = f \circ g$. Check this answer as follows:

$$(f \circ g)(x) = f(g(x)) = f(x - 3) = \sqrt{x - 3} = F(x)$$

b) Subtracting 6 from x can be accomplished by subtracting 3 from x and then subtracting 3 from that result, so $G = g \circ g$. Check as follows:

$$(g \circ g)(x) = g(g(x)) = g(x - 3) = (x - 3) - 3 = x - 6 = G(x)$$

c) For the function H, find the square root of x, then multiply by 2, and finally subtract 3. These three operations can be accomplished by composition, using f, then h, and then g. So $H = g \circ h \circ f$. Check as follows:

$$(g \circ h \circ f)(x) = g(h(f(x))) = g(h(\sqrt{x})) = g(2\sqrt{x}) = 2\sqrt{x} - 3 = H(x)$$ ◆

For Thought

?

True or false? Explain.

1. If $f = \{(2, 4)\}$ and $g = \{(1, 5)\}$, then $f + g = \{(3, 9)\}$.
2. If $f = \{(1, 6), (9, 5)\}$ and $g = \{(1, 3), (9, 0)\}$, then $f/g = \{(1, 2)\}$.
3. If $f = \{(1, 6), (9, 5)\}$ and $g = \{(1, 3), (9, 0)\}$, then $f \cdot g = \{(1, 18), (9, 0)\}$.
4. If $f(x) = x + 2$ and $g(x) = x - 3$, then $(f \cdot g)(5) = 14$.
5. If $s = P/4$ and $A = s^2$, then A is a function of P.
6. If $f(3) = 19$ and $g(19) = 99$, then $(g \circ f)(3) = 99$.
7. If $f(x) = \sqrt{x}$ and $g(x) = x - 2$, then $(f \circ g)(x) = \sqrt{x} - 2$.
8. If $f(x) = 5x$ and $g(x) = x/5$, then $(f \circ g)(x) = (g \circ f)(x) = x$.
9. If $F(x) = (x - 9)^2$, $g(x) = x^2$, and $h(x) = x - 9$, then $F = h \circ g$.
10. If $f(x) = x^3$, then $(f \circ f)(x) = x^6$.

3.5 Exercises

⊏▭⊐ **Tape 6**   💾 **Disk—5.25″: 2   3.5″: 1   Macintosh: 1**

Let $f(x) = x - 3$ and $g(x) = x^2 - x$. Find and simplify each expression.

1. $(f + g)(2)$
2. $(g + f)(3)$
3. $(f - g)(-2)$
4. $(g - f)(-6)$
5. $(f \cdot g)(-1)$
6. $(g \cdot f)(0)$
7. $(f/g)(4)$
8. $(g/f)(4)$
9. $(f + g)(a)$
10. $(f - g)(b)$
11. $(f \cdot g)(a)$
12. $(f/g)(b)$

Use the two given functions to write y as a function of x.

13. $y = 2a - 3, a = 3x + 1$
14. $y = -4d - 1, d = -3x - 2$
15. $y = w^2 - 2, w = x + 3$
16. $y = 3t^2 - 3, t = x - 1$
17. $y = 3m - 1, m = \dfrac{x + 1}{3}$
18. $y = 2z + 5, z = \dfrac{1}{2}x - \dfrac{5}{2}$
19. $y = 2k^3 - 1, k = \sqrt[3]{\dfrac{x + 1}{2}}$
20. $y = \dfrac{s^3 + 1}{5}, s = \sqrt[3]{5x - 1}$

Let $f = \{(-3, 1), (0, 4), (2, 0)\}$, $g = \{(-3, 2), (1, 2), (2, 6), (4, 0)\}$, and $h = \{(2, 4) (1, 0)\}$. Find each function.

21. $f + g$
22. $f - g$
23. g/f
24. f/g
25. $f \cdot g$
26. $g - f$
27. $f \circ g$
28. $g \circ f$
29. $f \circ h$
30. $h \circ f$
31. $h \circ g$
32. $g \circ h$

Let $f(x) = \sqrt{x}$, $g(x) = x - 4$, and $h(x) = \dfrac{1}{x - 2}$.

Find an equation defining each function and state the domain of the function.

33. $f + g$
34. $f + h$
35. $g \cdot h$
36. g/f
37. f/g
38. $f - h$
39. g/h
40. $f \cdot h$

Let $f(x) = 3x - 1$, $g(x) = x^2 + 1$, and $h(x) = \dfrac{x + 1}{3}$.

Find and simplify each expression.

41. $f(g(-1))$
42. $g(f(-1))$
43. $(f \circ h)(5)$
44. $(h \circ f)(-7)$
45. $(f \circ g)(4.39)$
46. $(g \circ h)(-9.87)$
47. $f(g(x))$
48. $g(f(x))$
49. $(g \circ f)(x)$
50. $(f \circ g)(x)$
51. $(h \circ g)(x)$
52. $(g \circ h)(x)$
53. $(f \circ f)(x)$
54. $(g \circ g)(x)$
55. $(h \circ g \circ f)(x)$
56. $(f \circ g \circ h)(x)$

Let $f(x) = \sqrt{x}$, $g(x) = 2x - 1$, and $h(x) = \dfrac{1}{x - 3}$.

Find an equation defining each function and state the domain of the function.

57. $f \circ g$ **58.** $h \circ g$ **59.** $h \circ f$ **60.** $f \circ h$

61. $h \circ h$ **62.** $g \circ h$ **63.** $f \circ f$ **64.** $f \circ h \circ g$

Show that $(f \circ g)(x) = x$ **and** $(g \circ f)(x) = x$ **for each given pair of functions.**

65. $f(x) = \sqrt[3]{4 - 3x}$, $g(x) = \dfrac{4 - x^3}{3}$

66. $f(x) = x^3 + 5$, $g(x) = \sqrt[3]{x - 5}$

67. $f(x) = \dfrac{x - 1}{x + 1}$, $g(x) = \dfrac{x + 1}{1 - x}$

68. $f(x) = \dfrac{x + 1}{x - 3}$, $g(x) = \dfrac{3x + 1}{x - 1}$

69. $f(x) = \sqrt[3]{2x^5 - 1}$, $g(x) = \sqrt[5]{\dfrac{x^3 + 1}{2}}$

70. $f(x) = x^{3/5} - 3$, $g(x) = (x + 3)^{5/3}$

Let $f(x) = |x|$, $g(x) = x - 7$, and $h(x) = x^2$. **Write each of the following functions as a composition of functions chosen from** f, g, **and** h.

71. $F(x) = x^2 - 7$ **72.** $G(x) = |x| - 7$

73. $H(x) = |x^2 - 7|$ **74.** $M(x) = x^2 - 14x + 49$

75. $N(x) = (|x| - 7)^2$ **76.** $R(x) = |x - 7|$

77. $P(x) = |x - 7| - 7$ **78.** $C(x) = (x^2 - 7)^2$

Solve each problem.

79. Write the area A of a square with a side of length s as a function of its diagonal d.

80. Write the perimeter of a square P as a function of the area A.

81. *Destruction of a rainforest* The function $f(x) = 0.1x$ gives the area used when a rainforest of area x is harvested. The function $g(x) = 0.45x$ found in Example 4 gives the area wasted as a function of the initial area x. Find and interpret the function $f + g$.

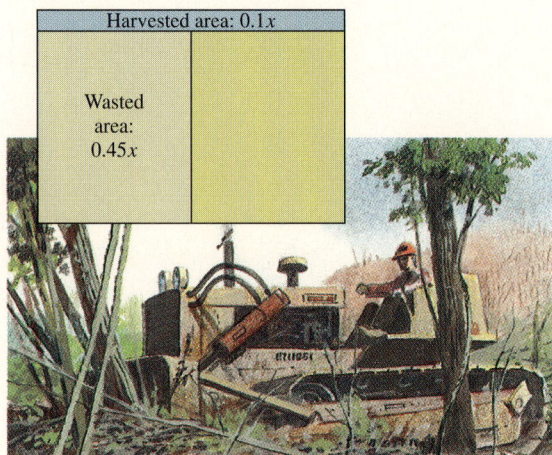

Figure for Exercise 81

82. *Deforestation in Nigeria* In Nigeria deforestation occurs at the rate of about 5.2% per year. If x is the total forest area of Nigeria at the start of 1994, then $f(x) = 0.948x$ is the amount of forest land in Nigeria at the start of 1995. Find and interpret $(f \circ f)(x)$. Find and interpret $(f \circ f \circ f)(x)$.

83. *Area of a window* A window is in the shape of a square with a side of length s, with a semicircle of diameter w adjoining the top of the square. Write the total area of the window W as a function of s.

84. *Area of a window* Using the window of Exercise 83, write the area of the square A as a function of the area of the semicircle S.

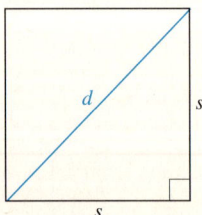

Figure for Exercises 79 and 80

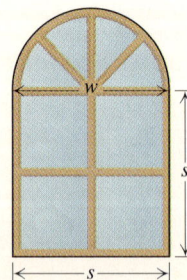

Figure for Exercises 83 and 84

85. *Packing a square piece of glass* A glass prism with a square cross section is to be shipped in a cylindrical cardboard tube that has an inside diameter of d inches. Given that s is the length of a side of the square and the glass fits snugly into the tube, write s as a function of d.

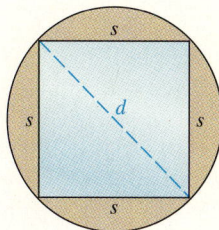

Figure for Exercise 85

86. *Packing a triangular piece of glass* A glass prism is to be shipped in a cylindrical cardboard tube that has an inside diameter of d inches. Given that the cross section of the prism is an equilateral triangle with side of length p and the prism fits snugly into the tube, write p as a function of d.

Figure for Exercise 86

87. *Spreading glue* Frank is designing a notched trowel for spreading glue. The notches on the trowel are in the shape of semicircles. The diameter of each notch and the space between consecutive notches is d inches. Write a formula expressing the number of square feet that a gallon of glue will cover as a function of the diameter of the notch d. Use the fact that 1 ft³ is approximately 7.5 gal. State any assumptions that you make in solving this problem.

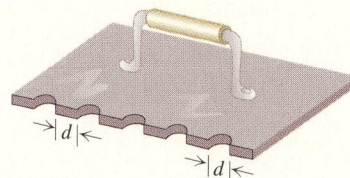

Figure for Exercise 87

88. *Saving energy* Caulking for sealing cracks around windows and doors is sold in a tube containing 20 in³ of caulking. The contents of one tube will make a cylindrical bead with diameter d that is n inches long. Write n as a function of d.

89. *Profit* The revenue in dollars that a company receives for installing x alarm systems per month is given by $R(x) = 3000x - 20x^2$, while the cost in dollars is given by $C(x) = 600x + 4000$. The function $P(x) = R(x) - C(x)$ gives the profit for installing x alarm systems per month. Find $P(x)$ and simplify it. Find $P(1)$, $P(5)$, and $P(122)$.

90. *Marginal profit* Use the functions $P(x)$, $R(x)$, and $C(x)$ from the last exercise to find the marginal profit function $MP(x)$, marginal revenue function $MR(x)$, and marginal cost function $MC(x)$. (See Exercises 99 and 100 of Section 3.2.) Is $MP(x) = MR(x) - MC(x)$?

Graphing Calculator Exercises

Graph each function without simplifying the given expression. Examine the graph and write a function for the graph. Then simplify the original function and see whether it is the same as your function. Is the domain of the original function equal to the domain of the function after it is simplified?

1. $y = ((x - 1)/(x + 1) + 1)/((x - 1)/(x + 1) - 1)$

2. $y = ((3x + 1)/(x - 1) + 1)/((3x + 1)/(x - 1) - 3)$

Most graphing calculators have the capability of combining functions. Define $y_1 = \sqrt{x + 1}$ and $y_2 = 3x - 4$ on your calculator. For each function y_3, defined in terms of y_1 and y_2, determine the domain and range of y_3 from its graph on your calculator and explain what each graph illustrates.

3. $y_3 = y_1 + y_2$ **4.** $y_3 = 3y_1 - 4$ **5.** $y_3 = \sqrt{y_2 + 1}$

Define $y_1 = \sqrt[3]{x}$, $y_2 = \sqrt{x}$, and $y_3 = x + 4$. For each function y_4, determine the domain and range of y_4 from its graph on your calculator and explain what each graph illustrates.

6. $y_4 = y_1 + y_2 + y_3$ **7.** $y_4 = \sqrt{y_1 + 4}$

3.6

Inverse Functions

It is possible for one function to undo what another function does. For example, squaring undoes the operation of taking a square root. The composition of two such functions is the identity function. In this section we explore this idea in detail.

Definitions

Consider again the medium pizza that costs $5 plus $2 per topping. Table 3.5 shows the ordered pairs of the function that determines the cost. If we know the cost of a pizza, can we determine how many toppings it has? We can, and Table 3.6 shows ordered pairs where the number of toppings is determined by the cost. Of course we could just read Table 3.5 backwards, but we make a new table because we want a function for finding the number of toppings depending on the cost. The function in Table 3.6 is the **inverse function** of the function in Table 3.5.

Toppings x	Cost y
0	$ 5
1	7
2	9
3	11
4	13

Table 3.5

Cost x	Toppings y
$ 5	0
7	1
9	2
11	3
13	4

Table 3.6

A function is a set of ordered pairs in which no two ordered pairs have the same first coordinates and different second coordinates. If we interchange the x- and y-coordinates in each ordered pair of a function, as in Tables 3.5 and 3.6, the resulting set of ordered pairs might or might not be a function. For example, if $f = \{(1, 3), (2, 4)\}$, then the set $\{(3, 1), (4, 2)\}$ is a function. However, if $f = \{(4, 6), (5, 6)\}$, then the set $\{(6, 4), (6, 5)\}$ is *not* a function. If the set obtained by interchanging the coordinates of each ordered pair of a function is a function, then the function is called an **invertible function.**

Definition: Inverse Function

The **inverse** of an invertible function f is the function f^{-1} (read "f inverse"), where the ordered pairs of f^{-1} are obtained by interchanging the coordinates in each ordered pair of f.

In this notation, the number -1 in f^{-1} does not represent a negative exponent. It is merely a symbol for denoting the inverse function.

Example 1   Finding an inverse function

Let $f = \{(1, 3), (2, 4), (5, 7)\}$. Find $f^{-1}, f^{-1}(3)$, and $(f^{-1} \circ f)(1)$.

Solution

Interchange the x- and y-coordinates of each ordered pair of f to find f^{-1}:

$$f^{-1} = \{(3, 1), (4, 2), (7, 5)\}$$

To find the value of $f^{-1}(3)$, notice that $f^{-1}(3)$ is the second coordinate when the first coordinate is 3 in the function f^{-1}. So $f^{-1}(3) = 1$. To find the composition, use the definition of composition of functions:

$$(f^{-1} \circ f)(1) = f^{-1}(f(1)) = f^{-1}(3) = 1$$

Since the coordinates in the ordered pairs are interchanged, the domain of f^{-1} is the range of f, and the range of f^{-1} is the domain of f. If f^{-1} is the inverse function of f, then certainly f is the inverse of f^{-1}. The functions f and f^{-1} are inverses of each other.

Whether a function is invertible depends on the ordered pairs of the function f. If f has ordered pairs with different first coordinates and the same second coordinate, then f is not invertible because the definition of function is violated when the ordered pairs are reversed.

Definition: One-to-One Function

If a function has no two ordered pairs with different first coordinates and the same second coordinate, then the function is called **one-to-one.**

In a one-to-one function, each member of the domain corresponds to a unique member of the range. The following theorem describes the invertible functions.

Theorem: Invertible Functions

A function is invertible if and only if it is a one-to-one function.

Example 2   Finding an inverse function

For each function, determine whether it is invertible. If it is invertible, then find the inverse.

a) $f = \{(-2, 3), (4, 5), (2, 3)\}$

b) $g = \{(3, 1), (5, 2), (7, 4), (9, 8)\}$

Solution

a) This function f is *not* one-to-one because of the ordered pairs $(-2, 3)$ and $(2, 3)$. So f is not invertible.

b) The function g is one-to-one, and so g is invertible. The inverse of g is the function $g^{-1} = \{(1, 3), (2, 5), (4, 7), (8, 9)\}$.

If the ordered pairs of an invertible function are defined by means of an equation, we can obtain the inverse function by simply interchanging x and y in the equation. Interchanging x and y in the equation causes the first and second coordinates of each ordered pair to be interchanged.

Example 3   Finding an inverse function

Given that $f = \{(x, y)\,|\,y = 2x\}$, find f^{-1}.

Solution

To interchange the coordinates in all of the ordered pairs, interchange x and y in the equation:

$$f^{-1} = \{(x, y)\,|\,x = 2y\} = \left\{(x, y)\,\Big|\,y = \frac{1}{2}x\right\}$$

For an ordered pair such as $(3, 6)$ in f, $(6, 3)$ is in f^{-1}.

Determining Whether a Function Is One-to-One

If a function is given as a short list of ordered pairs, then it is easy to determine whether the function is one-to-one. A graph of a function can also be used to determine whether a function is one-to-one using the **horizontal line test.**

Horizontal Line Test

If each horizontal line crosses the graph of a function at no more than one point, then the function is one-to-one.

The graph of a one-to-one function never has the same y-coordinate for two different x-coordinates on the graph. So if it is possible to draw a horizontal line that crosses the graph of a function two or more times, then the function is not one-to-one. The horizontal line test is a quick visual method that is limited by the accuracy of the graph.

Example 4   The horizontal line test

Use the horizontal line test to determine whether the functions shown in Fig. 3.60 are one-to-one.

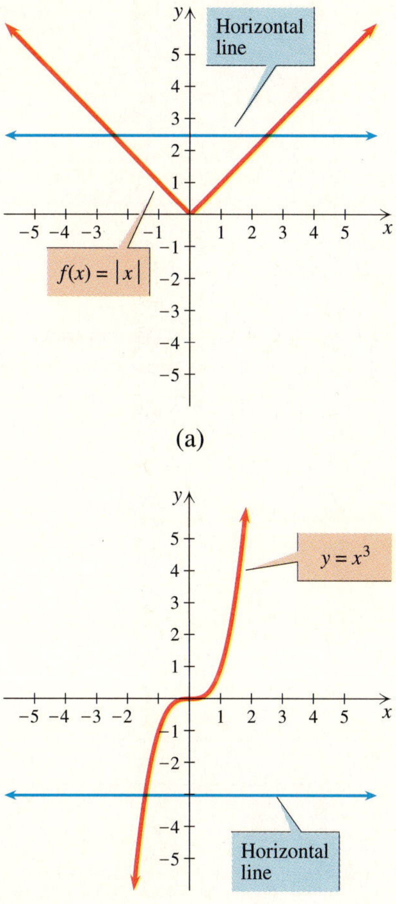

Solution

The function $f(x) = |x|$ is not one-to-one because it is possible to draw a horizontal line that crosses the graph twice as shown in Fig. 3.60(a). The function $y = x^3$ in Fig. 3.60(b) is one-to-one because it is not possible to draw a horizontal line that crosses the graph more than once.   ◆

Determining whether a function is one-to-one from an equation is a bit more involved than inspecting a graph. A function f is *not* one-to-one if it is possible to find two different numbers x_1 and x_2 such that $f(x_1) = f(x_2)$. For example, for the function $f(x) = x^2$, it is possible to find two different numbers, 2 and -2, such that $f(2) = f(-2)$. So $f(x) = x^2$ is not one-to-one. To prove that a function f *is* one-to-one, we must show that $f(x_1) = f(x_2)$ implies that $x_1 = x_2$.

Example 5   Using the definition of one-to-one

Determine whether each function is one-to-one.

a) $f(x) = \dfrac{2x + 1}{x - 3}$ **b)** $g(x) = |x|$

Solution

a) If $f(x_1) = f(x_2)$, then we have the following equation:

$$\frac{2x_1 + 1}{x_1 - 3} = \frac{2x_2 + 1}{x_2 - 3}$$

$$(2x_1 + 1)(x_2 - 3) = (2x_2 + 1)(x_1 - 3) \qquad \text{Multiply by the LCD.}$$

$$2x_1x_2 + x_2 - 6x_1 - 3 = 2x_2x_1 + x_1 - 6x_2 - 3$$

$$7x_2 = 7x_1$$

$$x_2 = x_1$$

Since $f(x_1) = f(x_2)$ implies that $x_1 = x_2$, $f(x)$ is a one-to-one function.

b) Since $g(3) = |3| = 3$ and $g(-3) = 3$, the function g is *not* one-to-one.   ◆

(a)

(b)

Figure 3.60

Inverse Functions Using *f*-Notation

The function $f(x) = 2x + 5$ gives the cost of a pizza where \$5 is the basic cost and x is the number of toppings at \$2 each. If we assume that x could be any real number, this function defines the set

$$f = \{(x, y) \mid y = 2x + 5\}.$$

Since the coordinates of each ordered pair are interchanged for f^{-1}, the ordered pairs of f^{-1} must satisfy $x = 2y + 5$:

$$f^{-1} = \{(x, y) \mid x = 2y + 5\}.$$

Since $x = 2y + 5$ is equivalent to $y = \dfrac{x - 5}{2}$,

$$f^{-1} = \left\{(x, y) \mid y = \frac{x - 5}{2}\right\}.$$

The function f^{-1} is described in *f*-notation as $f^{-1}(x) = \dfrac{x - 5}{2}$.

A one-to-one function f pairs members of the domain of f with members of the range of f, and its inverse function f^{-1} exactly reverses those pairings as shown in Fig. 3.61. The above example suggests the following steps for finding an inverse function using *f*-notation.

In the standard viewing window, $y = x^3 - 0.01x$ appears to be a one-to-one function and invertible.

This view of $y = x^3 - 0.01x$ leads us to believe that the function is not one-to-one. You should not make a conclusion based on a graph alone.

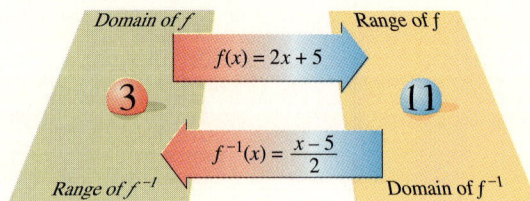

Domain of f Range of f

$f(x) = 2x + 5$

3   11

$f^{-1}(x) = \dfrac{x - 5}{2}$

Range of f^{-1} Domain of f^{-1}

Figure 3.61

Procedure: Finding $f^{-1}(x)$

To find the inverse of a one-to-one function given in *f*-notation:

1. **Replace $f(x)$ by y.**
2. **Interchange x and y.**
3. **Solve the equation for y.**
4. **Replace y by $f^{-1}(x)$.**
5. **Check that the domain of f is the range of f^{-1} and the range of f is the domain of f^{-1}.**

Example 6   Finding an inverse function

Find the inverse of the function $f(x) = \dfrac{2x + 1}{x - 3}$.

Solution

In Example 5(a) we showed that f is a one-to-one function. So we can find f^{-1} by interchanging x and y and solving for y:

$$y = \frac{2x + 1}{x - 3} \qquad \text{\color{blue}{Replace } } f(x) \text{\color{blue}{ by } } y.$$

$$x = \frac{2y + 1}{y - 3} \qquad \text{\color{blue}{Interchange } } x \text{\color{blue}{ and } } y.$$

$$x(y - 3) = 2y + 1 \qquad \text{\color{blue}{Solve for } } y.$$

$$xy - 3x = 2y + 1$$

$$xy - 2y = 3x + 1$$

$$y(x - 2) = 3x + 1$$

$$y = \frac{3x + 1}{x - 2}$$

Replace y by $f^{-1}(x)$ to get $f^{-1}(x) = \dfrac{3x + 1}{x - 2}$. The domain of f is all real numbers except 3. The range of f^{-1} is the set of all real numbers except 3. We exclude 3 because $\dfrac{3x + 1}{x - 2} = 3$ has no solution. Check that the range of f is equal to the domain of f^{-1}. ◆

To find an inverse function according to the definition, we simply interchange the coordinates in all of the ordered pairs of the given function. When a function is defined by a formula, it might not be easy to see that a "possible" formula for the inverse really does this. However, if the three conditions in the following theorem are satisfied, then we can conclude that g is the inverse of f (without checking all of the ordered pairs).

**Theorem: Determining
Whether g Is the Inverse of f**

A function g is the inverse of a function f if and only if

1. f is one-to-one,

2. the domain of g is equal to the range of f, and

3. $(g \circ f)(x) = x$ for any x in the domain of f.

The equation $(g \circ f)(x) = x$ means that the composition $g \circ f$ is the identity function. This condition is often used as the starting point for finding a formula for an inverse function. That is, we look for a function g that undoes the operations of f.

Example 7   Using composition to determine an inverse

For $f(x) = \sqrt{x}$ and $g(x) = x^2$, find $(g \circ f)(x)$ and determine whether g is the inverse of f.

Solution

The function $f(x) = \sqrt{x}$ is one-to-one because $\sqrt{x_1} = \sqrt{x_2}$ implies that $x_1 = x_2$. So f satisfies the first condition of the above theorem. The domain of f is $[0, \infty)$ and $(g \circ f)(x) = g(\sqrt{x}) = (\sqrt{x})^2 = x$ for any $x \geq 0$. So g satisfies the third condition of the theorem. Since the range of f is $[0, \infty)$ and the domain of g is $(-\infty, \infty)$, g fails the second condition of the theorem and g is *not* the inverse of f.   ◆

Even though $g(x) = x^2$ undoes what $f(x) = \sqrt{x}$ does, it is not the inverse of f because of its domain. This problem is easily corrected by specifying the domain of the squaring function. Since the inverse of $f(x) = \sqrt{x}$ must have domain $[0, \infty)$, we have $f^{-1}(x) = x^2$ for $x \geq 0$.

Graphs of f and f^{-1}

If a point (a, b) is on the graph of an invertible function f, then (b, a) is on the graph of f^{-1}. See Fig. 3.62. Since the points (a, b) and (b, a) are symmetric with

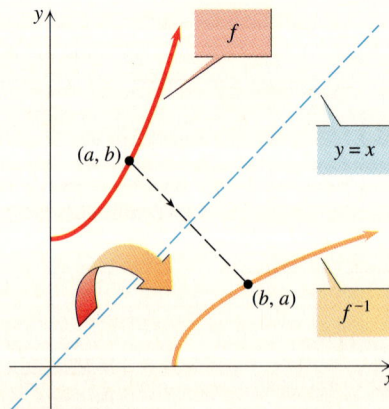

Figure 3.62

respect to the line $y = x$, the graph of f^{-1} is a reflection of the graph of f with respect to the line $y = x$.

$f^{-1}(x) = x^2 + 1$ for $x \geq 0$

$f(x) = \sqrt{x-1}$

$y = x$

Figure 3.63

Example 8   Graphing a function and its inverse

Find the inverse of the function $f(x) = \sqrt{x-1}$ and graph both f and f^{-1} on the same coordinate axes.

Solution

If $\sqrt{x_1 - 1} = \sqrt{x_2 - 1}$, then $x_1 = x_2$. So f is one-to-one. Since the range of $f(x) = \sqrt{x-1}$ is $[0, \infty)$, f^{-1} must be a function with domain $[0, \infty)$. To find a formula for the inverse, interchange x and y in $y = \sqrt{x-1}$ to get $x = \sqrt{y-1}$. Solving this equation for y gives $y = x^2 + 1$. So $f^{-1}(x) = x^2 + 1$ for $x \geq 0$. The graph of f is a translation one unit to the right of the graph of $y = \sqrt{x}$, and the graph of f^{-1} is a translation one unit upward of the graph of $y = x^2$ for $x \geq 0$. Both graphs are shown in Fig. 3.63 along with the line $y = x$. Notice that the graph of f^{-1} is a reflection of the graph of f.   ◆

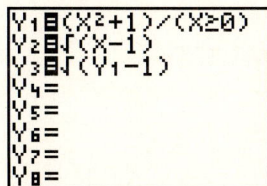

Using y_1 in place of x in y_3 makes y_3 the composition of y_1 and y_2.

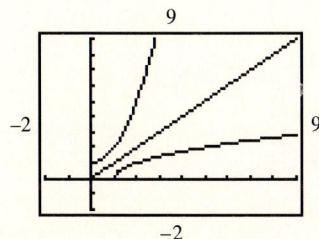

Since y_1 and y_2 are inverses, y_3 is the line $y = x$ for $x \geq 0$.

Finding Inverse Functions Mentally

It is no surprise that the inverse of $f(x) = x^2$ for $x \geq 0$ is the function $f^{-1}(x) = \sqrt{x}$. For nonnegative numbers, taking the square root undoes what squaring does. It is also no surprise that the inverse of $f(x) = 3x$ is $f^{-1}(x) = x/3$ or that the inverse of $f(x) = x + 9$ is $f^{-1}(x) = x - 9$. If an invertible function involves a single operation, it is usually easy to write the inverse function because for most operations there is an inverse operation. See Table 3.7. If an invertible function involves more than one operation, we find the inverse function by applying the inverse operations in the opposite order from the order in which they appear in the original function.

Function	Inverse
$f(x) = 2x$	$f^{-1}(x) = x/2$
$f(x) = x - 5$	$f^{-1}(x) = x + 5$
$f(x) = \sqrt{x}$	$f^{-1}(x) = x^2$ $\quad (x \geq 0)$
$f(x) = x^3$	$f^{-1}(x) = \sqrt[3]{x}$
$f(x) = 1/x$	$f^{-1}(x) = 1/x$
$f(x) = -x$	$f^{-1}(x) = -x$

Table 3.7

Example 9 Finding inverses mentally

Find the inverse of each function mentally.

a) $f(x) = 2x + 1$

b) $g(x) = \dfrac{x^3 + 5}{2}$

Solution

a) The function $f(x) = 2x + 1$ is a composition of multiplying x by 2 and then adding 1. So the inverse function is a composition of subtracting 1 and then dividing by 2: $f^{-1}(x) = (x - 1)/2$. See Fig. 3.64.

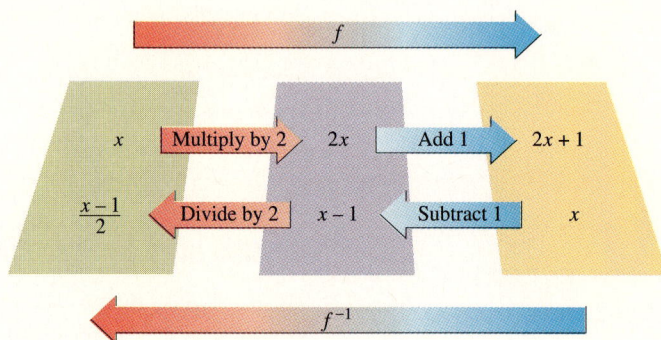

Figure 3.64

b) The function $g(x) = (x^3 + 5)/2$ is a composition of cubing x, adding 5, and dividing by 2. The inverse is a composition of multiplying by 2, subtracting 5, and then taking the cube root: $g^{-1}(x) = \sqrt[3]{2x - 5}$. See Fig. 3.65.

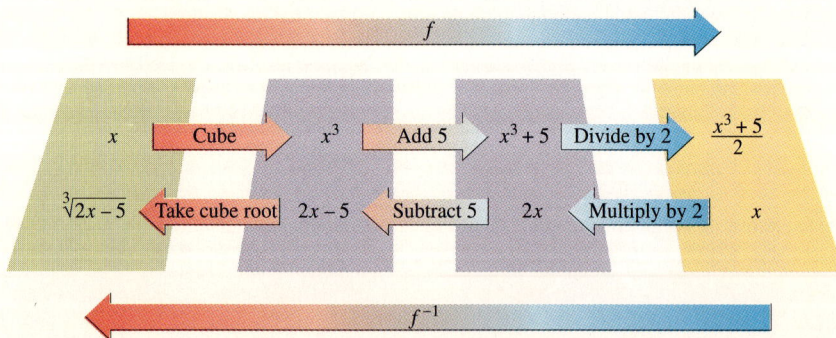

Figure 3.65

For Thought

True or false? Explain.

1. The inverse of the function $\{(2, 3), (5, 5)\}$ is $\{(5, 2), (5, 3)\}$.

2. The function $f(x) = -2$ is an invertible function.

3. If $g(x) = x^2$, then $g^{-1}(x) = \sqrt{x}$.

4. The only functions that are invertible are the one-to-one functions.

5. Every function has an inverse function.

6. The function $f(x) = x^4$ is invertible.

7. If $f(x) = 3\sqrt{x - 2}$, then $f^{-1}(x) = \dfrac{(x + 2)^2}{3}$ for $x \geq 0$.

8. If $f(x) = |x - 3|$, then $f^{-1}(x) = |x| + 3$.

9. According to the horizontal line test, $y = |x|$ is one-to-one.

10. The function $g(x) = -x$ is the inverse of the function $f(x) = -x$.

3.6 Exercises   Tape 7   Disk—5.25″: 2   3.5″: 1   Macintosh: 1

For each function f, find f^{-1}, $f^{-1}(5)$, and $(f^{-1} \circ f)(2)$.

1. $f = \{(2, 1), (3, 5)\}$

2. $f = \{(-1, 5), (0, 0), (2, 6)\}$

3. $f = \{(-3, -3), (0, 5), (2, -7)\}$

4. $f = \{(3.2, 5), (2, 1.99)\}$

Determine whether each function is invertible. If it is invertible, find the inverse.

5. $\{(-1, 0), (0, 0), (1, 0)\}$

6. $\{(0, 0)\}$

7. $\{(3, 0), (2, 5), (4, 6), (7, 9)\}$

8. $\{(-1, -\sqrt{3}), (1, -\sqrt{3})\}$

9. $\{(1, 1), (2, 2), (4.5, 4.5)\}$

10. $\{(-3, 2), (3, 0), (6, 0)\}$

11. $\{(1, 1), (2, 4), (3, 9), (4, 16)\}$

12. $\{(3, 1/3), (4, 1/4), (5, 1/6)\}$

13. $\{(x, y) | y = x + 2\}$

14. $\{(x, y) | y = 3x - 1\}$

15. $\{(x, y) | y = 2x + 7\}$

16. $\{(x, y) | y = 1/x^2\}$

17. $\{(x, y) | y = |x|\}$

18. $\{(x, y) | y = \sqrt{x}\}$

Use the horizontal line test to determine whether each function is one-to-one.

19. $f(x) = x^2 - 3x$

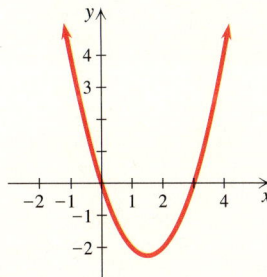

20. $g(x) = |x - 2| + 1$

21. $y = \sqrt[3]{x} + 2$

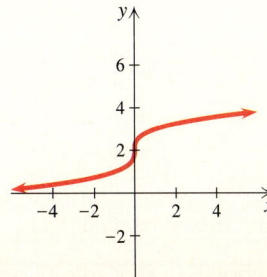

22. $y = (x - 2)^3$

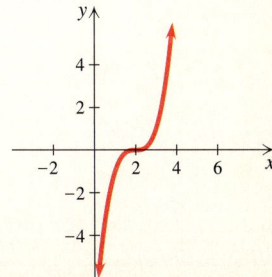

23. $y = x^3 - x$

24. $y = \dfrac{1}{x}$

Determine whether each function is one-to-one.

25. $\{(2, 3), (3, 3), (4, 6)\}$

26. $\{(5, \pi), (6, \pi/2), (7, \pi/3)\}$

27. $\{(x, y)|y = |x + 1|\}$

28. $\{(x, y)|y = x^6\}$

29. $\left\{(x, y)\Big|y = \dfrac{x + 1}{x - 9}\right\}$

30. $\left\{(x, y)\Big|y = 2 + \dfrac{3 - 4x}{x + 2}\right\}$

31. $y = 3 + \dfrac{1 - x}{x - 5}$

32. $y = \dfrac{-3x}{x + 7}$

33. $y = x^2 + 3$

34. $y = 2x^2 - 4$

35. $y = x^2 - x$

36. $f(x) = 5 + \sqrt[3]{x + 9}$

Find the inverse of each function.

37. $f(x) = 3x - 7$

38. $f(x) = -2x + 5$

39. $f(x) = 2 + \sqrt{x - 3}$

40. $f(x) = \sqrt{3x - 1}$

41. $f(x) = -x - 9$

42. $f(x) = -x + 3$

43. $f(x) = \dfrac{x + 3}{x - 5}$

44. $f(x) = \dfrac{2x - 1}{x - 6}$

45. $f(x) = \dfrac{2x - 3}{x - 2}$

46. $f(x) = \dfrac{-2x - 5}{x + 2}$

47. $f(x) = -\dfrac{1}{x}$

48. $f(x) = x$

49. $f(x) = \sqrt[3]{x - 9} + 5$

50. $f(x) = \sqrt[3]{\dfrac{x}{2}} + 5$

51. $f(x) = (x - 2)^2$ for $x \geq 2$

52. $f(x) = x^2$ for $x \leq 0$

In each case, find $(g \circ f)(x)$ and determine whether g is the inverse of f.

53. $f(x) = 4x + 4, g(x) = 0.25x - 1$

54. $f(x) = 20 - 5x, g(x) = -0.2x + 4$

55. $f(x) = x^2 + 1, g(x) = \sqrt{x - 1}$

56. $f(x) = \sqrt[4]{x}, g(x) = x^4$

57. $f(x) = \dfrac{1}{x} + 3, g(x) = \dfrac{1}{x - 3}$

58. $f(x) = 4 - \dfrac{1}{x}, g(x) = \dfrac{1}{4 - x}$

59. $f(x) = \dfrac{x + 1}{x - 2}, g(x) = \dfrac{2x + 1}{x - 1}$

60. $f(x) = \dfrac{3x - 2}{x + 2}, g(x) = \dfrac{2x + 2}{3 - x}$

61. $f(x) = \sqrt[3]{\dfrac{x - 2}{5}}, g(x) = 5x^3 + 2$

62. $f(x) = x^3 - 27, g(x) = \sqrt[3]{x} + 3$

Each of the following functions is invertible. Find the inverse mentally.

63. $f(x) = 5x$

64. $f(x) = 0.5x$

65. $f(x) = x - 88$

66. $f(x) = x + 99$

67. $f(x) = 3x - 7$

68. $f(x) = 5x + 1$

69. $f(x) = 4 - 3x$

70. $f(x) = 5 - 2x$

71. $f(x) = \dfrac{1}{2}x - 9$

72. $f(x) = \dfrac{x}{3} + 6$

73. $f(x) = -x$

74. $f(x) = \dfrac{1}{x}$

75. $f(x) = \sqrt[3]{x} - 9$

76. $f(x) = \sqrt[3]{x - 9}$

77. $f(x) = 2x^3 - 7$

78. $f(x) = -x^3 + 4$

Determine whether each pair of functions f and g are inverses of each other.

79.

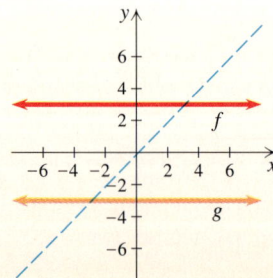

80.

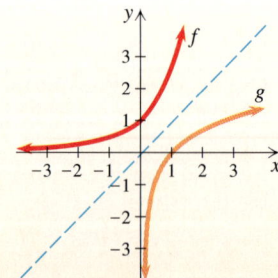

81.

82.

Find the inverse of each function and graph both f and f^{-1} on the same coordinate plane.

83. $f(x) = 3x + 2$

84. $f(x) = -x - 8$

85. $f(x) = x^2 - 4$ for $x \geq 0$

86. $f(x) = 1 - x^2$ for $x \geq 0$

87. $f(x) = x^3$

88. $f(x) = -x^3$

89. $f(x) = \sqrt{x} - 3$

90. $f(x) = \sqrt{x - 3}$

Solve each problem.

91. *Price of a car* The tax on a new car is 8% of the purchase price P. Express the total cost C as a function of the purchase price. Express the purchase price P as a function of the total cost C.

92. *Volume of a cube* Express the volume of a cube $V(x)$ as a function of the length of a side x. Express the length of a side of a cube $S(x)$ as a function of the volume x.

93. *Poiseuille's law* Under certain conditions, the velocity V of blood in a vessel at distance r from the center of the vessel is given by $V = 500(5.625 \times 10^{-5} - r^2)$ where $0 \leq r \leq 7.5 \times 10^{-3}$. Write r as a function of V.

94. *Expenditures versus income* The Consumer Expenditure Survey (1984) found that household expenditures E (in dollars) are related to income I (in dollars) by the equation $E = 0.4I + 10,000$. Write I as a function of E. At what income do expenditures equal income?

95. *Depreciation rate* A new Lexus 400 SC sells for about $40,000. The function $r = 1 - \left(\dfrac{V}{40,000}\right)^{1/5}$ expresses the depreciation rate r for a five-year-old car as a function of the value V of a five-year-old car. What is the depreciation rate for a car that is worth $18,000 after five years? Write V as a function of r.

96. *Annual growth rate* One measurement of the quality of a mutual fund is its average annual growth rate over the last 10 years. The function $r = \left(\dfrac{P}{10,000}\right)^{1/10} - 1$ expresses the

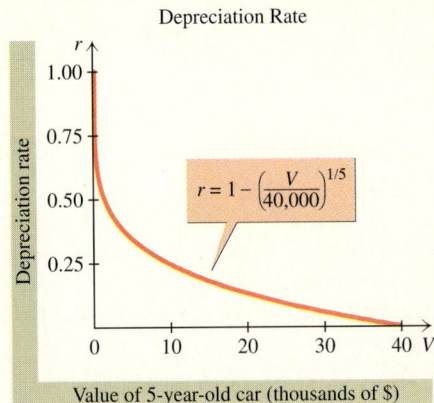

Depreciation Rate

$$r = 1 - \left(\frac{V}{40,000}\right)^{1/5}$$

Figure for Exercise 95

Annual Growth Rate

$$r = \left(\frac{P}{10,000}\right)^{1/10} - 1$$

Figure for Exercise 96

average annual growth rate r as a function of the present value P of an investment of $10,000 made 10 years ago. An investment of $10,000 in Fidelity's Contrafund in 1983 was worth $50,077 in 1993 (*Fidelity Contrafund Prospectus*). What was the annual growth rate for that period? Write P as a function of r.

For Writing/Discussion

97. Use a geometric argument to prove that if $a \neq b$, then (a, b) and (b, a) lie on a line perpendicular to the line $y = x$ and are equidistant from $y = x$.

98. Why is it difficult to find the inverse of a function such as $f(x) = (x - 3)/(x + 2)$ mentally?

99. Find an expression equivalent to $(x - 3)/(x + 2)$ in which x appears only once. Explain.

100. Use the result of Exercise 99 to find the inverse of $f(x) = (x - 3)/(x + 2)$ mentally. Explain.

101. If $f(x) = 2x + 1$ and $g(x) = 3x - 5$, verify that the equation $(f \circ g)^{-1} = g^{-1} \circ f^{-1}$.

102. Explain why $(f \circ g)^{-1} = g^{-1} \circ f^{-1}$ for any invertible functions f and g. Discuss any restrictions on the domains and ranges of f and g for this equation to be correct.

Graphing Calculator Exercises

For each exercise, graph the three functions on the same screen. Enter these functions as shown without simplifying any expressions. Explain what these exercises illustrate.

1. $y_1 = \sqrt[3]{x} - 1$, $y_2 = (x + 1)^3$, $y_3 = \left(\sqrt[3]{x} - 1 + 1\right)^3$

2. $y_1 = (2x - 1)^{1/3}$, $y_2 = (x^3 + 1)/2$, $y_3 = (2((x^3 + 1)/2) - 1)^{1/3}$

With most graphing calculators you can specify the domain of the function to be graphed. Check your calcula-

tor manual for how to do this and graph each pair of inverse functions. Observe the symmetry of the graphs.

3. $y_1 = \sqrt{x}$, $y_2 = x^2$ for $x \geq 0$

4. $y_1 = (x + 1)^4$ for $x \geq -1$, $y_2 = x^{1/4} - 1$

Determine whether each function is invertible by inspecting its graph.

5. $f(x) = (x + 0.01)(x + 0.02)(x + 0.03)$

6. $f(x) = x^3 - 0.6x^2 + 0.11x - 0.006$

3.7

Variation

The area of a circle is a function of the radius, $A = \pi r^2$. As the values of r change or vary, the values of A vary also. Certain relationships can be indicated by describing how one variable changes according to the values of another variable. In this section we will learn to write formulas for certain types of functions that are traditionally described in terms of variation.

Direct Variation

If we save 3 cubic yards of landfill space for every ton of paper that we recycle (Earth Works Group, 1989), then the volume V of space saved is a linear function of the number of tons recycled n, $V = 3n$. The volume is totally controlled by the number of tons recycled. For zero recycling, there is no saving in landfill space. The saving in landfill space is an example of **direct variation.** Compare the landfill saving to a pizza that costs $5 plus $2 per topping. The cost is a linear function of the number of toppings n: $C = 2n + 5$. But the number of toppings does not totally control the cost because of the $5 charge that is included whether we get any toppings or not.

Definition: Direct Variation

The statement **y varies directly as x** or **y is directly proportional to x** means that

$$y = kx$$

for some fixed nonzero real number k. The constant k is called the **variation constant** or **proportionality constant.**

Using these terms, we can say that the saving in landfill space varies directly as the number of tons of paper that is recycled, but the cost of the pizza is not directly proportional to the number of toppings. Note that if y varies directly as x ($y = kx$), then y is a linear function of x. We are merely introducing some new terms to describe an old idea.

Example 1   Direct variation

The cost, C, of a house in Wedgewood Estates is directly proportional to the size of the house, s. If a 2850-square-foot house costs \$182,400, then what is the cost of a 3640-square-foot house?

Solution

Because C is directly proportional to s, there is a constant k such that

$$C = ks.$$

We can find the constant by using $C = \$182{,}400$ when $s = 2850$:

$$182{,}400 = k(2850)$$

$$64 = k$$

The cost for a house is \$64 per square foot. To find the cost of a 3640-square-foot house, use the formula $C = 64s$:

$$C = 64(3640)$$

$$= 232{,}960$$

The 3640-square-foot house costs \$232,960.

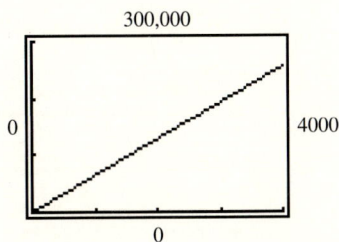

The graph of $y_1 = 64x$ shows how the cost of the house increases as the size increases.

Inverse Variation

In making the 3470-mile trip by air from New York to London, the time it takes is a function of the rate of the airplane. In fact, $T = 3470/R$ gives the time as a function of the rate. A Boeing 747 averaging 580 mph can make the trip in about 6 hours. But the Concorde, averaging 1350 mph, can make the same journey in about 2.5 hours. As the speed increases, the time decreases. This relationship between rate and time is an example of **inverse variation.**

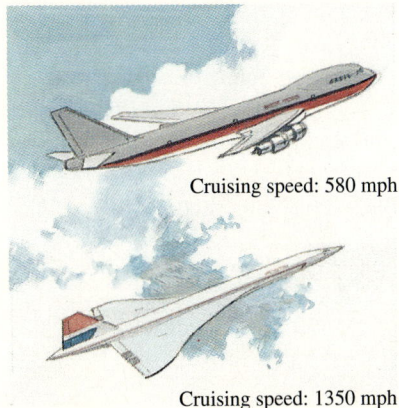

Cruising speed: 580 mph

Cruising speed: 1350 mph

Definition: Inverse Variation

The statement **y varies inversely as x** or **y is inversely proportional to x** means that

$$y = \frac{k}{x}$$

for fixed nonzero real number k.

Example 2 Inverse variation

The time it takes to remove all of the campaign signs in Bakersfield after an election varies inversely with the number of volunteers working on the job. If 12 volunteers could complete the job in 7 days, then how many days would it take for 15 volunteers?

Solution

Since the time T varies inversely as the number of volunteers n, we have

$$T = \frac{k}{n}.$$

Since $T = 7$ when $n = 12$, we can find k:

$$7 = \frac{k}{12}$$

$$84 = k$$

So the formula is $T = 84/n$. Now if $n = 15$, we get

$$T = \frac{84}{15} = 5.6.$$

It would take 15 volunteers 5.6 days to remove all of the political signs. ◆

The graph of $y_1 = 84/x$ shows how the time decreases as the number of volunteers increases.

Joint Variation

We have been studying functions of one variable, but situations often arise in which one variable depends on the values of two or more variables. For example, the area of a rectangular room depends on the length *and* the width, $A = LW$. If carpeting costs $25 per square yard, then the total cost is a function of the length and width, $C = 25LW$. This example illustrates **joint variation.**

Definition: Joint Variation

The statement **y varies jointly as x and z** or **y is jointly proportional to x and z** means that

$$y = kxz$$

for a fixed nonzero real number k.

Example 3 Joint variation

The cost of constructing a 9-foot by 12-foot patio is $734.40. If the cost varies jointly as the length and width, then what does an 8-foot by 14-foot patio cost?

Solution

Since the cost varies jointly as L and W, we have $C = kLW$ for some constant k. Use $W = 9$, $L = 12$, and $C = \$734.40$ to find k:

$$734.40 = k(12)(9)$$

$$6.80 = k$$

So the formula is $C = 6.80LW$. Use $W = 8$ and $L = 14$ to find C:

$$C = 6.80(14)(8)$$

$$= 761.60$$

The cost of constructing an 8-foot by 14-foot patio is \$761.60.

Combined Variation

The formula for the volume of a cylinder is $V = \pi r^2 h$. Using the language of variation, we can say that the volume of a can of tomato soup varies jointly as the height and the square of the radius. The next example gives more illustrations of combining direct, inverse, and joint variation involving powers and roots.

Example 4   Writing formulas for combined variation

Write each sentence as a formula involving a constant of variation k.

a) y varies directly as x and inversely as z.
b) y varies jointly as x and the square root of z.
c) y varies jointly as x and the square of z and inversely as the cube of w.

Solution

a) $y = \dfrac{kx}{z}$

b) $y = kx\sqrt{z}$

c) $y = \dfrac{kxz^2}{w^3}$

The language of variation evolved as a means of describing functions that involve only multiplication and/or division. The terms "varies directly," "directly proportional," and "jointly proportional" indicate multiplication. The term "inverse" indicates division. The term "variation" is *never* used in this text to refer to a formula involving addition or subtraction. Note that the formula for the cost of a pizza $C = 2n + 5$ is *not* the kind of formula that is described in terms of variation.

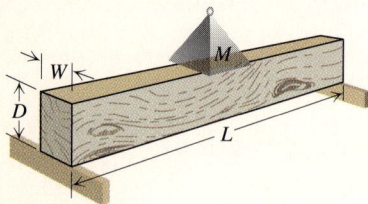

Figure 3.66

Example 5   Solving a combined variation problem

The recommended maximum load for the ceiling joist with rectangular cross section shown in Fig. 3.66 varies directly as the product of the width and the square of the depth of the cross section, and inversely as the length of the joist. If the recommended maximum load is 800 pounds for a 2-inch by 6-inch joist that is 12 feet long, then what is the recommended maximum load for a 2-inch by 12-inch joist that is 16 feet long?

Solution

The maximum load M varies directly as the product of the width W and the square of the depth D, and inversely as the length L:

$$M = \frac{kWD^2}{L}$$

Use $W = 2$, $D = 6$, $L = 12$, and $M = 800$ to find k:

$$800 = \frac{k \cdot 2 \cdot 6^2}{12}$$

$$800 = 6k$$

$$\frac{400}{3} = k$$

Now use $W = 2$, $D = 12$, $L = 16$, and $k = 400/3$ in $M = \dfrac{kWD^2}{L}$ to find M:

$$M = \frac{400 \cdot 2 \cdot 12^2}{3 \cdot 16} = 2400$$

The maximum load on the joist is 2400 pounds. Notice that feet and inches were used in the formula without converting one to the other. If we solve the problem using only feet or only inches, then the value of k is different but we get the same value for M. ◆

? For Thought

True or false? Explain.

1. If the cost of an 800 number is $29.95 per month plus 28 cents per minute, then the monthly cost varies directly as the number of minutes.

2. If bananas are $0.39 per pound, then your cost for a bunch of bananas varies directly with the number of bananas purchased.

3. The area of a circle varies directly as the square of the radius.

4. The area of a triangle varies jointly as the length of the base and the height.

5. The number π is a constant of proportionality.

6. If y is inversely proportional to x, then for $x = 0$, we get $y = 0$.

7. If the cost of potatoes varies directly with the number of pounds purchased, then the proportionality constant is the price per pound.

8. The height of a student in centimeters is directly proportional to the height of the same student in inches.

9. The surface area of a cube varies directly as the square of the length of an edge.

10. The surface area of a rectangular box varies jointly as its length, width, and height.

3.7 Exercises Tape 7 Disk—5.25": 2 3.5": 1 Macintosh: 1

Write a formula that expresses the relationship described by each statement. Use k as the constant of variation.

1. Your grade on the next test, G, varies directly with the number of hours, n, that you study for it.

2. The volume of a gas in a cylinder, V, is inversely proportional to the pressure on the gas, P.

3. For two lines that are perpendicular and have slopes, the slope of one is inversely proportional to the slope of the other.

4. The amount of sales tax on a new car is directly proportional to the purchase price of the car.

5. The cost of constructing a silo varies jointly as the height and the radius.

6. The volume of grain that a silo can hold varies jointly as the height and the square of the radius.

7. Y varies directly as x and inversely as the square root of z.

8. W varies directly with r and t and inversely with v.

For each formula that illustrates a type of variation, write the formula in words using the vocabulary of variation. The letters k and π represent constants, and all other letters represent variables.

9. $A = \pi r^2$

10. $C = \pi D$

11. $a = kzw$

12. $m_1 = \dfrac{-1}{m_2}$

13. $T = 3W + 5$

14. $y = k\sqrt{x - 3}$

15. $y = \dfrac{1}{x}$

16. $V = LWH$

17. $H = \dfrac{k\sqrt{t}}{s}$

18. $B = \dfrac{y^2}{2\sqrt{x}}$

19. $D = \dfrac{2LJ}{W}$

20. $E = mc^2$

Find the constant of variation and write the formula that is expressed by the indicated variation.

21. y varies directly as x, and $y = 5$ when $x = 9$.

22. h is directly proportional to z, and $h = 210$ when $z = 200$.

23. T is inversely proportional to y, and $T = -30$ when $y = 5$.

24. H is inversely proportional to n, and $H = 9$ when $n = -6$.

25. m varies directly as the square of t, and $m = 54$ when $t = 3\sqrt{2}$.

26. p varies directly as the cube root of w, and $p = \sqrt[3]{2}/2$ when $w = 4$.

27. y varies directly as x and inversely as the square root of z, and $y = 2.192$ when $x = 2.4$ and $z = 2.25$.

28. n is jointly proportional to x and the square root of b, and $n = -18.954$ when $x = -1.35$ and $b = 15.21$.

Solve each variation problem.

29. If y varies directly as x, and $y = 9$ when $x = 2$, what is y when $x = -3$?

30. If y is directly proportional to z, and $y = 6$ when $z = \sqrt{12}$, what is y when $z = \sqrt{75}$?

31. If P is inversely proportional to w, and $P = 2/3$ when $w = 1/4$, what is P when $w = 1/6$?

32. If H varies inversely as q, and $H = 0.03$ when $q = 0.01$, what is H when $q = 0.05$?

33. If A varies jointly as L and W, and $A = 30$ when $L = 3$ and $W = 5\sqrt{2}$, what is A when $L = 2\sqrt{3}$ and $W = 1/2$?

34. If J is jointly proportional to G and V, and $J = \sqrt{3}$ when $G = \sqrt{2}$ and $V = \sqrt{8}$, what is J when $G = \sqrt{6}$ and $V = 8$?

35. If y is directly proportional to u and inversely proportional to the square of v, and $y = 7$ when $u = 9$ and $v = 6$, what is y when $u = 4$ and $v = 8$?

36. If q is directly proportional to the square root of h and inversely proportional to the cube of j, and $q = 18$ when $h = 9$ and $j = 2$, what is q when $h = 16$ and $j = 1/2$?

In each case, determine whether the first variable varies directly or inversely with the other variable. Write a formula that describes the variation using the appropriate variation constant.

37. The length of a car in inches, the length of the same car in feet

38. The time in seconds to pop a bag of microwave popcorn, the time in minutes to pop the same bag of popcorn

39. The price for each person sharing a $20 pizza, the number of hungry people

40. The number of identical steel rods with a total weight of 40,000 lbs, the weight of one of those rods

41. The speed of a car in miles per hour, the speed of the same car at the same time in kilometers per hour

42. The weight of a sumo wrestler in pounds, the weight of the same sumo wrestler in kilograms. (Chad Rowan of the United States, at 455 lb or 206 kg, recently became the first-ever foreign grand champion in Tokyo's annual tournament.)

43. The temperature in degrees Celsius, the temperature at the same time and place in degrees Fahrenheit

44. The perimeter of a rectangle with a length of 4 ft, the width of the same rectangle

45. The area of a rectangle with a length of 30 in., the width of the same rectangle

46. The area of a triangle with a height of 10 cm, the base of the same triangle

47. The number of gallons of gasoline that you can buy for $5, the price per gallon of that same gasoline

48. The length of Yolanda's 40 ft² closet, the width of that closet

Solve each problem.

49. *Processing oysters* The time required to process a shipment of oysters varies directly with the number of pounds in the shipment and inversely with the number of workers assigned. If 3000 lb can be processed by 6 workers in 8 hr, then how long would it take 5 workers to process 4000 lb?

50. *View from an airplane* The view V from the air is directly proportional to the square root of the altitude A. If the view from horizon to horizon at an altitude of 16,000 ft is approximately 154 mi, then what is the view from 36,000 ft?

51. *Simple interest* The amount of simple interest on a deposit varies jointly with the principal and the time in days. If $20.80 is earned on a deposit of $4000 for 16 days, how much interest would be earned on a deposit of $6500 for 24 days?

52. *Free fall at Six Flags* Visitors at Six Flags in Atlanta get a brief thrill in the park's "free fall" ride. Passengers seated in a small cab drop from a 10-story tower down a track that eventually curves level with the ground. The distance the cab falls varies directly with the square of the time it is falling. If the cab falls 16 ft in the first second, then how far has it fallen after 2 sec?

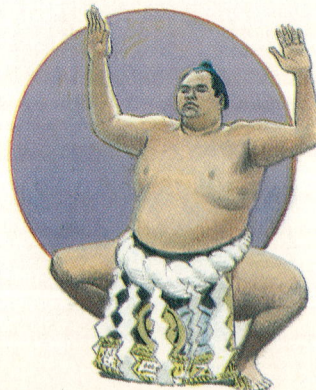

Figure for Exercise 42

53. *Cost of plastic sewer pipe* The cost of a plastic sewer pipe varies jointly as its diameter and its length. If a 20-ft pipe with a diameter of 6 in. costs $18.60, then what is the cost of a 16-ft pipe with a diameter of 8 in.?

54. *Cost of copper tubing* A plumber purchasing materials for a renovation observed that the cost of copper tubing varies jointly as the length and the diameter of the tubing. If 20 ft of $\frac{1}{2}$-in.-diameter tubing costs $36.60, then what is the cost of 100 ft of $\frac{3}{4}$-in.-diameter tubing?

55. *Weight of a can* The weight of a can of baked beans varies jointly with the height and the square of the diameter. If a 4-in.-high can with a 3-in. radius weighs 14.5 oz, then what is the weight of a 5-in.-high can with a diameter of 6 in.?

56. *Nitrogen gas shock absorber* The volume of gas in a nitrogen gas shock absorber varies directly with the temperature of the gas and inversely with the amount of weight on the piston. If the volume is 10 in^3 at 80° F with a weight of 600 lb, then what is the volume at 90° F with a weight of 800 lb?

57. *Velocity of underground water* Darcy's law states that the velocity V of underground water through sandstone varies directly as the head h and inversely as the length l of the flow. The head is the vertical distance between the point of intake into the rock and the point of discharge such as a spring, and the length is the length of the flow from intake to discharge. In a certain sandstone a velocity of 10 ft per year has been recorded with a head of 50 ft and length of 200 ft. What would we expect the velocity to be if the head is 60 ft and the length is 300 ft?

58. *Cross-sectional area of a well* The rate of discharge of a well, V, varies jointly as the hydraulic gradient, i, and the cross-sectional area of the well wall, A. Suppose that a well with a cross-sectional area of 10 ft^2 discharges 3 gal of water per minute in an area where the hydraulic gradient is 0.3. If we dig another well nearby where the hydraulic gradient is 0.4, and we want a discharge of 5 gal/min, then what should be the cross-sectional area for the well?

59. *Mass of Jupiter* A result of Kepler's harmonic law is that the mass of a planet with a satellite is directly proportional to the cube of the mean distance from the satellite to the planet, and inversely proportional to the square of the period of revolution. The period of revolution of the satellite Phobos around Mars is 7.65 hr, the mass of Mars is 6.424 × 10^{23} kg, and the mean distance between Mars and Phobos is 5800 mi. If the period of revolution of Io around Jupiter is 42.47 hr and the mean distance between Io and Jupiter is 262,000 mi, then what is the mass of Jupiter?

60. *Grade on an algebra test* Calvin believes that his grade on a college algebra test varies directly with the number of hours spent studying during the week prior to the test and inversely with the number of hours spent at the Beach Club playing volleyball during the week prior to the test. If he scored 76 on a test when he studied 12 hr and played 10 hr during the week prior to the test, then what score should he expect if he studies 9 hr and plays 15 hr?

61. *Carpeting a room* The cost of carpeting a room varies jointly as the length and width. Julie advertises a price of $263.40 to carpet a 9-ft by 12-ft room with Dupont Stainmaster. Julie gave a customer a price of $482.90 for carpeting a room with a width of 4 yd with that same carpeting, but she has forgotten the length of the room. What is the length of the room?

62. *Pole-vault principle* Due to increasingly flexible poles that return virtually all of the energy athletes put into them, the deciding factor in today's pole-vaulting competitions is speed down the runway (*Newsweek*, July 1992). The height a pole-vaulter attains is directly proportional to the square of his velocity on the runway. Given that a speed of 32 ft/sec will loft a vaulter to 16 ft, find the speed necessary for World Record Holder Sergei Bubka to reach a record height of 20 ft 2.5 in.

For Writing/Discussion

63. *Moving melons* To move more watermelons, a retailer decided to make the price of each watermelon purchased by a single customer inversely proportional to the number of watermelons purchased by the customer. Is this a good idea? Explain.

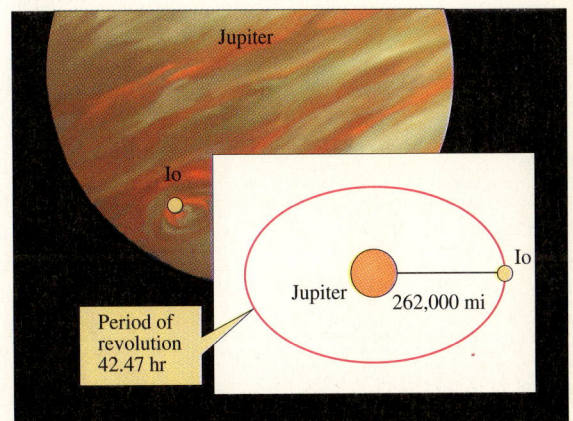

Figure for Exercise 59

64. *Common units* In general we do not perform computations involving inches and feet until they are converted to a common unit of measure. Explain how it is possible to use inches and feet in Example 5 without converting to a common unit of measure. Explain why a 2-in. by 6-in. ceiling joist is placed so that the 6-in. side is vertical.

65. *Cooperative learning* Each student in a small group should write in words a statement expressing a direct variation, a statement expressing inverse variation, a statement expressing joint variation, and a statement expressing combined variation. The group should write each statement in symbols and verify that it is of the required type.

◆ Highlights

Section 3.1   The Cartesian Coordinate System
1. The Cartesian coordinate system is a means of picturing sets of ordered pairs.
2. Any equation of the form $Ax + By = C$, where A and B are not both zero, has a graph that is a straight line.
3. The distance between (x_1, y_1) and (x_2, y_2) is $\sqrt{(x_2 - x_1)^2 + (y_2 - y_1)^2}$.
4. The midpoint for a line segment with endpoints (x_1, y_1) and (x_2, y_2) is $\left(\dfrac{x_1 + x_2}{2}, \dfrac{y_1 + y_2}{2}\right)$.

Section 3.2   Functions
1. Any set of ordered pairs is called a relation.
2. A function is a set of ordered pairs in which no two ordered pairs have the same first coordinate and different second coordinates.
3. The ordered pairs of a function may be defined by a table, a list, an equation, or a rule.
4. The average rate of change of a function is a measure of the rate of change of the dependent variable over some interval of the independent variable.
5. The difference quotient is the average rate of change of a function over the interval $(x, x + h)$.

Section 3.3   Graphs of Relations and Functions
1. The set of ordered pairs satisfying $(x - h)^2 + (y - k)^2 = r^2$ for $r > 0$ forms a circle centered at (h, k) with radius r.
2. The equation of a circle can be solved for y, giving the equations of two semicircles.
3. The vertical line test is a visual method for determining whether a graph is a graph of a function.
4. For piecewise functions, different formulas are used in different regions of the domain.
5. A function is increasing, decreasing, or constant on an interval, depending on whether its graph is rising, falling, or staying level as you move from left to right along the graph.

Section 3.4   Transformations and Symmetry of Graphs

1. If we know what some basic graphs look like, we can graph a wide variety of functions and equations by using the ideas of transformations and symmetry.
2. Any function of the form $f(x) = mx + b$ has a graph that is a straight line.
3. The graph of an even function ($f(-x) = f(x)$) is symmetric about the y-axis and the graph of an odd function ($f(-x) = -f(x)$) is symmetric about the origin.
4. The solution sets to certain inequalities can be determined by visually inspecting the graphs of functions.

Section 3.5   Operations with Functions

1. Two functions can be combined to form the sum, difference, product, and quotient functions.
2. In the composition $f \circ g$, the function g is followed by f (the input for f is the output of g).

Section 3.6   Inverse Functions

1. If a function has no two ordered pairs with different first coordinates and the same second coordinate, then the function is a one-to-one function.
2. It is only one-to-one functions that are invertible.
3. The horizontal line test is a visual method of determining whether a function is one-to-one.
4. The inverse function of a one-to-one function is the function formed by interchanging the coordinates in each ordered pair.
5. The graph of f^{-1} can be obtained by reflecting the graph of f about the line $y = x$.

Section 3.7   Variation

1. Certain functions of one and several variables are customarily described using the ideas of direct, inverse, and joint variation.
2. The terms "varies directly," "directly proportional," and "jointly proportional" indicate multiplication.
3. The terms "varies inversely" and "inversely proportional" indicate division.

Chapter 3 Review Exercises

Graph each set of ordered pairs. State the domain and range of each relation. Determine whether each set is a function.

1. $\{(x, y)\,|\,y = 3 - x\}$

2. $\{(0, 0), (1, 1), (-2, -2)\}$

3. $\{(0, -3), (1, -1), (2, 1), (2, 3)\}$

4. $\{(x, y)\,|\,2x + y = 5\}$

5. $\{(x, y)\,|\,x^2 + y^2 = 0.01\}$

6. $\{(x, y)\,|\,x^2 + y^2 = 2x - 4y\}$

7. $\{(x, y)\,|\,x = y^2 + 1\}$

8. $\{(x, y)\,|\,y = |x - 2|\}$

9. $\left\{(x, y)\,|\,y = \sqrt{x} - 3\right\}$

10. $\left\{(x, y)\,|\,x = \sqrt{y}\,\right\}$

Let $f(x) = x^2 + 3$ and $g(x) = 2x - 7$. **Find and simplify each expression.**

11. $f(-3)$ **12.** $g(3)$ **13.** $g(12)$

14. $f(-1)$ **15.** $(g \circ f)(-3)$ **16.** $(f \circ g)(3)$

17. $(f + g)(2)$ **18.** $(f - g)(-2)$ **19.** $(f \cdot g)(-1)$

20. $(f/g)(4)$ **21.** $f(g(2))$ **22.** $g(f(-2))$

23. $(f \circ f)(x)$ **24.** $(g \circ g)(x)$ **25.** $f(a + 1)$

26. $g(a + 2)$ **27.** $\dfrac{f(3 + h) - f(3)}{h}$

28. $\dfrac{g(5 + h) - g(5)}{h}$ **29.** $\dfrac{f(x + h) - f(x)}{h}$

30. $\dfrac{g(x + h) - g(x)}{h}$ **31.** $g\left(\dfrac{x + 7}{2}\right)$

32. $f\left(\sqrt{x - 3}\right)$ **33.** $g^{-1}(x)$

34. $g^{-1}(-3)$

Use transformations to graph each pair of functions on the same coordinate plane.

35. $f(x) = \sqrt{x}$, $g(x) = 2\sqrt{x + 3}$

36. $f(x) = \sqrt{x}$, $g(x) = -2\sqrt{x} + 3$

37. $f(x) = |x|$, $g(x) = -2|x + 2| + 4$

38. $f(x) = |x|$, $g(x) = |x - 1| - 3$

39. $f(x) = x^2$, $g(x) = (x - 2)^2 + 1$

40. $f(x) = x^2$, $g(x) = -2x^2 + 4$

Let $f(x) = \sqrt[3]{x}$, $g(x) = x - 4$, $h(x) = x/3$, and $j(x) = x^2$. **Write each function as a composition of the appropriate functions chosen from f, g, h, and j.**

41. $F(x) = \sqrt[3]{x - 4}$ **42.** $G(x) = -4 + \sqrt[3]{x}$

43. $H(x) = \sqrt[3]{\dfrac{x^2 - 4}{3}}$ **44.** $M(x) = \left(\dfrac{x - 4}{3}\right)^2$

45. $N(x) = \dfrac{1}{3}x^{2/3}$ **46.** $P(x) = x^2 - 16$

47. $R(x) = \dfrac{x^2}{3} - 4$ **48.** $Q(x) = x^2 - 8x + 16$

For each pair of points, find the distance between them and the midpoint of the line segment joining them.

49. $(-3, 5)$, $(2, -6)$ **50.** $(-1, 1)$, $(-2, -3)$

51. $(1/2, 1/3)$, $(1/4, 1)$ **52.** $(0.5, 0.2)$, $(-1.2, 2.1)$

Find the difference quotient for each function and simplify it.

53. $f(x) = -5x + 9$ **54.** $f(x) = \sqrt{x - 7}$

55. $f(x) = \dfrac{1}{2x}$ **56.** $f(x) = -5x^2 + x$

Sketch the graph of each function and state its domain and range. Determine the intervals on which the function is increasing, decreasing, or constant.

57. $f(x) = \sqrt{100 - x^2}$

58. $f(x) = -\sqrt{7 - x^2}$

59. $f(x) = \begin{cases} -x^2 & \text{for } x \le 0 \\ x^2 & \text{for } x > 0 \end{cases}$

60. $f(x) = \begin{cases} x^2 & \text{for } x \le 0 \\ x & \text{for } 0 < x \le 4 \end{cases}$

61. $f(x) = \begin{cases} -x - 4 & \text{for } x \le -2 \\ -|x| & \text{for } -2 < x < 2 \\ x - 4 & \text{for } x \ge 2 \end{cases}$

62. $f(x) = \begin{cases} (x + 1)^2 - 1 & \text{for } x \le -1 \\ -1 & \text{for } -1 < x < 1 \\ (x - 1)^2 - 1 & \text{for } x \ge 1 \end{cases}$

Each of the following graphs is the graph of a function involving absolute value. Write an equation for each graph and state the domain and range of the function.

63.

64.

65.

66.

67.

68.

Discuss the symmetry of the graph of each function.

69. $f(x) = x^4 - 9x^2$

70. $f(x) = |x| - 99$

71. $f(x) = -x^3 - 5x$

72. $f(x) = \dfrac{15}{x}$

73. $f(x) = -x + 1$

74. $f(x) = |x - 1|$

75. $f(x) = \sqrt{x^2}$

76. $f(x) = \sqrt{16 - x^2}$

Use the graph of an appropriate function to find the solution set to each inequality.

77. $|x - 3| \geq 1$

78. $|x + 2| + 1 < 0$

79. $-2|x| + 4 > 0$

80. $-\dfrac{1}{2}|x| + 2 \leq 0$

81. $-|x + 1| - 2 > 0$

82. $|x| - 3 < 0$

Graph each pair of functions on the same coordinate plane.

83. $f(x) = \sqrt{x + 3}$, $g(x) = x^2 - 3$ for $x \geq 0$

84. $f(x) = (x - 2)^3$, $g(x) = \sqrt[3]{x} + 2$

85. $f(x) = 2x - 4$, $g(x) = \dfrac{1}{2}x + 2$

86. $f(x) = -\dfrac{1}{2}x + 4$, $g(x) = -2x + 8$

Determine whether each function is invertible. If the function is invertible, then find the inverse function, and state the domain and range of the inverse function.

87. $\{(\pi, 0), (\pi/2, 1), (2\pi/3, 0)\}$

88. $\{(-2, 1/3), (-3, 1/4), (-4, 1/5)\}$

89. $f(x) = 3x - 21$

90. $f(x) = 3|x|$

91. $y = \sqrt{9 - x^2}$

92. $y = 7 - x$

93. $f(x) = \sqrt{x - 9}$

94. $f(x) = \sqrt{x} - 9$

95. $f(x) = \dfrac{x - 7}{x + 5}$

96. $f(x) = \dfrac{2x - 3}{5 - x}$

97. $f(x) = x^2 + 1$ for $x \leq 0$

98. $f(x) = (x + 3)^4$ for $x \geq 0$

Solve each problem.

99. Find the x- and y-intercepts for the graph of $3x - 4y = 9$.

100. Show that the points $(0, -2)$, $(6, 0)$, $(4, 6)$, and $(-2, 4)$ are the vertices of a square.

101. Given that D is directly proportional to W and $D = 9$ when $W = 25$, find D when $W = 100$.

102. Given that t varies directly as u and inversely as v, and $t = 6$ when $u = 8$ and $v = 2$, find t when $u = 19$ and $v = 3$.

103. Write in standard form the equation of the circle that has center $(-3, 5)$ and radius $\sqrt{3}$.

104. Use completing the square to write the equation $x^2 + y^2 = x - 2y + 1$ in the form $(x - h)^2 + (y - k)^2 = r^2$, and identify the center and radius of this circle.

105. Use completing the square to write the equation $y = -2x^2 + 4x + 5$ in the form $y = a(x - h)^2 + k$, and identify the vertex of the parabola.

106. Write the diameter of a circle, d, as a function of its area, A.

107. *Circle inscribed in a square* A cylindrical pipe with an outer radius r must fit snugly through a square hole in a wall. Write the area of the square as a function of the radius of the pipe.

108. *Load on a spring* When a load of 5 lb is placed on a spring, its length is 6 in., and when a load of 9 lb is placed on the spring, its length is 8 in. What is the average rate of change of the length of the spring as the load varies from 5 lb to 9 lb?

109. *Changing speed of a dragster* Suppose that 2 sec after starting, a dragster is traveling 40 mph, and 5 sec after starting, the dragster is traveling 130 mph. What is the average rate of change of the speed of the dragster over the time interval from 2 sec to 5 sec? What are the units for this measurement?

110. *Dinosaur speed* R. McNeill Alexander, a British paleontologist, uses observations of living animals to estimate the speed of dinosaurs (*Scientific American,* April 1991). Alexander believes that two animals of different sizes but geometrically similar shapes will move in a similar fashion. For animals of similar shapes, their velocity is directly proportional to the square root of their hip height. Alexander has compared a white rhinoceros, with a hip height of 1.5 m, and a member of the genus Triceratops, with a hip height of 2.8 m. If a white rhinoceros can move at 45 km/hr, then what is the estimated velocity of the dinosaur?

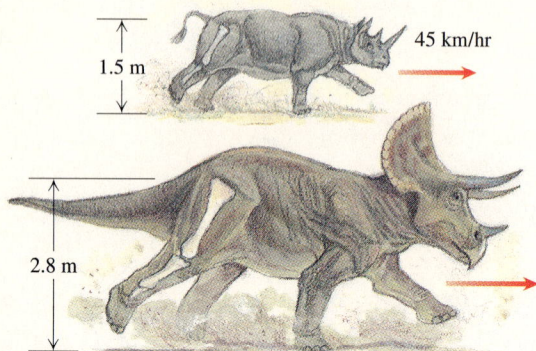

Figure for Exercise 110

111. *Arranging plants in a row* Timothy has a shipment of strawberry plants to plant in his field, which is prepared with rows of equal length. He knows that the number of rows that he will need varies inversely with the number of plants per row. If he plants 240 plants per row, then he needs 21 rows. How many rows does he need if he plants 224 plants per row?

112. *Handling charge for a globe* The shipping and handling charge for a globe from Mercator's Map Company varies directly as the square of the diameter. If the charge is $4.32 for a 6-in.-diameter globe, then what is the charge for a 16-in.-diameter globe?

113. *Newton's law of gravity* Newton's law of gravity states that the force of gravity acting between two objects varies directly as the product of the masses of the two objects and inversely as the square of the distance between their centers. Write an equation for Newton's law of gravity.

114. *Inefficiency of Ivory Coast mills* Not only are the tropical forests being consumed at an alarming rate, but the inefficiency of the lumber mills is contributing to the problem (*Scientific American,* April 1990). For example, Ivory Coast mills require 30% more logs to produce the same output as more efficient mills. Given that the output of an Ivory Coast mill varies directly with the input, and a mill produces 75 tons of lumber out of 130 tons of logs, write the output y as a function of the input x.

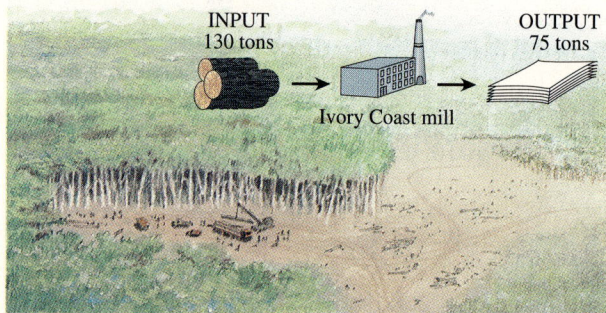

Figure for Exercise 114

For Writing/Discussion

115. Explain what the vertical and horizontal line tests are and how to apply them.

116. Is a vertical line the graph of a linear function? Explain.

Chapter 3 Test

Determine whether each equation defines y as a function of x.

1. $x^2 + y^2 = 25$

2. $3x - 5y = 20$

3. $x = |y + 2|$

4. $y = x^3 - 3x^2 + 2x - 1$

State the domain and range of each relation.

5. $\{(2, -3), (2, -4), (5, 7)\}$

6. $y = \sqrt{x - 9}$

7. $x = |y - 1|$

Sketch the graph of each function.

8. $3x - 4y = 12$ **9.** $y = 2x - 3$ **10.** $y = \sqrt{25 - x^2}$

11. $y = -(x - 2)^2 + 5$ **12.** $y = 2|x| - 4$

13. $f(x) = \begin{cases} x & \text{for } x < 0 \\ 2 & \text{for } x \geq 0 \end{cases}$

Let $f(x) = \sqrt{x + 2}$ and $g(x) = 3x - 1$. Find and simplify each of the following expressions.

14. $f(7)$ **15.** $(f \circ g)(2)$ **16.** $g^{-1}(x)$

17. $(f + g)(14)$

18. $\dfrac{g(x + h) - g(x)}{h}$

24. State the solution set to the inequality $|x - 1| - 3 < 0$ using interval notation.

25. Find the inverse of the function $g(x) = 3 + \sqrt[3]{x - 2}$.

Solve each problem.

19. Find the midpoint of the line segment whose endpoints are $(3, 5)$ and $(-3, 6)$.

20. Find the distance between the points $(-2, 4)$ and $(3, -1)$.

21. Write the equation in standard form for a circle with center $(-4, 1)$ and radius $\sqrt{7}$.

22. State the intervals on which $f(x) = (x - 3)^2$ is increasing and those on which $f(x)$ is decreasing.

23. Discuss the symmetry of the graph of the function $f(x) = x^4 - 3x^2 + 9$.

26. Jang's Postal Service charges \$35 for addressing 200 envelopes, and \$60 for addressing 400 envelopes. What is the average rate of change of the cost as the number of envelopes goes from 200 to 400?

27. The intensity of illumination I from a light source varies inversely as the square of the distance d from the source. If a flash on a camera has an intensity of 300 candlepower at a distance of 2 m, then what is the intensity of the flash at a distance of 10 m?

28. Write a formula expressing the volume V of a cube as a function of the length of the diagonal d of a side of the cube.

Tying It All Together
Chapters 1–3

Perform the indicated operations.

1. $2x + 3x$

2. $(3 - x)^2$

3. $\sqrt{x^2}$

4. $(x - 2)(x + 2)$

5. $x^2(x - 2)(x + 3)$

6. $(3 - 2i)(3 + 2i)$

7. $\dfrac{-4 \pm \sqrt{4^2 - 4(1)(2)}}{2}$

8. $\dfrac{-6 \pm \sqrt{(-6)^2 - 4(3)(5)}}{2(3)}$

Solve each equation.

9. $2x + 3x = 5x$

10. $\sqrt{9 - x^2} = 3 - x$

11. $|x| - 1 = 0$

12. $x^2 - 4 = 0$

13. $x^2 - 4x + 5 = 0$

14. $4x^2 - 4x - 1 = 0$

15. $x^3 - 9x = 0$

16. $x^4 - x^3 - 2x^2 = 0$

Graph each function.

17. $y = 5x$

18. $y = \sqrt{9 - x^2}$

19. $y = |x| - 1$

20. $y = x^2 - 4$

21. $y = 3 - x$

22. $y = \dfrac{1}{x}$

23. $y = 1 + \dfrac{1}{x}$

24. $y = (x - 1)(x^2 + x + 1)$

Write each expression as a single rational expression.

25. $x + \dfrac{1}{x}$

26. $x + 1 + \dfrac{1}{x - 1}$

27. $3 + \dfrac{2}{x - 4}$

28. $-2 + \dfrac{-1}{x + 3}$

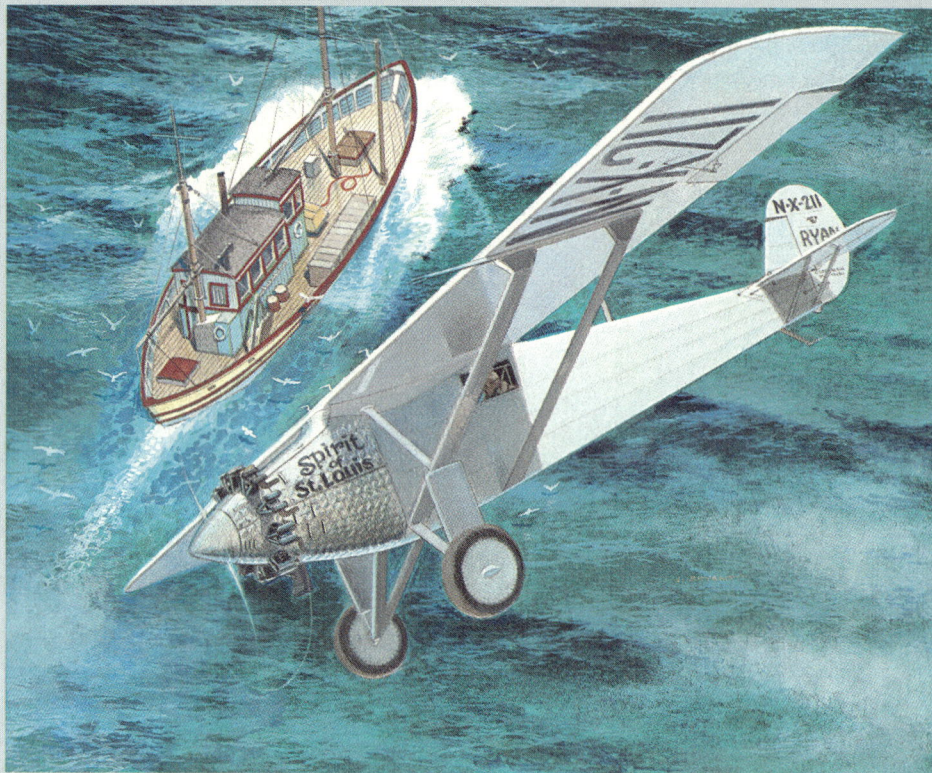

"Which way is Ireland?" A shout cut the air and hovered briefly over a fishing boat trawling the dark ocean waters below. The voice belonged to American aviator Charles Lindbergh, then in the 26th hour of his nonstop, solo flight across the Atlantic. A man stuck his head out of a porthole but did not reply. In exasperation the weary pilot gave up and headed his plane east.

Others had died attempting this aviation feat, and in May 1927, the $25,000 prize offered by a New York businessman was still unclaimed. Lindbergh, disoriented by hours of flying without sleep and by thick fog that obscured his view, desperately wanted to sight land. He feared he had drifted hundreds of miles off course. When no answer came from the boat, he knew he must make another vital determination of air speed: If he'd strayed too far south, he should throttle down to conserve fuel in order to fly the additional thousand miles it would take to reach Europe. But, if he was on course to Ireland and slowed his air speed, it was unlikely he'd sight land before dark.

For much of his flight, Lindbergh's calculations of the interrelationships among wind velocity, air speed, fuel consumption, and time elapsed held life-or-death implications. The longer he remained in the air, the wearier he became and the greater the risk of something going wrong. But the faster he flew, the

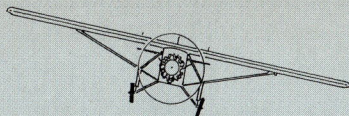

more rapidly the *Spirit of St. Louis* burned fuel and the greater the risk he'd fail to reach Paris, his ultimate destination.

In this chapter we explore polynomial and rational functions and their graphs. We'll learn how to use these basic functions of algebra to model many real-world situations, and we'll see how Lindbergh used them in planning several aspects of his 3610-mile flight between New York and Paris. In fact, as you'll discover, polynomial and rational functions have many common real-world applications. After reading this chapter, perhaps a better understanding of these functions will both extend your knowledge and—as with Lindbergh—expand your horizons.

4

Polynomial

and Rational

Functions

4.1

Linear Functions

In this chapter we will study the polynomial functions. The linear functions, first introduced and graphed in Section 3.1, are the simplest polynomial functions. In this section, we'll study different forms of their equations, see how graphs of lines are related to equations, and use linear functions in a variety of applications.

Definitions and Review

If a polynomial is used to define a function, then the function is called a **polynomial function.** Some examples of polynomial functions are

$$h(x) = 2, \qquad g(x) = 3x - 6, \qquad P(x) = 5x^2 - 2x + 6, \qquad \text{and} \qquad T(x) = x^3.$$

Definition: Polynomial Function

Let $a_n, a_{n-1}, \ldots, a_0$ be real numbers with $a_n \neq 0$. The function defined by

$$f(x) = a_n x^n + a_{n-1} x^{n-1} + \cdots + a_1 x + a_0$$

is called a **polynomial function of degree n. The leading coefficient** is a_n.

The degree of a polynomial function is the degree of the polynomial used to define it. The polynomial functions h, g, P, and T given as examples have degrees 0, 1, 2, and 3, respectively.

A linear function is a polynomial function of degree 1.

249

Definition: Linear Function

If m and b are real numbers with $m \neq 0$, then

$$f(x) = mx + b$$

is called a **linear function.**

If $m = 0$, then $f(x) = b$ has degree 0 (provided $b \neq 0$), and it is a **constant function.** The graph of $f(x) = b$ is a horizontal line. We learned in Chapter 3 that the graph of any linear function is a straight line.

Slope of a Line

In Section 3.2 we defined the average rate of change of a function. For a linear function the average rate of change is called the *slope* of the line.

Definition: Slope

The **slope** of a line through (x_1, y_1) and (x_2, y_2) with $x_1 \neq x_2$ is

$$\frac{y_2 - y_1}{x_2 - x_1}.$$

If (x_1, y_1) and (x_2, y_2) are two distinct points with $x_1 = x_2$, then the points lie on a vertical line. So slope is undefined for vertical lines. Equations of vertical lines were discussed in Chapter 3. Since a vertical line is not the graph of a function, it is not the graph of a linear function.

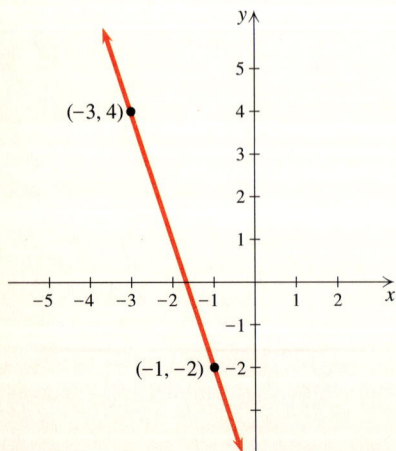

Figure 4.1

Example 1   Finding the slope of a line through two points

Find the slope of the line that contains the points $(-3, 4)$ and $(-1, -2)$.

Solution

The line through $(-3, 4)$ and $(-1, -2)$ is shown in Fig. 4.1. Use $(x_1, y_1) = (-3, 4)$ and $(x_2, y_2) = (-1, -2)$ in the expression for slope:

$$\text{slope} = \frac{y_2 - y_1}{x_2 - x_1} = \frac{-2 - 4}{-1 - (-3)} = \frac{-6}{2} = -3$$

The slope of the line is -3.

If (x_1, y_1) and (x_2, y_2) are *any* two points that satisfy $y = mx + b$, then $y_1 = mx_1 + b$ and $y_2 = mx_2 + b$. Using the definition of slope, we have

$$\text{slope} = \frac{y_2 - y_1}{x_2 - x_1} = \frac{(mx_2 + b) - (mx_1 + b)}{x_2 - x_1} = \frac{m(x_2 - x_1)}{x_2 - x_1} = m$$

provided that $x_1 \neq x_2$. This computation proves that the slope is the same no matter which two points of a line are used to calculate it and the slope actually appears in the formula $y = mx + b$ as m. Since $y = mx + b$ has y-intercept $(0, b)$ and slope m, $y = mx + b$ is called the **slope-intercept form** of the equation of a line.

Theorem: Slope-Intercept Form

The graph of $y = mx + b$ is a line with slope m and intercept $(0, b)$. Furthermore,

$$m = \frac{y_2 - y_1}{x_2 - x_1}$$

for any points (x_1, y_1) and (x_2, y_2) on the line $y = mx + b$ with $x_1 \neq x_2$.

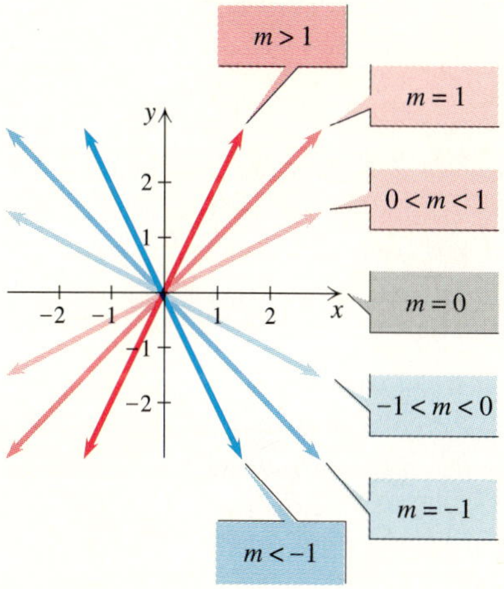

In Section 3.4 we considered $y = x$ as the basic linear function, with the graph of $y = mx$ obtained by transforming the graph of $y = x$. For example, if $m > 1$, each y-coordinate of $y = x$ is increased or stretched to a y-coordinate of $y = mx$. If $0 < m < 1$, each y-coordinate of $y = x$ is decreased or shrunk to a y-coordinate of $y = mx$. Graphs of $y = mx$ are shown in Fig. 4.2 for various values of m.

From Fig. 4.2 we can see that the slope is a measure of the steepness of the line. For a positive slope, the linear function is increasing on its entire domain, and it is called an *increasing function*. For a negative slope, $y = mx$ is a *decreasing function*. Since the average rate of change of a linear function is the constant m, the slope is the rate at which the linear function is increasing or decreasing.

Graphs of $y = mx$

Figure 4.2

Example 2   Slope-intercept form

Write the equation $2x - 3y = 6$ in slope-intercept form and identify the slope and y-intercept. Is the linear function increasing or decreasing?

Solution

Write $2x - 3y = 6$ in slope-intercept form by solving for y:

$$-3y = -2x + 6$$

$$y = \frac{2}{3}x - 2$$

The slope is 2/3 and the y-intercept is $(0, -2)$. The function $y = \frac{2}{3}x - 2$ is an increasing function because the slope is positive.

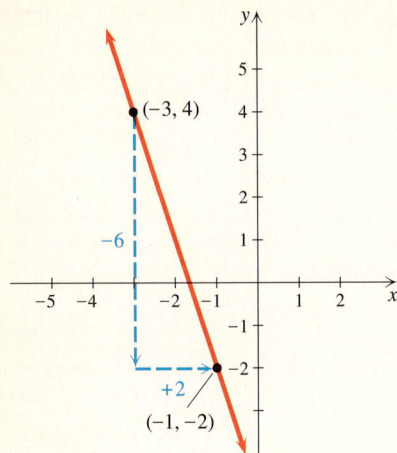

Figure 4.3

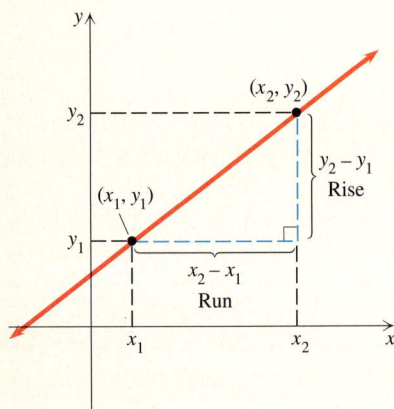

Figure 4.4

Using Slope in Graphing Lines

In Example 1 we found the slope of the line through $(-3, 4)$ and $(-1, -2)$. In this case, $y_2 - y_1 = -6$ and $x_2 - x_1 = 2$. We can move from $(-3, 4)$ to $(-1, -2)$ by going down six units and then moving two units to the right as shown in Fig. 4.3. The numerator in the slope formula, $y_2 - y_1$, is called the **rise,** and it gives the vertical change in going from (x_1, y_1) to (x_2, y_2) as shown in Fig. 4.4. The denominator $x_2 - x_1$ is called the **run,** and it gives the horizontal change. The slope of a line is the ratio of the rise to the run:

$$m = \frac{y_2 - y_1}{x_2 - x_1} = \frac{\text{rise}}{\text{run}}$$

Note that we can think of a positive rise as motion upward and a negative rise as motion downward. Likewise, a positive run indicates motion to the right and a negative run indicates motion to the left.

Example 3   Graphing a linear function using slope and y-intercept

Graph each linear function.

a) $f(x) = \dfrac{2}{3}x - 1$ **b)** $y = -3x + 4$

Solution

a) In the slope-intercept form $f(x) = mx + b$, the y-intercept is $(0, b)$ and the slope is m. So for $f(x) = \frac{2}{3}x - 1$, the y-intercept is $(0, -1)$ and the slope is $2/3$. The line goes through $(0, -1)$. To find a second point on the line, start at $(0, -1)$ and rise 2 and run 3 to get to the point $(3, 1)$ as shown in Fig. 4.5. The graph of $f(x)$ is a line through $(0, -1)$ and $(3, 1)$.

b) For $y = -3x + 4$, the y-intercept is $(0, 4)$ and the slope is -3. The graph goes through $(0, 4)$. The slope of -3 can be expressed as the rise/run ratio $-3/1$. To find a second point on the line, start at $(0, 4)$, move downward three units, and then one unit to the right. Draw a line through $(0, 4)$ and $(1, 1)$, as in Fig. 4.6.

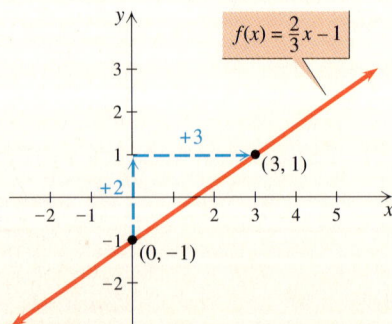

Figure 4.5   **Figure 4.6**

Point-Slope Form of a Line

In Example 3 we started with the equation of a line and graphed the line. We can also start with a description of a line or even a graph of the line and write the equation for the line. Suppose a line has slope m and goes through the point (x_1, y_1). Since the slope is a constant, any other point (x, y) on the line must satisfy

$$\frac{y - y_1}{x - x_1} = m.$$

Multiply each side of this equation by $x - x_1$ to get $y - y_1 = m(x - x_1)$, which is called the **point-slope form** of the equation of a line. We have proved the following theorem.

Theorem: Point-Slope Form

The equation of a line through (x_1, y_1) with slope m is

$$y - y_1 = m(x - x_1).$$

Example 4   Using the point-slope form

Find the equation of the line shown in Fig. 4.7 and write it in slope-intercept form. Assume that the line goes through the two points shown in Fig. 4.7.

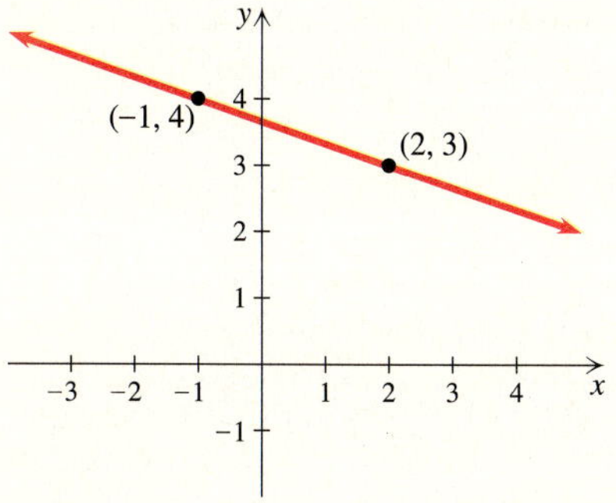

Figure 4.7

Solution

We can use any two points on the line to find its slope, and then use point-slope form for the equation of the line. The points in Fig. 4.7 are $(-1, 4)$ and $(2, 3)$. To get from $(-1, 4)$ to $(2, 3)$ we have a rise of -1 and a run of 3. So the slope is $-1/3$. Let $m = -1/3$ and $(x_1, y_1) = (-1, 4)$ in the equation $y - y_1 = m(x - x_1)$:

$$y - 4 = -\frac{1}{3}(x - (-1)) \qquad \text{The equation in point-slope form}$$

$$y - 4 = -\frac{1}{3}(x + 1)$$

$$y - 4 = -\frac{1}{3}x - \frac{1}{3}$$

$$y = -\frac{1}{3}x + \frac{11}{3} \qquad \text{The equation in slope-intercept form}$$

Check that $(-1, 4)$ and $(2, 3)$ both satisfy the last equation. All four of the equations are equations for the line in Fig. 4.7, but only the last one is in slope-intercept form as required.

Standard Form of the Equation of a Line

Since vertical lines are the only lines that have no slope and are not the graphs of functions, they are the only lines that cannot be described by the linear function $y = mx + b$. However, all lines including vertical lines can be described by an equation in standard form.

Definition: Standard Form

If A, B, and C are real numbers with A and B not both zero, then

$$Ax + By = C$$

is called the **standard form** of the equation of a line.

If $B = 0$ and $A = 1$ in the standard form, the graph of $x = C$ would be a vertical line with x-intercept $(C, 0)$.

Example 5   Equation of a line through two points

Write the standard form of the equation of the line through the given pair of points.

a) $(-1, 3)$ and $(2, 4)$

b) $(3, 5)$ and $(3, -6)$

Solution

a) First find the slope:

$$m = \frac{4 - 3}{2 - (-1)} = \frac{1}{3}$$

Use the point $(2, 4)$ and the slope $1/3$ in the point-slope form:

$$y - 4 = \frac{1}{3}(x - 2)$$

$$3y - 12 = x - 2 \qquad \text{Multiply each side by 3 to get integers.}$$

$$-x + 3y = 10$$

$$x - 3y = -10$$

The equation in standard form is $x - 3y = -10$. Both ordered pairs satisfy $x - 3y = -10$, because $-1 - 3(3) = -10$ and $2 - 3(4) = -10$.

b) The points $(3, 5)$ and $(3, -6)$ are two points on a vertical line through $(3, 0)$. Its equation is $x = 3$.

Parallel Lines

Two lines in a plane are said to be **parallel** if they have no points in common. Any two vertical lines are parallel, and slope can be used to determine whether non-vertical lines are parallel. For example, the lines $y = 3x - 4$ and $y = 3x + 1$ are parallel because their slopes are equal.

Theorem: Parallel Lines

> Two nonvertical lines in the coordinate plane are parallel if and only if their slopes are equal.

A proof to this theorem is outlined in Exercises 93 and 94.

Example 6   Parallel lines

Find the equation in slope-intercept form of the line through $(1, -4)$ that is parallel to $y = 3x + 2$.

Solution

Since $y = 3x + 2$ has slope 3, any line parallel to it also has slope 3. Write the equation of the line through $(1, -4)$ with slope 3 in point-slope form:

$$y - (-4) = 3(x - 1) \qquad \text{Point-slope form}$$

$$y + 4 = 3x - 3$$

$$y = 3x - 7 \qquad \text{Slope-intercept form}$$

The line $y = 3x - 7$ goes through $(1, -4)$ and is parallel to $y = 3x + 2$. ◆

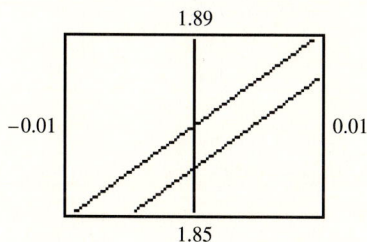

To graph $y = 2.11x + 1.86$ and $y = 2.11x + 1.87$, which are very close to each other, we must use a small viewing window.

Perpendicular Lines

Two lines are **perpendicular** if they intersect at a right angle. Slope can be used to determine whether lines are perpendicular. For example, lines with slopes such as 2/3 and $-3/2$ are perpendicular. The slope $-3/2$ is the opposite of the reciprocal of 2/3. In the following theorem we use the equivalent condition that the product of the slopes of two perpendicular lines is -1, provided they both have slopes.

Theorem: Perpendicular Lines

> Two lines with slopes m_1 and m_2 are perpendicular if and only if $m_1 m_2 = -1$.

Figure 4.8

Proof   The phrase "if and only if" means that there are two statements to prove. First we prove that if two lines are perpendicular, then the product of their slopes is -1. Suppose $y = m_1x + b_1$ and $y = m_2x + b_2$ are perpendicular as shown in Fig. 4.8. Assuming that $m_1 > 0$ and $m_2 < 0$, start at the intersection of the lines and form a triangle by a rise of m_1 and a run of 1 to indicate a slope of m_1. Use the slope m_2 as a negative rise of m_2 and a run of 1. By the Pythagorean theorem, $a = \sqrt{1 + m_1^2}$ and $b = \sqrt{1 + m_2^2}$. Applying the Pythagorean theorem to the third right triangle with $c = m_1 - m_2$, we get

$$(m_1 - m_2)^2 = \left(\sqrt{1 + m_1^2}\right)^2 + \left(\sqrt{1 + m_2^2}\right)^2 \qquad \text{Since } c^2 = a^2 + b^2$$

$$(m_1 - m_2)^2 = 1 + m_1^2 + 1 + m_2^2 \qquad \text{Simplify.}$$

$$m_1^2 - 2m_1m_2 + m_2^2 = 2 + m_1^2 + m_2^2$$

$$-2m_1m_2 = 2$$

$$m_1m_2 = -1$$

For the second half of the proof, we show that if $m_1m_2 = -1$, then the lines are perpendicular. If $m_1m_2 = -1$, we can reverse the order of the preceding equations to conclude that the triangle with sides a, b, and c is a right triangle and that the lines are perpendicular. Thus we have proved the theorem.   ◆

Example 7   Writing equations of perpendicular lines

Find the equation of the line perpendicular to the line $3x - 4y = 8$ and containing the point $(-2, 1)$. Write the answer in slope-intercept form.

Solution

Rewrite $3x - 4y = 8$ in slope-intercept form:

$$-4y = -3x + 8$$

$$y = \frac{3}{4}x - 2 \qquad \text{Slope of this line is 3/4.}$$

Since the product of the slopes of perpendicular lines is -1, the slope of the line that we seek is $-4/3$. Use the slope $-4/3$ and the point $(-2, 1)$ in the point-slope form:

$$y - 1 = -\frac{4}{3}(x - (-2))$$

$$y - 1 = -\frac{4}{3}x - \frac{8}{3}$$

$$y = -\frac{4}{3}x - \frac{5}{3}$$

The last equation is the required equation in slope-intercept form.   ◆

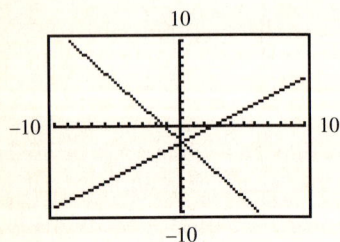

The graphs of $y_1 = \frac{3}{4}x - 2$ and $y_2 = -\frac{4}{3}x - \frac{5}{3}$ do not appear perpendicular in this viewing window, because each axis has a different unit length.

Geometric Application of Slope

In Section 3.1 the distance formula was used to establish some facts about geometric figures in the coordinate plane. We can also use slope to prove that lines are parallel or perpendicular.

Example 8   The diagonals of a rhombus are perpendicular

Given a rhombus with vertices $(0, 0)$, $(5, 0)$, $(3, 4)$, and $(8, 4)$, prove that the diagonals of this rhombus are perpendicular.

Solution

Plot the four points A, B, C, and D as shown in Fig. 4.9. You should show that each side of this figure has length 5, verifying that the figure is a rhombus. Find the slope of each diagonal and their product:

$$m_{AC} = \frac{4 - 0}{8 - 0} = \frac{1}{2} \qquad m_{BD} = \frac{0 - 4}{5 - 3} = -2$$

$$m_{AC} \cdot m_{BD} = \frac{1}{2}(-2) = -1$$

Since the product of the slopes of the diagonals is -1, the diagonals are perpendicular.

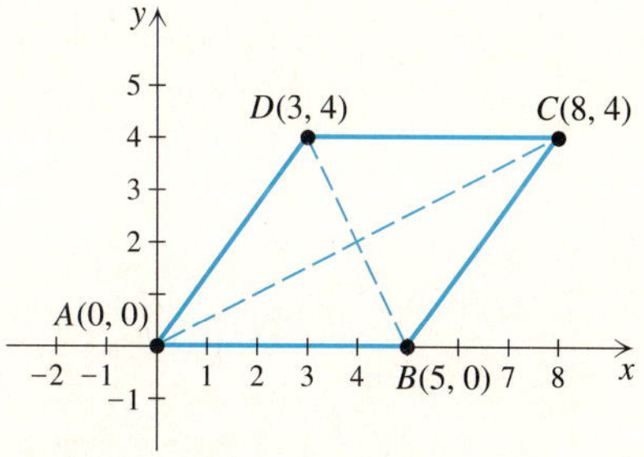

Figure 4.9

Applications of Linear Functions

If one variable is a linear function of another and we know two ordered pairs of that function, then we can write a formula for the linear function using the point-slope form of the equation of a line. In the next example we will see one of the linear functions that Charles Lindbergh used in planning his record-breaking flight across the Atlantic (from *The Spirit of St. Louis* by Charles Lindbergh).

Example 9   An application of linear functions

Lindbergh and Don Hall, chief engineer at Ryan Airlines, calculated that at the beginning of the flight the most economical air speed would be 97 mph, at which Lindbergh would get 1.2 miles per pound of fuel. At the flight's end, when the plane was almost out of fuel, the most economical air speed would be 67 mph, at which he would get 2.3 miles per pound of fuel. See Fig. 4.10. Through repeated testing, Lindbergh knew that the miles per pound M was a linear function of the most economical air speed A. Write a formula for that function.

Solution

The slope of the linear function through the points $(97, 1.2)$ and $(67, 2.3)$ is

$$m = \frac{2.3 - 1.2}{67 - 97} = \frac{1.1}{-30} = -\frac{11}{300}.$$

Use the point-slope form to get the equation of the line:

$$M - 1.2 = -\frac{11}{300}(A - 97)$$

$$M = -0.0366A + 4.7566$$

The number of miles per pound of fuel is determined from the most economical air speed A by the function $M = -0.0366A + 4.7566$, where $67 \leq A \leq 97$. ◆

Lindbergh planned to adjust his air speed to stay on the line shown in Fig. 4.10 from the start to the end of his flight. From knowledge of how much fuel he had used, he could estimate what portion of the flight he had completed and adjust his air speed accordingly.

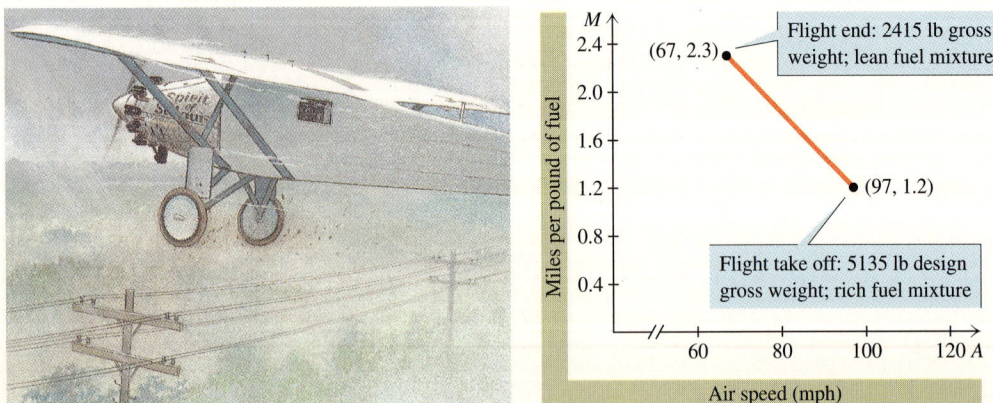

Figure 4.10

? For Thought

True or false? Explain.

1. The slope of the line through (2, 2) and (3, 3) is 3/2.
2. A linear function with positive slope is increasing on $(-\infty, \infty)$.
3. Any two distinct parallel lines have equal slopes.
4. The graph of $x = 3$ in the coordinate plane is the single point (3, 0).
5. Two lines with slopes m_1 and m_2 are perpendicular if $m_1 = -1/m_2$.
6. Every line in the coordinate plane has an equation in slope-intercept form.
7. The leading coefficient of the polynomial function $f(x) = 2x - x^3$ is 2.
8. Every line in the coordinate plane has an equation in standard form.
9. The line $y = 3x$ is parallel to the line $y = -3x$.
10. The line $x - 3y = 4$ contains the point $(1, -1)$ and has slope 1/3.

4.1 Exercises   ▭ **Tape 7**   ▣ **Disk—5.25″: 3   3.5″: 2   Macintosh: 1**

Find the slope of the line containing each pair of points.

1. $(-2, 3), (4, 5)$
2. $(-1, 2), (3, 6)$
3. $(1, 3), (3, -5)$
4. $(2, -1), (5, -3)$
5. $(5, 2), (-3, 2)$
6. $(0, 0), (5, 0)$
7. $(\sqrt{2}, 1), (3\sqrt{2}, -3)$
8. $(3, \sqrt{3}), (7, -\sqrt{3})$
9. $(3, \pi/2), (1, \pi/4)$
10. $(-3, \pi/2), (-5, \pi/3)$
11. $(1988, 3), (1990, 6)$
12. $(234, 188), (237, 156)$

Write each equation in slope-intercept form and identify the slope and y-intercept of the line.

13. $3x - 5y = 10$
14. $2x - 2y = 1$
15. $y - 3 = 2(x - 4)$
16. $y + 5 = -3(x - (-1))$
17. $y + 1 = \dfrac{1}{2}(x - (-3))$
18. $y - 2 = -\dfrac{3}{2}(x + 5)$
19. $y - \dfrac{1}{2} = -\dfrac{1}{3}\left(x - \dfrac{1}{4}\right)$
20. $y + \dfrac{3}{4} = \dfrac{2}{3}\left(x + \dfrac{3}{8}\right)$
21. $\dfrac{y - 6}{x - 8} = -\dfrac{1}{3}$
22. $\dfrac{y + 3}{x - 2} = \dfrac{3}{5}$
23. $y - 0.4 = 0.03(x - 100)$
24. $0.02x - y = 0.6$

Sketch the graph of each equation.

25. $y = \dfrac{1}{2}x - 200$
26. $y = \dfrac{2}{3}x + 100$
27. $f(x) = -3x + 1$
28. $f(x) = -x + 3$
29. $y = -\dfrac{3}{4}x - 1$
30. $y = -\dfrac{3}{2}x$
31. $x - y = 3$
32. $2x - 3y = 6$
33. $x = 5$
34. $x = -3$
35. $y - 5 = 0$
36. $6 - y = 0$
37. $y = \sqrt{2}x - \sqrt{2}$
38. $3x - 4y = 5$
39. $y = 3x - 4$
40. $y = 5x - 5$

Write an equation in slope-intercept form for each of the lines shown.

41.

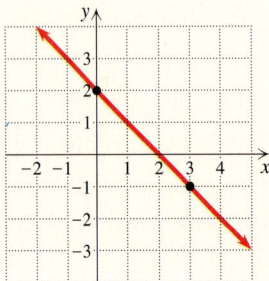

42.

43.

44.

45.

46.

Write the standard form of the equation of the line through the given pair of points.

47. (3, 0) and (0, −4)

48. (−2, 0) and (0, 3)

49. (2, 3) and (−3, −1)

50. (4, −1) and (−2, −6)

51. (−4, 2) and (−4, 5)

52. (−3, 6) and (9, 6)

Find the slope of each line described.

53. A line parallel to $y = 0.5x − 9$

54. A line parallel to $3x − 9y = 4$

55. A line perpendicular to $3y − 3x = 7$

56. A line perpendicular to $2x − 3y = 8$

57. A line perpendicular to the line $x = 4$

58. A line parallel to $y = 5$

Write an equation in standard form with integral coefficients for each of the lines described. In each case make a sketch.

59. The line with slope 2, going through $(1, −2)$

60. The line with slope $−3$, going through $(−3, 4)$

61. The line through $(1, 4)$, parallel to $y = −3x$

62. The line through $(−2, 3)$ parallel to $y = \dfrac{1}{2}x + 6$

63. The line perpendicular to $x − 2y = 3$ and containing $(−3, 1)$

64. The line perpendicular to $3x − y = 9$ and containing $(0, 0)$

65. The line perpendicular to $x = 4$ and containing $(2, 5)$

66. The line perpendicular to $y = 9$ and containing $(−1, 3)$

67. The line through $(0, 3)$ running parallel to the line through $(−1, −2)$ and $(4, 1)$

68. The line through $(0, −2)$ running perpendicular to the line through $(−2, 1)$ and $(3, 5)$

Find the value of a in each case.

69. The line through $(3, 4)$ and $(7, a)$ has slope 2/3.

70. The line through $(−2, a)$ and $(a, 3)$ has slope $−1/2$.

71. The line through $(−2, 3)$ and $(8, 5)$ is perpendicular to $y = ax + 2$.

72. The line through $(−1, a)$ and $(3, −4)$ is parallel to $y = ax$.

Either prove or disprove each statement. Use a graph only as a guide. Your proof should rely on algebraic calculations.

73. The points $(−1, 2)$, $(2, −1)$, $(3, 3)$, and $(−2, −2)$ are the vertices of a parallelogram.

74. The points $(−1, 1)$, $(−2, −5)$, $(2, −4)$, and $(3, 2)$ are the vertices of a parallelogram.

75. The points $(−5, −1)$, $(−3, −4)$, $(3, 0)$, and $(1, 3)$ are the vertices of a rectangle.

76. The points $(−5, −1)$, $(1, −4)$, $(4, 2)$, $(−1, 5)$ are the vertices of a square.

77. The points $(-5, 1)$, $(-2, -3)$, and $(4, 2)$ are the vertices of a right triangle.

78. The points $(-4, -3)$, $(1, -2)$, $(2, 3)$, and $(-3, 2)$ are the vertices of a rhombus.

79. The diagonals of the quadrilateral with vertices $(-3, 2)$, $(-1, -2)$, $(6, -1)$, and $(1, 4)$ are perpendicular.

80. The points $(-5, 1)$, $(-5, -4)$, $(4, 2)$, and $(-2, 3)$ are the vertices of a trapezoid.

Solve each problem.

81. *Celsius to Fahrenheit formula* Fahrenheit temperature F is a linear function of Celsius temperature C. The ordered pair $(0, 32)$ is an ordered pair of this function because $0°C$ is equivalent to $32°F$, the freezing point of water. The ordered pair $(100, 212)$ is also an ordered pair of this function because $100°C$ is equivalent to $212°F$, the boiling point of water. Use the two given points and the point-slope formula to write F as a function of C. Find the Fahrenheit temperature of an oven at $150°C$.

82. *Cost of business cards* Speedy Printing charges \$23 for 200 deluxe business cards and \$35 for 500 deluxe business cards. Given that the cost is a linear function of the number of cards printed, find a formula for that function and find the cost of 700 business cards.

83. *Annual bonus* The annual bonus for a technician at Hosaka Bio Tech was \$4560 in 1980 and \$5230 in 1990. Assuming that the bonus is a linear function of the year, find that function and predict the bonus in 1998.

84. *Shoe size formula* According to an old Sears catalog, a child with a $4\frac{1}{4}$ in. foot wears a size $2\frac{1}{2}$ shoe, and a child with a $5\frac{3}{4}$ in. foot wears a size 7 shoe. Assuming that shoe size is a linear function of foot size, find a formula for that function and the shoe size of a child with a $6\frac{1}{4}$ in. foot.

85. *Lindbergh's practical economical air speed* Lindbergh and Hall spent hours graphing variables such as flight speed, fuel consumption, and weight of the plane. In a graph like the one shown here, they experimented with the relationship between *practical economical air speed* and distance, as measured from the starting point of the flight. They estimated that at the starting point the practical economical air speed was 95 mph and at 4000 statute miles from the starting point it was 75 mph. Assume that the practical economical air speed S is a linear function of the distance D from the starting point and find a formula for that function. (The moral of the story: Start out fast when you have maximum load to lift, and decrease air speed as fuel burns off and load lightens.)

Figure for Exercise 85

86. *Lindbergh's air speed over Newfoundland* Lindbergh did not stick precisely to his calculations during his flight, but used them as a guide. By the time he reached Newfoundland, he had flown 11 hr and covered 1100 mi. Based on the formula in Exercise 85, what should his air speed have been at that point? (In his flying log, Lindbergh recorded his actual air speed over Newfoundland as 98 mph and estimated that with the wind at his tail he was making a mile every 30 sec.)

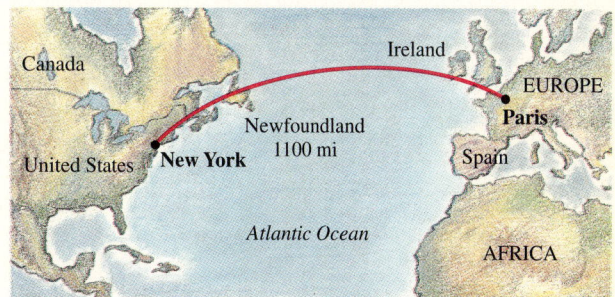

Figure for Exercise 86

87. *Negative income tax* One idea for income tax reform is that of a negative income tax. In this scheme families with an income below a certain level receive a payment from the government in addition to their income, while families above a certain level pay taxes. In one proposal a family with an earned income of \$6000 would receive a \$4000 payment, giving the family a disposable income of \$10,000. A family with an earned income of \$24,000 would pay \$2000 in taxes, giving the family a disposable income of \$22,000. Under this plan, a family's disposable income y is a linear function of the family's earned income x. Use the given data to write the disposable income y as a linear function of the earned income x.

Figure for Exercises 87 and 88

88. *Break-even point* In Exercise 87, what is the disposable income for a family with an earned income of $60,000? Find the break-even point, the income at which a family pays no taxes and receives no payment.

89. *Mean selling price of a car* In a 1994 survey the mean selling price for a new 1994 midsize car was found to be $16,400, while the mean price for a used 1988 midsize car was $3200. Assuming that the mean price is a linear function of the age of the car, find the function. What would you predict is the mean price for a used 1990 midsize car in 1994?

90. *Temperature and altitude* For an average day when the temperature at sea level is 59°F, the temperature at 1000 ft in the air is 56°F. Assume that the temperature T is a linear function of the altitude A and write a formula for T in terms of A. What is the temperature at 30,000 ft on an average day?

91. *Robot gap* The functions $J = 30n + 100$ and $U = 4n + 20$ can be used to model the growth of industrial robot usage in Japan and the United States (*Forbes,* April 16, 1990). The letter n represents the number of years since 1985, while J and U represent the number of robots in Japan and the United States in thousands, respectively. Which country has the more rapid growth in robot usage? In 1985 Japan had five times as many industrial robots as the United States. In what year will Japan have eight times as many?

92. *Marginal cost* A company estimates that its weekly cost $C(x)$ for manufacturing x items is $8000 plus $50 per item. Write a formula for $C(x)$. The marginal cost function $MC(x)$ is the difference quotient for the function $C(x)$ with $h = 1$. Find $MC(x)$. For what values of x is $MC(x)$ increasing, decreasing, or constant?

For Writing/Discussion

93. *Equal slopes* Show that if $y = mx + b_1$ and $y = mx + b_2$ are equations of lines with equal slopes, but $b_1 \neq b_2$, then they have no point in common. (Hint: Assume that they have a point in common and see that this assumption leads to a contradiction.)

94. *Unequal slopes* Show that the lines $y = m_1x + b_1$ and $y = m_2x + b_2$ intersect at a point with x-coordinate $(b_2 - b_1)/(m_1 - m_2)$ provided $m_1 \neq m_2$. Explain how this exercise and the previous exercise prove the theorem that two nonvertical lines are parallel if and only if they have equal slopes.

95. *Cooperative learning* Work in a small group to find and simplify the difference quotient for the function $f(x) = x^2$. Evaluate the difference quotient for $h = 0$ and $x = -1, 0, 1, 2$, and 3, and draw the graph of $f(x) = x^2$. Do these values give any indication of how fast the curve is increasing or decreasing at the points with x-coordinates $-1, 0, 1, 2$, and 3? Use the difference quotient to define the concept of ''slope'' of a curve $y = f(x)$ at any point on the curve.

Graphing Calculator Exercises

Using a "standard" viewing window, the graphs of two lines might be so close that they appear as one line or the graphs might lie entirely outside the viewing window. For example, try graphing $y = 2x$ and $y = 1.999x$ on the same screen. For each pair of lines given, choose an appropriate viewing window that allows you to see the point of intersection of the lines. Use the trace feature to estimate the point of intersection of the lines.

1. $y = 0.495x - 3$, $y = 0.5x - 4$

2. $y = \dfrac{2}{3}x - \dfrac{1}{3}$, $y = 0.67x - 0.33$

3. $y = 3x - 7$, $y = -2x + 20$

4. $y = 2.37x + 4.5$, $y = -1.4x - 2.9$

5. $y = 999x + 1$, $y = 0.0002x - 2$

6. $y = 9999x - 20000$, $y = 0.00001x + 3$

Each of the following lines is "tangent" to the parabola $y = x^2$. (The definition of "tangent to a curve" is given in calculus.) Determine the point of tangency by inspecting the graphs of $y = x^2$ and the line.

7. $y = -2x - 1$ **8.** $y = 6x - 9$

Write the equation of the line tangent to $y = x^2$ at each of the following points, using the fact (from calculus) that a line tangent to $y = x^2$ at (a, b) has slope $2a$. Graph $y = x^2$ and each tangent line to see whether the line appears to be tangent to the curve.

9. $(1, 1)$ **10.** $(-0.5, 0.25)$

4.2

Quadratic Functions

In Section 4.1 we studied linear functions, which are polynomial functions of degree 1. Quadratic functions are polynomial functions of degree 2.

Definition: Quadratic Function

A **quadratic function** is any function of the form

$$f(x) = ax^2 + bx + c,$$

where a, b, and c are real numbers and $a \neq 0$.

In Chapter 3 we graphed some quadratic functions, but we did not call them quadratic functions. In this section we will study quadratic functions in greater detail.

Using Transformations to Graph a Quadratic Function

The simplest quadratic function, $f(x) = x^2$, was graphed in Example 4(a) of Section 3.3. Its graph, shown again in Fig. 4.11, is called a parabola. We saw in Section 3.4 that the graph of any function of the form $f(x) = a(x - h)^2 + k$ is a transformation of the graph of $f(x) = x^2$. By completing the square, we can write any quadratic function in the form $f(x) = a(x - h)^2 + k$, and then quickly sketch the graph using transformations.

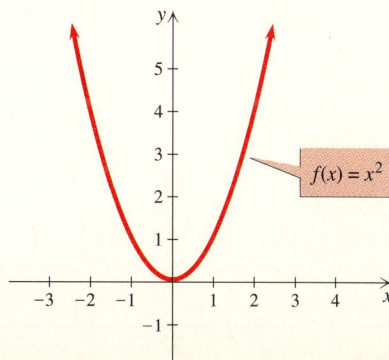

$f(x) = x^2$

Figure 4.11

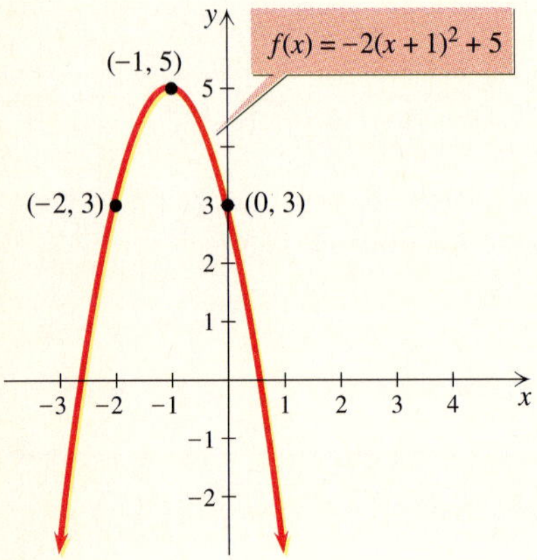

Figure 4.12

Example 1   Graphing a quadratic function

Rewrite $f(x) = -2x^2 - 4x + 3$ in the form $f(x) = a(x - h)^2 + k$ and sketch its graph.

Solution

Start by completing the square:

$$f(x) = -2(x^2 + 2x) + 3 \qquad \text{Factor out } -2 \text{ from first two terms.}$$
$$= -2(x^2 + 2x + 1 - 1) + 3 \qquad \text{Complete the square for } x^2 + 2x.$$
$$= -2(x^2 + 2x + 1) + 2 + 3 \qquad \text{Remove } -1 \text{ from the parentheses.}$$
$$= -2(x + 1)^2 + 5$$

The function is now in the form $f(x) = a(x - h)^2 + k$. The number 1 indicates that the graph of $f(x) = x^2$ is translated one unit to the left. Since $a = -2$, the graph is stretched by a factor of 2 and reflected below the x-axis. Finally, the graph is translated five units upward. The graph shown in Fig. 4.12 includes the points $(0, 3)$, $(-1, 5)$, and $(-2, 3)$. ◆

We can use completing the square as in Example 1 to write any quadratic function $f(x) = ax^2 + bx + c$ in the form $f(x) = a(x - h)^2 + k$:

$$f(x) = ax^2 + bx + c$$
$$= a\left(x^2 + \frac{b}{a}x\right) + c \qquad \text{Factor } a \text{ out of the first two terms.}$$

To complete the square for $x^2 + \dfrac{b}{a}x$, add and subtract $\dfrac{b^2}{4a^2}$ inside the parentheses:

$$f(x) = a\left(x^2 + \frac{b}{a}x + \frac{b^2}{4a^2} - \frac{b^2}{4a^2}\right) + c$$
$$= a\left(x^2 + \frac{b}{a}x + \frac{b^2}{4a^2}\right) - \frac{b^2}{4a} + c \qquad \text{Remove } -\frac{b^2}{4a^2} \text{ from the parentheses.}$$
$$= a\left(x + \frac{b}{2a}\right)^2 + \frac{4ac - b^2}{4a} \qquad \text{Factor and get a common denominator.}$$
$$= a(x - h)^2 + k \qquad \text{Let } h = -\frac{b}{2a} \text{ and } k = \frac{4ac - b^2}{4a}.$$

Since $y = ax^2 + bx + c$ is equivalent to $y = a(x - h)^2 + k$, *the graph of any quadratic function is a transformation of the graph of $y = x^2$.* The graph of any quadratic function is called a **parabola.**

Opening, Vertex, and Axis of Symmetry

If $a > 0$, the graph of $f(x) = a(x - h)^2 + k$ **opens upward;** if $a < 0$, the graph **opens downward** as shown in Fig. 4.13. Notice that h determines the amount of horizontal translation and k determines the amount of vertical translation of the graph of $f(x) = x^2$. Because of the translations, the point $(0, 0)$ on the graph of $f(x) = x^2$ moves to the point (h, k) on the graph of $f(x) = a(x - h)^2 + k$. Since $(0, 0)$ is the lowest point on the graph of $f(x) = x^2$, the lowest point on any parabola that opens upward is (h, k). Since $(0, 0)$ is the highest point on the graph of $f(x) = -x^2$, the highest point on any parabola that opens downward is (h, k). The point (h, k) is called the **vertex** of the parabola.

The graph of $y = 5.6(x - \sqrt{3})^2 + \pi$ opens up. See Appendix A for more examples.

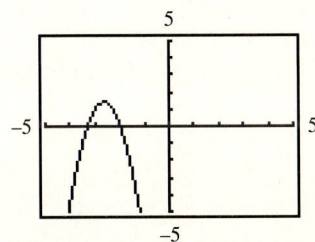

Opening upward

Opening downward

Figure 4.13

The graph of $y = -\pi(x + \sqrt{7})^2 + \sqrt{2}$ opens down. The parabolic shape occurs for any real choices of a, h, and k, provided $a \neq 0$.

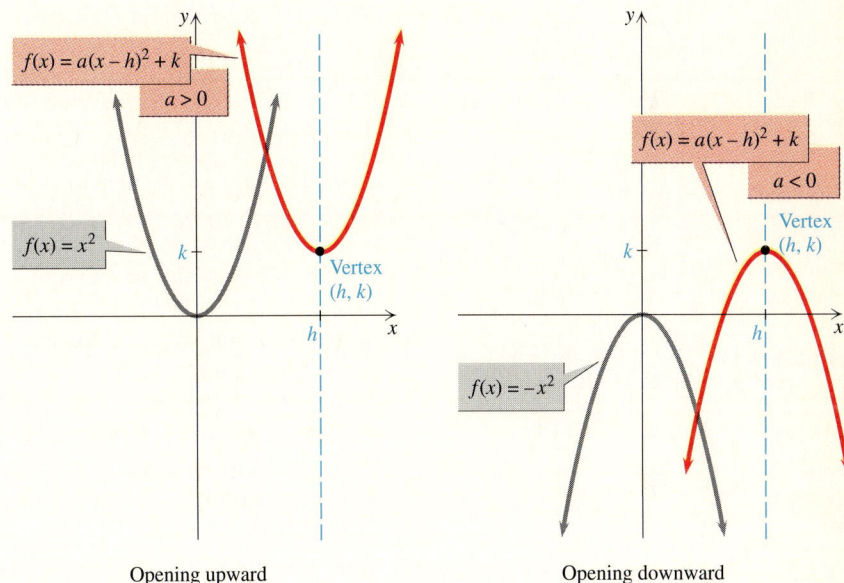

We can also find the vertex of the parabola when the function is written in the form $f(x) = ax^2 + bx + c$. The x-coordinate of the vertex is $-b/(2a)$ because we used $h = -b/(2a)$ when we obtained $y = a(x - h)^2 + k$ from $y = ax^2 + bx + c$. The y-coordinate is found by substitution.

Summary: Vertex of a Parabola

1. **For a quadratic function in the form $f(x) = a(x - h)^2 + k$, the vertex of the parabola is (h, k).**

2. **For the form $f(x) = ax^2 + bx + c$, the x-coordinate of the vertex is $-b/(2a)$. The y-coordinate is $f(-b/(2a))$.**

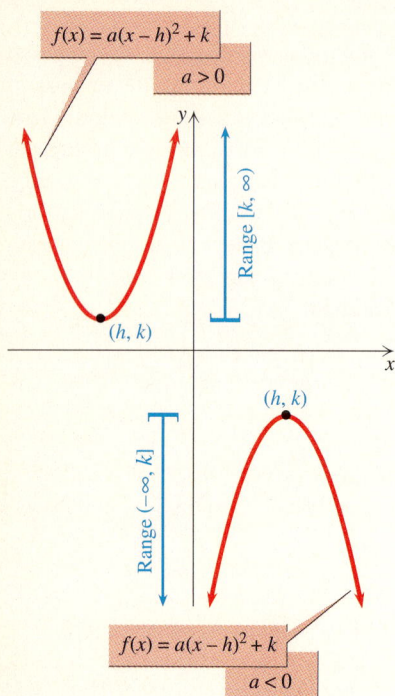

$f(x) = a(x - h)^2 + k$

$a > 0$

Range $[k, \infty)$

(h, k)

(h, k)

Range $(-\infty, k]$

$f(x) = a(x - h)^2 + k$

$a < 0$

Figure 4.14

Example 2   Finding the vertex

Find the vertex of the graph of $f(x) = -2x^2 - 4x + 3$.

Solution

Use $x = -b/(2a)$ to find the x-coordinate of the vertex:

$$x = -\frac{b}{2a} = -\frac{-4}{2(-2)} = -1$$

Now find $f(-1)$:

$$f(-1) = -2(-1)^2 - 4(-1) + 3 = 5$$

The vertex is $(-1, 5)$. In Example 1, $f(x) = -2x^2 - 4x + 3$ was rewritten as $f(x) = -2(x + 1)^2 + 5$. In this form we see immediately that the vertex is $(-1, 5)$. ◆

The domain of every quadratic function $f(x) = a(x - h)^2 + k$ is the set of real numbers, $(-\infty, \infty)$. The range of a quadratic function is determined from the second coordinate of the vertex. If $a > 0$, the range is $[k, \infty)$ and k is called the **minimum value of the function.** The function is decreasing on $(-\infty, h)$ and increasing on (h, ∞). See Fig. 4.14. If $a < 0$, the range is $(-\infty, k]$ and k is called the **maximum value of the function.** The function is increasing on $(-\infty, h)$ and decreasing on (h, ∞).

The graph of $f(x) = x^2$ is symmetric about the y-axis, which runs vertically through the vertex of the parabola. Since this symmetry is preserved in transformations, the graph of any quadratic function is symmetric about the vertical line through its vertex. The vertical line $x = -b/(2a)$ is called the **axis of symmetry** for the graph of $f(x) = ax^2 + bx + c$. Identifying these characteristics of a parabola before drawing a graph makes graphing easier and more accurate.

Example 3   Identifying the characteristics of a parabola

For each parabola, determine whether the parabola opens upward or downward, and find the vertex, axis of symmetry, and range of the function. Find the maximum or minimum value of the function and the intervals on which the function is increasing or decreasing.

a) $y = -2(x + 4)^2 - 8$ **b)** $y = 2x^2 - 4x - 9$

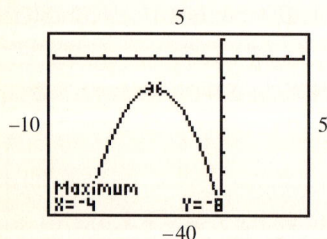

Many calculators have a maximum feature that can find the highest point on a graph.

Solution

a) Since $a = -2$, the parabola opens downward. In $y = a(x - h)^2 + k$, the vertex is (h, k). So the vertex is $(-4, -8)$. The axis of symmetry is the vertical line through the vertex, $x = -4$. Since the parabola opens downward from $(-4, -8)$, the range of the function is $(-\infty, -8]$. The maximum value of the function is -8, and the function is increasing on $(-\infty, -4)$ and decreasing on $(-4, \infty)$. This parabola is graphed in Example 4(a).

b) Since $a = 2$, the parabola opens upward. In the form $y = ax^2 + bx + c$, the x-coordinate of the vertex is $x = -b/(2a)$. In this case,

$$x = \frac{-b}{2a} = \frac{-(-4)}{2(2)} = 1.$$

Use $x = 1$ to find $y = 2(1)^2 - 4(1) - 9 = -11$. So the vertex is $(1, -11)$, and the axis of symmetry is the vertical line $x = 1$. Since the parabola opens upward, the vertex is the lowest point and the range is $[-11, \infty)$. The function is decreasing on $(-\infty, 1)$ and increasing on $(1, \infty)$. The minimum value of the function is -11. This parabola is graphed in Example 4(b). ◆

Intercepts

The x-intercepts and the y-intercept are important points on the graph of a parabola. The x-intercepts are used in solving quadratic inequalities and the y-intercept is the starting point on the graph of a function whose domain is the nonnegative real numbers. The y-intercept is easily found by letting $x = 0$. The x-intercepts are found by letting $y = 0$ and solving the resulting quadratic equation.

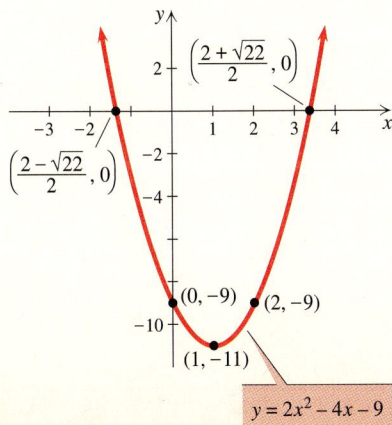

Figure 4.15

Figure 4.16

Example 4   Finding the intercepts

Find the y-intercept and the x-intercepts for each parabola and sketch the graph of each parabola.

a) $y = -2(x + 4)^2 - 8$ **b)** $y = 2x^2 - 4x - 9$

Solution

a) If $x = 0$, $y = -2(0 + 4)^2 - 8 = -40$. The y-intercept is $(0, -40)$. Because the graph is symmetric about the line $x = -4$, the point $(-8, -40)$ is also on the graph. Since the parabola opens downward from $(-4, -8)$, which is below the x-axis, there are no x-intercepts. If we try to solve $-2(x + 4)^2 - 8 = 0$ to find the x-intercepts, we get $(x + 4)^2 = -4$, which has no real solution. The graph is shown in Fig. 4.15.

b) If $x = 0$, $y = 2(0)^2 - 4(0) - 9 = -9$. The y-intercept is $(0, -9)$. From Example 3(b), the vertex is $(1, -11)$. Because the graph is symmetric about the line $x = 1$, the point $(2, -9)$ is also on the graph. The x-intercepts are found by solving $2x^2 - 4x - 9 = 0$:

$$x = \frac{4 \pm \sqrt{(-4)^2 - 4(2)(-9)}}{2(2)} = \frac{4 \pm \sqrt{88}}{4} = \frac{2 \pm \sqrt{22}}{2}$$

The x-intercepts are

$$\left(\frac{2 + \sqrt{22}}{2}, 0\right) \quad \text{and} \quad \left(\frac{2 - \sqrt{22}}{2}, 0\right).$$

See Fig. 4.16. ◆

Note that if $y = ax^2 + bx + c$ has x-intercepts, they can always be found by using the quadratic formula. The x-coordinates of the x-intercepts are

$$x = \frac{-b \pm \sqrt{b^2 - 4ac}}{2a} = \frac{-b}{2a} \pm \frac{\sqrt{b^2 - 4ac}}{2a}.$$

Note how the axis of symmetry, $x = -b/(2a)$, appears in this formula. The x-intercepts are on opposite sides of the graph and are equidistant from the axis of symmetry.

Solving Quadratic Inequalities

We first solved quadratic inequalities in Section 2.7. In Section 3.4 we saw that inequalities could be solved by graphing. Since we can graph any quadratic function, we can solve any quadratic inequality by graphing. As shown in the next example, the key points for solving a quadratic inequality are the x-intercepts.

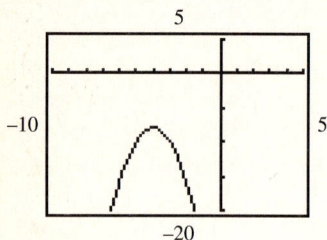

The graph of $y = -2(x + 4)^2 - 8$ supports the conclusion that no y-coordinate is greater than 0.

Example 5 Solving a quadratic inequality by graphing

Solve each inequality by graphing an appropriate function.

a) $-2(x + 4)^2 > 8$ **b)** $2x^2 - 4x - 9 < 0$

Solution

a) The inequality is equivalent to $-2(x + 4)^2 - 8 > 0$. The graph of $y = -2(x + 4)^2 - 8$ is shown in Fig. 4.15. Since no y-coordinate is greater than 0 on this graph, there is no solution to this inequality.

b) The graph of $y = 2x^2 - 4x - 9$ is shown in Fig. 4.16. The y-coordinates on the graph are less than zero provided that the x-coordinates are between the x-intercepts. So the solution to the inequality is the interval

$$\left(\frac{2 - \sqrt{22}}{2}, \frac{2 + \sqrt{22}}{2} \right).$$

The graph of $y = 2x^2 - 4x - 9$ supports the conclusion that $y < 0$ between the x-intercepts.

Applications of Maximum and Minimum

In applications we often seek to minimize cost, maximize profit, or maximize area. If one variable is a quadratic function of another, then the maximum or minimum value of the dependent variable occurs at the vertex of the parabola.

Example 6 Maximizing area of a rectangle

If 100 meters of fencing will be used to fence a rectangular region, then what dimensions for the rectangle will maximize the area of the region?

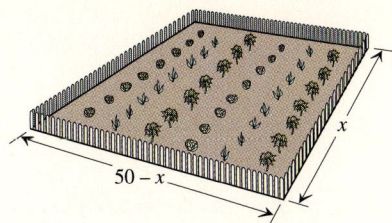

Solution

Since $P = 2(L + W)$ for a rectangle, the length plus the width is 50 meters. If we let x represent the length of the rectangular region, then $50 - x$ represents the width. See Fig. 4.17. Since $A = LW$ for a rectangle, $A = x(50 - x) = -x^2 + 50x$. So the area is a quadratic function of the length. The graph of this function is the parabola in Fig. 4.18.

Figure 4.17

$(25, 625)$

$A = -x^2 + 50x$

Area (m²)

Length of fenced region (m)

Figure 4.18

700

1

0   60

0

X=24.893617 Y=624.98868

0

Tracing to the highest point on the graph of $y = -x^2 + 50x$ supports the conclusion that the maximum area is 625 m².

Since the parabola opens downward, the maximum value of A occurs when

$$x = \frac{-b}{2a} = \frac{-50}{2(-1)} = 25.$$

So the length should be 25 meters and the width should be 25 meters. The rectangle that gives the maximum area is actually a square with an area of 625 m². ◆

? For Thought

True or false? Explain.

1. The domain and range of a quadratic function are $(-\infty, \infty)$.
2. The vertex of the graph of $y = 2(x - 3)^2 - 1$ is $(3, 1)$.
3. The graph of $y = -3(x + 2)^2 - 9$ has no x-intercepts.
4. The maximum value of y in the function $y = -4(x - 1)^2 + 9$ is 9.
5. For $y = 3x^2 - 6x + 7$, the value of y is at its minimum when $x = 1$.
6. The graph of $f(x) = 9x^2 + 12x + 4$ has one x-intercept and one y-intercept.
7. The graph of every quadratic function has exactly one y-intercept.
8. The inequality $\pi(x - \sqrt{3})^2 + \pi/2 \leq 0$ has no solution.
9. The maximum area of a rectangle with fixed perimeter p is $p^2/16$.
10. The function $f(x) = (x - 3)^2$ is increasing on the interval $[-3, \infty)$.

4.2 Exercises

Tape 8   Disk—5.25″: 3   3.5″: 2   Macintosh: 1

Write each quadratic function in the form
$y = a(x - h)^2 + k$ and sketch its graph.

1. $y = x^2 - 3x$

2. $y = -x^2 + x$

3. $y = 2x^2 - 12x + 22$

4. $y = -2x^2 - 4x - 5$

5. $y = -\frac{1}{2}x^2 + x + \frac{5}{2}$

6. $y = \frac{1}{2}x^2 - x + 1$

7. $y = x^2 + 3x + \frac{5}{2}$

8. $y = x^2 - x + 1$

9. $y = -2x^2 + 3x - 1$

10. $y = -2x^2 + x$

11. $y = -3x^2 + 2x$

12. $y = 3x^2 + 4x + 2$

Find the vertex of the graph of each quadratic function.

13. $f(x) = 3x^2 - 12x + 1$

14. $f(x) = -2x^2 - 8x + 9$

15. $f(x) = -3(x - 4)^2 + 1$

16. $f(x) = \frac{1}{2}(x + 6)^2 - \frac{1}{4}$

17. $y = -\frac{1}{2}x^2 - \frac{1}{3}x$

18. $y = \frac{1}{4}x^2 + \frac{1}{2}x - 1$

From the graph of each parabola, determine whether
the parabola opens upward or downward, and find the
vertex, axis of symmetry, and range of the function.
Find the maximum or minimum value of the function
and the intervals on which the function is increasing or
decreasing.

19.

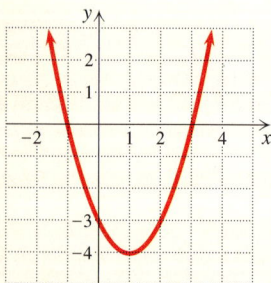

20.

Find the range of each quadratic function and the maximum or minimum value of the function. Identify the
intervals on which each function is increasing or decreasing.

21. $y = (x - 1)^2 - 1$

22. $y = (x + 3)^2 + 4$

23. $y = \sqrt{3} - x^2$

24. $y = \pi - x^2$

25. $f(x) = -2x^2 + 6x + 9$

26. $f(x) = -3x^2 - 9x + 4$

27. $y = \sqrt{2}x^2 + \sqrt{8}x - 2$

28. $f(x) = x^2 - \sqrt{2}x - 3$

29. $f(x) = \frac{1}{2}(x - 3)^2 + 4$

30. $f(x) = -\frac{1}{3}(x + 6)^2 + 37$

31. $y = -\frac{3}{4}\left(x - \frac{1}{2}\right)^2 + 9$

32. $y = \frac{3}{2}\left(x - \frac{1}{3}\right)^2 - 6$

Identify the vertex, axis of symmetry, y-intercept,
x-intercepts, and opening of each parabola, then
sketch the graph.

33. $y = x^2 - 3$

34. $y = 8 - x^2$

35. $f(x) = x^2 - \pi x$

36. $f(x) = 2\pi x - x^2$

37. $f(x) = x^2 + 6x + 9$

38. $f(x) = x^2 - 6x$

39. $f(x) = (x - 3)^2 - 4$

40. $f(x) = (x + 1)^2 - 9$

41. $y = -3(x - 2)^2 + 12$

42. $y = -2(x + 3)^2 + 8$

43. $y = -2x^2 + 4x + 1$

44. $y = -x^2 + 2x - 6$

Identify the solution set to each quadratic inequality
by inspecting the graphs of $y = x^2 - 2x - 3$ and
$y = -x^2 - 2x + 3$ shown here.

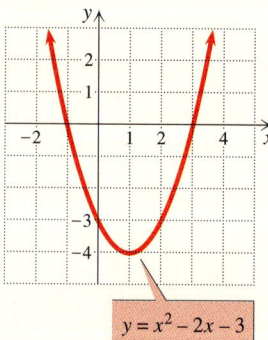

$y = x^2 - 2x - 3$

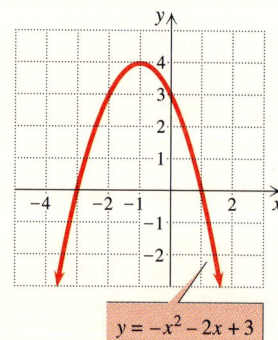

$y = -x^2 - 2x + 3$

45. $x^2 - 2x - 3 \geq 0$

46. $x^2 - 2x - 3 < 0$

47. $-x^2 - 2x + 3 > 0$

48. $-x^2 - 2x + 3 \leq 0$

49. $x^2 + 2x \leq 3$

50. $x^2 \leq 2x + 3$

Solve each quadratic inequality by graphing an appropriate quadratic function.

51. $x^2 - 5x + 6 \le 0$ **52.** $x^2 + 3x < 4$

53. $x^2 - 6x + 7 > 0$ **54.** $2x - x^2 \le -2$

55. $x^2 - 6x + 10 \ge 0$ **56.** $-2x^2 + 4x > 3$

Find the axis of symmetry for each parabola whose equation is given. Use the axis of symmetry to name a point on the parabola that has the same y-coordinate as the given point.

57. $y = x^2$, $(2, 4)$ **58.** $y = 5 - x^2$, $(3, -4)$

59. $y = 3(x - 1)^2 + 5$, $(2, 8)$

60. $y = -2(x + 3)^2 + 1$, $(0, -17)$

61. $y = 3x^2 - 6x + 7$, $(-1, 16)$

62. $y = 5x^2 - 5x + 12$, $(2, 22)$

63. $y = -6x^2 - 6x - 1$, $(-2, -13)$

64. $y = 2x^2 - x - 3$, $(-1/2, -2)$

Solve each problem.

65. Find the value of b so that the vertex of the parabola $y = -3x^2 + bx - 16$ is $(-2, -4)$.

66. Find the value of a so that the vertex of the parabola $y = ax^2 + 10x - 3$ is $(1, 2)$.

67. Find the value of a so that the parabola $y = a(x - 2)^2 + 3$ contains the point $(1, 6)$.

68. Find the value of a so that the parabola $y = a(x - 2)(x + 4)$ has vertex $(-1, 5)$.

69. Write the equation of the parabola that has vertex $(1, -3)$ and y-intercept $(0, 2)$.

70. Write the equation of the parabola that has x-intercepts $(1, 0)$ and $(5, 0)$ and y-intercept $(0, 6)$.

71. Write the equation of the parabola that has vertex $(-3, 2)$ and x-intercept $(-1, 0)$.

72. Write the equation of the parabola that has vertex $(3, -5)$ and goes through the point $(5, -2)$.

73. Find two numbers that have the maximum possible product and a sum of 7.

74. What is the maximum possible product that can be attained by two numbers with a sum of -8?

75. *Maximizing revenue* Mona Kalini gives a walking tour of Honolulu to one person for $49. To increase her business, she advertised at the National Orthodontist Convention that she would lower the price by $1 per person for each additional person, up to 49 people. Write her revenue as a function of the number of people on the tour. What number of people on the tour would maximize her revenue? What is the maximum revenue for her tour?

76. *Concert tickets* At $10 per ticket, Willie Williams and the Wranglers will fill all 8000 seats in the Assembly Center. The manager knows that for every $1 increase in the price, 500 tickets will go unsold. Write the total revenue as a function of the ticket price. What ticket price will maximize the revenue?

77. *Cross section of a gutter* Seth has a piece of aluminum that is 10 in. wide and 12 ft long. He plans to form a gutter with a rectangular cross section and an open top by folding up the sides as shown in the figure. What dimensions of the gutter would maximize the amount of water that it can hold?

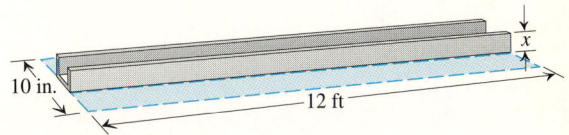

Figure for Exercise 77

78. *Maximum volume of a cage* Sharon has a 12-ft board that is 12 in. wide. She wants to cut it into five pieces to make a cage for two pigeons, as shown in the figure. The front and back will be covered with chicken wire. What should be the dimensions of the cage to maximize the volume and use all of the 12-ft board?

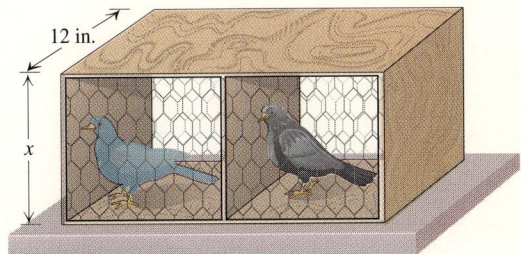

Figure for Exercise 78

79. *Lindbergh's most economical air speed* Lindbergh and Hall used a graph similar to the one shown here to determine the air speed at which the number of miles per pound of fuel, M, would be at a maximum for a fully loaded plane. The curve shown was determined by testing the loaded plane at Camp Kearney, San Diego, in 1927. Using empirical data, Lindbergh figured that the most economical air speed would occur at the highest point on the curve, at roughly 97 mph. In the figure, the curve appears to be a parabola. If we assume that M is a quadratic function of air speed A and is determined by the equation $M = -0.000653A^2 + 0.127A - 5.01$, then what value of A would maximize M?

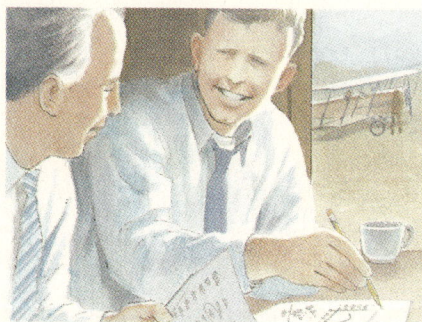

Miles per Pound of Fuel at Takeoff

Figure for Exercise 79

80. *Lindbergh's gas mileage* If Lindbergh flies at 97 mph and gets 1.2 mi per pound of fuel, then how many gallons of fuel is he burning per hour? Assume that gasoline weighs 6.12 lb/gal and see Exercise 79.

81. *Maximum height of a football* If a football is kicked straight up with an initial velocity of 128 ft/sec from a height of 5 ft, then its height above the earth is a function of time given by $h(t) = -16t^2 + 128t + 5$. What is the maximum height reached by this ball?

82. *Maximum height of a ball* If a juggler can toss a ball into the air at a velocity of 64 ft/sec from a height of 6 ft, then what is the maximum height reached by the ball?

83. *Minimizing distance* A caravan starting at $(0, 0)$ must travel in a line $(y = mx)$ to come as close as possible to outposts located at $(2, 3)$ and $(6, 5)$ as shown in the figure. One measure of the total distance from the caravan's path to the outposts is the sum of the squares of the vertical distances from the path to each outpost. In this case we get $d = (3 - 2m)^2 + (6m - 5)^2$. What value for m would minimize d?

Figure for Exercise 83

84. *Fitting a line to data points* In a test of a new type of brake pad, stopping distances from 30, 40, and 50 mph were measured at 40, 50, and 70 ft, respectively. Assuming that the stopping distance y is directly proportional to the speed of the vehicle x, find the equation of the form $y = mx$ that "best fits" the data points shown in the figure. Measure the fit of the line as in Exercise 83. Use your equation to predict the stopping distance from 65 mph.

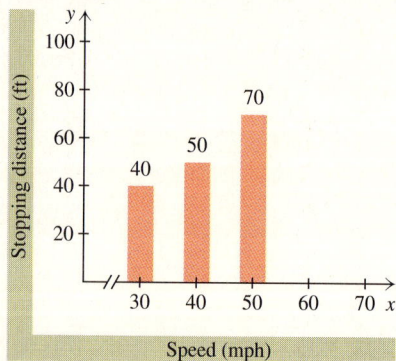

Figure for Exercise 84

85. *Variance of the number of smokers* If p is the probability that a randomly selected person in Chicago is a smoker, then $1 - p$ is the probability that the person is not a smoker. The variance of the number of smokers in a random sample of 50 Chicagoans is $50p(1 - p)$. What value of p maximizes the variance?

86. *Rate of flu infection* In a town of 5000 people the daily rate of infection with a flu virus varies directly with the product of the number of people who have been infected and the number of people not infected. When 1000 people have been infected, the flu is spreading at a rate of 40 new cases per day. For what number of people infected is the daily rate of infection at its maximum?

87. *Designing a paper clip*   Sylvia is developing a paper clip that will carry a company logo in the rectangular center section as shown in the figure. If the clip is to be made out of an 8-in. long piece of wire, then what dimensions for x and y will maximize the area of the rectangular center section? To simplify the problem, assume that the radii of the three semicircles on the ends are equal to $y/2$.

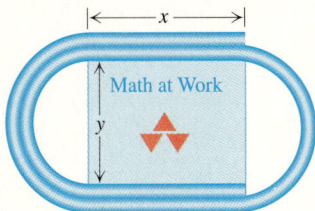

Math at Work

Figure for Exercise 87

88. *Designing a paper clip*   Repeat Exercise 87 assuming that the wire has a 1/16-in. diameter and do not assume that the radii of the semicircles are equal. What assumptions are you making?

89. *Increasing revenue*   A company's weekly revenue in dollars is given by $R(x) = 2000x - 2x^2$, where x is the number of items produced during a week. For what values of x is the revenue positive? For what values of x is the revenue increasing? Decreasing?

90. *Marginal revenue*   The marginal revenue function $MR(x)$ is the difference quotient for $R(x)$ with $h = 1$. Find $MR(x)$ for the revenue function of the previous exercise. For what values of x is the marginal revenue positive? Negative?

Graphing Calculator Exercises

To find the x-intercepts for $y = x^2 + 5x - 6$, we solve $x^2 + 5x - 6 = 0$ by factoring. We get $(x + 6)(x - 1) = 0$ and x-intercepts at $(-6, 0)$ and $(1, 0)$. Note that each x-intercept corresponds to a factor of the quadratic polynomial. Factor the quadratic polynomial that appears in each of the following functions by first locating the x-intercepts of the graph of the function.

1. $y = x^2 - 24x - 3456$ **2.** $y = x^2 - 0.56x - 0.192$

Solve each inequality by locating the x-intercepts of the graph of a quadratic function.

3. $x^2 + 0.1x - 0.021 < 0$ **4.** $x^2 + 60x + 899.5 < 0$

5. Graph the function $f(x) = (x^3 + 2x^2 - 13x + 10)/(x - 2)$. Find the x- and y-intercepts from the graphing calculator. Find the equation of a parabola with the x- and y-intercepts that you found. Check your answer by dividing $x^3 + 2x^2 - 13x + 10$ by $x - 2$.

6. Graph the parabola $y = \sqrt[3]{2}\, x^2 - (\pi/2)\, x - 3$. Use the trace feature to find the minimum value of the function accurate to 4 decimal places.

4.3

Zeros of Polynomial

Functions

In Sections 4.1 and 4.2 we studied polynomial functions of degree 2 or less. In this section we concentrate on polynomial functions in general, particularly those of degree 3 or higher.

The Remainder Theorem

If $y = P(x)$ is a polynomial function, then a value of x that satisfies $P(x) = 0$ is called a **zero** of the polynomial function or a zero of the polynomial. For example, 3 and -3 are zeros of the function $P(x) = x^2 - 9$, because $P(3) = 0$ and $P(-3) = 0$. Note that the zeros of $P(x) = x^2 - 9$ are the same as the solutions to the equation $x^2 - 9 = 0$. The zeros of a polynomial function appear on the graph of the function as the x-coordinates of the x-intercepts. The x-intercepts of the graph of $P(x) = x^2 - 9$ shown in Fig. 4.19 are $(-3, 0)$ and $(3, 0)$.

For polynomial functions of degree 2 or less, the zeros can be found by solving quadratic or linear equations. Our goal in this section is to find all of the zeros of a polynomial function when possible. For polynomials of degree higher than 2, the difficulty of this task ranges from easy to impossible, but we have some theorems to assist us. The remainder theorem relates evaluating a polynomial to division of polynomials.

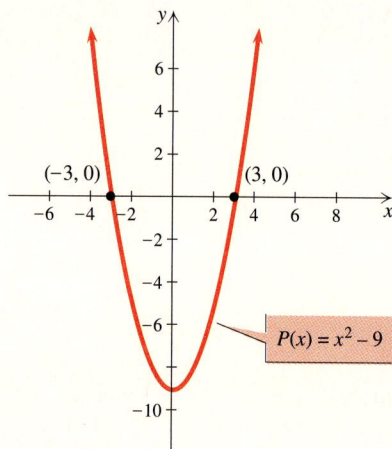

Figure 4.19

The Remainder Theorem

If R is the remainder when a polynomial $P(x)$ is divided by $x - c$, then $R = P(c)$.

Proof   Let $Q(x)$ be the quotient and R be the remainder when $P(x)$ is divided by $x - c$. Since the dividend is equal to the divisor times the quotient plus the remainder, we have

$$P(x) = (x - c)Q(x) + R$$

This statement is true for any value of x, and so it is also true for $x = c$:

$$P(c) = (c - c)Q(c) + R$$
$$= 0 \cdot Q(c) + R$$
$$= R$$

So $P(c)$ is equal to the remainder when $P(x)$ is divided by $x - c$. $\blacklozenge$

Example 1   Using the remainder theorem to evaluate a polynomial

Use the remainder theorem to find $P(3)$ if $P(x) = 2x^3 - 5x^2 + 4x - 6$.

Solution

By the remainder theorem $P(3)$ is the remainder when $P(x)$ is divided by $x - 3$:

$$
\begin{array}{r}
2x^2 + x + 7 \\
x - 3\overline{)2x^3 - 5x^2 + 4x - 6} \\
\underline{2x^3 - 6x^2} \\
x^2 + 4x \\
\underline{x^2 - 3x} \\
7x - 6 \\
\underline{7x - 21} \\
15
\end{array}
$$

The remainder is 15 and therefore $P(3) = 15$. We can check by finding $P(3)$ in the usual manner:

$$P(3) = 2 \cdot 3^3 - 5 \cdot 3^2 + 4 \cdot 3 - 6 = 54 - 45 + 12 - 6 = 15 \qquad \blacklozenge$$

Synthetic Division

In Example 1 we found $P(3) = 15$ in two different ways. Certainly, evaluating $P(x)$ for $x = 3$ in the usual manner is faster than dividing $P(x)$ by $x - 3$ using the ordinary method of dividing polynomials. However, there is a faster method, called **synthetic division,** for dividing by $x - 3$. Compare the two methods side by side, both showing $2x^3 - 5x^2 + 4x - 6$ divided by $x - 3$:

Ordinary Division of Polynomials

$$
\begin{array}{r}
2x^2 + x + 7 \leftarrow \textcolor{blue}{\textbf{Quotient}} \\
x - 3\overline{)2x^3 - 5x^2 + 4x - 6} \\
\underline{2x^3 - 6x^2} \\
x^2 + 4x \\
\underline{x^2 - 3x} \\
7x - 6 \\
\underline{7x - 21} \\
15 \leftarrow \textbf{\textcolor{blue}{Remainder}}
\end{array}
$$

Synthetic Division

$$
\begin{array}{r|rrrr}
3 & 2 & -5 & 4 & -6 \\
  &   & 6 & 3 & 21 \\
\hline
  & 2 & 1 & 7 & 15 \leftarrow \textbf{\textcolor{blue}{Remainder}}
\end{array}
$$

$\underbrace{}$
$\textcolor{blue}{\textbf{Quotient}}$

Synthetic division certainly looks easier than ordinary division, and in general it is faster than evaluating the polynomial by substitution. Synthetic division is used as a quick means of dividing a polynomial by a binomial of the form $x - c$.

In synthetic division we write just the necessary parts of the ordinary division. Instead of writing $2x^3 - 5x^2 + 4x - 6$, write the coefficients 2, -5, 4, and -6. For $x - 3$, write only the 3. The bottom row in synthetic division gives the coefficients of the quotient and the remainder.

To actually perform the synthetic division, start with the following arrangement of coefficients:

$$3 \,\underline{|\, 2 \quad -5 \quad 4 \quad -6}$$

Bring down the first coefficient, 2. Multiply 2 by 3 and write the answer beneath −5. Then add:

$$
\begin{array}{r|rrrr}
3 & 2 & -5 & 4 & -6 \\
  & \downarrow & 6 & & \\
\hline
\text{Multiply} \rightarrow & 2 & 1 & & \\
\end{array}
$$

$\uparrow$
Add

Using 3 rather than −3 when dividing by $x - 3$ allows us to multiply and add rather than multiply and subtract as in ordinary division. Now multiply 1 by 3 and write the answer beneath 4. Then add. Repeat the multiply-and-add step for the remaining column:

$$
\begin{array}{r|rrrr}
3 & 2 & -5 & 4 & -6 \\
  &   & 6 & 3 & 21 \\
\hline
  & 2 & 1 & 7 & 15 \\
\end{array}
$$

The quotient is $2x^2 + x + 7$, and the remainder is 15. Since the divisor in synthetic division is of the form $x - c$, the degree of the quotient is always one less than the degree of the dividend.

Example 2 Synthetic division

Use synthetic division to find the quotient and remainder when $x^4 - 14x^2 + 5x - 9$ is divided by $x + 4$.

Solution

Since $x + 4 = x - (-4)$, we use −4 in the synthetic division. Use 1, 0, −14, 5, and −9 as the coefficients of the polynomial. We use 0 for the coefficient of the missing x^3-term, as we would in ordinary division of polynomials.

$$
\begin{array}{r|rrrrr}
-4 & 1 & 0 & -14 & 5 & -9 \\
   &   & -4 & 16 & -8 & 12 \\
\hline
\text{Multiply} \rightarrow & 1 & -4 & 2 & -3 & 3 \\
\end{array}
$$

$\uparrow$
Add

The quotient is $x^3 - 4x^2 + 2x - 3$ and the remainder is 3. ◆

In the next example, we evaluate a polynomial using the remainder theorem, with synthetic division replacing the ordinary division used in Example 1.

Example 3   Using synthetic division to evaluate a polynomial

Let $f(x) = x^3$ and $g(x) = x^3 - 3x^2 + 5x - 12$. Use synthetic division to find the following function values.

a) $f(-2)$ **b)** $g(4)$

Solution

a) To find $f(-2)$, divide the polynomial x^3 by $x - (-2)$ or $x + 2$ using synthetic division. Write x^3 as $x^3 + 0x^2 + 0x + 0$, and use 1, 0, 0, and 0 as the coefficients. We use a zero for each power of x below x^3.

$$
\begin{array}{r|rrrr}
-2 & 1 & 0 & 0 & 0 \\
   &   & -2 & 4 & -8 \\
\hline
   & 1 & -2 & 4 & -8
\end{array}
$$

The remainder is -8, so $f(-2) = -8$. To check, find $f(-2) = (-2)^3 = -8$.

b) To find $g(4)$, use synthetic division to divide $x^3 - 3x^2 + 5x - 12$ by $x - 4$:

$$
\begin{array}{r|rrrr}
4 & 1 & -3 & 5 & -12 \\
  &   & 4 & 4 & 36 \\
\hline
  & 1 & 1 & 9 & 24
\end{array}
$$

The remainder is 24, so $g(4) = 24$. Check this answer by finding $g(4) = 4^3 - 3(4^2) + 5(4) - 12 = 24$. ◆

There are two ways to find the value of a polynomial $P(x)$ for $x = c$: Find the remainder when the polynomial is divided by $x - c$, or directly evaluate the polynomial by substituting $x = c$. In Example 2(a), it is easier to directly evaluate $(-2)^3$ than to do synthetic division. In Example 2(b), synthetic division is easier because there are fewer arithmetic operations in the synthetic division than in substituting $x = 4$ and computing the result. For polynomials of higher degree there may be an even bigger difference between the number of operations in the two methods.

The Factor Theorem

Consider the polynomial function $P(x) = x^2 - x - 6$. We can find the zeros of the function by solving $x^2 - x - 6 = 0$ by factoring:

$$(x - 3)(x + 2) = 0$$

$$x - 3 = 0 \quad \text{or} \quad x + 2 = 0$$

$$x = 3 \quad \text{or} \quad x = -2$$

Both 3 and -2 are zeros of the function $P(x) = x^2 - x - 6$. Note how each factor of the polynomial corresponds to a zero of the function. This example suggests the following theorem.

The Factor Theorem

> The number c is a zero of the polynomial function $y = P(x)$ if and only if $x - c$ is a factor of the polynomial $P(x)$.

Proof   If c is a zero of the polynomial function $y = P(x)$, then $P(c) = 0$. If $P(x)$ is divided by $x - c$, we get a quotient $Q(x)$ and a remainder R such that

$$P(x) = (x - c)Q(x) + R$$

By the remainder theorem, $R = P(c)$. Since $P(c) = 0$, we have $R = 0$ and $P(x) = (x - c)Q(x)$, which proves that $x - c$ is a factor of $P(x)$.

In Exercise 93 you will be asked to prove that if $x - c$ is a factor of $P(x)$, then c is a zero of the polynomial function. These two arguments together establish the truth of the factor theorem.   ◆

Synthetic division can be used in conjunction with the factor theorem. If the remainder of dividing $P(x)$ by $x - c$ is 0, then $P(c) = 0$ and c is a zero of the polynomial function. By the factor theorem, $x - c$ is a factor of $P(x)$.

Example 4   Using the factor theorem to factor a polynomial

Determine whether $x + 4$ is a factor of the polynomial $P(x) = x^3 - 13x + 12$. If it is a factor, then factor $P(x)$ completely.

Solution

By the factor theorem, $x + 4$ is a factor of $P(x)$ if and only if $P(-4) = 0$. We can find $P(-4)$ using synthetic division:

$$
\begin{array}{r|rrrr}
-4 & 1 & 0 & -13 & 12 \\
   &   & -4 & 16 & -12 \\
\hline
   & 1 & -4 & 3 & 0
\end{array}
$$

Since $P(-4)$ is equal to the remainder, $P(-4) = 0$ and -4 is a zero of $P(x)$. By the factor theorem, $x + 4$ is a factor of $P(x)$. Since the other factor is the quotient from the synthetic division, $P(x) = (x + 4)(x^2 - 4x + 3)$. Factor the quadratic polynomial to get $P(x) = (x + 4)(x - 1)(x - 3)$.   ◆

The graph of $y = x^3 - 13x + 12$ appears to cross the x-axis at -4, 1, and 3, which supports the conclusion that $P(x) = (x + 4)(x - 1)(x - 3)$.

The Fundamental Theorem of Algebra

Whether a number is a zero of a polynomial function can be determined by synthetic division. But does every polynomial function have a zero? This question was answered in the affirmative by Carl F. Gauss when he proved the fundamental theorem of algebra in his doctoral thesis in 1799 at the age of 22. Gauss also proved the n-root theorem of Section 4.4, which says that the number of zeros of a polynomial of degree n is at most n. These theorems do not tell us how to find the zeros of a polynomial function, they just indicate how many there are. Even though the proof of the fundamental theorem is beyond the scope of this text, it is certainly important to know how many zeros a polynomial function has before we begin searching for them.

The Fundamental Theorem of Algebra

> If $y = P(x)$ is a polynomial function of positive degree, then $y = P(x)$ has at least one zero in the set of complex numbers.

Note that the zero guaranteed by the fundamental theorem is in the set of complex numbers. So the zero might be real or imaginary. The theorem applies only to polynomial functions of degree 1 or more, because a polynomial function of zero degree might not have any zeros. For example, $P(x) = 7$ has no zeros. By definition, the coefficients of a polynomial in this text are real numbers, but the fundamental theorem is valid also for polynomials with complex coefficients.

We already know how to find the zeros of many polynomial functions. Since the zeros of $y = P(x)$ are found by solving $P(x) = 0$, the zeros of any linear or quadratic function are found by solving a linear or quadratic equation.

Example 5   Finding all zeros of a polynomial function

Find all real and imaginary zeros of each polynomial function.

a) $f(x) = 6x^2 - x - 35$ **b)** $g(x) = x^3 - 3x^2 + 4x - 12$

Solution

a) To find the zeros of the function, solve the equation $f(x) = 0$:

$$6x^2 - x - 35 = 0$$

$$(2x - 5)(3x + 7) = 0 \qquad \text{Factor.}$$

$$x = \frac{5}{2} \quad \text{or} \quad x = -\frac{7}{3}$$

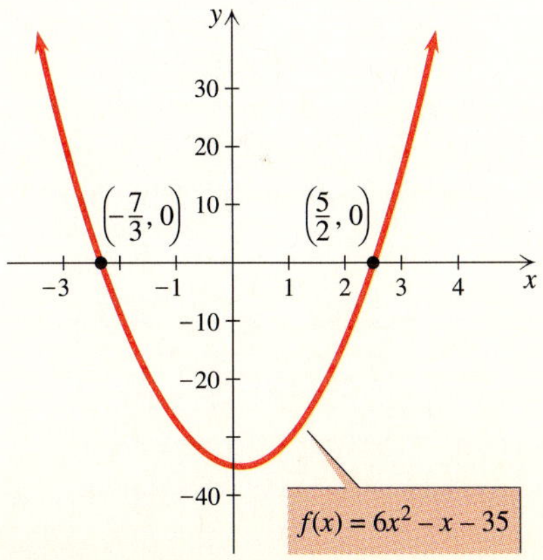

The only zeros of the function $f(x) = 6x^2 - x - 35$ are $5/2$ and $-7/3$. Keep in mind that the zeros of this function correspond to the x-intercepts of its graph shown in Fig. 4.20.

$$f(x) = 6x^2 - x - 35$$

Figure 4.20

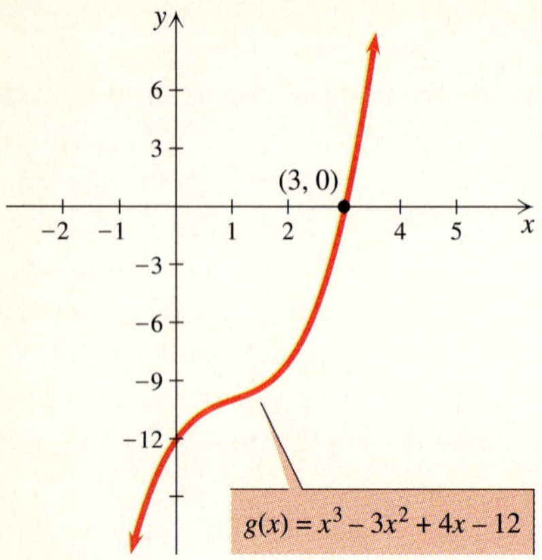

Figure 4.21

$g(x) = x^3 - 3x^2 + 4x - 12$

(3, 0)

b) To find the zeros of the function, solve $g(x) = 0$:

$$x^3 - 3x^2 + 4x - 12 = 0$$

$$x^2(x - 3) + 4(x - 3) = 0 \qquad \text{Factor by grouping.}$$

$$(x^2 + 4)(x - 3) = 0$$

$$x^2 + 4 = 0 \qquad \text{or} \qquad x - 3 = 0$$

$$x^2 = -4 \qquad \text{or} \qquad x = 3$$

$$x = \pm 2i \qquad \text{or} \qquad x = 3$$

The zeros of g are $-2i$, $2i$, and 3. Note that this function has only one real zero and only one x-intercept on its graph in Fig. 4.21. ◆

It was not necessary to graph the functions in Example 5 to find their zeros, but it is good to keep in mind that the real zeros of the function correspond to x-intercepts on the graph. We will study graphs of polynomial functions in Section 4.5.

The Rational Zero Theorem

The fundamental theorem of algebra guarantees that every polynomial function with positive degree has at least one zero, but does not tell us how to find the zeros. Zeros that are rational numbers, the **rational zeros,** are generally the easiest to find. The polynomial function

$$f(x) = 6x^2 - x - 35$$

in Example 5(a) has rational zeros 5/2 and $-7/3$. Note that in 5/2, 5 is a factor of 35 (the constant term) and 2 is a factor of 6 (the leading coefficient). For the zero $-7/3$, 7 is a factor of 35 and 3 is a factor of 6. Of course, these observations are not surprising, because we used these facts to factor the quadratic polynomial in the first place. Note that there are a lot of other factors of 35 and 6 for which the ratio is *not* a zero of this function. This example illustrates the rational zero theorem.

The Rational Zero Theorem

If $f(x) = a_n x^n + a_{n-1} x^{n-1} + a_{n-2} x^{n-2} + \cdots + a_1 x + a_0$ is a polynomial function with integral coefficients ($a_n \neq 0$ and $a_0 \neq 0$) and p/q (in lowest terms) is a rational zero of $f(x)$, then p is a factor of the constant term a_0 and q is a factor of the leading coefficient a_n.

Proof   If p/q is a zero of $f(x)$, then $f(p/q) = 0$:

$$a_n\left(\frac{p}{q}\right)^n + a_{n-1}\left(\frac{p}{q}\right)^{n-1} + a_{n-2}\left(\frac{p}{q}\right)^{n-2} + \cdots + a_1\frac{p}{q} + a_0 = 0$$

Subtract a_0 from each side of this equation.

$$a_n\left(\frac{p}{q}\right)^n + a_{n-1}\left(\frac{p}{q}\right)^{n-1} + a_{n-2}\left(\frac{p}{q}\right)^{n-2} + \cdots + a_1\frac{p}{q} = -a_0$$

Since p/q is in lowest terms, p is not a factor of q. So p appears at least once in each term on the left-hand side of the equation, and p is a factor of the left-hand side. Since the left-hand side is equal to the integer $-a_0$, p is also a factor of a_0.

To prove that q is a factor of a_n, multiply each side of the original equation by $(q/p)^n$, then subtract a_n from each side:

$$a_{n-1}\left(\frac{q}{p}\right) + a_{n-2}\left(\frac{q}{p}\right)^2 + a_{n-3}\left(\frac{q}{p}\right)^3 + \cdots + a_0\left(\frac{q}{p}\right)^n = -a_n$$

Now q is a factor of a_n by the same argument that we used above.   ◆

The rational zero theorem does not identify exactly which rational numbers are zeros of a function; it only gives *possibilities* for the rational zeros.

Example 6   Using the rational zero theorem

Find all possible rational zeros for each polynomial function.

a) $f(x) = 2x^3 - 3x^2 - 11x + 6$ **b)** $g(x) = 3x^3 - 8x^2 - 8x + 8$

Solution

a) If the rational number p/q is a zero of $f(x)$, then p is a factor of 6 and q is a factor of 2. The positive factors of 6 are 1, 2, 3, and 6. The positive factors of 2 are 1 and 2. Take each factor of 6 and divide by 1, to get $1/1$, $2/1$, $3/1$, and $6/1$. Take each factor of 6 and divide by 2, to get $1/2$, $2/2$, $3/2$, and $6/2$. Simplify the ratios, eliminate duplications, and put in the negative factors to get

$$\pm 1, \quad \pm 2, \quad \pm 3, \quad \pm 6, \quad \pm\frac{1}{2}, \quad \text{and} \quad \pm\frac{3}{2}$$

as the possible rational zeros to the function $f(x)$.

b) If the rational number p/q is a zero of $g(x)$, then p is a factor of 8 and q is a factor of 3. The factors of 8 are 1, 2, 4, and 8. The factors of 3 are 1 and 3. If we take all possible ratios of a factor of 8 over a factor of 3, we get

$$\pm 1, \quad \pm 2, \quad \pm 4, \quad \pm 8, \quad \pm\frac{1}{3}, \quad \pm\frac{2}{3}, \quad \pm\frac{4}{3}, \quad \text{and} \quad \pm\frac{8}{3}$$

as the possible rational zeros of the function $g(x)$.   ◆

A polynomial function might not have any rational zeros, but with the rational zero theorem we have a place to start in the problem of finding all of the zeros of a polynomial function. Synthetic division is used to determine whether a possible rational zero is actually a zero of the function.

Example 7   Finding all zeros of a polynomial function

Find all of the real and imaginary zeros for each polynomial function of Example 6.

a) $f(x) = 2x^3 - 3x^2 - 11x + 6$ **b)** $g(x) = 3x^3 - 8x^2 - 8x + 8$

Solution

a) The possible rational zeros of $f(x)$ are listed in Example 6(a). Use synthetic division to check each possible zero to see whether it is actually a zero. Try 1 first.

$$
\begin{array}{r|rrrr}
1 & 2 & -3 & -11 & 6 \\
  &   & 2 & -1 & -12 \\
\hline
  & 2 & -1 & -12 & -6
\end{array}
$$

Since the remainder is -6, 1 is not a zero of the function. Keep on trying numbers from the list of possible rational zeros. To save space, we will not show any more failures. So try 1/2 next.

$$
\begin{array}{r|rrrr}
\frac{1}{2} & 2 & -3 & -11 & 6 \\
  &   & 1 & -1 & -6 \\
\hline
  & 2 & -2 & -12 & 0
\end{array}
$$

Since the remainder is 0, 1/2 is a zero of $f(x)$. By the factor theorem, $x - 1/2$ is a factor of the polynomial. The quotient is the other factor.

$$2x^3 - 3x^2 - 11x + 6 = 0$$

$$\left(x - \frac{1}{2}\right)(2x^2 - 2x - 12) = 0 \qquad \text{\color{blue}Factor.}$$

$$(2x - 1)(x^2 - x - 6) = 0 \qquad \text{\color{blue}Factor 2 out of the second factor.}$$

$$(2x - 1)(x - 3)(x + 2) = 0 \qquad \text{\color{blue}Factor completely.}$$

$$2x - 1 = 0 \qquad \text{or} \qquad x - 3 = 0 \qquad \text{or} \qquad x + 2 = 0$$

$$x = \frac{1}{2} \qquad \text{or} \qquad x = 3 \qquad \text{or} \qquad x = -2$$

The zeros of the function f are 1/2, 3, and -2. Note that each zero of f corresponds to an x-intercept on the graph of f shown in Fig. 4.22.

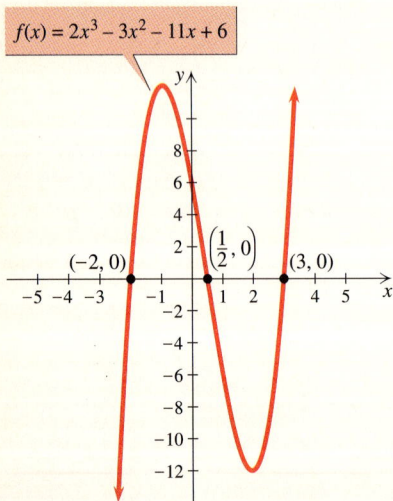

$f(x) = 2x^3 - 3x^2 - 11x + 6$

Figure 4.22

b) The possible rational zeros of $g(x)$ are listed in Example 6(b). First check $2/3$ to see whether it produces a remainder of 0.

$$
\begin{array}{r|rrrr}
\dfrac{2}{3} & 3 & -8 & -8 & 8 \\
 & & 2 & -4 & -8 \\
\hline
 & 3 & -6 & -12 & 0
\end{array}
$$

Since the remainder is 0, $2/3$ is a zero of $g(x)$. By the factor theorem, $x - 2/3$ is a factor of the polynomial. The quotient is the other factor.

$$3x^3 - 8x^2 - 8x + 8 = 0$$

$$\left(x - \frac{2}{3}\right)(3x^2 - 6x - 12) = 0$$

$$(3x - 2)(x^2 - 2x - 4) = 0$$

$$3x - 2 = 0 \qquad \text{or} \qquad x^2 - 2x - 4 = 0$$

$$x = \frac{2}{3} \qquad \text{or} \qquad x = \frac{2 \pm \sqrt{20}}{2}$$

$$x = \frac{2}{3} \qquad \text{or} \qquad x = 1 \pm \sqrt{5}$$

There are one rational and two irrational roots to the equation. So the zeros of the function g are $2/3$, $1 + \sqrt{5}$, and $1 - \sqrt{5}$. Each zero corresponds to an x-intercept on the graph of g shown in Fig. 4.23.

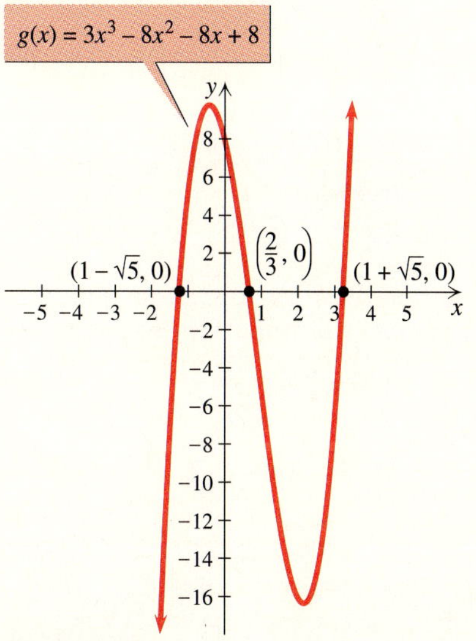

Figure 4.23

Note that in Example 7(a) all of the zeros were rational. All three could have been found by continuing to check the possible rational zeros using synthetic division. In Example 7(b) we would be wasting time if we continued to check the possible rational zeros, because there is only one. When we get to a quadratic polynomial, it is best to either factor the quadratic polynomial or use the quadratic formula to find the remaining zeros.

? For Thought

True or false? Explain.

1. The function $f(x) = 1/x$ has at least one zero.

2. If $P(x) = x^4 - 6x^2 - 8$ is divided by $x^2 - 2$, then the remainder is $P(\sqrt{2})$.

3. If $1 - 2i$ and $1 + 2i$ are zeros of $P(x) = x^3 - 5x^2 + 11x - 15$, then $x - 1 - 2i$ and $x - 1 + 2i$ are factors of $P(x)$.

4. If we divide $x^5 - 1$ by $x - 2$, then the remainder is 31.

5. Every polynomial function has at least one zero.

6. If $P(x) = x^3 - x^2 + 4x - 5$ and b is the remainder from division of $P(x)$ by $x - c$, then $b^3 - b^2 + 4b - 5 = c$.

7. If $P(x) = x^3 - 5x^2 + 4x - 15$, then $P(4) = 0$.

8. The equation $\pi^2 x^4 - \dfrac{1}{\sqrt{2}} x^3 + \dfrac{1}{\sqrt{7} + \pi} = 0$ has at least one complex solution.

9. The binomial $x - 1$ is a factor of $x^5 + x^4 - x^3 - x^2 - x + 1$.

10. The binomial $x + 3$ is a factor of $3x^4 - 5x^3 + 7x^2 - 9x - 2$.

4.3 Exercises   Tape 8   Disk—5.25″: 3   3.5″: 2   Macintosh: 2

Use ordinary division of polynomials to find the quotient and remainder when the first polynomial is divided by the second.

1. $x^2 - 5x + 7, x - 2$

2. $x^2 - 3x + 9, x - 4$

3. $-2x^3 + 4x - 9, x + 3$

4. $-4w^3 + 5w^2 - 7, w - 3$

5. $s^4 - 3s^2 + 6, s^2 - 5$

6. $h^4 + 3h^3 + h - 5, h^2 - 3$

Use synthetic division to find the quotient and remainder when the first polynomial is divided by the second.

7. $x^2 + 4x + 1, x - 2$

8. $2x^2 - 3x + 6, x - 5$

9. $-x^2 + 4x + 9, x + 3$

10. $-3x^2 + 4x - 1, x + 1$

11. $4x^3 - 5x + 2, x - \dfrac{1}{2}$

12. $-6x^3 + 25x^2 - 9, x - \dfrac{3}{2}$

13. $2a^3 - 3a^2 + 4a + 3, a + \dfrac{1}{2}$

14. $-3b^3 - b^2 - 3b - 1, b + \dfrac{1}{3}$

15. $x^4 - 3, x - 1$

16. $x^4 - 16, x - 2$

17. $x^2 - 3x + 1, x - \dfrac{1}{2}$

18. $x^3 - x^2 + x - 1, x - \dfrac{1}{2}$

Let $f(x) = x^5 - 1$, $g(x) = x^3 - 4x^2 + 8$, **and** $h(x) = 2x^4 + x^3 - x^2 + 3x + 3$. **Find the following function values by using synthetic division. Check by using substitution.**

19. $f(1)$ **20.** $f(-1)$ **21.** $f(-2)$ **22.** $f(3)$

23. $g(1)$ **24.** $g(-1)$ **25.** $g\left(-\dfrac{1}{2}\right)$ **26.** $g\left(\dfrac{1}{2}\right)$

27. $h(-1)$ **28.** $h(2)$ **29.** $h(1)$ **30.** $h(-3)$

Determine whether the given binomial is a factor of the polynomial following it. If it is a factor, then factor the polynomial completely.

31. $x + 3, x^3 + 4x^2 + x - 6$

32. $x + 5, x^3 + 8x^2 + 11x - 20$

33. $x - 4, x^3 + 4x^2 - 17x - 60$

34. $x - 2, x^3 - 12x^2 + 44x - 48$

Determine whether each given number is a zero of the polynomial function following the number.

35. $3, f(x) = 2x^3 - 5x^2 - 4x + 3$

36. $-2, g(x) = 3x^3 - 6x^2 - 3x - 19$

37. $-2, g(d) = d^3 + 2d^2 + 3d + 1$

38. $-1, w(x) = 3x^3 + 2x^2 - 2x - 1$

39. $-1, P(x) = x^4 + 2x^3 + 4x^2 + 6x + 3$

40. $3, G(r) = r^4 + 4r^3 + 5r^2 + 3r + 17$

41. $\dfrac{1}{2}, H(x) = x^3 + 3x^2 - 5x + 7$

42. $-\dfrac{1}{2}, T(x) = 2x^3 + 3x^2 - 3x - 2$

Find all real and imaginary zeros of each polynomial function.

43. $f(x) = 3x - 5$ **44.** $h(x) = -2x + 10$

45. $k(x) = x^2 + 5x + 6$ **46.** $m(x) = x^2 + x - 12$

47. $P(t) = t^2 + 5$ **48.** $W(t) = t^2 - 4t + 5$

49. $H(x) = x^3 - 3x^2 + x - 3$

50. $A(x) = x^3 + x^2 + 3x + 3$

51. $K(s) = 2s^3 + s^2 - 15s$ **52.** $J(s) = 4s^4 + 4s^3 - 15s^2$

53. $L(x) = -x^3 - x^2 + 6x + 6$

54. $Q(x) = -3x^3 + 3x^2 + 5x - 5$

55. $M(x) = x^3 - 125$ **56.** $N(x) = 8x^3 - 125$

57. $f(x) = x^4 - 1$

58. $g(x) = 16x^4 - 625$

59. $P(x) = x^6 - 1$

60. $M(x) = x^6 - 64$

Find all possible rational zeros for each polynomial function.

61. $f(x) = x^3 - 9x^2 + 26x - 24$

62. $g(x) = x^3 - 2x^2 - 5x + 6$

63. $h(x) = x^3 - x^2 - 7x + 15$

64. $m(x) = x^3 + 4x^2 + 4x + 3$

65. $P(x) = 8x^3 - 36x^2 + 46x - 15$

66. $T(x) = 18x^3 - 9x^2 - 5x + 2$

67. $M(x) = 18x^3 - 21x^2 + 10x - 2$

68. $N(x) = 4x^3 - 10x^2 + 4x + 5$

Find all of the real and imaginary zeros for each polynomial function.

69. $f(x) = x^3 - 9x^2 + 26x - 24$

70. $g(x) = x^3 - 2x^2 - 5x + 6$

71. $h(x) = x^3 - x^2 - 7x + 15$

72. $m(x) = x^3 + 4x^2 + 4x + 3$

73. $P(a) = 8a^3 - 36a^2 + 46a - 15$

74. $T(b) = 18b^3 - 9b^2 - 5b + 2$

75. $M(t) = 18t^3 - 21t^2 + 10t - 2$

76. $N(t) = 4t^3 - 10t^2 + 4t + 5$

77. $S(w) = w^4 + w^3 - w^2 + w - 2$

78. $W(v) = 2v^4 + 5v^3 + 3v^2 + 15v - 9$

79. $V(x) = x^4 + 2x^3 - x^2 - 4x - 2$

80. $U(x) = x^4 - 4x^3 + x^2 + 12x - 12$

Use division to write each rational expression in the form quotient + remainder/divisor. Use synthetic division when possible.

81. $\dfrac{2x + 1}{x - 2}$ **82.** $\dfrac{x - 1}{x + 3}$ **83.** $\dfrac{a^2 - 3a + 5}{a - 3}$

84. $\dfrac{2b^2 - 3b + 1}{b + 2}$ **85.** $\dfrac{c^2 - 3c - 4}{c^2 - 4}$ **86.** $\dfrac{2h^2 + h - 2}{h^2 - 1}$

87. $\dfrac{4t - 5}{2t + 1}$ **88.** $\dfrac{6y - 1}{3y - 1}$

Solve each problem.

89. *Open-top box* Joan intends to make an 18-in³ open-top box out of a 6 in. by 7 in. piece of copper by cutting equal squares (x in. by x in.) from the corners and folding up the sides. Write the difference between the intended volume and the actual volume as a function of x. For what value of x is there no difference between the intended volume and the actual volume?

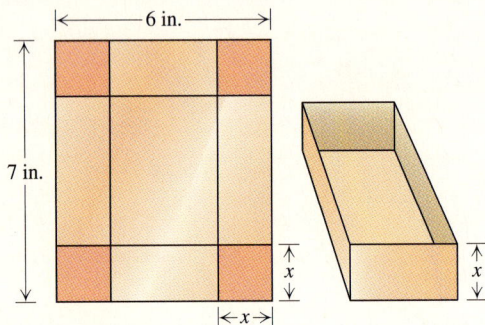

Figure for Exercise 89

90. *Drug testing* The concentration of a drug (in parts per million) in a patient's bloodstream t hours after administration of the drug is given by the function

$$P(t) = -t^4 + 12t^3 - 58t^2 + 132t.$$

How many hours after administration will the drug be totally eliminated from the bloodstream?

91. Verify that the equation $3x^4 + 5x^3 + 4x^2 + 3x + 1 = \{[(3x + 5)x + 4]x + 3\}x + 1$ is an identity.

92. Rewrite the polynomial $P(x) = 5x^4 - 3x^3 + 2x^2 - 7x + 4$ in the form shown in Exercise 91 and find $P(2)$ using that form. This method of evaluating a polynomial is known as Horner's method.

93. Prove that if $x - c$ is a factor of $P(x)$, then c is a zero of the polynomial function.

Figure for Exercise 90

94. For what value of c is the equation $\dfrac{x^2 - 2x + 7}{x - 3} = x + 1 + \dfrac{c}{x - 3}$ an identity?

For Writing/Discussion

95. *Horner's method* Write a detailed explanation of how to rewrite any polynomial in the new form of Exercise 91. Explain how the form of a polynomial shown in Exercises 91 and 92 relates to synthetic division and the remainder theorem.

96. *Synthetic division* Explain how synthetic division can be used to find the quotient and remainder when $3x^3 + 4x^2 + 2x - 4$ is divided by $3x - 2$.

97. *Cooperative learning* Write a brief report on the mathematician Carl F. Gauss and present it to your class.

Graphing Calculator Exercises

Find all rational zeros to each function by first estimating the zeros from its graph and then using your estimates to pick exact values from the list of all possible rational zeros. Use synthetic division to check.

1. $f(x) = 24x^3 - 26x^2 + 9x - 1$

2. $f(x) = 30x^3 - 47x^2 - x + 6$

3. $y = 16x^3 - 33x^2 + 82x - 5$

4. $y = 15x^3 - 37x^2 + 44x - 14$

5. $f(x) = 21x^4 - 31x^3 - 21x^2 - 31x - 42$

6. $f(x) = 119x^4 - 5x^3 + 214x^2 - 10x - 48$

4.4

The Theory of

Equations

One of the main goals in algebra is to keep expanding our knowledge of solving equations. The solutions (roots) of a polynomial equation $P(x) = 0$ are precisely the zeros of a polynomial function $y = P(x)$. Therefore the theorems of Section 4.3 concerning zeros of polynomial functions apply also to the roots of polynomial equations. In this section we study several additional theorems that are useful in solving polynomial equations.

The Number of Roots of a Polynomial Equation

When a polynomial equation is solved by factoring, a factor may occur more than once. For example, $x^2 - 10x + 25 = 0$ is equivalent to $(x - 5)^2 = 0$. Since the factor $x - 5$ occurs twice, we say that 5 is a root of the equation with *multiplicity* 2.

Definition: Multiplicity

> If the factor $x - c$ occurs k times in the complete factorization of the polynomial $P(x)$, then c is called a root of $P(x) = 0$ with **multiplicity** k.

If a quadratic equation has a single root, as in $x^2 - 10x + 25 = 0$, then that root has multiplicity 2. If a root with multiplicity 2 is counted as two roots, then every quadratic equation has two roots in the set of complex numbers. This situation is generalized in the following theorem, where the phrase "when multiplicity is considered" means that a root with multiplicity k is counted as k individual roots.

n-Root Theorem

> If $P(x) = 0$ is a polynomial equation with real or complex coefficients and positive degree n, then, when multiplicity is considered, $P(x) = 0$ has n roots.

Proof   By the fundamental theorem of algebra, the polynomial equation $P(x) = 0$ with degree n has at least one complex root c_1. By the factor theorem, $P(x) = 0$ is equivalent to

$$(x - c_1)Q_1(x) = 0,$$

where $Q_1(x)$ is a polynomial with degree $n - 1$ (the quotient when $P(x)$ is divided by $x - c_1$). Again, by the fundamental theorem of algebra, there is at least one complex root c_2 of $Q_1(x) = 0$. By the factor theorem, $P(x) = 0$ can be written as

$$(x - c_1)(x - c_2)Q_2(x) = 0,$$

where $Q_2(x)$ is a polynomial with degree $n - 2$. Reasoning in this manner n times, we get a quotient polynomial that has 0 degree, n factors for $P(x)$, and n complex roots, not necessarily all different.   ◆

Example 1   Finding all roots of a polynomial equation

State the degree of each polynomial equation. Find all real and imaginary roots of each equation, stating multiplicity when it is greater than one.

a) $6x^5 + 24x^3 = 0$

b) $(x - 3)^2(x + 4)^5 = 0$

Solution

a) This fifth-degree equation can be solved by factoring:

$$6x^3(x^2 + 4) = 0$$

$$6x^3 = 0 \quad \text{or} \quad x^2 + 4 = 0$$

$$x^3 = 0 \quad \text{or} \quad x^2 = -4$$

$$x = 0 \quad \text{or} \quad x = \pm 2i$$

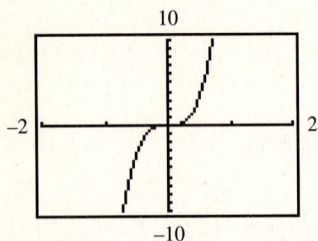

The graph of $y = 6x^5 + 24x^3$ has only one x-intercept at $(0, 0)$. See Appendix A for more examples.

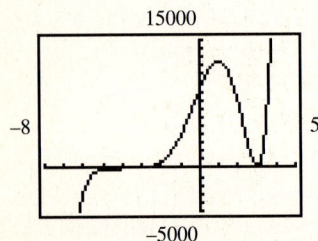

The roots are $\pm 2i$ and 0. Since there are two imaginary roots and 0 is a root with multiplicity 3, there are five roots when multiplicity is considered.

b) The highest power of x in $(x - 3)^2$ is 2, and in $(x + 4)^5$ is 5. By the product rule for exponents, the highest power of x in this equation is 7. The only roots of this seventh-degree equation are 3 and -4. The root 3 has multiplicity 2, and -4 has multiplicity 5. So there are seven roots when multiplicity is considered. ◆

The graph of $y = (x - 3)^2(x + 4)^5$ has x-intercepts $(3, 0)$ and $(-4, 0)$.

The Conjugate Pairs Theorem

For second-degree polynomial equations, we know that the roots occur in pairs. For example, the roots of $x^2 - 2x + 5 = 0$ are

$$x = \frac{2 \pm \sqrt{(-2)^2 - 4(1)(5)}}{2} = 1 \pm 2i.$$

The roots $1 - 2i$ and $1 + 2i$ are complex conjugates. The $\pm$ symbol in the quadratic formula causes the complex solutions of a quadratic equation with real coefficients to occur in conjugate pairs. The conjugate pairs theorem indicates that this situation occurs also for polynomial equations of higher degree.

Conjugate Pairs Theorem

If $P(x) = 0$ is a polynomial equation with real coefficients and the complex number $a + bi$ ($b \neq 0$) is a root, then $a - bi$ is also a root.

The proof for this theorem is left for the exercises.

Example 2 Using the conjugate pairs theorem

Find a polynomial equation with real coefficients that has 2 and $1 - i$ as roots.

Solution

If the polynomial has real coefficients, then its imaginary roots occur in conjugate pairs. So a polynomial with these two roots must actually have at least three roots: $2, 1 - i$, and $1 + i$. Since each root of the equation corresponds to a factor of the polynomial, we can write the following equation.

$$(x - 2)[x - (1 - i)][x - (1 + i)] = 0$$

$$(x - 2)(x^2 - 2x + 2) = 0 \qquad (1 - i)(1 + i) = 1 - i^2 = 2$$

$$x^3 - 4x^2 + 6x - 4 = 0$$

This equation has the required roots and the smallest degree. Any multiple of this equation would also have the required roots but would not be as simple. ◆

Descartes's Rule of Signs

None of the theorems in this chapter tells us how to find all of the n roots to a polynomial equation of degree n. However, the theorems and rules presented here add to our knowledge of polynomial equations and help us to predict the type and number of solutions to expect for a particular equation. Descartes's rule of signs is a method for determining the number of positive, negative, and imaginary solutions.

When a polynomial is written in descending order, a **variation of sign** occurs when the signs of consecutive terms change. For example, if

$$P(x) = 3x^5 - 7x^4 - 8x^3 - x^2 + 3x - 9,$$

there are sign changes in going from the first to the second term, from the fourth to the fifth term, and from the fifth to the sixth term. So there are three variations of sign for $P(x)$. This information determines the number of positive real solutions to $P(x) = 0$. Descartes's rule requires that we look at $P(-x)$ and also count the variations of sign after it is simplified:

$$P(-x) = 3(-x)^5 - 7(-x)^4 - 8(-x)^3 - (-x)^2 + 3(-x) - 9$$

$$= -3x^5 - 7x^4 + 8x^3 - x^2 - 3x - 9$$

In $P(-x)$ the signs of the terms change from the second to the third term and again from the third to the fourth term. So there are two variations of sign for $P(-x)$. This information determines the number of negative real solutions to $P(x) = 0$.

Descartes's Rule of Signs

Suppose $P(x) = 0$ is a polynomial equation with real coefficients and with terms written in descending order.

- The number of positive real roots of the equation is either equal to the number of variations of sign of $P(x)$ or less than that by an even number.
- The number of negative real roots of the equation is either equal to the number of variations of sign of $P(-x)$ or less than that by an even number.

The proof of Descartes's rule of signs is beyond the scope of this text, but we can apply the rule to polynomial equations. Descartes's rule of signs is especially helpful when the number of variations of sign is 0 or 1.

Example 3 Using Descartes's rule of signs

Discuss the possibilities for the roots to $2x^3 - 5x^2 - 6x + 4 = 0$.

Solution

The number of variations of sign in

$$P(x) = 2x^3 - 5x^2 - 6x + 4$$

is 2. By Descartes's rule, the number of positive real roots is either 2 or 0. Since

$$P(-x) = 2(-x)^3 - 5(-x)^2 - 6(-x) + 4$$
$$= -2x^3 - 5x^2 + 6x + 4$$

there is one variation of sign in $P(-x)$. So there is exactly one negative real root.

The equation must have three roots, because it is a third-degree polynomial equation. Since there must be three roots and one is negative, the other two roots must be either both imaginary numbers or both positive real numbers. Table 4.1 summarizes these two possibilities.

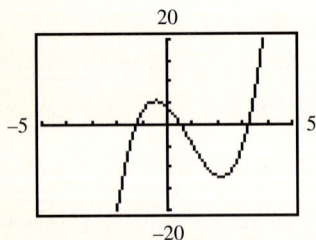

The graph of $y = 2x^3 - 5x^2 - 6x + 4$ crosses the positive x-axis twice and the negative x-axis once. So the first case in Table 4.1 is actually correct.

Positive	Negative	Imaginary
2	1	0
0	1	2

Table 4.1 Number of roots

Example 4 Using Descartes's rule of signs

Discuss the possibilities for the roots to $3x^4 - 5x^3 - x^2 - 8x + 4 = 0$.

Solution

There are two variations of sign in the polynomial

$$P(x) = 3x^4 - 5x^3 - x^2 - 8x + 4.$$

According to Descartes's rule, there are either two or zero positive real roots to the equation. Since

$$P(-x) = 3(-x)^4 - 5(-x)^3 - (-x)^2 - 8(-x) + 4$$

$$= 3x^4 + 5x^3 - x^2 + 8x + 4,$$

there are two variations of sign in $P(-x)$. So the number of negative real roots is either two or zero. Since the degree of the polynomial is 4, there must be four roots. Each line of Table 4.2 gives a possible distribution of the type of those four roots. Note that the number of imaginary roots is even in each case, as we would expect from the conjugate pairs theorem.

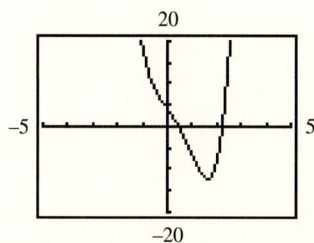

The graph of $y = 3x^4 - 5x^3 - x^2 - 8x + 4$ crosses the positive x-axis twice and does not cross the negative x-axis. So the second case in Table 4.2 is actually correct.

Positive	Negative	Imaginary
2	2	0
2	0	2
0	2	2
0	0	4

Table 4.2 Number of roots

Bounds on the Roots

If a polynomial equation has no roots greater than c, then c is called an **upper bound** for the roots. If there are no roots less than c, then c is called a **lower bound** for the roots. The next theorem is used to determine upper and lower bounds for the roots of a polynomial equation. We will not prove this theorem.

Theorem on Bounds

Suppose that $P(x)$ is a polynomial with real coefficients and a positive leading coefficient, and synthetic division with c is performed.

- If $c > 0$ and all terms in the bottom row are nonnegative, then c is an upper bound for the roots of $P(x) = 0$.

- If $c < 0$ and the terms in the bottom row alternate in sign, then c is a lower bound for the roots of $P(x) = 0$.

If 0 appears in the bottom row of the synthetic division, then it may be considered as a positive or negative term in determining whether the signs alternate. For example, the numbers 3, 0, 5, and -6 would be alternating in sign if we consider 0 as negative. The numbers -7, 5, -8, 0, and -2 would be alternating in sign if we consider 0 as positive.

Example 5   Finding bounds for the roots

Use the theorem on bounds to establish the best integral bounds for the roots of $2x^3 - 5x^2 - 6x + 4 = 0$.

Solution

Try synthetic division with the integers 1, 2, 3, and so on. The first integer for which all terms on the bottom row are nonnegative is the best upper bound for the roots according to the theorem on bounds.

$$
\begin{array}{r|rrrr}
1 & 2 & -5 & -6 & 4 \\
  &   & 2 & -3 & -9 \\
\hline
  & 2 & -3 & -9 & -5
\end{array}
\qquad
\begin{array}{r|rrrr}
2 & 2 & -5 & -6 & 4 \\
  &   & 4 & -2 & -16 \\
\hline
  & 2 & -1 & -8 & -12
\end{array}
$$

$$
\begin{array}{r|rrrr}
3 & 2 & -5 & -6 & 4 \\
  &   & 6 & 3 & -9 \\
\hline
  & 2 & 1 & -3 & -5
\end{array}
\qquad
\begin{array}{r|rrrr}
4 & 2 & -5 & -6 & 4 \\
  &   & 8 & 12 & 24 \\
\hline
  & 2 & 3 & 6 & 28
\end{array}
$$

By the theorem on bounds, no number greater than 4 can be a root to the equation. Now try synthetic division with the integers -1, -2, -3, and so on. The first negative integer for which the terms on the bottom row alternate in sign is the best lower bound for the roots.

$$
\begin{array}{r|rrrr}
-1 & 2 & -5 & -6 & 4 \\
   &   & -2 & 7 & -1 \\
\hline
   & 2 & -7 & 1 & 3
\end{array}
\qquad
\begin{array}{r|rrrr}
-2 & 2 & -5 & -6 & 4 \\
   &   & -4 & 18 & -24 \\
\hline
   & 2 & -9 & 12 & -20
\end{array}
$$

By the theorem on bounds, no number less than -2 can be a root to the equation. So all of the real roots to this equation are between -2 and 4.   ◆

In the next example we use all of the available information about roots.

All of the x-intercepts to $y = 2x^3 - 5x^2 - 6x + 4$ are between -2 and 4.

Example 6   Using all of the theorems about roots

Find all of the solutions to $2x^3 - 5x^2 - 6x + 4 = 0$.

Solution

In Example 3 we used Descartes's rule of signs on this equation to determine that it has either two positive roots and one negative root or one negative root and two imaginary roots. In Example 5 we used the theorem on bounds to determine that all of the real roots to this equation are between -2 and 4. From the rational zero theorem, the possible rational roots are ± 1, ± 2, ± 4, and $\pm 1/2$. Since there must be one negative root and it must be greater than -2, the only possible numbers from the list are -1 and $-1/2$. So start by checking $-1/2$ and -1 with synthetic division.

$$-\frac{1}{2} \;\bigg|\; \begin{array}{rrrr} 2 & -5 & -6 & 4 \\ & -1 & 3 & \dfrac{3}{2} \\ \hline 2 & -6 & -3 & \dfrac{11}{2} \end{array}$$

$$-1 \;\bigg|\; \begin{array}{rrrr} 2 & -5 & -6 & 4 \\ & -2 & 7 & -1 \\ \hline 2 & -7 & 1 & 3 \end{array}$$

Since neither -1 nor $-1/2$ is a root, the negative root must be irrational. Since there might be two positive roots smaller than 4, check $1/2$, 1, and 2.

$$\frac{1}{2} \;\bigg|\; \begin{array}{rrrr} 2 & -5 & -6 & 4 \\ & 1 & -2 & -4 \\ \hline 2 & -4 & -8 & 0 \end{array}$$

Since $1/2$ is a root of the equation, $x - 1/2$ is a factor of the polynomial. The last line in the synthetic division indicates that the other factor is $2x^2 - 4x - 8$.

$$\left(x - \frac{1}{2}\right)(2x^2 - 4x - 8) = 0$$

$$(2x - 1)(x^2 - 2x - 4) = 0$$

$$2x - 1 = 0 \quad \text{or} \quad x^2 - 2x - 4 = 0$$

$$x = \frac{1}{2} \quad \text{or} \quad x = \frac{2 \pm \sqrt{4 - 4(1)(-4)}}{2} = 1 \pm \sqrt{5}$$

There are two positive roots, $1/2$ and $1 + \sqrt{5}$. The negative root is $1 - \sqrt{5}$. Note that the roots guaranteed by Descartes's rule of signs are real numbers but not necessarily rational numbers. ◆

The graph of $y = 2x^3 - 5x^2 - 6x + 4$ has x-intercepts $(1 - \sqrt{5}, 0)$, $(1/2, 0)$, and $(1 + \sqrt{5}, 0)$.

For Thought

True or false? Explain.

1. The number 1 is a root of $x^3 - 1 = 0$ with multiplicity 3.

2. The equation $x^3 = 125$ has three complex number solutions.

3. For $(x + 1)^3(x^2 - 2x + 1) = 0$, -1 is a root with multiplicity 3.

4. For $(x - 5)^3(x^2 - 3x - 10) = 0$, 5 is a root with multiplicity 3.

5. If $4 - 5i$ is a solution to a polynomial equation with real coefficients, then $5i - 4$ is also a solution to the equation.

6. If $P(x) = 0$ is a polynomial equation with real coefficients and i, $2 - 3i$, and $5 + 7i$ are roots, then the degree of $P(x)$ is at least 6.

7. Both $-3 - i\sqrt{5}$ and $3 - i\sqrt{5}$ are solutions to $5x^3 - 9x^2 + 17x - 23 = 0$.

8. Both $3/2$ and 2 are solutions to $2x^5 - 4x^3 - 6x^2 - 3x - 6 = 0$.

9. The equation $x^3 - 5x^2 + 6x - 1 = 0$ has no negative roots.

10. The equation $5x^3 - 171 = 0$ has two imaginary solutions.

4.4 Exercises ▶ Tape 8 💾 Disk—5.25″: 3 3.5″: 2 Macintosh: 2

State the degree of each polynomial equation. Find all of the real and imaginary roots of each equation, stating multiplicity when it is greater than one.

1. $x^5 - 9x^3 = 0$

2. $x^6 + x^4 = 0$

3. $x^4 - 2x^3 + x^2 = 0$

4. $x^5 - 6x^4 + 9x^3 = 0$

5. $x^4 - 10x^2 + 25 = 0$

6. $x^4 - 18x^2 + 81 = 0$

7. $(2x - 3)^2(3x + 4)^2 = 0$

8. $(2x^2 + x)^2(3x - 1)^4 = 0$

9. $x^3 - 4x^2 - 6x = 0$

10. $-x^3 + 8x^2 - 14x = 0$

11. $x^3 + 3x^2 + x + 3 = 0$

12. $x^5 + x^3 - 2x = 0$

Find each product.

13. $(x - 3i)(x + 3i)$

14. $(x + 6i)(x - 6i)$

15. $\left[x - \left(1 + \sqrt{2}\right)\right]\left[x - \left(1 - \sqrt{2}\right)\right]$

16. $\left[x - \left(3 - \sqrt{5}\right)\right]\left[x - \left(3 + \sqrt{5}\right)\right]$

17. $[x - (3 + 2i)][x - (3 - 2i)]$

18. $[x - (3 - i)][x - (3 + i)]$

19. $(x - 2)[x - (3 + 4i)][x - (3 - 4i)]$

20. $(x + 1)\left(x - \dfrac{1 - i}{2}\right)\left(x - \dfrac{1 + i}{2}\right)$

Find a polynomial equation with real coefficients that has the given roots.

21. $-3, 5$

22. $6, -1$

23. $-4i, 4i$

24. $-9i, 9i$

25. $3 - i$

26. $4 + i$

27. $-2, i$

28. $4, -i$

29. $0, i\sqrt{3}$

30. $-2, i\sqrt{2}$

31. $3, 1 - i$

32. $5, 4 - 3i$

33. $1, 2, 3$

34. $-1, 2, -3$

35. $1, 2 - 3i$

36. $-1, 4 - 2i$

37. $\dfrac{1}{2}, \dfrac{1}{3}, \dfrac{1}{4}$

38. $-\dfrac{1}{2}, -\dfrac{1}{3}, 1$

39. $i, 1 + i$

40. $3i, 3 - i$

Use Descartes's rule of signs to discuss the possibilities for the roots of each equation. Do not solve the equation.

41. $x^3 + 5x^2 + 7x + 1 = 0$

42. $2x^3 - 3x^2 + 5x - 6 = 0$

43. $-x^3 - x^2 + 7x + 6 = 0$

44. $-x^4 - 5x^2 - x + 7 = 0$

45. $y^4 + 5y^2 + 7 = 0$

46. $-3y^4 - 6y^2 + 7 = 0$

47. $t^4 - 3t^3 + 2t^2 - 5t + 7 = 0$

48. $-5r^4 + 4r^3 + 7r - 16 = 0$

49. $x^5 + x^3 + 5x = 0$

50. $x^4 - x^2 + 1 = 0$

Use the theorem on bounds to establish the best integral bounds for the roots of each equation.

51. $2x^3 - 5x^2 + 6 = 0$

52. $2x^3 - x^2 - 5x + 3 = 0$

53. $4x^3 + 8x^2 - 11x - 15 = 0$

54. $6x^3 + 5x^2 - 36x - 35 = 0$

55. $w^4 - 5w^3 + 3w^2 + 2w - 1 = 0$

56. $3z^4 - 7z^2 + 5 = 0$

57. $-2x^3 + 5x^2 - 3x + 9 = 0$

58. $-x^3 + 8x - 12 = 0$

Use the rational zero theorem, Descartes's rule of signs, and the theorem on bounds as aids in finding all real and imaginary roots to each equation.

59. $x^3 - 4x^2 - 7x + 10 = 0$

60. $x^3 + 9x^2 + 26x + 24 = 0$

61. $x^3 - 10x - 3 = 0$

62. $2x^3 - 7x^2 - 16 = 0$

63. $x^4 + 2x^3 - 7x^2 + 2x - 8 = 0$

64. $x^4 - 4x^3 + 7x^2 - 16x + 12 = 0$

65. $6x^3 + 25x^2 - 24x + 5 = 0$

66. $6x^3 - 11x^2 - 46x - 24 = 0$

67. $x^4 + 2x^3 - 3x^2 - 4x + 4 = 0$

68. $x^5 + 3x^3 + 2x = 0$

69. $x^4 - 6x^3 + 12x^2 - 8x = 0$

70. $x^4 + 9x^3 + 27x^2 + 27x = 0$

71. $x^6 - x^5 - x^4 + x^3 - 12x^2 + 12x = 0$

72. $2x^7 - 2x^6 + 7x^5 - 7x^4 - 4x^3 + 4x^2 = 0$

Solve each problem.

73. *Growth rate for bacteria* The instantaneous growth rate r of a colony of bacteria t hours after the start of an experiment is given by the function $r = 0.01t^3 - 0.08t^2 + 0.11t + 0.20$ for $0 \le t \le 7$. Find the times for which the instantaneous growth rate is zero.

74. *Retail store profit* The manager of a retail store has figured that her monthly profit P (in thousands of dollars) is determined by her monthly advertising expense x (in tens of thousands of dollars) according to the formula
$$P = x^3 - 20x^2 + 100x \text{ for } 0 \le x \le 4.$$
For what value of x does she get \$147,000 in profit?

75. *Designing a crystal ball* The wizard Gandalf is creating a massive crystal ball. To achieve maximum power, the orb must be a perfect sphere of clear crystal mounted on a square solid silver base, as shown in the figure. The diameter d of the sphere must equal the length L of the side of the square. The thickness of the base must be π in. and the total amount of material used in the sphere and the base must be 1296π in^3. What is the radius of the sphere?

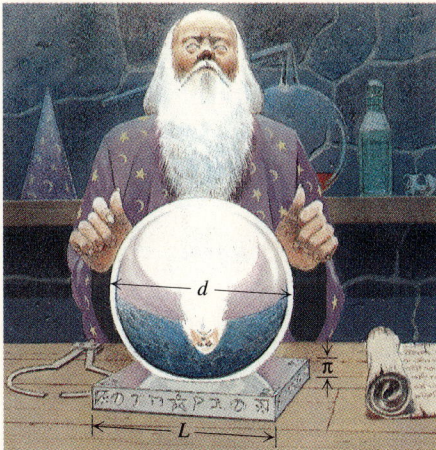

Figure for Exercise 75

76. *Designing fireworks* Marshall is designing a rocket for the Red Rocket Fireworks Company. The rocket will consist of a cardboard circular cylinder with a height that is four times as large as the radius. On top of the cylinder will be a cone with a height of 2 in. and a radius equal to the radius of the base as shown in the figure. If he wants to fill the cone and the cylinder with a total of 114π in^3 of powder, then what should be the radius of the cylinder?

Simplify each expression. (Recall that $\overline{a + bi}$ represents the conjugate of $a + bi$.)

77. $\overline{(3 - 2i) + (4 + 6i)}$

78. $\overline{(2 - i) - (3 + 4i)}$

79. $\overline{3 - 2i} + \overline{4 + 6i}$

80. $\overline{2 - i} - \overline{(3 + 4i)}$

81. $\overline{(3 - 2i)(4 + i)}$

82. $\overline{(3 - 2i)}\,\overline{(4 + i)}$

For Writing/Discussion

83. *Conjugate of a sum* Show that the conjugate of the sum of two complex numbers is equal to the sum of their conjugates.

84. *Conjugate of a product* Show that the conjugate of the product of two complex numbers is equal to the product of their conjugates.

85. *Conjugate of a real* Show that $\overline{a} = a$ for any real number a.

86. *Conjugate pairs* Assume that $a + bi$ is a root of $a_n x^n + a_{n-1}x^{n-1} + \cdots + a_1 x + a_0 = 0$ and substitute $a + bi$ for x. Take the conjugate of each side of the resulting equation and use the results of Exercises 83–85 to simplify it. Your final equation should show that $a - bi$ is a root of $a_n x^n + a_{n-1}x^{n-1} + \cdots + a_1 x + a_0 = 0$. This proves the conjugate pairs theorem.

87. *Missing polynomials* Find a third-degree polynomial function such that $f(0) = 3$ and whose zeros are 1, 2, and 3. Explain how you found it.

88. *Missing polynomials* Is there a third-degree polynomial function such that $f(0) = 6$ and $f(-1) = 12$ and whose zeros are 1, 2, and 3? Explain.

89. *Cooperative learning* Find out about the contributions of the French mathematician Évariste Galois to the theory of polynomial equations and tell your class what you have learned.

2 in.

Figure for Exercise 76

Graphing Calculator Exercises

For each of the following functions use synthetic division and the theorem on bounds to find integers a and b, such that the interval $[a, b]$ contains all real zeros of the function. This method does not necessarily give the shortest interval containing all real zeros. By inspecting the graph of each function, find the shortest interval $[c, d]$ that contains all real zeros of the function with c and d integers. The second interval should be a subinterval of the first.

1. $y = 2x^3 - 3x^2 - 50x + 18$

2. $f(x) = x^3 - 33x - 58$

3. $f(x) = x^4 - 26x^2 + 153$

4. $y = x^4 + x^3 - 16x^2 - 10x + 60$

5. $y = 4x^3 - 90x^2 - 2x + 45$

6. $f(x) = x^4 - 12x^3 + 27x^2 - 6x - 8$

4.5

Graphs of Polynomial

Functions

In Chapter 3 we learned that the graph of a polynomial function of degree 0 or 1 is a straight line. In Section 4.2 we learned that the graph of a second-degree polynomial function is a parabola. In this section we will concentrate on graphs of polynomial functions of degree greater than 2. If we plot enough points (as a graphing calculator does), we can usually discover what the graph looks like. Our goal in this section is to sketch a graph of a polynomial function without plotting a lot of points. To accomplish this goal we must understand symmetry, know the behavior of the graph at its x-intercepts, and know how the leading coefficient affects the shape of the graph.

Symmetry

Symmetry is a very special property of graphs of some functions but not others. Recognizing that the graph of a function has some symmetry usually cuts in half the work required to obtain the graph and also helps cut down on errors in graphing. So far we have discussed the following types of symmetry.

Summary: Types of Symmetry

1. The graph of a function $f(x)$ is *symmetric about the y-axis* if $f(-x) = f(x)$ for any value of x in the domain of the function. (Section 3.4)

2. The graph of a function $f(x)$ is *symmetric about the origin* if $f(-x) = -f(x)$ for any value of x in the domain of the function. (Section 3.4)

3. The graph of a quadratic function $f(x) = ax^2 + bx + c$ is symmetric about its *axis of symmetry*, $x = -b/(2a)$. (Section 4.2)

The graphs of $f(x) = x^2$ and $f(x) = x^3$ shown in Figs. 4.24 and 4.25 are nice examples of symmetry about the y-axis and symmetry about the origin, respectively. The axis of symmetry of $f(x) = x^2$ is the y-axis. The symmetry of the other quadratic functions comes from the fact that the graph of every quadratic function is a transformation of the graph of $f(x) = x^2$.

$f(x) = x^2$

$f(x) = x^3$

Symmetry about y-axis

Figure 4.24

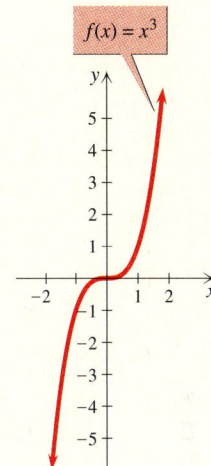

Symmetry about origin

Figure 4.25

Example 1   Determining the symmetry of a graph

Discuss the symmetry of the graph of each polynomial function.

a) $f(x) = 5x^3 - x$ **b)** $g(x) = 2x^4 - 3x^2$ **c)** $h(x) = x^2 - 3x + 6$

d) $j(x) = x^4 - x^3$

Solution

a) Replace x by $-x$ in $f(x) = 5x^3 - x$ and simplify: $f(-x) = 5(-x)^3 - (-x)$
$$= -5x^3 + x$$

Since $f(-x)$ is the opposite of $f(x)$, the graph is symmetric about the origin.

b) Replace x by $-x$ in $g(x) = 2x^4 - 3x^2$ and simplify:

$$g(-x) = 2(-x)^4 - 3(-x)^2$$
$$= 2x^4 - 3x^2$$

Since $g(-x) = g(x)$, the graph is symmetric about the y-axis.

c) Because h is a quadratic function, its graph is symmetric about the line $x = -b/(2a)$, which in this case is the line $x = 3/2$.

d) In this case

$$j(-x) = (-x)^4 - (-x)^3$$
$$= x^4 + x^3$$

So $j(-x) \neq j(x)$ and $j(-x) \neq -j(x)$. The graph of j has neither type of symmetry. ◆

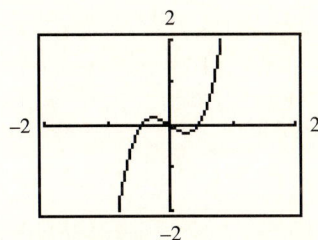

The graph of $y = 5x^3 - x$ is symmetric about the origin.

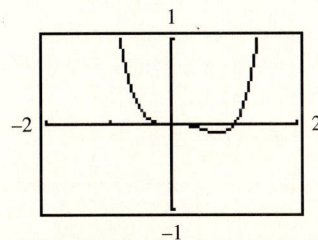

The graph of $y = x^4 - x^3$ has neither type of symmetry. See Appendix A for more examples.

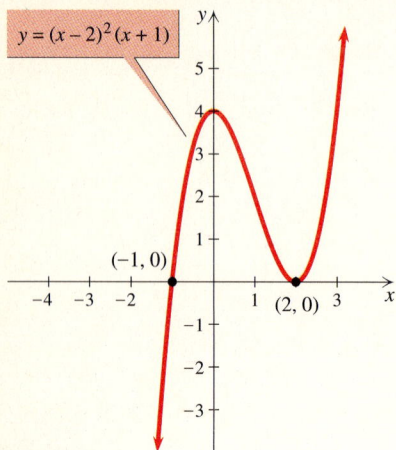

$y = (x - 2)^2(x + 1)$

(−1, 0)

(2, 0)

Figure 4.26

Behavior at the *x*-Intercepts

The *x*-intercepts are key points for the graph of a polynomial function, as they are for any function. Look at the *x*-intercepts $(2, 0)$ and $(-1, 0)$ shown on the graph of $y = (x - 2)^2(x + 1)$ in Fig. 4.26. Notice that the curve crosses the *x*-axis at $(-1, 0)$ but does not cross at $(2, 0)$. This behavior can be predicted from the equation $y = (x - 2)^2(x + 1)$.

Consider the value of the function near the *x*-intercepts. If *x* is chosen slightly to the right or slightly to the left of 2 (say, $x = 2.1$ or $x = 1.9$, respectively), then $(x - 2)^2$ is positive and $x + 1$ is positive. Since $y = (x - 2)^2(x + 1)$, the value of *y* is positive. See Table 4.3. For this reason the curve in Fig. 4.26 does not cross the *x*-axis at $(2, 0)$.

		x-intercepts				
x	− 1.1	−1	− 0.9	1.9	2	2.1
$y = (x - 2)^2(x + 1)$	− 0.961	0	0.841	0.029	0	0.031

↑ *y* changes sign; graph crosses *x*-axis   ↑ *y* does not change sign; graph does not cross *x*-axis

Table 4.3

If *x* is chosen slightly to the right of − 1 (say, $x = -0.9$), then $x + 1$ is positive, but if *x* is chosen slightly to the left of − 1 (say, $x = -1.1$), then $x + 1$ is negative. Since $(x - 2)^2$ is positive in either case, the value of *y* is negative slightly to the left of − 1 and positive slightly to the right of − 1. For this reason the curve in Fig. 4.26 crosses the *x*-axis at $(-1, 0)$.

The behavior of the graph of the polynomial function in Fig. 4.26 at its *x*-intercepts is not an isolated case. Every *x*-intercept corresponds to a factor of the polynomial. Whether that factor occurs an odd or even number of times determines the behavior of the graph at that intercept. *The graph crosses the x-axis at an x-intercept if the factor corresponding to that intercept is raised to an odd power, and the graph does not cross if the factor is raised to an even power.* As another example, consider the graphs of $f(x) = x^2$ and $f(x) = x^3$ shown in Figs. 4.24 and 4.25. Each has only one *x*-intercept. Note that $f(x) = x^2$ does not cross the *x*-axis at $(0, 0)$, but $f(x) = x^3$ does cross the *x*-axis at $(0, 0)$.

The graph of $y = x^2(x - 2)^2(x + 2)^2$ does not cross the *x*-axis at any of its three *x*-intercepts.

Example 2   Crossing at the *x*-intercepts

Find the *x*-intercepts and determine whether the graph of the function crosses the *x*-axis at each *x*-intercept.

a) $f(x) = (x - 1)^2(x - 3)$

b) $f(x) = x^3 + 2x^2 - 3x$

Solution

a) The *x*-intercepts are found by solving $(x - 1)^2(x - 3) = 0$. The *x*-intercepts are (1, 0) and (3, 0). The graph does not cross the *x*-axis at (1, 0) because the factor $x - 1$ occurs to an even power. The graph crosses the *x*-axis at (3, 0) because $x - 3$ occurs to an odd power.

b) The *x*-intercepts are found by solving $x^3 + 2x^2 - 3x = 0$. By factoring, we get $x(x + 3)(x - 1) = 0$. The *x*-intercepts are (0, 0), (−3, 0), and (1, 0). Since each factor occurs an odd number of times (once), the graph crosses the *x*-axis at each of the *x*-intercepts. ◆

The Leading Coefficient Test

Polynomial functions of degree 2 or less have graphs that are either parabolas or lines. Figure 4.27 shows some graphs of polynomial functions of degree 3 and higher.

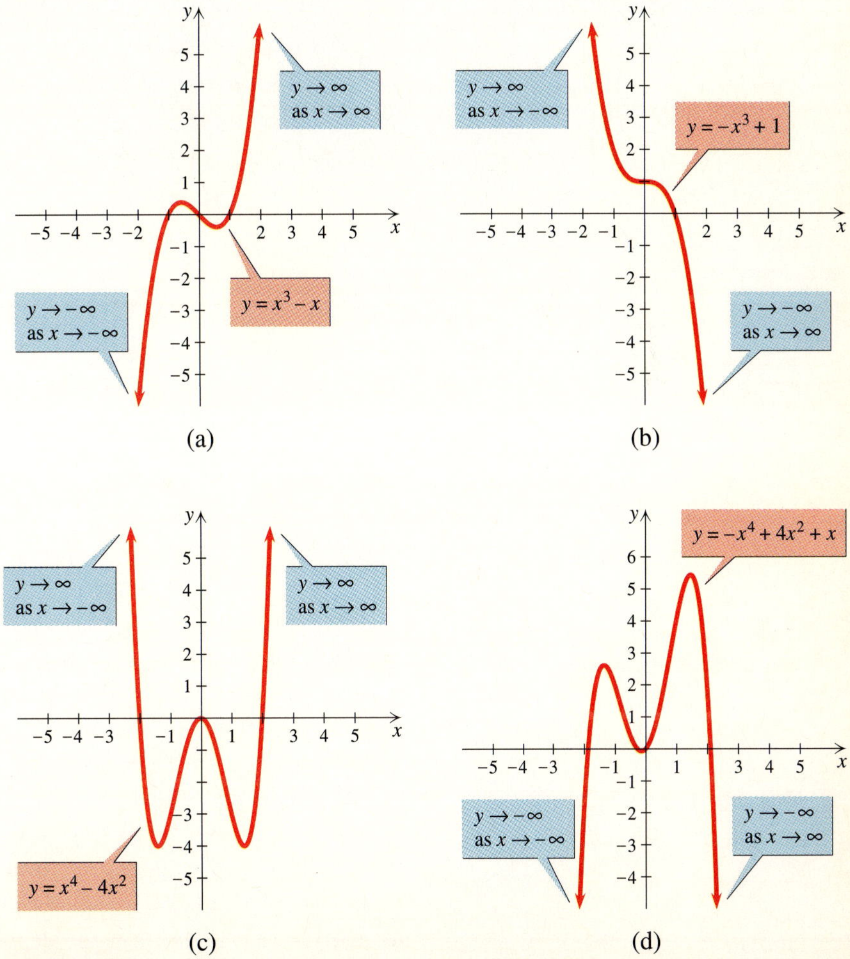

Figure 4.27

There is a lot of variety to the graphs in Fig. 4.27, but they do have similarities. All of these graphs are continuous curves with no breaks like those in the greatest integer function, and they are smooth curves without sharp corners like that in the graph of the absolute value function. We know why a curve crosses or does not cross the x-axis at an x-intercept. Now we see what determines the behavior of the curve as x takes on larger and larger values (or smaller and smaller values).

The graph of any polynomial function either rises or falls as x gets larger without bound, $x \to \infty$ (read "x goes to ∞"), or as x gets smaller without bound, $x \to -\infty$ (read "x goes to $-\infty$"). The curves in Fig. 4.27 are examples of the different possibilities. The **leading coefficient test** tells us how to determine the behavior of the graph as $x \to \infty$ or $x \to -\infty$ by observing the leading coefficient and the degree of the polynomial.

Leading Coefficient Test

If $f(x) = a_n x^n + a_{n-1} x^{n-1} + \cdots + a_1 x + a_0$, the behavior of the graph of f to the left and right is determined as follows:

For n odd and $a_n > 0$, $y \to \infty$ as $x \to \infty$, and $y \to -\infty$ as $x \to -\infty$.

For n odd and $a_n < 0$, $y \to -\infty$ as $x \to \infty$, and $y \to \infty$ as $x \to -\infty$.

For n even and $a_n > 0$, $y \to \infty$ as $x \to \infty$ or $x \to -\infty$.

For n even and $a_n < 0$, $y \to -\infty$ as $x \to \infty$ or $x \to -\infty$.

Example 3   The leading coefficient test

Apply the leading coefficient test to each function.

a) $y = x^3 - x$

b) $y = -x^3 + 1$

c) $y = x^4 - 4x^2$

d) $y = -x^4 + 4x^2 + x$

Solution

a) For $y = x^3 - x$ the degree is odd and the leading coefficient is positive. So as x gets larger and larger, y increases without bound; as x gets smaller and smaller, y decreases without bound as shown in Fig. 4.27(a).

b) For $y = -x^3 + 1$ the degree is odd and the leading coefficient is negative. So as x gets larger and larger, y decreases without bound; as x gets smaller and smaller, y increases without bound as shown in Fig. 4.27(b).

c) For $y = x^4 - 4x^2$ the degree is even and the leading coefficient is positive. So as x goes to ∞ or $-\infty$, y increases without bound as shown in Fig. 4.27(c).

d) For $y = -x^4 + 4x^2 + x$, the degree is even and the leading coefficient is negative. So as x goes to ∞ or $-\infty$, y goes to $-\infty$ as shown in Fig. 4.27(d).

Sketching Graphs of Polynomial Functions

Graphs of polynomial functions have some features that can be discovered through calculus but will not be discussed in this text. The following strategy will help you graph polynomial functions, but you might not be able to make the graphs as accurate as you would like without plotting an excessive number of points (which, of course, a graphing calculator can do for you).

Strategy: Graphing Polynomial Functions

1. **Check for symmetry of any type.**
2. **Find all real zeros of the polynomial function.**
3. **Determine the behavior at the corresponding x-intercepts.**
4. **Determine the behavior as $x \to \infty$ and as $x \to -\infty$.**
5. **Calculate several ordered pairs including the y-intercept to verify your suspicions about the shape of the graph.**
6. **Draw a smooth curve through the points to make the graph.**

Example 4 Graphing polynomial functions

Sketch the graph of each polynomial function.

a) $f(x) = x^3 - 5x^2 + 7x - 3$ **b)** $f(x) = x^4 - 2x^2 + 1$

Solution

a) First find $f(-x)$ to determine symmetry and the number of negative roots.

$$f(-x) = (-x)^3 - 5(-x)^2 + 7(-x) - 3$$
$$= -x^3 - 5x^2 - 7x - 3$$

From $f(-x)$, we see that the graph has neither type of symmetry. Because $f(-x)$ has no sign changes, $x^3 - 5x^2 + 7x - 3 = 0$ has no negative roots by Descartes's rule of signs. The only possible rational roots are 1 and 3.

$$
\begin{array}{r|rrrr}
1 & 1 & -5 & 7 & -3 \\
  &   & 1 & -4 & 3 \\
\hline
  & 1 & -4 & 3 & 0
\end{array}
$$

From the synthetic division we know that 1 is a root and we can factor $f(x)$:

$$f(x) = (x - 1)(x^2 - 4x + 3)$$
$$= (x - 1)^2(x - 3) \qquad \text{Factor completely.}$$

The x-intercepts are $(1, 0)$ and $(3, 0)$. The graph of f does not cross the x-axis at $(1, 0)$ because $x - 1$ occurs to an even power, while the graph crosses at $(3, 0)$ because $x - 3$ occurs to an odd power. The y-intercept is $(0, -3)$. Since the leading coefficient is positive and the degree is odd, $y \to \infty$ as $x \to \infty$ and $y \to -\infty$ as $x \to -\infty$. Calculate two more ordered pairs for accuracy, say $(2, -1)$ and $(4, 9)$. The graph is shown in Fig. 4.28.

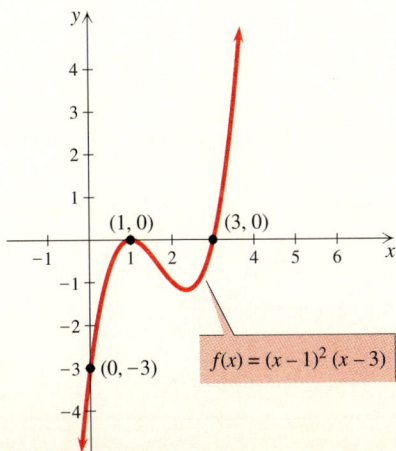

Figure 4.28

b) First find $f(-x)$:

$$f(-x) = (-x)^4 - 2(-x)^2 + 1$$
$$= x^4 - 2x^2 + 1$$

Since $f(x) = f(-x)$, the graph is symmetric about the y-axis. We can factor the polynomial as follows.

$$f(x) = x^4 - 2x^2 + 1$$
$$= (x^2 - 1)(x^2 - 1)$$
$$= (x - 1)(x + 1)(x - 1)(x + 1)$$
$$= (x - 1)^2(x + 1)^2$$

The x-intercepts are $(1, 0)$ and $(-1, 0)$. Since each factor for these intercepts has an even power, the graph does not cross the x-axis at the intercepts. The y-intercept is $(0, 1)$. Since the leading coefficient is positive and the degree is even, $y \to \infty$ as $x \to \infty$ or as $x \to -\infty$. Calculate two more ordered pairs for accuracy, say $(2, 9)$ and $(-2, 9)$. The graph is shown in Fig. 4.29.

The important features of $y = (x - 1)^2(x + 1)^2$ do not appear in this view.

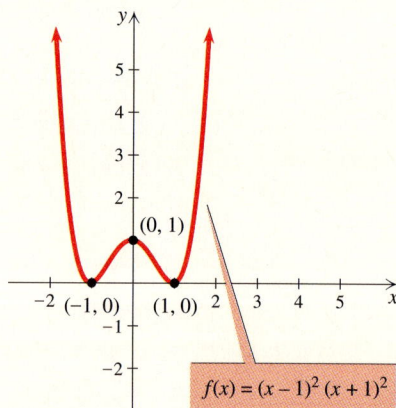

Figure 4.29

For Thought

True or false? Explain.

1. If P is a function for which $P(2) = 8$ and $P(-2) = -8$, then the graph of P is symmetric about the origin.

2. If $y = -3x^3 + 4x^2 - 6x + 9$, then $y \to -\infty$ as $x \to \infty$.

3. If the graph of $y = P(x)$ is symmetric about the origin and $P(8) = 4$, then $-P(-8) = 4$.

4. If $f(x) = x^3 - 3x$, then $f(x) = f(-x)$ for any value of x.

5. If $f(x) = x^4 - x^3 + x^2 - 6x + 7$,
   then $f(-x) = x^4 + x^3 + x^2 + 6x + 7$.

6. The graph of $f(x) = x^2 - 6x + 9$ has only one x-intercept.

7. The x-intercepts for $P(x) = (x - 1)^2(x + 1)$ are $(0, 1)$ and $(0, -1)$.

8. The y-intercept for $P(x) = 4(x - 3)^2 + 2$ is $(0, 2)$.

9. The graph of $f(x) = x^2(x + 8)^2$ has no points in quadrants III and IV.

10. The graph of $f(x) = x^3 - 1$ has three x-intercepts.

4.5 Exercises   Tape 9   Disk—5.25″: 3   3.5″: 2   Macintosh: 2

For each graph discuss its symmetry, indicate whether the graph crosses the x-axis at each x-intercept, and determine whether $y \to \infty$ or $y \to -\infty$ as $x \to \infty$ and $x \to -\infty$.

1.

2.

3.

4.

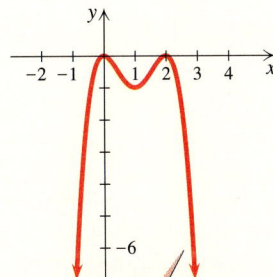

$y = x^4 - 2x^2 + 2$

Discuss the symmetry of the graph of each polynomial function.

5. $f(x) = x^6$

6. $f(x) = x^5 - x$

7. $f(x) = x^2 - 3x + 5$

8. $f(x) = 5x^2 + 10x + 1$

9. $f(x) = 3x^6 - 5x^2 + 3x$

10. $f(x) = x^6 - x^4 + x^2 - 8$

11. $f(x) = 4x^3 - x$

12. $f(x) = 7x^3 + x^2$

13. $f(x) = (x - 5)^2$

14. $f(x) = (x^2 - 1)^2$

15. $f(x) = -x$

16. $f(x) = 3x$

Find the x-intercepts and discuss the behavior of the graph of each polynomial function at its x-intercepts.

17. $f(x) = (x - 4)^2$

18. $f(x) = (x - 1)^2(x + 3)^2$

19. $f(x) = (2x - 1)^3$

20. $f(x) = x^6$

21. $f(x) = 4x - 1$

22. $f(x) = x^2 - 5x - 6$

23. $f(x) = x^2 - 3x + 10$

24. $f(x) = x^4 - 16$

25. $f(x) = x^3 - 3x^2$

26. $f(x) = x^3 - x^2 - x + 1$

27. $f(x) = 2x^3 - 5x^2 + 4x - 1$

28. $f(x) = x^3 - 3x^2 + 4$

For each function use the leading coefficient test to determine whether $y \to \infty$ or $y \to -\infty$ as $x \to \infty$.

29. $y = 2x^3 - x^2 + 9$

30. $y = -3x + 7$

31. $y = -3x^4 + 5$

32. $y = 6x^4 - 5x^2 - 1$

For each function use the leading coefficient test to determine whether $y \to \infty$ or $y \to -\infty$ as $x \to -\infty$.

33. $y = -2x^5 - 3x^2$

34. $y = x^3 + 8x + \sqrt{2}$

35. $y = 3x^6 - 999x^3$

36. $y = -12x^4 - 5x$

Match each polynomial function with its graph (a)–(h).

37. $f(x) = -2x + 1$

38. $f(x) = -2x^2 + 1$

39. $f(x) = -2x^3 + 1$

40. $f(x) = -2x^2 + 4x - 1$

41. $f(x) = -2x^4 + 6$

42. $f(x) = -2x^4 + 6x^2$

43. $f(x) = x^3 + 4x^2 - x - 4$

44. $f(x) = (x - 2)^4$

(a)

(b)

(c)

(d)

(e)

(f)

(g)

(h)

Sketch the graph of each polynomial function.

45. $f(x) = x - 3$

46. $f(x) = 4 - x$

47. $f(x) = (x - 3)^2$

48. $f(x) = (4 - x)^2$

49. $f(x) = x^3 - 4x^2$

50. $f(x) = x^3 - 9x$

51. $f(x) = (x - 1)^2(x + 1)^2$

52. $f(x) = (x - 2)^2(x + 1)$

53. $f(x) = -x^3 - x^2 + 5x - 3$

54. $f(x) = -x^4 + 6x^3 - 9x^2$

55. $f(x) = x^3 - x^2 - 6x$

56. $f(x) = -x^4 + 4x^3 - 4x^2$

57. $f(x) = x^3 - x + 6$

58. $f(x) = x^3 - 4x - 15$

59. $f(x) = -x^4 + x^2$

60. $f(x) = -x^4 + x^2 + 12$

61. $f(x) = x^3 + 3x^2 + 3x + 1$

62. $f(x) = -x^3 + 3x + 2$

Solve each problem.

63. *Economic forecast* The total annual profit (in thousands of dollars) for a department store chain is determined by $P = 90x - 6x^2 + 0.1x^3$, where x is the number of stores in the chain. The company now has 10 stores in operation and plans to pursue an aggressive expansion program. What happens to the total annual profit as the company opens more and more stores?

64. *Contaminated chicken* The number of bacteria of a certain type found on a chicken t minutes after processing in a contaminated plant is given by the function

$$N = 30t + 25t^2 + 44t^3 - 0.01t^4.$$

Find N for $t = 30$, 300, and 3000. What happens to N as t keeps increasing?

Graphing Calculator Exercises

Each of the following third-degree polynomial functions is increasing on $(-\infty, a)$, decreasing on (a, b), and increasing on (b, ∞), where $a < b$. There is no maximum or minimum value for this type of function, but we say that the function has a *local maximum* value of $f(a)$ at $x = a$ and a *local minimum* value of $f(b)$ at $x = b$. Graph each function and estimate the local maximum and local minimum values of the function.

1. $f(x) = x^3 + x^2 - 2x + 1$ 2. $y = x^3 - x^2 - 20x - 5$
3. $y = x^3 - 9x^2 - 12x + 20$
4. $f(x) = x^3 - x^2 - 4x + 6$
5. $f(x) = 2x^3 - 3x^2 - 4x + 20$
6. $y = x^3 - 7x^2 + 12x - 14$

Solve each problem.

7. *Maximum profit* A company's weekly profit (in thousands of dollars) is given by the function $P(x) = x^3 - 3x^2 + 2x + 3$, where x is the amount (in thousands of dollars) spent per week on advertising. Estimate the local maximum and the local minimum value for the profit. What amount must be spent on advertising to get the profit higher than the local maximum profit?

8. *Maximum volume* An open-top box is to be made from a 6 in. by 7 in. piece of copper by cutting equal squares (x in. by x in.) from each corner and folding up the sides. Write the volume of the box as a function of x and find the maximum possible volume to the nearest hundredth of a cubic inch.

4.6

Graphs of

Rational Functions

A rational expression was defined in Chapter 1 as a ratio of two polynomials. In this section we will study functions that are defined by rational expressions. Since polynomials are used to form rational functions, we will be able to use some of the facts that we have learned about polynomial functions.

Domain

If a ratio of two polynomials is used to define a function, then the function is called a rational function, provided the denominator is not the zero polynomial. Functions such as

$$y = \frac{1}{x}, \qquad f(x) = \frac{x - 3}{x - 1}, \qquad \text{and} \qquad g(x) = \frac{2x - 3}{x^2 - 4}$$

are rational functions.

Definition: Rational Function

If $P(x)$ and $Q(x)$ are polynomials, then a function of the form

$$f(x) = \frac{P(x)}{Q(x)}$$

is called a **rational function,** provided that $Q(x)$ is not the zero polynomial.

To simplify discussions of rational functions we will assume that $f(x)$ is in lowest terms ($P(x)$ and $Q(x)$ have no common factors) unless it is stated otherwise.

The domain of a polynomial function is the set of all real numbers, while the domain of a rational function is restricted to real numbers that do not cause the denominator to have a value of 0. The domain of $y = 1/x$ is the set of all real numbers except 0.

Example 1 The domain of a rational function

Find the domain of each rational function.

a) $f(x) = \dfrac{x - 3}{x - 1}$

b) $g(x) = \dfrac{2x - 3}{x^2 - 4}$

Solution

a) Since $x - 1 = 0$ only for $x = 1$, the domain of f is the set of all real numbers except 1. The domain is written in interval notation as $(-\infty, 1) \cup (1, \infty)$.

b) Since $x^2 - 4 = 0$ for $x = \pm 2$, any real number except 2 and -2 can be used for x. So the domain of g is $(-\infty, -2) \cup (-2, 2) \cup (2, \infty)$. ◆

Horizontal and Vertical Asymptotes

The graph of a rational function such as $f(x) = 1/x$ does not look like the graph of a polynomial function. The domain of $f(x) = 1/x$ is the set of all real numbers except 0. However, 0 is an important number for the graph of this function because of the behavior of the graph when x is close to 0. Table 4.4 shows ordered pairs in which $x > 0$ and x is close to 0. Notice that the closer x is to 0, the larger the y-coordinate. In symbols, $y \to \infty$ as $x \to 0$ from the right. Table 4.5 shows ordered pairs in which $x < 0$ and x is close to 0. Notice that the closer x is to 0, the smaller the y-coordinate. In symbols, $y \to -\infty$ as $x \to 0$ from the left. Plotting the ordered pairs of Tables 4.4 and 4.5 suggests a curve that gets closer and closer to the vertical line $x = 0$ (the y-axis) but never touches it. See Fig. 4.30. The y-axis is called a *vertical asymptote* for this curve.

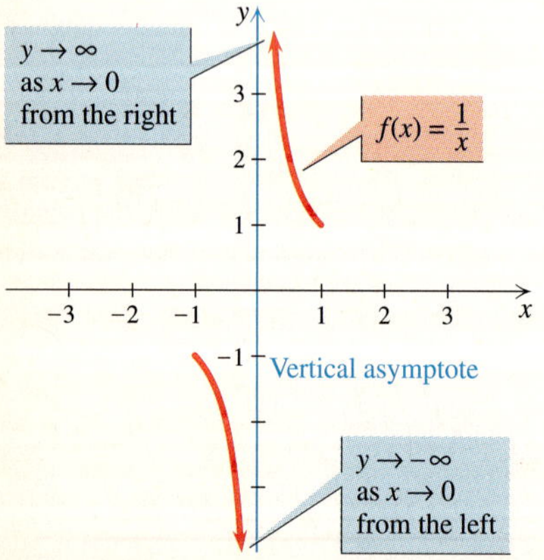

Figure 4.30

x	y
0.1	10
0.01	100
0.001	1,000
0.0001	10,000

Table 4.4 $y = \dfrac{1}{x}$ for $x > 0$

x	y
-0.1	-10
-0.01	-100
-0.001	$-1,000$
-0.0001	$-10,000$

Table 4.5 $y = \dfrac{1}{x}$ for $x < 0$

Table 4.6 shows ordered pairs in which $x > 0$ and x is large. Notice that as the x-coordinates get larger ($x \to \infty$), the y-coordinates get closer to 0 ($y \to 0$). The points from Table 4.6 lie just above the x-axis. Table 4.7 shows ordered pairs in which $x < 0$ and $|x|$ is large. Notice that as $|x|$ gets larger ($|x| \to \infty$), the y-coordinates are negative numbers that get closer to 0 ($y \to 0$). These points lie just below the x-axis. Plotting the ordered pairs of Tables 4.6 and 4.7 suggests a curve that gets closer and closer to the line $y = 0$ (the x-axis) but never touches it. The x-axis is called a *horizontal asymptote* for the graph of f. The complete graph of $f(x) = 1/x$ is shown in Fig. 4.31. Since $f(-x) = 1/(-x) = -f(x)$, the graph is symmetric about the origin.

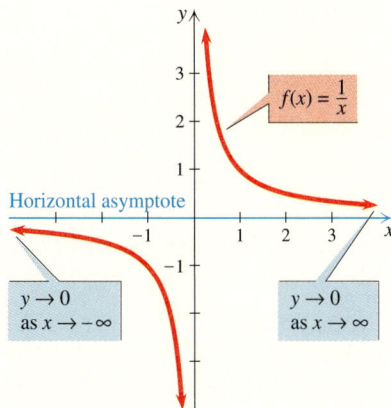

Figure 4.31

x	y
1	1
10	0.1
100	0.01
1000	0.001

Table 4.6 $\quad y = \dfrac{1}{x}$ for $x \to \infty$

x	y
-1	-1
-10	-0.1
-100	-0.01
-1000	-0.001

Table 4.7 $\quad y = \dfrac{1}{x}$ for $x \to -\infty$

For any rational function expressed in lowest terms, a horizontal asymptote is determined by the value approached by the rational expression as $|x| \to \infty$ ($x \to \infty$ or $x \to -\infty$). A vertical asymptote occurs for every number that causes the denominator of the function to have a value of 0. As we will learn shortly, not every rational function has a vertical and a horizontal asymptote. We can give a formal definition of asymptotes as follows.

Definition: Vertical and Horizontal Asymptotes

Let $f(x) = P(x)/Q(x)$ be a rational function written in lowest terms.

If $|f(x)| \to \infty$ as $x \to a$, then the vertical line $x = a$ is a **vertical asymptote.**

If $f(x) \to a$ as $|x| \to \infty$, then the horizontal line $y = a$ is a **horizontal asymptote.**

For a rational function, the asymptotes (if there are any) *must* be determined before the function is graphed. In the next example, we concentrate on finding the asymptotes of rational functions. Later, in Example 4 of this section, we will sketch the graphs of the functions of Example 2. If you wish, you may look ahead to the graphs in Figs. 4.32, 4.33, and 4.34 to see how important the asymptotes are to the graphs of these functions.

Example 2   Identifying horizontal and vertical asymptotes

Find the horizontal and vertical asymptotes for each rational function.

a) $f(x) = \dfrac{3}{x^2 - 1}$ **b)** $g(x) = \dfrac{x}{x^2 - 4}$

c) $h(x) = \dfrac{2x + 1}{x + 3}$

Solution

a) The denominator $x^2 - 1$ has a value of 0 if $x = \pm 1$. So the lines $x = 1$ and $x = -1$ are vertical asymptotes. As $x \to \infty$ or $x \to -\infty$, $x^2 - 1$ gets larger, making $3/(x^2 - 1)$ the ratio of 3 and a large number. Thus $3/(x^2 - 1) \to 0$ and the x-axis is a horizontal asymptote.

b) The denominator $x^2 - 4$ has a value of 0 if $x = \pm 2$. So the lines $x = 2$ and $x = -2$ are vertical asymptotes. As $x \to \infty$, $x/(x^2 - 4)$ is a ratio of two large numbers. The approximate value of this ratio is not clear. However, if we divide the numerator and denominator by x^2, the highest power of x, we can get a clearer picture of the value of this ratio:

$$g(x) = \frac{x}{x^2 - 4} = \frac{\dfrac{x}{x^2}}{\dfrac{x^2}{x^2} - \dfrac{4}{x^2}} = \frac{\dfrac{1}{x}}{1 - \dfrac{4}{x^2}}$$

As $|x| \to \infty$, the values of $1/x$ and $4/x^2$ go to 0. So

$$g(x) \to \frac{0}{1 - 0} = 0.$$

So the x-axis is a horizontal asymptote.

c) The denominator $x + 3$ has a value of 0 if $x = -3$. So the line $x = -3$ is a vertical asymptote. To find any horizontal asymptotes, rewrite the rational expression by dividing the numerator and denominator by the highest power of x:

$$h(x) = \frac{2x + 1}{x + 3} = \frac{\dfrac{2x}{x} + \dfrac{1}{x}}{\dfrac{x}{x} + \dfrac{3}{x}} = \frac{2 + \dfrac{1}{x}}{1 + \dfrac{3}{x}}$$

As $x \to \infty$ or $x \to -\infty$ the values of $1/x$ and $3/x$ approach 0. So

$$h(x) \to \frac{2 + 0}{1 + 0} = 2.$$

So the line $y = 2$ is a horizontal asymptote.

If the degree of the numerator of a rational function is less than the degree of the denominator, as in Examples 2(a) and 2(b), then the x-axis is a horizontal asymptote for the graph of the function. If the degree of the numerator is equal to the degree of the denominator, as in Example 2(c), then the x-axis is not a horizontal asymptote. We can see this clearly if we use division to rewrite the expression as quotient + remainder/divisor. For the function of Example 2(c), we get

$$h(x) = \frac{2x + 1}{x + 3} = 2 + \frac{-5}{x + 3}.$$

The graph of h is a translation two units upward of the graph of $y = -5/(x + 3)$, which has the x-axis as a horizontal asymptote. So $y = 2$ is a horizontal asymptote for h.

Oblique Asymptotes

Each rational function of Example 2 had one horizontal asymptote and had a vertical asymptote for each zero of the polynomial in the denominator. The horizontal asymptote $y = 0$ occurs because the y-coordinate gets closer and closer to 0 as $x \to \infty$ or $x \to -\infty$. Some rational functions have a nonhorizontal line for an asymptote. An asymptote that is neither horizontal nor vertical is called an **oblique asymptote** or **slant asymptote**.

Example 3   A rational function with an oblique asymptote

Determine all of the asymptotes for $g(x) = \dfrac{2x^2 + 3x - 5}{x + 2}$.

Solution

If $x + 2 = 0$, then $x = -2$. So the line $x = -2$ is a vertical asymptote. Because the degree of the numerator is larger than the degree of the denominator, we use long division to rewrite the function as quotient + remainder/divisor: (Dividing the numerator and denominator by x^2 will not work in this case.)

$$g(x) = \frac{2x^2 + 3x - 5}{x + 2} = 2x - 1 + \frac{-3}{x + 2}$$

If $|x| \to \infty$, then $-3/(x + 2) \to 0$. So the value of $g(x)$ approaches $2x - 1$ as $|x| \to \infty$. The line $y = 2x - 1$ is an oblique asymptote for the graph of g. You may look ahead to Fig. 4.35 to see the graph of this function with its oblique asymptote. ◆

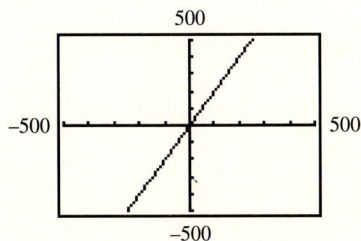

If the degree of $P(x)$ is 1 greater than the degree of $Q(x)$ and the degree of $Q(x)$ is at least 1, then the rational function has an oblique asymptote. In this case use division to rewrite the function as quotient + remainder/divisor. The graph of the equation formed by setting y equal to the quotient is an oblique asymptote. We conclude this discussion of asymptotes with a summary.

In this wide view, the graph of $y = 2 + \dfrac{-5}{x + 3}$ looks like its horizontal asymptote $y = 2$.

In a large viewing window, the graph of $y = \dfrac{2x^2 + 3x - 5}{x + 2}$ looks like the line $y = 2x - 1$, which is its oblique asymptote.

▼

Summary: Finding Asymptotes for a Rational Function

Let $f(x) = P(x)/Q(x)$ be a rational function in lowest terms with the degree of $Q(x)$ at least 1.

1. The graph of f has a vertical asymptote corresponding to each root of $Q(x) = 0$.
2. If the degree of $P(x)$ is less than the degree of $Q(x)$, then the x-axis is a horizontal asymptote.
3. If the degree of $P(x)$ is greater than or equal to the degree of $Q(x)$, then use division to rewrite the function as quotient + remainder/divisor. The graph of the equation formed by setting y equal to the quotient is an asymptote.

Sketching Graphs of Rational Functions

We now use asymptotes and symmetry to help us sketch the graphs of the rational functions discussed in Examples 2 and 3. Use the following steps to graph a rational function.

▼

Procedure: Graphing a Rational Function

To graph a rational function in lowest terms:

1. Determine the asymptotes and draw them as dashed lines.
2. Check for symmetry.
3. Find any intercepts.
4. Plot several selected points to determine how the graph approaches the asymptotes.
5. Draw curves through the selected points, approaching the asymptotes.

Example 4   Functions with horizontal and vertical asymptotes

Sketch the graph of each rational function.

a) $f(x) = \dfrac{3}{x^2 - 1}$ **b)** $g(x) = \dfrac{x}{x^2 - 4}$

c) $h(x) = \dfrac{2x + 1}{x + 3}$

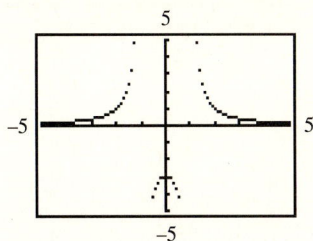

The graph of $y = \dfrac{3}{x^2 - 1}$ in dot mode

can be used as an aid in drawing a graph that shows the asymptotes as in Fig. 4.32.

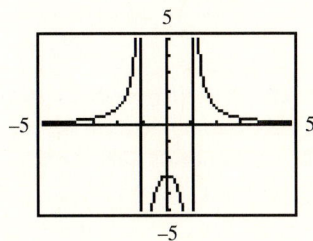

The graph of $y = \dfrac{3}{x^2 - 1}$ in connected

mode appears to show the vertical asymptotes, even though they are not part of the graph.

Solution

a) From Example 2(a), $x = 1$ and $x = -1$ are vertical asymptotes, and the x-axis is a horizontal asymptote. Draw the vertical asymptotes using dashed lines. Since all of the powers of x are even, $f(-x) = f(x)$ and the graph is symmetric about the y-axis. The y-intercept is $(0, -3)$. There are no x-intercepts because $f(x) = 0$ has no solution. Evaluate the function at $x = 0.9$ and $x = 1.1$ to see how the curve approaches the vertical asymptote at $x = 1$. Evaluate at $x = 2$ and $x = 3$ to see whether the curve approaches the horizontal asymptote from above or below. We get $(0.9, -15.789)$, $(1.1, 14.286)$, $(2, 1)$, and $(3, 3/8)$. Use the symmetry to sketch the graph shown in Fig. 4.32.

b) Draw the vertical asymptotes $x = 2$ and $x = -2$ from Example 2(b) as dashed lines. The x-axis is a horizontal asymptote. Because $f(-x) = -f(x)$, the graph is symmetric about the origin. The x-intercept is $(0, 0)$. Evaluate the function near the vertical asymptote $x = 2$, and for larger values of x to see how the curve approaches the horizontal asymptote. We get $(1.9, -4.872)$, $(2.1, 5.122)$, $(3, 3/5)$, and $(4, 1/3)$. Use the symmetry to get the graph shown in Fig. 4.33.

c) Draw the vertical asymptote $x = -3$ and the horizontal asymptote $y = 2$ from Example 2(c) as dashed lines. The x-intercept is $(-1/2, 0)$ and the y-intercept is $(0, 1/3)$. The points $(-2, -3)$, $(7, 1.5)$, $(-4, 7)$, and $(-13, 2.5)$ are also on the graph shown in Fig. 4.34.

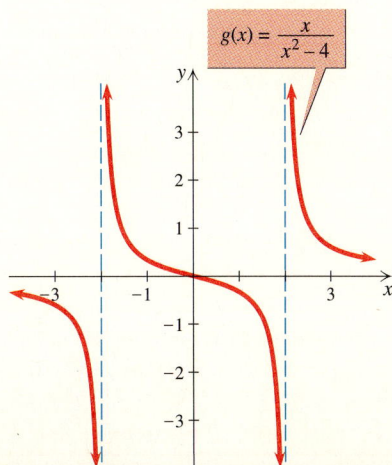

Figure 4.32

Figure 4.33

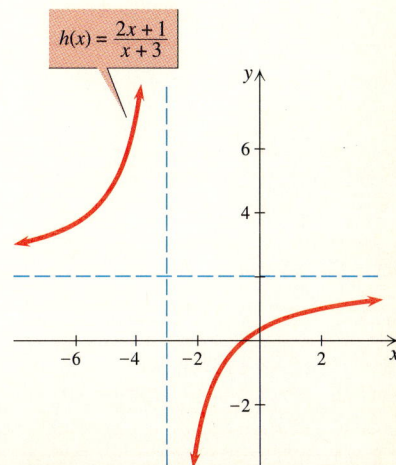

Figure 4.34

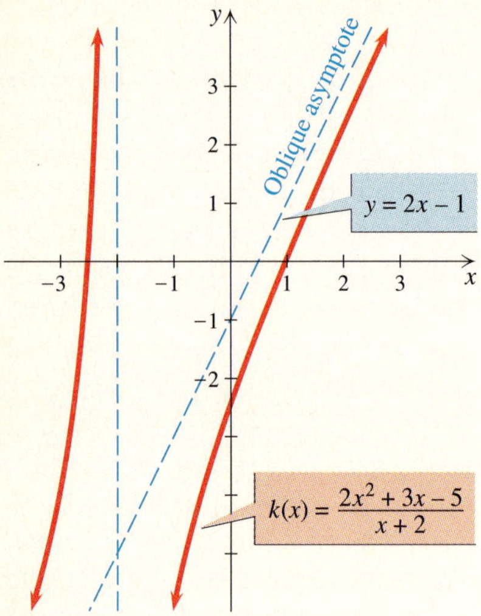

Figure 4.35

Example 5 Graphing a function with an oblique asymptote

Sketch the graph of $k(x) = \dfrac{2x^2 + 3x - 5}{x + 2}$.

Solution

Draw the vertical asymptote $x = -2$ and the oblique asymptote $y = 2x - 1$ determined in Example 3 as dashed lines. The x-intercepts, $(1, 0)$ and $(-2.5, 0)$, are found by solving $2x^2 + 3x - 5 = 0$. The y-intercept is $(0, -2.5)$. The points $(-1, -6)$, $(4, 6.5)$, and $(-3, -4)$ are also on the graph, which is shown in Fig. 4.35. ◆

The graph of a rational function cannot cross a vertical asymptote, but it can cross a nonvertical asymptote. The graph of a rational function gets closer and closer to a nonvertical asymptote as $|x| \to \infty$, but it can also cross a nonvertical asymptote at a "small" value of x, as illustrated in the next example.

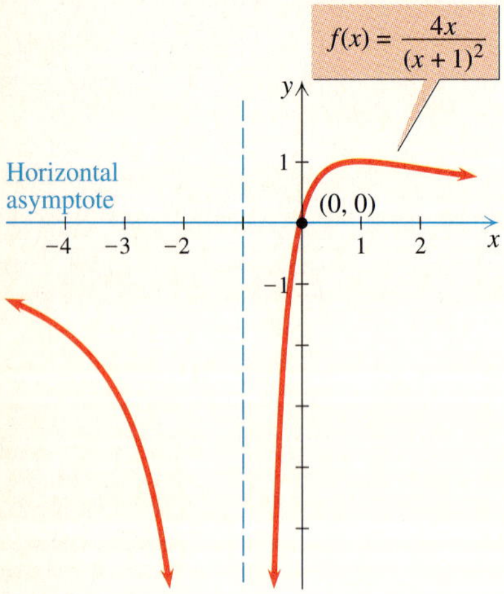

Figure 4.36

Example 6 A rational function that crosses an asymptote

Sketch the graph of $f(x) = \dfrac{4x}{(x + 1)^2}$.

Solution

The graph of f has a vertical asymptote at $x = -1$, and since the degree of the numerator is less than the degree of the denominator, the x-axis is a horizontal asymptote. If $x > 0$ then $f(x) > 0$, and if $x < 0$ then $f(x) < 0$. So the graph approaches the horizontal axis from above for $x > 0$, and from below for $x < 0$. However, the x-intercept is $(0, 0)$. So the graph crosses its horizontal asymptote at $(0, 0)$. The graph shown in Fig. 4.36 goes through $(1, 1)$, $(2, 8/9)$, $(-2, -8)$, and $(-3, -3)$. ◆

All rational functions so far have been given in lowest terms. In the next example the numerator and denominator have a common factor. In this case the graph does not have as many vertical asymptotes as you might expect. The graph is almost identical to the graph of the rational function obtained by reducing the expression to lowest terms.

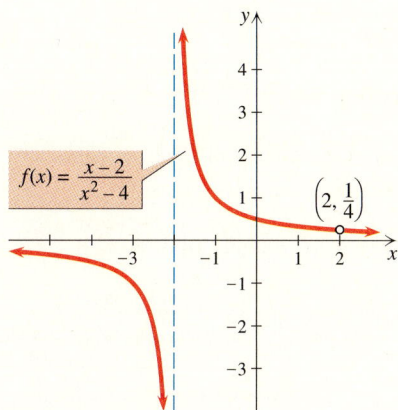

Figure 4.37

Example 7   A graph with a hole in it

Sketch the graph of $f(x) = \dfrac{x - 2}{x^2 - 4}$.

Solution

Since $x^2 - 4 = (x - 2)(x + 2)$, the domain of f is the set of all real numbers except 2 and -2. Since the rational expression can be reduced, the function f could also be defined as

$$f(x) = \frac{1}{x + 2} \qquad \text{for } x \neq 2 \text{ and } x \neq -2.$$

The graph of $y = 1/(x + 2)$ has only one vertical asymptote $x = -2$. In fact, the graph of $y = 1/(x + 2)$ is a translation two units to the left of $y = 1/x$. Since $f(2)$ is undefined, the point $(2, 1/4)$ that would normally be on the graph of $y = 1/(x + 2)$ is omitted. The missing point is indicated on the graph in Fig. 4.37 as a small open circle.   ◆

Applications

Rational functions can occur in many applied situations. A horizontal asymptote might indicate that the average cost of producing a product approaches a fixed value in the long run. A vertical asymptote might show that the cost of a project goes up astronomically as a certain barrier is approached.

Example 8   Average cost of a handbook

Eco Publishing spent $5000 in producing an environmental handbook and $8 each for printing. Write a function that gives the average cost to the company per printed handbook. Graph the function for $0 < x \leq 500$. What happens to the average cost if the book becomes very, very popular?

Solution

The total cost of producing and printing x handbooks is $8x + 5000$. To find the average cost per book, divide the total cost by the number of books:

$$C = \frac{8x + 5000}{x}$$

This rational function has a vertical asymptote at $x = 0$ and a horizontal asymptote $C = 8$. The graph is shown in Fig. 4.38. As x gets larger and larger, the $5000 production cost is spread out over more and more books, and the average cost per book approaches $8, as shown by the horizontal asymptote.   ◆

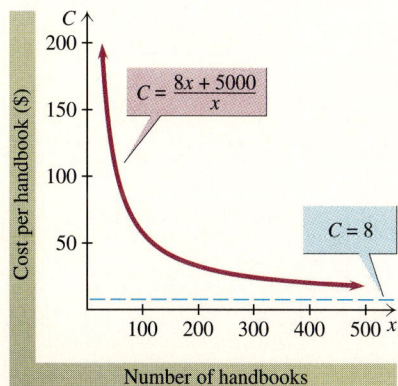

Figure 4.38

? For Thought

True or false? Explain.

1. The function $f(x) = \dfrac{1}{\sqrt{x} - 3}$ is a rational function.

2. The domain of $f(x) = \dfrac{x + 2}{x - 2}$ is $(-\infty, -2) \cup (-2, 2) \cup (2, \infty)$.

3. The number of vertical asymptotes for a rational function equals the degree of the denominator.

4. The graph of $f(x) = \dfrac{1}{x^4 - 4x^2}$ has four vertical asymptotes.

5. The x-axis is a horizontal asymptote for $f(x) = \dfrac{x^2 - 2x + 7}{x^3 - 4x}$.

6. The x-axis is a horizontal asymptote for the graph of $f(x) = \dfrac{5x - 1}{x + 2}$.

7. The line $y = x - 3$ is an asymptote for the graph of
$f(x) = x - 3 + \dfrac{1}{x - 1}$.

8. The graph of a rational function cannot intersect its asymptote.

9. The graph of $f(x) = \dfrac{4}{x^2 - 16}$ is symmetric about the y-axis.

10. The graph of $f(x) = \dfrac{2x + 6}{x^2 - 9}$ has only one vertical asymptote.

4.6 Exercises ▭ Tape 9 ▭ Disk—5.25″: 3   3.5″: 2   Macintosh: 2

Find the domain of each rational function.

1. $f(x) = \dfrac{4}{x + 2}$

2. $f(x) = \dfrac{-1}{x - 2}$

3. $f(x) = \dfrac{-x}{x^2 - 4}$

4. $f(x) = \dfrac{2}{x^2 - x - 2}$

5. $f(x) = \dfrac{2x + 3}{x - 3}$

6. $f(x) = \dfrac{4 - x}{x + 2}$

7. $f(x) = \dfrac{x^2 - 2x + 4}{x}$

8. $f(x) = \dfrac{x^3 + 2}{x^2}$

9. $f(x) = \dfrac{3x^2 - 1}{x^3 - x}$

10. $f(x) = \dfrac{x^2 + 1}{8x^3 - 2x}$

11. $f(x) = \dfrac{-x^2 + x}{x^2 + 5x + 6}$

12. $f(x) = \dfrac{-x^2 + 2x - 3}{x^2 + x - 12}$

Determine the domain and the equations of the asymptotes for the graph of each rational function.

13.

14.

15.

16.

Determine the equations of all asymptotes for the graph of each function.

17. $f(x) = \dfrac{5}{x - 2}$

18. $f(x) = \dfrac{-1}{x + 12}$

19. $f(x) = \dfrac{-x}{x^2 - 9}$

20. $f(x) = \dfrac{-2}{x^2 - 5x + 6}$

21. $f(x) = \dfrac{2x + 4}{x - 1}$

22. $f(x) = \dfrac{5 - x}{x + 5}$

23. $f(x) = \dfrac{x^2 - 2x + 1}{x}$

24. $f(x) = \dfrac{x^3 - 8}{x^2}$

25. $f(x) = \dfrac{3x^2 + 4}{x + 1}$

26. $f(x) = \dfrac{x^2}{x - 9}$

27. $f(x) = \dfrac{-x^2 + 4x}{x + 2}$

28. $f(x) = \dfrac{-x^2 + 3x - 7}{x - 3}$

Find all asymptotes, x-intercepts, and y-intercepts for the graph of each rational function and sketch the graph of the function.

29. $f(x) = \dfrac{-1}{x}$

30. $f(x) = \dfrac{1}{x^2}$

31. $f(x) = \dfrac{1}{x - 2}$

32. $f(x) = \dfrac{-1}{x + 1}$

33. $f(x) = \dfrac{1}{x^2 - 4}$

34. $f(x) = \dfrac{1}{x^2 - 2x + 1}$

35. $f(x) = \dfrac{-1}{(x + 1)^2}$

36. $f(x) = \dfrac{-2}{x^2 - 9}$

37. $f(x) = \dfrac{2x + 1}{x - 1}$

38. $f(x) = \dfrac{3x - 1}{x + 1}$

39. $f(x) = \dfrac{x - 3}{x + 2}$

40. $f(x) = \dfrac{2 - x}{x + 2}$

41. $f(x) = \dfrac{x}{x^2 - 1}$

42. $f(x) = \dfrac{-x}{x^2 - 9}$

43. $f(x) = \dfrac{4x}{x^2 - 2x + 1}$

44. $f(x) = \dfrac{-2x}{x^2 + 6x + 9}$

45. $f(x) = \dfrac{8 - x^2}{x^2 - 9}$

46. $f(x) = \dfrac{2x^2 + x - 8}{x^2 - 4}$

47. $f(x) = \dfrac{2x^2 + 8x + 2}{x^2 + 2x + 1}$

48. $f(x) = \dfrac{-x^2 + 7x - 9}{x^2 - 6x + 9}$

Find the oblique asymptote and sketch the graph of each rational function.

49. $f(x) = \dfrac{x^2 + 1}{x}$

50. $f(x) = \dfrac{x^2 - 1}{x}$

51. $f(x) = \dfrac{x^3 - 1}{x^2}$

52. $f(x) = \dfrac{x^3 + 1}{x}$

53. $f(x) = \dfrac{x^2}{x + 1}$

54. $f(x) = \dfrac{x^2}{x - 1}$

55. $f(x) = \dfrac{2x^2 - x}{x - 1}$

56. $f(x) = \dfrac{-x^2 + x + 1}{x + 1}$

Match each rational function with its graph (a)–(h).

57. $f(x) = -\dfrac{3}{x}$

58. $f(x) = \dfrac{1}{3 - x}$

59. $f(x) = \dfrac{x}{x - 3}$

60. $f(x) = \dfrac{x - 3}{x}$

61. $f(x) = \dfrac{1}{x^2 - 3x}$

62. $f(x) = \dfrac{x^2}{x^2 - 9}$

63. $f(x) = \dfrac{x^2 - 3}{x}$

64. $f(x) = \dfrac{-x^3 + 1}{x^2}$

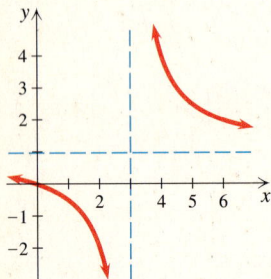

(a)

(b)

(c)

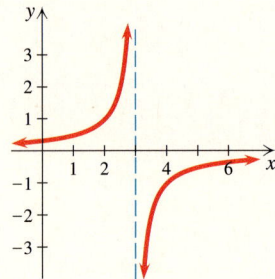

(d)

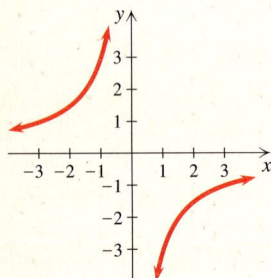

(e)

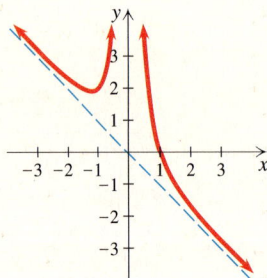

(f)

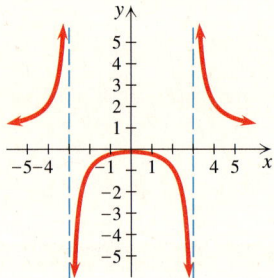

(g)

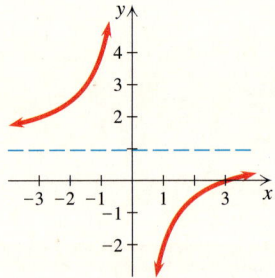

(h)

Sketch the graph of each rational function. Note that the functions are not in lowest terms.

65. $f(x) = \dfrac{x + 1}{x^2 - 1}$

66. $f(x) = \dfrac{x}{x^2 + 2x}$

67. $f(x) = \dfrac{x^2 - 1}{x - 1}$

68. $f(x) = \dfrac{x^2 - 5x + 6}{x - 2}$

Sketch the graph of each rational function.

69. $f(x) = \dfrac{2}{x^2 + 1}$

70. $f(x) = \dfrac{x}{x^2 + 1}$

71. $f(x) = \dfrac{x - 1}{x^3 - 9x}$

72. $f(x) = \dfrac{x^2 + 1}{x^3 - 4x}$

73. $f(x) = \dfrac{x + 1}{x^2}$

74. $f(x) = \dfrac{x - 1}{x^2}$

For each function, let $x = 5000$ and use a calculator to find the value of y to the nearest tenth.

75. $y = \dfrac{3x - 1}{2x - 9}$

76. $y = \dfrac{x - 9}{2 - x}$

77. $y = \dfrac{4}{3x + 40}$

78. $y = \dfrac{2x + 5}{x^2 - 1}$

For each function, let $x = -5000$ and use a calculator to find the value of y to the nearest tenth.

79. $y = \dfrac{9x^2 - 1}{3x^2 - 2}$

80. $y = \dfrac{5 - 4x^2}{10x^2 + 7}$

81. $y = \dfrac{-3x + 4}{2x^3 - 9x^2}$

82. $y = \dfrac{-7x + 9}{-10x^2 - 4x + 2}$

Solve each problem.

83. *Admission to the zoo* Winona paid $100 for a lifetime membership to Friends of the Zoo, so that she could gain admittance to the zoo for only $1. Write Winona's average cost per visit C as a function of the number of visits when she has visited x times. What is her average cost per visit when she has visited the zoo 100 times? Graph the function for $x > 0$. What happens to her average cost per visit if she starts when she is young and visits the zoo every day?

84. *Renting a car* The cost of renting a car for one day is $19 plus 30 cents per mile. Write the average cost per mile C as a function of the number of miles driven in one day x. Graph the function for $x > 0$. What happens to C as the number of miles gets very large?

85. *Average speed of an auto trip* A 200-mi trip by an electric car must be completed in 4 hr. Let x be the number of hours it takes to travel the first half of the distance and write the average speed for the second half of the trip as a function of x. Graph this function for $0 < x < 4$. What is the significance of the vertical asymptote?

86. *Billboard advertising* An Atlanta marketing agency figures that the monthly cost of a billboard advertising campaign depends on the fraction of the market p that the client wishes to reach. For $0 \le p < 1$ the cost in dollars is determined by the formula $C = (4p - 1200)/(p - 1)$. What is the monthly cost for a campaign intended to reach 95% of the market? Graph this function for $0 \le p < 1$. What happens to the cost for a client who wants to reach 100% of the market?

87. *Average cost* The cost to a company in dollars for producing x cellular telephones per day is given by the function $C(x) = 300x + 1000$. The function $AC(x) = C(x)/x$ gives the average cost of production per telephone. Find $AC(10)$ and $AC(50)$. What happens to the average cost as x gets large?

88. *Marginal average cost* The marginal average cost function, $MAC(x)$, is the difference quotient for the function $AC(x)$ with $h = 1$. Find $MAC(x)$ for the average cost function $AC(x)$ from the previous exercise. Find $MAC(5)$ and $MAC(20)$. What happens to the marginal average cost as x gets large?

For Writing/Discussion

89. *Approximate value* Given an arbitrary rational function $R(x)$, explain how you can find the approximate value of $R(x)$ when the absolute value of x is very large.

90. *Cooperative learning* Each student in your small group should write the equations of the horizontal, vertical, and/or oblique asymptotes of an unknown rational function. Then work as a group to find a rational function that has the specified asymptotes. Is there a rational function with any specified asymptotes?

Graphing Calculator Exercises

Use a graphing calculator to help you graph each function. Why do these graphs fail to have vertical asymptotes? What is the domain of each function? Estimate the range of each function from the graph. What is the equation of the horizontal asymptote?

1. $y = \dfrac{100}{x^2 + 1}$

2. $y = \dfrac{9x}{x^2 + 1}$

3. $y = \dfrac{5}{x^2 + 2x + 6}$

4. $y = \dfrac{-6x^2 + 12x - 29}{x^2 - 2x + 5}$

5. $f(x) = \dfrac{2x - 3}{x^2 + x + 1}$

6. $f(x) = \dfrac{x^3 + x + 5}{x^2 + 1}$

Highlights

Section 4.1   Linear Functions

1. The graph of every linear function $y = mx + b$ is a line with slope m and y-intercept $(0, b)$.

2. The slope of a line through (x_1, y_1) and (x_2, y_2) with $x_1 \neq x_2$ is $(y_2 - y_1)/(x_2 - x_1)$.

3. The slope-intercept form of the equation of a line through (x_1, y_1) with slope m is $y - y_1 = m(x - x_1)$.

4. Two nonvertical lines are parallel if and only if their slopes are equal.

5. Two lines with slopes m_1 and m_2 are perpendicular if and only if $m_1 m_2 = -1$.

Section 4.2   Quadratic Functions

1. The graph of every quadratic function $y = ax^2 + bx + c$ $(a \neq 0)$ is a parabola.

2. The graph of $y = ax^2 + bx + c$ $(a \neq 0)$ is a transformation of the graph of $y = x^2$.

3. The x-coordinate of the vertex for $y = ax^2 + bx + c$ is $-b/(2a)$.

4. The vertex for the parabola $y = a(x - h)^2 + k$ is (h, k).

5. The minimum or maximum value of a quadratic function occurs at the vertex of the parabola.

Section 4.3   Zeros of Polynomial Functions

1. The remainder when a polynomial $P(x)$ is divided by $x - c$ is $P(c)$.

2. Synthetic division provides a quick means of dividing $P(x)$ by $x - c$.

3. The number c is a zero of $y = P(x)$ if and only if $x - c$ is a factor of the polynomial $P(x)$.

4. A polynomial function of positive degree has at least one complex zero.

5. If p/q is a rational zero in lowest terms for a polynomial function with integral coefficients, then p is a factor of the constant term and q is a factor of the leading coefficient.

Section 4.4   The Theory of Equations

1. A polynomial equation with positive degree n has n roots, when multiplicity is counted.

2. The roots of an equation $P(x) = 0$ are the same as the zeros of the function $y = P(x)$.

3. In a polynomial equation with real coefficients, the imaginary roots occur in conjugate pairs.

4. Descartes's rule of signs and the theorem on bounds give us clues about the size and type of roots to a polynomial equation.

Section 4.5 Graphs of Polynomial Functions

1. The graph of a polynomial function can be symmetric about the y-axis $(f(-x) = f(x))$, symmetric about the origin $(f(-x) = -f(x))$, or have neither type of symmetry.

2. The graph of a quadratic function is symmetric about a vertical line through its vertex.

3. The graph of a polynomial function crosses the x-axis at an x-intercept if the factor corresponding to that intercept is raised to an odd power, and the graph does not cross if the factor is raised to an even power.

4. The leading coefficient and the degree of the function determine the behavior of a polynomial function as $x \to \infty$ or $x \to -\infty$.

Section 4.6 Graphs of Rational Functions

1. The graph of a rational function in lowest terms has a vertical asymptote for each value of x for which the denominator has a value of zero.

2. If the degree of the numerator of a rational function is less than the degree of the denominator, then the x-axis is a horizontal asymptote.

3. If the degree of the numerator of a rational function is greater than or equal to the degree of the denominator, then use division to discover the horizontal or oblique asymptote.

Chapter 4 Review Exercises

Solve each problem.

1. Find the slope of the line $3x - 5y = 8$.

2. Find the slope of the line through $(-3, 6)$ and $(5, -4)$.

3. Find the slope-intercept form of the equation of the line that goes through $(1, -2)$ and is perpendicular to the line $2x - 3y = 6$.

4. Find the equation of the line in slope-intercept form that goes through $(3, -4)$ and is parallel to the line through $(0, 2)$ and $(3, -1)$.

5. Write the function $f(x) = 3x^2 - 2x + 1$ in the form $f(x) = a(x - h)^2 + k$.

6. Write the function $f(x) = -4\left(x - \dfrac{1}{3}\right)^2 - \dfrac{1}{2}$ in the form $f(x) = ax^2 + bx + c$.

7. Find the vertex, axis of symmetry, x-intercepts, and y-intercept for the parabola $y = 2x^2 - 4x - 1$.

8. Find the maximum value of the function $y = -x^2 + 3x - 5$.

9. Write the equation of a parabola that has x-intercepts $(-1, 0)$ and $(3, 0)$ and y-intercept $(0, 6)$.

10. Write the equation of the parabola that has vertex $(1, 2)$ and y-intercept $(0, 5)$.

11. Solve the inequality $2x^2 + 5x < 3$.

12. Solve the inequality $-x^2 + 4x - 7 < 0$.

Find all real and imaginary zeros for each polynomial function.

13. $f(x) = 3x - 1$

14. $g(x) = 7$

15. $h(x) = x^2 - 8$

16. $m(x) = x^3 - 8$

17. $n(x) = 8x^3 - 1$

18. $C(x) = 3x^2 - 2$

19. $P(t) = t^4 - 100$

20. $S(t) = 25t^4 - 1$

21. $R(s) = 8s^3 - 4s^2 - 2s + 1$

22. $W(s) = s^3 + s^2 + s + 1$

23. $f(x) = x^3 + 2x^2 - 6x$

24. $f(x) = 2x^3 - 4x^2 + 3x$

For each polynomial, find the indicated value in two different ways.

25. $P(x) = 4x^3 - 3x^2 + x - 1, P(3)$

26. $P(x) = 2x^3 + 5x^2 - 3x - 2, P(-2)$

27. $P(x) = -8x^5 + 2x^3 - 6x + 2, P\left(-\dfrac{1}{2}\right)$

28. $P(x) = -4x^4 + 3x^2 + 1, P\left(\dfrac{1}{2}\right)$

Use the rational zero theorem to list all possible rational zeros for each polynomial function.

29. $f(x) = -3x^3 + 6x^2 + 5x - 2$

30. $f(x) = 2x^4 + 9x^2 - 8x - 3$

31. $f(x) = 6x^4 - x^2 - 9x + 3$

32. $f(x) = 4x^3 - 5x^2 - 13x - 8$

Find a polynomial equation with integral coefficients (and lowest degree) that has the given roots.

33. $-\dfrac{1}{2}, 3$

34. $\dfrac{1}{2}, -5$

35. $3 - 2i$

36. $4 + 2i$

37. $2, 1 - 2i$

38. $-3, 3 - 4i$

39. $2 - \sqrt{3}$

40. $1 + \sqrt{2}$

Use Descartes's rule of signs to discuss the possibilities for the roots to each equation. Do not solve the equation.

41. $x^8 + x^6 + 2x^2 = 0$ **42.** $-x^3 - x - 3 = 0$

43. $4x^3 - 3x^2 + 2x - 9 = 0$ **44.** $5x^5 + x^3 + 5x = 0$

45. $x^3 + 2x^2 + 2x + 1 = 0$

46. $-x^4 - x^3 + 3x^2 + 5x - 8 = 0$

Establish the best integral bounds for the roots of each equation according to the theorem on bounds.

47. $6x^2 + 5x - 50 = 0$

48. $4x^2 - 12x - 27 = 0$

49. $2x^3 - 15x^2 + 31x - 12 = 0$

50. $x^3 + 6x^2 + 11x + 7 = 0$

51. $12x^3 - 4x^2 - 3x + 1 = 0$

52. $4x^3 - 4x^2 - 9x + 10 = 0$

Find all real and imaginary solutions to each equation, stating multiplicity when it is greater than one.

53. $x^3 - 6x^2 + 11x - 6 = 0$

54. $x^3 + 7x^2 + 16x + 12 = 0$

55. $6x^4 - 5x^3 + 7x^2 - 5x + 1 = 0$

56. $6x^4 + 5x^3 + 25x^2 + 20x + 4 = 0$

57. $x^3 - 9x^2 + 28x - 30 = 0$

58. $x^3 - 4x^2 + 6x - 4 = 0$

59. $x^3 - 4x^2 + 7x - 6 = 0$

60. $2x^3 - 5x^2 + 10x - 4 = 0$

61. $2x^4 - 5x^3 - 2x^2 + 2x = 0$

62. $2x^5 - 15x^4 + 26x^3 - 12x^2 = 0$

Discuss the symmetry of the graph of each function.

63. $f(x) = 2x^2 - 3x + 9$

64. $f(x) = -3x^2 + 12x - 1$

65. $f(x) = -3x^4 - 2$

66. $f(x) = \dfrac{-x^3}{x^2 - 1}$

67. $f(x) = \dfrac{x}{x^2 + 1}$

68. $f(x) = 2x^4 + 3x^2 + 1$

Find the domain of each rational function.

69. $f(x) = \dfrac{x^2 - 4}{2x + 5}$ **70.** $f(x) = \dfrac{4x + 1}{x^2 - x - 6}$

71. $f(x) = \dfrac{1}{x^2 + 1}$ **72.** $f(x) = \dfrac{x - 9}{x^2 - 1}$

Find the *x*-intercepts, *y*-intercept, and asymptotes for the graph of each function and sketch the graph.

73. $f(x) = -\dfrac{1}{2}x + 4$ **74.** $f(x) = 3x - 5$

75. $f(x) = x^2 - x - 2$ **76.** $f(x) = -2(x - 1)^2 + 6$

77. $f(x) = x^3 - 3x - 2$

78. $f(x) = x^3 - 3x^2 + 4$

79. $f(x) = \dfrac{1}{2}x^3 - \dfrac{1}{2}x^2 - 2x + 2$

80. $f(x) = \dfrac{1}{2}x^3 - 3x^2 + 4x$

81. $f(x) = \dfrac{1}{4}x^4 - 2x^2 + 4$

82. $f(x) = \dfrac{1}{2}x^4 + 2x^3 + 2x^2$

83. $f(x) = \dfrac{2}{x + 3}$

84. $f(x) = \dfrac{1}{2 - x}$

85. $f(x) = \dfrac{2x}{x^2 - 4}$

86. $f(x) = \dfrac{2x^2}{x^2 - 4}$

87. $f(x) = \dfrac{x^2 - 2x + 1}{x - 2}$

88. $f(x) = \dfrac{-x^2 + x + 2}{x - 1}$

89. $f(x) = \dfrac{2x - 1}{2 - x}$

90. $f(x) = \dfrac{1 - x}{x + 1}$

Solve each problem.

91. Find the quotient and remainder when $x^3 - 6x^2 + 9x - 15$ is divided by $x - 3$.

92. Find the quotient and remainder when $3x^3 + 4x^2 + 2x - 4$ is divided by $3x - 2$.

93. *Placement test scores*   Dawn knows that the score on a mathematics placement test that she developed is a linear function of the student's high school grade point average. A student with a GPA of 2.4 scored 62, and a student with a GPA of 3.2 scored 78. Write the linear function that contains these two points. Use the function to predict the placement test score of a student with a GPA of 3.8.

94. *Altitude of a rocket*   If the altitude in feet of a model rocket is given by the equation $S = -16t^2 + 156t$, where t is the time in seconds after ignition, then what is the maximum height attained by the rocket?

95. *Antique saw*   Willard is making a reproduction of an antique saw. The handle consists of two pieces of wood joined at a right angle, with the blade being the hypotenuse of the right triangle as shown in the figure. If the total length of the handle is to be 36 in., then what length for each piece would minimize the square of the length of the blade?

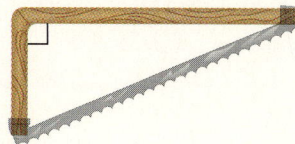

Figure for Exercise 95

96. *Physical fitness*   For a 25-year-old woman to be considered in the high fitness category, the maximum time allowed for a one-mile walk at a pulse rate of 110 is 19.1 min ("Monitoring Your Health," *Reader's Digest*, 1991). For walking at a pulse rate of 170, the maximum time allowed is 16.2 min. If the maximum time allowed is a linear function of the pulse rate and 25-year-old Tina walks one mile in 17.5 min with a pulse rate of 145, is she in the high fitness category?

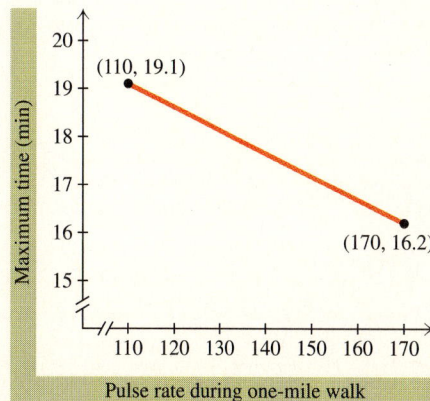

Figure for Exercise 96

Chapter 4 Test

Solve each problem.

1. Find the slope of the line that goes through $(3, -6)$ and $(-1, 2)$.

2. Find the slope of the line $3x - 4y = 9$.

3. Find the equation of the line through $(2, -4)$ that is perpendicular to the line $y = -3x - 5$. Write the answer in slope-intercept form.

4. Write $y = 3x^2 - 12x + 1$ in the form $y = a(x - h)^2 + k$.

5. Identify the vertex, axis of symmetry, y-intercept, x-intercepts, and range for $y = 3x^2 - 12x + 1$.

6. What is the minimum value of y in the function $y = 3x^2 - 12x + 1$?

7. Use synthetic division to find the quotient and remainder when $2x^3 - 4x + 5$ is divided by $x + 3$.

8. What is the remainder when
$$x^{98} - 19x^{73} + 17x^{44} - 12x^{23} + 2x^9 - 3$$
is divided by $x - 1$?

9. List the possible rational zeros for the function $f(x) = 3x^3 - 4x^2 + 5x - 6$.

10. Find a polynomial equation with real coefficients that has the roots -3 and $4i$.

11. Use Descartes's rule of signs to discuss the possibilities for the roots of $x^3 - 3x^2 + 5x + 7 = 0$.

12. The altitude in feet of a toy rocket t seconds after launch is given by the function $S(t) = -16t^2 + 128t$. Find the maximum altitude reached by the rocket.

Find all real and imaginary zeros for each polynomial function.

13. $f(x) = x^2 - 9$

14. $f(x) = x^4 - 16$

15. $f(x) = x^3 - 4x^2 - x + 10$

Find all real and imaginary roots of each equation. State the multiplicity of a root when it is greater than one.

16. $x^4 + 2x^2 + 1 = 0$

17. $(x^3 - 2x^2)(2x + 3)^3 = 0$

18. $2x^3 - 9x^2 + 14x - 5 = 0$

Sketch the graph of each function.

19. $y = -\dfrac{1}{2}x + 3$

20. $y = 2(x - 3)^2 + 1$

21. $y = (x - 2)^2(x + 1)$

22. $y = x^3 - 4x$

23. $y = \dfrac{1}{x - 2}$

24. $y = \dfrac{2x - 3}{x - 2}$

25. $f(x) = \dfrac{x^2 + 1}{x}$

26. $y = \dfrac{4}{x^2 - 4}$

Tying It All Together
Chapters 1–4

Find all real and imaginary solutions to each equation.

1. $2x + 1 = 3x - 1$

2. $(2x + 1)(3x - 1) = 0$

3. $\dfrac{2x + 1}{3x - 1} = 0$

4. $\dfrac{1}{2x + 1} = \dfrac{1}{3x - 1}$

5. $\dfrac{1}{3x - 1} - \dfrac{1}{2x + 1} = \dfrac{1}{6}$

6. $6x^4 + x^3 + 5x^2 + x - 1 = 0$

7. $\sqrt{2x + 1} = \sqrt{3x - 1}$

8. $3x^2 + 4 = 2x^2 + 1$

9. $\sqrt{3x + 4} - \sqrt{2x + 1} = 1$

10. $|2x + 1| = 3$

11. $(2x + 1)^{2/3} = 9$

12. $|2x + 1| = |3x - 1|$

Sketch the graph of each function.

13. $y = 3 - x$

14. $y = |3 - x|$

15. $y = 3 - x^2$

16. $y = \dfrac{x - 3}{|x - 3|}$

17. $y = 3 - \sqrt{x}$

18. $y = \dfrac{1}{3 - x}$

19. $y = x^2 - 6x + 9$

20. $y = \dfrac{1}{(x - 3)^2}$

21. $y = \dfrac{x^2 - 3}{x}$

22. $y = \sqrt{3 - x^2}$

23. $y = \sqrt{3 - x}$

24. $y = x^3 - 3x^2$

Simplify each expression.

25. 10^3

26. 3^0

27. $4^{-1/2}$

28. 10^{-3}

29. 3^{-2}

30. $9^{1/2}$

31. $27^{-2/3}$

32. $4^{3/2}$

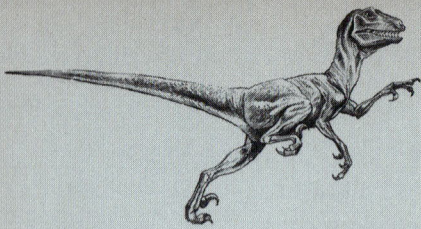

Vast, lonesome, remote, a no-man's land—the Montana badlands. In the summer of 1964, paleontologist John Ostrom was exploring this region when he spotted some weathered fragments of fossil bone protruding from a hillside. "Mine were the first human eyes to see those fragments," he wrote. "They had lain exposed there for centuries. But I knew at once that they were unique and that this was the most important discovery our expedition had made all summer."

Back at Yale University, as Ostrom carefully reassembled the skeletons, an extraordinary carnivore emerged. If fleshed out, this new dinosaur would have weighed about 170 pounds and measured 9 feet in length. Its structure suggested that it ran swiftly on its hind legs, much like a large bird. But the striking feature that inspired its name, *Deinonychus* or "terrible claw," was a 6-inch sickle-like claw on one toe of each hind foot. Ostrom believed these talons served as deadly weapons with which *Deinonychus* slashed its prey.

Later, when he unearthed several *Deinonychus* specimens entombed with the remains of a 1000-pound plant-eater, Ostrom deduced that the agile killer hunted in packs in order to overcome large quarry.

New information on dinosaur anatomy and ongoing studies of animal behavior are challenging old assumptions about what dinosaurs were like. Since imagination must often take over where scientific evidence ends, many of the new theories have sparked controversy.

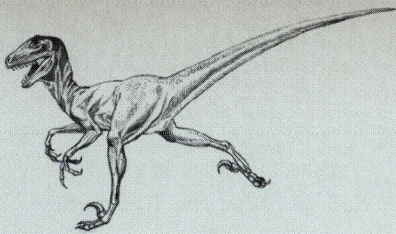

5

Exponential and Logarithmic Functions

Our search for the truth remains an ongoing detective hunt in which we interpret data extracted from mute bones and petrified footprints.

In this chapter we learn how scientists use exponential functions—which measure the rate of growth and decay over time—to date archaeological finds that are millions of years old. Through applications ranging from population growth to the growth of financial investments, we'll also see how exponential functions can help us predict the future as well as discover the past.

5.1

Exponential Functions

The functions that involve some combination of basic arithmetic operations, powers, or roots are called **algebraic functions.** Most of the functions studied so far are algebraic functions. In this chapter we turn to the exponential and logarithmic functions. These functions are used to describe phenomena ranging from growth of investments to the decay of radioactive materials, which cannot be described with algebraic functions. Since the exponential and logarithmic functions transcend what can be described with algebraic functions, they are called **transcendental functions.**

The Definition

In algebraic functions such as

$$j(x) = x^2, \qquad p(x) = x^5, \qquad \text{and} \qquad m(x) = x^{1/3}$$

the base is a variable and the exponent is constant. For the exponential functions the base is constant and the exponent is a variable. The functions

$$f(x) = 2^x, \qquad g(x) = 5^x, \qquad \text{and} \qquad h(x) = \left(\frac{1}{3}\right)^x$$

are exponential functions.

Definition: Exponential Function

> An **exponential function** with **base a** is a function of the form
>
> $$f(x) = a^x$$
>
> where a and x are real numbers such that $a > 0$ and $a \neq 1$.

We rule out the base $a = 1$ in the definition because $f(x) = 1^x$ is the constant function $f(x) = 1$. Negative numbers are not used as bases because powers such as $(-4)^{1/2}$ are not real numbers.

To evaluate an exponential function, we use our knowledge of exponents.

Define f, g, and h as y_1, y_2, and y_3.

Evaluating the functions for 2, 3, and -2 yields the same results as Example 1.

Example 1 Evaluating exponential functions

Let $f(x) = 4^x$, $g(x) = 5^{2-x}$, and $h(x) = \left(\dfrac{1}{3}\right)^x$. Find the following values.

a) $f(2)$ **b)** $f(3/2)$ **c)** $g(3)$ **d)** $h(-2)$

Solution

a) $f(2) = 4^2 = 16$

b) $f(3/2) = 4^{3/2} = \left(\sqrt{4}\right)^3 = 2^3 = 8$ *Take the square root, then cube.*

c) $g(3) = 5^{2-3} = 5^{-1} = \dfrac{1}{5}$

d) $h(-2) = \left(\dfrac{1}{3}\right)^{-2} = 3^2 = 9$ *Find the reciprocal, then square.* ◆

Domain of Exponential Functions

The domain of the exponential function $f(x) = 2^x$ is the set of all real numbers. To understand how this function can be evaluated for any *real* number, first note that the function is evaluated for any *rational* number using powers and roots. For example,

$$f(1.7) = f(17/10) = 2^{17/10} = \sqrt[10]{2^{17}} \approx 3.2490095985.$$

Until now, we have not considered *irrational* exponents.

Consider the expression $2^{\sqrt{3}}$ and recall that the irrational number $\sqrt{3}$ is an infinite nonterminating nonrepeating decimal number:

$$\sqrt{3} = 1.7320508075 \ldots$$

If we use rational approximations to $\sqrt{3}$ as exponents, we see a pattern:

$$2^{1.7} = 3.249009585 \ldots$$

$$2^{1.73} = 3.317278183 \ldots$$

$$2^{1.732} = 3.321880096 \ldots$$

$$2^{1.73205} = 3.321995226 \ldots$$

$$2^{1.7320508} = 3.321997068 \ldots$$

The $\wedge$ key is used for exponential expressions.

As the exponents get closer and closer to $\sqrt{3}$ we get results that are approaching some number. We define $2^{\sqrt{3}}$ to be that number. Of course, it is impossible to write the exact value of $\sqrt{3}$ or $2^{\sqrt{3}}$ as a decimal, but you can use a calculator to get $2^{\sqrt{3}} \approx 3.321997085$. Since any irrational number can be approximated by rational numbers in this same manner, 2^x is defined similarly for any irrational number. This idea extends to any exponential function.

Domain of an Exponential Function

The domain of $f(x) = a^x$ for $a > 0$ and $a \neq 1$ is the set of all real numbers.

Graphing Exponential Functions

Even though the domain of an exponential function is the set of real numbers, for ease of computation, we generally choose only rational numbers for x to find ordered pairs on the graph of the function.

Figure 5.1

Example 2   Graphing an exponential function ($a > 1$)

Sketch the graph of each exponential function and state the domain and range.

a) $f(x) = 2^x$ **b)** $g(x) = 10^x$

Solution

a) Find some ordered pairs satisfying $f(x) = 2^x$ as follows:

x	-2	-1	0	1	2	3
$y = 2^x$	1/4	1/2	1	2	4	8

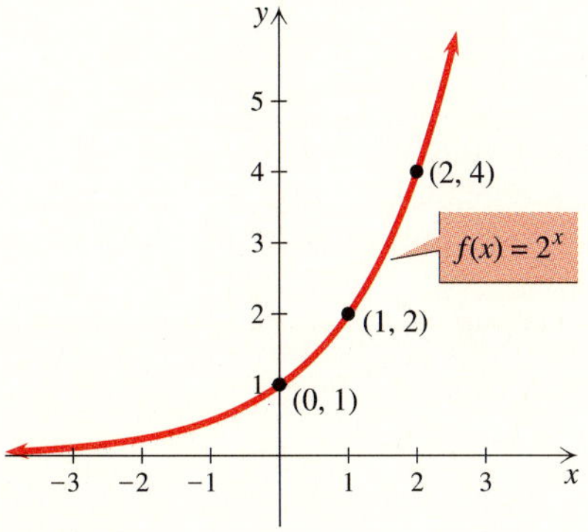

Plot these points. As x gets larger, so does 2^x. It can be shown that the graph of $f(x) = 2^x$ increases in a continuous manner, with no "jumps" or "breaks" in the graph. This fact allows us to draw a smooth curve through the points as shown in Fig. 5.1. The domain of $f(x) = 2^x$ is $(-\infty, \infty)$. From the graph in Fig. 5.1 we see that the range is $(0, \infty)$.

b) Find some ordered pairs satisfying $g(x) = 10^x$ as follows:

x	-2	-1	0	1	2	3
$y = 10^x$	0.01	0.1	1	10	100	1000

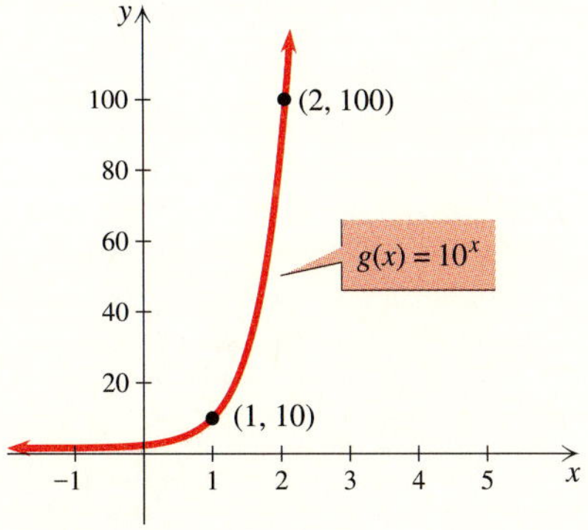

Figure 5.2

Plot these points and sketch the curve shown in Fig. 5.2. The graph of $g(x) = 10^x$ increases in the same manner as the graph of $f(x) = 2^x$. The domain of $g(x) = 10^x$ is $(-\infty, \infty)$. From the graph in Fig. 5.2 we see that the range is $(0, \infty)$.   ◆

Note the similarities between the graphs shown in Figs. 5.1 and 5.2. Both pass through the point $(0, 1)$, both functions are increasing, and both have the x-axis as a horizontal asymptote. Since a horizontal line can cross these graphs only once, the functions $f(x) = 2^x$ and $g(x) = 10^x$ are one-to-one by the horizontal line test. All functions of the form $f(x) = a^x$ for $a > 1$ have graphs similar to those in Figs. 5.1 and 5.2. In the next example we graph functions of the form $f(x) = a^x$ for $0 < a < 1$.

Example 3 Graphing an exponential function ($0 < a < 1$)

Sketch the graph of each function and state the domain and range of the function.

a) $f(x) = (1/2)^x$

b) $g(x) = 3^{-x}$

Solution

a) Find some ordered pairs satisfying $f(x) = (1/2)^x$ as follows:

x	-2	-1	0	1	2	3
$y = \left(\dfrac{1}{2}\right)^x$	4	2	1	1/2	1/4	1/8

Plot these points and draw a smooth curve through them as shown in Fig. 5.3. From the graph in Fig. 5.3 we see that the domain is $(-\infty, \infty)$ and the range is $(0, \infty)$. Note that the graph of $f(x) = (1/2)^x$ is a reflection in the y-axis of the graph of $y = 2^x$.

b) Since $3^{-x} = (1/3)^x$, this function is of the form $y = a^x$ for $0 < a < 1$. Find some ordered pairs satisfying $g(x) = 3^{-x}$ as follows:

x	-2	-1	0	1	2	3
$y = 3^{-x}$	9	3	1	1/3	1/9	1/27

Plot these points and draw a smooth curve through them as shown in Fig. 5.4. The domain is $(-\infty, \infty)$ and the range is $(0, \infty)$.

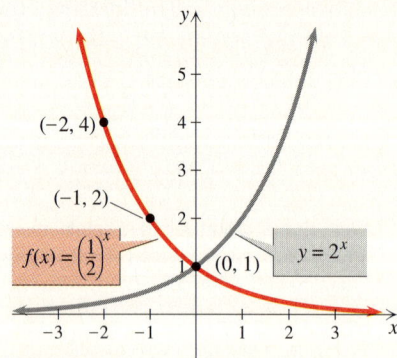

Figure 5.3

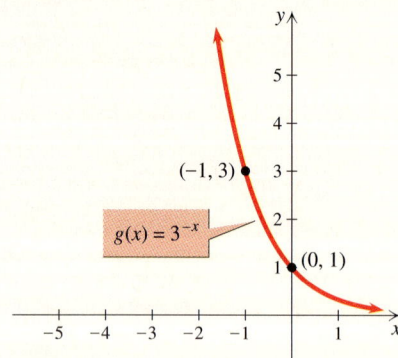

Figure 5.4

The only point on both $y_1 = (1/2)^x$ and $y_2 = 3^{-x}$ is $(0, 1)$.

Again note the similarities between the graphs in Figs. 5.3 and 5.4. Both pass through $(0, 1)$, both functions are decreasing, and both have the x-axis as a horizontal asymptote. By the horizontal line test, these functions are also one-to-one. The base determines whether the exponential function is increasing or decreasing. In general, we have the following properties of exponential functions.

▼

Properties of Exponential Functions

The exponential function $f(x) = a^x$ has the following properties:

1. The function f is increasing for $a > 1$ and decreasing for $0 < a < 1$.

2. The y-intercept of the graph of f is $(0, 1)$.

3. The graph has the x-axis as a horizontal asymptote.

4. The range of f is $(0, \infty)$.

5. The function f is one-to-one.

The phrase "growing exponentially" is often used to describe a population or other quantity that is increasing in the manner of an exponential function. For example, the earth's population is said to be growing exponentially because of the shape of the graph shown in Fig. 5.5. For an exponential function such as $f(x) = 2^x$, a small increase in x can cause a relatively large growth in y. If x increases from 9 to 10, $f(10) - f(9) = 512$. Compare this increase with the growth obtained in the algebraic function $g(x) = x^2$ when x increases from 9 to 10: $g(10) - g(9) = 19$.

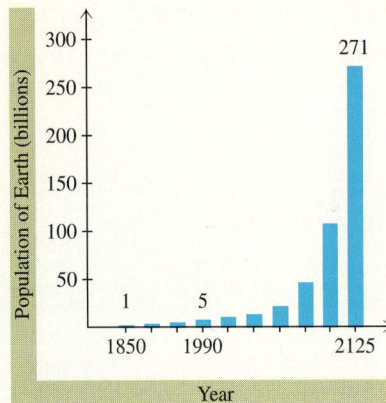

Figure 5.5

Transformation of Exponential Functions

The graphs of exponential functions can be translated, stretched, shrunk, or reflected using the ideas of Section 3.4. Just as $y = (x + 2)^2$ is a translation two units to the left of $y = x^2$, so $y = 3^{x+2}$ is a translation two units to the left of $y = 3^x$. If the graph is translated upward or downward, the horizontal asymptote will move also. If you have a graphing calculator, use it to graph the functions presented in the examples and exercises, and also experiment with it. Experimenting with a graphing calculator can lead you to an understanding of exponential functions much faster than doing all of the calculations by hand. Remember that the intention here is not simply to draw a picture, but to understand how the size of the base affects the shape of the graph and how transformations determine the location of the graph.

Example 4   Graphing an exponential function with translations

Sketch the graph of each function and state its domain and range.

a) $y = 2^{x+3}$

b) $f(x) = -4 + 3^{x+2}$

Solution

a) The graph of $y = 2^{x+3}$ is a translation three units to the left of the graph of $y = 2^x$ shown in Fig. 5.1. For accuracy, find a few ordered pairs that satisfy $y = 2^{x+3}$:

x	-3	-2	-1
$y = 2^{x+3}$	1	2	4

Draw the graph of $y = 2^{x+3}$ through these points as shown in Fig. 5.6. The domain of $y = 2^{x+3}$ is $(-\infty, \infty)$, and the range is $(0, \infty)$.

b) The graph of $y = 3^x$ goes through $(-1, 1/3)$, $(0, 1)$, and $(1, 3)$ as shown in Fig. 5.7. The graph of $f(x) = -4 + 3^{x+2}$ is obtained by translating $y = 3^x$ to the left two units and downward four units. For accuracy, find a few points on the graph of $f(x) = -4 + 3^{x+2}$:

x	-3	-2	-1	0
$y = -4 + 3^{x+2}$	$-11/3$	-3	-1	5

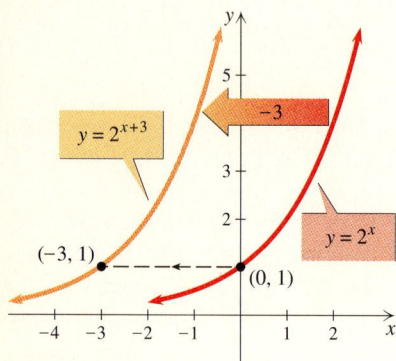

Figure 5.6

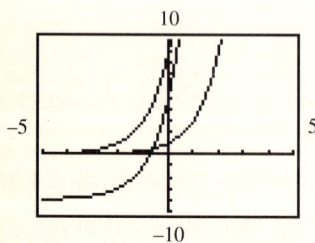

The graph of $y_2 = 3^{x+2}$ lies two units to the left of $y_1 = 3^x$. The graph of $y_3 = -4 + 3^{x+2}$ lies four units below y_2.

Figure 5.7

Sketch a smooth curve through these points as shown in Fig. 5.7. The horizontal asymptote is the line $y = -4$. The domain of f is $(-\infty, \infty)$, and the range is $(-4, \infty)$. ◆

If you recognize that a function is a transformation of a simpler function, then you know what its graph looks like. This information along with a few ordered pairs will help you to make accurate graphs.

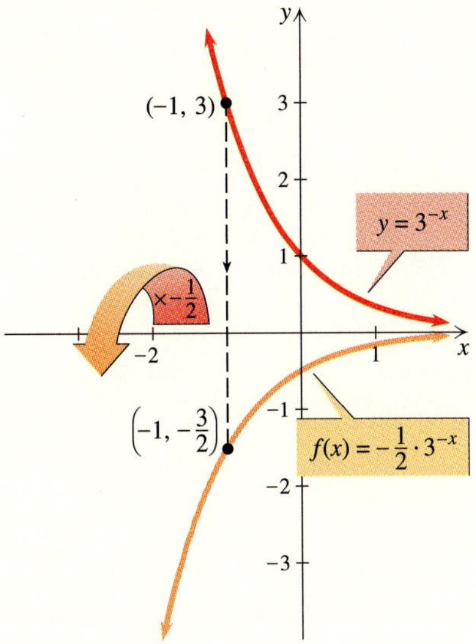

Figure 5.8

Example 5   Graphing an exponential function with a reflection

Sketch the graph of $f(x) = -\dfrac{1}{2} \cdot 3^{-x}$ and state the domain and range.

Solution

The graph of $y = 3^{-x}$ is shown in Fig. 5.4. Multiplying by $-1/2$ shrinks the graph and reflects it below the x-axis. The x-axis is the horizontal asymptote for the graph of f. Find a few ordered pairs as follows:

x	-2	-1	0	1	2
$y = -\dfrac{1}{2} \cdot 3^{-x}$	$-9/2$	$-3/2$	$-1/2$	$-1/6$	$-1/18$

Sketch the curve through these points as in Fig. 5.8. The domain is $(-\infty, \infty)$ and the range is $(-\infty, 0)$. ◆

Exponential Equations

From the graphs of exponential functions, we observed that they are one-to-one. For an exponential function, one-to-one means that if two exponential expressions with the same base are equal, then the exponents are equal.

One-to-One Property of Exponential Functions

For $a > 0$ and $a \neq 1$,

$$\text{if} \quad a^{x_1} = a^{x_2}, \quad \text{then} \quad x_1 = x_2.$$

The one-to-one property is used in solving simple exponential equations. For example, to solve $2^x = 8$, we recall that $8 = 2^3$. So the equation becomes $2^x = 2^3$. By the one-to-one property, $x = 3$ is the only solution. The one-to-one property applies only to equations in which each side is a power of the same base.

Example 6   Solving exponential equations

Solve each exponential equation.

a) $4^x = \dfrac{1}{4}$ **b)** $\left(\dfrac{1}{10}\right)^x = 100$

Solution

a) Since $1/4 = 4^{-1}$, we can write the right-hand side as a power of 4 and use the one-to-one property:

$$4^x = \frac{1}{4} = 4^{-1}$$

$$x = -1 \qquad \text{One-to-one property}$$

b) Since $(1/10)^x = 10^{-x}$ and $100 = 10^2$, we can write each side as a power of 10:

$$\left(\frac{1}{10}\right)^x = 100$$

$$10^{-x} = 10^2$$

$$-x = 2 \qquad \text{One-to-one property}$$

$$x = -2 \qquad \blacklozenge$$

The type of equation that we solved in Example 6 arises naturally when we try to find the first coordinate of an ordered pair of an exponential function when given the second coordinate.

Example 7   Finding the first coordinate given the second

Let $f(x) = 5^{2-x}$. Find x such that $f(x) = 125$.

Solution

To find x such that $f(x) = 125$, we must solve $5^{2-x} = 125$:

$$5^{2-x} = 5^3 \qquad \text{Since } 125 = 5^3$$

$$2 - x = 3 \qquad \text{One-to-one property}$$

$$x = -1 \qquad \blacklozenge$$

Compound Interest

Exponential functions are used to model phenomena such as population growth, radioactive decay, and compound interest. Here we will show how these functions are used to determine the amount of an investment earning compound interest.

If P dollars are deposited in an account with a simple annual interest rate r for t years, then the amount A in the account at the end of t years is found by using the formula $A = P + Prt$ or $A = P(1 + rt)$. If \$1000 is deposited at 6% annual rate for one quarter of a year, then at the end of three months the amount is

$$A = 1000\left(1 + 0.06 \cdot \frac{1}{4}\right) = 1000(1.015) = \$1015.$$

If the account begins the next quarter with \$1015, then at the end of the second quarter we again multiply by 1.015 to get the amount

$$A = 1000(1.015)^2 = \$1030.23.$$

This process is referred to as compound interest because interest is put back into the account and the interest also earns interest. At the end of 20 years, or 80 quarters, the amount is

$$A = 1000(1.015)^{80} = \$3290.66.$$

Compound interest can be thought of as simple interest computed over and over. The general **compound interest formula** follows.

Compound Interest Formula

If a principal P is invested for t years at an annual rate r compounded n times per year, then the amount A, or ending balance, is given by

$$A = P\left(1 + \frac{r}{n}\right)^{nt}.$$

The principal P is also called **present value** and the amount A is called **future value.**

The graph of
$y_1 = 20,000(1 + 0.06/365)^{365x}$
shows the growth of \$20,000 at 6%
compounded daily for x years.

Example 8   Using the compound interest formula

Find the amount of \$20,000 at 6% compounded daily for three years.

Solution

Use $P = \$20,000$, $r = 0.06$, $n = 365$, and $t = 3$ in the compound interest formula:

$$A = \$20,000\left(1 + \frac{0.06}{365}\right)^{365 \cdot 3} = \$23,943.99$$

It is also a common practice for banks to use 360 days per year.

Continuous Compounding and the Number *e*

The more often that interest is figured during the year, the more interest an investment will earn. The first five lines of Table 5.1 show the future value of $10,000 invested at 12% for one year for more and more frequent compounding. The last line shows the limiting amount $11,274.97, which we cannot exceed no matter how often we compound the interest on this investment for one year.

To better understand the last line of Table 5.1, we need to use a fact from calculus. The expression $[1 + (r/n)]^{nt}$ used in calculating the first five values in the table can be shown to approach e^{rt} as $n \to \infty$. The number e (like π) is an irrational number that occurs in many areas of mathematics. The factor e^{rt} is used in the last line of the table to find the limit to the values of the investment obtained by more and more frequent compounding. Using e^{rt} to find the future value is called **continuous compounding.**

Compounding	Future value in one year
Annually	$\$10{,}000\left(1 + \dfrac{0.12}{1}\right)^{1} = \$11{,}200$
Quarterly	$\$10{,}000\left(1 + \dfrac{0.12}{4}\right)^{4} = \$11{,}255.09$
Monthly	$\$10{,}000\left(1 + \dfrac{0.12}{12}\right)^{12} = \$11{,}268.25$
Daily	$\$10{,}000\left(1 + \dfrac{0.12}{365}\right)^{365} = \$11{,}274.75$
Hourly	$\$10{,}000\left(1 + \dfrac{0.12}{8760}\right)^{8760} = \$11{,}274.96$
Continuously	$\$10{,}000 e^{0.12(1)} = \$11{,}274.97$

Table 5.1

```
10000e^.12
         11274.96852
e^1
         2.718281828
```

The e^x key is used to find powers of e. See Appendix A for more examples.

You can find approximate values for powers of e using a calculator with an e^x key. To find an approximate value for e itself, find e^1 on a calculator:

$$e \approx 2.718281828459$$

▼

Continuous Compounding Formula

If a principal P is invested for t years at an annual rate r compounded continuously, then the amount A, or ending balance, is given by

$$A = P \cdot e^{rt}.$$

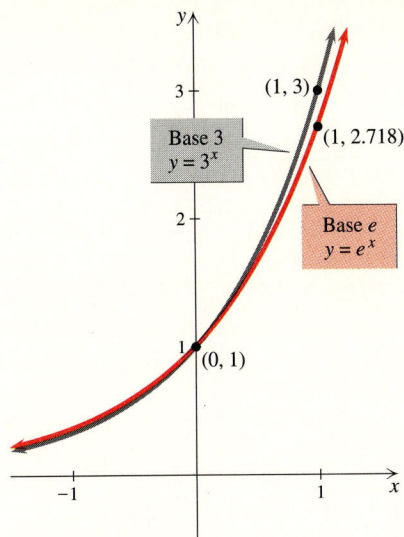

Exponential Functions

Figure 5.9

Example 9   Interest compounded continuously

Find the amount of \$5600 at $6\frac{1}{4}\%$ annual rate compounded continuously for 5 years and 9 months.

Solution

Convert 5 years and 9 months to 5.75 years. Use $r = 0.0625$, $t = 5.75$, and $P = \$5600$ in the continuous compounding formula:

$$A = 5600 \cdot e^{(0.0625)(5.75)} = \$8021.63$$

The function $f(x) = e^x$ is called the **base-e exponential function.** Variations of this function are used to model many types of growth and decay. The graph of $y = e^x$ looks like the graph of $y = 3^x$, because the value of e is close to 3. Use your calculator to check that the graph of $y = e^x$ shown in Fig. 5.9 goes approximately through the points $(-1, 0.368)$, $(0, 1)$, $(1, 2.718)$, and $(2, 7.389)$.

Radioactive Decay

The mathematical model of radioactive decay is based on the formula

$$A = A_0 e^{rt},$$

which gives the amount A of a radioactive substance remaining after t years, where A_0 is the initial amount present and r is the annual rate of decay. The only difference between this formula and the continuous compounding formula is that when decay is involved the rate r is negative.

Example 10   Radioactive decay of carbon-14

Finely-chipped spear points shaped by nomadic hunters in the Ice Age were found at the scene of a large kill in Lubbock, Texas. From the amount of decay of carbon-14 in the bone fragments of the giant bison, scientists determined that the kill took place about 9833 years ago (*National Geographic,* December 1955). Carbon-14 is a radioactive substance that is absorbed by an organism while it is alive but begins to decay upon the death of the organism. The number of grams of carbon-14 remaining in a fragment of charred bone from the giant bison after t years is given by the formula $A = 3.6e^{rt}$, where $r = -1.21 \times 10^{-4}$. Find the amount present initially and after 9833 years.

To see the decay on your calculator, enter the function as shown here.

This graph shows the decay of 3.6 grams of Carbon-14 over 25,000 years.

Solution

To find the initial amount, let $t = 0$ and $r = -1.21 \times 10^{-4}$ in the formula:

$$A = 3.6e^{(-1.21 \times 10^{-4})(0)} = 3.6 \text{ grams}$$

To find the amount present after 9833 years, let $t = 9833$ and $r = -1.21 \times 10^{-4}$:

$$A = 3.6e^{(-1.21 \times 10^{-4})(9833)}$$
$$\approx 1.1 \text{ grams}$$

The initial amount of carbon-14 was 3.6 grams, and after 9833 years, 1.1 grams of carbon-14 remained. See Fig. 5.10.

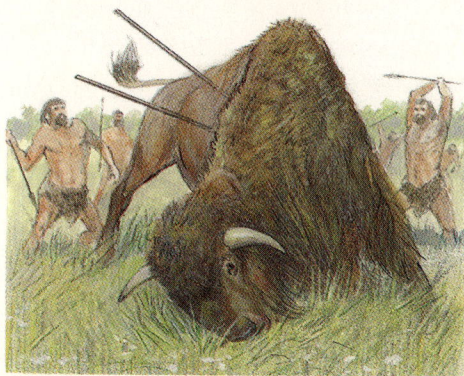

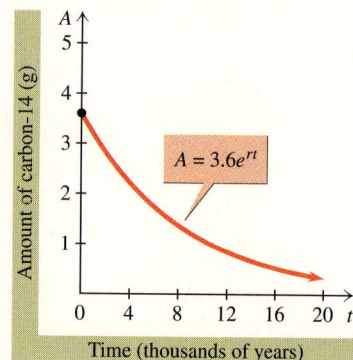

Figure 5.10

The formula for radioactive decay is used to determine the age of ancient objects such as bones, campfire charcoal, and meteorites. To date objects, we must solve the radioactive decay formula for t, which will be done in Section 5.4 after we have studied logarithms.

For Thought

True or false? Explain.

1. The function $f(x) = (-2)^x$ is an exponential function.
2. The function $f(x) = 2^x$ is invertible.
3. If $2^x = -\dfrac{1}{8}$, then $x = -3$.
4. If $f(x) = 3^x$, then $f(0.5) = \sqrt{3}$.
5. If $f(x) = e^x$, then $f(0) = 1$.
6. If $f(x) = e^x$ and $f(t) = e^2$, then $t = 2$.
7. The x-axis is a horizontal asymptote for the graph of $y = e^x$.
8. The function $f(x) = (0.5)^x$ is increasing.
9. The functions $f(x) = 4^{x-1}$ and $g(x) = (0.25)^{1-x}$ have the same graph.
10. $2^{1.73} = \sqrt[100]{2^{173}}$

5.1 Exercises [] Tape 9 [] Disk—5.25″: 3 3.5″: 2 Macintosh: 2

Let $f(x) = 3^x$, $g(x) = 2^{1-x}$, and $h(x) = (1/4)^x$. Find the following values.

1. $f(2)$ **2.** $f(4)$ **3.** $f(-2)$ **4.** $f(-3)$

5. $g(2)$ **6.** $g(1)$ **7.** $g(-2)$ **8.** $g(-3)$

9. $h(-1)$ **10.** $h(-2)$ **11.** $h(-1/2)$ **12.** $h(3/2)$

Sketch the graph of each function by finding at least three ordered pairs on the graph. State the domain, the range, and whether the function is increasing or decreasing.

13. $f(x) = 5^x$ **14.** $f(x) = 4^x$ **15.** $y = 10^{-x}$

16. $y = e^{-x}$ **17.** $f(x) = (1/4)^x$ **18.** $f(x) = (0.2)^x$

Use transformations to help you graph each function.

19. $f(x) = 2^x - 3$ **20.** $f(x) = 3^{-x} + 1$

21. $f(x) = 2^{x+3} - 5$ **22.** $f(x) = 3^{1-x} - 4$

23. $y = -2^{-x}$ **24.** $y = -10^{-x}$

25. $y = 1 - 2^x$ **26.** $y = -1 - 2^{-x}$

27. $f(x) = 0.5 \cdot 3^{x-2}$ **28.** $f(x) = -0.1 \cdot 5^{x+4}$

29. $y = 500(0.5)^x$ **30.** $y = 100 \cdot 2^x$

Solve each equation.

31. $2^x = 64$ **32.** $5^x = 1$ **33.** $10^x = 0.1$

34. $10^{2x} = 1000$ **35.** $-3^x = -27$ **36.** $-2^x = -\dfrac{1}{2}$

37. $3^{-x} = 9$ **38.** $2^x = \dfrac{1}{8}$ **39.** $8^x = 2$

40. $9^x = 3$ **41.** $e^x = \dfrac{1}{e^2}$ **42.** $e^{-x} = \dfrac{1}{e}$

43. $\left(\dfrac{1}{2}\right)^x = 8$ **44.** $\left(\dfrac{2}{3}\right)^x = \dfrac{9}{4}$

45. $10^{x-1} = 0.01$ **46.** $10^{|x|} = 1000$

Let $f(x) = 2^x$, $g(x) = (1/3)^x$, $h(x) = 10^x$, $m(x) = e^x$. Find the value of x in each equation.

47. $f(x) = 4$ **48.** $f(x) = 32$ **49.** $f(x) = \dfrac{1}{2}$

50. $f(x) = 1$ **51.** $g(x) = 1$ **52.** $g(x) = 9$

53. $g(x) = 27$ **54.** $g(x) = \dfrac{1}{9}$ **55.** $h(x) = 1000$

56. $h(x) = 10^5$ **57.** $h(x) = 0.1$ **58.** $h(x) = 0.0001$

59. $m(x) = e$ **60.** $m(x) = e^3$

61. $m(x) = \dfrac{1}{e}$ **62.** $m(x) = 1$

Fill in the missing coordinate in each ordered pair so that the pair is a solution to the given equation.

63. $y = 3^x$ $\quad (2, \ ), (\ , 3), (-1, \ ), (\ , 1/9)$

64. $y = 10^x$ $\quad (3, \ ), (\ , 1), (-2, \ ), (\ , 0.01)$

65. $f(x) = 5^{-x}$ $\quad (0, \ ), (\ , 25), (-1, \ ), (\ , 1/5)$

66. $f(x) = e^{-x}$ $\quad (1, \ ), (\ , e), (0, \ ), (\ , e^2)$

67. $f(x) = -2^x$ $\quad (4, \ ), (\ , -1/4), (-1, \ ), (\ , -32)$

68. $f(x) = -(1/4)^{x-1}$ $\quad (3, \ ), (\ , -4), (-1, \ ), (\ , -1/16)$

The following functions are not exponential functions, but they are similar to exponential functions. Sketch the graph of each function.

69. $f(x) = 2^{|x|}$ **70.** $f(x) = 3^{-|x|}$

71. $y = 2^{-x} + 2^x$ **72.** $y = x2^x$

Solve each problem.

73. *Interest compounded quarterly* If $4500 is deposited in a bank account paying 8% compounded quarterly, then what amount will be in the account at the end of 6 years? How much interest will be earned during the 6 years?

74. *Interest compounded monthly* If Melinda invests her $80,000 winnings from Publishers Clearing House at 9% compounded monthly, then what will the investment be worth at the end of 20 years? How much interest will be earned during the 20 years?

75. *Interest compounded daily* If a credit union pays 6.5% annual interest compounded daily, then what will a deposit of $2300 be worth after 5 years and 4 months? What assumptions are you making?

76. *High yields attract deposits* To attract funds, the financially troubled Commercial Federal Savings and Loan offered $9\frac{3}{4}\%$ annual interest compounded daily on certificates of deposit. At this rate how much interest would a deposit of $30,000 earn in 18 months?

77. *Interest compounded continuously* The Lakewood Savings Bank pays 5% annual interest compounded continuously on deposits. How much will a deposit of $100,000 amount to in 5 years and 45 days at 5% compounded continuously? Compounded daily?

78. *Interest compounded continuously* If one million dollars is deposited in an account paying 6% compounded continuously, then will it earn more or less than minimum wage in its first hour on deposit? How much interest is earned during its 500th hour on deposit?

79. *Population growth* The population of a small country in Central America is growing according to the function $P = 2,400,000e^{0.03t}$, where t is the number of years since 1990. What was the population in 1990? What will the population be in 2001 according to the formula?

80. *Radioactive decay* The number of grams of a certain radioactive substance present at time t is given by the formula $A = 200e^{-0.001t}$, where t is the number of years. How many grams are present at time $t = 0$? How many grams are present at time $t = 500$?

81. *Position of a football* The football is on the 10-yard line. Several penalties in a row are given, and each moves the ball half the distance to the closer goal line. Write a formula that gives the ball's position P after the nth such penalty.

82. *Cost of a parking ticket* The cost of a parking ticket on campus is $15 for the first offense. Given that the cost doubles for each additional offense, write a formula for the cost C as a function of the number of tickets n.

83. *The value of e* Make a table showing the values of $(1 + 1/n)^n$ for $n = 20, 200, 2000,$ and $20,000$. What is the difference between the last entry in your table and the value of e found on your calculator?

84. *A large investment* Make a table showing the amount of $80 million earning interest for one year at 9% compounded annually, monthly, daily, hourly, and every minute. What amount are the values in your table approaching?

85. *Challenger disaster* Using data on O-ring damage from 24 previous space shuttle launches, Professor Edward R. Tufte of Yale University concluded that the number of O-rings damaged per launch is an exponential function of the temperature at the time of the launch. If NASA had used a model such as $n = 644e^{-0.15t}$, where t is the Fahrenheit temperature at the time of launch and n is the number of O-rings damaged, then the tragic end to the flight of the space shuttle Challenger might have been avoided. Using this model, find the number of O-rings that would be expected to fail at 31°F, the temperature at the time of the Challenger launch on January 28, 1986.

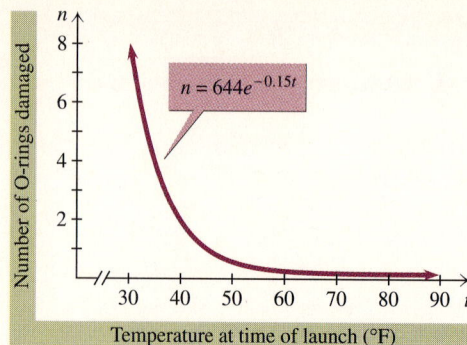

Figure for Exercise 85

86. *Manufacturing cost* The cost in dollars for manufacturing x units of a certain drug is given by the function $C(x) = xe^{0.001x}$. Find the cost for manufacturing 500 units. Find the function $AC(x)$ that gives the average cost per unit for manufacturing x units. What happens to the average cost per unit as x gets larger and larger?

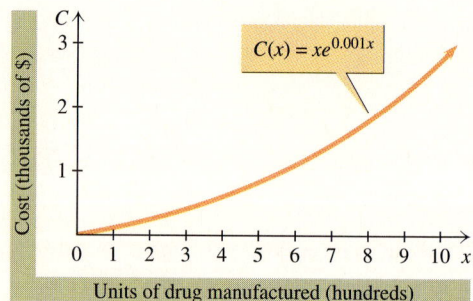

Figure for Exercise 86

For Writing/Discussion

87. *Inverse function* The function $f(x) = 10^x$ is a one-to-one function, and so it has an inverse function. What is the value of $f^{-1}(100)$? Explain your answer.

88. *Another inverse function* The function $f(x) = e^x$ is a one-to-one function, and so it has an inverse function. What is the value of $f^{-1}(1)$? Explain your answer.

89. *Cooperative learning* Work in a small group to write a summary (including drawings) of the types of graphs that can be obtained for exponential functions of the form $y = a^x$ for $a > 0$ and $a \neq 1$.

90. *Cooperative learning* Find an application of an exponential function that has not been mentioned in this text and present it to your class.

Graphing Calculator Exercises

Find an approximate solution to each equation by graphing an appropriate function on a graphing calculator and reading from the graph. Round answers to two decimal places.

1. $3^x - 5 = 0$

2. $2^{-x} - 3 = 0$

3. $2^x - 3^{x+1} = 0$

4. $e^x - 2^{-x+1} = 0$

5. $e^{x+1} = 9$

6. $2^x = 3^{-x}$

7. $200e^{0.06x} = 400$

8. $300(1.05)^x = 600$

The following functions are not exponential functions, but their formulas are similar to exponential functions. Functions like these occur in probability, statistics, and calculus. Sketch the graph of each function with the aid of a graphing calculator and determine the range of each function.

9. $y = 2^{-x^2}$

10. $y = \dfrac{e^x + e^{-x}}{2}$

11. $y = x^2 e^{-x}$

12. $y = x e^{-x}$

5.2

Logarithmic Functions

Since exponential functions are one-to-one functions (Section 5.1), they are invertible. In this section we will study the inverses of the exponential functions.

The Definition

The inverses of the exponential functions are called **logarithmic functions.** Since f^{-1} is a general name for an inverse function, we adopt a more descriptive notation for these inverses. If $f(x) = a^x$, then instead of $f^{-1}(x)$, we write $\log_a(x)$ for the inverse of the base-a exponential function. We read $\log_a(x)$ as "log of x with base a," and we call the expression $\log_a(x)$ a **logarithm.**

The meaning of $\log_a(x)$ will be clearer if we consider the exponential function

$$f(x) = 2^x$$

as an example. Since $f(3) = 2^3 = 8$, the base-2 exponential function pairs the exponent 3 with the value of the exponential expression 8. Since the function $\log_2(x)$ reverses that pairing, we have $\log_2(8) = 3$. So $\log_2(8)$ is the exponent that is used on the base 2 to obtain 8. *In general, $\log_a(x)$ is the exponent that is used on the base a to obtain the value x.*

▼

Definition: Logarithmic Function

For $a > 0$ and $a \neq 1$, the **logarithmic function with base a** is denoted $f(x) = \log_a(x)$, where

$$y = \log_a(x) \qquad \text{if and only if} \qquad a^y = x.$$

Example 1 Evaluating logarithmic functions

Find the indicated values of the logarithmic functions.

a) $\log_3(9)$ **b)** $\log_2\left(\dfrac{1}{4}\right)$ **c)** $\log_{1/2}(8)$

Solution

a) By the definition of logarithm, $\log_3(9)$ is the exponent that is used on the base 3 to obtain 9. Since $3^2 = 9$, we have $\log_3(9) = 2$.

b) Since $\log_2(1/4)$ is the exponent that is used on the base 2 to obtain 1/4, we try various powers of 2 until we find $2^{-2} = 1/4$. So $\log_2(1/4) = -2$.

c) Since $\log_{1/2}(8)$ is the exponent that is used on a base of 1/2 to obtain 8, we try various powers of 1/2 until we find $(1/2)^{-3} = 8$. So $\log_{1/2}(8) = -3$. ◆

Since the exponential function $f(x) = a^x$ has domain $(-\infty, \infty)$ and range $(0, \infty)$, the logarithmic function $f(x) = \log_a(x)$ has domain $(0, \infty)$ and range $(-\infty, \infty)$. So there are no logarithms of negative numbers or zero. Expressions such as $\log_2(-4)$ and $\log_3(0)$ are undefined. Note that $\log_a(1) = 0$ for any base a, because $a^0 = 1$ for any base a.

There are two bases that are used more frequently than the others; they are 10 and e. The notation $\log_{10}(x)$ is abbreviated $\log(x)$, and $\log_e(x)$ is abbreviated $\ln(x)$. Most scientific calculators have function keys for the exponential functions 10^x and e^x and their inverses $\log(x)$ and $\ln(x)$, which are called **common logarithms** and **natural logarithms,** respectively. Natural logarithms are also called **Napierian logarithms** after John Napier (1550–1617). You can use a calculator to find common or natural logarithms, but you should know how to find the values of logarithms such as those in Examples 1 and 2 without using a calculator.

Example 2 Evaluating logarithmic functions

Find the indicated values of the logarithmic functions.

a) $\log(1000)$ **b)** $\ln(1)$ **c)** $\ln(-6)$

Solution

a) To find $\log(1000)$, we must find the exponent that is used on the base 10 to obtain 1000. Since $10^3 = 1000$, we have $\log(1000) = 3$.

b) Since $e^0 = 1$, $\ln(1) = 0$.

c) The expression $\ln(-6)$ is undefined because -6 is not in the domain of the natural logarithm function. There is no power of e that results in -6. ◆

```
log 1000
               3
ln 1
               0
```

Use the LOG key for common logarithms and the LN key for natural logarithms.

Graphs of Logarithmic Functions

The functions $y = a^x$ and $y = \log_a(x)$ for $a > 0$ and $a \neq 1$ are inverse functions. So the graph of $y = \log_a(x)$ is a reflection about the line $y = x$ of the graph of

$y = a^x$. The graph of $y = a^x$ has the x-axis as its horizontal asymptote, while the graph of $y = \log_a(x)$ has the y-axis as its vertical asymptote.

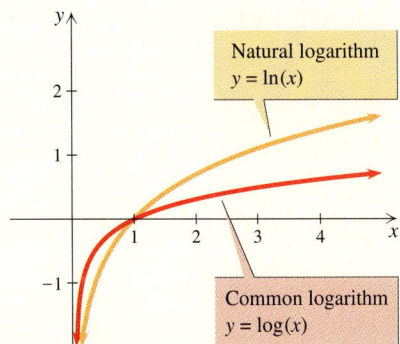

Figure 5.11

Example 3 Graph of a base-a logarithmic function with $a > 1$

Sketch the graphs of $y = 2^x$ and $y = \log_2(x)$ on the same coordinate system. State the domain and range of each function.

Solution

Since these two functions are inverses of each other, the graph $y = \log_2(x)$ is a reflection of the graph of $y = 2^x$ about the line $y = x$. Make a table of ordered pairs for each function:

x	-1	0	1	2
$y = 2^x$	$1/2$	1	2	4

x	$1/2$	1	2	4
$y = \log_2(x)$	-1	0	1	2

Sketch $y = 2^x$ as in Section 5.1. Then sketch $y = \log_2(x)$ through the points given in the table, keeping in mind that it is a reflection of $y = 2^x$. Both curves are shown in Fig. 5.11 along with the line $y = x$. The domain of $y = 2^x$ is $(-\infty, \infty)$, and its range is $(0, \infty)$. The domain of $y = \log_2(x)$ is $(0, \infty)$, and its range is $(-\infty, \infty)$.

Note that the function $y = \log_2(x)$ graphed in Fig. 5.11 is one-to-one by the horizontal line test and it is an increasing function. The graphs of the common logarithm function $y = \log(x)$ and the natural logarithm function $y = \ln(x)$ are shown in Fig. 5.12 and they are also increasing functions. The function $y = \log_a(x)$ is increasing if $a > 1$. By contrast, if $0 < a < 1$, the function $y = \log_a(x)$ is decreasing, as illustrated in the next example.

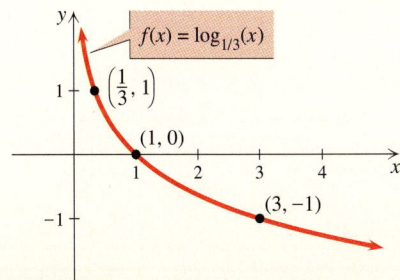

Figure 5.12

Example 4 Graph of a base-a logarithmic function with $0 < a < 1$

Sketch the graph of $f(x) = \log_{1/3}(x)$ and state its domain and range.

Solution

Make a table of ordered pairs for the function:

x	$1/9$	$1/3$	1	3	9
$y = \log_{1/3}(x)$	2	1	0	-1	-2

Sketch the curve through these points as shown in Fig. 5.13. The y-axis is a vertical asymptote for the curve. The domain of $f(x) = \log_{1/3}(x)$ is $(0, \infty)$, and its range is $(-\infty, \infty)$.

Figure 5.13

The logarithmic functions have properties corresponding to the properties of the exponential functions stated in Section 5.1.

Properties of Logarithmic Functions

The logarithmic function $f(x) = \log_a(x)$ has the following properties:

1. The function f is increasing for $a > 1$ and decreasing for $0 < a < 1$.
2. The x-intercept of the graph of f is $(1, 0)$.
3. The graph has the y-axis as a vertical asymptote.
4. The domain of f is $(0, \infty)$ and the range of f is $(-\infty, \infty)$.
5. The function f is one-to-one.

Transformation of Logarithmic Functions

The graphs of logarithmic functions can be transformed in the same manner as the graphs of other functions. For example, the graph of $y = \log_a(x + 2)$ lies two units to the left of the graph of $y = \log_a(x)$, while $y = -3 + \log_a(x)$ is three units below $y = \log_a(x)$, and $y = -\log_a(x)$ is a reflection of $y = \log_a(x)$.

Example 5 Transformation of the graph of a logarithmic function

Sketch the graph of each function and state its domain and range.

a) $y = \log_2(x - 1)$ **b)** $f(x) = -\dfrac{1}{2}\log_2(x + 3)$.

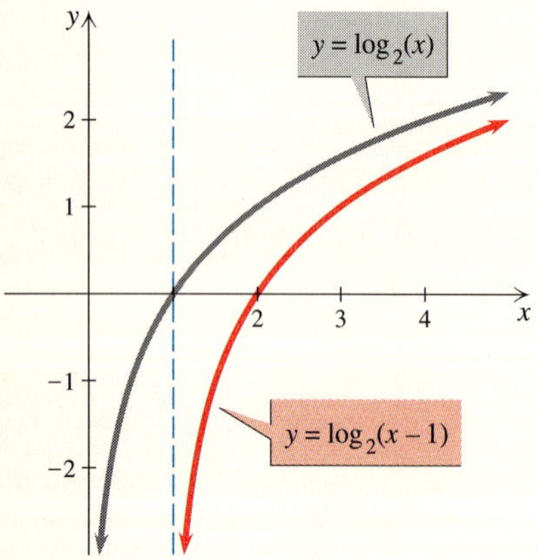

Solution

a) The graph of $y = \log_2(x - 1)$ is obtained by translating the graph of $y = \log_2(x)$ to the right one unit. Since the domain of $y = \log_2(x)$ is $(0, \infty)$, the domain of $y = \log_2(x - 1)$ is $(1, \infty)$. The line $x = 1$ is the vertical asymptote. Calculate a few ordered pairs to get an accurate graph.

x	1.5	2	3	5
$y = \log_2(x - 1)$	-1	0	1	2

Figure 5.14

Sketch the curve as shown in Fig. 5.14. The range is $(-\infty, \infty)$.

b) The graph of f is obtained by translating the graph of $y = \log_2(x)$ to the left three units. The vertical asymptote for f is the line $x = -3$. Multiplication by $-1/2$ shrinks the graph and reflects it in the x-axis. Calculate a few ordered pairs for accuracy.

x	-2	-1	1	5
$y = -\dfrac{1}{2}\log_2(x + 3)$	0	$-1/2$	-1	$-3/2$

Sketch the curve through these points as shown in Fig. 5.15. The domain of f is $(-3, \infty)$, and the range is $(-\infty, \infty)$.

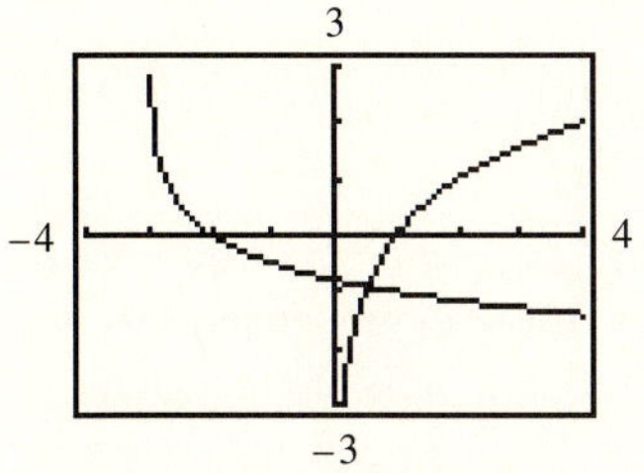

To graph $y = \log_2(x)$ and $y = -\frac{1}{2}\log_2(x + 3)$, you will need the base-change formula given in Section 5.3.

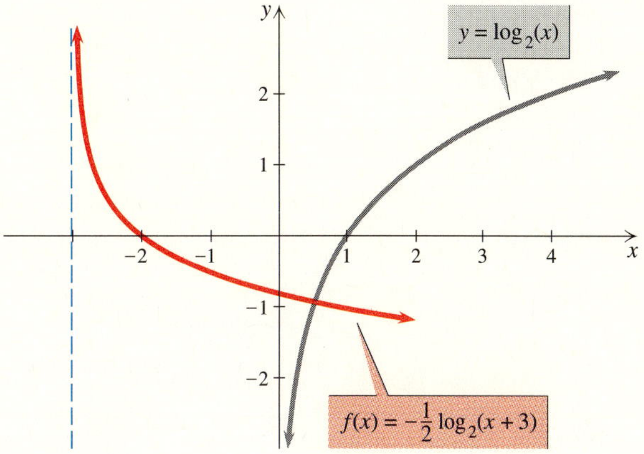

$y = \log_2(x)$

$f(x) = -\frac{1}{2}\log_2(x + 3)$

Figure 5.15

Logarithmic and Exponential Equations

Some equations involving logarithms can be solved by writing an equivalent exponential equation. On the other hand, some equations involving exponents can be solved by writing an equivalent logarithmic equation. Rewriting logarithmic and exponential equations is possible because of the definition of logarithms:

$$y = \log_a(x) \qquad \text{if and only if} \qquad a^y = x$$

Example 6   Rewriting logarithmic and exponential equations

Write each equation involving logarithms as an equivalent exponential equation, and write each equation involving exponents as an equivalent logarithmic equation.

a) $\log_5(625) = 4$ **b)** $\log_3(n - 1) = 5$ **c)** $3^{2x+3} = 50$ **d)** $e^{x-1} = 9$

Solution

a) $\log_5(625) = 4$ is equivalent to $5^4 = 625$.

b) $\log_3(n - 1) = 5$ is equivalent to $3^5 = n - 1$.

c) $3^{2x+3} = 50$ is equivalent to $2x + 3 = \log_3(50)$.

d) $e^{x-1} = 9$ is equivalent to $x - 1 = \ln(9)$.

The one-to-one property of exponential functions was used to solve exponential equations in Section 5.1. Likewise, the one-to-one property of logarithmic functions is used in solving logarithmic equations. The one-to-one property says that *if two quantities have the same base-a logarithm, then the quantities are equal.*

One-to-One Property of Logarithms

For $a > 0$ and $a \neq 1$,

$$\text{if} \quad \log_a(x_1) = \log_a(x_2), \quad \text{then} \quad x_1 = x_2.$$

The one-to-one properties and the definition of logarithm are at present the only new tools that we have for solving equations. Later in this chapter we will develop more properties of logarithms and solve more complicated equations.

Example 7   Solving equations involving logarithms

Solve each equation.

a) $\log_3(x) = -2$ **b)** $\log_x(5) = 2$ **c)** $5^x = 9$ **d)** $\ln(x^2) = \ln(3x)$

Solution

a) Use the definition of logarithm to write the equivalent exponential equation:

$$\log_3(x) = -2$$

$$x = 3^{-2} \qquad \text{Definition of logarithm}$$

$$= \frac{1}{9}$$

Since $\log_3(1/9) = -2$ is correct, the solution to the equation is $1/9$.

b) $\log_x(5) = 2$

$$x^2 = 5 \qquad \text{Definition of logarithm}$$

$$x = \pm\sqrt{5}$$

Since the base of a logarithm is always nonnegative, the only solution is $\sqrt{5}$.

c) $5^x = 9$

$$x = \log_5(9) \qquad \text{Definition of logarithm}$$

The exact solution to $5^x = 9$ is the irrational number $\log_5(9)$. In the next section we will learn how to find a rational approximation for $\log_5(9)$.

d)

$$\ln(x^2) = \ln(3x)$$

$$x^2 = 3x \qquad \text{One-to-one property of logarithms}$$

$$x^2 - 3x = 0$$

$$x(x - 3) = 0$$

$$x = 0 \quad \text{or} \quad x = 3$$

Checking $x = 0$ in the original equation, we get the undefined expression $\ln(0)$. So the only solution to the equation is 3.   ◆

Example 8   Equations involving common and natural logarithms

Use a scientific calculator to find the value of x rounded to four decimal places.

a) $e^x = 125$

b) $x \cdot \log(2) + 3x \cdot \log(17) = 6$

Solution

a) $e^x = 125$

$\quad\quad x = \ln(125)$ **Definition of logarithm**

$\quad\quad\quad \approx 4.8283$ **Use the natural logarithm key on a calculator.**

To check, find $e^{4.8283}$. You should get approximately 125.

b) $x \cdot \log(2) + 3x \cdot \log(17) = 6$

$\quad\quad x[\log(2) + 3 \cdot \log(17)] = 6$ **Factor out the common factor.**

$$x = \frac{6}{\log(2) + 3 \cdot \log(17)}$$

$\quad\quad\quad \approx 1.5029$ **Use the common logarithm key on a calculator.**

To check, find $1.5029 \cdot \log(2) + 3 \cdot 1.5029 \cdot \log(17)$. You should get approximately 6. ◆

```
6/(log 2+3log 17
)
        1.502864174
■
```

Be sure to enclose the denominator in parentheses to evaluate this expression.

Applications

We saw in Section 5.1 that if a principal of P dollars earns interest at an annual rate r compounded continuously, then the amount after t years is given by

$$A = Pe^{rt}.$$

We can solve this formula for t by rewriting the exponential equation as a logarithmic equation:

$$\frac{A}{P} = e^{rt}$$

$$rt = \ln\left(\frac{A}{P}\right) \quad\quad \textbf{Rewrite.}$$

$$t = \frac{1}{r}\ln\left(\frac{A}{P}\right) \quad\quad \textbf{Solve for } t.$$

Using this formula, we can find the time it would take for an investment to grow to a specified amount.

Example 9 Finding the time in a continuous compounding problem

If $8000 is invested at 9% compounded continuously, then how long will it take for the investment to grow to $20,000?

Solution

Use the formula $t = \dfrac{1}{r} \ln\left(\dfrac{A}{P}\right)$ with $r = 0.09$, $A = 20{,}000$, and $P = 8000$.

$$t = \frac{1}{0.09} \ln\left(\frac{20{,}000}{8000}\right)$$

$$= \frac{1}{0.09} \ln(2.5)$$

$$\approx 10.181 \text{ years} \qquad \text{Use the ln key on a scientific calculator.}$$

We can multiply 365 by 0.181 to get approximately 66 days. So the investment grows to $20,000 in approximately 10 years and 66 days. ◆

? For Thought

True or false? Explain.

1. The first coordinate of an ordered pair in an exponential function is a logarithm.

2. $\log_{100}(10) = 2$

3. If $f(x) = \log_3(x)$, then $f^{-1}(x) = 3^x$.

4. $10^{\log(1000)} = 1000$

5. The domain of $f(x) = \ln(x)$ is $(-\infty, \infty)$.

6. $\ln(e^{2.451}) = 2.451$

7. For any positive real number x, $e^{\ln(x)} = x$.

8. For any base a, where $a > 0$ and $a \neq 1$, $\log_a(0) = 1$.

9. $\log(10^3) + \log(10^5) = \log(10^8)$

10. $\log_2(32) - \log_2(8) = \log_2(4)$

5.2 Exercises ⬛ Tape 10 💾 Disk—5.25″: 3   3.5″: 2   **Macintosh: 2**

Let $f(x) = 5^x$. Evaluate each expression.

1. $f(1)$	**2.** $f(0)$
5. $f^{-1}(5)$	**6.** $f^{-1}(1)$

3. $f(-1)$ **4.** $f(-2)$

7. $f^{-1}(1/5)$ **8.** $f^{-1}(0.04)$

Find the indicated values of the logarithmic functions.

9. $\log_2(64)$ **10.** $\log_2(16)$ **11.** $\log_3(1/81)$

12. $\log_3(1)$ **13.** $\log_{16}(2)$ **14.** $\log_{16}(16)$

15. $\log_{1/5}(125)$ **16.** $\log_{1/5}(1/125)$ **17.** $\log(0.1)$

18. $\log(10^6)$ **19.** $\log(1)$ **20.** $\log(10)$

21. $\ln(e)$ **22.** $\ln(0)$

23. $\ln(e^{-5})$ **24.** $\ln(e^9)$

Sketch the graph of each function, and state the domain and range of each function.

25. $y = \log_3(x)$ **26.** $y = \log_4(x)$

27. $f(x) = \log_5(x)$ **28.** $g(x) = \log_8(x)$

29. $y = \log_{1/2}(x)$ **30.** $y = \log_{1/4}(x)$

31. $h(x) = \log_{1/5}(x)$ **32.** $s(x) = \log_{1/10}(x)$

33. $f(x) = \ln(x - 1)$ **34.** $f(x) = \log_3(x + 2)$

35. $f(x) = -3 + \log(x + 2)$ **36.** $f(x) = 4 - \log(x + 6)$

37. $f(x) = -\dfrac{1}{2}\log(x - 1)$ **38.** $f(x) = \log_2(-x)$

39. $y = -3 + 2 \cdot \log_2(x - 1)$

40. $y = \log_2(x^2)$

Write each equation involving logarithms as an equivalent exponential equation, and write each equation involving exponents as an equivalent logarithmic equation.

41. $\log(30) = y$ **42.** $\log(t) = 9$ **43.** $5^y = 7$

44. $10^s = 13$ **45.** $\log_2(1) = 0$ **46.** $\log_3(h) = k$

47. $e^{0.09t} = 2$ **48.** $e^{x-1} = y$

49. $\log_b(N/M) = 3$ **50.** $\log_a(NM) = t$

51. $b^{3x} = y$ **52.** $(1/2)^{5-y} = z$

Solve each equation. Find the exact solutions.

53. $\log_3(x) = \dfrac{1}{2}$ **54.** $\log_2(x) = 8$ **55.** $3^x = 77$

56. $\dfrac{1}{2^x} = 5$ **57.** $\ln(x - 3) = \ln(2x - 9)$

58. $\log_2(4x) = \log_2(x + 6)$ **59.** $\log_x(18) = 2$

60. $\log_x(9) = \dfrac{1}{2}$ **61.** $3^{x+1} = 7$ **62.** $5^{3-x} = 12$

63. $\log(x) = \log(6 - x^2)$ **64.** $\log_3(2x) = \log_3(24 - x^2)$

65. $100^{-3/2} = x$ **66.** $\left(\dfrac{1}{8}\right)^{-4/3} = x$

67. $\log_x\left(\dfrac{1}{9}\right) = -\dfrac{2}{3}$ **68.** $\log_x\left(\dfrac{1}{16}\right) = \dfrac{4}{3}$

69. $4^{2x-1} = \dfrac{1}{2}$ **70.** $e^{3x-4} = 1$

Use a scientific calculator to find the approximate solution rounded to four decimal places.

71. $10^x = 25$ **72.** $e^x = 2$

73. $x \cdot \ln(3) = \ln(7)$ **74.** $x \cdot \log(3) = \log(7)$

75. $x \cdot \ln(8) - 5 = \ln(20)$

76. $1.27 - x \cdot \log(23) = \log(54)$

77. $x \cdot \ln(2) - x \cdot \ln(3) = 5$

78. $x \cdot \log(17) + 3x \cdot \log(4) = 7$

79. $(x - 1)\log(5) = (x - 2)\log(9)$

80. $(x + 2)\ln(3) = (2x - 1)\ln(12)$

For each function, find f^{-1}.

81. $f(x) = 2^x$ **82.** $f(x) = 5^x$

83. $f(x) = \log_7(x)$ **84.** $f(x) = \log(x)$

85. $f(x) = \log_{1/8}(x)$

86. $f(x) = \log_{1/9}(x)$

87. $f(x) = \ln(x - 1)$

88. $f(x) = \log(x + 4)$

89. $f(x) = 3^{x+2}$

90. $f(x) = 6^{x-1}$

91. $f(x) = 3 + \log(x)$

92. $f(x) = 5 - \ln(x)$

Solve each problem.

93. *Becoming a millionaire* Find the amount of time, to the nearest day, that it would take for a deposit of $1000 to grow to $1 million at 14% compounded continuously.

94. *Doubling your money* How long does it take for a deposit of $1000 to double at 8% compounded continuously?

95. *Finding the rate* Solve the equation $A = Pe^{rt}$ for r, then find the rate at which a deposit of $1000 would double in 3 years compounded continuously.

96. *Finding the rate* At what interest rate would a deposit of $30,000 grow to $2,540,689 in 40 years with continuous compounding?

97. *Finding the time* Find the amount of time that it will take for an investment to double at 10% compounded continuously.

98. *Rule of 70* A popular rule for finding the time it takes for an investment to double is to simply divide 70 by the interest rate. For example, at 10% an investment will double in approximately 7 years. Compare this answer with the answer to Exercise 97 and explain why this rule works.

99. *Miracle in Boston* To illustrate the "miracle" of compound interest, Ben Franklin bequeathed $4000 to the city of Boston in 1790. The fund grew to $4.5 million in 200 years. Find the annual rate compounded continuously that would cause this "miracle" to happen.

100. *Miracle in Philadelphia* Ben Franklin's gift of $4000 to the city of Philadelphia in 1790 was not managed as well as his gift to Boston. The Philadelphia fund grew to only $2 million in 200 years. Find the annual rate compounded continuously that would yield this total value.

101. *Deforestation in Nigeria* In Nigeria, deforestation occurs at the rate of about 5.25% per year. Assuming that the amount of forest remaining is determined by the function

$$F = F_0 e^{-0.052t},$$

where F_0 is the present acreage of forest land and t is the time in years from the present. In how many years will there be only 60% of the present acreage remaining?

102. *Deforestation in El·Salvador* It is estimated that at the present rate of deforestation in El Salvador, in 20 years only 53% of the present forest will be remaining. Use the exponential model $F = F_0 e^{rt}$ to determine the annual rate of deforestation in El Salvador.

103. *World population* The population of the world doubled from 1950 to 1987, going from 2.5 billion to 5 billion people. Using the exponential model,

$$P = P_0 e^{rt},$$

find the annual growth rate r for that period. Although the annual growth rate has declined slightly to 1.63% annually, the population of the world is still growing at a tremendous rate. Using the initial population of 5 billion in 1987 and an annual rate of 1.63%, estimate the world population in the year 2000.

104. *Black Death* Because of the Black Death, or plague, the only substantial period in recorded history when the earth's population was not increasing was from 1348 to 1400. During that period the world population decreased by about 100 million people. Use the exponential model $P = P_0 e^{rt}$ and the data from the accompanying table to find the annual growth rate r_1 for the period 1400 to 1900 and the annual growth rate r_2 for the period 1900 to 1987. Now, assume that the Black Death had not occurred and the growth rate r_1 was in effect for the years 1348 through 1900. Under these assumptions, predict the no-plague population P_1 in 1900. Start with the no-plague population P_1 in 1900 and use the growth rate r_2 to predict a no-plague population P_2 for the year 2000. How much larger is P_2 than the population predicted for the year 2000 in Exercise 103?

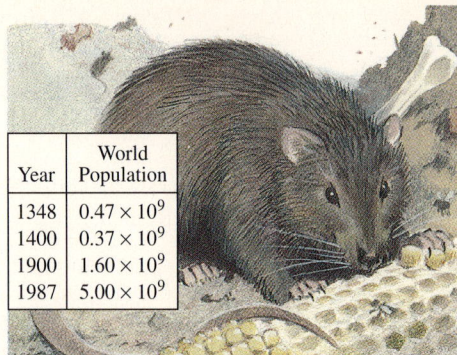

Year	World Population
1348	0.47×10^9
1400	0.37×10^9
1900	1.60×10^9
1987	5.00×10^9

Figure for Exercise 104

Please note the following for Exercises 105–108. In chemistry, the pH of a solution is defined to be

$$\text{pH} = -\log[H^+],$$

where H^+ is the hydrogen ion concentration of the solution in moles per liter. Distilled water has a pH of approximately 7. A substance with a pH under 7 is called an acid and one with a pH over 7 is called a base.

105. *Acidity of tomato juice* Tomato juice has a hydrogen ion concentration of $10^{-4.1}$ moles per liter. Find the pH of tomato juice.

106. *Acidity in your stomach* The gastric juices in your stomach have a hydrogen ion concentration of 10^{-1} moles per liter. Find the pH of your gastric juices.

107. *Acidity of orange juice* The hydrogen ion concentration of orange juice is $10^{-3.7}$ moles per liter. Find the pH of orange juice.

108. *Acidosis* A healthy body maintains the hydrogen ion concentration of human blood at $10^{-7.4}$ moles per liter. What is the pH of normal healthy blood? The condition of low blood pH is called acidosis. Its symptoms are sickly sweet breath, headache, and nausea.

For Writing/Discussion

109. *Increasing or decreasing* Which exponential and logarithmic functions are increasing? Decreasing? Is the inverse of an increasing function increasing or decreasing? Is the inverse of a decreasing function increasing or decreasing? Explain.

110. *Cooperative learning* Work in a small group to write a summary (including drawings) of the types of graphs that can be obtained for logarithmic functions of the form $y = \log_a(x)$ for $a > 0$ and $a \neq 1$.

Graphing Calculator Exercises

Find an approximate solution to each equation by graphing an appropriate function on a graphing calculator and reading from the graph. Round answers to two decimal places.

1. $\ln(x - 2) = 3.2$

2. $\log(x) = -\log(x + 2)$

3. $\log(x + 1) = -\ln(x + 2)$

4. $\log|x| = x^2 - 4$

Use your graphing calculator to help you sketch the graphs of each group of functions on the same coordinate plane.

5. $y_1 = \log(x)$, $y_2 = \log(10x)$, $y_3 = \log(100x)$

6. $y_1 = \log(x)$, $y_2 = \log(x/10)$, $y_3 = \log(x/100)$

7. $y_1 = \log(x)$, $y_2 = \log(x^3)$, $y_3 = \log(x^5)$

8. $y_1 = \log(x)$, $y_2 = \log(x^{1/3})$, $y_3 = \log(x^{1/5})$

Use the graphs drawn in Exercises 5–8 to answer the following questions.

9. How do the graphs of $y = \log(nx)$ and $y = \log(x/n)$ compare with the graph of $y = \log(x)$ for n a positive integral multiple of 10?

10. How do the graphs of $y = \log(x^n)$ and $y = \log(x^{1/n})$ compare with the graph of $y = \log(x)$ for n a positive odd integer?

11. Experiment with the graphs of $y = \log(x^n)$ and $y = \log(x^{1/n})$ for any nonzero integer n and write a summary of your findings.

5.3

Properties of

Logarithms

The properties of logarithms are closely related to the properties of exponents, because logarithms are exponents. The properties of exponents in Chapter 1 were given only for rational exponents. It can also be shown that the same properties hold if the exponents are real numbers. In this section we use the properties of exponents to develop some properties of logarithms. With these properties of logarithms we will be able to solve more equations involving exponents and logarithms.

The Logarithm of a Product

By the product rule for exponents, we add exponents when multiplying exponential expressions having the same base. To find a corresponding rule for logarithms, let's examine the equation $2^3 \cdot 2^2 = 2^5$. Notice that the exponents 3, 2, and 5 are the base-2 logarithms of 8, 4, and 32, respectively.

$$\overset{\displaystyle \log_2(8)\ \log_2(4)\ \ \log_2(32)}{2^3 \cdot 2^2 = 2^5}$$

When we add the exponents 3 and 2 to get 5, we are adding logarithms and getting another logarithm as the result. So the base-2 logarithm of 32 (the product of 8 and 4) is the sum of the base-2 logarithms of 8 and 4:

$$\log_2(8 \cdot 4) = \log_2(8) + \log_2(4)$$

This example suggests the **product rule for logarithms.**

Product Rule for Logarithms

For $M > 0$ and $N > 0$,

$$\log_a(MN) = \log_a(M) + \log_a(N).$$

Proof If $M = a^x$ and $N = a^y$, then

$$MN = a^x a^y = a^{x+y}.$$

By the definition of logarithm, $MN = a^{x+y}$ is equivalent to

$$\log_a(MN) = x + y.$$

Since $M = a^x$, we have $x = \log_a(M)$, and since $N = a^y$, we have $y = \log_a(N)$. So

$$\log_a(MN) = \log_a(M) + \log_a(N). \qquad \blacklozenge$$

```
log 36
       1.556302501
log 9+log 4
       1.556302501
∎
```

You can use a graphing calculator to illustrate the product rule.

The product rule for logarithms says that *the logarithm of a product of two numbers is equal to the sum of their logarithms,* provided that all of the logarithms are defined and all have the same base. There is no rule about the logarithm of a sum, and the logarithm of a sum is generally *not* equal to the sum of the logarithms. For example, $\log_2(8 + 8) \neq \log_2(8) + \log_2(8)$ because $\log_2(8 + 8) = 4$ while $\log_2(8) + \log_2(8) = 6$.

Example 1 Using the product rule for logarithms

Write each expression as a single logarithm.

a) $\log_3(x) + \log_3(6)$ **b)** $\ln(3) + \ln(x^2) + \ln(y)$

Solution

a) By the product rule for logarithms, $\log_3(x) + \log_3(6) = \log_3(6x)$.

b) By the product rule for logarithms, $\ln(3) + \ln(x^2) + \ln(y) = \ln(3x^2y)$. $\blacklozenge$

The Logarithm of a Quotient

By the quotient rule for exponents, we subtract the exponents when dividing exponential expressions having the same base. To find a corresponding rule for logarithms, examine the equation $2^5/2^2 = 2^3$. Notice that the exponents 5, 2, and 3 are the base-2 logarithms of 32, 4, and 8, respectively:

$$\frac{2^5}{2^2} = 2^3$$

$\log_2(32)$ $\log_2(8)$

$\log_2(4)$

When we subtract the exponents 5 and 2 to get 3, we are subtracting logarithms and getting another logarithm as the result. So the base-2 logarithm of 8 (the quotient of 32 and 4) is the difference of the base-2 logarithms of 32 and 4:

$$\log_2\left(\frac{32}{4}\right) = \log_2(32) - \log_2(4)$$

This example suggests the **quotient rule for logarithms.**

Quotient Rule for Logarithms

For $M > 0$ and $N > 0$,

$$\log_a\left(\frac{M}{N}\right) = \log_a(M) - \log_a(N).$$

The quotient rule for logarithms says that *the logarithm of a quotient of two numbers is equal to the difference of their logarithms,* provided that all logarithms are defined and all have the same base. (The proof of the quotient rule is similar to that of the product rule and so it is left as an exercise.) Note that the quotient rule does not apply to division of logarithms. For example,

$$\frac{\log_2(32)}{\log_2(4)} \neq \log_2(32) - \log_2(4),$$

because $\log_2(32)/\log_2(4) = 5/2$, while $\log_2(32) - \log_2(4) = 3$.

```
ln (7/9)
       -.2513144283
ln 7-ln 9
       -.2513144283
```

You can use a graphing calculator to illustrate the quotient rule.

Example 2   Using the quotient rule for logarithms

Write each expression as a single logarithm.

a) $\log_3(24) - \log_3(4)$ **b)** $\ln(x^6) - \ln(x^2)$

Solution

a) By the quotient rule, $\log_3(24) - \log_3(4) = \log_3(24/4) = \log_3(6)$.

b) $\ln(x^6) - \ln(x^2) = \ln\left(\dfrac{x^6}{x^2}\right)$ By the quotient rule for logarithms

$\qquad\qquad\qquad\quad = \ln(x^4)$ By the quotient rule for exponents

The Logarithm of a Power

By the power rule for exponents, we multiply the exponents when finding a power of an exponential expression. For example, $(2^3)^2 = 2^6$. Notice that the exponents 3 and 6 are the base-2 logarithms of 8 and 64, respectively.

$$\overset{\log_2(8)}{\underset{\downarrow}{}} \quad \overset{\log_2(64)}{\underset{\downarrow}{}}$$
$$(2^{3})^2 = 2^{6}$$

So the base-2 logarithm of 64 (the second power of 8) is twice the base-2 logarithm of 8:

$$\log_2(8^2) = 2 \cdot \log_2(8)$$

This example suggests the **power rule for logarithms.**

Power Rule for Logarithms

For $M > 0$ and any real number N,

$$\log_a(M^N) = N \cdot \log_a(M).$$

The power rule for logarithms says that *the logarithm of a power of a number is equal to the power times the logarithm of the number,* provided that all logarithms are defined and have the same base. The proof is left as an exercise.

Example 3   Using the power rule for logarithms

Rewrite each expression in terms of $\log(3)$.

a) $\log(3^8)$ **b)** $\log\left(\sqrt{3}\right)$

c) $\log\left(\dfrac{1}{3}\right)$

Solution

a) $\log(3^8) = 8 \cdot \log(3)$ By the power rule for logarithms

b) $\log\left(\sqrt{3}\right) = \log(3^{1/2}) = \dfrac{1}{2}\log(3)$ By the power rule for logarithms

c) $\log\left(\dfrac{1}{3}\right) = \log(3^{-1}) = -\log(3)$ By the power rule for logarithms

```
log (3^8)
       3.816970038
8log 3
       3.816970038
■
```

You can use a graphing calculator to illustrate the power rule.

The Inverse Properties

The definition of logarithms leads to two properties that are useful in solving equations. If $f(x) = a^x$ and $g(x) = \log_a(x)$, then

$$g(f(x)) = g(a^x) = \log_a(a^x) = x$$

for any real number x. The result of this composition is x because the functions are inverses. If we compose in the opposite order we get

$$f(g(x)) = f(\log_a(x)) = a^{\log_a(x)} = x$$

for any positive real number x. The results are called the **inverse properties.**

Inverse Properties

If $a > 0$ and $a \neq 1$, then

1. $\log_a(a^x) = x$ for any real number x
2. $a^{\log_a(x)} = x$ for $x > 0$.

```
e^ln 17.2
              17.2
ln e^6.1
               6.1
10^log 99
                99
■
```

Exponential functions and logarithmic functions are inverses of each other.

The inverse properties are easy to use if you remember that $\log_a(x)$ is the power of a that produces x. For example, $\log_2(67)$ is the power of 2 that produces 67. So $2^{\log_2(67)} = 67$. Similarly, $\ln(e^{99})$ is the power of e that produces e^{99}. So $\ln(e^{99}) = 99$.

Example 4   Using the inverse properties

Simplify each expression.

a) $e^{\ln(x^2)}$

b) $\log_7(7^{2x-1})$

Solution

By the inverse properties, $e^{\ln(x^2)} = x^2$ and $\log_7(7^{2x-1}) = 2x - 1$. ◆

Using the Rules and Properties

When simplifying or rewriting expressions, we often apply several rules or properties. In the following box we list all of the available rules and properties of logarithms. For simplicity, we refer to all of them as properties.

Properties of Logarithms

If M, N, and a are positive real numbers with $a \neq 1$, and x is any real number, then

1. $\log_a(a) = 1$ 2. $\log_a(1) = 0$
3. $\log_a(a^x) = x$ 4. $a^{\log_a(N)} = N$
5. $\log_a(MN) = \log_a(M) + \log_a(N)$
6. $\log_a(M/N) = \log_a(M) - \log_a(N)$
7. $\log_a(M^x) = x \cdot \log_a(M)$ 8. $\log_a(1/N) = -\log_a(N)$

Note that property 1 is a special case of property 3 with $x = 1$, property 2 follows from the fact that $a^0 = 1$ for any nonzero base a, and property 8 is a special case of property 6 with $M = 1$.

Example 5   Using the properties of logarithms

Rewrite each expression in terms of $\log(2)$ and $\log(3)$.

a) $\log(6)$ **b)** $\log\left(\dfrac{16}{3}\right)$ **c)** $\log\left(\dfrac{1}{3}\right)$

Solution

a) $\log(6) = \log(2 \cdot 3)$ Factor.

$\quad\quad\quad = \log(2) + \log(3)$ Property 5

b) $\log\left(\dfrac{16}{3}\right) = \log(16) - \log(3)$ Property 6

$\quad\quad\quad\quad = \log(2^4) - \log(3)$ Write 16 as a power of 2.

$\quad\quad\quad\quad = 4 \cdot \log(2) - \log(3)$ Property 7

c) $\log\left(\dfrac{1}{3}\right) = -\log(3)$ Property 8

```
log (16/3)
        .7269987279
4log 2-log 3
        .7269987279
```

A graphing calculator can be used to support the answer to Example 5(b).

Example 6   Rewriting a logarithmic expression

Rewrite each expression using a sum or difference of multiples of logarithms.

a) $\ln\left(\dfrac{3x^2}{yz}\right)$ **b)** $\log_3\left(\dfrac{(x-1)^2}{z^{3/2}}\right)$

Solution

a) $\ln\left(\dfrac{3x^2}{yz}\right) = \ln(3x^2) - \ln(yz)$ Quotient rule for logarithms

$\quad\quad\quad\quad = \ln(3) + \ln(x^2) - [\ln(y) + \ln(z)]$ Product rule for logarithms

$\quad\quad\quad\quad = \ln(3) + 2 \cdot \ln(x) - \ln(y) - \ln(z)$ Power rule for logarithms

b) $\log_3\left(\dfrac{(x-1)^2}{z^{3/2}}\right) = \log_3((x-1)^2) - \log_3(z^{3/2})$ Quotient rule for logarithms

$\quad\quad\quad\quad = 2 \cdot \log_3(x-1) - \dfrac{3}{2}\log_3(z)$ Power rule for logarithms

In example 7 we use the rules "in reverse" of the way we did in Example 6.

Example 7   Rewriting as a single logarithm

Rewrite each expression as a single logarithm.

a) $\ln(x-1) + \ln(3) - 3 \cdot \ln(x)$ **b)** $\dfrac{1}{2}\log(y) - \dfrac{1}{3}\log(z)$

Solution

a) Use the product rule, the power rule, and then the quotient rule for logarithms:

$$\ln(x - 1) + \ln(3) - 3 \cdot \ln(x) = \ln(3x - 3) - \ln(x^3)$$

$$= \ln\left(\frac{3x - 3}{x^3}\right)$$

b) $\dfrac{1}{2}\log(y) - \dfrac{1}{3}\log(z) = \log(\sqrt{y}) - \log(\sqrt[3]{z})$

$$= \log\left(\frac{\sqrt{y}}{\sqrt[3]{z}}\right)$$

The Base-Change Formula

The exact solution to an exponential equation is often expressed in terms of logarithms. For example, the exact solution to $(1.03)^x = 5$ is $x = \log_{1.03}(5)$. But how do we calculate $\log_{1.03}(5)$? The next example shows how to find a rational approximation for $\log_{1.03}(5)$ using properties of logarithms. In this example we also introduce a new idea in solving equations, *taking the logarithm of each side*. The base-a logarithms of two equal quantities are equal because logarithm is a function.

Example 8   A rational approximation for a logarithm

Find an approximate rational solution to $(1.03)^x = 5$. Round to four decimal places.

Solution

Take the logarithm of each side using one of the bases available on a calculator.

$$(1.03)^x = 5$$

$\log((1.03)^x) = \log(5)$	**Take the logarithm of each side.**
$x \cdot \log(1.03) = \log(5)$	**Power rule for logarithms**
$x = \dfrac{\log(5)}{\log(1.03)}$	**Divide each side by log(1.03).**
≈ 54.4487	**Divide the logarithms using a calculator.**

Check by evaluating $(1.03)^{54.4487}$ on a calculator.

```
log 5/log 1.03
          54.44868501
1.03^Ans
                    5
```

You can use the answer key to check the solution to Example 8.

The solution to the equation of Example 8 is a base-1.03 logarithm, but we obtained a rational approximation for it using base-10 logarithms. We can use the procedure of Example 8 to write a base-a logarithm in terms of a base-b logarithm for any bases a and b:

$$a^x = M \qquad \text{Equivalent equation: } x = \log_a(M)$$

$$\log_b(a^x) = \log_b(M) \qquad \text{Take the base-}b\text{ logarithm of each side}$$

$$x \cdot \log_b(a) = \log_b(M) \qquad \text{Power rule for logarithms}$$

$$x = \frac{\log_b(M)}{\log_b(a)} \qquad \text{Divide each side by } \log_b(a).$$

Since $x = \log_a(M)$, we have the following **base-change formula.**

Base-Change Formula

If $a > 0$, $b > 0$, $a \neq 1$, $b \neq 1$, and $M > 0$, then

$$\log_a(M) = \frac{\log_b(M)}{\log_b(a)}.$$

```
log 7/log 3
       1.771243749
ln 7/ln 3
       1.771243749
■
```

To find $\log_3(7)$, divide $\log(7)$ by $\log(3)$ or divide $\ln(7)$ by $\ln(3)$.

The base-change formula says that the logarithm of a number in one base is equal to the logarithm of the number in the new base divided by the logarithm of the old base. Using this formula, a logarithm such as $\log_3(7)$ can be easily found with a calculator. Let $a = 3$, $b = 10$, and $M = 7$ in the formula to get

$$\log_3(7) = \frac{\log(7)}{\log(3)} \approx 1.7712.$$

Note that you get the same value for $\log_3(7)$ using natural logarithms on your calculator. Find $\ln(7)/\ln(3)$ and compare.

Example 9   Using the base-change formula with compound interest

If $2500 is invested at 6% compounded daily, then how long (to the nearest day) would it take for the investment to double in value?

Solution

We want the number of years for $2500 to grow to $5000 at 6% compounded daily. Use $P = \$2500$, $A = \$5000$, $m = 365$, and $r = 0.06$ in the formula for compound interest:

$$A = P\left(1 + \frac{r}{n}\right)^{nt}$$

$$5000 = 2500\left(1 + \frac{0.06}{365}\right)^{365t}$$

$$2 = \left(1 + \frac{0.06}{365}\right)^{365t}$$

$$2 \approx (1.000164384)^{365t}$$

$$365t \approx \log_{1.000164384}(2) \qquad \text{Definition of logarithm}$$

$$t \approx \frac{1}{365} \cdot \frac{\ln(2)}{\ln(1.000164384)} \qquad \text{Base-change formula (Use either ln or log.)}$$

$$\approx 11.55337169 \text{ years}$$

The investment of \$2500 will double in about 11 years, 202 days.   ◆

Finding the Rate

If interest on an investment is compounded continuously, the rate appears as an exponent in the formula $A = Pe^{rt}$. We can solve this equation for r using natural logarithms and find the rate at which the money is growing. In the compound interest formula $A = P(1 + r/n)^{nt}$, the rate appears in the base and it can be found without using logarithms.

Example 10   Finding the rate with interest compounded periodically

For what annual percentage rate would \$1000 grow to \$3500 in 20 years compounded monthly?

Solution

Use $A = 3500$, $P = 1000$, $n = 12$, and $t = 20$ in the compound interest formula.

$$A = P\left(1 + \frac{r}{n}\right)^{nt}$$

$$3500 = 1000\left(1 + \frac{r}{12}\right)^{240}$$

$$3.5 = \left(1 + \frac{r}{12}\right)^{240} \qquad \text{Divide each side by 1000.}$$

$$(3.5)^{1/240} = \left(\left(1 + \frac{r}{12}\right)^{240}\right)^{1/240} \qquad \text{Raise each side to the power 1/240.}$$

$$1 + \frac{r}{12} = (3.5)^{1/240} \qquad \text{Write with } r \text{ on the left.}$$

$$\frac{r}{12} = (3.5)^{1/240} - 1$$

$$r = 12((3.5)^{1/240} - 1)$$

$$\approx 0.063$$

If \$1000 earns 6.3% compounded monthly, then it will grow to \$3500 in 20 years.   ◆

For Thought

True or false? Explain.

1. $\dfrac{\log(8)}{\log(3)} = \log(8) - \log(3)$

2. $\ln\left(\sqrt{3}\right) = \dfrac{\ln(3)}{2}$

3. $\dfrac{\log_{19}(8)}{\log_{19}(2)} = \log_3(27)$

4. $\dfrac{\log_2(7)}{\log_2(5)} = \dfrac{\log_3(7)}{\log_3(5)}$

5. $e^{\ln(x)} = x$ for any real number x.

6. The equations $\log(x - 2) = 4$ and $\log(x) - \log(2) = 4$ are equivalent.

7. The equations $x + 1 = 2x + 3$ and $\log(x + 1) = \log(2x + 3)$ are equivalent.

8. $\ln(e^x) = x$ for any real number x.

9. If $30 = x^{50}$, then x is between 1 and 2.

10. If $20 = a^4$, then $\ln(a) = \dfrac{1}{4}\ln(20)$.

5.3 Exercises

Tape 10   Disk—5.25″: 3   3.5″: 2   Macintosh: 2

Rewrite each expression as a single logarithm.

1. $\log(5) + \log(3)$
2. $\ln(6) + \ln(2)$
3. $\log_2(x - 1) + \log_2(x)$
4. $\log_3(x + 2) + \log_3(x - 1)$
5. $\log_4(12) - \log_4(2)$
6. $\log_2(25) - \log_2(5)$
7. $\ln(x^8) - \ln(x^3)$
8. $\log(x^2 - 4) - \log(x - 2)$

Rewrite each expression in terms of $\log(5)$.

9. $\log(25)$
10. $\log(1/5)$
11. $\log(1/125)$
12. $\log(5^x)$
13. $\log\left(\sqrt[3]{5}\right)$
14. $\log\left(\sqrt[4]{125}\right)$

Simplify each expression.

15. $e^{\ln\left(\sqrt{y}\right)}$
16. $10^{\log(3x + 1)}$
17. $\log(10^{y + 1})$
18. $\ln(e^{2k})$
19. $7^{\log_7(999)}$
20. $\log_4(2^{300})$

Rewrite each expression in terms of $\log(2)$ and $\log(5)$.

21. $\log(10)$
22. $\log(0.4)$
23. $\log(2.5)$
24. $\log(250)$
25. $\log\left(\sqrt{20}\right)$
26. $\log(0.0005)$
27. $\log(4/25)$
28. $\log(0.1)$

Rewrite each expression as a sum or difference of multiples of logarithms.

29. $\log_3(5x)$
30. $\log_2(xyz)$
31. $\log_2\left(\dfrac{5}{2y}\right)$
32. $\log\left(\dfrac{4a}{3b}\right)$
33. $\log\left(3\sqrt{x}\right)$
34. $\ln\left(\sqrt{x/4}\right)$
35. $\log(3 \cdot 2^{x-1})$
36. $\ln\left(\dfrac{5^{-x}}{2}\right)$
37. $\ln\left(\dfrac{\sqrt[3]{xy}}{t^{4/3}}\right)$
38. $\log\left(\dfrac{3x^2}{(ab)^{2/3}}\right)$
39. $\ln\left(\dfrac{6\sqrt{x - 1}}{5x^3}\right)$
40. $\log_4\left(\dfrac{3x\sqrt{y}}{\sqrt[3]{x - 1}}\right)$

Rewrite each expression as a single logarithm.

41. $\log_2(5) + 3 \cdot \log_2(x)$
42. $\log(x) + 5 \cdot \log(x)$
43. $\log_7(x^5) - 4 \cdot \log_7(x^2)$
44. $\dfrac{1}{3}\ln(6) - \dfrac{1}{3}\ln(2)$
45. $\log(2) + \log(x) + \log(y) - \log(z)$
46. $\ln(2) + \ln(3) + \ln(5) - \ln(7)$

47. $\frac{1}{2} \log(x) - \log(y) + \log(z) - \frac{1}{3} \log(w)$

48. $\frac{5}{6} \log_2(x) + \frac{2}{3} \log_2(y) - \frac{1}{2} \log_2(x) - \log_2(y)$

49. $3 \cdot \log_4(x^2) - 4 \cdot \log_4(x^{-3}) + 2 \cdot \log_4(x)$

50. $\frac{1}{2} [\log(x) + \log(y)] - \log(z)$

Find an approximate rational solution to each equation. Round answers to four decimal places.

51. $2^x = 9$ **52.** $3^x = 12$ **53.** $(0.56)^x = 8$

54. $(0.23)^x = 18.4$ **55.** $(1.06)^x = 2$ **56.** $(1.09)^x = 3$

57. $(0.73)^x = 0.5$ **58.** $(0.62)^x = 0.25$

Use a calculator and the base-change formula to find each logarithm to four decimal places.

59. $\log_4(9)$ **60.** $\log_3(4.78)$

61. $\log_{9.1}(2.3)$ **62.** $\log_{1.2}(13.7)$

Solve each equation. Round answers to four decimal places.

63. $(1.02)^{4t} = 3$ **64.** $(1.025)^{12t} = 3$

65. $(1.0001)^{365t} = 3.5$ **66.** $(1.00012)^{365t} = 2.4$

67. $(1 + r)^3 = 2.3$ **68.** $\left(1 + \dfrac{r}{4}\right)^{20} = 3$

69. $2\left(1 + \dfrac{r}{12}\right)^{360} = 8.4$ **70.** $5\left(1 + \dfrac{r}{360}\right)^{720} = 12$

Solve each equation. Round answers to four decimal places.

71. $\log_x(33.4) = 5$ **72.** $\log_x(12.33) = 2.3$

73. $\log_x(0.546) = -1.3$ **74.** $\log_x(0.915) = -3.2$

Use the properties of logarithms to determine whether each equation is true or false. Explain your answer.

75. $\log_3(81) = \log_3(9) \cdot \log_3(9)$

76. $\log(81) = \log(9) \cdot \log(9)$ **77.** $\ln(3^2) = (\ln(3))^2$

78. $\ln\left(\dfrac{5}{8}\right) = \dfrac{\ln(5)}{\ln(8)}$ **79.** $\log_2(8^4) = 12$

80. $\log(8.2 \times 10^{-9}) = -9 + \log(8.2)$

81. $\log(1006) = 3 + \log(6)$

82. $\dfrac{\log_2(16)}{\log_2(4)} = \log_2(16) - \log_2(4)$

83. $\dfrac{\log_2(8)}{\log_2(16)} = \log_2(8) - \log_2(16)$

84. $\ln\left(e^{(e^x)}\right) = e^x$ **85.** $\log_2(25) = 2 \cdot \log(5)$

86. $\log_3\left(\dfrac{5}{7}\right) = \log(5) - \log(7)$

87. $\log_2(7) = \dfrac{\log(7)}{\log(2)}$ **88.** $\dfrac{\ln(17)}{\ln(3)} = \dfrac{\log(17)}{\log(3)}$

For Exercises 89–94 find the time or rate required for each investment given in the table to grow to the specified amount. The letter W represents an unknown principal.

	Principal	Ending balance	Rate	Compounded	Time
89.	$800	$2000	8%	Daily	?
90.	$10,000	$1,000,000	7.75%	Annually	?
91.	$W	$3W	10%	Quarterly	?
92.	$W	$2W	12%	Monthly	?
93.	$500	$2000	?	Annually	25 yr
94.	$1000	$2500	?	Monthly	8 yr

95. *Ben's gift to Boston* Ben Franklin's gift of $4000 to Boston grew to $4.5 million in 200 years. At what interest rate compounded annually would this growth occur?

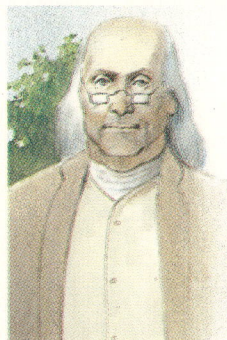

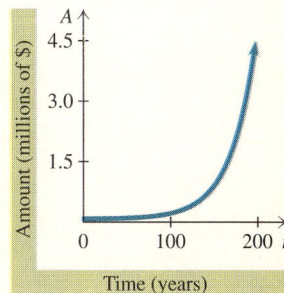

Figure for Exercise 95

96. *Ben's gift to Philadelphia* Ben Franklin's gift of $4000 to Philadelphia grew to $2 million in 200 years. At what interest rate compounded monthly would this growth occur?

97. *Richter scale* The common logarithm is used to measure the intensity of an earthquake on the Richter scale. The Richter scale rating of an earthquake of intensity I is given by $\log(I) - \log(I_0)$, where I_0 is the intensity of a small "benchmark" earthquake. Write the Richter scale rating as a single logarithm. What is the Richter scale rating of an earthquake for which $I = 1000 \cdot I_0$?

98. *Colombian earthquake of 1906* At 8.6 on the Richter scale, the Colombian earthquake of January 31, 1906, was one of the strongest earthquakes on record. Use the formula from Exercise 97 to write I as a multiple of I_0 for this earthquake.

99. *Time for growth* The time in years for a population of size P_0 to grow to size P at the annual growth rate r is given by $t = \ln((P/P_0)^{1/r})$. Use the properties of logarithms to express t in terms of $\ln(P)$ and $\ln(P_0)$.

100. *Formula for pH* The pH of a solution is given by pH = $\log(1/H^+)$, where H^+ is the hydrogen ion concentration of the solution. Use the properties of logarithms to express the pH in terms of $\log(H^+)$.

101. *Marginal revenue* The revenue in dollars from the sale of x items is given by the function $R(x) = 500 \cdot \log(x + 1)$. The marginal revenue function $MR(x)$ is the difference quotient for $R(x)$ when $h = 1$. Find $MR(x)$ and write it as a single logarithm. What happens to the marginal revenue as x gets larger and larger?

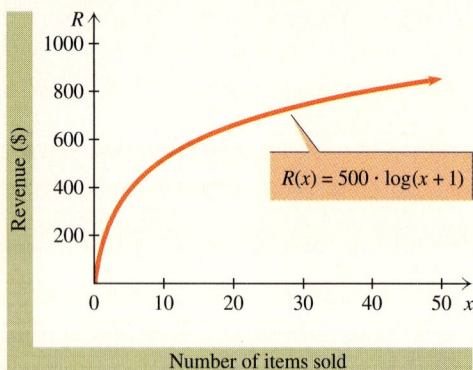

$R(x) = 500 \cdot \log(x + 1)$

Figure for Exercise 101

102. *Human memory model* A class of college algebra students was given a test on college algebra concepts every month for one year after completing a college algebra course. The mean score for the class t months after completing the course can be modeled by the function $m = \ln[e^{80}/(t + 1)^7]$ for $0 \le t \le 12$. Find the mean score of the class for $t = 0$, 5, and 12. Use the properties of logarithms to rewrite the function.

$m = \ln\left(\dfrac{e^{80}}{(t+1)^7}\right)$

Time after completing course (months)

Figure for Exercise 102

For Writing/Discussion

103. *Quotient rule* Write a proof for the quotient rule for logarithms that is similar to the proof given in the text for the product rule for logarithms.

104. *Cooperative learning* Work in a small group to write a proof for the power rule for logarithms.

Graphing Calculator Exercises

Logarithmic functions with bases other than e and 10 can be graphed on your graphing calculator if they are rewritten using the base-change formula. For example, the graph of $y = \log_2(x)$ is the same as the graph of $y = \ln(x)/\ln(2)$ or $y = \log(x)/\log(2)$. Graph each of the following pairs of functions in the same coordinate plane, and explain what each exercise illustrates.

1. $y_1 = \log_3(\sqrt{3}x)$, $y_2 = 0.5 + \log_3(x)$

2. $y_1 = \log_2(1/x)$, $y_2 = -\log_2(x)$

3. $y_1 = 3^{x-1}$, $y_2 = \log_3(x) + 1$

4. $y_1 = 3 + 2^{x-4}$, $y_2 = \log_2(x - 3) + 4$

5.4

More Equations

and Applications

The properties of Section 5.3 combined with the techniques that we have already used in Sections 5.1 and 5.2 allow us to solve several new types of equations involving exponents and logarithms.

Logarithmic Equations

An equation involving a single logarithm can usually be solved by using the definition of logarithm as we did in Section 5.2.

Example 1 An equation involving a single logarithm

Solve the equation $\log(x - 3) = 4$.

Solution

Write the equivalent equation using the definition of logarithm:

$$x - 3 = 10^4$$

$$x = 10{,}003$$

Check this number in the original equation. The solution is 10,003. ◆

When more than one logarithm is present, we can use the one-to-one property as in Section 5.2 or use the other properties of logarithms to combine logarithms.

Example 2 Equations involving more than one logarithm

Solve each equation.

a) $\log_2(x) + \log_2(x + 2) = \log_2(6x + 1)$ **b)** $\log(x) - \log(x - 1) = 2$

Solution

a) Since the sum of two logarithms is equal to the logarithm of a product, we can rewrite the left-hand side of the equation.

$$\log_2(x) + \log_2(x + 2) = \log_2(6x + 1)$$

$$\log_2(x(x + 2)) = \log_2(6x + 1) \qquad \text{Product rule for logarithms}$$

$$x^2 + 2x = 6x + 1 \qquad \text{One-to-one property of logarithms}$$

$$x^2 - 4x - 1 = 0 \qquad \text{Solve quadratic equation.}$$

$$x = \frac{4 \pm \sqrt{16 - 4(-1)}}{2} = 2 \pm \sqrt{5}$$

Since $2 - \sqrt{5}$ is a negative number, $\log_2(2 - \sqrt{5})$ is undefined and $2 - \sqrt{5}$ is not a solution. The only solution to the equation is $2 + \sqrt{5}$. Check this solution by using a calculator and the base-change formula.

```
2+√5
        4.236067977
log Ans/log 2+lo
g (Ans+2)/log 2
        4.723362395
```

```
2+√5
        4.236067977
log (6Ans+1)/log
 2
        4.723362395
■
```

To check $2 + \sqrt{5}$ in the original equation of Example 2(a), use the ANS key and the base-change formula.

b) Since the difference of two logarithms is equal to the logarithm of a quotient, we can rewrite the left-hand side of the equation:

$$\log(x) - \log(x - 1) = 2$$

$$\log\left(\frac{x}{x - 1}\right) = 2 \qquad \text{Quotient rule for logarithms}$$

$$\frac{x}{x - 1} = 10^2 \qquad \text{Definition of logarithm}$$

$$x = 100x - 100 \qquad \text{Solve for } x.$$

$$-99x = -100$$

$$x = \frac{100}{99}$$

Check $100/99$ in the original equation, using a calculator. ◆

All solutions to logarithmic and exponential equations should be checked in the original equation, because extraneous roots can occur, as in Example 2(a). Use your calculator to check every solution and you will increase your proficiency with your calculator.

Exponential Equations

An exponential equation with a single exponential expression can usually be solved by using the definition of logarithm, as in Section 5.2.

Example 3 Equations involving a single exponential expression

Solve the equation $(1.02)^{4t-1} = 5$.

Solution

Write an equivalent equation, using the definition of logarithm.

$$4t - 1 = \log_{1.02}(5)$$

$$t = \frac{1 + \log_{1.02}(5)}{4} \qquad \text{The exact solution}$$

$$= \frac{1 + \dfrac{\ln(5)}{\ln(1.02)}}{4} \qquad \text{Base-change formula}$$

$$\approx 20.5685$$

The approximate solution is 20.5685. Check this answer in the original equation, using a calculator. ◆

(Calculator screen)
```
log (100/99)-log
(100/99-1)
                2
```

Use a calculator to check the solution to Example 2(b).

(Calculator screen)
```
(1+ln 5/ln 1.02)
/4
     20.56848967
```

Use a calculator to check the solution to Example 3.

We could have solved Example 3 by taking the common or natural logarithm of each side and using the power rule for logarithms. If an equation has an exponential expression on each side, as in Example 4, then we must take the common or natural logarithm of each side to solve it.

Example 4   Equations involving two exponential expressions

Find the exact and approximate solutions to $3^{2x-1} = 5^x$.

Solution

$$\ln(3^{2x-1}) = \ln(5^x)$$ **Take the natural logarithm of each side.**

$$(2x - 1)\ln(3) = x \cdot \ln(5)$$ **Power rule for logarithms**

$$2x \cdot \ln(3) - \ln(3) = x \cdot \ln(5)$$ **Distributive property**

$$2x \cdot \ln(3) - x \cdot \ln(5) = \ln(3)$$

$$x[2 \cdot \ln(3) - \ln(5)] = \ln(3)$$

$$x = \frac{\ln(3)}{2 \cdot \ln(3) - \ln(5)}$$ **Exact solution**

$$\approx 1.8691$$

Had we used common logarithms, similar steps would give

$$x = \frac{\log(3)}{2 \cdot \log(3) - \log(5)}$$

$$\approx 1.8691$$

Check the approximate solution 1.8691 in the original equation. ◆

```
3^(2*1.8691-1)
          20.25132389
5^1.8691
          20.25092359
▪
```

The approximate solution does not exactly satisfy the equation in Example 4.

The technique of Example 4 can be used on any equation of the form $a^M = b^N$. Take the natural logarithm of each side and apply the power rule, to get an equation of the form $M \cdot \ln(a) = N \cdot \ln(b)$. This last equation usually has no exponents and is relatively easy to solve.

Strategy for Solving Equations

We solved equations involving exponential and logarithmic functions in Sections 5.1 through 5.4. There is no formula that will solve every exponential or logarithmic equation, but the following box summarizes the procedures we have been using.

▼

Strategy: Solving Exponential and Logarithmic Equations

1. If the equation involves a single logarithm or a single exponential expression, then use the definition of logarithm: $y = \log_a(x)$ if and only if $a^y = x$.

2. Use the one-to-one properties when applicable:
   a) if $a^M = a^N$, then $M = N$.
   b) if $\log_a(M) = \log_a(N)$, then $M = N$.

3. If an equation has several logarithms with the same base, then combine them using the product and quotient rules:
   a) $\log_a(M) + \log_a(N) = \log_a(MN)$
   b) $\log_a(M) - \log_a(N) = \log_a(M/N)$

4. If an equation has exponential expressions with different bases, then take the natural or common logarithm of each side and use the power rule: $a^M = b^N$ is equivalent to $\ln(a^M) = \ln(b^N)$ or $M \cdot \ln(a) = N \cdot \ln(b)$.

Radioactive Dating

In Section 5.1 we stated that the amount A of a radioactive substance remaining after t years is given by

$$A = A_0 e^{rt},$$

where A_0 is the initial amount present and r is the annual rate of decay for that particular substance. A standard measurement of the speed at which a radioactive substance decays is its **half-life.** The half-life of a radioactive substance is the amount of time that it takes for one-half of the substance to decay. Of course, when one-half has decayed, one-half remains.

Now that we have studied logarithms we can use the formula for radioactive decay to determine the age of an ancient object that contains a radioactive substance. One such substance is potassium-40, which is found in rocks. Once the rock is formed, the potassium-40 begins to decay. The amount of time that has passed since the formation of the rock can be determined by measuring the amount of potassium-40 that has decayed into argon-40. Dating rocks using potassium-40 is known as **potassium-argon dating.**

Example 5   Finding the age of *Deinonychus* ("terrible claw")

Our chapter opening case described the 1964 find of *Deinonychus* (*National Geographic,* August 1978). Since dinosaur bones are too old to contain enough organic material for radiocarbon dating, paleontologists often estimate the age of bones by dating volcanic debris in the surrounding rock. The age of *Deinonychus* was determined from the age of surrounding rocks by using potassium-argon dating. The half-life of potassium-40 is 1.31 billion years. If 94.5% of the original amount of potassium-40 is still present in the rock, then how old are the bones of *Deinonychus*?

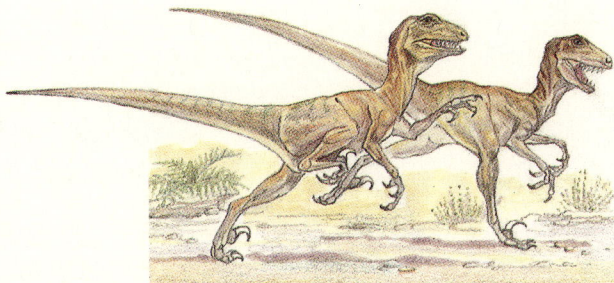

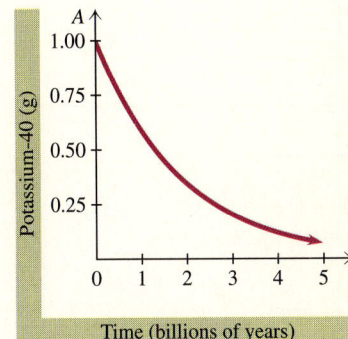

Solution

The half-life is the amount of time that it takes for 1 gram to decay to 0.5 gram. Use $A_0 = 1$, $A = 0.5$, and $t = 1.31 \times 10^9$ in the formula $A = A_0 e^{rt}$ to find r.

$$0.5 = 1 \cdot e^{(1.31 \times 10^9)(r)}$$

$$(1.31 \times 10^9)(r) = \ln(0.5) \qquad \text{Definition of logarithm}$$

$$r = \frac{\ln(0.5)}{1.31 \times 10^9}$$

$$\approx -5.29 \times 10^{-10}$$

Now we can find the amount of time that it takes for 1 gram to decay to 0.945 gram. Use $r \approx -5.29 \times 10^{-10}$, $A_0 = 1$, and $A = 0.945$ in the formula.

$$0.945 = 1 \cdot e^{(-5.29 \times 10^{-10})(t)}$$

$$(-5.29 \times 10^{-10})(t) = \ln(0.945) \qquad \text{Definition of logarithm}$$

$$t = \frac{\ln(0.945)}{-5.29 \times 10^{-10}}$$

$$\approx 107 \text{ million years}$$

The dinosaur *Deinonychus* lived about 107 million years ago. ◆

Newton's Law of Cooling

Newton's law of cooling states that when a warm object is placed in colder surroundings or a cold object is placed in warmer surroundings, then the difference between the two temperatures decreases in an exponential manner. If D_0 is the initial difference in temperature, then the difference D at time t is given by the formula

$$D = D_0 e^{kt},$$

where k is a constant that depends on the object and the surroundings. In the next example we use Newton's law of cooling to answer a question that you might have asked yourself as the appetizers were running low.

Example 6 Using Newton's law of cooling

A turkey with a temperature of 40°F is moved to a 350° oven. After 4 hours the internal temperature of the turkey is 170°F. If the turkey is done when its temperature reaches 185°, then how much longer must it cook?

Solution

The initial difference of 310° has dropped to a difference of 180° after 4 hours. See Fig. 5.16. With this information we can find k:

$$180 = 310e^{4k}$$

$$e^{4k} = \frac{180}{310} \qquad \text{Isolate } k \text{ by dividing by 310.}$$

$$4k = \ln(18/31) \qquad \text{Definition of logarithm}$$

$$k = \frac{\ln(18/31)}{4} \approx -0.1359$$

The turkey is done when the difference in temperature between the turkey and the oven is 165° (the oven temperature 350° minus the turkey temperature 185°). Now find the time for which the difference will be 165°:

$$165 = 310e^{-0.1359t}$$

$$e^{-0.1359t} = \frac{165}{310}$$

$$-0.1359t = \ln(165/310) \qquad \text{Definition of logarithm}$$

$$t = \frac{\ln(165/310)}{-0.1359} \approx 4.6404$$

The difference in temperature between the turkey and the oven will be 165° when the turkey has cooked 4.6404 hours. So the turkey must cook approximately 0.6404 hour (38.4 minutes) longer. ◆

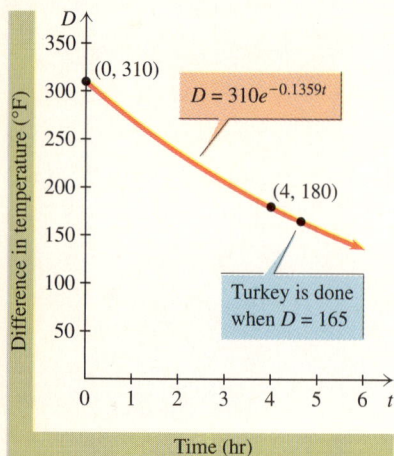

$D = 310e^{-0.1359t}$

(0, 310)

(4, 180)

Turkey is done when $D = 165$

Difference in temperature (°F)

Time (hr)

Figure 5.16

? For Thought

True or false? Explain.

1. The equation $3(1.02)^x = 21$ is equivalent to $x = \log_{1.02}(7)$.

2. If $x - x \cdot \ln(3) = 8$, then $x = \dfrac{8}{1 - \ln(3)}$.

3. The solution to $\ln(x) - \ln(x - 1) = 6$ is $1 - \sqrt{6}$.

4. If $2^{x-3} = 3^{2x+1}$, then $x - 3 = \log_2(3^{2x+1})$.

5. The exact solution to $3^x = 17$ is 2.5789.

6. The equation $\log(x) + \log(x - 3) = 1$ is equivalent to $\log(x^2 - 3x) = 1$.

7. The equation $4^x = 2^{x-1}$ is equivalent to $2x = x - 1$.

8. The equation $(1.09)^x = 2.3$ is equivalent to $x \cdot \ln(1.09) = \ln(2.3)$.

9. $\ln(2) \cdot \log(7) = \log(2) \cdot \ln(7)$

10. $\log(e) \cdot \ln(10) = 1$

5.4 Exercises

▭ **Tape 10**   ▭ **Disk—5.25″: 3   3.5″: 2   Macintosh: 2**

Solve each equation.

1. $\log(x + 20) = 2$

2. $\log(x^2 - 15) = 1$

3. $2 = \log_x(9)$

4. $-2 = \log_x(4)$

5. $\log_2(x + 2) + \log_2(x - 2) = 5$

6. $\log(x + 1) - \log(x) = 3$

7. $\log(5) = 2 - \log(x)$

8. $\log(4) = 1 + \log(x - 1)$

9. $\ln(x) + \ln(x + 2) = \ln(8)$

10. $\log_3(x) = \log_3(2) - \log_3(x - 2)$

11. $\log(4) + \log(x) = \log(5) - \log(x)$

12. $\ln(x) - \ln(x + 1) = \ln(x + 3) - \ln(x + 5)$

13. $\log_2(x) - \log_2(3x - 1) = 0$

14. $\log_3(x) + \log_3(1/x) = 0$

15. $x \cdot \ln(3) = 2 - x$

16. $x \cdot \log(5) + x \cdot \log(7) = \log(9)$

Solve each equation. If the exact solution is an irrational number, then find an approximate solution rounded to four decimal places.

17. $2^{x-1} = 7$

18. $5^{3x} = 29$

19. $(1.09)^{4x} = 3.4$

20. $(1.04)^{2x} = 2.5$

21. $3^{-x} = 30$

22. $10^{-x+3} = 102$

23. $9 = e^{-3x^2}$

24. $25 = 10^{-2x}$

25. $6^x = 3^{x+1}$

26. $2^x = 3^{x-1}$

27. $e^{x+1} = 10^x$

28. $e^x = 2^{x+1}$

29. $2^{x-1} = 4^{3x}$

30. $3^{3x-4} = 9^x$

31. $6^{x+1} = 12^x$

32. $2^x \cdot 2^{x+1} = 4^{x^2+x}$

33. $e^{-\ln(w)} = 3$

34. $10^{2 \cdot \log(y)} = 4$

35. $(\log(z))^2 = \log(z^2)$

36. $\ln(e^x) - \ln(e^6) = \ln(e^2)$

37. $4(1.02)^x = 3(1.03)^x$

38. $500(1.06)^x = 400(1.02)^{4x}$

39. $e^{3 \cdot \ln(x^2) - 2 \cdot \ln(x)} = \ln(e^{16})$

40. $\sqrt{\log(x) - 3} = \log(x) - 3$

41. $\left(\dfrac{1}{2}\right)^{2x-1} = \left(\dfrac{1}{4}\right)^{3x+2}$

42. $\left(\dfrac{2}{3}\right)^{x+1} = \left(\dfrac{9}{4}\right)^{x+2}$

Solve each problem.

43. *Dating a bone* A piece of bone is found to contain 10% of the carbon-14 that it contained when it was living. If the half-life of carbon-14 is 5730 years, then how long ago was the organism alive?

44. *Old clothes* If only 15% of the carbon-14 in a remnant of cloth has decayed, then how old is the cloth?

45. *Dating a tree* How long does it take for 12 g of carbon-14 in a tree trunk to be reduced to 10 g of carbon-14 by radioactive decay?

46. *Carbon-14 dating* How long does it take for 2.4 g of carbon-14 to be reduced to 1.3 g of carbon-14 by radioactive decay?

47. *Radioactive waste* If 25 g of radioactive waste reduces to 20 g of radioactive waste after 8000 years, then what is the half-life for this radioactive element?

48. *Finding the half-life* If 80% of a radioactive element remains radioactive after 250 million years, then what percent remains radioactive after 600 million years? What is the half-life of this element?

49. *Uranium-238 dating* Uranium-238 has a half-life of 4560 million years. When uranium-238 decays, it eventually turns into lead 206. If 60% of the original uranium-238 in a rock is still present in the rock, then how old is the rock?

50. *Dating a rock* If 23% of the uranium-238 that was present in a rock has decayed, then how old is the rock?

51. *Dead Sea Scrolls* Willard Libby, a nuclear chemist from the University of Chicago, developed radiocarbon dating in the 1940s. This dating method, effective on specimens up to about 40,000 years old, works best on objects like shells, charred bones, and plants that contain organic matter (carbon). Libby's first great success came in 1951 when he dated the Dead Sea Scrolls. Carbon-14 has a half-life of 5730 years. If Libby found 79.3% of the original carbon-14 still present, then in about what year were the scrolls made?

52. *Leakeys date Zinjanthropus* In 1959, archaeologists Louis and Mary Leakey were exploring Olduvai Gorge, Tanzania, when they uncovered the remains of *Zinjanthropus,* an early hominid with traits of both ape and man. Dating of the volcanic rock revealed that 91.2% of the original potassium had not decayed into argon. What age was assigned to the rock and the bones of *Zinjanthropus*? The radioactive decay of potassium-40 to argon-40 occurs with a half-life of 1.31 billion years.

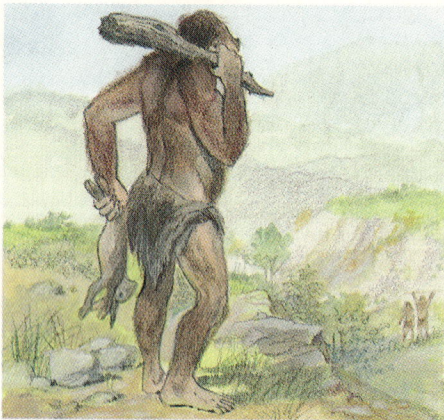

Figure for Exercise 52

53. *Cooking a roast* James knows that to get well-done beef, it should be brought to a temperature of 170°F. He placed a sirloin tip roast with a temperature of 35°F in an oven with a temperature of 325°, and after 3 hr the temperature of the roast was 140°. How much longer must the roast be in the oven to get it well done? If the oven temperature is set at 170°, how long will it take to get the roast well done?

54. *Room temperature* Marlene brought a can of polyurethane varnish that was stored at 40°F into her shop, where the temperature was 74°. After 2 hr the temperature of the varnish was 58°. If the varnish must be 68° for best results, then how much longer must Marlene wait until she uses the varnish?

55. *Time of death* A detective discovered a body in a vacant lot at 7 A.M. and found that the body temperature was 80°F. The county coroner examined the body at 8 A.M. and found that the body temperature was 72°. Assuming that the body temperature was 98° when the person died and that the air temperature was a constant 40° all night, what was the approximate time of death?

56. *Cooling hot steel* A blacksmith immersed a piece of steel at 600°F into a large bucket of water with a temperature of 65°. After 1 min the temperature of the steel was 200°. How much longer must the steel remain immersed to bring its temperature down to 100°?

57. *Poverty level* In a certain country the number of people below the poverty level, B, is currently 20 million and growing according to the formula $B = 20e^{0.07t}$, where B is in millions of people and t is the time in years from the present. The number of people above the poverty level, A, is 30 million and decreasing according to the formula $A = 30e^{-0.05t}$, where A is in millions of people and t is the time in years from the present. In how many years will the numbers of people above and below the poverty level be equal?

58. *Equality of investments* Tasha invested $400 at 5% compounded annually. At the same time, Eunice invested $500 at 4% compounded annually. In how many years will their investments be equal in value?

59. *Equality of investments* Fiona invested $1000 at 6% compounded continuously. At the same time, Maria invested $1100 at 6% compounded daily. How long will it take (to the nearest day) for their investments to be equal in value?

60. *Depreciation and inflation* Boris won a $35,000 luxury car on Wheel of Fortune. He plans to keep it until he can trade it evenly for a new compact car that currently costs $10,000. If the value of the luxury car decreases by 8% each year and the cost of the compact car increases by 5% each year, then in how many years will he be able to make the trade?

61. *Population of rabbits* The population of rabbits in a certain national forest appears to be growing according to the formula $P = 12,300 + 1000 \cdot \ln(t + 1)$, where t is the time in years from the present. How many rabbits are now in the forest? In how many years will there be 15,000 rabbits?

62. *Population of foxes* The population of foxes in the forest of Exercise 61 appears to be growing according to the formula $P = 400 + 50 \cdot \ln(90t + 1)$. When the population of foxes is equal to 5% of the population of rabbits, the system is considered to be out of ecological balance. In how many years will the system be out of balance?

63. *Visual magnitude of a star* If all stars were at the same distance, it would be a simple matter to compare their brightness. However, the brightness that we see, the apparent visual magnitude m, depends on a star's intrinsic brightness, or absolute visual magnitude M_V, and the distance d from the observer in parsecs (1 parsec $= 3.262$ light years), according to the formula $m = M_V - 5 + 5 \cdot \log(d)$. The values of M_V range from -8 for the intrinsically brightest stars to $+15$ for the intrinsically faintest stars. The nearest star to the sun, Alpha Centauri, has an apparent visual magnitude of 0 and an absolute visual magnitude of 4.39. Find the distance d in parsecs to Alpha Centauri.

64. *Visual magnitude of Deneb* The star Deneb is 490 parsecs away and has an apparent visual magnitude of 1.26, which means that it is harder to see than Alpha Centauri. Use the formula from Exercise 63 to find the absolute visual magnitude of Deneb. Is Deneb intrinsically very bright or very faint? If Deneb and Alpha Centauri were both 490 parsecs away, then which would appear brighter?

65. *Noise pollution* The level of a sound in decibels (db) is determined by the formula

$$\text{sound level} = 10 \cdot \log(I \times 10^{12}) \text{ db,}$$

where I is the intensity of the sound in watts per square meter. To combat noise pollution, a city has an ordinance prohibiting sounds above 90 db on a city street. What value of I gives a sound of 90 db?

66. *Doubling the sound level* A small stereo amplifier produces a sound of 50 db at a distance of 20 ft from the speakers. Use the formula from Exercise 65 to find the intensity of the sound at this point in the room. If the intensity just found is doubled, what happens to the sound level? What must the intensity be to double the sound level to 100 db?

67. *Present value of a CD* What amount (present value) must be deposited today in a certificate of deposit so that the investment will grow to $20,000 in 18 years at 6% compounded continuously?

68. *Present value of a bond* A $50 U.S. Savings Bond paying 6.22% compounded monthly matures in 11 years 2 months. What is the present value of the bond?

69. *Infinite series for e^x* The following formula from calculus is used to compute values of e^x:

$$e^x = 1 + x + \frac{x^2}{2!} + \frac{x^3}{3!} + \frac{x^4}{4!} + \cdots + \frac{x^n}{n!} + \cdots,$$

where $n! = 1 \cdot 2 \cdot 3 \cdot \cdots \cdot n$ for any positive integer n. The notation $n!$ is read "n factorial." For example, $3! = 1 \cdot 2 \cdot 3 = 6$. In calculating e^x, the more terms that we use from the formula, the closer we get to the true value of e^x. Use the first five terms of the formula to estimate the value of $e^{0.1}$ and compare your result to the value of $e^{0.1}$ obtained using the e^x-key on your calculator.

70. *Hyperbolic functions* The expression

$$\left(\frac{e^x + e^{-x}}{2}\right)^2 - \left(\frac{e^x - e^{-x}}{2}\right)^2$$

occurs in the study of hyperbolic functions in calculus. Simplify it.

71. *Logistic growth model* One student carrying a contagious virus enrolls at a school of 10,000 students. The virus spreads slowly at first, then more rapidly as more students are infected, and finally slows down as nearly all students become infected. To model this type of growth we can use the *logistic growth model* $n = 10,000/(1 + 9999e^{-0.3t})$, where t is the number of days that the virus is on campus and n is the number of infected students. See the accompanying graph of this function. How many students are infected after 20 days? In how many days will 95% of the students be infected?

$$n = \frac{10,000}{1 + 9999e^{-0.3t}}$$

Figure for Exercise 71

72. *Logistic growth model* If 10 students with the same virus enroll at the school described in the previous exercise, then the infection of the student body can be modeled by the logistic growth model $n = 10,000/(1 + 999e^{-0.3t})$, where t is the number of days that the virus is on campus and n is the number of infected students. In how many days will 95% of the students be infected? Find the number infected after 125 days. Use the formula to find t for $n = 10,000$.

For Writing/Discussion

73. *Cross examination* What assumptions are made in determining the time of death in Exercise 55? If a lawyer wants to refute the coroner's estimate of the time of death, what might the lawyer look for that the coroner may have overlooked?

74. *Cooperative learning* Write a brief report about the work of Willard Libby or Louis and Mary Leakey and present it to your class.

Graphing Calculator Exercises

Find the approximate solution to each equation by graphing an appropriate function on a graphing calculator and locating the *x*-intercept. Note that these equations cannot be solved by the techniques that we have learned in this chapter.

1. $2^x = 3^{x-1} + 5^{-x}$ **2.** $2^x = \log(x + 4)$

3. $\ln(x + 51) = \log(-48 - x)$

4. $2^x = 5 - 3^{x+1}$

Highlights

Section 5.1   Exponential Functions

1. If $a > 0$ and $a \neq 1$, then $f(x) = a^x$ is an exponential function. Its domain is all real numbers.

2. Exponential functions are one-to-one: if $a^{x_1} = a^{x_2}$ then $x_1 = x_2$.

3. If $a > 1$, then $f(x) = a^x$ is increasing, and if $0 < a < 1$, then $f(x) = a^x$ is decreasing.

4. The *x*-axis is a horizontal asymptote for the graph of $f(x) = a^x$.

5. *P* dollars invested at annual rate *r* compounded *n* times per year amounts to

$$A = P\left(1 + \frac{r}{n}\right)^{nt} \text{ in } t \text{ years.}$$

6. *P* dollars invested at an annual rate *r* compounded continuously amounts to $A = Pe^{rt}$ in *t* years, where $e \approx 2.718$.

Section 5.2   Logarithmic Functions

1. The inverse of $f(x) = a^x$ is $f^{-1}(x) = \log_a(x)$, where $y = \log_a(x)$ if and only if $a^y = x$.

2. The common logarithm function is $y = \log(x)$, and the natural logarithm function is $y = \ln(x)$.

3. The domain of $f(x) = \log_a(x)$ is $(0, \infty)$, and the *y*-axis is a vertical asymptote for its graph.

4. If $a > 1$, $f(x) = \log_a(x)$ is an increasing function, and if $0 < a < 1$, it is a decreasing function.

5. Logarithmic functions are one-to-one: if $\log_a(x_1) = \log_a(x_2)$, then $x_1 = x_2$.

Section 5.3   Properties of Logarithms

1. The logarithm of a product is the sum of two logarithms:
$\log_a(MN) = \log_a(M) + \log_a(N)$.

2. The logarithm of a quotient is the difference of two logarithms:
$\log_a(M/N) = \log_a(M) - \log_a(N)$.

3. The logarithm of a power of a number is the power times the logarithm of the number: $\log_a(M^N) = N \cdot \log_a(M)$.

4. Because exponential and logarithmic functions are inverses, $\log_a(a^x) = x$ and $a^{\log_a(x)} = x$.

5. The base-change formula $\log_a(M) = \dfrac{\log_b(M)}{\log_b(a)}$ is used to evaluate logarithms with a calculator.

Section 5.4   More Equations and Applications

1. The definition of logarithm, one-to-one properties, product rule, quotient rule, and power rule are all used in solving equations involving exponential functions and logarithmic functions.

2. Exponential and logarithmic functions have a wide variety of applications, including population growth, compound interest, radioactive dating, and Newton's law of cooling.

Chapter 5 Review Exercises

Simplify each expression.

1. 2^6 **2.** $\ln(e^2)$ **3.** $\log_2(64)$

4. $3 + 2 \cdot \log(10)$ **5.** $\log_9(1)$ **6.** $5^{\log_5(99)}$

7. $\log_2(2^{17})$ **8.** $\log_2(\log_2(16))$

Let $f(x) = 2^x$, $g(x) = 10^x$, and $h(x) = \log_2(x)$. Simplify each expression.

9. $f(5)$ **10.** $g(-1)$ **11.** $\log(g(3))$ **12.** $g(\log(5))$

13. $(h \circ f)(9)$ **14.** $(f \circ h)(7)$ **15.** $g^{-1}(1000)$

16. $g^{-1}(1)$ **17.** $h(1/8)$ **18.** $f(1/2)$

19. $f^{-1}(8)$ **20.** $(f^{-1} \circ f)(13)$

Rewrite each expression as a single logarithm.

21. $\log(x - 3) + \log(x)$ **22.** $(1/2)\ln(x) - 2 \cdot \ln(y)$

23. $2 \cdot \ln(x) + \ln(y) + \ln(3)$

24. $3 \cdot \log_2(x) - 2 \cdot \log_2(y) + \log_2(z)$

Rewrite each expression as a sum or difference of multiples of logarithms.

25. $\log(3x^4)$ **26.** $\ln\left(\dfrac{x^5}{y^3}\right)$

27. $\log_3\left(\dfrac{5\sqrt{x}}{y^4}\right)$ **28.** $\log_2\left(\sqrt{xy^3}\right)$

Rewrite each expression in terms of $\ln(2)$ and $\ln(5)$.

29. $\ln(10)$ **30.** $\ln(0.4)$ **31.** $\ln(50)$ **32.** $\ln\left(\sqrt{20}\right)$

Find the exact solution to each equation.

33. $\log(x) = 10$ **34.** $\log_3(x + 1) = -1$

35. $\log_x(81) = 4$ **36.** $\log_x(1) = 0$

37. $\log_{1/3}(27) = x + 2$ **38.** $\log_{1/2}(4) = x - 1$

39. $3^{x+2} = \dfrac{1}{9}$ **40.** $2^{x-1} = \dfrac{1}{4}$

41. $e^{x-2} = 9$ **42.** $\dfrac{1}{2^{1-x}} = 3$

43. $4^{x+3} = \dfrac{1}{2^x}$ **44.** $3^{2x-1} \cdot 9^x = 1$

45. $\log(x) + \log(2x) = 5$

46. $\log(x + 90) - \log(x) = 1$

47. $\log_2(x) + \log_2(x - 4) = \log_2(x + 24)$

48. $\log_5(x + 18) + \log_5(x - 6) = 2 \cdot \log_5(x)$

49. $2 \cdot \ln(x + 2) = 3 \cdot \ln(4)$

50. $x \cdot \log_2(12) = x \cdot \log_2(3) + 1$

51. $x \cdot \log(4) = 6 - x \cdot \log(25)$

52. $\log(\log(x)) = 1$

Find the missing coordinate so that each ordered pair satisfies the given equation.

53. $y = \left(\dfrac{1}{3}\right)^x$: $(-1, \ \ )$, $(\ \ , 27)$, $(-1/2, \ \ )$, $(\ \ , 1)$

54. $y = \log_9(x - 1)$: $(2, \ \ )$, $(4, \ \ )$, $(\ \ , 3/2)$, $(\ \ , -1)$

Match each equation to one of the graphs (a)–(h).

55. $y = 2^x$

56. $y = 2^{-x}$

57. $y = \log_2(x)$

58. $y = \log_{1/2}(x)$

59. $y = 2^{x+2}$

60. $y = \log_2(x + 2)$

61. $y = 2^x + 2$

62. $y = 2 + \log_2(x)$

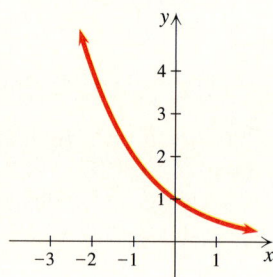

(a)   (b)   (c)   (d)

(e)   (f)   (g)   (h)

Sketch the graph of each function. State the domain, the range, and whether the function is increasing or decreasing.

63. $f(x) = 5^x$

64. $f(x) = e^x$

65. $f(x) = 10^{-x}$

66. $f(x) = (1/2)^x$

67. $y = \log_3(x)$

68. $y = \log_5(x)$

69. $y = 1 + \ln(x + 3)$

70. $y = 3 - \log_2(x)$

71. $f(x) = 1 + 2^{x-1}$

72. $f(x) = 3 - 2^{x+1}$

73. $y = \log_3(-x + 2)$

74. $y = 1 + \log_3(x + 2)$

For each function f, find f^{-1}.

75. $f(x) = 7^x$

76. $f(x) = \log_8(x)$

77. $f(x) = \log_5(x)$

78. $f(x) = 3^x$

Use a calculator to find an approximate solution to each equation. Round answers to four decimal places.

79. $3^x = 10$

80. $\log_3(x) = 1.876$

81. $5^x = 8^{x+1}$

82. $3^x = e^{x+1}$

Solve each problem.

83. *Comparing pH* If the hydrogen ion concentration of liquid A is 10 times the hydrogen ion concentration of liquid B, then how does the pH of A compare with the pH of B?

84. *Solving a formula* Solve the formula $A = P + Ce^{-kt}$ for t.

85. *Future value* If $50,000 is deposited in a bank account paying 5% compounded quarterly, then what will be the value of the account at the end of 18 years?

86. *Future value*   If $30,000 is deposited in First American Savings and Loan in an account paying 6.18% compounded continuously, then what will be the value of the account after 12 years and 3 months?

87. *Doubling time with quarterly compounding*   How long (to the nearest quarter) will it take for the investment of Exercise 85 to double?

88. *Doubling time with continuous compounding*   How long (to the nearest day) will it take for the investment of Exercise 86 to double?

89. *Finding the half-life*   The number of grams A of a certain radioactive substance present at time t is given by the formula $A = 25e^{-0.00032t}$, where t is the number of years from the present. How many grams are present initially? How many grams are present after 1000 years? What is the half-life of this substance?

90. *Comparing investments*   Shinichi invested $800 at 6% compounded continuously. At the same time Toshio invested $1000 at 5% compounded monthly. How long (to the nearest month) will it take for their investments to be equal in value?

91. *Learning curve*   According to an educational psychologist, the number of words learned by a student of a foreign language after t hours in the language laboratory is given by $f(t) = 40{,}000(1 - e^{-0.0001t})$. How many hours would it take to learn 10,000 words?

Chapter 5 Test

Simplify each expression.

1. $\log(1000)$ **2.** $\log_7(1/49)$ **3.** $3^{\log_3(6.47)}$ **4.** $\ln\!\left(e^{\sqrt{2}}\right)$

Find the inverse of each function.

5. $f(x) = \ln(x)$ **6.** $f(x) = 8^x$

Write each expression as a single logarithm.

7. $\log(x) + 3 \cdot \log(y)$ **8.** $\dfrac{1}{2} \cdot \ln(x - 1) - \ln(33)$

Write each expression in terms of $\log_a(2)$ and $\log_a(7)$.

9. $\log_a(28)$ **10.** $\log_a(3.5)$

Find the exact solution to each equation.

11. $\log_2(x) + \log_2(x - 2) = 3$

12. $\log(10x) - \log(x + 2) = 2 \cdot \log(3)$

Find the exact solution and an approximate solution (to four decimal places) for each equation.

13. $3^x = 5^{x-1}$ **14.** $\log_3(x - 1) = 5.46$

Graph each equation in the xy-plane. State the domain and range of the function and whether the function is increasing or decreasing. Identify any asymptotes.

15. $f(x) = 2^x + 1$ **16.** $y = \log_{1/2}(x - 1)$

Solve each problem.

17. What is the only ordered pair that satisfies both $y = \log_2(x)$ and $y = \ln(x)$?

18. A student invested $2000 at 8% annual percentage rate for 20 years. What is the amount of the investment if the interest is compounded quarterly? Compounded continuously?

19. A satellite has a radioisotope power supply. The power supply in watts is given by the formula $P = 50e^{-t/250}$, where t is the time in days. How much power is available at the end of 200 days? What is the half-life of the power supply? If the equipment aboard the satellite requires 9 watts of power to operate properly, then what is the operational life of the satellite?

20. If $4000 is invested at 6% compounded quarterly, then how long will it take for the investment to grow to $10,000?

21. An educational psychologist uses the model $t = -50 \cdot \ln(1 - p)$ for $0 \le p < 1$ to predict the number of hours t that it will take for a child to reach level p in the new video game Mario Goes to Mars. Level $p = 0$ means that the child knows nothing about MGM. What is the predicted level of a child for 100 hr of playing MGM? If $p = 1$ corresponds to mastery of MGM, then is it possible to master MGM?

Tying It All Together
Chapters 1–5

Solve each equation.

1. $(x - 3)^2 = 4$

2. $2 \cdot \log(x - 3) = \log(4)$

3. $\log_2(x - 3) = 4$

4. $2^{x-3} = 4$

5. $\sqrt{x - 3} = 4$

6. $|x - 3| = 4$

7. $x^2 - 4x = -2$

8. $2^{x-3} = 4^x$

9. $\sqrt{x} + \sqrt{x - 5} = 5$

10. $2^x = 3$

11. $\log(x - 3) + \log(4) = \log(x)$

12. $x^3 - 4x^2 + x + 6 = 0$

Sketch the graph of each function.

13. $y = x^2$

14. $y = (x - 2)^2$

15. $y = 2^x$

16. $y = x^{-2}$

17. $y = \log_2(x - 2)$

18. $y = x - 2$

19. $y = 2x$

20. $y = \log(2^x)$

21. $y = e^2$

22. $y = 2 - x^2$

23. $y = \dfrac{2}{x}$

24. $y = \dfrac{1}{x - 2}$

Find the inverse of each function.

25. $f(x) = \dfrac{1}{3}x$

26. $f(x) = \dfrac{1}{3^x}$

27. $f(x) = \sqrt{x - 2}$

28. $f(x) = 2 + (x - 5)^3$

29. $f(x) = \log(\sqrt{x} - 3)$

30. $f = \{(3, 1), (5, 4)\}$

31. $f(x) = 3 + \dfrac{1}{x - 5}$

32. $f(x) = 3 - e^{\sqrt{x}}$

Some Basic Functions of Algebra

$y = -x$

$y = x + 3$

$y = x$

Linear

$y = x^2$

$y = -x^2$

$y = (x - 3)^2$

Quadratic

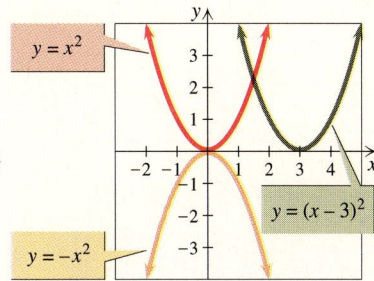

$y = -x^3$

$y = x^3$

$y = (x - 4)^3$

Cubic

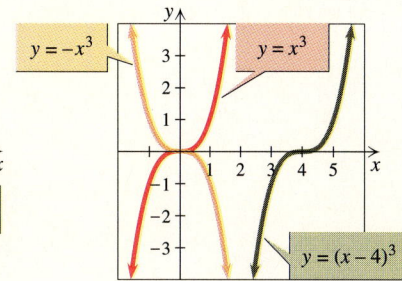

$y = |x|$

$y = |x - 2|$

$y = |x| - 3$

Absolute value

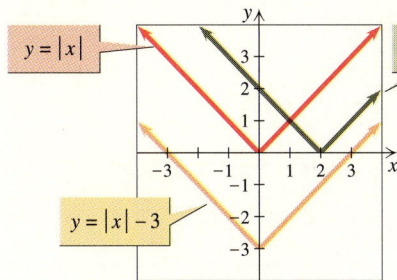

$y = 2^{-x}$

$y = 2^x$

$y = -2^x$

Exponential

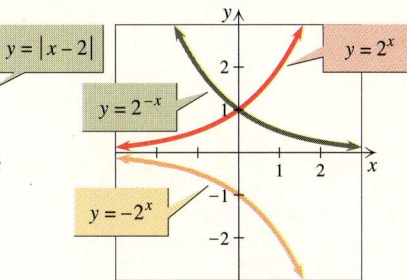

$y = \log_2(x + 4)$

$y = \log_2(x) + 3$

$y = \log_2(x)$

Logarithmic

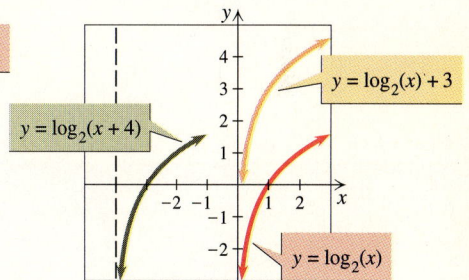

$y = \sqrt{x + 4}$

$y = \sqrt{x}$

$y = \sqrt{x + 3} - 4$

Square root

$y = \dfrac{1}{x}$

$y = -\dfrac{1}{x}$

Reciprocal

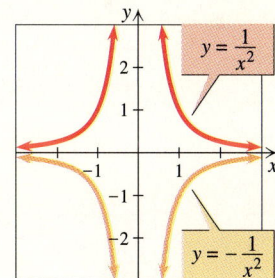

$y = \dfrac{1}{x^2}$

$y = -\dfrac{1}{x^2}$

Rational

$y = x^4$

$y = -x^4 - 1$

$y = (x - 4)^4$

Fourth degree

$y = \sqrt{4 - x^2}$

$y = -\sqrt{4 - x^2}$

Semicircle

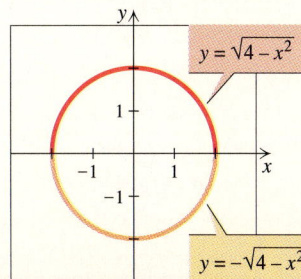

$y = [\![x]\!]$

$y = [\![x - 3]\!]$

Greatest integer

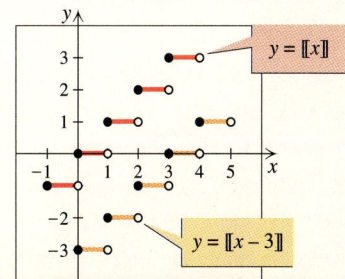

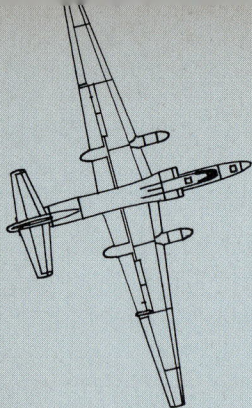

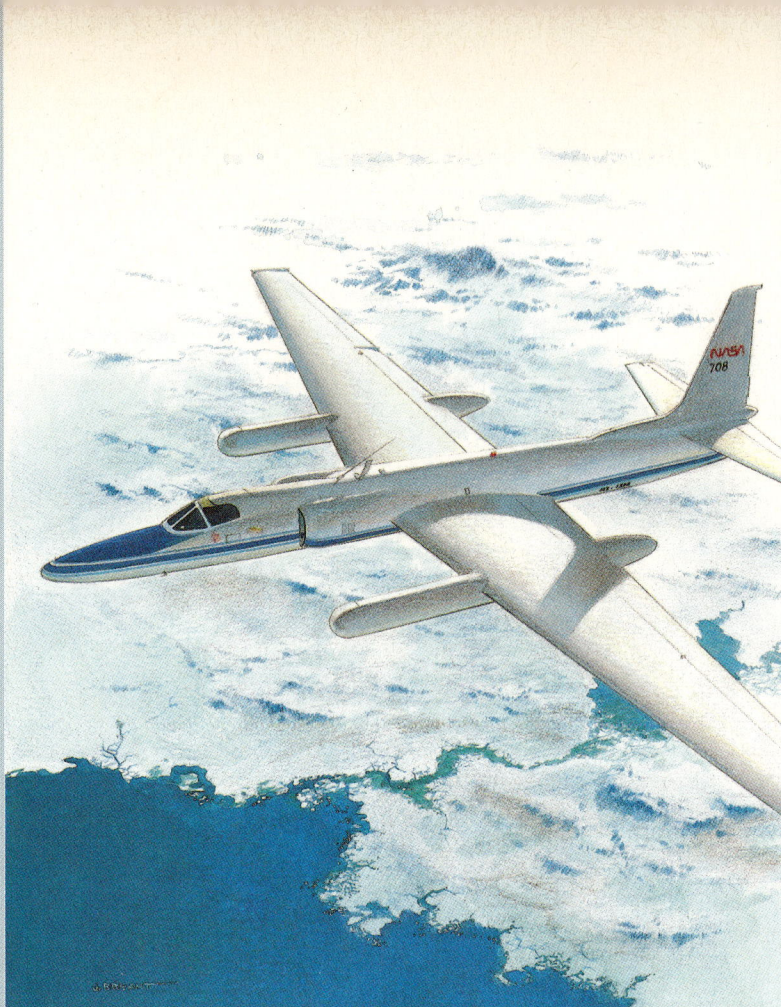

In the thin, icy air above Antarctica, a modified U-2 spy plane, now called ER2 (for Earth Resources), climbs to an altitude of 68,000 feet. Its mission is very different from the reconnaissance flights it once flew for the CIA decades ago, but just as dangerous. The year is 1988, and the same plane that gathered information on the Soviet Union and Cuba during the Cold War has been pressed into service gathering data on the ozone threat. In 1960, when the Soviets shot down and captured U-2 pilot Francis Gary Powers, the incident had international repercussions. Now, once again, the world community has much at stake in the U-2 missions.

Since the late 1980s, NASA pilots have risked their lives collecting air samples over Antarctica so that researchers can assess damage to ozone-rich gases in the stratosphere. Normally, Earth's thin ozone layer filters out destructive ultraviolet light from the sun, but increasing evidence indicates that this protective shield may be thinning at an alarming rate. Concerned about ozone loss, NASA-led teams continue to mount airborne expeditions to investigate the phenomenon.

Today, orbiting satellites have replaced the U-2 as spy aircraft because they can do a better job from their vantage points in outer space. From an altitude of 700 miles, satellites photograph scenes on Earth so clearly that observers can distinguish features the size of a baseball diamond. In addition to military applications, we employ satellites to locate valuable water and mineral deposits and to monitor the weather. The weather satellite Nimbus even tracks the movement of wildlife such as polar bears, which researchers have equipped with special monitors.

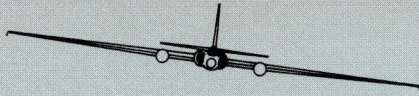

Determining the sizes of objects without measuring them physically is one of the principal uses of trigonometry. For example, using trigonometry, scientists calculated the distance to the moon long before we dreamed of going there. In this chapter we'll explore many applications of trigonometric functions, ranging from finding the velocity of a lawnmower blade to estimating the size of a building using an aerial photograph taken by a U-2 plane.

6

The

Trigonometric

Functions

6.1

Angles and Their

Measurements

Trigonometry was first studied by the Greeks, Egyptians, and Babylonians and used in surveying, navigation, and astronomy. Using trigonometry, they had a powerful tool for finding areas of triangular plots of land, as well as lengths of sides and measures of angles, without physically measuring them. We begin our study of trigonometry by studying angles and their measurements.

Degree Measure of Angles

In geometry a **ray** is defined as a point on a line together with all points of the line on one side of that point. Figure 6.1(a) shows ray $\overrightarrow{AB}$. An **angle** is defined as the union of two rays with a common endpoint, the **vertex.** The angle shown in Fig. 6.1(b) is named $\angle A$, $\angle BAC$, or $\angle CAB$. (Read the symbol $\angle$ as "angle.") Angles are also named using Greek letters such as α (alpha), β (beta), γ (gamma), or θ (theta).

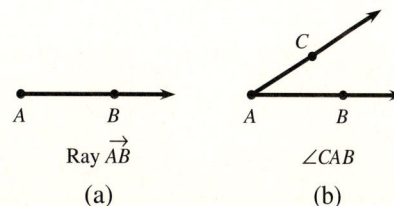

Ray $\overrightarrow{AB}$

(a)

$\angle CAB$

(b)

Figure 6.1

An angle is often thought of as being formed by rotating one ray away from a fixed ray as indicated by angle α and the arrow in Fig. 6.2(a). The fixed ray is the **initial side** and the rotated ray is the **terminal side.** An angle whose vertex is the center of a circle, as shown in Fig. 6.2(b), is a **central angle,** and the arc of the circle through which the terminal side moves is the **intercepted arc.** An angle in **standard position** is located in a rectangular coordinate system with the vertex at the origin and the initial side on the positive x-axis as shown in Fig. 6.2(c).

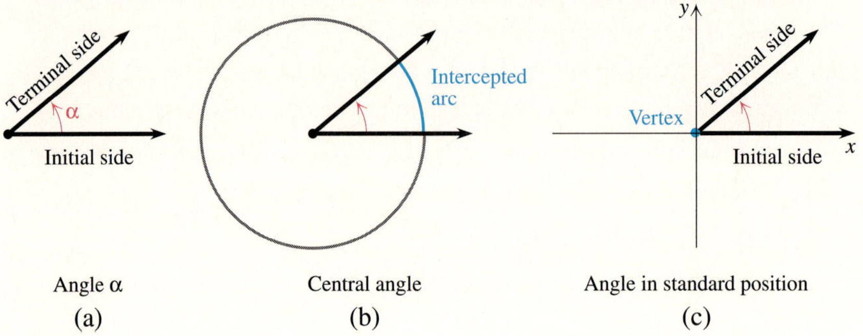

Angle α Central angle   Angle in standard position
(a)   (b)   (c)

Figure 6.2

The measure, $m(\alpha)$, of an angle α indicates the amount of rotation of the terminal side from the initial position. It is found using any circle centered at the vertex. The circle is divided into 360 equal arcs and each arc is one **degree** (1°).

Definition: Degree Measure

The **degree measure of an angle** is the number of degrees in the intercepted arc of a circle centered at the vertex. The degree measure is positive if the rotation is counterclockwise and negative if the rotation is clockwise.

Figure 6.3 shows the positions of the terminal sides of some angles in standard position with common positive measures between 0° and 360°.

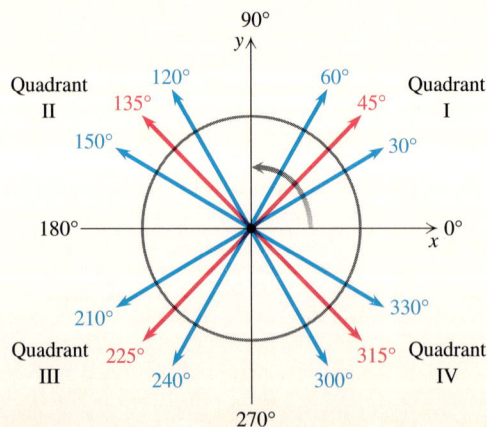

Figure 6.3

An angle with measure between 0° and 90° is an **acute angle.** An angle with measure between 90° and 180° is an **obtuse angle.** An angle of exactly 180° is a **straight angle.** See Fig. 6.4. A 90° angle is a **right angle.** An angle in standard position is said to lie in the quadrant where its terminal side lies. If the terminal side is on an axis, the angle is a **quadrantal angle.** We often think of the degree measure of an angle as the angle itself. For example, we write $m(\alpha) = 60°$ or $\alpha = 60°$, and we say that 60° is an acute angle.

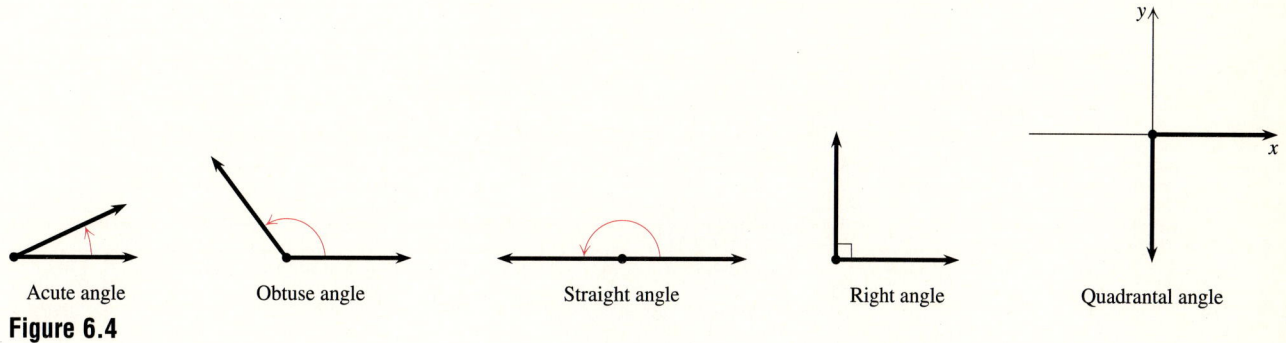

Acute angle   Obtuse angle   Straight angle   Right angle   Quadrantal angle

Figure 6.4

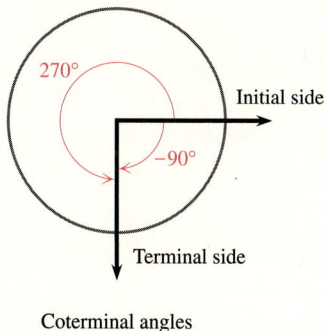

Coterminal angles

Figure 6.5

The initial side of an angle may be rotated in a positive or negative direction to get to the position of the terminal side. For example, if the initial side shown in Fig. 6.5 rotates clockwise for one-quarter of a revolution to get to the terminal position, then the measure of the angle is −90°. If the initial side had rotated counterclockwise to get to the position of the terminal side, then the measure of the angle would be 270°. If the initial side had rotated clockwise for one and a quarter revolutions to get to the terminal position, then the angle would be −450°. The angles −90°, −450°, and 270° are different angles in standard position with the same initial side and the same terminal side, and are called **coterminal angles.** Any two coterminal angles have degree measures that differ by a multiple of 360° (one complete revolution).

Coterminal Angles

> Angles α and β in standard position are coterminal if and only if there is an integer k such that $m(\beta) = m(\alpha) + k360°$.

Note that any angle formed by two rays can be thought of as one angle with infinitely many different measures, or infinitely many different coterminal angles.

Example 1   Finding coterminal angles

Find two positive angles and two negative angles that are coterminal with −50°.

Solution

Since any angle of the form $-50° + k360°$ is coterminal with $-50°$, there are infinitely many possible answers. For simplicity, we choose the positive integers 1 and 2 and the negative integers -1 and -2 for k to get the following angles:

$$-50° + 1 \cdot 360° = 310°$$

$$-50° + 2 \cdot 360° = 670°$$

$$-50° + (-1)360° = -410°$$

$$-50° + (-2)360° = -770°$$

The angles $310°$, $670°$, $-410°$, and $-770°$ are coterminal with $-50°$. ◆

Example 2   Determining whether angles are coterminal

Determine whether angles in standard position with the given measures are coterminal.

a) $m(\alpha) = 190°, m(\beta) = -170°$ **b)** $m(\alpha) = 150°, m(\beta) = 880°$

Solution

a) If there is an integer k such that $190 + 360k = -170$, then α and β are coterminal.

$$190 + 360k = -170$$

$$360k = -360$$

$$k = -1$$

Since the equation has an integral solution, α and β are coterminal.

b) If there is an integer k such that $150 + 360k = 880$, then α and β are coterminal.

$$150 + 360k = 880$$

$$360k = 730$$

$$k = \frac{73}{36}$$

Since there is no integral solution to the equation, α and β are not coterminal. ◆

The quadrantal angles, such as $90°$, $180°$, $270°$, and $360°$, have terminal sides on an axis and do not lie in any quadrant. Any angle that is not coterminal with a quadrantal angle lies in one of the four quadrants. To determine the quadrant in which an angle lies, add or subtract multiples of $360°$ (one revolution) to obtain a coterminal angle with a measure between $0°$ and $360°$.

Example 3   Determining in which quadrant an angle lies

Name the quadrant in which each angle lies.

a) 230°   **b)** −580°   **c)** 1380°

Solution

a) Since 180° < 230° < 270°, a 230° angle lies in quadrant III.

b) We must add 2(360°) to −580° to get an angle between 0° and 360°:

$$-580° + 2(360°) = 140°$$

So 140° and −580° are coterminal. Since 90° < 140° < 180°, 140° lies in quadrant II and so does a −580° angle.

c) From 1380° we must subtract 3(360°) to obtain an angle between 0° and 360°:

$$1380° - 3(360°) = 300°$$

So 1380° and 300° are coterminal. Since 270° < 300° < 360°, 300° lies in quadrant IV and so does 1380°.   ◆

Each degree is divided into 60 equal parts called **minutes,** and each minute is divided into 60 equal parts called **seconds.** A minute (min) is 1/60 of a degree (deg), and a second (sec) is 1/60 of a minute or 1/3600 of a degree. An angle with measure 44°12′30″ is an angle with a measure of 44 degrees, 12 minutes, and 30 seconds. Historically, angles were measured by using degrees-minutes-seconds, but with calculators it is convenient to have the fractional parts of a degree written as a decimal number such as 7.218°. Some calculators can handle angles in degrees-minutes-seconds and even convert them to decimal degrees.

```
44'12'30'
        44.20833333
■
```

A calculator can convert degrees/ minutes/seconds to decimal degrees. Note the different notation used here.

Example 4   Converting degrees-minutes-seconds to decimal degrees

Convert the measure 44°12′30″ to decimal degrees.

Solution

Since 1 degree = 60 minutes and 1 degree = 3600 seconds, we get

$$12 \text{ min} = 12 \text{ min} \cdot \frac{1 \text{ deg}}{60 \text{ min}} = \frac{12}{60} \text{ deg} \qquad \text{and}$$

$$30 \text{ sec} = 30 \text{ sec} \cdot \frac{1 \text{ deg}}{3600 \text{ sec}} = \frac{30}{3600} \text{ deg}.$$

So

$$44°12′30″ = \left(44 + \frac{12}{60} + \frac{30}{3600} \right)^° \approx 44.2083°.$$ ◆

Note that the conversion of Example 4 was done by "cancellation of units." Minutes in the numerator canceled with minutes in the denominator to give the result in degrees, and seconds in the numerator canceled with seconds in the denominator to give the result in degrees.

```
44.235▸DMS
          44°14'6"
```

A calculator can convert decimal degrees to degrees/minutes/seconds.

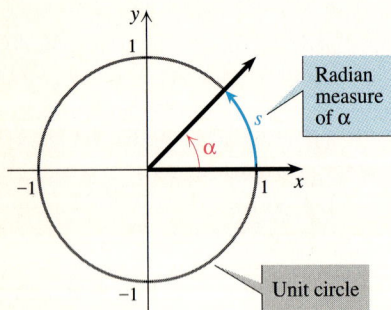

Example 5   Converting decimal degrees to degrees-minutes-seconds

Convert the measure 44.235° to degrees-minutes-seconds.

Solution

First convert 0.235° to minutes. Since 1 degree = 60 minutes,

$$0.235 \text{ deg} = 0.235 \text{ deg} \cdot \frac{60 \text{ min}}{1 \text{ deg}} = 14.1 \text{ min}.$$

So 44.235° = 44°14.1′. Now convert 0.1′ to seconds. Since 1 minute = 60 seconds,

$$0.1 \text{ min} = 0.1 \text{ min} \cdot \frac{60 \text{ sec}}{1 \text{ min}} = 6 \text{ sec}.$$

So 44.235° = 44°14′6″.   ◆

Note again how the original units canceled in Example 5, giving the result in the desired units of measurement.

Radian Measure of Angles

Degree measure of angles is used mostly in applied areas such as surveying, navigation, and engineering. Radian measure of angles is used more in scientific fields and results in simpler formulas in trigonometry and calculus.

For radian measure of angles we use a **unit circle** (a circle with radius 1) centered at the origin. The radian measure of an angle in standard position is simply the length of the intercepted arc on the unit circle. See Fig. 6.6. Since the radius of the unit circle is the real number 1 without any dimension (such as feet or inches), the length of an intercepted arc is a real number without any dimension and so the radian measure of an angle is also a real number without any dimension. One **radian** (abbreviated 1 rad) is the real number 1.

Figure 6.6

Definition: Radian Measure

To find the **radian measure** of the angle α in standard position, find the length of the intercepted arc on the unit circle. If the rotation is counterclockwise, the radian measure is the length of the arc. If the rotation is clockwise, the radian measure is the opposite of the length of the arc.

Radian measure is called a **directed length** because it is positive or negative depending on the direction of rotation of the initial side. If s is the length of the intercepted arc on the unit circle for an angle α as shown in Fig. 6.6, we write $m(\alpha) = s$. To emphasize that s is the length of an arc on a unit circle, we may write $m(\alpha) = s$ radians.

Because the circumference of a circle with radius r is $2\pi r$, the circumference of the unit circle is 2π. If the initial side rotates 360° (one complete revolution), then the length of the intercepted arc is 2π. So an angle of 360° has a radian measure of 2π radians. We express this relationship as 360° $= 2\pi$ rad or simply 360° $= 2\pi$. Dividing each side by 2 yields 180° $= \pi$, which is the basic relationship to remember for conversion of one unit of measurement to the other.

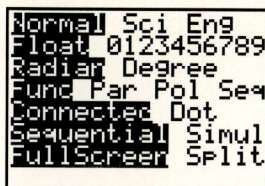

Your calculator operates in either radian or degree mode.

Degree-Radian Conversion

Conversion from degrees to radians or radians to degrees is based on

180 degrees = π radians.

To convert degrees to radians or radians to degrees we use 180 deg = π rad and cancellation of units. For example,

$$1 \text{ deg} = 1 \text{ deg} \cdot \frac{\pi \text{ rad}}{180 \text{ deg}} = \frac{\pi}{180} \text{ rad} \approx 0.01745 \text{ rad}$$

and

$$1 \text{ rad} = 1 \text{ rad} \cdot \frac{180 \text{ deg}}{\pi \text{ rad}} = \frac{180}{\pi} \text{ deg} \approx 57.3 \text{ deg}.$$

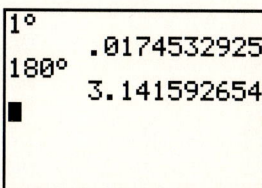

In radian mode, press ENTER to convert degrees to radians. See Appendix A for more examples.

Figure 6.7(a) shows an angle of 1° and Fig. 6.7(b) shows an angle of 1 radian. An angle of 1 radian intercepts an arc on the unit circle equal in length to the radius of the unit circle.

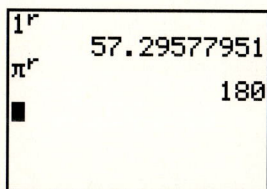

In degree mode, press ENTER to convert radians to degrees.

(a)

(b)

Figure 6.7

```
270°
         4.71238898
3π/2
         4.71238898
■
```

Use radian mode to convert 270° to radians. You can check Example 6(a) by evaluating $3\pi/2$ also.

Example 6   Converting from degrees to radians

Convert the degree measures to radians.

a) 270°   **b)** −23.6°

Solution

a) To convert degrees to radians, multiply the degree measure by π rad/180 deg:

$$270° = 270 \text{ deg} \cdot \frac{\pi \text{ rad}}{180 \text{ deg}} = \frac{3\pi}{2} \text{ rad}$$

The exact value, $3\pi/2$ rad, is approximately 4.71 rad; but when a measure in radians is a simple multiple of π, we usually write the exact value.

b) $-23.6° = -23.6 \text{ deg} \cdot \dfrac{\pi \text{ rad}}{180 \text{ deg}} \approx -0.412 \text{ rad}$ ◆

```
(7π/6)ʳ
              210
■
```

In degree mode, press ENTER to convert radians to degrees.

Example 7   Converting from radians to degrees

Convert the radian measures to degrees.

a) $\dfrac{7\pi}{6}$ **b)** 12.3

Solution

Multiply the radian measure by 180 deg/π rad:

a) $\dfrac{7\pi}{6} = \dfrac{7\pi}{6} \text{ rad} \cdot \dfrac{180 \text{ deg}}{\pi \text{ rad}} = 210°$ **b)** $12.3 = 12.3 \text{ rad} \cdot \dfrac{180 \text{ deg}}{\pi \text{ rad}} \approx 704.7°$ ◆

Figure 6.8 shows angles with common measures in standard position.

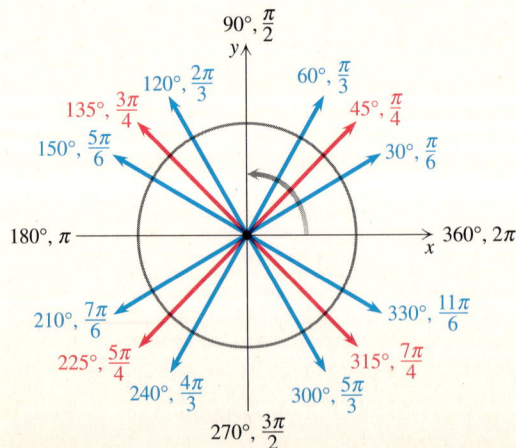

Figure 6.8

Coterminal angles in standard position have radian measures that differ by an integral multiple of 2π (their degree measures differ by an integral multiple of 360°).

Example 8   Finding coterminal angles using radian measure

Find two positive and two negative angles that are coterminal with $\pi/6$.

Solution

All angles coterminal with $\pi/6$ have a radian measure of the form $\pi/6 + k(2\pi)$, where k is an integer.

$$\frac{\pi}{6} + 1(2\pi) = \frac{\pi}{6} + \frac{12\pi}{6} = \frac{13\pi}{6}$$

$$\frac{\pi}{6} + 2(2\pi) = \frac{\pi}{6} + \frac{24\pi}{6} = \frac{25\pi}{6}$$

$$\frac{\pi}{6} + (-1)(2\pi) = \frac{\pi}{6} - \frac{12\pi}{6} = -\frac{11\pi}{6}$$

$$\frac{\pi}{6} + (-2)(2\pi) = \frac{\pi}{6} - \frac{24\pi}{6} = -\frac{23\pi}{6}$$

The angles $13\pi/6$, $25\pi/6$, $-11\pi/6$, and $-23\pi/6$ are coterminal with $\pi/6$. ◆

Arc Length

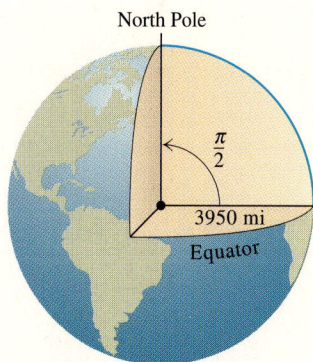

Figure 6.9

Radian measure of a central angle of a circle can be used to easily find the length of the intercepted arc of the circle. For example, an angle of $\pi/2$ radians positioned at the center of the earth, as shown in Fig. 6.9, intercepts an arc on the surface of the earth that runs from the Equator to the North Pole. Using 3950 miles as the approximate radius of the earth yields a circumference of $2\pi(3950) = 7900\pi$ miles. Since $\pi/2$ radians is $1/4$ of a complete circle, the length of the intercepted arc is $1/4$ of the circumference. So the distance from the Equator to the North Pole is $7900\pi/4$ miles, or about 6205 miles.

In general, a central angle α in a circle of radius r intercepts an arc whose length s is a fraction of the circumference of the circle, as shown in Fig. 6.10. Since a complete revolution is 2π radians, that fraction is $\alpha/(2\pi)$. Since the circumference is $2\pi r$, we get

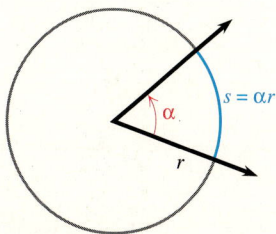

Figure 6.10

$$s = \frac{\alpha}{2\pi} \cdot 2\pi r = \alpha r.$$

Theorem: Length of an Arc

The length s of an arc intercepted by a central angle of α *radians* on a circle of radius r is given by

$$s = \alpha r.$$

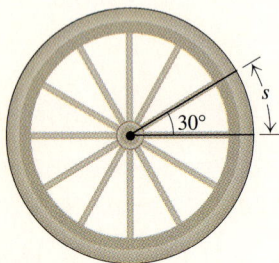

If α is negative, the formula $s = \alpha r$ gives a negative number for the length of the arc. So s is a directed length. Note that the formula $s = \alpha r$ applies only if α is in radians.

Example 9   Finding the length of an arc

The wagon wheel shown in Fig. 6.11 has a diameter of 28 inches and an angle of 30° between the spokes. What is the length of the arc s between two adjacent spokes?

Solution

First convert 30° to radians:

$$30° = 30 \text{ deg} \cdot \frac{\pi \text{ rad}}{180 \text{ deg}} = \frac{\pi}{6} \text{ rad}$$

Now use $r = 14$ inches and $\alpha = \pi/6$ in the formula for arc length $s = \alpha r$:

$$s = \frac{\pi}{6} \cdot 14 \text{ in.} = \frac{7\pi}{3} \text{ in.} \approx 7.33 \text{ in.}$$

Since the radian measure of an angle is a dimensionless real number, the product of radians and inches is given in inches. ◆

Example 10   Finding the central angle

For four years, U-2 spy planes flew reconnaissance over the Soviet Union, collecting information for the United States. The U-2, designed to fly at 70,000 feet —well out of range of Soviet guns—carried 12,000 feet of film in a camera that could photograph a path 2000 miles long and about 600 miles wide. Find the central angle, to the nearest tenth of a degree, that intercepts an arc of 600 miles on the surface of the earth (radius 3950 miles) as shown in Fig. 6.12.

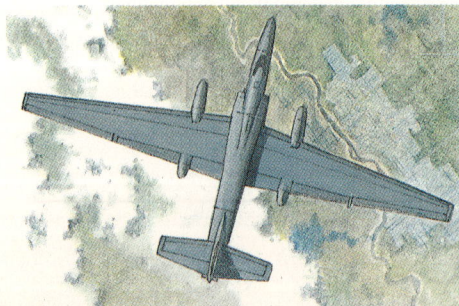

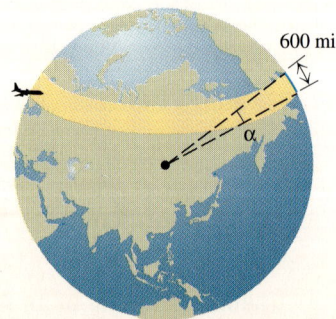

Figure 6.11

Figure 6.12

Solution

From the formula $s = \alpha r$ we get $\alpha = s/r$:

$$\alpha = \frac{600}{3950} \approx 0.1519 \text{ rad}$$

Multiply by 180 deg/π rad to find the angle in degrees:

$$\alpha = 0.1519 \text{ rad} \cdot \frac{180 \text{ deg}}{\pi \text{ rad}} \approx 8.7°$$

Linear and Angular Velocity

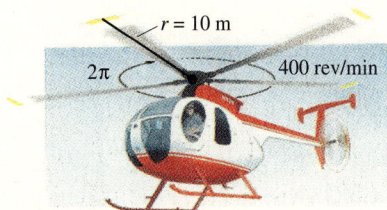

Figure 6.13

Suppose that a point is in motion on a circle. The distance traveled by the point in some unit of time is the length of an arc on the circle. The *linear velocity* of the point is the rate at which that distance is changing. For example, suppose that the helicopter blade of length 10 meters shown in Fig. 6.13 is rotating at 400 revolutions per minute about its center. A point on the tip of the blade (10 meters from the center) travels 20π meters for each revolution of the blade, because $C = 2\pi r$. Since the blade is rotating at 400 revolutions per minute, the point has a linear velocity of 8000π meters per minute.

If we draw a ray from the center of the circle through the point on the tip of the helicopter blade, then an angle is formed by the initial and terminal positions of the ray over some unit of time. The *angular velocity* of the point is the rate at which that angle is changing. For example, the ray shown in Fig. 6.13 rotates through an angle of 2π rad for each revolution of the blade. Since the blade is rotating at 400 revolutions per minute, the angular velocity of the point is 800π radians per minute. The angular velocity does not depend on the length of the blade (or the radius of the circular path), but only on the number of revolutions per unit of time.

We use the letter v to represent linear velocity for a point in circular motion and the Greek letter ω (omega) to represent angular velocity, and define them as follows.

Definition: Linear Velocity and Angular Velocity

If a point is in motion on a circle of radius r through an angle of α radians in time t, then its **linear velocity** is

$$v = \frac{s}{t},$$

where s is the arc length determined by $s = \alpha r$, and its **angular velocity** is

$$\omega = \frac{\alpha}{t}.$$

Since $v = s/t = \alpha r/t$ and $\omega = \alpha/t$, we get $v = r\omega$. We have proved the following theorem.

▼

Theorem: Linear Velocity in Terms of Angular Velocity

If v is the linear velocity of a point on a circle of radius r, and ω is its angular velocity, then

$$v = r\omega.$$

Example 11 Finding angular and linear velocity

What are the angular velocity in radians per second and the linear velocity in miles per hour of the tip of a 22-inch lawnmower blade that is rotating at 2500 revolutions per minute?

Solution

Use the fact that 2π radians = 1 revolution and 60 seconds = 1 minute to find the angular velocity:

$$\omega = \frac{2500 \text{ rev}}{\text{min}} = \frac{2500 \text{ rev}}{\text{min}} \cdot \frac{2\pi \text{ rad}}{1 \text{ rev}} \cdot \frac{1 \text{ min}}{60 \text{ sec}} \approx 261.799 \text{ rad/sec}$$

To find the linear velocity, use $v = r\omega$ with $r = 11$ inches:

$$v = 11 \text{ in.} \cdot \frac{261.799 \text{ rad}}{\text{sec}} = 2879.789 \text{ in./sec}$$

Convert to miles per hour:

$$v = \frac{2879.789 \text{ in.}}{\text{sec}} \cdot \frac{1 \text{ ft}}{12 \text{ in.}} \cdot \frac{1 \text{ mi}}{5280 \text{ ft}} \cdot \frac{3600 \text{ s}}{1 \text{ hr}} \approx 163.624 \text{ mi/hr}$$ ◆

Any point on the surface of the earth (except at the poles) makes one revolution about the axis of the earth in 24 hours. So the angular velocity of a point on the earth is $\pi/12$ radians per hour. The linear velocity of a point on the surface of the earth depends on its distance from the axis of the earth.

Example 12 Linear velocity on the surface of the earth

What is the linear velocity in miles per hour of a point on the equator?

Solution

A point on the equator has an angular velocity of $\pi/12$ radians per hour on a circle of radius 3950 miles. Using the formula $v = \omega r$, we get

$$v = \frac{\pi}{12} \text{ rad/hr} \cdot 3950 \text{ mi} = \frac{3950\pi}{12} \text{ mi/hr} \approx 1034 \text{ mi/hr}.$$

Note that we write angular velocity as radians per hour, but radians are omitted from the answer in miles per hour because radians are simply real numbers. ◆

For Thought

True or false? Explain.

1. An angle is a union of two rays with a common endpoint.
2. The lengths of the rays of $\angle A$ determine the degree measure of $\angle A$.
3. Angles of $5°$ and $-365°$ are coterminal.
4. The radian measure of an angle cannot be negative.
5. An angle of $38\pi/4$ radians is a quadrantal angle.
6. If $m(\angle A) = 210°$, then $m(\angle A) = 5\pi/6$ radians.
7. $25°20'40'' = 25.34°$
8. The angular velocity of Seattle is $\pi/12$ radians per hour.
9. Seattle and Los Angeles have the same linear velocity.
10. A central angle of 1 rad in a circle of radius r intercepts an arc of length r.

6.1 Exercises   Tape 11   Disk—5.25": 4   3.5": 3   Macintosh: 3

Find the missing degree or radian measure for each position of the terminal side shown. For Exercise 1, use degrees between $0°$ and $360°$ and radians between 0 and 2π. For Exercise 2, use degrees between $-360°$ and $0°$ and radians between -2π and 0. Practice these two exercises until you have memorized the degree and radian measures corresponding to these common angles.

1.

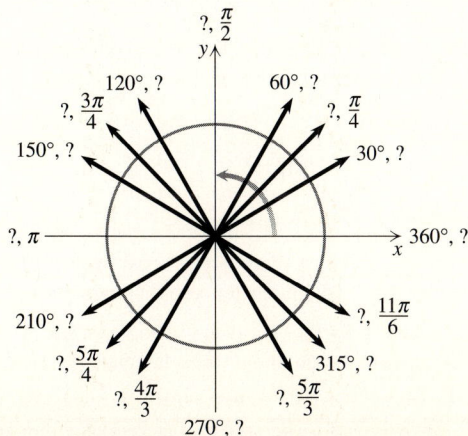

2.

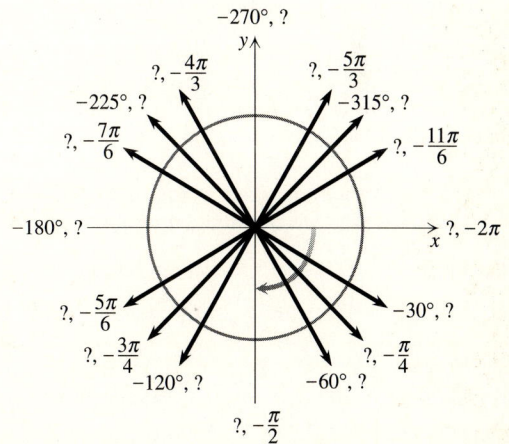

Find two positive angles and two negative angles that are coterminal with each given angle. For angles given in degrees, find coterminal angles in degrees. For angles given in radians, find coterminal angles in radians.

3. $60°$ **4.** $45°$ **5.** $-16°$ **6.** $-90°$

7. $\dfrac{\pi}{3}$ **8.** $\dfrac{\pi}{4}$ **9.** -1.2 **10.** 2.3

Determine whether the angles in each given pair are co-terminal.

11. $123.4°, -236.6°$ **12.** $744°, -336°$

13. $1055°, 155°$ **14.** $0°, 359.9°$

15. $\dfrac{3\pi}{4}, \dfrac{29\pi}{4}$ **16.** $-\dfrac{\pi}{3}, \dfrac{5\pi}{3}$

17. $\dfrac{7\pi}{6}, -\dfrac{5\pi}{6}$ **18.** $1, 361$

Name the quadrant in which each angle lies.

19. $85°$ **20.** $110°$ **21.** $-125°$

22. $-200°$ **23.** $300°$ **24.** $205°$

25. $\dfrac{5\pi}{12}$ **26.** $\dfrac{13\pi}{12}$ **27.** $-\dfrac{6\pi}{7}$

28. $-\dfrac{39\pi}{20}$ **29.** $\dfrac{13\pi}{8}$ **30.** $-\dfrac{11\pi}{8}$

31. $750°$ **32.** $-980°$ **33.** -7.3

34. 23.1 **35.** 3 **36.** -2

Determine which angle of the three given angles is not coterminal with the other two.

37. $150°, -\dfrac{7\pi}{6}, -120°$ **38.** $\dfrac{\pi}{4}, 400°, \dfrac{17\pi}{4}$

39. $\dfrac{9\pi}{4}, \dfrac{5\pi}{4}, -\dfrac{3\pi}{4}$ **40.** $840°, -600°, 1100°$

41. $\dfrac{3\pi}{2}, \dfrac{\pi}{2}, -\dfrac{11\pi}{2}$ **42.** $-35°, -325°, 685°$

Find the measure in degrees of the least positive angle that is coterminal with each given angle.

43. $400°$ **44.** $540°$ **45.** $-340°$ **46.** $-180°$

47. $-1100°$ **48.** $-840°$ **49.** $900.54°$ **50.** $1235.6°$

51. $-209°32'14''$ **52.** $-105°15'43''$

Convert each angle to decimal degrees. When necessary, round to four decimal places.

53. $13°12'$ **54.** $45°6'$ **55.** $-8°30'18''$

56. $-5°45'30''$ **57.** $28°5'9''$ **58.** $44°19'32''$

Convert each angle to degrees-minutes-seconds. Round to the nearest whole number of seconds.

59. $75.5°$ **60.** $39.4°$ **61.** $-17.33°$

62. $-9.12°$ **63.** $18.123°$ **64.** $122.786°$

Convert each degree measure to radian measure. Give exact answers.

65. $18°$ **66.** $48°$ **67.** $-67.5°$

68. $-105°$ **69.** $630°$ **70.** $495°$

Convert each radian measure to degree measure. Use the value of π found on a scientific calculator and round approximate answers to three decimal places.

71. $\dfrac{5\pi}{12}$ **72.** $\dfrac{17\pi}{12}$ **73.** -6π

74. -9π **75.** 2.39 **76.** 0.452

Convert each degree measure to radian measure. Use the value of π found on a scientific calculator and round answers to three decimal places.

77. $37.4°$ **78.** $125.3°$ **79.** $-13°47'$

80. $-99°15'$ **81.** $53°37'6''$ **82.** $187°49'36''$

Find the measure in radians of the least positive angle that is coterminal with each given angle.

83. 3π **84.** 6π **85.** $\dfrac{9\pi}{2}$ **86.** $\dfrac{19\pi}{2}$

87. $-\dfrac{5\pi}{3}$ **88.** $-\dfrac{7\pi}{6}$ **89.** $-\dfrac{13\pi}{3}$ **90.** $-\dfrac{19\pi}{4}$

91. 8.32 **92.** -23.55

Find the length of the arc intercepted by the given central angle α in a circle of radius r.

93. $\alpha = \dfrac{\pi}{4}, r = 12$ ft   **94.** $\alpha = 1, r = 4$ cm

95. $\alpha = 3°, r = 4000$ mi   **96.** $\alpha = 60°, r = 2$ m

Find the radius of the circle in which the given central angle α intercepts an arc of the given length s.

97. $\alpha = 1, s = 1$ mi   **98.** $\alpha = 0.004, s = 99$ km

99. $\alpha = 180°, s = 10$ km   **100.** $\alpha = 360°, s = 8$ m

Solve each problem.

101. *Linear velocity* What is the linear velocity for any edge point of a 12-in.-diameter record spinning at $33\frac{1}{3}$ rev/min?

102. *Angular velocity* What is the angular velocity in radians per minute for any point on a 45-rpm record?

103. *Lawnmower blade* What is the linear velocity of the tip of a lawnmower blade spinning at 2800 rev/min for a lawnmower that cuts a 20-in.-wide path?

104. *Router bit* A router bit makes 45,000 rev/min. What is the linear velocity in miles per hour of the outside edge of a bit that cuts a 1-in.-wide path?

105. *Table saw* The blade on a table saw rotates at 3450 revolutions per minute. What is the difference in the linear velocity in miles per hour for a point on the edge of a 12-in.-diameter blade and a 10-in.-diameter blade?

106. *Automobile tire* If a car runs over a nail at 55 mph and the nail is lodged in the tire tread 13 in. from the center of the wheel, then what is the angular velocity of the nail in radians per hour?

107. *Distance to North Pole* Peshtigo, Wisconsin, is on the 45th parallel. This means that an arc from Peshtigo to the North Pole subtends a central angle of 45° as shown in the figure. If the radius of the earth is 3950 mi, then how far is it from Peshtigo to the North Pole?

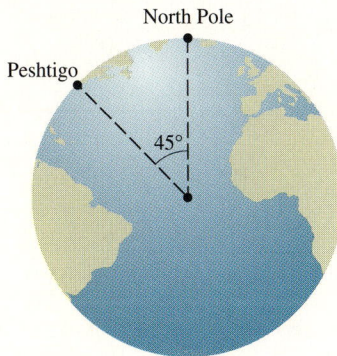

Figure for Exercise 107

108. *Distance to the helper* A surveyor sights her 6-ft 2-in. helper on a nearby hill as shown in the figure. If the angle of sight between the helper's feet and head is 0°37′, then approximately how far away is the helper? What assumption are you making about the helper?

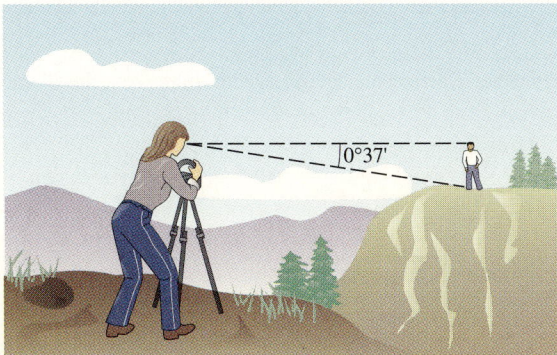

Figure for Exercise 108

109. *Photographing the earth* From an altitude of 161 mi, the Orbiting Geophysical Observatory (OGO-1) can photograph a path on the surface of the earth that is approximately 2000 mi wide. Find the central angle, to the nearest tenth of a degree, that intercepts an arc of 2000 mi on the surface of the earth (radius 3950 mi).

110. *Margin of error in a field goal* Morten Andersen of the New Orleans Saints has kicked many game-winning field goals from over 50 yd. Assume that the distance between the goal posts, 18.5 ft, is the length of an arc on a circle of radius 50 yd as shown in the figure. Andersen aims to kick the ball midway between the uprights. To score a field goal, what is the maximum number of degrees that the actual trajectory can deviate from the intended trajectory? What is the maximum number of degrees for a 20-yd field goal?

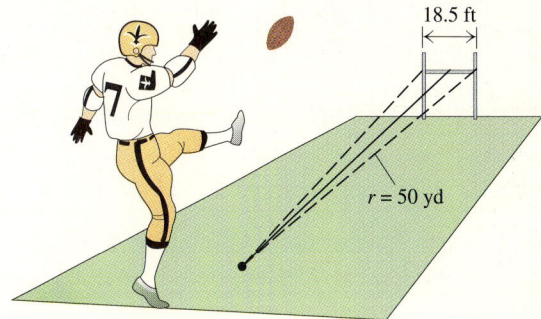

Figure for Exercise 110

111. *Linear velocity near the North Pole* Find the linear velocity for a point on the surface of the earth that is one mile from the North Pole.

112. *Linear velocity of Peshtigo* What are the linear and angular velocities for Peshtigo, Wisconsin (on the 45th parallel) with respect to its rotation around the axis of the earth? (See Exercise 107.)

113. *Eratosthenes measures the earth* Over 2200 years ago Eratosthenes read in the Alexandria library that at noon on June 21 a vertical stick in Syene cast no shadow. So on June 21 at noon Eratosthenes set out a vertical stick in Alexandria and found an angle of 7° in the position shown in the drawing. Eratosthenes reasoned that since the sun is so far away, sunlight must be arriving at the earth in parallel rays. With this assumption he concluded that the earth is round and the central angle in the drawing must also be 7°. He then paid a man to pace off the distance between Syene and Alexandria and found it to be 800 km. From these facts, calculate the circumference of the earth as Eratosthenes did and compare his answer with the circumference calculated by using the currently accepted radius of 6378 km.

Figure for Exercise 113

114. *Achieving synchronous orbit* The space shuttle orbits the earth in about 90 min at an altitude of about 125 mi, but a communication satellite must always remain above a fixed location on the earth, in synchronous orbit with the earth. Since the earth rotates 15° per hour, the angular velocity of a communication satellite must be 15° per hour. To achieve a synchronous orbit with the earth, the radius of the orbit of

a satellite must be 6.5 times the radius of the earth (3950 mi). What is the linear velocity in miles per hour of such a satellite?

115. *Area of a slice of pizza* If a 16-in.-diameter pizza is cut into six slices of the same size, then what is the area of each piece?

116. *Area of a sector of a circle* If a slice with central angle α radians is cut from a pizza of radius r, then what is the area of the slice?

For Writing/Discussion

117. *Swinging buckets* When the author was a child, he was fascinated by the fact that he could swing a bucket of water over his head fast enough so that the water did not spill even though the bucket was momentarily upside down. Experiment with a small can of water tied to a cord. Swing the can in a circle using radii of 1, 2, 3, and 4 ft. For each radius, determine the minimum angular and linear velocity of the can for which the water stays in the can. What conclusions can you draw from your data?

118. *Other units* Angles are measured in units other than degrees and radians. Find another unit that is used for measuring angles. Explain how to use it and indicate who might use it.

119. *Cooperative learning* In a small group, discuss the difference between radian measure and degree measure of an angle. Exactly how big is a unit circle? Why is a radian a real number?

6.2

The Sine and

Cosine Functions

In Section 6.1 we learned that angles can be measured in degrees or radians. Now we define two trigonometric functions whose domain is the set of all angles (measured in degrees or radians). These two functions are unlike any functions defined in algebra, but they form the foundation of trigonometry.

Definition

If α is an angle in standard position whose terminal side intersects the unit circle at point (x, y) as shown in Fig. 6.14, then **sine of α**—abbreviated $\sin(\alpha)$ or $\sin \alpha$—is the y-coordinate of that point, and **cosine of α**—abbreviated $\cos(\alpha)$ or $\cos \alpha$—is the x-coordinate. Sine and cosine are called **trigonometric functions.**

Definition: Sine and Cosine

If α is an angle in standard position and (x, y) is the point of intersection of the terminal side and the unit circle, then

$$\sin \alpha = y \quad \text{and} \quad \cos \alpha = x.$$

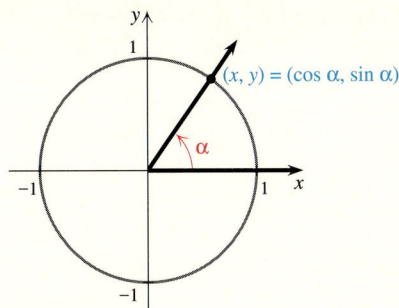

Figure 6.14

The domain of the sine function and the cosine function is the set of angles in standard position, but since each angle has a measure in degrees or radians, we generally use the set of degree measures or the set of radian measures as the domain. If (x, y) is on the unit circle, then $-1 \leq x \leq 1$ and $-1 \leq y \leq 1$, so the range of each of these functions is the interval $[-1, 1]$.

Example 1   Evaluating sine and cosine

Find the exact values of the sine and cosine functions for each angle.

a) $90°$

b) $\pi/2$

c) $180°$

d) $7\pi/2$

e) $720°$

Solution

a) Consider the unit circle shown in Fig. 6.15. Since the terminal side of $90°$ intersects the unit circle at $(0, 1)$, $\sin(90°) = 1$ and $\cos(90°) = 0$.

b) Since $\pi/2$ is coterminal with $90°$, we have $\sin(\pi/2) = 1$ and $\cos(\pi/2) = 0$.

c) The terminal side for $180°$ intersects the unit circle at $(-1, 0)$. So $\sin(180°) = 0$ and $\cos(180°) = -1$.

d) Since $7\pi/2$ is coterminal with $3\pi/2$, the terminal side of $7\pi/2$ lies on the negative y-axis and intersects the unit circle at $(0, -1)$. So $\sin(7\pi/2) = -1$ and $\cos(7\pi/2) = 0$.

e) Since $720°$ is coterminal with $0°$, the terminal side of $720°$ lies on the positive x-axis and intersects the unit circle at $(1, 0)$. So $\sin(720°) = 0$ and $\cos(720°) = 1$.   ◆

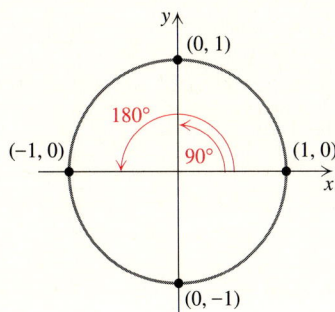

Figure 6.15

The *signs* of the sine and cosine functions depend on the quadrant in which the angle lies. For any point (x, y) on the unit circle in quadrant I, the x- and y-coordinates are positive. So if α is an angle in quadrant I, $\sin \alpha > 0$ and $\cos \alpha > 0$. Since the x-coordinate of any point in quadrant II is negative, the cosine of any angle in quadrant II is negative. Figure 6.16 gives the signs of sine and cosine for each of the four quadrants.

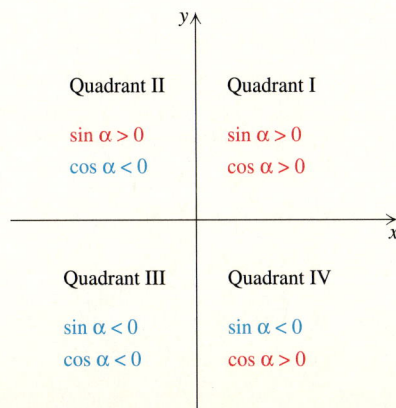

Figure 6.16

Sine and Cosine of a Multiple of 45°

In Example 1, exact values of the sine and cosine functions were found for some angles that were multiples of $90°$ (quadrantal angles). We can also find the exact values of $\sin(45°)$ and $\cos(45°)$ and then use that information to find the exact values of these functions for any multiple of $45°$.

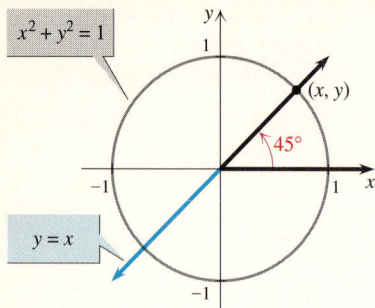

Figure 6.17

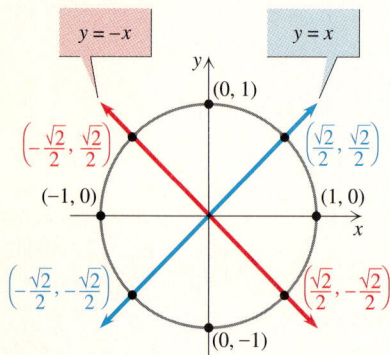

Figure 6.18

Since the terminal side for 45° lies on the line $y = x$, as shown in Fig. 6.17, the x- and y-coordinates at the point of intersection with the unit circle are equal. From Section 3.3, the equation of the unit circle is $x^2 + y^2 = 1$. Because $y = x$, we can write $x^2 + x^2 = 1$ and solve for x:

$$2x^2 = 1$$

$$x^2 = \frac{1}{2}$$

$$x = \pm\sqrt{\frac{1}{2}} = \pm\frac{\sqrt{2}}{2}$$

For a 45° angle in quadrant I, x and y are both positive numbers. So $\sin(45°) = \cos(45°) = \sqrt{2}/2$.

There are four points where the lines $y = x$ and $y = -x$ intersect the unit circle. The coordinates of these points, shown in Fig. 6.18, can all be found as above or by the symmetry of the unit circle. Figure 6.18 shows the coordinates of the key points for determining the sine and cosine of any angle that is an integral multiple of 45°.

Example 2   Evaluating sine and cosine at a multiple of 45°

Find the exact value of each expression.

a) $\sin(135°)$

b) $\sin\left(\dfrac{5\pi}{4}\right)$

c) $2\sin\left(-\dfrac{9\pi}{4}\right)\cos\left(-\dfrac{9\pi}{4}\right)$

Solution

a) The terminal side for 135° lies in quadrant II, halfway between 90° and 180°. As shown in Fig. 6.18, the point on the unit circle halfway between 90° and 180° is $\left(-\sqrt{2}/2, \sqrt{2}/2\right)$. So $\sin(135°) = \sqrt{2}/2$.

b) The terminal side for $5\pi/4$ is in quadrant III, halfway between $4\pi/4$ and $6\pi/4$. From Fig. 6.18, $\sin(5\pi/4)$ is the y-coordinate of $\left(-\sqrt{2}/2, -\sqrt{2}/2\right)$. So $\sin(5\pi/4) = -\sqrt{2}/2$.

c) Since $-8\pi/4$ is one clockwise revolution, $-9\pi/4$ is coterminal with $-\pi/4$. From Fig. 6.18, we have $\cos(-9\pi/4) = \cos(-\pi/4) = \sqrt{2}/2$ and $\sin(-9\pi/4) = -\sqrt{2}/2$. So

$$2\sin\left(-\frac{9\pi}{4}\right)\cos\left(-\frac{9\pi}{4}\right) = 2 \cdot \left(-\frac{\sqrt{2}}{2}\right) \cdot \frac{\sqrt{2}}{2} = -1.$$

Sine and Cosine of a Multiple of 30°

Exact values for the sine and cosine of 60° and the sine and cosine of 30° are found with a little help from geometry. The terminal side of a 60° angle intersects the unit circle at a point (x, y), as shown in Fig. 6.19(a), where x and y are the lengths of the legs of a 30-60-90 triangle whose hypotenuse is length 1.

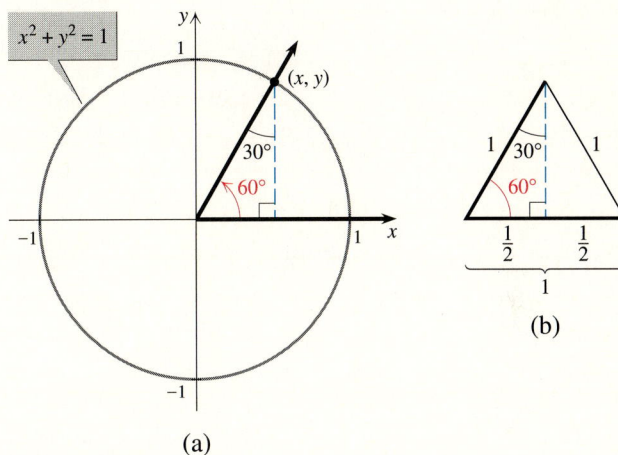

(b)

(a)

Figure 6.19

Since two congruent 30-60-90 triangles are formed by the altitude of an equilateral triangle, as shown in Fig. 6.19(b), the side opposite the 30° angle is half the length of the hypotenuse. Since the length of the hypotenuse is 1, the side opposite 30° is 1/2, that is, $x = 1/2$. Use the fact that $x^2 + y^2 = 1$ for the unit circle to find y:

$$\left(\frac{1}{2}\right)^2 + y^2 = 1$$

$$y^2 = \frac{3}{4}$$

$$y = \pm\sqrt{\frac{3}{4}} = \pm\frac{\sqrt{3}}{2}$$

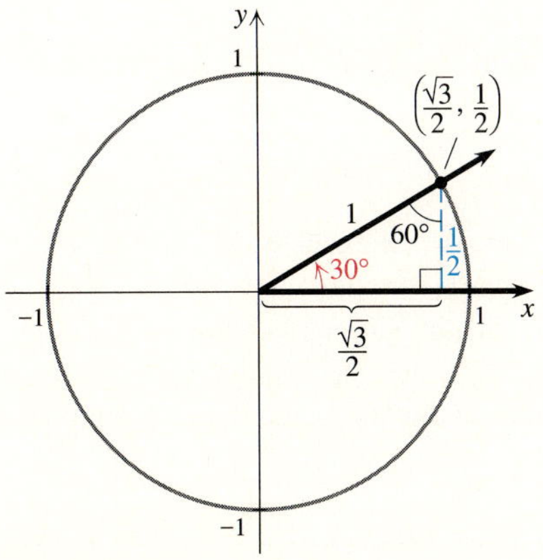

Figure 6.20

Since $y > 0$ (in quadrant I), $\cos(60°) = 1/2$ and $\sin(60°) = \sqrt{3}/2$.

The point of intersection of the terminal side for a 30° angle with the unit circle determines a 30-60-90 triangle exactly the same size as the one described above. In this case the longer leg is on the x-axis, as shown in Fig. 6.20, and the point on the unit circle is $\left(\sqrt{3}/2, 1/2\right)$. So $\cos(30°) = \sqrt{3}/2$, and $\sin(30°) = 1/2$.

Using the symmetry of the unit circle and the values just found for the sine and cosine of 30° and 60°, we can label points on the unit circle as shown in Fig. 6.21 on the next page. This figure shows the coordinates of every point on the unit circle where the terminal side of a multiple of 30° intersects the unit circle. For

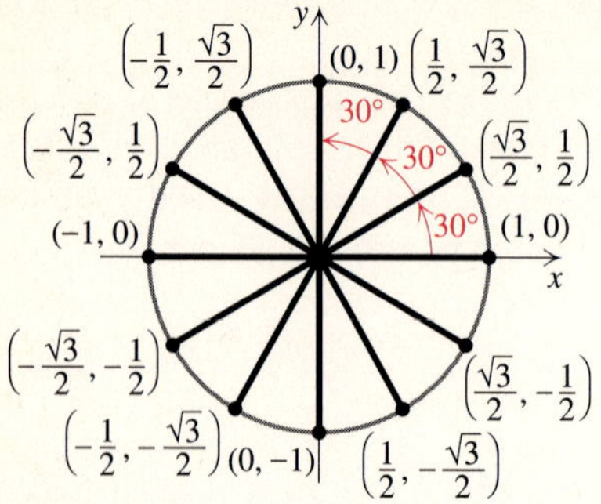

Figure 6.21

example, the terminal side of a 120° angle in standard position intersects the unit circle at $\left(-1/2, \sqrt{3}/2\right)$, because 120° = 4 · 30°.

In the next example, we use Fig. 6.21 to evaluate expressions involving multiples of 30° ($\pi/6$). Note that the square of sin α could be written as $(\sin \alpha)^2$, but for simplicity it is written as $\sin^2 \alpha$. Likewise, $\cos^2 \alpha$ is used for $(\cos \alpha)^2$.

Example 3 Evaluating sine and cosine at a multiple of 30°

Find the exact value of each expression.

a) $\sin\left(\dfrac{7\pi}{6}\right)$

b) $\cos^2(-240°) - \sin^2(-240°)$

Solution

a) Since $\pi/6 = 30°$, we have $7\pi/6 = 210°$. To determine the location of the terminal side of $7\pi/6$, notice that $7\pi/6$ is 30° larger than the straight angle 180° and so it lies in quadrant III. From Fig. 6.21, $7\pi/6$ intersects the unit circle at $\left(-\sqrt{3}/2, -1/2\right)$ and $\sin(7\pi/6) = -1/2$.

b) Since $-240°$ is coterminal with 120°, the terminal side for $-240°$ lies in quadrant II and intersects the unit circle at $\left(-1/2, \sqrt{3}/2\right)$, as indicated in Fig. 6.21. So $\cos(-240°) = -1/2$ and $\sin(-240°) = \sqrt{3}/2$. Now use these values in the original expression:

$$\cos^2(-240°) - \sin^2(-240°) = \left(-\frac{1}{2}\right)^2 - \left(\frac{\sqrt{3}}{2}\right)^2 = \frac{1}{4} - \frac{3}{4} = -\frac{1}{2}$$

Reference Angles

One way to find sin α and cos α is to determine the coordinates of the intersection of the terminal side of α and the unit circle, as we did in Examples 2 and 3. Another way to find sin α and cos α is to find the sine and cosine of the corresponding *reference angle,* which is in the interval [0, $\pi/2$], and attach the appropriate sign. Reference angles are essential for evaluating trigonometric functions if you are using tables that give values only for angles in the interval [0, $\pi/2$].

Definition: Reference Angle

If θ is a nonquadrantal angle in standard position, then the **reference angle** for θ is the positive acute angle θ' (read ''theta prime'') formed by the terminal side of θ and the positive or negative x-axis.

Figure 6.22 shows the reference angle θ' for an angle θ with terminal side in each of the four quadrants. Notice that in each case θ' is the acute angle formed by the terminal side of θ and the x-axis.

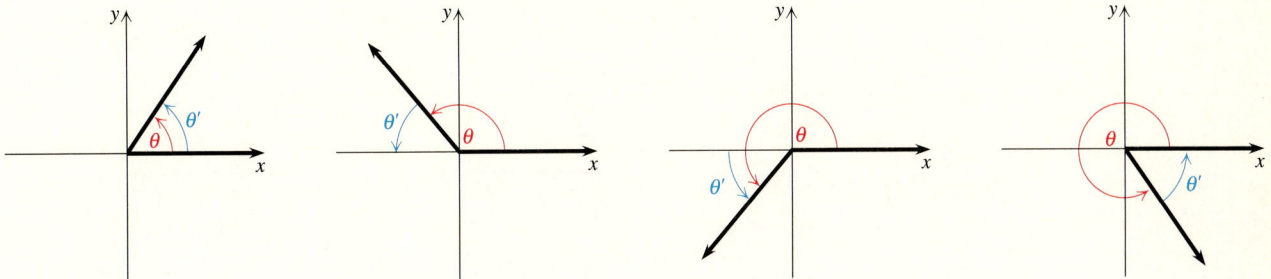

Figure 6.22

Example 4   Finding reference angles

For each given angle θ, sketch the reference angle θ' and give the measure of θ' in both radians and degrees.

a) $\theta = 120°$ **b)** $\theta = 7\pi/6$ **c)** $\theta = 690°$ **d)** $\theta = -7\pi/4$

Solution

a) The terminal side of $120°$ is in quadrant II, as shown in Fig. 6.23(a). The reference angle θ' is $180° - 120° = 60°$ or $\pi/3$.

b) The terminal side of $7\pi/6$ is in quadrant III, as shown in Fig. 6.23(b). The reference angle θ' is $7\pi/6 - \pi = \pi/6$ or $30°$.

c) Since $690°$ is coterminal with $330°$, the terminal side is in quadrant IV, as shown in Fig. 6.23(c). The reference angle θ' is $360° - 330° = 30°$ or $\pi/6$.

d) An angle of $-7\pi/4$ radians has a terminal side in quadrant I, as shown in Fig. 6.23(d), and the reference angle is $\pi/4$ or $45°$.

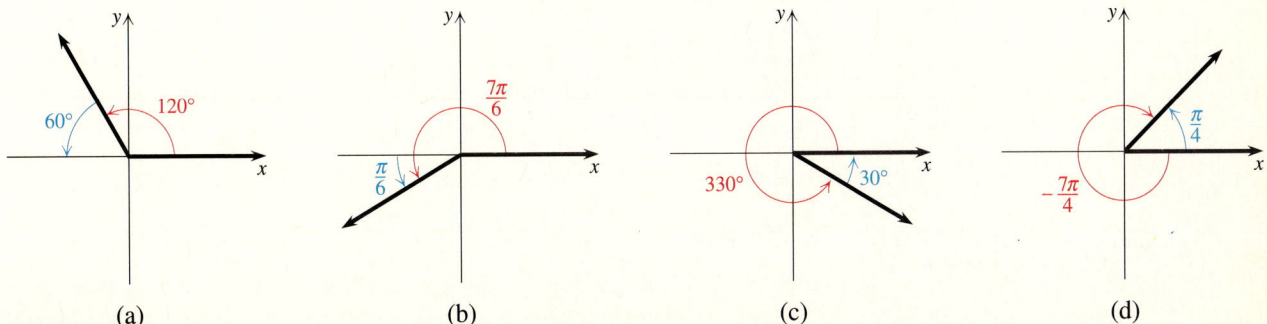

(a)   (b)   (c)   (d)

Figure 6.23

Because the measure of any reference angle is between 0 and $\pi/2$, any reference angle can be positioned in the first quadrant. Figure 6.24(a) shows a typical reference angle θ' in standard position with its terminal side intersecting the unit circle at (x, y). So $\cos \theta' = x$ and $\sin \theta' = y$.

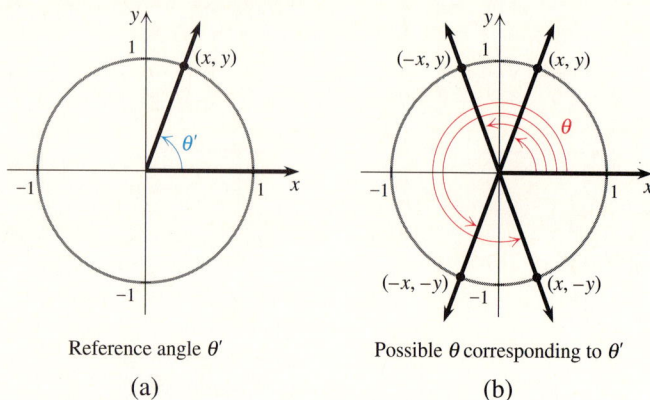

Reference angle θ'

(a)

Possible θ corresponding to θ'

(b)

Figure 6.24

For what angle θ could θ' be the reference angle? The angle θ (in standard position) that corresponds to θ' could have terminated in quadrant I, II, III, or IV, where it would intersect the unit circle at (x, y), $(-x, y)$, $(-x, -y)$, or $(x, -y)$ as shown in Fig. 6.24(b). So $\cos \theta = \pm x$ and $\sin \theta = \pm y$. Therefore, we can find $\cos \theta'$ and $\sin \theta'$ and prefix the appropriate sign to get $\cos \theta$ and $\sin \theta$. The appropriate sign is determined by the quadrant in which the terminal side of θ lies. The signs of $\sin \theta$ and $\cos \theta$ for the four quadrants were summarized in Fig. 6.16.

Theorem: Evaluating Trigonometric Functions Using Reference Angles

For an angle θ in standard position that is not a quadrantal angle, the value of a trigonometric function of θ can be found by finding the value of the function for its reference angle θ' and prefixing the appropriate sign.

Example 5 Trigonometric functions using reference angles

Find the sine and cosine for each angle using reference angles.

a) $\theta = 120°$ **b)** $\theta = 7\pi/6$ **c)** $\theta = 690°$ **d)** $\theta = -7\pi/4$

Solution

a) Figure 6.23(a) shows the terminal side of $120°$ in quadrant II and its reference angle $60°$. Since $\sin \theta > 0$ and $\cos \theta < 0$ for any angle θ in quadrant II, $\sin(120°) > 0$ and $\cos(120°) < 0$. So to get $\cos(120°)$ from $\cos(60°)$ we must prefix a negative sign. We have

$$\sin(120°) = \sin(60°) = \frac{\sqrt{3}}{2} \qquad \text{and} \qquad \cos(120°) = -\cos(60°) = -\frac{1}{2}.$$

b) Figure 6.23(b) shows the terminal side of $7\pi/6$ in quadrant III and its reference angle $\pi/6$. Since $\sin\theta < 0$ and $\cos\theta < 0$ for any angle θ in quadrant III, we must prefix a negative sign to both $\sin(\pi/6)$ and $\cos(\pi/6)$. So

$$\sin\left(\frac{7\pi}{6}\right) = -\sin\left(\frac{\pi}{6}\right) = -\frac{1}{2} \quad \text{and} \quad \cos\left(\frac{7\pi}{6}\right) = -\cos\left(\frac{\pi}{6}\right) = -\frac{\sqrt{3}}{2}.$$

c) The angle $690°$ terminates in quadrant IV (where $\sin\theta < 0$ and $\cos\theta > 0$) and its reference angle is $30°$, as shown in Fig. 6.23(c). So

$$\sin(690°) = -\sin(30°) = -\frac{1}{2} \quad \text{and} \quad \cos(690°) = \cos(30°) = \frac{\sqrt{3}}{2}.$$

d) The angle $-7\pi/4$ terminates in quadrant I (where $\sin\theta > 0$ and $\cos\theta > 0$) and its reference angle is $\pi/4$, as shown in Fig. 6.23(d). So

$$\sin\left(-\frac{7\pi}{4}\right) = \sin\left(\frac{\pi}{4}\right) = \frac{\sqrt{2}}{2} \quad \text{and} \quad \cos\left(-\frac{7\pi}{4}\right) = \cos\left(\frac{\pi}{4}\right) = \frac{\sqrt{2}}{2}.$$

The Sine and Cosine of a Real Number

The domain of sine and cosine can be thought of as the set of all angles, degree measures of angles, or radian measures of angles. However, since radian measures are just real numbers, the domain of sine and cosine then becomes the set of real numbers. The set of real numbers is often visualized as the domain of sine and cosine by placing a number line (the set of real numbers) next to the unit circle as shown in Fig. 6.25(a). To find $\sin s$ or $\cos s$ for $s > 0$, wrap the top half of the number line around the unit circle as shown in Fig. 6.25(b). The real number s then corresponds to an arc on the unit circle of length s, ending at (x, y). Of course, $\sin s = y$ and $\cos s = x$. Arcs corresponding to negative real numbers are found

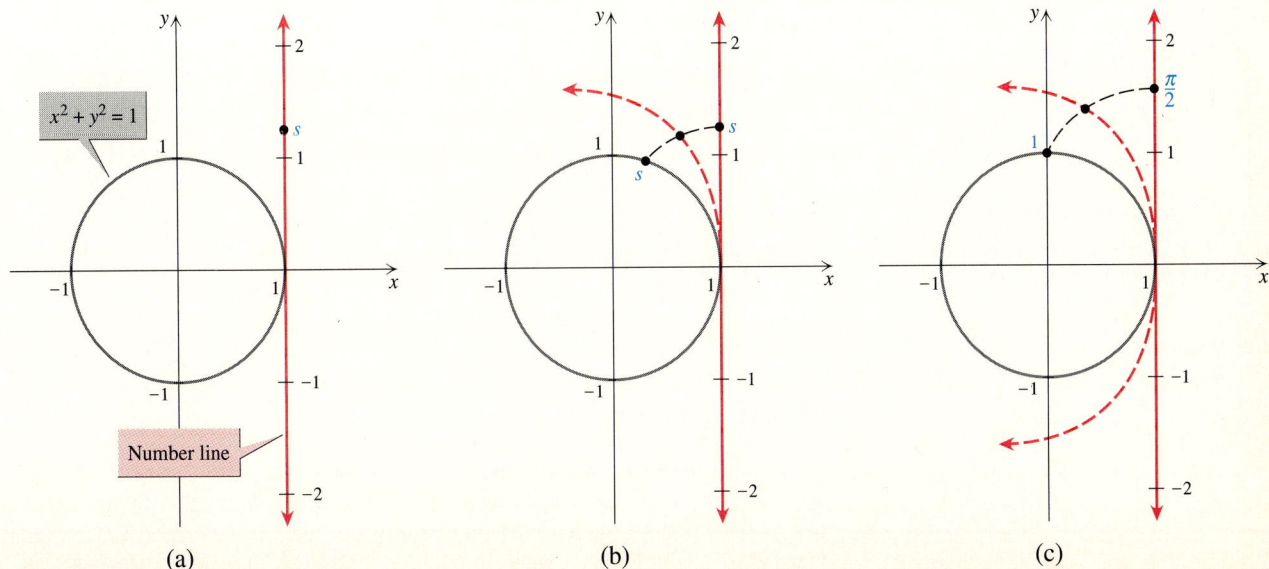

(a)   (b)   (c)

Figure 6.25

by wrapping the negative half of the number line around the unit circle in a clockwise direction as shown in Fig. 6.25(c). Because of their relationship to the unit circle, sine and cosine are often called **circular functions.**

Example 6   Sine and cosine of a real number

Find each function value.

a) $\sin(\pi/2)$ **b)** $\cos(-\pi)$

Solution

a) When wrapping the number line around the unit circle, $\pi/2$ on the number line corresponds to the point $(0, 1)$, which is also the point of intersection of the angle $\pi/2$ and the unit circle. So $\sin(\pi/2) = 1$.

b) When wrapping the number line around the unit circle, $-\pi$ on the number line wraps around to $(-1, 0)$. So $\cos(-\pi) = -1$. ◆

Approximate Values for Sine and Cosine

The sine and cosine for any angle that is a multiple of $30°$ or $45°$ can be found exactly. These angles are so common that it is important to know these exact values. However, for most other angles or real numbers we use approximate values for the sine and cosine, found with the help of a scientific calculator or a graphing calculator. Calculators can evaluate $\sin \alpha$ or $\cos \alpha$ if α is a real number (radian) or α is the degree measure of an angle. Generally, there is a mode key that sets the calculator to degree mode or radian mode. Consult your calculator manual.

```
sin (4.27r)
      -.9037315214
cos (-39.46°)
      .7720684617
```

If you indicate that the angle is radians or degrees, then it is not necessary to change the mode.

Example 7   Evaluating sine and cosine with a calculator

Find each function value rounded to four decimal places.

a) $\sin(4.27)$ **b)** $\cos(-39.46°)$

Solution

a) Make sure the calculator is in radian mode. We get $\sin(4.27) \approx -0.9037$.

b) Make sure the calculator is in degree mode. We get $\cos(-39.46°) \approx 0.7721$. ◆

The Fundamental Identity

An identity is an equation that is satisfied for all values of the variable for which both sides are defined. There are identities involving functions of trigonometry just as there are identities involving functions of algebra. The most fundamental identity in trigonometry comes from the definition of sine and cosine. For any angle, α, $\sin \alpha = y$ and $\cos \alpha = x$, where (x, y) is the point of intersection of the terminal side of α and the unit circle. Since every point on the unit circle satisfies the equation $x^2 + y^2 = 1$, we get the following identity.

The Fundamental Identity of Trigonometry

If α is any angle or real number,

$$\sin^2 \alpha + \cos^2 \alpha = 1.$$

If we know the value of the sine or cosine of an angle, then we can use the fundamental identity to find the value of the other function of the angle.

```
(sin 5.9)²+(cos
5.9)²
                 1
(sin 48)²+(cos 4
8)²
                 1
■
```

You can illustrate the fundamental identity for any angle or real number.

Example 8   Using the fundamental identity

Find $\cos \alpha$, given that $\sin \alpha = 3/5$ and α is an angle in quadrant II.

Solution

Use the fundamental identity $\sin^2 \alpha + \cos^2 \alpha = 1$ to find $\cos \alpha$:

$$\left(\frac{3}{5}\right)^2 + \cos^2 \alpha = 1$$

$$\cos^2 \alpha = \frac{16}{25}$$

$$\cos \alpha = \pm\sqrt{\frac{16}{25}} = \pm\frac{4}{5}$$

Since $\cos \alpha < 0$ for any α in quadrant II, we choose the negative sign and get $\cos \alpha = -4/5$. ◆

Motion of a Spring

The sine and cosine functions are used in modeling the motion of a spring. If a weight is at rest while hanging from a spring, as shown in Fig. 6.26, then it is at the **equilibrium** position, or 0 on a vertical number line. If the weight is set in motion with an initial velocity v_0 from location x_0, then the location at time t is given by

$$x = \frac{v_0}{\omega} \sin(\omega t) + x_0 \cos(\omega t).$$

The letter ω (omega) is a constant that depends on the stiffness of the spring and the amount of weight on the spring. For positive values of x the weight is below equilibrium, and for negative values it is above equilibrium. The initial velocity is considered to be positive if it is in the downward direction and negative if it is upward. Note that the domain of the sine and cosine functions in this formula is the nonnegative real numbers, and this application has nothing to do with angles.

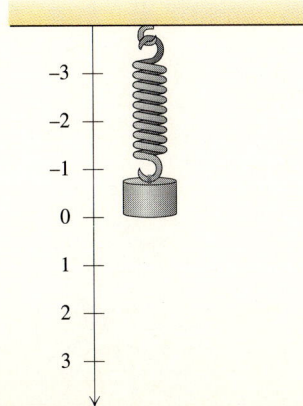

Figure 6.26

Example 9   Motion of a spring

A weight on a certain spring is set in motion with an upward velocity of 3 centimeters per second from a position 2 centimeters below equilibrium. Assume that for this spring and weight combination the constant ω has a value of 1. Write a formula that gives the location of the weight in centimeters as a function of the time t in seconds, and find the location of the weight 2 seconds after the weight is set in motion.

Solution

Since upward velocity is negative and locations below equilibrium are positive, use $v_0 = -3$, $x_0 = 2$, and $\omega = 1$ in the formula for the motion of a spring:

$$x = -3 \sin t + 2 \cos t$$

If $t = 2$ seconds, then $x = -3 \sin(2) + 2 \cos(2) \approx -3.6$ centimeters, which is 3.6 centimeters above the equilibrium position.

? For Thought

True or false? Explain.

1. $\cos(90°) = 1$
2. $\cos(90) = 0$
3. $\sin(45°) = 1/\sqrt{2}$
4. $\sin(-\pi/3) = \sin(\pi/3)$
5. $\cos(-\pi/3) = -\cos(\pi/3)$
6. $\sin(390°) = \sin(30°)$
7. If $\sin(\alpha) < 0$ and $\cos(\alpha) > 0$, then α is an angle in quadrant III.
8. If $\sin^2(\alpha) = 1/4$ and α is in quadrant III, then $\sin(\alpha) = 1/2$.
9. If $\cos^2(\alpha) = 3/4$ and α is an angle in quadrant I, then $\alpha = \pi/6$.
10. For any angle α, $\cos^2(\alpha) = (1 - \sin(\alpha))(1 + \sin(\alpha))$.

6.2 Exercises   Tape 11   Disk—5.25″: 4   3.5″: 3   Macintosh: 3

Redraw each diagram and label the indicated points with the proper coordinates. Exercise 1 shows the points where the terminal side of every multiple of 45° intersects the unit circle. Exercise 2 shows the points where the terminal side of every multiple of 30° intersects the unit circle. Repeat Exercises 1 and 2 until you can do them from memory.

1.

2.

Use the diagrams drawn in Exercises 1 and 2 to help you find the exact value of each function. Do not use a calculator.

3. $\sin(30°)$ **4.** $\sin(120°)$ **5.** $\cos(-60°)$

6. $\cos(-120°)$ **7.** $\sin(135°)$ **8.** $\sin(180°)$

9. $\cos(-135°)$ **10.** $\cos(-90°)$ **11.** $\cos(\pi)$

12. $\cos(2\pi)$ **13.** $\sin(-\pi/4)$ **14.** $\sin(3\pi/4)$

15. $\cos(7\pi/6)$ **16.** $\cos(-2\pi/3)$ **17.** $\sin(-4\pi/3)$

18. $\sin(5\pi/6)$ **19.** $\sin(390°)$ **20.** $\sin(765°)$

21. $\cos(-420°)$ **22.** $\cos(-450°)$ **23.** $\cos(13\pi/6)$

24. $\cos(-7\pi/3)$ **25.** $\sin(-11\pi/2)$ **26.** $\sin(13\pi/4)$

Sketch the given angle in standard position and find its reference angle in degrees and radians.

27. $210°$ **28.** $330°$ **29.** $5\pi/3$ **30.** $7\pi/6$

31. $-300°$ **32.** $-210°$ **33.** $5\pi/6$ **34.** $-13\pi/6$

35. $405°$ **36.** $390°$ **37.** $-3\pi/4$ **38.** $-2\pi/3$

Use reference angles to find the exact value of each expression.

39. $\sin(135°)$ **40.** $\sin(420°)$ **41.** $\cos(5\pi/3)$

42. $\cos(11\pi/6)$ **43.** $\sin(7\pi/4)$ **44.** $\sin(-13\pi/4)$

45. $\cos(-17\pi/6)$ **46.** $\cos(-5\pi/3)$ **47.** $\sin(-45°)$

48. $\cos(-120°)$ **49.** $\cos(-240°)$ **50.** $\sin(-225°)$

Find the exact value of each expression without using a calculator. Check your answer with a calculator.

51. $\dfrac{\cos(\pi/3)}{\sin(\pi/3)}$ **52.** $\dfrac{\sin(-5\pi/6)}{\cos(-5\pi/6)}$ **53.** $\dfrac{\sin(7\pi/4)}{\cos(7\pi/4)}$

54. $\dfrac{\sin(-3\pi/4)}{\cos(-3\pi/4)}$ **55.** $\sin\left(\dfrac{\pi}{3}+\dfrac{\pi}{6}\right)$ **56.** $\cos\left(\dfrac{\pi}{3}-\dfrac{\pi}{6}\right)$

57. $\dfrac{1-\cos(5\pi/6)}{\sin(5\pi/6)}$ **58.** $\dfrac{\sin(5\pi/6)}{1+\cos(5\pi/6)}$

59. $\sin(\pi/4)+\cos(\pi/4)$ **60.** $\sin^2(\pi/6)+\cos^2(\pi/6)$

61. $\cos(45°)\cos(60°)-\sin(45°)\sin(60°)$

62. $\sin(30°)\cos(135°)+\cos(30°)\sin(135°)$

63. $\dfrac{1-\cos\left(2\cdot\dfrac{\pi}{6}\right)}{2}$ **64.** $\dfrac{1+\cos\left(2\cdot\dfrac{\pi}{8}\right)}{2}$

65. $2\cos^2(210°)-1$

66. $\cos^2(-270°)-\sin^2(-270°)$

Use a calculator to find the value of each function. Round answers to four decimal places.

67. $\sin(43°)$ **68.** $\cos(55°)$ **69.** $\cos(230.46°)$

70. $\sin(344.1°)$ **71.** $\cos(-359.4°)$ **72.** $\sin(-269.5°)$

73. $\sin(23°48')$ **74.** $\cos(49°13')$ **75.** $\sin(-48°3'12'')$

76. $\cos(-9°4'7'')$ **77.** $\sin(\pi/4)$ **78.** $\cos(\pi/4)$

79. $\sin(1.57)$ **80.** $\cos(3.14)$ **81.** $\sin(14.6)$

82. $\cos(-18.9)$ **83.** $\cos(7\pi/12)$ **84.** $\sin(-13\pi/8)$

85. $\sin(0.99\pi)$ **86.** $\cos(0.163\pi)$

Find the approximate value of each expression. Round answers to four decimal places.

87. $\dfrac{1+\cos(44.3°)}{2}$ **88.** $\dfrac{1-\cos(98.6°)}{\sin(98.6°)}$

89. $\dfrac{\sin(\pi/12)}{\cos(\pi/12)}$ **90.** $\dfrac{\sin(9.2)}{\cos(9.2)}$

91. $\sin^2(4)+\cos^2(4)$ **92.** $\sin^2(3)-\cos^2(3)$

93. $1-2\sin^2(-88.4°)$

94. $2\cos^2(-9.5°)-1$

Solve each problem.

95. Find $\cos(\alpha)$, given that $\sin(\alpha)=5/13$ and α is in quadrant II.

96. Find $\sin(\alpha)$, given that $\cos(\alpha)=-4/5$ and α is in quadrant III.

97. Find $\sin(\alpha)$, given that $\cos(\alpha)=3/5$ and α is in quadrant IV.

98. Find $\cos(\alpha)$, given that $\sin(\alpha)=-12/13$ and α is in quadrant IV.

99. Find $\cos(\alpha)$, given that $\sin(\alpha)=1/3$ and $\cos(\alpha)>0$.

100. Find $\sin(\alpha)$, given that $\cos(\alpha)=2/5$ and $\sin(\alpha)<0$.

Solve each problem.

101. *Motion of a spring* A weight on a vertical spring is given an initial downward velocity of 4 cm/sec from a point 3 cm above equilibrium. Assuming that the constant ω has a value of 1, write the formula for the location of the weight at time t, and find its location 3 sec after it is set in motion.

102. *Motion of a spring* A weight on a vertical spring is given an initial upward velocity of 3 in./sec from a point 1 in. below equilibrium. Assuming that the constant ω has a value of $\sqrt{3}$, write the formula for the location of the weight at time t, and find its location 2 sec after it is set in motion.

103. *Spacing between teeth* The length of an arc intercepted by a central angle of θ radians in a circle of radius r is $r\theta$. The length of the chord, c, joining the endpoints of that arc is given by $c = r\sqrt{2 - 2\cos\theta}$. Find the actual distance between the tips of two adjacent teeth on a 12-in.-diameter carbide-tipped circular saw blade with 22 equally spaced teeth. Compare your answer with the length of a circular arc joining two adjacent teeth on a circle 12 in. in diameter.

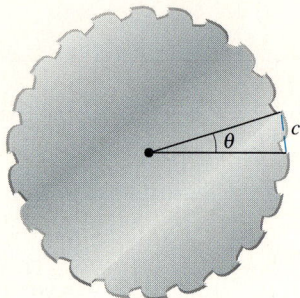

Figure for Exercise 103

104. *Shipping stop signs* A manufacturer of stop signs ships 100 signs in a circular drum. Use the formula from Exercise 103 to find the radius of the inside of the circular drum, given that the length of an edge of the stop sign is 10 in. and the signs just fit as shown in the figure.

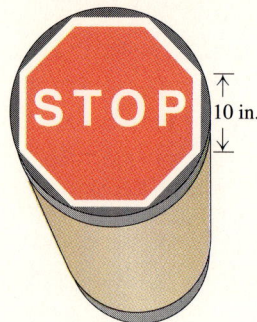

10 in.

Figure for Exercise 104

For Writing/Discussion

105. *Cosine in terms of sine* Use the fundamental identity to write two formulas for $\cos\alpha$ in terms of $\sin\alpha$. Indicate which formula to use for a given value of α.

106. *Cooperative learning* Work in a small group to find $\sin(-19\pi/4)$ without a calculator, using the two different methods presented in the text. Which method do you prefer and why?

Graphing Calculator Exercises

The sine and cosine functions can be evaluated by using formulas called infinite series. It is shown in calculus that

$$\sin(x) = x - \frac{x^3}{3!} + \frac{x^5}{5!} - \frac{x^7}{7!} + \cdots \quad \text{and}$$

$$\cos(x) = 1 - \frac{x^2}{2!} + \frac{x^4}{4!} - \frac{x^6}{6!} + \cdots$$

for any real number x. You (or your calculator) can evaluate only a finite number of terms of the series, but the more terms you use, the closer the computation gets to the true value of the function. For any positive integer n, the notation $n!$ is read "n factorial" and $n! = 1 \cdot 2 \cdot 3 \cdot \cdots \cdot n$, the product of the positive integers from 1 through n inclusive.

1. Evaluate the expressions x, $x - x^3/3!$, $x - x^3/3! + x^5/5!$, $x - x^3/3! + x^5/5! - x^7/7!$, and so on, for $x = \pi/6$. Theoretically, none of these expressions has a value that is exactly $\sin(\pi/6)$. According to your calculator, which expression is the first one to have the same value as your calculator's value for $\sin(\pi/6)$? Repeat this exercise with $x = 13\pi/6$.

2. Evaluate the expressions $1 - x^2/2!$, $1 - x^2/2! + x^4/4!$, $1 - x^2/2! + x^4/4! - x^6/6!$, and so on, for $x = \pi/4$. Theoretically, none of these expressions gives the exact value of $\cos(\pi/4)$. According to your calculator, which expression is the first one to have the same value as your calculator's value for $\cos(\pi/4)$? Repeat this exercise with $x = 9\pi/4$.

6.3

The Graphs

of the Sine and

Cosine Functions

In Section 6.2 we studied the sine and cosine functions. In this section we will study their graphs. The graphs of trigonometric functions are important for understanding their use in modeling physical phenomena such as radio, sound, and light waves, and the motion of a spring or a pendulum.

The Graph of $y = \sin(x)$

Until now, s or α has been used as the independent variable in writing $\sin(s)$ or $\sin(\alpha)$. When graphing functions in an xy-coordinate system, it is customary to use x as the independent variable and y as the dependent variable, as in $y = \sin(x)$. We assume that x is a real number or radian measure unless it is stated that x is the degree measure of an angle. We often omit the parentheses and write $\sin x$ for $\sin(x)$.

Consider some ordered pairs that satisfy $y = \sin x$ for x in $[0, 2\pi]$:

x	0	$\pi/6$	$\pi/4$	$\pi/3$	$\pi/2$	$3\pi/4$	π	$5\pi/4$	$3\pi/2$	$7\pi/4$	2π
$y = \sin x$	0	0.5	0.707	0.866	1	0.707	0	-0.707	-1	-0.707	0

y-values increasing · *y*-values decreasing · *y*-values increasing

Maximum *y*-value · Minimum *y*-value

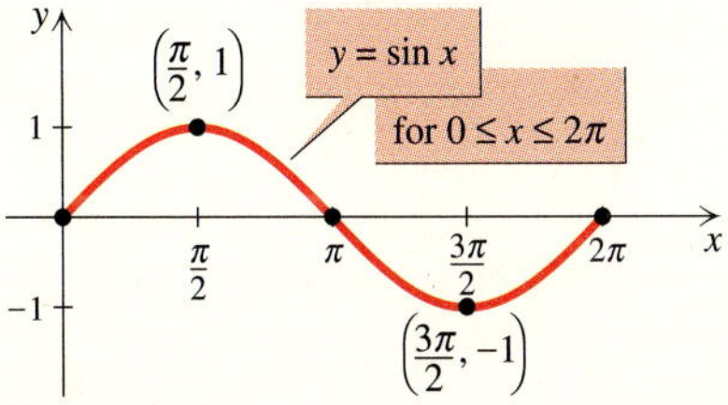

Figure 6.27

Figure 6.28

From this table, we see that the graph of $y = \sin x$ has x-intercepts at $(0, 0)$, $(\pi, 0)$, and $(2\pi, 0)$. On the interval $(0, \pi/2)$, y is increasing to reach its maximum value of 1 at $x = \pi/2$. On the interval $(\pi/2, 3\pi/2)$, y is decreasing to reach its minimum value of -1 at $3\pi/2$. On the interval $(3\pi/2, 2\pi)$, y is increasing up to 0. The graph of $y = \sin x$ for x in the interval $[0, 2\pi]$ is shown in Fig. 6.27. Since the x-intercepts and the maximum and minimum values of the function occur at multiples of π, we usually label the x-axis with multiples of π as in Fig. 6.27.

The graph of $y = \sin x$ can be better understood by recalling that $\sin x$ is the second coordinate of the terminal point on the unit circle for an arc of length x, as shown in Fig. 6.28. As the arc length x increases from length 0 to $\pi/2$; the second coordinate of the endpoint increases to 1. As the length of x increases from $\pi/2$ to π, the second coordinate of the endpoint decreases to 0. As the arc length x increases from π to 2π, the second coordinate decreases to -1 and then increases to 0.

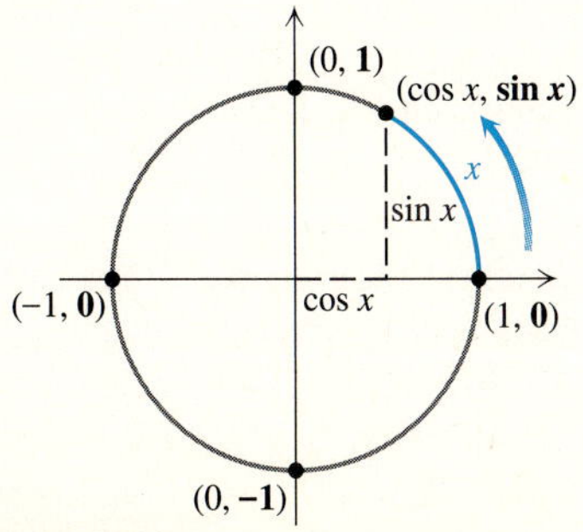

Since the domain of $y = \sin x$ is the set of all real numbers (or radian measures of angles), we must consider values of x outside the interval $[0, 2\pi]$. As x increases from 2π to 4π, the values of $\sin x$ again increase from 0 to 1, decrease to -1, then increase to 0. Because $\sin(x + 2\pi) = \sin x$, the exact shape that we saw in Fig. 6.27 is repeated for x in intervals such as $[2\pi, 4\pi]$, $[-2\pi, 0]$,

$[-4\pi, -2\pi]$, and so on. So the curve shown in Fig. 6.29 continues indefinitely to the left and right. The range of $y = \sin x$ is $[-1, 1]$. The graph $y = \sin x$ or any transformation of $y = \sin x$ is called a **sine wave**, a **sinusoidal wave**, or a **sinusoid**.

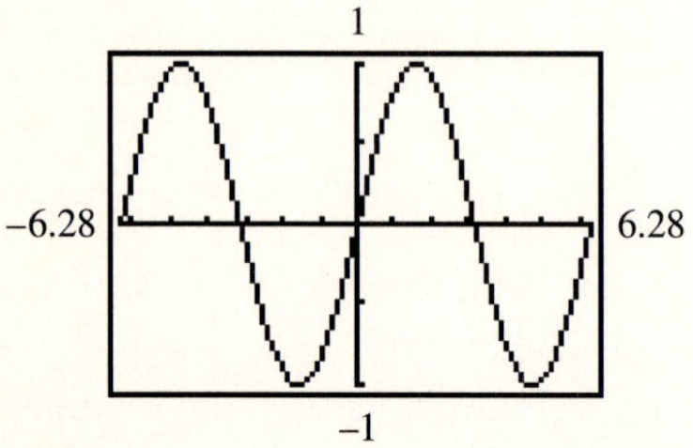

Graph $y = \sin x$ in radian mode. Use 6.28 for 2π.

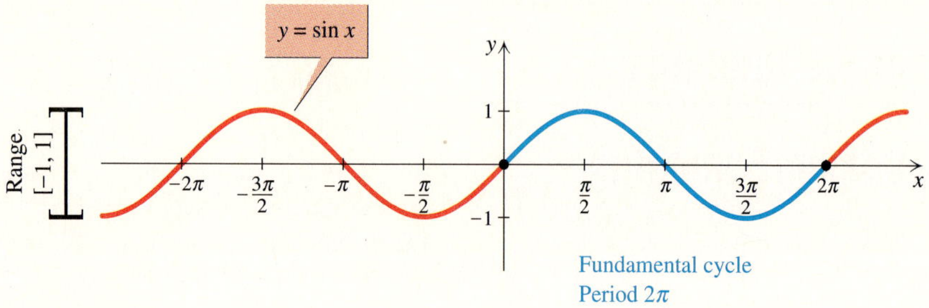

Figure 6.29

Since the shape of $y = \sin x$ for x in $[0, 2\pi]$ is repeated infinitely often, $y = \sin x$ is called a *periodic function*.

Definition: Periodic Function

If $y = f(x)$ is a function and a is a nonzero constant such that $f(x) = f(x + a)$ for every x in the domain of f, then f is called a **periodic function.** The smallest such positive constant a is the **period** of the function.

For the sine function, the smallest value of a such that $\sin x = \sin(x + a)$ is $a = 2\pi$, and so the period of $y = \sin x$ is 2π. To prove that 2π is actually the *smallest* value of a for which $\sin x = \sin(x + a)$ is a bit complicated, and we will omit the proof. The graph of $y = \sin x$ over any interval of length 2π is called a **cycle** of the sine wave. The graph of $y = \sin x$ over $[0, 2\pi]$ is called the **fundamental cycle** of $y = \sin x$.

Example 1 Graphing a periodic function

Sketch the graph of $y = 2 \sin x$ for x in the interval $[-2\pi, 2\pi]$.

Solution

The y-coordinates for $y = 2 \sin x$ can be obtained by doubling the y-coordinates in the table for $y = \sin x$. The five most important ordered pairs for sketching the graph between 0 and 2π are listed in the following table. Note that these five x-values divide the interval $[0, 2\pi]$ into four equal parts.

x	0	$\pi/2$	π	$3\pi/2$	2π
$y = 2 \sin x$	0	2	0	-2	0

The x-intercepts for $y = 2 \sin x$ on $[0, 2\pi]$ are $(0, 0)$, $(\pi, 0)$, and $(2\pi, 0)$. Midway between the intercepts are found the highest point $(\pi/2, 2)$ and the lowest point $(3\pi/2, -2)$. Draw one cycle of $y = 2 \sin x$ through these five points as shown in Fig. 6.30. Draw another cycle of the sine wave for x in $[-2\pi, 0]$ to complete the graph of $y = 2 \sin x$ for x in $[-2\pi, 2\pi]$. ◆

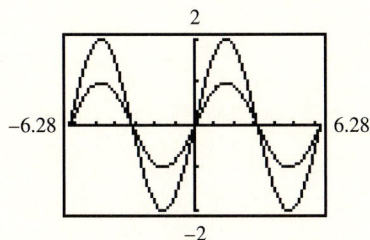

Compare the amplitudes of $y_1 = \sin x$ and $y_2 = 2 \sin x$.

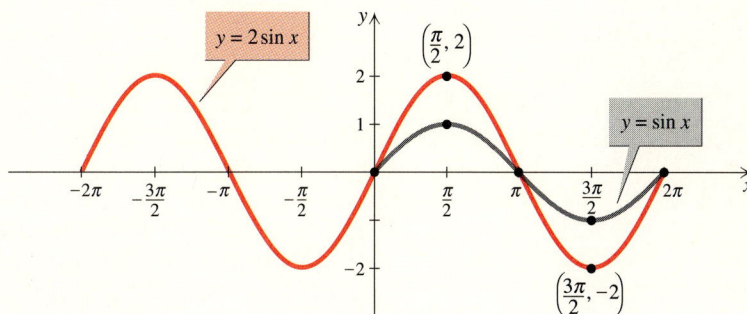

Figure 6.30

The amplitude of a sine wave is a measure of the "height" of the wave. When an oscilloscope is used to get a picture of the sine wave corresponding to a sound, the amplitude of the sine wave corresponds to the intensity or loudness of the sound.

Definition: Amplitude

The **amplitude** of a sine wave, or the amplitude of the function, is the absolute value of half the difference between the maximum and minimum y-coordinates on the wave.

Example 2 Finding amplitude

Find the amplitude of the functions $y = \sin x$ and $y = 2 \sin x$.

Solution

For $y = \sin x$ the maximum y-coordinate is 1 and the minimum is -1. So the amplitude is

$$\left| \frac{1}{2} [1 - (-1)] \right| = 1.$$

For $y = 2 \sin x$, the maximum y-coordinate is 2 and the minimum is -2. So the amplitude is

$$\left| \frac{1}{2} [2 - (-2)] \right| = 2.$$

◆

The Graph of $y = \cos(x)$

Consider some ordered pairs that satisfy $y = \cos x$ for x in $[0, 2\pi]$:

x	0	$\pi/4$	$\pi/2$	$3\pi/4$	π	$5\pi/4$	$3\pi/2$	$7\pi/4$	2π
$y = \cos x$	1	0.707	0	-0.707	-1	-0.707	0	0.707	1

$\underbrace{\qquad\qquad\qquad}_{\text{y-values decreasing}}$ $\underbrace{\qquad\qquad\qquad}_{\text{y-values increasing}}$

↑ Maximum y-value   ↑ Minimum y-value   ↑ Maximum y-value

On the interval $[0, 2\pi]$, the curve starts at the high point $(0, 1)$ and decreases to its lowest point $(\pi, -1)$ and then increases up to another high point $(2\pi, 1)$. Draw a curve passing through these points and the x-intercepts $(\pi/2, 0)$ and $(3\pi/2, 0)$ as shown in Fig. 6.31. The graph of $y = \cos x$ over $[0, 2\pi]$ is called the **fundamental cycle** of $y = \cos x$. Since $\cos x = \cos(x + 2\pi)$, the fundamental cycle of $y = \cos x$ is repeated on $[2\pi, 4\pi]$, $[-2\pi, 0]$, and so on. The graph of $y = \cos x$ is shown in Fig. 6.31.

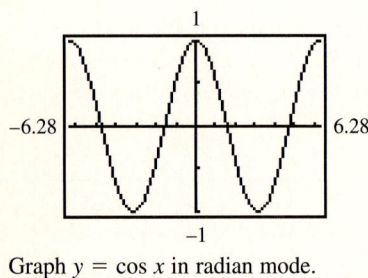

Graph $y = \cos x$ in radian mode.

Figure 6.31

The graph of $y = \cos x$ can be better understood by recalling that $\cos x$ is the first coordinate of the terminal point on the unit circle for an arc of length x, as shown in Fig. 6.28. As the arc length x increases from length 0 to $\pi/2$, the first coordinate of the endpoint decreases from 1 to 0. As the length of x increases from $\pi/2$ to π, the first coordinate decreases from 0 to -1. As the length increases from π to 2π, the first coordinate increases from -1 to 1. The graph of $y = \cos x$ has exactly the same shape as the graph of $y = \sin x$. If the graph of $y = \sin x$ is shifted a distance of $\pi/2$ to the left, the graphs would coincide. For this reason the graph of $y = \cos x$ is also called a sine wave with amplitude 1 and period 2π.

Example 3   Graphing another periodic function

Sketch the graph of $y = -3 \cos x$ for x in the interval $[-2\pi, 2\pi]$ and find its amplitude.

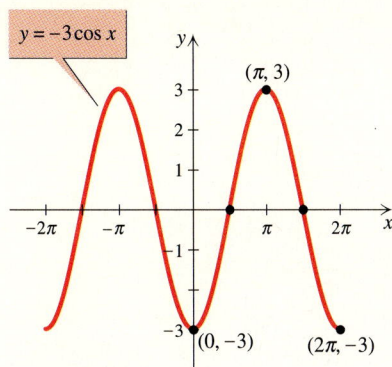

$y = -3\cos x$

$(\pi, 3)$

$(0, -3)$ $(2\pi, -3)$

Figure 6.32

Solution

Make a table of ordered pairs for x in $[0, 2\pi]$ to get one cycle of the graph. Note that the five x-coordinates in the table divide the interval $[0, 2\pi]$ into four equal parts.

x	0	$\pi/2$	π	$3\pi/2$	2π
$y = -3\cos x$	-3	0	3	0	-3

Draw one cycle of $y = -3\cos x$ through these five points as shown in Fig. 6.32. Repeat the same shape for x in the interval $[-2\pi, 0]$ to get the graph of $y = -3\cos x$ for x in $[-2\pi, 2\pi]$. The amplitude is $|0.5(3 - (-3))| = 3$. ◆

Transformations of Sine and Cosine

In Section 3.4 we discussed how various changes in a formula affect the graph of the function. We know the changes that cause horizontal or vertical translations, reflections, and stretching or shrinking. The graph of $y = 2\sin x$ from Example 1 can be obtained by stretching the graph of $y = \sin x$. The graph of $y = -3\cos x$ in Example 3 can be obtained by stretching and reflecting the graph of $y = \cos x$. The amplitude of $y = 2\sin x$ is 2 and the amplitude of $y = -3\sin x$ is 3. In general, the amplitude is determined by the coefficient of the sine or cosine function.

▼

Theorem: Amplitude

> The amplitude of $y = A\sin x$ or $y = A\cos x$ is $|A|$.

In Section 3.4 we saw that the graph of $y = f(x - C)$ is a horizontal translation of the graph of $y = f(x)$, to the right if $C > 0$ and to the left if $C < 0$. So the graphs of $y = \sin(x - C)$ and $y = \cos(x - C)$ are horizontal translations of $y = \sin x$ and $y = \cos x$, respectively, to the right if $C > 0$ and to the left if $C < 0$.

▼

Definition: Phase Shift

> The **phase shift** of the graph of $y = \sin(x - C)$ or $y = \cos(x - C)$ is $|C|$.

Example 4   Horizontal translation

Graph two cycles of $y = \sin(x + \pi/6)$, and determine the phase shift of the graph.

Solution

Since $x + \pi/6 = x - (-\pi/6)$ and $C < 0$, the graph of $y = \sin(x + \pi/6)$ is obtained by moving $y = \sin x$ a distance of $\pi/6$ to the left. Since the phase shift is

$\pi/6$, label the x-axis with multiples of $\pi/6$ as shown in Fig. 6.33. Concentrate on moving the fundamental cycle of $y = \sin x$. The three x-intercepts $(0, 0)$, $(\pi, 0)$, and $(2\pi, 0)$ move to $(-\pi/6, 0)$, $(5\pi/6, 0)$, and $(11\pi/6, 0)$. The high and low points, $(\pi/2, 1)$ and $(3\pi/2, -1)$, move to $(\pi/3, 1)$ and $(4\pi/3, -1)$. Draw one cycle through these five points and continue the pattern for another cycle as shown in Fig. 6.33.

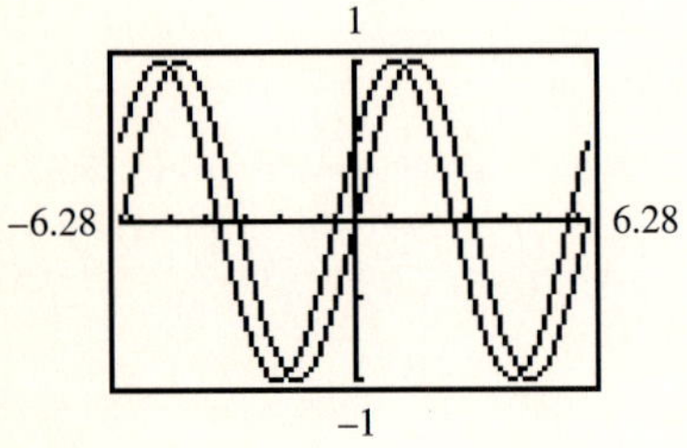

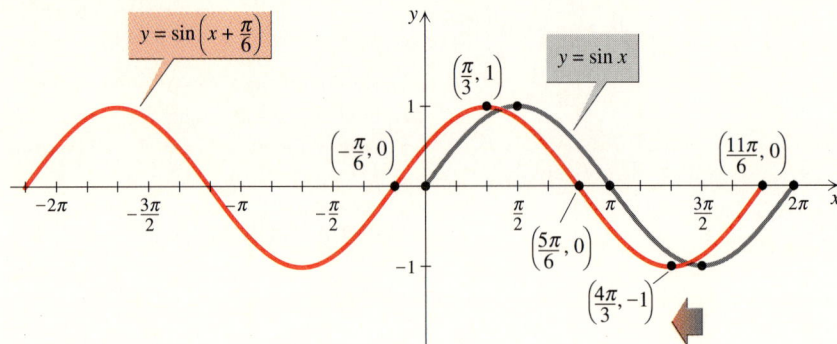

Graph $y_1 = \sin x$ and $y_2 = \sin(x + \pi/6)$ to see the phase shift.

Figure 6.33

The graphs of $y = \sin(x) + D$ and $y = \cos(x) + D$ are vertical translations of $y = \sin x$ and $y = \cos x$, respectively. The translation is upward for $D > 0$ and downward for $D < 0$. The next example combines a phase shift and a vertical translation.

Example 5   Horizontal and vertical translation

Graph two cycles of $y = \cos(x - \pi/4) + 2$, and determine the phase shift of the graph.

Solution

The graph of $y = \cos(x - \pi/4) + 2$ is obtained by moving $y = \cos x$ a distance of $\pi/4$ to the right and two units upward. Since the phase shift is $\pi/4$, label the x-axis with multiples of $\pi/4$ as shown in Fig. 6.34.

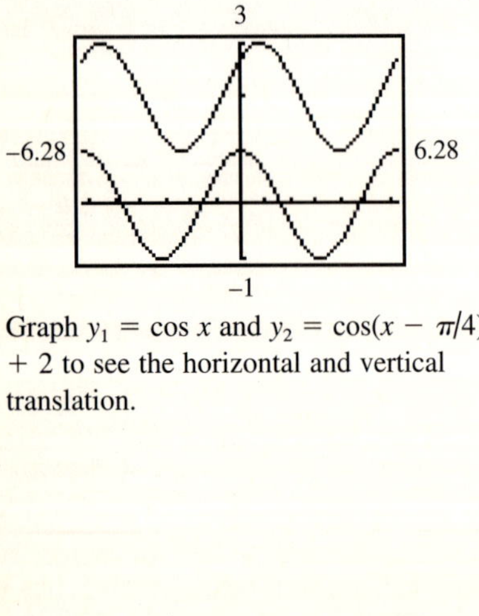

Graph $y_1 = \cos x$ and $y_2 = \cos(x - \pi/4)$ $+ 2$ to see the horizontal and vertical translation.

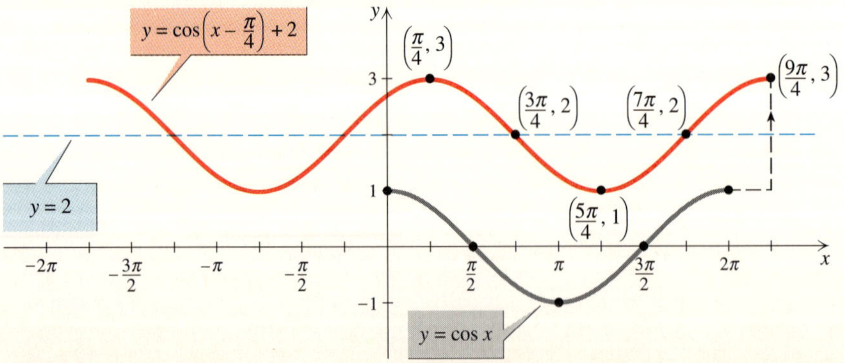

Figure 6.34

Concentrate on moving the fundamental cycle of $y = \cos x$. The points $(0, 1)$, $(\pi, -1)$, and $(2\pi, 1)$ move to $(\pi/4, 3)$, $(5\pi/4, 1)$, and $(9\pi/4, 3)$. The x-intercepts $(\pi/2, 0)$ and $(3\pi/2, 0)$ move to $(3\pi/4, 2)$ and $(7\pi/4, 2)$. Draw one cycle through these five points and continue the pattern for another cycle as shown in Fig. 6.34.   ◆

Changing the Period

We can alter the sine and cosine functions in a way that we did not alter the functions of algebra in Chapter 3. The period of a periodic function can be changed by replacing x by a multiple of x.

Example 6   Changing the period

Graph two cycles of $y = \sin(2x)$ and determine the period of the function.

Solution

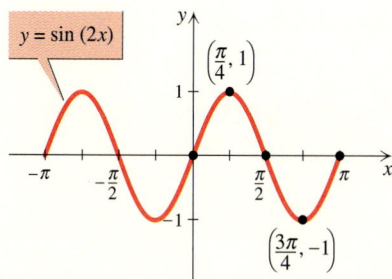

Figure 6.35

The graph of $y = \sin x$ completes its fundamental cycle for $0 \le x \le 2\pi$. So $y = \sin(2x)$ completes one cycle for $0 \le 2x \le 2\pi$, or $0 \le x \le \pi$. So the period is π. For $0 \le x \le \pi$, the x-intercepts are $(0, 0)$, $(\pi/2, 0)$, and $(\pi, 0)$. Midway between 0 and $\pi/2$, at $x = \pi/4$, the function reaches a maximum value of 1, and the function attains its minimum value of -1 at $x = 3\pi/4$. Draw one cycle of the graph through the x-intercepts and through $(\pi/4, 1)$ and $(3\pi/4, -1)$. Then graph another cycle of $y = \sin(2x)$ as shown in Fig. 6.35.   ◆

Note that in Example 6 the period of $y = \sin(2x)$ is the period of $y = \sin x$ divided by 2. In general, one complete cycle of $y = \sin(Bx)$ or $y = \cos(Bx)$ for $B > 0$ occurs for $0 \le Bx \le 2\pi$, or $0 \le x \le 2\pi/B$.

Theorem: Period of $y = \sin(Bx)$ and $y = \cos(Bx)$

The period P of $y = \sin(Bx)$ and $y = \cos(Bx)$ is given by

$$P = \frac{2\pi}{B}.$$

Note that the period is a natural number (that is, not a multiple of π) when B is a multiple of π.

Example 7   A period that is not a multiple of π

Determine the period of $y = \cos\left(\frac{\pi}{2} x\right)$ and graph two cycles of the function.

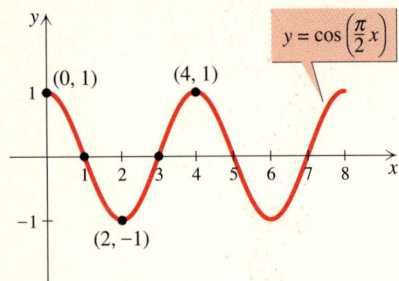

Figure 6.36

Solution

For this function $B = \pi/2$. To find the period, use $P = 2\pi/B$:

$$P = \frac{2\pi}{\pi/2} = 4$$

So one cycle of $y = \cos\left(\frac{\pi}{2}x\right)$ is completed for $0 \leq x \leq 4$. The cycle starts at $(0, 1)$ and ends at $(4, 1)$. A minimum point occurs halfway in between, at $(2, -1)$. The x-intercepts are $(1, 0)$ and $(3, 0)$. Draw a curve through these five points to get one cycle of the graph. Continue this pattern from 4 to 8 to get a second cycle as shown in Fig. 6.36.   ◆

The General Sine Wave

We can use any combination of translating, reflecting, phase shifting, stretching, shrinking, or period changing in a single trigonometric function.

The General Sine Wave

The graph of

$$y = A\,\sin(B[x - C]) + D \qquad \text{or} \qquad y = A\,\cos(B[x - C]) + D$$

is a sine wave with an amplitude $|A|$, period $2\pi/B$ ($B > 0$), phase shift $|C|$, and vertical translation D.

We assume that $B > 0$, because any general sine or cosine function can be rewritten with $B > 0$ using identities from Chapter 7. Notice that A and B affect the shape of the curve, while C and D determine its location.

Procedure: Graphing a Sine Wave

To graph $y = A\,\sin(B[x - C]) + D$ or $y = A\,\cos(B[x - C]) + D$:

1. **Sketch one cycle of $y = \sin Bx$ or $y = \cos Bx$ on $[0, 2\pi/B]$.**
2. **Change the amplitude of the cycle according to the value of A.**
3. **If $A < 0$, reflect the curve in the x-axis.**
4. **Translate the cycle $|C|$ units to the right if $C > 0$ and to the left if $C < 0$.**
5. **Translate the cycle $|D|$ units upward if $D > 0$ and downward if $D < 0$.**

Example 8   Stretching, translating, and period changing

Determine amplitude, period, and phase shift, and sketch two cycles of $y = 2 \sin(3x + \pi) + 1$.

Solution

First we rewrite the function in the form $y = A \sin(B[x - C]) + D$ by factoring 3 out of $3x + \pi$:

$$y = 2 \sin\left(3\left[x + \frac{\pi}{3}\right]\right) + 1$$

From this equation we get $A = 2$, $B = 3$, and $C = -\pi/3$. So the amplitude is 2, the period is $2\pi/3$, and the phase shift is $\pi/3$ to the left. The period change causes the fundamental cycle of $y = \sin x$ on $[0, 2\pi]$ to shrink to the interval $[0, 2\pi/3]$. Now draw one cycle of $y = \sin 3x$ on $[0, 2\pi/3]$ as shown in Fig. 6.37. Stretch the cycle vertically so that it has an amplitude of 2. The numbers $\pi/3$ and 1 shift the cycle a distance of $\pi/3$ to the left and up one unit. So one cycle of the function occurs on $[-\pi/3, \pi/3]$. Check by evaluating the function at the endpoints and midpoint of the interval $[-\pi/3, \pi/3]$ to get $(-\pi/3, 1)$, $(0, 1)$, and $(\pi/3, 1)$. Since the graph is shifted one unit upward, these points are the points where the curve intersects the line $y = 1$. Evaluate the function midway between these points to get $(-\pi/6, 3)$ and $(\pi/6, -1)$, the highest and lowest points of this cycle. One cycle is drawn through these five points and continued for another cycle, as shown in Fig. 6.37.

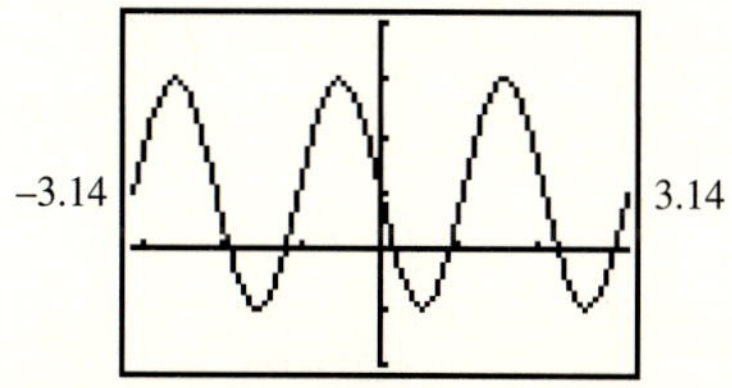

Graph $y = 2 \sin(3x + \pi) + 1$ to check the amplitude, period, and phase shift found in Example 8.

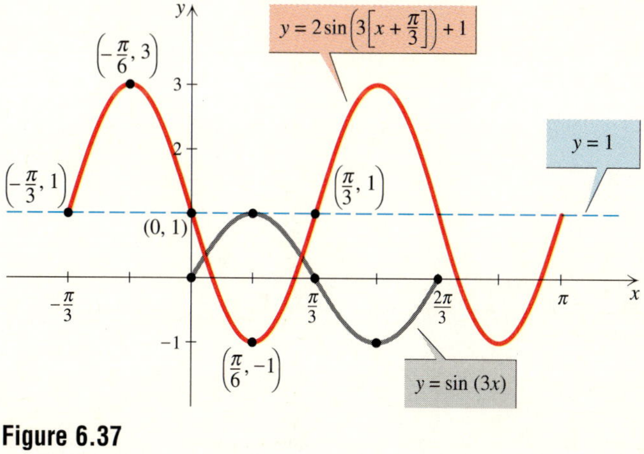

Figure 6.37

In all of the examples given we have used the real numbers or the set of radian measures of angles as the domain of the functions. If we use degree measure as the domain, then we must make some adjustments. For example, the period of $y = \sin x$ is $360°$ rather than 2π when x is the degree measure of an angle.

Example 9   Degree measures as the domain

Determine amplitude, period, and phase shift, and sketch one cycle of $y = -3 \cos(2x - 180°) - 1$, where x is the degree measure of an angle.

Solution

Rewrite the function in the general form as

$$y = -3 \cos(2[x - 90°]) - 1.$$

The amplitude is 3. Since the period in degree measure is $360°/B$, the period is $360°/2$ or $180°$. The fundamental cycle is shrunk to the interval $[0°, 180°]$. Sketch one cycle of $y = \cos 2x$ on $[0°, 180°]$, as shown in Fig. 6.38. This cycle is reflected in the x-axis and stretched by a factor of 3. A shift of $90°$ to the right means that one cycle of the original function occurs for $90° \le x \le 270°$. Evaluate the function at the endpoints and midpoint of the interval $[90°, 270°]$ to get $(90°, -4)$, $(180°, 2)$, and $(270°, -4)$ for the endpoints and midpoint of this cycle. Since the graph is translated downward one unit, midway between these maximum and minimum points we get points where the graph intersects the line $y = -1$. These points are $(135°, -1)$ and $(225°, -1)$. Draw one cycle of the graph through these five points as shown in Fig. 6.38.

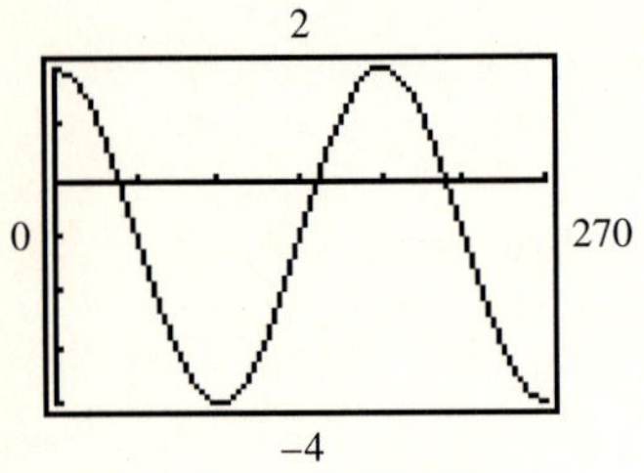

Use degree mode to graph
$y = -3 \cos(2x - 180) - 1.$

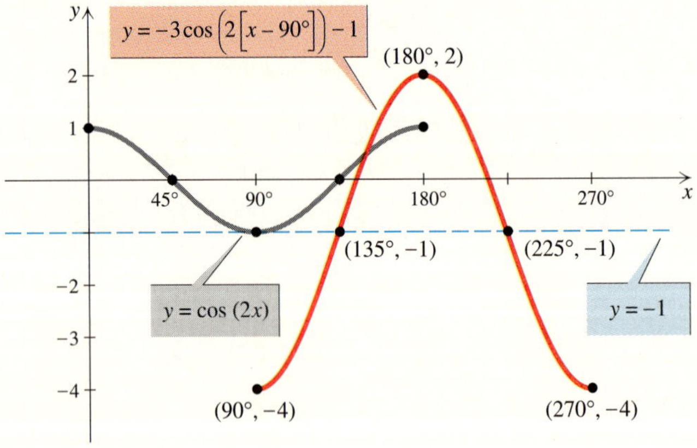

Figure 6.38

Frequency

Sine waves are used to model physical phenomena such as radio, sound, or light waves. A high-frequency radio wave is a wave that has a large number of cycles per second. If we think of the x-axis as a time axis, then the period of a sine wave is the amount of time required for the wave to complete one cycle, and the reciprocal of the period is the number of cycles per unit of time. For example, the sound wave for middle C on a piano completes 264 cycles per second. The period of the wave is 1/264 second, which means that one cycle is completed in 1/264 second.

Definition: Frequency

The **frequency** F of a sine wave with period P is defined by $F = 1/P$.

Example 10   Frequency of a sine wave

Find the frequency of the sine wave given by $y = \sin(528\pi x)$.

Solution

First find the period:

$$P = \frac{2\pi}{528\pi} = \frac{1}{264}$$

Since $F = 1/P$, the frequency is 264. A sine wave with this frequency completes 264 cycles for x in [0, 1] (or x in any interval of length one).   ◆

Note that if B is a large positive number in $y = \sin(Bx)$ or $y = \cos(Bx)$, then the period is short and the frequency is high.

? For Thought

True or false? Explain.

1. The period of $y = \cos(2\pi x)$ is π.

2. The range of $y = 4\sin(x) + 3$ is $[-4, 4]$.

3. The graph of $y = \sin(2x + \pi/6)$ has a period of π and a phase shift $\pi/6$.

4. The points $(5\pi/6, 0)$ and $(11\pi/6, 0)$ are on the graph of $y = \cos(x - \pi/3)$.

5. The frequency of the sine wave $y = \sin x$ is $1/(2\pi)$.

6. The period for $y = \sin(0.1\pi x)$ is 20.

7. The graphs of $y = \sin x$ and $y = \cos(x + \pi/2)$ are identical.

8. If x is measured in degrees, then the period of $y = \cos(4x)$ is $90°$.

9. The maximum value of the function $y = -2\cos(3x) + 4$ is 6.

10. The range of the function $y = 3\sin(5x - \pi) + 2$ is $[-1, 5]$.

6.3 Exercises ◖▭◗ Tape 11 🖫 Disk—5.25": 4   3.5": 3   Macintosh: 3

Determine the amplitude, period, and phase shift for each function.

1. $y = -2 \cos x$ **2.** $y = -4 \cos x$

3. $f(x) = \cos(x - \pi/2)$ **4.** $f(x) = \sin(x + \pi/2)$

5. $y = -2 \sin(x + \pi/3)$ **6.** $y = -3 \sin(x - \pi/6)$

Determine the amplitude and the phase shift for each function, and sketch the graph of each function for $-2\pi \le x \le 2\pi$.

7. $y = -\sin x$ **8.** $y = -\cos x$ **9.** $y = -3 \sin x$

10. $y = 4 \sin x$ **11.** $y = \dfrac{1}{2} \cos x$ **12.** $y = \dfrac{1}{3} \cos x$

13. $y = \sin(x + \pi)$ **14.** $y = \cos(x - \pi)$

15. $y = \cos(x - \pi/3)$ **16.** $y = \cos(x + \pi/4)$

17. $f(x) = \cos(x) + 2$ **18.** $f(x) = \cos(x) - 3$

19. $y = -\sin(x) - 1$ **20.** $y = -\sin(x) + 2$

21. $y = \sin(x + \pi/4) + 2$ **22.** $y = \sin(x - \pi/2) - 2$

23. $y = 2 \cos(x + \pi/6) + 1$ **24.** $y = 3 \cos(x + 2\pi/3) - 2$

25. $f(x) = -2 \sin(x - \pi/3) + 1$

26. $f(x) = -3 \cos(x + \pi/3) - 1$

Sketch the graph of each function, where x is the degree measure of an angle and $-360° \le x \le 360°$. Determine the amplitude and phase shift for each function.

27. $y = \sin x$ **28.** $y = \cos x$

29. $f(x) = 3 \cos(x) + 1$ **30.** $f(x) = 2 \sin(x) - 2$

31. $y = \sin(x - 45°)$ **32.** $y = \sin(x + 30°)$

33. $y = -2 \cos(x + 90°)$ **34.** $y = -\cos(x - 60°)$

35. $y = 4 \sin(x - 90°) + 1$ **36.** $y = 3 \sin(x + 180°) - 1$

37. $f(x) = \dfrac{1}{2} \cos(x - 30°) - 2$

38. $f(x) = -\cos(x - 90°) - 1$

Determine amplitude, period, and phase shift for each function.

39. $y = 3 \sin(4x)$ **40.** $y = -\cos(x/2) + 3$

41. $y = -2 \cos\left(2x + \dfrac{\pi}{2}\right) - 1$

42. $y = 4 \cos(3x - 2\pi)$

43. $y = -2 \sin(\pi x - \pi)$

44. $y = \sin\left(\dfrac{\pi}{2}x + \pi\right)$

Sketch at least two cycles of the graph of each function, where x is a real number. Determine the period, the phase shift, and the range of the function.

45. $y = \sin(3x)$ **46.** $y = \cos(x/3)$ **47.** $y = -\sin(2x)$

48. $y = -\cos(3x)$ **49.** $y = \cos(4x) + 2$

50. $y = \sin(3x) - 1$ **51.** $y = 2 - \sin(x/4)$

52. $y = 3 - \cos(x/5)$ **53.** $y = \sin\left(\dfrac{\pi}{3}x\right)$

54. $y = \sin\left(\dfrac{\pi}{4}x\right)$ **55.** $f(x) = \sin\left(2\left[x - \dfrac{\pi}{2}\right]\right)$

56. $f(x) = \sin\left(3\left[x + \dfrac{\pi}{3}\right]\right)$

57. $f(x) = \sin\left(\dfrac{\pi}{2}x + \dfrac{3\pi}{2}\right)$ **58.** $f(x) = \cos\left(\dfrac{\pi}{3}x - \dfrac{\pi}{3}\right)$

59. $y = 2 \cos\left(2\left[x + \dfrac{\pi}{6}\right]\right) + 1$

60. $y = 3 \cos\left(4\left[x - \dfrac{\pi}{2}\right]\right) - 1$

61. $y = -\dfrac{1}{2} \sin\left(3\left[x - \dfrac{\pi}{6}\right]\right) - 1$

62. $y = -\dfrac{1}{2} \sin\left(4\left[x + \dfrac{\pi}{4}\right]\right) + 1$

Sketch at least two cycles of the graph of each function, where x is the degree measure of an angle in standard position. Determine the period, the phase shift, and the range of the function.

63. $y = \sin(18x)$ **64.** $y = \sin(12x)$

65. $y = \cos(3x + 90°)$ **66.** $y = \cos(4x - 180°)$

67. $y = -2 \sin(2[x - 60°]) + 1$

68. $y = 3 \sin(3[x + 120°])$

Write an equation of the form $y = A \sin(B[x - C]) + D$
whose graph is the given sine wave.

69.

70.

71.

72.

Solve each problem.

73. What is the frequency of the sine wave determined by $y = \sin(200\pi x)$, where x is time in seconds?

74. What is the frequency of the sine wave determined by $y = \cos(0.001\pi x)$, where x is time in seconds?

75. If the period of a sine wave is 0.025 hr, then what is the frequency?

76. If the frequency of a sine wave is 40,000 cycles per second, then what is the period?

77. *Motion of a spring* A weight hanging on a vertical spring is set in motion with a downward velocity of 6 cm/sec from its equilibrium position. Assume that the constant ω for this particular spring and weight combination is 2. Write the formula that gives the location of the weight in centimeters as a function of the time t in seconds. Find the amplitude and period of the function and sketch its graph for t in the interval $[0, 2\pi]$. (See Section 6.2 for the general formula that describes the motion of a spring.)

78. *Motion of a spring* A weight hanging on a vertical spring is set in motion with an upward velocity of 4 cm/sec from its

equilibrium position. Assume that the constant ω for this particular spring and weight combination is π. Write the formula that gives the location of the weight in centimeters as a function of the time t in seconds. Find the period of the function and sketch its graph for t in the interval $[0, 4]$.

79. *Sun spots* Astronomers have been recording sunspot activity for over 130 years. The number of sunspots per year varies periodically over time as shown in the graph. What is the approximate period of this periodic function?

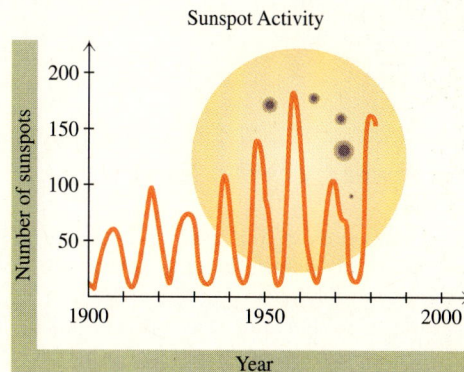

Figure for Exercise 79

80. *First pulsar* In 1967, Jocelyn Bell, a graduate student at Cambridge University, England, found the peculiar pattern shown in the graph on a paper chart from a radio telescope. She had made the first discovery of a pulsar, a very small neutron star that emits beams of radiation as it rotates as fast as 1000 times per second. From the graph shown here, estimate the period of the first discovered pulsar, now known as CP 1919.

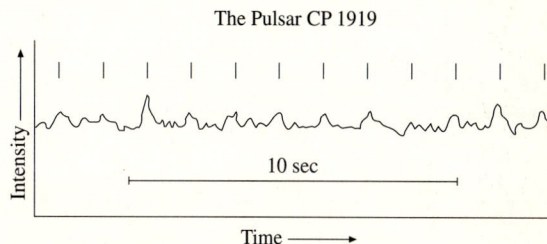

Figure for Exercise 80

81. *Lung capacity* The volume of air v in cubic centimeters in the lungs of a certain long distance runner is modeled by the equation $v = 400 \sin(60\pi t) + 900$, where t is time in minutes.

a) What are the maximum and minimum volumes of air in the runner's lungs at any time?

b) How many breaths does the runner take per minute?

82. *Blood velocity* The velocity v of blood at a valve in the heart of a certain rodent is modeled by the equation $v = -4 \cos(6\pi t) + 4$, where v is in centimeters per second and t is time in seconds.

a) What are the maximum and minimum velocities of the blood at this valve?

b) What is the rodent's heart rate in beats per minute?

For Writing/Discussion

83. *Periodic temperature* Air temperature T generally varies in a periodic manner, with highs during the day and lows during the night. Assume that no drastic changes in the weather are expected and let t be time in hours with $t = 0$ at midnight tonight. Graph T for t in the interval $[0, 48]$ and write a function of the form $T = A \sin(B[t - C]) + D$ for your graph. Explain your choices for A, B, C, and D.

84. *Even or odd* Determine whether $y = \sin x$ and $y = \cos x$ are even or odd functions and explain your answers.

Graphing Calculator Exercises

In calculus it is proved that the sine function can be approximated to any degree of accuracy by polynomial functions called Taylor polynomials. The polynomials in the following exercises are Taylor polynomials.

1. Graph the functions

$$y_1 = x - \frac{x^3}{3!}, \qquad y_2 = x - \frac{x^3}{3!} + \frac{x^5}{5!},$$

$$y_3 = x - \frac{x^3}{3!} + \frac{x^5}{5!} - \frac{x^7}{7!}, \qquad \text{and} \qquad y_4 = \sin x.$$

Which of the three polynomials approximates the sine curve the best for $-\pi \le x \le \pi$?

2. Graph the functions

$$y_1 = x - \frac{x^3}{3!} + \frac{x^5}{5!} - \frac{x^7}{7!} + \frac{x^9}{9!} - \frac{x^{11}}{11!} + \frac{x^{13}}{13!} \qquad \text{and}$$

$$y_2 = \sin x.$$

For what values of x do the y-coordinates on these curves differ by less than 0.4?

6.4

The Other

Trigonometric

Functions and

Their Graphs

So far we have studied two of the six trigonometric functions. In this section we will define the remaining four functions.

Definitions

The tangent, cotangent, secant, and cosecant functions are all defined in terms of the sine and cosine functions as shown at the top of the next page. We use the abbreviation tan for tangent, cot for cotangent, sec for secant, and csc for cosecant. As usual, we may think of α as an angle, the measure of an angle in degrees, the measure of an angle in radians, or a real number.

The domain of the tangent and secant function is the set of angles except those for which $\cos \alpha = 0$. The only points on the unit circle where the first coordinate is 0 are $(0, 1)$ and $(0, -1)$. Angles such as $\pi/2, 3\pi/2, 5\pi/2$, and so on, have terminal sides through either $(0, 1)$ or $(0, -1)$. These angles are of the form $\pi/2 + k\pi$, where k is any integer. So $\cos(\pi/2 + k\pi) = 0$ for any integer k.

Definition: Tangent, Cotangent, Secant, and Cosecant Functions

If α is an angle in standard position, we define the **tangent, cotangent, secant,** and **cosecant** functions as

$$\tan \alpha = \frac{\sin \alpha}{\cos \alpha}, \quad \cot \alpha = \frac{\cos \alpha}{\sin \alpha}, \quad \sec \alpha = \frac{1}{\cos \alpha}, \quad \text{and} \quad \csc \alpha = \frac{1}{\sin \alpha}.$$

We exclude from the domain of each function any value of α for which the denominator is 0.

By definition, the zeros of the cosine function are excluded from the domain of the tangent and secant. So the domain of tangent and secant is

$$\left\{ \alpha \,\middle|\, \alpha \neq \frac{\pi}{2} + k\pi, \text{ where } k \text{ is an integer} \right\}.$$

The only points on the unit circle where the second coordinate is 0 are $(1, 0)$ and $(-1, 0)$. Angles that are multiples of π, such as 0, π, 2π, and so on, have terminal sides that go through one of these points. Any angle of the form $k\pi$, where k is any integer, has a terminal side that goes through either $(1, 0)$ or $(-1, 0)$. So $\sin(k\pi) = 0$ for any integer k. By definition, the zeros of the sine function are excluded from the domain of cotangent and cosecant; thus the domain of cotangent and cosecant is

$$\{ \alpha \,|\, \alpha \neq k\pi, \text{ where } k \text{ is an integer} \}.$$

To find the values of the six trigonometric functions for an angle α, first find $\sin \alpha$, $\cos \alpha$, and $\tan \alpha$ (by dividing $\sin \alpha$ by $\cos \alpha$). By definition, $\csc \alpha$, $\sec \alpha$, and $\cot \alpha$ are the reciprocals of $\sin \alpha$, $\cos \alpha$, and $\tan \alpha$, respectively. The signs of the six trigonometric functions for angles in each quadrant are shown in Fig. 6.39. It is not necessary to memorize these signs, because they can be easily obtained by knowing the signs of the sine and cosine functions in each quadrant.

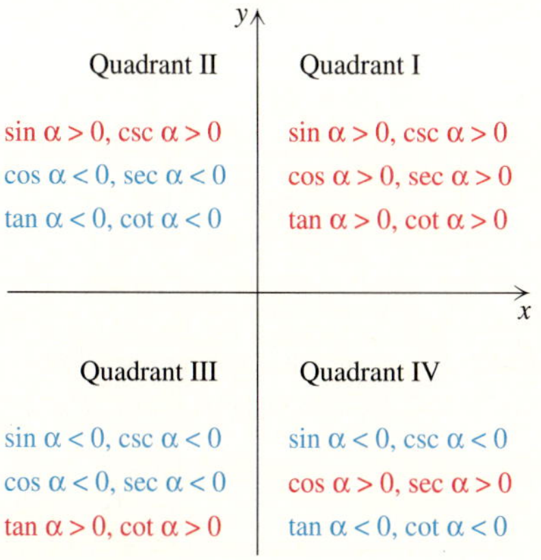

Quadrant II	Quadrant I
$\sin \alpha > 0$, $\csc \alpha > 0$	$\sin \alpha > 0$, $\csc \alpha > 0$
$\cos \alpha < 0$, $\sec \alpha < 0$	$\cos \alpha > 0$, $\sec \alpha > 0$
$\tan \alpha < 0$, $\cot \alpha < 0$	$\tan \alpha > 0$, $\cot \alpha > 0$
Quadrant III	Quadrant IV
$\sin \alpha < 0$, $\csc \alpha < 0$	$\sin \alpha < 0$, $\csc \alpha < 0$
$\cos \alpha < 0$, $\sec \alpha < 0$	$\cos \alpha > 0$, $\sec \alpha > 0$
$\tan \alpha > 0$, $\cot \alpha > 0$	$\tan \alpha < 0$, $\cot \alpha < 0$

Figure 6.39

Example 1   Evaluating the trigonometric functions

Find the values of all six trigonometric functions for each angle.

a) $\pi/4$ **b)** $150°$

Solution

a) Use $\sin(\pi/4) = \sqrt{2}/2$ and $\cos(\pi/4) = \sqrt{2}/2$ to find the other values:

$$\tan(\pi/4) = \frac{\sin(\pi/4)}{\cos(\pi/4)} = \frac{\sqrt{2}/2}{\sqrt{2}/2} = 1 \qquad \cot(\pi/4) = \frac{1}{\tan(\pi/4)} = 1$$

$$\sec(\pi/4) = \frac{1}{\cos(\pi/4)} = \frac{2}{\sqrt{2}} = \sqrt{2} \qquad \csc(\pi/4) = \frac{1}{\sin(\pi/4)} = \frac{2}{\sqrt{2}} = \sqrt{2}$$

b) The reference angle for $150°$ is $30°$. So $\sin(150°) = \sin(30°) = 1/2$ and $\cos(150°) = -\cos(30°) = -\sqrt{3}/2$. Use these values to find the other four values:

$$\tan(150°) = \frac{1/2}{-\sqrt{3}/2} = -\frac{1}{\sqrt{3}} = -\frac{\sqrt{3}}{3} \qquad \cot(150°) = -\frac{3}{\sqrt{3}} = -\sqrt{3}$$

$$\sec(150°) = \frac{1}{\cos(150°)} = -\frac{2}{\sqrt{3}} = -\frac{2\sqrt{3}}{3} \qquad \csc(150°) = \frac{1}{\sin(150°)} = 2$$

◆

Most scientific calculators have keys for the sine, cosine, and tangent functions only. To find values for the other three functions, we use the definitions. Keys labeled $\sin^{-1}$, $\cos^{-1}$, and $\tan^{-1}$ on a calculator are for the inverse trigonometric functions, which we will study in the next section. These keys do *not* give the values of $1/\sin x$, $1/\cos x$, or $1/\tan x$.

Example 2   Evaluating with a calculator

Use a calculator to find approximate values rounded to four decimal places.

a) $\sec(\pi/12)$

b) $\csc(123°)$

c) $\cot(-12.4)$

Solution

a) Make sure that your calculator is in radian mode. Since $\sec x = 1/\cos x$,

$$\sec(\pi/12) = \frac{1}{\cos(\pi/12)} \approx 1.0353.$$

One way to perform this computation on scientific calculators is to find $\cos(\pi/12)$ and then use the reciprocal function key, which is labeled $1/x$.

b) Change your calculator to degree mode. Since $\csc x = 1/\sin x$, we have

$$\csc(123°) = \frac{1}{\sin(123°)} \approx 1.1924.$$

c) Change your calculator to radian mode. Since $\cot x = 1/\tan x$, we have

$$\cot(-12.4) = \frac{1}{\tan(-12.4)} \approx 5.9551.$$

◆

```
1/cos (π/12)
           1.03527618
1/sin (123°)
           1.192363293
1/tan -12.4
           5.955117373
```

In radian mode, you can use the degree symbol to find $1/\sin(123°)$.

Graph of $y = \tan(x)$

Consider some ordered pairs that satisfy $y = \tan x$ for x in $(-\pi/2, \pi/2)$. Note that $-\pi/2$ and $\pi/2$ are not in the domain of $y = \tan x$, but x can be chosen close to $\pm \pi/2$.

x	-1.56	-1.5	-1.2	$-\pi/4$	0	$\pi/4$	1.2	1.5	1.56
$y = \tan x$	-92.6	-14.1	-2.6	-1	0	1	2.6	14.1	92.6

$y \to -\infty$ as ↑
$x \to -\pi/2$ from the right

$y \to \infty$ as ↑
$x \to \pi/2$ from the left

The graph of $y = \tan x$ includes the points $(-\pi/4, -1)$, $(0, 0)$, and $(\pi/4, 1)$. Since $\tan x = \sin x/\cos x$, the graph of $y = \tan x$ has a vertical asymptote for every zero of the cosine function. So the vertical lines $x = \pi/2 + k\pi$ for any integer k are the **vertical asymptotes.** The behavior of $y = \tan x$ near the asymptotes $x = \pm \pi/2$ can be seen from the table of ordered pairs. As x approaches $\pi/2$ (approximately 1.57) from the left, $\tan x \to \infty$. As x approaches $-\pi/2$ from the right, $\tan x \to -\infty$. In the interval $(-\pi/2, \pi/2)$, the function is increasing. The graph of $y = \tan x$ is shown in Fig. 6.40. The shape of the tangent function between $-\pi/2$ and $\pi/2$ is repeated between each pair of consecutive asymptotes, as is shown in the figure. The period of $y = \tan(x)$ is π, and the fundamental cycle is the portion of the graph between $-\pi/2$ and $\pi/2$. Since the range of $y = \tan x$ is $(-\infty, \infty)$, the concept of amplitude is not defined.

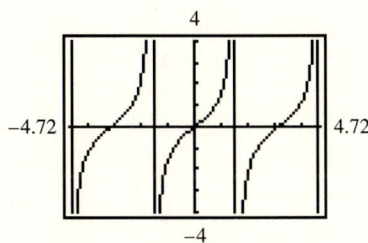

The graph $y = \tan x$ in connected mode appears to include the vertical asymptotes. Points on those lines do not satisfy $y = \tan x$.

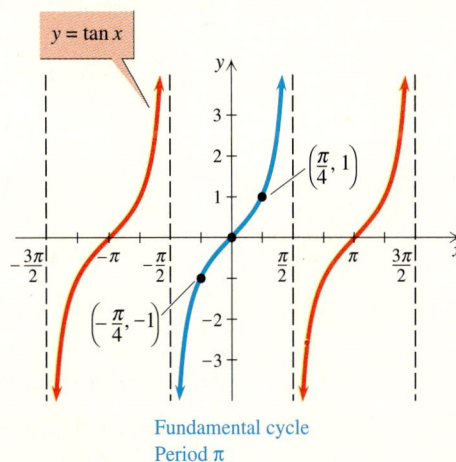

Figure 6.40

We can transform the graph of the tangent function by using the same techniques that we used for the sine and cosine functions.

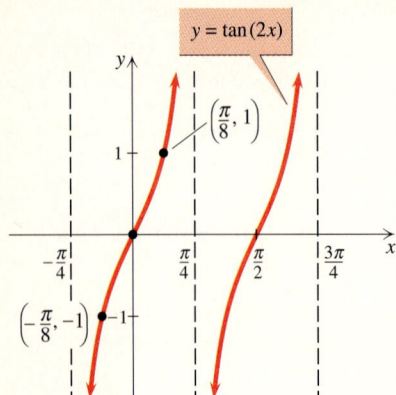

Figure 6.41

Example 3 A tangent function with a transformation

Sketch two cycles of the function $y = \tan(2x)$ and determine the period.

Solution

The graph of $y = \tan x$ completes one cycle for $-\pi/2 < x < \pi/2$. So the graph of $y = \tan(2x)$ completes one cycle for $-\pi/2 < 2x < \pi/2$, or $-\pi/4 < x < \pi/4$. The graph includes the points $(-\pi/8, -1)$, $(0, 0)$, and $(\pi/8, 1)$. The graph is similar to the graph of $y = \tan x$, but the asymptotes are $x = \pi/4 + k\pi/2$ for any integer k, and the period of $y = \tan(2x)$ is $\pi/2$. Two cycles of the graph are shown in Fig. 6.41. The period is $\pi/2$. ◆

By understanding what happens to the fundamental cycles of the sine, cosine, and tangent functions when the period changes, we can easily determine the location of the new function and sketch its graph quickly. We start with the fundamental cycles of $y = \sin x$, $y = \cos x$, and $y = \tan x$, which occur over the intervals $[0, 2\pi]$, $[0, 2\pi]$, and $[-\pi/2, \pi/2]$, respectively. For $y = \sin Bx$, $y = \cos Bx$, and $y = \tan Bx$ (for $B > 0$) these fundamental cycles move to the intervals $[0, 2\pi/B]$, $[0, 2\pi/B]$, and $[-\pi/(2B), \pi/(2B)]$, respectively. In every case, divide the old period by B to get the new period. Note that only the nonzero endpoints of the intervals are changed. Figure 6.42 shows these fundamental cycles for $B > 1$.

Fundamental cycles

Figure 6.42

Graph of $y = \cot(x)$

By definition, $\cot x = 1/\tan x$. So we can use $y = \tan x$ to graph $y = \cot x$. Since $\tan x = 0$ for $x = k\pi$, where k is any integer, the vertical lines $x = k\pi$ for any integer k are the vertical asymptotes of $y = \cot x$. Consider some ordered pairs that satisfy $y = \cot x$ between the asymptotes $x = 0$ and $x = \pi$:

x	0.02	$\pi/4$	$\pi/2$	$3\pi/4$	3.1
$y = \cot x$	49.99	1	0	-1	-24.0

$y \to \infty$ as
$x \to 0$ from the right

$y \to -\infty$ as
$x \to \pi$ from the left

As $x \to 0$ from the right, $y \to \infty$, and as $x \to \pi$ from the left, $y \to -\infty$. In the interval $(0, \pi)$, the function is decreasing. The graph of $y = \cot x$ is shown in Fig. 6.43. The shape of the curve between $x = 0$ and $x = \pi$ is repeated between each pair of consecutive asymptotes. The period of $y = \cot x$ is π, and its fundamental cycle is the portion of the graph between 0 and π. The range of $y = \cot x$ is $(-\infty, \infty)$.

Because $\cot x = 1/\tan x$, $\cot x$ is large when $\tan x$ is small, and vice versa. The graph of $y = \cot x$ has an x-intercept wherever $y = \tan x$ has a vertical asymptote, and a vertical asymptote wherever $y = \tan x$ has an x-intercept. So for every integer k, $(\pi/2 + k\pi, 0)$ is an x-intercept of $y = \cot x$, and the vertical line $x = k\pi$ is an asymptote, as shown in Fig. 6.43.

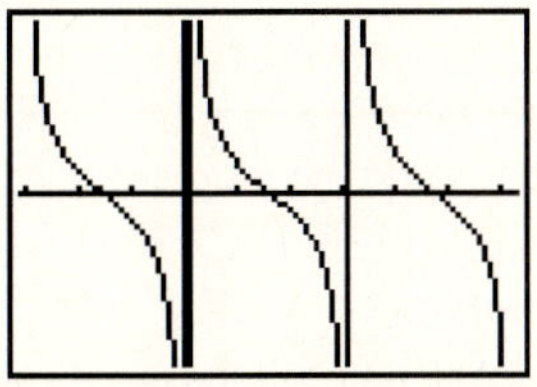

To graph $y = \cot x$, define y_1 as $y_1 = 1/\tan x$.

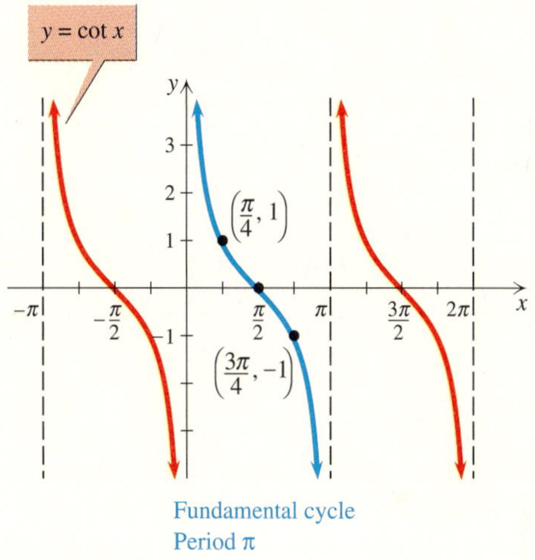

Figure 6.43

Example 4   A cotangent function with a transformation

Sketch two cycles of the function $y = 0.3 \cot(2x + \pi/2)$, and determine the period.

Solution

Factor out 2 to write the function as

$$y = 0.3 \cot\left(2\left[x + \frac{\pi}{4}\right]\right).$$

Since $y = \cot x$ completes one cycle for $0 < x < \pi$, the graph of $y = 0.3 \cot(2[x + \pi/4])$ completes one cycle for

$$0 < 2\left[x + \frac{\pi}{4}\right] < \pi, \text{ or } 0 < x + \frac{\pi}{4} < \frac{\pi}{2}, \text{ or } -\frac{\pi}{4} < x < \frac{\pi}{4}.$$

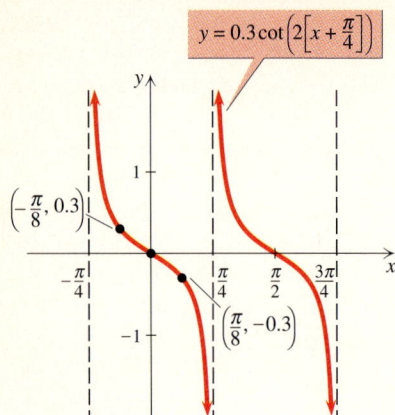

$$y = 0.3\cot\left(2\left[x + \frac{\pi}{4}\right]\right)$$

Figure 6.44

The interval $(-\pi/4, \pi/4)$ for the fundamental cycle can also be obtained by dividing the period π of $y = \cot x$ by 2 to get $\pi/2$ as the period, and then shifting the interval $(0, \pi/2)$ a distance of $\pi/4$ to the left. The factor 0.3 shrinks the y-coordinates. The graph goes through $(-\pi/8, 0.3)$, $(0, 0)$, and $(\pi/8, -0.3)$ as it approaches its vertical asymptotes $x = \pm\pi/4$. The graph for two cycles is shown in Fig. 6.44. ◆

Since the period for $y = \cot x$ is π, the period for $y = \cot(Bx)$ with $B > 0$ is π/B and $y = \cot(Bx)$ completes one cycle on the interval $(0, \pi/B)$. The left-hand asymptote for the fundamental cycle of $y = \cot x$ remains fixed at $x = 0$, while the right-hand asymptote changes to $x = \pi/B$.

Graph of $y = \sec(x)$

Since $\sec x = 1/\cos x$, the values of $\sec x$ are large when the values of $\cos x$ are small. For any x such that $\cos x$ is 0, $\sec x$ is undefined and the graph of $y = \sec x$ has a vertical asymptote. Because of the reciprocal relationship between $\sec x$ and $\cos x$, we first draw the graph of $y = \cos x$ for reference when graphing $y = \sec x$. At every x-intercept of $y = \cos x$, we draw a vertical asymptote, as shown in Fig. 6.45. If $\cos x = \pm 1$, then $\sec x = \pm 1$. So every maximum or minimum point on the graph of $y = \cos x$ is also on the graph of $y = \sec x$. If $\cos x > 0$ and $\cos x \to 0$, then $\sec x \to \infty$. If $\cos x < 0$ and $\cos x \to 0$, then $\sec x \to -\infty$. These two facts cause the graph of $y = \sec x$ to approach its asymptotes in the manner shown in Fig. 6.45. The period of $y = \sec x$ is 2π, the same as the period for $y = \cos x$. The range of $y = \sec x$ is $(-\infty, -1] \cup [1, \infty)$.

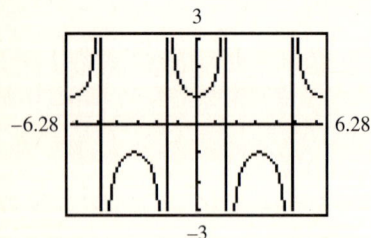

The graph of $y = 1/\cos x$ in connected mode appears to show the vertical asymptotes. Points on these lines do not satisfy $y = \sec x$.

$\sec x \to \infty$ because $\cos x > 0$ and $\cos x \to 0$

$y = \sec x$

$\sec x \to -\infty$ because $\cos x < 0$ and $\cos x \to 0$

$y = \cos x$

Figure 6.45

Example 5   A secant function with a transformation

Sketch two cycles of the function $y = 2\sec(x - \pi/2)$, and determine the period and the range of the function.

Solution

Since
$$y = 2 \sec\left(x - \frac{\pi}{2}\right) = \frac{2}{\cos(x - \pi/2)},$$

we first graph $y = \cos(x - \pi/2)$ as shown in Fig. 6.46. The function $y = \sec(x - \pi/2)$ goes through the maximum and minimum points on the graph of $y = \cos(x - \pi/2)$, but $y = 2 \sec(x - \pi/2)$ stretches $y = \sec(x - \pi/2)$ by a factor of 2. So the portions of the curve that open up do not go lower than 2, and the portions that open down do not go higher than -2, as shown in the figure. The period is 2π, and the range is $(-\infty, -2] \cup [2, \infty)$.

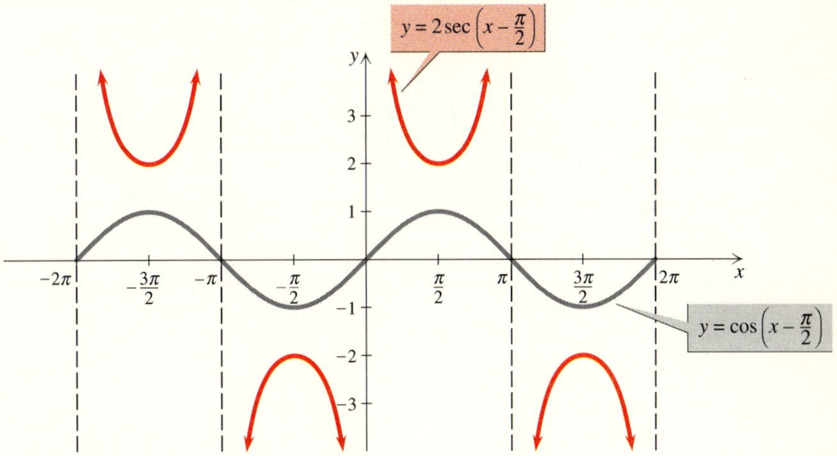

Figure 6.46

Graph of $y = \csc(x)$

Since $\csc x = 1/\sin x$, the graph of $y = \csc x$ is related to the graph of $y = \sin x$ in the same way that the graphs of $y = \sec x$ and $y = \cos x$ are related. To graph $y = \csc x$, first draw the graph of $y = \sin x$ and a vertical asymptote at each x-intercept. Since the graph of $y = \sin x$ can be obtained by shifting $y = \cos x$ a distance of $\pi/2$ to the right, the graph of $y = \csc x$ is obtained from $y = \sec x$ by shifting a distance of $\pi/2$ to the right as shown in Fig. 6.47.

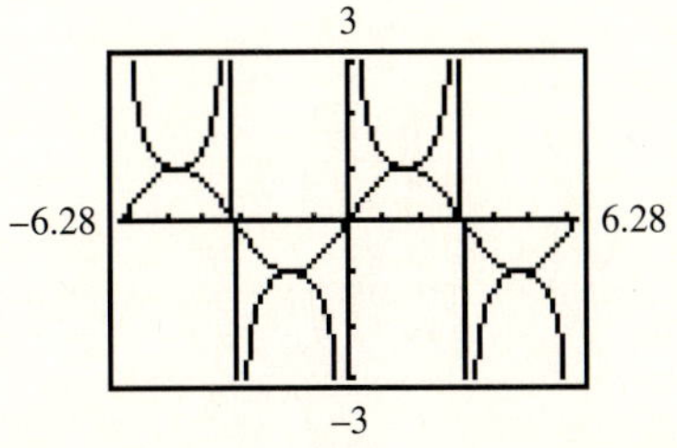

Graph $y_1 = \sin x$ and $y_2 = 1/\sin x$ to see the relationship between the sine and the cosecant functions.

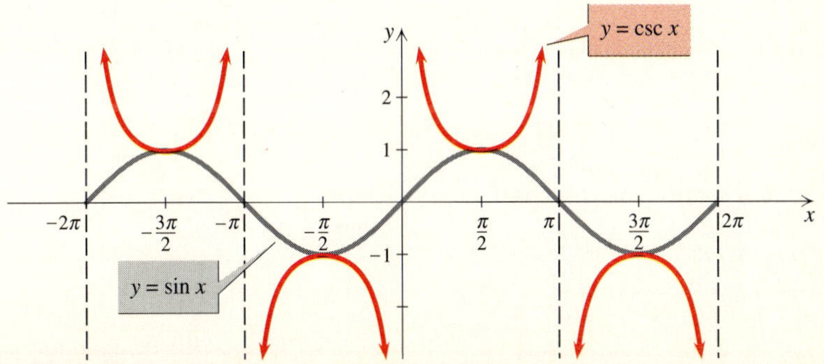

Figure 6.47

The graph in Fig. 6.47 also shows us that the period of $y = \csc x$ is 2π, the same as the period for $y = \sin x$.

Example 6   A cosecant function with a transformation

Sketch two cycles of the graph of $y = \csc(2x - 120°)$, where x is the degree measure of an angle, and determine the period and the range of the function.

Solution

Since

$$y = \csc(2x - 120°)$$

$$= \frac{1}{\sin(2[x - 60°])},$$

we first graph $y = \sin(2[x - 60°])$. The period for $y = \sin(2[x - 60°])$ is $180°$ with phase shift of $60°$ to the right. So the fundamental cycle of $y = \sin x$ is transformed to occur on the interval $[60°, 240°]$. Draw at least two cycles of $y = \sin(2[x - 60°])$ with a vertical asymptote (for the cosecant function) at every x-intercept, as shown in Fig. 6.48. Each portion of $y = \csc(2[x - 60°])$ that opens up has a minimum value of 1, and each portion that opens down has a maximum value of -1. The period of the function is $180°$, and the range is $(-\infty, -1] \cup [1, \infty)$.

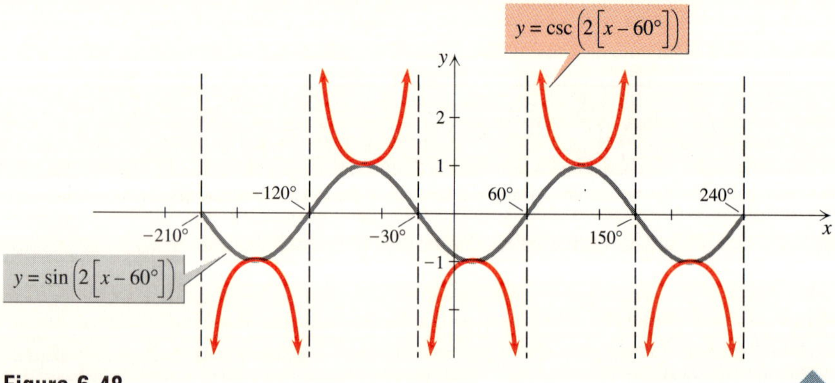

Figure 6.48

The table on the next page summarizes some of the facts that we have learned about the six trigonometric functions. Also, one cycle of the graph of each trigonometric function is shown.

▼

Summary: Trigonometric Functions

Function	Domain (k any integer)	Range	Period	Fundamental cycle	Graph
$y = \sin x$	$(-\infty, \infty)$	$[-1, 1]$	2π	$[0, 2\pi]$	
$y = \cos x$	$(-\infty, \infty)$	$[-1, 1]$	2π	$[0, 2\pi]$	
$y = \tan x$	$x \neq \pi/2 + k\pi$	$(-\infty, \infty)$	π	$(-\pi/2, \pi/2)$	
$y = \csc x$	$x \neq k\pi$	$(-\infty, -1] \cup [1, \infty)$	2π	$(0, 2\pi)$	
$y = \sec x$	$x \neq \pi/2 + k\pi$	$(-\infty, -1] \cup [1, \infty)$	2π	$(0, 2\pi)$	
$y = \cot x$	$x \neq k\pi$	$(-\infty, \infty)$	π	$(0, \pi)$	

? For Thought

True or false? Explain.

1. $\sec(\pi/4) = 1/\sin(\pi/4)$
2. $\cot(1.24) = 1/\tan(1.24)$
3. $\csc(60°) = 2\sqrt{3}/3$
4. $\tan(5\pi/2) = 0$
5. $\sec(95°) = \sqrt{5}$
6. $\csc(2[90° - 30°]) = 2/\sqrt{3}$
7. The graphs of $y = 2 \csc x$ and $y = 1/(2 \sin x)$ are identical.
8. The range of $y = 0.5 \csc(13x - 5\pi)$ is $(-\infty, -0.5] \cup [0.5, \infty)$.
9. The graph of $y = \tan(3x)$ has vertical asymptotes at $x = \pm\pi/6$.
10. The graph of $y = \cot(4x)$ has vertical asymptotes at $x = \pm\pi/4$.

6.4 Exercises

Tape 12   Disk—5.25": 4   3.5": 3   Macintosh: 3

Redraw each unit circle and label each indicated point with the proper value of the tangent function. The points in Exercise 1 are the terminal points for arcs with lengths that are multiples of $\pi/4$. The points in Exercise 2 are the terminal points for arcs with lengths that are multiples of $\pi/6$. Repeat Exercises 1 and 2 until you can do them from memory.

1.

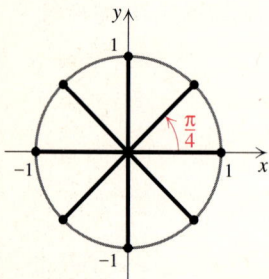

2.

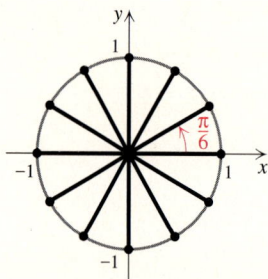

Find the exact value of each of the following expressions.

3. $\tan(\pi/3)$

4. $\tan(\pi/4)$

5. $\tan(-\pi/4)$

6. $\tan(\pi/6)$

7. $\cot(\pi/2)$

8. $\cot(2\pi/3)$

9. $\cot(-\pi/3)$

10. $\cot(0)$

11. $\sec(\pi/6)$

12. $\sec(\pi/3)$

13. $\sec(\pi/2)$

14. $\sec(\pi)$

15. $\csc(-\pi)$

16. $\csc(\pi/6)$

17. $\csc(3\pi/4)$

18. $\csc(-\pi/3)$

19. $\tan(135°)$

20. $\tan(270°)$

21. $\cot(210°)$

22. $\cot(120°)$

23. $\sec(-120°)$

24. $\sec(-90°)$

25. $\csc(315°)$

26. $\csc(240°)$

Find the approximate value of each expression. Round to four decimal places.

27. $\tan(1.55)$

28. $\tan(1.6)$

29. $\cot(-3.48)$

30. $\cot(22.4)$

31. $\csc(0.002)$

32. $\csc(1.54)$

33. $\sec(\pi/12)$

34. $\sec(-\pi/8)$

35. $\cot(0.09°)$

36. $\cot(179.4°)$

37. $\csc(-44.3°)$

38. $\csc(-124.5°)$

39. $\sec(89.2°)$

40. $\sec(-0.024°)$

41. $\tan(-44.6°)$

42. $\tan(138°)$

Sketch two cycles of the graph of each function and determine the period. Assume that x is a real number.

43. $y = \tan(3x)$

44. $y = \tan(4x)$

45. $y = \cot(x + \pi/4)$

46. $y = \cot(x - \pi/6)$

47. $y = \cot(x/2)$

48. $y = \cot(x/3)$

49. $y = \tan(\pi x)$

50. $y = \tan(\pi x/2)$

51. $y = -2\tan x$

52. $y = -\tan(x - \pi/2)$

53. $y = -\cot(x + \pi/2)$

54. $y = 2 + \cot x$

55. $y = \cot(2x - \pi/2)$

56. $y = \cot(3x + \pi)$

57. $y = \tan\left(\dfrac{\pi}{2}x - \dfrac{\pi}{2}\right)$

58. $y = \tan\left(\dfrac{\pi}{4}x + \dfrac{3\pi}{4}\right)$

Sketch two cycles of the graph of each function and determine the period and the range. Assume that x is a real number.

59. $y = \sec(2x)$

60. $y = \sec(3x)$

61. $y = \csc(x - \pi/2)$

62. $y = \csc(x + \pi/4)$

63. $y = \csc(x/2)$

64. $y = \csc(x/4)$

65. $y = \sec(\pi x/2)$

66. $y = \sec(\pi x)$

67. $y = 2\sec x$

68. $y = \dfrac{1}{2}\sec x$

69. $y = \csc(2x - \pi/2)$

70. $y = \csc(3x + \pi)$

71. $y = -\csc\left(\dfrac{\pi}{2}x + \dfrac{\pi}{2}\right)$

72. $y = -2\csc(\pi x - \pi)$

73. $y = 2 + 2\sec(2x)$

74. $y = 2 - 2\sec\left(\dfrac{x}{2}\right)$

Sketch two cycles of the graph of each function. Assume that x is the degree measure of an angle in standard position.

75. $y = -\cot(3x)$ **76.** $y = -\tan(4x)$

77. $y = \sec(2x)$ **78.** $y = \csc(5x)$

79. $y = \tan(x/2)$ **80.** $y = \cot(x/3)$

81. $y = -\csc(x/2)$ **82.** $y = -\sec(x/3)$

83. $y = \cot(2x - 180°)$ **84.** $y = \tan(x - 30°)$

85. $y = -\sec(x + 45°)$ **86.** $y = \csc(2x + 60°)$

87. $y = 2 + \csc(x - 30°)$ **88.** $y = \sec(2x) - 2$

89. $y = \tan(90x + 90°)$ **90.** $y = \cot(45x - 45°)$

Consider a nonvertical line through the point (x_1, y_1). Let r be the ray that extends horizontally from (x_1, y_1) in the direction of the positive x-axis. Let θ be the angle of inclination formed by rotating r either counterclockwise or clockwise until it first meets the line, so that $-\pi/2 < \theta < \pi/2$. It can be shown that the equation of the line is $y = (x - x_1)\tan\theta + y_1$. **Find the equation of each line described below and write it in the form $y = mx + b$.**

91. The line through $(2, 3)$ with angle of inclination $\pi/4$

92. The line through $(-1, 2)$ with angle of inclination $-\pi/4$

93. The line through $(3, -1)$ with angle of inclination $-80°$

94. The line through $(-2, -1)$ with angle of inclination $17°$

For Writing/Discussion

95. *Addition of ordinates*   The function $y = \sin x + \cos x$ can be graphed by a method called *addition of ordinates*. First graph $y = \sin x$ and $y = \cos x$ on the same coordinate system. Then select a value for x such as $x = 0$ and observe the approximate values of y on each curve. Add these ordinates to get the approximate location of a point on the curve $y = \sin x + \cos x$ with x-coordinate 0. Repeat this procedure for $x = \pi/4, \pi/2, 3\pi/4$, and so on, until you get enough points on $y = \sin x + \cos x$ so that you can sketch the curve. Determine the period and range of $y = \sin x + \cos x$.

96. *Addition of ordinates again*   Explain how to sketch the curve $y = \sin x + \cos x$ by addition of ordinates without actually doing any arithmetic, but by simply looking at the curves $y = \sin x$ and $y = \cos x$. Graph $y = \sin x - \cos x$ and $y = 2 \sin x + \sin(2x)$ by using addition of ordinates.

97. *Cooperative learning*   Work in a small group to combine some basic trigonometric functions using addition, subtraction, multiplication, or division and graph the combined function. (For instance, graph $y = \sin x + \cos x - \tan x$.) A graphing calculator would be useful here. Are these combined functions periodic? Can you determine the period of a combined function from the periods of the basic functions that are used to form it?

Graphing Calculator Exercises

The function $y = \sin x$ oscillates between the two horizontal lines $y = 1$ and $y = -1$. By multiplying $\sin x$ by another function $g(x)$ we can get curves that oscillate between the graphs of $y = g(x)$ and $y = -g(x)$.

1. Graph $y_1 = x \sin x$, $y_2 = x$, and $y_3 = -x$ for $0 \le x \le 100$ and $-100 \le y \le 100$. For what values of x is $x = x \sin x$?

2. Graph $y_1 = x^2 \sin x$, $y_2 = x^2$, and $y_3 = -x^2$ for $0 \le x \le 20$ and $-400 \le y \le 400$. For what values of x is $x^2 = x^2 \sin x$?

3. Graph $y_1 = \dfrac{1}{x} \sin x$ for $0 \le x \le 10$ and $-2 \le y \le 2$. Is it true that for $x > 0$, $-\dfrac{1}{x} < \dfrac{1}{x} \sin x < \dfrac{1}{x}$?

4. Graph $f(x) = \dfrac{1}{x} \sin x$ for $-2 \le x \le 2$ and $-2 \le y \le 2$. What is the value of $f(0)$? Is it true that for all x in the interval $[-0.1, 0.1]$ for which $f(x)$ is defined, $f(x)$ satisfies $0.99 < f(x) < 1$?

5. Graph $y = x + \sin x$ for $-100 \le x \le 100$ and $-100 \le y \le 100$. Explain your results.

6. Graph $y = x + \tan x$ for $-6 \le x \le 6$ and $-10 \le y \le 10$. Explain your results.

6.5

The Inverse

Trigonometric

Functions

We have learned how to find the values of the trigonometric functions for angles or real numbers, but to make the trigonometric functions really useful we must be able to reverse this process. In this section we define the inverses of the trigonometric functions.

The Inverse of the Sine Function

In Chapter 3 we learned that only one-to-one functions are invertible. Since $y = \sin x$ with domain $(-\infty, \infty)$ is a periodic function, it is certainly not one-to-one. However, if we restrict the domain to the interval $[-\pi/2, \pi/2]$, then the restricted function is one-to-one and invertible. Other intervals could be used, but this interval is chosen to keep the inverse function as simple as possible.

The graph of the sine function with domain $[-\pi/2, \pi/2]$ is shown in Fig. 6.49(a). Its range is $[-1, 1]$.

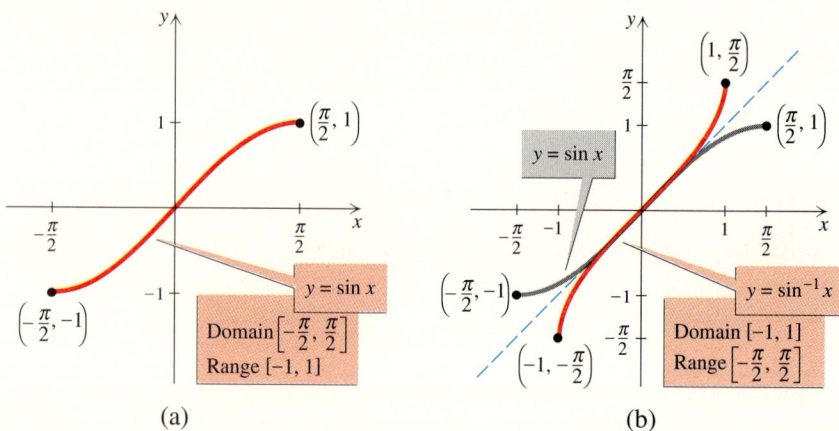

(a)

(b)

Figure 6.49

We now define the inverse sine function and denote it as $f^{-1}(x) = \sin^{-1} x$ (read "inverse sine of x") or $f^{-1}(x) = \arcsin x$ (read "arc sine of x").

Definition: The Inverse Sine Function

The function $y = \sin^{-1} x$ or $y = \arcsin x$ is the inverse of the function $y = \sin x$ restricted to $[-\pi/2, \pi/2]$. The domain of $y = \sin^{-1} x$ is $[-1, 1]$ and its range is $[-\pi/2, \pi/2]$.

The graph of $y = \sin^{-1} x$ is a reflection about the line $y = x$ of the graph of $y = \sin x$ on $[-\pi/2, \pi/2]$ as shown in Fig. 6.49(b).

If $y = \sin^{-1} x$, then y is the real number such that $-\pi/2 \le y \le \pi/2$ and $\sin y = x$. Depending on the context, $\sin^{-1} x$ might also be an angle, a measure of an angle in degrees or radians, or the length of an arc of the unit circle. The expression $\sin^{-1} x$ can be read as "the angle whose sine is x" or "the arc length whose sine is x." The notation $y = \arcsin x$ reminds us that y is the arc length whose sine is x. For example, $\arcsin(1)$ is the arc length in $[-\pi/2, \pi/2]$ whose sine is 1. Since we know that $\sin(\pi/2) = 1$, we have $\arcsin(1) = \pi/2$. We will assume that $\sin^{-1} x$ is a real number unless indicated otherwise.

Note that the nth power of the sine function is usually written as $\sin^n(x)$ as a shorthand notation for $(\sin x)^n$, provided $n \neq -1$. The -1 used in $\sin^{-1} x$ indicates the inverse function and does *not* mean reciprocal. To write $1/\sin x$ using exponents, we must write $(\sin x)^{-1}$.

Example 1   Evaluating the inverse sine function

Find the exact value of each expression without using a table or a calculator.

a) $\sin^{-1}(1/2)$ **b)** $\arcsin\left(-\sqrt{3}/2\right)$

Solution

a) The value of $\sin^{-1}(1/2)$ is the number α in the interval $[-\pi/2, \pi/2]$ such that $\sin(\alpha) = 1/2$. We recall that $\sin(\pi/6) = 1/2$, and so $\sin^{-1}(1/2) = \pi/6$. Note that $\pi/6$ is the only value of α in $[-\pi/2, \pi/2]$ for which $\sin(\alpha) = 1/2$.

b) The value of $\arcsin\left(-\sqrt{3}/2\right)$ is the number α in $[-\pi/2, \pi/2]$ such that $\sin(\alpha) = -\sqrt{3}/2$. Since $\sin(-\pi/3) = -\sqrt{3}/2$, we have $\arcsin\left(-\sqrt{3}/2\right) = -\pi/3$. Note that $-\pi/3$ is the only value of α in $[-\pi/2, \pi/2]$ for which $\sin(\alpha) = -\sqrt{3}/2$. ◆

Example 2   Evaluating the inverse sine function

Find the exact value of each expression in degrees without using a table or a calculator.

a) $\sin^{-1}\left(\sqrt{2}/2\right)$ **b)** $\arcsin(0)$

Solution

a) The value of $\sin^{-1}\left(\sqrt{2}/2\right)$ in degrees is the angle α in the interval $[-90°, 90°]$ such that $\sin(\alpha) = \sqrt{2}/2$. We recall that $\sin(45°) = \sqrt{2}/2$, and so $\sin^{-1}\left(\sqrt{2}/2\right) = 45°$.

b) The value of $\arcsin(0)$ in degrees is the angle α in the interval $[-90°, 90°]$ for which $\sin(\alpha) = 0$. Since $\sin(0°) = 0$, we have $\arcsin(0) = 0°$. ◆

In the next example we use a calculator to find the degree measure of an angle whose sine is given. To obtain degree measure make sure the calculator is in degree mode. Scientific calculators usually have a key labeled $\sin^{-1}$ that gives values for the inverse sine function.

```
sin⁻¹ .88
        61.64236342
sin Ans
              .88
```

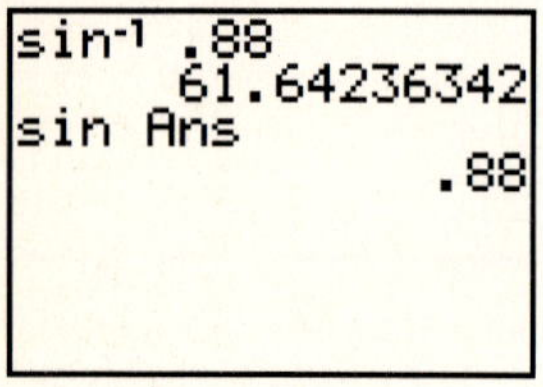

Use degree mode to find $\sin^{-1}(0.88)$.
Check using the ANS feature.

Example 3 Finding an angle given its sine

Given that α is an angle such that $0° < \alpha < 90°$ and $\sin \alpha = 0.88$, find α to the nearest tenth of a degree.

Solution

The value of $\sin^{-1}(0.88)$ is the only angle in $[-90°, 90°]$ with a sine of 0.88. Use a scientific calculator in degree mode to get

$$\alpha = \sin^{-1}(0.88) \approx 61.6°.$$

Checking by finding $\sin(61.6°) \approx 0.88$. ◆

The Inverse Cosine Function

Since the cosine function is not one-to-one on $(-\infty, \infty)$, we restrict the domain to $[0, \pi]$, where the cosine function is one-to-one and invertible. The graph of the cosine function with this restricted domain is shown in Fig. 6.50(a). Note that the range of the restricted function is $[-1, 1]$.

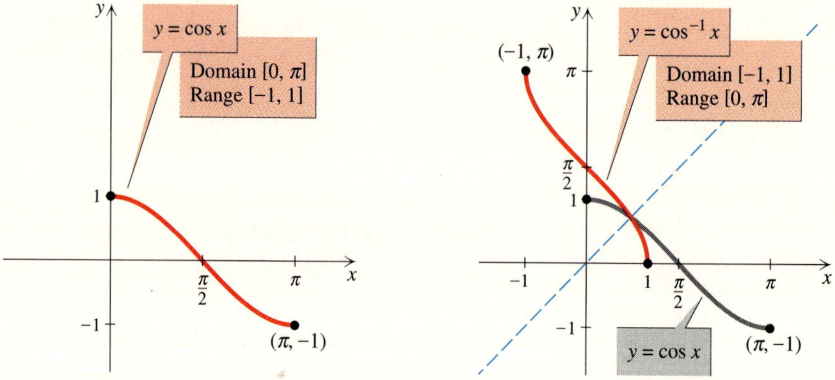

(a)   (b)

Figure 6.50

We now define the inverse of $f(x) = \cos x$ for x in $[0, \pi]$ and denote it as $f^{-1}(x) = \cos^{-1} x$ or $f^{-1}(x) = \arccos x$.

Definition: The Inverse Cosine Function

The function $y = \cos^{-1} x$ or $y = \arccos x$ is the inverse of the function $y = \cos x$ restricted to $[0, \pi]$. The domain of $y = \cos^{-1} x$ is $[-1, 1]$ and its range is $[0, \pi]$.

If $y = \cos^{-1} x$, then y is the real number in $[0, \pi]$ such that $\cos y = x$. The expression $\cos^{-1} x$ can be read as "the angle whose cosine is x" or "the arc length whose cosine is x." The graph of $y = \cos^{-1} x$, shown in Fig. 6.50(b), is obtained by reflecting the graph of $y = \cos x$ (restricted to $[0, \pi]$) about the line $y = x$. We will assume that $\cos^{-1} x$ is a real number unless indicated otherwise.

Example 4   Evaluating the inverse cosine function

Find the exact value of each expression without using a table or a calculator.

a) $\cos^{-1}(-1)$

b) $\arccos(-1/2)$

Solution

a) The value of $\cos^{-1}(-1)$ is the number α in $[0, \pi]$ such that $\cos(\alpha) = -1$. We recall that $\cos(\pi) = -1$, and so $\cos^{-1}(-1) = \pi$.

b) The value of $\arccos(-1/2)$ is the number α in $[0, \pi]$ such that $\cos(\alpha) = -1/2$. We recall that $\cos(2\pi/3) = -1/2$, and so $\arccos(-1/2) = 2\pi/3$. ◆

In the next example we use a calculator to find the degree measure of an angle whose cosine is given. Most scientific calculators have a key labeled $\cos^{-1}$ that gives values for the inverse cosine function. To get the degree measure, make sure the calculator is in degree mode.

Example 5   Finding an angle given its cosine

Find the angle α to the nearest tenth of a degree, given that $0° < \alpha < 90°$ and $\cos \alpha = 0.23$.

Solution

Since $\cos^{-1}(0.23)$ is the unique angle in $[0°, 180°]$ with a cosine of 0.23, $\alpha = \cos^{-1}(0.23) \approx 76.7°$. ◆

```
cos⁻¹ .23
        76.70292825
cos Ans
              .23
```

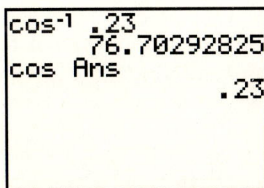

Use degree mode to find $\cos^{-1}(0.23)$ and use the ANS feature to check.

Inverses of Tangent, Cotangent, Secant, and Cosecant

Since all of the trigonometric functions are periodic, they must all be restricted to a domain where they are one-to-one before their inverse functions can be defined. There is more than one way to choose a domain to get a one-to-one function, but we will use the most common restrictions. The restricted domain for $y = \tan x$ is $(-\pi/2, \pi/2)$, for $y = \csc x$ it is $[-\pi/2, 0) \cup (0, \pi/2]$, for $y = \sec x$ it is $[0, \pi/2) \cup (\pi/2, \pi]$, and for $y = \cot x$ it is $(0, \pi)$. The functions $\tan^{-1}$, $\cot^{-1}$,

$\sec^{-1}$, and $\csc^{-1}$ are defined to be the inverses of these restricted functions. The notations arctan, arccot, arcsec, and arccsc are also used for these inverse functions. Figure 6.51 shows the graphs of all six inverse functions, with their domains and ranges.

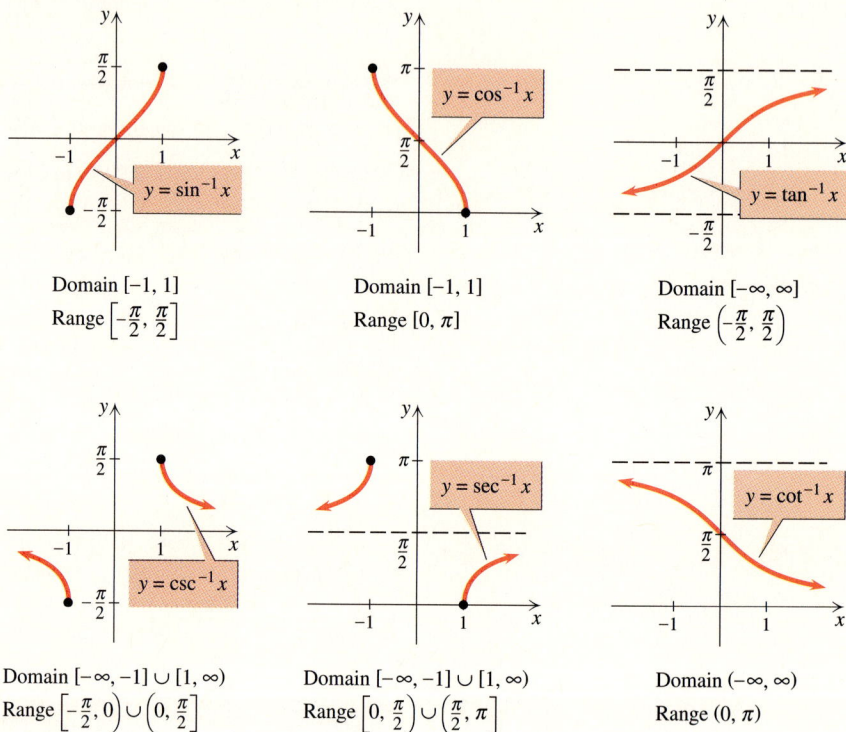

Domain $[-1, 1]$
Range $\left[-\frac{\pi}{2}, \frac{\pi}{2}\right]$

Domain $[-1, 1]$
Range $[0, \pi]$

Domain $[-\infty, \infty]$
Range $\left(-\frac{\pi}{2}, \frac{\pi}{2}\right)$

Domain $[-\infty, -1] \cup [1, \infty)$
Range $\left[-\frac{\pi}{2}, 0\right) \cup \left(0, \frac{\pi}{2}\right]$

Domain $[-\infty, -1] \cup [1, \infty)$
Range $\left[0, \frac{\pi}{2}\right) \cup \left(\frac{\pi}{2}, \pi\right]$

Domain $(-\infty, \infty)$
Range $(0, \pi)$

Figure 6.51

When studying inverse trigonometric functions, you should first learn to evaluate $\sin^{-1}$, $\cos^{-1}$, and $\tan^{-1}$. Those values found can be used along with the identities

$$\csc \alpha = \frac{1}{\sin \alpha}, \qquad \sec \alpha = \frac{1}{\cos \alpha}, \qquad \text{and} \qquad \cot \alpha = \frac{1}{\tan \alpha}$$

to evaluate $\csc^{-1}$, $\sec^{-1}$, and $\cot^{-1}$. For example, $\sin(\pi/6) = 1/2$ and $\csc(\pi/6) = 2$. So the angle whose cosecant is 2 is the same as the angle whose sine is $1/2$. In symbols,

$$\csc^{-1}(2) = \sin^{-1}\left(\frac{1}{2}\right) = \frac{\pi}{6}.$$

In general, $\csc^{-1} x = \sin^{-1}(1/x)$. Likewise, $\sec^{-1} x = \cos^{-1}(1/x)$. For the inverse cotangent, $\cot^{-1} x = \tan^{-1}(1/x)$ only for positive values of x, because of the choice of $(0, \pi)$ as the range of the inverse cotangent. We have $\cot^{-1}(0) = \pi/2$ and, if x is negative, $\cot^{-1}(x) = \tan^{-1}(1/x) + \pi$. These identities are summarized.

Identities for the Inverse Functions

$$\csc^{-1}(x) = \sin^{-1}(1/x) \text{ for } |x| \geq 1$$
$$\sec^{-1}(x) = \cos^{-1}(1/x) \text{ for } |x| \geq 1$$
$$\cot^{-1}(x) = \begin{cases} \tan^{-1}(1/x) & \text{for } x > 0 \\ \tan^{-1}(1/x) + \pi & \text{for } x < 0 \\ \pi/2 & \text{for } x = 0 \end{cases}$$

We can see another relationship between $\cot^{-1}$ and $\tan^{-1}$ from their graphs in Fig. 6.51. We get the graph of $y = \cot^{-1}(x)$ by reflecting the graph of $y = \tan^{-1}(x)$ with respect to the x-axis and translating it up $\pi/2$ units. Thus,

$$\cot^{-1}(x) = \frac{\pi}{2} - \tan^{-1}(x).$$

Example 6 Evaluating the inverse functions

Find the exact value of each expression without using a table or a calculator.

a) $\tan^{-1}(1)$ **b)** $\text{arcsec}(-2)$ **c)** $\csc^{-1}(\sqrt{2})$ **d)** $\text{arccot}(-1/\sqrt{3})$

Solution

a) Since $\tan(\pi/4) = 1$ and since $\pi/4$ is in the range of $\tan^{-1}$, we have

$$\tan^{-1}(1) = \frac{\pi}{4}.$$

b) To evaluate the inverse secant we use the identity $\sec^{-1}(x) = \cos^{-1}(1/x)$. In this case, the arc whose secant is -2 is the same as the arc whose cosine is $-1/2$. So we must find $\cos^{-1}(-1/2)$. Since $\cos(2\pi/3) = -1/2$ and since $2\pi/3$ is in the range of arccos, we have

$$\text{arcsec}(-2) = \arccos\left(-\frac{1}{2}\right) = \frac{2\pi}{3}.$$

c) To evaluate $\csc^{-1}(\sqrt{2})$, we use the identity $\csc^{-1}(x) = \sin^{-1}(1/x)$ with $x = \sqrt{2}$. So we must find $\sin^{-1}(1/\sqrt{2})$. Since $\sin(\pi/4) = 1/\sqrt{2}$ and since $\pi/4$ is in the range of $\csc^{-1}$,

$$\csc^{-1}(\sqrt{2}) = \sin^{-1}\left(\frac{1}{\sqrt{2}}\right) = \frac{\pi}{4}.$$

d) If x is negative we use the identity $\cot^{-1}(x) = \tan^{-1}(1/x) + \pi$. Since $x = -1/\sqrt{3}$, we must find $\tan^{-1}(-\sqrt{3})$. Since $\tan^{-1}(-\sqrt{3}) = -\pi/3$, we get

$$\text{arccot}\left(\frac{-1}{\sqrt{3}}\right) = \tan^{-1}(-\sqrt{3}) + \pi = -\frac{\pi}{3} + \pi = \frac{2\pi}{3}.$$

Note that $\cot(-\pi/3) = \cot(2\pi/3) = -1/\sqrt{3}$, but $\text{arccot}(-1/\sqrt{3}) = 2\pi/3$ because $2\pi/3$ is in the range of the function arccot.

The functions $\sin^{-1}$, $\cos^{-1}$, and $\tan^{-1}$ are available on scientific and graphing calculators. The calculator values of the inverse functions are given in degrees or radians, depending on the mode setting. We will assume that the values of the inverse functions are to be in radians unless indicated otherwise. Calculators use the domains and ranges of these functions defined here. The functions $\sec^{-1}$, $\csc^{-1}$, and $\cot^{-1}$ are generally not available on a calculator, and expressions involving these functions must be written in terms of $\sin^{-1}$, $\cos^{-1}$, and $\tan^{-1}$ by using the identities.

Example 7   Evaluating the inverse functions with a calculator

Find the approximate value of each expression rounded to four decimal places.

a) $\sin^{-1}(0.88)$ **b)** $\text{arccot}(2.4)$

c) $\csc^{-1}(4)$ **d)** $\cot^{-1}(0)$

Solution

a) Use the inverse sine function in radian mode to get $\sin^{-1}(0.88) \approx 1.0759$.

b) Use the identity $\cot^{-1}(x) = \tan^{-1}(1/x)$ because $2.4 > 0$.

$$\text{arccot}(2.4) = \tan^{-1}\left(\frac{1}{2.4}\right) \approx 0.3948$$

c) $\csc^{-1}(4) = \sin^{-1}(0.25) \approx 0.2527$

d) The value of $\cot^{-1}(0)$ cannot be found on a calculator. But we know that $\cot(\pi/2) = 0$ and $\pi/2$ is in the range $(0, \pi)$ for $\cot^{-1}$, so we have $\cot^{-1}(0) = \pi/2$. ◆

```
sin⁻¹ .88
         1.0758622
tan⁻¹ (1/2.4)
         .3947911197
sin⁻¹ .25
         .2526802551
```

Use radian mode to evaluate these inverse functions.

Compositions of Functions

One trigonometric function can be followed by another to form a composition of functions. We can evaluate an expression such as $\sin(\sin(\alpha))$ because the sine of the real number $\sin(\alpha)$ is defined. However, it is more common to have a composition of a trigonometric function and an inverse trigonometric function. For example, $\tan^{-1}(\alpha)$ is the angle whose tangent is α, and so $\sin(\tan^{-1}(\alpha))$ is the sine of the angle whose tangent is α.

Example 8   Evaluating compositions of functions

Find the exact value of each composition without using a table or a calculator.

a) $\sin(\tan^{-1}(0))$ **b)** $\arcsin(\cos(\pi/6))$

c) $\tan\left(\sec^{-1}\left(\sqrt{2}\right)\right)$

Solution

a) Since $\tan(0) = 0$, $\tan^{-1}(0) = 0$. Therefore

$$\sin(\tan^{-1}(0)) = \sin(0) = 0.$$

b) Since $\cos(\pi/6) = \sqrt{3}/2$, we have

$$\arcsin\left(\cos\left(\frac{\pi}{6}\right)\right) = \arcsin\left(\frac{\sqrt{3}}{2}\right) = \frac{\pi}{3}.$$

c) To find $\sec^{-1}(\sqrt{2})$, we use the identity $\sec^{-1}(x) = \cos^{-1}(1/x)$. Since $\cos(\pi/4) = 1/\sqrt{2}$, we have $\cos^{-1}(1/\sqrt{2}) = \pi/4$ and $\sec^{-1}(\sqrt{2}) = \pi/4$. Therefore,

$$\tan\left(\sec^{-1}(\sqrt{2})\right) = \tan\left(\frac{\pi}{4}\right) = 1. \qquad \blacklozenge$$

The composition of $y = \sin x$ restricted to $[-\pi/2, \pi/2]$ and $y = \sin^{-1} x$ is the identity function, because they are inverse functions of each other. So

$$\sin^{-1}(\sin x) = x \qquad \text{for } x \text{ in } [-\pi/2, \pi/2]$$

and

$$\sin(\sin^{-1} x) = x \qquad \text{for } x \text{ in } [-1, 1].$$

Note that if x is not in $[-\pi/2, \pi/2]$, then the sine function followed by the inverse sine function is not the identity function. For example, $\sin^{-1}(\sin(2\pi/3)) = \pi/3$.

Check the answers to Example 8.

```
sin tan-1 0
                    0
sin-1 cos (π/6)
         1.047197551
tan cos-1 (1/√2)
                    1
■
```

? For Thought

True or false? Explain.

1. $\sin^{-1}(0) = \sin(0)$

2. $\sin(3\pi/4) = 1/\sqrt{2}$

3. $\cos^{-1}(0) = 1$

4. $\sin^{-1}(\sqrt{2}/2) = 135°$

5. $\cot^{-1}(5) = \dfrac{1}{\tan^{-1}(5)}$

6. $\sec^{-1}(5) = \cos^{-1}(0.2)$

7. $\sin\left(\cos^{-1}(\sqrt{2}/2)\right) = 1/\sqrt{2}$

8. $\sec(\sec^{-1}(2)) = 2$

9. The functions $f(x) = \sin^{-1} x$ and $f^{-1}(x) = \sin x$ are inverse functions.

10. The secant and cosecant functions are inverses of each other.

6.5 Exercises ▭ Tape 12 ▦ Disk—5.25″: 4 3.5″: 3 Macintosh: 3

Find the exact value of each expression without using a calculator or table.

1. $\sin^{-1}(-1/2)$
2. $\sin^{-1}(0)$
3. $\arcsin(1/2)$
4. $\arcsin(\sqrt{3}/2)$
5. $\cos^{-1}(\sqrt{2}/2)$
6. $\cos^{-1}(1)$
7. $\arccos(1/2)$
8. $\arccos(-\sqrt{3}/2)$

Find the exact value of each expression in degrees without using a calculator or table.

9. $\arcsin(-1)$
10. $\sin^{-1}(1/\sqrt{2})$
11. $\cos^{-1}(-\sqrt{2}/2)$
12. $\cos^{-1}(\sqrt{3}/2)$
13. $\arcsin(0.5)$
14. $\sin^{-1}(\sqrt{3}/2)$
15. $\arccos(-1)$
16. $\arccos(0)$

Find the exact value of each expression without using a calculator or table.

17. $\tan^{-1}(-1)$
18. $\cot^{-1}(1/\sqrt{3})$
19. $\sec^{-1}(2)$
20. $\csc^{-1}(2/\sqrt{3})$
21. $\operatorname{arcsec}(\sqrt{2})$
22. $\arctan(-1/\sqrt{3})$
23. $\operatorname{arccsc}(-2)$
24. $\operatorname{arccot}(-\sqrt{3})$
25. $\tan^{-1}(0)$
26. $\sec^{-1}(1)$
27. $\csc^{-1}(1)$
28. $\csc^{-1}(-1)$
29. $\cot^{-1}(-1)$
30. $\cot^{-1}(0)$
31. $\cot^{-1}(-\sqrt{3}/3)$
32. $\cot^{-1}(1)$

Find the approximate value of each expression with a calculator. Round answers to two decimal places.

33. $\arcsin(0.5682)$
34. $\sin^{-1}(-0.4138)$
35. $\cos^{-1}(-0.993)$
36. $\cos^{-1}(0.7392)$
37. $\tan^{-1}(-0.1396)$
38. $\cot^{-1}(4.32)$
39. $\sec^{-1}(-3.44)$
40. $\csc^{-1}(6.8212)$
41. $\operatorname{arcsec}(\sqrt{6})$
42. $\arctan(-2\sqrt{7})$
43. $\operatorname{arccsc}(-2\sqrt{2})$
44. $\operatorname{arccot}(-\sqrt{5})$
45. $\operatorname{arccot}(-12)$
46. $\operatorname{arccot}(0.001)$
47. $\cot^{-1}(15.6)$
48. $\cot^{-1}(-1.01)$

Find the exact value of each composition without using a calculator or table.

49. $\tan(\arccos(1/2))$
50. $\sec(\arcsin(1/\sqrt{2}))$
51. $\sin^{-1}(\cos(2\pi/3))$
52. $\tan^{-1}(\sin(\pi/2))$
53. $\cot^{-1}(\cot(\pi/6))$
54. $\sec^{-1}(\sec(\pi/3))$
55. $\arcsin(\sin(3\pi/4))$
56. $\arccos(\cos(-\pi/3))$
57. $\tan(\arctan(1))$
58. $\cot(\operatorname{arccot}(0))$
59. $\cos^{-1}(\cos(3\pi/2))$
60. $\sin(\csc^{-1}(-2))$
61. $\cos(2\sin^{-1}(\sqrt{2}/2))$
62. $\tan(2\cos^{-1}(1/2))$
63. $\sin^{-1}(2\sin(\pi/6))$
64. $\cos^{-1}(0.5\tan(\pi/4))$

Use a calculator to find the approximate value of each composition. Round answers to four decimal places. Some of these expressions are undefined.

65. $\sin(\cos^{-1}(0.45))$
66. $\cos(\tan^{-1}(44.33))$
67. $\sec^{-1}(\cos(\pi/9))$
68. $\csc^{-1}(\sec(\pi/8))$
69. $\tan(\sin^{-1}(-0.7))$
70. $\csc^{-1}(\sin(3\pi/7))$
71. $\cot(\cos^{-1}(-1/\sqrt{7}))$
72. $\sin^{-1}(\csc(1.08))$
73. $\cot(\csc^{-1}(3.6))$
74. $\csc(\sec^{-1}(-2.4))$
75. $\sec(\cot^{-1}(5.2))$
76. $\csc(\cot^{-1}(-3.1))$

In each case α is an angle such that $0° < \alpha < 90°$. Find α to the nearest tenth of a degree.

77. $\sin\alpha = 0.557$
78. $\sin\alpha = 0.93$
79. $\cos\alpha = 0.06$
80. $\cos\alpha = 0.71$
81. $\csc\alpha = 1.3$
82. $\sec\alpha = 1.1$
83. $\tan\alpha = 6.4$
84. $\cot\alpha = 9.8$

Sketch the graph of each function using degree measures for the range.

85. $y = \sin^{-1}(x)$
86. $y = \arccos x$
87. $f(x) = \sec^{-1}(x)$
88. $f(x) = \csc^{-1}(x)$
89. $y = \arctan x$
90. $f(x) = \operatorname{arccot} x$

In a circle with radius r, a central angle θ intercepts a chord of length c, where $\theta = \cos^{-1}\left(1 - \dfrac{c^2}{2r^2}\right)$. In each case, find the central angle θ.

91. $r = 10$, $c = 10\sqrt{2}$ **92.** $r = c$

93. $r = 12$ m, $c = 3$ m   **94.** $r = 3.2$ m, $c = 6.1$ m

For Writing/Discussion

95. *Redefining inverse sine* Find an interval other than $[-\pi/2, \pi/2]$ on which $y = \sin x$ is a one-to-one function. Restrict $y = \sin x$ to this interval and define its inverse function. Find $\sin^{-1} \alpha$ for several values of α using your definition of the inverse function. Which inverse sine function is simpler, the new one or the old one?

96. *Redefining inverse cotangent* Restrict $y = \cot x$ to the domain $(-\pi/2, 0) \cup (0, \pi/2]$. Define the inverse of this restricted function and sketch its graph. Write an identity that gives the values of the new $\cot^{-1}$ function in terms of the function $\tan^{-1}$. Does the identity $\cot^{-1}(x) = \pi/2 - \tan^{-1}(x)$ hold for the new $\cot^{-1}$? Which inverse cotangent function is simpler, the new one or the old one?

Graphing Calculator Exercises

1. Graph the function $y = \sin(\sin^{-1} x)$ for $-2\pi \leq x \leq 2\pi$ and explain your result.

2. Graph the function $y = \sin^{-1}(\sin x)$ for $-2\pi \leq x \leq 2\pi$ and explain your result.

3. Graph $y = \sin^{-1}(1/x)$ and explain why the graph looks like the graph of $y = \csc^{-1} x$ shown in Fig. 6.51.

4. Graph $y = \tan^{-1}(1/x)$ and explain why the graph does *not* look like the graph of $y = \cot^{-1} x$ shown in Fig. 6.51.

6.6

Right Triangle

Trigonometry

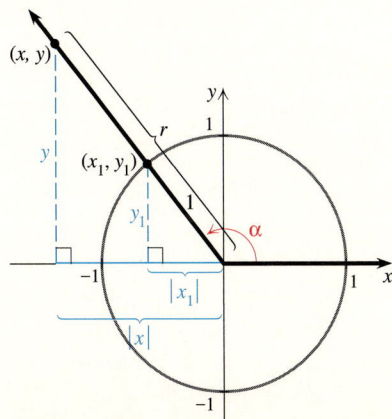

Figure 6.52

One reason trigonometry was invented was to determine the measures of sides and angles of geometric figures without actually measuring them. In this section we study right triangles (triangles that have a 90° angle) and see what information is needed to determine the measures of all unknown sides and angles of a right triangle.

Trigonometric Ratios

We defined the sine and cosine functions for an angle in standard position in terms of the point at which the terminal side intersects the unit circle. However, it is often the case that we do not know that point but we do know a different point on the terminal side. When this situation occurs, we can find the values for the sine and cosine of the angle using trigonometric ratios.

Figure 6.52 shows an angle α in quadrant II with the terminal side passing through the point (x, y) and intersecting the unit circle at (x_1, y_1). If we draw vertical line segments down to the x-axis, we form two similar right triangles as shown in Fig. 6.52. If r is the distance from (x, y) to the origin, then $r = \sqrt{x^2 + y^2}$. The lengths of the legs in the larger triangle are y and $|x|$, and its hypotenuse is r. The lengths of the legs in the smaller triangle are y_1 and $|x_1|$, and its hypotenuse is 1. Since ratios of the lengths of corresponding sides of similar triangles are equal, we have

$$\frac{y_1}{y} = \frac{1}{r} \quad \text{and} \quad \frac{|x_1|}{|x|} = \frac{1}{r}.$$

The first equation can be written as $y_1 = y/r$ and the second as $x_1 = x/r$. (Since x_1 and x are both negative in this case, the absolute value symbols can be omitted.) Since $y_1 = \sin \alpha$ and $x_1 = \cos \alpha$, we get

$$\sin \alpha = \frac{y}{r} \qquad \text{and} \qquad \cos \alpha = \frac{x}{r}.$$

This argument can be repeated in each quadrant with (x, y) chosen inside, outside, or on the unit circle. These ratios also give the correct sine or cosine if α is a quadrantal angle. Since all other trigonometric functions are defined in terms of sine and cosine, their values can also be obtained from x, y, and r.

Theorem: Trigonometric Ratios

If (x, y) is any point other than the origin on the terminal side of an angle α in standard position and $r = \sqrt{x^2 + y^2}$, then

$$\sin \alpha = \frac{y}{r}, \qquad \cos \alpha = \frac{x}{r}, \qquad \text{and} \qquad \tan \alpha = \frac{y}{x} \quad (x \neq 0).$$

Example 1   Trigonometric ratios

Find the values of the six trigonometric functions of the angle α in standard position whose terminal side passes through $(4, -2)$.

Solution

Use $x = 4$, $y = -2$, and $r = \sqrt{4^2 + (-2)^2} = \sqrt{20} = 2\sqrt{5}$ to get

$$\sin \alpha = \frac{-2}{2\sqrt{5}} = -\frac{\sqrt{5}}{5}, \quad \cos \alpha = \frac{4}{2\sqrt{5}} = \frac{2\sqrt{5}}{5}, \quad \text{and} \quad \tan \alpha = \frac{-2}{4} = -\frac{1}{2}.$$

Since cosecant, secant, and cotangent are the reciprocals of sine, cosine, and tangent,

$$\csc \alpha = \frac{1}{\sin \alpha} = -\frac{5}{\sqrt{5}} = -\sqrt{5}, \qquad \sec \alpha = \frac{1}{\cos \alpha} = \frac{5}{2\sqrt{5}} = \frac{\sqrt{5}}{2}, \qquad \text{and}$$

$$\cot \alpha = \frac{1}{\tan \alpha} = -2.$$

Right Triangles

So far the trigonometric functions have been tied to a coordinate system and an angle in standard position. Trigonometric ratios can also be used to evaluate the trigonometric functions for an acute angle of a right triangle without having the angle or the triangle located in a coordinate system.

Consider a right triangle with acute angle α, legs of length x and y, and hypotenuse r, as shown in Fig. 6.53(a).

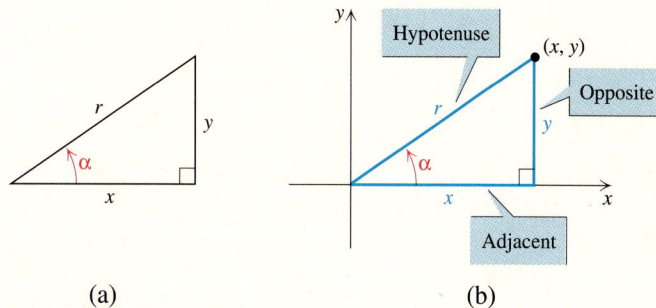

(a)   (b)

Figure 6.53

If this triangle is positioned in a coordinate system as in Fig. 6.53(b), then (x, y) is a point on the terminal side of α and

$$\sin(\alpha) = \frac{y}{r}, \qquad \cos(\alpha) = \frac{x}{r}, \qquad \text{and} \qquad \tan(\alpha) = \frac{y}{x}.$$

However, the values of the trigonometric functions are simply ratios of the lengths of the sides of the right triangle and it is not necessary to move the triangle to a coordinate system to find them. Notice that y is the length of the side **opposite** the angle α, x is the length of the side **adjacent** to α, and r is the length of the **hypotenuse.** We use the abbreviations opp, adj, and hyp to represent the lengths of these sides in the following theorem.

Theorem: Trigonometric Functions of an Acute Angle of a Right Triangle

If α is an acute angle of a right triangle, then

$$\sin \alpha = \frac{\text{opp}}{\text{hyp}}, \qquad \cos \alpha = \frac{\text{adj}}{\text{hyp}}, \qquad \text{and} \qquad \tan \alpha = \frac{\text{opp}}{\text{adj}}.$$

Since the cosecant, secant, and cotangent are the reciprocals of the sine, cosine, and tangent, respectively, the values of all six trigonometric functions can be found for an acute angle of a right triangle.

Example 2   Trigonometric functions in a right triangle

Find the values of all six trigonometric functions for the angle α of the right triangle with legs of length 1 and 4 as shown in Fig. 6.54 on the next page.

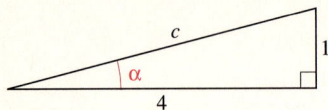

Figure 6.54

Solution

The length of the hypotenuse is $c = \sqrt{4^2 + 1^2} = \sqrt{17}$. Since the length of the side opposite α is 1 and the length of the adjacent side is 4, we have

$$\sin \alpha = \frac{\text{opp}}{\text{hyp}} = \frac{1}{\sqrt{17}} = \frac{\sqrt{17}}{17}, \qquad \cos \alpha = \frac{\text{adj}}{\text{hyp}} = \frac{4}{\sqrt{17}} = \frac{4\sqrt{17}}{17},$$

$$\tan \alpha = \frac{\text{opp}}{\text{adj}} = \frac{1}{4}, \qquad \csc \alpha = \frac{1}{\sin \alpha} = \sqrt{17},$$

$$\sec \alpha = \frac{1}{\cos \alpha} = \frac{\sqrt{17}}{4}, \qquad \cot \alpha = \frac{1}{\tan \alpha} = 4.$$

Solving a Right Triangle

The value of the trigonometric functions for an acute angle of a right triangle are determined by ratios of lengths of sides of the triangle. We can use those ratios along with the inverse trigonometric functions to find missing parts of a right triangle in which some of the measures of angles or lengths of sides are known. Finding all of the unknown lengths of sides or measures of angles is called **solving the triangle.** A triangle can be solved only if enough information is given to determine a unique triangle. For example, the lengths of the sides in a 30–60–90 triangle cannot be found because there are infinitely many such triangles of different sizes. However, the lengths of the missing sides in a 30–60–90 triangle with a hypotenuse of 6 will be found in Example 3.

In solving right triangles, we usually name the acute angles α and β and the lengths of the sides opposite those angles a and b. The 90° angle is γ, and the length of the side opposite γ is c.

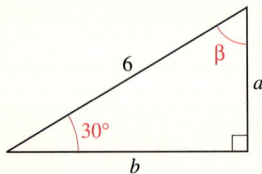

Figure 6.55

Example 3 Solving a right triangle

Solve the right triangle in which $\alpha = 30°$ and $c = 6$.

Solution

The triangle is shown in Fig. 6.55. Since $\alpha = 30°$, $\gamma = 90°$, and the sum of the measures of the angles of any triangle is 180°, we have $\beta = 60°$. Since $\sin \alpha = \text{opp}/\text{hyp}$, we get $\sin 30° = a/6$ and

$$a = 6 \cdot \sin 30° = 6 \cdot \frac{1}{2} = 3.$$

Since $\cos \alpha = \text{adj}/\text{hyp}$, we get $\cos 30° = b/6$ and

$$b = 6 \cdot \cos 30° = 6 \cdot \frac{\sqrt{3}}{2} = 3\sqrt{3}.$$

The angles of the right triangle are 30°, 60°, and 90°, and the sides opposite those angles are 3, $3\sqrt{3}$, and 6, respectively.

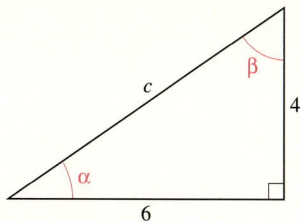

Figure 6.56

Example 4   Solving a right triangle

Solve the right triangle in which $a = 4$ and $b = 6$. Find the acute angles to the nearest tenth of a degree.

Solution

The triangle is shown in Fig. 6.56. By the Pythagorean theorem, $c^2 = 4^2 + 6^2$, or $c = \sqrt{52} = 2\sqrt{13}$. To find α, first find $\sin \alpha$:

$$\sin \alpha = \frac{\text{opp}}{\text{hyp}} = \frac{4}{2\sqrt{13}} = \frac{2}{\sqrt{13}}$$

Now, α is the angle whose sine is $2/\sqrt{13}$:

$$\alpha = \sin^{-1}\left(\frac{2}{\sqrt{13}}\right) \approx 33.7°$$

Since $\alpha + \beta = 90°$, $\beta = 90° - 33.7° = 56.3°$. The angles of the triangle are $33.7°$, $56.3°$, and $90°$, and the sides opposite those angles are 4, 6, and $2\sqrt{13}$, respectively.

In Example 4 we could have found α by using $\alpha = \tan^{-1}(4/6) \approx 33.7°$. We could then have found c by using $\cos(33.7°) = 6/c$, or $c = 6/\cos(33.7°) \approx 7.2$. There are many ways to solve a right triangle, but the basic strategy is always the same.

Strategy: Solving a Right Triangle

To solve a right triangle:

1. **Use the Pythagorean theorem to find the length of a third side when the lengths of two sides are known.**
2. **Use the trigonometric ratios to find missing sides or angles.**
3. **Use the fact that the sum of the measures of the angles of a triangle is 180° to determine a third angle when two are known.**

Significant Digits

In solving right triangles and in most other problems in algebra and trigonometry, we usually assume that the given numbers are exact. Starting with exact numbers, exact answers or approximations to any degree of accuracy desired can be obtained. However, in real applications, lengths of sides in triangles or measures of angles are determined by actual measurement. In such cases, we must be concerned with the accuracy of our measurements and the values obtained from those measurements using trigonometry. For example, an angle of 21.75° at the center of a circle with a one-mile radius intercepts an arc of length 2004.3 feet, but if the same angle is given to the nearest tenth of a degree, 21.8°, then the length of the intercepted arc is 2008.9 feet, a difference of 4.6 feet.

To say that an angle is measured at 21.8° means that the angle is actually between 21.75° and 21.85°. Since all digits in 21.8° are meaningful, we say that 21.8° has three **significant digits.** Zeros that follow the decimal point but precede nonzero digits are not significant. A measurement of 0.00034 centimeters has two significant digits. In the measurement 44.0 pounds, the zero means that the measurement was made to the nearest tenth, and so it is significant. Zeros that follow the decimal point but are not followed by nonzero digits are significant. For whole-number measurements without decimal points, we assume that all digits are significant. So measurements of 4782 feet and 8000 miles both have four significant digits. In scientific notation, only significant digits are written. For example, 8×10^3 has only one significant digit, while 8.000×10^3 has four.

When arithmetic operations or trigonometric functions are applied to measurements, the answers are only as accurate as the *least accurate* measurement involved. For example, if it is 23 miles from your house to the airport and 8×10^3 miles (to the nearest thousand) to Hong Kong, then to say that it is 8023 miles to Hong Kong from your house implies an accuracy that is not justified. Since 8×10^3 miles is a less accurate measurement than 23 miles and 8×10^3 miles has one significant digit, any answer obtained from these two measurements has only one significant digit. So only the first digit of 8023 is significant, and it is 8×10^3 miles (to the nearest thousand) from your house to Hong Kong. A "rule of thumb" that works well in most cases is that an answer should be given with the same number of significant digits as is in the least accurate measurement in the computation.

Example 5   Significant digits

Assume that the given numbers are measurements. Perform each computation, and give the answer with the appropriate number of significant digits.

a) $\sin(123.4°)$ **b)** $\dfrac{\cos(3°)}{\sin(1.9°)}$

c) $544 \cdot \sin(25°)$

Solution

a) Since 123.4° has four significant digits, we round the calculator value to four decimal places. So $\sin(123.4°) \approx 0.8348$.

b) Since the least number of significant digits in 3° and 1.9° is one, round the value obtained on a calculator to one significant digit:

$$\frac{\cos(3°)}{\sin(1.9°)} \approx 3 \times 10^1 \text{ (to the nearest ten)}$$

c) The least number of significant digits in 544 and 25° is two. So

$$544 \cdot \sin(25°) \approx 230 \text{ (to the nearest ten)}.$$

In scientific notation, $544 \cdot \sin(25°) \approx 2.3 \times 10^2$.   ◆

Applications

Using trigonometry, we can find the size of an object without actually measuring the object but by measuring an angle. Two common terms used in this regard are **angle of elevation** and **angle of depression.** The angle of elevation α for a point above a horizontal line is the angle formed by the horizontal line and the observer's line of sight through the point as shown in Fig. 6.57. The angle of depression β for a point below a horizontal line is the angle formed by the horizontal line and the observer's line of sight through the point as shown in Fig. 6.57. We use these angles and our skills in solving triangles to find the sizes of objects that would be inconvenient to measure.

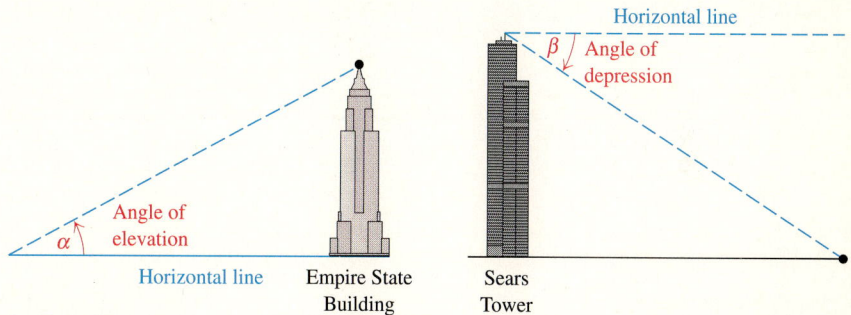

Figure 6.57

Example 6   Finding the height of an object

A guy wire of length 108 meters runs from the top of an antenna to the ground. If the angle of elevation of the top of the antenna, sighting along the guy wire, is 42.3°, then what is the height of the antenna?

Solution

Let y represent the height of the antenna as shown in Fig. 6.58. Since $\sin(42.3°) = y/108$,

$$y = 108 \cdot \sin(42.3°) \approx 72.7 \text{ meters.}$$

Three significant digits are used in the answer because both of the given measurements contain three significant digits.

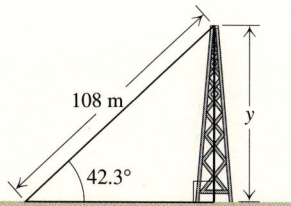

Figure 6.58

In Example 6 we knew the distance to the top of the antenna, and we found the height of the antenna. If we knew the distance on the ground to the base of the antenna and the angle of elevation of the guy wire, we could still have found the height of the antenna. Both cases involve knowing the distance to the antenna either on the ground or through the air. However, one of the biggest triumphs of trigonometry is being able to find the size of an object or the distance to an object (such as the moon) without going to the object. The next example shows one way to find the height of an object without actually going to it.

Example 7 Finding the height of an object without going there

The angle of elevation of the top of a water tower from point A on the ground is 19.9°. From point B, 50.0 feet closer to the tower, the angle of elevation is 21.8°. What is the height of the tower?

Solution

Let y represent the height of the tower and x represent the distance from point B to the base of the tower as shown in Fig. 6.59.

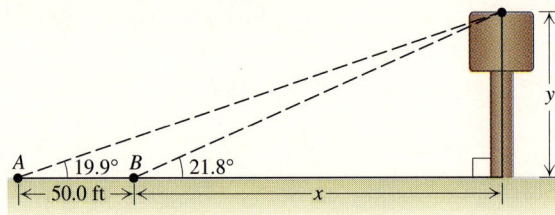

Figure 6.59

At each point where the angles of elevation are given, the tangent of the angle of elevation is the ratio of the opposite and adjacent sides of a right triangle. At point B, $\tan 21.8° = y/x$ or

$$x = \frac{y}{\tan 21.8°}.$$

Since the distance to the base of the tower from point A is $x + 50$,

$$\tan 19.9° = \frac{y}{x + 50}$$

or

$$y = (x + 50) \tan 19.9°.$$

To find the value of y we must write an equation that involves only y. Since $x = y/\tan 21.8°$, we can substitute $y/\tan 21.8°$ for x in the last equation:

$$y = \left(\frac{y}{\tan 21.8°} + 50 \right) \tan 19.9°$$

$$y = \frac{y \cdot \tan 19.9°}{\tan 21.8°} + 50 \tan 19.9° \qquad \text{\textcolor{blue}{\textbf{Distributive property}}}$$

$$y - \frac{y \cdot \tan 19.9°}{\tan 21.8°} = 50 \tan 19.9°$$

$$y \left(1 - \frac{\tan 19.9°}{\tan 21.8°} \right) = 50 \tan 19.9° \qquad \text{\textcolor{blue}{\textbf{Factor out y.}}}$$

$$y = \frac{50 \tan 19.9°}{1 - \dfrac{\tan 19.9°}{\tan 21.8°}} \approx 191 \text{ feet}$$

```
50tan 19.9/(1-ta
n 19.9/tan 21.8)
          190.6278641
```

Use degree mode to find y in Example 7.

In the next example we combine the solution of a right triangle with the arc length of a circle from Section 6.1 to solve a problem of aerial photography.

Example 8   Photography from a spy plane

In the late 1950s, the Soviets labored to develop a missile that could stop the U-2 spy plane. On May 1, 1960, Nikita S. Khrushchev announced to the world that the Soviets had shot down Francis Gary Powers while Powers was photographing the Soviet Union from a U-2 at an altitude of 14 miles. How wide a path on the earth's surface could Powers see from that altitude? (Use 3950 miles as the earth's radius.)

Solution

Figure 6.60 shows the line of sight to the horizon on the left-hand side and right-hand side of the airplane while flying at the altitude of 14 miles.

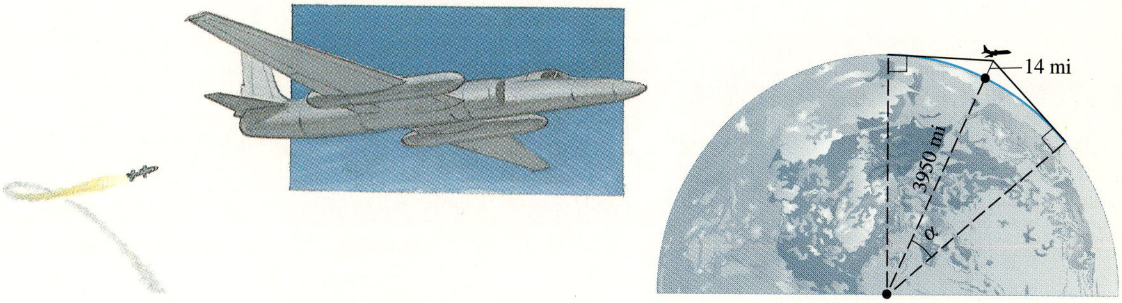

Figure 6.60

Since a line tangent to a circle (the line of sight) is perpendicular to the radius at the point of tangency, the angle α at the center of the earth in Fig. 6.60 is an acute angle of a right triangle with hypotenuse $3950 + 14$ or 3964. So we have

$$\cos \alpha = \frac{3950}{3964}$$

$$\alpha = \cos^{-1}\left(\frac{3950}{3964}\right) \approx 4.8°.$$

The width of the path seen by Powers is the length of the arc intercepted by the central angle 2α or $9.6°$. Using the formula $s = \alpha r$ from Section 6.1, where α is in radians, we get

$$s = 9.6 \text{ deg} \cdot \frac{\pi \text{ rad}}{180 \text{ deg}} \cdot 3950 \text{ miles} \approx 661.8 \text{ miles}.$$

From an altitude of 14 miles, Powers could see a path that was 661.8 miles wide. Actually, he photographed a path that was somewhat narrower, because parts of the photographs near the horizon were not usable.   ◆

For Thought

?

True or false? Explain. For Exercises 1–4, α is an angle in standard position.

1. If the terminal side of α goes through $(5, -10)$, then $\sin \alpha = 10/\sqrt{125}$.
2. If the terminal side of α goes through $(-1, 2)$, then $\sec \alpha = -\sqrt{5}$.
3. If the terminal side of α goes through $(-2, 3)$, then $\alpha = \sin^{-1}(3/\sqrt{13})$.
4. If the terminal side of α goes through $(3, 1)$, then $\alpha = \cos^{-1}(3/\sqrt{10})$.
5. In a right triangle, $\sin \alpha = \cos \beta$, $\sec \alpha = \csc \beta$, and $\tan \alpha = \cot \beta$.
6. If $a = 4$ and $b = 2$ in a right triangle, then $c = \sqrt{6}$.
7. If $a = 6$ and $b = 2$ in a right triangle, then $\beta = \tan^{-1}(3)$.
8. If $a = 8$ and $\alpha = 55°$ in a right triangle, then $b = 8/\tan(55°)$.
9. In a right triangle with sides of length 3, 4, and 5, the smallest angle is $\cos^{-1}(0.8)$.
10. In a right triangle, $\sin(90°) = \text{hyp/adj}$.

6.6 Exercises

⬛ **Tape 12**   💾 **Disk—5.25″: 4   3.5″: 3   Macintosh: 3**

Assume that α is an angle in standard position whose terminal side contains the given point. Find the exact values of $\sin \alpha$, $\cos \alpha$, $\tan \alpha$, $\csc \alpha$, $\sec \alpha$, and $\cot \alpha$.

1. $(3, 4)$ 2. $(4, 4)$ 3. $(-2, 6)$
4. $(-3, 6)$ 5. $(-2, -\sqrt{2})$ 6. $(-1, -\sqrt{3})$
7. $(\sqrt{6}, -\sqrt{2})$ 8. $(2\sqrt{3}, -2)$

Assume that α is an angle in standard position whose terminal side contains the given point and that $0° < \alpha < 90°$. Find the degree measure of α to the nearest tenth of a degree.

9. $(1.5, 9)$ 10. $(4, 5)$ 11. $(\sqrt{2}, \sqrt{6})$ 12. $(4.3, 6.9)$

Assume that α is an angle in standard position whose terminal side contains the given point and that $0 < \alpha < \pi/2$. Find the radian measure of α to the nearest tenth of a radian.

13. $(4, 6.3)$ 14. $(1/3, 1/2)$
15. $(\sqrt{5}, 1)$ 16. $(\sqrt{7}, \sqrt{3})$

For Exercises 17–22 find exact values of $\sin \alpha$, $\cos \alpha$, $\tan \alpha$, $\sin \beta$, $\cos \beta$, and $\tan \beta$ for the given right triangle.

17.

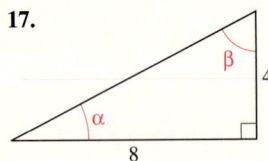

18.

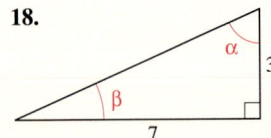

19.

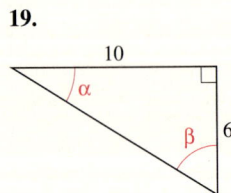

20.

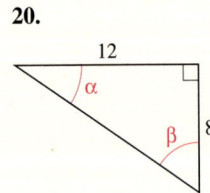

21.

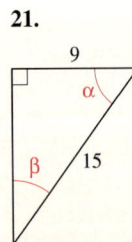

22.

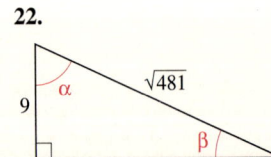

Solve each right triangle with the given sides and angles. In each case, make a sketch.

23. $a = 6$, $b = 8$

24. $a = 10$, $c = 12$

25. $b = 6$, $c = 8.3$

26. $\alpha = 32.4°$, $b = 10$

27. $\alpha = 16°$, $c = 20$

28. $\beta = 47°$, $a = 3$

29. $\alpha = 39°9'$, $a = 9$

30. $\beta = 19°12'$, $b = 60$

Perform each computation with the given measurements, and give your answers with the appropriate number of significant digits.

31. $456 \cdot \tan 3.2°$

32. $123.32 \cdot \tan 36.3°$

33. $\sin^2 65.7°$

34. $4 + 6.33 \sin 79.7°$

35. $7.50/\sin 23.559°$

36. $12.3/\cot 0.03°$

37. $500/\sec 125.3°$

38. $16.3 - 8.31 \cdot \csc 8.76°$

Solve each problem.

39. *Aerial photography* An aerial photograph from a U-2 spy plane is taken of a building suspected of housing nuclear warheads. The photograph is made when the angle of elevation of the sun is 32°. By comparing the shadow cast by the building to objects of known size in the photograph, analysts determine that the shadow is 80 ft long. How tall is the building?

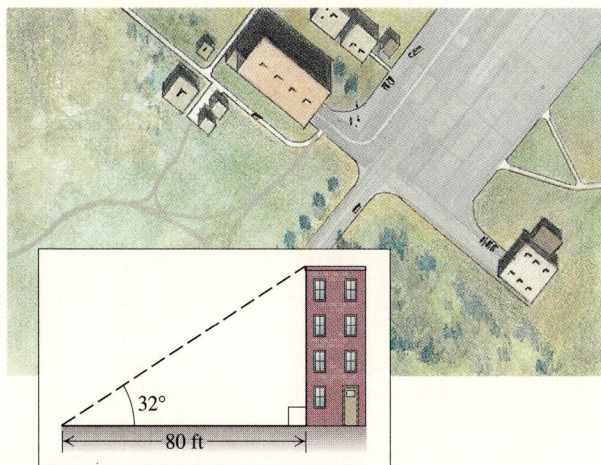

Figure for Exercise 39

40. *Avoiding a swamp* Muriel was hiking directly toward a long, straight road when she encountered a swamp. She turned 65° to the right and hiked 4 mi in that direction to reach the road. How far was she from the road when she encountered the swamp?

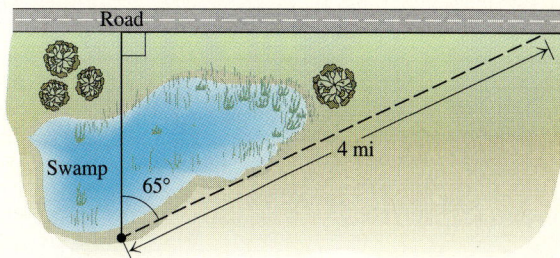

Figure for Exercise 40

41. *Angle of depression* From a highway overpass, 14.3 m above the road, the angle of depression of an oncoming car is measured at 18.3°. How far is the car from a point on the highway directly below the observer?

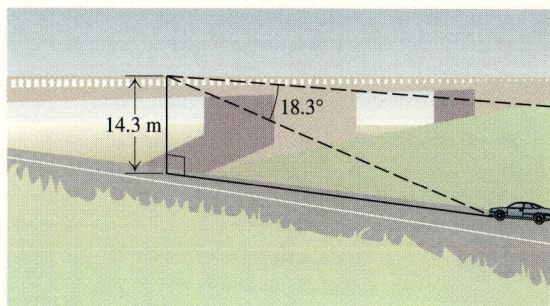

Figure for Exercise 41

42. *Length of a tunnel* A tunnel under a river is 196.8 ft below the surface at its lowest point as shown in the drawing. If the angle of depression of the tunnel is 4.962°, then how far apart on the surface are the entrances to the tunnel? How long is the tunnel?

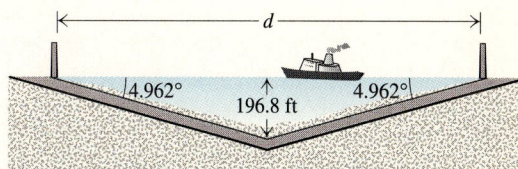

Figure for Exercise 42

43. *Height of a crosswalk* The angle of elevation of a pedestrian crosswalk over a busy highway is 8.34° as shown in the drawing. If the distance between the ends of the crosswalk measured on the ground is 342 ft, then what is the height *h* of the crosswalk at the center?

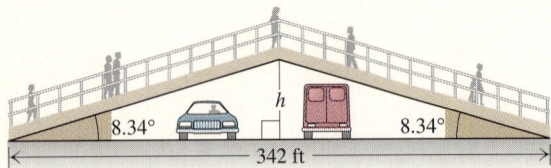

Figure for Exercise 43

44. *Shortcut to Snyder*   To get from Muleshoe to Snyder, Harry drives 50 mph for 178 mi south on route 214 to Seminole, then goes east on route 180 to Snyder. Harriet leaves Muleshoe one hour later at 55 mph, but takes US 84, which goes straight from Muleshoe to Snyder through Lubbock. If US 84 intersects route 180 at a 50° angle, then how many more miles does Harry drive?

45. *Installing a guy wire*   A 41-m guy wire is attached to the top of a 34.6-m antenna and to a point on the ground. How far is the point on the ground from the base of the antenna, and what angle does the guy wire make with the ground?

46. *Robin and Marian*   Robin Hood plans to use a 30-ft ladder to reach the castle window of Maid Marian. Little John, who made the ladder, advised Robin that the angle of elevation of the ladder must be between 55° and 70° for safety. What are the minimum and maximum heights that can safely be reached by the top of the ladder when it is placed against the 50-ft castle wall?

47. *Detecting a speeder*   A policewoman has positioned herself 500 ft from the intersection of two roads. She has carefully measured the angles of the lines of sight to points *A* and *B* as shown in the drawing. If a car passes from *A* to *B* in 1.75 s and the speed limit is 55 mph, is the car speeding?

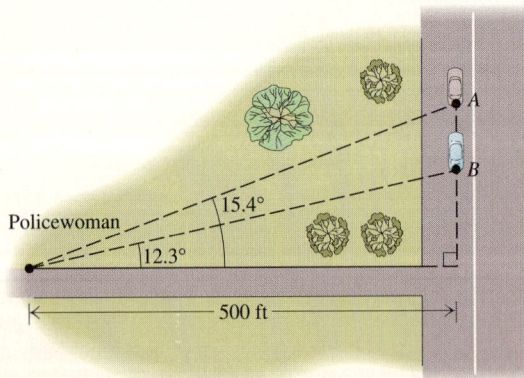

Figure for Exercise 47

48. *Progress of a forest fire*   A forest ranger atop a 3248-ft mesa is watching the progress of a forest fire spreading in her direction. In five minutes the angle of depression of the leading edge of the fire changed from 11.34° to 13.51°. At what speed in miles per hour is the fire spreading in the direction of the ranger?

49. *Height of a rock*   Inscription Rock rises almost straight upward from the valley floor. From one point the angle of elevation of the top of the rock is 16.7°. From a point 168 m closer to the rock, the angle of elevation of the top of the rock is 24.1°. How high is Inscription Rock?

50. *Height of a balloon*   A hot air balloon is between two spotters who are 1.2 mi apart. One spotter reports that the angle of elevation of the balloon is 76°, and the other reports that it is 68°. What is the altitude of the balloon in miles?

51. *Passing in the night*   A boat sailing north sights a lighthouse to the east at an angle of 32° from the north as shown in the drawing. After the boat travels one more kilometer, the angle of the lighthouse from the north is 36°. If the boat continues to sail north, then how close will the boat come to the lighthouse?

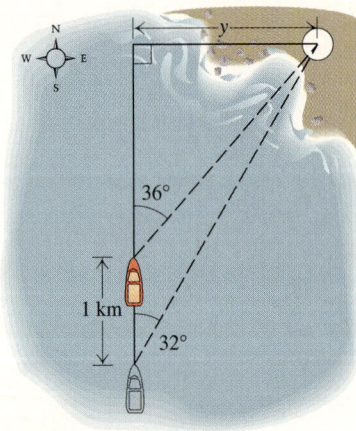

Figure for Exercise 51

52. *Height of a skyscraper*   For years the Woolworth skyscraper in New York held the record for the world's tallest office building. If the length of the shadow of the Woolworth building increases by 17.4 m as the angle of elevation of the sun changes from 44° to 42°, then how tall is the building?

53. *Parsecs*   In astronomy the light year (abbreviated ly) and the parsec (abbreviated pc) are the two units used to measure distances. The distance in space at which a line from the earth to the sun subtends an angle of 1 second is 1 parsec. Find the number of miles in 1 parsec by using a right triangle positioned in space with its right angle at the sun as shown in the figure.

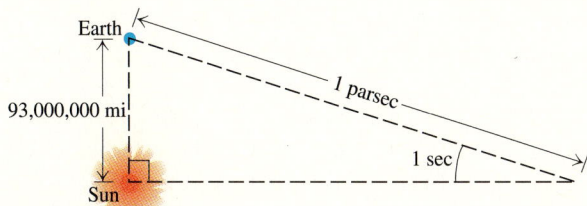

Earth

93,000,000 mi

1 parsec

1 sec

Sun

Figure for Exercise 53

54. *Light years*   In astronomy, 1 astronomical unit is defined as the distance from the earth to the sun, or 1 AU = 93,000,000 mi. A light year (ly) is the distance that light travels in one year. If 1 ly = 63,240 AU, then how many years does it take light to travel 1 parsec? (See Exercise 53.)

55. *View from Landsat*   The satellite Landsat orbits the earth at an altitude of 700 mi, as shown in the figure. What is the width of the path on the surface of the earth that can be seen by the cameras of Landsat?

700 mi

Figure for Exercise 55

56. *Communicating via satellite*   A communication satellite is usually put into a synchronous orbit with the earth, which means that it stays above a fixed point on the surface of the earth at all times. The radius of the orbit of such a satellite is 6.5 times the radius of the earth (3950 mi). The satellite is used to relay a signal from one point on the earth to another point on the earth. The sender and receiver of a signal must be in a line of sight with the satellite, as shown in the figure. What is the maximum distance on the surface of the earth between the sender and receiver for this type of satellite?

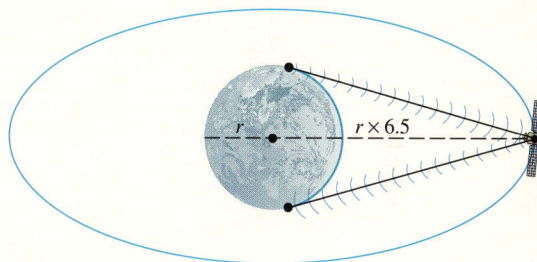

r $r \times 6.5$

Figure for Exercise 56

For Writing/Discussion

57. *Cooperative learning*   Work in a group to select a local building and find its height by measuring the length of its shadow and the angle of elevation of the sun. Explain how you determined the angle of elevation of the sun.

58. *Cooperative learning*   Work in a group to select a local building and find its height without going up to the building. Use the difference in the lengths of its shadow at two different times of day and the technique of Example 7.

Highlights

Section 6.1   Angles and Their Measurements

1. If the circumference of a circle is divided into 360 equal arcs (degrees), then the degree measure of a central angle is the number of degrees through which the initial side must rotate to get to the terminal side. The degree measure is positive if the rotation is counterclockwise, negative if clockwise.

2. The radian measure of an angle is the length of the arc on the unit circle through which the initial side rotates to get to the terminal side. The radian measure is positive if the rotation is counterclockwise, negative if clockwise.

3. To convert from radians to degrees or degrees to radians use π rad $= 180°$.

4. The arc length s intercepted by a central angle of α radians on a circle of radius r is $s = \alpha r$.

5. The angular velocity ω of a point in motion on a circle of radius r through an angle of α radians in time t is $\omega = \alpha/t$. The linear velocity v is $v = s/t$, where s is the arc length determined by $s = \alpha r$. The relationship between v and ω is $v = r\omega$.

Section 6.2 The Sine and Cosine Functions

1. If α is an angle in standard position and (x, y) is the point of intersection of the terminal side and the unit circle, then $\sin \alpha = x$ and $\cos \alpha = y$.

2. Exact values for $\sin \alpha$ and $\cos \alpha$ are known if α is an integral multiple of $30°$ or $45°$.

3. To find the sine (or cosine) of any angle, first find the sine (or cosine) of its reference angle in the first quadrant and prefix the appropriate sign.

4. For any angle α, $\sin^2 \alpha + \cos^2 \alpha = 1$.

5. The domain for sine or cosine can be the set of angles in standard position, the measures of those angles in degrees or radians, or the set of real numbers.

6. A radian measure is a real number because it is the length of an arc on the unit circle.

Section 6.3 The Graphs of the Sine and Cosine Functions

1. The graphs of $y = \sin x$ and $y = \cos x$ are sine waves, each with period 2π.

2. The graph of $y = A \sin(B[x - C]) + D$ or $y = A \cos(B[x - C]) + D$ is a sine wave with amplitude $|A|$, period $2\pi/|B|$, phase shift $|C|$, and vertical translation D.

3. To graph a sine wave for $B > 0$, first draw one cycle of $y = \sin Bx$ or $y = \cos Bx$ on $[0, 2\pi/B]$, then perform any necessary translation, reflection, or amplitude change on that cycle.

4. If P is the period of a sine wave, then the frequency F is defined by $F = 1/P$.

Section 6.4 The Other Trigonometric Functions and Their Graphs

1. For an angle α in standard position we define four new functions: $\tan(\alpha) = \sin \alpha/\cos \alpha$, $\cot \alpha = \cos \alpha/\sin \alpha$, $\sec \alpha = 1/\cos \alpha$, and $\csc \alpha = 1/\sin \alpha$ provided no denominator is 0.

2. The tangent and cotangent functions have period π, while the secant and cosecant have period 2π.

3. The fundamental cycle of $y = \tan Bx$ for $B > 0$ occurs on the interval $[-\pi/B, \pi/B]$.

4. The vertical asymptotes for tangent and secant occur at the x-intercepts of $y = \cos x$, and the vertical asymptotes for cotangent and cosecant occur at the x-intercepts for $y = \sin x$.

Section 6.5   The Inverse Trigonometric Functions

1. Each of the six trigonometric functions is invertible when restricted to a suitable domain.

2. If $y = \sin^{-1} x$ for x in $[-1, 1]$, then y is the real number in $[-\pi/2, \pi/2]$ such that $\sin y = x$.

3. If $y = \cos^{-1} x$ for x in $[-1, 1]$, then y is the real number in $[0, \pi]$ such that $\cos y = x$.

4. If $y = \tan^{-1} x$ for x in $(-\infty, \infty)$, then y is the real number in $[-\pi/2, \pi/2]$ such that $\tan y = x$.

5. Identities are used to evaluate the functions $\csc^{-1}$, $\sec^{-1}$, and $\cot^{-1}$.

Section 6.6   Right Triangle Trigonometry

1. If (x, y) is any point other than the origin on the terminal side of an angle α in standard position and $r = \sqrt{x^2 + y^2}$, then $\sin \alpha = y/r$, $\cos \alpha = x/r$, and $\tan \alpha = y/x$ $(x \neq 0)$.

2. If α is an acute angle of a right triangle, then $\sin \alpha = $ opp/hyp, $\cos \alpha = $ adj/hyp, and $\tan \alpha = $ opp/adj.

3. Solving a right triangle means finding all unknown sides or angles.

4. In computing with numbers obtained from measurements, answers are only as accurate as the least accurate measurement in the computation.

Chapter 6 Review Exercises

Find the measure in degrees of the least positive angle that is coterminal with each given angle.

1. $388°$

2. $-840°$

3. $-153°14'27''$

4. $455°39'24''$

5. $-\pi$

6. $-35\pi/6$

7. $13\pi/5$

8. $29\pi/12$

Convert each radian measure to degree measure. Do not use a calculator.

9. $5\pi/3$ **10.** $-3\pi/4$ **11.** $3\pi/2$ **12.** $5\pi/6$

Convert each degree measure to radian measure. Do not use a calculator.

13. $330°$

14. $405°$

15. $-300°$

16. $-210°$

Give the exact values of each of the following expressions. Do not use a calculator.

17. $\sin(-\pi/4)$

18. $\cos(-2\pi/3)$

19. $\tan(\pi/3)$

20. $\sec(\pi/6)$

21. $\csc(-120°)$

22. $\cot(135°)$

23. $\sin(180°)$

24. $\tan(0°)$

25. $\cos(3\pi/2)$

26. $\csc(5\pi/6)$

27. $\sec(-\pi)$

28. $\cot(-4\pi/3)$

29. $\cot(420°)$

30. $\sin(390°)$

31. $\cos(-135°)$

32. $\tan(225°)$

33. $\sec(2\pi/3)$

34. $\csc(-3\pi/4)$

35. $\tan(5\pi/6)$

36. $\sin(7\pi/6)$

For each triangle shown below, find the exact values of sin α, cos α, tan α, csc α, sec α, and cot α.

37.

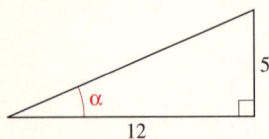

38.

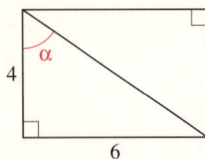

Find an approximate value for each expression. Round to four decimal places.

39. $\sin(44°)$

40. $\cos(-205°)$

41. $\cos(4.62)$

42. $\sin(3.14)$

43. $\tan(\pi/17)$

44. $\sec(2.33)$

45. $\csc(105°4')$

46. $\cot(55°3'12'')$

Find the exact value of each expression.

47. $\sin^{-1}(-0.5)$

48. $\cos^{-1}(-0.5)$

49. $\arctan(-1)$

50. $\text{arccot}(1/\sqrt{3})$

51. $\sec^{-1}(\sqrt{2})$

52. $\csc^{-1}(\sec(\pi/3))$

53. $\sin^{-1}(\sin(5\pi/6))$

54. $\cos(\cos^{-1}(-\sqrt{3}/2))$

Find the exact value of each expression in degrees.

55. $\sin^{-1}(1)$

56. $\tan^{-1}(1)$

57. $\arccos(-1/\sqrt{2})$

58. $\text{arcsec}(2)$

59. $\cot^{-1}(\sqrt{3})$

60. $\cot^{-1}(-\sqrt{3})$

61. $\text{arccot}(0)$

62. $\text{arccot}(-\sqrt{3}/3)$

Solve each right triangle with the given parts.

63. $a = 2, b = 3$

64. $a = 3, c = 7$

65. $a = 3.2, \alpha = 21.3°$

66. $\alpha = 34.6°, c = 9.4$

Sketch two cycles of the graph of each function and determine the period and range.

67. $f(x) = 2\sin(3x)$

68. $f(x) = 1 + \cos(x + \pi/4)$

69. $y = \tan(2x + \pi)$

70. $y = \cot(x - \pi/4)$

71. $y = \sec\left(\frac{1}{2}x\right)$

72. $y = \csc\left(\frac{\pi}{2}x\right)$

Sketch the graph of each function for $-360° \le x \le 360°$, and determine the period and range.

73. $y = \dfrac{1}{2}\cos(2x)$

74. $y = 1 - \sin(x - 60°)$

75. $y = \cot(2x + 60°)$

76. $y = 2\tan(x + 45°)$

77. $y = \dfrac{1}{3}\csc(2x + 180°)$

78. $y = 1 + 2\sec(x - 45°)$

Solve each problem.

79. Find sin α, given that cos $\alpha = 1/5$ and α is in quadrant IV.

80. Find tan α, given that sin $\alpha = 1/3$ and α is in quadrant II.

81. *Broadcasting the oldies* If radio station Q92 is broadcasting its oldies at 92.3 FM, then it is broadcasting at a frequency of 92.3 megahertz, or 92.3×10^6 cycles per second. What is the period of a wave with this frequency? (FM stands for frequency modulation.)

82. *AM radio* If WLS in Chicago is broadcasting at 890 AM (amplitude modulation), then its signal has a frequency of 890 kilohertz, or 890×10^3 cycles per second. What is the period of a wave with this frequency?

83. *Irrigation* A center-pivot irrigation system waters a circular field by rotating about the center of the field. If the system makes one revolution in 8 hr, what is the linear velocity in feet per hour of a nozzle that is 120 ft from the center?

84. *Angular velocity* If a bicycle with a 26-in.-diameter wheel is traveling 16 mph, then what is the angular velocity of the valve stem in radians per hour?

85. *Crooked man*   A man is standing 1000 ft from a surveyor. The surveyor measures an angle of 0.4° sighting from the man's feet to his head. Assume that the man can be represented by the arc intercepted by a central angle of 0.4° in a circle of radius 1000 ft. Find the height of the man.

86. *Straight man*   Assume that the man of Exercise 85 can be represented by the side a of a right triangle with angle $\alpha = 0.4°$ and $b = 1000$ ft. Find the height of the man, and compare your answer with the answer of the previous exercise.

87. *Shooting a target*   Judy is standing 200 ft from a circular target with a radius of 3 in. To hit the center of the circle, she must hold the gun perfectly level, as shown in the figure. Will she hit the target if her aim is off by one-tenth of a degree in any direction?

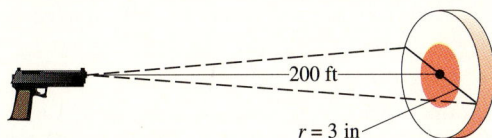

200 ft

$r = 3$ in

Figure for Exercise 87

88. *Height of buildings*   Two buildings are 300 ft apart. From the top of the shorter building the angle of elevation of the top of the taller building is 23°, and the angle of depression of the base of the taller building is 36°, as shown in the figure. How tall is each building?

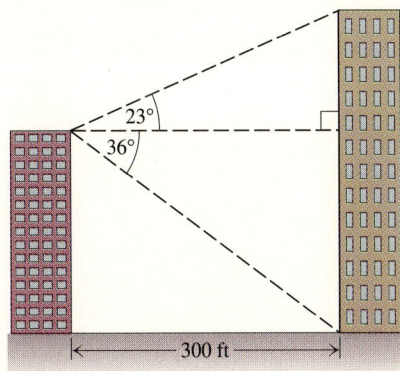

23°
36°
300 ft

Figure for Exercise 88

89. *Rotating space station*   One design for a space station is a giant rotating "doughnut" where humans live and work on the outside wall, as shown in the figure. The inhabitants would have an artificial gravity equal to the normal gravity

on earth when the time T in seconds for one revolution is related to the radius r in meters by the equation $T^2g = 4\pi^2r$. If g is acceleration of gravity (9.8 m/sec²) and the station is built with $r = 90$ m, then what angular velocity in radians per second would be necessary to produce artificial gravity?

$r = 90$ m

$T^2g = 4\pi^2r$

Figure for Exercise 89

90. *Cloud height*   Visual Flight Rules require that the height of the clouds be more than 1000 ft for a pilot to fly without instrumentation. At night, cloud height can be determined from the ground by using a search light and an observer as shown in the figure. If the beam of light is aimed straight upward and the observer 500 ft away sights the cloud with an angle of elevation of 55°, then what is the height of the cloud cover?

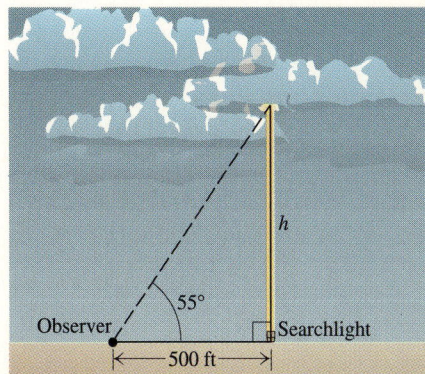

h

Observer   55°   Searchlight

500 ft

Figure for Exercise 90

Chapter 6 Test

Find the exact value of each expression.

1. $\cos 420°$

2. $\sin(-390°)$

3. $\tan(3\pi/4)$

4. $\sec(-\pi/3)$

5. $\csc(7\pi/6)$

6. $\cot(-2\pi/3)$

7. $\sin^{-1}(-1/2)$

8. $\cos^{-1}(-1/2)$

9. $\arctan(-1)$

10. $\sec^{-1}(\sqrt{3}/2)$

11. $\csc^{-1}(-\sqrt{2}/2)$

12. $\sin(\cos^{-1}(-1/3))$

Sketch the graph of each function for $-2\pi \le x \le 2\pi$. Determine the period, range, and amplitude for each function.

13. $y = \sin(3x) - 2$

14. $y = \cos(x + \pi/2)$

15. $y = \tan(\pi x/2)$

16. $y = 2\sin(2x + \pi)$

17. $y = 2\sec(x - \pi)$

18. $y = \csc(x - \pi/2)$

19. $y = \cot(2x)$

20. $y = -\cos(x - \pi/2)$

Solve each problem.

21. Find the arc length intercepted by a central angle of $46°24'6''$ in a circle with a radius of 35.62 m.

22. Find the degree measure of an angle of 2.34 radians.

23. Find exact value of $\cos\alpha$, given that $\sin\alpha = 1/4$ and α is in quadrant II.

24. Find the exact values of all six trigonometric functions for an angle α in standard position whose terminal side contains the point $(5, -2)$.

25. If a bicycle wheel with a 26-in. diameter is making 103 revolutions per minute, then what is the angular velocity in radians per minute for a point on the tire?

26. At what speed in miles per hour will a bicycle travel if the rider can cause the 26-in.-diameter wheel to rotate 103 revolutions per minute?

27. To estimate the blood pressure of *Brachiosaurus,* Professor Ostrom wanted to estimate the height of the head of the famed *Brachiosaurus* skeleton at Humboldt University in Berlin. From a distance of 11 m from a point directly below the head, the angle of elevation of the head was approximately 48°. Use this information to find the height of the head of *Brachiosaurus.*

28. From a point on the street the angle of elevation of the top of the John Hancock Building is 65.7°. From a point on the street that is 100 ft closer to the building the angle of elevation is 70.1°. Find the height of the building.

Tying It All Together
Chapters 1–6

Sketch the graph of each function. State the domain and range of each function.

1. $y = 2 + e^x$ **2.** $y = 2 + x^2$ **3.** $y = 2 + \sin x$ **4.** $y = 2 + \ln(x)$

5. $y = \ln(x - \pi/4)$ **6.** $y = \sin(x - \pi/4)$ **7.** $y = \log_2(2x)$ **8.** $y = \cos(2x)$

Determine whether each function is even, odd, or neither.

9. $f(x) = e^x$ **10.** $f(x) = \sin x$ **11.** $f(x) = \cos x$ **12.** $f(x) = x^4 - x^2 + 1$

13. $f(x) = \tan x$ **14.** $f(x) = \sec x$ **15.** $f(x) = \ln(x)$ **16.** $f(x) = x^3 - 3x$

Determine whether each function is increasing or decreasing on the given interval.

17. $f(x) = \sin x, \ (0, \pi/2)$ **18.** $f(x) = 2^x, \ (-\infty, \infty)$ **19.** $f(x) = \tan x, \ (-\pi/2, \pi/2)$

20. $f(x) = \cos x, \ (0, \pi)$ **21.** $f(x) = x^2, \ (-\infty, 0)$ **22.** $f(x) = \sec x, \ (0, \pi/2)$

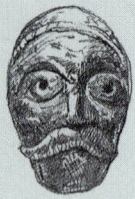

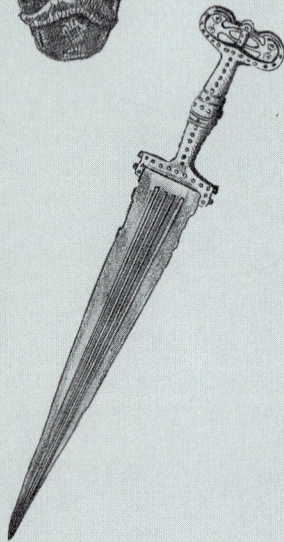

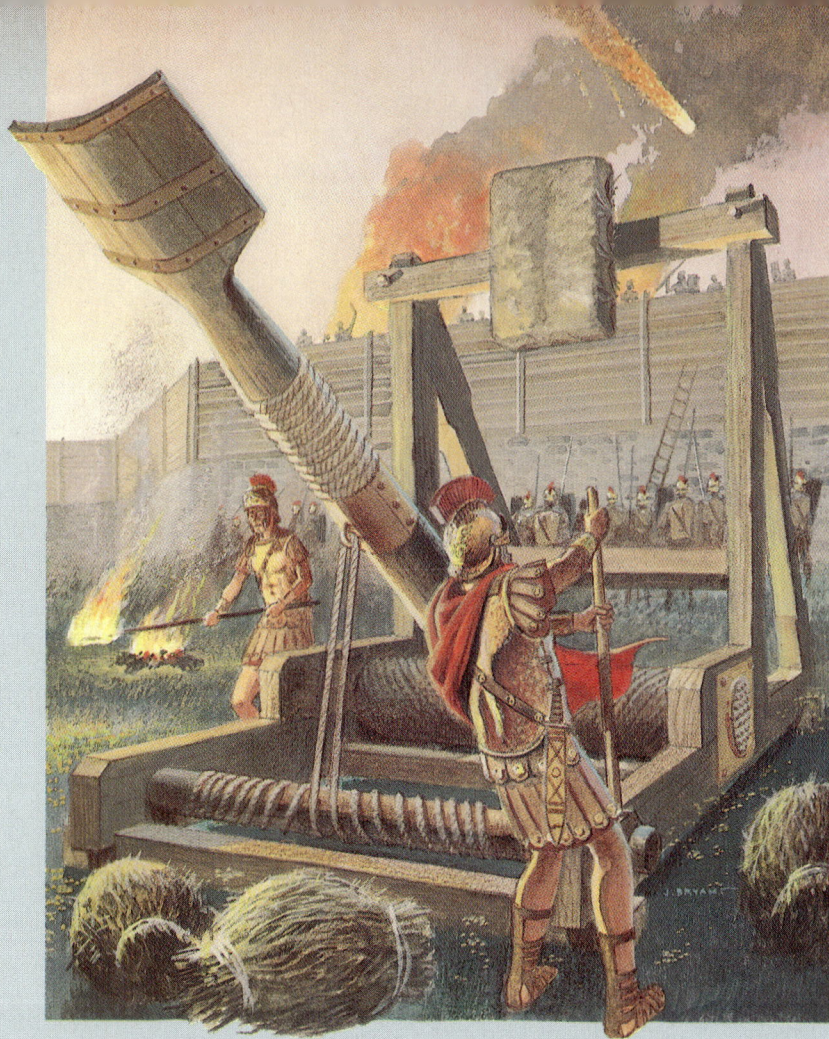

A Roman centurion orders his men to load boulders and burning bales of hay onto their ready-made catapults. At his command, an artillery barrage hurls missiles at deadly speed toward the British fortress. This hill-fort is a cluster of huts, shrines, and animal pens, enclosed by a maze of ditches and earth-packed ramparts. Atop the towering walls, Celtic warriors brandish their weapons and shout harsh cries of defiance. Backed by every last man, woman, and child, the defenders hold firm, ready to fight to the death. But, under cover of the catapults, the Roman infantry advances. Using the famous tortoise formation—tightly interlocking shields held high above their heads—they forge toward the vulnerable eastern gate.

From the time of Caesar, spring-powered catapults had been part of Rome's battering arm. These precursors of the modern mortar were capable of accurately flinging stones of over a hundred pounds 300 to 400 meters. In 44 A.D., during Rome's second invasion of the British Isles, future emperor Vespasian employed artillery pieces and other engines of war to subdue tribes of native Celts. Hot-headed, flamboyant, fearsome in appearance, the Celtic "barbarians" resisted courageously. However, in the end, their impetuous approach to war proved no match for the cold calculations and disciplined fighting of the Roman troops.

7

Trigonometric Identities and Conditional Equations

People have been hurling objects since the Ice Age. Whether a ball, rock, or missile, humans have learned through trial and error that a projectile's path is determined by the initial velocity and the angle at which it is launched. In the 20th century, athletes achieve a balance between accuracy and distance through years of practice, while operators of modern artillery achieve accuracy by calculating velocity and angle of trajectory with mathematical precision.

In this chapter, we'll see how trigonometric equations involving initial velocity and angle of trajectory determine a projectile's flight. Using these equations, we'll determine the velocity and angle of trajectory required to launch a projectile a given distance. Using an identity, we'll find the proper angle of trajectory to achieve maximum distance for the projectile.

7.1

Basic Identities

All of the trigonometric functions are related to each other because they are all defined in terms of the coordinates on a unit circle. For this reason, any expression involving them can be written in many different forms. Identities are used to simplify expressions and determine whether expressions are equivalent. Recall that an identity is an equation that is satisfied by *every* number for which both sides are defined. There are infinitely many trigonometric identities, but only the most common identities should be memorized. Here we review the most basic ones.

Identities from Definitions

The following identities include the definitions of tangent, cotangent, secant, and cosecant, and four identities that follow directly from these definitions. Each of these equations is satisfied by every real number for which both sides are defined.

Some Basic Identities

Definitions:

$$\tan x = \frac{\sin x}{\cos x} \qquad \cot x = \frac{1}{\tan x} \qquad \sec x = \frac{1}{\cos x} \qquad \csc x = \frac{1}{\sin x}$$

Related identities:

$$\cot x = \frac{\cos x}{\sin x} \qquad \tan x = \frac{1}{\cot x} \qquad \cos x = \frac{1}{\sec x} \qquad \sin x = \frac{1}{\csc x}$$

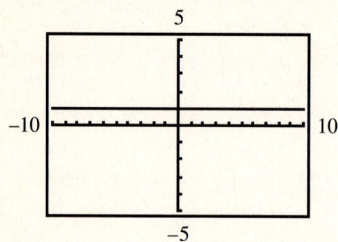

The graph of $y = (\sin x)^2 + (\cos x)^2$ is the horizontal line $y = 1$.

The fundamental identity $\sin^2 x + \cos^2 x = 1$ from Section 6.2 is based on the definitions of $\sin x$ and $\cos x$ as coordinates of a point on the unit circle. If we divide each side of this identity by $\sin^2 x$, we get a new identity:

$$\frac{\sin^2 x}{\sin^2 x} + \frac{\cos^2 x}{\sin^2 x} = \frac{1}{\sin^2 x}$$

$$1 + \left(\frac{\cos x}{\sin x}\right)^2 = \left(\frac{1}{\sin x}\right)^2$$

$$1 + \cot^2 x = \csc^2 x$$

If we divide each side of the fundamental identity by $\cos^2 x$, we get another new identity:

$$\frac{\sin^2 x}{\cos^2 x} + \frac{\cos^2 x}{\cos^2 x} = \frac{1}{\cos^2 x}$$

$$\tan^2 x + 1 = \sec^2 x$$

Since these three identities come from the Pythagorean theorem, they are called the **Pythagorean identities.**

Pythagorean Identities

$$\sin^2 x + \cos^2 x = 1 \qquad 1 + \cot^2 x = \csc^2 x \qquad \tan^2 x + 1 = \sec^2 x$$

Using Identities

Identities are used in many ways. One way is to simplify expressions involving trigonometric functions. In the next example we follow the strategy of writing each expression in terms of sines and cosines only and then simplifying the expression.

Example 1   Using identities to simplify

Write each expression in terms of sines and/or cosines, and then simplify.

a) $\dfrac{\tan x}{\sec x}$ b) $\sin x + \cot x \cos x$

Solution

a) Rewrite each trigonometric function in terms of sine and cosine:

$$\frac{\tan x}{\sec x} = \frac{\dfrac{\sin x}{\cos x}}{\dfrac{1}{\cos x}} = \frac{\sin x}{\cos x} \cdot \frac{\cos x}{1} \qquad \text{Invert and multiply.}$$

$$= \sin x$$

b) $\sin x + \cot x \cos x = \sin x + \dfrac{\cos x}{\sin x} \cdot \cos x$ **Rewrite using sines and cosines.**

$\qquad\qquad\qquad\qquad = \sin x + \dfrac{\cos^2 x}{\sin x}$

$\qquad\qquad\qquad\qquad = \dfrac{\sin^2 x}{\sin x} + \dfrac{\cos^2 x}{\sin x}$ **Multiply sin x by sin x/sin x.**

$\qquad\qquad\qquad\qquad = \dfrac{\sin^2 x + \cos^2 x}{\sin x}$

$\qquad\qquad\qquad\qquad = \dfrac{1}{\sin x}$ **Since $\sin^2 x + \cos^2 x = 1$:**

$\qquad\qquad\qquad\qquad = \csc x$ **Use a basic identity.**   ◆

Using identities we can write any one of the six trigonometric functions in terms of any other. For example, the identity

$$\sin x = \frac{1}{\csc x}$$

expresses the sine function in terms of the cosecant. To write sine in terms of cosine, we rewrite the identity $\sin^2 x + \cos^2 x = 1$:

$$\sin^2 x = 1 - \cos^2 x$$
$$\sin x = \pm\sqrt{1 - \cos^2 x}$$

The $\pm$ symbol in this identity means that if $\sin x > 0$, then the identity is $\sin x = \sqrt{1 - \cos^2 x}$, and if $\sin x < 0$, then the identity is $\sin x = -\sqrt{1 - \cos^2 x}$.

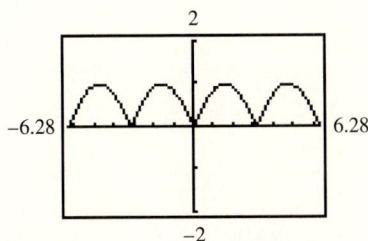

The graph of $y = \sqrt{1 - (\cos x)^2}$ shown here is the same as the graph of $y = \sin x$ only when $\sin x > 0$.

Example 2   Writing one function in terms of another

Write an identity that expresses the tangent function in terms of the sine function.

Solution

The Pythagorean identity $\sin^2 x + \cos^2 x = 1$ yields $\cos x = \pm\sqrt{1 - \sin^2 x}$. Since $\tan x = \sin x/\cos x$, we can write

$$\tan x = \pm\frac{\sin x}{\sqrt{1 - \sin^2 x}}.$$

We use the positive or negative sign depending on the value of x.   ◆

Because all of the trigonometric functions are related by identities, we can find their values for an angle if we know the value of any one of them and the quadrant of the angle.

Example 3   Using identities to find function values

Given that $\tan \alpha = -2/3$ and α is in quadrant IV, find the values of the remaining five trigonometric functions at α by using identities.

Solution

Use $\tan \alpha = -2/3$ in the identity $\sec^2 \alpha = 1 + \tan^2 \alpha$:

$$\sec^2 \alpha = 1 + \left(-\frac{2}{3}\right)^2 = \frac{13}{9}$$

$$\sec \alpha = \pm \frac{\sqrt{13}}{3}$$

Since α is in quadrant IV, $\cos \alpha > 0$ and $\sec \alpha > 0$. So $\sec \alpha = \sqrt{13}/3$. Since cosine is the reciprocal of secant and cotangent is the reciprocal of tangent,

$$\cos \alpha = \frac{3}{\sqrt{13}} = \frac{3\sqrt{13}}{13} \quad \text{and} \quad \cot \alpha = -\frac{3}{2}.$$

Since $\sin \alpha$ is negative in quadrant IV, $\sin \alpha = -\sqrt{1 - \cos^2 \alpha}$:

$$\sin \alpha = -\sqrt{1 - \frac{9}{13}} = -\sqrt{\frac{4}{13}} = -\frac{2\sqrt{13}}{13}$$

Since cosecant is the reciprocal of sine, $\csc \alpha = -\sqrt{13}/2$. ◆

The identities discussed so far give relationships between the different trigonometric functions. The next identities show how the value of each trigonometric function for $-x$ is related to its value for x.

Odd and Even Identities

In Section 3.4 we defined an **odd function** as one for which $f(-x) = -f(x)$ and an **even function** as one for which $f(-x) = f(x)$. In algebra most of the odd functions have odd exponents and the even functions have even exponents. We can also classify each of the trigonometric functions as either odd or even, but we cannot make the determination based on exponents. Instead, we examine the definitions of the functions in terms of the unit circle.

Figure 7.1 shows the real numbers s and $-s$ as arcs on the unit circle. If the terminal point of s is (x, y), then the terminal point of $-s$ is $(x, -y)$. The y-coordinate for the terminal point of $-s$ is the opposite of the y-coordinate for the terminal point of s. So $\sin(-s) = -\sin(s)$, and sine is an odd function. Since the x-coordinates for s and $-s$ are equal, $\cos(-s) = \cos(s)$, and cosine is an even function. Because sine is an odd function, cosecant is also an odd function:

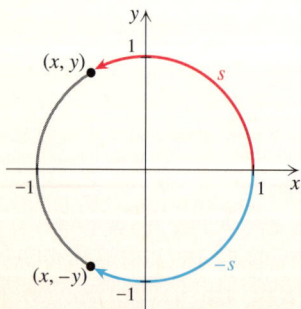

Figure 7.1

$$\csc(-s) = \frac{1}{\sin(-s)} = \frac{1}{-\sin(s)} = -\csc(s)$$

Likewise we can establish that secant is an even function and tangent and cotangent are both odd functions. These arguments establish the following identities.

▼

Odd and Even Identities

Odd: $\sin(-x) = -\sin(x)$ $\csc(-x) = -\csc(x)$
         $\tan(-x) = -\tan(x)$ $\cot(-x) = -\cot(x)$
Even: $\cos(-x) = \cos(x)$ $\sec(-x) = \sec(x)$

To help you remember which of the trigonometric functions are odd and which are even, remember that the graph of an odd function is symmetric about the origin and the graph of an even function is symmetric about the y-axis. The graphs of $y = \sin x$, $y = \csc x$, $y = \tan x$, and $y = \cot x$ are symmetric about the origin, while $y = \cos x$ and $y = \sec x$ are symmetric about the y-axis.

Example 4   Using odd and even identities

Simplify each expression.

a) $\sin(-x)\cot(-x)$

b) $\dfrac{1}{1 + \cos(-x)} + \dfrac{1}{1 - \cos x}$

Solution

a) $\sin(-x)\cot(-x) = (-\sin x)(-\cot x)$

$$= \sin x \cdot \cot x = \sin x \cdot \frac{\cos x}{\sin x} = \cos x$$

b) To add these expressions, we need a common denominator.

$$\frac{1}{1 + \cos(-x)} + \frac{1}{1 - \cos x} = \frac{1}{1 + \cos x} + \frac{1}{1 - \cos x} \qquad \text{\color{blue}Because } \cos(-x) = \\ \text{\color{blue}$\cos x$}$$

$$= \frac{1(1 - \cos x)}{(1 + \cos x)(1 - \cos x)} + \frac{1(1 + \cos x)}{(1 - \cos x)(1 + \cos x)}$$

$$= \frac{1 - \cos x}{1 - \cos^2 x} + \frac{1 + \cos x}{1 - \cos^2 x}$$

$$= \frac{1 - \cos x}{\sin^2 x} + \frac{1 + \cos x}{\sin^2 x} \qquad \text{\color{blue}Pythagorean} \\ \text{\color{blue}identity}$$

$$= \frac{1 - \cos x + 1 + \cos x}{\sin^2 x} = \frac{2}{\sin^2 x}$$

$$= 2 \cdot \frac{1}{\sin^2 x} = 2 \csc^2 x$$

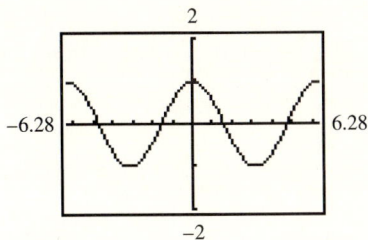

Graph $y_1 = \sin(-x)/\tan(-x)$ to check the answer to Example 4(a). The graph looks like the graph of the cosine function.

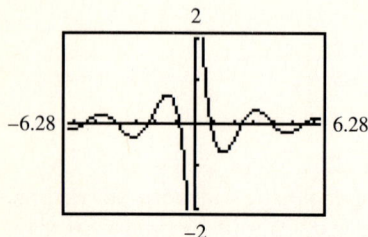

The graph of $y = \cos(2x)/x$ supports the conclusion that the function is odd and the graph is symmetric about the origin.

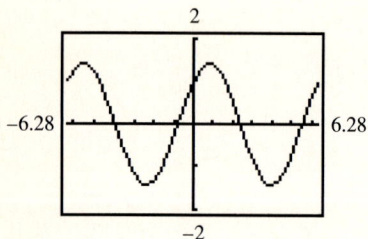

The graph of $y = \sin x + \cos x$ is neither symmetric about the origin nor symmetric about the y-axis.

Example 5 Odd or even functions

Determine whether each function is odd, even, or neither.

a) $f(x) = \dfrac{\cos(2x)}{x}$ **b)** $g(t) = \sin t + \cos t$

Solution

a) We replace x by $-x$ and see whether $f(-x)$ is equal to $f(x)$ or $-f(x)$:

$$f(-x) = \frac{\cos(2(-x))}{-x} \qquad \text{Replace } x \text{ by } -x.$$

$$= \frac{\cos(-2x)}{-x}$$

$$= \frac{\cos(2x)}{-x} \qquad \text{Since cosine is an even function}$$

$$= -\frac{\cos(2x)}{x}$$

Since $f(-x) = -f(x)$, the function f is an odd function.

b) $g(-t) = \sin(-t) + \cos(-t)$ Replace t by $-t$.
$\qquad = -\sin t + \cos t$ Sine is odd and cosine is even.

Since $g(-t) \neq g(t)$ and $g(-t) \neq -g(t)$, the function is neither odd nor even. ◆

Identity or Not?

In this section we have seen the basic identities, the Pythagorean identities, and the odd-even identities and we are just getting started. With so many identities in trigonometry it might seem difficult to determine whether a given equation is an identity. Remember that an identity is satisfied by every value of the variable for which both sides are defined. If you can find one number for which the left-hand side of the equation has a value different from the right-hand side, then the equation is not an identity.

Example 6 Proving that an equation is not an identity

Show that $\sin(2t) = 2\sin(t)$ is not an identity.

Solution

If $t = \pi/4$, then

$$\sin(2t) = \sin\left(2 \cdot \frac{\pi}{4}\right) = \sin\left(\frac{\pi}{2}\right) = 1$$

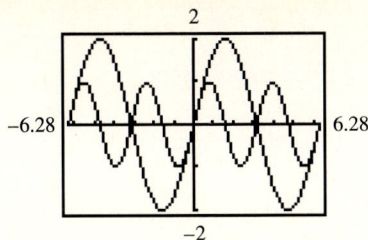

The graphs of $y_1 = \sin(2x)$ and $y_2 = 2 \sin x$ support the conclusion that $\sin(2t) = 2 \sin t$ is not an identity.

and

$$2 \sin(t) = 2 \sin\left(\frac{\pi}{4}\right) = 2 \cdot \frac{\sqrt{2}}{2} = \sqrt{2}.$$

Since the values of $\sin(2t)$ and $2 \sin(t)$ are unequal for $t = \pi/4$, the equation is not an identity.   ◆

An equation that is not an identity may have many solutions. So you might have to try more than one value for the variable to find one that fails to satisfy the equation. Section 7.2 will be devoted to proving that given equations are identities.

? For Thought

True or false? Explain.

1. The equation $\sin x = \cos x$ is an identity.

2. If we simplify the expression $(\tan x)(\cot x)$, we get 1.

3. The function $f(x) = \sin^2 x$ is an even function.

4. The function $f(x) = \cos^3 x$ is an odd function.

5. If $\sin(2x) = 2 \sin(x)\cos(x)$ is an identity, then so is $\dfrac{\sin(2x)}{\sin(x)} = 2 \cos(x)$.

6. The equation $(\sin x + \cos x)^2 = \sin^2 x + \cos^2 x$ is an identity.

7. Tangent is written in terms of secant as $\tan x = \pm\sqrt{1 - \sec^2 x}$.

8. $\sin(-3)\cos(-3)\tan(-3)\sec(3)\csc(-3)\cot(3) = -1$

9. $\sin^2(-\pi/9) + \cos^2(-\pi/9) = -1$

10. $(1 - \sin(\pi/7))(1 + \sin(\pi/7)) = \cos^2(-\pi/7)$

7.1 Exercises   ▧ Tape 13   ▨ Disk—5.25": 4   3.5": 3   Macintosh: 3

Write each expression in terms of sines and/or cosines, and then simplify.

1. $\dfrac{\sec x}{\tan x}$

2. $\dfrac{\cot x}{\csc x}$

3. $\dfrac{\sin x}{\csc x} + \dfrac{\cos x}{\sec x}$

4. $\dfrac{1}{\sin^2 x} - \dfrac{1}{\tan^2 x}$

5. $(1 - \sin \alpha)(1 + \sin \alpha)$

6. $(\sec \alpha - 1)(\sec \alpha + 1)$

7. $(\cos \beta \tan \beta + 1)(\sin \beta - 1)$

8. $(1 + \cos \beta)(1 - \cot \beta \sin \beta)$

9. $\dfrac{1 + \cos \alpha \tan \alpha \csc \alpha}{\csc \alpha}$

10. $\dfrac{(\cos \alpha \tan \alpha + 1)(\sin \alpha - 1)}{\cos^2 \alpha}$

Write an identity that expresses the first function in terms of the second.

11. cotangent, in terms of cosecant

12. secant, in terms of tangent

13. sine, in terms of cotangent

14. cosine, in terms of tangent

15. tangent, in terms of cosecant

16. cotangent, in terms of secant

In each exercise, use identities to find the exact values at α for the remaining five trigonometric functions.

17. $\tan \alpha = 1/2$ and $0 < \alpha < \pi/2$

18. $\sin \alpha = 3/4$ and $\pi/2 < \alpha < \pi$

19. $\cos \alpha = -\sqrt{3}/5$ and α is in quadrant III

20. $\sec \alpha = -4\sqrt{5}/5$ and α is in quadrant II

21. $\cot \alpha = -1/3$ and $-\pi/2 < \alpha < 0$

22. $\csc \alpha = \sqrt{3}$ and $0 < \alpha < \pi/2$

Simplify each expression.

23. $\sin(-x)\cot(-x)$

24. $\sec(-x) - \sec(x)$

25. $\sin(y) + \sin(-y)$

26. $\cos(y) + \cos(-y)$

27. $\dfrac{\sin(x)}{\cos(-x)} + \dfrac{\sin(-x)}{\cos(x)}$

28. $\dfrac{\cos(-x)}{\sin(-x)} - \dfrac{\cos(-x)}{\sin(x)}$

29. $(1 + \sin(\alpha))(1 + \sin(-\alpha))$

30. $(1 - \cos(-\alpha))(1 + \cos(\alpha))$

31. $\sin(-\beta)\cos(-\beta)\csc(\beta)$

32. $\tan(-\beta)\csc(-\beta)\cos(\beta)$

Determine whether each function is odd, even, or neither.

33. $f(x) = \sin(2x)$

34. $f(x) = \cos(2x)$

35. $f(x) = \cos x + \sin x$

36. $f(x) = 2 \sin x \cos x$

37. $f(t) = \sec^2(t) - 1$

38. $f(t) = 2 + \tan(t)$

39. $f(\alpha) = 1 + \sec \alpha$

40. $f(\beta) = 1 + \csc \beta$

41. $f(x) = \dfrac{\sin x}{x}$

42. $f(x) = x \cos x$

43. $f(x) = x + \sin x$

44. $f(x) = \csc(x^2)$

Show that each equation is not an identity. Write your explanation in paragraph form.

45. $(\sin \gamma + \cos \gamma)^2 = \sin^2 \gamma + \cos^2 \gamma$

46. $\tan^2 x - 1 = \sec^2 x$

47. $(1 + \sin \beta)^2 = 1 + \sin^2 \beta$

48. $\sin(2\alpha) = \sin \alpha \cos \alpha$

49. $\sin \alpha = \sqrt{1 - \cos^2 \alpha}$

50. $\tan \alpha = \sqrt{\sec^2 \alpha - 1}$

51. $\sin(y) = \sin(-y)$

52. $\cos(-y) = -\cos y$

53. $\cos^2 y - \sin^2 y = \sin(2y)$

54. $\cos(2x) = 2 \cos x \sin x$

Use identities to simplify each expression.

55. $1 - \dfrac{1}{\cos^2 x}$

56. $\dfrac{\sin^4 x - \sin^2 x}{\sec x}$

57. $\dfrac{-\tan^2 t - 1}{\sec^2 t}$

58. $\dfrac{\cos w \sin^2 w + \cos^3 w}{\sec w}$

59. $\dfrac{\sin^2 \alpha - \cos^2 \alpha}{1 - 2 \cos^2 \alpha}$

60. $\dfrac{\sin^3(-\theta)}{\sin^3 \theta - \sin \theta}$

61. $\dfrac{\tan^3 x - \sec^2 x \tan x}{\cot(-x)}$

62. $\sin x + \dfrac{\cos^2 x}{\sin x}$

63. $\dfrac{1}{\sin^3 x} - \dfrac{\cot^2 x}{\sin x}$

64. $1 - \dfrac{\sec^2 x}{\tan^2 x}$

65. $\sin^4 x - \cos^4 x$

66. $\csc^4 x - \cot^4 x$

A projectile is fired with initial velocity of v_0 feet per second. The projectile can be pictured as being fired from the origin into the first quadrant, making an angle θ with the positive x-axis, as shown in the figure. If there is no air resistance, then at time t the projectile is at the point (x, y), where $x = v_0 t \cos \theta$ and $y = -16t^2 + v_0 t \sin \theta$. The time t is in seconds and x and y are in feet.

67. Given that v_0 is 4 ft/sec, write y as a function of x and use identities to simplify.

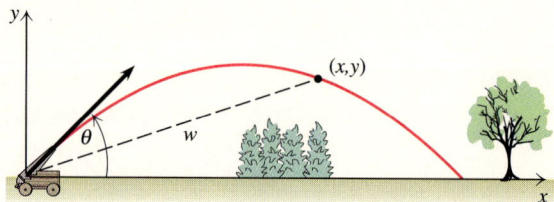

Figure for Exercises 67 and 68

68. Let w be the distance of the projectile from the origin at time t and let $v_0 = 4$ ft/sec. Write w as a function of t and simplify.

For Writing/Discussion

69. *Algebraic identities* List as many algebraic identities as you can and explain why each one is an identity.

70. *Trigonometric identities* List as many trigonometric identities as you can and explain why each one is an identity.

71. *Cooperative learning* Each student in your small group should "randomly" write an equation involving trigonometric functions (without consulting the text). Determine whether each equation is an identity. Which are more "common," identities or conditional equations?

7.2

Verifying Identities

In Section 7.1 we saw that any trigonometric function can be expressed in terms of any other trigonometric function. So an expression involving one or more trigonometric functions could be written in many different equivalent forms. In trigonometry and calculus, where trigonometric functions are used, we must often decide whether two expressions are equivalent. The fact that two expressions are equivalent is expressed as an identity. In this section we concentrate on techniques for verifying that a given equation is an identity.

Developing a Strategy

In simplifying expressions in Section 7.1, we used known identities and properties of algebra to rewrite a complicated expression as a simpler equivalent expression. It may not always be clear that one expression is simpler than another, but an expression that involves fewer symbols is generally considered simpler. To prove or verify that an equation is an identity, we start with the expression on one side of the equation and use known identities and properties of algebra to convert it into the expression on the other side of the equation. It is the same process as simplifying expressions, except that in this case we know exactly what we want as the final expression. If one side appears to be more complicated than the other, we usually start with the more complicated side and simplify it.

Example 1   Verifying an identity

Prove that the following equation is an identity:

$$1 + \sec x \sin x \tan x = \sec^2 x$$

Solution

We start with the left-hand side, the more complicated side, and write it in terms of sine and cosine. Our goal is to get $\sec^2 x$ as the final simplified expression.

$$1 + \sec x \sin x \tan x = 1 + \frac{1}{\cos x} \sin x \frac{\sin x}{\cos x}$$

$$= 1 + \frac{\sin^2 x}{\cos^2 x}$$

$$= 1 + \tan^2 x$$

$$= \sec^2 x \qquad \text{\color{blue}{Pythagorean identity}} \qquad \blacklozenge$$

In verifying identities it is often necessary to multiply binomials or factor trinomials involving trigonometric functions.

Example 2   Multiplying Binomials

Find each product.

a) $(1 + \tan x)(1 - \tan x)$

b) $(2 \sin \alpha + 1)^2$

Solution

a) The product of a sum and a difference is equal to the difference of two squares:

$$(1 + \tan x)(1 - \tan x) = 1 - \tan^2 x$$

b) Use $(a + b)^2 = a^2 + 2ab + b^2$ with $a = 2 \sin \alpha$ and $b = 1$:

$$(2 \sin \alpha + 1)^2 = 4 \sin^2 \alpha + 4 \sin \alpha + 1 \qquad \blacklozenge$$

Example 3   Factoring with trigonometric functions

Factor each expression.

a) $\sec^2 x - \tan^2 x$

b) $\sin^2 \beta + \sin \beta - 2$

Solution

a) A difference of two squares is equal to a product of a sum and a difference:

$$\sec^2 x - \tan^2 x = (\sec x + \tan x)(\sec x - \tan x)$$

b) Factor this expression as you would factor $x^2 + x - 2 = (x + 2)(x - 1)$:

$$\sin^2 \beta + \sin \beta - 2 = (\sin \beta + 2)(\sin \beta - 1) \qquad \blacklozenge$$

The equations that appear in Examples 2 and 3 are identities. They are not the kind of identities that are memorized because they can be obtained at any time by the principles of multiplying or factoring. The next example uses the identity $(1 - \sin \alpha)(1 + \sin \alpha) = 1 - \sin^2 \alpha$.

Example 4   An identity with fractions

Prove that the following equation is an identity:

$$\frac{\cos \alpha}{1 - \sin \alpha} = \frac{1 + \sin \alpha}{\cos \alpha}$$

Solution

Start with the left-hand side and multiply the numerator and denominator by $1 + \sin \alpha$ because $1 + \sin \alpha$ appears in the numerator of the right-hand side.

$$\frac{\cos \alpha}{1 - \sin \alpha} = \frac{\cos \alpha(1 + \sin \alpha)}{(1 - \sin \alpha)(1 + \sin \alpha)}$$

$$= \frac{\cos \alpha(1 + \sin \alpha)}{1 - \sin^2 \alpha} \qquad \text{\color{blue}{Leave numerator in factored form.}}$$

$$= \frac{\cos \alpha(1 + \sin \alpha)}{\cos^2 \alpha} \qquad \text{\color{blue}{Pythagorean identity}}$$

$$= \frac{1 + \sin \alpha}{\cos \alpha} \qquad \text{\color{blue}{Reduce.}}$$

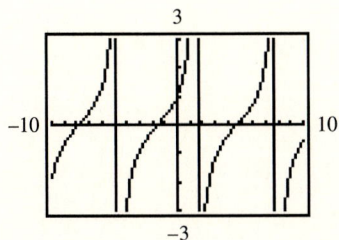

The graphs of $y_1 = \cos x/(1 - \sin x)$ and $y_2 = (1 + \sin x)/\cos x$ coincide.

In Example 4 the numerator and denominator were multiplied by $1 + \sin \alpha$ because the expression on the right-hand side has $1 + \sin \alpha$ in its numerator. We could have multiplied the numerator and denominator of the left-hand side by $\cos \alpha$ because the expression on the right-hand side has $\cos \alpha$ in its denominator. You should prove that the equation of Example 4 is an identity by multiplying the numerator and denominator by $\cos \alpha$.

Another strategy to use when a fraction has a sum or difference in its numerator is to write the fraction as a sum or difference of two fractions, as in the next example.

Example 5   Verifying an identity

Prove that the following equation is an identity:

$$\frac{\csc x - \sin x}{\sin x} = \cot^2 x$$

Solution

First rewrite the left-hand side as a difference of two rational expressions.

$$\frac{\csc x - \sin x}{\sin x} = \frac{\csc x}{\sin x} - \frac{\sin x}{\sin x}$$

$$= \csc x \cdot \frac{1}{\sin x} - 1$$

$$= \csc^2 x - 1 \qquad \text{\color{blue}{Since $1/\sin x = \csc x$}}$$

$$= \cot^2 x \qquad \text{\color{blue}{Pythagorean identity}}$$

In verifying an identity, you can start with the expression on either side of the equation and use known identities and properties of algebra to get the expression on the other side. Note that we do *not* start with the given equation and work on it to get simpler and simpler equations as we do when solving an equation, because we do not yet know that the given equation is actually a correct identity. So when an expression involves the sum or difference of fractions, as in the next example, get a common denominator and combine the fractions. Do *not* multiply each side by the LCD as is done when solving an equation involving fractions.

Example 6   Starting with the right-hand side

Prove that the following equation is an identity:

$$2 \tan^2 x = \frac{1}{\csc x - 1} - \frac{1}{\csc x + 1}$$

Solution

In this equation, the right-hand side is the more complicated one. We will simplify the right-hand side by getting a common denominator and combining the fractions, keeping $2 \tan^2 x$ in mind as our goal.

$$\frac{1}{\csc x - 1} - \frac{1}{\csc x + 1} = \frac{1(\csc x + 1)}{(\csc x - 1)(\csc x + 1)} - \frac{1(\csc x - 1)}{(\csc x + 1)(\csc x - 1)}$$

$$= \frac{\csc x + 1}{\csc^2 x - 1} - \frac{\csc x - 1}{\csc^2 x - 1}$$

$$= \frac{2}{\csc^2 x - 1} \qquad \text{\color{blue}Subtract the numerators.}$$

$$= \frac{2}{\cot^2 x} \qquad \text{\color{blue}Pythagorean identity}$$

$$= 2 \cdot \frac{1}{\cot^2 x} = 2 \tan^2 x \qquad \blacklozenge$$

In every example so far, we started with one side and converted it into the other side. Although all identities can be verified by that method, sometimes it is simpler to convert both sides into the same expression. If both sides are shown to be equivalent to a common expression, then the identity is certainly verified. This technique works best when both sides are rather complicated and we simply set out to simplify them.

Example 7   Verifying an identity

Prove that the following equation is an identity:

$$\frac{1 - \sin^2 t}{1 - \csc(-t)} = \frac{1 + \sin(-t)}{\csc t}$$

Solution

Use $\csc(-t) = -\csc t$ to simplify the left-hand side:

$$\frac{1 - \sin^2 t}{1 - \csc(-t)} = \frac{1 - \sin^2 t}{1 + \csc t} = \frac{\cos^2 t}{1 + \csc t}$$

Now look for a way to get the right-hand side of the original equation equivalent to $\cos^2 t/(1 + \csc t)$. Since $\sin(-t) = -\sin t$, we can write $1 + \sin(-t)$ as $1 - \sin t$ and then get $\cos^2 t$ in the numerator by multiplying by $1 + \sin t$.

$$\frac{1 + \sin(-t)}{\csc t} = \frac{1 - \sin t}{\csc t} = \frac{(1 - \sin t)(1 + \sin t)}{\csc t(1 + \sin t)}$$

$$= \frac{1 - \sin^2 t}{\csc t + \csc t \sin t} = \frac{\cos^2 t}{\csc t + \dfrac{1}{\sin t} \sin t}$$

$$= \frac{\cos^2 t}{1 + \csc t}$$

Since both sides of the equation are equivalent to the same expression, the equation is an identity.

The Strategy

Verifying identities takes practice. There are many different ways to verify a particular identity, so just go ahead and get started. If you do not seem to be getting anywhere, try another approach. The strategy that we used in the examples is stated below. Other techniques are sometimes needed to prove identities, but you can prove most of the identities in this text using the following strategy.

Strategy: Verifying Identities

1. **Work on one side of the equation (usually the more complicated side), keeping in mind the expression on the other side as your goal.**

2. **Some expressions can be simplified quickly if they are rewritten in terms of sines and cosines only. (See Example 1.)**

3. **To convert one rational expression into another, multiply the numerator and denominator of the first by either the numerator or the denominator of the desired expression. (See Example 4.)**

4. **If the numerator of a rational expression is a sum or difference, convert the rational expression into a sum or difference of two rational expressions. (See Example 5.)**

5. **If a sum or difference of two rational expressions occurs on one side of the equation, then find a common denominator and combine them into one rational expression. (See Example 6.)**

? For Thought

True or false? Answer true if the equation is an identity and false if it is not. If false, find a value of x for which the two sides have different values.

1. $\dfrac{\sin x}{\csc x} = \sin^2 x$

2. $\dfrac{\cot x}{\tan x} = \tan^2 x$

3. $\dfrac{\sec x}{\csc x} = \tan x$

4. $\sin x \sec x = \tan x$

5. $\dfrac{\cos x + \sin x}{\cos x} = 1 + \tan x$

6. $\sec x + \dfrac{\sin x}{\cos x} = \dfrac{1 + \sin x \cos x}{\cos x}$

7. $\dfrac{1}{1 - \sin x} = \dfrac{1 + \sin x}{\cos^2 x}$

8. $\tan x \cot x = 1$

9. $(1 - \cos x)(1 - \cos x) = \sin^2 x$

10. $(1 - \csc x)(1 + \csc x) = \cot^2 x$

7.2 Exercises   Tape 13   Disk—5.25": 4   3.5": 3   Macintosh: 3

Match each expression on the left with one on the right that completes each equation as an identity.

1. $\cos x \tan x = $?

2. $\sec x \cot x = $?

3. $(\csc x - \cot x)(\csc x + \cot x) = $?

4. $\dfrac{\sin x + \cos x}{\sin x} = $?

5. $(1 - \sec x)(1 + \sec x) = $?

6. $(\csc x - 1)(\csc x + 1) = $?

7. $\dfrac{\csc x - \sin x}{\csc x} = $?

8. $\dfrac{\cos x - \sec x}{\sec x} = $?

9. $\dfrac{\csc x}{\sin x} = $?

10. $\dfrac{\sin x}{\cos x} + \dfrac{\cos x}{\sin x} = $?

A. 1

B. $-\tan^2 x$

C. $\cot^2 x$

D. $\sin x$

E. $-\sin^2 x$

F. $\dfrac{1}{\sin x \cos x}$

G. $1 + \cot^2 x$

H. $\cos^2 x$

I. $\csc x$

J. $1 + \cot x$

Find the products.

11. $(\sin \alpha + 1)(\sin \alpha - 1)$

12. $(\tan \alpha + 2)(\tan \alpha - 2)$

13. $(2 \cos \beta + 1)(\cos \beta - 1)$

14. $(2 \csc \beta - 1)(\csc \beta - 3)$

15. $(\csc x + \sin x)^2$

16. $(2 \cos x - \sec x)^2$

17. $(2 \sin \theta - 1)(2 \sin \theta + 1)$

18. $(3 \sec \theta - 2)(3 \sec \theta + 2)$

19. $(3 \sin \theta + 2)^2$

20. $(3 \cos \theta - 2)^2$

21. $(2 \sin^2 y - \csc^2 y)^2$

22. $(\tan^2 y + \cot^2 y)^2$

Factor each trigonometric expression.

23. $2 \sin^2 \gamma - 5 \sin \gamma - 3$

24. $\cos^2 \gamma - \cos \gamma - 6$

25. $\tan^2 \alpha - 6 \tan \alpha + 8$

26. $2 \cot^2 \alpha + \cot \alpha - 3$

27. $4 \sec^2 \beta + 4 \sec \beta + 1$

28. $9 \csc^2 \theta - 12 \csc \theta + 4$

29. $\tan^2 \alpha - \sec^2 \beta$

30. $\sin^4 y - \cos^4 x$

31. $\sin^2 \beta \cos \beta + \sin \beta \cos \beta - 2 \cos \beta$

32. $\cos^2 \theta \tan \theta - 2 \cos \theta \tan \theta - 3 \tan \theta$

33. $4 \sec^4 x - 4 \sec^2 x + 1$

34. $\cos^4 x - 2 \cos^2 x + 1$

35. $\sin \alpha \cos \alpha + \cos \alpha + \sin \alpha + 1$

36. $2 \sin^2 \theta + \sin \theta - 2 \sin \theta \cos \theta - \cos \theta$

Prove that each of the following equations is an identity.

37. $\tan x \cos x + \csc x \sin^2 x = 2 \sin x$

38. $\cot x \sin x - \cos^2 x \sec x = 0$

39. $(1 + \sin \alpha)^2 + \cos^2 \alpha = 2 + 2 \sin \alpha$

40. $(1 + \cot \alpha)^2 - 2 \cot \alpha = \dfrac{1}{(1 - \cos \alpha)(1 + \cos \alpha)}$

41. $2 - \csc \beta \sin \beta = \sin^2 \beta + \cos^2 \beta$

42. $(1 - \sin^2 \beta)(1 + \sin^2 \beta) = 2 \cos^2 \beta - \cos^4 \beta$

43. $\tan x + \cot x = \sec x \csc x$

44. $\dfrac{\csc x}{\cot x} - \dfrac{\cot x}{\csc x} = \dfrac{\tan x}{\csc x}$

45. $\dfrac{\sec x}{\tan x} - \dfrac{\tan x}{\sec x} = \cos x \cot x$

46. $\dfrac{1 - \sin^2 x}{1 - \sin x} = \dfrac{\csc x + 1}{\csc x}$

47. $\sec^2 x = \dfrac{\csc x}{\csc x - \sin x}$

48. $\dfrac{\sin x}{\sin x + 1} = \dfrac{\csc x - 1}{\cot^2 x}$

49. $1 + \csc x \sec x = \dfrac{\cos(-x) - \csc(-x)}{\cos(x)}$

50. $\tan^2(-x) - \dfrac{\sin(-x)}{\sin x} = \sec^2 x$

51. $\dfrac{1}{\csc \theta - \cot \theta} = \dfrac{1 + \cos \theta}{\sin \theta}$

52. $\dfrac{-1}{\tan \theta - \sec \theta} = \dfrac{1 + \sin \theta}{\cos \theta}$

53. $\dfrac{\csc y + 1}{\csc y - 1} = \dfrac{1 + \sin y}{1 - \sin y}$

54. $\dfrac{1 - 2 \cos^2 y}{1 - 2 \cos y \sin y} = \dfrac{\sin y + \cos y}{\sin y - \cos y}$

55. $\ln(\sec \theta) = -\ln(\cos \theta)$

56. $\ln(\tan \theta) = \ln(\sin \theta) + \ln(\sec \theta)$

57. $\ln|\sec \alpha + \tan \alpha| = -\ln|\sec \alpha - \tan \alpha|$

58. $\ln|\csc \alpha + \cot \alpha| = -\ln|\csc \alpha - \cot \alpha|$

59. $\dfrac{\cot x + \tan x}{\csc x} = \dfrac{1}{\cos x}$

60. $\dfrac{\sin(-x)}{-x} = \dfrac{\sin x}{x}$

61. $\dfrac{1 - \sin^2(-x)}{1 - \sin(-x)} = 1 - \sin x$

62. $\dfrac{1 - \sin^2 x \csc^2 x + \sin^2 x}{\cos^2 x} = \tan^2 x$

63. $\dfrac{1 - \cot^2 w + \cos^2 w \cot^2 w}{\csc^2 w} = \sin^4 w$

64. $\dfrac{\sec^2 z - \csc^2 z + \csc^2 z \cos^2 z}{\cot^2 z} = \tan^4 z$

For each equation either prove that it is an identity or prove that it is not an identity.

65. $\dfrac{\sin \theta + \cos \theta}{\sin \theta} = 1 + \cot \theta$

66. $\dfrac{\sin \theta + \cos \theta}{\cos \theta} = \cot \theta + 1$

67. $(\sin x + \csc x)^2 = \sin^2 x + \csc^2 x$

68. $\tan x + \sec x = \dfrac{\sin^2 x + 1}{\cos x}$

69. $\cot x + \sin x = \dfrac{1 + \cos x - \cos^2 x}{\sin x}$

70. $1 - 2 \cos^2 x + \cos^4 x = \sin^4 x$

For Writing/Discussion

71. Find functions $f_1(x)$ and $f_2(x)$ such that $f_1(x) = f_2(x)$ for infinitely many values of x, but yet $f_1(x) = f_2(x)$ is not an identity. Explain your example.

72. Find functions $f_1(x)$ and $f_2(x)$ such that $f_1(x) \neq f_2(x)$ for infinitely many values of x, but yet $f_1(x) = f_2(x)$ is an identity. Explain your example.

Graphing Calculator Exercises

The equation $f_1(x) = f_2(x)$ is an identity if and only if the graphs of $y = f_1(x)$ and $y = f_2(x)$ coincide at all values of x for which *both* sides are defined. Graph $y = f_1(x)$ and $y = f_2(x)$ on the same screen of your calculator for each of the following equations. From the graphs, make a conjecture as to whether each equation is an identity, then prove your conjecture.

1. $\dfrac{\sin x}{\cos x} - \dfrac{\cos x}{\sin x} = \dfrac{2\cos^2 x - 1}{\sin x \cos x}$

2. $\dfrac{1}{1 - \sin x} + \dfrac{1}{1 + \sin x} = \dfrac{2}{\cos^2 x}$

3. $\dfrac{\cos(-x)}{1 - \sin x} = \dfrac{1 - \sin(-x)}{\cos x}$

4. $\dfrac{\sin^2 x}{1 - \cos x} = 0.99 + \cos x$

The graphs on a graphing calculator corresponding to $f_1(x) = f_2(x)$ will usually appear to be identical if $f_1(x) = f_2(x)$ is an identity and different if it is not an identity.

5. Explain how the graphs on a calculator of $y = f_1(x)$ and $y = f_2(x)$ could appear identical when $f_1(x) = f_2(x)$ is not an identity.

6. Explain how the graphs on a calculator of $y = f_1(x)$ and $y = f_2(x)$ could appear different when $f_1(x) = f_2(x)$ is an identity.

7.3

Sum and Difference

Identities

In every identity discussed in Sections 7.1 and 7.2, the trigonometric functions were functions of only a single variable, such as θ or x. In this section we establish several new identities involving sine, cosine, and tangent of a sum or difference of two variables. These identities are used in solving equations and in simplifying expressions. They do not follow from the known identities but rather from the geometry of the unit circle.

The Cosine of a Sum

With so many identities showing relationships between the trigonometric functions, we might be fooled into thinking that almost any equation is an identity. For example, consider the equation

$$\cos(\alpha + \beta) = \cos \alpha + \cos \beta.$$

This equation looks nice, but is it an identity? It is easy to check with a calculator that if $\alpha = 30°$ and $\beta = 45°$, we get

$$\cos(30° + 45°) = \cos(75°) \approx 0.2588 \quad \text{and} \quad \cos(30°) + \cos(45°) \approx 1.573.$$

So the equation $\cos(\alpha + \beta) = \cos \alpha + \cos \beta$ is *not* an identity. However, there is an identity for the cosine of a sum of two angles.

To derive an identity for $\cos(\alpha + \beta)$, consider angles α, $\alpha + \beta$, and $-\beta$ in standard position, as shown in Fig. 7.2. Assume that α and β are acute angles, although any angles α and β may be used. The terminal side of α intersects the unit circle at the point $A(\cos \alpha, \sin \alpha)$. The terminal side of $\alpha + \beta$ intersects the unit circle at $B(\cos(\alpha + \beta), \sin(\alpha + \beta))$. The terminal side of $-\beta$ intersects the unit circle at $C(\cos(-\beta), \sin(-\beta))$ or $C(\cos \beta, -\sin \beta)$. Let D be

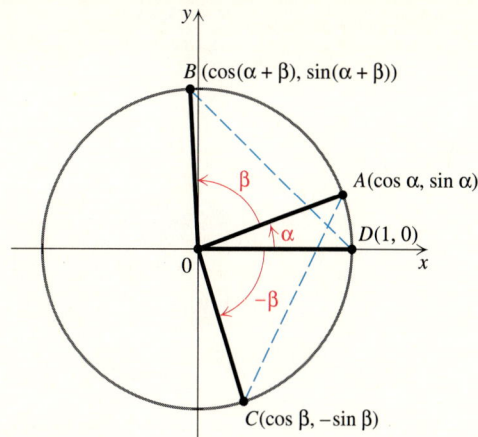

Figure 7.2

the point $(1, 0)$. Since $\angle BOD = \alpha + \beta$ and $\angle AOC = \alpha + \beta$, the chords $\overline{BD}$ and $\overline{AC}$ are equal in length. Using the distance formula to find the lengths, we get the following equation:

$$\sqrt{(\cos(\alpha + \beta) - 1)^2 + (\sin(\alpha + \beta) - 0)^2} = \\ \sqrt{(\cos\alpha - \cos\beta)^2 + (\sin\alpha - (-\sin\beta))^2}$$

Square each side and simplify:

$$(\cos(\alpha + \beta) - 1)^2 + \sin^2(\alpha + \beta) = (\cos\alpha - \cos\beta)^2 + (\sin\alpha + \sin\beta)^2$$

Square the binomials:

$$\cos^2(\alpha + \beta) - 2\cos(\alpha + \beta) + 1 + \sin^2(\alpha + \beta) = \\ \cos^2\alpha - 2\cos\alpha\cos\beta + \cos^2\beta + \sin^2\alpha + 2\sin\alpha\sin\beta + \sin^2\beta$$

Using the identity $\cos^2 x + \sin^2 x = 1$ once on the left-hand side and twice on the right-hand side, we get the following equation:

$$2 - 2\cos(\alpha + \beta) = 2 - 2\cos\alpha\cos\beta + 2\sin\alpha\sin\beta$$

Subtract 2 from each side and divide each side by -2 to get the identity for the cosine of a sum:

Identity: Cosine of a Sum

$$\cos(\alpha + \beta) = \cos\alpha\cos\beta - \sin\alpha\sin\beta$$

If an angle is the sum of two angles for which we know the exact values of the trigonometric functions, we can use this new identity to find the exact value of the cosine of the sum. For example, we can find the exact values of expressions such as $\cos(75°)$, $\cos(7\pi/12)$, and $\cos(195°)$ because

$$75° = 45° + 30°, \qquad \frac{7\pi}{12} = \frac{\pi}{3} + \frac{\pi}{4}, \qquad \text{and} \qquad 195° = 150° + 45°.$$

Example 1   The cosine of a sum

Find the exact value of $\cos(75°)$.

Solution

Use $75° = 30° + 45°$ and the identity for the cosine of a sum.

$$\begin{aligned}
\cos 75° &= \cos(30° + 45°) \\
&= \cos(30°)\cos(45°) - \sin(30°)\sin(45°) \\
&= \frac{\sqrt{3}}{2} \cdot \frac{\sqrt{2}}{2} - \frac{1}{2} \cdot \frac{\sqrt{2}}{2} \\
&= \frac{\sqrt{6} - \sqrt{2}}{4}
\end{aligned}$$

To check, evaluate $\cos 75°$ and $(\sqrt{6} - \sqrt{2})/4$ using a calculator. ◆

Note that we found the exact value of $\cos(75°)$ only to illustrate an identity. If we need $\cos(75°)$ in an application, we can use the calculator's value for it.

The Cosine of a Difference

To derive an identity for the cosine of a difference we write $\alpha - \beta = \alpha + (-\beta)$ and use the identity for the cosine of a sum:

$$\begin{aligned}
\cos(\alpha - \beta) &= \cos(\alpha + (-\beta)) \\
&= \cos\alpha\cos(-\beta) - \sin\alpha\sin(-\beta)
\end{aligned}$$

Now use the identities $\cos(-\beta) = \cos\beta$ and $\sin(-\beta) = -\sin\beta$ to get the identity for the cosine of a difference:

Identity: Cosine of a Difference

$$\cos(\alpha - \beta) = \cos\alpha\cos\beta + \sin\alpha\sin\beta$$

If an angle is the difference of two angles for which we know the exact values of sine and cosine, then we can find the exact value for the cosine of the angle.

Example 2   The cosine of a difference

Find the exact value of $\cos(\pi/12)$.

Solution

Since $\dfrac{\pi}{3} = \dfrac{4\pi}{12}$ and $\dfrac{\pi}{4} = \dfrac{3\pi}{12}$, we can use $\dfrac{\pi}{12} = \dfrac{\pi}{3} - \dfrac{\pi}{4}$ and the identity for the cosine of a difference:

$$\cos\left(\frac{\pi}{12}\right) = \cos\left(\frac{\pi}{3} - \frac{\pi}{4}\right)$$

$$= \cos\frac{\pi}{3}\cos\frac{\pi}{4} + \sin\frac{\pi}{3}\sin\frac{\pi}{4}$$

$$= \frac{1}{2}\cdot\frac{\sqrt{2}}{2} + \frac{\sqrt{3}}{2}\cdot\frac{\sqrt{2}}{2}$$

$$= \frac{\sqrt{2} + \sqrt{6}}{4}$$

To check, evaluate $\cos(\pi/12)$ and $(\sqrt{2} + \sqrt{6})/4$ using a calculator.   ◆

The identities for the cosine of a sum or a difference can be used to find identities for the sine of a sum or difference. To develop those identities we first find identities that relate each trigonometric function and its **cofunction.** Sine and cosine are cofunctions, tangent and cotangent are cofunctions, and secant and cosecant are cofunctions.

Cofunction Identities

If $\alpha = \pi/2$ is substituted into the identity for $\cos(\alpha - \beta)$, we get another identity:

$$\cos\left(\frac{\pi}{2} - \beta\right) = \cos\frac{\pi}{2}\cos\beta + \sin\frac{\pi}{2}\sin\beta$$

$$= 0\cdot\cos\beta + 1\cdot\sin\beta$$

$$= \sin\beta$$

The identity $\cos(\pi/2 - \beta) = \sin\beta$ is a **cofunction identity.**

If we let $u = \pi/2 - \beta$ or $\beta = \pi/2 - u$ in the identity $\sin\beta = \cos(\pi/2 - \beta)$, we get the identity

$$\sin\left(\frac{\pi}{2} - u\right) = \cos u.$$

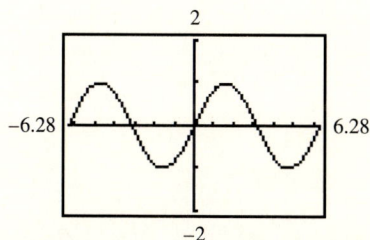

The graphs of $y_1 = \cos(\pi/2 - x)$ and $y_2 = \sin x$ coincide.

Using these identities, we can establish the cofunction identities for secant, cosecant, tangent, and cotangent. For the cofunction identities *the value of any trigonometric function at u is equal to the value of its cofunction at* $(\pi/2 - u)$.

Cofunction Identities

$$\sin\left(\frac{\pi}{2} - u\right) = \cos u \qquad \cos\left(\frac{\pi}{2} - u\right) = \sin u$$

$$\tan\left(\frac{\pi}{2} - u\right) = \cot u \qquad \cot\left(\frac{\pi}{2} - u\right) = \tan u$$

$$\sec\left(\frac{\pi}{2} - u\right) = \csc u \qquad \csc\left(\frac{\pi}{2} - u\right) = \sec u$$

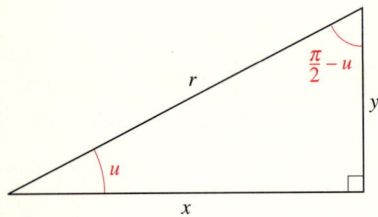

Figure 7.3

If $0 < u < \pi/2$, then u and $(\pi/2 - u)$ are the measures of the acute angles of a right triangle as shown in Fig. 7.3. The term *cofunction* comes from the fact that these angles are complementary. The cofunction identities indicate that the value of a trigonometric function of one acute angle of a right triangle is equal to the value of the cofunction of the other. Since the cofunction identities hold also for complementary angles measured in degrees, we can write equations such as

$$\sin(20°) = \cos(70°), \qquad \cot(89°) = \tan(1°), \qquad \text{and} \qquad \sec(50°) = \csc(40°).$$

The cofunction identities are consistent with the values obtained for the trigonometric functions for an acute angle using the ratios of the sides of a right triangle. For example, the cofunction identity $\sin(u) = \cos(\pi/2 - u)$ is correct for angles u and $(\pi/2 - u)$ shown in Fig. 7.3, because

$$\sin(u) = \frac{\text{opp}}{\text{hyp}} = \frac{y}{r}$$

and

$$\cos\left(\frac{\pi}{2} - u\right) = \frac{\text{adj}}{\text{hyp}} = \frac{y}{r}.$$

Example 3   Using the cofunction identities

Use a cofunction identity to find the exact value of $\sin(5\pi/12)$.

Solution

Since $\sin(u) = \cos(\pi/2 - u)$,

$$\sin\left(\frac{5\pi}{12}\right) = \cos\left(\frac{\pi}{2} - \frac{5\pi}{12}\right)$$

$$= \cos\left(\frac{\pi}{12}\right).$$

From Example 2, $\cos(\pi/12) = \left(\sqrt{2} + \sqrt{6}\right)/4$. So $\sin(5\pi/12) = \left(\sqrt{2} + \sqrt{6}\right)/4$.

Sine of a Sum or a Difference

We now use a cofunction identity to find an identity for $\sin(\alpha + \beta)$:

$$\sin(\alpha + \beta) = \cos\left(\frac{\pi}{2} - (\alpha + \beta)\right)$$

$$= \cos\left(\left(\frac{\pi}{2} - \alpha\right) - \beta\right)$$

$$= \cos\left(\frac{\pi}{2} - \alpha\right)\cos\beta + \sin\left(\frac{\pi}{2} - \alpha\right)\sin\beta$$

$$= \sin\alpha\cos\beta + \cos\alpha\sin\beta$$

To find an identity for $\sin(\alpha - \beta)$, use the identity for $\sin(\alpha + \beta)$:

$$\sin(\alpha - \beta) = \sin(\alpha + (-\beta))$$

$$= \sin\alpha\cos(-\beta) + \cos\alpha\sin(-\beta)$$

$$= \sin\alpha\cos\beta - \cos\alpha\sin\beta$$

The identities for the sine of a sum or difference are stated as follows.

Identities: Sine of a Sum or Difference

$$\sin(\alpha + \beta) = \sin\alpha\cos\beta + \cos\alpha\sin\beta$$

$$\sin(\alpha - \beta) = \sin\alpha\cos\beta - \cos\alpha\sin\beta$$

Example 4   The sine of a sum

Find the exact value of $\sin(195°)$.

Solution

Use $195° = 150° + 45°$ and the identity for the sine of a sum:

$$\sin(195°) = \sin(150° + 45°)$$

$$= \sin(150°)\cos(45°) + \cos(150°)\sin(45°)$$

$$= \frac{1}{2} \cdot \frac{\sqrt{2}}{2} + \left(-\frac{\sqrt{3}}{2}\right) \cdot \frac{\sqrt{2}}{2}$$

$$= \frac{\sqrt{2} - \sqrt{6}}{4}$$

To check, evaluate $\sin(195°)$ and $(\sqrt{2} - \sqrt{6})/4$ using a calculator.   ◆

Tangent of a Sum or Difference

We can use the identities for $\sin(\alpha + \beta)$ and $\cos(\alpha + \beta)$ to find an identity for $\tan(\alpha + \beta)$:

$$\tan(\alpha + \beta) = \frac{\sin(\alpha + \beta)}{\cos(\alpha + \beta)}$$

$$= \frac{\sin \alpha \cos \beta + \cos \alpha \sin \beta}{\cos \alpha \cos \beta - \sin \alpha \sin \beta}$$

To express the right-hand side in terms of tangent, multiply the numerator and denominator by $1/(\cos \alpha \cos \beta)$ and use the identity $\tan x = \sin x / \cos x$:

$$\tan(\alpha + \beta) = \frac{\dfrac{\sin \alpha \cos \beta}{\cos \alpha \cos \beta} + \dfrac{\cos \alpha \sin \beta}{\cos \alpha \cos \beta}}{\dfrac{\cos \alpha \cos \beta}{\cos \alpha \cos \beta} - \dfrac{\sin \alpha \sin \beta}{\cos \alpha \cos \beta}}$$

$$= \frac{\tan \alpha + \tan \beta}{1 - \tan \alpha \tan \beta}$$

Use the identity $\tan(-\beta) = -\tan \beta$ to get a similar identity for $\tan(\alpha - \beta)$.

Identities: Tangent of a Sum or Difference

$$\tan(\alpha + \beta) = \frac{\tan \alpha + \tan \beta}{1 - \tan \alpha \tan \beta} \qquad \tan(\alpha - \beta) = \frac{\tan \alpha - \tan \beta}{1 + \tan \alpha \tan \beta}$$

Example 5   The tangent of a difference

Find the exact value of $\tan(\pi/12)$.

Solution

Use $\pi/12 = \pi/3 - \pi/4$ and the identity for the tangent of a difference:

$$\tan\left(\frac{\pi}{12}\right) = \tan\left(\frac{\pi}{3} - \frac{\pi}{4}\right)$$

$$= \frac{\tan \dfrac{\pi}{3} - \tan \dfrac{\pi}{4}}{1 + \tan \dfrac{\pi}{3} \tan \dfrac{\pi}{4}} = \frac{\sqrt{3} - 1}{1 + \sqrt{3} \cdot 1}$$

$$= \frac{(\sqrt{3} - 1)(\sqrt{3} - 1)}{(\sqrt{3} + 1)(\sqrt{3} - 1)}$$

$$= \frac{3 - 2\sqrt{3} + 1}{2} = \frac{4 - 2\sqrt{3}}{2} = 2 - \sqrt{3}$$

In the next example we use sum and difference identities in reverse to simplify an expression.

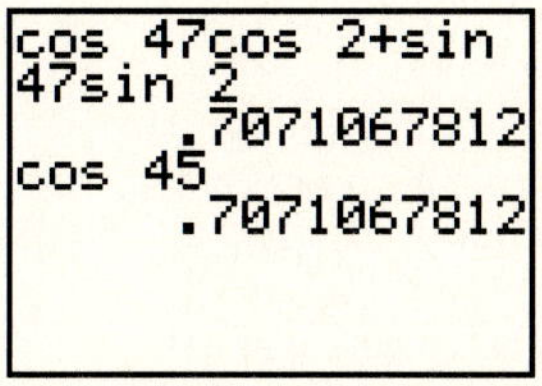

Example 6 Simplifying with sum and difference identities

Use an appropriate identity to simplify each expression.

a) $\cos(47°)\cos(2°) + \sin(47°)\sin(2°)$ **b)** $\sin t \cos 2t - \cos t \sin 2t$

Solution

a) This expression is the right-hand side of the identity for the cosine of a difference, $\cos(\alpha - \beta) = \cos \alpha \cos \beta + \sin \alpha \sin \beta$, with $\alpha = 47°$ and $\beta = 2°$:

$$\cos(47°)\cos(2°) + \sin(47°)\sin(2°) = \cos(47° - 2°) = \cos(45°) = \frac{\sqrt{2}}{2}.$$

b) We recognize this expression as the right-hand side of the identity for the sine of a difference, $\sin(\alpha - \beta) = \sin \alpha \cos \beta - \cos \alpha \sin \beta$. So

$$\sin t \cos 2t - \cos t \sin 2t = \sin(t - 2t) = \sin(-t) = -\sin t. \qquad \blacklozenge$$

As shown next we use identities to find $\sin(\alpha + \beta)$ without knowing α or β.

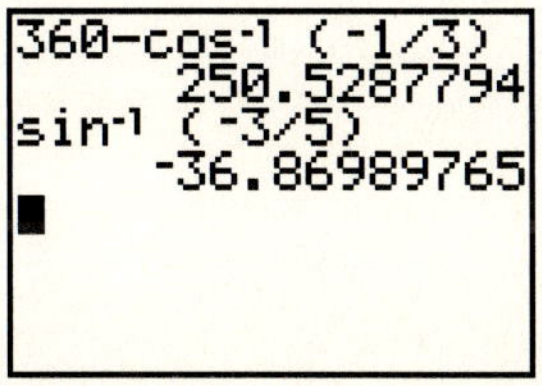

Example 7 Using identities

Find the exact value of $\sin(\alpha + \beta)$, given that $\sin \alpha = -3/5$ and $\cos \beta = -1/3$, with α in quadrant IV and β in quadrant III.

Solution

To use the identity for $\sin(\alpha + \beta)$, we need $\cos \alpha$ and $\sin \beta$ in addition to the given values. Use $\sin \alpha = -3/5$ in the identity $\sin^2 x + \cos^2 x = 1$ to find $\cos \alpha$:

$$\left(-\frac{3}{5}\right)^2 + \cos^2 \alpha = 1, \qquad \cos^2 \alpha = \frac{16}{25}, \qquad \cos \alpha = \pm \frac{4}{5}.$$

Since cosine is positive in quadrant IV, $\cos \alpha = 4/5$. Use $\cos \beta = -1/3$ in $\sin^2 x + \cos^2 x = 1$ to find $\sin \beta$:

$$\sin^2 \beta + \left(-\frac{1}{3}\right)^2 = 1, \qquad \sin^2 \beta = \frac{8}{9}, \qquad \sin \beta = \pm \frac{2\sqrt{2}}{3}.$$

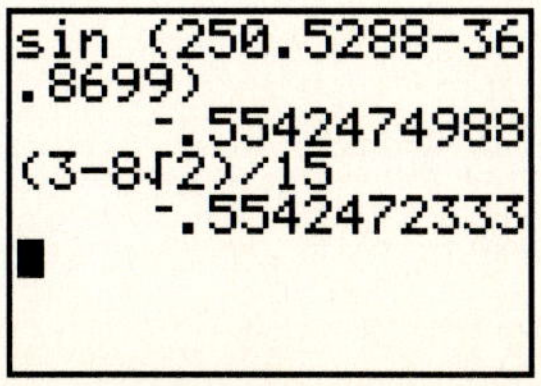

Since sine is negative in quadrant III, $\sin \beta = -2\sqrt{2}/3$. Now use the appropriate values in the identity $\sin(\alpha + \beta) = \sin \alpha \cos \beta + \cos \alpha \sin \beta$:

$$\sin(\alpha + \beta) = -\frac{3}{5}\left(-\frac{1}{3}\right) + \frac{4}{5}\left(-\frac{2\sqrt{2}}{3}\right) = \frac{3}{15} - \frac{8\sqrt{2}}{15} = \frac{3 - 8\sqrt{2}}{15}.$$

Check: Use a calculator to find approximate values for α, β, and $\sin(\alpha + \beta)$. $\qquad \blacklozenge$

Sidebar (left column):

```
cos 47cos 2+sin
47sin 2
        .7071067812
cos 45
        .7071067812
```

Use a calculator to check Example 6(a).

```
360-cos⁻¹ (-1/3)
        250.5287794
sin⁻¹ (-3/5)
        -36.86989765
■
```

Since $\cos^{-1}(-1/3)$ is in quadrant II, α is $360 - \cos^{-1}(-1/3)$. The angle β is $\sin^{-1}(-3/5)$.

```
sin (250.5288-36
.8699)
        -.5542474988
(3-8√2)/15
        -.5542472333
■
```

Use α and β that were found above to check that $\sin(\alpha + \beta)$ is equal to $(3 - 8\sqrt{2})/15$.

For Thought

?

True or false? Explain.

1. $\cos(3°)\cos(2°) - \sin(3°)\sin(2°) = \cos(1°)$

2. $\cos(4)\cos(5) + \sin(4)\sin(5) = \cos(-1)$

3. For any real number t, $\cos(t - \pi/2) = \sin t$.

4. For any radian measure α, $\sin(\alpha - \pi/2) = \cos \alpha$.

5. $\sec(\pi/3) = \csc(\pi/6)$

6. $\sin(5\pi/6) = \sin(2\pi/3) + \sin(\pi/6)$

7. $\sin(5\pi/12) = \sin(\pi/3)\cos(\pi/4) + \cos(\pi/3)\sin(\pi/4)$

8. $\tan(21°30'5'') = \cot(68°29'55'')$

9. For any real number x, $\sec(x) - \csc(\pi/2 - x) = \tan(x) - \cot(\pi/2 - x)$.

10. $\dfrac{\tan(13°) + \tan(-20°)}{1 - \tan(13°)\tan(-20°)} = \dfrac{\tan(45°) + \tan(-52°)}{1 - \tan(45°)\tan(-52°)}$

7.3 Exercises

Tape 13   Disk—5.25": 4   3.5": 3   Macintosh: 3

Find the following sums or differences.

1. $\dfrac{\pi}{4} + \dfrac{\pi}{3}$

2. $\dfrac{2\pi}{3} - \dfrac{\pi}{4}$

3. $\dfrac{3\pi}{4} + \dfrac{\pi}{3}$

4. $\dfrac{\pi}{6} + \dfrac{\pi}{4}$

Express each given angle as $\alpha + \beta$ or $\alpha - \beta$, where $\sin \alpha$ and $\sin \beta$ are known exactly.

5. $75°$ 6. $15°$ 7. $165°$ 8. $195°$

Use appropriate identities to find the exact value of each expression.

9. $\cos(5\pi/12)$

10. $\cos(7\pi/12)$

11. $\sin(7\pi/12)$

12. $\sin(5\pi/12)$

13. $\tan(75°)$

14. $\tan(-15°)$

15. $\sin(-15°)$

16. $\sin(165°)$

17. $\cos(195°)$

18. $\cos(-75°)$

19. $\tan(-13\pi/12)$

20. $\tan(7\pi/12)$

Simplify each expression by using appropriate identities. Do not use a calculator.

21. $\sin(23°)\cos(67°) + \cos(23°)\sin(67°)$

22. $\sin(34°)\cos(13°) + \cos(-34°)\sin(-13°)$

23. $\cos(-\pi/2)\cos(\pi/5) + \sin(\pi/2)\sin(\pi/5)$

24. $\cos(12°)\cos(-3°) - \sin(12°)\sin(-3°)$

25. $\dfrac{\tan(\pi/9) + \tan(\pi/6)}{1 - \tan(\pi/9)\tan(\pi/6)}$

26. $\dfrac{\tan(\pi/3) - \tan(\pi/5)}{1 + \tan(\pi/3)\tan(\pi/5)}$

27. $\dfrac{\tan(\pi/7) + \cot(\pi/2 - \pi/6)}{1 + \tan(-\pi/7)\tan(\pi/6)}$

28. $\dfrac{\cot(\pi/2 - \pi/3) + \tan(-\pi/6)}{1 + \tan(\pi/3)\cot(\pi/2 - \pi/6)}$

29. $\sin(14°)\cos(35°) + \cos(14°)\cos(55°)$

30. $\cos(10°)\cos(20°) + \cos(-80°)\sin(-20°)$

31. $\cos(-\pi/5)\cos(\pi/3) + \sin(-\pi/5)\sin(-\pi/3)$

32. $\cos(-6)\cos(4) - \sin(6)\sin(-4)$

33. $\sin 2k \cos k + \cos 2k \sin k$

34. $\cos 3y \cos y - \sin 3y \sin y$

Match each expression on the left with an expression on the right that has exactly the same value. Do not use a calculator.

35. $\cos(44°)$

A. $\cos(0)$

36. $\sin(-46°)$

B. $-\cos(44°)$

37. $\cos(46°)$

C. $\cos(36°)$

38. $\sin(136°)$

D. $\cot\left(\dfrac{5\pi}{14}\right)$

39. $\sec(1)$

E. $-\cos(46°)$

40. $\tan\left(\dfrac{\pi}{7}\right)$

F. $\csc\left(\dfrac{\pi - 2}{2}\right)$

41. $\csc\left(\dfrac{\pi}{2}\right)$

G. $\sin(46°)$

42. $\sin(-44°)$

H. $\sin(44°)$

Solve each problem.

43. Find the exact value of $\sin(\alpha + \beta)$, given that $\sin \alpha = 3/5$ and $\sin \beta = 5/13$, with α in quadrant II and β in quadrant I.

44. Find the exact value of $\sin(\alpha - \beta)$, given that $\sin \alpha = -4/5$ and $\cos \beta = 12/13$, with α in quadrant III and β in quadrant IV.

45. Find the exact value of $\cos(\alpha + \beta)$, given that $\sin \alpha = 2/3$ and $\sin \beta = -1/2$, with α in quadrant I and β in quadrant III.

46. Find the exact value of $\cos(\alpha - \beta)$, given that $\cos \alpha = \sqrt{3}/4$ and $\cos \beta = -\sqrt{2}/3$, with α in quadrant I and β in quadrant II.

Write each expression as a function of α alone.

47. $\cos(\pi/2 + \alpha)$

48. $\sin(\alpha - \pi)$

49. $\cos(180° - \alpha)$

50. $\sin(180° - \alpha)$

51. $\sin(360° - \alpha)$

52. $\cos(\alpha - \pi)$

53. $\sin(90° + \alpha)$

54. $\cos(360° - \alpha)$

55. $\tan(\pi/4 + \alpha)$

56. $\tan(\pi/4 - \alpha)$

57. $\tan(180° + \alpha)$

58. $\tan(360° - \alpha)$

Verify that each equation is an identity.

59. $\sin(180° - \alpha) = \dfrac{1 - \cos^2 \alpha}{\sin \alpha}$

60. $\cos(x - \pi/2) = \cos x \tan x$

61. $\dfrac{\cos(x + y)}{\cos x \cos y} = 1 - \tan x \tan y$

62. $\dfrac{\sin(x + y)}{\sin x \cos y} = 1 + \cot x \tan y$

63. $\sin(\alpha + \beta)\sin(\alpha - \beta) = \sin^2 \alpha - \sin^2 \beta$

64. $\cos(\alpha + \beta)\cos(\alpha - \beta) = \cos^2 \beta - \sin^2 \alpha$

65. $\cos(2x) = \cos^2 x - \sin^2 x$

66. $\sin(2x) = 2 \sin x \cos x$

67. $\sin(x - y) - \sin(y - x) = 2 \sin x \cos y - 2 \cos x \sin y$

68. $\cos(x - y) + \cos(y - x) = 2 \cos x \cos y + 2 \sin x \sin y$

69. $\tan(s + t)\tan(s - t) = \dfrac{\tan^2 s - \tan^2 t}{1 - \tan^2 s \tan^2 t}$

70. $\tan(\pi/4 + x) = \cot(\pi/4 - x)$

71. $\dfrac{\cos(\alpha + \beta)}{\sin(\alpha - \beta)} = \dfrac{1 - \tan \alpha \tan \beta}{\tan \alpha - \tan \beta}$

72. $\dfrac{\cos(\alpha - \beta)}{\sin(\alpha + \beta)} = \dfrac{1 + \tan \alpha \tan \beta}{\tan \alpha + \tan \beta}$

73. $\sec(v + t) = \dfrac{\cos v \cos t + \sin v \sin t}{\cos^2 v - \sin^2 t}$

74. $\csc(v - t) = \dfrac{\sin v \cos t + \cos v \sin t}{\sin^2 v - \sin^2 t}$

75. $\dfrac{\cos(x + y)}{\cos(x - y)} = \dfrac{\cot y - \tan x}{\cot y + \tan x}$

76. $\dfrac{\sin(x + y)}{\sin(x - y)} = \dfrac{\cot y + \cot x}{\cot y - \cot x}$

77. $\dfrac{\sin(\alpha + \beta)}{\sin \alpha + \sin \beta} = \dfrac{\sin \alpha - \sin \beta}{\sin(\alpha - \beta)}$

78. $\dfrac{\cos(\alpha + \beta)}{\cos \alpha + \sin \beta} = \dfrac{\cos \alpha - \sin \beta}{\cos(\beta - \alpha)}$

For Writing/Discussion

79. Verify the four cofunction identities that were not proved in the text.

80. Verify the identity for the tangent of a difference.

81. Show that $\sin(\alpha + \beta) = \sin \alpha + \sin \beta$ is not an identity.

82. Show that $\cos(\alpha - \beta) = \cos \alpha - \cos \beta$ is not an identity.

83. Explain why $\sin(180° - \alpha) = \sin \alpha$ using the unit circle.

84. Explain why $\cos(180° - \alpha) = -\cos \alpha$ using the unit circle.

7.4

Double-Angle and

Half-Angle Identities

The graphs of $y_1 = \sin(2x)$ and $y_2 = 2 \sin x \cos x$ coincide.

In Section 7.3 we studied identities for functions of sums and differences of angles. The double-angle and half-angle identities, which we develop next, are special cases of those identities. These special cases occur so often that they are remembered as separate identities.

Double-Angle Identities

To get an identity for $\sin 2x$, replace both α and β by x in the identity for $\sin(\alpha + \beta)$:

$$\sin 2x = \sin(x + x) = \sin x \cos x + \cos x \sin x = 2 \sin x \cos x$$

To find an identity for $\cos 2x$, replace both α and β by x in the identity for $\cos(\alpha + \beta)$:

$$\cos 2x = \cos(x + x) = \cos x \cos x - \sin x \sin x = \cos^2 x - \sin^2 x$$

To get a second form of the identity for $\cos 2x$, replace $\sin^2 x$ with $1 - \cos^2 x$:

$$\cos(2x) = \cos^2 x - \sin^2 x = \cos^2 - (1 - \cos^2 x) = 2 \cos^2 x - 1$$

Replacing $\cos^2 x$ with $1 - \sin^2 x$ produces a third form of the identity for $\cos 2x$:

$$\cos(2x) = 1 - 2 \sin^2 x$$

To get an identity for $\tan 2x$, we can replace both α and β by x in the identity for $\tan(\alpha + \beta)$:

$$\tan 2x = \tan(x + x) = \frac{\tan x + \tan x}{1 - \tan x \tan x} = \frac{2 \tan x}{1 - \tan^2 x}$$

These identities, which are summarized below, are known as the **double-angle identities.**

▼

Double-Angle Identities

$$\sin 2x = 2 \sin x \cos x \qquad \tan 2x = \frac{2 \tan x}{1 - \tan^2 x}$$

$$\cos 2x = \cos^2 x - \sin^2 x$$

$$\cos 2x = 2 \cos^2 x - 1$$

$$\cos 2x = 1 - 2 \sin^2 x$$

Be careful to learn the double-angle identities exactly as they are written. A "nice looking" equation such as $\cos 2x = 2 \cos x$ could be mistaken for an identity if you are not careful. [Since $\cos(\pi/2) \neq 2 \cos(\pi/4)$, the "nice looking" equation is not an identity.] Remember that an equation is not an identity if at least one permissible value of the variable fails to satisfy the equation.

Example 1   Using the double-angle identities

Find $\sin(120°)$, $\cos(120°)$, and $\tan(120°)$ by using double-angle identities.

Solution

Note that $120° = 2 \cdot 60°$ and use the values $\sin(60°) = \sqrt{3}/2$, $\cos(60°) = 1/2$, and $\tan(60°) = \sqrt{3}$ in the appropriate identities:

$$\sin(120°) = 2\sin(60°)\cos(60°) = 2 \cdot \frac{\sqrt{3}}{2} \cdot \frac{1}{2} = \frac{\sqrt{3}}{2}$$

$$\cos(120°) = \cos^2(60°) - \sin^2(60°) = \left(\frac{1}{2}\right)^2 - \left(\frac{\sqrt{3}}{2}\right)^2 = -\frac{1}{2}$$

$$\tan(120°) = \frac{2\tan(60°)}{1 - \tan^2(60°)} = \frac{2 \cdot \sqrt{3}}{1 - (\sqrt{3})^2} = \frac{2\sqrt{3}}{-2} = -\sqrt{3}$$

These results are the well-known values of $\sin(120°)$, $\cos(120°)$, and $\tan(120°)$.   ◆

Example 2   A triple-angle identity

Prove that the following equation is an identity:

$$\sin(3x) = \sin x(3\cos^2 x - \sin^2 x)$$

Solution

Write $3x$ as $2x + x$ and use the identity for the sine of a sum:

$\sin(3x) = \sin(2x + x)$

$\qquad = \sin 2x \cos x + \cos 2x \sin x$ **Sine of a sum identity**

$\qquad = 2\sin x \cos x \cos x + (\cos^2 x - \sin^2 x)\sin x$ **Double-angle identities**

$\qquad = 2\sin x \cos^2 x + \sin x \cos^2 x - \sin^3 x$ **Distributive property**

$\qquad = 3\sin x \cos^2 x - \sin^3 x$

$\qquad = \sin x(3\cos^2 x - \sin^2 x)$ **Factor.**   ◆

Note that in Example 2 we used identities to expand the simpler side of the equation. In this case, simplifying the more complicated side is more difficult.

The double-angle identities can be used to get identities for $\sin(x/2)$, $\cos(x/2)$, and $\tan(x/2)$. These identities, which are discussed on the next page, are known as **half-angle identities.**

Half-Angle Identities

To get an identity for $\cos(x/2)$, start by solving the double-angle identity $\cos 2x = 2\cos^2 x - 1$ for $\cos x$:

$$2\cos^2 x - 1 = \cos 2x$$

$$\cos^2 x = \frac{1 + \cos 2x}{2}$$

$$\cos x = \pm\sqrt{\frac{1 + \cos 2x}{2}}$$

Replacing x by $x/2$ yields the half-angle identity for cosine:

$$\cos\frac{x}{2} = \pm\sqrt{\frac{1 + \cos x}{2}}$$

For any value of x, $\cos(x/2)$ is equal to either the positive or the negative square root of $(1 + \cos x)/2$. Whether the positive or negative sign is chosen depends on the value of $x/2$.

Example 3 Using half-angle identities

Use the half-angle identity to find the exact value of $\cos(\pi/8)$.

Solution

Use $x = \pi/4$ in the half-angle identity for cosine:

$$\cos\frac{\pi}{8} = \cos\frac{\pi/4}{2} = \sqrt{\frac{1 + \cos \pi/4}{2}} = \sqrt{\frac{1 + \frac{\sqrt{2}}{2}}{2}}$$

$$= \sqrt{\frac{2 + \sqrt{2}}{4}} = \frac{\sqrt{2 + \sqrt{2}}}{2}$$

The positive square root is used because the angle $\pi/8$ is in the first quadrant, where cosine is positive.

To get an identity for $\sin(x/2)$, solve $1 - 2\sin^2 x = \cos 2x$ for $\sin x$:

$$1 - 2\sin^2 x = \cos 2x$$

$$-2\sin^2 x = \cos 2x - 1$$

$$\sin^2 x = \frac{1 - \cos 2x}{2}$$

$$\sin x = \pm\sqrt{\frac{1 - \cos 2x}{2}}$$

Replacing x by $x/2$ yields the half-angle identity for sine:

$$\sin\frac{x}{2} = \pm\sqrt{\frac{1 - \cos x}{2}}$$

The half-angle identity for sine can be used to find the exact value for any angle that is one-half of an angle for which we know the exact value of cosine.

Example 4   Using half-angle identities

Use the half-angle identity to find the exact value of $\sin(75°)$.

Solution

Use $x = 150°$ in the half-angle identity for sine:

$$\sin(75°) = \sin\left(\frac{150°}{2}\right) = \pm\sqrt{\frac{1 - \cos(150°)}{2}}$$

Since 75° is in quadrant I, $\sin(75°)$ is positive. The reference angle for 150° is 30° and $\cos(150°)$ is negative, because 150° is in quadrant II. So

$$\cos(150°) = -\cos(30°) = -\frac{\sqrt{3}}{2}$$

and

$$\sin(75°) = \sqrt{\frac{1 + \frac{\sqrt{3}}{2}}{2}} = \sqrt{\frac{2 + \sqrt{3}}{4}} = \frac{\sqrt{2 + \sqrt{3}}}{2}.$$ ◆

To get a half-angle identity for tangent, use the half-angle identities for sine and cosine:

$$\tan\frac{x}{2} = \frac{\sin\dfrac{x}{2}}{\cos\dfrac{x}{2}} = \frac{\pm\sqrt{\dfrac{1 - \cos x}{2}}}{\pm\sqrt{\dfrac{1 + \cos x}{2}}} = \pm\sqrt{\frac{1 - \cos x}{1 + \cos x}}$$

To get another form of this identity, multiply the numerator and denominator inside the radical by $1 + \cos x$. Use the fact that $\sqrt{\sin^2 x} = |\sin x|$ but $\sqrt{(1 + \cos x)^2} = 1 + \cos x$ because $1 + \cos x$ is nonnegative:

$$\tan\frac{x}{2} = \pm\sqrt{\frac{(1 - \cos x)(1 + \cos x)}{(1 + \cos x)^2}} = \pm\sqrt{\frac{\sin^2 x}{(1 + \cos x)^2}} = \pm\frac{|\sin x|}{1 + \cos x}$$

We use the positive or negative sign depending on the sign of $\tan(x/2)$. It can be shown that $\sin x$ has the same sign as $\tan(x/2)$. (See Exercise 67.) With this fact we can omit the absolute value and the $\pm$ symbol and write the identity as

$$\tan \frac{x}{2} = \frac{\sin x}{1 + \cos x}.$$

To get a third form of the identity for $\tan(x/2)$, multiply the numerator and denominator of the right-hand side by $1 - \cos x$:

$$\tan \frac{x}{2} = \frac{\sin x(1 - \cos x)}{(1 + \cos x)(1 - \cos x)} = \frac{\sin x(1 - \cos x)}{\sin^2 x} = \frac{1 - \cos x}{\sin x}$$

The half-angle identities are summarized as follows.

Half-Angle Identities

$$\sin \frac{x}{2} = \pm \sqrt{\frac{1 - \cos x}{2}} \qquad \cos \frac{x}{2} = \pm \sqrt{\frac{1 + \cos x}{2}}$$

$$\tan \frac{x}{2} = \pm \sqrt{\frac{1 - \cos x}{1 + \cos x}} \qquad \tan \frac{x}{2} = \frac{\sin x}{1 + \cos x} \qquad \tan \frac{x}{2} = \frac{1 - \cos x}{\sin x}$$

Example 5   Using half-angle identities

Use the half-angle identity to find the exact value of $\tan(-15°)$.

Solution

Use $x = -30°$ in the half-angle identity $\tan \dfrac{x}{2} = \dfrac{\sin x}{1 + \cos x}$.

$$\tan(-15°) = \tan\left(\frac{-30°}{2}\right)$$

$$= \frac{\sin(-30°)}{1 + \cos(-30°)}$$

$$= \frac{-\dfrac{1}{2}}{1 + \dfrac{\sqrt{3}}{2}} = \frac{-1}{2 + \sqrt{3}} = \frac{-1(2 - \sqrt{3})}{(2 + \sqrt{3})(2 - \sqrt{3})}$$

$$= -2 + \sqrt{3}$$

Example 6   Verifying identities involving half-angles

Prove that the following equation is an identity:

$$\tan \frac{x}{2} + \cot \frac{x}{2} = 2 \csc x$$

Solution

Write the left-hand side in terms of $\tan(x/2)$ and then use two different half-angle identities for $\tan(x/2)$:

$$\tan\frac{x}{2} + \cot\frac{x}{2} = \tan\frac{x}{2} + \frac{1}{\tan\dfrac{x}{2}}$$

$$= \frac{\sin x}{1 + \cos x} + \frac{\sin x}{1 - \cos x}$$

$$= \frac{\sin x(1 - \cos x)}{(1 + \cos x)(1 - \cos x)} + \frac{\sin x(1 + \cos x)}{(1 - \cos x)(1 + \cos x)}$$

$$= \frac{2\sin x}{1 - \cos^2 x}$$

$$= \frac{2\sin x}{\sin^2 x}$$

$$= 2\frac{1}{\sin x}$$

$$= 2\csc x$$

Example 7   Using the identities

Find $\sin(\alpha/2)$, given that $\sin\alpha = 3/5$ and $0 < \alpha < \pi/2$.

Solution

Use the identity $\sin^2\alpha + \cos^2\alpha = 1$ to find $\cos\alpha$:

$$\left(\frac{3}{5}\right)^2 + \cos^2\alpha = 1$$

$$\cos^2\alpha = \frac{16}{25}$$

$$\cos\alpha = \pm\frac{4}{5}$$

For $0 < \alpha < \pi/2$, we have $\cos\alpha > 0$. So $\cos\alpha = 4/5$. Now use the half-angle identity for sine:

$$\sin\frac{\alpha}{2} = \pm\sqrt{\frac{1 - \cos\alpha}{2}} = \pm\sqrt{\frac{1 - \dfrac{4}{5}}{2}} = \pm\sqrt{\frac{1}{10}} = \pm\frac{\sqrt{10}}{10}$$

If $0 < \alpha < \pi/2$, then $0 < \alpha/2 < \pi/4$. So $\sin(\alpha/2) > 0$ and $\sin(\alpha/2) = \sqrt{10}/10$.

For Thought

True or false? Explain.

1. $\dfrac{\sin 42°}{2} = \sin 21° \cos 21°$

2. $\cos(\sqrt{8}) = 2\cos^2(\sqrt{2}) - 1$

3. $\sin 150° = \sqrt{\dfrac{1 - \cos 75°}{2}}$

4. $\sin 200° = -\sqrt{\dfrac{1 - \cos 40°}{2}}$

5. $\tan\dfrac{7\pi}{8} = \sqrt{\dfrac{1 - \cos(7\pi/4)}{1 + \cos(7\pi/4)}}$

6. $\tan(-\pi/8) = \dfrac{1 - \cos(\pi/4)}{\sin(-\pi/4)}$

7. For any real number x, $\dfrac{\sin 2x}{2} = \sin x$.

8. The equation $\cos x = \sqrt{\dfrac{1 + \cos 2x}{2}}$ is an identity.

9. The equation $\sqrt{(1 - \cos x)^2} = 1 - \cos x$ is an identity.

10. If $180° < \alpha < 360°$, then $\sin \alpha < 0$ and $\tan(\alpha/2) < 0$.

7.4 Exercises

Tape 14   Disk—5.25": 4   3.5": 3   Macintosh: 3

Find the exact value of each expression using double-angle identities.

1. $\sin(90°)$ 2. $\cos(60°)$ 3. $\tan(60°)$ 4. $\cos(180°)$

5. $\sin(3\pi/2)$ 6. $\cos(4\pi/3)$ 7. $\tan(4\pi/3)$ 8. $\sin(2\pi/3)$

Find the exact value of each expression using the half-angle identities.

9. $\sin(15°)$ 10. $\cos(15°)$ 11. $\tan(15°)$

12. $\sin(-\pi/6)$ 13. $\cos(\pi/8)$ 14. $\tan(3\pi/8)$

15. $\sin(22.5°)$ 16. $\tan(75°)$

For each equation determine whether the positive or negative sign makes the equation correct. Do not use a calculator.

17. $\sin 118.5° = \pm\sqrt{\dfrac{1 - \cos 237°}{2}}$

18. $\sin 222.5° = \pm\sqrt{\dfrac{1 - \cos 445°}{2}}$

19. $\cos 100° = \pm\sqrt{\dfrac{1 + \cos 200°}{2}}$

20. $\cos\dfrac{9\pi}{7} = \pm\sqrt{\dfrac{1 + \cos(18\pi/7)}{2}}$

21. $\tan\dfrac{-5\pi}{12} = \pm\sqrt{\dfrac{1 - \cos(-5\pi/6)}{1 + \cos(-5\pi/6)}}$

22. $\tan\dfrac{17\pi}{12} = \pm\sqrt{\dfrac{1 - \cos(17\pi/6)}{1 + \cos(17\pi/6)}}$

Use identities to simplify each expression. Do not use a calculator.

23. $2\sin 13° \cos 13°$

24. $\sin^2\left(\dfrac{\pi}{5}\right) - \cos^2\left(\dfrac{\pi}{5}\right)$

25. $2\cos^2(22.5°) - 1$

26. $1 - 2\sin^2\left(-\dfrac{\pi}{8}\right)$

27. $\dfrac{\tan 15°}{1 - \tan^2(15°)}$

28. $\dfrac{2}{\cot 5(1 - \tan^2 5)}$

29. $\dfrac{\tan 30°}{1 - \tan^2(30°)}$

30. $\cos^2\left(\dfrac{\pi}{9}\right) - \sin^2\left(\dfrac{\pi}{9}\right)$

31. $2 \sin\left(\dfrac{\pi}{9} - \dfrac{\pi}{2}\right)\cos\left(\dfrac{\pi}{2} - \dfrac{\pi}{9}\right)$

32. $2 \cos^2\left(\dfrac{\pi}{5} - \dfrac{\pi}{2}\right) - 1$

33. $\dfrac{\sin 12°}{1 + \cos 12°}$

34. $\csc 8°(1 - \cos 8°)$

Verify that each equation is an identity.

35. $\cos^4 s - \sin^4 s = \cos 2s$

36. $\sin 2s = -2 \sin s \sin(s - \pi/2)$

37. $\dfrac{\sin 4t}{4} = \cos^3 t \sin t - \sin^3 t \cos t$

38. $\cos 3t = \cos^3 t - 3 \sin^2 t \cos t$

39. $\dfrac{\cos 2x + \cos 2y}{\sin x + \cos y} = 2 \cos y - 2 \sin x$

40. $(\sin \alpha - \cos \alpha)^2 = 1 - \sin 2\alpha$

41. $\dfrac{\cos 2x}{\sin^2 x} = \csc^2 x - 2$

42. $\dfrac{\cos 2s}{\cos^2 s} = \sec^2 s - 2 \tan^2 s$

43. $2 \sin^2\left(\dfrac{u}{2}\right) = \dfrac{\sin^2 u}{1 + \cos u}$

44. $\cos 2y = \dfrac{1 - \tan^2 y}{1 + \tan^2 y}$

45. $\tan^2\left(\dfrac{x}{2}\right) = \dfrac{\sec x + \cos x - 2}{\sec x - \cos x}$

46. $\sec^2\left(\dfrac{x}{2}\right) = \dfrac{2 \sec x + 2}{\sec x + 2 + \cos x}$

47. $\dfrac{1 - \sin^2\left(\dfrac{x}{2}\right)}{1 + \sin^2\left(\dfrac{x}{2}\right)} = \dfrac{1 + \cos x}{3 - \cos x}$

48. $\dfrac{1 - \cos^2\left(\dfrac{x}{2}\right)}{1 - \sin^2\left(\dfrac{x}{2}\right)} = \dfrac{1 - \cos x}{1 + \cos x}$

In each case, find $\sin \alpha$, $\cos \alpha$, $\tan \alpha$, $\csc \alpha$, $\sec \alpha$, **and** $\cot \alpha$.

49. $\cos 2\alpha = 3/5$ and $0° < 2\alpha < 90°$

50. $\cos 2\alpha = 1/3$ and $360° < 2\alpha < 450°$

51. $\cos(\alpha/2) = -1/4$ and $\pi/2 < \alpha/2 < 3\pi/4$

52. $\sin(\alpha/2) = -1/3$ and $7\pi/4 < \alpha/2 < 2\pi$

53. $\sin(\alpha/2) = 4/5$ and $\alpha/2$ is in quadrant II

54. $\sin(\alpha/2) = 1/5$ and $\alpha/2$ is in quadrant II

Solve each problem.

55. In the figure, $\angle B$ is a right angle and the line segment $\overline{AD}$ bisects $\angle A$. If $AB = 5$ and $BC = 3$, what is the exact value of BD?

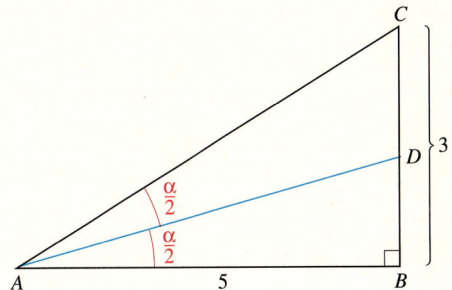

Figure for Exercise 55

56. In the figure, $\angle B$ is a right angle and the line segment $\overline{AD}$ bisects $\angle A$. If $AB = 10$ and $BD = 2$, what is the exact value of CD?

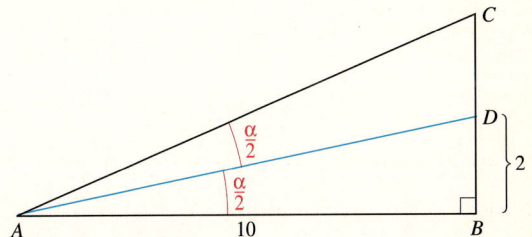

Figure for Exercise 56

Assuming no air resistance, the distance in feet that a projectile (such as a football) travels when launched from the ground with an initial velocity of v_0 ft/sec is given by $x = (v_0^2 \sin \theta \cos \theta)/16$, where θ is the initial angle between the trajectory and the ground.

57. *Maximum distance of a projectile* A field-goal kicker knows instinctively the angle at which to kick a football to achieve the maximum distance, but we can figure it out. Use an identity to write the distance x as a function of 2θ. Graph the function and determine the value of θ that maximizes x.

Figure for Exercises 57 and 58

58. *Initial velocity of a projectile* When Morten Anderson of the New Orleans Saints kicks a football 55 yd using the angle determined in the previous exercise, what is the initial velocity of the football in miles per hour? (To overcome the air resistance and travel 55 yd, the football must actually have a greater initial velocity than that determined here.)

For each equation, either prove that it is an identity or prove that it is not an identity.

59. $\sin(2x) = 2 \sin x$

60. $\dfrac{\cos 2x}{2} = \cos x$

61. $\tan\left(\dfrac{x}{2}\right) = \dfrac{1}{2} \tan x$

62. $\tan\left(\dfrac{x}{2}\right) = \sqrt{\dfrac{1 - \cos x}{1 + \cos x}}$

63. $\sin(2x) \cdot \sin\left(\dfrac{x}{2}\right) = \sin^2 x$

64. $\tan x + \tan x = \tan 2x$

65. $\cot \dfrac{x}{2} - \tan \dfrac{x}{2} = \dfrac{\sin 2x}{\sin^2 x}$

66. $\csc^2\left(\dfrac{x}{2}\right) + \sec^2\left(\dfrac{x}{2}\right) = 4 \csc^2 x$

For Writing/Discussion

67. Show that $\tan(x/2)$ has the same sign as $\sin x$ for any real number x.

68. Explain why $1 + \cos x \geq 0$ for any real number x.

69. Explain why $\sin 2x = 2 \sin x$ is not an identity by using graphs and by using the definition of the sine function.

7.5

Product and

Sum Identities

In Section 6.2 we saw that the location x at time t for an object on a spring, given an initial velocity v_0 from an initial location x_0, is given by

$$x = \frac{v_0}{\omega} \sin(\omega t) + x_0 \cos(\omega t),$$

where ω is a constant. Even though x is determined by a combination of sine and cosine, an object in motion on a spring oscillates in a "periodic" fashion. In this section we will develop identities that involve sums of trigonometric functions. These identities are used in solving equations, graphing functions, and explaining how the combination of the sine and cosine functions in the spring equation can work together to produce a periodic motion.

Product-to-Sum Identities

In Section 7.3 we learned identities for sine and cosine of a sum or difference:

$$\sin(A + B) = \sin A \cos B + \cos A \sin B$$

$$\sin(A - B) = \sin A \cos B - \cos A \sin B$$

$$\cos(A + B) = \cos A \cos B - \sin A \sin B$$

$$\cos(A - B) = \cos A \cos B + \sin A \sin B$$

Now we can add the first two of these identities to get a new identity:

$$\sin(A + B) = \sin A \cos B + \cos A \sin B$$
$$\underline{\sin(A - B) = \sin A \cos B - \cos A \sin B}$$
$$\sin(A + B) + \sin(A - B) = 2 \sin A \cos B$$

When we divide each side by 2, we get an identity expressing a product in terms of a sum:

$$\sin A \cos B = \frac{1}{2}[\sin(A + B) + \sin(A - B)]$$

We can produce three other, similar identities from the sum and difference identities. The four identities, known as the **product-to-sum identities** are listed below. These identities are not used as often as the other identities that we study in this chapter. It is not necessary to memorize these identities. Just remember them by name and look them up as necessary.

Product-to-Sum Identities

$$\sin A \cos B = \frac{1}{2}[\sin(A + B) + \sin(A - B)]$$

$$\sin A \sin B = \frac{1}{2}[\cos(A - B) - \cos(A + B)]$$

$$\cos A \sin B = \frac{1}{2}[\sin(A + B) - \sin(A - B)]$$

$$\cos A \cos B = \frac{1}{2}[\cos(A - B) + \cos(A + B)]$$

Example 1   Expressing a product as a sum

Use the product-to-sum identities to rewrite each expression:

a) $\sin 12° \cos 9°$

b) $\sin(\pi/12) \sin(\pi/8)$

Solution

a) Use the product-to-sum identity for $\sin A \cos B$:

$$\sin 12° \cos 9° = \frac{1}{2}[\sin(12° + 9°) + \sin(12° - 9°)]$$

$$= \frac{1}{2}[\sin 21° + \sin 3°]$$

b) Use the product-to-sum identity for $\sin A \sin B$:

$$\sin\left(\frac{\pi}{12}\right)\sin\left(\frac{\pi}{8}\right) = \frac{1}{2}\left[\cos\left(\frac{\pi}{12} - \frac{\pi}{8}\right) - \cos\left(\frac{\pi}{12} + \frac{\pi}{8}\right)\right]$$

$$= \frac{1}{2}\left[\cos\left(-\frac{\pi}{24}\right) - \cos\left(\frac{5\pi}{24}\right)\right]$$

$$= \frac{1}{2}\left[\cos\frac{\pi}{24} - \cos\frac{5\pi}{24}\right]$$

◆

Example 2 Evaluating a product

Use a product-to-sum identity to find the exact value of $\cos(67.5°) \sin(112.5°)$.

Solution

Use the identity $\cos A \sin B = \dfrac{1}{2}[\sin(A + B) - \sin(A - B)]$:

$$\cos(67.5°) \sin(112.5°) = \frac{1}{2}[\sin(67.5° + 112.5°) - \sin(67.5° - 112.5°)]$$

$$= \frac{1}{2}[\sin(180°) - \sin(-45°)]$$

$$= \frac{1}{2}[0 + \sin(45°)]$$

$$= \frac{1}{2} \cdot \frac{\sqrt{2}}{2} = \frac{\sqrt{2}}{4}$$

◆

Sum-to-Product Identities

It is sometimes useful to write a sum of two trigonometric functions as a product. Consider the product-to-sum formula

$$\sin A \cos B = \frac{1}{2}[\sin(A + B) + \sin(A - B)].$$

To make the right-hand side look simpler, we let $A + B = x$ and $A - B = y$. From these equations we get $x + y = 2A$ and $x - y = 2B$, or

$$A = \frac{x + y}{2} \qquad \text{and} \qquad B = \frac{x - y}{2}.$$

Substitute these values into the identity for $\sin A \cos B$:

$$\sin\left(\frac{x + y}{2}\right)\cos\left(\frac{x - y}{2}\right) = \frac{1}{2}[\sin x + \sin y]$$

Multiply each side by 2 to get

$$\sin x + \sin y = 2 \sin\left(\frac{x + y}{2}\right)\cos\left(\frac{x - y}{2}\right).$$

The other three product-to-sum formulas can also be rewritten as sum-to-product identities by using similar procedures.

Sum-to-Product Identities

$$\sin x + \sin y = 2 \sin\left(\frac{x + y}{2}\right)\cos\left(\frac{x - y}{2}\right)$$

$$\sin x - \sin y = 2 \cos\left(\frac{x + y}{2}\right)\sin\left(\frac{x - y}{2}\right)$$

$$\cos x + \cos y = 2 \cos\left(\frac{x + y}{2}\right)\cos\left(\frac{x - y}{2}\right)$$

$$\cos x - \cos y = -2 \sin\left(\frac{x + y}{2}\right)\sin\left(\frac{x - y}{2}\right)$$

Example 3   Expressing a sum or difference as a product

Use the sum-to-product identities to rewrite each expression.

a) $\sin 8° + \sin 6°$ **b)** $\cos(\pi/5) - \cos(\pi/8)$ **c)** $\sin(6t) - \sin(4t)$

Solution

a) Use the sum-to-product identity for $\sin x + \sin y$:

$$\sin 8° + \sin 6° = 2 \sin\left(\frac{8° + 6°}{2}\right)\cos\left(\frac{8° - 6°}{2}\right) = 2 \sin 7° \cos 1°$$

b) Use the sum-to-product identity for $\cos x - \cos y$:

$$\cos\left(\frac{\pi}{5}\right) - \cos\left(\frac{\pi}{8}\right) = -2 \sin\left(\frac{\pi/5 + \pi/8}{2}\right)\sin\left(\frac{\pi/5 - \pi/8}{2}\right)$$

$$= -2 \sin\left(\frac{13\pi}{80}\right)\sin\left(\frac{3\pi}{80}\right)$$

c) Use the sum-to-product identity for $\sin x - \sin y$:

$$\sin 6t - \sin 4t = 2 \cos\left(\frac{6t + 4t}{2}\right)\sin\left(\frac{6t - 4t}{2}\right)$$

$$= 2 \cos 5t \sin t$$

Example 4 Evaluating a sum

Use a sum-to-product identity to find the exact value of $\cos(112.5°) + \cos(67.5°)$.

Solution

Use the sum-to-product identity $\cos x + \cos y = 2 \cos\left(\dfrac{x+y}{2}\right)\cos\left(\dfrac{x-y}{2}\right)$:

$$\cos(112.5°) + \cos(67.5°) = 2 \cos\left(\frac{112.5° + 67.5°}{2}\right)\cos\left(\frac{112.5° - 67.5°}{2}\right)$$

$$= 2 \cos(90°)\cos(22.5°)$$

$$= 0 \qquad \text{Because } \cos(90°) = 0.$$

The Function $y = a \sin x + b \cos x$

The function $y = a \sin x + b \cos x$ involves an expression similar to those in the sum-to-product identities, but it is not covered by the sum-to-product identities. Functions of this type occur in applications such as the position of a weight in motion due to the force of a spring, the position of a swinging pendulum, and the current in an electrical circuit. In these applications it is important to express this function in terms of a single trigonometric function.

To get an identity for $a \sin x + b \cos x$ for nonzero values of a and b, we let α be an angle in standard position whose terminal side contains the point (a, b). Using trigonometric ratios from Section 6.6, we get

$$\sin \alpha = \frac{b}{r} = \frac{b}{\sqrt{a^2 + b^2}} \qquad \text{and} \qquad \cos \alpha = \frac{a}{r} = \frac{a}{\sqrt{a^2 + b^2}}.$$

Now write $a \sin x + b \cos x$ with the trigonometric ratios, and then replace the ratios by $\sin \alpha$ and $\cos \alpha$:

$$a \sin x + b \cos x = \sqrt{a^2 + b^2}\left((\sin x)\frac{a}{\sqrt{a^2 + b^2}} + (\cos x)\frac{b}{\sqrt{a^2 + b^2}}\right)$$

$$= \sqrt{a^2 + b^2}\,(\sin x \cos \alpha + \cos x \sin \alpha)$$

$$= \sqrt{a^2 + b^2}\,\sin(x + \alpha) \qquad \text{By the identity for the sine of a sum}$$

This identity is called the **reduction formula** because it reduces two trigonometric functions to one.

Theorem: Reduction Formula

If α is an angle in standard position whose terminal side contains (a, b), then

$$a \sin x + b \cos x = \sqrt{a^2 + b^2}\,\sin(x + \alpha) \qquad \text{for any real number } x.$$

To rewrite an expression of the form $a \sin x + b \cos x$ by using the reduction formula, we need to find α so that the terminal side of α goes through (a, b). By using trigonometric ratios, we have

$$\sin \alpha = \frac{b}{\sqrt{a^2 + b^2}}, \qquad \cos \alpha = \frac{a}{\sqrt{a^2 + b^2}}, \qquad \text{and} \qquad \tan \alpha = \frac{a}{b}.$$

Since we know a and b, we can find $\sin \alpha$, $\cos \alpha$, or $\tan \alpha$, and then use an inverse trigonometric function to find α. However, because of the ranges of the inverse functions, the angle obtained from an inverse function might not have its terminal side through (a, b) as required.

Example 5 Using the reduction formula

Use the reduction formula to rewrite $-3 \sin x - 3 \cos x$ in the form $A \sin(x + C)$.

Solution

Because $a = -3$ and $b = -3$, we have

$$\sqrt{a^2 + b^2} = \sqrt{18} = 3\sqrt{2}.$$

Since the terminal side of α must go through $(-3, -3)$, we have

$$\cos \alpha = \frac{-3}{3\sqrt{2}} = -\frac{\sqrt{2}}{2}.$$

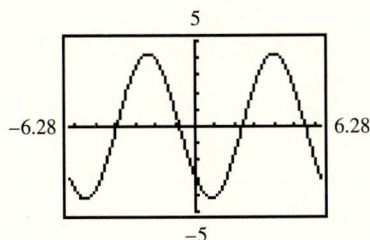

The graph of $y = -3 \sin x - 3 \cos x$ shown here is a sine wave because of the reduction formula.

Now $\cos^{-1}\left(-\sqrt{2}/2\right) = 3\pi/4$, but the terminal side of $3\pi/4$ is in quadrant II, as shown in Fig. 7.4. However, we also have $\cos(5\pi/4) = -\sqrt{2}/2$ and the terminal side for $5\pi/4$ does pass through $(-3, -3)$ in quadrant III. So $\alpha = 5\pi/4$. By the reduction formula

$$-3 \sin x - 3 \cos x = 3\sqrt{2} \sin\left(x + \frac{5\pi}{4}\right).$$

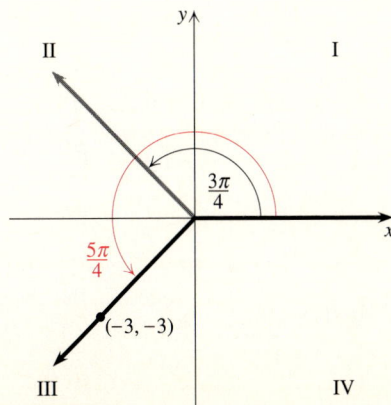

Figure 7.4

In the next example, we see that the reduction formula can make it easier to graph a function of the form $y = a \sin x + b \cos x$.

Example 6   Using the reduction formula in graphing

Graph one cycle of the function

$$y = \sqrt{3} \sin x + \cos x$$

and state the amplitude, period, and phase shift.

Solution

Because $a = \sqrt{3}$ and $b = 1$, we have $\sqrt{a^2 + b^2} = \sqrt{4} = 2$. The terminal side of α must go through $(\sqrt{3}, 1)$. So $\sin \alpha = 1/2$, and $\sin^{-1}(1/2) = \pi/6$. Since the terminal side of $\pi/6$ is in quadrant I and goes through $(\sqrt{3}, 1)$, we can rewrite the function as

$$y = \sqrt{3} \sin x + \cos x = 2 \sin\left(x + \frac{\pi}{6}\right).$$

From the new form, we see that the graph is a sine wave with amplitude 2, period 2π, and phase shift $\pi/6$ to the left, as shown in Fig. 7.5.

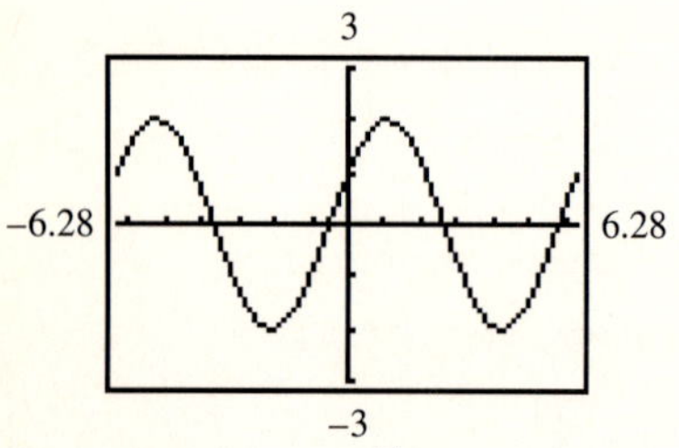

You can graph $y = \sqrt{3} \sin x + \cos x$ on your calculator, but you need the form $y = 2 \sin(x + \pi/6)$ to find the amplitude, period, and phase shift.

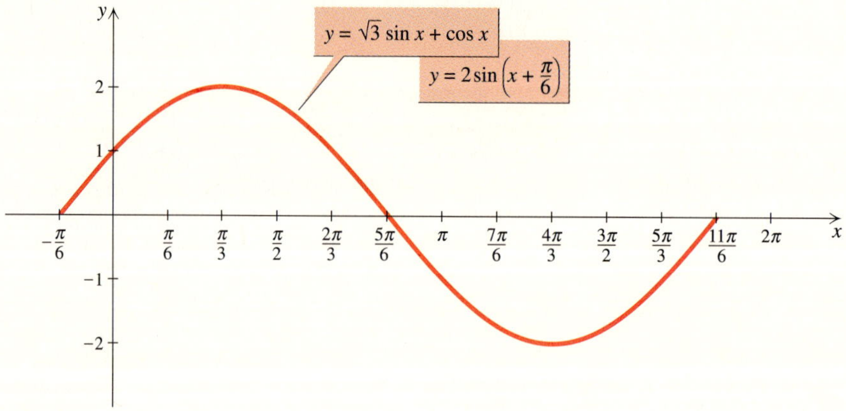

Figure 7.5

Since $y = a \sin x + b \cos x$ and $y = \sqrt{a^2 + b^2} \sin(x + \alpha)$ are the same function, the graph of $y = a \sin x + b \cos x$ is a transformation of $y = \sin x$. So, for nonzero a and b, the graph of $y = a \sin x + b \cos x$ is a sine wave with amplitude $\sqrt{a^2 + b^2}$, period 2π, and phase shift $|\alpha|$.

Applications

In Section 6.2 the function

$$x = \frac{v_0}{\omega} \sin(\omega t) + x_0 \cos(\omega t)$$

was used to give the position x at time t for a weight in motion on a vertical spring. In this formula, v_0 is the initial velocity, x_0 is the initial position, and ω is a constant. The same equation is used for a weight attached to a horizontal spring and set in motion on a frictionless surface, as shown in Fig. 7.6.

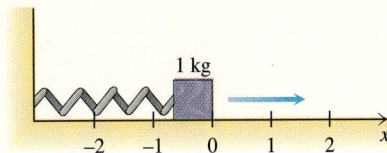

Figure 7.6

We use the reduction formula on the spring equation to determine how far a spring actually stretches and compresses when the block is set in motion with a particular velocity.

Example 7   Using the reduction formula with springs

At time $t = 0$ the 1-kilogram block shown in Fig. 7.6 is moved (and the spring compressed) to a position 1 meter to the left of the resting position. From this position the block is given a velocity of 2 meters per second to the right. Use the reduction formula to find the maximum distance reached by the block from rest. Assume that $\omega = 1$.

Solution

Since $v_0 = 2$, $x_0 = -1$, and $\omega = 1$, the position of the block at time t is given by

$$x = 2 \sin t - \cos t.$$

Use the reduction formula to rewrite this equation. If $a = 2$ and $b = -1$, then

$$\sqrt{a^2 + b^2} = \sqrt{2^2 + (-1)^2} = \sqrt{5}.$$

If the terminal side of α goes through $(2, -1)$, then $\tan \alpha = -1/2$ and

$$\tan^{-1}(-1/2) \approx -0.46.$$

An angle of -0.46 rad lies in quadrant IV and goes through $(2, -1)$. Use $\alpha = -0.46$ in the reduction formula to get

$$x = \sqrt{5} \sin(t - 0.46).$$

Since the amplitude of this function is $\sqrt{5}$, the block oscillates between $x = \sqrt{5}$ meters and $x = -\sqrt{5}$ meters. The maximum distance from $x = 0$ is $\sqrt{5}$ meters. ◆

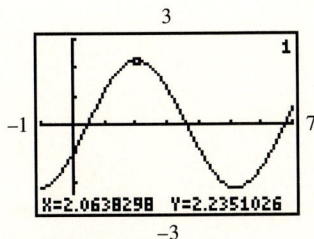

Use TRACE on the graph of $y = 2 \sin x - \cos x$ to see that the amplitude is about 2.2.

? For Thought

True or false? Explain.

1. $\sin 45° \cos 15° = 0.5[\sin 60° + \sin 30°]$

2. $\cos(\pi/8)\sin(\pi/4) = 0.5[\sin(3\pi/8) - \sin(\pi/8)]$

3. $2 \cos 6° \cos 8° = \cos 2° + \cos 14°$

4. $\sin 5° - \sin 9° = 2 \cos 7° \sin 2°$

5. $\cos 4 + \cos 12 = 2 \cos 8 \cos 4$

6. $\cos(\pi/3) - \cos(\pi/2) = -2 \sin(5\pi/12)\sin(\pi/12)$

7. $\sin(\pi/6) + \cos(\pi/6) = \sqrt{2} \sin(\pi/6 + \pi/4)$

8. $\dfrac{1}{2} \sin(\pi/6) + \dfrac{\sqrt{3}}{2} \cos(\pi/6) = \sin(\pi/2)$

9. The graph of $y = \dfrac{1}{2} \sin x + \dfrac{\sqrt{3}}{2} \cos x$ is a sine wave with amplitude 1.

10. The equation $\dfrac{1}{\sqrt{2}} \sin x + \dfrac{1}{\sqrt{2}} \cos x = \sin\left(x + \dfrac{\pi}{4}\right)$ is an identity.

7.5 Exercises

Tape 14   Disk—5.25″: 5   3.5″: 3   Macintosh: 3

Use the product-to-sum identities to rewrite each expression.

1. $\sin 13° \sin 9°$ 2. $\cos 34° \cos 39°$ 3. $\sin 16° \cos 20°$

4. $\cos 9° \sin 8°$

5. $\cos\left(\dfrac{\pi}{6}\right)\cos\left(\dfrac{\pi}{5}\right)$

6. $\sin\left(\dfrac{2\pi}{9}\right)\sin\left(\dfrac{3\pi}{4}\right)$

7. $\cos(5y^2)\cos(7y^2)$

8. $\cos 3t \sin 5t$

9. $\sin(2s - 1)\cos(s + 1)$

10. $\sin(3t - 1)\sin(2t + 3)$

Find the exact value of each product.

11. $\sin(52.5°)\sin(7.5°)$ 12. $\cos(105°)\cos(75°)$

13. $\sin\left(\dfrac{13\pi}{24}\right)\cos\left(\dfrac{5\pi}{24}\right)$ 14. $\cos\left(\dfrac{5\pi}{24}\right)\sin\left(-\dfrac{\pi}{24}\right)$

Use the sum-to-product identities to rewrite each expression.

15. $\sin 12° - \sin 8°$ 16. $\sin 7° + \sin 11°$

17. $\cos 80° - \cos 87°$ 18. $\cos 44° + \cos 31°$

19. $\sin 3.6 - \sin 4.8$ 20. $\sin 5.1 + \sin 6.3$

21. $\cos(5y - 3) - \cos(3y + 9)$

22. $\cos(6t^2 - 1) + \cos(4t^2 - 1)$

23. $\sin 5\alpha - \sin 8\alpha$ 24. $\sin 3s + \sin 5s$

25. $\cos\left(\dfrac{\pi}{3}\right) - \cos\left(\dfrac{\pi}{5}\right)$

26. $\cos\left(\dfrac{1}{2}\right) + \cos\left(\dfrac{2}{3}\right)$

Find the exact value of each sum.

27. $\sin(75°) + \sin(15°)$ 28. $\sin(285°) - \sin(15°)$

29. $\cos\left(-\dfrac{\pi}{24}\right) - \cos\left(\dfrac{7\pi}{24}\right)$

30. $\cos\left(\dfrac{5\pi}{24}\right) + \cos\left(\dfrac{\pi}{24}\right)$

Rewrite each expression in the form $A \sin(x + C)$.

31. $\sin x - \cos x$ 32. $2 \sin x + 2 \cos x$

33. $-\dfrac{1}{2}\sin x + \dfrac{\sqrt{3}}{2}\cos x$

34. $\dfrac{\sqrt{2}}{2}\sin x - \dfrac{\sqrt{2}}{2}\cos x$

35. $\dfrac{\sqrt{3}}{2}\sin x - \dfrac{1}{2}\cos x$

36. $-\dfrac{\sqrt{3}}{2}\sin x - \dfrac{1}{2}\cos x$

Write each function in the form $y = A\sin(x + C)$. Then graph the function for $-2\pi \le x \le 2\pi$ and state the amplitude, period, and phase shift.

37. $y = -\sin x + \cos x$

38. $y = \sin x + \sqrt{3}\cos x$

39. $y = \sqrt{2}\sin x - \sqrt{2}\cos x$

40. $y = 2\sin x - 2\cos x$

41. $y = -\sqrt{3}\sin x - \cos x$

42. $y = -\dfrac{1}{2}\sin x - \dfrac{\sqrt{3}}{2}\cos x$

For each function, determine the exact amplitude and find the phase shift in radians (to the nearest tenth).

43. $y = 3\sin x + 4\cos x$

44. $y = \sin x + 5\cos x$

45. $y = -6\sin x + \cos x$

46. $y = -\sqrt{5}\sin x + 2\cos x$

47. $y = -3\sin x - 5\cos x$

48. $y = -\sqrt{2}\sin x - \sqrt{7}\cos x$

Prove that each equation is an identity.

49. $\dfrac{\sin 3t - \sin t}{\cos 3t + \cos t} = \tan t$

50. $\dfrac{\sin 3x + \sin 5x}{\sin 3x - \sin 5x} = -\dfrac{\tan 4x}{\tan x}$

51. $\dfrac{\cos x - \cos 3x}{\cos x + \cos 3x} = \tan 2x \tan x$

52. $\dfrac{\cos 5y + \cos 3y}{\cos 5y - \cos 3y} = -\cot 4y \cot y$

53. $\cos^2 x - \cos^2 y = -\sin(x + y)\sin(x - y)$

54. $\sin^2 x - \sin^2 y = \sin(x + y)\sin(x - y)$

55. $\left(\sin\dfrac{x + y}{2} + \cos\dfrac{x + y}{2}\right)\left(\sin\dfrac{x - y}{2} + \cos\dfrac{x - y}{2}\right)$
$= \sin x + \cos y$

56. $\sin 2A \sin 2B = \sin^2(A + B) - \sin^2(A - B)$

57. $\cos^2(A - B) - \cos^2(A + B) = \sin^2(A + B) - \sin^2(A - B)$

58. $(\sin A + \cos A)(\sin B + \cos B)$
$= \sin(A + B) + \cos(A - B)$

Solve each problem.

59. *Motion of a spring* A block is attached to a spring and set in motion, as in Example 7. For this block and spring, the location on the surface at any time t in seconds is given in meters by $x = \sqrt{3}\sin t + \cos t$. Use the reduction formula to rewrite this function and find the maximum distance reached by the block from the resting position.

60. *Motion of a spring* A block hanging from a spring, as shown in the figure, oscillates in the same manner as the block of Example 7. If a 1-kg block attached to a certain spring is given an upward velocity of 0.3 m/sec from a point 0.5 m below its resting position, then its position at any time t in seconds is given in meters by $x = -0.3\sin t + 0.5\cos t$. Use the reduction formula to find the maximum distance that the block travels from the resting position.

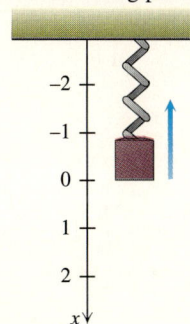

Figure for Exercise 60

For Writing/Discussion

61. Derive the identity $\cos(2x) = \cos^2 x - \sin^2 x$ from a product-to-sum identity.

62. Derive the identity $\sin(2x) = 2\sin x \cos x$ from a product-to-sum identity.

63. Prove the three product-to-sum identities that were not proved in the text.

64. Prove the three sum-to-product identities that were not proved in the text.

7.6

Conditional Equations

An identity is satisfied by *all* values of the variable for which both sides are defined. In this section we investigate **conditional equations,** those equations that are not identities but that have *at least one* solution. For example, the equation $\sin x = 0$ is a conditional equation. Because the trigonometric functions are periodic, conditional equations involving trigonometric functions usually have infinitely many solutions. The equation $\sin x = 0$ is satisfied by $x = 0$, $\pm\pi$, $\pm2\pi$, $\pm3\pi$, and so on. All of these solutions are of the form $k\pi$, where k is an integer. In this section we will solve conditional equations and see that identities play a fundamental role in their solution.

Cosine Equations

The most basic conditional equation involving cosine is of the form $\cos x = a$, where a is a number in the interval $[-1, 1]$. If a is not in $[-1, 1]$, $\cos x = a$ has no solution. For a in $[-1, 1]$, the equation $x = \cos^{-1} a$ provides one solution in the interval $[0, \pi]$. From this single solution we can determine all of the solutions because of the periodic nature of the cosine function. We must remember also that the cosine of an arc is the x-coordinate of its terminal point on the unit circle. So arcs that terminate at opposite ends of a vertical chord in the unit circle have the same cosine.

Example 1   Solving a cosine equation

Find all real numbers that satisfy each equation.

a) $\cos x = 1$ **b)** $\cos x = 0$ **c)** $\cos x = -1/2$

Solution

a) One solution to $\cos x = 1$ is

$$x = \cos^{-1}(1) = 0.$$

Since the period of cosine is 2π, any integral multiple of 2π can be added to this solution to get additional solutions. So the equation is satisfied by 0, $\pm2\pi$, $\pm4\pi$, and so on. Now $\cos x = 1$ is satisfied only if the arc of length x on the unit circle has terminal point $(1, 0)$, as shown in Fig. 7.7. So there are no more solutions. The solution set is written as

$$\{x \mid x = 2k\pi\},$$

where k is any integer.

b) One solution to $\cos x = 0$ is

$$x = \cos^{-1}(0) = \frac{\pi}{2}.$$

The terminal point for the arc of length $\pi/2$ is $(0, 1)$. So any arc length of the form $\pi/2 + 2k\pi$ is a solution to $\cos x = 0$. However, an arc of length x that

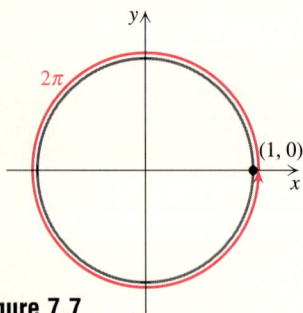

Figure 7.7

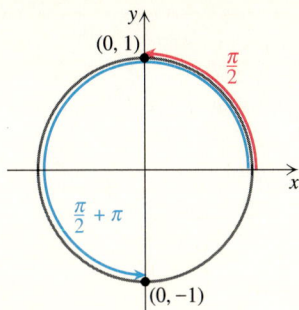

Figure 7.8

terminates at $(0, -1)$ also satisfies $\cos x = 0$, as shown in Fig. 7.8, and these arcs are not included in the form $\pi/2 + 2k\pi$. Since the distance between $(0, 1)$ and $(0, -1)$ on the unit circle is π, all arcs that terminate at these points are of the form $\pi/2 + k\pi$. So the solution set is

$$\left\{ x \mid x = \frac{\pi}{2} + k\pi \right\},$$

where k is any integer.

c) One solution to the equation $\cos x = -1/2$ is

$$x = \cos^{-1}\left(-\frac{1}{2}\right) = \frac{2\pi}{3}.$$

Since the period of cosine is 2π, all arcs of length $2\pi/3 + 2k\pi$ (where k is any integer) satisfy the equation. The arc of length $4\pi/3$ also terminates at a point with x-coordinate $-1/2$ as shown in Fig. 7.9.

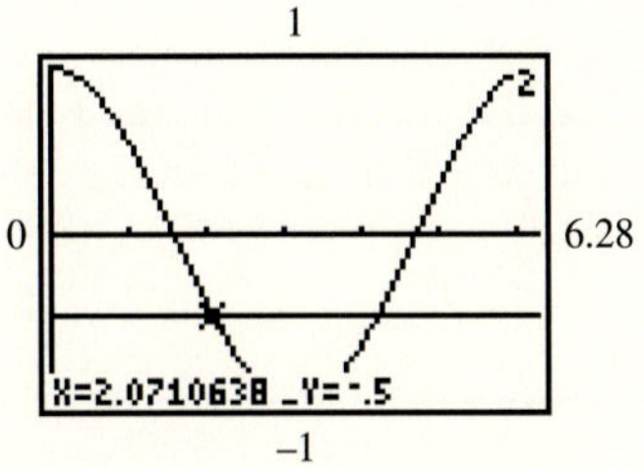

The graphs of $y_1 = \cos x$ and $y_2 = -1/2$ show that $\cos x = -1/2$ twice in the interval $[0, 2\pi]$, at $2\pi/3$ and $4\pi/3$.

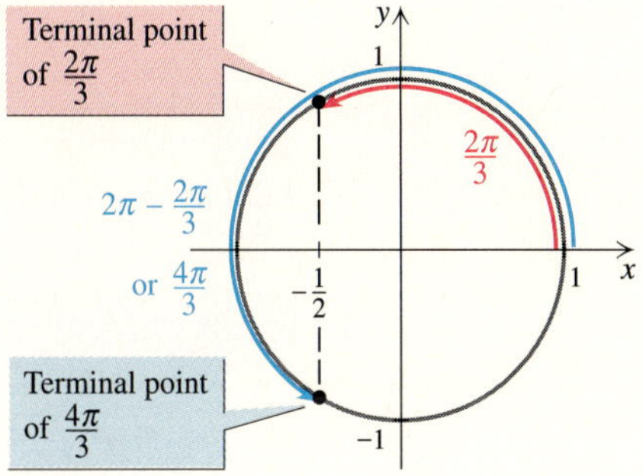

Figure 7.9

So $4\pi/3$ is also a solution to $\cos x = -1/2$, but it is not included in the form $2\pi/3 + 2k\pi$. So the solution set to $\cos x = -1/2$ is

$$\left\{ x \mid x = \frac{2\pi}{3} + 2k\pi \quad \text{or} \quad x = \frac{4\pi}{3} + 2k\pi \right\},$$

where k is any integer. Note that the arcs of length $2\pi/3$ and $4\pi/3$ terminate at opposite ends of a vertical chord in the unit circle.

The procedure used in Example 1(c) can be used to solve $\cos x = a$ for any nonzero a between -1 and 1. First find two values of x between 0 and 2π that satisfy the equation. (In general, $s = \cos^{-1}(a)$ and $2\pi - s$ are the values.) Next write all solutions by adding $2k\pi$ to the first two solutions. The solution sets for $a = -1$, 0, and 1 are easier to find and remember. All of the cases for solving $\cos x = a$ are summarized on the next page. The letter k is used to represent any integer. You should not simply memorize this summary, but you should be able to solve $\cos x = a$ for any real number a with the help of a unit circle.

▼

> **Summary: Solving cos $x = a$**
>
> 1. If $-1 < a < 1$ and $a \neq 0$, then the solution set to $\cos x = a$ is $\{x | x = s + 2k\pi \text{ or } x = 2\pi - s + 2k\pi\}$, where $s = \cos^{-1} a$.
> 2. The solution set to $\cos x = 1$ is $\{x | x = 2k\pi\}$.
> 3. The solution set to $\cos x = 0$ is $\{x | x = \pi/2 + k\pi\}$.
> 4. The solution set to $\cos x = -1$ is $\{x | x = \pi + 2k\pi\}$.
> 5. If $|a| > 1$, then $\cos x = a$ has no solution.

Sine Equations

To solve $\sin x = a$, we use the same strategy that we used to solve $\cos x = a$. We look for the smallest nonnegative solution and then build on it to obtain all solutions. However, the "first" solution obtained from $s = \sin^{-1} a$ might be negative because the range for the function $\sin^{-1}$ is $[-\pi/2, \pi/2]$. In this case $s + 2\pi$ is positive and we build on it to find all solutions. Remember that the sine of an arc on the unit circle is the y-coordinate of the terminal point of the arc. So arcs that terminate at opposite ends of a horizontal chord in the unit circle have the same y-coordinate for the terminal point and the same sine. This fact is the key to finding all solutions of a sine equation.

▬▬▬▬

Example 2   Solving a sine equation

Find all real numbers that satisfy $\sin x = -1/2$.

Solution

One solution to $\sin x = -1/2$ is

$$x = \sin^{-1}\left(-\frac{1}{2}\right) = -\frac{\pi}{6}.$$

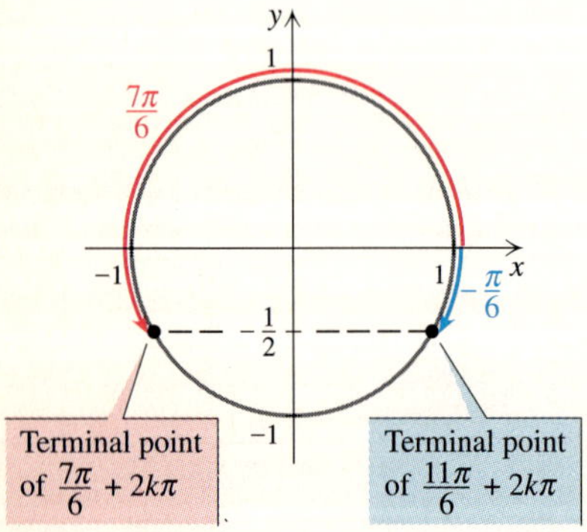

See Fig. 7.10. The smallest positive arc with the same terminal point as $-\pi/6$ is $-\pi/6 + 2\pi = 11\pi/6$. Since the period of the sine function is 2π, all arcs of the form $11\pi/6 + 2k\pi$ have the same terminal point and satisfy $\sin x = -1/2$. The smallest positive arc that terminates at the other end of the horizontal chord shown in Fig. 7.10 is $\pi - (-\pi/6)$ or $7\pi/6$. So $7\pi/6$ also satisfies the equation but it is not included in the form $11\pi/6 + 2k\pi$. So the solution set is

$$\left\{ x \,\middle|\, x = \frac{7\pi}{6} + 2k\pi \quad \text{or} \quad x = \frac{11\pi}{6} + 2k\pi \right\},$$

where k is any integer.   ◆

$\frac{7\pi}{6}$

Terminal point of $\frac{7\pi}{6} + 2k\pi$

Terminal point of $\frac{11\pi}{6} + 2k\pi$

Figure 7.10

The procedure used in Example 2 can be used to solve $\sin x = a$ for any nonzero a between -1 and 1. First find two values between 0 and 2π that satisfy the equation. (In general, $s = \sin^{-1} a$ and $\pi - s$ work when s is positive; $s + 2\pi$ and $\pi - s$ work when s is negative.) Next write all solutions by adding $2k\pi$ to the first two solutions. The equations $\sin x = 0$, $\sin x = 1$, and $\sin x = -1$ have solutions that are similar to the corresponding cosine equations. We can summarize the different solution sets to $\sin x = a$ as follows. You should not simply memorize this summary, but you should be able to solve $\sin x = a$ for any real number a with the help of a unit circle.

Summary: Solving $\sin x = a$

1. **If $-1 < a < 1$, $a \neq 0$, and $s = \sin^{-1} a$, then the solution set to $\sin x = a$ is**

    $\{x \mid x = s + 2k\pi \text{ or } x = \pi - s + 2k\pi\}$ **for $s > 0$,**

    $\{x \mid x = s + 2\pi + 2k\pi \text{ or } x = \pi - s + 2k\pi\}$ **for $s < 0$.**

2. **The solution set to $\sin x = 1$ is $\{x \mid x = \pi/2 + 2k\pi\}$.**
3. **The solution set to $\sin x = 0$ is $\{x \mid x = k\pi\}$.**
4. **The solution set to $\sin x = -1$ is $\{x \mid x = 3\pi/2 + 2k\pi\}$.**
5. **If $|a| > 1$, then $\sin x = a$ has no solution.**

Tangent Equations

Tangent equations are a little simpler than sine and cosine equations, because the tangent function is one-to-one in its fundamental cycle, while sine and cosine are not one-to-one in their fundamental cycles. In the next example, we see that the solution set to $\tan x = a$ consists of any single solution plus multiples of π.

Example 3   Solving a tangent equation

Find all solutions, in degrees.

a) $\tan \alpha = 1$ **b)** $\tan \alpha = -1.34$

Solution

a) Since $\tan^{-1}(1) = 45°$ and the period of tangent is $180°$, all angles of the form $45° + k180°$ satisfy the equation. There are no additional angles that satisfy the equation. The solution set to $\tan \alpha = 1$ is

$$\{\alpha \mid \alpha = 45° + k180°\}.$$

b) Since $\tan^{-1}(-1.34) = -53.5°$, one solution is $\alpha = -53.3°$. Since all solutions to $\tan \alpha = -1.34$ differ by a multiple of $180°$, $-53.3° + 180° = 126.7°$ is the smallest positive solution. So the solution set is

$$\{\alpha \mid \alpha = 126.7° + k180°\}. \quad \blacklozenge$$

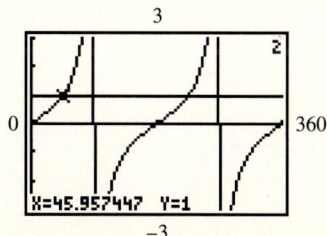

The graphs of $y_1 = \tan x$ and $y_2 = 1$ show that the solutions to $\tan x = 1$ are $180°$ apart.

To solve $\tan x = a$ for any real number a, we first find the smallest nonnegative solution. (In general, $s = \tan^{-1} a$ works if $s > 0$ and $s + \pi$ works if $s < 0$.) Next add on all integral multiples of π. The solution to the equation $\tan x = a$ is summarized as follows.

Summary: Solving $\tan x = a$

If a is any real number and $s = \tan^{-1} a$, then the solution set to $\tan x = a$ is $\{x | x = s + k\pi\}$ for $s \geq 0$, and $\{x | x = s + \pi + k\pi\}$ for $s < 0$.

In the summaries for solving sine, cosine, and tangent equations, the domain of x is the set of real numbers. Similar summaries can be made if the domain of x is the set of degree measures. Start by finding $\sin^{-1} a$, $\cos^{-1} a$, or $\tan^{-1} a$ in degrees, then write the solution set in terms of multiples of $180°$ or $360°$.

Equations Involving Multiple Angles

Equations can involve expressions such as $\sin 2x$, $\cos 3\alpha$, or $\tan(s/2)$. These expressions involve a multiple of the variable rather than a single variable such as x, α, or s. In this case we solve for the multiple just as we would solve for a single variable and then find the value of the single variable in the last step of the solution.

Example 4   A sine equation involving a double angle

Find all solutions in degrees to $\sin 2\alpha = 1/\sqrt{2}$.

Solution

The only values for 2α between $0°$ and $360°$ that satisfy the equation are $45°$ and $135°$. (Note that $\sin^{-1}(1/\sqrt{2}) = 45°$ and $135° = 180° - 45°$.) See Fig. 7.11.

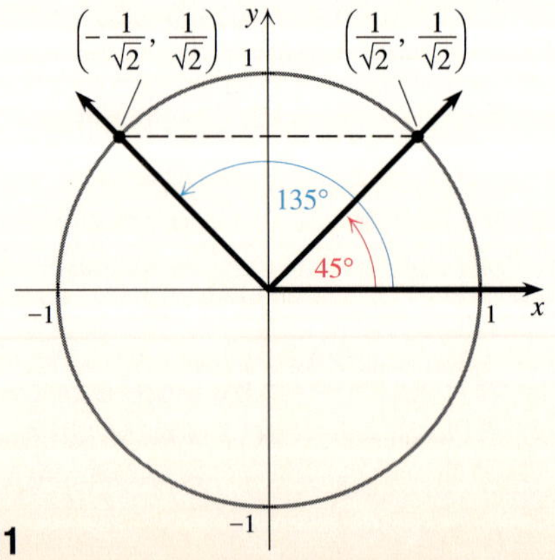

Figure 7.11

So proceed as follows:

$$\sin 2\alpha = \frac{1}{\sqrt{2}}$$

$$2\alpha = 45° + k360° \qquad \text{or} \qquad 2\alpha = 135° + k360°$$

$$\alpha = 22.5° + k180° \qquad \text{or} \qquad \alpha = 67.5° + k180° \qquad \color{blue}{\text{Divide each side by 2.}}$$

The solution set is $\{\alpha \mid \alpha = 22.5° + k180°$ or $\alpha = 67.5° + k180°\}$, where k is any integer.   ◆

Note that in Example 4, all possible values for 2α are found and *then* divided by 2 to get all possible values for α. Finding $\alpha = 22.5°$ and then adding on multiples of $360°$ will not produce the same solutions. Observe the same procedure in the next example, where we wait until the final step to divide each side by 3.

Example 5   A tangent equation involving angle multiples

Find all solutions to $\tan 3x = -\sqrt{3}$ in the interval $(0, 2\pi)$.

Solution

First find the smallest positive value for $3x$ that satisfies the equation. Then form all of the solutions by adding on multiples of π. Since $\tan^{-1}(-\sqrt{3}) = -\pi/3$, the smallest positive value for $3x$ that satisfies the equation is $-\pi/3 + \pi$, or $2\pi/3$. So proceed as follows:

$$\tan 3x = -\sqrt{3}$$

$$3x = \frac{2\pi}{3} + k\pi$$

$$x = \frac{2\pi}{9} + \frac{k\pi}{3} \qquad \color{blue}{\text{Divide each side by 3.}}$$

The solutions between 0 and 2π occur if $k = 0, 1, 2, 3, 4$, and 5. If $k = 6$, then x is greater than 2π. So the solution set is

$$\left\{ \frac{2\pi}{9}, \frac{5\pi}{9}, \frac{8\pi}{9}, \frac{11\pi}{9}, \frac{14\pi}{9}, \frac{17\pi}{9} \right\}.$$ ◆

The graphs of $y_1 = \tan(3x)$ and $y_2 = -\sqrt{3}$ show the six solutions to $\tan(3x) = -\sqrt{3}$ in $[0, 2\pi]$.

More Complicated Equations

More complicated equations involving trigonometric functions are solved by first solving for $\sin x$, $\cos x$, or $\tan x$, and then solving for x. In solving for the trigonometric functions we may use trigonometric identities or properties of algebra, such as factoring or the quadratic formula. In stating formulas for solutions to equations, we will continue to use the letter k to represent any integer.

Example 6   An equation solved by factoring

Find all solutions in the interval $[0, 2\pi)$ to the equation

$$\sin 2x = \sin x.$$

Solution

First we use the identity $\sin 2x = 2 \sin x \cos x$ to get all of the trigonometric functions written in terms of the variable x alone. Then we rearrange and factor:

$$\sin 2x = \sin x$$

$$2 \sin x \cos x = \sin x \qquad \text{By the double-angle identity}$$

$$2 \sin x \cos x - \sin x = 0 \qquad \text{Subtract } \sin x \text{ from each side.}$$

$$\sin x(2 \cos x - 1) = 0 \qquad \text{Factor.}$$

$$\sin x = 0 \quad \text{or} \quad 2 \cos x - 1 = 0 \qquad \text{Set each factor equal to 0.}$$

$$x = k\pi \quad \text{or} \quad 2 \cos x = 1$$

$$\cos x = \frac{1}{2}$$

$$x = \frac{\pi}{3} + 2k\pi \quad \text{or} \quad x = \frac{5\pi}{3} + 2k\pi$$

The only solutions in the interval $[0, 2\pi)$ are 0, $\pi/3$, π, and $5\pi/3$. ◆

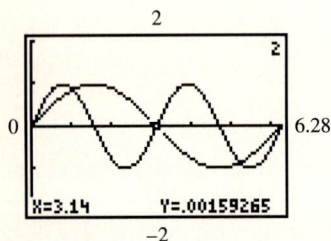

The graphs of $y_1 = \sin(2x)$ and $y_2 = \sin x$ show the four solutions to $\sin(2x) = \sin x$ in $[0, 2\pi)$.

In algebraic equations, we generally do not divide each side by any expression that involves a variable, and the same rule holds for trigonometric equations. In Example 6 we did not divide by $\sin x$ when it appeared on opposite sides of the equation. If we had, the solutions 0 and π would have been lost.

In the next example, an equation of quadratic type is solved by factoring.

Example 7   An equation of quadratic type

Find all solutions to the equation

$$6 \cos^2\left(\frac{x}{2}\right) - 7 \cos\left(\frac{x}{2}\right) + 2 = 0$$

in the interval $[0, 2\pi)$. Round approximate answers to four decimal places.

Solution

Let $y = \cos(x/2)$ to get a quadratic equation:

$$6y^2 - 7y + 2 = 0$$

$$(2y - 1)(3y - 2) = 0 \qquad \text{Factor.}$$

$$2y - 1 = 0 \qquad\qquad \text{or} \qquad\qquad 3y - 2 = 0$$

$$y = \frac{1}{2} \qquad\qquad \text{or} \qquad\qquad y = \frac{2}{3}$$

$$\cos\frac{x}{2} = \frac{1}{2} \qquad \text{or} \qquad \cos\frac{x}{2} = \frac{2}{3} \qquad \text{Replace } y \text{ by } \cos(x/2).$$

$$\frac{x}{2} = \frac{\pi}{3} + 2k\pi \text{ or } \frac{4\pi}{3} + 2k\pi \qquad \text{or} \qquad \frac{x}{2} \approx 0.8411 + 2k\pi \text{ or } 5.4421 + 2k\pi$$

Now multiply by 2 to solve for x:

$$x = \frac{2\pi}{3} + 4k\pi \text{ or } \frac{8\pi}{3} + 4k\pi \qquad \text{or} \qquad x \approx 1.6822 + 4k\pi \text{ or } 10.822 + 4k\pi$$

The solutions in the interval $[0, 2\pi)$ are $2\pi/3$ and 1.6822. ◆

The graph of
$y = 6(\cos(x/2))^2 - 7\cos(x/2) + 2$
has x-intercepts at $2\pi/3$ and 1.6822.

For a trigonometric equation to be of quadratic type it must be written in terms of a trigonometric function and the square of that function. In the next example an identity is used to convert an equation involving sine and cosine to one with only sine. This equation is of quadratic type like Example 7, but it does not factor.

Example 8   An equation solved by the quadratic formula

Find all solutions to the equation

$$\cos^2 \alpha - 0.2 \sin \alpha = 0.9$$

in the interval $[0°, 360°)$. Round answers to the nearest tenth of a degree.

Solution

$$\cos^2 \alpha - 0.2 \sin \alpha = 0.9$$

$$1 - \sin^2 \alpha - 0.2 \sin \alpha = 0.9 \qquad \text{Replace } \cos^2 \alpha \text{ with } 1 - \sin^2 \alpha.$$

$$\sin^2 \alpha + 0.2 \sin \alpha - 0.1 = 0 \qquad \text{An equation of quadratic type.}$$

$$\sin \alpha = \frac{-0.2 \pm \sqrt{(0.2)^2 - 4(1)(-0.1)}}{2} \qquad \text{Use } a = 1, b = 0.2, \text{ and } c = -0.1 \text{ in the quadratic formula.}$$

$$\sin \alpha \approx 0.2317 \qquad \text{or} \qquad \sin \alpha \approx -0.4317$$

Find two positive solutions to $\sin \alpha = 0.2317$ in $[0, 360°)$ by using a calculator to get $\alpha = \sin^{-1}(0.2317) = 13.4°$ and $180° - \alpha = 166.6°$. Now find two positive solutions to $\sin \alpha = -0.4317$ in $[0, 360°)$. Using a calculator, we get $\alpha = \sin^{-1}(-0.4317) = -25.6°$ and $180° - \alpha = 205.6°$. Since $-25.6°$ is negative, we use $-25.6° + 360°$ or $334.4°$ along with $205.6°$ as the two solutions. We list all possible solutions as

$$\alpha \approx 13.4°, \ 166.6°, \ 205.6°, \ \text{or} \ 334.4° \qquad (+k360° \text{ in each case}).$$

The solutions in $[0°, 360°)$ are $13.4°$, $166.6°$, $205.6°$, and $334.4°$.   ◆

The Pythagorean identities $\sin^2 x = 1 - \cos^2 x$, $\csc^2 x = 1 + \cot^2 x$, and $\sec^2 x = 1 + \tan^2 x$ are frequently used to replace one function by the other. If an equation involves sine and cosine, cosecant and cotangent, or secant and tangent, we might be able to square each side and then use these identities.

Example 9   Square each side of the equation

Find all values of y in the interval $[0°, 360°)$ that satisfy the equation

$$\tan 3y + 1 = \sqrt{2} \sec 3y.$$

Solution

$$(\tan 3y + 1)^2 = \left(\sqrt{2} \sec 3y\right)^2 \qquad \text{\color{blue}{Square each side.}}$$

$$\tan^2 3y + 2 \tan 3y + 1 = 2 \sec^2 3y$$

$$\tan^2 3y + 2 \tan 3y + 1 = 2\left(\tan^2 3y + 1\right) \qquad \text{\color{blue}{Since } \sec^2 x = \tan^2 x + 1}$$

$$-\tan^2 3y + 2 \tan 3y - 1 = 0 \qquad \text{\color{blue}{Subtract 2} \tan^2 3y + 2}$$
$$\text{\color{blue}{from both sides.}}$$

$$\tan^2 3y - 2 \tan 3y + 1 = 0$$

$$(\tan 3y - 1)^2 = 0$$

$$\tan 3y - 1 = 0$$

$$\tan 3y = 1$$

$$3y = 45° + k180°$$

Because we squared each side, we must check for extraneous roots. First check $3y = 45°$. If $3y = 45°$, then $\tan 3y = 1$ and $\sec 3y = \sqrt{2}$. Substituting these values into $\tan 3y + 1 = \sqrt{2} \sec 3y$ gives us

$$1 + 1 = \sqrt{2} \cdot \sqrt{2}. \qquad \text{\color{blue}{Correct.}}$$

If we add any *even* multiple of $180°$, or any multiple of $360°$, to $45°$ we get the same values for $\tan 3y$ and $\sec 3y$. So for any k, $3y = 45° + k360°$ satisfies the original equation. Now check $45°$ plus *odd* multiples of $180°$. If $3y = 225°$ $(k = 1)$, then

$$\tan 3y = 1 \qquad \text{and} \qquad \sec 3y = -\sqrt{2}.$$

These values do not satisfy the original equation. Since $\tan 3y$ and $\sec 3y$ have these same values for $3y = 45° + k180°$ for any odd k, the only solutions are of the form $3y = 45° + k360°$, or $y = 15° + k120°$. The solutions in the interval $[0°, 360°)$ are $15°$, $135°$, and $255°$. ◆

There is no single method that applies to all trigonometric equations, but the following strategy will help you to solve them.

▼

Strategy: Solving Trigonometric Equations

1. **Know the solutions to $\sin x = a$, $\cos x = a$, and $\tan x = a$.**

2. **Solve an equation involving multiple angles as if the equation had a single variable.**

3. **Simplify complicated equations by using identities. Try to get an equation involving only one trigonometric function.**

4. **If possible, factor to get different trigonometric functions into separate factors.**

5. **For equations of quadratic type, solve by factoring or the quadratic formula.**

6. **Square each side of the equation, if necessary, so that identities involving squares can be applied. (Check for extraneous roots.)**

Applications

Next, we consider the motion of a weight on a spring, discussed in Section 7.5.

Example 10 Solving a spring equation

In Example 7 of Section 7.5, a weight in motion attached to a spring had location x given by $x = 2 \sin t - \cos t$. For what values of t is the weight at position $x = 0$?

Solution

To solve $2 \sin t - \cos t = 0$, divide each side by $\cos t$. We can divide by $\cos t$ because the values of t for which $\cos t$ is zero do not satisfy the original equation.

$$2 \sin t = \cos t$$

$$\frac{\sin t}{\cos t} = \frac{1}{2}$$

$$\tan t = \frac{1}{2}$$

Since the weight is set in motion at time $t = 0$, the values of t are positive. Since $\tan^{-1}(1/2) = 0.46$, the weight is at position $x = 0$ for $t = 0.46 + k\pi$ for k, a nonnegative integer. ◆

The distance d (in feet) traveled by a projectile fired from the ground with an angle of elevation θ is related to the initial velocity v_0 (in feet per second) by the equation $v_0^2 \sin 2\theta = 32d$. If the projectile is pictured as being fired from the origin into the first quadrant, then the x- and y-coordinates (in feet) of the projectile at time t (in seconds) are given by $x = v_0 t \cos \theta$ and $y = -16t^2 + v_0 t \sin \theta$.

Example 11   The path of a projectile

A catapult is placed 100 feet from the castle wall, which is 35 feet high. The soldier wants the burning bale of hay to clear the top of the wall and land 50 feet inside the castle wall. If the initial velocity of the bale is 70 feet per second, then at what angle should the bale of hay be launched so that it will travel 150 feet and pass over the castle wall?

Solution

Use the equation $v_0^2 \sin 2\theta = 32d$ with $v_0 = 70$ feet per second and $d = 150$ feet to find θ:

$$70^2 \sin 2\theta = 32(150)$$

$$\sin 2\theta = \frac{32(150)}{70^2} \approx 0.97959$$

The launch angle θ must be in the interval $(0, 90°)$, so we look for values of 2θ in the interval $(0, 180°)$. Since $\sin^{-1}(0.97959) \approx 78.4°$, both $78.4°$ and $180° - 78.4° = 101.6°$ are possible values for 2θ. So possible values for θ are $39.2°$ and $50.8°$. See Fig. 7.12.

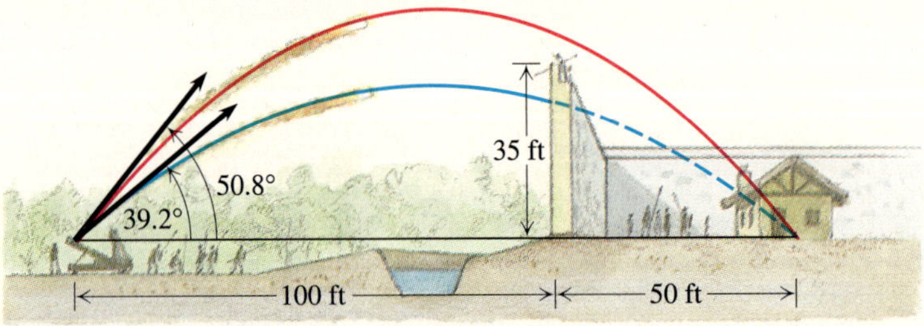

Figure 7.12

Use the equation $x = v_0 t \cos \theta$ to find the time at which the bale is 100 feet from the catapult (measured horizontally) by using each of the possible values for θ:

$100 = 70t \cos(39.2°)$ $100 = 70t \cos(50.8°)$

$t = \dfrac{100}{70 \cos(39.2°)} \approx 1.84$ seconds $t = \dfrac{100}{70 \cos(50.8°)} \approx 2.26$ seconds

Use the equation $y = -16t^2 + v_0t \sin \theta$ to find the altitude of the bale at time $t = 1.84$ seconds and $t = 2.26$ seconds:

$$y = -16(1.84)^2 + 70(1.84)\sin 39.2° \qquad y = -16(2.26)^2 + 70(2.26)\sin 50.8°$$
$$\approx 27.2 \text{ feet} \qquad\qquad\qquad \approx 40.9 \text{ feet}$$

If the burning bale is launched on a trajectory with an angle of 39.2°, then it will have an altitude of only 27.2 feet when it reaches the castle wall. If it is launched with an angle of 50.8°, then it will have an altitude of 40.9 feet when it reaches the castle wall. Since the castle wall is 35 feet tall, the 50.8° angle must be used for the bale to reach its intended target. ◆

For Thought

True or false? Explain.

1. The only solutions to $\cos \alpha = 1/\sqrt{2}$ in $[0°, 360°)$ are 45° and 135°.

2. The only solution to $\sin x = -0.55$ in $[0, \pi)$ is $\sin^{-1}(-0.55)$.

3. $\{x|x = -29° + k360°\} = \{x|x = 331° + k360°\}$, where k is any integer.

4. The solution set to $\tan x = -1$ is $\left\{ x \middle| x = \dfrac{7\pi}{4} + k\pi \right\}$, where k is any integer.

5. The equation $2\cos^2 x + \cos x - 1 = (2\cos x - 1)(\cos x + 1)$ is an identity.

6. The equation $\sin^2 x = \sin x \cos x$ is equivalent to $\sin x = \cos x$.

7. One solution to $\sec x = 2$ is $\dfrac{1}{\cos^{-1}(2)}$.

8. The solution set to $\cot x = 3$ for x in $[0, \pi)$ is $\{\tan^{-1}(1/3)\}$.

9. The equation $\sin x = \cos x$ is equivalent to $\sin^2 x = \cos^2 x$.

10. $\left\{ x \middle| 3x = \dfrac{\pi}{2} + 2k\pi \right\} = \left\{ x \middle| x = \dfrac{\pi}{6} + 2k\pi \right\}$, where k is any integer.

7.6 Exercises ▣ Tape 14 ▣ Disk—5.25″: 5   3.5″: 3   Macintosh: 3

Find all real numbers that satisfy each equation.

1. $\cos x = 1/2$
2. $\cos x = \sqrt{2}/2$
3. $\sin x = \sqrt{2}/2$
4. $\sin x = \sqrt{3}/2$
5. $\tan x = 1$
6. $\tan x = \sqrt{3}/3$
7. $\cos x = -\sqrt{3}/2$
8. $\cos x = -\sqrt{2}/2$
9. $\sin x = -\sqrt{2}/2$
10. $\sin x = -\sqrt{3}/2$
11. $\tan x = -1$
12. $\tan x = -\sqrt{3}$

Find all angles in degrees that satisfy each equation. Round approximate answers to the nearest tenth of a degree.

13. $\cos \alpha = 0$
14. $\cos \alpha = -1$
15. $\sin \alpha = 1$

16. $\sin \alpha = -1$ **17.** $\tan \alpha = 0$ **18.** $\tan \alpha = -1$

19. $\cos \alpha = 0.873$ **20.** $\cos \alpha = -0.158$

21. $\sin \alpha = -0.244$ **22.** $\sin \alpha = 0.551$

23. $\tan \alpha = 5.42$ **24.** $\tan \alpha = -2.31$

Find all real numbers that satisfy each equation.

25. $\cos(x/2) = 1/2$ **26.** $2 \cos 2x = -\sqrt{2}$

27. $\cos 3x = 1$ **28.** $\cos 2x = 0$

29. $2 \sin(x/2) - 1 = 0$ **30.** $\sin 2x = 0$

31. $2 \sin 2x = -\sqrt{2}$ **32.** $\sin(x/3) + 1 = 0$

33. $\tan 2x = \sqrt{3}$ **34.** $\sqrt{3} \tan(3x) + 1 = 0$

35. $\tan 4x = 0$ **36.** $\tan 3x = -1$

37. $\sin(\pi x) = 1/2$ **38.** $\tan(\pi x/4) = 1$

39. $\cos(2\pi x) = 0$ **40.** $\sin(3\pi x) = 1$

Find all values of α in $[0°, 360°)$ that satisfy each equation.

41. $2 \sin \alpha = -\sqrt{3}$ **42.** $\tan \alpha = -\sqrt{3}$

43. $\sqrt{2} \cos 2\alpha - 1 = 0$ **44.** $\sin 6\alpha = 1$

45. $\sec 3\alpha = -\sqrt{2}$ **46.** $\csc(5\alpha) + 2 = 0$

47. $\cot(\alpha/2) = \sqrt{3}$ **48.** $\sec(\alpha/2) = \sqrt{2}$

Find all values of α in degrees that satisfy each equation. Round approximate answers to the nearest tenth of a degree.

49. $\sin 3\alpha = 0.34$ **50.** $\cos 2\alpha = -0.22$

51. $\sin 3\alpha = -0.6$ **52.** $\tan 4\alpha = -3.2$

53. $\sec 2\alpha = 4.5$ **54.** $\csc 3\alpha = -1.4$

55. $\csc(\alpha/2) = -2.3$ **56.** $\cot(\alpha/2) = 4.7$

Find all real numbers in the interval $[0, 2\pi)$ that satisfy each equation. Round approximate answers to the nearest tenth.

57. $3 \sin^2 x = \sin x$ **58.** $2 \tan^2 x = \tan x$

59. $2 \cos^2 x + 3 \cos x = -1$ **60.** $2 \sin^2 x + \sin x = 1$

61. $\tan x = \sec x - \sqrt{3}$ **62.** $\csc x - \sqrt{3} = \cot x$

63. $\sin x + \sqrt{3} = 3\sqrt{3} \cos x$ **64.** $6 \sin^2 x - 2 \cos x = 5$

65. $5 \sin^2 x - 2 \sin x = \cos^2 x$

66. $\sin^2 x - \cos^2 x = 0$

67. $\tan x \sin 2x = 0$

68. $3 \sec^2 x \tan x = 4 \tan x$

69. $\sin 2x - \sin x \cos x = \cos x$

70. $2 \cos^2 2x - 8 \sin^2 x \cos^2 x = -1$

71. $\sin x \cos(\pi/4) + \cos x \sin(\pi/4) = 1/2$

72. $\sin(\pi/6)\cos x - \cos(\pi/6)\sin x = -1/2$

73. $\sin 2x \cos x - \cos 2x \sin x = -1/2$

74. $\cos 2x \cos x - \sin 2x \sin x = 1/2$

Find all values of θ in the interval $[0°, 360°)$ that satisfy each equation. Round approximate answers to the nearest tenth of a degree.

75. $\cos^2\left(\dfrac{\theta}{2}\right) = \sec \theta$

76. $2 \sin^2\left(\dfrac{\theta}{2}\right) = \cos \theta$

77. $2 \sin \theta = \cos \theta$

78. $3 \sin 2\theta = \cos 2\theta$

79. $\sin 3\theta = \csc 3\theta$

80. $\tan^2 \theta - \cot^2 \theta = 0$

81. $\tan^2 \theta - 2 \tan \theta - 1 = 0$

82. $\cot^2 \theta - 4 \cot \theta + 2 = 0$

83. $9 \sin^2 \theta + 12 \sin \theta + 4 = 0$

84. $12 \cos^2 \theta + \cos \theta - 6 = 0$

85. $\dfrac{\tan 3\theta - \tan \theta}{1 + \tan 3\theta \tan \theta} = \sqrt{3}$

86. $\dfrac{\tan 3\theta + \tan 2\theta}{1 - \tan 3\theta \tan 2\theta} = 1$

87. $8 \cos^4 \theta - 10 \cos^2 \theta + 3 = 0$

88. $4 \sin^4 \theta - 5 \sin^2 \theta + 1 = 0$

89. $\sec^4 \theta - 5 \sec^2 \theta + 4 = 0$

90. $\cot^4 \theta - 4 \cot^2 \theta + 3 = 0$

Solve each equation. (These equations are types that will arise in Chapter 8.)

91. $\dfrac{\sin 33.2°}{a} = \dfrac{\sin 45.6°}{13.7}$

92. $\dfrac{\sin 49.6°}{55.1} = \dfrac{\sin 88.2°}{b}$

93. $\dfrac{\sin \alpha}{23.4} = \dfrac{\sin 67.2°}{25.9}$ for $0° < \alpha < 90°$

94. $\dfrac{\sin 9.7°}{15.4} = \dfrac{\sin \beta}{52.9}$ for $90° < \beta < 180°$

95. $(3.6)^2 = (5.4)^2 + (8.2)^2 - 2(5.4)(8.2)\cos \alpha$
for $0° < \alpha < 90°$

96. $(6.8)^2 = (3.2)^2 + (4.6)^2 - 2(3.2)(4.6)\cos \alpha$
for $90° < \alpha < 180°$

Solve each problem.

97. *Motion of a spring* A block is attached to a spring and set in motion on a frictionless plane. Its location on the surface at any time t in seconds is given in meters by $x = \sqrt{3} \sin 2t + \cos 2t$. For what values of t is the block at its resting position $x = 0$?

98. *Motion of a spring* A block is set in motion hanging from a spring and oscillates about its resting position $x = 0$ according to the function $x = -0.3 \sin 3t + 0.5 \cos 3t$. For what values of t is the block at its resting position $x = 0$?

99. *Wave action* The vertical position of a floating ball in an experimental wave tank is given by the equation $x = 2 \sin(\pi t/3)$, where x is the number of feet above sea level and t is the time in seconds. For what values of t is the ball $\sqrt{3}$ ft above sea level?

100. *Periodic sales* The number of car stereos sold by a national department store chain varies seasonally and is a function of the month of the year. The function

$$x = 6.2 + 3.1 \sin\left(\frac{\pi}{6}(t - 9)\right)$$

gives the anticipated sales (in thousands of units) as a function of the number of the month ($t = 1, 2, \ldots, 12$). In what month does the store anticipate selling 9300 units?

101. *Firing an M-16* A soldier is accused of breaking a window 3300 ft away during target practice. If the muzzle velocity for an M-16 is 325 ft/sec, then at what angle would it have to be aimed for the bullet to travel 3300 ft? The distance d (in feet) traveled by a projectile fired at an angle θ is related to the initial velocity v_0 (in feet per second) by the equation $v_0^2 \sin 2\theta = 32d$.

102. *Firing an M-16* If you were accused of firing an M-16 into the air and breaking a window 4000 ft away, what would be your defense?

103. *Choosing the right angle* Seattle Mariners centerfielder Ken Griffey, Jr., fields a ground ball and attempts to make a 230-ft throw to home plate. Given that Griffey commonly makes long throws at 90 mph, find the two possible angles at which he can throw the ball to home plate. Find the time saved by choosing the smaller angle.

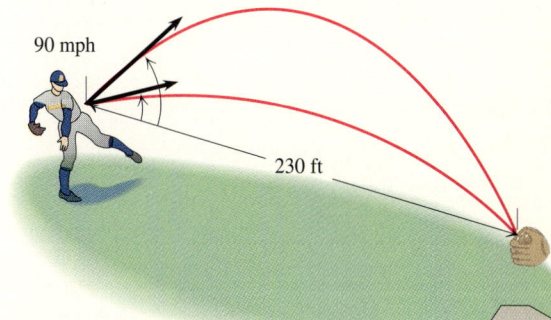

Figure for Exercise 103

104. *Muzzle velocity* The 8-in. (diameter) howitzer on the U.S. Army's M-110 can propel a projectile a distance of 18,500 yd. If the angle of elevation of the barrel is 45°, then what muzzle velocity (in feet per second) is required to achieve this distance?

Graphing Calculator Exercises

One way to solve an equation with a graphing calculator is to rewrite the equation with 0 on the right-hand side, then graph the function that is on the left-hand side. The x-coordinate of each x-intercept of the graph is a solution to the original equation. For each equation, find all real solutions (to the nearest tenth) in the interval $[0, 2\pi)$.

1. $\sin(x/2) = \cos 3x$

2. $2 \sin x = \csc(x + 0.2)$

3. $\dfrac{x}{2} - \dfrac{\pi}{6} + \dfrac{\sqrt{3}}{2} = \sin x$

4. $x^2 = \sin x$

Highlights

Section 7.1 Basic Identities

1. An identity is an equation that is satisfied by every number for which both sides are defined.

2. The simplest identities follow directly from the definitions of the trigonometric functions.

3. Cos x and sec x are even functions; sin x, csc x, tan x, and cot x are odd.

4. Identities are used to get equivalent (possibly simpler) expressions.

5. To show that an equation is not an identity, find a value for the variable that does not satisfy the equation.

Section 7.2 Verifying Identities

1. To verify or prove an identity, we start with the expression on one side of the equation and use known identities and properties of algebra to convert it into the expression on the other side of the equation. An identity can also be proved by converting each expression into the same expression.

2. There is no set method to verify every identity, but there are guidelines.

Section 7.3 Sum and Difference Identities

1. The identities for $\cos(\alpha \pm \beta)$, $\sin(\alpha \pm \beta)$, and $\tan(\alpha \pm \beta)$ are developed.

2. Any trigonometric function value at s equals the cofunction value at $(\pi/2 - s)$.

Section 7.4 Double-Angle and Half-Angle Identities

1. The double-angle identities for sin $2x$, cos $2x$, and tan $2x$ are developed.

2. The half-angle identities for $\sin(x/2)$, $\cos(x/2)$, and $\tan(x/2)$ follow from the double-angle identities.

Section 7.5 Product and Sum Identities

1. The product-to-sum identities allow us to write products of sines and/or cosines in terms of sums or differences of sines and/or cosines.

2. The sum-to-product identities allow us to write sums or differences of sines or cosines as products of sines and/or cosines.

3. If α is an angle in standard position with terminal side through (a, b), then $a \sin x + b \cos x = \sqrt{a^2 + b^2} \sin(x + \alpha)$ (the reduction formula).

4. By the reduction formula, the function $y = a \sin x + b \cos x$ is a sine wave with amplitude $\sqrt{a^2 + b^2}$, period 2π, and phase shift $|\alpha|$.

Section 7.6 Conditional Equations

1. A conditional equation has at least one solution but is not an identity.

2. The solutions to $\sin x = a$, $\cos x = a$, and $\tan x = a$ provide the basis for solving more complicated conditional equations. More complicated equations are solved by first using identities or techniques of algebra to simplify the equations.

Chapter 7 Review Exercises

Simplify each expression.

1. $(1 - \sin \alpha)(1 + \sin \alpha)$

2. $\csc x \tan x + \sec(-x)$

3. $(1 - \csc x)(1 - \csc(-x))$

4. $\dfrac{\cos^2 x - \sin^2 x}{\sin 2x}$

5. $\dfrac{1}{1 + \sin \alpha} - \dfrac{\sin(-\alpha)}{\cos^2 \alpha}$

6. $2 \sin\left(\dfrac{\pi}{2} - \alpha\right)\cos\left(\dfrac{\pi}{2} - \alpha\right)$

7. $\dfrac{2 \tan 2s}{1 - \tan^2 2s}$

8. $\dfrac{\tan 2w - \tan 4w}{1 + \tan 2w \tan 4w}$

9. $\sin 3\theta \cos 6\theta - \cos 3\theta \sin 6\theta$

10. $\dfrac{\sin 2y}{1 + \cos 2y}$

11. $\dfrac{1 - \cos 2z}{\sin 2z}$

12. $\cos^2\left(\dfrac{x}{2}\right) - \sin^2\left(\dfrac{x}{2}\right)$

Use identities to find the exact values of the remaining five trigonometric functions at α.

13. $\cos \alpha = -5/13$ and $\pi/2 < \alpha < \pi$

14. $\tan \alpha = 5/12$ and $\pi < \alpha < 3\pi/2$

15. $\sin\left(\dfrac{\pi}{2} - \alpha\right) = -3/5$ and $\pi < \alpha < 3\pi/2$

16. $\csc\left(\dfrac{\pi}{2} - \alpha\right) = 3$ and $0 < \alpha < \pi/2$

17. $\sin(\alpha/2) = 3/5$ and $3\pi/4 < \alpha/2 < \pi$

18. $\cos(\alpha/2) = -1/3$ and $\pi/2 < \alpha/2 < 3\pi/4$

Determine whether each equation is an identity. Prove your answer.

19. $(\sin x + \cos x)^2 = 1 + \sin 2x$

20. $\cos(A - B) = \cos A \cos B - \sin A \sin B$

21. $\csc^2 x - \cot^2 x = \tan^2 x - \sec^2 x$

22. $\sin^2\left(\dfrac{x}{2}\right) = \dfrac{\sin^2 x}{2 + \sin 2x \csc x}$

Determine whether each function is odd, even, or neither.

23. $f(x) = \dfrac{\sin x - \tan x}{\cos x}$

24. $f(x) = 1 + \sin^2 x$

25. $f(x) = \dfrac{\cos x - \sin x}{\sec x}$

26. $f(x) = \csc^3 x - \tan^3 x$

27. $f(x) = \dfrac{\sin x \tan x}{\cos x + \sec x}$

28. $f(x) = \sin x + \cos x$

Prove that each of the following equations is an identity.

29. $\sec 2\theta = \dfrac{1 + \tan^2 \theta}{1 - \tan^2 \theta}$

30. $\tan^2 \theta = \dfrac{1 - \cos 2\theta}{1 + \cos 2\theta}$

31. $\sin^2\left(\dfrac{x}{2}\right) = \dfrac{\csc^2 x - \cot^2 x}{2 \csc^2 x + 2 \csc x \cot x}$

32. $\cot(-x) = \dfrac{1 - \sin^2 x}{\cos(-x)\sin(-x)}$

33. $\cot(\alpha - 45°) = \dfrac{1 + \tan \alpha}{\tan \alpha - 1}$

34. $\cos(\alpha + 45°) = \dfrac{\cos \alpha - \sin \alpha}{\sqrt{2}}$

35. $\dfrac{\sin 2\beta}{2 \csc \beta} = \sin^2 \beta \cos \beta$

36. $\sin(45° - \beta) = \dfrac{\cos 2\beta}{\sqrt{2}(\cos \beta + \sin \beta)}$

37. $\dfrac{\cot^3 y - \tan^3 y}{\sec^2 y + \cot^2 y} = 2 \cot 2y$

38. $\dfrac{\sin^3 y - \cos^3 y}{\sin y - \cos y} = \dfrac{2 + \sin 2y}{2}$

39. $\cos 4x = 8 \sin^4 x - 8 \sin^2 x + 1$

40. $\cos 3x = \cos x(1 - 4 \sin^2 x)$

41. $\sin^4 2x = 16 \sin^4 x - 32 \sin^6 x + 16 \sin^8 x$

42. $1 - \cos^6 x = 3 \sin^2 x - 3 \sin^4 x + \sin^6 x$

Use an appropriate identity to find the exact value of each expression.

43. $\tan(-\pi/12)$

44. $\sin(-\pi/8)$

45. $\sin(-75°)$

46. $\cos(105°)$

Write each function in the form $y = A \sin(x + C)$, and graph the function for $-2\pi \le x \le 2\pi$. Determine the amplitude and phase shift.

47. $y = 4 \sin x + 4 \cos x$ **48.** $y = \sqrt{3} \sin x + 3 \cos x$

49. $y = -2 \sin x + \cos x$ **50.** $y = -2 \sin x - \cos x$

Find all real numbers that satisfy each equation.

51. $2 \cos 2x + 1 = 0$ **52.** $2 \sin 2x + \sqrt{3} = 0$

53. $(\sqrt{3} \csc x - 2)(\csc x - 2) = 0$

54. $(\sec x - \sqrt{2})(\sqrt{3} \sec x + 2) = 0$

55. $2 \sin^2 x + 1 = 3 \sin x$ **56.** $4 \sin^2 x = \sin x + 3$

57. $-8\sqrt{3} \sin \dfrac{x}{2} = -12$ **58.** $-\cos \dfrac{x}{2} = \sqrt{2} + \cos \dfrac{x}{2}$

59. $\cos \dfrac{x}{2} - \sin x = 0$ **60.** $\sin 2x = \tan x$

61. $\cos 2x + \sin^2 x = 0$ **62.** $\tan \dfrac{x}{2} = \sin x$

63. $\sin x \cos x + \sin x + \cos x + 1 = 0$

64. $\sin 2x \cos 2x - \cos 2x + \sin 2x - 1 = 0$

Find all angles α in $[0°, 360°)$ that satisfy each equation.

65. $\sin \alpha \cos \alpha = \dfrac{1}{2}$ **66.** $\cos 2\alpha = \cos \alpha$

67. $\sin \alpha = \cos \alpha + 1$ **68.** $\cos \alpha \csc \alpha = \cot^2 \alpha$

69. $\sin^2 \alpha + \cos^2 \alpha = \dfrac{1}{2}$ **70.** $\sec^2 \alpha - \tan^2 \alpha = 0$

71. $4 \sin^4 2\alpha = 1$ **72.** $\sin 2\alpha = \tan \alpha$

73. $\tan 2\alpha = \tan \alpha$ **74.** $\tan \alpha = \cot \alpha$

75. $\sin 2\alpha \cos \alpha + \cos 2\alpha \sin \alpha = \cos 3\alpha$

76. $\cos 2\alpha \cos \alpha - \sin 2\alpha \sin \alpha = \cot 3\alpha$

Use the sum-to-product identities to rewrite each expression as a product.

77. $\cos 15° + \cos 19°$ **78.** $\cos 4° - \cos 6°$

79. $\sin(\pi/4) - \sin(-\pi/8)$ **80.** $\sin(-\pi/6) + \sin(\pi/12)$

Use the product-to-sum identities to write each expression as a sum or difference.

81. $2 \sin 11° \cos 13°$ **82.** $2 \sin 8° \sin 12°$

83. $2 \cos(x/4)\cos(x/3)$ **84.** $2 \cos s \sin 3s$

Solve each problem.

85. *Motion of a spring* A block is set in motion hanging from a spring and oscillates about its resting position $x = 0$ according to the function $x = 0.6 \sin 2t + 0.4 \cos 2t$, where x is in centimeters and t is in seconds. For what values of t in the interval $[0, 3]$ is the block at its resting position $x = 0$?

86. *Battle of Gettysburg* The Confederates had at least one 24-lb mortar at the battle of Gettysburg in the Civil War. If the muzzle velocity of a projectile was 400 ft/sec, then at what angles could the cannon be aimed to hit the Union Army 3000 ft away? The distance d (in feet) traveled by a projectile fired at an angle θ is related to the initial velocity v_0 (in feet per second) by the equation $v_0^2 \sin 2\theta = 32d$.

Chapter 7 Test

Use identities to simplify each expression.

1. $\sec x \cot x \sin 2x$

2. $\sin 2t \cos 5t + \cos 2t \sin 5t$

3. $\dfrac{1}{1 - \cos y} + \dfrac{1}{1 + \cos y}$ **4.** $\dfrac{\tan(\pi/5) + \tan(\pi/10)}{1 - \tan(\pi/5)\tan(\pi/10)}$

Prove that each of the following equations is an identity.

5. $\dfrac{\sin \beta \cos \beta}{\tan \beta} = 1 - \sin^2 \beta$

6. $\dfrac{1}{\sec \theta - 1} - \dfrac{1}{\sec \theta + 1} = 2 \cot^2 \theta$

7. $\cos\left(\dfrac{\pi}{2} - x\right)\cos(-x) = \dfrac{\sin(2x)}{2}$

8. $\tan \dfrac{t}{2} \cos^2 t - \tan \dfrac{t}{2} = \dfrac{\sin t}{\sec t} - \sin t$

Find all solutions to each equation.

9. $\sin(-\theta) = 1$ **10.** $\cos 3s = \dfrac{1}{2}$

11. $\tan 2t = -\sqrt{3}$ **12.** $\sin 2\theta = \cos \theta$

Find all values of α in $[0, 360°)$ that satisfy each equation. Round approximate answers to the nearest tenth of a degree.

13. $3 \sin^2 \alpha - 4 \sin \alpha + 1 = 0$

14. $\dfrac{\tan 2\alpha - \tan 7\alpha}{1 + \tan 2\alpha \tan 7\alpha} = 1$

Solve each problem.

15. Write $y = \sin x - \sqrt{3} \cos x$ in the form $y = A \sin(x + C)$ and graph the function. Determine the period, amplitude, and phase shift.

16. Given that $\csc \alpha = 2$ and α is in quadrant II, find the exact values at α for the remaining five trigonometric functions.

17. Determine whether the function $f(x) = x \sin x$ is odd, even, or neither.

18. Use an appropriate identity to find the exact value of $\sin(-\pi/12)$.

19. Prove that $\tan x + \tan y = \tan(x + y)$ is not an identity.

20. A car with worn shock absorbers hits a pothole and oscillates about its normal riding position. At time t (in seconds) the front bumper is distance d (in inches) above or below its normal position, where $d = 2 \sin 3t - 4 \cos 3t$. For what values of t (to the nearest tenth of a second) in the interval $[0, 4]$ is the front bumper at its normal position $d = 0$?

Tying It All Together
Chapters 1–7

Determine whether each function is even or odd.

1. $f(x) = 3x^3 - 2x$ **2.** $f(x) = 2|x|$ **3.** $f(x) = x^3 + \sin x$ **4.** $f(x) = x^3 \sin x$

5. $f(x) = x^4 - x^2 + 1$ **6.** $f(x) = 1/x$ **7.** $f(x) = \dfrac{\sin x}{x}$ **8.** $f(x) = |\sin x|$

9. $f(x) = \cos^5 x + \cos^3 x - 2 \cos x$ **10.** $f(x) = x^3 \sin^4 x + x \sin^2 x$

Determine whether each equation is an identity. Prove your answer.

11. $\sin(\alpha + \beta) = \sin \alpha + \sin \beta$ **12.** $(\alpha + \beta)^2 = \alpha^2 + \beta^2$ **13.** $3x + 5x = 8x$

14. $(x + 3)^2 = x^2 + 6x + 9$ **15.** $\sin^{-1}(x) = \dfrac{1}{\sin x}$ **16.** $\sin^2 x = \sin(x^2)$

Solve each right triangle that has the given parts.

17. $\alpha = 30°$ and $a = 4$ **18.** $a = \sqrt{3}$ and $b = 1$ **19.** $\cos \beta = 0.3$ and $b = 5$ **20.** $\sin \alpha = 0.6$ and $a = 2$

In Chapter 8 we will study expressions like the following, which combine trigonometric functions and complex numbers. Write each expression in the form $a + bi$, where a and b are real numbers.

21. $\cos(\pi/3) + i \sin(\pi/3)$ **22.** $2\left(\cos \dfrac{2\pi}{3} + i \sin \dfrac{2\pi}{3}\right)$ **23.** $(\cos 225° + i \sin 225°)^2$

24. $(\cos 3° + i \sin 3°)(\cos 3° - i \sin 3°)$ **25.** $(2 + i)^3$ **26.** $\left(\sqrt{2} - \sqrt{2}i\right)^4$

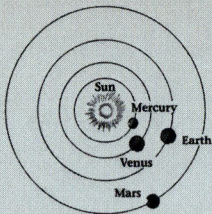

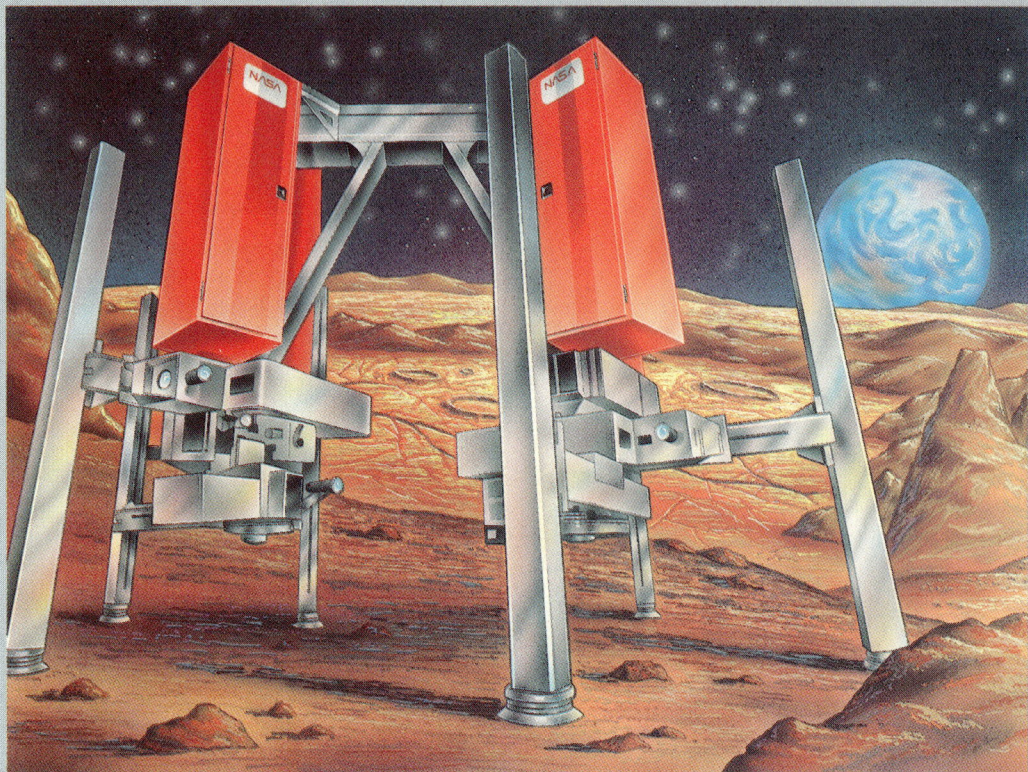

Martian winds swirl flurries of red dust around the legs of an 18-foot tall explorer striding across the barren landscape. It is 60 days into the first Mars Rover Sample Return Mission, and NASA's explorer is an autonomous walking robot called Ambler.

Demonstrating the agility of a trained athlete, the robot has neatly sidestepped miles of sharp boulders and traversed dune-like deposits of fine soil. Walking slowly but steadily on six legs that not only telescope up and down, but also rotate forward and backward, Ambler (for Autonomous Mobile Exploration Robot) surveys the scene with powerful sensor "eyes." Coming into view are the great Valles Marineris canyons, so large that our Grand Canyon would literally disappear into their depths. Ambler is programmed to collect soil and rock samples from the rim to be delivered via rocket back to Earth. With six-axis force sensors embedded in each foot, Ambler doesn't make a false move.

This scenario is only one of several that NASA is developing for future missions to Mars. With no life-support requirements and numerous technical capabilities, robots are likely candidates to replace astronauts in space exploration. And without the need for paychecks, coffee breaks, or vacations, robots are also rapidly becoming the worker bees of industry.

Mechanical devices that can perform manipulative tasks under their own power are as old as recorded history. As early as 3000 B.C., Egyptians built waterclocks and articulated figures. The first industrial robot joined GM's production line in 1961. In the 21st century, robotic "steel collar" workers will increasingly perform boring, repetitive jobs associated with a high degree of human error. They will assemble parts, mix chemicals, spot weld, inspect, test, and enter contaminated areas deadly to their human counterparts.

In this chapter we'll see that trigonometry is the mathematics you need to model repetitive motion. For example, in robotics—one of the most exciting industrial applications—we use trigonometry to position a robot arm so that it can perform a series of repetitive tasks. We will also see how trigonometry is used to design machines and find forces acting on parts.

8
Applications of
Trigonometry

8.1

The Law of Sines

In Chapter 6 we used trigonometry to solve right triangles. In this section and the next we will learn to solve any triangle for which we have enough information to determine the shape of the triangle.

Oblique Triangles

Any triangle without a right angle is called an **oblique triangle.** As usual, we use α, β, and γ for the angles of a triangle and a, b, and c, respectively, for the lengths of the sides opposite those angles, as shown in Fig. 8.1. The vertices at angles α, β, and γ are labeled A, B, and C, respectively. To solve an oblique triangle, we must know at least three parts of the triangle, at least one of which must be the length of a side. We can classify the different cases for the three known parts as follows:

1. One side and any two angles (ASA or AAS)
2. Two sides and a nonincluded angle (SSA)
3. Two sides and an included angle (SAS)
4. Three sides (SSS)

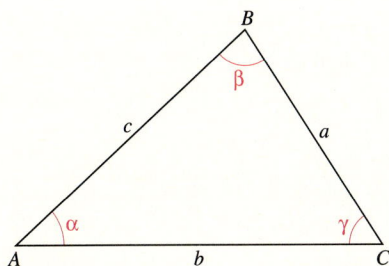

Figure 8.1

We can actually solve all of these cases by dividing the triangles into right triangles and using right triangle trigonometry. Since that method is quite tedious, we develop the *law of sines* and the *law of cosines*. The first two cases can be handled with the law of sines. We will discuss the last two cases in Section 8.2, when we develop the law of cosines.

The Law of Sines

The **law of sines** says that *the ratio of the sine of an angle and the length of the side opposite the angle is the same for each angle of a triangle.*

The Law of Sines

In any triangle,

$$\frac{\sin \alpha}{a} = \frac{\sin \beta}{b} = \frac{\sin \gamma}{c}.$$

Proof   Either the triangle is an acute triangle (all acute angles) or it is an obtuse triangle (one obtuse angle). We consider the case of the obtuse triangle here and leave the case of the acute triangle as Exercise 37. Triangle *ABC* is shown in Fig. 8.2 with an altitude of length h_1 drawn from point *C* to the opposite side and an altitude of length h_2 drawn from point *B* to the extension of the opposite side.

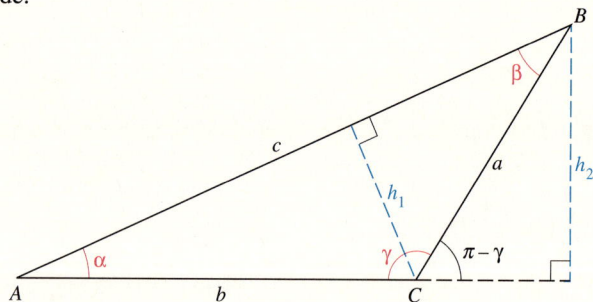

Figure 8.2

Since α and β are now in right triangles, we have

$$\sin \alpha = \frac{h_1}{b} \qquad \text{or} \qquad h_1 = b \sin \alpha$$

and

$$\sin \beta = \frac{h_1}{a} \qquad \text{or} \qquad h_1 = a \sin \beta.$$

Replace h_1 by $b \sin \alpha$ in the equation $h_1 = a \sin \beta$ to get

$$b \sin \alpha = a \sin \beta.$$

Dividing each side of the last equation by ab yields

$$\frac{\sin \alpha}{a} = \frac{\sin \beta}{b}.$$

Using the largest right triangle in Fig. 8.2, we have

$$\sin \alpha = \frac{h_2}{c} \qquad \text{or} \qquad h_2 = c \sin \alpha.$$

Note that the acute angle $\pi - \gamma$ in Fig. 8.2 is the reference angle for the obtuse angle γ. We know that sine is positive for both acute and obtuse angles, so $\sin \gamma = \sin(\pi - \gamma)$. Since $\pi - \gamma$ is an angle of a right triangle, we can write

$$\sin \gamma = \sin(\pi - \gamma) = \frac{h_2}{a} \qquad \text{or} \qquad h_2 = a \sin \gamma.$$

From $h_2 = c \sin \alpha$ and $h_2 = a \sin \gamma$ we get $c \sin \alpha = a \sin \gamma$ or

$$\frac{\sin \alpha}{a} = \frac{\sin \gamma}{c}$$

and we have proved the law of sines.   ◆

The law of sines can also be written in the form

$$\frac{a}{\sin \alpha} = \frac{b}{\sin \beta} = \frac{c}{\sin \gamma}.$$

In solving triangles, it is usually simplest to use the form in which the unknown quantity appears in the numerator.

In our first example we use the law of sines to solve a triangle for which we are given two angles and an included side. Remember that if the dimensions given for triangles are measurements, then any answers obtained from those measurements are only as accurate as the least accurate of the measurements. (See the discussion of significant digits in Chapter 6.)

Example 1   Given two angles and an included side (ASA)

Given $\beta = 34°$, $\gamma = 64°$, and $a = 5.3$, solve the triangle.

Solution

To sketch the triangle, first draw side a, then draw angles of approximately $34°$ and $64°$ on opposite ends of a. Label all parts (see Fig. 8.3). Since the sum of the three angles of a triangle is $180°$, the third angle α is $82°$. By the law of sines,

$$\frac{5.3}{\sin 82°} = \frac{b}{\sin 34°}$$

$$b = \frac{5.3 \sin 34°}{\sin 82°} \approx 3.0.$$

Again, by the law of sines,

$$\frac{c}{\sin 64°} = \frac{5.3}{\sin 82°}$$

$$c = \frac{5.3 \sin 64°}{\sin 82°} \approx 4.8.$$

So $\alpha = 82°$, $b \approx 3.0$, and $c \approx 4.8$ solves the triangle.   ◆

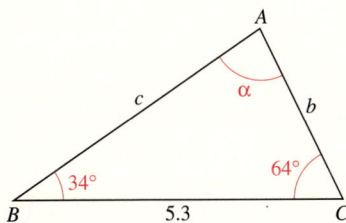

Figure 8.3

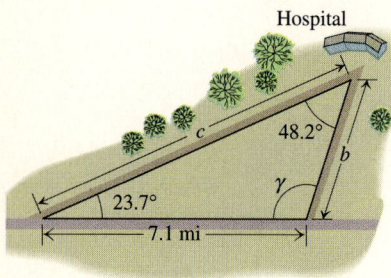

Figure 8.4

Example 2 Given two angles and a nonincluded side (AAS)

A corporation is planning to build a hospital to the north of an east-west highway. Two straight roads intersect at the hospital site at an angle of 48.2° and cross the east-west highway 7.1 miles apart, as shown in Fig. 8.4. If one of the roads intersects the highway at a 23.7° angle, then how far is it along each road from the hospital site to the highway?

Solution

Let b and c be the distances from the hospital to the highway on each road and γ be the unknown angle in the triangle shown in Fig. 8.4. Since the sum of the angles of the triangle is 180°, the measure of the third angle γ is 108.1°. By the law of sines,

$$\frac{b}{\sin 23.7°} = \frac{7.1}{\sin 48.2°}$$

or

$$b = \frac{7.1 \sin 23.7°}{\sin 48.2°} \approx 3.8 \text{ miles.}$$

Again by the law of sines,

$$\frac{c}{\sin 108.1°} = \frac{7.1}{\sin 48.2°}$$

or

$$c = \frac{7.1 \sin 108.1°}{\sin 48.2°} \approx 9.1 \text{ miles.}$$

So the distance from the hospital site to the highway is approximately 3.8 miles on one road and 9.1 miles on the other. ◆

Bearing

The measure of an angle that describes the direction of a ray is called the **bearing** of the ray. In air navigation, bearing is given as a nonnegative angle less than 360° measured in a clockwise direction from a ray pointing due north. So in Fig. 8.5 the bearing of ray $\overrightarrow{OA}$ is 60°, the bearing of ray $\overrightarrow{OB}$ is 150°, and the bearing of $\overrightarrow{OC}$ is 225°.

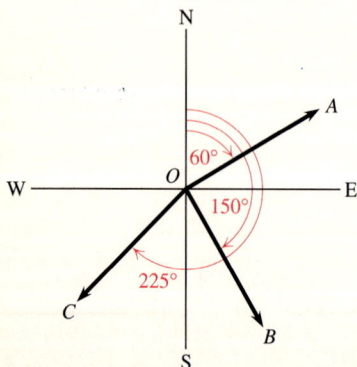

Figure 8.5

Example 3 Using bearing in solving triangles

A bush pilot left the Fairbanks Airport in a light plane and flew 100 miles toward Fort Yukon in still air on a course with a bearing of 18°. She then flew due east (bearing 90°) for some time to drop supplies to a snowbound family. After the drop, her course to return to Fairbanks had a bearing of 225°. What was her maximum distance from Fairbanks?

Solution

Figure 8.6 shows the course of her flight. To change course from bearing 18° to bearing 90° at point B, the pilot must add 72° to the bearing. So $\angle ABC$ is 108°. A bearing of 225° at point C means that $\angle BCA$ is 45°. Finally, we obtain $\angle BAC = 27°$ by subtracting 18° and 45° from 90°.

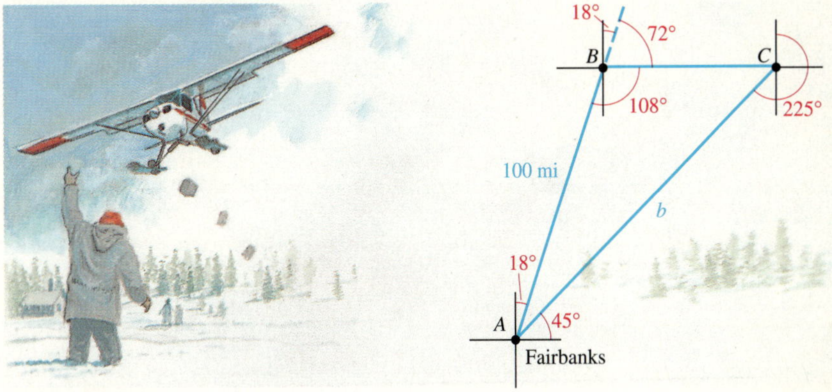

Figure 8.6

We can find the length of $\overline{AC}$ (the maximum distance from Fairbanks) by using the law of sines:

$$\frac{b}{\sin 108°} = \frac{100}{\sin 45°}$$

$$b = \frac{100 \cdot \sin 108°}{\sin 45°} \approx 134.5$$

So the pilot's maximum distance from Fairbanks was 134.5 miles.

In marine navigation and surveying, the bearing of a ray is the acute angle the ray makes with a ray pointing due north or due south. Along with the acute angle, directions are given that indicate in which quadrant the ray lies. For example, in Fig. 8.5, $\overrightarrow{OA}$ has a bearing 60° east of north (N60°E), $\overrightarrow{OB}$ has a bearing 30° east of south (S30°E), and $\overrightarrow{OC}$ has a bearing 45° west of south (S45°W).

The Ambiguous Case (SSA)

In the AAS and ASA cases we are given any two angles with positive measures and the length of any side. If the total measure of the two angles is less than 180°, then a unique triangle is determined. However, for two sides and a *nonincluded* angle (SSA), there are several possibilities. So the SSA case is called the **ambiguous case.** Drawing the diagram in the proper order will help you in understanding the ambiguous case.

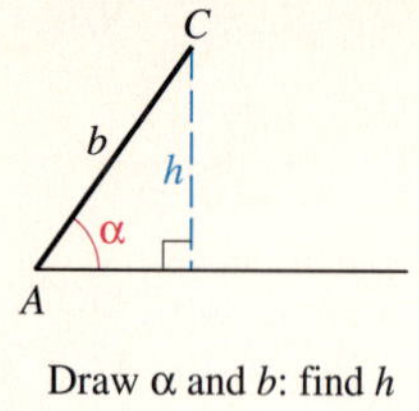

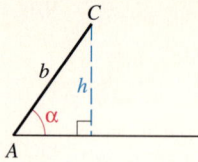

Draw α and b: find h

Figure 8.7

Suppose we are given an acute angle α ($0° < \alpha < 90°$) and sides a and b. Side a is opposite the angle α and side b is adjacent to it. Draw an angle of approximate size α in standard position with terminal side of length b as shown in Fig. 8.7. Don't draw in side a yet. Let h be the distance from C to the initial side of α. Since $\sin \alpha = h/b$, we have $h = b \sin \alpha$. Now we are ready to draw in side a, but there are four possibilities for its location.

1. If $a < h$, then no triangle can be formed, because a cannot reach from point C to the initial side of α. This situation is shown in Fig. 8.8(a).

2. If $a = h$, then exactly one right triangle is formed, as in Fig. 8.8(b).

3. If $h < a < b$, then exactly two triangles are formed, because a reaches to the initial side in two places, as in Fig. 8.8(c).

4. If $a \geq b$, then only one triangle is formed, as in Fig. 8.8(d).

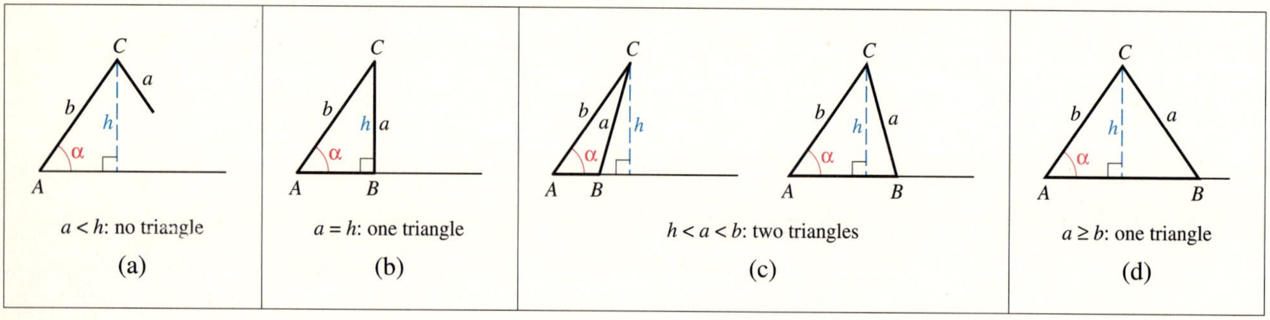

$a < h$: no triangle	$a = h$: one triangle	$h < a < b$: two triangles	$a \geq b$: one triangle
(a)	(b)	(c)	(d)

Figure 8.8

If we start with α, a, and b, where $90° \leq \alpha < 180°$, then there are only two possibilities for the number of triangles determined. If $a \leq b$, then no triangle is formed, as in Fig. 8.9(a). If $a > b$, one triangle is formed, as in Fig. 8.9(b).

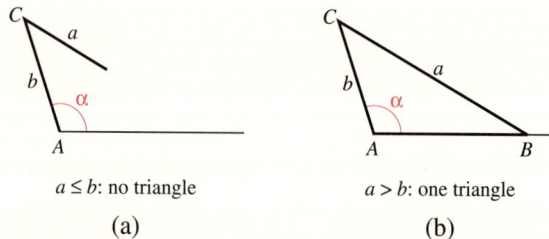

$a \leq b$: no triangle $a > b$: one triangle

(a)   (b)

Figure 8.9

It is not necessary to memorize all of the SSA cases shown in Figs. 8.8 and 8.9. If you draw the triangle for a given problem in the order suggested, then it will be clear how many triangles are possible with the given parts. We must decide how many triangles are possible, before we can solve the triangle(s).

Example 4   Given two sides and a nonincluded angle (SSA)

Given $\alpha = 41°$, $a = 3.3$, and $b = 5.4$, solve the triangle.

Solution

Draw the 41° angle and label its terminal side 5.4 as shown in Fig. 8.10. Side a must go opposite α, but do not put it in yet. Find the length of the altitude h from C. Using a trigonometric ratio, we get $\sin 41° = h/5.4$, or $h = 5.4 \sin 41° \approx 3.5$. Since $a = 3.3$, a is *shorter* than the altitude h, and side a will not reach from point C to the initial side of α. So there is no triangle with the given parts.

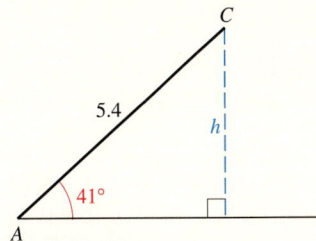

Figure 8.10

Example 5   Given two sides and a nonincluded angle (SSA)

Given $\gamma = 125°$, $b = 5.7$, and $c = 8.6$, solve the triangle.

Solution

Draw the 125° angle and label its terminal side 5.7. Since $8.6 > 5.7$, side c will reach from point A to the initial side of γ as shown in Fig. 8.11, and form a single triangle. To solve this triangle, we use the law of sines:

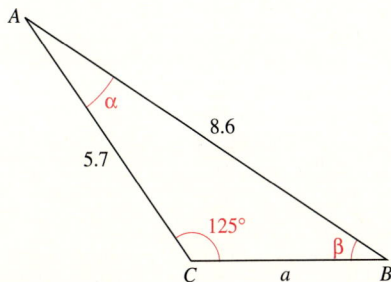

Figure 8.11

$$\frac{\sin 125°}{8.6} = \frac{\sin \beta}{5.7}$$

$$\sin \beta = \frac{5.7 \sin 125°}{8.6}$$

$$\beta = \sin^{-1}\left(\frac{5.7 \sin 125°}{8.6}\right) \approx 32.9°$$

Since the sum of α, β, and γ is 180°, $\alpha = 22.1°$. Now use α and the law of sines to find a:

$$\frac{a}{\sin 22.1°} = \frac{8.6}{\sin 125°}$$

$$a = \frac{8.6 \sin 22.1°}{\sin 125°} \approx 3.9$$

Example 6 Given two sides and a nonincluded angle (SSA)

Given $\beta = 56.3°$, $a = 8.3$, and $b = 7.6$, solve the triangle.

Solution

Draw an angle of approximately 56.3°, and label its terminal side 8.3. Side b must go opposite β, but do not put it in yet. Find the length of the altitude h from point C to the initial side of β, as shown in Fig. 8.12(a). Since $\sin 56.3° = h/8.3$, we get $h = 8.3 \sin 56.3° \approx 6.9$. Because b is longer than h but shorter than a, there are two triangles with the given parts, as shown in parts (b) and (c) of the figure. In either triangle we have

$$\frac{\sin \alpha}{8.3} = \frac{\sin 56.3°}{7.6}$$

$$\sin \alpha = 0.9086.$$

This equation has two solutions in $[0°, 180°]$. For part (c) we get $\alpha = \sin^{-1}(0.9086) = 65.3°$ and for part (b) we get $\alpha = 180° - 65.3° = 114.7°$. Using the law of sines, we get $\gamma = 9.0°$ and $c = 1.4$ for part (b), and we get $\gamma = 58.4°$ and $c = 7.8$ for part (c).

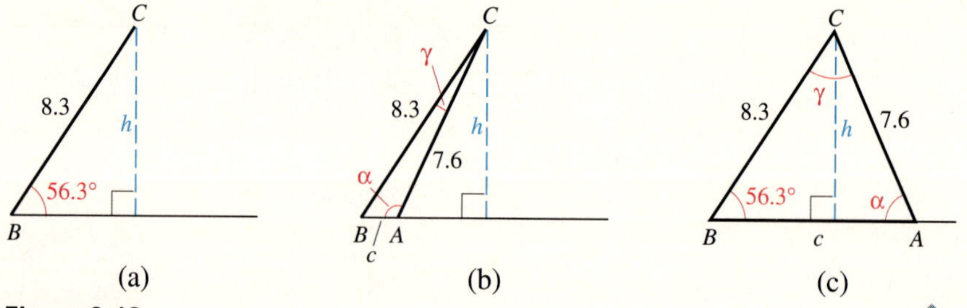

(a)   (b)   (c)

Figure 8.12

Area of a Triangle

The formula $A = \frac{1}{2}bh$ gives the area of a triangle in terms of a side and the altitude to that side. We can rewrite this formula in terms of two sides and their included angle. Consider the triangle shown in Fig. 8.13, in which α is an acute angle and h is inside the triangle. Since $\sin \alpha = h/c$, we get $h = c \sin \alpha$. Substitute $c \sin \alpha$ for h in the formula to get

$$A = \frac{1}{2} bc \sin \alpha.$$

Figure 8.13

This formula is also correct in all other possible cases: α is acute and h is the side of the triangle opposite α, α is acute and h lies outside the triangle, α is a right

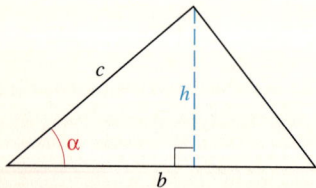

angle, and α is obtuse. (See Exercise 38.) Likewise, the formula can be written using the angles β or γ. We have the following theorem.

Theorem: Area of a Triangle

The area of a triangle is given by

$$A = \frac{1}{2} bc \sin \alpha, \qquad A = \frac{1}{2} ac \sin \beta, \qquad \text{and} \qquad A = \frac{1}{2} ab \sin \gamma.$$

Surveyors describe property in terms of the lengths of the sides and the angles between those sides. In the next example we find the area of a piece of property, using the new area formula.

Example 7   Area of a quadrilateral

The town of Hammond is considering the purchase of a four-sided piece of property to be used for a playground. The dimensions of the property are given in Fig. 8.14. Find the area of the property in square feet.

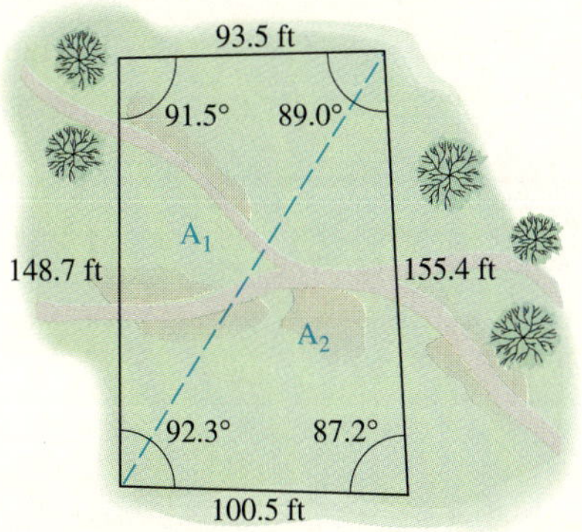

Figure 8.14

Solution

Divide the property into two triangles as shown in Fig. 8.14, and find the area of each:

$$A_1 = \frac{1}{2} (148.7)(93.5)\sin 91.5° \approx 6949.3 \text{ square feet}$$

$$A_2 = \frac{1}{2} (100.5)(155.4)\sin 87.2° \approx 7799.5 \text{ square feet}$$

The total area of the property is 14,748.8 square feet.

? For Thought

True or false? Explain.

1. If we know the measures of two angles of a triangle, then the measure of the third angle is determined.

2. If $\dfrac{\sin 9°}{a} = \dfrac{\sin 17°}{88}$, then $a = \dfrac{\sin 17°}{88 \sin 9°}$.

3. The equation $\dfrac{\sin \alpha}{5} = \dfrac{\sin 44°}{18}$ has exactly one solution in $[0, 180°]$.

4. One solution to $\dfrac{\sin \beta}{2.3} = \dfrac{\sin 39°}{1.6}$ is $\beta = \sin^{-1}\left(\dfrac{2.3 \sin 39°}{1.6}\right)$.

5. $\dfrac{\sin 60°}{\sqrt{3}} = \dfrac{\sin 30°}{1}$.

6. No triangle exists with $\alpha = 60°$, $b = 10$ feet, and $a = 500$ feet.

7. There is exactly one triangle with $\beta = 30°$, $c = 20$, and $b = 10$.

8. There are two triangles with $\alpha = 135°$, $b = 17$, and $a = 19$.

9. The area of a triangle is determined by the lengths of two sides and the measure of the included angle.

10. A right triangle's area is one-half the product of the lengths of its legs.

8.1 Exercises   Tape 15   Disk—5.25″: 5   3.5″: 3   Macintosh: 3

Solve each triangle.

1.

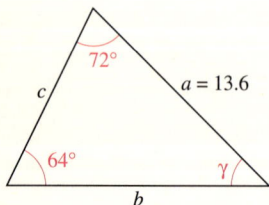

$72°$ $a = 13.6$

c

$64°$ γ

b

2.

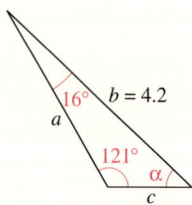

$16°$ $b = 4.2$

a

$121°$ α

c

3.

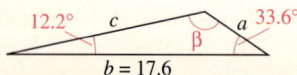

$12.2°$ c $33.6°$

β a

$b = 17.6$

4.

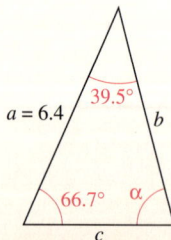

$a = 6.4$ $39.5°$ b

$66.7°$ α

c

Solve each triangle with the given parts.

5. $\alpha = 10.3°$, $\gamma = 143.7°$, $c = 48.3$

6. $\beta = 94.7°$, $\alpha = 30.6°$, $b = 3.9$

7. $\beta = 120.7°$, $\gamma = 13.6°$, $a = 489.3$

8. $\alpha = 39.7°$, $\gamma = 91.6°$, $b = 16.4$

Determine the number of triangles with the given parts and solve each triangle.

9. $\alpha = 39.6°$, $c = 18.4$, $a = 3.7$

10. $\beta = 28.6°$, $a = 40.7$, $b = 52.5$

11. $\gamma = 60°$, $b = 20$, $c = 10\sqrt{3}$

12. $\alpha = 41.2°$, $a = 8.1$, $b = 10.6$

13. $\beta = 138.1°$, $c = 6.3$, $b = 15.6$

14. $\gamma = 128.6°$, $a = 9.6$, $c = 8.2$

15. $\beta = 32.7°$, $a = 37.5$, $b = 28.6$

16. $\alpha = 30°$, $c = 40$, $a = 20$

17. $\gamma = 99.6°$, $b = 10.3$, $c = 12.4$

18. $\alpha = 75.3$, $a = 12.4$, $b = 9.8$

Find the area of each triangle with the given parts.

19. $a = 12.9$, $b = 6.4$, $\gamma = 13.7°$

20. $b = 42.7$, $c = 64.1$, $\alpha = 74.2°$

21. $\alpha = 39.4°$, $b = 12.6$, $a = 13.7$

22. $\beta = 74.2°$, $c = 19.7$, $b = 23.5$

23. $\alpha = 42.3°$, $\beta = 62.1°$, $c = 14.7$

24. $\gamma = 98.6°$, $\beta = 32.4°$, $a = 24.2$

25. $\alpha = 56.3°$, $\beta = 41.2°$, $a = 9.8$

26. $\beta = 25.6°$, $\gamma = 74.3°$, $b = 17.3$

Find the area of each region.

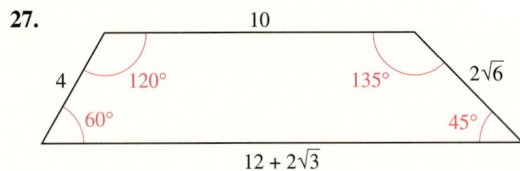

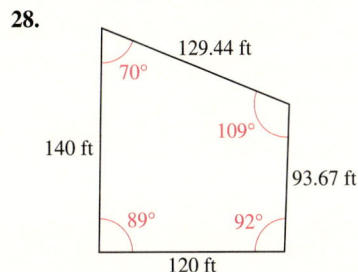

27.

28.

Solve each problem.

29. *Designing an addition* A 40-ft-wide house has a roof with a 6-12 pitch (the roof rises 6 ft for a run of 12 ft). The owner plans a 14-ft-wide addition that will have a 3-12 pitch to its roof. Find the lengths of $\overline{AB}$ and $\overline{BC}$ shown below.

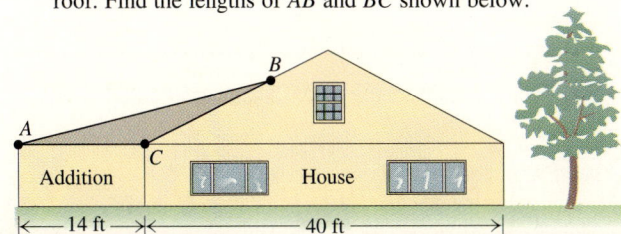

Figure for Exercise 29

30. *Saving an endangered hawk* A hill has an angle of inclination of 36° as shown in the figure. A study completed by a state's highway commission showed that the placement of a highway requires that 400 ft of the hill, measured horizontally, be removed. The engineers plan to leave a slope alongside the highway with an angle of inclination of 62°, as shown in the figure. Located 750 ft up the hill measured from the base is a tree containing the nest of an endangered hawk. Will this tree be removed in the excavation?

Figure for Exercise 30

31. *Observing traffic* A traffic report helicopter left the WKPR studios on a course with a bearing of 210°. After flying 12 mi to reach interstate highway 20, the helicopter flew due east along I-20 for some time. The helicopter headed back to WKPR on a course with a bearing of 310° and reported no accidents along I-20. For how many miles did the helicopter fly along I-20?

32. *Course of a fighter plane* During an important NATO exercise, an F-14 Tomcat left the carrier Nimitz on a course with a bearing of 34° and flew 400 mi. Then the F-14 flew for some distance on a course with a bearing of 162°. Finally, the plane flew back to its starting point on a course with a bearing of 308°. What distance did the plane fly on the final leg of the journey?

33. *Surveying property* A surveyor locating the corners of a triangular piece of property started at one corner and walked 480 ft in the direction N36°W to reach the next corner. The surveyor turned and walked S21°W to get to the next corner of the property. Finally, the surveyor walked in the direction N82°E to get back to the starting point. What is the area of the property in square feet?

34. *Sailing* Joe and Jill set sail from the same point, with Joe sailing in the direction S4°E and Jill sailing in the direction S9°W. After 4 hr, Jill was 2 mi due west of Joe. How far had Jill sailed?

35. *Making a kite* A kite is made in the shape shown in the figure. Find the surface area of the kite in square inches.

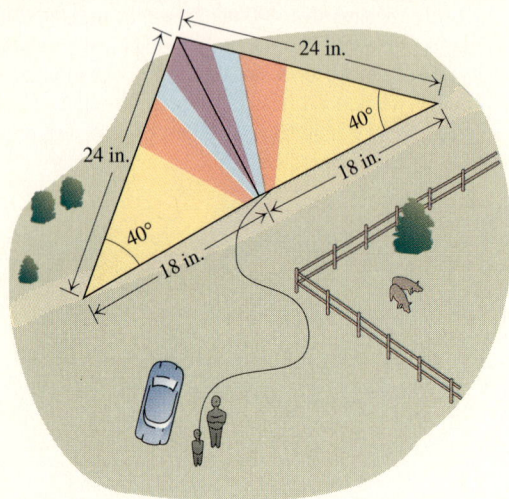

Figure for Exercise 35

36. *Area of a wing* The F-106 Delta Dart once held a world speed record of Mach 2.3. Its sweptback triangular wings have the dimensions given in the figure. Find the area of one wing in square feet.

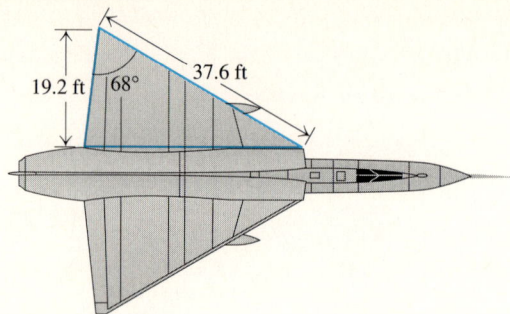

Figure for Exercise 36

For Writing/Discussion

37. *Law of sines* Prove the law of sines for the case in which the triangle is an acute triangle.

38. *Area of a triangle* Prove the trigonometric formula for the area of a triangle in the cases mentioned in the text but not proved in the text.

39. *Cooperative learning* Find a map or a description of a piece of property in terms of sides and angles. Show it to your class and explain how to find the area of the property in square feet.

8.2

The Law of Cosines

In Section 8.1 we discussed the AAS, ASA, and SSA cases for solving triangles using the law of sines. For the SAS and SSS cases, which cannot be handled with the law of sines, we develop the law of cosines. This law is a generalization of the Pythagorean theorem and applies to any triangle. However, it is usually stated and used only for oblique triangles because it is not needed for right triangles.

The Law of Cosines

The **law of cosines** gives a formula for the square of any side of an oblique triangle in terms of the other two sides and their included angle.

Theorem: Law of Cosines

If triangle ABC is an oblique triangle with sides a, b, and c and angles α, β, and γ, then

$$a^2 = b^2 + c^2 - 2bc \cos \alpha,$$

$$b^2 = a^2 + c^2 - 2ac \cos \beta,$$

$$c^2 = a^2 + b^2 - 2ab \cos \gamma.$$

Proof Given triangle ABC, position the triangle as shown in Fig. 8.15. The vertex C is in the first quadrant if α is acute and in the second if α is obtuse. Both cases are shown in Fig. 8.15.

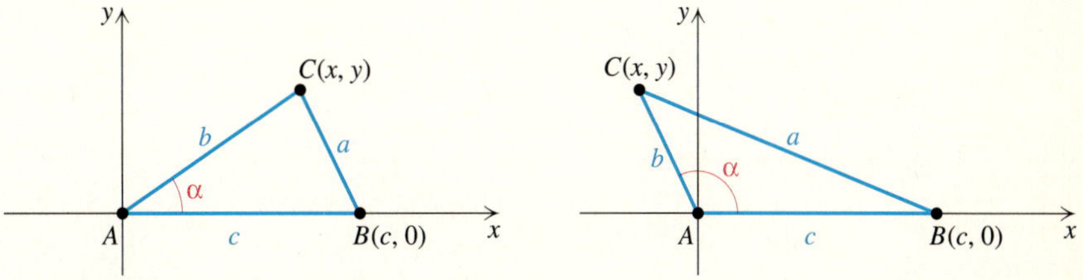

Figure 8.15

In either case, the x-coordinate of C is $x = b\cos\alpha$, and the y-coordinate of C is $y = b\sin\alpha$. The distance from C to B is a, but we can also find that distance by using the distance formula:

$$\begin{aligned} a &= \sqrt{(b\cos\alpha - c)^2 + (b\sin\alpha - 0)^2} \\ a^2 &= (b\cos\alpha - c)^2 + (b\sin\alpha)^2 \\ &= b^2\cos^2\alpha - 2bc\cos\alpha + c^2 + b^2\sin^2\alpha \\ &= b^2(\cos^2\alpha + \sin^2\alpha) + c^2 - 2bc\cos\alpha \end{aligned}$$

Using $\cos^2 x + \sin^2 x = 1$, we get the first equation of the theorem:

$$a^2 = b^2 + c^2 - 2bc\cos\alpha$$

Similar arguments with B and C at (0, 0) produce the other two equations. ◆

In any triangle with unequal sides, the largest angle is opposite the largest side and the smallest angle is opposite the smallest side. We need this fact in the first example.

Example 1 Given three sides of a triangle (SSS)

Given $a = 8.2$, $b = 3.7$, and $c = 10.8$, solve the triangle.

Solution

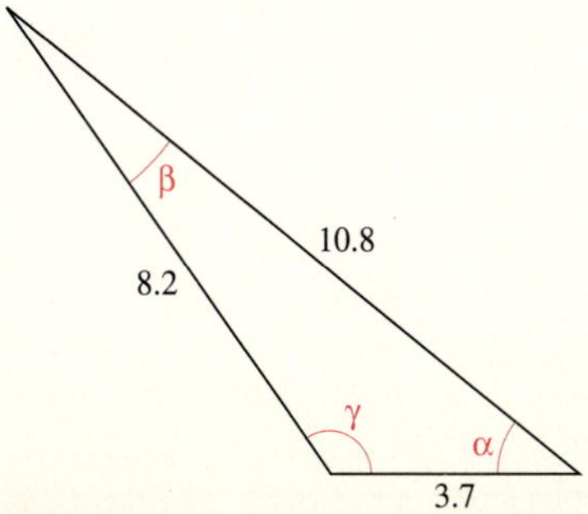

Figure 8.16

Draw the triangle and label it as in Fig. 8.16. Since c is the longest side, use

$$c^2 = a^2 + b^2 - 2ab\cos\gamma$$

to find the largest angle γ:

$$\begin{aligned} -2ab\cos\gamma &= c^2 - a^2 - b^2 \\ \cos\gamma &= \frac{c^2 - a^2 - b^2}{-2ab} = \frac{(10.8)^2 - (8.2)^2 - (3.7)^2}{-2(8.2)(3.7)} \approx -0.5885 \\ \gamma &= \cos^{-1}(-0.5885) \approx 126.1^\circ \end{aligned}$$

We could finish with the law of cosines, but in this case it is simpler to use the law of sines, which involves fewer computations:

$$\frac{\sin \beta}{b} = \frac{\sin \gamma}{c}$$

$$\frac{\sin \beta}{3.7} = \frac{\sin 126.1°}{10.8}$$

$$\sin \beta \approx 0.2768$$

There are two solutions to $\sin \beta = 0.2768$ in $[0°, 180°]$. However, β must be less than $90°$, because γ is $126.1°$. So $\beta = \sin^{-1}(0.2768) \approx 16.1°$. Finally, $\alpha = 180° - 126.1° - 16.1° = 37.8°$. ◆

In solving the SSS case, we always *find the largest angle first,* using the law of cosines. The remaining two angles must be acute angles. So when using the law of sines to find another angle, we need only find an acute solution to the equation.

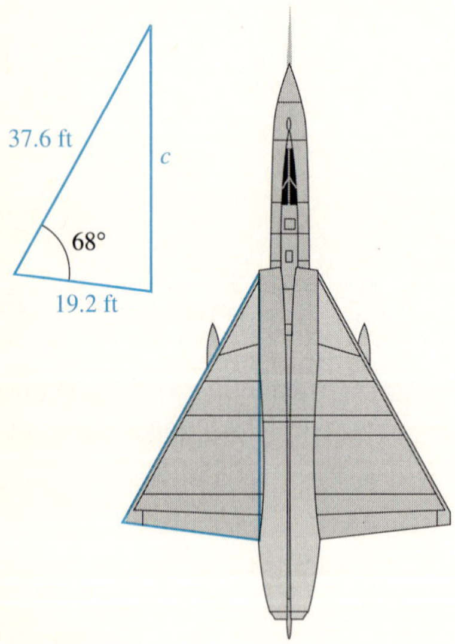

Example 2   Given two sides and an included angle (SAS)

The wing of the F-106 Delta Dart is triangular in shape, with the dimensions given in Fig. 8.17. Find the length of the side labeled c in Fig. 8.17.

Solution

Using the law of cosines, we find c as follows:

$$c^2 = (19.2)^2 + (37.6)^2 - 2(19.2)(37.6)\cos 68° \approx 1241.5$$
$$c \approx \sqrt{1241.5} \approx 35.2 \text{ ft}$$

◆

In Examples 1 and 2, the law of cosines was used to find the length of a side of a triangle. Next, we use it to write a formula for the length of a chord in terms of the radius of the circle and the central angle subtended by the chord.

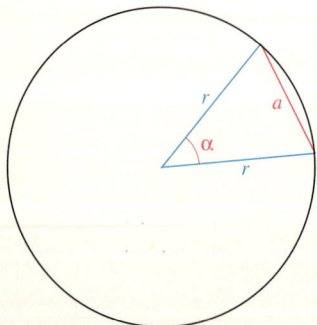

Figure 8.17

Example 3   A formula for the length of a chord

A central angle α in a circle of radius r intercepts a chord of length a. Write a formula for a in terms of α and r.

Solution

We can apply the law of cosines to the triangle formed by the cord and the radii at the endpoints of the chord, as shown in Fig. 8.18.

$$a^2 = r^2 + r^2 - 2r^2 \cos \alpha$$
$$= r^2(2 - 2 \cos \alpha)$$
$$a = r\sqrt{2 - 2 \cos \alpha}$$

◆

Figure 8.18

Area of a Triangle by Heron's Formula

In Section 8.1 we saw the formulas $A = \frac{1}{2}bh$ and $A = \frac{1}{2}bc \sin \alpha$ for the area of a triangle. For one formula we must know a side and the altitude to that side, and for the other we need two sides and the measure of their included angle. Using the law of cosines, we can get a formula for the area of a triangle that involves only the lengths of the sides. This formula is known as **Heron's area formula,** named after Heron of Alexandria, who is believed to have discovered it in about A.D. 75.

Heron's Area Formula

> The area of a triangle with sides a, b, and c is given by the formula
> $$A = \sqrt{S(S - a)(S - b)(S - c)},$$
> where $S = (a + b + c)/2$.

Proof   First rewrite the equation $a^2 = b^2 + c^2 - 2bc \cos \alpha$ as follows:

$$2bc \cos \alpha = b^2 + c^2 - a^2$$
$$4b^2c^2 \cos^2 \alpha = (b^2 + c^2 - a^2)^2$$

Now write the area formula $A = \frac{1}{2}bc \sin \alpha$ in terms of $4b^2c^2 \cos \alpha$:

$$A = \frac{1}{2}bc \sin \alpha$$

$$4A = 2bc \sin \alpha$$

$$16A^2 = 4b^2c^2 \sin^2 \alpha \qquad \text{Square both sides.}$$

$$16A^2 = 4b^2c^2(1 - \cos^2 \alpha) \qquad \text{Pythagorean identity}$$

$$16A^2 = 4b^2c^2 - 4b^2c^2 \cos^2 \alpha$$

Using the expression for $4b^2c^2 \cos^2 \alpha$ obtained above, we get

$$16A^2 = 4b^2c^2 - (b^2 + c^2 - a^2)^2.$$

The last equation could be solved for A in terms of the lengths of the three sides, but it would not be Heron's formula. To get Heron's formula, let

$$S = \frac{(a + b + c)}{2} \qquad \text{or} \qquad 2S = a + b + c.$$

It can be shown (with some effort) that

$$4b^2c^2 - (b^2 + c^2 - a^2)^2 = 2S(2S - 2a)(2S - 2b)(2S - 2c).$$

Using this fact in the above equation yields Heron's formula:

$$16A^2 = 16S(S - a)(S - b)(S - c)$$
$$A^2 = S(S - a)(S - b)(S - c)$$
$$A = \sqrt{S(S - a)(S - b)(S - c)} \qquad \blacklozenge$$

Example 4   Area of a triangle using only the sides

A piece of property in downtown Houston is advertised for sale at $45 per square foot. If the lengths of the sides of the triangular lot are 220 feet, 234 feet, and 160 feet, then what is the asking price for the lot?

Solution

Using Heron's formula with

$$S = \frac{220 + 234 + 160}{2} = 307,$$

we get

$$A = \sqrt{307(307 - 220)(307 - 234)(307 - 160)} \approx 16{,}929.69 \text{ square feet.}$$

At $45 per square foot, the asking price is $761,836.   ◆

Applications

When you reach for a light switch, your brain controls the angles at your elbow and your shoulder that put your hand at the location of the switch. An arm on a robot works in much the same manner. In the next example we see how the law of sines and the law of cosines are used to determine the proper angles at the joints so that a robot's hand can be moved to a position given in the coordinates of the workspace.

Example 5   Positioning a robotic arm

A robotic arm with a 0.5-meter segment and a 0.3-meter segment is attached at the origin as shown in Fig. 8.19. The computer controlled arm is positioned by rotating each segment through angles θ_1 and θ_2 as shown in Fig. 8.19. Given that we want to have the end of the arm at the point (0.7, 0.2), find θ_1 and θ_2.

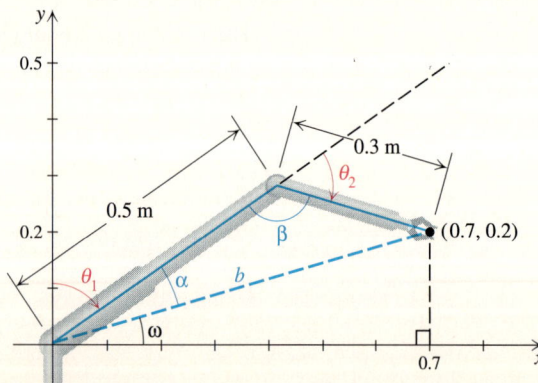

Figure 8.19

Solution

In the right triangle shown in Fig. 8.19, we have $\tan \omega = 0.2/0.7$ and

$$\omega = \tan^{-1}\left(\frac{0.2}{0.7}\right) \approx 15.95°.$$

Find b using the Pythagorean theorem:

$$b = \sqrt{0.7^2 + 0.2^2} \approx 0.73$$

Find β using the law of cosines:

$$(0.73)^2 = 0.5^2 + 0.3^2 - 2(0.5)(0.3) \cos \beta$$

$$\cos \beta = \frac{0.73^2 - 0.5^2 - 0.3^2}{-2(0.5)(0.3)} = -0.643$$

$$\beta = \cos^{-1}(-0.643) \approx 130.02°$$

Find α using the law of sines:

$$\frac{\sin \alpha}{0.3} = \frac{\sin 130.02°}{0.73}$$

$$\sin \alpha = \frac{0.3 \cdot \sin 130.02°}{0.73} = 0.3147$$

$$\alpha = \sin^{-1}(0.3147) \approx 18.34°$$

Since $\theta_1 = 90° - \alpha - \omega$,

$$\theta_1 = 90° - 18.34° - 15.95° = 55.71°.$$

Since β and θ_2 are supplementary,

$$\theta_2 = 180° - 130.02° = 49.98°.$$

So the longer segment of the arm is rotated 55.71° and the shorter segment is rotated 49.98°. ◆

Since the angles in Example 4 describe a clockwise rotation, the angles could be given negative signs to indicate the direction of rotation. In robotics, the direction of rotation is important, because there may be more than one way to position a robotic arm at a desired location. In fact, Example 4 has infinitely many solutions. See if you can find another one.

? For Thought

True or false? Explain.

1. If $\gamma = 90°$ in triangle ABC, then $c^2 = a^2 + b^2$.
2. If a, b, and c are the sides of a triangle, then $a = \sqrt{c^2 + b^2 - 2bc \cos \gamma}$.
3. If a, b, and c are the sides of any triangle, then $c^2 = a^2 + b^2$.
4. The smallest angle of a triangle lies opposite the shortest side.

5. The equation $\cos \alpha = -0.3421$ has two solutions in $[0°, 180°]$.

6. If the largest angle of a triangle is obtuse, then the other two are acute.

7. The equation $\sin \beta = 0.1235$ has two solutions in $[0°, 180°]$.

8. In the SSS case of solving a triangle it is best to find the largest angle first.

9. There is no triangle with sides $a = 3.4$, $b = 4.2$, and $c = 8.1$.

10. There is no triangle with $\alpha = 179.9°$, $b = 1$ inch, and $c = 1$ mile.

8.2 Exercises   Tape 15   Disk—5.25″: 5   3.5″: 3   Macintosh: 3

Solve each triangle.

1.

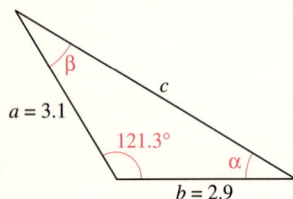

2.

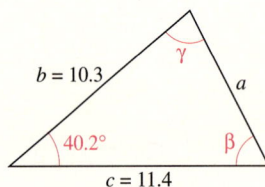

3.

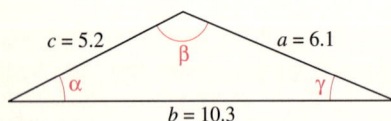

4.

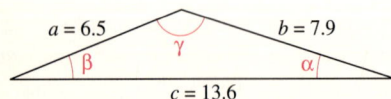

Solve each triangle with the given information.

5. $a = 6.8$, $c = 2.4$, $\beta = 10.5°$

6. $a = 1.3$, $b = 14.9$, $\gamma = 9.8°$

7. $a = 18.5$, $b = 12.2$, $c = 8.1$

8. $a = 30.4$, $b = 28.9$, $c = 31.6$

9. $b = 9.3$, $c = 12.2$, $\alpha = 30°$

10. $a = 10.3$, $c = 8.4$, $\beta = 88°$

11. $a = 6.3$, $b = 7.1$, $c = 6.8$

12. $a = 4.1$, $b = 9.8$, $c = 6.2$

Find the area of each triangle using Heron's formula.

13. $a = 16$, $b = 9$, $c = 10$

14. $a = 12$, $b = 8$, $c = 17$

15. $a = 3.6$, $b = 9.8$, $c = 8.1$

16. $a = 5.4$, $b = 8.2$, $c = 12.0$

17. $a = 346$, $b = 234$, $c = 422$

18. $a = 124.8$, $b = 86.4$, $c = 154.2$

Use the most appropriate formula for the area of a triangle to find the area of each triangle.

19.

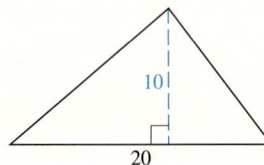

20.

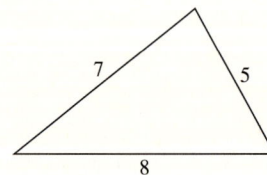

21.

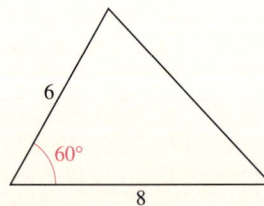

22.

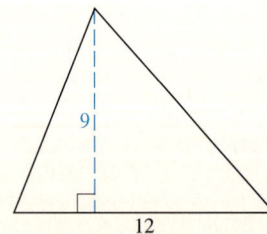

23.

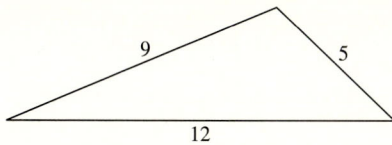

24.

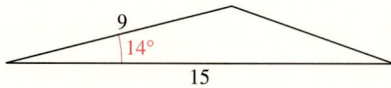

Solve each problem.

25. *Hiking*   Jan and Dean started hiking from the same location at the same time. Jan hiked at 4 mph with bearing N12°E, and Dean hiked at 5 mph with bearing N31°W. How far apart were they after 6 hr?

26. *Flying*   Andrea and Carlos left the airport at the same time. Andrea flew at 180 mph on a course with bearing 80°, and Carlos flew at 240 mph on a course with bearing 210°. How far apart were they after 3 hr?

27. *Positioning a solar panel*   A solar panel with a width of 1.2 m is positioned on a flat roof as shown in the figure. What is the angle of elevation α of the solar panel?

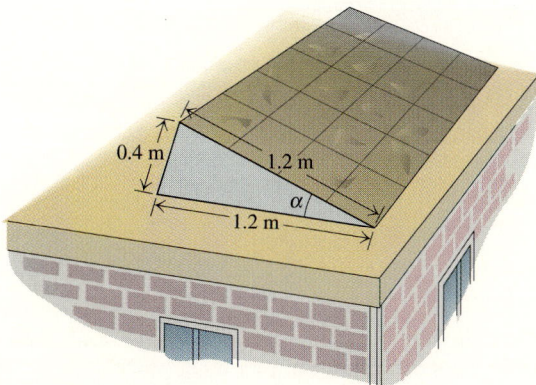

Figure for Exercise 27

28. *Installing an antenna*   A 6-ft antenna is installed at the top of a roof as shown in the figure. A guy wire is to be attached to the top of the antenna and to a point 10 ft down the roof. If the angle of elevation of the roof is 28°, then what length guy wire is needed?

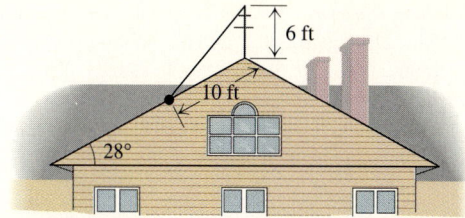

Figure for Exercise 28

29. *Rotating gears*   An inventor wants to mount three gears at the vertices of a triangle as shown in the figure. If gears A, B, and C have radii 2 in., 3 in., and 4 in., respectively, then what are the measures of the angles of triangle ABC? If the angular velocity of gear A is 12π radians per second in a clockwise direction, then what are the angular velocity and direction of rotation of gears B and C?

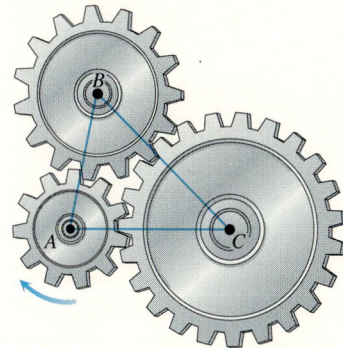

Figure for Exercise 29

30. *Firing a torpedo*   A submarine sights a moving target at a distance of 820 m. A torpedo is fired 9° ahead of the target as shown in the drawing and travels 924 m in a straight line to hit the target. How far has the target moved from the time the torpedo is fired to the time of the hit?

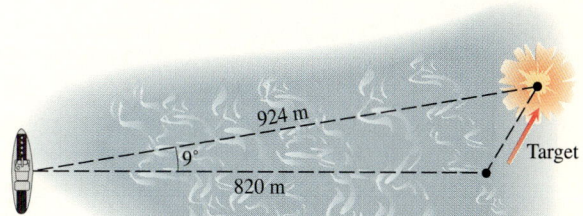

Figure for Exercise 30

31. *Planning a tunnel*   A tunnel is planned through a mountain to connect points A and B on two existing roads as shown in the figure. If the angle between the roads at point C is 28°, what is the distance from point A to B?

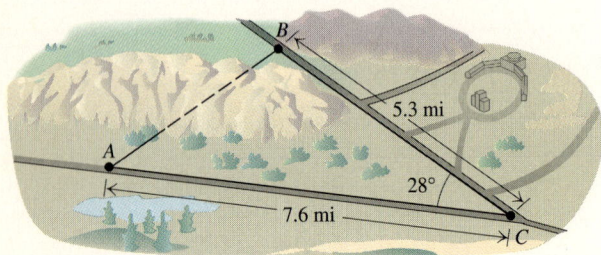

Figure for Exercise 31

32. *Digging a tunnel*   To simultaneously dig the tunnel in Exercise 31 from both sides of the mountain, it is important to know the measures of $\angle CBA$ and $\angle CAB$. Find them to the nearest tenth of a degree.

33. *Side of a pentagon*   A regular pentagon is inscribed in a circle of radius 10 m. Find the length of a side of the pentagon.

34. *Central angle*   A central angle α in a circle of radius 5 m intercepts a chord of length 1 m. What is the measure of α?

35. *Making a shaft*   The end of a steel shaft for an electric motor is to be machined so that it has three flat sides of equal widths as shown in the figure. Given that the radius of the shaft is 10 mm and the length of the arc between each pair of flat sides must be 2.5 mm, find the width s of each flat side.

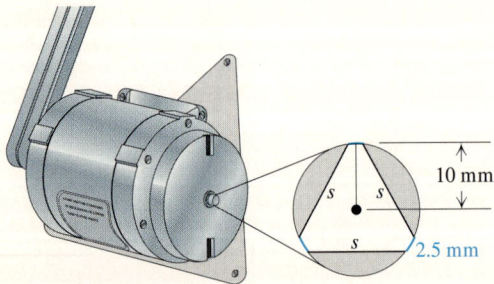

Figure for Exercise 35

36. *Positioning a human arm*   A human arm consists of an upper arm of 30 cm and a lower arm of 30 cm as shown in the figure. To move the hand to the point (36, 8), the human brain chooses angle θ_1 and θ_2 as shown in the figure. Find θ_1 and θ_2 to the nearest tenth of a degree.

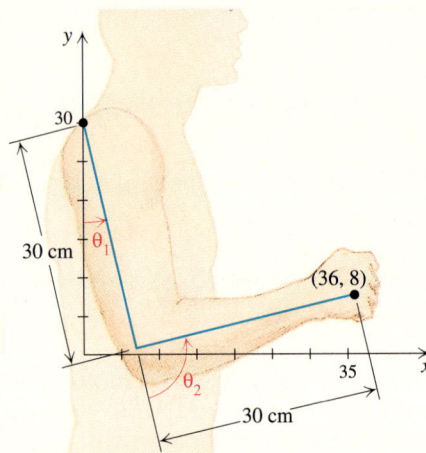

Figure for Exercise 36

For Writing/Discussion

37. A central angle θ in a circle of radius r intercepts a chord of length a, where $0° \leq \theta \leq 180°$. Show that $a = 2r \sin(\theta/2)$.

38. Explain why the second largest side in a triangle with unequal sides is opposite an acute angle.

39. Explain why the Pythagorean theorem is a special case of the law of cosines.

40. Find the area of the triangle with sides 37, 48, and 86, using Heron's formula. Explain your result.

41. Find the area of the triangle with sides 31, 87, and 56, using Heron's formula. Explain your result.

42. Find the area of the triangle with sides of length 6 ft, 9 ft, and 13 ft by using the formula

$$A = \frac{1}{4}\sqrt{4b^2c^2 - (b^2 + c^2 - a^2)^2},$$

and check your result using a different formula for the area of a triangle. Prove that this formula gives the area of any triangle with sides a, b, and c.

8.3

Vectors

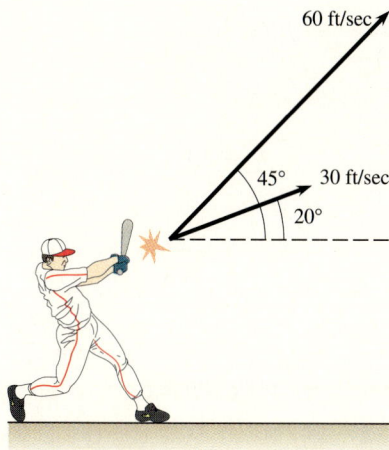

Figure 8.20

By using the law of sines and the law of cosines we can find all of the missing parts of any triangle for which we have enough information to determine its shape. In this section we will use these and many other tools of trigonometry in the study of vectors.

Definitions

Quantities such as length, area, volume, temperature, and time have magnitude only and are completely characterized by a single real number with appropriate units (such as feet, degrees, or hours). Such quantities are called **scalar quantities,** and the corresponding real numbers are **scalars.** Quantities that involve both a magnitude and a direction, such as velocity, acceleration, and force, are **vector quantities,** and they can be represented by **directed line segments.** These directed line segments are called **vectors.**

For an example of vectors, consider two baseballs hit into the air as shown in Fig. 8.20. One is hit with an initial velocity of 30 feet per second at an angle of 20° from the horizontal and the other is hit with an initial velocity of 60 feet per second at an angle of 45° from the horizontal. The length of a vector represents the **magnitude** of the vector quantity. So the vector representing the faster baseball is drawn twice as long as the other vector. The **direction** is indicated by the position of the vector and the arrowhead at one end.

We have used the notation $\overline{AB}$ to name a line segment with endpoints A and B and $\overrightarrow{AB}$ to name a ray with initial point A and passing through B. When studying vectors, the notation $\overrightarrow{AB}$ is used to name a vector with **initial point** A and **terminal point** B. The vector $\overrightarrow{AB}$ terminates at B, while the ray $\overrightarrow{AB}$ goes beyond B. Since the direction of the vector is the same as the order of the letters, to indicate a vector in print we will omit the arrow and use boldface type. In handwritten work an arrow over the letter or letters indicates a vector. Thus **AB** and $\overrightarrow{AB}$ represent the same vector. The vector **BA** is a vector with initial point B and terminal point A, and is not the same as **AB**. A vector whose endpoints are not specified is named by a single uppercase or lowercase letter. For example, **b**, **B**, $\overrightarrow{b}$, and $\overrightarrow{B}$ are names of vectors. The magnitude of vector **A** is written as $|\mathbf{A}|$.

Two vectors are **equal** if they have the same magnitude and the same direction. Equal vectors may be in different locations. They do not necessarily coincide. In Fig. 8.21, **A** = **B** because they have the same direction and the same magnitude. Vector **B** is *not* equal to vector **C** because they have the same magnitude but opposite direction. **C** ≠ **D** because they have different magnitudes and different directions. The **zero vector** is a vector that has no magnitude and no direction. It is denoted by a boldface zero, **0**.

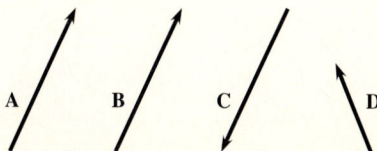

Figure 8.21

Scalar Multiplication and Addition

Suppose two tugboats are pulling on a barge that has run aground in the Mississippi River as shown in Fig. 8.22.

Figure 8.22

The smaller boat exerts a force of 2000 pounds in an easterly direction and its force is represented by the vector **A**. If the larger boat exerts a force of 4000 pounds in the same direction, then its force can be represented by the vector 2**A**. This example illustrates the operation of **scalar multiplication.**

Definition: Scalar Multiplication

For any scalar k and vector **A**, k**A** is a vector with magnitude $|k|$ times the magnitude of **A**. If $k > 0$, then the direction of k**A** is the same as the direction of **A**. If $k < 0$, the direction of k**A** is opposite to the direction of **A**. If $k = 0$, then k**A** $= \mathbf{0}$.

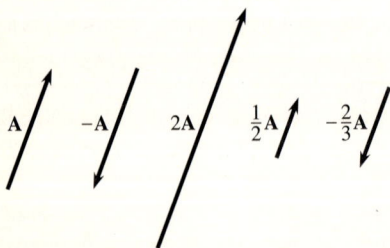

The vector 2**A** has twice the magnitude of **A**, and $-\frac{2}{3}$**A** has two-thirds of the magnitude. For -1**A** we write $-$**A**. Some examples of scalar multiplication are shown in Fig. 8.23.

Suppose two draft horses are pulling on a tree stump with forces of 200 pounds and 300 pounds as shown in Fig. 8.24, with an angle of 65° between the forces.

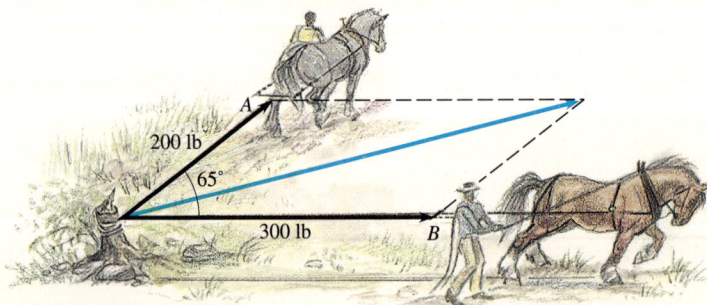

Figure 8.23

Figure 8.24

The two forces are represented by vectors **A** and **B**. If **A** and **B** had the same direction, then there would be a total force of 500 pounds acting on the stump. However, a total force of 500 pounds is not achieved because of the angle between the forces. It can be shown by experimentation that one force acting along the diagonal of the parallelogram shown in Fig. 8.24, with a magnitude equal to the length of the diagonal, has the same effect on the stump as the two forces **A** and **B**. In physics, this result is known as the **parallelogram law.** The single force **A** + **B** acting along the diagonal is called the **sum** or **resultant** of **A** and **B**. The parallelogram law motivates our definition of vector addition.

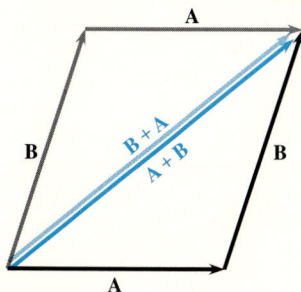

Definition: Vector Addition

Figure 8.25

To find the resultant or sum **A** + **B** of any vectors **A** and **B**, position **B** (without changing its magnitude or direction) so that the initial point of **B** coincides with the terminal point of **A**, as in Fig. 8.25. The vector that begins at the initial point of **A** and terminates at the terminal point of **B** is the vector **A** + **B**.

Note that the vector **A** + **B** in Fig. 8.25 coincides with the diagonal of a parallelogram whose adjacent sides are **A** and **B**. We can see from Fig. 8.25 that **A** + **B** = **B** + **A**, and vector addition is commutative. If **A** and **B** have the same direction or opposite directions, then no parallelogram is formed, but **A** + **B** can be found by the procedure given in the definition. Each vector in the sum **A** + **B** is called a **component** of the sum.

For every vector **A** there is a vector −**A**, the **opposite** of **A**, having the same magnitude as **A** but opposite direction. The sum of a vector and its opposite is the zero vector, **A** + (−**A**) = **0**. Vector subtraction is defined like subtraction of real numbers. For any two vectors **A** and **B**, **A** − **B** = **A** + (−**B**). Fig. 8.26 shows **A** − **B**.

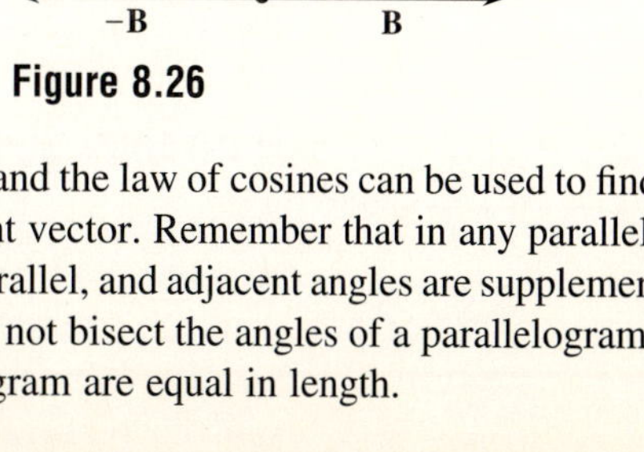

Figure 8.26

The law of sines and the law of cosines can be used to find the magnitude and direction of a resultant vector. Remember that in any parallelogram the opposite sides are equal and parallel, and adjacent angles are supplementary. The diagonals of a parallelogram do not bisect the angles of a parallelogram unless the adjacent sides of the parallelogram are equal in length.

Example 1   Magnitude and direction of a resultant

Two draft horses are pulling on a tree stump with forces of 200 pounds and 300 pounds, as shown in Fig. 8.27. If the angle between the forces is 65°, then what is the magnitude of the resultant force? What is the angle between the resultant and the 300-pound force?

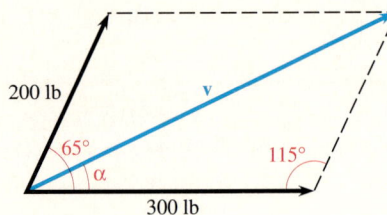

Figure 8.27

Solution

The resultant coincides with the diagonal of the parallelogram shown in Fig. 8.27. We can view the resultant as a side of a triangle in which the other two sides are 200 and 300, and their included angle is 115°. The law of cosines can be used to find the magnitude of the resultant vector $\mathbf{v}$:

$$|\mathbf{v}|^2 = 300^2 + 200^2 - 2(300)(200) \cos 115° = 180,714.19$$

$$|\mathbf{v}| = 425.1$$

We can use the law of sines to find the angle α between the resultant and the 300-pound force:

$$\frac{\sin \alpha}{200} = \frac{\sin 115°}{425.1}$$

$$\sin \alpha = 0.4264$$

Since $\sin^{-1}(0.4264) = 25.2°$, both 25.2° and 154.8° are solutions to $\sin \alpha = 0.4264$. However, α must be smaller than 65°, so the angle between the resultant and the 300-pound force is 25.2°. So one horse pulling in the direction of the resultant with a force of 425.1 pounds would have the same effect on the stump as the two horses pulling at an angle of 65°. ◆

Horizontal and Vertical Components

When a peregrine falcon dives at its prey with a velocity of 120 feet per second at an angle of 40° from the horizontal, its location is changing both horizontally and vertically. So its velocity vector can be thought of as a sum of a horizontal velocity vector and a vertical velocity vector. In fact, any nonzero vector $\mathbf{w}$ is the sum of a **vertical component** and a **horizontal component** as shown in Fig. 8.28.

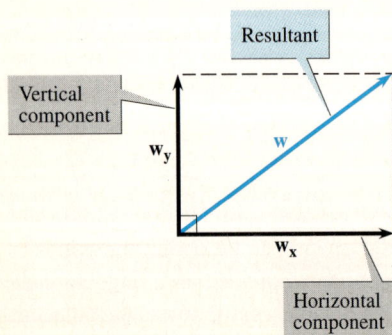

Figure 8.28

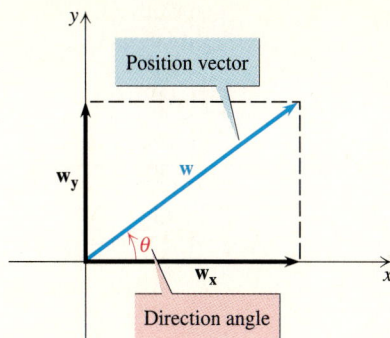

Figure 8.29

The horizontal component is denoted $\mathbf{w_x}$ and the vertical component is denoted $\mathbf{w_y}$. The vector $\mathbf{w}$ is the diagonal of the rectangle formed by the vertical and horizontal components.

If a vector $\mathbf{w}$ is placed in a rectangular coordinate system so that its initial point is the origin (as in Fig. 8.29), then $\mathbf{w}$ is called a **position vector** or **radius vector.** A position vector is a convenient representative to focus on when considering all vectors that are equal to a certain vector. The angle θ ($0° \leq \theta < 360°$) formed by the positive x-axis and a position vector is the **direction angle** for the position vector (or any other vector that is equal to the position vector).

If the vector $\mathbf{w}$ has magnitude r, direction angle θ, horizontal component $\mathbf{w_x}$, and vertical component $\mathbf{w_y}$, as shown in Fig. 8.29, then by using trigonometric ratios we get

$$\cos \theta = \frac{|\mathbf{w_x}|}{r} \qquad \text{and} \qquad \sin \theta = \frac{|\mathbf{w_y}|}{r}$$

or

$$|\mathbf{w_x}| = r \cos \theta \qquad \text{and} \qquad |\mathbf{w_y}| = r \sin \theta.$$

If the direction of $\mathbf{w}$ is such that $\sin \theta$ or $\cos \theta$ is negative, then we can write

$$|\mathbf{w_x}| = |r \cos \theta| \qquad \text{and} \qquad |\mathbf{w_y}| = |r \sin \theta|.$$

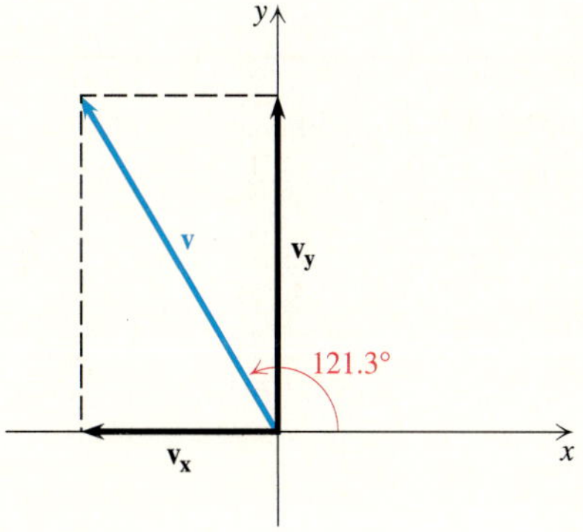

Example 2   Finding horizontal and vertical components

Find the magnitude of the horizontal and vertical components for a vector $\mathbf{v}$ with magnitude 8.3 and direction angle 121.3°.

Solution

Figure 8.30

The vector $\mathbf{v}$ and its horizontal and vertical components $\mathbf{v_x}$ and $\mathbf{v_y}$ are shown in Fig. 8.30. The magnitudes of $\mathbf{v_x}$ and $\mathbf{v_y}$ are found as follows:

$$|\mathbf{v_x}| = |8.3 \cos 121.3°| = 4.3$$
$$|\mathbf{v_y}| = |8.3 \sin 121.3°| = 7.1$$

The direction angle for $\mathbf{v_x}$ is 180°, and the direction angle for $\mathbf{v_y}$ is 90°. ◆

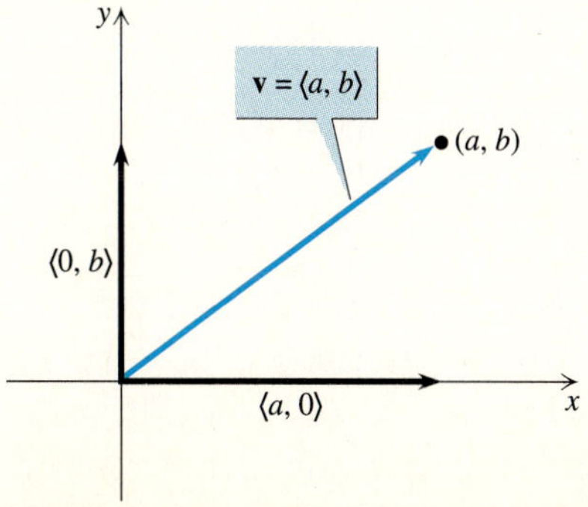

Component Form of a Vector

Any vector is the resultant of its horizontal and vertical components. Since the horizontal and vertical components of a vector determine the vector, it is convenient to use a notation for vectors that involves them. The notation $\langle a, b \rangle$ is used for the position vector with terminal point (a, b) as shown in Fig. 8.31. The form $\langle a, b \rangle$ is called **component form** because its horizontal component is $\langle a, 0 \rangle$ and its vertical component is $\langle 0, b \rangle$. Since the vector $\mathbf{v} = \langle a, b \rangle$ extends from $(0, 0)$ to (a, b), its magnitude is the distance between these points:

$$|\mathbf{v}| = \sqrt{a^2 + b^2}$$

Figure 8.31

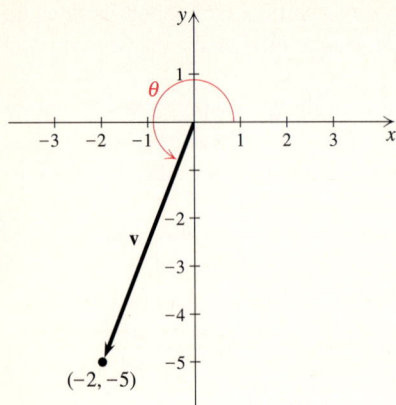

Figure 8.32

In Example 3 we find the magnitude and direction for a vector given in component form and in Example 4 we find the component form for a vector given as a directed line segment.

Example 3 Finding magnitude and direction from component form

Find the magnitude and direction angle of the vector $\mathbf{v} = \langle -2, -5 \rangle$.

Solution

The vector $\mathbf{v}$ shown in Fig. 8.32 has magnitude

$$|\mathbf{v}| = \sqrt{(-2)^2 + (-5)^2} = \sqrt{29}.$$

To find the direction angle θ, use trigonometric ratios to get $\sin \theta = -5/\sqrt{29}$. Since $\sin^{-1}(-5/\sqrt{29}) = -68.2°$, we have $\theta = 180° - (-68.2°) = 248.2°$.

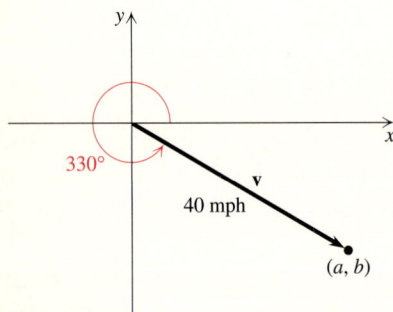

Figure 8.33

Example 4 Finding the component form given magnitude and direction

Find the component form for a vector of magnitude 40 mph with direction angle 330°.

Solution

To find the component form, we need the terminal point (a, b) for a vector of magnitude 40 mph positioned as shown in Fig. 8.33. Since $a = r \cos \theta$ and $b = r \sin \theta$, we have

$$a = 40 \cos 330° = 40 \, \frac{\sqrt{3}}{2} = 20\sqrt{3} \qquad \text{and}$$

$$b = 40 \sin 330° = 40\left(-\frac{1}{2}\right) = -20.$$

So the component form of the vector is $\langle 20\sqrt{3}, -20 \rangle$.

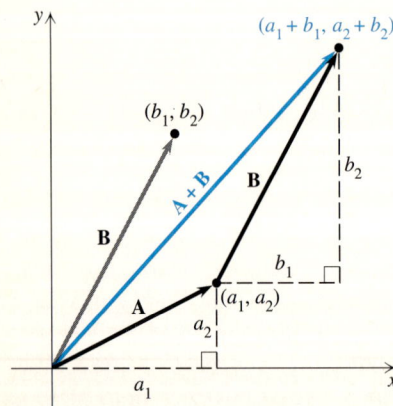

Figure 8.34

We originally defined vectors as directed line segments and performed operations with these directed line segments. When vectors are written in component form, operations with vectors are easier to perform. Figure 8.34 illustrates the addition of $\mathbf{A} = \langle a_1, a_2 \rangle$ and $\mathbf{B} = \langle b_1, b_2 \rangle$, where (a_1, a_2) and (b_1, b_2) are in the first quadrant. It is easy to see that the endpoint of $\mathbf{A} + \mathbf{B}$ is $(a_1 + b_1, a_2 + b_2)$ and so $\mathbf{A} + \mathbf{B} = \langle a_1 + b_1, a_2 + b_2 \rangle$. The sum can be found in component form by adding the components instead of drawing directed line segments. We leave it to the reader to show that the other operations can be performed according to the following rules.

Rules for Scalar Product, Vector Sum, Vector Difference

If $\mathbf{A} = \langle a_1, a_2 \rangle$, $\mathbf{B} = \langle b_1, b_2 \rangle$, and k is a scalar, then

1. $k\mathbf{A} = \langle ka_1, ka_2 \rangle$ **Scalar product**
2. $\mathbf{A} + \mathbf{B} = \langle a_1 + b_1, a_2 + b_2 \rangle$ **Vector sum**
3. $\mathbf{A} - \mathbf{B} = \langle a_1 - b_1, a_2 - b_2 \rangle$ **Vector difference**

Example 5   Operations with vectors in component form

Let $\mathbf{w} = \langle -3, 2 \rangle$ and $\mathbf{z} = \langle 5, -1 \rangle$. Perform the operations indicated.

a) $\mathbf{w} - \mathbf{z}$ b) $-8\mathbf{z}$ c) $3\mathbf{w} + 4\mathbf{z}$

Solution

a) $\mathbf{w} - \mathbf{z} = \langle -3, 2 \rangle - \langle 5, -1 \rangle = \langle -8, 3 \rangle$

b) $-8\mathbf{z} = -8\langle 5, -1 \rangle = \langle -40, 8 \rangle$

c) $3\mathbf{w} + 4\mathbf{z} = 3\langle -3, 2 \rangle + 4\langle 5, -1 \rangle = \langle -9, 6 \rangle + \langle 20, -4 \rangle = \langle 11, 2 \rangle$ ◆

The vectors $\mathbf{i} = \langle 1, 0 \rangle$ and $\mathbf{j} = \langle 0, 1 \rangle$ are called **unit vectors** because each has magnitude one. For any vector $\langle a_1, a_2 \rangle$ as shown in Fig. 8.35, we have

$$\langle a_1, a_2 \rangle = a_1 \langle 1, 0 \rangle + a_2 \langle 0, 1 \rangle = a_1 \mathbf{i} + a_2 \mathbf{j}.$$

The form $a_1 \mathbf{i} + a_2 \mathbf{j}$ is called a **linear combination** of the vectors $\mathbf{i}$ and $\mathbf{j}$. These unit vectors are thought of as fundamental vectors, because any vector can be expressed as a linear combination of them.

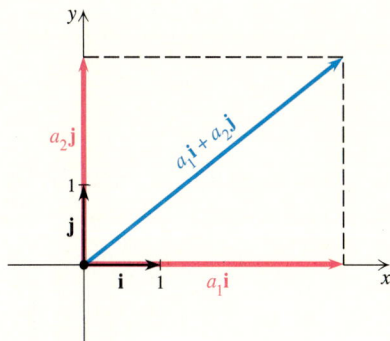

Figure 8.35

Example 6   Unit vectors

Write each vector as a linear combination of the unit vectors $\mathbf{i}$ and $\mathbf{j}$.

a) $\mathbf{A} = \langle -2, 6 \rangle$ b) $\mathbf{B} = \langle -4, -1 \rangle$

Solution

a) $\mathbf{A} = \langle -2, 6 \rangle = -2\mathbf{i} + 6\mathbf{j}$

b) $\mathbf{B} = \langle -4, -1 \rangle = -4\mathbf{i} - \mathbf{j}$ ◆

Applications

The applications of vectors usually involve triangles and trigonometry. When forces are acting on an object, we generally neglect friction and assume that all forces are acting on a single point.

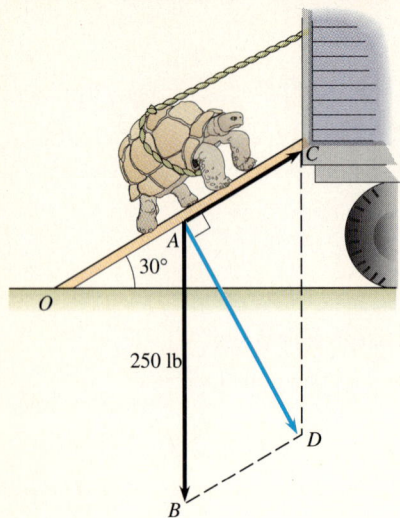

Figure 8.36

Example 7 Finding a force

Workers at the Audubon Zoo must move a giant tortoise to his new home. Find the amount of force required to pull a 250-pound tortoise up a ramp leading into a truck. The angle of elevation of the ramp is 30°. See Fig. 8.36.

Solution

The weight of the tortoise is a 250-pound force in a downward direction, shown as vector **AB** in Fig. 8.36. The tortoise exerts a force against the ramp at a 90° angle with the ramp, shown as vector **AD**. The magnitude of **AD** is less than the magnitude of **AB** because of the incline of the ramp. The force required to pull the tortoise up the ramp is vector **AC** in Fig. 8.36. The force against the ramp, **AD**, is the resultant of **AB** and **AC**. Since $\angle O = 30°$, we have $\angle OAB = 60°$ and $\angle BAD = 30°$. Because $\angle CAD = 90°$, we have $\angle ADB = 90°$. We can find any of the missing parts to the right triangle ABD. Since

$$\sin 30° = \frac{\text{opposite}}{\text{hypotenuse}} = \frac{|\mathbf{BD}|}{|\mathbf{AB}|},$$

we have

$$|\mathbf{BD}| = |\mathbf{AB}|\sin 30° = 250 \cdot \frac{1}{2} = 125 \text{ pounds.}$$

Since the opposite sides of a parallelogram are equal, the magnitude of **AC** is also 125 pounds. ◆

If an airplane heads directly against the wind or with the wind, the wind speed is subtracted from or added to the air speed of the plane to get the ground speed of the plane. When the wind is at some other angle to the direction of the plane, some portion of the wind speed will be subtracted from or added to the air speed to determine the ground speed. In addition, the wind causes the plane to travel on a course different from where it is headed. We can use the vector $\mathbf{v_1}$ to represent the heading and air speed of the plane, as shown in Fig. 8.37.

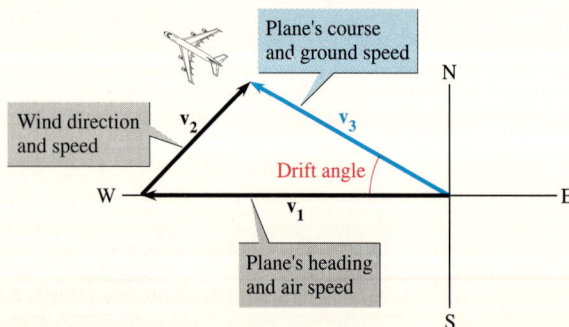

Figure 8.37

The vector $\mathbf{v_2}$ represents the wind direction and speed. The resultant of $\mathbf{v_1}$ and $\mathbf{v_2}$ is the vector $\mathbf{v_3}$, where $\mathbf{v_3}$ represents the course and ground speed of the plane. The angle between the heading and the course is the **drift angle.** Recall that the bearing of a vector used to describe direction in air navigation is a nonnegative angle smaller than 360° measured in a clockwise direction from due north. For example, the wind in Fig. 8.37 is out of the southwest and has a bearing of 45°. The vector $\mathbf{v_1}$ describing the heading of the plane points due west and has a bearing of 270°.

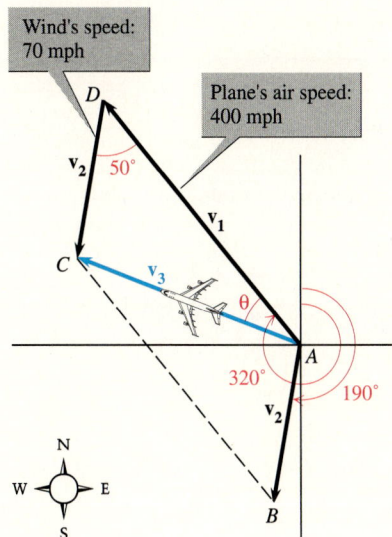

Wind's speed: 70 mph

Plane's air speed: 400 mph

Figure 8.38

Example 8   Finding course and ground speed of an airplane

The heading of an executive's Lear jet has a bearing of 320°. The wind is 70 mph with a bearing of 190°. Given that the air speed of the plane is 400 mph, find the drift angle, the ground speed, and the course of the airplane.

Solution

Figure 8.38 shows vectors for the heading $\mathbf{v_1}$, the wind $\mathbf{v_2}$, and the course $\mathbf{v_3}$. Subtract the bearing of the wind from that of the heading to get $\angle DAB = 320° - 190° = 130°$. Since $ABCD$ is a parallelogram, $\angle CDA = 50°$. Apply the law of cosines to triangle ACD to get

$$|\mathbf{v_3}|^2 = 70^2 + 400^2 - 2(70)(400)\cos 50° = 128{,}903.9$$

$$|\mathbf{v_3}| \approx 359.0$$

The ground speed is approximately 359.0 mph. The drift angle θ is found by using the law of sines:

$$\frac{\sin \theta}{70} = \frac{\sin 50°}{359.0}$$

$$\sin \theta = 0.1494$$

Since θ is an acute angle, $\theta = \sin^{-1}(0.1494) = 8.6°$. The course $\mathbf{v_3}$ has a bearing of $320° - 8.6° = 311.4°$. ◆

? For Thought

True or false? Explain.

1.   The vector $2\mathbf{v}$ has the same direction as $\mathbf{v}$ but twice the magnitude of $\mathbf{v}$.

2.   The magnitude of $\mathbf{A} + \mathbf{B}$ is the sum of the magnitudes of $\mathbf{A}$ and $\mathbf{B}$.

3.   The magnitude of $-\mathbf{A}$ is equal to the magnitude of $\mathbf{A}$.

4.   For any vector $\mathbf{A}$, $\mathbf{A} + (-\mathbf{A}) = \mathbf{0}$.

5.   The parallelogram law says that the opposite sides of a parallelogram are equal in length.

(continued on next page)

6. In the coordinate plane, the direction angle is the angle formed by the y-axis and a vector with initial point at the origin.

7. If v has magnitude r and direction angle θ, then the horizontal component of v is a vector with direction angle $0°$ and magnitude $r \cos \theta$.

8. The magnitude of $\langle 3, -4 \rangle$ is 5.

9. The vectors $\langle -2, 3 \rangle$ and $\langle -6, 9 \rangle$ have the same direction angle.

10. The direction angle of $\langle -2, 2 \rangle$ is $\cos^{-1}\left(-2/\sqrt{8}\right)$.

8.3 Exercises 🖭 Tape 15 💾 Disk—5.25″: 5   3.5″: 3   Macintosh: 3

Find the magnitude of the horizontal and vertical components for each vector v with the given magnitude and given direction angle θ.

1. $|v| = 4.5$, $\theta = 65.2°$ 2. $|v| = 6000$, $\theta = 13.1°$

3. $|v| = 8000$, $\theta = 155.1°$ 4. $|v| = 445$, $\theta = 211.1°$

5. $|v| = 234$, $\theta = 248°$ 6. $|v| = 48.3$, $\theta = 349°$

Find the magnitude and direction angle of each vector.

7. $\langle \sqrt{3}, 1 \rangle$ 8. $\langle -1, \sqrt{3} \rangle$

9. $\langle -\sqrt{2}, \sqrt{2} \rangle$ 10. $\langle \sqrt{2}, -\sqrt{2} \rangle$

11. $\langle 8, -8\sqrt{3} \rangle$ 12. $\langle -1/2, -\sqrt{3}/2 \rangle$

13. $\langle 5, 0 \rangle$ 14. $\langle 0, -6 \rangle$

15. $\langle -3, 2 \rangle$ 16. $\langle -4, -2 \rangle$

17. $\langle 3, -1 \rangle$ 18. $\langle 2, -6 \rangle$

Let $r = \langle 3, -2 \rangle$, $s = \langle -1, 5 \rangle$, and $t = \langle 4, -6 \rangle$. Perform the operations indicated. Write the answers in the form $\langle a, b \rangle$.

19. $5r$ 20. $-4s$ 21. $2r + 3t$

22. $r - t$ 23. $s + 3t$ 24. $\dfrac{r + s}{2}$

25. $r - (s + t)$ 26. $r - s - t$

Find the component form for each vector v with the given magnitude and direction angle θ.

27. $|v| = 8$, $\theta = 45°$ 28. $|v| = 12$, $\theta = 120°$

29. $|v| = 290$, $\theta = 145°$ 30. $|v| = 5.3$, $\theta = 321°$

31. $|v| = 18$, $\theta = 347°$ 32. $|v| = 3000$, $\theta = 209.1°$

Write each vector as a linear combination of the unit vectors i and j.

33. $\langle 2, 1 \rangle$ 34. $\langle 1, 5 \rangle$

35. $\langle -3, \sqrt{2} \rangle$ 36. $\langle \sqrt{2}, -5 \rangle$

37. $\langle 0, -9 \rangle$ 38. $\langle -1/2, 0 \rangle$

39. $\langle -7, -1 \rangle$ 40. $\langle 1, 1 \rangle$

Given that $A = \langle 3, 1 \rangle$ and $B = \langle -2, 3 \rangle$, find the magnitude and direction angle for each of the following vectors.

41. $A + B$ 42. $A - B$

43. $-3A$ 44. $5B$

45. $B - A$ 46. $B + A$

47. $-A + \dfrac{1}{2}B$ 48. $\dfrac{1}{2}A - 2B$

For each given pair of vectors, find the magnitude and direction angle of the resultant.

49.

50.

51.

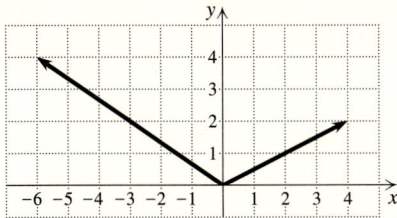

52.

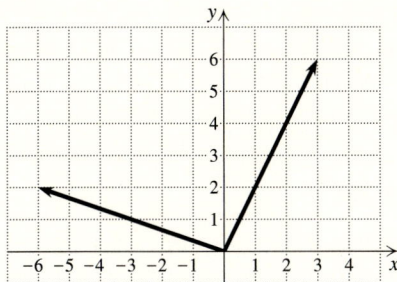

53.

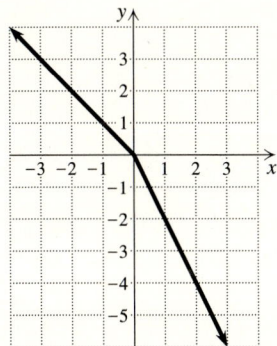

54.

In each case, find the magnitude of the resultant force and the angle between the resultant and each force.

55. Forces of 3 lb and 8 lb act at an angle of 90° to each other.

56. Forces of 2 lb and 12 lb act at an angle of 60° to each other.

57. Forces of 4.2 newtons (a unit of force from physics) and 10.3 newtons act at an angle of 130° to each other.

58. Forces of 34 newtons and 23 newtons act at an angle of 100° to each other.

Solve each problem.

59. *Magnitude of a force* The resultant of a 10-lb force and another force has a magnitude of 12.3 lb at an angle of 23.4° with the 10-lb force. Find the magnitude of the other force and the angle between the two forces.

60. *Magnitude of a force* The resultant of a 15-lb force and another force has a magnitude of 9.8 lb at an angle of 31° with the 15-lb force. Find the magnitude of the other force and the angle between the other force and the resultant.

61. *Moving a donkey* Two prospectors are pulling on ropes attached around the neck of a donkey that does not want to move. One prospector pulls with a force of 55 lb, and the other pulls with a force of 75 lb. If the angle between the ropes is 25°, as shown in the figure, then how much force must the donkey use in order to stay put? (The donkey knows the proper direction in which to apply his force.)

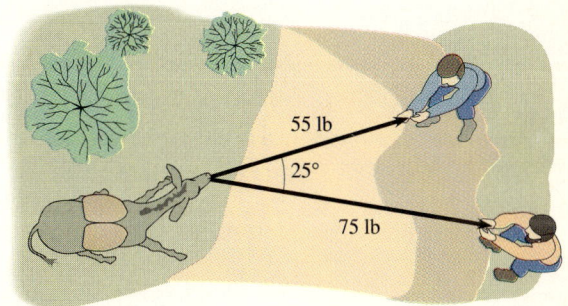

Figure for Exercise 61

62. *Pushing a shopping cart* Ronnie, Phyllis, and Ted are conducting a vector experiment in a Wal-Mart parking lot. Ronnie is pushing a cart containing Phyllis to the east at 5 mph while Ted is pushing it to the north at 3 mph. What is Phyllis's speed and in what direction (measured from north) is she moving?

63. *Gaining altitude* An airplane with an air speed of 520 mph is climbing at an angle of 30° from the horizontal. What are the magnitudes of the horizontal and vertical components of the speed vector?

64. *Acceleration of a missile* A missile is fired with an angle of elevation of 22°, with an acceleration of 30 m/sec². What are the magnitudes of the horizontal and vertical components of the acceleration vector?

65. *Rock and roll* In Roman mythology, Sisyphus, king of Corinth, revealed a secret of Zeus and thus incurred the god's wrath. As punishment, Zeus banished him to Hades, where he was doomed for eternity to roll a rock uphill, only to have it roll back on him. If Sisyphus stands in front of a 4000-lb spherical rock on a 20° incline, as shown in the figure, then what force applied in the direction of the incline would keep the rock from rolling down the incline?

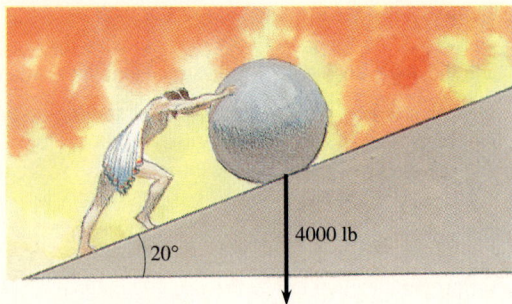

Figure for Exercise 65

66. *Weight of a ball* A solid steel ball is placed on a 10° incline. If a force of 3.2 lb in the direction of the incline is required to keep the ball in place, then what is the weight of the ball?

67. *Course of an airplane* An airplane is heading on a bearing of 102° with an air speed of 480 mph. If the wind is out of the northeast (bearing 225°) at 58 mph, then what are the bearing of the course and the ground speed of the airplane?

68. *Course of a helicopter* The heading of a helicopter has a bearing of 240°. If the 70-mph wind has a bearing of 185° and the air speed of the helicopter is 195 mph, then what are the bearing of the course and the ground speed of the helicopter?

69. *Going with the flow* A river is 2000 ft wide and flowing at 6 mph from north to south. A woman in a canoe starts on the eastern shore and heads west at her normal paddling speed of 2 mph. In what direction (measured clockwise from north) will she actually be traveling? How far downstream from a point directly across the river will she land?

70. *Crossing a river* If the woman in Exercise 69 wants to go directly across the river and she paddles at 8 mph, then in what direction (measured clockwise from north) must she aim her canoe? How long will it take her to go directly across the river?

For Writing/Discussion

71. *Distributive* Prove that scalar multiplication is distributive over vector addition, first using the component form and then using a geometric argument.

72. *Associative* Prove that vector addition is associative, first using the component form and then using a geometric argument.

73. *Cooperative learning* Working with a classmate, find a scale of the type that is used to weigh objects that are suspended from the scale. Weigh an object with this scale. Use vectors to predict the force that would be needed to keep the object from sliding down an inclined plane. Then use the scale to actually measure this force. Repeat this experiment for several different inclines and write a summary of your findings.

8.4

Trigonometric Form

of Complex Numbers

Until now we have concentrated on applying the trigonometric functions to real-life situations. So the trigonometric form of complex numbers, introduced in this section, may be somewhat surprising. However, with this form we can expand our knowledge about complex numbers and find powers and roots of complex numbers that we could not find without it.

The Complex Plane

A complex number $a + bi$ can be thought of as an ordered pair (a, b), where a is the real part and b is the imaginary part of the complex number. Ordered pairs representing complex numbers are found in a coordinate system just like ordered pairs of real numbers. The horizontal axis is called the **real axis,** and the vertical axis is called the **imaginary axis,** as shown in Fig. 8.39. This coordinate system is called the **complex plane.** The complex plane provides an order to the complex number system and allows us to treat complex numbers like vectors.

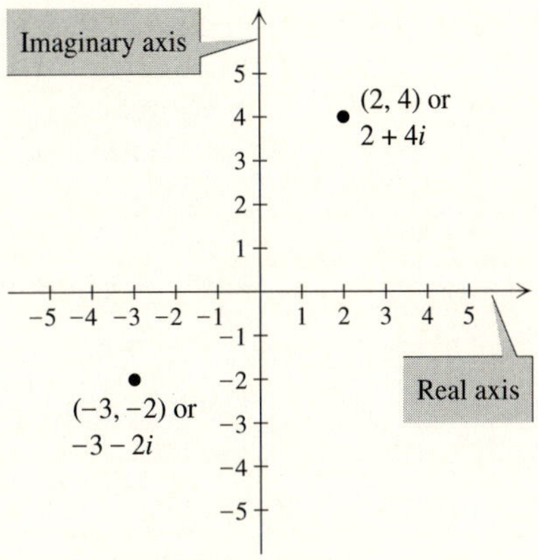

The Complex Plane

Figure 8.39

The absolute value of a real number is the distance on the number line between the number and the origin. The **absolute value of a complex number** is the distance between the complex number and the origin in the complex plane.

Definition: Absolute Value of *a* + *bi*

The absolute value of the complex number $a + bi$ is defined by

$$|a + bi| = \sqrt{a^2 + b^2}.$$

The absolute value of $a + bi$ is the same as the magnitude of the vector $\langle a, b \rangle$.

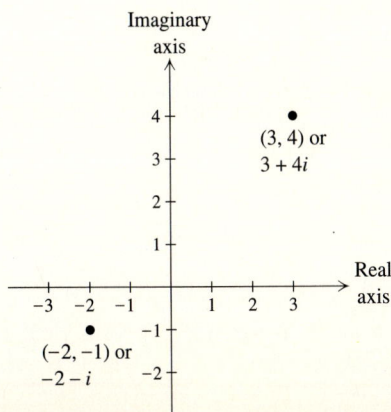

Figure 8.40

Example 1 Graphing complex numbers

Graph each complex number and find its absolute value.

a) $3 + 4i$ **b)** $-2 - i$

Solution

a) The complex number $3 + 4i$ is located in the first quadrant, as shown in Fig. 8.40, three units to the right of the origin and four units upward. Using the definition of absolute value of a complex number, we have

$$|3 + 4i| = \sqrt{3^2 + 4^2} = 5.$$

b) The complex number $-2 - i$ is located in the third quadrant (Fig. 8.40) and

$$|-2 - i| = \sqrt{(-2)^2 + (-1)^2} = \sqrt{5}.$$

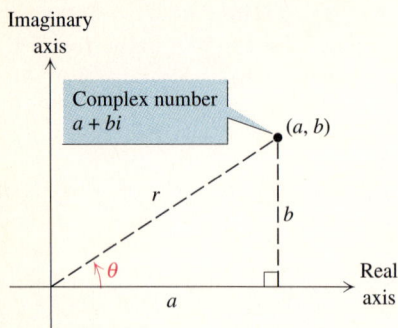

Imaginary axis

Complex number $a + bi$

(a, b)

r

b

θ

a

Real axis

Figure 8.41

Trigonometric Form a Complex Number

Multiplication and division of complex numbers in the standard form $a + bi$ is rather complicated, as is finding powers and roots, but in trigonometric form these operations are simpler and more natural. A complex number is written in **trigonometric form** by a process similar to finding the horizontal and vertical components of a vector.

Consider the complex number $a + bi$ shown in Fig. 8.41, where

$$r = |a + bi| = \sqrt{a^2 + b^2}.$$

Since (a, b) is a point on the terminal side of θ in standard position, we have

$$a = r \cos \theta \quad \text{and} \quad b = r \sin \theta.$$

Substituting for a and b, we get

$$a + bi = (r \cos \theta) + (r \sin \theta)i = r(\cos \theta + i \sin \theta).$$

Definition: Trigonometric Form of a Complex Number

If $z = a + bi$ is a complex number, then the **trigonometric form** of z is

$$z = r(\cos \theta + i \sin \theta),$$

where $r = \sqrt{a^2 + b^2}$ and θ is an angle in standard position whose terminal side contains the point (a, b).

The number r is called the **modulus** of $a + bi$, and θ is called the **argument** of $a + bi$. Since there are infinitely many angles whose terminal side contains (a, b), the trigonometric form of a complex number is not unique. However, we usually use the smallest nonnegative value for θ that satisfies

$$a = r \cos \theta \quad \text{or} \quad b = r \sin \theta.$$

In the next two examples, we convert from standard to trigonometric form and vice versa.

Imaginary axis

$(-2\sqrt{3}, 2)$

r

θ

Real axis

Figure 8.42

Example 2 Writing a complex number in trigonometric form

Write each complex number in trigonometric form.

a) $-2\sqrt{3} + 2i$ **b)** $3 - 2i$

Solution

a) First we locate $-2\sqrt{3} + 2i$ in the complex plane, as shown in Fig. 8.42. Find r using $a = -2\sqrt{3}$, $b = 2$, and $r = \sqrt{a^2 + b^2}$:

$$r = \sqrt{(-2\sqrt{3})^2 + 2^2} = 4$$

Since $a = r \cos \theta$, we have $-2\sqrt{3} = 4 \cos \theta$ or

$$\cos \theta = -\frac{\sqrt{3}}{2}.$$

The smallest nonnegative solution to this equation is $\theta = 5\pi/6$. Since the terminal side of $5\pi/6$ goes through $\left(-2\sqrt{3}, 2\right)$, we get

$$-2\sqrt{3} + 2i = 4\left(\cos \frac{5\pi}{6} + i \sin \frac{5\pi}{6}\right).$$

b) Locate $3 - 2i$ in the complex plane, as shown in Fig. 8.43. Next, find r by using $a = 3$, $b = -2$, and $r = \sqrt{a^2 + b^2}$:

$$r = \sqrt{(3)^2 + (-2)^2} = \sqrt{13}$$

Since $a = r \cos \theta$, we have $3 = \sqrt{13} \cos \theta$ or

$$\cos \theta = \frac{3}{\sqrt{13}}.$$

Use a calculator to get

$$\cos^{-1}\left(\frac{3}{\sqrt{13}}\right) = 33.7°.$$

Since θ is an angle whose terminal side contains $(3, -2)$, choose $\theta = 360° - 33.7° = 326.3°$. So

$$3 - 2i = \sqrt{13}(\cos 326.3° + i \sin 326.3°). \qquad \blacklozenge$$

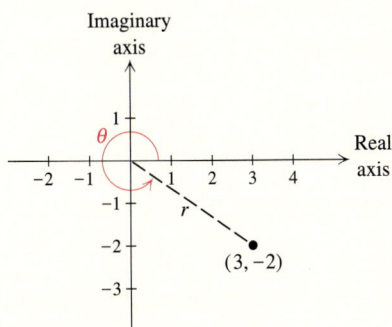

Imaginary axis

Real axis

$(3, -2)$

θ

r

Figure 8.43

Example 3 Writing a complex number in standard form

Write each complex number in the form $a + bi$.

a) $\sqrt{2}\left(\cos \frac{\pi}{4} + i \sin \frac{\pi}{4}\right)$

b) $3.6(\cos 143° + i \sin 143°)$

Solution

a) Since $\cos(\pi/4) = \sqrt{2}/2$ and $\sin(\pi/4) = \sqrt{2}/2$, we have

$$\sqrt{2}\left(\cos \frac{\pi}{4} + i \sin \frac{\pi}{4}\right) = \sqrt{2}\left(\frac{\sqrt{2}}{2} + i \frac{\sqrt{2}}{2}\right)$$

$$= 1 + i.$$

b) Use a calculator to find $\cos 143°$ and $\sin 143°$:

$$3.6(\cos 143° + i \sin 143°) \approx -2.875 + 2.167i \qquad \blacklozenge$$

Products and Quotients in Trigonometric Form

Multiplication and division of complex numbers in standard form are rather complicated, but in trigonometric form these operations are simpler to perform. We multiply the complex numbers

$$z_1 = r_1(\cos \theta_1 + i \sin \theta_1) \qquad \text{and} \qquad z_2 = r_2(\cos \theta_2 + i \sin \theta_2)$$

as follows:

$$
\begin{aligned}
z_1 z_2 &= r_1(\cos \theta_1 + i \sin \theta_1) \cdot r_2(\cos \theta_2 + i \sin \theta_2) \\
&= r_1 r_2(\cos \theta_1 + i \sin \theta_1)(\cos \theta_2 + i \sin \theta_2) \\
&= r_1 r_2[(\cos \theta_1 \cos \theta_2 - \sin \theta_1 \sin \theta_2) + i(\sin \theta_1 \cos \theta_2 + \cos \theta_1 \sin \theta_2)]
\end{aligned}
$$

Using the identities for cosine and sine of a sum, we get

$$z_1 z_2 = r_1 r_2[\cos(\theta_1 + \theta_2) + i \sin(\theta_1 + \theta_2)].$$

The product of z_1 and z_2 can be found by multiplying their moduli r_1 and r_2 and adding their arguments θ_1 and θ_2.

We find the quotient of z_1 and z_2 as follows:

$$
\begin{aligned}
\frac{z_1}{z_2} &= \frac{r_1(\cos \theta_1 + i \sin \theta_1)}{r_2(\cos \theta_2 + i \sin \theta_2)} \\
&= \frac{r_1(\cos \theta_1 + i \sin \theta_1)(\cos \theta_2 - i \sin \theta_2)}{r_2(\cos \theta_2 + i \sin \theta_2)(\cos \theta_2 - i \sin \theta_2)} \\
&= \frac{r_1}{r_2}[(\cos \theta_1 \cos \theta_2 + \sin \theta_1 \sin \theta_2) + i(\sin \theta_1 \cos \theta_2 - \cos \theta_1 \sin \theta_2)] \\
&= \frac{r_1}{r_2}[\cos(\theta_1 - \theta_2) + i \sin(\theta_1 - \theta_2)]
\end{aligned}
$$

The quotient of z_1 and z_2 can be found by dividing their moduli r_1 and r_2 and subtracting their arguments θ_1 and θ_2.

These two results are stated in the following theorem.

Theorem: Product and Quotient of Complex Numbers

If $z_1 = r_1(\cos \theta_1 + i \sin \theta_1)$ and $z_2 = r_2(\cos \theta_2 + i \sin \theta_2)$, then

$$z_1 z_2 = r_1 r_2[\cos(\theta_1 + \theta_2) + i \sin(\theta_1 + \theta_2)]$$

and

$$\frac{z_1}{z_2} = \frac{r_1}{r_2}[\cos(\theta_1 - \theta_2) + i \sin(\theta_1 - \theta_2)].$$

Example 4   A product and quotient using trigonometric form

Find $z_1 z_2$ and z_1/z_2 for $z_1 = 6(\cos 2.4 + i \sin 2.4)$ and $z_2 = 2(\cos 1.8 + i \sin 1.8)$. Express the answers in the form $a + bi$.

Solution

Find the product by multiplying the moduli and adding the arguments:

$$z_1 z_2 = 6 \cdot 2(\cos(2.4 + 1.8) + i \sin(2.4 + 1.8))$$
$$= 12(\cos 4.2 + i \sin 4.2)$$
$$= 12 \cos 4.2 + 12i \sin 4.2$$
$$\approx -5.88 - 10.46i$$

Find the quotient by dividing the moduli and subtracting the arguments:

$$\frac{z_1}{z_2} = \frac{6}{2}(\cos(2.4 - 1.8) + i \sin(2.4 - 1.8))$$
$$= 3(\cos(0.6) + i \sin(0.6))$$
$$= 3 \cos 0.6 + 3i \sin 0.6$$
$$\approx 2.48 + 1.69i$$

Example 5   A product and quotient using trigonometric form

Use trigonometric form to find $z_1 z_2$ and z_1/z_2, given that $z_1 = -2 + 2i\sqrt{3}$ and $z_2 = \sqrt{3} + i$.

Solution

First convert z_1 and z_2 into trigonometric form. Since $\sqrt{(-2)^2 + (2\sqrt{3})^2} = 4$ and $120°$ is a positive angle whose terminal side contains $(-2, 2\sqrt{3})$, we have

$$z_1 = 4(\cos 120° + i \sin 120°).$$

Since $\sqrt{(\sqrt{3})^2 + 1^2} = 2$ and $30°$ is a positive angle whose terminal side contains $(\sqrt{3}, 1)$, we have

$$z_2 = 2(\cos 30° + i \sin 30°).$$

To find the product we multiply the moduli and add the arguments:

$$z_1 z_2 = 2 \cdot 4[\cos(120° + 30°) + i \sin(120° + 30°)]$$
$$= 8(\cos 150° + i \sin 150°)$$

Using $\cos 150° = \sqrt{3}/2$ and $\sin 150° = 1/2$, we get

$$z_1 z_2 = 8\left(\frac{\sqrt{3}}{2} + i\frac{1}{2}\right) = 4\sqrt{3} + 4i.$$

To find the quotient, we divide the moduli and subtract the arguments:

$$\frac{z_1}{z_2} = \frac{4}{2}[\cos(120° - 30°) + i \sin(120° - 30°)]$$
$$= 2(\cos 90° + i \sin 90°)$$
$$= 2(0 + i) = 2i$$

Check by performing the operations with z_1 and z_2 in standard form.

1. The complex number $3 - 4i$ lies in quadrant IV of the complex plane.
2. The absolute value of $-2 - 5i$ is $2 + 5i$.
3. If θ is an angle whose terminal side contains $(1, -3)$, then $\cos \theta = 1/\sqrt{10}$.
4. If θ is an angle whose terminal side contains $(-2, -3)$, then $\tan \theta = 2/3$.
5. $i = 1(\cos 0° + i \sin 0°)$
6. The smallest positive solution to $\sqrt{3} = 2 \cos \theta$ is $\theta = 30°$.
7. The modulus of $2 - 5i$ is $\sqrt{29}$.
8. An argument for $2 - 4i$ is $\theta = 360° - \cos^{-1}(1/\sqrt{5})$.
9. $3(\cos \pi/4 + i \sin \pi/4) \cdot 2(\cos \pi/2 + i \sin \pi/2) = 6(\cos 3\pi/4 + i \sin 3\pi/4)$
10. $\dfrac{3(\cos \pi/4 + i \sin \pi/4)}{2(\cos \pi/2 + i \sin \pi/2)} = 1.5(\cos \pi/4 + i \sin \pi/4)$

8.4 Exercises

Tape 16   Disk—5.25″: 5   3.5″: 3   Macintosh: 3

Graph each complex number, and find its absolute value.

1. $2 - 6i$
2. $-1 - i$
3. $-2 + 2i\sqrt{3}$
4. $-\sqrt{3} - i$
5. $8i$
6. $-3i$
7. -9
8. $-\sqrt{6}$
9. $\dfrac{1}{\sqrt{2}} - \dfrac{i}{\sqrt{2}}$
10. $\dfrac{\sqrt{3}}{2} + \dfrac{i}{2}$
11. $3 + 3i$
12. $-4 - 4i$

Write each complex number in trigonometric form, using degree measure for the argument.

13. $-3 + 3i$
14. $4 - 4i$
15. $-\dfrac{3}{\sqrt{2}} + \dfrac{3i}{\sqrt{2}}$
16. $\dfrac{\sqrt{3}}{6} + \dfrac{i}{6}$
17. 8
18. -7
19. $i\sqrt{3}$
20. $-5i$
21. $-\sqrt{3} + i$
22. $-2 - 2i\sqrt{3}$
23. $3 + 4i$
24. $-2 + i$
25. $-3 + 5i$
26. $-2 - 4i$
27. $3 - 6i$
28. $5 - 10i$

Write each complex number in the form $a + bi$.

29. $\sqrt{2}(\cos 45° + i \sin 45°)$
30. $6(\cos 30° + i \sin 30°)$
31. $\dfrac{\sqrt{3}}{2}(\cos 150° + i \sin 150°)$
32. $12(\cos(\pi/10) + i \sin(\pi/10))$
33. $\dfrac{1}{2}(\cos 3.7 + i \sin 3.7)$
34. $4.3(\cos(\pi/9) + i \sin(\pi/9))$
35. $3(\cos 90° + i \sin 90°)$
36. $4(\cos 180° + i \sin 180°)$
37. $\sqrt{3}(\cos(3\pi/2) + i \sin(3\pi/2))$
38. $8.1(\cos \pi + i \sin \pi)$
39. $\sqrt{6}(\cos 60° + i \sin 60°)$
40. $0.5(\cos(5\pi/6) + i \sin(5\pi/6))$

Perform the indicated operations. Write the answer in the form $a + bi$.

41. $2(\cos 150° + i \sin 150°) \cdot 3(\cos 300° + i \sin 300°)$
42. $\sqrt{3}(\cos 45° + i \sin 45°) \cdot \sqrt{2}(\cos 315° + i \sin 315°)$
43. $\sqrt{3}(\cos 10° + i \sin 10°) \cdot \sqrt{2}(\cos 20° + i \sin 20°)$
44. $8(\cos 100° + i \sin 100°) \cdot 3(\cos 35° + i \sin 35°)$

45. $[3(\cos 45° + i \sin 45°)]^2$

46. $\left[\sqrt{5}(\cos(\pi/12) + i \sin(\pi/12))\right]^2$

47. $\dfrac{4(\cos(\pi/3) + i \sin(\pi/3))}{2(\cos(\pi/6) + i \sin(\pi/6))}$

48. $\dfrac{9(\cos(\pi/4) + i \sin(\pi/4))}{3(\cos(5\pi/4) + i \sin(5\pi/4))}$

49. $\dfrac{4.1(\cos 36.7° + i \sin 36.7°)}{8.2(\cos 84.2° + i \sin 84.2°)}$

50. $\dfrac{18(\cos 121.9° + i \sin 121.9°)}{2(\cos 325.6° + i \sin 325.6°)}$

Find $z_1 z_2$ and z_1/z_2 for each pair of complex numbers, using trigonometric form. Write the answer in the form $a + bi$.

51. $z_1 = 4 + 4i, z_2 = -5 - 5i$

52. $z_1 = -3 + 3i, z_2 = -2 - 2i$

53. $z_1 = \sqrt{3} + i, z_2 = 2 + 2i\sqrt{3}$

54. $z_1 = -\sqrt{3} + i, z_2 = 4\sqrt{3} - 4i$

55. $z_1 = 2 + 2i, z_2 = \sqrt{2} - i\sqrt{2}$

56. $z_1 = \sqrt{5} + i\sqrt{5}, z_2 = -\sqrt{6} - i\sqrt{6}$

57. $z_1 = 3 + 4i, z_2 = -5 - 2i$

58. $z_1 = 3 - 4i, z_2 = -1 + 3i$

59. $z_1 = 2 - 6i, z_2 = -3 - 2i$

60. $z_1 = 1 + 4i, z_2 = -4 - 2i$

61. $z_1 = 3i, z_2 = 1 + i$ **62.** $z_1 = 4, z_2 = -3 + i$

Solve each problem.

63. Given that $z = 3 + 3i$, find z^3 by writing z in trigonometric form and computing the product $z \cdot z \cdot z$.

64. Given that $z = \sqrt{3} + i$, find z^4 by writing z in trigonometric form and computing the product $z \cdot z \cdot z \cdot z$.

For Writing/Discussion

65. Show that the complex conjugate of $z = r(\cos \theta + i \sin \theta)$ is $\bar{z} = r(\cos(-\theta) + i \sin(-\theta))$.

66. Find the product of z and $\bar{z}$ using the trigonometric forms from Exercise 65.

67. Show that the reciprocal of $z = r(\cos \theta + i \sin \theta)$ is $z^{-1} = r^{-1}(\cos \theta - i \sin \theta)$ provided $r \neq 0$.

68. Given that $z = r(\cos \theta + i \sin \theta)$, find z^2 and z^{-2} provided $r \neq 0$.

8.5

Powers and Roots of

Complex Numbers

We know that the square root of a negative number is an imaginary number, but we have not yet encountered roots of imaginary numbers. In this section we will use the trigonometric form of a complex number to find any power or root of a complex number.

De Moivre's Theorem

Powers of a complex number can be found by repeated multiplication. Repeated multiplication can be done with standard form or trigonometric form. However, in trigonometric form a product is found by simply multiplying the moduli and adding the arguments. So, if $z = r(\cos \theta + i \sin \theta)$, then

$$z^2 = r(\cos \theta + i \sin \theta) \cdot r(\cos \theta + i \sin \theta) = r^2(\cos 2\theta + i \sin 2\theta).$$

We can easily find z^3 because $z^3 = z^2 \cdot z$:

$$z^3 = r^2(\cos 2\theta + i \sin 2\theta) \cdot r(\cos \theta + i \sin \theta) = r^3(\cos 3\theta + i \sin 3\theta)$$

Multiplying z^3 by z gives

$$z^4 = r^4(\cos 4\theta + i \sin 4\theta).$$

If you examine the results of z^2, z^3, and z^4, you will see a pattern, which can be stated as shown on the following page.

De Moivre's Theorem

If $z = r(\cos\theta + i\sin\theta)$ is a complex number and n is any positive integer, then

$$z^n = r^n(\cos n\theta + i\sin n\theta).$$

De Moivre's theorem, named after the French mathematician Abraham De Moivre (1667–1754), can be proved by using mathematical induction, which is presented in Chapter 12.

Example 1   A power of a complex number

Use De Moivre's theorem to simplify $(-\sqrt{3} + i)^8$. Write the answer in the form $a + bi$.

Solution

First write $-\sqrt{3} + i$ in trigonometric form. The modulus is 2, and the argument is 150°. So

$$-\sqrt{3} + i = 2(\cos 150° + i\sin 150°).$$

Use De Moivre's theorem to find the eighth power:

$$\begin{aligned}
(-\sqrt{3} + i)^8 &= [2(\cos 150° + i\sin 150°)]^8 \\
&= 2^8[\cos(8 \cdot 150°) + i\sin(8 \cdot 150°)] \\
&= 256[\cos 1200° + i\sin 1200°]
\end{aligned}$$

Since $1200° = 3 \cdot 360° + 120°$,

$$\cos 1200° = \cos 120° = -\frac{1}{2} \quad\text{and}\quad \sin 1200° = \sin 120° = \frac{\sqrt{3}}{2}.$$

Use these values to simplify the trigonometric form of $(-\sqrt{3} + i)^8$:

$$(-\sqrt{3} + i)^8 = 256\left(-\frac{1}{2} + i\frac{\sqrt{3}}{2}\right) = -128 + 128i\sqrt{3}$$

We have seen how De Moivre's theorem is used to find a positive integral power of a complex number. It is also used for finding roots of complex numbers.

Roots of a Complex Number

In the real number system, a is an nth root of b if $a^n = b$. We have a similar definition of nth root in the complex number system.

Definition: *n*th Root of a Complex Number

The complex number $w = a + bi$ is an *n*th root of the complex number z if

$$(a + bi)^n = z.$$

In Example 1, we saw that $\left(-\sqrt{3} + i\right)^8 = -128 + 128i\sqrt{3}$. So $-\sqrt{3} + i$ is an eighth root of $-128 + 128i\sqrt{3}$.

The definition of *n*th root does not indicate how to find an *n*th root, but a formula for all *n*th roots can be found by using trigonometric form. Suppose that the complex number w is an *n*th root of the complex number z, where the trigonometric forms of w and z are

$$w = s(\cos \alpha + i \sin \alpha) \qquad \text{and} \qquad z = r(\cos \theta + i \sin \theta).$$

By De Moivre's theorem, $w^n = s^n(\cos n\alpha + i \sin n\alpha)$. Since $w^n = z$, we have

$$w^n = s^n(\cos n\alpha + i \sin n\alpha) = r(\cos \theta + i \sin \theta).$$

This last equation gives two different trigonometric forms for w^n. The absolute value of w^n in one form is s^n and in the other it is r. So $s^n = r$, or $s = r^{1/n}$. Because the argument of a complex number is any angle in the complex plane whose terminal side goes through the point representing the complex number, any two angles used for the argument must differ by a multiple of 2π. Since the argument for w^n is either $n\alpha$ or θ, we have $n\alpha = \theta + 2k\pi$ or

$$\alpha = \frac{\theta + 2k\pi}{n}.$$

We can get n different values for α from this formula by using $k = 0, 1, 2, \ldots, n - 1$. These results are summarized in the following theorem.

Theorem: *n*th Roots of a Complex Number

For any positive integer n, the complex number $z = r(\cos \theta + i \sin \theta)$ has exactly n distinct *n*th roots, given by the expression

$$r^{1/n}\left[\cos\left(\frac{\theta + 2k\pi}{n}\right) + i \sin\left(\frac{\theta + 2k\pi}{n}\right) \right]$$

for $k = 0, 1, 2, \ldots, n - 1$.

The n distinct *n*th roots of a complex number all have the same modulus, $r^{1/n}$. What distinguishes between these roots is the n different arguments determined by

$$\alpha = \frac{\theta + 2k\pi}{n}$$

for $k = 0, 1, 2, \ldots, n - 1$. If θ is measured in degrees, then the n different arguments are determined by

$$\alpha = \frac{\theta + k360°}{n}$$

for $k = 0, 1, 2, \ldots, n - 1$. Note that if $k \geq n$, we get new values for α but no new nth roots, because the values of $\sin \alpha$ and $\cos \alpha$ have already occurred for some $k < n$. For example, if $k = n$, then $\alpha = \theta/n + 2\pi$, and the values for $\cos \alpha$ and $\sin \alpha$ are the same as they were for $\alpha = \theta/n$, which corresponds to $k = 0$.

Example 2   Finding nth roots

Find all of the fourth roots of the complex number $-8 + 8i\sqrt{3}$.

Solution

First find the modulus of $-8 + 8i\sqrt{3}$:

$$\sqrt{(-8)^2 + (8\sqrt{3})^2} = 16$$

Since the terminal side of θ contains $(-8, 8\sqrt{3})$, we have $\cos \theta = -8/16 = -1/2$. Choose $\theta = 120°$. So the fourth roots are generated by the expression

$$16^{1/4}\left[\cos\left(\frac{120° + k360°}{4}\right) + i \sin\left(\frac{120° + k360°}{4}\right)\right].$$

Evaluating this expression for $k = 0, 1, 2,$ and 3 (because $n = 4, n - 1 = 3$), we get

$$2[\cos 30° + i \sin 30°] = \sqrt{3} + i$$
$$2[\cos 120° + i \sin 120°] = -1 + i\sqrt{3}$$
$$2[\cos 210° + i \sin 210°] = -\sqrt{3} - i$$
$$2[\cos 300° + i \sin 300°] = 1 - i\sqrt{3}.$$

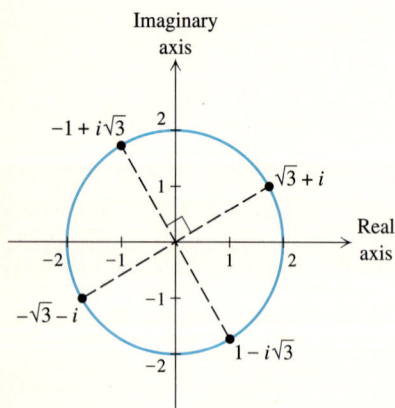

Fourth Roots of $-8 + 8i\sqrt{3}$

Figure 8.44

The graph of the original complex number in Example 2 and its four fourth roots reveals an interesting pattern. Since the modulus of $-8 + 8i\sqrt{3}$ is 16, $-8 + 8i\sqrt{3}$ lies on a circle of radius 16 centered at the origin. Since the modulus of each fourth root is $16^{1/4} = 2$, the fourth roots lie on a circle of radius 2 centered at the origin as shown in Fig. 8.44. Note how the fourth roots in Fig. 8.44 divide the circumference of the circle into four equal parts. In general, the nth roots of $z = r(\cos \theta + i \sin \theta)$ divide the circumference of a circle of radius $r^{1/n}$ into n equal parts. This symmetric arrangement of the nth roots on a circle provides a simple check on whether the nth roots are correct.

Consider the equation $x^6 - 1 = 0$. By the fundamental theorem of algebra, this equation has six solutions in the complex number system if multiplicity is considered. Since this equation is equivalent to $x^6 = 1$, each solution is a sixth root of 1. The roots of 1 are called the **roots of unity.** The fundamental theorem of algebra does not guarantee that all of the roots are unique, but the theorem on the nth roots of a complex number does.

Example 3　The six sixth roots of unity

Solve the equation $x^6 - 1 = 0$.

Solution

The solutions to the equation are the sixth roots of 1. Since $1 = 1(\cos 0° + i \sin 0°)$, the expression for the sixth roots is

$$1^{1/6}\left[\cos\left(\frac{0° + k360°}{6}\right) + i \sin\left(\frac{0° + k360°}{6}\right)\right].$$

Evaluating this expression for $k = 0, 1, 2, \ldots, 5$ gives the six roots:

$$\cos 0° + i \sin 0° = 1 \qquad\qquad \cos 60° + i \sin 60° = \frac{1}{2} + i\frac{\sqrt{3}}{2}$$

$$\cos 120° + i \sin 120° = -\frac{1}{2} + i\frac{\sqrt{3}}{2} \qquad \cos 180° + i \sin 180° = -1$$

$$\cos 240° + i \sin 240° = -\frac{1}{2} - i\frac{\sqrt{3}}{2} \qquad \cos 300° + i \sin 300° = \frac{1}{2} - i\frac{\sqrt{3}}{2}$$

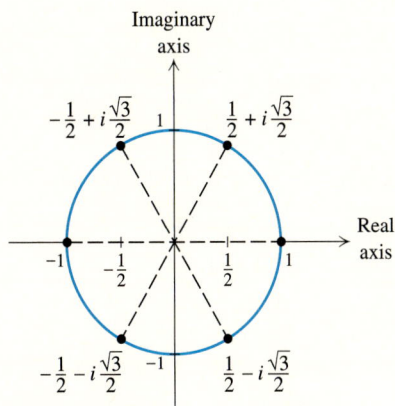

Sixth Roots of Unity

Figure 8.45

A graph of the sixth roots of unity found in Example 3 is shown in Fig. 8.45. Notice that they divide the circle of radius 1 into six equal parts. According to the conjugate pairs theorem of Chapter 4, the complex solutions to a polynomial equation with real coefficients occur in conjugate pairs. In Example 3, the complex solutions of $x^6 - 1 = 0$ did occur in conjugate pairs. This fact gives us another method of checking whether the six roots are correct. In general, the nth roots of any real number occur in conjugate pairs. Note that the fourth roots of the imaginary number in Example 2 did not occur in conjugate pairs.

For Thought

True or false? Explain.

1. $(2 + 3i)^2 = 4 + 9i^2$

2. If $z = 2(\cos 120° + i \sin 120°)$, then $z^3 = 8i$.

3. If z is a complex number with modulus r, then the modulus of z^4 is r^4.

4. If the argument of z is θ, then the argument of z^4 is $\cos 4\theta$.

5. $[\cos(\pi/3) + i \sin(\pi/6)]^2 = \cos(2\pi/3) + i \sin(\pi/3)$

6. If $\cos \alpha = \cos \beta$, then $\alpha = \beta + 2k\pi$ for some integer k.

7. One of the fifth roots of 32 is $2(\cos 72° + i \sin 72°)$.

8. All solutions to $x^8 = 1$ in the complex plane lie on the unit circle.

9. All solutions to $x^8 = 1$ lie on the axes or the lines $y = \pm x$.

10. The equation $x^4 + 81 = 0$ has two real and two imaginary solutions.

8.5 Exercises   ▭ Tape 16   ▣ Disk—5.25″: 5   3.5″: 3   Macintosh: 3

Use De Moivre's theorem to simplify each expression. Write the answer in the form $a + bi$.

1. $[3(\cos 30° + i \sin 30°)]^3$

2. $[2(\cos 45° + i \sin 45°)]^5$

3. $\left[\sqrt{2}(\cos 120° + i \sin 120°)\right]^4$

4. $\left[\sqrt{3}(\cos 150° + i \sin 150°)\right]^3$

5. $[\cos(\pi/12) + i \sin(\pi/12)]^8$

6. $[\cos(\pi/6) + i \sin(\pi/6)]^9$

7. $\left[\sqrt{6}(\cos(2\pi/3) + i \sin(2\pi/3))\right]^4$

8. $\left[\sqrt{18}(\cos(5\pi/6) + i \sin(5\pi/6))\right]^3$

9. $[4.3(\cos 12.3° + i \sin 12.3°)]^5$

10. $[4.9(\cos 37.4° + i \sin 37.4°)]^6$

Simplify each expression, by using trigonometric form and De Moivre's theorem.

11. $(2 + 2i)^3$

12. $(1 - i)^3$

13. $(\sqrt{3} - i)^4$

14. $\left(-2 + 2i\sqrt{3}\right)^4$

15. $(-3 - 3i\sqrt{3})^5$

16. $\left(2\sqrt{3} - 2i\right)^5$

17. $(2 + 3i)^4$

18. $(4 - i)^5$

19. $(2 - i)^4$

20. $(-1 - 2i)^6$

21. $(1.2 + 3.6i)^3$

22. $(-2.3 - i)^3$

Find the indicated roots. Express answers in trigonometric form.

23. The square roots of $4(\cos 90° + i \sin 90°)$

24. The cube roots of $8(\cos 30° + i \sin 30°)$

25. The fourth roots of $\cos 120° + i \sin 120°$

26. The fifth roots of $32(\cos 300° + i \sin 300°)$

27. The sixth roots of $64(\cos \pi + i \sin \pi)$

28. The fourth roots of $16[\cos(3\pi/2) + i \sin(3\pi/2)]$

Find the indicated roots in the form $a + bi$. Check by graphing the roots in the complex plane.

29. The cube roots of 1

30. The cube roots of 8

31. The fourth roots of 16

32. The fourth roots of 1

33. The fourth roots of -1

34. The fourth roots of -16

35. The cube roots of i

36. The cube roots of $-8i$

37. The square roots of $-2 + 2i\sqrt{3}$

38. The square roots of $-4i$

39. The square roots of $1 + 2i$

40. The cube roots of $-1 + 3i$

Solve each equation. Express answers in the form $a + bi$.

41. $x^3 + 1 = 0$

42. $x^3 + 125 = 0$

43. $x^4 - 81 = 0$

44. $x^4 + 81 = 0$

45. $x^2 + 2i = 0$

46. $ix^2 + 3 = 0$

47. $x^7 - 64x = 0$

48. $x^9 - x = 0$

Solve each equation. Express answers in trigonometric form.

49. $x^5 - 2 = 0$

50. $x^5 + 3 = 0$

51. $x^4 + 3 - i = 0$

52. $ix^3 + 2 - i = 0$

Solve each problem.

53. Write the expression $[\cos(\pi/3) + i \sin(\pi/6)]^3$ in the form $a + bi$.

54. Solve the equation $x^6 - 1 = 0$ by factoring and the quadratic formula. Compare your answers to Example 3 of this section.

55. Use the quadratic formula and De Moivre's theorem to solve $x^2 + (-1 + i)x - i = 0$.

56. Use the quadratic formula and De Moivre's theorem to solve $x^2 + (-1 - 3i)x + (-2 + 2i) = 0$.

For Writing/Discussion

57. Explain why $x^6 - 2x^3 + 1 = 0$ has three distinct solutions, $x^6 - 2x^3 = 0$ has four distinct solutions, and $x^6 - 2x = 0$ has six distinct solutions.

58. Find all real numbers a and b for which it is true that $\sqrt{a + bi} = \sqrt{a} + i\sqrt{b}$.

8.6

Polar Equations

Figure 8.46

The Cartesian coordinate system is a system for describing the location of points in a plane. It is not the only coordinate system available. In this section we will study the **polar coordinate system.** In this coordinate system, a point is located by using a *directed distance* and an *angle* in a manner that will remind you of the magnitude and direction angle of a vector and the modulus and argument for a complex number.

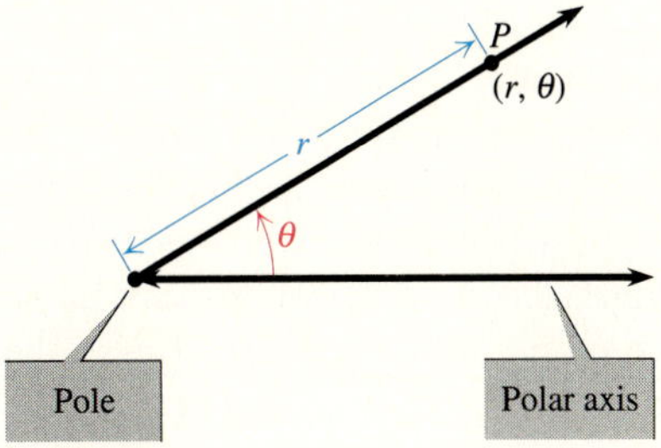

Polar Coordinates

In the rectangular coordinate system, points are named according to their position with respect to an x-axis and a y-axis. In the polar coordinate system, we have a fixed point called the **pole** and a fixed ray called the **polar axis.** A point P has coordinates (r, θ), where r is the directed distance from the pole to P and θ is an angle whose initial side is the polar axis and whose terminal side contains the point. See Fig. 8.46.

Since we are so familiar with rectangular coordinates, we retain the x- and y-axes when using polar coordinates. The pole is placed at the origin, and the polar axis is placed along the positive x-axis. The angle θ is any angle (in degrees or radians) in standard position whose terminal side contains the point. As usual, θ is positive for a counterclockwise rotation and negative for a clockwise rotation. In polar coordinates, r can be any real number. For example, the ordered pair $(2, \pi/4)$ represents the point that lies two units from the origin on the terminal side of the angle $\pi/4$. The point $(0, \pi/4)$ is at the origin. The point $(-2, \pi/4)$ lies two units from the origin on the line through the terminal side of $\pi/4$ but in the direction opposite to $(2, \pi/4)$. See Fig. 8.47. Any ordered pair in polar coordinates names a single point, but the coordinates of a point in polar coordinates are not unique. For example, $(-2, \pi/4)$, $(2, 5\pi/4)$, and $(2, -3\pi/4)$ all name the same point.

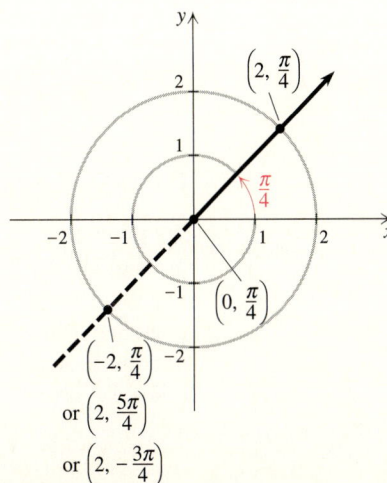

Figure 8.47

Example 1 Plotting points in polar coordinates

Plot the points whose polar coordinates are $(2, 5\pi/6)$, $(-3, \pi)$, $(1, -\pi/2)$, and $(-1, 450°)$.

Solution

The terminal side of $5\pi/6$ lies in the second quadrant, so $(2, 5\pi/6)$ is two units from the origin along this ray. The terminal side of the angle π points in the direction of the negative x-axis, but since the first coordinate of $(-3, \pi)$ is negative, the point is located three units in the direction opposite to the direction of the ray. So $(-3, \pi)$ lies three units from the origin on the positive x-axis. The point $(-3, \pi)$ is the same point as $(3, 0)$. The terminal side of $-\pi/2$ lies on the negative y-axis. So $(1, -\pi/2)$ lies one unit from the origin on the negative y-axis. The terminal side of $450°$ lies on the positive y-axis. Since the first coordinate of $(-1, 450°)$ is negative, the point is located one unit in the opposite direction. The point $(-1, 450°)$ is the same as $(1, -\pi/2)$. All points are shown in Fig. 8.48.

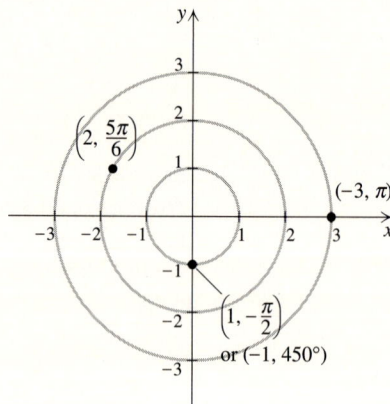

Figure 8.48

Polar-Rectangular Conversions

To graph equations in polar coordinates, we must be able to switch the coordinates of a point in either system to the other. Suppose that (r, θ) is a point in the first quadrant with $r > 0$ and θ acute as shown in Fig. 8.49. If (x, y) is the same point in rectangular coordinates, then x and y are the lengths of the legs of the right triangle shown in Fig. 8.49. Using the Pythagorean theorem and trigonometric ratios we have

$$x^2 + y^2 = r^2, \qquad x = r\cos\theta, \qquad y = r\sin\theta, \qquad \text{and} \qquad \tan\theta = \frac{y}{x}.$$

It can be shown that these equations hold for any point (r, θ), except that $\tan\theta$ is undefined if $x = 0$. The following rules for converting from one system to the other follow from these relationships.

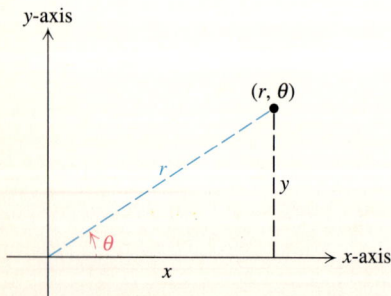

Figure 8.49

**Polar-Rectangular
Conversion Rules**

To convert (r, θ) to rectangular coordinates (x, y), use

$$x = r \cos \theta \qquad \text{and} \qquad y = r \sin \theta.$$

To convert (x, y) to polar coordinates (r, θ), use

$$r = \sqrt{x^2 + y^2}$$

and any angle θ in standard position whose terminal side contains (x, y).

There are several ways to find an angle θ whose terminal side contains (x, y). One way to find θ (provided $x \neq 0$) is to find an angle that satisfies $\tan \theta = y/x$ and goes through (x, y). Remember that the angle $\tan^{-1}(y/x)$ is between $-\pi/2$ and $\pi/2$ and will go through (x, y) only if $x > 0$. If $x = 0$, then the point (x, y) is on the y-axis, and you can use $\theta = \pi/2$ or $\theta = -\pi/2$ as appropriate.

Example 2   Polar-rectangular conversion

a) Convert $(6, 210°)$ to rectangular coordinates.

b) Convert $(-3, 6)$ to polar coordinates.

Solution

```
P▶Rx(6,210)
          -5.196152423
-3√3
          -5.196152423
P▶Ry(6,210)
                     -3
■
```

Use degree mode to check the polar-to-rectangular conversion of Example 2(a).

a) Use $r = 6$, $\cos 210° = -\sqrt{3}/2$, and $\sin 210° = -1/2$ in the formulas $x = r \cos \theta$ and $y = r \sin \theta$:

$$x = 6\left(-\frac{\sqrt{3}}{2}\right) = -3\sqrt{3} \qquad \text{and} \qquad y = 6\left(-\frac{1}{2}\right) = -3$$

So $(6, 210°)$ in polar coordinates is $\left(-3\sqrt{3}, -3\right)$ in rectangular coordinates.

b) To convert $(-3, 6)$ to polar coordinates, find r:

$$r = \sqrt{(-3)^2 + 6^2} = 3\sqrt{5}$$

```
R▶Pr(-3,6)
           6.708203932
3√5
           6.708203932
R▶Pθ(-3,6)
           116.5650512
```

Use degree mode to check the rectangular-to-polar conversion of Example 2(b).

Use a calculator to find $\tan^{-1}(-2) \approx -63.4°$. To get an angle whose terminal side contains $(-3, 6)$, use $\theta = 180° - 63.4° = 116.6°$. So $(-3, 6)$ in polar coordinates is $\left(3\sqrt{5}, 116.6°\right)$. Since there are infinitely many representations for any point in polar coordinates, this answer is not unique. In fact, another possibility is $\left(-3\sqrt{5}, -63.4°\right)$. ◆

Polar Equations

An equation in two variables (typically x and y) that is graphed in the rectangular coordinate system is called a **rectangular** or **Cartesian equation.** An equation in two variables (typically r and θ) that is graphed in the polar coordinate system is called a **polar equation.** Certain polar equations are easier to graph than the equivalent Cartesian equations. We can graph a polar equation in the same way that we graph a rectangular equation, that is, we can simply plot enough points to get the shape of the graph. However, since most of our polar equations involve trigonometric functions, finding points on these curves can be tedious. A graphing calculator can be used to great advantage here.

Example 3   Graphing a polar equation

Sketch the graph of the polar equation $r = 2 \cos \theta$.

Solution

If $\theta = 0°$, then $r = 2 \cos 0° = 2$. So the ordered pair $(2, 0°)$ is on the graph. If $\theta = 30°$, then $r = 2 \cos 30° = \sqrt{3}$, and $(\sqrt{3}, 30°)$ is on the graph. These ordered pairs and several others that satisfy the equation are listed in the following table:

θ	0°	30°	45°	60°	90°	120°	135°	150°	180°
r	2	$\sqrt{3}$	$\sqrt{2}$	1	0	-1	$-\sqrt{2}$	$-\sqrt{3}$	-2

Plot these points and draw a smooth curve through them to get the graph shown in Fig. 8.50. If θ is larger than 180° or smaller than 0°, we get different ordered pairs, but they are all located on the curve drawn in Fig. 8.50. For example, $(-\sqrt{3}, 210°)$ satisfies $r = 2 \cos \theta$, but it has the same location as $(\sqrt{3}, 30°)$.

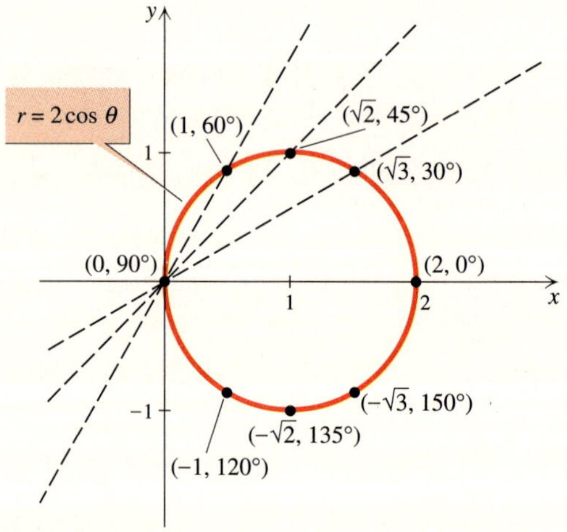

Figure 8.50

The graph $r = 2 \cos \theta$ in Fig. 8.50 looks like a circle. To verify that it is a circle, we can convert the polar equation to an equivalent rectangular equation because we know the form of the equation of a circle in rectangular coordinates. This conversion is done in Example 6.

In the next example, a simple polar equation produces a curve that is not usually graphed when studying rectangular equations because the equivalent rectangular equation is quite complicated.

Example 4   Graphing a polar equation

Sketch the graph of the polar equation $r = 3 \sin 2\theta$.

Solution

The ordered pairs in the following table satisfy the equation $r = 3 \sin 2\theta$. The values of r are rounded to the nearest tenth.

θ	0°	15°	30°	45°	60°	90°	135°	180°	225°	270°	315°	360°
r	0	1.5	2.6	3	2.6	0	−3	0	3	0	−3	0

As θ varies from 0° to 90°, the value of r goes from 0 to 3, then back to 0, creating a loop in quadrant I. As θ varies from 90° to 180°, the value of r goes from 0 to −3 then back to 0, creating a loop in quadrant IV. As θ varies from 180° to 270°, a loop in quadrant III is formed; and as θ varies from 270° to 360°, a loop in quadrant II is formed. If θ is chosen greater than 360° or less than 0°, we get the same points over and over because the sine function is periodic. The graph of $r = 3 \sin 2\theta$ is shown in Fig. 8.51. The graph is called a **four-leaf rose.**

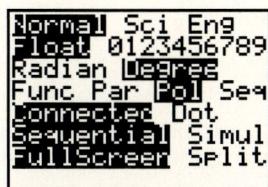

Use this mode setting to graph a polar equation with the angles in degrees.

In polar mode, the Y= key gives r as a function of θ.

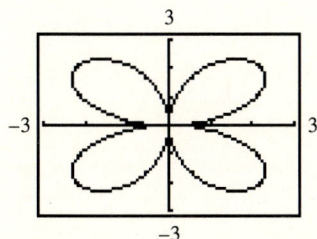

Graph $r = 3 \sin(2\theta)$ for θ ranging from 0° to 360°.

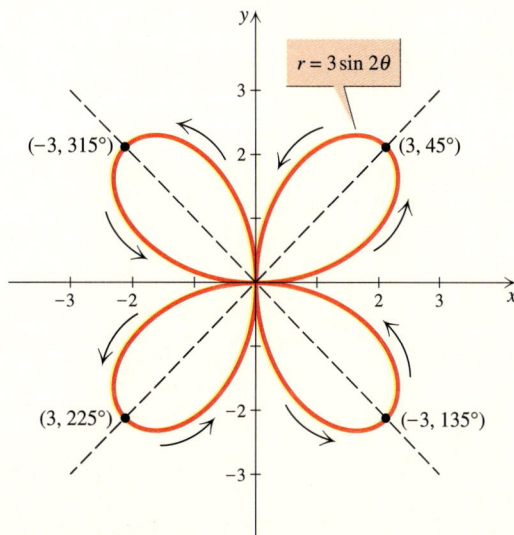

Figure 8.51

Example 5   Graphing a polar equation

Sketch the graph of the polar equation $r = \theta$, where θ is in radians and $\theta \geq 0$.

Solution

The ordered pairs in the following table satisfy $r = \theta$. The values of r are rounded to the nearest tenth.

θ	r
0	0
$\pi/4$	0.8
$\pi/2$	1.6
$3\pi/4$	2.4

θ	r
π	3.1
$3\pi/2$	4.7
2π	6.3
3π	9.4

The graph of $r = \theta$, called **the spiral of Archimedes,** is shown in Fig. 8.52. As the value of θ increases, the value of r increases, causing the graph to spiral out from the pole. There is no repetition of points as there was in Examples 3 and 4, because no periodic function is involved.

Graph $r = \theta$ in radian mode for θ ranging from 0 to 40 to see the spiral.

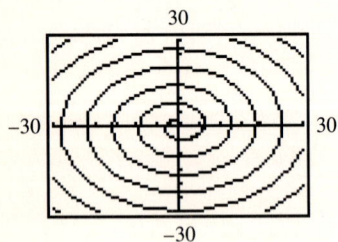

Figure 8.52

Figure 8.53 on the next page shows the graphs of several types of polar equations.

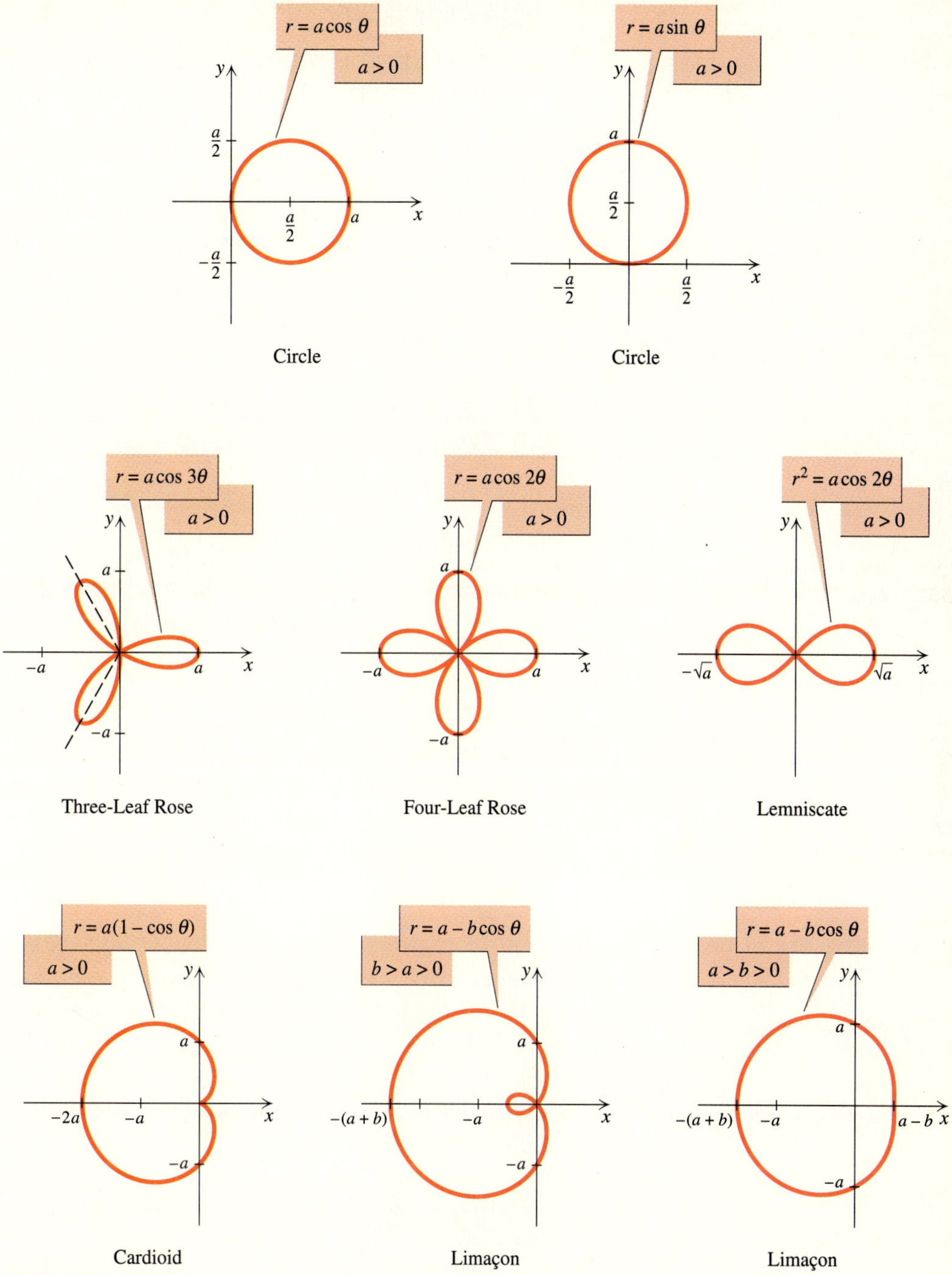
$r = a\cos\theta$
$a > 0$
Circle
$r = a\sin\theta$
$a > 0$
Circle
$r = a\cos 3\theta$
$a > 0$
Three-Leaf Rose
$r = a\cos 2\theta$
$a > 0$
Four-Leaf Rose
$r^2 = a\cos 2\theta$
$a > 0$
$-\sqrt{a}$
$\sqrt{a}$
Lemniscate
$r = a(1 - \cos\theta)$
$a > 0$
$-2a$
Cardioid
$r = a - b\cos\theta$
$b > a > 0$
$-(a + b)$
Limaçon
$r = a - b\cos\theta$
$a > b > 0$
$-(a + b)$
$a - b$
Limaçon

Figure 8.53

Converting Equations

We know that certain types of rectangular equations have graphs that are particular geometric shapes such as lines, circles, and parabolas. We can use our knowledge of equations in rectangular coordinates with equations in polar coordinates (and vice versa) by converting the equations from one system to the other. For example, we can determine whether the graph of $r = 2 \cos \theta$ in Example 3 is a circle by finding the equivalent Cartesian equation and deciding whether it is the equation of a circle. When converting from one system to another, we use the relationships

$$x^2 + y^2 = r^2, \qquad x = r \cos \theta, \qquad \text{and} \qquad y = r \sin \theta.$$

Example 6   Converting a polar equation to a rectangular equation

Write an equivalent rectangular equation for the polar equation $r = 2 \cos \theta$.

Solution

First multiply each side of $r = 2 \cos \theta$ by r to get $r^2 = 2r \cos \theta$. Now eliminate r and θ by making substitutions using $r^2 = x^2 + y^2$ and $x = r \cos \theta$:

$$r^2 = 2r \cos \theta$$
$$x^2 + y^2 = 2x$$

From Section 3.3, the standard equation of a circle with center (h, k) and radius r is $(x - h)^2 + (y - k)^2 = r^2$. Complete the square to get $x^2 + y^2 = 2x$ into the standard form of the equation of a circle:

$$x^2 - 2x + y^2 = 0$$
$$x^2 - 2x + 1 + y^2 = 1$$
$$(x - 1)^2 + y^2 = 1$$

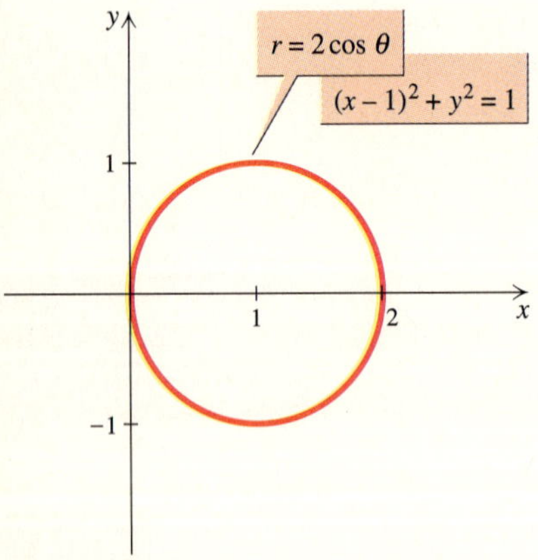

Since we recognize the rectangular equation as the equation of a circle centered at $(1, 0)$ with radius 1, the graph of $r = 2 \cos \theta$ shown in Fig. 8.54 is a circle centered at $(1, 0)$ with radius 1. ◆

$r = 2 \cos \theta$

$(x - 1)^2 + y^2 = 1$

Figure 8.54

In the next example we convert the rectangular equation of a line and circle into polar coordinates.

Example 7   Converting a rectangular equation to a polar equation

For each rectangular equation, write an equivalent polar equation.

a) $y = 3x - 2$ **b)** $x^2 + y^2 = 9$

Solution

a) Substitute $x = r \cos \theta$ and $y = r \sin \theta$, and then solve for r:

$$y = 3x - 2$$
$$r \sin \theta = 3r \cos \theta - 2$$
$$r \sin \theta - 3r \cos \theta = -2$$
$$r(\sin \theta - 3 \cos \theta) = -2$$
$$r = \frac{-2}{\sin \theta - 3 \cos \theta}$$

b) Substitute $r^2 = x^2 + y^2$ to get polar coordinates:

$$x^2 + y^2 = 9$$
$$r^2 = 9$$
$$r = \pm 3$$

A polar equation for a circle of radius 3 centered at the origin is $r = 3$. The equation $r = -3$ is also a polar equation for the same circle.   ◆

In Example 7 we saw that a straight line has a rather simple equation in rectangular coordinates but a more complicated equation in polar coordinates. A circle centered at the origin has a very simple equation in polar coordinates but a more complicated equation in rectangular coordinates. The graphs of simple polar equations are typically circular or somehow "centered" at the origin.

? For Thought

True or false? Explain.

1. The distance of the point (r, θ) from the origin depends only on r.

2. The distance of the point (r, θ) from the origin is r.

3. The ordered pairs $(2, \pi/4)$, $(2, -3\pi/4)$, and $(-2, 5\pi/4)$ all represent the same point in polar coordinates.

4. The equations relating rectangular and polar coordinates are $x = r \sin \theta$, $y = r \cos \theta$, and $x^2 + y^2 = r^2$.

5. The point $(-4, 225°)$ in polar coordinates is $\left(2\sqrt{2}, 2\sqrt{2}\right)$ in rectangular coordinates.

6. The graph of $0 \cdot r + \theta = \pi/4$ in polar coordinates is a straight line.

7. The graphs of $r = 5$ and $r = -5$ are identical.

8. The ordered pairs $\left(-\sqrt{2}/2, \pi/3\right)$ and $\left(\sqrt{2}/2, \pi/3\right)$ satisfy $r^2 = \cos 2\theta$.

9. The graph of $r = 1/\sin \theta$ is a vertical line.

10. The graphs of $r = \theta$ and $r = -\theta$ are identical.

8.6 Exercises   Tape 16   Disk—5.25″: 5   3.5″: 3   Macintosh: 3

Find polar coordinates for each given point, using radian measure for the angle.

1.

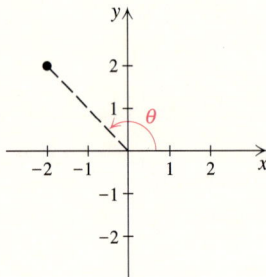

2.

3.

4.

Plot the points whose polar coordinates are given.

5. $(3, \pi/6)$

6. $(2, \pi/4)$

7. $(-2, 2\pi/3)$

8. $(-1, \pi/6)$

9. $(2, -\pi/4)$

10. $(1, -2\pi/3)$

11. $(3, -225°)$

12. $(2, -180°)$

13. $(-2, 45°)$

14. $(-3, 30°)$

15. $(4, 390°)$

16. $(3, 13\pi/6)$

Convert the polar coordinates of each point to rectangular coordinates.

17. $(1, \pi/6)$

18. $(2, \pi/4)$

19. $(-3, 3\pi/2)$

20. $(-2, 2\pi)$

21. $(\sqrt{2}, 135°)$

22. $(\sqrt{3}, 150°)$

23. $(-\sqrt{6}, -60°)$

24. $(-\sqrt{2}/2, -45°)$

Convert the rectangular coordinates of each point to polar coordinates. Use degrees for θ.

25. $(\sqrt{3}, 3)$

26. $(4, 4)$

27. $(-2, 2)$

28. $(-2, 2\sqrt{3})$

29. $(0, 2)$

30. $(-2, 0)$

31. $(-3, -3)$

32. $(2, -2)$

33. $(1, 4)$

34. $(-2, 3)$

35. $(\sqrt{2}, -2)$

36. $(-2, -\sqrt{3})$

Sketch the graph of each polar equation.

37. $r = 2 \sin \theta$

38. $r = 3 \cos \theta$

39. $r = 3 \cos 2\theta$

40. $r = -2 \sin 2\theta$

41. $r = 2\theta$ for θ in radians

42. $r = \theta$ for $\theta \le 0$ and θ in radians

43. $r = 1 + \cos \theta$ (cardioid)

44. $r = 1 - \cos \theta$ (cardioid)

45. $r^2 = 9 \cos 2\theta$ (lemniscate)

46. $r^2 = 4 \sin 2\theta$ (lemniscate)

47. $r = 4 \cos 2\theta$ (four-leaf rose)

48. $r = 3 \sin 2\theta$ (four-leaf rose)

49. $r = 2 \sin 3\theta$ (three-leaf rose)

50. $r = 4 \cos 3\theta$ (three-leaf rose)

51. $r = 1 + 2 \cos \theta$ (limaçon)

52. $r = 2 + \cos \theta$ (limaçon)

53. $r = 3.5$

54. $r = -5$

55. $\theta = 30°$

56. $\theta = 3\pi/4$

For each polar equation, write an equivalent rectangular equation.

57. $r = 4 \cos \theta$

58. $r = 2 \sin \theta$

59. $r = \dfrac{3}{\sin \theta}$

60. $r = \dfrac{-2}{\cos \theta}$

61. $r = 3 \sec \theta$

62. $r = 2 \csc \theta$

63. $r = 5$

64. $r = -3$

65. $\theta = \dfrac{\pi}{4}$

66. $\theta = 0$

67. $r = \dfrac{2}{1 - \sin \theta}$

68. $r = \dfrac{3}{1 + \cos \theta}$

For each rectangular equation, write an equivalent polar equation.

69. $x = 4$

70. $y = -6$

71. $y = -x$

72. $y = x\sqrt{3}$

73. $x^2 = 4y$

74. $y^2 = 2x$

75. $x^2 + y^2 = 4$

76. $2x^2 + y^2 = 1$

77. $y = 2x - 1$

78. $y = -3x + 5$

79. $x^2 + (y - 1)^2 = 1$

80. $(x + 1)^2 + y^2 = 4$

For Writing/Discussion

81. Explain why θ must be radian measure for the equation $r = 2\theta$, but θ can be in radians or degrees for the equation $r = 2 \cos \theta$.

82. Show that a polar equation for the straight line $y = mx$ is $\theta = \tan^{-1} m$.

Graphing Calculator Exercises

Graph each pair of polar equations on the same screen of your calculator and use the trace feature to estimate the polar coordinates of all points of intersection of the curves. Check your calculator manual to see how to graph polar equations on your calculator.

1. $r = 1$, $r = 2 \sin 3\theta$

2. $r = \sin \theta$, $r = \sin 2\theta$

3. $r = 3 \sin 2\theta$, $r = 1 - \cos \theta$

Highlights

Section 8.1   The Law of Sines

1. To solve an oblique triangle, we must know at least three parts of the triangle, at least one of which must be the length of a side.

2. The ASA and AAS cases are solved by using the law of sines.

3. The ratio of the sine of an angle and the length of the opposite side is the same for each angle of a triangle (the law of sines).

4. In the ambiguous SSA case it is the lengths of the sides and the size of the angle that determine the number of triangles formed (0, 1, or 2).

5. The area of a triangle is one-half the product of the lengths of any two sides and the sine of their included angle.

Section 8.2   The Law of Cosines

1. The SAS and SSS cases are solved by using the law of cosines.

2. The law of cosines expresses the square of the length of one side in terms of the other two sides and the cosine of their included angle. For side c, we have $c^2 = a^2 + b^2 - 2ab \cos \gamma$.

3. Heron's area formula for the area of triangle ABC is $A = \sqrt{S(S - a)(S - b)(S - c)}$, where $S = (a + b + c)/2$.

Section 8.3   Vectors

1. A vector quantity **A** has magnitude and direction.

2. The magnitude of the scalar product $k\mathbf{A}$ is $|k|$ times the magnitude of **A**. The direction of $k\mathbf{A}$ is the same as the direction of **A** if $k > 0$ and opposite to the direction of **A** if $k < 0$.

3. The sum or resultant of **A** and **B** is found by placing **B** so that its initial point coincides with the terminal point of **A** and drawing a vector from the initial point of **A** to the terminal point of **B**.

4. If the vector **w** has magnitude r and direction angle θ, then the horizontal component **u** has magnitude $|r \cos \theta|$, and the vertical component **v** has magnitude $|r \sin \theta|$.

5. The form $\langle a, b \rangle$ is the component form for the vector with initial point $(0, 0)$ and terminal point (a, b). Operations are easily performed with vectors in component form.

Section 8.4   Trigonometric Form of Complex Numbers

1. Complex numbers are graphed as ordered pairs in the complex plane, where the horizontal axis is the real axis and the vertical axis is the imaginary axis.

2. The absolute value of a complex number $a + bi$ is defined by $|a + bi| = \sqrt{a^2 + b^2}$.

3. The trigonometric form of the complex number $z = a + bi$ is $z = r(\cos\theta + i\sin\theta)$, where $a = r\cos\theta$ and $b = r\sin\theta$. The modulus r is given by $r = \sqrt{a^2 + b^2}$ and the argument θ is an angle in standard position whose terminal side contains (a, b).

4. In trigonometric form, the product of two complex numbers is found by multiplying their moduli and adding their arguments; the quotient is found by dividing their moduli and subtracting their arguments.

Section 8.5   Powers and Roots of Complex Numbers

1. If $z = r(\cos\theta + i\sin\theta)$ and n is any positive integer, then $z^n = r^n(\cos n\theta + i\sin n\theta)$ (De Moivre's theorem).

2. The n distinct nth roots of $r(\cos\theta + i\sin\theta)$ are given by

$$r^{1/n}\left[\cos\left(\frac{\theta + 2k\pi}{n}\right) + i\sin\left(\frac{\theta + 2k\pi}{n}\right)\right]$$

for $k = 0, 1, 2, \ldots, n - 1$.

Section 8.6   Polar Equations

1. In polar coordinates, if $r > 0$, then the point (r, θ) is r units from the origin (pole) on the terminal side of the angle θ in standard position. If $r < 0$, then (r, θ) is $|r|$ units from the origin on the extension of the terminal side of the angle θ.

2. To convert (r, θ) to rectangular coordinates (x, y), use $x = r\cos\theta$ and $y = r\sin\theta$.

3. To convert (x, y) to polar coordinates (r, θ), use $r = \sqrt{x^2 + y^2}$ and an angle θ whose terminal side contains (x, y).

4. The graphs of some simple polar equations are curves such as the cardioid, lemniscate, and four-leaf rose that are not usually encountered when studying rectangular coordinates.

Chapter 8 Review Exercises

Solve each triangle that exists with the given parts. If there is more than one triangle with the given parts, then solve each one.

1. $\gamma = 48°$, $a = 3.4$, $b = 2.6$

2. $a = 6$, $b = 8$, $c = 10$

3. $\alpha = 13°$, $\beta = 64°$, $c = 20$

4. $\alpha = 50°$, $a = 3.2$, $b = 8.4$

5. $a = 3.6$, $b = 10.2$, $c = 5.9$

6. $\beta = 36.2°$, $\gamma = 48.1°$, $a = 10.6$

7. $a = 30.6$, $b = 12.9$, $c = 24.1$

8. $\alpha = 30°$, $a = \sqrt{3}$, $b = 2\sqrt{3}$

9. $\beta = 22°$, $c = 4.9$, $b = 2.5$

10. $\beta = 121°$, $a = 5.2$, $c = 7.1$

Find the area of each triangle.

11.

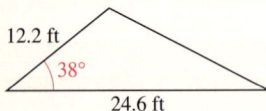

12.2 ft, 38°, 24.6 ft

12.

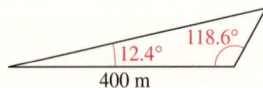

118.6°, 12.4°, 400 m

13.

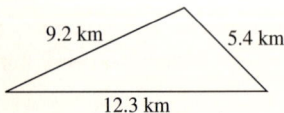

9.2 km, 5.4 km, 12.3 km

14.

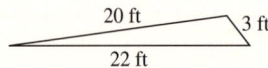

20 ft, 3 ft, 22 ft

Find the magnitude of the horizontal and vertical components for each vector **v** with the given magnitude and given direction angle θ.

15. $|\mathbf{v}| = 6$, $\theta = 23.3°$

16. $|\mathbf{v}| = 4.5$, $\theta = 156°$

17. $|\mathbf{v}| = 3.2$, $\theta = 231.4°$

18. $|\mathbf{v}| = 7.3$, $\theta = 344°$

Find the magnitude and direction for each vector.

19. $\langle 2, 3 \rangle$

20. $\langle -4, 3 \rangle$

21. $\langle -3.2, -5.1 \rangle$

22. $\langle 2.1, -3.8 \rangle$

Find the component form for each vector **v** with the given magnitude and direction angle.

23. $|\mathbf{v}| = \sqrt{2}$, $\theta = 45°$

24. $|\mathbf{v}| = 6$, $\theta = 60°$

25. $|\mathbf{v}| = 9.1$, $\theta = 109.3°$

26. $|\mathbf{v}| = 5.5$, $\theta = 344.6°$

Perform the vector operations. Write your answer in the form $\langle a, b \rangle$.

27. $2\langle -3, 4 \rangle$

28. $-3\langle 4, -1 \rangle$

29. $\langle 2, -5 \rangle - 2\langle 1, 6 \rangle$

30. $3\langle 1, 2 \rangle + 4\langle -1, -2 \rangle$

Rewrite each vector **v** in the form $a_1 \mathbf{i} + a_2 \mathbf{j}$, where $\mathbf{i} = \langle 1, 0 \rangle$ and $\mathbf{j} = \langle 0, 1 \rangle$.

31. In component form, $\mathbf{v} = \langle -4, 8 \rangle$.

32. In component form, $\mathbf{v} = \langle 3.2, -4.1 \rangle$.

33. The direction angle for **v** is $30°$ and its magnitude is 7.2.

34. The magnitude of **v** is 6 and it has the same direction as the vector $\langle 2, 5 \rangle$.

Find the absolute value of each complex number.

35. $3 - 5i$

36. $3.6 + 4.8i$

37. $\sqrt{5} + i\sqrt{3}$

38. $-2\sqrt{2} + 3i\sqrt{5}$

Write each complex number in trigonometric form, using degree measure for the argument.

39. $-4.2 + 4.2i$

40. $3 - i\sqrt{3}$

41. $-2.3 - 7.2i$

42. $4 + 9.2i$

Write each complex number in the form $a + bi$.

43. $\sqrt{3}(\cos 150° + i \sin 150°)$

44. $\sqrt{2}(\cos 225° + i \sin 225°)$

45. $6.5(\cos 33.1° + i \sin 33.1°)$

46. $14.9(\cos 289.4° + i \sin 289.4°)$

Find the product and quotient of each pair of complex numbers, using trigonometric form.

47. $z_1 = 2.5 + 2.5i, z_2 = -3 - 3i$

48. $z_1 = -\sqrt{3} + i, z_2 = -2 - 2i\sqrt{3}$

49. $z_1 = 2 + i, z_2 = 3 - 2i$

50. $z_1 = -3 + i, z_2 = 2 - i$

Use De Moivre's theorem to simplify each expression. Write the answer in the form $a + bi$.

51. $[2(\cos 45° + i \sin 45°)]^3$

52. $\left[\sqrt{3}(\cos 210° + i \sin 210°)\right]^4$

53. $(4 + 4i)^3$

54. $\left(1 - i\sqrt{3}\right)^4$

Find the indicated roots. Express answers in the form $a + bi$.

55. The square roots of i

56. The cube roots of $-i$

57. The cube roots of $\sqrt{3} + i$

58. The square roots of $3 + 3i$

59. The cube roots of $2 + i$

60. The cube roots of $3 - i$

61. The fourth roots of $625i$

62. The fourth roots of $-625i$

Convert the polar coordinates of each point to rectangular coordinates.

63. $(5, 60°)$

64. $(-4, 30°)$

65. $\left(\sqrt{3}, 100°\right)$

66. $\left(\sqrt{5}, 230°\right)$

Convert the rectangular coordinates of each point to polar coordinates. Use radians for θ.

67. $\left(-2, -2\sqrt{3}\right)$

68. $\left(-3\sqrt{2}, 3\sqrt{2}\right)$

69. $(2, -3)$

70. $(-4, -5)$

Sketch the graph of each polar equation.

71. $r = -2 \sin \theta$

72. $r = 5 \sin 3\theta$

73. $r = 2 \cos 2\theta$

74. $r = 1.1 - \cos \theta$

75. $r = 500 + \cos \theta$

76. $r = 500$

77. $r = \dfrac{1}{\sin \theta}$

78. $r = \dfrac{-2}{\cos \theta}$

For each polar equation, write an equivalent rectangular equation.

79. $r = \dfrac{1}{\sin \theta + \cos \theta}$

80. $r = -6 \cos \theta$

81. $r = -5$

82. $r = \dfrac{1}{1 + \sin \theta}$

For each rectangular equation, write an equivalent polar equation.

83. $y = 3$

84. $x^2 + (y + 1)^2 = 1$

85. $x^2 + y^2 = 49$

86. $2x + 3y = 6$

Solve each problem.

87. *Resultant force* Forces of 12 lb and 7 lb act at a 30° angle to each other. Find the magnitude of the resultant force and the angle that the resultant makes with each force.

88. *Course of a Cessna* A twin-engine Cessna is heading on a bearing of 35° with an air speed of 180 mph. If the wind is out of the west (bearing 90°) at 40 mph, then what are the bearing of the course and the ground speed of the airplane?

89. *Dividing property* Mrs. White Eagle gave each of her children approximately half of her four-sided lot in Gallup by dividing it on a diagonal. If Susan's piece is 482 ft by 364 ft by 241 ft and Seth's piece is 482 ft by 369 ft by 238 ft, then which child got the larger piece?

90. *Area, area, area* A surveyor found that two sides of a triangular lot were 135.4 ft and 164.1 ft, with an included angle of 86.4°. Find the area of this lot, using each of the three area formulas.

91. *Pipeline detour* A pipeline was planned to go from A to B as shown in the figure. However, Mr. Smith would not give permission for the pipeline to cross his property. The pipeline was laid 431 ft from A to C and then 562 feet from C to B. If ∠C is 122° and the cost of the pipeline was $21.60/ft, then how much extra was spent to go around Mr. Smith's property?

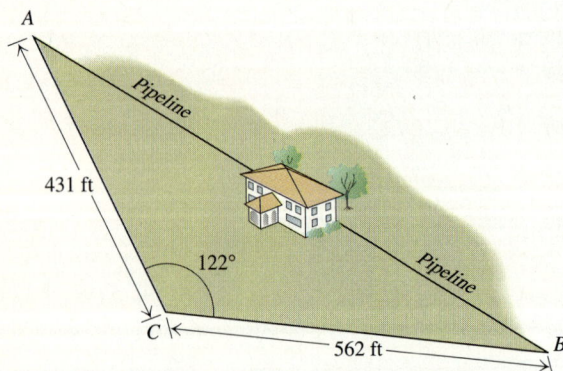

Figure for Exercise 91

92. *In the wrong place* In a lawsuit filed against a crane operator, a pedestrian of average height claims that he was struck by a wrecking ball. At the time of the accident, the operator had the ball extended 40 ft from the end of the 60-ft boom as shown in the figure, and the angle of elevation of the boom was 53°. How far from the crane would the pedestrian have to stand to be struck by this wrecking ball?

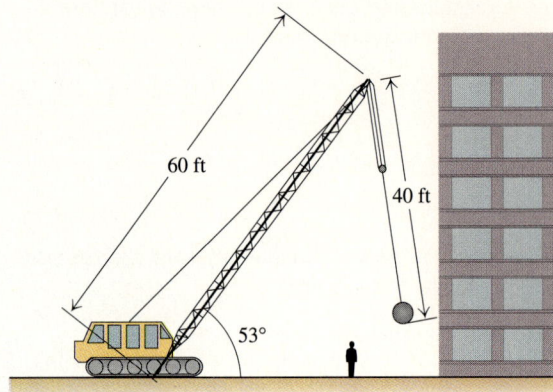

Figure for Exercise 92

Chapter 8 Test

Determine the number of triangles with the given parts and solve each triangle.

1. $\alpha = 30°, b = 4, a = 2$

2. $\alpha = 60°, b = 4.2, a = 3.9$

3. $a = 3.6, \alpha = 20.3°, \beta = 14.1°$

4. $a = 2.8, b = 3.9, \gamma = 17°$

5. $a = 4.1, b = 8.6, c = 7.3$

Given $\mathbf{A} = \langle -3, 2 \rangle$ and $\mathbf{B} = \langle 1, 4 \rangle$, find the magnitude and direction angle for each of the following vectors.

6. $\mathbf{A} + \mathbf{B}$

7. $\mathbf{A} - \mathbf{B}$

8. $3\mathbf{B}$

Write each complex number in trigonometric form, using degree measure for the argument.

9. $3 + 3i$

10. $-1 + i\sqrt{3}$

11. $-4 - 2i$

Perform the indicated operations. Write the answer in the form $a + bi$.

12. $3(\cos 20° + i \sin 20°) \cdot 2(\cos 25° + i \sin 25°)$

13. $[2(\cos 10° + i \sin 10°)]^9$

14. $\dfrac{3(\cos 63° + i \sin 63°)}{2(\cos 18° + i \sin 18°)}$

Give the rectangular coordinates for each of the following points in the polar coordinate system.

15. $(5, 30°)$

16. $(-3, -\pi/4)$

17. $(33, 217°)$

Sketch the graph of each equation in polar coordinates.

18. $r = 5 \cos \theta$

19. $r = 3 \cos 2\theta$

Solve each problem.

20. Find the area of the triangle in which $a = 4.1$ m, $b = 6.8$ m, and $c = 9.5$ m.

21. A vector $\mathbf{v}$ in the coordinate plane has direction angle $\theta = 37.2°$ and $|\mathbf{v}| = 4.6$. Find real numbers a_1 and a_2 such that $\mathbf{v} = a_1\mathbf{i} + a_2\mathbf{j}$, where $\mathbf{i} = \langle 1, 0 \rangle$ and $\mathbf{j} = \langle 0, 1 \rangle$.

22. Find all of the fourth roots of -81.

23. Write an equation equivalent to $x^2 + y^2 + 5y = 0$ in polar coordinates.

24. Write an equation equivalent to $r = 5 \sin 2\theta$ in rectangular coordinates.

25. The bearing of an airplane is 40° with an air speed of 240 mph. If the wind is out of the northwest (bearing 135°) at 30 mph, then what are the bearing of the course and the ground speed of the airplane?

Tying It All Together
Chapters 1–8

Find all real and imaginary solutions to each equation.

1. $x^4 - x = 0$

2. $x^3 - 2x^2 - 5x + 6 = 0$

3. $x^6 + 2x^5 - x - 2 = 0$

4. $x^7 - x^4 + 2x^3 - 2 = 0$

5. $2 \sin 2x - 2 \cos x + 2 \sin x = 1$

6. $4x \sin x + 2 \sin x - 2x = 1$

7. $e^{\sin x} = 1$

8. $\sin(e^x) = 1/2$

9. $2^{2x-3} = 32$

10. $\log(x - 1) - \log(x + 2) = 2$

Sketch the graph of each function using rectangular or polar coordinates as appropriate.

11. $y = \sin x$

12. $y = e^x$

13. $r = \sin \theta$

14. $r = \theta$

15. $y = \sqrt{\sin x}$

16. $y = \ln(\sin x)$

17. $r = \sin(\pi/3)$

18. $y = x^{1/3}$

Evaluate each expression.

19. $\log(\sin(\pi/2))$

20. $\sin(\log(1))$

21. $\cos(\ln(e^\pi))$

22. $\ln(\cos(2\pi))$

23. $\sin^{-1}\left(\log_2\left(\sqrt{2}\right)\right)$

24. $\cos^{-1}\left(\ln\left(\sqrt{e}\right)\right)$

25. $\tan^{-1}(\log(0.1))$

26. $\tan^{-1}(\ln(e))$

Perform the indicated operations and simplify.

27. $\dfrac{1}{x + 2} + \dfrac{1}{x - 2}$

28. $\dfrac{1}{1 - \sin x} + \dfrac{1}{1 + \sin x}$

29. $\dfrac{1}{2 + \sqrt{3}} + \dfrac{1}{2 - \sqrt{3}}$

30. $\log\left(\dfrac{1}{x + 2}\right) + \log\left(\dfrac{1}{x - 2}\right)$

31. $\dfrac{x^2 - 9}{2x - 6} \cdot \dfrac{4x + 12}{x^2 + 6x + 9}$

32. $\dfrac{1 - \sin^2(x)}{2\cos^2 x - \sin(2x)} \cdot \dfrac{\cos^2 x - \sin^2 x}{2\cos^2 x + \sin(2x)}$

33. $\dfrac{\sqrt{8}}{2 - \sqrt{3}} \cdot \dfrac{\sqrt{2}}{4 + \sqrt{12}}$

34. $\log\left(\dfrac{x^2 + 3x + 2}{x^2 + 5x + 6}\right) - \log\left(\dfrac{x + 1}{x + 3}\right)$

35. $\dfrac{\dfrac{1}{x - 2} + \dfrac{1}{x + 2}}{\dfrac{1}{x^2 - 4} - \dfrac{1}{x + 2}}$

36. $\dfrac{\dfrac{1}{\sin(2x)} - \tan(x)}{\dfrac{\cot(x)}{2} - \dfrac{\tan(x)}{2}}$

Some Basic Functions of Trigonometry

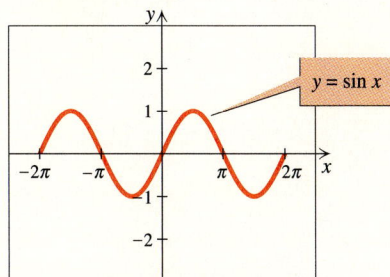

$y = \sin x$

Period 2π
Amplitude 1

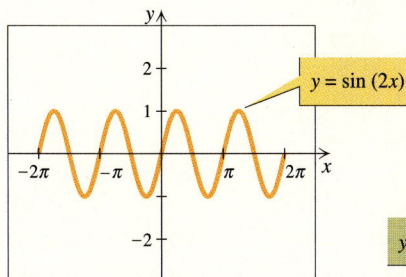

$y = \sin(2x)$

Period π
Amplitude 1

$y = 2\sin x$

Period 2π
Amplitude 2

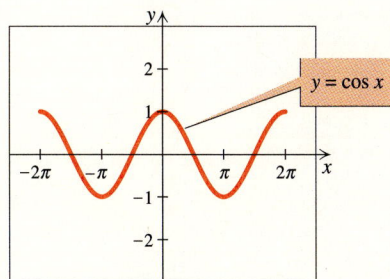

$y = \cos x$

Period 2π
Amplitude 1

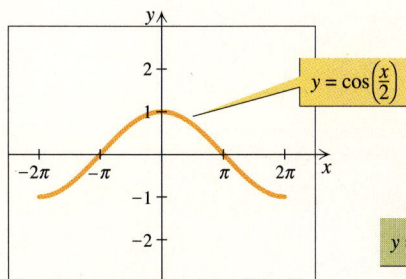

$y = \cos\left(\dfrac{x}{2}\right)$

Period 4π
Amplitude 1

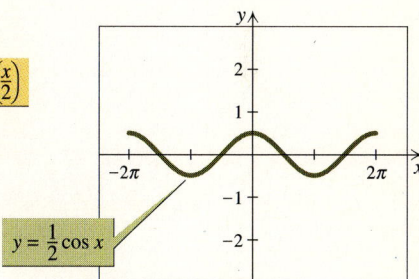

$y = \dfrac{1}{2}\cos x$

Period 2π
Amplitude $\dfrac{1}{2}$

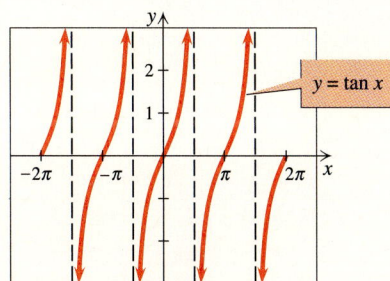

$y = \tan x$

Period π

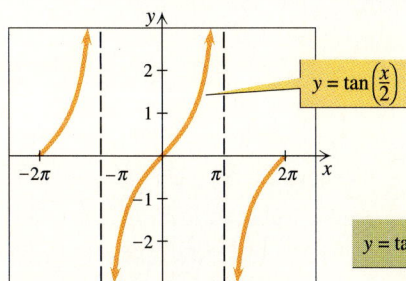

$y = \tan\left(\dfrac{x}{2}\right)$

Period 2π

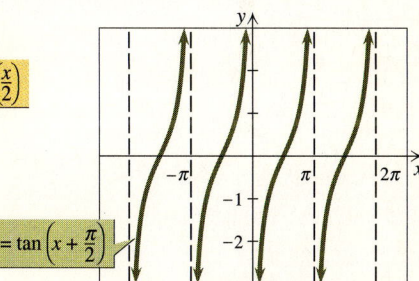

$y = \tan\left(x + \dfrac{\pi}{2}\right)$

Period π

$y = \csc x$

Period 2π

$y = \sec x$

Period 2π

$y = \cot x$

Period π

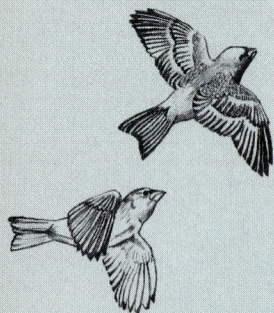

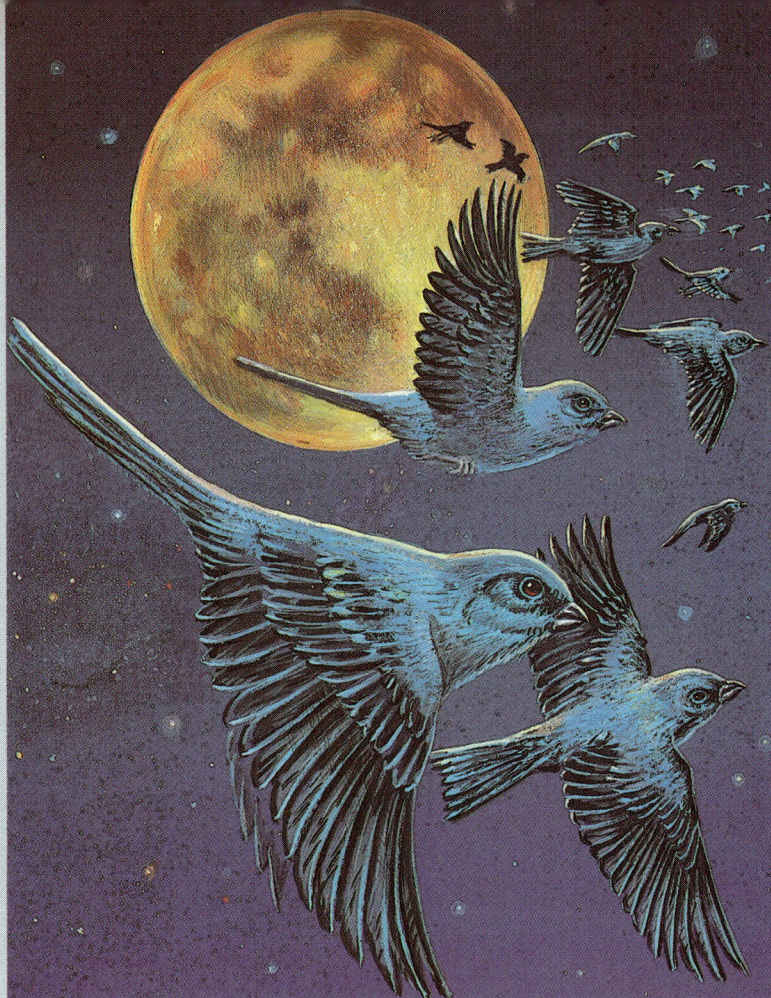

It's a moonlit night in early October. With only stars to guide them, a flock of indigo buntings wings silently over New England. Somehow these small midnight-blue finches can identify the stationary Pole Star around which other stars move and, orienting their flight away from it, they head south with unerring accuracy.

Scientists have been investigating the mysteries of bird migration for decades. In the East alone over ten million birds may fly from Cape Cod on a single fall night, heading for winter quarters in Mexico, the West Indies, and Central America. This 3000-kilometer ocean flight is one of the longest for small birds. Even the tiny ruby-throated hummingbird flies nonstop across the Gulf of Mexico to Central America—an incredible journey of 600 miles in 20 hours.

How can birds fly such long distances with neither food nor water? We know they navigate by smells, sounds, and visual landmarks and set their course by the sun, the stars, and Earth's magnetic field. Experiments indicate their stamina comes from an ability to efficiently burn body fat for fuel. At Duke University, researcher Vance Tucker set up a wind tunnel experiment. He trained parakeets and gulls to fly into a stream of air, then measured the rate at which they converted fats, carbohydrates, and proteins into usable energy. His data show that birds are experts at choosing a flight speed that will minimize the number of calories they burn. In this chapter we'll learn how to pinpoint the most economic flight speed by using systems of equations to fit a parabolic curve to Tucker's data.

In earlier chapters we solved equations containing one variable. However, many business, engineering, and science applications require solving equations or inequalities containing two or more variables. As you study the techniques for solving systems of equations and inequalities, consider the advantages and disadvantages of each method.

9

Systems of

Equations

and

Inequalities

9.1

Systems of Linear

Equations in Two

Variables

In Section 3.1 we defined a linear equation in two variables as an equation of the form

$$Ax + By = C$$

where A and B are not both zero, and we discussed numerous applications of linear equations. There are infinitely many ordered pairs that satisfy a single linear equation. In applications, however, we are often interested in finding a single ordered pair that satisfies a *pair* of linear equations. In this section we discuss several methods for solving that problem.

Solving a System by Graphing

Any collection of two or more equations is called a **system of equations.** For example, the system of equations consisting of $x + 2y = 6$ and $2x - y = -8$ is written as follows:

$$x + 2y = 6$$
$$2x - y = -8$$

The **solution set** of a system of two linear equations in two variables is the set of all ordered pairs that satisfy *both* equations of the system. The graph of an equation shows all ordered pairs that satisfy it, so we can solve some systems by graphing the equations and observing which points (if any) satisfy all of the equations.

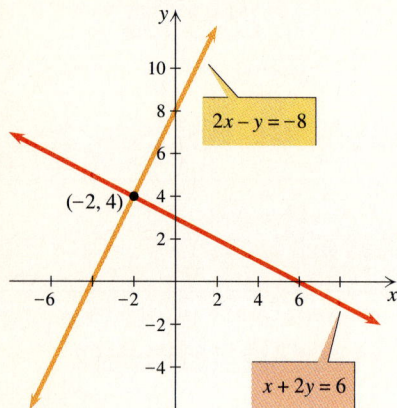

Figure 9.1

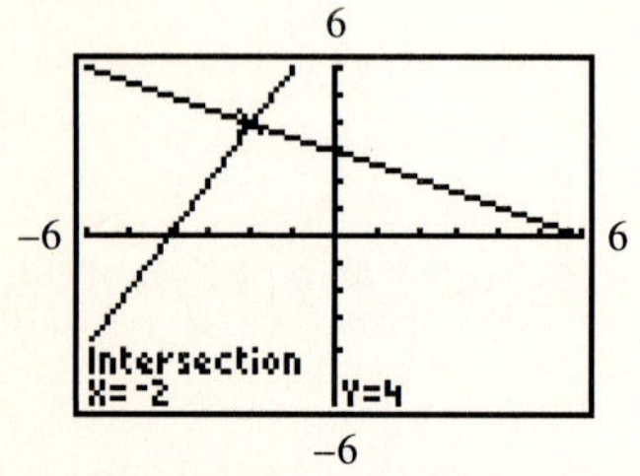

Some calculators can find the intersection of the lines in Example 1(a).

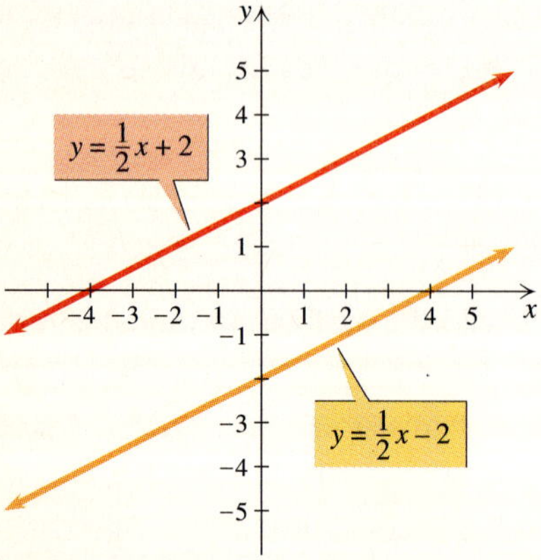

Figure 9.2

Example 1 Solving a system by graphing

Solve each system by graphing.

a) $x + 2y = 6$
  $2x - y = -8$

b) $y = \dfrac{1}{2}x + 2$
  $x - 2y = 4$

Solution

a) Graph the straight line

$$x + 2y = 6$$

by using its intercepts, $(0, 3)$ and $(6, 0)$. Graph the straight line

$$2x - y = -8$$

by using its intercepts, $(0, 8)$ and $(-4, 0)$. The graphs are shown in Fig. 9.1. The lines appear to intersect at $(-2, 4)$. Check $(-2, 4)$ in both equations. Since

$$-2 + 2(4) = 6 \qquad \text{and} \qquad 2(-2) - 4 = -8$$

are both correct, we can be certain that $(-2, 4)$ satisfies both equations. The solution set of the system is $\{(-2, 4)\}$.

b) Use the y-intercept $(0, 2)$ and the slope $1/2$ to graph

$$y = \frac{1}{2}x + 2$$

as shown in Fig. 9.2. Since $x - 2y = 4$ is equivalent to

$$y = \frac{1}{2}x - 2,$$

its graph has y-intercept $(0, -2)$ and is parallel to the first line. Since the lines are parallel, there is no point that satisfies both equations of the system. In fact, if you substitute any value of x in the two equations, the corresponding y-values will differ by 4.

Independent, Inconsistent, and Dependent Equations

Most of the systems of equations that occur in applications correspond to pairs of lines that intersect in a single point, as in Fig. 9.1. In this case the equations are called **independent** or the system is called an independent system. If the two lines corresponding to the equations are parallel, as in Fig. 9.2, then there is no solution to the system and the equations are called **inconsistent** or the system is called inconsistent. The third possibility is that two equations are equivalent and have the same graph. In this case the equations are called **dependent.** For example, the equations of the system

$$x + y = 5$$
$$2x + 2y = 10$$

have the same graph (a line) and the system is dependent. Any ordered pair that satisfies one of these equations, satisfies both of these equations. So the solution set to the system consists of all ordered pairs that satisfy one of the equations, described in set notation as

$$\{(x, y) \,|\, x + y = 5\}.$$

Since the equations are equivalent, we could use either equation in this set notation, but we usually use the simpler one. Figure 9.3 illustrates typical graphs for independent, inconsistent, and dependent systems.

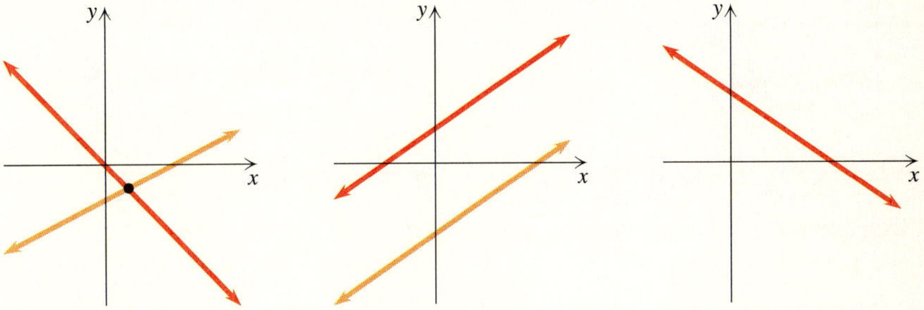

Independent system: one solution

Inconsistent system: no solution

Dependent system: infinitely many solutions

Figure 9.3

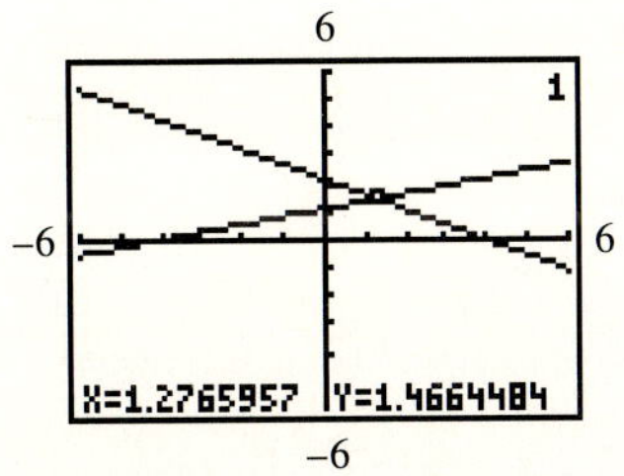

The Substitution Method

Graphing the equations of a system helps us to visualize the system and determine how many solutions it has. However, solving systems of linear equations by graphing is not very accurate unless the solution is fairly simple. The accuracy of graphing can be improved with a graphing calculator, but even with a graphing calculator, we generally get only approximate solutions. However, by using an algebraic technique such as the substitution method, we can get exact solutions quickly. In this method, shown in Example 2, we eliminate a variable from one equation by substituting an expression for that variable from the other equation.

The solution to $y_1 = (28 - 7x)/13$ and $y_2 = (29 + 7x)/26$ is $(9/7, 19/13)$, but it is difficult to find exactly with a graphing calculator.

Example 2   Solving a system by substitution

Solve each system by substitution.

a) $3x - y = 6$
$6x + 5y = -23$

b) $\quad y = 2x + 1000$
$0.05x + 0.06y = 400$

Solution

a) Since y occurs with coefficient -1 in $3x - y = 6$, it is simpler to isolate y in this equation than to isolate any other variable in the system.

$$-y = -3x + 6$$

$$y = 3x - 6$$

Use $3x - 6$ in place of y in the equation $6x + 5y = -23$:

$$6x + 5(3x - 6) = -23 \qquad \text{\color{blue}Substitution}$$

$$6x + 15x - 30 = -23$$

$$21x = 7$$

$$x = \frac{1}{3}$$

The x-coordinate of the solution is $\frac{1}{3}$. To find y, use $x = \frac{1}{3}$ in $y = 3x - 6$:

$$y = 3\left(\frac{1}{3}\right) - 6$$

$$y = -5$$

Check that $(\frac{1}{3}, -5)$ satisfies both of the original equations. The solution set is $\{(\frac{1}{3}, -5)\}$.

b) The first equation, $y = 2x + 1000$, already has one variable isolated. So we can replace y by $2x + 1000$ in $0.05x + 0.06y = 400$:

$$0.05x + 0.06(2x + 1000) = 400 \qquad \text{\color{blue}Substitution}$$

$$0.05x + 0.12x + 60 = 400$$

$$0.17x = 340$$

$$x = 2000$$

Use $x = 2000$ in $y = 2x + 1000$ to find y:

$$y = 2(2000) + 1000$$

$$y = 5000$$

Check $(2000, 5000)$ in the original equations. The solution set is $\{(2000, 5000)\}$. ◆

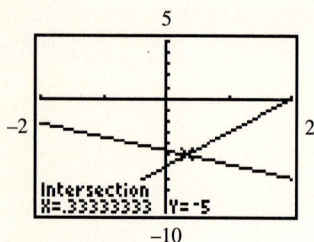

The graphs of $y_1 = 3x - 6$ and $y_2 = (-23 - 6x)/5$ support the solution to Example 2(a).

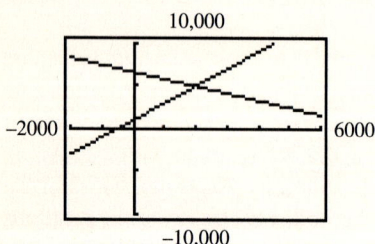

The graphs of $y_1 = 2x + 1000$ and $y_2 = (400 - 0.05x)/0.06$ support the solution in Example 2(b).

If substitution results in a false statement, then the system is inconsistent. If substitution results in an identity, then the system is dependent. In the next example on the following page we solve an inconsistent system and a dependent system by substitution.

Example 3   Inconsistent and dependent systems

Solve each system by substitution.

a) $3x - y = 9$ **b)** $\dfrac{1}{2}x - \dfrac{2}{3}y = -2$
$2y - 6x = 7$

$4y = 3x + 12$

Solution

a) Solve $3x - y = 9$ for y to get $y = 3x - 9$. Replace y by $3x - 9$ in the equation $2y - 6x = 7$:

$$2(3x - 9) - 6x = 7$$

$$6x - 18 - 6x = 7$$

$$-18 = 7$$

Since the last statement is false, the system is inconsistent and has *no solution*.

b) Solve $4y = 3x + 12$ for y to get $y = \frac{3}{4}x + 3$. Replace y by $\frac{3}{4}x + 3$ in the first equation:

$$\frac{1}{2}x - \frac{2}{3}\left(\frac{3}{4}x + 3\right) = -2$$

$$\frac{1}{2}x - \frac{1}{2}x - 2 = -2$$

$$-2 = -2$$

Since the last statement is an identity, the system is dependent. The solution set is $\{(x, y) \mid 4y = 3x + 12\}$. ◆

Note that there are many ways of writing the solution set in Example 3(b). Since $4y = 3x + 12$ is equivalent to $y = \frac{3}{4}x + 3$ or to $x = \frac{4}{3}y - 4$, we could write the solution set as

$$\left\{\left(x, \frac{3}{4}x + 3\right) \middle| x \text{ is any real number}\right\} \quad \text{or}$$

$$\left\{\left(\frac{4}{3}y - 4, y\right) \middle| y \text{ is any real number}\right\}.$$

The graphs of $y_1 = 3x - 9$ and $y_2 = (6x + 7)/2$ are parallel lines because the system in Example 3(a) is inconsistent.

The Addition Method

In the substitution method we eliminate a variable in one equation by substituting from the other equation. In the addition method we eliminate a variable by adding the two equations. It might be necessary to multiply each equation by an appropriate number so that a variable will be eliminated by the addition.

Example 4   Solving systems by addition

Solve each system by addition.

a) $3x - y = 9$ b) $2x - 3y = -2$
   $2x + y = 1$ $3x - 2y = 12$

Solution

a) Add the equations to eliminate the y-variable:

$$\begin{array}{rl} 3x - y &= 9 \\ 2x + y &= 1 \\ \hline 5x &= 10 \\ x &= 2 \end{array}$$

Use $x = 2$ in $2x + y = 1$ to find y: $2(2) + y = 1$

$$y = -3$$

Substituting the values $x = 2$ and $y = -3$ in the original equations yields $3(2) - (-3) = 9$ and $2(2) + (-3) = 1$, which are both correct. So $(2, -3)$ satisfies both equations and the solution set to the system is $\{(2, -3)\}$.

b) To eliminate x upon addition, we multiply the first equation by 3 and the second equation by -2:

$$3(2x - 3y) = 3(-2)$$
$$-2(3x - 2y) = -2(12)$$

This multiplication produces $6x$ in one equation and $-6x$ in the other. So the x-variable is eliminated upon addition of the equations.

$$\begin{array}{rl} 6x - 9y &= -6 \\ -6x + 4y &= -24 \\ \hline -5y &= -30 \\ y &= 6 \end{array}$$

Use $y = 6$ in $2x - 3y = -2$ to find x: $2x - 3(6) = -2$

$$2x - 18 = -2$$
$$2x = 16$$
$$x = 8$$

If $y = 6$ is used in the other equation, $3x - 2y = 12$, we would also get $x = 8$. Substituting $x = 8$ and $y = 6$ in both of the original equations yields $2(8) - 3(6) = -2$ and $3(8) - 2(6) = 12$, which are both correct. So the solution set to the system is $\{(8, 6)\}$.   ◆

In Example 4(b), we started with the given system and multiplied the first equation by 3 and the second equation by -2 to get

$$6x - 9y = -6$$
$$-6x + 4y = -24.$$

Since each equation of the new system is equivalent to an equation of the old system, the solution sets to these systems are identical. Two systems with the same solution set are **equivalent systems.** If we had multiplied the first equation by 2 and the second by -3, we would have obtained the equivalent system

$$4x - 6y = -4$$
$$-9x + 6y = -36$$

and we would have eliminated y by adding the equations.

When we have a choice of which method to use for solving a system, we generally avoid graphing because it is often inaccurate. Substitution or addition both yield exact solutions, but sometimes one method is easier to apply than the other. Substitution is usually used when one equation gives one variable in terms of the other as in Example 2. Addition is usually used when both equations are in the form $Ax + By = C$ as in Example 4. By doing the exercises, you will soon discover which method works best on a given system.

When a system is solved by the addition method, an inconsistent system results in a false statement and a dependent system results in an identity, just as they did for the substitution method.

Example 5 Inconsistent and dependent systems

Solve each system by addition.

a) $0.2x - 0.4y = 0.5$
$x - 2y = 1.3$

b) $\dfrac{1}{2}x - \dfrac{2}{3}y = -2$
$-3x + 4y = 12$

Solution

a) It is usually a good idea to eliminate the decimals in the coefficients, so we multiply the first equation by 10:

$$2x - 4y = 5 \qquad \text{\color{blue}First equation multiplied by 10}$$
$$x - 2y = 1.3$$

Now multiply the second equation by -2 and add to eliminate x:

$$2x - 4y = 5$$
$$\underline{-2x + 4y = -2.6}$$
$$0 = 2.4$$

Since $0 = 2.4$ is false, there is no solution to the system.

b) To eliminate fractions in the coefficients, multiply the first equation by the LCD 6:

$$3x - 4y = -12$$
$$\underline{-3x + 4y = 12}$$
$$0 = 0$$

Since $0 = 0$ is an identity, the solution set is $\{(x, y) \mid -3x + 4y = 12\}$. ◆

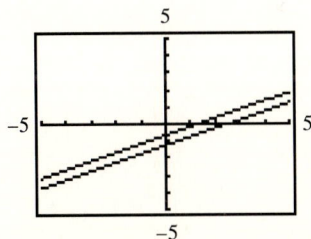

The graphs of $y_1 = (0.5 - 0.2x)/(-0.4)$ and $y_2 = (1.3 - x)/(-2)$ are parallel lines because the system in Example 5(a) is inconsistent.

Note that there are many ways to solve a system by addition. In Example 5(a), we could have multiplied the first equation by -5 or the second equation by -0.2. In either case, x would be eliminated upon addition. Try this for yourself.

Applications

We solved many problems involving linear equations in the past, but we always wrote all unknown quantities in terms of a single variable. Now that we can solve systems of equations, we can solve problems involving two unknown quantities by using two variables and a system of at least two equations.

Example 6 Problem solving using a system of equations

Ahmed has observed that every day at 9 A.M., five English professors come into the cafeteria and buy three doughnuts and five coffees for $3.30. At 9:05 A.M. three mathematics professors buy four doughnuts and three coffees for $2.75. At 9:10 A.M. the provost comes in alone and gets a doughnut and a cup of coffee. How much is the provost's bill?

Solution

Let x be the cost of one doughnut and y be the cost of one cup of coffee. We can write a system of equations about x and y:

$$3x + 5y = 3.30$$
$$4x + 3y = 2.75$$

To eliminate x, multiply the first equation by -4 and the second by 3:

$$-4(3x + 5y) = -4(3.30)$$
$$3(4x + 3y) = 3(2.75)$$

Add the two resulting equations:

$$-12x - 20y = -13.20$$
$$\underline{12x + 9y = 8.25}$$
$$-11y = -4.95$$
$$y = 0.45$$

Use $y = 0.45$ in $3x + 5y = 3.30$ to find x:

$$3x + 5(0.45) = 3.30$$
$$3x + 2.25 = 3.30$$
$$3x = 1.05$$
$$x = 0.35$$

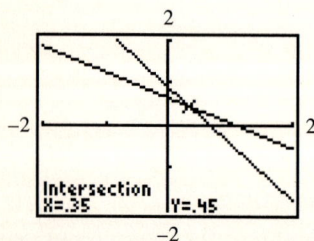

Graph $y_1 = (3.30 - 3x)/5$ and $y_2 = (2.75 - 4x)/3$ and use the intersection feature to find $(0.35, 0.45)$.

Check that 35 cents for a doughnut and 45 cents for a cup of coffee satisfy the statements in the original problem. Given these prices, the provost's bill is 80 cents.   ◆

? For Thought

True or false? Explain.

The following systems are referenced in these statements.

a) $x + y = 5$ **b)** $x - 2y = 4$ **c)** $x = 5 + 3y$
      $x - y = 1$ $3x - 6y = 8$ $9y - 3x = -15$

1. The ordered pair (2, 3) is in the solution set to $x + y = 5$.

2. The ordered pair (2, 3) is in the solution set to system (a).

3. System (a) is inconsistent.

4. There is no solution to system (b).

5. Adding the equations in system (a) would eliminate y.

6. To solve system (c) we could substitute $5 + 3y$ for x in $9y - 3x = -15$.

7. System (c) is inconsistent.

8. The solution set to system (c) is the set of all real numbers.

9. The graphs of the equations of system (c) intersect at a single point.

10. The graphs of the equations of system (b) are parallel.

9.1 Exercises   📼 **Tape 17**   💾 **Disk—5.25″: 6   3.5″: 4   Macintosh: 4**

Solve each system by inspecting the graphs of the equations.

1. $2x - 3y = -4$
   $y = -2x + 4$

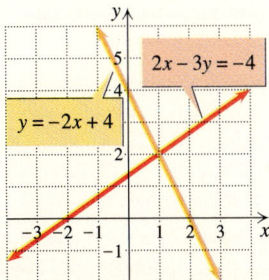

2. $x + 2y = -1$
   $2x + 3y = -3$

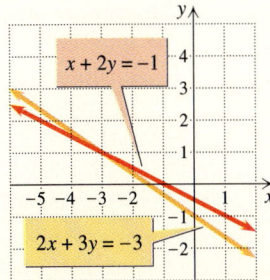

3. $3x - 4y = 0$
   $y = \dfrac{3}{4}x + 2$

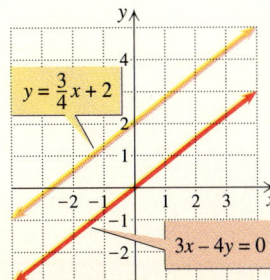

4. $x - 2y = -3$
   $y = \dfrac{1}{2}x + \dfrac{3}{2}$

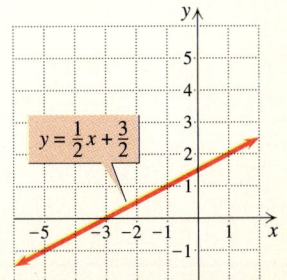

Solve each system by graphing.

5. $x + y = 5$
$x - y = 1$

6. $2x + y = -1$
$y - x = 5$

7. $y = x - 2$
$y = -x + 4$

8. $y = -3x$
$x - 2y = 7$

9. $3y + 2x = 6$
$y = -\dfrac{2}{3}x - 1$

10. $2x + 4y = 12$
$2y = 6 - x$

11. $y = \dfrac{1}{2}x - 3$
$2x - 4y = 12$

12. $y = -2x + 6$
$4x + 2y = 8$

Solve each system by substitution. Determine whether each system is independent, inconsistent, or dependent.

13. $y = 2x + 1$
$3x - 4y = 1$

14. $5x - 6y = 23$
$x = 6 - 3y$

15. $x + y = 1$
$2x - 3y = 8$

16. $x + 2y = 3$
$2x + y = 5$

17. $y - 3x = 5$
$3(x + 1) = y - 2$

18. $2y = 1 - 4x$
$2x + y = 0$

19. $2y = 6 - 3x$
$\dfrac{1}{2}x + \dfrac{1}{3}y = 3$

20. $2x = 10 - 5y$
$\dfrac{1}{5}x + \dfrac{1}{2}y = 1$

21. $x + y = 200$
$0.05x + 0.06y = 10.50$

22. $2x + y = 300$
$\dfrac{1}{2}x + \dfrac{1}{3}y = 80$

23. $y = 3x + 1$
$y = 3x - 7$

24. $2x + y = 9$
$4x + 2y = 10$

25. $\dfrac{1}{2}x - \dfrac{1}{3}y = 12$
$\dfrac{1}{4}x - \dfrac{1}{2}y = 1$

26. $0.05x + 0.1y = 10$
$0.06x + 0.2y = 16$

Solve each system by addition. Determine whether each system is independent, inconsistent, or dependent.

27. $x + y = 20$
$x - y = 6$

28. $3x - 2y = 7$
$-3x + y = 5$

29. $x - y = 5$
$3x + 2y = 10$

30. $x - 4y = -3$
$-3x + 5y = 2$

31. $x - y = 7$
$y - x = 5$

32. $2x - y = 6$
$-4x + 2y = 9$

33. $2x + 3y = 1$
$3x - 5y = -8$

34. $-2x + 5y = 14$
$7x + 6y = -2$

35. $0.05x + 0.1y = 0.6$
$x + 2y = 12$

36. $0.02x - 0.04y = 0.08$
$x - 2y = 4$

37. $\dfrac{x}{2} + \dfrac{y}{2} = 5$
$\dfrac{3x}{2} - \dfrac{2y}{3} = 2$

38. $\dfrac{x}{4} + \dfrac{y}{3} = 0$
$\dfrac{x}{8} - \dfrac{y}{6} = 2$

39. $3x - 2.5y = -4.2$
$0.12x + 0.09y = 0.4932$

40. $1.5x - 2y = 8.5$
$3x + 1.5y = 6$

Classify each system as independent, inconsistent, or dependent without doing any written work.

41. $y = 5x - 6$
$y = -5x - 6$

42. $y = 5x - 6$
$y = 5x + 4$

43. $5x - y = 6$
$y = 5x - 6$

44. $5x - y = 6$
$y = -5x + 6$

Solve each problem using two variables and a system of two equations. Solve the system by the method of your choice.

45. *Income on investments* Carmen made $25,000 profit on the sale of her condominium. She lent part of the profit to Jim's Orange Grove at 10% interest and the remainder to Ricky's Used Cars at 8% interest. If she received $2200 in interest after one year, then how much did she lend to each business?

46. *Stock market losses* In 1993 Gerhart lost twice as much in the futures market as he did in the stock market. If his losses totaled $18,630, then how much did he lose in each market?

47. *Zoo admission prices* The Lincoln Park Zoo has different admission prices for adults and children. When Mr. and Mrs. Weaver went with their five children, the bill was $33. If Mrs. Wong and her three children got in for $18.50, then what is the price of an adult's ticket and what is the price of a child's ticket?

48. *Book prices* At the Book Exchange, all paperbacks sell for one price and all hardbacks sell for another price. Tanya got six paperbacks and three hardbacks for $8.25, while Gretta got four paperbacks and five hardbacks for $9.25. What was Todd's bill for seven paperbacks and nine hardbacks?

49. *Political party preference* The results of a survey of students at Central High School concerning political party preference are given in the accompanying table. If 230 students preferred the Democratic party and 260 students preferred the Republican party, then how many students are there at CHS?

	Male	Female
Democratic	50%	30%
Republican	20%	60%
Other	30%	10%

Table for Exercise 49

50. *Distribution of coin types*   Isabelle paid for her $1.75 lunch with 87 coins. If all of the coins were nickels and pennies, then how many were there of each type?

51. *Coin collecting*   Theodore has a collection of 166 old coins consisting of quarters and dimes. If he figures that each coin is worth two and a half times its face value, then his collection is worth $61.75. How many of each type of coin does he have?

52. *Real estate development*   During one week a real estate developer gave either a Florida vacation coupon or a complete home stereo system to 120 people who listened to a one-hour presentation on the advantages of owning a lot inside the Arctic Circle. It costs the developer $8 for the complete home stereo and $22 for the Florida vacation coupon. If she sold 14 lots and her cost for prizes that week was $1688, then how many of each did she give away?

53. *Protein and carbohydrates*   Nutritional information for Rice Krispies and Grape-nuts is given in the accompanying table. How many servings of each would it take to get exactly 23 g of protein and 215 g of carbohydrates?

	Rice Krispies	Grape-nuts
Protein (g/serving)	2	3
Carbohydrates (g/serving)	25	23

Table for Exercise 53

54. *Doubles and singles*   The Executive Inn rents a double room for $10 more per night than a single. One night the motel took in $2159 by renting 15 doubles and 26 singles. What is the rental price for each type of room?

55. *Furniture rental*   A civil engineer has a choice of two plans for renting furniture for her new office. Under Plan A, she pays $800 plus $150 per month, while under Plan B she pays $200 plus $200 per month. For each plan, write the cost as a function of the number of months. Which plan is cheaper in the long run? For what number of months do the two plans cost the same?

56. *Equilibrium price*   At Milton's Gallery, the weekly demand d for oil paintings is determined from the average price p in his gallery by the equation $d = 1614 - 15p$. Milton's supply of paintings s also depends on the average price at which they are sold, according to the function $s = 2 + 0.5p$. What is the equilibrium price, the price at which the supply equals the demand?

A system of equations can be used to find the equation of a line that goes through two points. For example, if $y = ax + b$ goes through (3, 5), then a and b must satisfy $3a + b = 5$. For each given pair of points, find the equation of the line $y = ax + b$ that goes through the points by solving a system of equations.

57. $(-3, 9), (2, -1)$

58. $(1, -1), (3, 7)$

59. $(-2, 3), (4, -7)$

60. $(-3, -1), (4, 9)$

For Writing/Discussion

61. *Number of solutions*   Explain how you can tell (without graphing) whether a system of linear equations has one solution, no solutions, or infinitely many solutions. Be sure to account for linear equations that are not functions.

62. *Cooperative learning*   Write a step-by-step procedure (or algorithm) based on the addition method that will solve any system of two equations of the form $Ax + By = C$. Ask a classmate to solve a system using your procedure.

63. *Cooperative learning*   Write an independent system of two linear equations for which $(2, -3)$ is the solution. Ask a classmate to solve your system.

64. *Cooperative learning*   Write a dependent system of two linear equations for which $\{(t, t + 5) | t$ is any real number$\}$ is the solution set. Ask a classmate to solve your system.

Graphing Calculator Exercises

Solve each system by graphing the equations on a graphing calculator and estimating the point of intersection.

1. $y = 0.5x + 3$
$y = 0.499x + 2$

2. $y = 2x - 3$
$y = 1.9999x - 2$

3. $0.23x + 0.32y = 1.25$
$0.47x - 1.26y = 3.58$

4. $342x - 78y = 474$
$123x + 145y = 397$

9.2

Systems of Linear

Equations in

Three Variables

Figure 9.4

Systems of many linear equations in many variables are used to model a variety of situations ranging from airline scheduling to allocating resources in manufacturing. The same techniques that we are studying with small systems can be extended to much larger systems. In this section we use the techniques of substitution and addition from Section 9.1 to solve systems of linear equations in three variables.

Definitions

A **linear equation in three variables** x, y, and z is an equation of the form

$$Ax + By + Cz = D$$

where A, B, C, and D are real numbers with A, B, and C not all equal to zero. For example,

$$x + y + 2z = 9$$

is a linear equation in three variables. The equation is called *linear* because its form is similar to that of a linear equation in two variables. A solution to a linear equation in three variables is an **ordered triple** of real numbers in the form (x, y, z) that satisfies the equation. For instance, the ordered triple $(1, 2, 3)$ is a solution to

$$x + y + 2z = 9$$

because $1 + 2 + 2(3) = 9$. Other ordered triples such as $(4, 5, 0)$ or $(3, 4, 1)$ are also in the solution set to $x + y + 2z = 9$. In fact, there are infinitely many ordered triples in the solution set to a linear equation in three variables.

The graph of the solution set of a linear equation in three variables requires a three-dimensional coordinate system. A three-dimensional coordinate system has a z-axis through the origin of the xy-plane, as shown in Fig. 9.4. The third coordinate of a point indicates its distance above or below the xy-plane. The point $(1, 2, 3)$ is shown in Fig. 9.4.

You will not be required to graph equations in three variables in this text. However, it is helpful to know that the graph of a linear equation in three variables

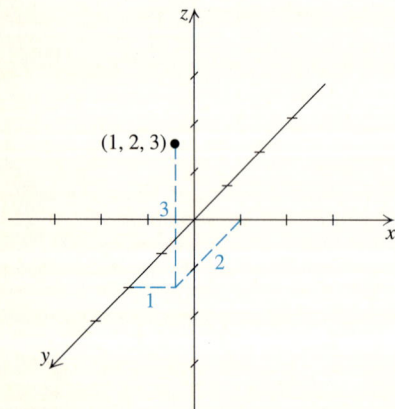

is a plane and not a line. For example, the graph of

$$x + y + 2z = 9$$

is shown in Fig. 9.5. We will refer to graphs as an aid to understanding the solutions to linear systems, but the systems will be solved without graphs by using addition, substitution, or both.

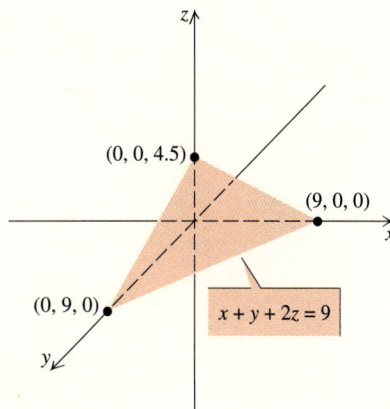

Figure 9.5

Systems with Infinite Solution Sets

Two planes in three-dimensional space either are parallel or intersect along a line as shown in Fig. 9.6. If the planes are parallel, there is no common point and no solution to the system. If the two planes intersect along a line, then there are infinitely many points on that line that satisfy both equations of the system. In our first example we solve such a system.

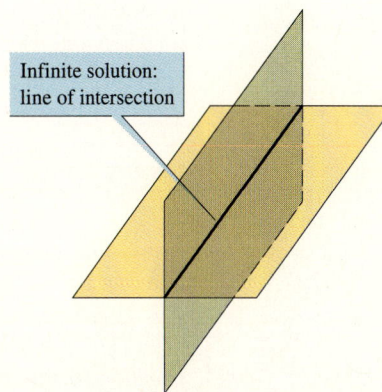

Dependent System of Two Equations

Figure 9.6

Example 1   Two linear equations in three variables

Solve the system.

$$(1) \qquad -2x + 3y - z = -1$$

$$(2) \qquad x - 2y + z = 3$$

Solution

Add the equations to eliminate z:

$$-2x + 3y - z = -1$$
$$\underline{x - 2y + z = 3}$$
$$-x + y = 2$$
$$(3) \qquad\qquad\qquad y = x + 2$$

Equation (3) indicates that (x, y, z) satisfies both equations if and only if $y = x + 2$. Write Eq. (2) as $z = 3 - x + 2y$ and substitute $y = x + 2$ into this equation:

$$z = 3 - x + 2(x + 2)$$
$$z = x + 7$$

Now (x, y, z) satisfies (1) and (2) if and only if $y = x + 2$ and $z = x + 7$. So the solution set to the system could be written as

$$\{(x, y, z) \mid y = x + 2 \text{ and } z = x + 7\}$$

or more simply

$$\{(x, x + 2, x + 7) \mid x \text{ is any real number}\}.$$

This system has infinitely many solutions. Note that we can write the solution set in terms of x, y, or z, depending on which variable we choose to eliminate in the first step and which variable we solve for after that. For example, the solution set could also be written as

$$\{(z - 7, z - 5, z) \mid z \text{ is any real number}\},$$

because $z = x + 7$.   ◆

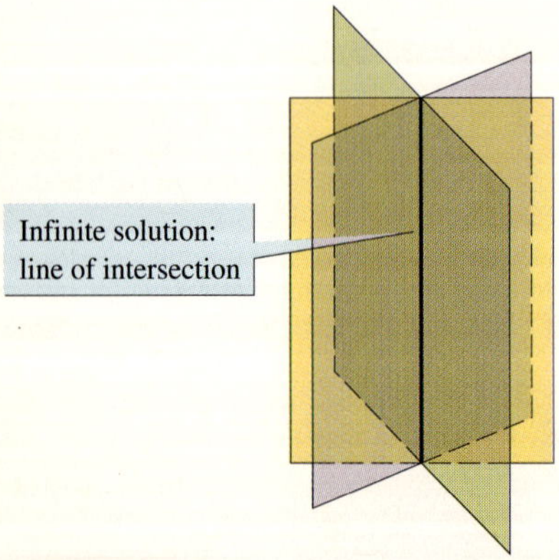

Infinite solution: line of intersection

Dependent System of Three Equations

Figure 9.7

A line in three-dimensional space does not have a simple equation like a line in a two-dimensional coordinate system. In Example 1, the triple $(x, x + 2, x + 7)$ is a point on the line of intersection of the two planes, for any real number x. For example, the points $(0, 2, 7)$, $(1, 3, 8)$, and $(-2, 0, 5)$ all satisfy both equations of the system and lie on the line of intersection of the planes.

In the next example we solve a system consisting of three planes that intersect along a single line, as shown in Fig. 9.7. This example is similar to Example 1.

Example 2   A dependent system of three equations

Solve the system.

$$\begin{aligned}
(1) \qquad 2x + y - z &= 1 \\
(2) \qquad -3x - 3y + 2z &= 1 \\
(3) \qquad -10x - 14y + 8z &= 10
\end{aligned}$$

Solution

Examine the system and decide which variable to eliminate. If Eq. (1) is multiplied by 2, then $-2z$ will appear in Eq. (1) and $2z$ in Eq. (2). So, multiply Eq. (1) by 2 and add the result to Eq. (2) to eliminate z:

$$\begin{aligned}
4x + 2y - 2z &= 2 \qquad &&\text{Eq. (1) multiplied by 2} \\
-3x - 3y + 2z &= 1 \qquad &&\text{Eq. (2)} \\
\hline
(4) \qquad x - y &= 3
\end{aligned}$$

Now eliminate z from Eqs. (2) and (3) by multiplying Eq. (2) by -4 and adding the result to Eq. (3):

$$\begin{aligned}
12x + 12y - 8z &= -4 \qquad &&\text{Eq. (2) multiplied by } -4 \\
-10x - 14y + 8z &= 10 \qquad &&\text{Eq. (3)} \\
\hline
(5) \qquad 2x - 2y &= 6
\end{aligned}$$

Note that Eq. (5) is a multiple of Eq. (4). Since Eqs. (4) and (5) are dependent, the original system has infinitely many solutions. We can describe all solutions in terms of the single variable x. To do this, get $y = x - 3$ from Eq. (4). Then substitute $x - 3$ for y in Eq. (1) to find z in terms of x:

$$\begin{aligned}
z &= 2x + y - 1 \qquad &&\text{Eq. (1) solved for } z \\
z &= 2x + (x - 3) - 1 \qquad &&\text{Replace } y \text{ by } x - 3. \\
z &= 3x - 4
\end{aligned}$$

An ordered triple (x, y, z) satisfies the system provided $y = x - 3$ and $z = 3x - 4$. So the solution set to the system is $\{(x, x - 3, 3x - 4) \mid x \text{ is any real number}\}$. ◆

If all of the original equations are equivalent, then the solution set to the system is the set of all points that satisfy one of the equations. For example, the system

$$\begin{aligned}
x + y + z &= 1 \\
2x + 2y + 2z &= 2 \\
3x + 3y + 3z &= 3
\end{aligned}$$

has solution set $\{(x, y, z) \mid x + y + z = 1\}$.

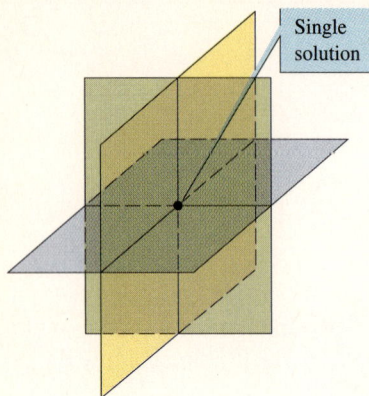

Single solution

Independent System of Three Equations

Figure 9.8

Independent Systems

As with linear systems in two variables, a linear system in three variables can have one, infinitely many, or zero solutions. If the three equations correspond to three planes that intersect at a single point, as in Fig. 9.8, then the solution to the system is a single ordered triple. In this case, the system is called **independent**.

Example 3   An independent system of equations

Solve the system.

$$(1) \qquad x + y - z = 0$$
$$(2) \qquad 3x - y + 3z = -2$$
$$(3) \qquad x + 2y - 3z = -1$$

Solution

Look for a variable that is easy to eliminate by addition. Since y occurs in Eq. (1) and $-y$ occurs in Eq. (2), we can eliminate y by adding Eqs. (1) and (2):

$$x + y - z = 0$$
$$\underline{3x - y + 3z = -2}$$
$$(4) \qquad 4x + 2z = -2$$

Now repeat the process to eliminate y from Eqs. (1) and (3). Multiply Eq. (1) by -2 and add the result to Eq. (3):

$$-2x - 2y + 2z = 0 \qquad \text{Eq. (1) multiplied by } -2$$
$$\underline{x + 2y - 3z = -1} \qquad \text{Eq. (3)}$$
$$(5) \qquad -x - z = -1$$

Equations (4) and (5) are a system of two linear equations in two variables. We could solve this system by substitution or addition. To solve by addition, multiply Eq. (5) by 2 and add to Eq. (4) to eliminate z:

$$4x + 2z = -2 \qquad \text{Eq. (4)}$$
$$\underline{-2x - 2z = -2} \qquad \text{Eq. (5) multiplied by 2}$$
$$2x = -4$$
$$x = -2$$

Use $x = -2$ in $4x + 2z = -2$ to find z:

$$4(-2) + 2z = -2$$
$$2z = 6$$
$$z = 3$$

To find y, use $x = -2$ and $z = 3$ in $x + y - z = 0$ (Eq. 1):

$$-2 + y - 3 = 0$$
$$y = 5$$

Check that the ordered triple $(-2, 5, 3)$ satisfies all three of the original equations:

(1) $-2 + 5 - 3 = 0$ *Correct.*
(2) $3(-2) - 5 + 3(3) = -2$ *Correct.*
(3) $-2 + 2(5) - 3(3) = -1$ *Correct.*

The solution set is $\{(-2, 5, 3)\}$.   ◆

When solving a system of three equations in three variables, we try to get a system of two equations in two variables by eliminating one of the variables. Any one of the three variables can be eliminated first, but we usually look for the easiest one. We may eliminate that variable from the first and second, the first and third, or the second and third equations. For example, we could have started Example 3 by adding Eqs. (2) and (3) to eliminate z. We then repeat the process, eliminating the same variable from another pair of equations in the system. After getting two equations in two unknowns, we solve that system using methods for systems involving two variables.

Inconsistent Systems

There are several ways that three planes can be positioned so that there is no single point that is on all three planes. For example, Fig. 9.9 corresponds to a system where there are ordered triples that satisfy two equations, but no ordered triple that satisfies all three equations. Whatever the configuration of the planes, if there are no points in common to all three, the corresponding system is called **inconsistent** and has no solution. It is easy to identify a system that has no solution because a false statement will occur when we try to solve the system.

No solution: no point on all three planes

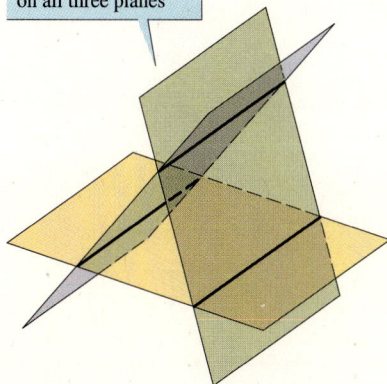

Inconsistent System of Three Equations

Figure 9.9

Example 4   A system with no solution

Solve the system.

(1) $x + y - z = 9$
(2) $x + y - z = 7$
(3) $x - y + z = 3$

Solution

Multiply Eq. (1) by -1 and add the result to Eq. (2):

$$-x - y + z = -9 \qquad \text{Eq. (1) multiplied by } -1$$
$$\underline{x + y - z = 7} \qquad \text{Eq. (2)}$$
$$0 = -2$$

Since $0 = -2$ is false, the system is inconsistent. There is no solution.   ◆

Applications

It is a fairly simple matter to write the equation of a line when given two points on the line, but it is a bit more complicated to write the equation of a parabola when given three points on the parabola. Surprisingly, the equation of a parabola through three points can be found by solving a system of three linear equations. The process of finding a curve that goes through some given points is called **curve fitting.**

In the next example, we fit a parabola to some given data points to determine the most economic flight speed for a parakeet in level flight.

Example 5   Power expenditure of birds in flight

In studying how migratory birds can fly thousands of miles without rest, Vance Tucker gathered data on parakeets and sea gulls in a wind tunnel (*Scientific American,* May 1969). Tucker found that a parakeet in level flight at 12 mph has a *power expenditure* of 152. (That is, it requires 152 calories to move each gram of body weight at that speed for one hour.) At 22 mph the parakeet's power expenditure is 105, and at 30 mph its power expenditure is 165. These points are graphed in Fig. 9.10. Tucker's data suggest that power expenditure is a quadratic function of the flying speed. Find the quadratic function whose graph goes through the three given points and find the speed at which the power expenditure is a minimum.

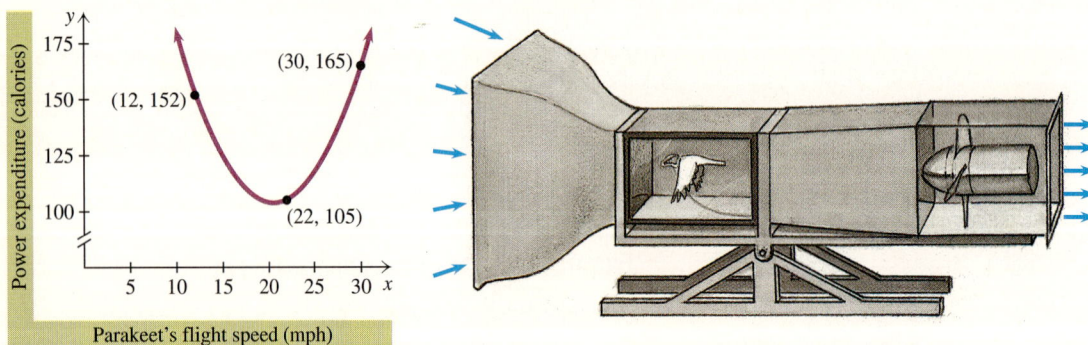

Figure 9.10

Solution

The ordered pairs (12, 152), (22, 105), and (30, 165) satisfy $y = ax^2 + bx + c$, where x is the speed and y is the power expenditure. We get three equations by using values of x and y from the ordered pairs in the equation $y = ax^2 + bx + c$.

(1) $144a + 12b + c = 152$ Use $x = 12$ and $y = 152$.

(2) $484a + 22b + c = 105$ Use $x = 22$ and $y = 105$.

(3) $900a + 30b + c = 165$ Use $x = 30$ and $y = 165$.

The easiest variable to eliminate in this system is c. Multiplying Eq. (1) by -1 and adding it to Eq. (2) yields

(4) $340a + 10b = -47.$

Multiplying Eq. (2) by -1 and adding it to Eq. (3) yields

$$416a + 8b = 60.$$

Dividing each side of this equation by 4 (to simplify it) yields

(5) $104a + 2b = 15.$

We now solve the system consisting of Eqs. (4) and (5). Multiplying Eq. (5) by -5 and adding it to Eq. (4), we get

$$-180a = -122$$
$$a = 0.6778.$$

Substituting $a = 0.6778$ into Eq. (5) to find b, we get

$$b = -27.7446.$$

Using the values of a and b in Eq. (1) gives

$$c = 387.3349.$$

The equation

$$y = 0.6778x^2 - 27.7446x + 387.3349$$

expresses power expenditure y as a function of speed x. The minimum value of y occurs at the vertex of the parabola, at $x = -b/(2a)$:

$$x = \frac{27.7446}{2(0.6778)} = 20.5$$

So the parakeet expends the least amount of energy flying at about 20.5 mph. In fact, other research has confirmed that many migrating birds fly (without the influence of the wind) at speeds of 20 to 25 mph.   ◆

The next example involves three unknown prices, but we are given enough information to write three linear equations concerning those prices.

```
QuadReg
 y=ax²+bx+c
 a=.6777777778
 b=-27.74444444
 c=387.3333333
```

A calculator can find the quadratic equation through three given points using quadratic regression, which is not the same as the method in Example 5.

Example 6   A problem involving three unknowns

Lionel delivers milk, bread, and eggs to Marcie's Camp Store. On Monday the bill for eight half-gallons of milk, four loaves of bread, and six dozen eggs was $14.20. On Tuesday the bill for five half-gallons of milk, ten loaves of bread, and three dozen eggs was $14.50. On Wednesday the bill for two half-gallons of milk, five loaves of bread, and seven dozen eggs was $9.50. How much is Thursday's bill for one half-gallon of milk, two loaves of bread, and one dozen eggs?

Solution

Let x represent the price of a half-gallon of milk, y represent the price of a loaf of bread, and z represent the price of a dozen eggs. We can write an equation for the bill on each of the three days:

$$(1) \qquad 8x + 4y + 6z = 14.20$$

$$(2) \qquad 5x + 10y + 3z = 14.50$$

$$(3) \qquad 2x + 5y + 7z = 9.50$$

Multiply Eq. (2) by -2 and add the result to Eq. (1):

$$8x + 4y + 6z = 14.20 \qquad \text{Eq. (1)}$$

$$\underline{-10x - 20y - 6z = -29} \qquad \text{Eq. (2) multiplied by } -2$$

$$-2x - 16y \qquad = -14.80$$

$$(4) \qquad x + 8y = 7.40$$

Multiply Eq. (2) by -7, multiply Eq. (3) by 3, and add the results:

$$-35x - 70y - 21z = -101.50 \qquad \text{Eq. (2) multiplied by } -7$$

$$\underline{6x + 15y + 21z = 28.50} \qquad \text{Eq. (3) multiplied by 3}$$

$$(5) \qquad -29x - 55y \qquad = -73$$

Multiply Eq. (4) by 29 and add the result to Eq. (5) to eliminate x:

$$29x + 232y = 214.60 \qquad \text{Eq. (4) multiplied by 29}$$

$$\underline{-29x - 55y = -73} \qquad \text{Eq. (5)}$$

$$177y = 141.60$$

$$y = 0.80$$

Use $y = 0.80$ in $x + 8y = 7.40$ (Eq. 4):

$$x + 8(0.80) = 7.40$$

$$x + 6.40 = 7.40$$

$$x = 1$$

Use $x = 1$ and $y = 0.80$ in $8x + 4y + 6z = 14.20$ (Eq. 1) to find z:

$$8(1) + 4(0.80) + 6z = 14.20$$

$$6z = 3$$

$$z = 0.50$$

So milk is \$1.00 per half-gallon, bread is 80 cents per loaf, and eggs are 50 cents per dozen. Thursday's bill should be \$3.10. ◆

For Thought

True or false? Explain.

The following systems are referenced in statements 1–7:

(a)	(b)	(c)
$x + y - z = 2$	$x + y - z = 6$	$x - y + z = 1$
$-x - y + z = 4$	$x + y + z = 4$	$-x + y - z = -1$
$x - 2y + 3z = 9$	$x - y - z = 8$	$2x - 2y + 2z = 2$

1. The point $(1, 1, 0)$ is in the solution set to $x + y - z = 2$.

2. The point $(1, 1, 0)$ is in the solution set to system (a).

3. System (a) is inconsistent.

4. The point $(2, 3, -1)$ satisfies all equations of system (b).

5. The point $(6, -1, -1)$ satisfies all equations of system (b).

6. The solution set to system (c) is $\{(x, y, z) \mid x - y + z = 1\}$.

7. System (c) is dependent.

8. The solution set to $y = 2x + 3$ is $\{(x, 2x + 3) \mid x$ is any real number$\}$.

9. $(3, 1, 0) \in \{(x + 2, x, x - 1) \mid x$ is any real number$\}$.

10. x nickels, y dimes, and z quarters are worth $5x + 10y + 25z$ dollars.

9.2 Exercises   Tape 17   Disk—5.25″: 6   3.5″: 4   Macintosh: 4

Solve each system of equations.

1. $x + y + z = 6$
   $2x - 2y - z = -5$
   $3x + y - z = 2$

2. $3x - y + 2z = 14$
   $x + y - z = 0$
   $2x - y + 3z = 18$

3. $3x + 2y + z = 1$
   $x + y - 2z = -4$
   $2x - 3y + 3z = 1$

4. $4x - 2y + z = 13$
   $3x - y + 2z = 13$
   $x + 3y - 3z = -10$

5. $2x + y - 2z = -15$
   $4x - 2y + z = 15$
   $x + 3y + 2z = -5$

6. $x - 2y - 3z = 4$
   $2x - 4y + 5z = -3$
   $5x - 6y + 4z = -7$

7. $x - 2y + 3z = 5$
   $2x - 4y + 6z = 3$
   $2x - 3y + z = 9$

8. $-2x + y - 3z = 6$
   $4x - y + z = 2$
   $2x - y + 3z = 1$

9. $x + 2y - 3z = -17$
   $3x - 2y - z = -3$

10. $x + 2y + z = 4$
    $2x - y - z = 3$

11. $x + 2y - 3z = 5$
    $-x - 2y + 3z = -5$
    $2x + 4y - 6z = 10$

12. $2x - 6y + 4z = 8$
    $3x - 9y + 6z = 12$
    $5x - 15y + 10z = 20$

13. $x + y - z = 2$
    $2x - y + z = 4$

14. $-2x + 2y - z = 4$
    $2x - y + z = 1$

15. $x + y = 5$
    $y - z = 2$
    $x + z = 3$

16. $2x - y = -1$
    $-2x + z = 1$
    $y - z = 0$

17. $x - y + z = 7$
    $2y - 3z = -13$
    $3x - 2z = -3$

18. $2x + y - z = 5$
    $2y + 3z = -14$
    $-3y - 2z = 11$

19. $x + y + 2z = 7.5$
    $3x + 4y + z = 12$
    $5x + 2y + 5z = 21$

20. $100x + 200y + 500z = 47$
    $350x + 5y + 250z = 33.9$
    $200x + 80y + 100z = 23.4$

21. $x + y + z = 9000$
    $0.05x + 0.06y + 0.09z = 710$
    $z = 3y$

22.
$$x + y + z = 200{,}000$$
$$0.09x + 0.08y + 0.12z = 20{,}200$$
$$z = x + y$$

23.
$$x = 2y - 1$$
$$y = 3z + 2$$
$$z = 2x - 3$$

24.
$$x + 2y - 3z = 0$$
$$2x - y + z = 0$$
$$3x + y - 4z = 0$$

Use a system of equations to find the parabola of the form $y = ax^2 + bx + c$ that goes through the three given points.

25. $(-1, -2), (2, 1), (-2, 1)$ **26.** $(1, 2), (2, 3), (3, 6)$

27. $(0, 0), (1, 3), (2, 2)$ **28.** $(0, -6), (1, -3), (2, 6)$

29. $(0, 4), (-2, 0), (-3, 1)$ **30.** $(0, 6), (3, 0), (-1, 12)$

Solve each problem by using a system of three linear equations in three variables.

31. *Efficiency for descending flight*   In studying flight of birds (*Scientific American,* May 1969), Vance Tucker measured the efficiency (the relationship between power input and power output) for parakeets flying at various speeds in a descending flight pattern. He recorded an efficiency of 0.18 at 12 mph, 0.23 at 22 mph, and 0.14 at 30 mph. Tucker's measurements suggest that efficiency E is a quadratic function of the speed s. Find the quadratic function whose graph goes through the three given ordered pairs, and find the speed that gives the maximum efficiency for descending flight.

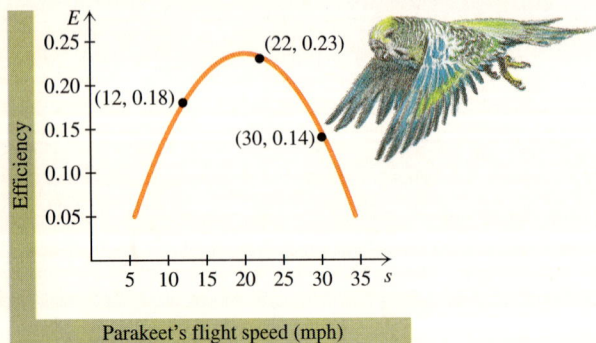

Figure for Exercise 31

32. *Path of a missile*   A missile is fired from the origin in the direction of the positive x-axis as shown in the drawing. Radar shows the missile at 4000 ft when it is 1 mi from the origin and 7000 ft when it is 2 mi from the origin. Assuming that the path of the missile is parabolic, find the equation of the parabola. What is the highest altitude reached by the missile? How many miles from the origin will the missile strike?

Figure for Exercise 32

33. *Stocks, bonds, and a mutual fund*   Marita invested a total of $25,000 in stocks, bonds, and a mutual fund. In one year she earned 8% on her stock investment, 10% on her bond investment, and 6% on her mutual fund, with a total return of $1860. Unfortunately, the amount she invested in the mutual fund was twice as large as the amount she invested in the bonds. How much did she invest in each?

34. *Age groups*   In 1980 the population of Springfield was 1911. In 1990 the number of people under 20 years old increased 10%, while the number of people in the 20 to 60 category decreased by 8%, and the number of people over 60 increased by one-third. If the 1990 population was 2136 and in 1990 the number of people over 60 was equal to the number of people 60 and under, then how many were in each age group in 1980?

35. *Fast food inflation*   Last year you could get a hamburger, fries, and a Coke at Francisco's Drive-In for $1.90. Since the price of a hamburger has increased 10%, the price of fries has increased 20%, and the price of a Coke has increased 50%, the same meal now costs $2.37. If the price of a Coke is now 9 cents more than that of a hamburger, then what was the price of each item last year?

36. *Misplaced house numbers*   Angelo on Elm Street removed his three-digit house number for painting and noticed that the sum of the digits was 9 and that the units digit was three times as large as the hundreds digit. When the painters put the house number back up, they reversed the digits. The new house number was now 396 larger than the correct house number. What is Angelo's correct address?

37. *Distribution of coins* Emma paid the $10.36 bill for her lunch with 232 coins consisting of pennies, nickels, and dimes. If the number of nickels plus the number of dimes was equal to the number of pennies, then how many coins of each type did she use?

38. *Students, teachers, and pickup trucks* Among the 564 students and teachers at Jefferson High School, 128 drive to school each day. One-fourth of the male students, one-sixth of the female students, and three-fourths of the teachers drive. Among those who drive to school, there are 41 who drive pickup trucks. If one-half of the driving male students, one-tenth of the driving female students, and one-third of the driving teachers drive pickups, then how many male students, female students, and teachers are there?

39. *Milk, coffee, and doughnuts* The employees from maintenance go for coffee together every day at 9 A.M. On Monday, Hector paid $5.45 for three cartons of milk, four cups of coffee, and seven doughnuts. On Tuesday, Guillermo paid $5.30 for four milks, two coffees, and eight doughnuts. On Wednesday, Anna paid $5.15 for two milks, five coffees, and six doughnuts. On Thursday, Alphonse had only $6.00. Did he have enough money to pay for five milks, two coffees, and nine doughnuts?

40. *Average age of vehicles* The average age of the Johnsons' cars is eight years. Three years ago the Toyota was twice as old as the Ford. Two years ago the sum of the Buick's and the Ford's ages was equal to the age of the Toyota. How old is each car now?

For Writing/Discussion

41. *Cooperative learning* Write a system of three linear equations in three unknowns for which $(1/2, 1/3, 1/4)$ is the only solution. Ask a classmate to solve the system.

42. *Cooperative learning* Write a word problem for which a system of three linear equations in three unknowns can be used to find the solution. Ask a classmate to solve the problem.

9.3

Nonlinear Systems

of Equations

In Sections 9.1 and 9.2 we solved linear systems of equations, but we have seen many equations in two variables that are not linear. Equations such as

$$y = 3x^2, \qquad y = \sqrt{x}, \qquad y = |x|, \qquad y = 10^x, \qquad \text{and} \qquad y = \log(x)$$

are called *nonlinear equations,* because their graphs are not straight lines. If a system has at least one nonlinear equation, it is called a **nonlinear system.** Systems of nonlinear equations arise in applications just as systems of linear equations do. In this section we will use the techniques that we learned for linear systems to solve nonlinear systems.

Solving by Elimination of Variables

To solve nonlinear systems, we combine equations to eliminate variables just as we did for linear systems. However, since the graphs of nonlinear equations are not straight lines, the graphs might intersect at more than one point, and the solution set might contain more than one point. The next examples show systems whose solution sets contain two points, four points, and one point.

Example 1 A parabola and a line

Solve the system of equations and sketch the graph of each equation on the same coordinate plane.

$$y = x^2 + 1$$
$$y - x = 2$$

Solution

The graph of $y = x^2 + 1$ is a parabola opening upward. The graph of $y = x + 2$ is a line with y-intercept $(0, 2)$ and slope 1. The graphs are shown in Fig. 9.11. To find the exact coordinates of the points of intersection of the graphs, we solve the system by substitution. Substitute $y = x^2 + 1$ into $y - x = 2$:

$$x^2 + 1 - x = 2$$
$$x^2 - x - 1 = 0$$

Solve this quadratic equation using the quadratic formula:

$$x = \frac{1 \pm \sqrt{1 - 4(1)(-1)}}{2(1)} = \frac{1 \pm \sqrt{5}}{2}$$

Use $x = \left(1 \pm \sqrt{5}\right)/2$ in $y = x + 2$ to find y:

$$y = \frac{1 \pm \sqrt{5}}{2} + 2 = \frac{5 \pm \sqrt{5}}{2}$$

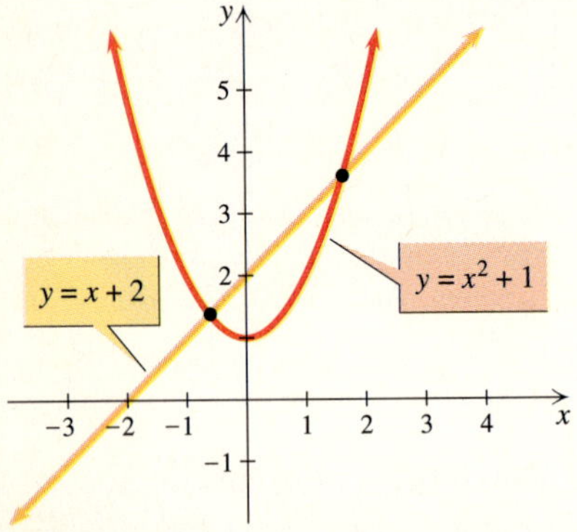

The solution set to the system is

$$\left\{ \left(\frac{1 + \sqrt{5}}{2}, \frac{5 + \sqrt{5}}{2} \right), \left(\frac{1 - \sqrt{5}}{2}, \frac{5 - \sqrt{5}}{2} \right) \right\}.$$

To check the solution, use a calculator to find the approximations $(1.62, 3.62)$ and $(-0.62, 1.38)$. These points of intersection are consistent with Fig. 9.11. You should also check the decimal approximations in the original equations. Note that we could have solved this system by solving the second equation for x (or for y) and then substituting into the first. ◆

In the next example we find the points of intersection for the graph of an absolute value function and a parabola.

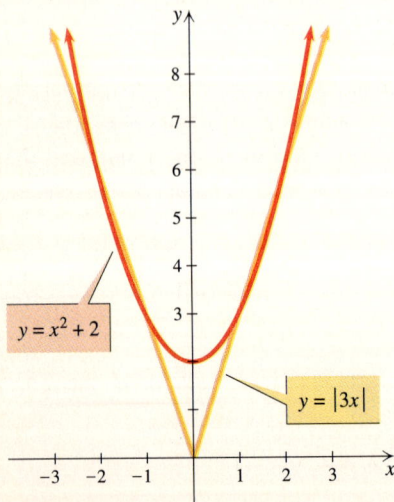

Figure 9.11

Example 2 An absolute value function and a parabola

Solve the system of equations and sketch the graph of each equation on the same coordinate plane.

$$y = |3x|$$
$$y = x^2 + 2$$

Solution

The graph of $y = x^2 + 2$ is a parabola opening upward with vertex at $(0, 2)$. The graph of $y = |3x|$ is V-shaped and passes through $(0, 0)$ and $(\pm 1, 3)$. Both graphs are shown in Fig. 9.12. To eliminate y, we can substitute $|3x|$ for y in the second equation:

$$|3x| = x^2 + 2$$

Figure 9.12

Write an equivalent compound equation without absolute value and solve:

$$3x = x^2 + 2 \qquad \text{or} \qquad 3x = -(x^2 + 2)$$
$$x^2 - 3x + 2 = 0 \qquad \text{or} \qquad x^2 + 3x + 2 = 0$$
$$(x - 2)(x - 1) = 0 \qquad \text{or} \qquad (x + 2)(x + 1) = 0$$
$$x = 2 \ \text{ or } \ x = 1 \qquad \text{or} \qquad x = -2 \ \text{ or } \ x = -1$$

Using each of these values for x in the equation $y = |3x|$ yields the solution set $\{(-2, 6), (-1, 3), (1, 3), (2, 6)\}$. This solution set is consistent with what we see on the graphs in Fig. 9.12.   ◆

It is not necessary to draw the graphs of the equations of a nonlinear system to solve it. However, the graphs give us an idea of how many solutions to expect for the system. In the next two examples we solve systems without graphing.

Example 3   Solving a nonlinear system

Solve the system of equations.

$$(1) \qquad x^2 + y^2 = 25$$

$$(2) \qquad \frac{x^2}{18} + \frac{y^2}{32} = 1$$

Solution

Write Eq. (1) as $y^2 = 25 - x^2$ and substitute into Eq. (2):

$$\frac{x^2}{18} + \frac{25 - x^2}{32} = 1$$

$$288\left(\frac{x^2}{18} + \frac{25 - x^2}{32}\right) = 288 \cdot 1 \qquad \text{The LCD for 18 and 32 is 288.}$$

$$16x^2 + 9(25 - x^2) = 288$$

$$7x^2 + 225 = 288$$

$$7x^2 = 63$$

$$x^2 = 9$$

$$x = \pm 3$$

Use $x = 3$ in $y^2 = 25 - x^2$:
$$y^2 = 25 - 3^2$$
$$y^2 = 16$$
$$y = \pm 4$$

Use $x = -3$ in $y^2 = 25 - x^2$:
$$y^2 = 25 - (-3)^2$$
$$y^2 = 16$$
$$y = \pm 4$$

The solution set is $\{(-3, 4), (-3, -4), (3, 4), (3, -4)\}$. Check that all four ordered pairs satisfy both equations of the original system.   ◆

Sidebar (left margin):

```
Y₁⊟√(25-X²)
Y₂⊟-√(25-X²)
Y₃⊟√(32-32X²/18)
Y₄⊟-√(32-32X²/18
)
Y₅=
Y₆=
```

Solve each equation for y.

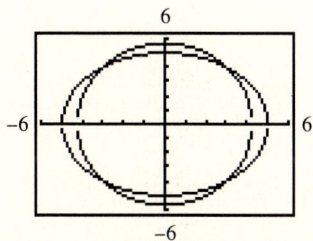

The curves intersect at four points.

Example 4   Solving a nonlinear system

Solve the system of equations.

$$(1) \quad \frac{2}{x} + \frac{3}{y} = \frac{1}{2}$$

$$(2) \quad \frac{4}{x} + \frac{1}{y} = \frac{2}{3}$$

Solution

Since these equations have the same form, we can use addition to eliminate a variable. Multiply Eq. (2) by -3 and add the result to Eq. (1):

$$\frac{2}{x} + \frac{3}{y} = \frac{1}{2}$$

$$\frac{-12}{x} + \frac{-3}{y} = -2$$

$$\overline{\frac{-10}{x} \qquad = -\frac{3}{2}}$$

$$3x = 20$$

$$x = \frac{20}{3}$$

Use $x = 20/3$ in Eq. (1) to find y:

$$\frac{2}{20/3} + \frac{3}{y} = \frac{1}{2}$$

$$\frac{3}{10} + \frac{3}{y} = \frac{1}{2}$$

$$3y + 30 = 5y \qquad \text{\textcolor{blue}{\textbf{Multiply each side by 10y.}}}$$

$$30 = 2y$$

$$15 = y$$

The solution set is $\{(20/3, 15)\}$. Check this solution in the original system. ◆

Applications

In the next example we solve a problem using a nonlinear system.

Example 5   Application of a nonlinear system

A 10-inch diagonal-measure television is advertised as having a viewing area of 48 square inches. What are the width and height of the screen?

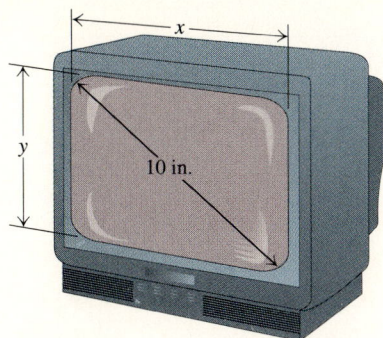

Figure 9.13

Solution

Let x be the width and y be the height as shown in Fig. 9.13. Use the Pythagorean theorem to write the first equation and the formula $A = LW$ to write the second:

$$x^2 + y^2 = 10^2$$
$$xy = 48$$

Write $xy = 48$ as $y = 48/x$ and replace y by $48/x$ in the first equation:

$$x^2 + \left(\frac{48}{x}\right)^2 = 100$$

$$x^2 + \frac{2304}{x^2} = 100$$

$$x^4 + 2304 = 100x^2$$

$x^4 - 100x^2 + 2304 = 0$ An equation of quadratic type

$(x^2 - 36)(x^2 - 64) = 0$ Solve by factoring.

$x^2 = 36$ or $x^2 = 64$

$x = \pm 6$ or $x = \pm 8$

Since x cannot be negative in this situation, x is either 6 or 8. If $x = 6$ then $y = 8$; and if $x = 8$, then $y = 6$. The width of a television screen is usually larger than the height, so we conclude that the screen is 8 inches wide and 6 inches high. ◆

❓ For Thought

True or false? Explain.

1. The line $y = x$ intersects the circle $x^2 + y^2 = 1$ at two points.

2. A line and a circle intersect at two points or not at all.

3. A parabola and a circle can intersect at three points.

4. The parabolas $y = x^2 - 1$ and $y = 1 - x^2$ do not intersect.

5. In a 30-60-90 triangle, the length of the side opposite the 30° angle is half the length of the hypotenuse.

6. Two distinct circles can intersect at more than two points.

7. The area of a right triangle is half the product of the lengths of its legs.

8. The surface area of a rectangular solid with length L, width W, and height H is $2LW + 2LH + 2WH$.

9. Two numbers with a sum of 6 and a product of 7 are $3 - \sqrt{2}$ and $3 + \sqrt{2}$.

10. It is impossible to find two numbers with a sum of 7 and a product of 1.

9.3 Exercises

Tape 17   Disk—5.25″: 6   3.5″: 4   Macintosh: 4

Graph both equations on the same coordinate system and determine the points of intersection of the graphs.

1. $5x - y = 6$
$y = x^2$

2. $2x^2 - y = 8$
$7x + y = -4$

3. $y = |x| - 1$
$2y - x = 1$

4. $y = x + 3$
$y = |x|$

5. $y = \sqrt{x}$
$y = 2x$

6. $y = \sqrt{x + 3}$
$x - 4y = -7$

7. $y = x^3$
$y = 4x$

8. $y = x^2$
$x = y^2$

9. $x^2 + (y - 2)^2 = 4$
$2y = x^2$

10. $y = 2^x$
$y = 3^{-x}$

11. $y = \log_2(x)$
$y = \log_{1/3}(x)$

12. $y = x^3 - x$
$y = x$

13. $y = x^4 - x^2$
$y = x^2$

14. $x^2 + y^2 = 25$
$2x - 3y = -6$

Solve each system.

15. $x + y = -4$
$xy = 1$

16. $x + y = 10$
$xy = 21$

17. $2x^2 - y^2 = 1$
$x^2 - 2y^2 = -1$

18. $xy - 2x = 2$
$2x - y = 1$

19. $\dfrac{3}{x} - \dfrac{1}{y} = \dfrac{13}{10}$
$\dfrac{1}{x} + \dfrac{2}{y} = \dfrac{9}{10}$

20. $\dfrac{2}{x} + \dfrac{3}{2y} = \dfrac{11}{4}$
$\dfrac{5}{2x} - \dfrac{2}{y} = \dfrac{3}{2}$

21. $x^2 + xy - y^2 = -5$
$x + y = 1$

22. $x^2 + xy + y^2 = 12$
$x + y = 2$

23. $x^2 + 2xy - 2y^2 = -11$
$-x^2 - xy + 2y^2 = 9$

24. $-3x^2 + 2xy - y^2 = -9$
$3x^2 - xy + y^2 = 15$

Solve each system of exponential or logarithmic equations.

25. $y = 2^{x+1}$
$y = 4^{-x}$

26. $y = 3^{2x+1}$
$y = 9^{-x}$

27. $y = \log_2(x)$
$y = \log_4(x + 2)$

28. $y = \log_2(-x)$
$y = \log_2(x + 4)$

29. $y = \log_2(x + 2)$
$y = 3 - \log_2(x)$

30. $y = \log(2x + 4)$
$y = 1 + \log(x - 2)$

31. $y = 3^x$
$y = 2^x$

32. $y = 6^{x-1}$
$y = 2^{x+1}$

Solve each problem using a system of two equations in two unknowns.

33. *Legs of a right triangle*   Find the lengths of the legs of a right triangle whose hypotenuse is 15 m and whose area is 54 m².

34. *Sides of a rectangle*   What are the length and width of a rectangle that has a perimeter of 98 cm and a diagonal of 35 cm?

35. *Sides of a triangle*   Find the lengths of the sides of a triangle whose perimeter is 12 ft and whose angles are 30°, 60°, and 90°.

36. *Size of a vent*   Kwan is constructing a triangular vent in the gable end of a house, as shown in the diagram. If the pitch of the roof is 6-12 (run 12 ft and rise 6 ft) and the vent must have an area of 4.5 ft², then what size should he make the base and height of the triangle?

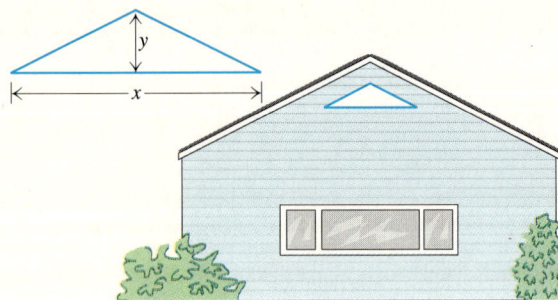

Figure for Exercise 36

37. *Pumping tomato soup*   At the Acme Soup Company a large vat of tomato soup can be filled in 8 min by pump A and pump B working together. Pump B can be reversed so that it empties the vat at the same rate at which it fills the vat. One day a worker filled the vat in 12 min using pumps A and B, but accidentally ran pump B in reverse. How long does it take each pump to fill the vat working alone?

38. *Planting strawberries* Blanche and Morris can plant an acre of strawberries in 8 hr working together. Morris takes 2 hr longer to plant an acre of strawberries working alone than it takes Blanche working alone. How long does it take Morris to plant an acre by himself?

39. *Lost numbers* Find two complex numbers whose sum is 6 and whose product is 10.

40. *More lost numbers* Find two complex numbers whose sum is 1 and whose product is 5.

41. *Voyage of the whales* In one of Captain James Kirk's most challenging missions, he returned to late twentieth century San Francisco to bring back a pair of humpback whales. Chief Engineer Scotty built a tank of transparent aluminum to hold the time-traveling cetaceans. If Scotty's 20-ft-high tank had a volume of 36,000 ft^3, and it took 7,200 ft^2 of transparent aluminum to cover all six sides, then what were the length and width of the tank?

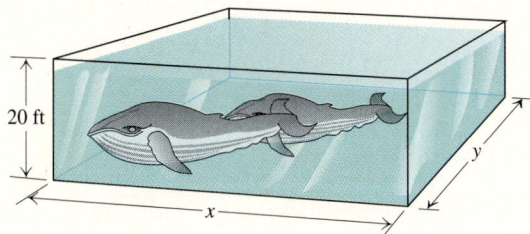

Figure for Exercise 41

42. *Dimensions of a laundry room* The plans call for a rectangular laundry room in the Wilsons' new house. Connie wants to increase the width by 1 ft and the length by 2 ft, which will increase the area by 30 ft^2. Christopher wants to increase the width by 2 ft and decrease the length by 3 ft, which will decrease the area by 6 ft^2. What are the original dimensions for the laundry room?

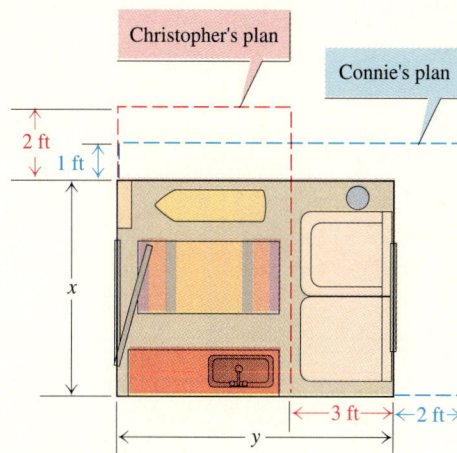

Figure for Exercise 42

For Writing/Discussion

43. *Cooperative learning* Write a nonlinear system of two equations that has three points in its solution set and a nonlinear system of two equations that has no solution. Ask a classmate to solve your systems.

44. *How many?* Explain why the system
$$y = 3 + \log_2(x - 1)$$
$$8x - 8 = 2^y$$
has infinitely many solutions.

45. *Line and circle* For what values of b does the solution set to $y = x + b$ and $x^2 + y^2 = 1$ consist of one point, two points, and no points? Explain.

Graphing Calculator Exercises

Using a graphing calculator, we can solve systems that are too difficult to solve algebraically. Solve each system by graphing its equations on a graphing calculator and finding points of intersection to the nearest tenth.

1. $y = \log_2(x)$
   $y = 2^x - 3$

2. $x^2 - y^2 = 1$
   $y = 2^{x-3}$

3. $x^2 + y^2 = 4$
   $y = \log_3(x)$

4. $y = \dfrac{x^3}{6} + \dfrac{x^2}{2} + x + 1$
   $y = e^x$

5. One demographer believes that the population growth of a certain country is best modeled by the function $P(t) = 20e^{0.07t}$, while a second demographer believes that the population growth of that same country is best modeled by the function $P(t) = 20 + 2t$. In each case t is the number of years from the present and $P(t)$ is given in millions of people. For what values of t do these two models give the same population? In how many years is the population predicted by the exponential model twice as large as the population predicted by the linear model?

9.4

Partial Fractions

In algebra we usually learn a process and then learn to reverse it. For example, we multiply two binomials, and then we factor trinomials into a product of two binomials. We solve polynomial equations, and we write polynomial equations with given solutions. After we studied functions, we learned about their inverses. In Section 1.6 we learned how to add rational expressions. Now we will reverse the process of addition. We start with a rational expression and write it as a sum of two or more simpler rational expressions. This technique is useful in calculus, and it also shows a nice application of systems of equations.

The Basic Idea

Before trying to reverse addition of rational expressions, recall how to add them.

Example 1 Adding rational expressions

Perform the indicated operation.

$$\frac{3}{x - 3} + \frac{-2}{x + 1}$$

Solution

The least common denominator (LCD) for $x - 3$ and $x + 1$ is $(x - 3)(x + 1)$. We convert each rational expression or fraction into an equivalent fraction with this denominator:

$$\frac{3}{x - 3} + \frac{-2}{x + 1} = \frac{3(x + 1)}{(x - 3)(x + 1)} + \frac{-2(x - 3)}{(x + 1)(x - 3)}$$

$$= \frac{3x + 3}{(x - 3)(x + 1)} + \frac{-2x + 6}{(x - 3)(x + 1)}$$

$$= \frac{x + 9}{(x - 3)(x + 1)} \qquad \blacklozenge$$

The following rational expression is similar to the result in Example 1.

$$\frac{8x - 7}{(x + 1)(x - 2)}$$

Thus, it is possible that this fraction is the sum of two fractions with denominators $x + 1$ and $x - 2$. In the next example, we will find those two fractions.

Example 2 Reversing the addition of rational expressions

Write the following rational expression as a sum of two rational expressions.

$$\frac{8x - 7}{(x + 1)(x - 2)}$$

Solution

To write the given expression as a sum, we need numbers A and B such that

$$\frac{8x - 7}{(x + 1)(x - 2)} = \frac{A}{x + 1} + \frac{B}{x - 2}.$$

Simplify this equation by multiplying each side by the LCD, $(x + 1)(x - 2)$:

$$(x + 1)(x - 2)\frac{8x - 7}{(x + 1)(x - 2)} = (x + 1)(x - 2)\left(\frac{A}{x + 1} + \frac{B}{x - 2}\right)$$

$$8x - 7 = A(x - 2) + B(x + 1)$$

$$8x - 7 = Ax - 2A + Bx + B$$

$$8x - 7 = (A + B)x - 2A + B \qquad \text{Combine like terms.}$$

Since the last equation is an identity, the coefficient of x on one side equals the coefficient of x on the other, and the constant on one side equals the constant on the other. So A and B satisfy the following two equations.

$$A + B = 8$$

$$-2A + B = -7$$

We can solve this system of two linear equations in two unknowns by addition:

$$\begin{aligned} A + B &= 8 \\ \underline{2A - B} &= \underline{7} \qquad \text{Second equation multiplied by } -1 \\ 3A &= 15 \\ A &= 5 \end{aligned}$$

If $A = 5$, then $B = 3$, and we have

$$\frac{8x - 7}{(x + 1)(x - 2)} = \frac{5}{x + 1} + \frac{3}{x - 2}.$$

Check by adding the fractions on the right-hand side of the equation. ◆

Each of the two fractions on the right-hand side of the equation

$$\frac{8x - 7}{(x + 1)(x - 2)} = \frac{5}{x + 1} + \frac{3}{x - 2}$$

is called a **partial fraction.** This equation shows the **partial fraction decomposition** of the rational expression on the left-hand side.

General Decomposition

In general, let $N(x)$ be the polynomial in the numerator and $D(x)$ be the polynomial in the denominator of the fraction that is to be decomposed. *We will decompose only fractions for which the degree of the numerator is smaller than the degree of the denominator.* If the degree of $N(x)$ is not smaller than the degree of

$D(x)$, we can use long division to write the rational expression as quotient + remainder/divisor. For example,

$$\frac{x^3 + x^2 - x + 5}{x^2 + x - 6} = x + \frac{5x + 5}{x^2 + x - 6} = x + \frac{A}{x + 3} + \frac{B}{x - 2}.$$

You should find the values of A and B in the above equation as we did in Example 2, and check.

If a factor of $D(x)$ is repeated n times, then all powers of the factor from 1 through n might occur as denominators in the partial fractions. To understand the reason for this statement, look at

$$\frac{7}{8} = \frac{1}{2} + \frac{1}{4} + \frac{1}{8} = \frac{1}{2} + \frac{1}{2^2} + \frac{1}{2^3}.$$

Here the factor 2 is repeated three times in $D(x) = 8$. Notice that each of the powers of 2 (2^1, 2^2, 2^3) occurs in the denominators of the partial fractions.

Now consider a fraction $N(x)/D(x)$, where $D(x) = (x - 1)(x + 3)^2$. Since the factor $x + 3$ occurs to the second power, both $x + 3$ and $(x + 3)^2$ occur in the partial fraction decomposition of $N(x)/D(x)$. For example, we write the partial fraction decomposition for $(3x^2 + 17x + 12)/D(x)$ as follows:

$$\frac{3x^2 + 17x + 12}{(x - 1)(x + 3)^2} = \frac{A}{x - 1} + \frac{B}{x + 3} + \frac{C}{(x + 3)^2}$$

It is possible that we do not need the fraction $B/(x + 3)$, but we do not know this until we find the value of B. If $B = 0$, then the decomposition does not include a fraction with denominator $x + 3$. This decomposition is completed in Example 3.

To find the partial fraction decomposition of a rational expression, the denominator must be factored into a product of prime polynomials. *If a quadratic prime polynomial occurs in the denominator, then the numerator of the partial fraction for that polynomial is of the form $Ax + B$.* For example, if $D(x) = x^3 + x^2 + 4x + 4$, then $D(x) = (x^2 + 4)(x + 1)$. The partial fraction decomposition for $(5x^2 + 3x + 13)/D(x)$ is written as follows:

$$\frac{5x^2 + 3x + 13}{x^3 + x^2 + 4x + 4} = \frac{Ax + B}{x^2 + 4} + \frac{C}{x + 1}$$

This decomposition is completed in Example 4.

The main points to remember for partial fraction decomposition are summarized as follows.

Strategy: Decomposition into Partial Fractions

To decompose a rational expression $N(x)/D(x)$ into partial fractions, use the following strategies:

1. If the degree of the numerator $N(x)$ is greater than or equal to the degree of the denominator $D(x)$, use division to express $N(x)/D(x)$ as quotient + remainder/divisor and decompose the resulting fraction.

2. **If the degree of $N(x)$ is less than the degree of $D(x)$, factor the denominator completely into prime factors that are either linear $(ax + b)$ or quadratic $(ax^2 + bx + c)$.**

3. **For each linear factor of the form $(ax + b)^n$, the partial fraction decomposition must include the following fractions:**

$$\frac{A_1}{ax + b} + \frac{A_2}{(ax + b)^2} + \cdots + \frac{A_n}{(ax + b)^n}$$

4. **For each quadratic factor of the form $(ax^2 + bx + c)^m$, the partial fraction decomposition must include the following fractions:**

$$\frac{B_1x + C_1}{ax^2 + bx + c} + \frac{B_2x + C_2}{(ax^2 + bx + c)^2} + \cdots + \frac{B_mx + C_m}{(ax^2 + bx + c)^m}$$

5. **Set up and solve a system of equations involving the A's, B's, and/or C's.**

The strategy for decomposition applies to very complicated rational expressions. To actually carry out the decomposition, we must be able to solve the system of equations that arises. Theoretically, we can solve systems of many equations in many unknowns. Practically, we are limited to fairly simple systems of equations. Large systems of equations are generally solved by computers, using techniques that we will develop in the next chapter.

Example 3   Repeated linear factor

Find the partial fraction decomposition for the rational expression

$$\frac{3x^2 + 17x + 12}{(x - 1)(x + 3)^2}.$$

Solution

Since the factor $x + 3$ occurs twice in the original denominator, it might occur in the partial fractions with powers 1 and 2:

$$\frac{3x^2 + 17x + 12}{(x - 1)(x + 3)^2} = \frac{A}{x - 1} + \frac{B}{x + 3} + \frac{C}{(x + 3)^2}$$

Multiply each side of the equation by the LCD, $(x - 1)(x + 3)^2$.

$$3x^2 + 17x + 12 = A(x + 3)^2 + B(x - 1)(x + 3) + C(x - 1)$$
$$= Ax^2 + 6Ax + 9A + Bx^2 + 2Bx - 3B + Cx - C$$
$$= (A + B)x^2 + (6A + 2B + C)x + 9A - 3B - C$$

Next, write a system of equations by equating the coefficients of like terms from opposite sides of the equation. The corresponding coefficients are highlighted on the previous page.

$$A + B \qquad = 3$$
$$6A + 2B + C = 17$$
$$9A - 3B - C = 12$$

We can solve this system of three equations in the variables A, B, and C by first eliminating C. Add the last two equations to get $15A - B = 29$. Add this equation to $A + B = 3$:

$$15A - B = 29$$
$$\underline{A + B = 3}$$
$$16A \qquad = 32$$
$$A = 2$$

If $A = 2$ and $A + B = 3$, then $B = 1$. Use $A = 2$ and $B = 1$ in the equation $6A + 2B + C = 17$:

$$6(2) + 2(1) + C = 17$$
$$C = 3$$

The partial fraction decomposition is written as follows:

$$\frac{3x^2 + 17x + 12}{(x - 1)(x + 3)^2} = \frac{2}{x - 1} + \frac{1}{x + 3} + \frac{3}{(x + 3)^2}$$
◆

Example 4 Single prime quadratic factor

Find the partial fraction decomposition for the rational expression

$$\frac{5x^2 + 3x + 13}{x^3 + x^2 + 4x + 4}.$$

Solution

Factor the denominator by grouping:

$$x^3 + x^2 + 4x + 4 = x^2(x + 1) + 4(x + 1) = (x^2 + 4)(x + 1)$$

Write the partial fraction decomposition:

$$\frac{5x^2 + 3x + 13}{x^3 + x^2 + 4x + 4} = \frac{Ax + B}{x^2 + 4} + \frac{C}{x + 1}$$

Note that $Ax + B$ is used over the prime quadratic polynomial $x^2 + 4$. Multiply each side of this equation by the LCD, $(x^2 + 4)(x + 1)$:

$$5x^2 + 3x + 13 = (Ax + B)(x + 1) + C(x^2 + 4)$$
$$= Ax^2 + Bx + Ax + B + Cx^2 + 4C$$
$$= (A + C)x^2 + (A + B)x + B + 4C$$

Write a system of equations by equating the coefficients of like terms from opposite sides of the last equation:

$$A + C = 5$$
$$A + B = 3$$
$$B + 4C = 13$$

Solve the system by substituting $C = 5 - A$ and $B = 3 - A$ into $B + 4C = 13$:

$$3 - A + 4(5 - A) = 13$$
$$3 - A + 20 - 4A = 13$$
$$23 - 5A = 13$$
$$-5A = -10$$
$$A = 2$$

Since $A = 2$, we get $C = 5 - 2 = 3$ and $B = 3 - 2 = 1$. So the partial fraction decomposition is written as follows:

$$\frac{5x^2 + 3x + 13}{x^3 + x^2 + 4x + 4} = \frac{2x + 1}{x^2 + 4} + \frac{3}{x + 1}$$

◆

Example 5   Repeated prime quadratic factor

Find the partial fraction decomposition for the rational expression

$$\frac{4x^3 - 2x^2 + 7x - 6}{4x^4 + 12x^2 + 9}.$$

Solution

The denominator factors as $(2x^2 + 3)^2$, and $2x^2 + 3$ is prime. Write the partial fractions using denominators $2x^2 + 3$ and $(2x^2 + 3)^2$.

$$\frac{4x^3 - 2x^2 + 7x - 6}{(2x^2 + 3)^2} = \frac{Ax + B}{2x^2 + 3} + \frac{Cx + D}{(2x^2 + 3)^2}$$

Multiply each side of the equation by the LCD, $(2x^2 + 3)^2$, to get the following equation:

$$4x^3 - 2x^2 + 7x - 6 = (Ax + B)(2x^2 + 3) + Cx + D$$
$$= 2Ax^3 + 2Bx^2 + 3Ax + 3B + Cx + D$$
$$= 2Ax^3 + 2Bx^2 + (3A + C)x + 3B + D$$

Equating the coefficients produces the following system of equations:

$$2A = 4$$
$$2B = -2$$
$$3A + C = 7$$
$$3B + D = -6$$

From the first two equations $2A = 4$ and $2B = -2$, we get $A = 2$ and $B = -1$. Using $A = 2$ in $3A + C = 7$ gives $C = 1$. Using $B = -1$ in $3B + D = -6$ gives $D = -3$. So the partial fraction decomposition is written as follows:

$$\frac{4x^3 - 2x^2 + 7x - 6}{(2x^2 + 3)^2} = \frac{2x - 1}{2x^2 + 3} + \frac{x - 3}{(2x^2 + 3)^2}$$

◆

For Thought

True or false? Explain.

1. $\dfrac{1}{x} + \dfrac{3}{x + 1} = \dfrac{4x + 1}{x^2 + x}$ for any real number x except 0 and -1.

2. $x + \dfrac{3x}{x^2 - 1} = \dfrac{x^3 + 2x}{x^2 - 1}$ for any real number x except -1 and 1.

3. The partial fraction decomposition of $\dfrac{x^2}{x^2 - 9}$ is $\dfrac{x^2}{x^2 - 9} = \dfrac{A}{x - 3} + \dfrac{B}{x + 3}$.

4. In the decomposition $\dfrac{5}{8} = \dfrac{A}{2} + \dfrac{B}{2^2} + \dfrac{C}{2^3}$, $A = 1$, $B = 0$, and $C = 1$.

5. The partial fraction decomposition of $\dfrac{3x - 1}{x^3 + x}$ is $\dfrac{3x - 1}{x^3 + x} = \dfrac{A}{x} + \dfrac{B}{x^2 + 1}$.

6. $\dfrac{1}{x^2 - 1} = \dfrac{1}{x - 1} + \dfrac{1}{x + 1}$ for any real number except 1 and -1.

7. $\dfrac{x^3 + 1}{x^2 + x - 2} = x - 1 + \dfrac{3x - 1}{x^2 + x - 2}$ for any real number except 1 and -2.

8. $x^3 - 8 = (x - 2)(x^2 + 4x + 4)$ for any real number x.

9. $\dfrac{x^2 + 2x}{x^3 - 1} = \dfrac{1}{x - 1} + \dfrac{1}{x^2 + x + 1}$ for any real number except 1.

10. There is no partial fraction decomposition for $\dfrac{2x}{x^2 + 9}$.

9.4 Exercises

Tape 18   Disk—5.25": 6   3.5": 4   Macintosh: 4

Perform the indicated operations.

1. $\dfrac{3}{x - 2} + \dfrac{4}{x + 1}$

2. $\dfrac{-1}{x + 5} + \dfrac{-3}{x - 4}$

3. $\dfrac{1}{x - 1} + \dfrac{-3}{x^2 + 2}$

4. $\dfrac{x + 3}{x^2 + x + 1} + \dfrac{1}{x - 1}$

5. $\dfrac{2x + 1}{x^2 + 3} + \dfrac{x^3 + 2x + 2}{(x^2 + 3)^2}$

6. $\dfrac{3x - 1}{x^2 + x - 3} + \dfrac{x^3 + x - 1}{(x^2 + x - 3)^2}$

7. $\dfrac{1}{x - 1} + \dfrac{2x + 3}{(x - 1)^2} + \dfrac{x^2 + 1}{(x - 1)^3}$

8. $\dfrac{3}{x + 2} + \dfrac{x - 1}{(x + 2)^2} + \dfrac{1}{x^2 + 2}$

Find A and B for each partial fraction decomposition.

9. $\dfrac{12}{x^2 - 9} = \dfrac{A}{x - 3} + \dfrac{B}{x + 3}$

10. $\dfrac{5x + 2}{x^2 - 4} = \dfrac{A}{x - 2} + \dfrac{B}{x + 2}$

Find the partial fraction decomposition for each rational expression.

11. $\dfrac{5x - 1}{(x + 1)(x - 2)}$

12. $\dfrac{-3x - 5}{(x + 3)(x - 1)}$

13. $\dfrac{2x + 5}{x^2 + 6x + 8}$

14. $\dfrac{x + 2}{x^2 + 12x + 32}$

15. $\dfrac{2}{x^2 - 9}$

16. $\dfrac{1}{9x^2 - 1}$

17. $\dfrac{1}{x^2 - x}$

18. $\dfrac{2}{x^2 - 2x}$

Find A, B, and C for each partial fraction decomposition.

19. $\dfrac{x^2 + x - 31}{(x + 3)^2(x - 2)} = \dfrac{A}{x + 3} + \dfrac{B}{(x + 3)^2} + \dfrac{C}{x - 2}$

20. $\dfrac{x^2 - x - 7}{(x + 1)(x^2 + 4)} = \dfrac{A}{x + 1} + \dfrac{Bx + C}{x^2 + 4}$

Find each partial fraction decomposition.

21. $\dfrac{-2x - 7}{x^2 + 4x + 4}$

22. $\dfrac{-3x + 2}{x^2 - 2x + 1}$

23. $\dfrac{6x^2 - x + 1}{x^3 + x^2 + x + 1}$

24. $\dfrac{3x^2 - 2x + 8}{x^3 + 2x^2 + 4x + 8}$

25. $\dfrac{3x^3 - x^2 + 19x - 9}{x^4 + 18x^2 + 81}$

26. $\dfrac{-x^3 - 10x - 3}{x^4 + 10x^2 + 25}$

27. $\dfrac{3x^2 + 17x + 14}{x^3 - 8}$

28. $\dfrac{2x^2 + 17x - 21}{x^3 + 27}$

29. $\dfrac{2x^3 + x^2 + 3x - 2}{x^2 - 1}$

30. $\dfrac{2x^3 - 19x - 9}{x^2 - 9}$

31. $\dfrac{3x^3 - 2x^2 + x - 2}{(x^2 + x + 1)^2}$

32. $\dfrac{x^3 - 8x^2 - 5x - 33}{(x^2 + x + 4)^2}$

33. $\dfrac{3x^3 + 4x^2 - 12x + 16}{x^4 - 16}$

34. $\dfrac{9x^2 + 3}{81x^4 - 1}$

35. $\dfrac{5x^3 + x^2 + x - 3}{x^4 - x^3}$

36. $\dfrac{2x^3 + 3x^2 - 8x + 4}{x^4 - 4x^2}$

37. $\dfrac{6x^2 - 28x + 33}{(x - 2)^2(x - 3)}$

38. $\dfrac{7x^2 + 45x + 58}{(x + 3)^2(x + 1)}$

39. $\dfrac{9x^2 + 21x - 24}{x^3 + 4x^2 - 11x - 30}$

40. $\dfrac{3x^2 + 24x + 6}{2x^3 - x^2 - 13x - 6}$

41. $\dfrac{x^2 - 2}{x^3 - 3x^2 + 3x - 1}$

42. $\dfrac{2x^2}{x^3 + 3x^2 + 3x + 1}$

Find the partial fraction decomposition for each rational expression. Assume that a, b, and c are nonzero constants.

43. $\dfrac{x}{(ax + b)^2}$

44. $\dfrac{x^2}{(ax + b)^3}$

45. $\dfrac{x + c}{ax^2 + bx}$

46. $\dfrac{1}{x^3(ax + b)}$

47. $\dfrac{1}{x^2(ax + b)}$

48. $\dfrac{1}{x^2(ax + b)^2}$

For Writing/Discussion

49. *Togetherness* Suppose Andrew can paint a fence by himself in a hours and Betty can paint the same fence by herself in b hours, where a and b are positive integers. If they paint the fence together in 2 hr 24 min, then what are the values of a and b? Explain your solution.

50. *More help* Suppose the time that it takes for Andrew, Betty, or Carl to paint a fence working alone is a, b, or c hours respectively, where a, b, and c are positive integers. If they paint the fence together in 1 hr 15 min, then what are the values of a, b, and c? Explain your solution.

9.5

Inequalities and

Systems of

Inequalities in

Two Variables

In Sections 2.6 and 2.7 we discussed linear and quadratic inequalities in one variable. Earlier in this chapter we solved systems of equations in two variables. In this section we turn to linear and nonlinear inequalities in two variables and systems of inequalities in two variables. In the next section systems of linear inequalities will be used to solve linear programming problems.

Linear Inequalities

Three hamburgers and two Cokes cost at least $5.50. If a hamburger costs x dollars and a Coke costs y dollars, then this sentence can be written as the linear inequality

$$3x + 2y \geq 5.50.$$

A linear inequality in two variables is simply a linear equation in two variables with the equal sign replaced by an inequality symbol.

Definition: Linear Inequality

If A, B, and C are real numbers with A and B not both zero, then

$$Ax + By < C$$

is called a **linear inequality in two variables.** In place of $<$ we can also use the symbols $\leq$, $>$, or $\geq$.

Some examples of linear inequalities are

$$y > x - 3, \qquad 2x - y \geq 1, \qquad x + y < 0, \qquad \text{and} \qquad y \leq 3.$$

The inequality $y \leq 3$ is considered an inequality in two variables because it can be written in the form $0 \cdot x + y \leq 3$.

An ordered pair (a, b) is a **solution to an inequality** if the inequality is true when x is replaced by a and y is replaced by b. Ordered pairs such as $(9, 1)$ and $(7, 3)$ satisfy $y \leq 3$. Ordered pairs such as $(4, 2)$ and $(-1, -2)$ satisfy $y > x - 3$.

The ordered pairs satisfying $y = x - 3$ form a line in the rectangular coordinate system, but what does the solution set to $y > x - 3$ look like? An ordered pair satisfies $y > x - 3$ whenever the y-coordinate is *greater than* the x-coordinate minus 3. For example, $(4, 1)$ satisfies the equation $y = x - 3$ and is on the line, while $(4, 1.001)$ satisfies $y > x - 3$ and is above the line. In fact, any point in the coordinate plane directly above $(4, 1)$ satisfies $y > x - 3$ (and any point directly below $(4, 1)$ satisfies $y < x - 3$). Since this fact holds true for every point on the line $y = x - 3$, the solution set to the inequality is the set of all points *above* the line. The graph of $y > x - 3$ is indicated by shading the region above the line as shown in Fig. 9.14. We draw a dashed line for $y = x - 3$ because it is not part of the solution set to $y > x - 3$. However, the graph of $y \geq x - 3$ in Fig.

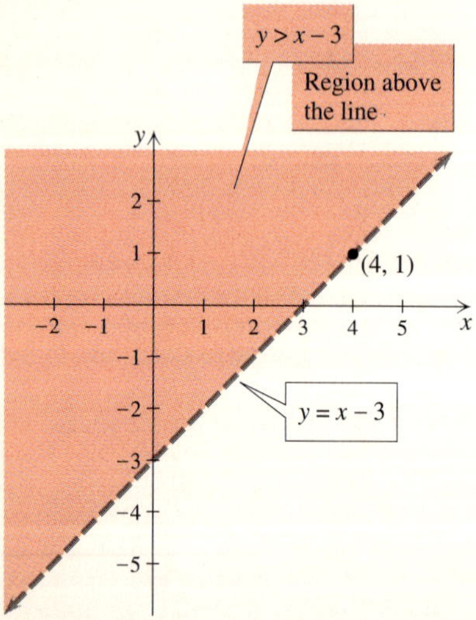

$y > x - 3$

Region above the line

$(4, 1)$

$y = x - 3$

Figure 9.14

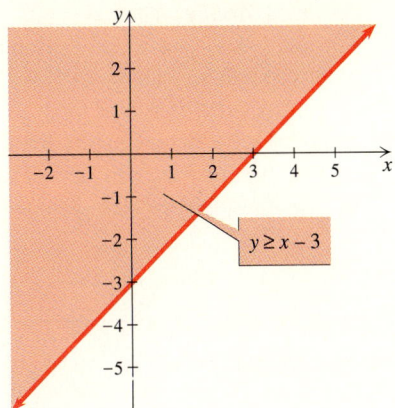

Figure 9.15

9.15 has a solid boundary line because the line is included in its solution set. Notice also that the region below the line is the solution set to $y < x - 3$.

Example 1   Graphing linear inequalities

Graph the solution set to each inequality.

a) $y < -\dfrac{1}{2}x + 2$ **b)** $2x - y \geq 1$ **c)** $y > 2$ **d)** $x \leq 3$

Solution

a) Graph the corresponding equation

$$y = -\frac{1}{2}x + 2$$

as a dashed line by using its intercept $(0, 2)$ and slope $-\frac{1}{2}$. Since the inequality symbol is $<$, the solution set to the inequality is the region below this line, as shown in Fig. 9.16.

b) Solve the inequality for y:

$$2x - y \geq 1$$
$$-y \geq -2x + 1$$
$$y \leq 2x - 1 \qquad \text{Multiply by } -1 \text{ and reverse the inequality.}$$

Graph the line $y = 2x - 1$ by using its intercept $(0, -1)$ and its slope 2. Because of the symbol $\leq$, we use a solid line and shade the region below it, as shown in Fig. 9.17.

c) Every point in the region above the horizontal line $y = 2$ satisfies $y > 2$. See Fig. 9.18.

d) Every point on or to the left of the vertical line $x = 3$ satisfies $x \leq 3$. See Fig. 9.19.

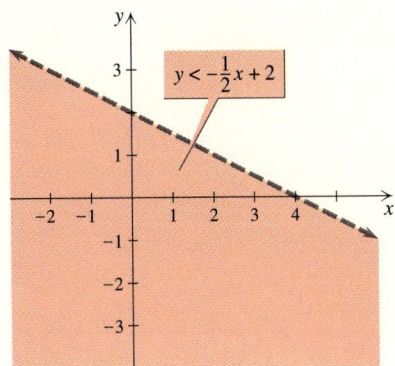

Figure 9.16

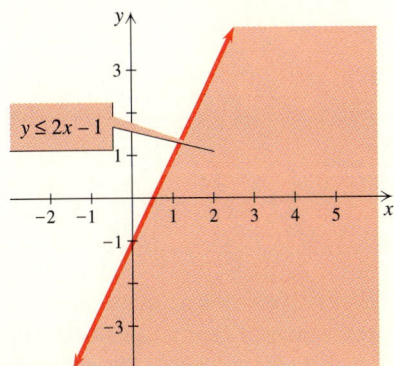

Figure 9.17

Figure 9.18

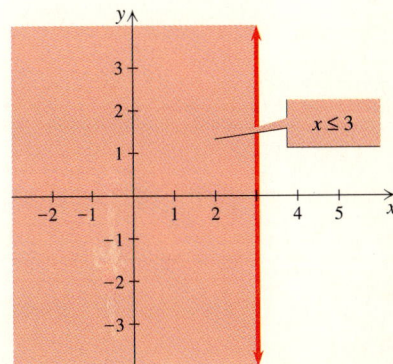

Figure 9.19

The graph in Example 1(b) is the region below the line because $2x - y \geq 1$ is equivalent to $y \leq 2x - 1$. The graph of $y > mx + b$ is above the line and $y < mx + b$ is below the line. The symbol $>$ corresponds to "above the line" and $<$ corresponds to "below the line" *only if the inequality is solved for y.*

With the **test point method** it is not necessary to solve the inequality for y. The graph of $Ax + By = C$ divides the plane into two regions. On one side, $Ax + By > C$; and on the other, $Ax + By < C$. We can simply test a single point in one of the regions to see which is which.

Example 2 Graphing an inequality using test points

Use the test point method to graph $3x - 6y > 9$.

Solution

First graph $3x - 6y = 9$, using a dashed line going through its intercepts $(0, -\frac{3}{2})$ and $(3, 0)$. Select a test point that is not on the line, say, $(0, 0)$. Test $(0, 0)$ in $3x - 6y > 9$:

$$3 \cdot 0 - 6 \cdot 0 > 9$$

$$0 > 9 \qquad \text{Incorrect.}$$

Since $(0, 0)$ does not satisfy $3x - 6y > 9$, any point on the *other* side of the line satisfies $3x - 6y > 9$. So we shade the region that does not contain $(0, 0)$, as shown in Fig. 9.20.

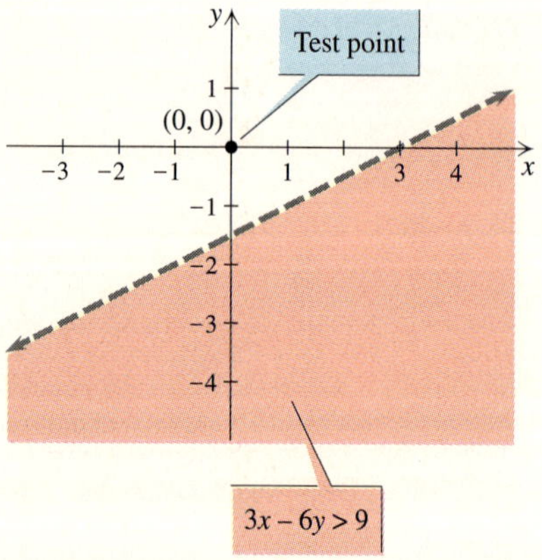

Figure 9.20

Nonlinear Inequalities

Any nonlinear equation in two variables becomes a nonlinear inequality in two variables when the equal sign is replaced by an inequality symbol. The test point method is generally the easiest to use for graphing nonlinear inequalities in two variables.

Example 3   Graphing nonlinear inequalities in two variables

Graph the solution set to each nonlinear inequality.

a) $y > x^2$ **b)** $x^2 + y^2 \leq 4$ **c)** $y < \log_2(x)$

Solution

a) The graph of $y = x^2$ is a parabola opening upward with vertex at $(0, 0)$. Draw the parabola dashed, as shown in Fig. 9.21, and select a point that is not on the parabola, say, $(0, 5)$. Since $5 > 0^2$ is correct, the region of the plane containing $(0, 5)$ is shaded.

b) The graph of $x^2 + y^2 = 4$ is a circle of radius 2 centered at $(0, 0)$. Select $(0, 0)$ as a test point. Since $0^2 + 0^2 \leq 4$ is correct, we shade the region inside the circle, as shown in Fig. 9.22.

c) The graph of $y = \log_2(x)$ is a curve through the points $(1, 0)$, $(2, 1)$, and $(4, 2)$ as shown in Fig. 9.23. Select $(4, 0)$ as a test point. Since $0 < \log_2(4)$ is correct, shade the region shown in Fig. 9.23.

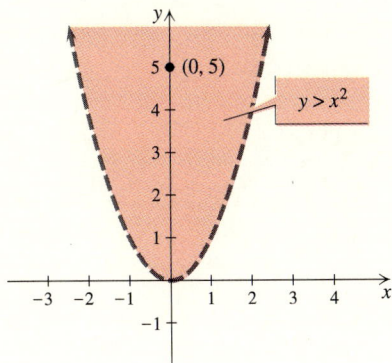

Figure 9.21

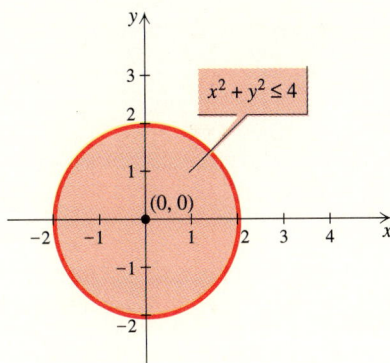

Figure 9.22

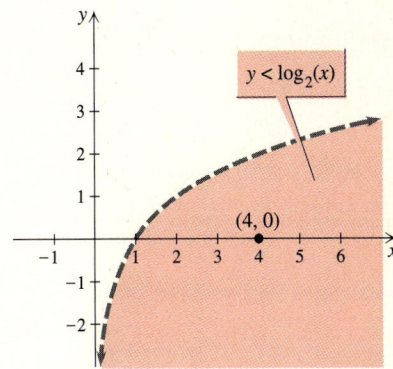

Figure 9.23

Systems of Inequalities

The solution set to a system of inequalities in two variables consists of all ordered pairs that satisfy *all* of the inequalities in the system. For example, the system

$$x + y > 5$$
$$x - y < 9$$

has $(4, 2)$ as a solution because $4 + 2 > 5$ and $4 - 2 < 9$. There are infinitely many solutions to this system.

The solution set to a system is generally a region of the coordinate plane. It is the intersection of the solution sets to the individual inequalities. To find the solution set to a system, we graph the equation corresponding to each inequality in the system and then test a point in each region to see whether it satisfies all inequalities of the system.

Example 4   Solving a system of linear inequalities

Graph the solution set to the system.

$$x + 2y \le 4$$
$$y \ge x - 3$$

Solution

The graph of $x + 2y = 4$ is a line through $(0, 2)$ and $(4, 0)$. The graph of $y = x - 3$ is a line through $(0, -3)$ with slope 1. These two lines divide the plane into four regions, as shown in Fig. 9.24. Select a test point in each of the four regions. We use $(0, 0)$, $(0, 5)$, $(0, -5)$, and $(5, 0)$ as test points. Only $(0, 0)$ satisfies both of the inequalities of the system. So we shade the region containing $(0, 0)$, as shown in Fig. 9.25.

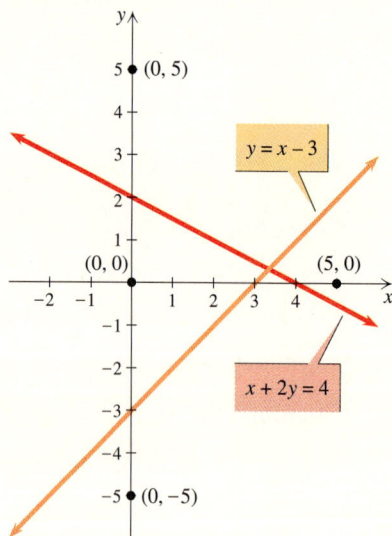

Figure 9.24

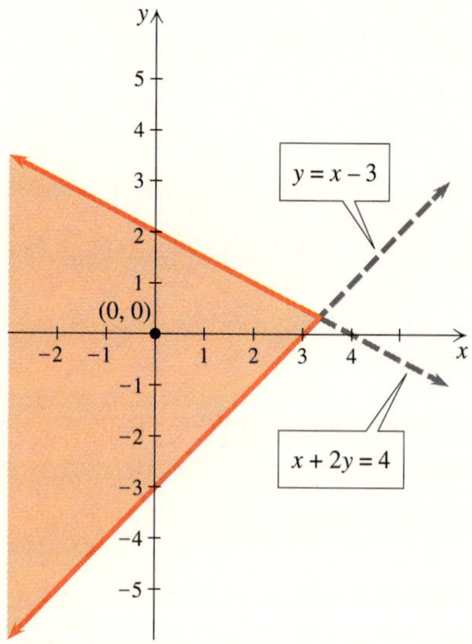

Figure 9.25

Note that the set of points indicated in Fig. 9.25 is the intersection of the set of points on or below the line $x + 2y = 4$ with the set of points on or above the line $y = x - 3$. If you can visualize how these regions intersect, then you can find the solution set without test points.

Example 5   Solving a system of nonlinear inequalities

Graph the solution set to the system.

$$x^2 + y^2 \le 16$$
$$y > 2^x$$

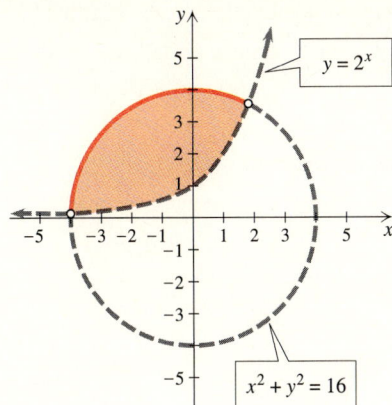

Figure 9.26

Solution

Points that satisfy $x^2 + y^2 \leq 16$ are on or inside the circle of radius 4 centered at $(0, 0)$. Points that satisfy $y > 2^x$ are above the curve $y = 2^x$. Points that satisfy both inequalities are on or inside the circle and above the curve $y = 2^x$ as shown in Fig. 9.26. Note that the circular boundary of the solution set is drawn as a solid curve because of the $\leq$ symbol. We could also find the solution set by using test points. ◆

Example 6   Solving a system of three inequalities

Graph the solution set to the system.

$$y > x^2$$
$$y < x + 6$$
$$y < -x + 6$$

Solution

Points that satisfy $y > x^2$ are above the parabola $y = x^2$. Points that satisfy $y < x + 6$ are below the line $y = x + 6$. Points that satisfy $y < -x + 6$ are below the line $y = -x + 6$. Points that satisfy all three inequalities lie in the region shown in Fig. 9.27.

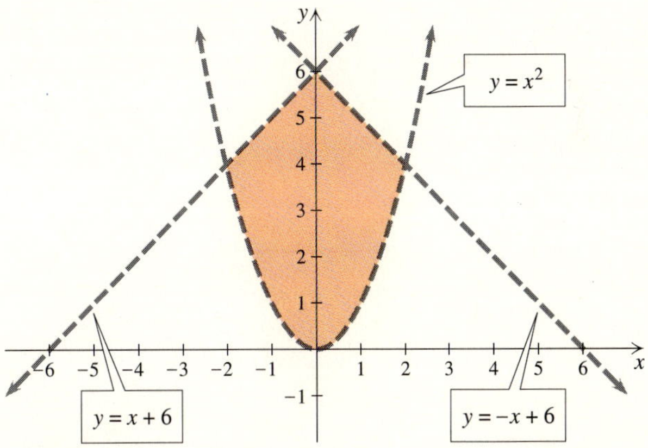

Figure 9.27   ◆

For Thought

True or false? Explain.

1. The point $(1, 3)$ satisfies the inequality $y > x + 2$.
2. The graph of $x - y < 2$ is the region below the line $x - y = 2$.
3. The graph of $x + y > 2$ is the region above the line $x + y = 2$.

4. The graph of $x^2 + y^2 > 5$ is the region outside the circle of radius 5.
The following systems are referenced in statements 5–10:

a) $x - 2y < 3$ **b)** $x^2 + y^2 > 9$ **c)** $y \geq x^2 - 4$
$\quad\;\; y - 3x > 5$ $y < x + 2$ $y < x + 2$

5. The point $(-2, 1)$ is in the solution set to system (a).

6. The solution to system (b) consists of points outside a circle and below a line.

7. The point $(-2, 0)$ is in the solution set to system (c).

8. The point $(-1, 2)$ is a test point for system (a).

9. The origin is in the solution set to system (c).

10. The point $(2, 3)$ is in the solution set to system (b).

9.5 Exercises   Tape 18   Disk—5.25″: 6   3.5″: 4   **Macintosh: 4**

Match each inequality with one of the graphs (a)–(d).

1. $y > x - 2$ **2.** $x < 2 - y$

3. $x - y > 2$ **4.** $x + y > 2$

(a)

(b)

(c)

(d)

Sketch the graph of the solution set to each linear inequality in the rectangular coordinate system.

5. $x + y > 3$ **6.** $2x + y < 1$

7. $2x - y \leq 4$ **8.** $x - 2y \geq 6$

9. $y < -3x - 4$ **10.** $y > \frac{2}{3}x - 3$

11. $x - 3 \geq 0$ **12.** $y + 1 \leq 0$

13. $20x - 30y \leq 6000$ **14.** $30y + 40x > 1200$

15. $y < 3$ **16.** $x > 0$

Sketch the graph of each nonlinear inequality.

17. $y > -x^2$ **18.** $y < 4 - x^2$ **19.** $x^2 + y^2 \geq 1$

20. $x^2 + y^2 < 36$ **21.** $x > |y|$ **22.** $y < x^{1/3}$

23. $x \geq y^2$ **24.** $y^2 > x - 1$ **25.** $y \geq x^3$

26. $y < |x - 1|$ **27.** $y > 2^x$ **28.** $y < \log_2(x)$

Sketch the graph of the solution set to each system of inequalities.

29. $y > x - 4$ **30.** $y < \frac{1}{2}x + 1$
$\quad\;\; y < -x - 2$ $y < -\frac{1}{3}x + 1$

31. $3x - 4y \leq 12$ **32.** $2x + y \geq -1$
$\quad\;\; x + y \geq -3$ $y - 2x \leq -3$

33. $3x - y < 4$
$y < 3x + 5$

34. $x - y > 0$
$y + 4 > x$

35. $y + x < 0$
$y > 3 - x$

36. $3x - 2y \leq 6$
$2y - 3x \leq -8$

37. $x + y < 5$
$y \geq 2$

38. $x \leq 2$
$y > -2$

39. $y < x - 3$
$x \leq 4$

40. $y > 0$
$y \leq x$

Sketch the graph of the solution set to each nonlinear system of inequalities.

41. $y > x^2 - 3$
$y < x + 1$

42. $y < 5 - x^2$
$y > (x - 1)^2$

43. $x^2 + y^2 \geq 4$
$x^2 + y^2 \leq 16$

44. $x^2 + y^2 \leq 9$
$y \geq x - 1$

45. $(x - 3)^2 + y^2 < 25$
$(x + 3)^2 + y^2 < 25$

46. $x^2 + y^2 \geq 64$
$x^2 + y^2 \leq 16$

47. $x^2 + y^2 > 4$
$|x| \leq 4$

48. $x^2 + y^2 < 36$
$|y| < 3$

49. $y > |2x| - 4$
$y \leq \sqrt{4 - x^2}$

50. $y < 4 - |x|$
$y \geq |x| - 4$

51. $|x - 1| < 2$
$|y - 1| < 4$

52. $x \geq y^2 - 1$
$(x + 1)^2 + y^2 \geq 4$

Solve each system of inequalities.

53. $x \geq 0$
$y \geq 0$
$x + y \leq 4$

54. $x \geq 0$
$y \geq 0$
$y \geq -\dfrac{1}{2}x + 2$

55. $x \geq 0, y \geq 0$
$x + y \geq 4$
$y \geq -2x + 6$

56. $x \geq 0, y \geq 0$
$4x + 3y \leq 12$
$3x + 4y \leq 12$

57. $x^2 + y^2 \geq 9$
$x^2 + y^2 \leq 25$
$y \geq |x|$

58. $x - 2 < y < x + 2$
$x^2 + y^2 < 16$
$x > 0, y > 0$

59. $y > (x - 1)^3$
$y > 1$
$x + y > -2$

60. $x \geq |y|$
$y \geq -3$
$2y - x \leq 4$

61. $y > 2^x$
$y < 6 - x^2$
$x + y > 0$

62. $x^2 + y \leq 5$
$y \geq x^3 - x$
$y \leq 4$

Write a system of inequalities whose solution set is the region shown.

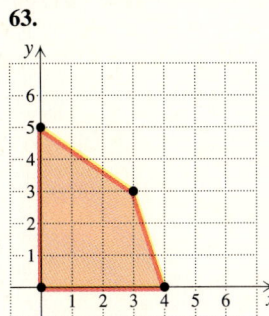

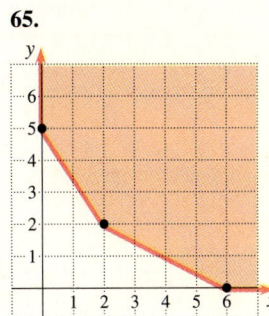

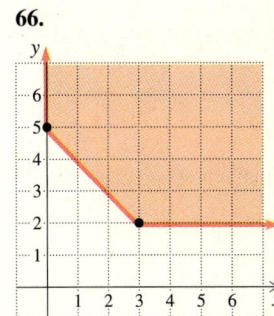

63.

64.

65.

66.

Find a system of inequalities to describe the given region.

67. Points inside the square that has vertices $(2, 2)$, $(-2, 2)$, $(-2, -2)$, and $(2, -2)$.

68. Points inside the triangle that has vertices $(0, 0)$, $(0, 6)$, and $(3, 0)$.

69. Points in the first quadrant less than nine units from the origin.

70. Points that are closer to the x-axis than they are to the y-axis.

Write a system of inequalities that describes the possible solutions to each problem and graph the solution set to the system.

71. *Inventory control* A car dealer stocks mid-size and full-size cars on her lot, which cannot hold more than 110 cars. On the average, she borrows $10,000 to purchase a mid-size car and $15,000 to purchase a full-size car. How many cars of each type should she stock if her total debt cannot exceed $1.5 million?

72. *Delicate balance* A fast food restaurant must have a minimum of 30 employees and a maximum of 50. To avoid charges of sexual bias, the company has a policy that the number of employees of one sex must never exceed the number of employees of the other sex by more than six. How many persons of each sex could be employed at this restaurant?

73. *Political correctness*   A political party is selling $50 tickets and $100 tickets for a fund-raising banquet in a hall that cannot hold more than 500 people. To show that the party represents the people, the number of $100 tickets must not be greater than 20% of the total number of tickets sold. How many tickets of each type can be sold?

74. *Mixing alloys*   A metallurgist has two alloys available. Alloy A is 20% zinc and 80% copper, while alloy B is 60% zinc and 40% copper. He wants to melt down and mix x ounces of alloy A with y ounces of alloy B to get a metal that is at most 50% zinc and at most 60% copper. How many ounces of each alloy should he use if the new piece of metal must not weigh more than 20 ounces?

Graphing Calculator Exercises

Use a graphing calculator to graph the equation corresponding to each inequality. From the display of the graphing calculator, locate one ordered pair in the solution set to the system and check that it satisfies both inequalities of the system.

1. $y > 2^{x+2}$
$y < x^3 - 3x$

2. $y < -0.5x^2 + 150x - 11226.6$
$y > -0.11x + 38$

3. $y > e^{x-0.8}$
$y < \log(x + 2.5)$

9.6

Linear Programming

In this section we apply our knowledge of systems of linear inequalities to solving linear programming problems. Linear programming is a method that can be used to solve many practical business problems. Linear programming can tell us how to allocate resources to achieve a maximum profit, minimum labor cost, or a most nutritious meal.

Graphing the Constraints

In the simplest linear programming applications we have two variables that must satisfy several linear inequalities. These inequalities are called the **constraints** because they restrict the variables to only certain values. The graph of the solution set to the system is used to indicate the points that satisfy all of the constraints. Any point that satisfies all of the constraints is called a **feasible solution** to the problem.

Example 1   Graphing the constraints

Graph the solution set to the system of inequalities and identify each vertex of the region.

$$x \geq 0, \qquad y \geq 0$$
$$2x + y \leq 6$$
$$x + y \leq 4$$

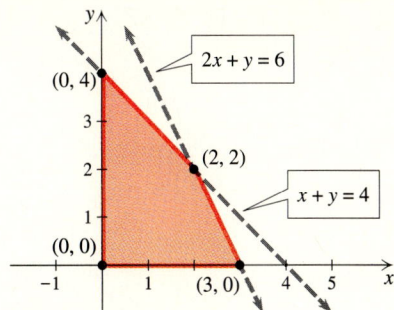

Figure 9.28

Solution

The points on or to the right of the y-axis satisfy $x \geq 0$. The points on or above the x-axis satisfy $y \geq 0$. The points on or below the line $2x + y = 6$ satisfy $2x + y \leq 6$. The points on or below the line $x + y = 4$ satisfy $x + y \leq 4$. Graph each straight line and shade the region that satisfies all four inequalities, as shown in Fig. 9.28. Three of the vertices are easily identified as $(0, 0)$, $(0, 4)$, and $(3, 0)$. The fourth vertex is at the intersection of $x + y = 4$ and $2x + y = 6$. Multiply $x + y = 4$ by -1 and add the result to $2x + y = 6$:

$$
\begin{array}{r}
-x - y = -4 \\
\underline{2x + y = 6} \\
x \quad\;\; = 2
\end{array}
$$

If $x = 2$ and $x + y = 4$, then $y = 2$. So the fourth vertex is $(2, 2)$.   ◆

In linear programming, the constraints usually come from physical limitations of resources described within the specific problem. Constraints that are always satisfied are called **natural constraints.** For example, the requirement that the number of employees be greater than or equal to zero is a natural constraint. In the next example we write the constraints and then graph the points in the coordinate plane that satisfy all of the constraints.

Example 2   Finding and graphing the constraints

Bruce builds portable storage buildings. He uses 10 sheets of plywood and 15 studs in a small building, and he uses 15 sheets of plywood and 45 studs in a large building. Bruce has available only 60 sheets of plywood and 135 studs. Write the constraints on the number of small and large portable buildings that he can build with the available supplies, and graph the solution set to the system of constraints.

Solution

Let x represent the number of small buildings and y represent the number of large buildings. The natural constraints are

$$x \geq 0 \quad \text{and} \quad y \geq 0$$

because he cannot build a negative number of buildings. Since he has only 60 sheets of plywood available, we must have

$$10x + 15y \leq 60$$
$$2x + 3y \leq 12 \qquad \text{Divide each side by 5.}$$

Since he has only 135 studs available, we must have

$$15x + 45y \leq 135$$
$$x + 3y \leq 9 \qquad \text{Divide each side by 15.}$$

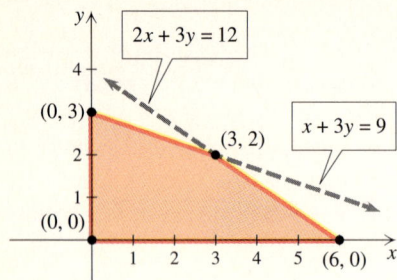

Figure 9.29

The conditions stated lead to the following system of constraints:

$$x \geq 0, \qquad y \geq 0 \qquad \text{Natural constraints}$$

$$2x + 3y \leq 12 \qquad \text{Constraint on amount of plywood}$$

$$x + 3y \leq 9 \qquad \text{Constraint on number of studs}$$

The graph of the solution set to this system is shown in Fig. 9.29.

Maximizing or Minimizing a Linear Function

In Example 2, any ordered pair within the shaded region of Fig. 9.29 is a feasible solution to the problem of deciding how many buildings of each type could be built. For each feasible solution within the shaded region Bruce makes some amount of profit. Of course, Bruce wants to find a feasible solution that will yield the maximum possible profit. In general, the function that we wish to maximize or minimize, subject to the constraints, is called the **objective function.**

If Bruce makes a profit of \$400 on a small building and \$500 on a large building, then the total profit from x small and y large buildings is

$$P = 400x + 500y.$$

Since the profit is a function of x and y, we write the objective function as

$$P(x, y) = 400x + 500y.$$

The function P is a linear function of x and y. The domain of P is the region graphed in Fig. 9.29.

Definition: Linear Function in Two Variables

A **linear function in two variables** is a function of the form

$$f(x, y) = Ax + By + C,$$

where A, B, and C are real numbers such that A and B are not both zero.

Bruce is interested in the maximum profit, subject to the constraints on x and y. Suppose $x = 1$ and $y = 1$; then the profit is

$$P(1, 1) = 400(1) + 500(1) = \$900.$$

In fact, the profit is \$900 at any point on the line

$$400x + 500y = 900.$$

The profit is \$1300 at any point on the line

$$400x + 500y = 1300,$$

and the profit is \$1800 at any point on the line

$$400x + 500y = 1800.$$

The graphs of these lines are shown in Fig. 9.30. Notice that the larger profit is found on the higher *profit line* and all of the profit lines are parallel. Bruce wants the highest profit line that intersects the region of feasible solutions. You can see in Fig. 9.30 that the highest line that intersects the region and is parallel to the other profit lines will intersect the region at the vertex (6, 0). So he should build six small buildings and no large buildings to maximize the profit. The maximum profit is $P(6, 0) = 400(6) + 500(0) = \$2400$.

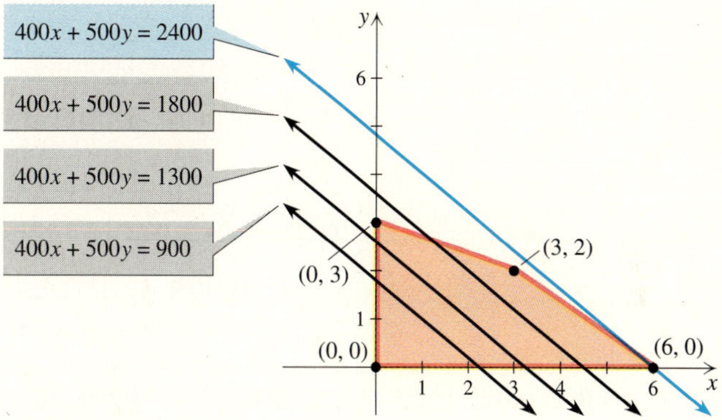

$400x + 500y = 2400$

$400x + 500y = 1800$

$400x + 500y = 1300$

$400x + 500y = 900$

(0, 3)

(3, 2)

(0, 0)

(6, 0)

Figure 9.30

In another case we may be looking for the point that would *minimize* a linear function of the two variables. *In general, if the maximum or minimum value exists, then the maximum or minimum value of a linear function subject to linear constraints occurs at a vertex of the region determined by the constraints.* The minimum profit for the portable buildings occurs when no buildings are built, at the vertex (0, 0). It is possible that the maximum or minimum value occurs at two adjacent vertices and at every point along the line segment joining them.

It is a bit cumbersome and possibly inaccurate to graph parallel lines and then find the highest (or lowest) one that intersects the region determined by the constraints. Instead, we can use the following procedure for linear programming, which does not depend as much on graphing.

Procedure: Linear Programming

Use the following steps to find the maximum or minimum value of a linear function subject to linear constraints.

1. **Graph the region that satisfies all of the constraints.**
2. **Determine the coordinates of each vertex of the region.**
3. **Evaluate the function at each vertex of the region.**
4. **Identify which vertex gives the maximum or minimum value of the function.**

To use the new procedure on Bruce's buildings, note that in Fig. 9.30 the vertices are $(0, 0)$, $(6, 0)$, $(0, 3)$, and $(3, 2)$. Compute the profit at each vertex:

$$P(0, 0) = 400(0) + 500(0) = \$0 \qquad \text{Minimum profit}$$
$$P(6, 0) = 400(6) + 500(0) = \$2400 \qquad \text{Maximum profit}$$
$$P(0, 3) = 400(0) + 500(3) = \$1500$$
$$P(3, 2) = 400(3) + 500(2) = \$2200$$

From this list, we see that the maximum profit is $2400, when six small buildings and no large buildings are built, and the minimum profit is $0, when no buildings of either type are built.

In the next example we use the linear programming technique to find the minimum value of a linear function subject to a system of constraints.

Example 3 Finding the minimum value of a linear function

One serving of Muesli breakfast cereal contains 4 grams of protein and 30 grams of carbohydrates. One serving of Multi Bran Chex contains 2 grams of protein and 25 grams of carbohydrates. A dietitian wants to mix these two cereals to make a batch that contains at least 44 grams of protein and at least 450 grams of carbohydrates. If the cost of Muesli is 21 cents per serving and the cost of Multi Bran Chex is 14 cents per serving, then how many servings of each cereal would minimize the cost and satisfy the constraints?

Solution

Let x = the number of servings of Muesli and y = the number of servings of Multi Bran Chex. If the batch is to contain at least 44 grams of protein, then

$$4x + 2y \geq 44.$$

If the batch is to contain at least 450 grams of carbohydrates, then

$$30x + 25y \geq 450.$$

Simplify each inequality and use the two natural constraints to get the following system:

$$x \geq 0, \qquad y \geq 0$$
$$2x + y \geq 22$$
$$6x + 5y \geq 90$$

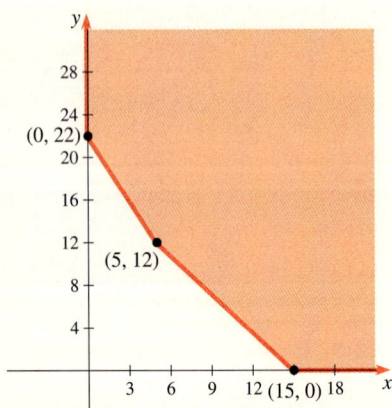

Figure 9.31

The graph of the constraints is shown in Fig. 9.31. The vertices are $(0, 22)$, $(5, 12)$, and $(15, 0)$. The cost in dollars for x servings of Muesli and y servings of Multi Bran Chex is $C(x, y) = 0.21x + 0.14y$. Evaluate the cost at each vertex.

$$C(0, 22) = 0.21(0) + 0.14(22) = \$3.08$$
$$C(5, 12) = 0.21(5) + 0.14(12) = \$2.73 \qquad \text{Minimum cost}$$
$$C(15, 0) = 0.21(15) + 0.14(0) = \$3.15$$

The minimum cost of $2.73 is attained by using 5 servings of Muesli and 12 servings of Multi Bran Chex. Note that $C(x, y)$ does not have a maximum value on this region. The cost increases without bound as x and y increase. ◆

The examples of linear programming given in this text are simple examples. Problems in linear programming in business can involve a hundred or more variables subject to as many inequalities. These problems are solved by computers using matrix methods, but the basic idea is the same as we have seen in this section.

? For Thought

True or false? Explain.

1. The graph of $x \geq 0$ in the coordinate plane consists only of points on the x-axis that are at or to the right of the origin.

2. The graph of $y \geq 2$ in the coordinate plane consists of the points on or to the right of the line $y = 2$.

3. The graph of $x + y \leq 5$ does not include the origin.

4. The graph of $2x + 3y = 12$ has x-intercept $(0, 4)$ and y-intercept $(6, 0)$.

5. The graph of a system of inequalities is the intersection of their individual solution sets.

6. In linear programming, constraints are inequalities that restrict the values of the variables.

7. The function $f(x, y) = 4x^2 + 9y^2 + 36$ is a linear function of x and y.

8. The value of $R(x, y) = 30x + 15y$ at the point $(1, 3)$ is 75.

9. If $C(x, y) = 7x + 9y + 3$, then $C(0, 5) = 45$.

10. To solve a linear programming problem, we evaluate the objective function at the vertices of the region determined by the constraints.

9.6 Exercises 📼 Tape 18 💾 Disk—5.25″: 6   3.5″: 4   Macintosh: 4

Graph the solution set to each system of inequalities and identify each vertex of the region.

1. $x \geq 0, y \geq 0$
$x + y \leq 4$

2. $x \geq 0, y \geq 0$
$2x + y \leq 4$

3. $x \geq 0, y \geq 0$
$x \leq 1, y \leq 3$

4. $x \geq 0, y \geq 0$
$y \leq x, x \leq 3$

5. $x \geq 0, y \geq 0$
$x + y \leq 4$
$2x + y \leq 6$

6. $x \geq 0, y \geq 0$
$50x + 40y \leq 200$
$10x + 20y \leq 60$

7. $x \geq 0, y \geq 0$
$2x + y \geq 4$
$x + y \geq 3$

8. $x \geq 0, y \geq 0$
$20x + 10y \geq 40$
$5x + 5y \geq 15$

9. $x \geq 0, y \geq 0$
$3x + y \geq 6$
$x + y \geq 4$

10. $x \geq 0, y \geq 0$
$2x + y \geq 6$
$x + 2y \geq 6$

11. $x \geq 0, y \geq 0$
$3x + y \geq 8$
$x + y \geq 6$

12. $x \geq 0, y \geq 0$
$x + 4y \leq 20$
$4x + y \leq 64$

Find the maximum value of the objective function
$T(x, y) = 2x + 3y$ **on each given region.**

13.

14.

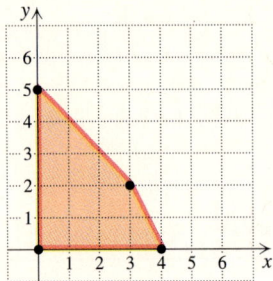

Find the minimum value of the objective function
$H(x, y) = 2x + 2y$ **on each given region.**

15.

16.

Find the maximum or minimum value of each objective function subject to the given constraints.

17. Maximize $P(x, y) = 5x + 9y$ subject to $x \geq 0$, $y \geq 0$, and $x + 2y \leq 6$.

18. Maximize $P(x, y) = 25x + 31y$ subject to $x \geq 0$, $y \geq 0$, and $5x + 6y \leq 30$.

19. Minimize $C(x, y) = 10x + 20y$ subject to $x \geq 0$, $y \geq 0$, $x + y \geq 8$, and $3x + 5y \geq 30$.

20. Maximize $R(x, y) = 50x + 20y$ subject to $x \geq 0$, $y \geq 0$, $3x + y \leq 18$, and $2x + y \leq 14$.

Solve each problem by linear programming.

21. *Maximizing revenue* Bob and Betty make door harps and mailboxes in their craft shop near Gatlinburg. Each door harp requires 3 hr work from Bob and 1 hr from Betty. Each mailbox requires 4 hr work from Bob and 2 hr work from Betty. Bob cannot work more than 48 hr per week and Betty cannot work more than 20 hr per week. If each door harp sells for $12 and each mailbox sells for $20, then how many of each should they make to maximize their revenue?

22. *Maximizing revenue* At Taco Town a taco contains 2 oz of ground beef and 1 oz of chopped tomatoes. A burrito contains 1 oz of ground beef and 3 oz of chopped tomatoes. Near closing time the cook discovers that they have only 22 oz of ground beef and 36 oz of tomatoes left. The manager directs the cook to use the available resources to maximize their revenue for the remainder of the shift. If a taco sells for 40 cents and a burrito sells for 65 cents, then how many of each should they make to maximize their revenue?

23. *Door harps and mailboxes* If a door harp sells for $18 and a mailbox for $20, then how many of each should Bob and Betty build to maximize their revenue, subject to the constraints of Exercise 21?

24. *Tacos and burritos* If a taco sells for 20 cents and a burrito for 65 cents, then how many of each should be made to maximize the revenue, subject to the constraints of Exercise 22?

25. *Minimizing operating costs* Kimo's Material Company hauls gravel to a construction site, using a small truck and a large truck. The carrying capacity and operating cost per load are given in the accompanying table. Kimo must deliver a minimum of 120 yd³ per day to satisfy his contract with the builder. The union contract with his drivers requires that the total number of loads per day be a minimum of 8. How many loads should be made in each truck per day to minimize the total cost?

	Small truck	**Large truck**
Capacity (yd³)	20	40
Cost/load	$70	$60

Table for Exercise 25

26. *Minimizing labor costs* Tina's Telemarketing employs part-time and full-time workers. The number of hours worked per week and the pay per hour for each is given in the accompanying table. Tina needs at least 1200 hr of work done per week. To qualify for certain tax breaks, she must have at least 45 employees. How many part-time and full-time employees should be hired to minimize Tina's weekly labor cost?

	Part-time	**Full-time**
Hr/wk	20	40
Pay/hr	$6	$8

Table for Exercise 26

27. *Small trucks and large trucks* If it costs $70 per load to operate the small truck and $75 per load to operate the large truck, then how many loads should be made in each truck per day to minimize the total cost, subject to the constraints of Exercise 25?

28. *Part-time and full-time workers* If the labor cost for a part-timer is $9/hr and the labor cost for a full-timer is $8/hr, then how many of each should be employed to minimize the weekly labor cost, subject to the constraints of Exercise 26?

Highlights

Section 9.1   Systems of Linear Equations in Two Variables
1. A linear equation in two variables has the form $Ax + By = C$, where A and B are not both 0.
2. The solution set to a system of equations in two variables is the intersection of the graphs of the equations, or the set of ordered pairs that satisfy all of the equations.
3. The methods of addition and substitution can be used on linear systems to eliminate variables.
4. A system of linear equations in two variables may have one solution (independent system), no solution (inconsistent system), or infinitely many solutions (dependent system).

Section 9.2   Systems of Linear Equations in Three Variables
1. A linear equation in three variables has the form $Ax + By + Cz = D$, where A, B, and C are not all 0.
2. A system of linear equations in three variables can have a single solution, no solution, or infinitely many solutions.

Section 9.3   Nonlinear Systems of Equations
1. A system with at least one nonlinear equation is a nonlinear system.
2. The graphs of a nonlinear system may intersect at more than one point.
3. The methods of addition and substitution can be used on nonlinear systems to eliminate variables.

Section 9.4   Partial Fractions
1. In partial fraction decomposition we reverse the process of addition of rational expressions.
2. The numerators of the partial fractions are found by solving a system of linear equations.

Section 9.5   Inequalities and Systems of Inequalities in Two Variables

1. The solution set to $y > mx + b$ is the region above the line $y = mx + b$, and the solution set to $y < mx + b$ is the region below the line $y = mx + b$.

2. The solution set in the coordinate plane to $x > k$ (for any real number k) is the region to the right of the vertical line $x = k$, and the solution set to $x < k$ is the region to the left of $x = k$.

3. The solution set to a $\leq$ or $\geq$ inequality includes the boundary (drawn solid), while the solution set to a $<$ or $>$ inequality excludes the boundary (drawn dashed).

4. A system of inequalities is solved by testing a point from each region formed by the boundary lines or curves.

Section 9.6   Linear Programming

1. Inequalities concerning the variables in linear programming are called constraints.

2. A system of constraints determines a region of possible values for the variables in a linear programming problem.

3. The maximum or minimum value of a linear objective function subject to linear constraints occurs at a vertex of the region determined by the constraints.

Chapter 9 Review Exercises

Solve each system by graphing.

1. $2x - 3y = -9$
    $3x + y = 14$

2. $3x - 2y = 0$
    $y = -2x - 7$

3. $x + y = 2$
    $2y - 3x = 9$

4. $x - y = 30$
    $2x + 3y = 10$

Solve each system by the method of your choice. Indicate whether each system is independent, dependent, or inconsistent.

5. $3x - 5y = 19$
    $y = x$

6. $x + y = 9$
    $y = x - 3$

7. $4x - 3y = 6$
    $3x + 2y = 9$

8. $3x - 2y = 4$
    $5x + 7y = 1$

9. $6x + 2y = 2$
    $y = -3x + 1$

10. $x - y = 9$
    $2y - 2x = -18$

11. $3x - 4y = 12$
    $8y - 6x = 9$

12. $y = -5x + 3$
    $5x + y = 6$

Solve each system.

13. $x + y - z = 8$
    $2x + y + z = 1$
    $x + 2y + 3z = -5$

14. $2x + 3y - 2z = 8$
    $3x - y + 4z = -20$
    $x + y - z = 3$

15. $x + y + z = 1$
    $2x - y + 2z = 2$
    $2x + 2y + 2z = 2$

16. $x - y - z = 9$
    $x + y + 2z = -9$
    $-2x + 2y + 2z = -18$

17. $x + y + z = 1$
    $2x - y + 3z = 5$
    $x + y + z = 4$

18. $2x - y + z = 4$
    $x - y + z = -1$
    $-x + y - z = 0$

Solve each nonlinear system of equations. Find real solutions only.

19. $x^2 + y^2 = 4$
    $x = y^2$

20. $x^2 - y^2 = 9$
    $x^2 + y^2 = 7$

21. $y = |x|$
    $y = x^2$

22. $y = 2x^2 + x - 3$
    $6x + y = 12$

Find the partial fraction decomposition for each rational expression.

23. $\dfrac{7x - 7}{(x - 3)(x + 4)}$

24. $\dfrac{x - 13}{x^2 - 6x + 5}$

25. $\dfrac{7x^2 - 7x + 23}{x^3 - 3x^2 + 4x - 12}$

26. $\dfrac{10x^2 - 6x + 2}{(x - 1)^2(x + 2)}$

Graph the solution set to each inequality.

27. $x^2 + (y - 3)^2 < 9$

28. $2x - 9y \leq 18$

29. $x \leq (y - 1)^2$

30. $y < 6 - 2x^2$

Graph the solution set to each system of inequalities.

31. $2x - 3y \geq 6$
    $x \leq 2$

32. $x \leq 3, y \geq 1$
    $x - y \geq -5$

33. $y \geq 2x^2 - 6$
    $x^2 + y^2 \leq 9$

34. $x^2 + y^2 \geq 16$
    $2y \geq x^2 - 16$

35. $x \geq 0, y \geq 1$
    $x + 2y \leq 10$
    $3x + 4y \leq 24$

36. $x \geq 0, y \geq 0$
    $30x + 60y \leq 1200$
    $x + \quad y \leq 30$

37. $x \geq 0, y \geq 0$
    $x + 6y \geq 60$
    $x + \quad y \geq 35$

38. $x \geq 0$
    $y \geq x + 1$
    $x + y \leq 5$

Solve each problem, using a system of equations.

39. Find the equation of the line through $(-2, 3)$ and $(4, -1)$.

40. Find the equation of the line through $(4, 7)$ and $(-2, -3)$.

41. Find the equation of the parabola through $(1, 4)$, $(3, 20)$, and $(-2, 25)$.

42. Find the equation of the parabola through $(-1, 10)$, $(2, -5)$, and $(3, -18)$.

43. *Tacos and burritos* At Taco Town a taco contains 1 oz of meat and 2 oz of cheese, while a burrito contains 2 oz of meat and 3 oz of cheese. In one hour the cook used 181 oz of meat and 300 oz of cheese making tacos and burritos. How many of each were made?

44. *Imported and domestic cars* Nicholas had 10% imports on his used car lot, and Seymour had 30% imports on his used car lot. After Nicholas bought Seymour's entire stock, Nicholas had 300 cars, of which 22% were imports. How many cars were on each lot originally?

45. *Daisies, carnations, and roses* Esther's Flower Shop sells a bouquet containing five daisies, three carnations, and two roses for $3.05. Esther also sells a bouquet containing three daisies, one carnation, and four roses for $2.75. Her Valentine's Day Special contains four daisies, two carnations, and one rose for $2.10. How much should she charge for her Economy Special, which contains one daisy, one carnation, and one rose?

46. *Peppers, tomatoes, and eggplants* Ngan planted 81 plants in his garden at a total cost of $23.85. The 81 plants consisted of peppers at 20 cents each, tomatoes at 35 cents each, and eggplants at 30 cents each. If the total number of peppers and tomatoes was only half the number of eggplants, then how many of each did he plant?

Solve each linear programming problem.

47. *Minimum* Find the minimum value of the function $C(x, y) = 0.42x + 0.84y$ subject to the constraints $x \geq 0$, $y \geq 0$, $x + 6y \geq 60$, and $x + y \geq 35$.

48. *Maximum* Find the maximum value of the function $P(x, y) = 1.23x + 1.64y$ subject to the constraints $x \geq 0$, $y \geq 0$, $x + 2y \leq 40$, and $x + y \leq 30$.

49. *Pipeline or barge* A refinery gets its oil either from a pipeline or from barges. The refinery needs at least 12 million barrels per day. The maximum capacity of the pipeline is 12 million barrels per day, and the maximum that can be delivered by barge is 8 million barrels per day. The refinery has a contract to buy at least 6 million barrels per day through the pipeline. If the cost of oil by barge is $18 per barrel and the cost of oil from the pipeline is $20 per barrel, then how much oil should be purchased from each source to minimize the cost and satisfy the constraints?

50. *Fluctuating costs* If the cost of oil by barge goes up to $21 per barrel while the cost of oil by pipeline stays at $20 per barrel, then how much oil should be purchased from each source to minimize the cost and satisfy the constraints of Exercise 49?

Chapter 9 Test

Solve the system by the indicated method.

1. Graphing:
$$2x + 3y = 6$$
$$y = \frac{1}{3}x + 5$$

2. Substitution:
$$2x + y = 4$$
$$3x - 4y = 9$$

3. Addition:
$$10x - 3y = 22$$
$$7x + 2y = 40$$

Determine whether each of the following systems is independent, inconsistent, or dependent.

4. $x = 6 - y$
$3x + 3y = 4$

5. $y = \frac{1}{2}x + 3$
$x - 2y = -6$

6. $y = 2x - 1$
$y = 3x + 20$

7. $y = -x + 2$
$y = -x + 5$

Find the solution set to each system of equations in three variables.

8. $2x - y + z = 4$
$-x + 2y - z = 6$

9. $x - 2y - z = 2$
$2x + 3y + z = -1$
$3x - y - 3z = -4$

10. $x + y + z = 1$
$x + y - z = 4$
$-x - y + z = 2$

Solve each system.

11. $x^2 + y^2 = 16$
$x^2 - 4y^2 = 16$

12. $x + y = -2$
$y = x^2 - 5x$

Find the partial fraction decomposition for each rational expression.

13. $\dfrac{2x + 10}{x^2 - 2x - 8}$

14. $\dfrac{4x^2 + x - 2}{x^3 - x^2}$

Graph the solution set to each inequality or system of inequalities.

15. $2x - y < 8$

16. $x + y \le 5$
$x - y < 0$

17. $x^2 + y^2 \le 9$
$y \le 1 - x^2$

Solve each problem.

18. In a survey of 52 students in the cafeteria, it was found that 15 were commuters. If one-quarter of the female students and one-third of the male students in the survey were commuters, then how many of each sex were surveyed?

19. General Hospital is planning an aggressive advertising campaign to bolster the hospital's image in the community. Each television commercial reaches 14,000 people and costs $9000, while each newspaper ad reaches 6000 people and costs $3000. The advertising budget for the campaign is limited to $99,000 and the advertising agency has the capability of producing a maximum of 23 ads and/or commercials during the time allotted for the campaign. What mix of television commercials and newspaper ads will maximize the audience exposure, subject to the given constraints?

Tying It All Together
Chapters 1–9

Solve each equation.

1. $\dfrac{x-2}{x+5} = \dfrac{11}{24}$

2. $\dfrac{1}{x} + \dfrac{x-2}{x+5} = \dfrac{11}{24}$

3. $5 - 3(x+2) - 2(x-2) = 7$

4. $|3 - 2x| = 5$

5. $\sqrt{3 - 2x} = 5$

6. $3x^2 - 4 = 0$

7. $\dfrac{(x-2)^2}{x^2} = 1$

8. $2^{x-1} = 9$

9. $\log(x+1) + \log(x+4) = 1$

10. $x^{-2/3} = 0.25$

11. $x^2 - 3x = 6$

12. $2(x-3)^2 - 1 = 0$

Solve each inequality and graph the solution set on the number line.

13. $3 - 2x > 0$

14. $|3 - 2x| > 0$

15. $x^2 \geq 9$

16. $(x-2)(x+4) \leq 27$

Graph the solution set to each inequality in the coordinate plane.

17. $3 - 2x > y$

18. $|3 - 2x| > y$

19. $x^2 \geq 9$

20. $(x-2)(x+4) \leq y$

In the year 2010, traffic gridlocks on Los Angeles freeways will be history. In response to electronic commands, great platoons of linked vehicles will cruise the once congested lanes at an unvarying 75 miles per hour. Safely locked inside their "smart cars," commuters will read, use laptop computers, even catch a few extra winks on the way to work. They'll do almost anything, it seems, except keep their eyes on the road.

This is the vision of Intelligent Vehicle/Highway Systems (IVHS), a nonprofit group whose members foresee a cluster of emerging technologies transforming modern highways. They are already testing some on-board navigation systems along a 13-mile "smart corridor" of southern California's Santa Monica Freeway. In the new interactive systems, powerful computers will pinpoint travelers' locations, select routes, display maps, warn of impending collisions, and even issue weather reports. At last, driving will come up to speed.

To perform such complex feats, IVHS systems must instantly solve huge problems in linear programming involving hundreds of thousands of variables. That degree of complexity is echoed in AT&T's domestic long-distance communication network, which manipulates 800,000 variables! Problems of this scope were once impossible to solve, even with the help of supercomputers. However, in

10

Matrices and

Determinants

1984, a Bell Lab employee discovered a new matrix technique that allowed computers to solve highly complex problems in a short period of time. By implementing this method, AT&T solved one of its unmanageable linear programming problems in less than half an hour and obtained another 10 percent capacity out of its $15 billion system.

We covered basic linear programming in Chapter 9. But as AT&T discovered, huge problems require more efficient techniques. In this chapter we'll study matrices, along with several methods for using them to solve systems of linear equations. These methods lie at the core of the large-scale computer programs being used today to manage traffic flow, schedule airline flights, and track data in inventory control, accounting, and purchasing.

10.1

Solving Linear

Systems Using

Matrices

In this section we learn a method for solving systems of linear equations that is an improved version of the addition method of Section 9.1. The new method requires some new terminology.

Matrices

Twenty-six female students and twenty-four male students responded to a survey on income in a college algebra class. Among the female students, 5 classified themselves as low-income, 10 as middle-income, and 11 as high-income. Among the male students, 9 were low-income, 2 were middle-income, and 13 were high-income. Each student is classified in two ways, according to sex and income. This information can be written in a *matrix:*

$$\begin{array}{c} \\ \text{Female} \\ \text{Male} \end{array} \begin{array}{ccc} \text{L} & \text{M} & \text{H} \\ \left[\begin{array}{ccc} 5 & 10 & 11 \\ 9 & 2 & 13 \end{array}\right] \end{array}$$

In this matrix we can see the class makeup according to sex and income. A matrix provides a convenient way to organize a two-way classification of data.

A **matrix** is a rectangular array of real numbers. The **rows** of a matrix run horizontally, and the **columns** run vertically. A **row matrix** is a matrix with only one row, and a **column matrix** is a matrix with only one column. A matrix with m rows and n columns has **order** $m \times n$ (read "m by n") or **dimension** $m \times n$. The number of rows is always given first. For example, the matrix used to classify the students is a 2×3 matrix.

Example 1   Finding the order of a matrix

Determine the order of each matrix.

a) $[-4 \quad 3 \quad -2]$ b) $\begin{bmatrix} 3 & -1 \\ 4 & 2 \end{bmatrix}$ c) $\begin{bmatrix} -5 & 19 \\ 14 & 2 \\ 0 & -1 \end{bmatrix}$ d) $\begin{bmatrix} -4 & 46 & 8 \\ 1 & 0 & 13 \\ -12 & -5 & 2 \end{bmatrix}$

Solution

Matrix (a) is a row matrix with order 1×3. Matrix (b) has order 2×2, matrix (c) has order 3×2, and matrix (d) has order 3×3. ◆

A **square** matrix has an equal number of rows and columns. Matrices (b) and (d) of Example 1 are square matrices. Each number in a matrix is called an **entry** or an **element.** The matrix [5] is a 1×1 matrix with only one entry, 5.

The Augmented Matrix

We now see how matrices are used to represent systems of linear equations. The solution to a system of linear equations such as

$$x - 3y = 11$$
$$2x + y = 1$$

depends on the coefficients of x and y and the constants on the right-hand side of the equation. The **coefficient matrix** for this system is the matrix

$$\begin{bmatrix} 1 & -3 \\ 2 & 1 \end{bmatrix},$$

whose entries are the coefficients of the variables. (The coefficient of y in $x - 3y = 11$ is -3.) The constants from the right-hand side of the system are inserted into the matrix of coefficients, to form the **augmented matrix** of the system:

$$\begin{bmatrix} 1 & -3 & | & 11 \\ 2 & 1 & | & 1 \end{bmatrix}$$

Each row of the augmented matrix represents an equation of the system, while the columns represent the coefficients of x, the coefficients of y, and the constants, respectively. The vertical line represents the equal signs. Two systems of linear equations are **equivalent** if they have the same solution set, while two augmented matrices are **equivalent** if the systems they represent are equivalent.

Example 2   Determining the augmented matrix

Write the augmented matrix for each system of equations.

a) $x = y + 3$ b) $x + 2y - z = 1$ c) $x - y = 2$
   $y = 4 - x$ $2x + 3z = 5$ $y - z = 3$
    $3x - 2y + z = 0$

Solution

a) To write the augmented matrix, the equations must be in standard form with the variables on the left-hand side and constants on the right-hand side:

$$x - y = 3$$
$$x + y = 4$$

We write the augmented matrix using the coefficients of the variables and the constants:

$$\left[\begin{array}{cc|c} 1 & -1 & 3 \\ 1 & 1 & 4 \end{array}\right]$$

b) Use the coefficient 0 for each variable that is missing:

$$\left[\begin{array}{ccc|c} 1 & 2 & -1 & 1 \\ 2 & 0 & 3 & 5 \\ 3 & -2 & 1 & 0 \end{array}\right]$$

c) The augmented matrix for this system is a 2 × 4 matrix:

$$\left[\begin{array}{ccc|c} 1 & -1 & 0 & 2 \\ 0 & 1 & -1 & 3 \end{array}\right]$$

Example 3 Writing a system for an augmented matrix

Write the system of equations represented by each augmented matrix.

a) $\left[\begin{array}{cc|c} 1 & 3 & -5 \\ 2 & -3 & 1 \end{array}\right]$ **b)** $\left[\begin{array}{cc|c} 1 & 0 & 7 \\ 0 & 1 & 3 \end{array}\right]$ **c)** $\left[\begin{array}{ccc|c} 2 & 5 & 1 & 2 \\ -3 & 0 & 4 & -1 \\ 4 & -5 & 2 & 3 \end{array}\right]$

Solution

a) Use the first two numbers in each row as the coefficients of x and y and the last number as the constant to get the following system:

$$x + 3y = -5$$
$$2x - 3y = 1$$

b) The augmented matrix represents the following system:

$$x = 7$$
$$y = 3$$

c) Use the first three numbers in each row as the coefficients of x, y, and z and the last number as the constant to get the following system:

$$2x + 5y + z = 2$$
$$-3x + 4z = -1$$
$$4x - 5y + 2z = 3$$

Recall that to solve a single equation, we write simpler and simpler equivalent equations to get an equation whose solution is obvious. Similarly, to solve a system of equations, we write simpler and simpler equivalent systems to get a system whose solution is obvious. We now look at operations that can be performed on augmented matrices to obtain simpler equivalent augmented matrices.

The Gauss-Jordan Method

The rows of an augmented matrix represent the equations of a system. Since the equations of a system can be written in any order, two rows of an augmented matrix can be interchanged if necessary. Since multiplication of both sides of an equation by the same nonzero number produces an equivalent equation, multiplying each entry in a row of the augmented matrix by a nonzero number produces an equivalent augmented matrix. In Section 9.1, two equations were added to eliminate a variable. In the augmented matrix, elimination of variables is accomplished by adding the entries in one row to the corresponding entries in another row. These two row operations can be combined to add a multiple of one row to another, just as was done in solving systems by addition. The three **row operations** for an augmented matrix are summarized as follows.

> ### Summary: Row Operations
>
> **Any of the following row operations on an augmented matrix gives an equivalent augmented matrix:**
>
> 1. **Interchanging two rows of the matrix.**
> 2. **Multiplying every entry in a row by the same nonzero real number.**
> 3. **Adding to a row a multiple of another row.**

To solve a system of two linear equations in two variables using the **Gauss-Jordan method,** we use row operations to obtain simpler and simpler augmented matrices. We want to get an augmented matrix that corresponds to a system whose solution is obvious. An augmented matrix of the following form is the simplest:

$$\left[\begin{array}{cc|c} 1 & 0 & a \\ 0 & 1 & b \end{array}\right]$$

Notice that this augmented matrix corresponds to the system $x = a$ and $y = b$, for which the solution set is $\{(a, b)\}$.

The **diagonal** of a matrix consists of the entries in the first row first column, second row second column, third row third column, and so on. The goal of the Gauss-Jordan method is to convert the original augmented matrix into an equivalent augmented matrix that has 1's on its diagonal and 0's above and below the 1's. If the system has a unique solution, then it will appear in the right-most column of the final augmented matrix.

Example 4   Using the Gauss-Jordan method

Use row operations to solve the system.

$$2x - 4y = 16$$
$$3x + \ y = 3$$

Solution

Start with the augmented matrix:

$$\begin{bmatrix} 2 & -4 & | & 16 \\ 3 & 1 & | & 3 \end{bmatrix}$$

The first step is to multiply the first row R_1 by $\frac{1}{2}$ to get a 1 in the first position on the diagonal. Think of this step as replacing R_1 by $\frac{1}{2}R_1$. We show this in symbols as $\frac{1}{2}R_1 \rightarrow R_1$.

$$\begin{bmatrix} 1 & -2 & | & 8 \\ 3 & 1 & | & 3 \end{bmatrix} \qquad \frac{1}{2}R_1 \rightarrow R_1$$

To get a 0 in the first position of R_2, multiply R_1 by -3 and add the result to R_2. Since $-3R_1 = [-3, 6, -24]$ and $R_2 = [3, 1, 3]$, $-3R_1 + R_2 = [0, 7, -21]$. We are replacing R_2 by $-3R_1 + R_2$:

$$\begin{bmatrix} 1 & -2 & | & 8 \\ 0 & 7 & | & -21 \end{bmatrix} \qquad -3R_1 + R_2 \rightarrow R_2$$

To get a 1 in the second position on the diagonal, multiply R_2 by $\frac{1}{7}$:

$$\begin{bmatrix} 1 & -2 & | & 8 \\ 0 & 1 & | & -3 \end{bmatrix} \qquad \frac{1}{7}R_2 \rightarrow R_2$$

Now row 2 is in the form needed to solve the system. We next get a 0 as the second entry in R_1. Multiply row 2 by 2 and add the result to row 1. Since $2R_2 = [0, 2, -6]$ and $R_1 = [1, -2, 8]$, $2R_2 + R_1 = [1, 0, 2]$. We get the following matrix.

$$\begin{bmatrix} 1 & 0 & | & 2 \\ 0 & 1 & | & -3 \end{bmatrix} \qquad 2R_2 + R_1 \rightarrow R_1$$

Note that the coefficient of y in the first equation is now 0. The system associated with the last augmented matrix is $x = 2$ and $y = -3$. So the solution set to the system is $\{(2, -3)\}$. Check in the original system.   ◆

The procedure used in Example 4 to solve a system that has a unique solution is summarized below. For inconsistent and dependent systems see Examples 6, 7, and 8.

▼

Procedure: The Gauss-Jordan Method for an Independent System of Two Equations

To solve a system of two linear equations in two variables using the Gauss-Jordan method, perform the following row operations on the augmented matrix.

1. If necessary, interchange R_1 and R_2 so that R_1 begins with a non-zero entry.
2. Get a 1 in the first position on the diagonal by multiplying R_1 by the reciprocal of the first entry in R_1.
3. Add an appropriate multiple of R_1 to R_2 to get 0 below the first 1.
4. Get a 1 in the second position on the diagonal by multiplying R_2 by the reciprocal of the second entry in R_2.
5. Add an appropriate multiple of R_2 to R_1 to get 0 above the second 1.
6. Read the unique solution from the last column of the final augmented matrix.

In the next example, the Gauss-Jordan method is used on a system involving three variables. For three linear equations in three variables, x, y, and z, we try to get the augmented matrix into the form

$$\left[\begin{array}{ccc|c} 1 & 0 & 0 & a \\ 0 & 1 & 0 & b \\ 0 & 0 & 1 & c \end{array}\right],$$

from which we conclude that $x = a$, $y = b$, and $z = c$.

Example 5   Gauss-Jordan with three variables

Use the Gauss-Jordan method to solve the following system:

$$2x - y + z = 1$$
$$x + y - 2z = 5$$
$$3x - y - z = 8$$

Solution

Write the augmented matrix:

$$\left[\begin{array}{ccc|c} 2 & -1 & 1 & 1 \\ 1 & 1 & -2 & 5 \\ 3 & -1 & -1 & 8 \end{array}\right]$$

To get the first 1 on the diagonal, we could multiply R_1 by $\frac{1}{2}$ or interchange R_1 and R_2. Interchanging R_1 and R_2 is simpler because it avoids getting fractions in the entries:

$$\left[\begin{array}{ccc|c} 1 & 1 & -2 & 5 \\ 2 & -1 & 1 & 1 \\ 3 & -1 & -1 & 8 \end{array}\right] \qquad R_1 \leftrightarrow R_2$$

Now use the first row to get 0's below the first 1 on the diagonal. First, to get a 0 in the first position of the second row, multiply the first row by -2 and add the result to the second row. Then, to get a 0 in the first position of the third row, multiply the first row by -3 and add the result to the third row. These two steps eliminate the variable x from the second and third rows.

$$\left[\begin{array}{ccc|c} 1 & 1 & -2 & 5 \\ 0 & -3 & 5 & -9 \\ 0 & -4 & 5 & -7 \end{array}\right] \qquad \begin{array}{l} -2R_1 + R_2 \rightarrow R_2 \\ -3R_1 + R_3 \rightarrow R_3 \end{array}$$

To get a 1 in the second position on the diagonal, multiply the second row by $-\frac{1}{3}$:

$$\left[\begin{array}{ccc|c} 1 & 1 & -2 & 5 \\ 0 & 1 & -\frac{5}{3} & 3 \\ 0 & -4 & 5 & -7 \end{array}\right] \qquad -\frac{1}{3}R_2 \rightarrow R_2$$

To get a 0 above the second 1 on the diagonal, multiply R_2 by -1 and add the result to R_1. To get a 0 below the second 1 on the diagonal, multiply R_2 by 4 and add the result to the third row:

$$\left[\begin{array}{ccc|c} 1 & 0 & -\frac{1}{3} & 2 \\ 0 & 1 & -\frac{5}{3} & 3 \\ 0 & 0 & -\frac{5}{3} & 5 \end{array}\right] \qquad \begin{array}{l} -1R_2 + R_1 \rightarrow R_1 \\ 4R_2 + R_3 \rightarrow R_3 \end{array}$$

To get a 1 in the third position on the diagonal, multiply R_3 by $-\frac{3}{5}$:

$$\left[\begin{array}{ccc|c} 1 & 0 & -\frac{1}{3} & 2 \\ 0 & 1 & -\frac{5}{3} & 3 \\ 0 & 0 & 1 & -3 \end{array}\right] \qquad -\frac{3}{5}R_3 \rightarrow R_3$$

Use the third row to get 0's above the third 1 on the diagonal:

$$\left[\begin{array}{ccc|c} 1 & 0 & 0 & 1 \\ 0 & 1 & 0 & -2 \\ 0 & 0 & 1 & -3 \end{array}\right] \qquad \begin{array}{l} \frac{1}{3}R_3 + R_1 \rightarrow R_1 \\ \frac{5}{3}R_3 + R_2 \rightarrow R_2 \end{array}$$

This last matrix corresponds to $x = 1$, $y = -2$, and $z = -3$. So the solution set to the system is $\{(1, -2, -3)\}$. Check in the original system. ◆

In Example 5, there were two different row operations that would produce the first 1 on the diagonal. When doing the Gauss-Jordan method by hand, you should choose the row operations that make the computations the simplest. When this method is performed by a computer, the same sequence of steps is used on every system. Any system of three equations that has a unique solution can be solved with the following sequence of steps.

> **Procedure: The Gauss-Jordan Method for an Independent System of Three Equations**
>
> To solve a system of three linear equations in three variables using the Gauss-Jordan method, perform the following row operations on the augmented matrix.
>
> 1. Get a 1 in the first position on the diagonal by multiplying R_1 by the reciprocal of the first entry in R_1. (First interchange rows if necessary.)
> 2. Add appropriate multiples of R_1 to R_2 and R_3 to get 0's below the first 1.
> 3. Get a 1 in the second position on the diagonal by multiplying R_2 by the reciprocal of the second entry in R_2. (First interchange rows if necessary.)
> 4. Add appropriate multiples of R_2 to R_1 and R_3 to get 0's above and below the second 1.
> 5. Get a 1 in the third position on the diagonal by multiplying R_3 by the reciprocal of the third entry in R_3.
> 6. Add appropriate multiples of R_3 to R_1 and R_2 to get 0's above the third 1.
> 7. Read the unique solution from the last column of the final augmented matrix.

Inconsistent and Dependent Equations

Inconsistent and dependent systems can also be solved by using the Gauss-Jordan method, but in these cases it is impossible to get the augmented matrix into the desired form.

Example 6   An inconsistent system in two variables

Solve the system.

$$x = y + 3$$
$$2y = 2x + 5$$

Solution

Write both equations in the form $Ax + By = C$:

$$x - y = 3$$
$$-2x + 2y = 5$$

Start with the augmented matrix:

$$\begin{bmatrix} 1 & -1 & | & 3 \\ -2 & 2 & | & 5 \end{bmatrix}$$

To get a 0 in the first position of R_2, multiply R_1 by 2 and add the result to R_2:

$$\begin{bmatrix} 1 & -1 & | & 3 \\ 0 & 0 & | & 11 \end{bmatrix} \qquad 2R_1 + R_2 \rightarrow R_2$$

It is impossible to convert this augmented matrix to the desired form of the Gauss-Jordan method. However, we can obtain the solution by observing that the second row corresponds to the equation $0 = 11$. So the system is inconsistent, and there is no solution.   ◆

Applying the Gauss-Jordan method to an inconsistent system (as in Example 6) causes a row to appear with 0 as the entry for each coefficient but a nonzero entry for the constant. For a dependent system of two equations in two variables (as in the next example), a 0 will appear in every entry of some row.

Example 7   A dependent system in two variables

Solve the system.

$$2x + y = 4$$
$$4x + 2y = 8$$

Solution

Start with the augmented matrix:

$$\begin{bmatrix} 2 & 1 & | & 4 \\ 4 & 2 & | & 8 \end{bmatrix}$$

To get a 0 in the first position of R_2, multiply R_1 by -2 and add the result to R_2:

$$\begin{bmatrix} 2 & 1 & | & 4 \\ 0 & 0 & | & 0 \end{bmatrix} \qquad -2R_1 + R_2 \rightarrow R_2$$

The second row of this augmented matrix gives us the equation $0 = 0$. So the system is dependent. Every ordered pair that satisfies the first equation satisfies both equations. The solution set is $\{(x, y) | 2x + y = 4\}$.   ◆

Solving a dependent system that has fewer equations than variables does not necessarily result in a row in which all entries are 0. The next example shows a dependent system of two equations in three variables solved by the Gauss-Jordan method.

Example 8   A dependent system in three variables

Solve the system.

$$x - y + z = 2$$
$$-2x + y + 2z = 5$$

Solution

Start with the augmented matrix:

$$\begin{bmatrix} 1 & -1 & 1 & | & 2 \\ -2 & 1 & 2 & | & 5 \end{bmatrix}$$

Perform the following row operations to get 1's and 0's in the first two columns as you would for a 2×2 matrix:

$$\begin{bmatrix} 1 & -1 & 1 & | & 2 \\ 0 & -1 & 4 & | & 9 \end{bmatrix} \qquad 2R_1 + R_2 \rightarrow R_2$$

$$\begin{bmatrix} 1 & -1 & 1 & | & 2 \\ 0 & 1 & -4 & | & -9 \end{bmatrix} \qquad -1R_2 \rightarrow R_2$$

$$\begin{bmatrix} 1 & 0 & -3 & | & -7 \\ 0 & 1 & -4 & | & -9 \end{bmatrix} \qquad R_2 + R_1 \rightarrow R_1$$

The last matrix corresponds to the system

$$x - 3z = -7$$
$$y - 4z = -9$$

or $x = 3z - 7$ and $y = 4z - 9$. The system is dependent, and the solution set is

$$\{(3z - 7, 4z - 9, z) | z \text{ is any real number}\}.$$

Replace x by $3z - 7$ and y by $4z - 9$ in the original equations to check.   ◆

The Gauss-Jordan method can be applied to a system of n linear equations in m unknowns. However, it is a rather tedious method to perform when n and m are greater than 2, especially when fractions are involved. Computers can be programmed to work with matrices, and the Gauss-Jordan method is frequently used for computer solutions.

Applications

There is much discussion among city planners and transportation experts about Intelligent Vehicle/Highway Systems. (See *Omni*, January 1992.) Such a system would use a central computer and computers within vehicles to manage traffic flow by controlling traffic lights, rerouting traffic away from congested areas, and perhaps even driving the vehicle for you. It is easy to say that computers will control the complex traffic system of the future, but a computer does only what it is programmed to do. The next example shows one type of problem that a computer would be continually solving in order to control traffic flow.

Example 9   Traffic control in the future

Figure 10.1 shows the intersections of four one-way streets. The numbers on the arrows represent the number of cars per hour that desire to enter each intersection and leave each intersection. For example, 400 cars per hour want to enter intersection P from the north on First Ave, while 300 cars per hour want to head east from intersection Q on Elm Street. The letters w, x, y, and z represent the number of cars per hour passing the four points between these four intersections, as shown in Fig. 10.1.

a) Find values for w, x, y, and z that would realize this desired traffic flow.

b) If construction on Oak Street limits z to 300 cars per hour, then how many cars per hour would have to pass w, x, and y?

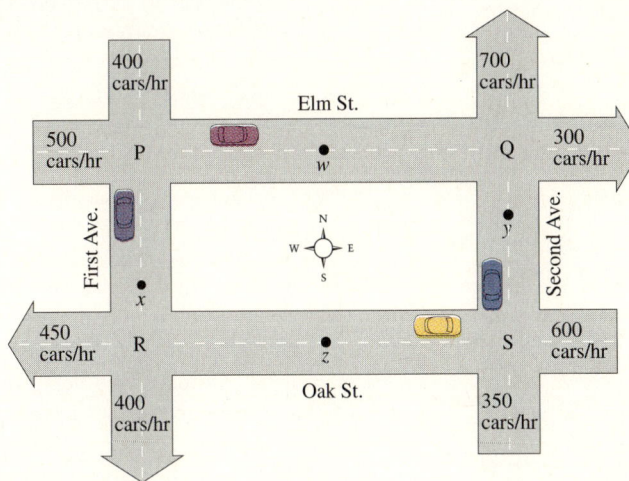

Figure 10.1

Solution

a) The solution to the problem is based on the fact that the number of cars entering an intersection per hour must equal the number leaving that intersection per hour, if the traffic is to keep flowing. Since 900 cars (400 + 500) enter intersection P, 900 must leave, $x + w = 900$. Writing a similar equation for each intersection yields the system shown at the top of the next page.

$$w + x = 900$$
$$w + y = 1000$$
$$x + z = 850$$
$$y + z = 950$$

For this system, the augmented matrix is

$$\left[\begin{array}{cccc|c} 1 & 1 & 0 & 0 & 900 \\ 1 & 0 & 1 & 0 & 1000 \\ 0 & 1 & 0 & 1 & 850 \\ 0 & 0 & 1 & 1 & 950 \end{array}\right].$$

Use row operations to get the equivalent matrix:

$$\left[\begin{array}{cccc|c} 1 & 1 & 0 & 0 & 900 \\ 0 & 1 & 0 & 1 & 850 \\ 0 & 0 & 1 & 1 & 950 \\ 0 & 0 & 0 & 0 & 0 \end{array}\right].$$

The system is a dependent system and does not have a unique solution. From this matrix we get $w = 50 + z$, $x = 850 - z$, and $y = 950 - z$. The problem is solved by any ordered 4-tuple of the form $(50 + z, 850 - z, 950 - z, z)$ where z is a nonnegative integer.

b) If z is limited to 300 because of construction, then the solution is (350, 550, 650, 300). To keep traffic flowing when $z = 300$, the system must route 350 cars past w, 550 past x, and 650 past y. ◆

One-way streets were used in Example 9 to keep the problem simple, but you can imagine two-way streets where at each intersection there are three ways to route traffic. You can also imagine many streets and many intersections all subject to systems of equations, with computers continually controlling flow to keep all traffic moving. Computerized traffic control may be a few years away, but similar problems are being solved today at AT&T to route long-distance calls and at American Airlines to schedule flight crews and equipment.

? For Thought

True or false? Explain.

1. The augmented matrix for a system of two linear equations in two unknowns is a 2×2 matrix.

2. The augmented matrix for $\begin{array}{l} x - y = 4 \\ 3x + y = 5 \end{array}$ is $\left[\begin{array}{cc|c} x & -y & 4 \\ 3x & y & 5 \end{array}\right].$

3. The augmented matrix for $\begin{array}{l} 3x - 2y = 4 \\ x + y = 6 \end{array}$ is $\left[\begin{array}{cc|c} 3 & -2 & 4 \\ 1 & 1 & 6 \end{array}\right].$

4. The matrix $\begin{bmatrix} 1 & 0 & | & 2 \\ 1 & -1 & | & -3 \end{bmatrix}$ corresponds to the system $\begin{array}{l} x = 2 \\ x - y = -3. \end{array}$

5. The matrix $\begin{bmatrix} 1 & 3 & | & -2 \\ -1 & -5 & | & 4 \end{bmatrix}$ is equivalent to $\begin{bmatrix} 1 & 3 & | & -2 \\ 0 & -2 & | & 2 \end{bmatrix}$.

6. The matrix $\begin{bmatrix} 1 & 2 & | & 3 \\ 1 & -3 & | & 2 \end{bmatrix}$ is equivalent to $\begin{bmatrix} 1 & 2 & | & 3 \\ 0 & -5 & | & -1 \end{bmatrix}$.

7. The matrix $\begin{bmatrix} 1 & 0 & | & 2 \\ 0 & 1 & | & 7 \end{bmatrix}$ corresponds to the system $\begin{array}{l} x + y = 2 \\ x + y = 7. \end{array}$

8. The system corresponding to $\begin{bmatrix} 1 & 3 & | & 5 \\ 0 & 0 & | & 7 \end{bmatrix}$ is inconsistent.

9. The system corresponding to $\begin{bmatrix} -1 & 2 & | & -3 \\ 0 & 0 & | & 0 \end{bmatrix}$ is inconsistent.

10. The notation $2R_1 + R_3 \rightarrow R_3$ means to replace R_3 by $2R_1 + R_3$.

10.1 Exercises Tape 19   Disk—5.25″: 6   3.5″: 4   Macintosh: 4

Determine the order of each matrix.

1. $[1 \quad 5 \quad 8]$

2. $\begin{bmatrix} -3 \\ y \\ 5 \end{bmatrix}$

3. $[7]$

4. $\begin{bmatrix} x & y \\ z & w \end{bmatrix}$

5. $\begin{bmatrix} -5 & 12 \\ 99 & 6 \\ 0 & 0 \end{bmatrix}$

6. $\begin{bmatrix} 1 & 5 & 7 \\ 3 & 0 & 5 \\ 2 & -6 & -3 \end{bmatrix}$

Write the augmented matrix for each system of equations.

7. $\begin{array}{l} x - 2y = 4 \\ 3x + 2y = -5 \end{array}$

8. $\begin{array}{l} 4x - y = 1 \\ x + 3y = 5 \end{array}$

9. $\begin{array}{l} x - y - z = 4 \\ x + 3y - z = 1 \\ 2y - 5z = -6 \end{array}$

10. $\begin{array}{l} x + 3y \quad = 5 \\ y - 4z = 8 \\ -2x \quad + 5z = 7 \end{array}$

Write the system of equations represented by each augmented matrix.

11. $\begin{bmatrix} 3 & 4 & | & -2 \\ 3 & -5 & | & 0 \end{bmatrix}$

12. $\begin{bmatrix} 1 & 0 & | & -7 \\ 0 & 1 & | & 5 \end{bmatrix}$

13. $\begin{bmatrix} 5 & 0 & 0 & | & 6 \\ -4 & 0 & 2 & | & -1 \\ 4 & 4 & 0 & | & 7 \end{bmatrix}$

14. $\begin{bmatrix} 1 & 0 & 1 & | & 2 \\ 0 & 1 & -1 & | & -6 \\ 1 & -1 & 1 & | & 5 \end{bmatrix}$

Describe in words the row operation(s) that will convert the first augmented matrix into the second.

15. $\begin{bmatrix} 2 & 4 & | & 14 \\ 5 & 4 & | & 5 \end{bmatrix}, \begin{bmatrix} 1 & 2 & | & 7 \\ 5 & 4 & | & 5 \end{bmatrix}$

16. $\begin{bmatrix} 1 & 2 & | & 7 \\ 5 & 4 & | & 5 \end{bmatrix}, \begin{bmatrix} 1 & 2 & | & 7 \\ 0 & -6 & | & -30 \end{bmatrix}$

17. $\begin{bmatrix} 1 & 2 & | & 7 \\ 0 & -6 & | & -30 \end{bmatrix}, \begin{bmatrix} 1 & 2 & | & 7 \\ 0 & 1 & | & 5 \end{bmatrix}$

18. $\begin{bmatrix} 1 & 2 & | & 7 \\ 0 & 1 & | & 5 \end{bmatrix}, \begin{bmatrix} 1 & 0 & | & -3 \\ 0 & 1 & | & 5 \end{bmatrix}$

Solve each system using the Gauss-Jordan method.

19. $\begin{array}{l} x + y = 5 \\ -2x + y = -1 \end{array}$

20. $\begin{array}{l} x - y = 2 \\ 3x - y = 12 \end{array}$

21. $\begin{array}{l} y = 4 - 2x \\ x = 8 + y \end{array}$

22. $\begin{array}{l} 3x = 1 + 2y \\ y = 2 - x \end{array}$

23. $\begin{array}{l} 2x - y = 3 \\ 3x + 2y = 15 \end{array}$

24. $\begin{array}{l} 2x - 3y = -1 \\ 3x - 2y = 1 \end{array}$

25. $\begin{array}{l} 0.4x - 0.2y = 0 \\ x + 1.5y = 2 \end{array}$

26. $\begin{array}{l} 0.2x + 0.6y = 0.7 \\ 0.5x - y = 0.5 \end{array}$

27. $\begin{array}{l} 3a - 5b = 7 \\ -3a + 5b = 4 \end{array}$

28. $\begin{array}{l} 2s - 3t = 9 \\ 4s - 6t = 1 \end{array}$

29. $\begin{array}{l} 0.5u + 1.5v = 2 \\ 3u + 9v = 12 \end{array}$

30. $\begin{array}{l} m - 2.5n = 0.5 \\ -4m + 10n = -2 \end{array}$

31. $x + y + z = 6$
$x - y - z = 0$
$2y - z = 3$

32. $x - y + z = 2$
$-x + y + z = 4$
$-x \quad + z = 2$

33. $2x + y = 2 + z$
$x + 2y = 2 + z$
$x + 2z = 2 + y$

34. $3x = 4 + y$
$x + y = z - 1$
$2z = 3 - x$

35. $2a - 2b + c = -2$
$a + b - 3c = 3$
$a - 3b + c = -5$

36. $r - 3s - t = -3$
$-r - s + 2t = 1$
$-r + 2s - t = -2$

37. $3y = x + z$
$x - y - 3z = 4$
$x + y + 2z = -1$

38. $z = 2 + x$
$2x - y = 1$
$y + 3z = 15$

39. $x - 2y + 3z = 1$
$2x - 4y + 6z = 2$
$-3x + 6y - 9z = -3$

40. $4x - 2y + 6z = 4$
$2x - y + 3z = 2$
$-2x + y - 3z = -2$

41. $x - y + z = 2$
$2x + y - z = 1$
$2x - 2y + 2z = 5$

42. $x - y + z = 4$
$x + y - z = 1$
$x + y - z = 3$

43. $x + y - z = 3$
$3x + y + z = 7$
$x - y + 3z = 1$

44. $x + 2y + 2z = 4$
$2x + y + z = 1$
$-x + y + z = 3$

45. $2x - y + 3z = 1$
$x + y - z = 4$

46. $x + 3y + z = 6$
$-x + y - z = 2$

47. $x - y + z - w = 2$
$-x + 2y - z - w = -1$
$2x - y - z + w = 4$
$x + 3y - 2z - 3w = 6$

48. $3a - 2b + c + d = 0$
$a - b + c - d = -4$
$-2a + b + 3c - 2d = 5$
$2a + 3b - c - d = -3$

Write a system of linear equations for each problem and solve the system using the Gauss-Jordan method.

49. *Wages from two jobs* Mike works a total of 60 hr per week at his two jobs. He makes \$5 per hour at Burgers-R-Us and \$5.50 per hour at the Soap Opera Laundromat. If his total pay for one week is \$311 before taxes, then how many hours does he work at each job?

50. *Postal rates* Mary Ellen spent \$15.40 on postage inviting a total of 60 boys and girls to her 10th birthday party. Each boy was sent an invitation on a postcard, while each girl was invited with a letter. If she put a 19-cent stamp on each postcard and a 29-cent stamp on each letter, then how many kids of each sex were invited?

51. *Investment portfolio* Petula invested a total of \$40,000 in a no-load mutual fund, treasury bills, and municipal bonds. Her total return of \$3660 came from an 8% return on her investment in the no-load mutual fund, a 9% return on the treasury bills, and 12% return on the municipal bonds. If her total investment in treasury bills and municipal bonds was equal to her investment in the no-load mutual fund, then how much did she invest in each?

52. *Nutrition* The table shows the percentage of U.S. Recommended Daily Allowances (RDA) for phosphorus, magnesium, and calcium in one ounce of three breakfast cereals (without milk). If Hulk Hogan got 98% of the RDA of phosphorus, 84% of the RDA of magnesium, and 38% of the RDA of calcium by eating a large bowl of each (without milk), then how many ounces of each cereal did he eat?

Percentage of U.S. RDA

	Phosphorus	Magnesium	Calcium
Kix	4%	2%	4%
Quick Oats	10%	10%	0%
Muesli	8%	8%	2%

Table for Exercise 52

53. *Curve fitting* Find a, b, and c such that the graph of $y = ax^3 + bx + c$ goes through the points $(-1, 4)$, $(1, 2)$, and $(2, 7)$.

54. *Curve fitting* Find a, b, and c such that the graph of $y = ax^2 + bx + c$ goes through the points $(-1, 0)$, $(1, 0)$, and $(3, 0)$.

55. *Traffic control* The diagram shows the number of cars that desire to enter and leave each of three intersections on three one-way streets in a 30-minute period. The letters x, y, and z represent the number of cars passing the three points between these three intersections as shown in the diagram. Find values for x, y, and z that would realize this desired traffic flow. If construction on JFK Boulevard limits the value of z to 50, then what values for x and y would keep the traffic flowing?

56. *Traffic control* Southbound, M. L. King Drive in the figure for Exercise 55 leads to another intersection that cannot always handle 700 cars in a 30-minute period. Change 700 to 600 in the figure and write a system of three equations in x, y, and z. What is the solution to this system? If you had control over the 400 cars coming from the north into the first intersection on M. L. King Drive, what number would you use in place of 400 to get the system flowing again?

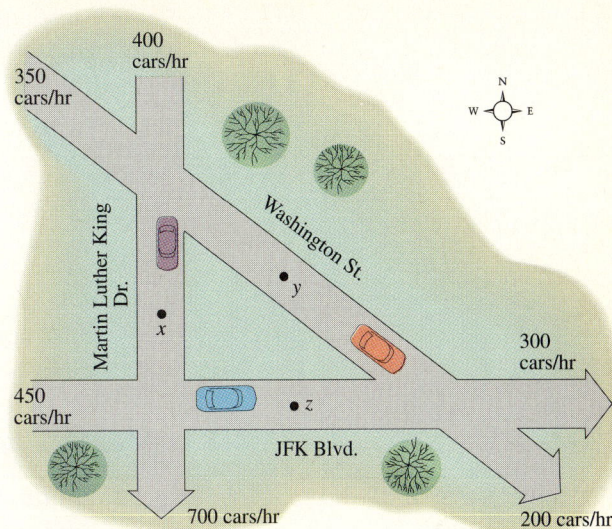

Figure for Exercise 55

10.2

Operations

with Matrices

In Section 10.1, matrices were used to keep track of the coefficients and constants in systems of equations. Matrices are also useful for simplifying and organizing information ranging from inventories to win-loss records of sports teams. In this section we study matrices in more detail and learn how operations with matrices are used in applications.

Notation

In Section 10.1 a matrix was defined as a rectangular array of numbers. Capital letters are used to name matrices and lowercase letters to name their entries. A general $m \times n$ matrix with m rows and n columns is given as follows:

$$A = \begin{bmatrix} a_{11} & a_{12} & a_{13} & \cdots & a_{1n} \\ a_{21} & a_{22} & a_{23} & \cdots & a_{2n} \\ a_{31} & a_{32} & a_{33} & \cdots & a_{3n} \\ \vdots & \vdots & \vdots & & \vdots \\ a_{m1} & a_{m2} & a_{m3} & \cdots & a_{mn} \end{bmatrix}$$

The subscripts indicate the position of each entry. For example, a_{32} is the entry in the third row and second column. The entry in the ith row and jth column is denoted by a_{ij}.

Two matrices are **equal** if they have the same order and the corresponding entries are equal. We write

$$\begin{bmatrix} 0.5 \\ 0.25 \end{bmatrix} = \begin{bmatrix} \frac{1}{2} \\ \frac{1}{4} \end{bmatrix}$$

because these matrices have the same order and their corresponding entries are equal. The matrices

$$\begin{bmatrix} 4 & 3 \\ 2 & 1 \end{bmatrix} \quad \text{and} \quad \begin{bmatrix} 1 & 3 \\ 2 & 1 \end{bmatrix}$$

have the same order, but they are not equal because the entries in the first row and first column are not equal. The matrices

$$[3 \quad 5] \quad \text{and} \quad \begin{bmatrix} 3 \\ 5 \end{bmatrix}$$

are not equal because the first has order 1×2 (a row matrix) and the second has order 2×1 (a column matrix).

Example 1 Equal matrices

Determine the values of x, y, and z that make the following matrix equation true:

$$\begin{bmatrix} x & y - 1 \\ z & 7 \end{bmatrix} = \begin{bmatrix} 1 & 3 \\ 2 & 7 \end{bmatrix}$$

Solution

If these matrices are equal, then the corresponding entries are equal. So $x = 1$, $y = 4$, and $z = 2$. ◆

Addition and Subtraction of Matrices

Matrices are rectangular arrays of real numbers, and in many ways they behave like real numbers. We can define matrix operations that have many properties similar to the properties of the real numbers.

Definition: Matrix Addition

The sum of two $m \times n$ matrices A and B is the $m \times n$ matrix denoted $A + B$ whose entries are the sums of the corresponding entries of A and B.

Note that only matrices that have the same order can be added. There is no definition for the sum of matrices of different orders.

Example 2 Sum of matrices

Find $A + B$ given that $A = \begin{bmatrix} -4 & 3 \\ 5 & -2 \end{bmatrix}$ and $B = \begin{bmatrix} 7 & -3 \\ 2 & -5 \end{bmatrix}$.

```
MATRIX[A] 2 ×2
[ -4    3    ]
[ 5    -2    ]
```

```
MATRIX[B] 2 ×2■
[ 7    -3    ]
[ 2    -5    ]
```

Define matrices A and B using the MATRX EDIT feature. See Appendix A for more details.

```
[A]+[B]
        [[3  0 ]
         [7  -7]]
```

Use MATRX NAMES to display $A + B$ and find the sum of the two matrices.

Solution

To find $A + B$, add the corresponding entries of A and B:

$$A + B = \begin{bmatrix} -4 & 3 \\ 5 & -2 \end{bmatrix} + \begin{bmatrix} 7 & -3 \\ 2 & -5 \end{bmatrix} = \begin{bmatrix} -4 + 7 & 3 + (-3) \\ 5 + 2 & -2 + (-5) \end{bmatrix} = \begin{bmatrix} 3 & 0 \\ 7 & -7 \end{bmatrix}$$

If all of the entries of a matrix are zero, the matrix is called a **zero matrix.** There is a zero matrix for every order. In matrix addition, the zero matrix behaves just like the additive identity 0 in the set of real numbers. For example,

$$\begin{bmatrix} 5 & -2 \\ 3 & -4 \end{bmatrix} + \begin{bmatrix} 0 & 0 \\ 0 & 0 \end{bmatrix} = \begin{bmatrix} 5 & -2 \\ 3 & -4 \end{bmatrix}.$$

In general, an $n \times n$ zero matrix is called the **additive identity** for $n \times n$ matrices.

For any matrix A, the **additive inverse** of A, denoted $-A$, is the matrix of the same order as A such that each entry of $-A$ is the opposite of the corresponding entry of A. Since corresponding entries are added in matrix addition, $A + (-A)$ is a zero matrix.

Example 3   Additive inverses of matrices

Find $-A$ and $A + (-A)$ for $A = \begin{bmatrix} -1 & 2 & 0 \\ 4 & -3 & 5 \\ 2 & 0 & -9 \end{bmatrix}$.

Solution

To find $-A$, find the opposite of every entry of A:

$$-A = \begin{bmatrix} 1 & -2 & 0 \\ -4 & 3 & -5 \\ -2 & 0 & 9 \end{bmatrix}$$

$$A + (-A) = \begin{bmatrix} -1 & 2 & 0 \\ 4 & -3 & 5 \\ 2 & 0 & -9 \end{bmatrix} + \begin{bmatrix} 1 & -2 & 0 \\ -4 & 3 & -5 \\ -2 & 0 & 9 \end{bmatrix} = \begin{bmatrix} 0 & 0 & 0 \\ 0 & 0 & 0 \\ 0 & 0 & 0 \end{bmatrix}$$

Therefore, the sum of A and $-A$ is the additive identity for 3×3 matrices.

The difference of two real numbers a and b is defined by $a - b = a + (-b)$. The difference of two matrices of the same order is defined similarly.

Definition: Matrix Subtraction

The difference of two $m \times n$ matrices A and B is the $m \times n$ matrix denoted $A - B$, where $A - B = A + (-B)$.

Even though subtraction is defined as addition of the additive inverse, we can certainly find the difference for two matrices by subtracting their corresponding entries. Note that we can subtract corresponding entries only if the matrices have the same order.

Example 4 Subtraction of matrices

Let $A = [3 \quad 5 \quad 8]$, $B = [3 \quad -1 \quad 6]$, $C = \begin{bmatrix} -3 \\ 5 \\ 6 \end{bmatrix}$, and $D = \begin{bmatrix} 4 \\ 7 \\ 2 \end{bmatrix}$. Find the following matrices.

a) $A - B$ **b)** $C - D$ **c)** $A - C$

Solution

a) To find $A - B$, subtract the corresponding entries of the matrices:

$$A - B = [3 \quad 5 \quad 8] - [3 \quad -1 \quad 6] = [0 \quad 6 \quad 2]$$

b) To find $C - D$, subtract the corresponding entries:

$$C - D = \begin{bmatrix} -3 \\ 5 \\ 6 \end{bmatrix} - \begin{bmatrix} 4 \\ 7 \\ 2 \end{bmatrix} = \begin{bmatrix} -7 \\ -2 \\ 4 \end{bmatrix}$$

c) Since A and C do not have the same order, $A - C$ is not defined. ◆

Scalar Multiplication

A matrix of order 1×1 is a matrix with only one entry. To distinguish a 1×1 matrix from a real number, a real number is called a **scalar** when we are dealing with matrices. We define multiplication of a matrix by a scalar as follows.

Definition: Scalar Multiplication

If A is an $m \times n$ matrix and b is a scalar, then the matrix bA is the $m \times n$ matrix obtained by multiplying each entry of A by the real number b.

Example 5 Scalar multiplication

Given that $A = [-2 \quad 3 \quad 5]$ and $B = \begin{bmatrix} -3 & 4 \\ 2 & -6 \end{bmatrix}$, find the following matrices.

a) $3A$

b) $-2B$

c) $-1A$

Solution

a) The matrix $3A$ is the 1×3 matrix formed by multiplying each entry of A by 3:

$$3A = [-6 \quad 9 \quad 15]$$

b) Multiply each entry of B by -2:

$$-2B = \begin{bmatrix} 6 & -8 \\ -4 & 12 \end{bmatrix}$$

c) Multiply each entry of A by -1:

$$-1A = [2 \quad -3 \quad -5]$$

The scalar product of A and -1 is the additive inverse of A, $-1A = -A$.

◆

```
[A]
       [[-2 3 5]]
3[A]
       [[-6 9 15]]
```

You can define the 1×3 matrix A and perform scalar multiplication on your calculator.

In Section 10.3 multiplication of matrices will be defined in a manner that is very different from scalar multiplication.

Applications

In Section 10.1 we saw how a matrix is used to represent a system of equations. Just the essential parts of the system, the coefficients, are listed in the matrix. Matrices are also very useful in two-way classifications of data. The matrix just contains the essentials, and we must remember what the entries represent. For example, an electronics manufacturer buys transistors and resistors from suppliers A and B. The following table shows the number of transistors and resistors supplied by each for the first week of January.

	A	B
Transistors	400	800
Resistors	600	500

This supply information can be listed in a 2×2 matrix S_1:

$$S_1 = \begin{bmatrix} 400 & 800 \\ 600 & 500 \end{bmatrix}$$

If the manufacturer wants to increase the orders by 10% for the second week, the supply matrix for the second week is $S_1 + 0.1S_1$ or $1.1S_1$, which is found by scalar multiplication:

$$S_2 = 1.1S_1 = (1.1)\begin{bmatrix} 400 & 800 \\ 600 & 500 \end{bmatrix} = \begin{bmatrix} 440 & 880 \\ 660 & 550 \end{bmatrix}$$

The matrix $T = S_1 + S_2$ gives the total supplies for the first two weeks.

$$T = S_1 + S_2 = \begin{bmatrix} 400 & 800 \\ 600 & 500 \end{bmatrix} + \begin{bmatrix} 440 & 880 \\ 660 & 550 \end{bmatrix} = \begin{bmatrix} 840 & 1680 \\ 1260 & 1050 \end{bmatrix}$$

This simple example illustrates the operations of addition and scalar multiplication, but you can imagine the same operations with matrices showing amounts of many different items ordered from many different suppliers. In applications involving large matrices, computers are used to store the matrices and perform the matrix operations.

? For Thought

True or false? Explain.

The following statements refer to the matrices

$$A = \begin{bmatrix} 1 \\ 3 \end{bmatrix}, \qquad B = \begin{bmatrix} 1 \\ 3 \end{bmatrix}, \qquad C = \begin{bmatrix} 1 & 1 \\ 3 & 3 \end{bmatrix},$$

$$D = \begin{bmatrix} -3 & 5 \\ 1 & -2 \end{bmatrix}, \qquad \text{and} \qquad E = \begin{bmatrix} -2 & 6 \\ 4 & 1 \end{bmatrix}.$$

1. $A = B$ **2.** $A = C$ **3.** $A + B = C$ **4.** $C + D = E$

5. $A - B = \begin{bmatrix} 0 \\ 0 \end{bmatrix}$ **6.** $3B = \begin{bmatrix} 3 \\ 3 \end{bmatrix}$ **7.** $-A = \begin{bmatrix} 3 \\ 1 \end{bmatrix}$

8. $A + C = \begin{bmatrix} 2 & 1 \\ 6 & 3 \end{bmatrix}$ **9.** $C - A = \begin{bmatrix} 1 & 0 \\ 3 & 0 \end{bmatrix}$

10. $C + 2D = \begin{bmatrix} -5 & 11 \\ 4 & 1 \end{bmatrix}$

10.2 Exercises   Tape 19   Disk—5.25": 6   3.5": 4   Macintosh: 4

Determine the values of x, y, and z that make each matrix equation true.

1. $\begin{bmatrix} x \\ 5 \end{bmatrix} = \begin{bmatrix} 2 \\ y \end{bmatrix}$ **2.** $\begin{bmatrix} 2z \\ 5x \end{bmatrix} = \begin{bmatrix} -1 \\ 2 \end{bmatrix}$

3. $\begin{bmatrix} 2x & 4y \\ 3z & 8 \end{bmatrix} = \begin{bmatrix} 6 & 10 \\ z+y & 8 \end{bmatrix}$

4. $\begin{bmatrix} -x & 2y \\ 3 & x+y \end{bmatrix} = \begin{bmatrix} 3 & -6 \\ 3 & 4z \end{bmatrix}$

Perform the following operations. If it is not possible to perform an operation, explain.

5. $\begin{bmatrix} 3 \\ 5 \end{bmatrix} + \begin{bmatrix} 2 \\ 1 \end{bmatrix}$ **6.** $\begin{bmatrix} -1 \\ 2 \end{bmatrix} + \begin{bmatrix} 0 \\ 3 \end{bmatrix}$

7. $\begin{bmatrix} 0.2 & 0.1 \\ 0.4 & 0.3 \end{bmatrix} + \begin{bmatrix} 0.2 & 0.05 \\ 0.3 & 0.8 \end{bmatrix}$

8. $\begin{bmatrix} \frac{1}{2} & \frac{1}{3} \\ \frac{1}{4} & 1 \end{bmatrix} - \begin{bmatrix} -\frac{1}{2} & \frac{1}{6} \\ 1 & \frac{1}{3} \end{bmatrix}$ **9.** $3\begin{bmatrix} \frac{1}{6} & \frac{1}{2} \\ 1 & -4 \end{bmatrix}$

10. $-2\begin{bmatrix} -1 & 4 \\ 3 & 1 \end{bmatrix}$

11. $\begin{bmatrix} -2 & 4 \\ 6 & 8 \end{bmatrix} - 4\begin{bmatrix} -3 & 1 \\ 2 & -2 \end{bmatrix}$

12. $2\begin{bmatrix} \frac{1}{4} \\ \frac{1}{3} \end{bmatrix} + 3\begin{bmatrix} \frac{1}{4} \\ \frac{1}{6} \end{bmatrix}$ **13.** $\begin{bmatrix} -1 & 3 \\ 2 & 5 \end{bmatrix} + 5\begin{bmatrix} -2 \\ 4 \end{bmatrix}$

14. $\begin{bmatrix} \sqrt{2} & 4 & \sqrt{12} \end{bmatrix} + \begin{bmatrix} \sqrt{8} & -2 & \sqrt{3} \end{bmatrix}$

15. $2\begin{bmatrix} -\sqrt{2} \\ \sqrt{5} \\ \sqrt{27} \end{bmatrix} - \begin{bmatrix} -\sqrt{8} \\ \sqrt{20} \\ -\sqrt{3} \end{bmatrix}$ **16.** $\begin{bmatrix} 1 & 3 & 7 \end{bmatrix} + \begin{bmatrix} -1 \\ 2 \\ 3 \end{bmatrix}$

17. $2\begin{bmatrix} a \\ b \end{bmatrix} + 3\begin{bmatrix} 2a \\ 4b \end{bmatrix} - 5\begin{bmatrix} -a \\ 3b \end{bmatrix}$

18. $2\begin{bmatrix} -a \\ b \\ c \end{bmatrix} - \begin{bmatrix} -3a \\ 4b \\ -c \end{bmatrix} - 6\begin{bmatrix} a \\ -b \\ 2c \end{bmatrix}$

19. $0.4\begin{bmatrix} -x & y \\ 2x & 8y \end{bmatrix} - 0.3\begin{bmatrix} 2x & 3y \\ 5x & -y \end{bmatrix}$

20. $a\begin{bmatrix} a & b \\ c & 2b \end{bmatrix} - b\begin{bmatrix} b & a \\ 0 & 3a \end{bmatrix}$

21. $\begin{bmatrix} -5 & 4 \\ -2 & 3 \\ 0 & 1 \end{bmatrix} - \begin{bmatrix} -4 & -9 \\ 7 & 0 \\ -6 & 3 \end{bmatrix}$

22. $\begin{bmatrix} 2 & 3 & 4 \\ 4 & 6 & 8 \\ 6 & 3 & 1 \end{bmatrix} + \begin{bmatrix} 1 & 0 & 0 \\ 0 & 1 & 0 \\ 0 & 0 & 1 \end{bmatrix}$

23. $2\begin{bmatrix} x & y & z \\ -x & 2y & 3z \\ x & -y & -3z \end{bmatrix} - \begin{bmatrix} -x & 0 & 3z \\ 4x & y & -z \\ 2x & 5y & z \end{bmatrix}$

24. $\dfrac{1}{2}\begin{bmatrix} 2 & -8 & -4 \\ 14 & 16 & 10 \\ 4 & -6 & 2 \end{bmatrix} + \dfrac{1}{3}\begin{bmatrix} 6 & -9 & 0 \\ 0 & 12 & -6 \\ 21 & -18 & -3 \end{bmatrix}$

Let $A = \begin{bmatrix} -4 & 1 \\ 3 & 0 \end{bmatrix}$, $B = \begin{bmatrix} -1 & -2 \\ 7 & 4 \end{bmatrix}$, $C = \begin{bmatrix} -3 & -4 \\ 2 & -5 \end{bmatrix}$,

$D = \begin{bmatrix} -4 \\ 5 \end{bmatrix}$, and $E = \begin{bmatrix} -1 \\ 2 \end{bmatrix}$. Find each of the following matrices, if possible.

25. $A + B$ **26.** $A - B$ **27.** $A - D$

28. $E - D$ **29.** $(A + B) + C$ **30.** $3A + 3C$

31. $A + (B + C)$ **32.** $3(A + C)$ **33.** $D + E$

34. $D + A$ **35.** $E + D$ **36.** $B + A$

37. $2A - B$ **38.** $-C - 3B$ **39.** $2D - 3E$

40. $4D + 0E$

Each of the following matrix equations corresponds to a system of linear equations. Write the system of equations and solve it by the method of your choice.

41. $\begin{bmatrix} x + y \\ x - y \end{bmatrix} = \begin{bmatrix} 5 \\ 1 \end{bmatrix}$ **42.** $\begin{bmatrix} x - y \\ 2x + y \end{bmatrix} = \begin{bmatrix} -1 \\ 4 \end{bmatrix}$

43. $\begin{bmatrix} 2x + 3y \\ x - 4y \end{bmatrix} = \begin{bmatrix} 7 \\ -13 \end{bmatrix}$ **44.** $\begin{bmatrix} x - 3y \\ 2x + y \end{bmatrix} = \begin{bmatrix} 1 \\ -5 \end{bmatrix}$

45. $\begin{bmatrix} x + y + z \\ x - y - z \\ x - y + z \end{bmatrix} = \begin{bmatrix} 8 \\ -7 \\ 2 \end{bmatrix}$

46. $\begin{bmatrix} 2x + y + z \\ x - 2y - z \\ x - y + z \end{bmatrix} = \begin{bmatrix} 7 \\ -6 \\ 2 \end{bmatrix}$

Solve each problem.

47. *Budgeting* In January, Terry spent $120 on food, $30 on clothing, and $40 on utilities. In February she spent $130 on food, $70 on clothing, and $50 on utilities. In March she spent $140 on food, $60 on clothing, and $45 on utilities. Write a 3×1 matrix for each month's expenditures and find the sum of the three matrices. What do the entries in the sum represent?

48. *Nutritional content* According to manufacturers' labels, one serving of Kix contains 110 calories, 2 g of protein, and 40 mg of potassium. One serving of Quick Oats contains 100 calories, 4 g of protein, and 100 mg of potassium. One serving of Muesli contains 120 calories, 3 g of protein, and 115 mg of potassium. Write 1×3 matrices K, Q, and M, which express the nutritional content of each cereal. Find $2K + 2Q + 3M$ and indicate what the entries represent.

49. *Inventories* The first week of July, farmer Thomas supplied Tiffany's Farm Market with 40 cantaloupes, 30 watermelons, and 80 tomatoes, and farmer Wilson supplied Tiffany's with 80 cantaloupes, 90 watermelons, and 200 tomatoes. Write this information as a 3×2 matrix. The next week Tiffany increased her orders in each category from each supplier by 50%. Write a 3×2 supply matrix for the second week of July.

50. *Recommended Daily Allowances* The percentages of the U.S. Recommended Daily Allowances for phosphorus, magnesium, and calcium for a 1-oz serving of Kix are 4, 2, and 4, respectively. The percentages of the U.S. Recommended Daily Allowances for phosphorus, magnesium, and calcium in 1/2 cup of milk are 11, 4, and 16, respectively. Write a 1×3 matrix K giving the percentages for 1 oz of Kix and a 1×3 matrix M giving the percentages for 1/2 cup of milk. Find the matrix $K + 2M$; indicate what its entries represent.

For Writing/Discussion

Let $A = \begin{bmatrix} a_{11} & a_{12} \\ a_{21} & a_{22} \end{bmatrix}$, $B = \begin{bmatrix} b_{11} & b_{12} \\ b_{21} & b_{22} \end{bmatrix}$, and $C = \begin{bmatrix} c_{11} & c_{12} \\ c_{21} & c_{22} \end{bmatrix}$ for the following problems.

51. Is $A + B = B + A$? Is addition of 2×2 matrices commutative? Explain.

52. Is addition of 3×3 matrices commutative? Explain.

53. Is $(A + B) + C = A + (B + C)$? Is addition of 2×2 matrices associative? Explain.

54. Is addition of 3×3 matrices associative? Explain.

55. Is $k(A + B) = kA + kB$ for any constant k? Is scalar multiplication distributive over addition of 2×2 matrices?

56. Is scalar multiplication distributive over addition of $n \times n$ matrices for each natural number n?

57. In the set of 2×2 matrices, which matrix is the additive identity?

58. Does every 2×2 matrix have an additive inverse with respect to the appropriate additive identity?

10.3

Multiplication of

Matrices

In Sections 10.1 and 10.2 we saw how matrices are used for solving equations and for representing two-way classifications of data. We saw how addition and scalar multiplication of matrices could be useful in applications. In this section you will learn to find the product of two matrices. Matrix multiplication is more complicated than addition or subtraction, but it is also very useful in applications.

An Application

Before presenting the general definition of multiplication, let us look at an example where multiplication of matrices is useful. Table 10.1 shows the number of economy, mid-size, and large cars rented by individuals and corporations at a rental agency in a single day. Table 10.2 shows the number of bonus points and free miles given in a promotional program for each of the three car types.

	Econo	Mid	Large
Individuals	3	2	6
Corporations	5	2	4

Table 10.1

	Bonus points	Free miles
Econo	20	50
Mid	30	100
Large	40	150

Table 10.2

The 1×3 row matrix [3 2 6] from Table 10.1 represents the number of economy, mid-size, and large cars rented by individuals. The 3×1 column matrix
$$\begin{bmatrix} 20 \\ 30 \\ 40 \end{bmatrix}$$
from Table 10.2 represents the bonus points given for each economy, mid-

size, and large car that is rented. The product of these two matrices is a 1×1 matrix whose entry is the sum of the products of the corresponding entries:

$$[3 \quad 2 \quad 6] \begin{bmatrix} 20 \\ 30 \\ 40 \end{bmatrix} = [3(20) + 2(30) + 6(40)] = [360]$$

The product of this row matrix and this column matrix gives the total number of bonus points given to individuals on the rental of the 11 cars.

Now write Table 10.1 as a 2×3 matrix giving the number of cars of each type rented by individuals and corporations and write Table 10.2 as a 3×2 matrix giving the number of bonus points and free miles for each type of car rented.

$$\text{Individuals} \to \begin{bmatrix} 3 & 2 & 6 \\ 5 & 2 & 4 \end{bmatrix} \quad \begin{array}{c} \text{Economy} \to \\ \text{Mid-size} \to \\ \text{Large} \to \end{array} \begin{bmatrix} 20 & 50 \\ 30 & 100 \\ 40 & 150 \end{bmatrix}$$

(Economy, Mid-size, Large columns for first matrix; Bonus points, Free miles columns for second matrix)

The product of these two matrices is a 2×2 matrix that gives the total bonus points and free miles both for individuals and corporations.

$$\begin{bmatrix} 3 & 2 & 6 \\ 5 & 2 & 4 \end{bmatrix} \begin{bmatrix} 20 & 50 \\ 30 & 100 \\ 40 & 150 \end{bmatrix} = \begin{bmatrix} 360 & 1250 \\ 320 & 1050 \end{bmatrix} \begin{array}{l} \leftarrow \text{Individuals} \\ \leftarrow \text{Corporations} \end{array}$$

(columns: Bonus points, Free miles)

The product of the 2×3 matrix and the 3×2 matrix is a 2×2 matrix. Take a careful look at where the entries in the 2×2 matrix come from:

$3(20) + 2(30) + 6(40) = 360$	**Total bonus points for individuals**
$3(50) + 2(100) + 6(150) = 1250$	**Total free miles for individuals**
$5(20) + 2(30) + 4(40) = 320$	**Total bonus points for corporations**
$5(50) + 2(100) + 4(150) = 1050$	**Total free miles for corporations**

Each entry in the 2×2 matrix is found by multiplying the entries of a *row* of the first matrix by the corresponding entries of a *column* of the second matrix and adding the results. To multiply any matrices, the number of entries in a row of the first matrix must equal the number of entries in a column of the second matrix.

Matrix Multiplication

The product of two matrices is illustrated by the previous example of the rental cars. The general definition of matrix multiplication follows.

Definition: Matrix Multiplication

The product of an $m \times n$ matrix A and an $n \times p$ matrix B is an $m \times p$ matrix AB whose entries are found as follows. The entry in the ith row and jth column of AB is found by multiplying each entry in the ith row of A by the corresponding entry in the jth column of B and adding the results.

Note that by the definition we can multiply only an $m \times n$ matrix and an $n \times p$ matrix to get an $m \times p$ matrix. To find a product AB, *each row of A must have the same number of entries as each column of B*. The entry c_{ij} in AB comes from the ith row of A and the jth column of B as shown below.

ith row of A jth column of B ijth entry of AB

$$\begin{bmatrix} * & * & * \\ & & \end{bmatrix} \begin{bmatrix} * \\ * \\ * \end{bmatrix} = \begin{bmatrix} & \\ & c_{ij} \end{bmatrix}$$

Example 1   Multiplying matrices

Find the following products.

a) $\begin{bmatrix} 2 & 3 \end{bmatrix} \begin{bmatrix} -3 \\ 4 \end{bmatrix}$ b) $\begin{bmatrix} 1 & 2 \\ 3 & 4 \end{bmatrix} \begin{bmatrix} -2 \\ 5 \end{bmatrix}$ c) $\begin{bmatrix} 1 & 3 \\ 5 & 7 \end{bmatrix} \begin{bmatrix} 2 & 4 \\ 6 & 8 \end{bmatrix}$

Solution

a) The product of a 1×2 matrix and a 2×1 matrix is a 1×1 matrix. The only entry in the product is found by multiplying the first row of the first matrix by the first column of the second matrix:

$$\begin{bmatrix} 2 & 3 \end{bmatrix} \begin{bmatrix} -3 \\ 4 \end{bmatrix} = [2(-3) + 3(4)] = [6]$$

b) The product of a 2×2 matrix and a 2×1 matrix is a 2×1 matrix. Multiply the corresponding entries in each row of the first matrix and the only column of the second matrix:

$$\begin{bmatrix} 1 & 2 \\ 3 & 4 \end{bmatrix} \begin{bmatrix} -2 \\ 5 \end{bmatrix} = \begin{bmatrix} 8 \\ 14 \end{bmatrix} \qquad \begin{array}{l} 1(-2) + 2(5) = 8 \\ 3(-2) + 4(5) = 14 \end{array}$$

c) The product of a 2×2 matrix and a 2×2 matrix is a 2×2 matrix. Multiply the corresponding entries in each row of the first matrix and each column of the second matrix.

$$\begin{bmatrix} 1 & 3 \\ 5 & 7 \end{bmatrix} \begin{bmatrix} 2 & 4 \\ 6 & 8 \end{bmatrix} = \begin{bmatrix} 20 & 28 \\ 52 & 76 \end{bmatrix} \qquad \begin{array}{l} 1 \cdot 2 + 3 \cdot 6 = 20 \\ 1 \cdot 4 + 3 \cdot 8 = 28 \\ 5 \cdot 2 + 7 \cdot 6 = 52 \\ 5 \cdot 4 + 7 \cdot 8 = 76 \end{array}$$

Example 2   Multiplying matrices

Find AB and BA in each case.

a) $A = \begin{bmatrix} 1 & 3 \\ 5 & 7 \\ 8 & 2 \end{bmatrix}$, $B = \begin{bmatrix} 2 & 4 & -1 \\ 6 & -3 & 2 \end{bmatrix}$

b) $A = \begin{bmatrix} 1 & 3 & 4 \\ 2 & 5 & 6 \\ 7 & 8 & 9 \end{bmatrix}$, $B = \begin{bmatrix} 1 & 0 & 1 \\ 0 & 1 & 0 \\ 0 & 1 & 1 \end{bmatrix}$

Solution

a) The product of 3×2 matrix A and 2×3 matrix B is the 3×3 matrix AB. The first row of AB is found by multiplying the corresponding entries in the first row of A and each column of B:

$$1 \cdot 2 + 3 \cdot 6 = 20, \quad 1 \cdot 4 + 3(-3) = -5, \quad \text{and} \quad 1(-1) + 3 \cdot 2 = 5$$

So 20, -5, and 5 form the first row of AB. The second row of AB is formed from multiplying the second row of A and each column from B:

$$5 \cdot 2 + 7 \cdot 6 = 52, \quad 5 \cdot 4 + 7(-3) = -1, \quad \text{and} \quad 5(-1) + 7 \cdot 2 = 9$$

So 52, -1, and 9 form the second row of AB. The third row of AB is formed by multiplying corresponding entries in the third row of A and each column of B.

$$AB = \begin{bmatrix} 1 & 3 \\ 5 & 7 \\ 8 & 2 \end{bmatrix} \begin{bmatrix} 2 & 4 & -1 \\ 6 & -3 & 2 \end{bmatrix} = \begin{bmatrix} 20 & -5 & 5 \\ 52 & -1 & 9 \\ 28 & 26 & -4 \end{bmatrix}$$

The product of 2×3 matrix B and 3×2 matrix A is the 2×2 matrix BA:

$$BA = \begin{bmatrix} 2 & 4 & -1 \\ 6 & -3 & 2 \end{bmatrix} \begin{bmatrix} 1 & 3 \\ 5 & 7 \\ 8 & 2 \end{bmatrix}$$

$$= \begin{bmatrix} 14 & 32 \\ 7 & 1 \end{bmatrix} \quad \begin{aligned} 2(1) + 4(5) + (-1)(8) &= 14 \\ 2(3) + 4(7) + (-1)(2) &= 32 \\ 6(1) + (-3)(5) + 2(8) &= 7 \\ 6(3) + (-3)(7) + 2(2) &= 1 \end{aligned}$$

b) The product of 3×3 matrix A and 3×3 matrix B is the 3×3 matrix AB:

$$AB = \begin{bmatrix} 1 & 3 & 4 \\ 2 & 5 & 6 \\ 7 & 8 & 9 \end{bmatrix} \begin{bmatrix} 1 & 0 & 1 \\ 0 & 1 & 0 \\ 0 & 1 & 1 \end{bmatrix} = \begin{bmatrix} 1 & 7 & 5 \\ 2 & 11 & 8 \\ 7 & 17 & 16 \end{bmatrix}$$

The product of 3×3 matrix B and 3×3 matrix A is the 3×3 matrix BA:

$$BA = \begin{bmatrix} 1 & 0 & 1 \\ 0 & 1 & 0 \\ 0 & 1 & 1 \end{bmatrix} \begin{bmatrix} 1 & 3 & 4 \\ 2 & 5 & 6 \\ 7 & 8 & 9 \end{bmatrix} = \begin{bmatrix} 8 & 11 & 13 \\ 2 & 5 & 6 \\ 9 & 13 & 15 \end{bmatrix}$$

◆

```
[A][B]
  [[20 -5  5 ]
   [52 -1  9 ]
   [28 26 -4]]
[B][A]
     [[14 32]
      [7   1 ]]
```

Once the matrices A and B are defined as given in Example 2(a), the products AB and BA are easily found with a calculator.

Operations with matrices have some properties similar to the properties of operations with real numbers. For example, multiplication of matrices is associative, multiplicative inverses exist for some matrices, and there is a multiplicative identity. However, multiplication of matrices is generally not commutative. Example 2 shows matrices A and B where $AB \neq BA$. In Example 2(a), AB and BA are not even the same order. It can also happen that AB is defined but BA is undefined because of the orders of A and B. We will not study all of the properties of the operations with matrices in detail in this text, but some properties for 2×2 matrices are discussed in the exercises.

Matrix Equations

Recall that two matrices are equal provided they are the same order and their corresponding entries are equal. This definition is used to solve matrix equations.

Example 3   Solving a matrix equation

Find the values of x and y that satisfy the matrix equation

$$\begin{bmatrix} 3 & 2 \\ 4 & -1 \end{bmatrix} \begin{bmatrix} x \\ y \end{bmatrix} = \begin{bmatrix} 1 \\ -6 \end{bmatrix}.$$

Solution

Multiply the matrices on the left-hand side to get the matrix equation

$$\begin{bmatrix} 3x + 2y \\ 4x - y \end{bmatrix} = \begin{bmatrix} 1 \\ -6 \end{bmatrix}.$$

Since matrices of the same order are equal only when all of their corresponding entries are equal, we have the following system of equations:

$$(1) \qquad 3x + 2y = 1$$
$$(2) \qquad 4x - y = -6$$

Solve by eliminating y:

$$\begin{array}{ll} 3x + 2y = 1 & \\ \underline{8x - 2y = -12} & \text{Eq. (2) multiplied by 2} \\ 11x \qquad = -11 & \\ x = -1 & \end{array}$$

Now use $x = -1$ in Eq. (1):

$$3(-1) + 2y = 1$$
$$2y = 4$$
$$y = 2$$

You should check that the matrix equation is satisfied if $x = -1$ and $y = 2$. ◆

In Example 3, we rewrote a matrix equation as a system of equations. We will now write a system of equations as a matrix equation. In fact, any system of linear equations can be written as a matrix equation in the form $AX = B$, where A is a matrix of coefficients, X is a column matrix of variables, and B is a column matrix of constants. In the next example we write a system of two equations as a matrix equation. In Section 10.4 we will solve systems of equations by using matrix operations on the corresponding matrix equations.

Example 4   Writing a matrix equation

Write the following system as an equivalent matrix equation in the form $AX = B$.

$$2x + y = 5$$
$$x - y = 4$$

Solution

Let $A = \begin{bmatrix} 2 & 1 \\ 1 & -1 \end{bmatrix}$, $X = \begin{bmatrix} x \\ y \end{bmatrix}$, and $B = \begin{bmatrix} 5 \\ 4 \end{bmatrix}$. The system of equations is equivalent to the matrix equation

$$\begin{bmatrix} 2 & 1 \\ 1 & -1 \end{bmatrix} \begin{bmatrix} x \\ y \end{bmatrix} = \begin{bmatrix} 5 \\ 4 \end{bmatrix},$$

which is of the form $AX = B$. Check by multiplying the two matrices on the left-hand side to get

$$\begin{bmatrix} 2x + y \\ x - y \end{bmatrix} = \begin{bmatrix} 5 \\ 4 \end{bmatrix}.$$

By the definition of equal matrices, this matrix equation is correct provided that $2x + y = 5$ and $x - y = 4$, which is the original system. ◆

? For Thought

True or false? Explain.

The following statements refer to the matrices

$$A = \begin{bmatrix} 1 \\ 6 \end{bmatrix}, \quad B = [7 \ 9], \quad C = \begin{bmatrix} 2 & 3 \\ 4 & 5 \end{bmatrix}, \quad D = \begin{bmatrix} 1 & 0 \\ 0 & 1 \end{bmatrix}, \quad \text{and} \quad E = \begin{bmatrix} 2 & -1 \\ 0 & 3 \end{bmatrix}.$$

1. The order of AB is 2×2.
2. The order of BA is 1×1.
3. The order of AC is 1×2.
4. The order of CA is 2×2.
5. $DC = C$ and $CD = C$.
6. $BC = [50 \quad 66]$
7. $AB = \begin{bmatrix} 7 & 9 \\ 42 & 54 \end{bmatrix}$
8. $CE = \begin{bmatrix} 4 & 7 \\ 8 & 11 \end{bmatrix}$
9. $BA = [61]$
10. $CE = EC$

10.3 Exercises

Tape 19   Disk—5.25″: 6   3.5″: 4   Macintosh: 4

Find the order of AB in each case if the matrices can be multiplied.

1. A has order 3×2, B has order 2×5

2. A has order 3×1, B has order 1×3

3. A has order 1×4, B has order 4×1

4. A has order 4×2, B has order 2×5

5. A has order 5×1, B has order 1×5

6. A has order 2×1, B has order 1×6

7. A has order 3×3, B has order 3×3

8. A has order 4×4, B has order 4×1

9. A has order 3×4, B has order 3×4

10. A has order 4×2, B has order 3×4

Let $A = \begin{bmatrix} 2 \\ -3 \\ 1 \end{bmatrix}$, $B = [2 \quad 3 \quad 4]$, $C = \begin{bmatrix} 2 & 3 \\ 4 & 5 \\ 1 & 0 \end{bmatrix}$,

$D = \begin{bmatrix} 2 & -1 & 1 \\ 0 & 3 & 2 \end{bmatrix}$, and $E = \begin{bmatrix} 1 & 1 & 1 \\ 0 & 1 & 1 \\ 0 & 0 & 1 \end{bmatrix}$.

Find each product if possible.

11. AB **12.** BC **13.** BE **14.** BA

15. CE **16.** DE **17.** EC **18.** BD

19. DC **20.** CD **21.** ED **22.** EE

23. EA **24.** DA **25.** AE **26.** EB

Find each product if possible.

27. $\begin{bmatrix} 2 & 0 \\ 3 & 1 \end{bmatrix}\begin{bmatrix} 1 & 1 \\ 0 & 1 \end{bmatrix}$

28. $\begin{bmatrix} -1 & 2 \\ 3 & 4 \end{bmatrix}\begin{bmatrix} 2 & -1 \\ 5 & 2 \end{bmatrix}$

29. $\begin{bmatrix} 7 & 4 \\ 5 & 3 \end{bmatrix}\begin{bmatrix} 3 & -4 \\ -5 & 7 \end{bmatrix}$

30. $\begin{bmatrix} -2 & -3 \\ 5 & 8 \end{bmatrix}\begin{bmatrix} -8 & -3 \\ 5 & 2 \end{bmatrix}$

31. $\begin{bmatrix} -0.5 & 4 \\ 9 & 0.7 \end{bmatrix}\begin{bmatrix} 1 & 0 \\ 0 & 1 \end{bmatrix}$

32. $\begin{bmatrix} 1 & 0 \\ 0 & 1 \end{bmatrix}\begin{bmatrix} -0.7 & 1.2 \\ 3 & 1.1 \end{bmatrix}$

33. $[-2 \quad 3]\begin{bmatrix} a & 3b \\ 2a & b \end{bmatrix}$

34. $\begin{bmatrix} -2 & 5 \\ 6 & 4 \end{bmatrix}\begin{bmatrix} x \\ y \end{bmatrix}$

35. $\begin{bmatrix} a & 0 \\ 0 & b \end{bmatrix}\begin{bmatrix} 2 & 5 & 3 \\ 1 & 4 & 6 \end{bmatrix}$

36. $\begin{bmatrix} 1 & 1 \\ 1 & -1 \end{bmatrix}\begin{bmatrix} x \\ y \end{bmatrix}$

37. $[1 \quad 2 \quad 3]\begin{bmatrix} 1 & 0 & 1 \\ 0 & 1 & 1 \\ 1 & 0 & 1 \end{bmatrix}$

38. $\begin{bmatrix} 1 & 1 & 0 \\ 1 & 0 & 1 \\ 0 & 1 & 1 \end{bmatrix}\begin{bmatrix} -2 \\ 3 \\ 5 \end{bmatrix}$

39. $[-1 \quad 0 \quad 3]\begin{bmatrix} -5 \\ 1 \\ 4 \end{bmatrix}$

40. $\begin{bmatrix} -5 \\ 1 \\ 4 \end{bmatrix}[-1 \quad 0 \quad 3]$

41. $\begin{bmatrix} x \\ y \end{bmatrix}[x \quad y]$

42. $[x \quad y]\begin{bmatrix} x \\ y \end{bmatrix}$

43. $\begin{bmatrix} -1 & 2 & 3 \\ 3 & 4 & 4 \end{bmatrix}\begin{bmatrix} \sqrt{2} \\ 0 \\ \sqrt{2} \end{bmatrix}$

44. $[2 \quad 3]\begin{bmatrix} 0 & \sqrt{2} & 5 \\ -1 & \sqrt{8} & 0 \end{bmatrix}$

45. $\begin{bmatrix} \frac{1}{2} & \frac{1}{3} \\ \frac{1}{4} & \frac{1}{5} \end{bmatrix}\begin{bmatrix} -8 & 12 \\ -5 & 15 \end{bmatrix}$

46. $\begin{bmatrix} \frac{1}{4} & \frac{1}{2} \\ \frac{1}{8} & -\frac{1}{2} \end{bmatrix}\begin{bmatrix} -\frac{1}{2} & \frac{1}{4} \\ \frac{1}{2} & \frac{1}{4} \end{bmatrix}$

47. $[3 \quad 0 \quad 3][2 \quad 4 \quad 6]$

48. $\begin{bmatrix} 2 \\ 5 \end{bmatrix}\begin{bmatrix} 7 & 2 \\ 3 & 1 \end{bmatrix}$

49. $\begin{bmatrix} 1 & 0 & -1 \\ 0 & 1 & 0 \\ 1 & 1 & 1 \end{bmatrix}\begin{bmatrix} 9 & 8 & 10 \\ 3 & 5 & 2 \\ 7 & 8 & 4 \end{bmatrix}$

50. $\begin{bmatrix} 1 & 1 & 1 \\ 0 & 1 & 1 \\ 0 & 0 & 1 \end{bmatrix}\begin{bmatrix} -2 & 3 & -4 \\ 2 & 5 & 7 \\ -3 & 0 & -6 \end{bmatrix}$

51. $\begin{bmatrix} 5 & -3 & -2 \\ 4 & 2 & 6 \\ 2 & 3 & -8 \end{bmatrix}\begin{bmatrix} 0.2 & 0.3 \\ 0.2 & -0.4 \\ 0.3 & 0.5 \end{bmatrix}$

52. $\begin{bmatrix} 0.2 & 0.1 & 0.7 \\ 0.3 & 0.3 & 0.4 \end{bmatrix}\begin{bmatrix} 20 & 30 & 40 \\ 10 & 20 & 50 \\ 60 & 50 & 40 \end{bmatrix}$

Write each matrix equation as a system of equations and solve the system by the method of your choice.

53. $\begin{bmatrix} 2 & -3 \\ 1 & 2 \end{bmatrix}\begin{bmatrix} x \\ y \end{bmatrix} = \begin{bmatrix} 0 \\ 7 \end{bmatrix}$

54. $\begin{bmatrix} 1 & 5 \\ -2 & 4 \end{bmatrix}\begin{bmatrix} x \\ y \end{bmatrix} = \begin{bmatrix} 2 \\ 10 \end{bmatrix}$

55. $\begin{bmatrix} 2 & 3 \\ 4 & 6 \end{bmatrix}\begin{bmatrix} x \\ y \end{bmatrix} = \begin{bmatrix} 5 \\ 9 \end{bmatrix}$

56. $\begin{bmatrix} 1 & -3 \\ -2 & 6 \end{bmatrix}\begin{bmatrix} x \\ y \end{bmatrix} = \begin{bmatrix} 1 \\ -2 \end{bmatrix}$

57. $\begin{bmatrix} 1 & 1 & 1 \\ 0 & 1 & 1 \\ 0 & 0 & 1 \end{bmatrix}\begin{bmatrix} x \\ y \\ z \end{bmatrix} = \begin{bmatrix} 4 \\ 5 \\ 6 \end{bmatrix}$

58. $\begin{bmatrix} 2 & 3 & 1 \\ 0 & 1 & 4 \\ 0 & 0 & 2 \end{bmatrix}\begin{bmatrix} x \\ y \\ z \end{bmatrix} = \begin{bmatrix} 0 \\ 3 \\ 6 \end{bmatrix}$

Write a matrix equation of the form $AX = B$ that corresponds to each system of equations.

59. $2x + 3y = 9$
$4x - y = 6$

60. $x - y = -7$
$x + 2y = 8$

61. $x + 2y - z = 3$
$3x - y + 3z = 1$
$2x + y - 4z = 0$

62. $x + y + z = 1$
$2x + y - z = 4$
$x - y - 3z = 2$

Solve each problem.

63. *Building costs* A contractor builds two types of houses. The costs for labor and materials for the economy model and the deluxe model are shown in the table. Write the information in the table as a matrix A. Suppose that the contractor built four economy models and seven deluxe models. Write a matrix Q of the appropriate order containing the quantity of each type. Find the product matrix AQ. What do the entries of AQ represent?

	Economy	**Deluxe**
Labor	$24,000	$40,000
Materials	$38,000	$70,000

Table for Exercise 63

64. *Nutritional information* According to the manufacturers, the breakfast cereals Almond Delight and Basic 4 contain the numbers of grams of protein, carbohydrates, and fat per serving listed in the table. Write the information in the table as a matrix A. In one week Julia ate four servings of Almond Delight and three servings of Basic 4. Write a matrix Q of the appropriate order expressing the quantity of each type that Julia ate. Find the product AQ. What do the entries of AQ represent?

	Almond Delight	**Basic 4**
Protein	2	3
Carbohydrates	23	28
Fat	2	2

Table for Exercise 64

65. *Ranking soccer teams* The table shown here gives the records of all four teams in a soccer league. An entry of 1 indicates that the row team has defeated the column team. (There are no ties.) The teams are ranked, not by their percentage of victories, but by the number of points received under a ranking scheme that gives a team credit for the quality of the teams it defeats. Since team A defeated B and C, A gets two points. Since B defeated C and D, and C defeated D, A gets 3 more secondary points for a total of 5 points. Since D's only victory is over A, D gets 1 point for that victory plus 2 secondary points for A's defeats of B and C, giving D a total of 3 points. Write the table as a 4×4 matrix M and find M^2. Explain what the entries of M^2 represent.

	A	B	C	D
A	0	1	1	0
B	0	0	1	1
C	0	0	0	1
D	1	0	0	0

Table for Exercise 65

66. *Ranking soccer teams* Let M be the 4×4 matrix from Exercise 65 and let T be a 4×1 matrix with a 1 in every entry. Find $(M + M^2)T$ and explain what its entries represent. Explain how it is possible for one team to have a better win-loss record than another, but end up ranked lower than the other because of this point scheme. (This system, known as the Harbin Football-Team Rating System, is used in Ohio to determine which high school teams are eligible to participate in playoffs.)

For Writing/Discussion

Let $A = \begin{bmatrix} a_{11} & a_{12} \\ a_{21} & a_{22} \end{bmatrix}$, $B = \begin{bmatrix} b_{11} & b_{12} \\ b_{21} & b_{22} \end{bmatrix}$, and $C = \begin{bmatrix} c_{11} & c_{12} \\ c_{21} & c_{22} \end{bmatrix}$.
Determine whether each of the following statements is true, and explain your answer.

67. $AB = BA$ (commutative)

68. $(AB)C = A(BC)$ (associative)

69. For any real number k, $k(A + B) = kA + kB$.

70. $A(B + C) = AB + AC$ (distributive)

71. For any real numbers s and t, $sA + tA = (s + t)A$.

72. Multiplication of 1×1 matrices is commutative.

Graphing Calculator Exercises

1. Make up a win-loss table like the table for Exercise 65 for a six-team soccer league. Assume that there are no ties. Use a graphing calculator or other device that can do matrix operations to find $(M + M^2)T$. Compare the percentage of games won by each team with its ranking by this scheme.

2. Find $(2M + M^2)T$, using M from the previous exercise and explain what its entries mean.

3. Compare the ranking of the six teams, using $(2M + M^2)T$ and $(M + M^2)T$. Is it possible that $(2M + M^2)T$ could change the order of the teams?

10.4

Inverses of Matrices

In previous sections we learned to add, subtract, and multiply matrices. There is no definition for division of matrices. However, there is an identity matrix for multiplication that behaves like the multiplicative identity 1 in the real number system. (The number 1 is called the multiplicative identity because $1 \cdot a = a$ and $a \cdot 1 = a$ for any real number a.) In this section we will see that for certain matrices there are inverse matrices such that the product of a matrix and its inverse matrix is the identity matrix.

The Identity Matrix

Consider the product of a 2×2 matrix $A = \begin{bmatrix} a_{11} & a_{12} \\ a_{21} & a_{22} \end{bmatrix}$ and $I = \begin{bmatrix} 1 & 0 \\ 0 & 1 \end{bmatrix}$:

$$\begin{bmatrix} a_{11} & a_{12} \\ a_{21} & a_{22} \end{bmatrix} \begin{bmatrix} 1 & 0 \\ 0 & 1 \end{bmatrix} = \begin{bmatrix} a_{11} & a_{12} \\ a_{21} & a_{22} \end{bmatrix} \quad \text{and} \quad \begin{bmatrix} 1 & 0 \\ 0 & 1 \end{bmatrix} \begin{bmatrix} a_{11} & a_{12} \\ a_{21} & a_{22} \end{bmatrix} = \begin{bmatrix} a_{11} & a_{12} \\ a_{21} & a_{22} \end{bmatrix}$$

Since $AI = A$ and $IA = A$, the matrix I is called the 2×2 **identity matrix.** The matrix I has 1's on its diagonal and 0's elsewhere. The 2×2 identity matrix is not an identity matrix for 3×3 matrices, but a 3×3 matrix with 1's on the diagonal and 0's elsewhere is the identity matrix for 3×3 matrices.

Definition: Identity Matrix

For each positive integer n, the $n \times n$ **identity matrix** I is an $n \times n$ matrix with 1's on the diagonal and 0's elsewhere. In symbols,

$$I = \begin{bmatrix} 1 & 0 & 0 & \dots & 0 \\ 0 & 1 & 0 & \dots & 0 \\ 0 & 0 & 1 & \dots & 0 \\ \vdots & \vdots & \vdots & & \vdots \\ 0 & 0 & 0 & \dots & 1 \end{bmatrix}.$$

We use the letter I for the identity matrix for any order, but the order of I should be clear from the context.

Example 1   Using an identity matrix

Find a matrix I such that $BI = B$ and $IB = B$ for

$$B = \begin{bmatrix} 2 & 3 & 5 \\ 1 & 0 & 4 \\ 5 & 7 & 2 \end{bmatrix}.$$

Solution

The matrix I must be the 3×3 identity matrix

$$I = \begin{bmatrix} 1 & 0 & 0 \\ 0 & 1 & 0 \\ 0 & 0 & 1 \end{bmatrix}.$$

Check that $BI = B$:

$$\begin{bmatrix} 2 & 3 & 5 \\ 1 & 0 & 4 \\ 5 & 7 & 2 \end{bmatrix} \begin{bmatrix} 1 & 0 & 0 \\ 0 & 1 & 0 \\ 0 & 0 & 1 \end{bmatrix} = \begin{bmatrix} 2 & 3 & 5 \\ 1 & 0 & 4 \\ 5 & 7 & 2 \end{bmatrix} \qquad \begin{array}{l} 2 \cdot 1 + 3 \cdot 0 + 5 \cdot 0 = 2 \\ 2 \cdot 0 + 3 \cdot 1 + 5 \cdot 0 = 3 \\ 2 \cdot 0 + 3 \cdot 0 + 5 \cdot 1 = 5 \end{array}$$

The computations at the right show that 2, 3, and 5 form the first row of BI. The second and third rows are found similarly. Now check that $IB = B$:

$$\begin{bmatrix} 1 & 0 & 0 \\ 0 & 1 & 0 \\ 0 & 0 & 1 \end{bmatrix} \begin{bmatrix} 2 & 3 & 5 \\ 1 & 0 & 4 \\ 5 & 7 & 2 \end{bmatrix} = \begin{bmatrix} 2 & 3 & 5 \\ 1 & 0 & 4 \\ 5 & 7 & 2 \end{bmatrix} \qquad \blacklozenge$$

The Inverse of a Matrix

Every nonzero real number a has a multiplicative inverse $1/a$ such that $a \cdot (1/a) = 1$. The product of a real number and its multiplicative inverse is the multiplicative identity. The situation is similar for matrices.

Definition: Inverse of a Matrix

> The **inverse** of an $n \times n$ matrix A is an $n \times n$ matrix A^{-1} (if it exists) such that $AA^{-1} = I$ and $A^{-1}A = I$. (Read A^{-1} as "A inverse.")

If A has an inverse, then A is **invertible.** Before we learn how to find the inverse of a matrix, we use the definition to determine whether two given matrices are inverses.

Example 2   Using the definition of inverse matrices

Determine whether $A = \begin{bmatrix} 3 & 4 \\ 5 & 7 \end{bmatrix}$ and $B = \begin{bmatrix} 7 & -4 \\ -5 & 3 \end{bmatrix}$ are inverses of each other.

Solution

Find the products AB and BA:

$$AB = \begin{bmatrix} 3 & 4 \\ 5 & 7 \end{bmatrix} \begin{bmatrix} 7 & -4 \\ -5 & 3 \end{bmatrix} = \begin{bmatrix} 1 & 0 \\ 0 & 1 \end{bmatrix} \qquad BA = \begin{bmatrix} 7 & -4 \\ -5 & 3 \end{bmatrix} \begin{bmatrix} 3 & 4 \\ 5 & 7 \end{bmatrix} = \begin{bmatrix} 1 & 0 \\ 0 & 1 \end{bmatrix}$$

A and B are inverses because $AB = BA = I$, where I is the 2×2 identity matrix. $\quad\blacklozenge$

If we are given the matrix

$$A = \begin{bmatrix} 3 & 4 \\ 5 & 7 \end{bmatrix}$$

from Example 2, how do we find its inverse if it is not already known? According to the definition, A^{-1} is a 2×2 matrix such that $AA^{-1} = I$ and $A^{-1}A = I$. So if

$$A^{-1} = \begin{bmatrix} x & y \\ z & w \end{bmatrix},$$

then

$$\begin{bmatrix} 3 & 4 \\ 5 & 7 \end{bmatrix} \begin{bmatrix} x & y \\ z & w \end{bmatrix} = \begin{bmatrix} 1 & 0 \\ 0 & 1 \end{bmatrix} \qquad \text{and} \qquad \begin{bmatrix} x & y \\ z & w \end{bmatrix} \begin{bmatrix} 3 & 4 \\ 5 & 7 \end{bmatrix} = \begin{bmatrix} 1 & 0 \\ 0 & 1 \end{bmatrix}.$$

To find A^{-1} we solve these matrix equations. If A is invertible, both equations will have the same solution. We will work with the first one. Multiply the two matrices on the left-hand side of the first equation to get the following equation:

$$\begin{bmatrix} 3x + 4z & 3y + 4w \\ 5x + 7z & 5y + 7w \end{bmatrix} = \begin{bmatrix} 1 & 0 \\ 0 & 1 \end{bmatrix}$$

Equate the corresponding terms from these equal matrices to get the following two systems:

$$\begin{aligned} 3x + 4z &= 1 & 3y + 4w &= 0 \\ 5x + 7z &= 0 & 5y + 7w &= 1 \end{aligned}$$

We can solve these two systems by using the Gauss-Jordan method from Section 10.1. The augmented matrices for these systems are

$$\begin{bmatrix} 3 & 4 & | & 1 \\ 5 & 7 & | & 0 \end{bmatrix} \qquad \text{and} \qquad \begin{bmatrix} 3 & 4 & | & 0 \\ 5 & 7 & | & 1 \end{bmatrix}.$$

Note that the two augmented matrices have the same coefficient matrix. Since we would use the same row operations on each of them, we can solve the systems simultaneously by combining the two systems into one augmented matrix denoted $[A\,|\,I]$:

$$[A\,|\,I] = \begin{bmatrix} 3 & 4 & | & 1 & 0 \\ 5 & 7 & | & 0 & 1 \end{bmatrix}$$

So the problem of finding A^{-1} is equivalent to the problem of solving two systems by the Gauss-Jordan method. A is invertible if and only if these systems have a solution. Multiply the first row of $[A|I]$ by $\frac{1}{3}$ to get a 1 in the first row, first column:

$$\begin{bmatrix} 1 & \frac{4}{3} & \frac{1}{3} & 0 \\ 5 & 7 & 0 & 1 \end{bmatrix} \qquad \frac{1}{3}R_1 \rightarrow R_1$$

Now multiply row 1 by -5 and add the result to row 2:

$$\begin{bmatrix} 1 & \frac{4}{3} & \frac{1}{3} & 0 \\ 0 & \frac{1}{3} & -\frac{5}{3} & 1 \end{bmatrix} \qquad -5R_1 + R_2 \rightarrow R_2$$

Multiply row 2 by 3:

$$\begin{bmatrix} 1 & \frac{4}{3} & \frac{1}{3} & 0 \\ 0 & 1 & -5 & 3 \end{bmatrix} \qquad 3R_2 \rightarrow R_2$$

Multiply row 2 by $-\frac{4}{3}$ and add the result to row 1:

$$\begin{bmatrix} 1 & 0 & 7 & -4 \\ 0 & 1 & -5 & 3 \end{bmatrix} \qquad -\frac{4}{3}R_2 + R_1 \rightarrow R_1$$

The numbers in the first column to the right of the bar give the values of x and z, while the numbers in the second column give the values of y and w. So $x = 7$, $y = -4$, $z = -5$, and $w = 3$ give the solutions to the two systems, and

$$A^{-1} = \begin{bmatrix} 7 & -4 \\ -5 & 3 \end{bmatrix}.$$

Since A^{-1} is the same matrix that was called B in Example 2, we can be sure that $AA^{-1} = I$ and $A^{-1}A = I$. Note that the matrix A^{-1} actually appeared on the right-hand side of the final augmented matrix, while the 2×2 identity matrix I appeared on the left-hand side. So A^{-1} is found by simply using row operations to convert the matrix $[A|I]$ into the matrix $[I|A^{-1}]$.

The essential steps for finding the inverse of a matrix are listed as follows.

▼ **Procedure: Finding A^{-1}**

Use the following steps to find the inverse of a square matrix A.

1. **Write the augmented matrix $[A|I]$, where I is the identity matrix of the same order as A.**

2. **Use row operations (the Gauss-Jordan method) to convert the left-hand side of the augmented matrix into I.**

3. **If the left-hand side can be converted to I, then $[A|I]$ becomes $[I|A^{-1}]$, and A^{-1} appears on the right-hand side of the augmented matrix.**

4. **If the left-hand side cannot be converted to I, then A is not invertible.**

Example 3 Finding the inverse of a matrix

Find the inverse of the matrix

$$A = \begin{bmatrix} 2 & -3 \\ 1 & 1 \end{bmatrix}.$$

Solution

Write the augmented matrix $[A|I]$:

$$\begin{bmatrix} 2 & -3 & | & 1 & 0 \\ 1 & 1 & | & 0 & 1 \end{bmatrix}$$

Use row operations to convert $[A|I]$ into $[I|A^{-1}]$:

$$\begin{bmatrix} 1 & 1 & | & 0 & 1 \\ 2 & -3 & | & 1 & 0 \end{bmatrix} \qquad R_1 \leftrightarrow R_2$$

$$\begin{bmatrix} 1 & 1 & | & 0 & 1 \\ 0 & -5 & | & 1 & -2 \end{bmatrix} \qquad -2R_1 + R_2 \rightarrow R_2$$

$$\begin{bmatrix} 1 & 1 & | & 0 & 1 \\ 0 & 1 & | & -\frac{1}{5} & \frac{2}{5} \end{bmatrix} \qquad -\frac{1}{5}R_2 \rightarrow R_2$$

$$\begin{bmatrix} 1 & 0 & | & \frac{1}{5} & \frac{3}{5} \\ 0 & 1 & | & -\frac{1}{5} & \frac{2}{5} \end{bmatrix} \qquad -R_2 + R_1 \rightarrow R_1$$

Since the last matrix is in the form $[I|A^{-1}]$, we get

$$A^{-1} = \begin{bmatrix} \frac{1}{5} & \frac{3}{5} \\ -\frac{1}{5} & \frac{2}{5} \end{bmatrix}.$$

Check that $AA^{-1} = A^{-1}A = I$. ◆

Example 4 A noninvertible matrix

Find the inverse of the matrix

$$A = \begin{bmatrix} 1 & -3 \\ -1 & 3 \end{bmatrix}.$$

Solution

Use row operations on the augmented matrix $[A|I]$:

$$\begin{bmatrix} 1 & -3 & | & 1 & 0 \\ -1 & 3 & | & 0 & 1 \end{bmatrix}$$

$$\begin{bmatrix} 1 & -3 & | & 1 & 0 \\ 0 & 0 & | & 1 & 1 \end{bmatrix} \qquad R_1 + R_2 \rightarrow R_2$$

The row containing all zeros on the left-hand side of the augmented matrix indicates that the left-hand side (the matrix A) cannot be converted to I using row operations. So A is not invertible. Note that the row of 0's in the augmented matrix means that there is no solution to the systems that must be solved to find A^{-1}.

◆

Example 5   The inverse of a 3×3 matrix

Find the inverse of the matrix

$$A = \begin{bmatrix} 0 & 1 & 2 \\ 1 & 0 & 3 \\ 0 & 1 & 4 \end{bmatrix}.$$

Solution

Perform row operations to convert $[A\,|\,I]$ into $[I\,|\,A^{-1}]$, where I is the 3×3 identity matrix.

$$\begin{bmatrix} 0 & 1 & 2 & | & 1 & 0 & 0 \\ 1 & 0 & 3 & | & 0 & 1 & 0 \\ 0 & 1 & 4 & | & 0 & 0 & 1 \end{bmatrix} \quad \text{The augmented matrix}$$

Interchange the first and second rows to get the first 1 on the diagonal:

$$\begin{bmatrix} 1 & 0 & 3 & | & 0 & 1 & 0 \\ 0 & 1 & 2 & | & 1 & 0 & 0 \\ 0 & 1 & 4 & | & 0 & 0 & 1 \end{bmatrix} \quad R_1 \leftrightarrow R_2$$

Since the first column is now in the desired form, we work on the second column:

$$\begin{bmatrix} 1 & 0 & 3 & | & 0 & 1 & 0 \\ 0 & 1 & 2 & | & 1 & 0 & 0 \\ 0 & 0 & 2 & | & -1 & 0 & 1 \end{bmatrix} \quad -R_2 + R_3 \rightarrow R_3$$

We now get a 1 in the last position on the diagonal and 0's above it:

$$\begin{bmatrix} 1 & 0 & 3 & | & 0 & 1 & 0 \\ 0 & 1 & 2 & | & 1 & 0 & 0 \\ 0 & 0 & 1 & | & -0.5 & 0 & 0.5 \end{bmatrix} \quad \tfrac{1}{2} R_3 \rightarrow R_3$$

$$\begin{bmatrix} 1 & 0 & 0 & | & 1.5 & 1 & -1.5 \\ 0 & 1 & 0 & | & 2 & 0 & -1 \\ 0 & 0 & 1 & | & -0.5 & 0 & 0.5 \end{bmatrix} \quad -3R_3 + R_1 \rightarrow R_1 \text{ and } -2R_3 + R_2 \rightarrow R_2$$

So

$$A^{-1} = \begin{bmatrix} 1.5 & 1 & -1.5 \\ 2 & 0 & -1 \\ -0.5 & 0 & 0.5 \end{bmatrix}.$$

Check that $AA^{-1} = I$ and $A^{-1}A = I$.

◆

[A]⁻¹
  [[1.5 1 -1.5]
   [2   0 -1]
   [-.5 0 .5]]

Define A as given in Example 5. Then use the x^{-1} key to find A^{-1}.

Finding the inverse of a matrix is a rather tedious process unless the matrix is designed to have a simple inverse. However, if we find the inverse of the matrix of coefficients for a system of linear equations, we can use it to solve the system.

Solving Systems of Equations Using Matrix Inverses

In Section 10.3 we saw that a system of n linear equations in n unknowns could be written as a matrix equation of the form $AX = B$, where A is a matrix of coefficients, X is a matrix of variables, and B is a matrix of constants. If A^{-1} exists, then we can multiply each side of this equation by A^{-1}. Since matrix multiplication is not commutative in general, A^{-1} is placed to the left of the matrices on each side:

$$AX = B$$
$$A^{-1}(AX) = A^{-1}B \qquad \text{Multiply each side by } A^{-1}.$$
$$(A^{-1}A)X = A^{-1}B \qquad \text{Matrix multiplication is associative}$$
$$IX = A^{-1}B \qquad \text{Since } A^{-1}A = I$$
$$X = A^{-1}B \qquad \text{Since } I \text{ is the identity matrix}$$

The last equation indicates that the values of the variables in the matrix X are equal to the entries in the matrix $A^{-1}B$. So solving the system is equivalent to finding $A^{-1}B$. This result is summarized in the following theorem.

Theorem: Solving a System Using A^{-1}

If a system of n linear equations in n variables has a unique solution, then the solution is given by

$$X = A^{-1}B,$$

where A is the matrix of coefficients, B is the matrix of constants, and X is the matrix of variables.

If there is no solution or there are infinitely many solutions, the matrix of coefficients is not invertible.

Example 6   Using the inverse of a matrix to solve a system

Solve the system by using A^{-1}.

$$2x - 3y = 1$$
$$x + y = 8$$

Solution

For this system,

$$A = \begin{bmatrix} 2 & -3 \\ 1 & 1 \end{bmatrix}, \qquad X = \begin{bmatrix} x \\ y \end{bmatrix}, \qquad \text{and} \qquad B = \begin{bmatrix} 1 \\ 8 \end{bmatrix}.$$

Since the matrix A is the same as in Example 3, use A^{-1} from Example 3. Multiply A^{-1} and B to obtain

$$\begin{bmatrix} \frac{1}{5} & \frac{3}{5} \\ -\frac{1}{5} & \frac{2}{5} \end{bmatrix} \begin{bmatrix} 1 \\ 8 \end{bmatrix} = \begin{bmatrix} 5 \\ 3 \end{bmatrix}.$$

Since $X = A^{-1}B$, we have

$$X = \begin{bmatrix} x \\ y \end{bmatrix} = \begin{bmatrix} 5 \\ 3 \end{bmatrix}.$$

So $x = 5$ and $y = 3$. Check this solution in the original system.   ◆

Define A and B as in Example 6. Then find $A^{-1}B$ to solve the equation.

[A]⁻¹[B]
[[5]
[3]]

Since computers and even pocket calculators can find inverses of matrices, the method of solving an equation by first finding A^{-1} is a popular one for use with machines. By hand, this method may seem somewhat tedious. However, it is useful when there are many systems to solve with the same coefficients. The following system has the same coefficients as the system of Example 6:

$$2x - 3y = -1$$
$$x + \ y = -3$$

So

$$X = A^{-1}B = \begin{bmatrix} \frac{1}{5} & \frac{3}{5} \\ -\frac{1}{5} & \frac{2}{5} \end{bmatrix} \begin{bmatrix} -1 \\ -3 \end{bmatrix} = \begin{bmatrix} -2 \\ -1 \end{bmatrix}$$

or $x = -2$ and $y = -1$. As long as the coefficients of x and y are unchanged, the same A^{-1} is used to solve the system.

Application of Matrices to Coding

There are many ways of encoding a message so that no one other than the intended recipient can understand the message. One way is to use a matrix to encode the message and the inverse matrix to decode the message. Assume that a space is 0, A is 1, B is 2, C is 3, and so on. The numerical equivalent of the word HELP is **8, 5, 12, 16.** List these numbers in two 2×1 matrices. Then multiply the matrices by a coding matrix. We will use the 2×2 matrix A from Example 6:

$$\begin{bmatrix} 2 & -3 \\ 1 & 1 \end{bmatrix} \begin{bmatrix} 8 \\ 5 \end{bmatrix} = \begin{bmatrix} 1 \\ 13 \end{bmatrix} \qquad \begin{bmatrix} 2 & -3 \\ 1 & 1 \end{bmatrix} \begin{bmatrix} 12 \\ 16 \end{bmatrix} = \begin{bmatrix} -24 \\ 28 \end{bmatrix}$$

So the encoded message sent is $1, 13, -24, 28$. The person receiving the message must know that

$$A^{-1} = \begin{bmatrix} \frac{1}{5} & \frac{3}{5} \\ -\frac{1}{5} & \frac{2}{5} \end{bmatrix}.$$

To decode the message, we find

$$\begin{bmatrix} \frac{1}{5} & \frac{3}{5} \\ -\frac{1}{5} & \frac{2}{5} \end{bmatrix} \begin{bmatrix} 1 \\ 13 \end{bmatrix} = \begin{bmatrix} 8 \\ 5 \end{bmatrix} \quad \text{and} \quad \begin{bmatrix} \frac{1}{5} & \frac{3}{5} \\ -\frac{1}{5} & \frac{2}{5} \end{bmatrix} \begin{bmatrix} -24 \\ 28 \end{bmatrix} = \begin{bmatrix} 12 \\ 16 \end{bmatrix}.$$

So the message is 8, 5, 12, 16, or HELP. Of course, any invertible matrix of any order and its inverse could be used for this coding scheme, and all of the encoding and decoding could be done by a computer. If a person or computer didn't know A or A^{-1} or even their order, then how hard do you think it would be to break this code?

For Thought

True or false? Explain.

The following statements refer to the matrices

$$A = \begin{bmatrix} 2 & 3 \\ 3 & 5 \end{bmatrix}, \qquad B = \begin{bmatrix} 5 & -3 \\ -3 & 2 \end{bmatrix}, \qquad C = \begin{bmatrix} 4 & 6 \\ 3 & 5 \\ 2 & 1 \end{bmatrix},$$

$$D = \begin{bmatrix} 11 \\ 19 \end{bmatrix}, \qquad \text{and} \qquad I = \begin{bmatrix} 1 & 0 \\ 0 & 1 \end{bmatrix}.$$

1. $AB = BA = I$ **2.** $B = A^{-1}$ **3.** B is an invertible matrix.

4. $AC = CA$ **5.** $CI = C$ **6.** C is an invertible matrix.

7. The system $\begin{aligned} 2x + 3y &= 11 \\ 3x + y &= 19 \end{aligned}$ is equivalent to $A \begin{bmatrix} x \\ y \end{bmatrix} = D$.

8. $A^{-1}D = \begin{bmatrix} -2 \\ 5 \end{bmatrix}$

9. The solution set to the system $\begin{aligned} 2x + 3y &= 11 \\ 3x + y &= 19 \end{aligned}$ is $\{(-2, 5)\}$.

10. The solution set to the system $\begin{aligned} 2x + 3y &= 3 \\ 3x + y &= -7 \end{aligned}$ is obtained from $A^{-1} \begin{bmatrix} 3 \\ -7 \end{bmatrix}$.

10.4 Exercises Tape 20 Disk—5.25": 6 3.5": 4 Macintosh: 4

Find the following products.

1. $\begin{bmatrix} 1 & 0 \\ 0 & 1 \end{bmatrix} \begin{bmatrix} -3 & 5 \\ 12 & 6 \end{bmatrix}$

2. $\begin{bmatrix} 4 & 8 \\ 9 & -2 \end{bmatrix} \begin{bmatrix} 1 & 0 \\ 0 & 1 \end{bmatrix}$

3. $\begin{bmatrix} -2 & -3 \\ 3 & 4 \end{bmatrix} \begin{bmatrix} 4 & 3 \\ -3 & -2 \end{bmatrix}$

4. $\begin{bmatrix} 3 & 2 \\ 3 & 3 \end{bmatrix} \begin{bmatrix} 1 & -\frac{2}{3} \\ -1 & 1 \end{bmatrix}$

5. $\begin{bmatrix} 3 & 4 \\ 3 & 5 \end{bmatrix} \begin{bmatrix} \frac{5}{3} & -\frac{4}{3} \\ -1 & 1 \end{bmatrix}$

6. $\begin{bmatrix} 1 & 2 \\ 4 & 5 \end{bmatrix} \begin{bmatrix} -\frac{5}{3} & \frac{2}{3} \\ \frac{4}{3} & -\frac{1}{3} \end{bmatrix}$

7. $\begin{bmatrix} 3 & 5 & 1 \\ 4 & 5 & 7 \\ 4 & 9 & 2 \end{bmatrix} \begin{bmatrix} 1 & 0 & 0 \\ 0 & 1 & 0 \\ 0 & 0 & 1 \end{bmatrix}$

8. $\begin{bmatrix} 1 & 0 & 0 \\ 0 & 1 & 0 \\ 0 & 0 & 1 \end{bmatrix} \begin{bmatrix} 4 & 0 & 5 \\ 0 & 7 & 9 \\ 3 & 1 & 2 \end{bmatrix}$

9. $\begin{bmatrix} 1 & 0 & 2 \\ 1 & 3 & 0 \\ 0 & 1 & 0 \end{bmatrix} \begin{bmatrix} 0 & 1 & -3 \\ 0 & 0 & 1 \\ 0.5 & -0.5 & 1.5 \end{bmatrix}$

10. $\begin{bmatrix} 2 & 1 & 0 \\ 1 & 0 & 2 \\ 0 & 1 & 1 \end{bmatrix} \begin{bmatrix} 0.4 & 0.2 & -0.4 \\ 0.2 & -0.4 & 0.8 \\ -0.2 & 0.4 & 0.2 \end{bmatrix}$

11. $\begin{bmatrix} 1 & 1 & 0 \\ 0 & 1 & 1 \\ 1 & 0 & 1 \end{bmatrix} \begin{bmatrix} 0.5 & -0.5 & 0.5 \\ 0.5 & 0.5 & -0.5 \\ -0.5 & 0.5 & 0.5 \end{bmatrix}$

12. $\begin{bmatrix} 0.4 & 0.2 & -0.4 \\ 0.2 & -0.4 & 0.8 \\ -0.2 & 0.4 & 0.2 \end{bmatrix} \begin{bmatrix} 2 & 1 & 0 \\ 1 & 0 & 2 \\ 0 & 1 & 1 \end{bmatrix}$

Determine whether the matrices in each pair are inverses of each other.

13. $\begin{bmatrix} 3 & 1 \\ 11 & 4 \end{bmatrix}, \begin{bmatrix} 4 & -1 \\ -11 & 3 \end{bmatrix}$ **14.** $\begin{bmatrix} \frac{1}{2} & 0 \\ 0 & \frac{1}{2} \end{bmatrix}, \begin{bmatrix} 2 & 0 \\ 0 & 2 \end{bmatrix}$

15. $\begin{bmatrix} \frac{1}{2} & -1 \\ 3 & -12 \end{bmatrix}, \begin{bmatrix} 4 & 2 \\ 1 & 1 \end{bmatrix}$

16. $\begin{bmatrix} 1 & 2 & 3 \\ 0 & 1 & 2 \\ 0 & 0 & 1 \end{bmatrix}, \begin{bmatrix} 1 & -2 & 1 \\ 0 & 1 & -2 \\ 0 & 0 & 1 \end{bmatrix}$

17. $\begin{bmatrix} 1 & 0 & 0 \\ 0 & \frac{1}{2} & 0 \end{bmatrix}, \begin{bmatrix} 1 & 0 \\ 0 & 2 \\ 3 & 4 \end{bmatrix}$ **18.** $\begin{bmatrix} 1 & 2 \\ 3 & 4 \\ 5 & 6 \end{bmatrix}, \begin{bmatrix} 1 & \frac{1}{2} \\ \frac{1}{4} & \frac{1}{4} \\ \frac{1}{5} & \frac{1}{6} \end{bmatrix}$

Find the inverse of each matrix A if possible. Check that $AA^{-1} = I$ and $A^{-1}A = I$.

19. $\begin{bmatrix} 1 & 4 \\ 0 & 2 \end{bmatrix}$ **20.** $\begin{bmatrix} 1 & 3 \\ 0 & -1 \end{bmatrix}$

21. $\begin{bmatrix} 1 & 6 \\ 1 & 9 \end{bmatrix}$ **22.** $\begin{bmatrix} 1 & 4 \\ 3 & 8 \end{bmatrix}$

23. $\begin{bmatrix} -2 & -3 \\ 3 & 4 \end{bmatrix}$ **24.** $\begin{bmatrix} 3 & 4 \\ 4 & 5 \end{bmatrix}$

25. $\begin{bmatrix} 1 & -5 \\ -1 & 3 \end{bmatrix}$ **26.** $\begin{bmatrix} 4 & 3 \\ -3 & -2 \end{bmatrix}$

27. $\begin{bmatrix} -1 & 5 \\ 2 & -10 \end{bmatrix}$ **28.** $\begin{bmatrix} 2 & 6 \\ 1 & 3 \end{bmatrix}$

29. $\begin{bmatrix} 1 & 1 & 0 \\ 0 & -1 & -1 \\ 1 & 0 & -1 \end{bmatrix}$ **30.** $\begin{bmatrix} 1 & -1 & 2 \\ 1 & 2 & 3 \\ 2 & 1 & 5 \end{bmatrix}$

31. $\begin{bmatrix} 1 & 1 & 1 \\ 1 & -1 & -1 \\ 1 & -1 & 1 \end{bmatrix}$ **32.** $\begin{bmatrix} 1 & 0 & 2 \\ 0 & 2 & 0 \\ 1 & 3 & 0 \end{bmatrix}$

33. $\begin{bmatrix} 0 & 2 & 0 \\ 3 & 3 & 2 \\ 2 & 5 & 1 \end{bmatrix}$ **34.** $\begin{bmatrix} 4 & 1 & -3 \\ 0 & 1 & 0 \\ -3 & 1 & 2 \end{bmatrix}$

Solve each system of equations by using A^{-1}. Note that the matrix of coefficients in each system is a matrix from Exercises 19–34.

35. $x + 6y = -3$
$\quad\ x + 9y = -6$

36. $x + 4y = 5$
$\quad 3x + 8y = 7$

37. $x + 6y = 4$
$\quad\ x + 9y = 5$

38. $x + 4y = 1$
$\quad 3x + 8y = 5$

39. $-2x - 3y = 1$
$\quad\ \ 3x + 4y = -1$

40. $3x + 4y = 1$
$\quad 4x + 5y = 2$

41. $x - 5y = -5$
$\quad -x + 3y = 1$

42. $4x + 3y = 2$
$\quad -3x - 2y = -1$

43. $x + y + z = 3$
$\quad x - y - z = -1$
$\quad x - y + z = 5$

44. $x + 2z = -4$
$\quad\quad 2y = 6$
$\quad x + 3y = 7$

45. $\quad\quad\ 2y = 6$
$\quad 3x + 3y + 2z = 16$
$\quad 2x + 5y + \ z = 19$

46. $4x + y - 3z = 3$
$\quad\quad\quad\ y = -2$
$\quad -3x + y + 2z = -5$

Solve each system of equations by using A^{-1} if possible.

47. $0.3x = 3 - 0.1y$
$\quad 4y = 7 - 2x$

48. $2x = 3y - 7$
$\quad y = x + 4$

49. $x - y + z = 5$
$\quad 2x - y + 3z = 1$
$\quad\quad y + z = -9$

50. $x + y - z = 4$
$\quad 2x - 3y + z = 2$
$\quad 4x - y - z = 6$

51. $x + y + z = 1$
$\quad 2x + 4y + z = 2$
$\quad x + 3y + 6z = 3$

52. $0.5x - 0.25y + 0.1z = 3$
$\quad 0.2x - 0.5y + 0.2z = -2$
$\quad 0.1x + 0.3y - 0.5z = -8$

Determine whether each matrix is invertible. If so, state the inverse.

53. $\begin{bmatrix} 1 & 0 & 1 \\ 0 & 2 & 2 \\ 2 & 1 & 0 \end{bmatrix}$ **54.** $\begin{bmatrix} 1 & 3 & 0 \\ 0 & 3 & -2 \\ 0 & -5 & 3 \end{bmatrix}$

55. $\begin{bmatrix} 0 & 4 & 2 \\ 0 & 3 & 2 \\ 1 & -1 & 1 \end{bmatrix}$ **56.** $\begin{bmatrix} 1 & 2 & 0 \\ 2 & 3 & -1 \\ 0 & 1 & 2 \end{bmatrix}$

57. $\begin{bmatrix} 1 & 2 & 3 & 4 \\ 0 & 1 & 2 & 3 \\ 0 & 0 & 1 & 2 \\ 0 & 0 & 0 & 1 \end{bmatrix}$ **58.** $\begin{bmatrix} 1 & 0 & 0 & 0 \\ -2 & 1 & 0 & 0 \\ 3 & -2 & 1 & 0 \\ 5 & 3 & -2 & 1 \end{bmatrix}$

Solve each problem.

59. Find all matrices A such that $A = \begin{bmatrix} a & 7 \\ 3 & b \end{bmatrix}$, $A^{-1} = \begin{bmatrix} -b & 7 \\ 3 & -a \end{bmatrix}$, and a and b are positive integers.

60. Find all matrices of the form $A = \begin{bmatrix} a & a \\ 0 & c \end{bmatrix}$ such that $A^2 = I$.

Write a system of equations for each problem. Solve the system using an inverse matrix.

61. *Eggs and magazines* Stephanie bought a dozen eggs and a magazine at the Handy Mart. Her bill including tax was

$2.70. If groceries are taxed at 8% and magazines at 5% and she paid 15 cents in tax on the purchase, then what was the price of each item?

62. *Dogs and suds*   The French Club sold 48 hot dogs and 120 soft drinks at the game on Saturday for a total of $103.20. If the price of a hot dog was 40 cents more than the price of a soft drink, then what was the price of each item?

63. *Plywood and insulation*   A contractor purchased four loads of plywood and six loads of insulation on Monday for $2500, and three loads of plywood and five loads of insulation on Tuesday for $1950. Find the cost of one load of plywood and the cost of one load of insulation.

64. *Adjusting for inflation*   The contractor from Exercise 63 made identical purchases on Monday and Tuesday of the following week, but paid $2770 on Monday and $2165 on Tuesday as a result of price increases. What were the new costs of one load of plywood and one load of insulation?

The following messages were encoded by using the matrix $A = \begin{bmatrix} 3 & 1 \\ 5 & 2 \end{bmatrix}$ **and the coding scheme described in this section. Find** A^{-1} **and use it to decode the messages.**

65. 36, 65, 49, 83, 12, 24, 66, 111, 33, 55

66. 15, 29, 26, 45, 24, 46, 3, 5, 46, 83, 6, 12, 77, 133

Graphing Calculator Exercises

Most graphing calculators can perform operations with matrices, including matrix inversion and multiplication. Solve the following systems, using a graphing calculator to find A^{-1} and the product $A^{-1}B$.

1. $0.1x + 0.2y + 0.1z = 27$
$0.5x + 0.2y + 0.3z = 9$
$0.4x + 0.8y + 0.1z = 36$

2. $3x + 6y + 4z = 9$
$x + 2y - 2z = -18$
$-x + 4y + 3z = 54$

3. $1.5x - 5y + 3z = 16$
$2.25x - 4y + z = 24$
$2x + 3.5y - 3z = -8$

4. $2.1x - 3.4y + 5z = 100$
$1.3x + 2y - 8z = 250$
$2.5x + 3y - 9.1z = 300$

Write a system of equations for each of the following problems and solve the system using matrix inversion and matrix multiplication on a graphing calculator.

5. Fernando works as a shipping clerk in an office supply warehouse. His first shipment on Monday was for 24 reams of typing paper, 33 packs of pens, and 12 boxes of paper clips for a total price of $202.23. His second shipment was for 19 reams of typing paper, 40 packs of pens, and 22 boxes of paper clips for a total price of $209.38. His third shipment was for 30 reams of typing paper, 9 packs of pens, and 19 boxes of paper clips for a total price of $167.66. For the fourth shipment the computer was down, and Fernando did not know the price of each item. What is the price of each item?

6. A-Bear's Catering Service charges its customers according to the number of servings of each item that is supplied at the party. The table shows the number of servings of jambalaya, crawfish pie, filé gumbo, iced tea, and dessert for the last five customers, along with the total cost of each party. What amount does A-Bear's charge per serving of each item?

	Jambalaya	Crawfish pie	Filé gumbo	Iced tea	Dessert	Cost
Boudreaux	36	28	35	90	68	$344.35
Thibodeaux	37	19	56	84	75	$369.10
Fontenot	49	55	70	150	125	$588.90
Arceneaux	58	34	52	122	132	$529.50
Gautreaux	44	65	39	133	120	$521.65

Table for Graphing Calculator Exercise 6

10.5

Solution of Linear

Systems in Two

Variables Using

Determinants

We have solved linear systems of equations by graphing, substitution, addition, the Gauss-Jordan method, and inverse matrices. Graphing, substitution, and addition are feasible only with relatively simple systems. By contrast, the Gauss-Jordan and inverse matrix methods are readily performed by computers or even hand-held calculators. With a machine doing the work, they can be applied to complicated systems such as those in the Graphing Calculator Exercises of Section 10.4. Determinants, which we now discuss, can also be used by computers and calculators, and give us another method that is not limited to simple systems.

The Determinant of a 2 × 2 Matrix

Before we can solve a system of equations by using determinants, we need to learn what a determinant is and how to find it. The **determinant** of a square matrix is a real number associated with the matrix. Every square matrix has a determinant. The determinant of a 1×1 matrix is the single entry of the matrix. For a 2×2 matrix the determinant is defined as follows.

Definition: Determinant of a 2 × 2 Matrix

The **determinant** of the matrix $\begin{bmatrix} a_{11} & a_{12} \\ a_{21} & a_{22} \end{bmatrix}$ is the real number $a_{11}a_{22} - a_{21}a_{12}$. In symbols,

$$\begin{vmatrix} a_{11} & a_{12} \\ a_{21} & a_{22} \end{vmatrix} = a_{11}a_{22} - a_{21}a_{12}.$$

If a matrix is named A, then the determinant of that matrix is denoted as $|A|$. Even though the symbol for determinant looks like the absolute value symbol, the value of a determinant may be any real number. For a 2×2 matrix, that number is found by subtracting the products of the diagonal entries:

$$\begin{vmatrix} a_{11} & a_{12} \\ a_{21} & a_{22} \end{vmatrix} = a_{11}a_{22} - a_{21}a_{12}$$

```
det [A]
              -11
det [B]
              0
```

Define A and B as in Example 1. Then use the determinant function to find the determinants.

Example 1   The determinant of a 2 × 2 matrix

Find the determinant of each matrix.

a) $\begin{bmatrix} 3 & -1 \\ 4 & -5 \end{bmatrix}$ **b)** $\begin{bmatrix} 4 & -6 \\ 2 & -3 \end{bmatrix}$

Solution

a) $\begin{vmatrix} 3 & -1 \\ 4 & -5 \end{vmatrix} = 3(-5) - (4)(-1) = -15 + 4 = -11$

b) $\begin{vmatrix} 4 & -6 \\ 2 & -3 \end{vmatrix} = 4(-3) - (2)(-6) = -12 + 12 = 0$

Cramer's Rule for Systems in Two Variables

We will now see how determinants arise in the solution of a system of two linear equations in two unknowns. Consider a general system of two linear equations in two unknowns, x and y,

$$(1) \qquad a_1x + b_1y = c_1$$

$$(2) \qquad a_2x + b_2y = c_2$$

where a_1, a_2, b_1, b_2, c_1, and c_2 are real numbers.

To eliminate y, multiply Eq. (1) by b_2 and Eq. (2) by $-b_1$:

$$
\begin{array}{ll}
b_2a_1x + b_2b_1y = b_2c_1 & \text{Eq. (1) multiplied by } b_2 \\
\underline{-b_1a_2x - b_1b_2y = -b_1c_2} & \text{Eq. (2) multiplied by } -b_1 \\
a_1b_2x - a_2b_1x \qquad\quad = c_1b_2 - c_2b_1 & \text{Add.}
\end{array}
$$

$$
(a_1b_2 - a_2b_1)x = c_1b_2 - c_2b_1 \qquad \text{Factor out } x.
$$

$$
x = \frac{c_1b_2 - c_2b_1}{a_1b_2 - a_2b_1} \qquad \text{Provided that } a_1b_2 - a_2b_1 \neq 0
$$

This formula for x can be written using determinants as

$$
x = \frac{\begin{vmatrix} c_1 & b_1 \\ c_2 & b_2 \end{vmatrix}}{\begin{vmatrix} a_1 & b_1 \\ a_2 & b_2 \end{vmatrix}}, \qquad \text{provided } a_1b_2 - a_2b_1 \neq 0.
$$

The same procedure is used to eliminate x and get the following formula for y in terms of determinants:

$$
y = \frac{\begin{vmatrix} a_1 & c_1 \\ a_2 & c_2 \end{vmatrix}}{\begin{vmatrix} a_1 & b_1 \\ a_2 & b_2 \end{vmatrix}}, \qquad \text{provided } a_1b_2 - a_2b_1 \neq 0
$$

Notice that there are three determinants involved in solving for x or y. Let

$$
D = \begin{vmatrix} a_1 & b_1 \\ a_2 & b_2 \end{vmatrix}, \quad D_x = \begin{vmatrix} c_1 & b_1 \\ c_2 & b_2 \end{vmatrix}, \quad \text{and} \quad D_y = \begin{vmatrix} a_1 & c_1 \\ a_2 & c_2 \end{vmatrix}.
$$

Note that D is the determinant of the original matrix of coefficients of x and y. D appears in the denominator for both x and y. D_x is the determinant D with the constants c_1 and c_2 replacing the first column of D. D_y is the determinant D with the constants c_1 and c_2 replacing the second column of D. These formulas for solving a system of two linear equations in two variables are known as **Cramer's rule**.

Cramer's Rule for Systems in Two Variables

The solution to the system

$$a_1x + b_1y = c_1$$

$$a_2x + b_2y = c_2$$

is given by $x = \dfrac{D_x}{D}$ and $y = \dfrac{D_y}{D}$, where

$$D = \begin{vmatrix} a_1 & b_1 \\ a_2 & b_2 \end{vmatrix}, \quad D_x = \begin{vmatrix} c_1 & b_1 \\ c_2 & b_2 \end{vmatrix}, \quad \text{and} \quad D_y = \begin{vmatrix} a_1 & c_1 \\ a_2 & c_2 \end{vmatrix},$$

provided that $D \neq 0$.

Example 2   Applying Cramer's rule

Use Cramer's rule to solve the system.

$$3x = 2y + 9$$

$$3y = x + 3$$

Solution

To apply Cramer's rule, rewrite both equations in the form $Ax + By = C$:

$$3x - 2y = 9$$

$$-x + 3y = 3$$

First find the determinant of the coefficient matrix using the coefficients of x and y:

$$D = \begin{vmatrix} 3 & -2 \\ -1 & 3 \end{vmatrix} = 3(3) - (-1)(-2) = 7$$

Next we find the determinants D_x and D_y. For D_x, use 9 and 3 in the x-column, and for D_y, use 9 and 3 in the y-column:

$$D_x = \begin{vmatrix} 9 & -2 \\ 3 & 3 \end{vmatrix} = 33 \quad \text{and} \quad D_y = \begin{vmatrix} 3 & 9 \\ -1 & 3 \end{vmatrix} = 18.$$

By Cramer's rule,

$$x = \frac{D_x}{D} = \frac{33}{7} \quad \text{and} \quad y = \frac{D_y}{D} = \frac{18}{7}.$$

Check that $x = 33/7$ and $y = 18/7$ satisfy the original system.

[D]
```
        [[3  -2]
         [-1  3 ]]
```
[A]
```
        [[9  -2]
         [3  3 ]]
```

Define the matrices D and A as shown.

```
det [A]/det [D]
        4.714285714
Ans▶Frac
            33/7
■
```

Find x by using Cramer's rule. You can find y in a similar manner.

A system of two linear equations in two unknowns may have a unique solution, no solution, or infinitely many solutions. Cramer's rule works only on systems that have a unique solution. For inconsistent or dependent systems, $D = 0$ and Cramer's rule will not give the solution. If $D = 0$, then another method must be used to determine the solution set.

Example 3   Inconsistent and dependent systems

Use Cramer's rule to solve each system.

a) $2x - 4y = 8$ **b)** $2x - 4y = 8$
 $-x + 2y = -4$ $-x + 2y = 6$

Solution

The matrix D is the same for both systems:

$$D = \begin{vmatrix} 2 & -4 \\ -1 & 2 \end{vmatrix} = 0$$

So Cramer's rule does not apply to either system. Multiply the second equation in each system by 2 and add the equations:

a) $2x - 4y = 8$ **b)** $2x - 4y = 8$
 $\underline{-2x + 4y = -8}$ $\underline{-2x + 4y = 12}$
 $\ 0 = 0$ $\ 0 = 20$

System (a) is dependent, and the solution set is $\{(x, y) \mid -x + 2y = -4\}$. System (b) is inconsistent and has no solution. ◆

A system of two linear equations in two variables has a unique solution if and only if the determinant of the matrix of coefficients is nonzero. We have not yet seen how to find a determinant of a larger matrix, but the same result is true for a system of n linear equations in n variables. In Section 10.4 we learned that a system of n linear equations in n variables has a unique solution if and only if the matrix of coefficients is invertible. These two results are combined in the following theorem to give a means of identifying whether a matrix is invertible.

Theorem: Invertible Matrices

A matrix is invertible if and only if it has a nonzero determinant.

Example 4   Determinants and inverse matrices

Are the matrices $A = \begin{bmatrix} 2 & -3 \\ 4 & 5 \end{bmatrix}$ and $B = \begin{bmatrix} 3 & 5 \\ 6 & 10 \end{bmatrix}$ invertible?

Solution

Since $|A| = 10 - (-12) = 22$, A is an invertible matrix. However, because $|B| = 30 - 30 = 0$, B is not an invertible matrix. ◆

For Thought

True or false? Explain.

The following statements refer to the matrices

$$A = \begin{bmatrix} 3 & -5 \\ 1 & 4 \end{bmatrix}, \quad B = \begin{bmatrix} 4 & -2 \\ -10 & 5 \end{bmatrix}, \quad C = \begin{bmatrix} 2 & -5 \\ 6 & 4 \end{bmatrix}, \quad \text{and} \quad E = \begin{bmatrix} 3 & 2 \\ 1 & 6 \end{bmatrix}.$$

1. $|A| = 7$ **2.** A is invertible.

3. $|B| = 0$ **4.** B is invertible.

5. The system $\begin{array}{l} 3x - 5y = 2 \\ x + 4y = 6 \end{array}$ is independent.   **6.** $|CE| = |C| \cdot |E|$

7. The solution to the system $\begin{array}{l} 3x^2 - 5y^2 = 2 \\ x^2 + 4y = 6 \end{array}$ is $x = \dfrac{|C|}{|A|}$ and $y = \dfrac{|E|}{|A|}$.

8. The determinant of the 2×2 identity matrix I is 1.

9. The matrix $\begin{bmatrix} 2 & 0.1 \\ 100 & 5 \end{bmatrix}$ is invertible.

10. $\begin{bmatrix} 5 & 3 \\ 1 & 6 \end{bmatrix} = 27$

10.5 Exercises   Tape 20   Disk—5.25": 7   3.5": 4   Macintosh: 4

Find the determinant of each matrix.

1. $\begin{bmatrix} 1 & 3 \\ 0 & 2 \end{bmatrix}$ **2.** $\begin{bmatrix} 0 & 4 \\ 2 & -1 \end{bmatrix}$

3. $\begin{bmatrix} 3 & 4 \\ 2 & 9 \end{bmatrix}$ **4.** $\begin{bmatrix} 7 & 2 \\ 3 & -2 \end{bmatrix}$

5. $\begin{bmatrix} -0.3 & -0.5 \\ -0.7 & 0.2 \end{bmatrix}$ **6.** $\begin{bmatrix} -\frac{1}{3} & \frac{4}{3} \\ -3 & \frac{2}{3} \end{bmatrix}$

7. $\begin{bmatrix} \frac{1}{8} & -\frac{3}{8} \\ 2 & -\frac{1}{4} \end{bmatrix}$ **8.** $\begin{bmatrix} -1 & -3 \\ -5 & -8 \end{bmatrix}$

9. $\begin{bmatrix} 0.02 & 0.4 \\ 1 & 20 \end{bmatrix}$ **10.** $\begin{bmatrix} -0.3 & 0.4 \\ 3 & -4 \end{bmatrix}$

11. $\begin{bmatrix} 3 & -5 \\ -9 & 15 \end{bmatrix}$ **12.** $\begin{bmatrix} -6 & 2 \\ 3 & -1 \end{bmatrix}$

Solve each system, using Cramer's rule when possible.

13. $\begin{array}{l} 2x - y = -11 \\ x + 3y = 12 \end{array}$ **14.** $\begin{array}{l} 3x - 2y = -4 \\ -5x + 4y = -1 \end{array}$

15. $\begin{array}{l} x = y + 6 \\ x + y = 5 \end{array}$ **16.** $\begin{array}{l} 3x + y = 7 \\ 4x = y - 4 \end{array}$

17. $\begin{array}{l} \frac{1}{2}x - \frac{1}{3}y = 4 \\ \frac{1}{4}x + \frac{1}{2}y = 6 \end{array}$ **18.** $\begin{array}{l} \frac{1}{4}x + \frac{2}{3}y = 25 \\ \frac{3}{5}x - \frac{1}{10}y = 12 \end{array}$

19. $\begin{array}{l} 0.2x + 0.12y = 148 \\ x + y = 900 \end{array}$ **20.** $\begin{array}{l} 0.08x + 0.05y = 72 \\ 2x - y = 0 \end{array}$

21. $\begin{array}{l} 3x + y = 6 \\ -6x - 2y = -12 \end{array}$ **22.** $\begin{array}{l} 8x - 4y = 2 \\ 4x - 2y = 1 \end{array}$

23. $8x - y = 9$
$-8x + y = 10$

24. $12x + 3y = 9$
$4x + y = 6$

25. $y = x - 3$
$y = 3x + 9$

26. $y = \dfrac{x - 3}{2}$
$x + 2y = 15$

27. $\sqrt{2}x + \sqrt{3}y = 4$
$\sqrt{18}x - \sqrt{12}y = -3$

28. $\dfrac{\sqrt{3}x}{3} + y = 1$
$x - \sqrt{3}y = 0$

29. $x^2 + y^2 = 25$
$x^2 - y = 5$

30. $x^2 + y = 8$
$x^2 - y = 4$

31. $x - 2y = y^2$
$\dfrac{1}{2}x - y = 2$

32. $y = 0.15x$
$x + y = 736$

Determine whether each matrix is invertible by finding the determinant of the matrix.

33. $\begin{bmatrix} 4 & 0.5 \\ 2 & 3 \end{bmatrix}$

34. $\begin{bmatrix} -5 & 2 \\ 4 & -1 \end{bmatrix}$

35. $\begin{bmatrix} 3 & -4 \\ 9 & -12 \end{bmatrix}$

36. $\begin{bmatrix} \frac{1}{2} & 12 \\ \frac{1}{3} & 8 \end{bmatrix}$

Solve each problem, using two linear equations in two variables and Cramer's rule.

37. *The survey says* A survey of 615 teenagers found that 44% of the boys and 35% of the girls would like to be taller. If altogether 231 teenagers in the survey wished they were taller, how many boys and how many girls were in the survey?

38. *Average weight* The average weight of the starting quarterback and the backup quarterback for the Hudson Hornets is

148 lb. If the starting quarterback is 18 lb heavier than the backup, then how much does each weigh?

39. *Modern maturity* One morning Sarah awoke to find that the digits in her age had reversed, and she was 72 years older than she was when she went to bed. If the sum of the digits in her age is 10, then how old was she when she went to bed?

40. *Acute angles* One acute angle of a right triangle is 1° larger than twice the other acute angle. What are the measures of the acute angles?

41. *An isosceles triangle* If the smallest angle of an isosceles triangle is 2° smaller than any other angle, then what is the measure of each angle?

42. *A losing situation* Morton Motor Express lost a full truckload of TVs and VCRs. The truck carrying the shipment had a capacity of 2350 ft³. On the insurance claim the TVs were valued at $400 each and the VCRs were valued at $225 each, for a total value of $147,500. If each TV was in a carton with a volume of 8 ft³ and each VCR was in a carton of 2.5 ft³, then how many TVs and VCRs were in the shipment?

For Writing/Discussion

The following exercises investigate some of the properties of determinants. For these exercises let
$M = \begin{bmatrix} 3 & 2 \\ 5 & 4 \end{bmatrix}$ and $N = \begin{bmatrix} 2 & 7 \\ 1 & 5 \end{bmatrix}$.

43. Find $|M|$, $|N|$, and $|MN|$. Is $|MN| = |M| \cdot |N|$?

44. Find M^{-1} and $|M^{-1}|$. Is $|M^{-1}| = 1/|M|$?

45. Prove that the determinant of a product of two 2 × 2 matrices is equal to the product of their determinants.

46. Prove that if A is any 2 × 2 invertible matrix, then the determinant of A^{-1} is the reciprocal of the determinant of A.

47. Find $|-2M|$. Is $|-2M| = -2 \cdot |M|$?

48. Prove that if k is any scalar and A is any 2 × 2 matrix, then $|kA| = k^2 \cdot |A|$.

Graphing Calculator Exercises

Most graphing calculators can perform operations with matrices and calculate the determinant of a matrix. Solve the following systems using Cramer's rule and a graphing calculator.

1. $3.47x + 23.09y = 5978.95$
$12.48x + 3.98y = 2765.34$

2. $0.0875x + 0.1625y = 564.40$
$x + y = 4232$

10.6

Solution of Linear

Systems in Three

Variables Using

Determinants

The determinant can be defined for any square matrix. In this section we define the determinant of a 3×3 matrix by extending the definition of the determinant for 2×2 matrices. We can then solve linear systems of three equations in three unknowns, using an extended version of Cramer's rule. The first step is to define a determinant of a certain part of a matrix, a *minor*.

Minors

To each entry of a 3×3 matrix there corresponds a 2×2 matrix, which is obtained by deleting the row and column in which that entry appears. The determinant of this 2×2 matrix is called the **minor** of that entry.

Example 1   Finding the minor of an entry

Find the minors for the entries -2, 5, and 6 in the 3×3 matrix

$$\begin{bmatrix} -2 & -3 & -1 \\ -7 & 4 & 5 \\ 0 & 6 & 1 \end{bmatrix}.$$

Solution

To find the minor for the entry -2, delete the first row and first column.

$$\begin{bmatrix} -2 & -3 & -1 \\ -7 & 4 & 5 \\ 0 & 6 & 1 \end{bmatrix} \qquad \text{The minor for } -2 \text{ is } \begin{vmatrix} 4 & 5 \\ 6 & 1 \end{vmatrix} = 4 - (30) = -26.$$

To find the minor for the entry 5, delete the second row and third column.

$$\begin{bmatrix} -2 & -3 & -1 \\ -7 & 4 & 5 \\ 0 & 6 & 1 \end{bmatrix} \qquad \text{The minor for } 5 \text{ is } \begin{vmatrix} -2 & -3 \\ 0 & 6 \end{vmatrix} = -12 - (0) = -12.$$

To find the minor for the entry 6, delete the third row and second column.

$$\begin{bmatrix} -2 & -3 & -1 \\ -7 & 4 & 5 \\ 0 & 6 & 1 \end{bmatrix} \qquad \text{The minor for } 6 \text{ is } \begin{vmatrix} -2 & -1 \\ -7 & 5 \end{vmatrix} = -10 - (7) = -17.$$

The Determinant of a 3 × 3 Matrix

The determinant of a 3×3 matrix is defined in terms of minors. Let M_{ij} be the 2×2 matrix obtained from M by deleting the ith row and jth column. The determinant of M_{ij}, $|M_{ij}|$, is the minor for a_{ij}.

Definition: Determinant of a 3 × 3 Matrix

If $A = \begin{bmatrix} a_{11} & a_{12} & a_{13} \\ a_{21} & a_{22} & a_{23} \\ a_{31} & a_{32} & a_{33} \end{bmatrix}$, then the determinant of A, $|A|$, is defined as

$$|A| = a_{11} |M_{11}| - a_{21} |M_{21}| + a_{31} |M_{31}|.$$

To find the determinant of A, each entry in the first column of A is multiplied by its minor. This process is referred to as **expansion by minors** about the first column. Note the sign change on the middle term in the expansion.

Example 2   The determinant of a 3 × 3 matrix

Find $|A|$, given that $A = \begin{bmatrix} -2 & -3 & -1 \\ -7 & 4 & 5 \\ 0 & 6 & 1 \end{bmatrix}$.

Solution

Use the definition to expand by minors about the first column.

$$|A| = -2 \cdot \begin{vmatrix} 4 & 5 \\ 6 & 1 \end{vmatrix} - (-7) \cdot \begin{vmatrix} -3 & -1 \\ 6 & 1 \end{vmatrix} + 0 \cdot \begin{vmatrix} -3 & -1 \\ 4 & 5 \end{vmatrix}$$

$$= -2(-26) + 7(3) + 0(-11)$$

$$= 73$$

The value of the determinant of a 3 × 3 matrix can be found by expansion by minors about any row or column. However, you must use alternating plus and minus signs to precede the coefficients of the minors according to the following **sign array:**

$$\begin{bmatrix} + & - & + \\ - & + & - \\ + & - & + \end{bmatrix}$$

The signs in the sign array are used for the determinant of *any* 3 × 3 matrix and they are independent of the signs of the entries in the matrix. Notice that in Example 2, when we expanded by minors about the first column, we used the signs "+ − +" from the first column of the sign array. These signs were used in addition to the signs that appear on the entries themselves (-2, -7, and 0).

Example 3   Expansion by minors about the second column

Expand by minors using the second column to find $|A|$, given that

$A = \begin{bmatrix} -2 & -3 & -1 \\ -7 & 4 & 5 \\ 0 & 6 & 1 \end{bmatrix}$.

Solution

Use the signs "$- + -$" from the second column of the sign array:

$$\begin{bmatrix} + & - & + \\ - & + & - \\ + & - & + \end{bmatrix}$$

The coefficients $-3, 4$, and 6 from the second column of A are preceded by the signs from the second column of the sign array:

From the sign array

$$|A| = -(-3) \cdot \begin{vmatrix} -7 & 5 \\ 0 & 1 \end{vmatrix} + (4) \cdot \begin{vmatrix} -2 & -1 \\ 0 & 1 \end{vmatrix} - (6) \cdot \begin{vmatrix} -2 & -1 \\ -7 & 5 \end{vmatrix}$$

From second column of A

$$= 3(-7) + 4(-2) - 6(-17)$$

$$= 73$$

[A]
    [[-2 -3 -1]
     [-7   4   5]
     [0   6   1]]
det [A]
                73

You can define the matrix A and find its determinant with your calculator.

In Examples 2 and 3 we got the same value for $|A|$ by using two different expansions. Expanding about any row or column gives the same result, but the computations can be easier if we examine the matrix and choose the row or column that contains the most 0's. Using 0's for the coefficients of one or more minors simplifies the work, because we do not have to evaluate the minors that are multiplied by zero. If a row or column of a matrix contains all 0's, then the determinant of the matrix is 0.

Example 4   Expansion by minors using the simplest row or column

Find $|B|$, given that $B = \begin{bmatrix} 4 & 2 & 1 \\ -6 & 3 & 5 \\ 0 & 0 & -7 \end{bmatrix}$.

Solution

Since the third row has two 0's, we expand by minors about the third row. Use the signs "$+ - +$" from the third row of the sign array and the coefficients $0, 0$, and -7 from the third row of B:

$$|B| = 0 \cdot \begin{vmatrix} 2 & 1 \\ 3 & 5 \end{vmatrix} - 0 \cdot \begin{vmatrix} 4 & 1 \\ -6 & 5 \end{vmatrix} + (-7) \cdot \begin{vmatrix} 4 & 2 \\ -6 & 3 \end{vmatrix}$$

$$= -7(24)$$

$$= -168$$

Determinant of a 4 × 4 Matrix

The determinant of a 4 × 4 matrix is also found by expanding by minors about a row or column. The following 4 × 4 sign array of alternating + and − signs (starting with + in the upper-left position) is used for the signs in the expansion:

$$\begin{bmatrix} + & - & + & - \\ - & + & - & + \\ + & - & + & - \\ - & + & - & + \end{bmatrix}$$

The minor for an entry of a 4 × 4 matrix is the determinant of the 3 × 3 matrix found by deleting the row and column of that entry. In general, the determinant of an $n \times n$ matrix is defined in terms of determinants of $(n - 1) \times (n - 1)$ matrices in the same manner.

Example 5   Determinant of a 4 × 4 matrix

Find $|A|$, given that $A = \begin{bmatrix} -2 & -3 & 0 & 4 \\ 1 & -6 & 1 & -1 \\ 2 & 0 & 1 & 5 \\ 4 & 0 & 3 & 1 \end{bmatrix}$

Solution

Since the second column has two zeros, we expand by minors about the second column, using the signs "− + − +" from the second column of the sign array:

$$|A| = -(-3)\begin{vmatrix} 1 & 1 & -1 \\ 2 & 1 & 5 \\ 4 & 3 & 1 \end{vmatrix} + (-6)\begin{vmatrix} -2 & 0 & 4 \\ 2 & 1 & 5 \\ 4 & 3 & 1 \end{vmatrix}$$

$$- 0\begin{vmatrix} -2 & 0 & 4 \\ 1 & 1 & -1 \\ 4 & 3 & 1 \end{vmatrix} + 0\begin{vmatrix} -2 & 0 & 4 \\ 1 & 1 & -1 \\ 2 & 1 & 5 \end{vmatrix}$$

Evaluate the determinant of the first two 3 × 3 matrices to get

$$|A| = 3(2) - 6(36) = -210.$$

```
[A]
 [[-2 -3 0 4 ]
  [1  -6 1 -1]
  [2  0  1 5 ]
  [4  0  3 1 ]]
det [A]
              -210
■
```

You can use a calculator to find the determinant of a large matrix and to solve systems by Cramer's rule.

Cramer's Rule for Systems in Three Variables

Cramer's rule for solving a system of three linear equations in three variables is similar to Cramer's rule for two variables. The rule consists of formulas for x, y, and z in terms of determinants. The development of Cramer's rule for three variables is similar to the development for two variables that was presented in Section 10.5, and so we will omit it.

Cramer's Rule for Systems in Three Variables

The solution to the system

$$a_1 x + b_1 y + c_1 z = d_1$$

$$a_2 x + b_2 y + c_2 z = d_2$$

$$a_3 x + b_3 y + c_3 z = d_3$$

is given by $x = \dfrac{D_x}{D}$, $y = \dfrac{D_y}{D}$, and $z = \dfrac{D_z}{D}$, where

$$D = \begin{vmatrix} a_1 & b_1 & c_1 \\ a_2 & b_2 & c_2 \\ a_3 & b_3 & c_3 \end{vmatrix}, \qquad D_x = \begin{vmatrix} d_1 & b_1 & c_1 \\ d_2 & b_2 & c_2 \\ d_3 & b_3 & c_3 \end{vmatrix},$$

$$D_y = \begin{vmatrix} a_1 & d_1 & c_1 \\ a_2 & d_2 & c_2 \\ a_3 & d_3 & c_3 \end{vmatrix}, \quad \text{and} \quad D_z = \begin{vmatrix} a_1 & b_1 & d_1 \\ a_2 & b_2 & d_2 \\ a_3 & b_3 & d_3 \end{vmatrix}, \quad \text{for } D \neq 0.$$

Note that D_x, D_y, and D_z are obtained by replacing the first, second, and third columns of D by the constants d_1, d_2, and d_3, respectively.

Example 6   Solving a system using Cramer's rule

Use Cramer's rule to solve the system

$$x + y + z = 0$$

$$2x - y + z = -1$$

$$-x + 3y - z = -8.$$

Solution

To use Cramer's rule, we first evaluate D, D_x, D_y, and D_z:

$$D = \begin{vmatrix} 1 & 1 & 1 \\ 2 & -1 & 1 \\ -1 & 3 & -1 \end{vmatrix} = 1 \cdot \begin{vmatrix} -1 & 1 \\ 3 & -1 \end{vmatrix} - 2 \cdot \begin{vmatrix} 1 & 1 \\ 3 & -1 \end{vmatrix} + (-1) \cdot \begin{vmatrix} 1 & 1 \\ -1 & 1 \end{vmatrix}$$

$$= 1(-2) - 2(-4) - 1(2)$$

$$= 4$$

To find D_x, D_y, or D_z, expand by minors about the first row because the first row contains a zero in each case:

$$D_x = \begin{vmatrix} 0 & 1 & 1 \\ -1 & -1 & 1 \\ -8 & 3 & -1 \end{vmatrix} = 0 \cdot \begin{vmatrix} -1 & 1 \\ 3 & -1 \end{vmatrix} - (1) \cdot \begin{vmatrix} -1 & 1 \\ -8 & -1 \end{vmatrix} + (1) \cdot \begin{vmatrix} -1 & -1 \\ -8 & 3 \end{vmatrix}$$

$$= -1(9) + 1(-11)$$

$$= -20$$

$$D_y = \begin{vmatrix} 1 & 0 & 1 \\ 2 & -1 & 1 \\ -1 & -8 & -1 \end{vmatrix} = 1 \cdot \begin{vmatrix} -1 & 1 \\ -8 & -1 \end{vmatrix} - (0) \cdot \begin{vmatrix} 2 & 1 \\ -1 & -1 \end{vmatrix} + (1) \cdot \begin{vmatrix} 2 & -1 \\ -1 & -8 \end{vmatrix}$$

$$= 1(9) + 1(-17)$$
$$= -8$$

$$D_z = \begin{vmatrix} 1 & 1 & 0 \\ 2 & -1 & -1 \\ -1 & 3 & -8 \end{vmatrix} = 1 \cdot \begin{vmatrix} -1 & -1 \\ 3 & -8 \end{vmatrix} - (1) \cdot \begin{vmatrix} 2 & -1 \\ -1 & -8 \end{vmatrix} + (0) \cdot \begin{vmatrix} 2 & -1 \\ -1 & 3 \end{vmatrix}$$

$$= 1(11) - 1(-17)$$
$$= 28$$

Now, by Cramer's rule,

$$x = \frac{D_x}{D} = \frac{-20}{4} = -5, \quad y = \frac{D_y}{D} = \frac{-8}{4} = -2, \quad \text{and} \quad z = \frac{D_z}{D} = \frac{28}{4} = 7.$$

Check that the ordered triple $(-5, -2, 7)$ satisfies all three equations. The solution set to the system is $\{(-5, -2, 7)\}$. ◆

Cramer's rule can provide the solution to any system of three linear equations in three variables that has a unique solution. Its advantage is that it can give the value of any one of the variables without having to solve for the others. If $D = 0$, then Cramer's rule does not give the solution to the system, but it does indicate that the system is either dependent or inconsistent. If $D = 0$, then we must use another method to complete the solution to the system.

Example 7 Solving a system with $D = 0$

Use Cramer's rule to solve the system.

$$(1) \qquad 2x + y - z = 3$$
$$(2) \qquad 4x + 2y - 2z = 6$$
$$(3) \qquad 6x + 3y - 3z = 9$$

Solution

To use Cramer's rule, we first evaluate D:

$$D = \begin{vmatrix} 2 & 1 & -1 \\ 4 & 2 & -2 \\ 6 & 3 & -3 \end{vmatrix} = 2 \cdot \begin{vmatrix} 2 & -2 \\ 3 & -3 \end{vmatrix} - 4 \cdot \begin{vmatrix} 1 & -1 \\ 3 & -3 \end{vmatrix} + 6 \cdot \begin{vmatrix} 1 & -1 \\ 2 & -2 \end{vmatrix}$$

$$= 2(0) - 4(0) + 6(0)$$
$$= 0$$

Because $D = 0$, Cramer's rule cannot be used to solve the system. We could use the Gauss-Jordan method, but note that Eqs. (2) and (3) are obtained by multiplying Eq. (1) by 2 and 3, respectively. Since all three equations are equivalent, the solution set to the system is $\{(x, y, z) \mid 2x + y - z = 3\}$. ◆

In the last two chapters we have discussed several different methods for solving systems of linear equations. Studying different methods increases our understanding of systems of equations. A small system can usually be solved by any of these methods, but for large systems that are solved with computers, the most efficient and popular method is probably the Gauss-Jordan method or a variation of it. Since you can find determinants, invert matrices, and perform operations with them on a graphing calculator, you can use either Cramer's rule or inverse matrices with a graphing calculator.

? For Thought

True or false? Explain.

Statements 1–5 reference the matrix $A = \begin{bmatrix} 2 & -3 & 1 \\ 3 & 4 & 2 \\ 0 & 0 & 1 \end{bmatrix}$.

1. The sign array is used to determine whether $|A|$ is positive or negative.

2. $|A| = 2 \cdot \begin{vmatrix} 4 & 2 \\ 0 & 1 \end{vmatrix} - (-3) \cdot \begin{vmatrix} 3 & 2 \\ 0 & 1 \end{vmatrix} + 1 \cdot \begin{vmatrix} 2 & -3 \\ 3 & 4 \end{vmatrix}$

3. We can find $|A|$ by expanding about any row or column.

4. $|A| = \begin{vmatrix} 2 & -3 \\ 3 & 4 \end{vmatrix}$

5. We can find $|A|$ by expanding by minors about the diagonal.

6. A minor is a 2×2 matrix.

7. By Cramer's rule, the value of x is D/D_x.

8. If a matrix has a row in which all entries are zero, then the determinant of the matrix is 0.

9. By Cramer's rule, there is no solution to a system for which $D = 0$.

10. Cramer's rule works on any system of nonlinear equations.

10.6 Exercises   Tape 20   Disk—5.25″: 7   3.5″: 4   Macintosh: 4

Find the indicated minors, using the matrix
$\begin{bmatrix} 2 & -3 & 1 \\ 4 & 5 & -6 \\ 7 & 9 & -8 \end{bmatrix}$.

1. Minor for 2   2. Minor for −3   3. Minor for 1

4. Minor for 4   5. Minor for 5   6. Minor for −6

7. Minor for 9   8. Minor for −8

Find the determinant of each 3×3 matrix, using expansion by minors about the first column.

9. $\begin{bmatrix} 1 & -4 & 0 \\ -3 & 1 & -2 \\ 3 & -1 & 5 \end{bmatrix}$ 10. $\begin{bmatrix} 1 & -3 & 2 \\ 3 & 1 & -4 \\ 2 & 3 & 6 \end{bmatrix}$

11. $\begin{bmatrix} 3 & -1 & 2 \\ 0 & 4 & -1 \\ 5 & 1 & -2 \end{bmatrix}$ 12. $\begin{bmatrix} -1 & 3 & -1 \\ 0 & 2 & -3 \\ 2 & 6 & -9 \end{bmatrix}$

13. $\begin{bmatrix} -2 & 5 & 1 \\ -3 & 0 & -1 \\ 0 & 2 & -7 \end{bmatrix}$ **14.** $\begin{bmatrix} 0 & -6 & 2 \\ -1 & 4 & -2 \\ 5 & 3 & -1 \end{bmatrix}$

15. $\begin{bmatrix} 0.1 & 30 & 1 \\ 0.4 & 20 & 6 \\ 0.7 & 90 & 8 \end{bmatrix}$ **16.** $\begin{bmatrix} 3 & 0.3 & 10 \\ 5 & 0.5 & 30 \\ 8 & 0.1 & 80 \end{bmatrix}$

Evaluate the following determinants, using expansion by minors about the row or column of your choice.

17. $\begin{vmatrix} -1 & 3 & 5 \\ -2 & 0 & 0 \\ 4 & 3 & -4 \end{vmatrix}$ **18.** $\begin{vmatrix} 8 & -9 & 1 \\ 3 & 4 & 0 \\ -2 & 1 & 0 \end{vmatrix}$

19. $\begin{vmatrix} 1 & 1 & 1 \\ 2 & 2 & 2 \\ 4 & 4 & 4 \end{vmatrix}$ **20.** $\begin{vmatrix} 4 & -1 & 3 \\ 4 & -1 & 3 \\ 4 & -1 & 3 \end{vmatrix}$

21. $\begin{vmatrix} 0 & -1 & 0 \\ 3 & 4 & 6 \\ -2 & 3 & -5 \end{vmatrix}$ **22.** $\begin{vmatrix} 2 & 0 & 0 \\ 56 & 3 & -4 \\ 88 & 5 & -2 \end{vmatrix}$

23. $\begin{vmatrix} 2 & 0 & 1 \\ 4 & 0 & 6 \\ -7 & 9 & -8 \end{vmatrix}$ **24.** $\begin{vmatrix} 2 & -3 & 1 \\ -2 & 5 & -6 \\ 0 & 0 & 0 \end{vmatrix}$

25. $\begin{vmatrix} 3 & 0 & 1 & 5 \\ 2 & -3 & 2 & 0 \\ -2 & 3 & 1 & 2 \\ 2 & -4 & 1 & 3 \end{vmatrix}$ **26.** $\begin{vmatrix} 1 & -4 & 2 & 0 \\ -2 & -1 & 0 & -3 \\ 2 & 2 & 4 & 1 \\ 3 & 0 & -3 & 1 \end{vmatrix}$

27. $\begin{vmatrix} 2 & -3 & 4 & 6 \\ 1 & -5 & 0 & 0 \\ 1 & 3 & 1 & -3 \\ -2 & 0 & 2 & 1 \end{vmatrix}$ **28.** $\begin{vmatrix} -2 & 4 & 0 & 5 \\ 2 & -1 & 0 & 7 \\ 3 & 2 & 0 & -1 \\ 2 & 2 & -3 & 4 \end{vmatrix}$

Solve each system, using Cramer's rule where possible.

29. $\begin{aligned} x + y + z &= 6 \\ x - y + z &= 2 \\ 2x + y + z &= 7 \end{aligned}$ **30.** $\begin{aligned} 2x - 2y + 3z &= 7 \\ x + y - z &= -2 \\ 3x + y - 2z &= 5 \end{aligned}$

31. $\begin{aligned} x + 2y &= 8 \\ x - 3y + z &= -2 \\ 2x - y &= 1 \end{aligned}$ **32.** $\begin{aligned} 2x + y &= -4 \\ 3y - z &= -1 \\ x + 3z &= -16 \end{aligned}$

33. $\begin{aligned} 2x - 3y + z &= 1 \\ x + 4y - z &= 0 \\ 3x - y + 2z &= 0 \end{aligned}$ **34.** $\begin{aligned} -2x + y - z &= 0 \\ x - y + 3z &= 1 \\ 3x + 3y + 2z &= 0 \end{aligned}$

35. $\begin{aligned} x + y + z &= 2 \\ 2x - y + 3z &= 0 \\ 3x + y - z &= 0 \end{aligned}$ **36.** $\begin{aligned} x - 2y - z &= 0 \\ -x + y + 3z &= 0 \\ x + 3y + z &= 3 \end{aligned}$

37. $\begin{aligned} x + y - 2z &= 1 \\ x - 2y + z &= 2 \\ 2x - y - z &= 3 \end{aligned}$ **38.** $\begin{aligned} x + y + z &= 4 \\ -2x - y + 3z &= 1 \\ y + 5z &= 9 \end{aligned}$

39. $\begin{aligned} x - y + z &= 5 \\ x + 2y + 3z &= 8 \\ 2x - 2y + 2z &= 16 \end{aligned}$

40. $\begin{aligned} 3x + 6y + 9z &= 12 \\ x + 2y + 3z &= 0 \\ x - y - 3z &= 0 \end{aligned}$

Solve each problem, using a system of three equations in three unknowns and Cramer's rule.

41. *Age disclosure*   Jackie, Rochelle, and Alisha will not disclose their ages. However, the average of the ages of Jackie and Rochelle is 33, the average for Rochelle and Alisha is 25, and the average for Jackie and Alisha is 19. How old is each?

42. *Bennie's coins*   Bennie emptied his pocket of 49 coins to pay for his \$5.50 lunch. He used only nickels, dimes, and quarters, and the total number of dimes and quarters was one more than the number of nickels. How many of each type of coin did he use?

43. *What a difference a weight makes*   A sociology professor gave two one-hour exams and a final exam. Ian was distressed with his average score of 60 for the three tests and went to see the professor. Because of Ian's improvement during the semester, the professor offered to count the final exam as 60% of the grade and the two tests equally, giving Ian a weighted average of 76. Ian countered that since he improved steadily during the semester, the first test should count 10%, the second 20%, and the final 70% of the grade, giving a weighted average of 83. What were Ian's actual scores on the two tests and the final exam?

44. *Cookie time*   Cheryl, of Cheryl's Famous Cookies, set out 18 cups of flour, 14 cups of sugar, and 13 cups of shortening for her employees to use in making some batches of chocolate chip, oatmeal, and peanut butter cookies. She left for the day without telling them how many batches of each type to bake. The table gives the number of cups of each ingredient required for one batch of each type of cookie. How many batches of each were they supposed to bake?

	Flour	Sugar	Shortening
Chocolate chip	2 cups	2 cups	1 cup
Oatmeal	1 cup	1 cup	2 cups
Peanut butter	4 cups	2 cups	1 cup

Table for Exercise 44

The equation of a line through two points can be expressed as an equation involving a determinant.

45. Show that the following equation is equivalent to the equation of the line through $(3, -5)$ and $(-2, 6)$.

$$\begin{vmatrix} x & y & 1 \\ 3 & -5 & 1 \\ -2 & 6 & 1 \end{vmatrix} = 0$$

46. Show that the following equation is equivalent to the equation of the line through (x_1, y_1) and (x_2, y_2).

$$\begin{vmatrix} x & y & 1 \\ x_1 & y_1 & 1 \\ x_2 & y_2 & 1 \end{vmatrix} = 0$$

For Writing/Discussion

Prove each of the following statements for any 3×3 matrix A.

47. If all entries in any row or column of A are zero, then $|A| = 0$.

48. If A has two identical rows (or columns), then $|A| = 0$.

49. If all entries in a row (or column) of A are multiplied by a constant k, then the determinant of the new matrix is $k \cdot |A|$.

50. If two rows (or columns) of A are interchanged, then the determinant of the new matrix is $-|A|$.

Graphing Calculator Exercises

Use the determinant feature of a graphing calculator to solve each system by Cramer's rule.

1. $0.2x - 0.3y + 1.2z = 13.11$
$0.25x + 0.35y - 0.9z = -1.575$
$2.4x - y + 1.25z = 42.02$

2. $3.6x + 4.5y + 6.8z = 45,300$
$0.09x + 0.05y + 0.04z = 474$
$x + y - z = 0$

Solve each problem, using Cramer's rule and a graphing calculator.

3. *Gasoline sales* The Runway Deli sells regular unleaded, plus unleaded, and supreme unleaded Chevron gasoline. The number of gallons of each grade and the total receipts for gasoline are shown in the table for the first three weeks of February. What was the price per gallon for each grade?

	Regular	Plus	Supreme	Receipts
Week 1	1270	980	890	$3728.66
Week 2	1450	1280	1050	$4496.82
Week 3	1340	1190	1060	$4279.01

4. *Gasoline sales* John's Curb Market sells regular unleaded, plus unleaded, and supreme unleaded Citgo gasoline. John was experimenting with prices during the first three weeks of February. The prices that he charged during each week for each grade and the total receipts for each week are given in the table. John unexpectedly found that, regardless of price, the number of gallons sold in each grade was the same for all three weeks. How many gallons of each grade did he sell every week?

	Regular	Plus	Supreme	Receipts
Week 1	$1.099	$1.209	$1.259	$5457.47
Week 2	$1.069	$1.219	$1.289	$5455.17
Week 3	$1.029	$1.239	$1.299	$5422.47

Extend Cramer's rule to four linear equations in four unknowns. Solve each system, using the extended Cramer's rule.

5. $w + x + y + z = 4$
$2w - x + y + 3z = 13$
$w + 2x - y + 2z = -2$
$w - x - y + 4z = 8$

6. $2w + 2x - 2y + z = 9$
$w + x + y + z = 7$
$4w - 3x + 2y - 5z = -13$
$w + 3x - y + 9z = 10$

Highlights

Section 10.1 Solving Linear Systems Using Matrices

1. An $m \times n$ matrix is a rectangular array of numbers with m rows and n columns.

2. For a system of linear equations, the augmented matrix is a matrix whose entries are the coefficients of the variables of the system together with the constants.

3. In the Gauss-Jordan method for solving a system of linear equations, there are three row operations that can be used on the augmented matrix to simplify it.

4. Applying the Gauss-Jordan method to a dependent system causes an entire row to appear with zero in each entry.

5. Applying the Gauss-Jordan method to an inconsistent system causes a row to appear with zero as the entry for each coefficient but a nonzero entry for the constant.

Section 10.2 Operations with Matrices

1. Two matrices are equal if they have the same order and all corresponding entries are equal.

2. Two matrices of the same order can be added or subtracted by adding or subtracting the corresponding entries.

3. Matrices of a given order have an additive identity and each matrix has an additive inverse.

4. The product of a real number (scalar) and a matrix is found by multiplying each entry of the matrix by the real number.

Section 10.3 Multiplication of Matrices

1. The product of an $m \times n$ matrix A and an $n \times p$ matrix B is an $m \times p$ matrix AB, where the ijth entry of AB is the sum of the products of the corresponding entries in the ith row of A and the jth column of B.

2. The product AB is defined only if the number of columns of A is equal to the number of rows of B. AB is not necessarily equal to BA.

3. A linear system of equations can be written as a matrix equation in the form $AX = B$, where A is the matrix of coefficients, X is a column matrix containing the variables, and B is a column matrix containing the constants.

Section 10.4 Inverses of Matrices

1. The $n \times n$ identity matrix has 1's on the diagonal and 0's elsewhere.

2. If A is an $n \times n$ matrix and I is the $n \times n$ identity matrix, then $AI = IA = A$.

3. If A is an $n \times n$ matrix for which there is a matrix A^{-1} such that $AA^{-1} = A^{-1}A = I$, then A is an invertible matrix and A and A^{-1} are inverses of each other.

4. If A is an invertible matrix, then the solution to $AX = B$ is $X = A^{-1}B$.

Section 10.5   Solution of Linear Systems in Two Variables Using Determinants

1. The determinant of a square matrix is a real number associated with the matrix.
2. Cramer's rule gives formulas involving determinants for finding the solution to an independent system of two linear equations in two unknowns or three linear equations in three unknowns.
3. A matrix has an inverse if and only if its determinant is nonzero.

Section 10.6   Solution of Linear Systems in Three Variables Using Determinants

1. The determinant of a 3×3 matrix is defined in terms of determinants of 2×2 matrices, using the idea of minors and the sign array.
2. Cramer's rule will not work if the determinant of the matrix of coefficients is zero.

Chapter 10 Review Exercises

Let $A = \begin{bmatrix} 2 & -3 \\ -2 & 4 \end{bmatrix}$, $B = \begin{bmatrix} 3 & 7 \\ 1 & 2 \end{bmatrix}$, $C = \begin{bmatrix} -1 \\ 3 \end{bmatrix}$,

$D = \begin{bmatrix} 5 \\ -3 \end{bmatrix}$, $E = \begin{bmatrix} 1 \\ -4 \\ 3 \end{bmatrix}$, $F = [3 \quad 2 \quad -1]$, and

$G = \begin{bmatrix} -1 & 0 & 0 \\ 1 & 1 & 0 \\ -2 & 3 & 1 \end{bmatrix}$. Find each of the following matrices

or determinants if possible.

1. $A + B$ 2. $A - B$ 3. $2A - B$ 4. $2A + 3B$
5. AB 6. BA 7. $D + E$ 8. $F + G$
9. AC 10. BD 11. EF 12. FE
13. FG 14. GE 15. GF 16. EG
17. A^{-1} 18. B^{-1} 19. G^{-1} 20. $A^{-1}C$
21. $(AB)^{-1}$ 22. $A^{-1}B^{-1}$ 23. AA^{-1} 24. GG^{-1}
25. $|A|$ 26. $|B|$ 27. $|G|$ 28. $|C|$

Solve each of the following systems by all three methods: Gauss-Jordan, matrix inversion, and Cramer's rule.

29. $x + y = 9$
$2x - y = 1$

30. $x - 2y = 3$
$x + 2y = 2$

31. $2x + y = -1$
$3x + 2y = 0$

32. $3x - y = 1$
$-2x + y = 1$

33. $x - 5y = 9$
$-2x + 10y = -18$

34. $3x - y = 4$
$6x - 2y = 6$

35. $0.05x + 0.1y = 1$
$10x + 20y = 20$

36. $0.04x - 0.2y = 3$
$2x - 10y = 150$

37. $x + y - 2z = -3$
$-x + 2y - z = 0$
$-x - y + 3z = 6$

38. $x - y + z = 5$
$x + y + 3z = 11$
$-x + 2y - z = -5$

39. $y - 3z = 1$
$x + 2y = 5$
$x + 4z = 1$

40. $3x + y - 2z = 0$
$-2y + z = 0$
$y + 3z = 14$

41. $x - y + z = 2$
$x - 2y - z = 1$
$2x - 3y = 3$

42. $x + 2y + z = 1$
$2x + 4y + 2z = 0$
$-x - 2y - z = 2$

43. $x - 3y - z = 2$
$x - 3y - z = 1$
$x - 3y - z = 0$

44. $2x - y - z = 0$
$x + y + z = 3$
$3x = 3$

Find the values of x, y, and z that make each of the equations true.

45. $\begin{bmatrix} x \\ x + y \end{bmatrix} = \begin{bmatrix} 9 \\ -3 \end{bmatrix}$

46. $\begin{bmatrix} x^2 \\ x - y \end{bmatrix} = \begin{bmatrix} 4 \\ 1 \end{bmatrix}$

47. $\begin{bmatrix} 1 & 1 \\ 2 & 1 \end{bmatrix} \begin{bmatrix} x \\ y \end{bmatrix} = \begin{bmatrix} 6 \\ 8 \end{bmatrix}$

48. $\begin{bmatrix} 0 & 0.5 \\ 1 & 1 \end{bmatrix} \begin{bmatrix} x \\ y \end{bmatrix} = \begin{bmatrix} 7 \\ 9 \end{bmatrix}$

49. $\begin{bmatrix} x \\ y \end{bmatrix} + \begin{bmatrix} y \\ -x \end{bmatrix} = \begin{bmatrix} -3 \\ y \end{bmatrix}$ **50.** $\begin{bmatrix} y \\ x \end{bmatrix} - \begin{bmatrix} x \\ y \end{bmatrix} = \begin{bmatrix} 4 \\ 5 \end{bmatrix}$

51. $\begin{bmatrix} x+y & 0 & 0 \\ 0 & y+z & 0 \\ 0 & 0 & x+z \end{bmatrix} = \begin{bmatrix} 1 & 0 & 0 \\ 0 & 1 & 0 \\ 0 & 0 & 1 \end{bmatrix}$

52. $\begin{bmatrix} 0 & x-y & 2z \\ 0 & 0 & y-z \\ 0 & 0 & 0 \end{bmatrix} = \begin{bmatrix} 0 & 2 & 5 \\ 0 & 0 & 3 \\ 0 & 0 & 0 \end{bmatrix}$

53. $\begin{bmatrix} 1 & 1 & 0 \\ 0 & 1 & 2 \\ 1 & 0 & 3 \end{bmatrix} \begin{bmatrix} x \\ y \\ z \end{bmatrix} = \begin{bmatrix} -1 \\ 7 \\ 17 \end{bmatrix}$

54. $\begin{bmatrix} 1 & 1 & 1 \\ -1 & 1 & -1 \\ 1 & 0 & 2 \end{bmatrix} \begin{bmatrix} x \\ y \\ z \end{bmatrix} = \begin{bmatrix} 0 \\ 0 \\ 0 \end{bmatrix}$

Solve each problem, using a system of linear equations in two or three variables. Use the method of your choice from this chapter.

55. *Fine for polluting* A small manufacturing plant must pay a fine of $10 for each gallon of pollutant A and $6 for each gallon of pollutant B per day that it discharges into a nearby stream. If the manufacturing process produces three gallons of pollutant A for every four gallons of pollutant B and the daily fine is $4060, then how many gallons of each are discharged each day?

56. *Spending money* Heiko, Seth, and André spent a total of $216 on entertainment last month. Heiko's and Seth's expenses totaled only half as much as André's. If Seth spent $12 more than Heiko, then how much did each spend?

57. *Utility bills* Bette's total expense for water, gas, and electricity for one month including tax was $189.83. There is a 6% state tax on electricity, a 5% city tax on gas, and a 4% county tax on water. If her total expenses included $9.83 in taxes and her electric bill including tax was twice the gas bill including tax, then how much was each bill including tax?

58. *Predicting car sales* In the first three months of the year, West Coast Cadillac sold 38, 42, and 49 new cars, respectively. Find the equation of the parabola of the form $y = ax^2 + bx + c$ that passes through $(1, 38)$, $(2, 42)$, and $(3, 49)$ as shown in the figure. (The fact that each point satisfies $y = ax^2 + bx + c$ gives three linear equations in a, b, and c.) Assuming that the fourth month's sales will fall on that same parabola, what would be the predicted sales for the fourth month?

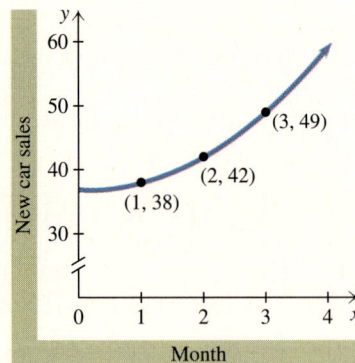

Figure for Exercise 58

Chapter 10 Test

Solve each system, using the Gauss-Jordan method.

1. $2x - 3y = 1$
$x + 9y = 4$

2. $2x - y + z = 5$
$x - 2y - z = -2$
$3x - y - z = 6$

3. $x - y - z = 1$
$2x + y - z = 0$
$5x - 2y - 4z = 3$

Let $A = \begin{bmatrix} 1 & -1 \\ -2 & 4 \end{bmatrix}$, $B = \begin{bmatrix} 2 & -3 \\ -4 & 6 \end{bmatrix}$, $C = \begin{bmatrix} -2 \\ 1 \end{bmatrix}$,

$D = \begin{bmatrix} 3 \\ -2 \end{bmatrix}$, $E = \begin{bmatrix} 2 \\ 3 \\ -1 \end{bmatrix}$, $F = [1 \ \ 0 \ \ -1]$, **and**

$G = \begin{bmatrix} -2 & 3 & 1 \\ -3 & 1 & 3 \\ 0 & 2 & -1 \end{bmatrix}$. **Find each of the following matrices or determinants if possible.**

4. $A + B$ **5.** $2A - B$

6. AB **7.** AC

8. CB **9.** FG

10. EF **11.** A^{-1}

12. G^{-1} **13.** $|A|$

14. $|B|$ **15.** $|G|$

Solve each system, using Cramer's rule.

16. $\begin{aligned} x - y &= 2 \\ -2x + 4y &= 2 \end{aligned}$ **17.** $\begin{aligned} 2x - 3y &= 6 \\ -4x + 6y &= 1 \end{aligned}$

18. $\begin{aligned} -2x + 3y + z &= -2 \\ -3x + y + 3z &= -4 \\ 2y - z &= 0 \end{aligned}$

Solve each system by using inverse matrices.

19. $\begin{aligned} x - y &= 1 \\ -2x + 4y &= -8 \end{aligned}$ **20.** $\begin{aligned} -2x + 3y + z &= 1 \\ -3x + y + 3z &= 0 \\ 2y - z &= -1 \end{aligned}$

Solve by using a method from this chapter.

21. The manager of a computer store bought x copies of the program Math Skillbuilder for $10 each at the beginning of the year and sold y copies of the program for $35 each. At the end of the year the program was obsolete and she destroyed 12 unsold copies. If her net profit for the year was $730, then how many were bought and how many were sold?

22. Find a, b, and c such that the graph of $y = ax^2 + b\sqrt{x} + c$ goes through the points $(0, 3)$, $(1, -1/2)$, and $(4, 3)$.

Tying It All Together
Chapters 1–10

Solve each equation.

1. $2(x + 3) - 5x = 7$ **2.** $\dfrac{1}{2}\left(x - \dfrac{1}{3}\right) + \dfrac{1}{5} = 1$ **3.** $\dfrac{1}{2}(2x - 2)(6x - 8) = 4$ **4.** $1 - \dfrac{1}{2}(8x - 4) = 9$

Solve each system of equations by the specified method.

5. Graphing:
$\begin{aligned} 2x + y &= 6 \\ x - 2y &= 8 \end{aligned}$

6. Substitution:
$\begin{aligned} y + 2x &= 1 \\ 2x + 6y &= 2 \end{aligned}$

7. Addition:
$\begin{aligned} 2x - 0.06y &= 20 \\ 3x + 0.01y &= 20 \end{aligned}$

8. Gauss-Jordan:
$\begin{aligned} 2x - y &= -1 \\ x + 3y &= -11 \end{aligned}$

9. Matrix inversion:
$\begin{aligned} 3x - 5y &= -7 \\ -x + y &= 1 \end{aligned}$

10. Cramer's rule:
$\begin{aligned} 4x - 3y &= 5 \\ 3x - 5y &= 1 \end{aligned}$

11. Your choice:
$\begin{aligned} x^2 + y^2 &= 25 \\ x - y &= -1 \end{aligned}$

12. Your choice:
$\begin{aligned} x^2 - y &= 1 \\ x + y &= 1 \end{aligned}$

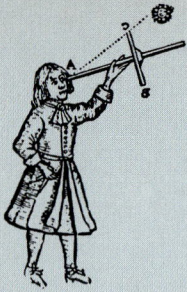

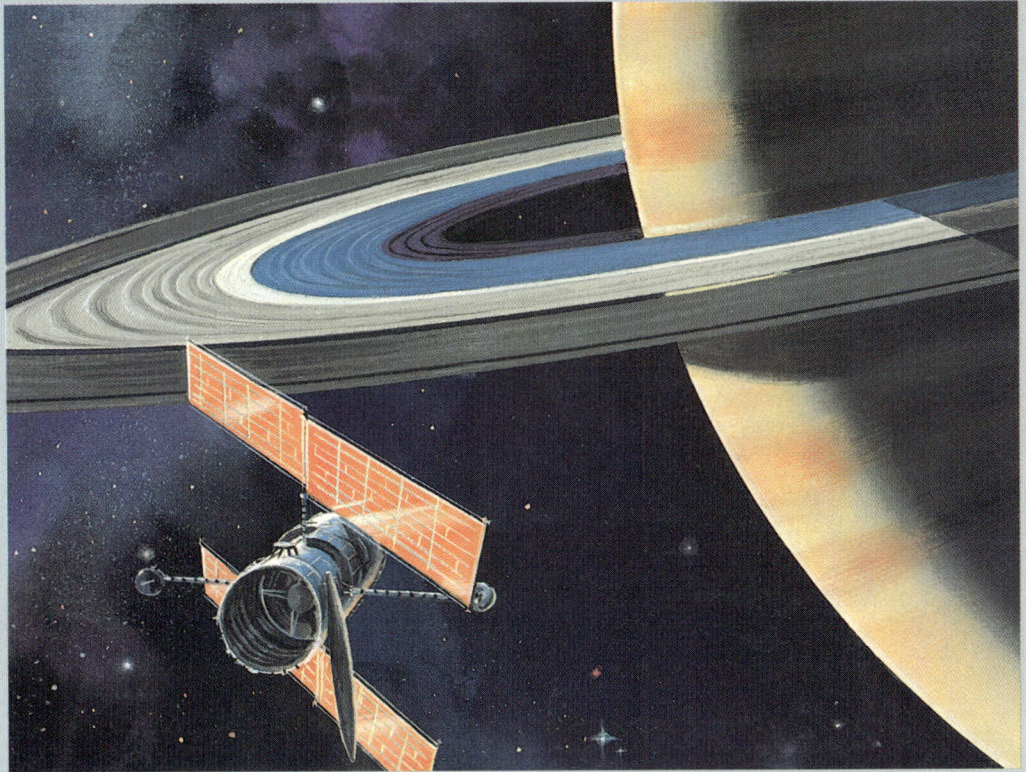

The images of Saturn were spectacular! Orbiting far beyond Earth's distorting atmosphere, the Hubble space telescope had recorded a 50,000-kilometer-wide storm of ammonia ice crystals, which appears only once every 60 years. Hubble also revealed changes in Jupiter's cloud bands and famed Great Red Spot— possibly a perpetual hurricane twice the diameter of Earth. Then, turning its gaze beyond the solar system, Hubble sent back information on glowing nebulae, the birth and death of stars, and the possible presence of a massive black hole—quite a record of accomplishment for what at first appeared to be a failed mission!

Hubble's odyssey began on April 25, 1990, amid the popping of champagne corks, as scientists and technicians celebrated its successful launch from the shuttle Discovery. Initial enthusiasm for the space telescope fizzled, however, when tests revealed an inherent focusing problem: Hubble's main mirror had been ground to the wrong shape. Although the error was only two millionths of a meter smaller than design specifications, it was enough to dim astronomers' hopes.

Refusing to concede defeat, scientists used computers to enhance transmitted images until in December 1993, astronauts succeeded in repairing the telescope's defects. Through Hubble's newly sharpened pictures, scientists hope to refine the big bang

11

The Conic

Sections

theory, to better understand the evolution of galaxies, and perhaps even to predict the fate of the cosmos.

What gives Hubble and other optical instruments the power to record distant objects? The answer lies in the precisely calculated curves of mirrors and lenses that gather light rays and focus them into dazzling images. The concave mirror at the heart of Hubble has a parabolic shape. And a parabola is but one of several conic sections found useful in exploring space and describing planetary motion. In this chapter we'll explore the amazing properties of parabolas, ellipses, circles, and hyperbolas. We begin with a discussion of the parabola.

11.1

The Parabola

The parabola, circle, ellipse, and hyperbola can be defined as the four curves that are obtained by intersecting a right circular cone and a plane as shown in Fig. 11.1. That is why these curves are known as **conic sections.** If the plane passes through the vertex of the cone, then the intersection of the cone and the plane is called a **degenerate conic.** The conic sections can also be defined as the graphs of certain equations (as we did for the parabola in Section 4.2). However, the useful properties of the curves are not apparent in either of these approaches. In this chapter we give geometric definitions from which we derive equations for the conic sections. This approach allows us to better understand the properties of the conic sections.

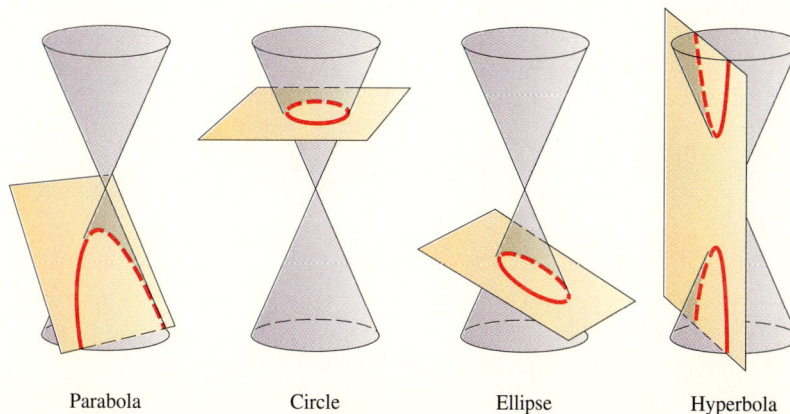

Parabola   Circle   Ellipse   Hyperbola

Figure 11.1

Definition

Previously in Section 4.2 we defined a parabola algebraically as the graph of $y = ax^2 + bx + c$ for $a \neq 0$. This equation can be expressed also in the form $y = a(x - h)^2 + k$. No equation is mentioned in our geometric definition of a parabola.

Definition: Parabola

A **parabola** is the set of all points in the plane that are equidistant from a fixed line (the **directrix**) and a fixed point not on the line (the **focus**).

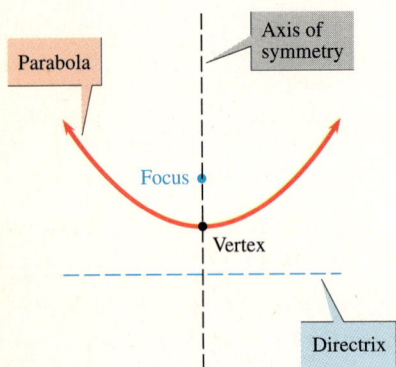

Figure 11.2

Figure 11.2 shows a parabola with its directrix, focus, axis of symmetry, and vertex. In terms of the directrix and focus, the **axis of symmetry** can be described as the line perpendicular to the directrix and containing the focus. The **vertex** is the point on the axis of symmetry that is equidistant from the focus and directrix. If we position the directrix and focus in a coordinate plane with the directrix horizontal, we can find an equation that is satisfied by all points of the parabola.

Developing the Equation

Start with a focus and a horizontal directrix as shown in Fig. 11.3. If we use the coordinates (h, k) for the vertex, then the focus is $(h, k + p)$ and the directrix is $y = k - p$, where p (the **focal length**) is the directed distance from the vertex to the focus. If the focus is above the vertex, then $p > 0$, and if the focus is below the vertex, then $p < 0$. The distance from the vertex to the focus or the vertex to the directrix is $|p|$.

The distance d_1 from an arbitrary point (x, y) on the parabola to the directrix is the distance from (x, y) to $(x, k - p)$, as shown in Fig. 11.3. We use the distance formula from Section 3.1 to find d_1:

$$d_1 = \sqrt{(x - x)^2 + (y - (k - p))^2} = \sqrt{y^2 - 2(k - p)y + (k - p)^2}$$

Now we find the distance d_2 between (x, y) and the focus $(h, k + p)$:

$$d_2 = \sqrt{(x - h)^2 + (y - (k + p))^2} = \sqrt{(x - h)^2 + y^2 - 2(k + p)y + (k + p)^2}$$

Since $d_1 = d_2$ for every point (x, y) on the parabola, we have the following equation.

$$\sqrt{y^2 - 2(k - p)y + (k - p)^2} = \sqrt{(x - h)^2 + y^2 - 2(k + p)y + (k + p)^2}$$

You should verify that squaring each side and simplifying yields

$$y = \frac{1}{4p}(x - h)^2 + k.$$

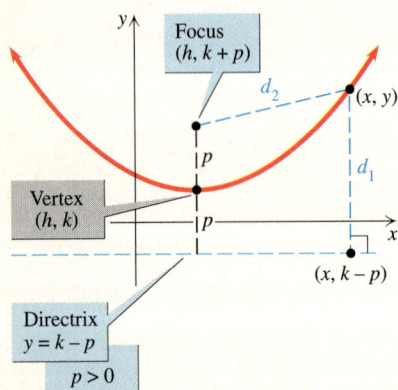

Figure 11.3

This equation is of the form $y = a(x - h)^2 + k$, where $a = 1/(4p)$. So the curve determined by the geometric definition has an equation that is an equation of a parabola according to the algebraic definition. We state these results as follows.

Theorem: The Equation of a Parabola

The equation of a parabola with focus $(h, k + p)$ and directrix $y = k - p$ is

$$y = a(x - h)^2 + k,$$

where $a = 1/(4p)$ and (h, k) is the vertex.

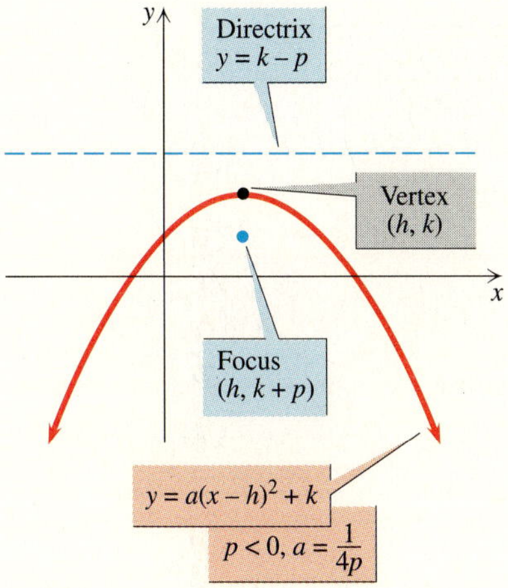

Figure with labels: Directrix $y = k - p$; Vertex (h, k); Focus $(h, k + p)$; $y = a(x - h)^2 + k$; $p < 0$, $a = \frac{1}{4p}$

Figure 11.4

The link between the geometric definition and the equation of a parabola is

$$a = \frac{1}{4p}.$$

For any particular parabola, a and p have the same sign. If they are both positive, the parabola opens upward and the focus is above the directrix. If they are both negative, the parabola opens downward and the focus is below the directrix. Figure 11.4 shows the positions of the focus, directrix, and vertex for a parabola with $p < 0$. Since a is inversely proportional to p, smaller values of $|p|$ correspond to larger values of $|a|$ and to "narrower" parabolas.

Example 1   Writing the equation from the focus and directrix

Find the equation of the parabola with focus $(-1, 3)$ and directrix $y = 2$.

Solution

The focus is one unit above the directrix as shown in Fig. 11.5. So $p = 1/2$. Therefore $a = 1/(4p) = 1/2$. Since the vertex is halfway between the focus and directrix, the y-coordinate of the vertex is $(3 + 2)/2$ and the vertex is $(-1, 5/2)$. Use $a = 1/2$, $h = -1$, and $k = 5/2$ in the formula $y = a(x - h)^2 + k$ to get the equation

$$y = \frac{1}{2}(x - (-1))^2 + \frac{5}{2}.$$

Simplify to get the equation $y = \frac{1}{2}x^2 + x + 3$. ◆

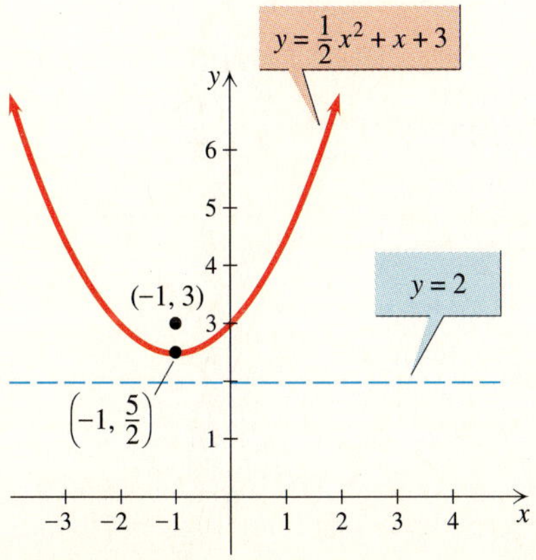

Figure with labels: $y = \frac{1}{2}x^2 + x + 3$; $(-1, 3)$; $y = 2$; $\left(-1, \frac{5}{2}\right)$

Figure 11.5

The Standard Equation of a Parabola

If we start with the standard equation of a parabola, $y = ax^2 + bx + c$, we can identify the vertex, focus, and directrix by rewriting it in the form $y = a(x - h)^2 + k$.

Example 2 Finding the vertex, focus, and directrix

Find the vertex, focus, and directrix of the graph of $y = -3x^2 - 6x + 2$.

Solution

Use completing the square to write the equation in the form $y = a(x - h)^2 + k$:

$$\begin{aligned} y &= -3(x^2 + 2x) + 2 \\ &= -3(x^2 + 2x + 1 - 1) + 2 \\ &= -3(x^2 + 2x + 1) + 2 + 3 \\ &= -3(x + 1)^2 + 5 \end{aligned}$$

Graph $y_1 = -3x^2 - 6x + 2$ and $y_2 = 61/12$ to see how close the directrix is to the vertex.

The vertex is $(-1, 5)$, and the parabola opens downward because $a = -3$. Since $a = 1/(4p)$, we have $1/(4p) = -3$, or $p = -1/12$. Because the parabola opens downward, the focus is $1/12$ unit below the vertex $(-1, 5)$ and the directrix is a horizontal line $1/12$ unit above the vertex. The focus is $(-1, 59/12)$, and the directrix is $y = 61/12$. ◆

In Section 4.2 we learned that the x-coordinate of the vertex of the parabola $y = ax^2 + bx + c$ is $-b/(2a)$. We can use $x = -b/(2a)$ and $a = 1/(4p)$ to determine the focus and directrix without completing the square.

Example 3 Finding the vertex, focus, and directrix

Find the vertex, focus, and directrix of the parabola $y = 2x^2 + 6x - 7$ without completing the square, and determine whether the parabola opens upward or downward.

Solution

First use $x = -b/(2a)$ to find the x-coordinate of the vertex:

$$x = \frac{-b}{2a} = \frac{-6}{2 \cdot 2} = -\frac{3}{2}$$

The graph of $y_2 = -93/8$ is close to the vertex of $y_1 = 2x^2 + 6x - 7$.

To find the y-coordinate of the vertex, let $x = -3/2$ in $y = 2x^2 + 6x - 7$:

$$y = 2\left(-\frac{3}{2}\right)^2 + 6\left(-\frac{3}{2}\right) - 7 = \frac{9}{2} - 9 - 7 = -\frac{23}{2}$$

The vertex is $(-3/2, -23/2)$. Use $2 = 1/(4p)$ to get $p = 1/8$. Since the parabola opens upward, the directrix is $1/8$ unit below the vertex and the focus is $1/8$ unit above the vertex. The directrix is $y = -93/8$, and the focus is $(-3/2, -91/8)$. ◆

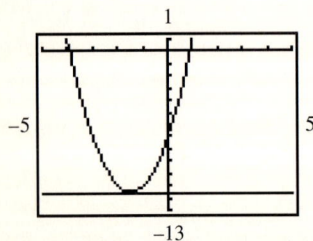

Graphing a Parabola

According to the geometric definition of a parabola, every point on the parabola is equidistant from its focus and directrix. However, it is not easy to find points satisfying that condition. The German mathematician Johannes Kepler (1571–1630) devised a method for drawing a parabola with a given focus and directrix: A piece of string, with length equal to the length of the T-square, is attached to the end of the T-square and the focus, as shown in Fig. 11.6. A pencil is moved down the edge of the T-square, holding the string against it, while the T-square is moved along the directrix toward the focus. While the pencil is aligning the string with the edge of the T-square, it remains equidistant from the focus and directrix.

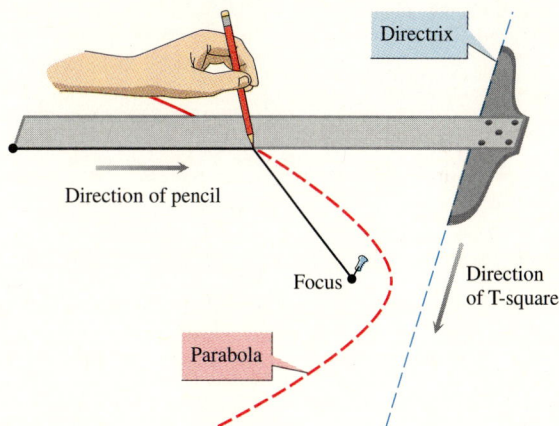

Figure 11.6

Although Kepler's method does give the graph of the parabola from the focus and directrix, you may not want to use it too often. Next, we find the equation of a parabola with a given focus and directrix and then graph it as in Section 4.2.

Example 4   Graphing a parabola given its focus and directrix

Find the vertex, axis of symmetry, x-intercepts, and y-intercept of the parabola that has focus $(3, 7/4)$ and directrix $y = 9/4$. Sketch the graph, showing the focus and directrix.

Solution

First draw the focus and directrix on the graph as shown in Fig. 11.7. Since the vertex is midway between the focus and directrix, the vertex is $(3, 2)$. The distance between the focus and vertex is $1/4$. Since the directrix is above the focus, the parabola opens downward, and we have $p = -1/4$. Use $a = 1/(4p)$ to get $a = -1$. Use $a = -1$ and the vertex $(3, 2)$ in the equation $y = a(x - h)^2 + k$ to get

$$y = -1(x - 3)^2 + 2.$$

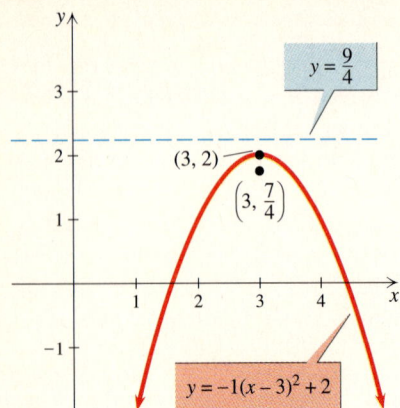

Figure 11.7

The axis of symmetry is the vertical line $x = 3$. If $x = 0$, then $y = -1(0 - 3)^2 + 2 = -7$. So the y-intercept is $(0, -7)$. Find the x-intercepts by setting y equal to 0 in the equation:

$$-1(x - 3)^2 + 2 = 0$$
$$(x - 3)^2 = 2$$
$$x - 3 = \pm\sqrt{2}$$
$$x = 3 \pm \sqrt{2}$$

The x-intercepts are $\left(3 - \sqrt{2}, 0\right)$ and $\left(3 + \sqrt{2}, 0\right)$. Two additional points that satisfy $y = -1(x - 3)^2 + 2$ are $(1, -2)$ and $(4, 1)$. Using all of this information, we get the graph shown in Fig. 11.7. ◆

Parabolas Opening to the Left or Right

If we interchange x and y in $y = a(x - h)^2 + k$, we get

$$x = a(y - h)^2 + k.$$

The graph of this equation is also a parabola. By interchanging the variables, the roles of the x- and y-axes are interchanged. For example, the graph of $y = 2x^2$ opens upward, while the graph of $x = 2y^2$ opens to the right. For parabolas opening right or left, the directrix is a vertical line. If the focus is to the right of the directrix, then the parabola opens to the right, and if the focus is to the left of the directrix, then the parabola opens to the left. Figure 11.8 shows the relative locations of the vertex, focus, and directrix for parabolas of the form $x = a(y - h)^2 + k$.

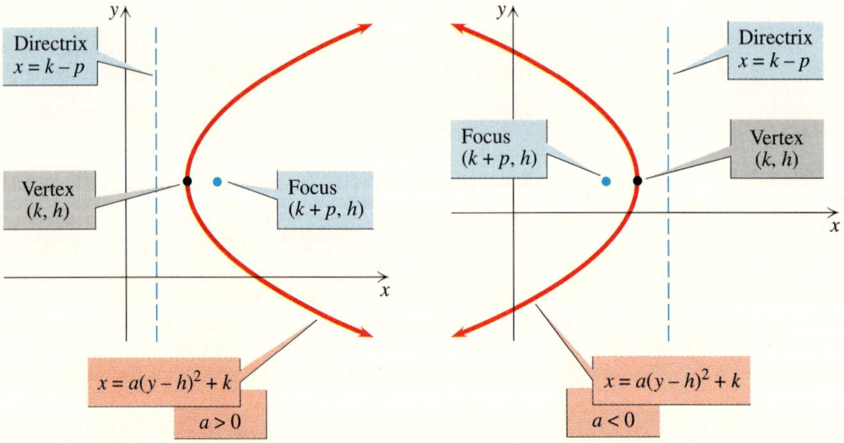

Figure 11.8

The equation $x = a(y - h)^2 + k$ can be written as $x = ay^2 + by + c$. So the graph of $x = ay^2 + by + c$ is also a parabola opening to the right for $a > 0$ and to the left for $a < 0$. Because the roles of x and y are interchanged, $-b/(2a)$ is now the y-coordinate of the vertex, and the axis of symmetry is the horizontal line $y = -b/(2a)$.

Example 5   Graphing a parabola with a vertical directrix

Find the vertex, axis of symmetry, y-intercepts, focus, and directrix for the parabola $x = y^2 - 2y$. Find several other points on the parabola and sketch the graph.

Solution

Because $a = 1$, the parabola opens to the right. The y-coordinate of the vertex is

$$y = \frac{-b}{2a} = \frac{-(-2)}{2(1)} = 1.$$

If $y = 1$, then $x = (1)^2 - 2(1) = -1$ and the vertex is $(-1, 1)$. The axis of symmetry is the horizontal line $y = 1$. To find the y-intercepts, we solve $y^2 - 2y = 0$ by factoring:

$$y(y - 2) = 0$$

$$y = 0 \quad \text{or} \quad y = 2$$

The y-intercepts are $(0, 0)$ and $(0, 2)$. Using all of this information and the additional points $(3, 3)$ and $(3, -1)$, we get the graph shown in Fig. 11.9. Because $a = 1$ and $a = 1/(4p)$, we have $p = 1/4$. So the focus is $1/4$ unit to the right of the vertex at $(-3/4, 1)$. The directrix is the vertical line $1/4$ unit to the left of the vertex, $x = -5/4$.   ◆

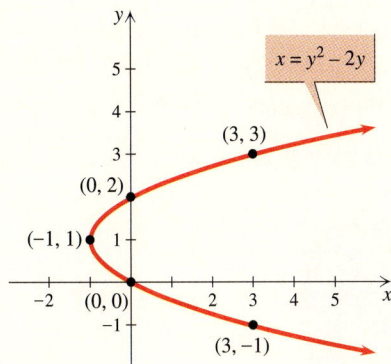

Figure 11.9

Applications

In Section 4.2 we saw an important application of a parabola. Because of the shape of a parabola, a quadratic function has a maximum value or a minimum value at the vertex of the parabola. However, parabolas are important for another totally different reason. When a ray of light, traveling parallel to the axis of symmetry, hits a parabolic reflector, it is reflected toward the focus of the parabola. See Fig. 11.10. This property is used in telescopes to magnify the light from distant stars. For spotlights, in which the light source is at the focus, the reflecting property is used in reverse. Light originating at the focus is reflected off the parabolic reflector and projected outward in a narrow beam. The reflecting property is used also in telephoto camera lenses, radio antennas, satellite dishes, eavesdropping devices, and flashlights.

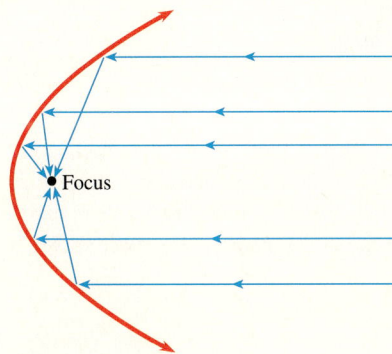

Figure 11.10

Example 6   The Hubble telescope

The Hubble space telescope uses a glass mirror with a parabolic cross section and a diameter of 2.4 meters, as shown in Fig. 11.11 on the following page. If the focus of the parabola is 57.6 meters from the vertex, then what is the equation of the parabola used for the mirror? How much thicker is the mirror at the edge than at its center?

Solution

If we position the parabola with the vertex at the origin and opening upward as shown in Fig. 11.11, then its equation is of the form $y = ax^2$. Since the distance between the focus and vertex is 57.6 meters, $p = 57.6$. Using $a = 1/(4p)$, we get $a = 1/(4 \cdot 57.6) \approx 0.004340$. The equation of the parabola is

$$y = 0.004340x^2.$$

To find the difference in thickness at the edge, let $x = 1.2$ in the equation of the parabola:

$$y = 0.004340(1.2)^2 \approx 0.006250 \text{ meter}$$

The mirror is 0.006250 meter thicker at the edge than at the center.

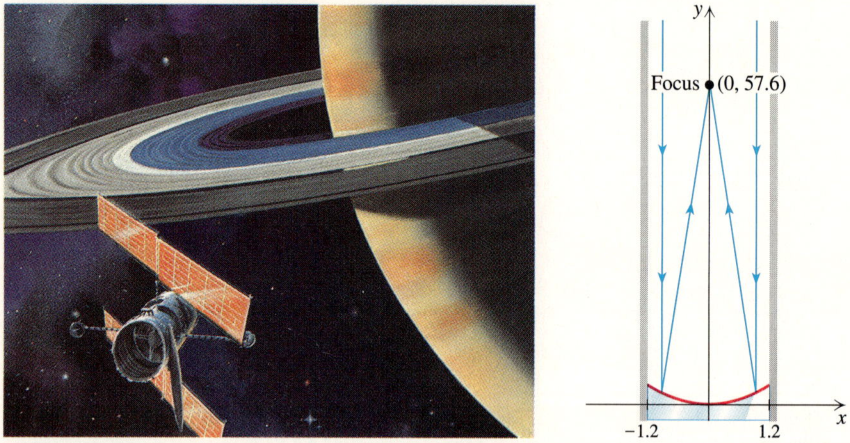

Figure 11.11

For a telescope to work properly, the glass mirror must be ground with great precision. Prior to the 1993 repair, the Hubble space telescope did not work as well as hoped (*Scientific American,* June 1992), because the mirror was actually ground to be only 0.006248 meter thicker at the outside edge, two millionths of a meter smaller than it should have been!

? **For Thought**

True or false? Explain.

1. A parabola with focus (0, 0) and directrix $y = -1$ has vertex (0, 1).
2. A parabola with focus (3, 0) and directrix $x = -1$ opens to the right.
3. A parabola with focus (4, 5) and directrix $x = 1$ has vertex (5/2, 5).
4. For $y = x^2$, the focus is (0, −1/4).
5. For $x = y^2$, the focus is (1/4, 0).

6. A parabola with focus (2, 3) and directrix $y = -5$ has no x-intercepts.

7. The parabola $y = 4x^2 - 9x$ has no y-intercept.

8. The vertex of the parabola $x = 2(y + 3)^2 - 5$ is $(-5, -3)$.

9. The parabola $y = (x - 5)^2 + 4$ has its focus at (5, 4).

10. The parabola $x = -3y^2 + 7y - 5$ opens downward.

11.1 Exercises   Tape 21   Disk—5.25″: 7   3.5″: 4   Macintosh: 4

Find the equation of the parabola with the given focus and directrix.

1.

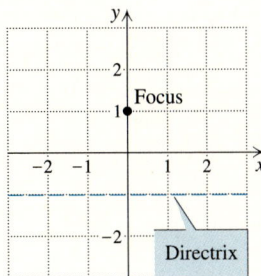

2.

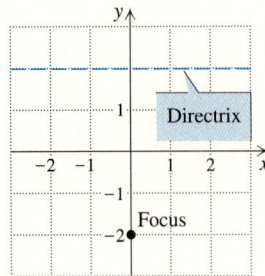

3.

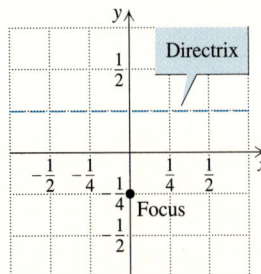

4.

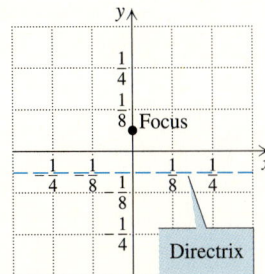

Find the equation of the parabola with the given focus and directrix.

5. Focus (3, 5), directrix $y = 2$

6. Focus (−1, 5), directrix $y = 3$

7. Focus (1, −3), directrix $y = 2$

8. Focus (1, −4), directrix $y = 0$

9. Focus (−2, 1.2), directrix $y = 0.8$

10. Focus (3, 9/8), directrix $y = 7/8$

Use completing the square to write each equation in the form $y = a(x - h)^2 + k$. Identify the vertex, focus, and directrix.

11. $y = x^2 - 8x + 3$

12. $y = x^2 + 2x - 5$

13. $y = 2x^2 + 12x + 5$

14. $y = 3x^2 + 12x + 1$

15. $y = -2x^2 + 6x + 1$

16. $y = -3x^2 - 6x + 5$

17. $y = 5x^2 + 30x$

18. $y = -2x^2 + 12x$

19. $y = \dfrac{1}{8} x^2 - \dfrac{1}{2} x + \dfrac{9}{2}$

20. $y = \dfrac{1}{4} x^2 + \dfrac{1}{2} x - \dfrac{7}{4}$

Find the vertex, focus, and directrix of each parabola without completing the square, and determine whether the parabola opens upward or downward.

21. $y = x^2 - 4x + 3$

22. $y = x^2 - 6x - 7$

23. $y = -x^2 + 2x - 5$

24. $y = -x^2 + 4x + 3$

25. $y = 3x^2 - 6x + 1$

26. $y = 2x^2 + 4x - 1$

27. $y = -\dfrac{1}{2} x^2 - 3x + 2$

28. $y = -\dfrac{1}{2} x^2 + 3x - 1$

29. $y = \dfrac{1}{4} x^2 + 5$

30. $y = -\dfrac{1}{8} x^2 - 6$

Find the vertex, axis of symmetry, x-intercepts, and y-intercept of the parabola that has the given focus and directrix. Sketch the graph, showing the focus and directrix.

31. Focus (1/2, −2), directrix $y = -5/2$

32. Focus (1, −35/4), directrix $y = -37/4$

33. Focus (−1/2, 6), directrix $y = 13/2$

34. Focus (−1, 35/4), directrix $y = 37/4$

Find the vertex, axis of symmetry, x-intercepts, y-intercept, focus, and directrix for each parabola. Sketch the graph, showing the focus and directrix.

35. $y = \dfrac{1}{2}(x + 2)^2 + 2$ **36.** $y = \dfrac{1}{2}(x - 4)^2 + 1$

37. $y = -\dfrac{1}{4}(x + 4)^2 + 2$

38. $y = -\dfrac{1}{4}(x - 2)^2 + 4$

39. $y = \dfrac{1}{2}x^2 - 2$

40. $y = -\dfrac{1}{4}x^2 + 4$

41. $y = x^2 - 4x + 4$

42. $y = (x - 4)^2$

43. $y = \dfrac{1}{3}x^2 - x$

44. $y = \dfrac{1}{5}x^2 + x$

Find the vertex, axis of symmetry, x-intercept, y-intercepts, focus, and directrix for each parabola. Sketch the graph, showing the focus and directrix.

45. $x = -y^2$ **46.** $x = y^2 - 2$

47. $x = -\dfrac{1}{4}y^2 + 1$ **48.** $x = \dfrac{1}{2}(y - 1)^2$

49. $x = y^2 + y - 6$ **50.** $x = y^2 + y - 2$

51. $x = -\dfrac{1}{2}y^2 - y - 4$ **52.** $x = -\dfrac{1}{2}y^2 + 3y + 4$

53. $x = 2(y - 1)^2 + 3$ **54.** $x = 3(y + 1)^2 - 2$

55. $x = -\dfrac{1}{2}(y + 2)^2 + 1$ **56.** $x = -\dfrac{1}{4}(y - 2)^2 - 1$

Find the equation of the parabola determined by the given information.

57. Focus $(1, 5)$, vertex $(1, 4)$

58. Directrix $y = 5$, vertex $(2, 3)$

59. Vertex $(0, 0)$, directrix $x = -2$

60. Focus $(-2, 3)$, vertex $(-9/4, 3)$

Solve each problem.

61. *The Hale telescope*   The focus of the Hale telescope on Palomar Mountain in California is 55 ft above the mirror (at the vertex). The Pyrex glass mirror is 200 in. in diameter and 23 in. thick at the center. Find the equation for the parabola that was used to shape the glass. How thick is the glass on the outside edge?

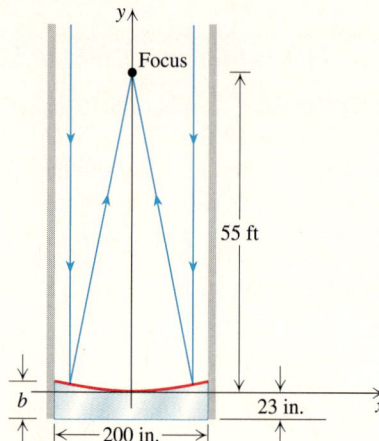

Figure for Exercise 61

62. *Eavesdropping*   From the Edmund Scientific catalog you can buy a device that will "pull in voices up to three-quarters of a mile away with our electronic parabolic microphone." The 18.75-in.-diameter plastic shield reflects sound waves to a microphone located at the focus. Given that the microphone is located 6 in. from the vertex of the parabolic shield, find the equation for a cross section of the shield. What is the depth of the shield?

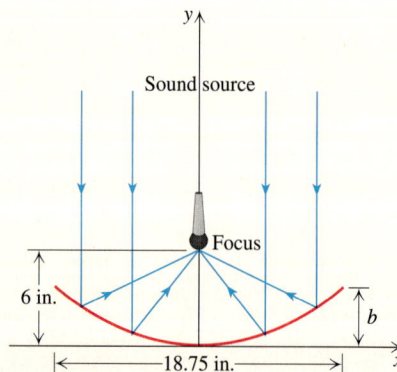

Figure for Exercise 62

For Writing/Discussion

63. *Derive the equation*   In this section we derived the equation of a parabola with a given focus and a horizontal directrix. Use the same technique to derive the equation of a parabola with a given focus and a vertical directrix.

64. *Cooperative learning*   With the help of a partner or two, design and build a portable parabolic pool table. The table should have a parabolic bumper with one pocket at the focus. To demonstrate the reflecting property of the parabola, shoot a hard rubber ball into the pocket by first bouncing it off the bumper.

Graphing Calculator Exercises

1. Graph $y = x^2$ using the viewing window with $-1 \leq x \leq 1$ and $0 \leq y \leq 1$. Graph $y = 2x^2 - 4x + 5$ using the viewing window with $-1 \leq x \leq 3$ and $3 \leq y \leq 11$. What can you say about the two graphs?

2. Find two different viewing windows in which the graph of $y = 3x^2 + 30x + 71$ looks just like the graph of $y = x^2$ in the viewing window with $-1 \leq x \leq 1$ and $0 \leq y \leq 1$.

3. The graph of $x = -y^2$ is a parabola opening to the left with vertex at the origin. Find two functions whose graphs

will together form this parabola and graph them on your calculator.

4. You can illustrate the reflective property of the parabola $x = -y^2$ on the screen of your calculator. First graph $x = -y^2$ as in the previous exercise. Next, by using the graphs of two line segments, make it appear that a particle coming in horizontally from the left toward $(-1, 1)$ is reflected off the parabola and heads toward the focus $(-1/4, 0)$. Consult your manual to see how to graph line segments.

11.2

The Ellipse and

the Circle

Ellipses and circles can be obtained by intersecting planes and cones as shown in Fig. 11.1. Here we will develop general equations for ellipses, and then circles, by starting with geometric definitions as we did with the parabola.

Definition of Ellipse

An easy way to draw an ellipse is illustrated in Fig. 11.12. A string is attached at two fixed points, and a pencil is used to take up the slack. As the pencil is moved around the paper, the sum of the distances of the pencil point from the two fixed points remains constant, because the length of the string is constant. This method of drawing an ellipse illustrates the geometric definition for an ellipse.

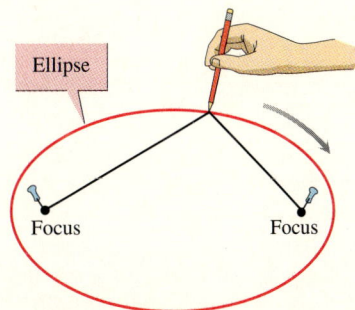

Figure 11.12

Definition: Ellipse

An **ellipse** is the set of points in a plane such that the sum of their distances from two fixed points is a constant. Each fixed point is called a **focus** (plural: foci) of the ellipse.

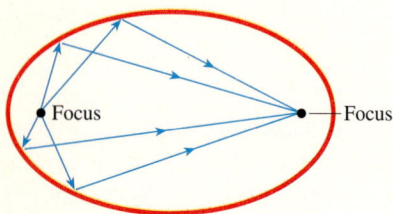

Figure 11.13

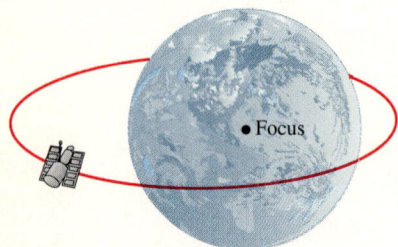

Figure 11.14

The ellipse, like the parabola, has interesting reflecting properties. All light or sound waves emitted from one focus are reflected off the ellipse to concentrate at the other focus as shown in Fig. 11.13. This property is used in light fixtures such as a dentist's light, for which a concentration of light at a point is desired, and in a whispering gallery like the one in the U.S. Capitol Building. In a whispering gallery, a whisper emitted at one focus is reflected off the elliptical ceiling and is amplified so that it can be heard at the other focus, but not anywhere in between.

The orbits of the planets around the sun and satellites around the earth are elliptical. For the orbit of the earth, the sun is at one focus of the elliptical path. For the orbit of a satellite such as the Hubble space telescope, the center of the earth is one focus. See Fig. 11.14.

The Equation of the Ellipse

The ellipse shown in Fig. 11.15 has foci at $(c, 0)$ and $(-c, 0)$, and y-intercepts $(0, b)$ and $(0, -b)$, where $c > 0$ and $b > 0$. The line segment $\overline{V_1 V_2}$ is the **major axis,** and the line segment $\overline{B_1 B_2}$ is the **minor axis.** For any ellipse, the major axis is longer than the minor axis and the foci are on the major axis. The **center** of an ellipse is the midpoint of the major (or minor) axis. The ellipse in Fig. 11.15 is centered at the origin. The **vertices** of an ellipse are the endpoints of the major axis. The vertices of the ellipse in Fig. 11.15 are the x-intercepts.

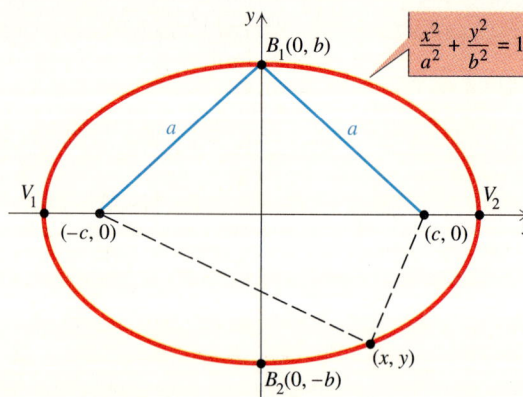

$$\frac{x^2}{a^2} + \frac{y^2}{b^2} = 1$$

Figure 11.15

Let a be the distance between $(c, 0)$ and the y-intercept $(0, b)$ as shown in Fig. 11.15. The sum of the distances from the two foci to $(0, b)$ is $2a$. So for any point (x, y) on the ellipse, the distance from (x, y) to $(c, 0)$ plus the distance from

(x, y) to $(-c, 0)$ is equal to $2a$. Writing this last statement as an equation (using the distance formula) gives the equation of the ellipse shown in Fig. 11.15:

$$\sqrt{(x - c)^2 + (y - 0)^2} + \sqrt{(x - (-c))^2 + (y - 0)^2} = 2a$$

With some effort, this equation can be greatly simplified. We will provide the major steps in simplifying it, and leave the details as an exercise. First simplify inside the radicals and isolate them to get

$$\sqrt{x^2 - 2xc + c^2 + y^2} = 2a - \sqrt{x^2 + 2xc + c^2 + y^2}.$$

Next, square each side and simplify again to get

$$a\sqrt{x^2 + 2xc + c^2 + y^2} = a^2 + xc.$$

Squaring each side again yields

$$a^2x^2 - c^2x^2 + a^2y^2 = a^4 - a^2c^2$$

$$(a^2 - c^2)x^2 + a^2y^2 = a^2(a^2 - c^2) \qquad \text{Factor.}$$

Since $a^2 = b^2 + c^2$, or $a^2 - c^2 = b^2$ (from Fig. 11.5), replace $a^2 - c^2$ by b^2:

$$b^2x^2 + a^2y^2 = a^2b^2$$

$$\frac{x^2}{a^2} + \frac{y^2}{b^2} = 1 \qquad \text{Divide each side by } a^2b^2.$$

If we start with the foci on the y-axis and x-intercepts $(b, 0)$ and $(-b, 0)$, then we can develop a similar equation, which is stated in the following theorem. If the foci are on the y-axis, then the y-intercepts are the vertices and the major axis is vertical. We have proved the following theorem.

Theorem: Equation of an Ellipse with Center (0, 0)

The equation of an ellipse centered at the origin with foci $(c, 0)$ and $(-c, 0)$ and y-intercepts $(0, b)$ and $(0, -b)$ is

$$\frac{x^2}{a^2} + \frac{y^2}{b^2} = 1, \qquad \text{where } a^2 = b^2 + c^2.$$

The equation of an ellipse centered at the origin with foci $(0, c)$ and $(0, -c)$ and x-intercepts $(b, 0)$ and $(-b, 0)$ is

$$\frac{x^2}{b^2} + \frac{y^2}{a^2} = 1, \qquad \text{where } a^2 = b^2 + c^2.$$

Consider the ellipse in Fig. 11.16 on the next page with foci $(c, 0)$ and $(-c, 0)$ and equation

$$\frac{x^2}{a^2} + \frac{y^2}{b^2} = 1,$$

where $a > b > 0$. If $y = 0$ in this equation, then $x = \pm a$. So the vertices (or x-intercepts) of the ellipse are $(a, 0)$ and $(-a, 0)$, and a is the distance from the center to a vertex. The distance from a focus to an endpoint of the minor axis is a also. *So in any ellipse the distance from the focus to an endpoint of the minor axis is the same as the distance from the center to a vertex.* The graphs in Fig. 11.16 will help you remember the relationship between a, b, and c.

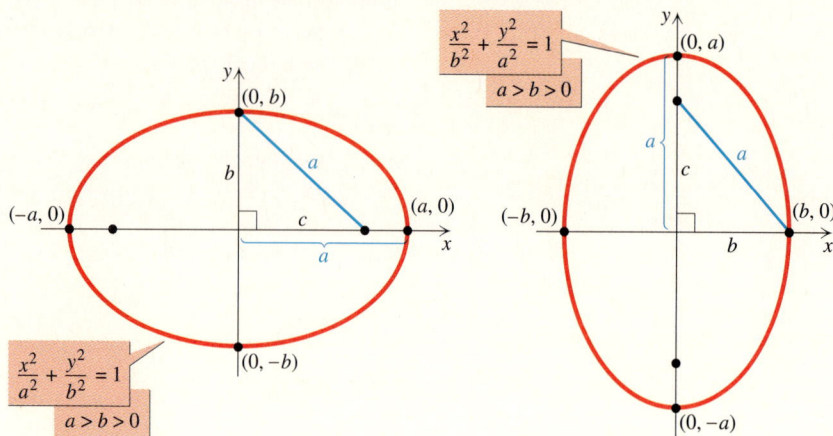

Figure 11.16

Example 1   Writing the equation of an ellipse

Sketch an ellipse with foci at $(0, 4)$ and $(0, -4)$ and vertices $(0, 6)$ and $(0, -6)$, and find the equation for this ellipse.

Solution

Since the vertices are on the y-axis, the major axis is vertical and the ellipse is elongated in the direction of the y-axis. A sketch of the ellipse appears in Fig. 11.17. Because a is the distance from the center to a vertex, $a = 6$. Because c is the distance from the center to a focus, $c = 4$. To write the equation of the ellipse, we need the value of b^2. Use $a = 6$ and $c = 4$ in $a^2 = b^2 + c^2$ to get

$$6^2 = b^2 + 4^2$$

$$b^2 = 20$$

So the x-intercepts are $\left(\sqrt{20}, 0\right)$ and $\left(-\sqrt{20}, 0\right)$, and the equation of the ellipse is

$$\frac{x^2}{20} + \frac{y^2}{36} = 1.$$

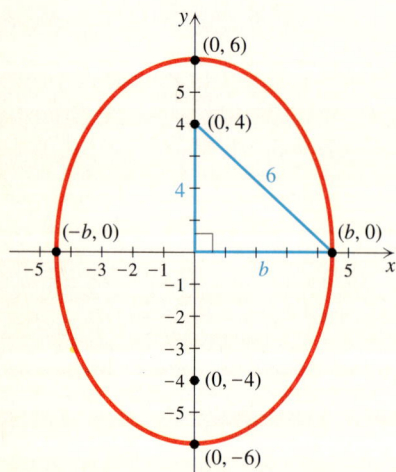

Figure 11.17

Graphing an Ellipse Centered at the Origin

For $a > b > 0$, the graph of

$$\frac{x^2}{a^2} + \frac{y^2}{b^2} = 1$$

is an ellipse centered at the origin with a horizontal major axis, x-intercepts $(a, 0)$ and $(-a, 0)$, and y-intercepts $(0, b)$ and $(0, -b)$, as shown in Fig. 11.16. The foci are on the major axis and are determined by $a^2 = b^2 + c^2$ or $c^2 = a^2 - b^2$. Remember that when the denominator for x^2 is larger than the denominator for y^2, the major axis is horizontal. *To sketch the graph of an ellipse centered at the origin, simply locate the four intercepts and draw an ellipse through them.*

Example 2   Graphing an ellipse with foci on the x-axis

Sketch the graph and identify the foci of the ellipse

$$\frac{x^2}{9} + \frac{y^2}{4} = 1.$$

Solution

To sketch the ellipse, we find the x-intercepts and the y-intercepts. If $x = 0$, then $y^2 = 4$ or $y = \pm 2$. So the y-intercepts are $(0, 2)$ and $(0, -2)$. If $y = 0$, then $x = \pm 3$. So the x-intercepts are $(3, 0)$ and $(-3, 0)$. To make a rough sketch of an ellipse, plot only the intercepts and draw an ellipse through them, as shown in Fig. 11.18. Since this ellipse is elongated in the direction of the x-axis, the foci are on the x-axis. Use $a = 3$ and $b = 2$ in $c^2 = a^2 - b^2$, to get $c^2 = 9 - 4 = 5$. So $c = \pm\sqrt{5}$, and the foci are $(\sqrt{5}, 0)$ and $(-\sqrt{5}, 0)$.

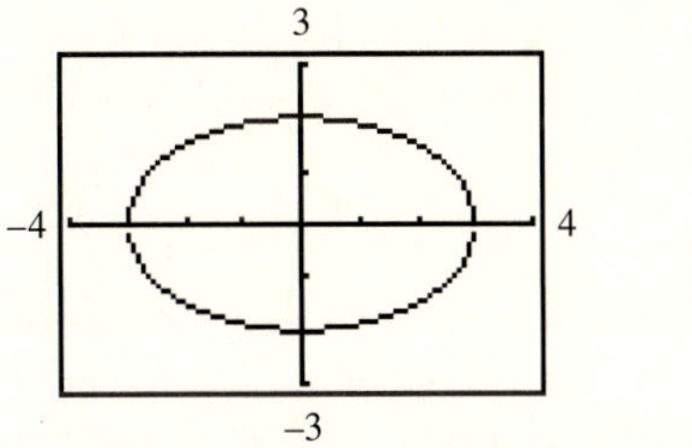

To graph the ellipse in Example 2 on a calculator, solve the equation for y and graph $y_1 = \sqrt{4 - 4x^2/9}$ and $y_2 = -y_1$.

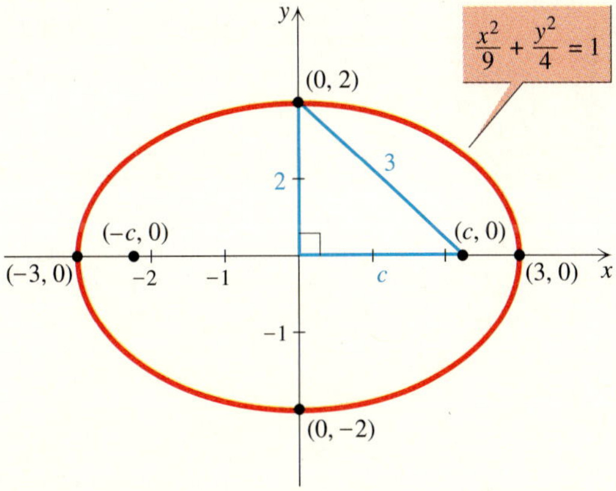

Figure 11.18

For $a > b > 0$, the graph of

$$\frac{x^2}{b^2} + \frac{y^2}{a^2} = 1$$

is an ellipse with a vertical major axis, x-intercepts $(b, 0)$ and $(-b, 0)$, and y-intercepts $(0, a)$ and $(0, -a)$, as shown in Fig. 11.16. When the denominator for y^2 is larger than the denominator for x^2, the major axis is vertical. The foci are always on the major axis and are determined by $a^2 = b^2 + c^2$ or $c^2 = a^2 - b^2$.

Remember that this relationship between a, b, and c is determined by the Pythagorean theorem, because a, b, and c are the lengths of sides of a right triangle, as shown in Fig. 11.16.

Example 3   Graphing an ellipse with foci on the y-axis

Sketch the graph of $11x^2 + 3y^2 = 66$ and identify the foci.

Solution

We first divide each side of the equation by 66 to get the standard equation:

$$\frac{x^2}{6} + \frac{y^2}{22} = 1$$

If $x = 0$, then $y^2 = 22$ or $y = \pm\sqrt{22}$. So the y-intercepts are $\left(0, \sqrt{22}\right)$ and $\left(0, -\sqrt{22}\right)$. If $y = 0$, then $x = \pm\sqrt{6}$ and the x-intercepts are $\left(\sqrt{6}, 0\right)$ and $\left(-\sqrt{6}, 0\right)$. Plot these four points and draw an ellipse through them as shown in Fig. 11.19. Since this ellipse is elongated in the direction of the y-axis, the foci are on the y-axis. Use $a^2 = 22$ and $b^2 = 6$ in $c^2 = a^2 - b^2$ to get $c^2 = 22 - 6 = 16$. So $c = \pm 4$, and the foci are $(0, 4)$ and $(0, -4)$.

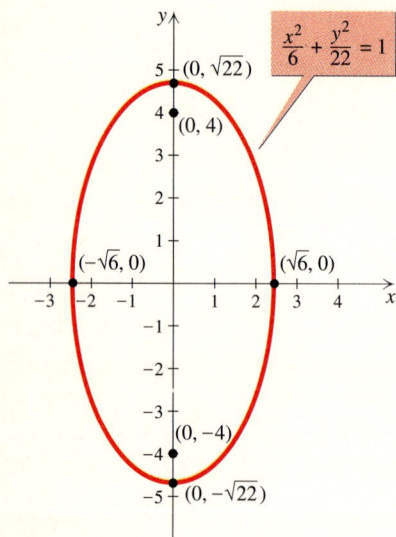

Figure 11.19

Translations of Ellipses

Although an ellipse is not the graph of a function, its graph can be translated in the same manner. Figure 11.20 shows the graphs of

$$\frac{(x - h)^2}{a^2} + \frac{(y - k)^2}{b^2} = 1 \qquad \text{and} \qquad \frac{x^2}{a^2} + \frac{y^2}{b^2} = 1.$$

They have the same size and shape, but the graph of the first equation is centered at (h, k) rather than at the origin. So the graph of the first equation is obtained by translating the graph of the second horizontally h units and vertically k units.

Figure 11.20

Example 4   Graphing an ellipse centered at (h, k)

Sketch the graph and identify the foci of the ellipse

$$\frac{(x-3)^2}{25} + \frac{(y+1)^2}{9} = 1.$$

Solution

The graph of this equation is a translation of the graph of

$$\frac{x^2}{25} + \frac{y^2}{9} = 1,$$

three units to the right and one unit downward. The center of the ellipse is $(3, -1)$. Since $a^2 = 25$, the vertices lie five units to the right and five units to the left of $(3, -1)$. So the ellipse goes through $(8, -1)$ and $(-2, -1)$. Since $b^2 = 9$, the graph includes points three units above and three units below the center. So the ellipse goes through $(3, 2)$ and $(3, -4)$, as shown in Fig. 11.21. Since $c^2 = 25 - 9 = 16$, $c = \pm 4$. The major axis is on the horizontal line $y = -1$. So the foci are found four units to the right and four units to the left of the center at $(7, -1)$ and $(-1, -1)$.

To graph the ellipse in Example 4, graph
$y_1 = -1 + \sqrt{9 - 9(x-3)^2/25}$ and
$y_2 = -1 - \sqrt{9 - 9(x-3)^2/25}$.

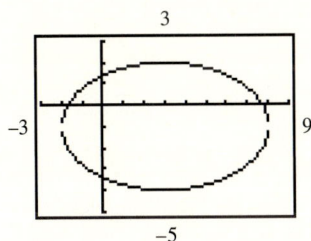

Figure 11.21

The Circle

The circle is the simplest curve of the four conic sections. In keeping with our approach to the conic sections, we give a geometric definition of a circle and then use the distance formula to derive the standard equation for a circle. The standard equation was derived in this way in Section 3.3. For completeness, we repeat the definition and derivation on the following page.

▼

Definition: Circle

A **circle** is the set of all points in a plane such that their distance from a fixed point (the **center**) is a constant (the **radius**).

As shown in Fig. 11.22, a point (x, y) is on a circle with center (h, k) and radius r if and only if

$$\sqrt{(x - h)^2 + (y - k)^2} = r.$$

If we square both sides of this equation, we get the standard equation of a circle.

▼

Theorem: Standard Equation of a Circle

The standard equation of a circle with center (h, k) and radius r ($r > 0$) is

$$(x - h)^2 + (y - k)^2 = r^2.$$

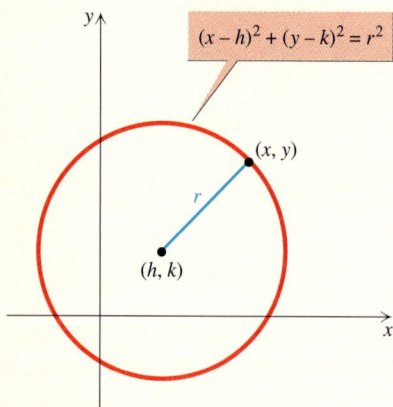

Figure 11.22

The equation $(x - h)^2 + (y - k)^2 = a$ is a circle of radius $\sqrt{a}$ if $a > 0$. If $a = 0$, only (h, k) satisfies $(x - h)^2 + (y - k)^2 = 0$ and the point (h, k) is a degenerate circle. If $a < 0$, then no ordered pair satisfies $(x - h)^2 + (y - k)^2 = a$. If h and k are zero, then we get the standard equation of a circle centered at the origin, $x^2 + y^2 = r^2$.

A circle can be thought of as an ellipse in which the two foci coincide at the center. If the foci are identical, then $a = b$ and the equation for an ellipse becomes the equation for a circle with radius a.

Example 5 Finding the equation for a circle

Write the equation for the circle that has center $(4, 5)$ and passes through $(-1, 2)$.

Solution

The radius is the distance from $(4, 5)$ to $(-1, 2)$:

$$r = \sqrt{(4 - (-1))^2 + (5 - 2)^2} = \sqrt{25 + 9} = \sqrt{34}$$

Use $h = 4$, $k = 5$, and $r = \sqrt{34}$ in $(x - h)^2 + (y - k)^2 = r^2$ to get the equation

$$(x - 4)^2 + (y - 5)^2 = 34. \qquad \blacklozenge$$

To graph a circle, we must know the center and radius. A compass or a string can be used to keep the pencil at a fixed distance from the center.

Example 6 Finding the center and radius

Determine the center and radius, and sketch the graph of $x^2 + 4x + y^2 - 2y = 11$.

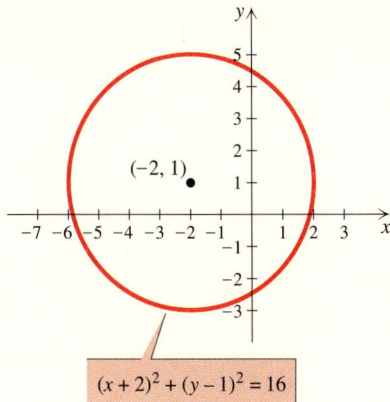

$(x + 2)^2 + (y - 1)^2 = 16$

Figure 11.23

Solution

Use completing the square to get the equation into the standard form:

$$x^2 + 4x + 4 + y^2 - 2y + 1 = 11 + 4 + 1$$

$$(x + 2)^2 + (y - 1)^2 = 16$$

From the standard form we recognize that the center is $(-2, 1)$ and the radius is 4. The graph is shown in Fig. 11.23.   ◆

Applications

The **eccentricity** e of an ellipse is defined by $e = c/a$, where c is the distance from the center to a focus and a is one-half the length of the major axis. Since $0 < c < a$, we have $0 < e < 1$. For an ellipse that appears circular, the foci are close to the center and c is small compared with a. So its eccentricity is near 0. For an ellipse that is very elongated, the foci are close to the vertices and c is nearly equal to a. So its eccentricity is near 1. See Fig. 11.24. A satellite that orbits the earth has an elliptical orbit that is nearly circular, with eccentricity close to 0. On the other hand, Halley's comet is in a very elongated elliptical orbit around the sun, with eccentricity close to 1.

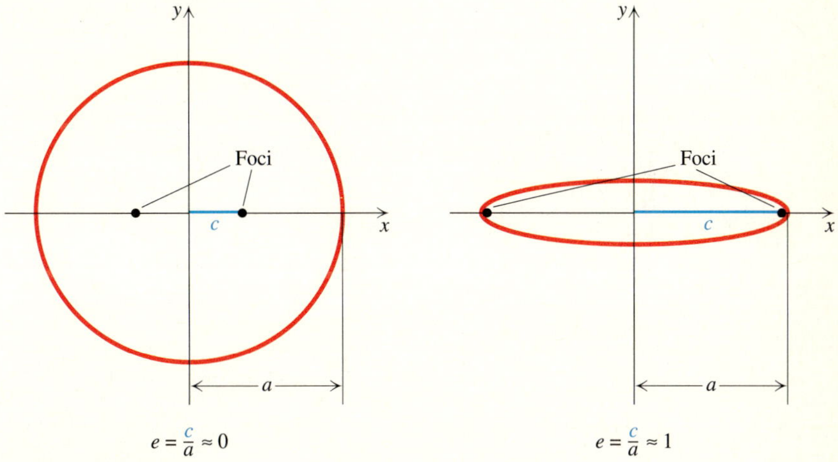

$e = \dfrac{c}{a} \approx 0$ $e = \dfrac{c}{a} \approx 1$

Figure 11.24

Example 7   Eccentricity of an orbit

The first artificial satellite to orbit the earth was Sputnik I, launched by the Soviet Union in 1957. The altitude of Sputnik varied from 132 miles to 583 miles above the surface of the earth. If the center of the earth is one focus of its elliptical orbit and the radius of the earth is 3950 miles, then what was the eccentricity of the orbit?

Solution

The center of the earth is one focus F_1 of the elliptical orbit as shown in Fig. 11.25. When Sputnik I was 132 miles above the earth, it was at a vertex V_1, 4082 miles from F_1 (4082 = 3950 + 132). When Sputnik I was 583 miles above the earth, it was at the other vertex V_2, 4533 miles from F_1 (4533 = 3950 + 583). So the length of the major axis is 4082 + 4533 = 8615 miles. Since the length of the major axis is $2a$, we get $a = 4307.5$ miles. Since c is the distance from the center of the ellipse C to F_1, we get $c = 4307.5 - 4082 = 225.5$. Use $e = c/a$ to get

$$e = \frac{225.5}{4307.5} \approx 0.052.$$

So the eccentricity of the orbit was approximately 0.052.

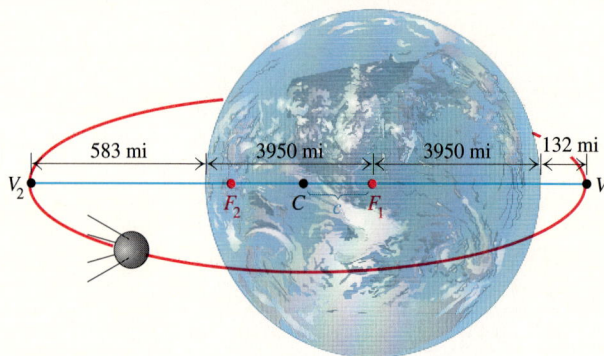

Figure 11.25

For Thought

True or false? Explain.

1. The x-intercepts for $\dfrac{x^2}{9} + \dfrac{y^2}{4} = 1$ are (9, 0) and (−9, 0).

2. The graph of $2x^2 + y^2 = 1$ is an ellipse.

3. The ellipse $\dfrac{x^2}{16} + \dfrac{y^2}{25} = 1$ has a major axis of length 10.

4. The x-intercepts for $0.5x^2 + y^2 = 1$ are $\left(\sqrt{2}, 0\right)$ and $\left(-\sqrt{2}, 0\right)$.

5. The y-intercepts for $x^2 + \dfrac{y^2}{3} = 1$ are $\left(0, \sqrt{3}\right)$ and $\left(0, -\sqrt{3}\right)$.

6. A circle is a set of points, and the center is one of those points.

7. If the foci of an ellipse coincide, then the ellipse is a circle.

8. No ordered pair satisfies the equation $(x - 3)^2 + (y + 1)^2 = 0$.

9. The graph of $(x - 1)^2 + (y + 2)^2 + 9 = 0$ is a circle of radius 3.

10. The radius of the circle $x^2 - 4x + y^2 + y = 9$ is 3.

11.2 Exercises

Tape 21   Disk—5.25": 7   3.5": 4   Macintosh: 4

Find the equation of each ellipse described below and sketch its graph.

1. Foci $(-2, 0)$ and $(2, 0)$, and y-intercepts $(0, -3)$ and $(0, 3)$

2. Foci $(-3, 0)$ and $(3, 0)$, and y-intercepts $(0, -4)$ and $(0, 4)$

3. Foci $(-4, 0)$ and $(4, 0)$, and x-intercepts $(-5, 0)$ and $(5, 0)$

4. Foci $(-1, 0)$ and $(1, 0)$, and x-intercepts $(-3, 0)$ and $(3, 0)$

5. Foci $(0, 2)$ and $(0, -2)$, and x-intercepts $(2, 0)$ and $(-2, 0)$

6. Foci $(0, 6)$ and $(0, -6)$, and x-intercepts $(2, 0)$ and $(-2, 0)$

7. Foci $(0, 4)$ and $(0, -4)$, and y-intercepts $(0, 7)$ and $(0, -7)$

8. Foci $(0, 3)$ and $(0, -3)$, and y-intercepts $(0, 4)$ and $(0, -4)$

Sketch the graph of each ellipse and identify the foci.

9. $\dfrac{x^2}{16} + \dfrac{y^2}{4} = 1$

10. $\dfrac{x^2}{16} + \dfrac{y^2}{9} = 1$

11. $\dfrac{x^2}{9} + \dfrac{y^2}{36} = 1$

12. $x^2 + \dfrac{y^2}{4} = 1$

13. $\dfrac{x^2}{25} + y^2 = 1$

14. $\dfrac{x^2}{6} + \dfrac{y^2}{10} = 1$

15. $\dfrac{y^2}{25} + \dfrac{x^2}{9} = 1$

16. $\dfrac{y^2}{9} + \dfrac{x^2}{4} = 1$

17. $9x^2 + y^2 = 9$

18. $x^2 + 4y^2 = 4$

19. $4x^2 + 9y^2 = 36$

20. $9x^2 + 25y^2 = 225$

Sketch the graph of each ellipse and identify the foci.

21. $\dfrac{(x - 1)^2}{16} + \dfrac{(y + 3)^2}{9} = 1$

22. $\dfrac{(x + 2)^2}{16} + \dfrac{(y + 1)^2}{4} = 1$

23. $\dfrac{(x - 3)^2}{9} + \dfrac{(y + 2)^2}{25} = 1$

24. $(x - 5)^2 + \dfrac{(y - 3)^2}{9} = 1$

25. $(x + 4)^2 + 36(y + 3)^2 = 36$

26. $9(x - 1)^2 + 4(y + 3)^2 = 36$

Find the equation of each ellipse and identify its foci.

27.

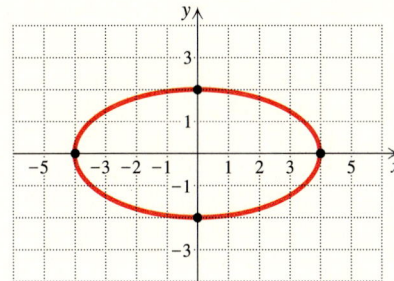

28.

29.

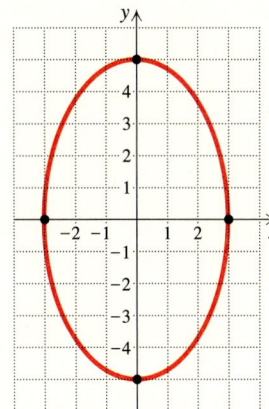

30.

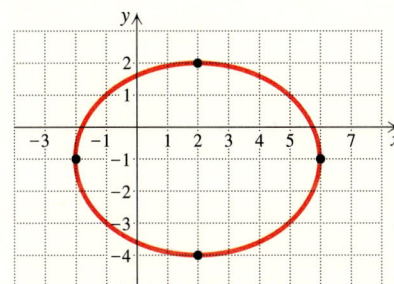

For Exercises 31–38, write the equation for each circle described.

31. Center $(0, 0)$ and radius 2

32. Center $(0, 0)$ and radius $\sqrt{5}$

33. Center $(0, 0)$ and passing through $(4, 5)$

34. Center $(0, 0)$ and passing through $(-3, -4)$

35. Center $(2, -3)$ and passing through $(4, 1)$

36. Center $(-2, -4)$ and passing through $(1, -1)$

37. Diameter has endpoints $(3, 4)$ and $(-1, 2)$.

38. Diameter has endpoints $(3, -1)$ and $(-4, 2)$.

Determine the center and radius of each circle and sketch its graph.

39. $x^2 + y^2 = 100$

40. $x^2 + y^2 = 25$

41. $(x - 1)^2 + (y - 2)^2 = 4$

42. $(x + 2)^2 + (y - 3)^2 = 9$

43. $(x + 2)^2 + (y + 2)^2 = 8$

44. $x^2 + (y - 3)^2 = 9$

Find the center and radius of each circle.

45. $x^2 + y^2 + 2y = 8$

46. $x^2 - 6x + y^2 = 1$

47. $x^2 + 8x + y^2 = 10y$

48. $x^2 + y^2 = 12x - 12y$

49. $x^2 + 4x + y^2 = 5$

50. $x^2 + y^2 - 6y = 0$

51. $x^2 - x + y^2 + y = \dfrac{1}{2}$

52. $x^2 + 5x + y^2 + 3y = \dfrac{1}{2}$

53. $x^2 + \dfrac{2}{3}x + y^2 + \dfrac{1}{3}y = \dfrac{1}{9}$

54. $x^2 + \dfrac{1}{2}x + y^2 + \dfrac{1}{2}y = \dfrac{1}{8}$

55. $2x^2 + 4x + 2y^2 = 1$

56. $x^2 + y^2 = \dfrac{3}{2}y$

Write each of the following equations in one of the forms:
$$y = a(x - h)^2 + k, \qquad x = a(y - h)^2 + k,$$
$$\frac{(x - h)^2}{a^2} + \frac{(y - k)^2}{b^2} = 1, \quad \text{or} \quad (x - h)^2 + (y - k)^2 = r^2.$$

Then identify each equation as the equation of a parabola, an ellipse, or a circle.

57. $y = x^2 + y^2$

58. $4x = x^2 + y^2$

59. $4x^2 + 12y^2 = 4$

60. $2x^2 + 2y^2 = 4 - y$

61. $2x^2 + 4x = 4 - y$

62. $2x^2 + 4y^2 = 4 - y$

63. $2 - x = (2 - y)^2$

64. $3(3 - x)^2 = 9 - y$

65. $2(4 - x)^2 = 4 - y^2$

66. $\dfrac{x^2}{4} + \dfrac{y^2}{4} = 1$

67. $9x^2 = 1 - 9y^2$

68. $9x^2 = 1 - 9y$

Solve each problem.

69. *Foci and a point* Find the equation of the ellipse that goes through the point $(2, 3)$ and has foci $(2, 0)$ and $(-2, 0)$.

70. *Foci and a point* Find the equation of the ellipse that goes through $(6, 5)$ and has foci $(6, 0)$ and $(-6, 0)$.

71. *Focus of elliptical reflector* An elliptical reflector is 10 in. in diameter and 3 in. deep. A light source is at one focus $(0, 0)$, 3 in. from the vertex $(-3, 0)$, as shown in the figure. Find the location of the other focus.

Figure for Exercise 71

72. *Constructing an elliptical arch* A mason is constructing an elliptical form for an arch out of a 4-ft by 12-ft sheet of plywood, as shown in the figure. What length string is needed to draw the ellipse on the plywood (by attaching the ends at the foci as discussed at the beginning of this section)? Where are the foci for this ellipse?

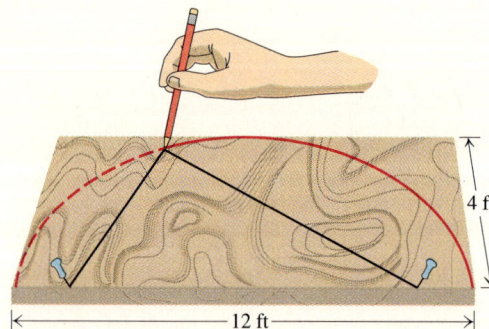

Figure for Exercise 72

73. *Foci and eccentricity* Find the equation of an ellipse with foci $(\pm 5, 0)$ and eccentricity $1/2$.

74. *Orbit of the moon*   The moon travels on an elliptical path with the earth at one focus. If the maximum distance from the moon to the earth is 405,500 km and the minimum distance is 363,300 km, then what is the eccentricity of the orbit?

75. *Halley's comet*   The comet Halley, last seen in 1986, travels in an elliptical orbit with the sun at one focus. Its minimum distance to the sun is 8×10^7 km, and the eccentricity of its orbit is 0.97. What is its maximum distance from the sun? Comet Halley will next be visible from the earth in 2062.

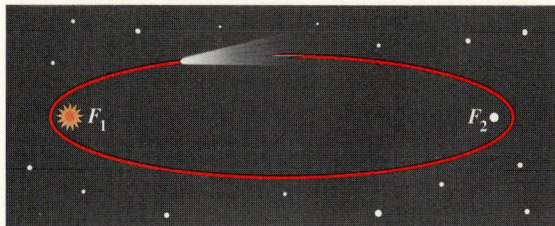

Figure for Exercise 75

76. *Adjacent circles*   A 13-in.-diameter mag wheel and a 16-in.-diameter mag wheel are placed in the first quadrant as shown in the figure. Write an equation for the circular boundary of each wheel.

Figure for Exercise 76

For Writing/Discussion

77. *Best reflector*   Both parabolic and elliptical reflectors reflect sound waves to a focal point where they are amplified. To eavesdrop on the conversation of the quarterback on a football field, which type of reflector is preferable and why?

78. *Details*   Fill in the details in the development of the equation of the ellipse that is outlined in this section.

79. *Development*   Develop the equation of the ellipse with foci at $(0, \pm c)$ and x-intercepts $(\pm b, 0)$ from the definition of ellipse.

80. *Cooperative learning*   To make drapes gather properly, they must be made 2.5 times as wide as the window. Discuss with your classmates the problem of making drapes for a semicircular window. Explain in detail how to cut the material.

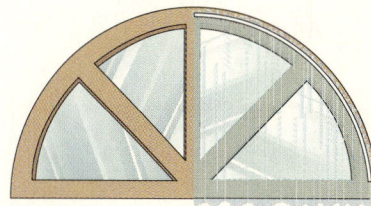

Figure for Exercise 80

81. *Cooperative learning*   With the help of a partner or two, design and build a portable elliptical pool table. The table should have an elliptical bumper with a pocket at one focus. A hard rubber ball shot from the other focus in any direction should go into the pocket.

Graphing Calculator Exercises

1. Solve $x^2 + y^2 = 3950^2$ for y and graph the resulting two equations to get a picture of the earth (a circle of radius 3950 mi) as seen from space. The center of the earth is at the origin. Use $-4500 \leq x \leq 4500$ and $-4500 \leq y \leq 4500$.

2. The elliptical orbit of Sputnik I had the equation

$$\frac{(x + 225.5)^2}{4307.5^2} + \frac{y^2}{4301.6^2} = 1,$$

where one of the foci was the center of the earth. Solve this equation for y and graph the resulting two functions on the same screen as the graph obtained in the previous exercise. The graph of the circle and ellipse together will give you an idea of the altitude of the orbit of Sputnik compared with the size of the earth.

3. Use the information in Example 7 to verify that the equation given in the previous exercise for the orbit of Sputnik I is correct when the earth is centered at the origin.

11.3

The Hyperbola

The last of the four conic sections, the hyperbola, has two branches and each one has a focus as shown in Fig. 11.26. The hyperbola also has a useful reflecting property. A light ray aimed at one focus is reflected toward the other focus as shown in the figure. This reflecting property is used in telescopes, as shown in Fig. 11.27. Within the telescope, a small hyperbolic mirror with the same focus as the large parabolic mirror reflects the light to a more convenient location for viewing.

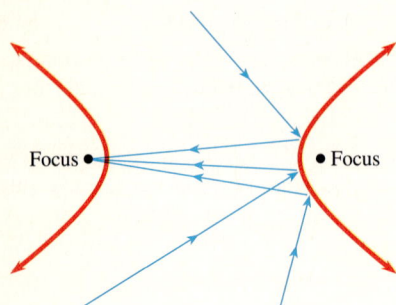

Figure 11.26

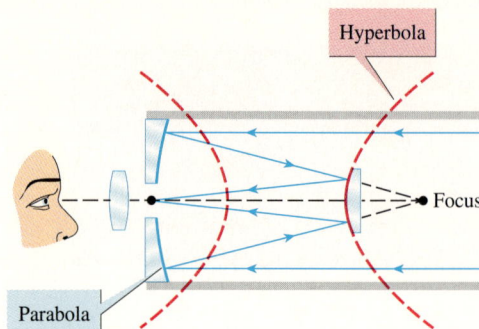

Figure 11.27

Hyperbolas also occur in the context of supersonic noise pollution. The sudden change in air pressure from an aircraft traveling at supersonic speed creates a cone-shaped wave through the air. The plane of the ground and this cone intersect along one branch of a hyperbola as shown in Fig. 11.28. A sonic boom is heard on the ground along this branch of the hyperbola. Since this curve where the sonic boom is heard travels along the ground with the aircraft, supersonic planes such as the Concorde are restricted from flying across the continental United States.

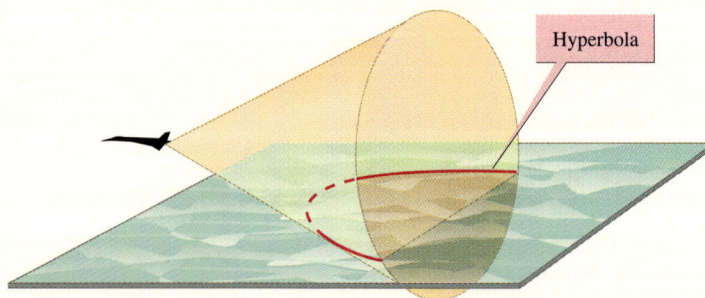

Figure 11.28

A hyperbola may also occur as the path of a moving object such as a spacecraft traveling past the moon on its way toward Venus, a comet passing in the neighborhood of the sun, or an alpha particle passing by the nucleus of an atom.

The Definition

A hyperbola can be defined as the intersection of a cone and a plane, as shown in Fig. 11.1. As we did for the other conic sections, we will give a geometric definition of a hyperbola and use the distance formula to derive its equation.

Definition: Hyperbola

A **hyperbola** is the set of points in a plane such that the difference between the distances from two fixed points (foci) is constant.

For a point on a hyperbola the *difference* between the distances from two fixed points is constant, and for a point on an ellipse the *sum* of the distances from two fixed points is constant. The definitions of a hyperbola and an ellipse are similar, and we will see that their equations are similar also. Their graphs, however, are very different. In the hyperbola shown in Fig. 11.29, the branches look like parabolas, *but they are not parabolas.*

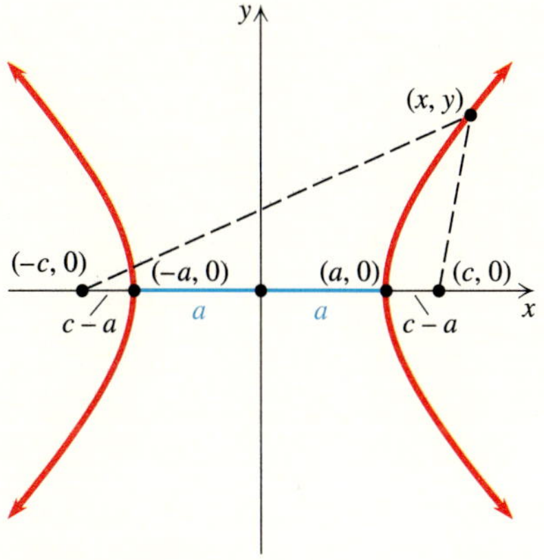

Figure 11.29

Developing the Equation

The hyperbola shown in Fig. 11.29 has foci at $(c, 0)$ and $(-c, 0)$ and x-intercepts or **vertices** at $(a, 0)$ and $(-a, 0)$, where $a > 0$ and $c > 0$. The line segment between the vertices is the **transverse axis.** The point $(0, 0)$, halfway between the foci, is the **center.** The point $(a, 0)$ is on the hyperbola. The distance from $(a, 0)$ to the focus $(c, 0)$ is $c - a$. The distance from $(a, 0)$ to $(-c, 0)$ is $2a + (c - a)$. So the constant difference is $2a$. For an arbitrary point (x, y), the distance to $(c, 0)$ is subtracted from the distance to $(-c, 0)$ to get the constant $2a$:

$$\sqrt{(x - (-c))^2 + (y - 0)^2} - \sqrt{(x - c)^2 + (y - 0)^2} = 2a$$

For (x, y) on the other branch we would subtract in the opposite order, but we get the same simplified form. The simplified form for this equation is similar to that for the ellipse. We will provide the major steps for simplifying it and leave the details as an exercise. First simplify inside the radicals and isolate them to get

$$\sqrt{x^2 + 2xc + c^2 + y^2} = 2a + \sqrt{x^2 - 2xc + c^2 + y^2}.$$

Squaring each side and simplifying yields

$$xc - a^2 = a\sqrt{x^2 - 2xc + c^2 + y^2}.$$

Square each side again and simplify to get

$$c^2x^2 - a^2x^2 - a^2y^2 = c^2a^2 - a^4$$

$$(c^2 - a^2)x^2 - a^2y^2 = a^2(c^2 - a^2) \qquad \text{Factor.}$$

Now $c^2 - a^2$ is positive, because $c > a$. So let $b^2 = c^2 - a^2$ and substitute:

$$b^2x^2 - a^2y^2 = a^2b^2$$

$$\frac{x^2}{a^2} - \frac{y^2}{b^2} = 1 \qquad \text{Divide each side by } a^2b^2.$$

If the foci are on the x-axis, we say that the hyperbola opens to the left and to the right. If the foci are on the y-axis, x and y are interchanged in the equation, and we say that the hyperbola opens up and down. These results are summarized in the following theorem.

Theorem: Equation of a Hyperbola Centered at (0, 0)

The equation of a hyperbola centered at the origin with foci $(c, 0)$ and $(-c, 0)$ and x-intercepts $(a, 0)$ and $(-a, 0)$ is

$$\frac{x^2}{a^2} - \frac{y^2}{b^2} = 1, \qquad \text{where } b^2 = c^2 - a^2.$$

The equation of a hyperbola centered at the origin with foci $(0, c)$ and $(0, -c)$ and y-intercepts $(0, a)$ and $(0, -a)$ is

$$\frac{y^2}{a^2} - \frac{x^2}{b^2} = 1, \qquad \text{where } b^2 = c^2 - a^2.$$

Graphing a Hyperbola Centered at (0, 0)

If we solve the equation

$$\frac{x^2}{a^2} - \frac{y^2}{b^2} = 1$$

for y, we get the following:

$$\frac{y^2}{b^2} = \frac{x^2}{a^2} - 1$$

$$y^2 = \frac{b^2 x^2}{a^2} - b^2 \qquad\qquad \text{Multiply each side by } b^2.$$

$$y^2 = \frac{b^2 x^2}{a^2} - \frac{a^2 b^2 x^2}{a^2 x^2} \qquad\qquad \text{Write } b^2 \text{ as } \frac{a^2 b^2 x^2}{a^2 x^2}.$$

$$y^2 = \frac{b^2 x^2}{a^2} \left(1 - \frac{a^2}{x^2} \right) \qquad\qquad \text{Factor out } \frac{b^2 x^2}{a^2}.$$

$$y = \pm \frac{b}{a} x \sqrt{1 - \frac{a^2}{x^2}}$$

As $x \rightarrow \infty$, the value of $(a^2/x^2) \rightarrow 0$ and the value of y can be approximated by $y = \pm(b/a)x$. So the lines

$$y = \frac{b}{a} x \qquad \text{and} \qquad y = -\frac{b}{a} x$$

are oblique asymptotes for the graph of the hyperbola. The graph of this hyperbola is shown with its asymptotes in Fig. 11.30. The asymptotes are essential for determining the proper shape of the hyperbola. The asymptotes go through the points (a, b), $(a, -b)$, $(-a, b)$, and $(-a, -b)$. The rectangle with these four points as vertices is called the **fundamental rectangle.** The line segment with endpoints $(0, b)$ and $(0, -b)$ is called the **conjugate axis.** The location of the fundamental rectangle is determined by the conjugate axis and the transverse axis.

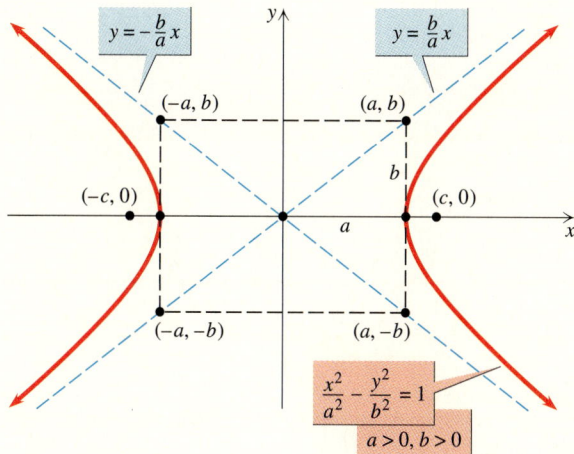

Figure 11.30

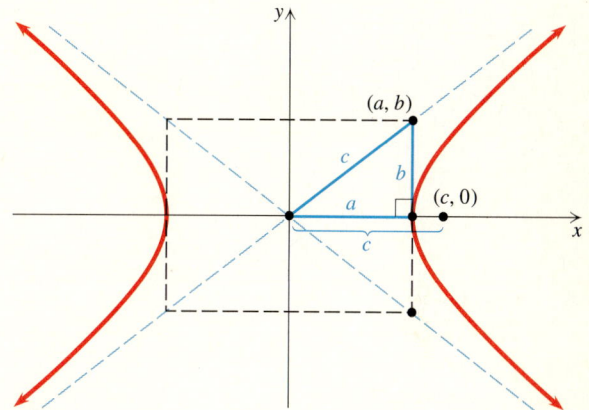

Figure 11.31

Note that since $c^2 = a^2 + b^2$, the distance c from the center to a focus is equal to the distance from the center to (a, b), as shown in Fig. 11.31.

The following steps will help you graph hyperbolas opening left and right.

Procedure: Graphing the Hyperbola $\dfrac{x^2}{a^2} - \dfrac{y^2}{b^2} = 1$

To graph $\dfrac{x^2}{a^2} - \dfrac{y^2}{b^2} = 1$ for $a > 0$ and $b > 0$, do the following:

1. Locate the x-intercepts $(a, 0)$ and $(-a, 0)$.
2. Draw the rectangle through $(\pm a, 0)$ and through $(0, \pm b)$.
3. Extend the diagonals of the rectangle to get the asymptotes.
4. Draw a hyperbola opening to the left and right from the x-intercepts approaching the asymptotes.

Example 1   Graphing a hyperbola opening left and right

Determine the foci and the equations of the asymptotes, and sketch the graph of

$$\frac{x^2}{36} - \frac{y^2}{9} = 1.$$

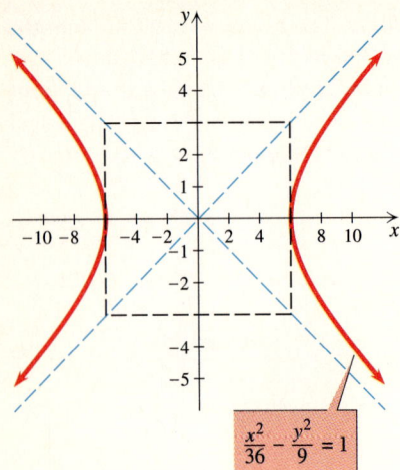

$$\frac{x^2}{36} - \frac{y^2}{9} = 1$$

Figure 11.32

Solution

If $y = 0$, then $x = \pm 6$. The x-intercepts are $(6, 0)$ and $(-6, 0)$. Since $b^2 = 9$, the fundamental rectangle goes through $(0, 3)$ and $(0, -3)$ and through the x-intercepts. Next draw the fundamental rectangle and extend its diagonals to get the asymptotes. Draw the hyperbola opening to the left and right, as shown in Fig. 11.32. To find the foci, use $c^2 = a^2 + b^2$:

$$c^2 = 36 + 9 = 45$$

$$c = \pm\sqrt{45}$$

So the foci are $\left(\sqrt{45}, 0\right)$ and $\left(-\sqrt{45}, 0\right)$. From the graph we get the asymptotes

$$y = \frac{1}{2}x \quad \text{and} \quad y = -\frac{1}{2}x.$$

We can show that hyperbolas opening up and down have asymptotes just as hyperbolas opening right and left have. The lines

$$y = \frac{a}{b}x \quad \text{and} \quad y = -\frac{a}{b}x$$

are asymptotes for the graph of

$$\frac{y^2}{a^2} - \frac{x^2}{b^2} = 1,$$

as shown in Fig. 11.33. Note that the asymptotes are essential for determining the shape of the hyperbola.

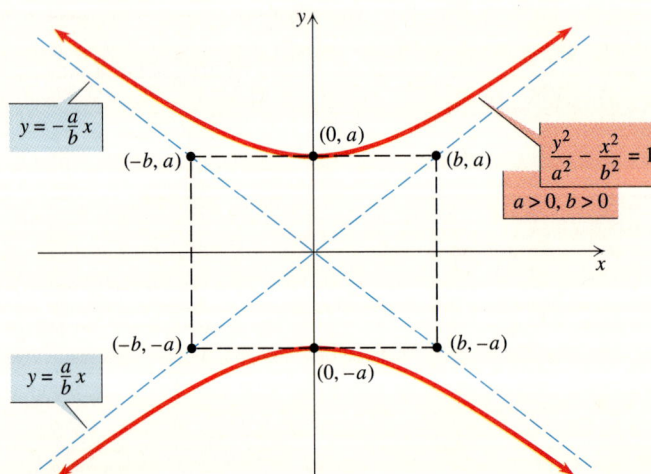

$$y = -\frac{a}{b}x$$

$$\frac{y^2}{a^2} - \frac{x^2}{b^2} = 1$$

$$a > 0, b > 0$$

$$y = \frac{a}{b}x$$

Figure 11.33

The following steps will help you graph hyperbolas opening up and down.

Procedure: Graphing the Hyperbola $\dfrac{y^2}{a^2} - \dfrac{x^2}{b^2} = 1$

To graph $\dfrac{y^2}{a^2} - \dfrac{x^2}{b^2} = 1$ for $a > 0$ and $b > 0$, do the following:

1. Locate the y-intercepts $(0, a)$ and $(0, -a)$.
2. Draw the rectangle through $(0, \pm a)$ and through $(b, 0)$ and $(-b, 0)$.
3. Extend the diagonals of the rectangle to get the asymptotes.
4. Draw a hyperbola opening up and down from the y-intercepts approaching the asymptotes.

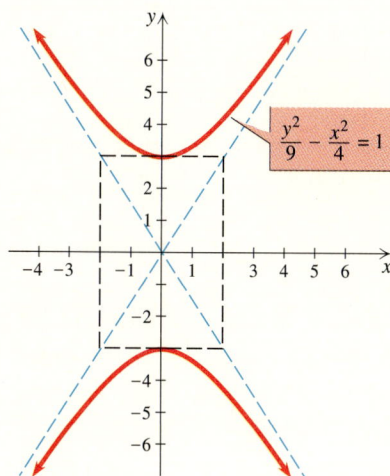

Example 2   Graphing a hyperbola opening up and down

Determine the foci and the equations of the asymptotes, and sketch the graph of

$$4y^2 - 9x^2 = 36.$$

Solution

Divide each side of the equation by 36 to get the equation into the standard form for the equation of the hyperbola:

$$\frac{y^2}{9} - \frac{x^2}{4} = 1$$

If $x = 0$, then $y = \pm 3$. The y-intercepts are $(0, 3)$ and $(0, -3)$. Since $b^2 = 4$, the fundamental rectangle goes through the y-intercepts and through $(2, 0)$ and $(-2, 0)$. Draw the fundamental rectangle and extend its diagonals to get the asymptotes. Draw a hyperbola opening up and down from the y-intercepts approaching the asymptotes as in Fig. 11.34. To find the foci, use $c^2 = a^2 + b^2$:

$$c^2 = 9 + 4 = 13$$

$$c = \pm \sqrt{13}$$

The foci are $\left(0, \sqrt{13}\right)$ and $\left(0, -\sqrt{13}\right)$. From the fundamental rectangle we can see that the equations of the asymptotes are

$$y = \frac{3}{2}x \quad \text{and} \quad y = -\frac{3}{2}x.$$

Figure 11.34

To graph the hyperbola and its asymptotes in Example 2, graph $y_1 = \sqrt{9 + 9x^2/4}$, $y_2 = -y_1$, $y_3 = 1.5x$, and $y_4 = -y_3$.

Hyperbolas Centered at (h, k)

The graph of a hyperbola can be translated horizontally and vertically by replacing x by $x - h$ and y by $y - k$.

▼

Theorem: Hyperbolas Centered at (h, k)

A hyperbola centered at (h, k), opening left and right, has a horizontal transverse axis and equation

$$\frac{(x - h)^2}{a^2} - \frac{(y - k)^2}{b^2} = 1.$$

A hyperbola centered at (h, k), opening up and down, has a vertical transverse axis and equation

$$\frac{(y - k)^2}{a^2} - \frac{(x - h)^2}{b^2} = 1.$$

Example 3   Graphing a hyperbola centered at (h, k)

Determine the foci and the equations of the asymptotes, and sketch the graph of

$$\frac{(x - 2)^2}{9} - \frac{(y + 1)^2}{4} = 1.$$

Solution

The graph that we seek is the graph of

$$\frac{x^2}{9} - \frac{y^2}{4} = 1$$

translated so that its center is (2, −1). Since $a = 3$, the vertices are three units from (2, −1) at (5, −1) and (−1, −1). Because $b = 2$, the fundamental rectangle passes through the vertices and points that are two units above and below (2, −1). Draw the fundamental rectangle through the vertices and through (2, 1) and (2, −3). Extend the diagonals of the fundamental rectangle for the asymptotes, and draw the hyperbola opening to the left and right as shown in Fig. 11.35.

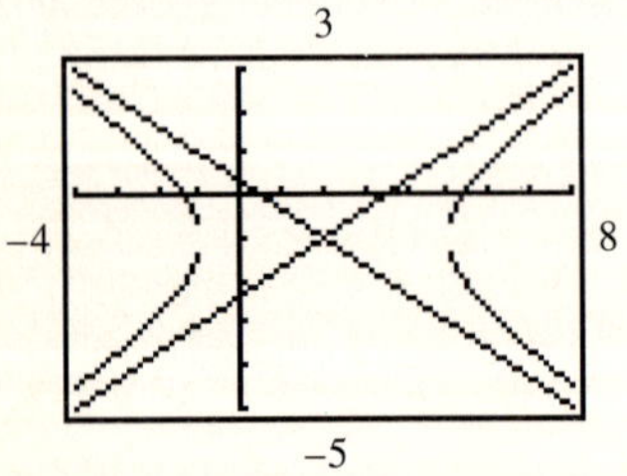

To graph the hyperbola and its asymptotes in Example 3, graph

$y_1 = -1 + \sqrt{4(x - 2)^2/9 - 4}$,

$y_2 = -1 - \sqrt{4(x - 2)^2/9 - 4}$,

$y_3 = (2x - 7)/3$, and

$y_4 = (-2x + 1)/3$.

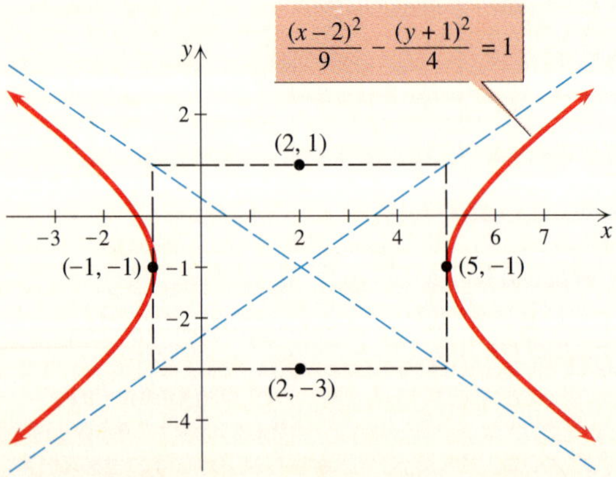

Figure 11.35

Use $c^2 = a^2 + b^2$ to get $c = \sqrt{13}$. The foci are on the transverse axis, $\sqrt{13}$ units from the center $(2, -1)$. So the foci are $\left(2 + \sqrt{13}, -1\right)$ and $\left(2 - \sqrt{13}, -1\right)$. The asymptotes have slopes $\pm 2/3$ and pass through $(2, -1)$. Using the point-slope form for the equation of the line, we get the equations

$$y = \frac{2}{3}x - \frac{7}{3} \qquad \text{and} \qquad y = -\frac{2}{3}x + \frac{1}{3}$$

as the equations of the asymptotes.

Finding the Equation of a Hyperbola

The equation of a hyperbola depends on the location of the foci, center, vertices, transverse axis, conjugate axis, and asymptotes. However, it is not necessary to have all of this information to write the equation. In the next example, we find the equation of a hyperbola given only its transverse axis and conjugate axis.

Example 4   Writing the equation of a hyperbola

Find the equation of the hyperbola whose transverse axis has endpoints $(0, \pm 4)$ and whose conjugate axis has endpoints $(\pm 2, 0)$.

Solution

Since the vertices are the endpoints of the transverse axis, the vertices are $(0, \pm 4)$ and the hyperbola opens up and down. Since the fundamental rectangle goes through the endpoints of the transverse axis and the endpoints of the conjugate axis, we can sketch the hyperbola shown in Fig. 11.36. Since the hyperbola opens up and down and is centered at the origin, its equation is of the form

$$\frac{y^2}{a^2} - \frac{x^2}{b^2} = 1,$$

for $a > 0$ and $b > 0$. From the fundamental rectangle we get $a = 4$ and $b = 2$. So the equation of the hyperbola is

$$\frac{y^2}{16} - \frac{x^2}{4} = 1.$$

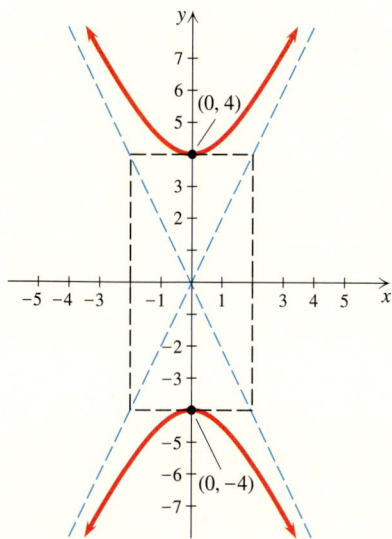

Figure 11.36

For Thought

True or false? Explain.

1. The graph of $\dfrac{x^2}{4} - \dfrac{y}{9} = 1$ is a hyperbola.

2. The graph of $\dfrac{x^2}{16} - \dfrac{y^2}{9} = 1$ has y-intercepts at $(0, 3)$ and $(0, -3)$.

3. The hyperbola $y^2 - x^2 = 1$ opens up and down.

4. The graph of $y^2 = 4 + 16x^2$ is a hyperbola.

5. Every point that satisfies $y = \dfrac{b}{a} x$ must satisfy $\dfrac{x^2}{a^2} - \dfrac{y^2}{b^2} = 1$.

6. The asymptotes for $x^2 - \dfrac{y^2}{4} = 1$ are $y = 2x$ and $y = -2x$.

7. The foci for $\dfrac{x^2}{16} - \dfrac{y^2}{9} = 1$ are $(5, 0)$ and $(-5, 0)$.

8. The points $\left(0, \sqrt{8}\right)$ and $\left(0, -\sqrt{8}\right)$ are the foci for $\dfrac{y^2}{3} - \dfrac{x^2}{5} = 1$.

9. The graph of $\dfrac{x^2}{9} - \dfrac{y^2}{4} = 1$ intersects the line $y = \dfrac{2}{3} x$.

10. The graph of $y^2 = 1 - x^2$ is a hyperbola centered at the origin.

11.3 Exercises

Tape 21　　Disk—5.25″: 7　　3.5″: 4　　**Macintosh: 4**

Determine the foci and the equations of the asymptotes, and sketch the graph of each hyperbola.

1. $\dfrac{x^2}{4} - \dfrac{y^2}{9} = 1$

2. $\dfrac{x^2}{16} - \dfrac{y^2}{9} = 1$

3. $\dfrac{y^2}{4} - \dfrac{x^2}{25} = 1$

4. $\dfrac{y^2}{9} - \dfrac{x^2}{16} = 1$

5. $\dfrac{x^2}{4} - y^2 = 1$

6. $x^2 - \dfrac{y^2}{4} = 1$

7. $x^2 - \dfrac{y^2}{9} = 1$

8. $\dfrac{x^2}{9} - y^2 = 1$

9. $16x^2 - 9y^2 = 144$

10. $9x^2 - 25y^2 = 225$

11. $x^2 - y^2 = 1$

12. $y^2 - x^2 = 1$

Sketch the graph of each hyperbola. Determine the foci and the equations of the asymptotes.

13. $\dfrac{(x + 1)^2}{4} - \dfrac{(y - 2)^2}{9} = 1$

14. $\dfrac{(x + 3)^2}{16} - \dfrac{(y + 2)^2}{25} = 1$

15. $\dfrac{(y - 1)^2}{4} - (x + 2)^2 = 1$

16. $\dfrac{(y - 2)^2}{4} - \dfrac{(x - 1)^2}{9} = 1$

17. $\dfrac{(x + 2)^2}{16} - \dfrac{(y - 3)^2}{9} = 1$

18. $\dfrac{(x + 1)^2}{16} - \dfrac{(y + 2)^2}{25} = 1$

19. $(y - 3)^2 - (x - 3)^2 = 1$

20. $(y + 2)^2 - (x + 2)^2 = 1$

Find the equation of each hyperbola described below.

21. Asymptotes $y = \dfrac{1}{2} x$ and $y = -\dfrac{1}{2} x$, and x-intercepts $(6, 0)$ and $(-6, 0)$

22. Asymptotes $y = x$ and $y = -x$, and y-intercepts $(0, 2)$ and $(0, -2)$

23. Foci $(5, 0)$ and $(-5, 0)$, and x-intercepts $(3, 0)$ and $(-3, 0)$

24. Foci $(0, 5)$ and $(0, -5)$, and y-intercepts $(0, 4)$ and $(0, -4)$

25. Vertices of the fundamental rectangle $(3, \pm 5)$ and $(-3, \pm 5)$, and opening left and right

26. Vertices of the fundamental rectangle $(1, \pm 7)$ and $(-1, \pm 7)$, and opening up and down

Find the equation of each hyperbola.

27.

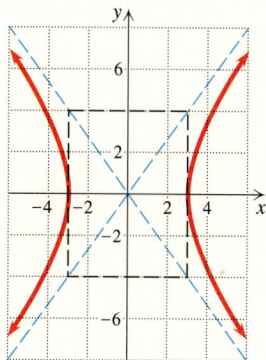

28.

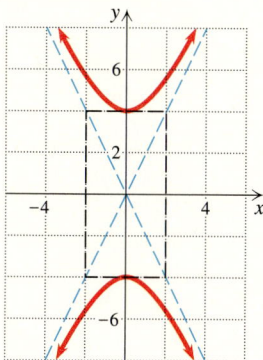

29.

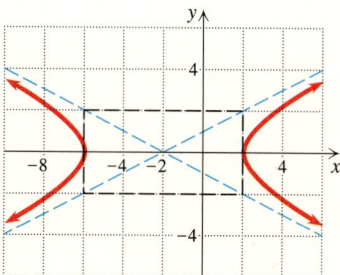

30.

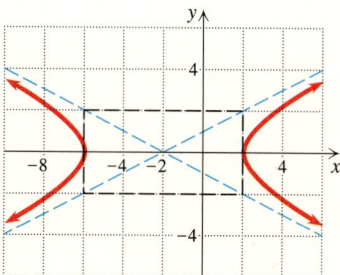

Determine which conic section each equation represents by rewriting each equation in one of the standard forms of the conic sections.

31. $y^2 - x^2 + 2x = 2$

32. $4y^2 + x^2 - 2x = 15$

33. $y - x^2 = 2x$

34. $y^2 + x^2 - 2x = 0$

35. $25x^2 = 2500 - 25y^2$

36. $100x^2 = 25y^2 + 2500$

37. $100y^2 = 2500 - 25x$

38. $100y^2 = 2500 - 25x^2$

39. $2x^2 - 4x + 2y^2 - 8y = -9$

40. $2x^2 + 4x + y = -7$

41. $2x^2 + 4x + y^2 + 6y = -7$

42. $9x^2 - 18x + 4y^2 + 16y = 11$

Solve each problem.

43. *Telephoto lens*   The focus of the main parabolic mirror of a telephoto lens is 10 in. above the vertex as shown in the drawing. A hyperbolic mirror is placed so that its vertex is at $(0, 8)$. If the hyperbola has center $(0, 0)$ and foci $(0, \pm 10)$, then what is the equation of the hyperbola?

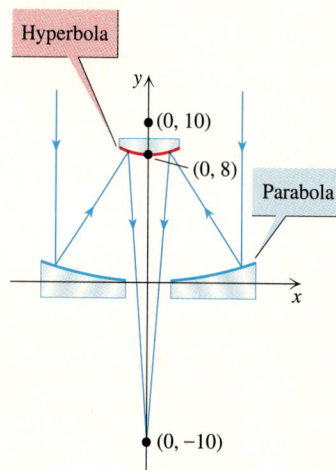

Figure for Exercise 43

44. *Parabolic mirror*   Find the equation of the cross section of the parabolic mirror described in Exercise 43.

45. *Marine navigation*   In 1990 the loran (long range navigation) system had about 500,000 users (*Popular Mechanics,*

March 1990). A $300 loran unit measures the difference in time that it takes for radio signals from pairs of fixed points to reach a ship. The unit then finds the equations of two hyperbolas that pass through the location of the ship and determines the location of the ship. Suppose that the hyperbolas $9x^2 - 4y^2 = 36$ and $16y^2 - x^2 = 16$ pass through the location of a ship in the first quadrant. Find the exact location of the ship.

46. *Air navigation* A pilot is flying in the coordinate system shown in the figure. Using radio signals emitted from $(4, 0)$ and $(-4, 0)$, he knows that he is four units closer to $(4, 0)$ than he is to $(-4, 0)$. Find the equation of the hyperbola with foci $(\pm 4, 0)$ that passes through his location. He also knows that he is two units closer to $(0, 4)$ than he is to $(0, -4)$. Find the equation of the hyperbola with foci $(0, \pm 4)$ that passes through his location. If the pilot is in the first quadrant, then what is his exact location?

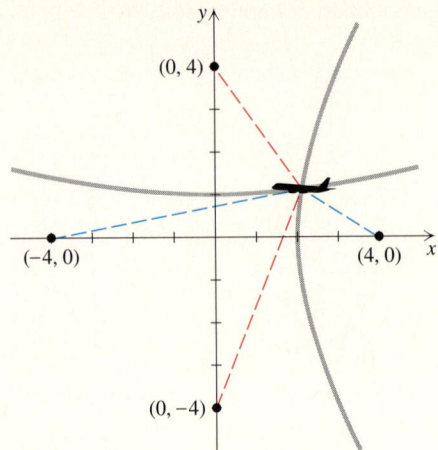

Figure for Exercise 46

For Writing/Discussion

47. *Details* Fill in the details in the development of the equation of a hyperbola that is outlined at the beginning of this section.

48. *Cooperative learning* Work in groups to find conditions on A, B, C, D, and E that will determine whether $Ax^2 + Bx + Cy^2 + Dy = E$ represents a parabola, an ellipse, a circle, or a hyperbola (without rewriting the equation). Test your conditions on the equations of Exercises 31–42.

Graphing Calculator Exercises

Graph each hyperbola on a graphing calculator along with its asymptotes. Observe how close the hyperbola gets to its asymptotes as x gets larger and larger. If $x = 50$, what is the difference between the y-value on the asymptotes and the y-value on the hyperbola?

1. $x^2 - y^2 = 1$

2. $\dfrac{y^2}{4} - \dfrac{x^2}{9} = 1$

Highlights

Section 11.1   The Parabola

1. A parabola is the set of all points in the plane that are equidistant from a fixed line and a fixed point not on the line.

2. The graph of $y = ax^2 + bx + c$ is a parabola opening up if $a > 0$ and down if $a < 0$, and the x-coordinate of the vertex is $x = -b/(2a)$.

3. The graph of $x = ay^2 + by + c$ is a parabola opening right if $a > 0$ and left if $a < 0$, and the y-coordinate of the vertex is $y = -b/(2a)$.

4. The vertex of a parabola is halfway between the focus and directrix.

5. The parabola $y = a(x - h)^2 + k$ has vertex (h, k), focus $(h, k + p)$, and directrix $y = k - p$, where $a = 1/(4p)$.

6. The parabola $x = a(y - h)^2 + k$ has vertex (k, h), focus $(k + p, h)$, and directrix $x = k - p$, where $a = 1/(4p)$.

Section 11.2   The Ellipse and Circle

1. An ellipse is the set of points in the plane such that the sum of their distances from two fixed points is a constant.

2. If $a > b > 0$, the ellipse $\dfrac{x^2}{a^2} + \dfrac{y^2}{b^2} = 1$ has intercepts $(\pm a, 0)$ and $(0, \pm b)$ and foci $(\pm c, 0)$, where $c^2 = a^2 - b^2$.

3. If $a > b > 0$, the ellipse $\dfrac{x^2}{b^2} + \dfrac{y^2}{a^2} = 1$ has intercepts $(\pm b, 0)$ and $(0, \pm a)$ and foci $(0, \pm c)$, where $c^2 = a^2 - b^2$.

4. A circle is the set of points in a plane such that their distance from a fixed point is a constant.

5. The graph of $(x - h)^2 + (y - k)^2 = r^2$ for $r > 0$ is a circle with radius r and center (h, k).

6. The graph of $x^2 + y^2 = r^2$ for $r > 0$ is a circle with radius r centered at $(0, 0)$.

Section 11.3   The Hyperbola

1. A hyperbola is the set of points in a plane such that the difference between the distances from two fixed points is constant.

2. The equation of the hyperbola with foci $(\pm c, 0)$ and x-intercepts $(\pm a, 0)$ is $\dfrac{x^2}{a^2} - \dfrac{y^2}{b^2} = 1$, where $b^2 = c^2 - a^2$.

3. The equation of the hyperbola with foci $(0, \pm c)$ and y-intercepts $(0, \pm a)$ is $\dfrac{y^2}{a^2} - \dfrac{x^2}{b^2} = 1$, where $b^2 = c^2 - a^2$.

4. Draw the fundamental rectangle and the asymptotes to determine the shape of the hyperbola.

5. By replacing x by $x - h$ and y by $y - k$, an ellipse, circle, or hyperbola with center $(0, 0)$ is translated to one with center at (h, k).

Chapter 11 Review Exercises

For Exercises 1–6, sketch the graph of each parabola. Determine the x- and y-intercepts, vertex, axis of symmetry, focus, and directrix for each.

1. $y = x^2 + 4x - 12$

2. $y = 4x - x^2$

3. $y = 6x - 2x^2$

4. $y = 2x^2 - 4x + 2$

5. $x = y^2 + 4y - 6$ **6.** $x = -y^2 + 6y - 9$

Sketch the graph of each ellipse, and determine its foci.

7. $\dfrac{x^2}{16} + \dfrac{y^2}{36} = 1$ **8.** $\dfrac{x^2}{64} + \dfrac{y^2}{16} = 1$

9. $\dfrac{(x-1)^2}{8} + \dfrac{(y-1)^2}{24} = 1$ **10.** $\dfrac{(x+2)^2}{16} + \dfrac{(y-1)^2}{7} = 1$

11. $5x^2 - 10x + 4y^2 + 24y = -1$

12. $16x^2 - 16x + 4y^2 + 4y = 59$

Determine the center and radius of each circle, and sketch its graph.

13. $x^2 + y^2 = 81$ **14.** $6x^2 + 6y^2 = 36$

15. $(x+1)^2 + y^2 = 4$ **16.** $(x-2)^2 + (y+3)^2 = 9$

17. $x^2 + 5x + y^2 + \dfrac{1}{4} = 0$ **18.** $x^2 + 3x + y^2 + 5y = \dfrac{1}{2}$

Write the standard equation for each circle with the given center and radius.

19. Center $(0, -4)$, radius 3

20. Center $(-2, -5)$, radius 1

21. Center $(-2, -7)$, radius $\sqrt{6}$

22. Center $\left(\dfrac{1}{2}, -\dfrac{1}{4}\right)$, radius $\dfrac{\sqrt{2}}{2}$

23. Center $\left(1, -\dfrac{1}{3}\right)$, radius $\dfrac{\sqrt{3}}{2}$

24. Center $(0.2, 0.4)$, radius 0.1

Sketch the graph of each hyperbola. Determine the foci and the equations of the asymptotes.

25. $\dfrac{x^2}{64} - \dfrac{y^2}{36} = 1$

26. $\dfrac{y^2}{100} - \dfrac{x^2}{64} = 1$

27. $\dfrac{(y-2)^2}{64} - \dfrac{(x-4)^2}{16} = 1$

28. $\dfrac{(x-5)^2}{100} - \dfrac{(y-10)^2}{225} = 1$

29. $x^2 - 4x - 4y^2 + 32y = 64$

30. $y^2 - 6y - 4x^2 + 48x = 279$

Identify each equation as the equation of a parabola, ellipse, circle, or hyperbola. Try to do these problems without rewriting the equations.

31. $x^2 = y^2 + 1$ **32.** $x^2 + y^2 = 1$

33. $x^2 = 1 - 4y^2$ **34.** $x^2 + 4x + y^2 = 0$

35. $x^2 + y = 1$ **36.** $y^2 = 1 - x$

37. $x^2 + 4x = y^2$ **38.** $9x^2 + 7y^2 = 63$

Write each equation in standard form, then sketch the graph of the equation.

39. $x^2 = 4 - y^2$ **40.** $x^2 = 4y^2 + 4$

41. $x^2 = 4y + 4$ **42.** $y^2 = 4x - 4$

43. $x^2 = 4 - 4y^2$ **44.** $x^2 = 4y - y^2$

45. $4y^2 = 4x - x^2$

46. $x^2 - 4x = y^2 + 4y + 4$

Determine the equation of each conic section described below.

47. A parabola with focus $(1, 3)$ and directrix $x = 1/2$

48. A circle centered at the origin and passing through $(2, 8)$

49. An ellipse with foci $(\pm 4, 0)$ and vertices $(\pm 6, 0)$

50. A hyperbola with asymptotes $y = \pm 3x$ and y-intercepts $(0, \pm 3)$

51. A circle with center $(1, 3)$ and passing through $(-1, -1)$

52. An ellipse with foci $(3, 2)$ and $(-1, 2)$, and vertices $(5, 2)$ and $(-3, 2)$

53. A hyperbola with foci $(\pm 3, 0)$ and x-intercepts $(\pm 2, 0)$

54. A parabola with focus $(0, 3)$ and directrix $y = 1$

Assume that each of the following graphs is the graph of a parabola, ellipse, circle, or hyperbola. Find the equation for each graph.

55.

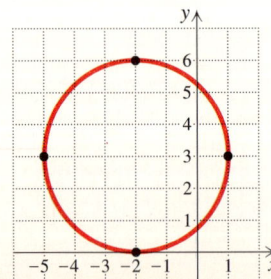

56.

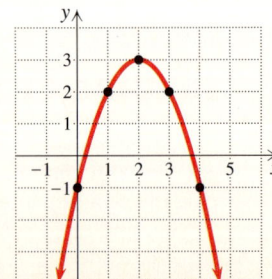

57.

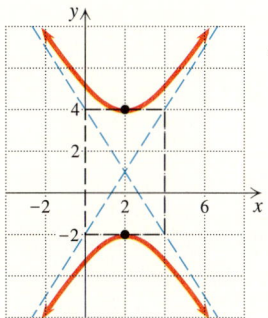

58.

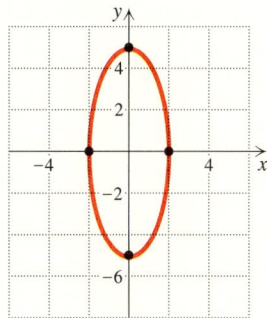

59.

60.

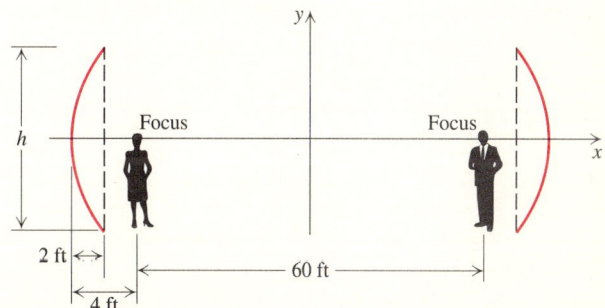

Figure for Exercise 63

Figure for Exercise 64

Solve each problem.

61. *Points on an ellipse*   Find all points of the ellipse $9x^2 + 25y^2 = 225$ that are twice as far from one focus as they are from the other focus.

62. *Points on a hyperbola*   Find all points of the hyperbola $x^2 - y^2 = 1$ that are twice as far from one focus as they are from the other focus.

63. *Nuclear power*   A cooling tower for a nuclear power plant has a hyperbolic cross section as shown in the figure. The diameter of the tower at the top and bottom is 240 ft, while the diameter at the middle is 200 ft. The height of the tower is $48\sqrt{11}$ ft. Find the equation of the hyperbola, using the coordinate system shown in the figure.

64. *Searchlight*   The bulb in a searchlight is positioned 10 in. above the vertex of its parabolic reflector, as shown in the figure. The width of the reflector is 30 in. Using the coordinate system given in the figure, find the equation of the parabola and the thickness t of the reflector at its outside edge.

65. *Whispering gallery*   In the whispering gallery shown in the figure, the foci of the ellipse are 60 ft apart. Each focus is 4 ft from the vertex of an elliptical reflector. Using the coordinate system given in the figure, find the equation of the ellipse that is used to make the elliptical reflectors and the dimension marked h in the figure.

Figure for Exercise 65

Chapter 11 Test

Sketch the graph of each equation.

1. $x^2 + y^2 = 8$

2. $100x^2 + 9y^2 = 900$

3. $x^2 + 6x - y = -8$

4. $\dfrac{y^2}{25} - \dfrac{x^2}{9} = 1$

5. $x^2 + 6x + y^2 - 2y = 0$

6. $\dfrac{(x-2)^2}{9} - \dfrac{(y+3)^2}{4} = 1$

Determine whether each equation is the equation of a parabola, an ellipse, a circle, or a hyperbola.

7. $x^2 - 8x = y^2$

8. $x^2 - 8x = y$

9. $8x - x^2 = y^2$

10. $8x - x^2 = 8y^2$

Determine the equation of each conic section.

11. A circle with center $(-3, 4)$ and radius $2\sqrt{3}$

12. A parabola with focus $(2, 0)$ and directrix $x = -2$

13. An ellipse with focus $\left(0, \pm\sqrt{6}\right)$ and x-intercepts $(\pm 2, 0)$

14. A hyperbola with foci $(\pm 8, 0)$ and vertices $(\pm 6, 0)$

Solve each problem.

15. Find the focus, directrix, vertex, and axis of symmetry for the parabola $y = x^2 - 4x$.

16. Find the foci, length of the major axis, and length of the minor axis for the ellipse $x^2 + 4y^2 = 16$.

17. Find the foci, vertices, equations of the asymptotes, length of the transverse axis, and length of the conjugate axis for the hyperbola $16y^2 - x^2 = 16$.

18. Find the center and radius of the circle

$$x^2 + x + y^2 - 3y = -\frac{1}{4}.$$

19. A lithotripter is used to disintegrate kidney stones by bombarding them with high-energy shock waves generated at one focus of an elliptical reflector. The lithotripter is positioned so that the kidney stone is at the other focus of the reflector. If the equation $81x^2 + 225y^2 = 18{,}225$ is used for the cross section of the elliptical reflector, with centimeters as the unit of measurement, then how far from the point of generation will the waves be focused?

Tying It All Together
Chapters 1–11

Sketch the graph of each equation.

1. $y = 6x - x^2$ **2.** $y = 6x$ **3.** $y = 6 - x^2$ **4.** $y^2 = 6 - x^2$

5. $y = 6 + x$ **6.** $y^2 = 6 - x$ **7.** $y = (6 - x)^2$ **8.** $y = |x + 6|$

9. $y = 6^x$ **10.** $y = \log_6(x)$ **11.** $y = \dfrac{1}{x^2 - 6}$ **12.** $4x^2 + 9y^2 = 36$

13. $4x^2 - 9y^2 = 36$ **14.** $y = 6x - x^3$ **15.** $2x + 3y = 6$ **16.** $x^2 - 6x = y^2 - 6$

Solve each equation.

17. $3(x - 3) + 5 = -9$ **18.** $5x - 4(2x - 3) = 17$ **19.** $2\left(x - \dfrac{1}{3}\right) - 3\left(\dfrac{1}{2} - x\right) = \dfrac{3}{2}$

20. $-4\left(x - \dfrac{1}{2}\right) = 3\left(x + \dfrac{2}{3}\right)$ **21.** $\dfrac{1}{2}x - \dfrac{1}{3} = \dfrac{1}{4}x + \dfrac{3}{2}$

22. $0.05(x - 20) + 0.02(x + 10) = 2.7$

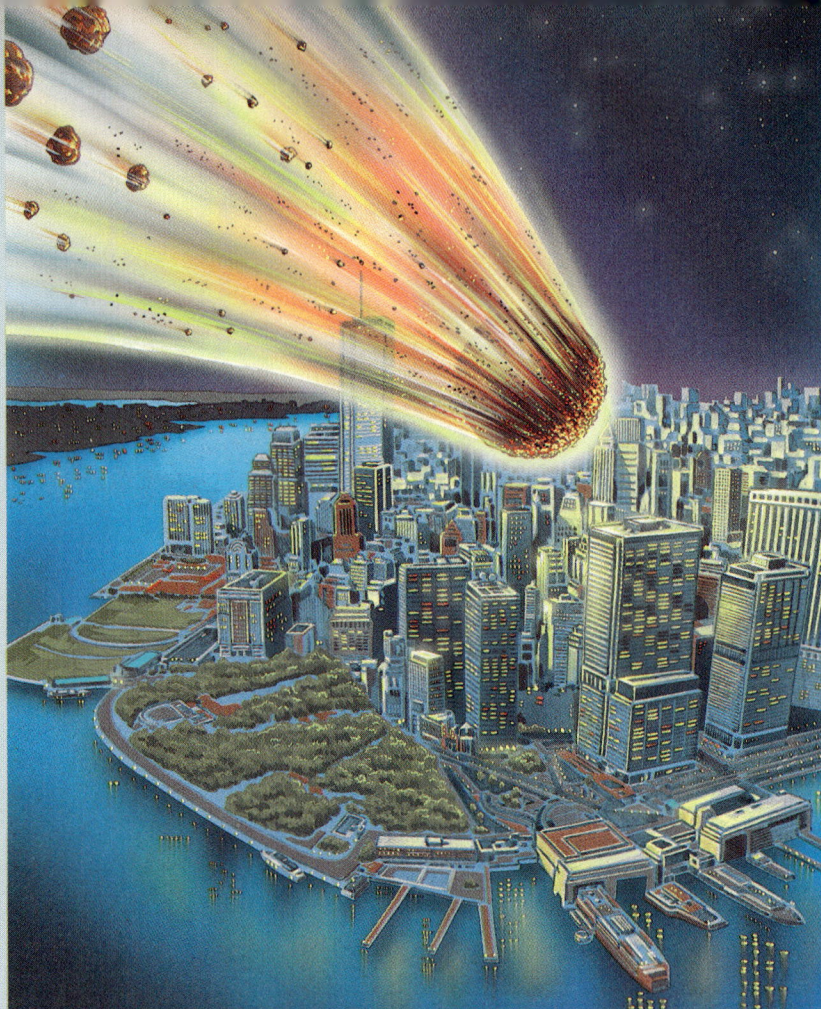

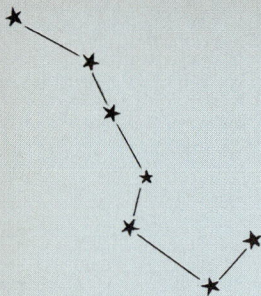

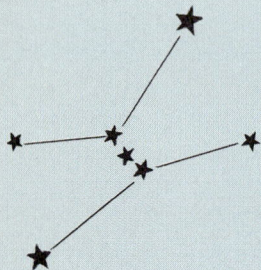

Across the eastern United States, people watched the spectacular meteor shower light up the night sky. Fiery trails cast an eerie glow over the spectators below, yet few onlookers showed signs of fear. After all, what was the chance of a direct hit? Then suddenly, a giant fireball appeared over New York City and began descending rapidly. Crowds panicked as a mighty blast rocked skyscrapers and tore apart concrete canyons. The asteroid's impact, which resembled a several-megaton nuclear blast, triggered an earthquake that leveled buildings and plunged the streets into darkness. In a matter of minutes, the great city had been reduced to rubble.

Science fiction? Perhaps, but chunks of rock and ice have been tumbling about in space since the creation of the solar system. And small meteoroids, packing the wallop of a thousand tons of TNT, do enter Earth's orbit every year or so. Usually they disintegrate upon contact with the atmosphere. However, in 1908, a comet possibly the size of a football field struck Siberia. And in July 1994, the Hubble Telescope captured images of a series of collisions between fragments of the comet Shoemaker-Levy 9 and the planet Jupiter. Said one planetary scientist, "We can be very glad that this comet was heading for Jupiter and not Earth." Even one of the smaller fragments released enough explosive

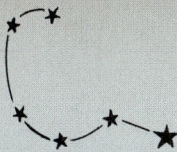

energy to have instantly vaporized the city of Los Angeles.

Today scientists are debating the probability that a "doomsday rock" will collide with Earth. In 1992 a NASA team called for an international watch for these unwelcome visitors.

In this chapter we'll see how discrete mathematics can help us calculate accumulated monthly savings, the economic impact of a $2,000,000 company payroll on a community, and various ways to select lottery numbers. We'll also see how, based on estimates of past collisions, we can determine the probability of an asteroid strike during the next 100 years — an event that could have a lasting impact on your future!

12

Sequences,

Series, and

Probability

12.1

Sequences

In everyday life, we hear the term *sequence* in many contexts. We describe a sequence of events, we make a sequence of car payments, or we get something out of sequence. In this section we will give a mathematical definition of the term and explore several applications.

Definition

We can think of a sequence of numbers as *an ordered list* of numbers. For example, your grades on the first four algebra tests can be listed to form a finite sequence. The sequence

$$10, 20, 30, 40, 50 \ldots$$

is an infinite sequence that lists the positive multiples of 10.

We can think of a sequence as a list, but saying that a sequence is a list is too vague for mathematics. The definition of sequence can be clearly stated by using the terminology of functions.

▼

Definition: Sequence

A **finite sequence** is a function whose domain is $\{1, 2, 3, \ldots, n\}$, the positive integers less than or equal to a fixed positive integer n.

An **infinite sequence** is a function whose domain is the set of all positive integers.

When the domain is apparent, we will refer to either a finite or an infinite sequence as a sequence.

The function $f(n) = n^2$ with domain $\{1, 2, 3, 4, 5\}$ is a finite sequence. For the independent variable of a sequence, we usually use n (for natural number) rather than x, and assume that only natural numbers can be used in place of n. For the dependent variable $f(n)$, we generally write a_n (read "a sub n"). So this finite sequence is also defined by

$$a_n = n^2 \qquad \text{for } 1 \leq n \leq 5.$$

The **terms** of the sequence are the values of the dependent variable a_n. We call a_n the **nth term** or the **general term** of the sequence. The equation $a_n = n^2$ provides a formula for finding the nth term. The five terms of this sequence are $a_1 = 1$, $a_2 = 4$, $a_3 = 9$, $a_4 = 16$, and $a_5 = 25$. We refer to a listing of the terms as the sequence. So 1, 4, 9, 16, 25 is a finite sequence with five terms.

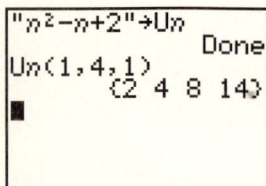

```
"n²-n+2"→Un
                Done
Un(1,4,1)
        {2 4 8 14}
■
```

Some calculators have a sequence feature. Here u_n is defined as $n^2 - n + 2$, and u_n is found for $n = 1$ through $n = 4$ in increments of 1. For more details see Appendix A.

Example 1   Listing terms of a finite sequence

Find all terms of the sequence

$$a_n = n^2 - n + 2 \qquad \text{for } 1 \leq n \leq 4.$$

Solution

Replace n in the formula by each integer from 1 through 4:

$$a_1 = 1^2 - 1 + 2 = 2, \qquad a_2 = 2^2 - 2 + 2 = 4, \qquad a_3 = 8, \qquad a_4 = 14$$

The four terms of the sequence are 2, 4, 8, and 14. ◆

Example 2   Listing terms of an infinite sequence

Find the first four terms of the infinite sequence

$$a_n = \frac{(-1)^{n-1}2^n}{n}.$$

Solution

Replace n in the formula by each integer from 1 through 4:

$$a_1 = \frac{(-1)^{1-1}2^1}{1} = 2 \qquad a_2 = \frac{(-1)^1 2^2}{2} = -2$$

$$a_3 = \frac{(-1)^2 2^3}{3} = \frac{8}{3} \qquad a_4 = \frac{(-1)^3 2^4}{4} = -4$$

The first four terms of the infinite sequence are 2, -2, 8/3, and -4. ◆

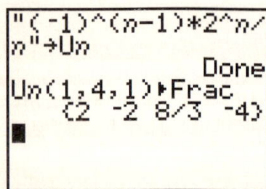

```
"(-1)^(n-1)*2^n/
n"→Un
                Done
Un(1,4,1)▶Frac
        {2 -2 8/3 -4}
■
```

You can define the sequence in Example 2 on your calculator and find the first four terms.

Factorial Notation

Products of consecutive positive integers occur often in sequences and other functions discussed in this chapter. For example, consider the product

$$5 \cdot 4 \cdot 3 \cdot 2 \cdot 1 = 120.$$

The notation 5! (read "five factorial") is used to represent the product of the positive integers from 1 through 5. So $5! = 5 \cdot 4 \cdot 3 \cdot 2 \cdot 1 = 120$. In general, $n!$ is the product of the positive integers from 1 through n. We will find it convenient when writing formulas to have a meaning for 0! even though it does not represent a product of positive integers. The value given to 0! is 1.

Definition: Factorial Notation

> For any positive integer n, the notation $n!$ (read "**n factorial**") is defined by
>
> $$n! = 1 \cdot 2 \cdot 3 \cdot \; \cdots \; \cdot (n - 1) \cdot n.$$
>
> The symbol 0! is defined to be 1, $0! = 1$.

Example 3   A sequence involving factorial notation

Find the first five terms of the infinite sequence whose nth term is

$$a_n = \frac{(-1)^n}{(n - 1)!}.$$

Solution

You can use the factorial function and the fraction feature to find a_3, a_4, and a_5 in Example 3.

Replace n in the formula by each integer from 1 through 5. Use $0! = 1$, $1! = 1$, $2! = 2 \cdot 1 = 2$, $3! = 3 \cdot 2 \cdot 1 = 6$, and $4! = 4 \cdot 3 \cdot 2 \cdot 1 = 24$.

$$a_1 = \frac{(-1)^1}{(1 - 1)!} = \frac{-1}{0!} = -1 \qquad a_2 = \frac{(-1)^2}{1!} = 1$$

$$a_3 = \frac{(-1)^3}{2!} = -\frac{1}{2} \qquad a_4 = \frac{(-1)^4}{3!} = \frac{1}{6} \qquad a_5 = \frac{(-1)^5}{4!} = -\frac{1}{24}$$

Using these five terms, we write the infinite sequence as follows:

$$-1, 1, -\frac{1}{2}, \frac{1}{6}, -\frac{1}{24}, \cdots$$

Most scientific calculators have a factorial key, which might be labeled $x!$ or $n!$. The factorial key is used to find values such as 12!, which is 479,001,600. The value of 69! is the largest factorial that most calculators can calculate, because 70! is larger than 10^{100}.

Finding a Formula for the nth Term

We often have the terms of a sequence and want to write a formula for the sequence. For example, consider your two parents, four grandparents, eight great-grandparents, and so on. The sequence 2, 4, 8, ... lists the number of ancestors you have in each generation going backward in time. Continuing the sequence, the number of great-great-grandparents is 16. Therefore, we want a formula in which each term is twice the one preceding it. The formula

$$a_n = 2^n$$

is the correct formula for the sequence of ancestors. Note that the formula of Example 1,

$$a_n = n^2 - n + 2,$$

gives the same first three terms as this formula but does not have 16 as the fourth term. So if a sequence such as 2, 4, 8, ... is given out of context, many formulas might be found that will produce the same given terms. When attempting to write a formula for a given sequence of terms, look for an "obvious" pattern. Of course, some sequences do not have an obvious pattern. For example, there is no known pattern to the infinite sequence 2, 3, 5, 7, 11, 13, 17, ..., the sequence of prime numbers.

Example 4   Finding a formula for a sequence

Write a formula for the general term of each infinite sequence.

a) 6, 8, 10, 12, ...

b) 3, 5, 7, 9, ...

c) $1, -\dfrac{1}{4}, \dfrac{1}{9}, -\dfrac{1}{16}, \ldots$

Solution

a) Assuming that all terms of the sequence are multiples of 2, we might try the expression $2n$ for the general term. Since the domain of a sequence is the set of positive integers, the expression $2n$ gives the values 2, 4, 6, 8, and so on. But 2 and 4 are not part of this sequence. To get the desired sequence, use the formula $a_n = 2n + 4$. Check by using $a_n = 2n + 4$ to find

$$a_1 = 2(1) + 4 = 6, \qquad a_2 = 2(2) + 4 = 8,$$
$$a_3 = 2(3) + 4 = 10, \qquad a_4 = 2(4) + 4 = 12.$$

b) Since every odd number is an even number plus 1, let $a_n = 2n + 1$. Using $a_n = 2n + 1$, we get $a_1 = 3$, $a_2 = 5$, $a_3 = 7$, and so on. So the general term of the given sequence is $a_n = 2n + 1$.

```
"(-1)^(n+1)/n²"→
Un
              Done
Un(1,4,1)▶Frac
{1 -1/4 1/9 -1/...
■
```

You can use a calculator to check the answer to Example 4(c).

c) To obtain alternating signs, use a power of -1. The expression $(-1)^{n+1}$ for the numerator will give a value of 1 in the numerator when n is odd and a value of -1 in the numerator when n is even. Since the denominators are the squares of the positive integers, we use n^2 to produce the denominators. So the nth term of the sequence is

$$a_n = \frac{(-1)^{n+1}}{n^2}.$$

Check the formula by finding a_1, a_2, a_3, and a_4.   ◆

Recursion Formulas

So far, the formulas used for the nth term of a sequence have expressed the nth term as a function of n, the number of the term. In another approach, a **recursion formula** gives the nth term as a function *of the previous term*. If the first term is known, then a recursion formula determines the remaining terms of the sequence.

```
Un⦿Un-1²-5
Vn=■
```

A recursion formula is defined in sequence mode using the Y= key.

Example 5   A recursion formula

Find the first four terms of the infinite sequence in which $a_1 = 3$ and $a_n = (a_{n-1})^2 - 5$ for $n \geq 2$.

Solution

We are given $a_1 = 3$. If $n = 2$, the recursion formula is $a_2 = (a_1)^2 - 5$. Since $a_1 = 3$, we get $a_2 = 3^2 - 5 = 4$. To find the next two terms, we let $n = 3$ and $n = 4$ in the recursion formula:

$$a_3 = (a_2)^2 - 5 = 4^2 - 5 = 11 \qquad \text{Since } a_2 = 4$$
$$a_4 = (a_3)^2 - 5 = 11^2 - 5 = 116 \qquad \text{Since } a_3 = 11$$

```
Un(1,4,1)
      {3 4 11 116}
■
```

Set $u_1 = 3$ in the WINDOW menu. Then find the first four terms of the sequence in Example 5.

So the first four terms of the infinite sequence are 3, 4, 11, and 116.   ◆

Arithmetic Sequences

In the sequence

$$4, 9, 14, 19, 24, \ldots$$

each term is 5 larger than the previous term. So the sequence could be written as follows:

$$4, \quad 4 + 5, \quad 4 + 2 \cdot 5, \quad 4 + 3 \cdot 5, \quad 4 + 4 \cdot 5, \ldots$$

A formula for the general term of this sequence is

$$a_n = 4 + (n - 1)5.$$

This sequence is an example of an arithmetic sequence.

Definition: Arithmetic Sequence

A sequence that has an nth term of the form

$$a_n = a_1 + (n - 1)d,$$

where a_1 and d are any real numbers, is called an **arithmetic sequence.**

In an arithmetic sequence, each term after the first is obtained by adding a constant to the previous term. The first term is a_1, the second term is $a_1 + d$, the third term is $a_1 + 2d$, the fourth term is $a_1 + 3d$, and so on. The number d is called the **common difference.** A sequence is an arithmetic sequence if and only if there is a common difference between consecutive terms.

Example 6   Finding a formula for an arithmetic sequence

Determine whether each sequence is arithmetic. If it is, then write a formula for the general term of the sequence.

a) 3, 7, 11, 15, 19, . . .

b) 5, 2, −1, −4, . . .

c) 2, 6, 18, 54, . . .

Solution

a) Since each term of the sequence is 4 larger than the previous term, the sequence is arithmetic and $d = 4$. Since $a_1 = 3$, the formula for the nth term is

$$a_n = 3 + (n - 1)4.$$

This formula can be simplified to

$$a_n = 4n - 1.$$

b) Since each term of this sequence is 3 smaller than the previous term, the sequence is arithmetic and $d = -3$. Since $a_1 = 5$, we have

$$a_n = 5 + (n - 1)(-3).$$

This formula can be simplified to

$$a_n = -3n + 8.$$

c) For 2, 6, 18, 54, . . . , we get $6 - 2 = 4$ and $18 - 6 = 12$. Since the difference between consecutive terms is not constant, the sequence is not arithmetic. ◆

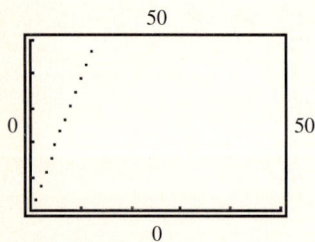

Points that satisfy $u_n = 4n - 1$ lie in a straight line because an arithmetic sequence has the same form as a linear function.

Note that when the general term of an arithmetic sequence is simplified, it has the form of a linear function.

Example 7   Finding the terms of an arithmetic sequence

Find the first four terms and the 30th term of each arithmetic sequence.

a) $a_n = -7 + (n - 1)6$

b) $a_n = -\dfrac{1}{2} n + 8$

Solution

a) Let n take the values from 1 through 4:

$$a_1 = -7 + (1 - 1)6 = -7$$
$$a_2 = -7 + (2 - 1)6 = -1$$
$$a_3 = -7 + (3 - 1)6 = 5$$
$$a_4 = -7 + (4 - 1)6 = 11$$

The first four terms of the sequence are $-7, -1, 5, 11$. The 30th term is

$$a_{30} = -7 + (30 - 1)6 = 167.$$

b) Let n take the values from 1 through 4:

$$a_1 = -\frac{1}{2}(1) + 8 = \frac{15}{2}$$

$$a_2 = -\frac{1}{2}(2) + 8 = 7$$

$$a_3 = -\frac{1}{2}(3) + 8 = \frac{13}{2}$$

$$a_4 = -\frac{1}{2}(4) + 8 = 6$$

The first four terms of the sequence are $15/2, 7, 13/2, 6$. The 30th term is

$$a_{30} = -\frac{1}{2}(30) + 8 = -7.$$ ◆

The formula $a_n = a_1 + (n - 1)d$ involves four quantities, a_n, a_1, n, and d. If any three of them are known, the fourth can be found.

Example 8   Finding a term of an arithmetic sequence

An insurance representative made \$30,000 her first year and \$60,000 her seventh year. Assume that her annual salary figures form an arithmetic sequence and predict what she will make in her tenth year.

Solution

The seventh term is $a_7 = a_1 + (7 - 1)d$. Use $a_7 = 60,000$ and $a_1 = 30,000$ in this equation to find d:

$$60,000 = 30,000 + (7 - 1)d$$
$$30,000 = 6d$$
$$5000 = d$$

Now use the formula $a_n = a_1 + (n - 1)(5000)$ to find a_{10}:

$$a_{10} = 30,000 + (10 - 1)(5000)$$
$$a_{10} = 75,000$$

So the predicted salary for her tenth year is $75,000.   ◆

Using a recursion formula, an arithmetic sequence with first term a_1 and constant difference d is defined by $a_n = a_{n-1} + d$ for $n \geq 2$.

Example 9   A recursion formula for an arithmetic sequence

Find the first four terms of the sequence in which $a_1 = -7$ and $a_n = a_{n-1} + 6$ for $n \geq 2$.

Solution

The recursion formula indicates that each term after the first is obtained by adding 6 to the previous term:

$$a_1 = -7$$
$$a_2 = a_1 + 6 = -7 + 6 = -1$$
$$a_3 = a_2 + 6 = -1 + 6 = 5$$
$$a_4 = a_3 + 6 = 5 + 6 = 11$$

The first four terms are -7, -1, 5, and 11. Note that this recursion formula produces the same sequence as the formula in Example 7(a).   ◆

```
Un=Un-1+6
Un=
```

Define u_n using the Y= key in sequence mode. Set UnStart $= -7$ in the WINDOW menu.

```
Un(1,4,1)
    {-7 -1 5 11}
■
```

Find the first four terms of the sequence in Example 9.

? For Thought

True or false? Explain.

1.   The equation $a_n = e^n$ for n a natural number defines a sequence.

2.   The domain of a finite sequence is the set of positive integers.

3.   We can think of a sequence as a list of the values of the dependent variable.

4. The letter n is used to represent the dependent variable.

5. The first four terms of $a_n = (-1)^{n-1}n^3$ are $-1, 8, -27, 81$.

6. The fifth term of $a_n = -3 + (n-1)6$ is 5.

7. The common difference in the arithmetic sequence $7, 4, 1, -2, \ldots$ is 3.

8. The sequence $1, 4, 9, 16, 25, 36, \ldots$ is an arithmetic sequence.

9. If the first term of an arithmetic sequence is 4 and the third term is 14, then the fourth term is 24.

10. The sequence $a_n = 5 + 2n$ is an arithmetic sequence.

12.1 Exercises ▭ Tape 22 ▭ Disk—5.25″: 7   3.5″: 4   Macintosh: 4

Find all terms of each finite sequence.

1. $a_n = n^2, 1 \le n \le 7$

2. $a_n = (n-1)^2, 1 \le n \le 5$

3. $b_n = \dfrac{(-1)^{n+1}}{n+1}, 1 \le n \le 8$

4. $b_n = (-1)^n 3n, 1 \le n \le 4$

5. $c_n = (-2)^{n-1}, 1 \le n \le 6$

6. $c_n = (-3)^{n-2}, 1 \le n \le 6$

7. $a_n = 2^{2-n}, 1 \le n \le 5$

8. $a_n = \left(\dfrac{1}{2}\right)^{3-n}, 1 \le n \le 7$

9. $a_n = -6 + (n-1)(-4), 1 \le n \le 5$

10. $a_n = -2 + (n-1)4, 1 \le n \le 8$

11. $b_n = 5 + (n-1)(0.5), 1 \le n \le 7$

12. $b_n = \dfrac{1}{4} + (n-1)\left(-\dfrac{1}{2}\right), 1 \le n \le 5$

13. $c_n = \dfrac{n^2}{n!}, 1 \le n \le 5$

14. $c_n = \dfrac{n!}{(n-2)!}, 2 \le n \le 9$

15. $a_n = (n-1)!, 1 \le n \le 7$

16. $t_n = (2n)!, 1 \le n \le 4$

Find the first four terms and the 10th term of each infinite sequence whose nth term is given.

17. $a_n = 8 + (n-1)(-3)$

18. $a_n = -7 + (n-1)(0.5)$

19. $a_n = \dfrac{4}{2n+1}$

20. $a_n = \dfrac{2}{n^2+1}$

21. $a_n = \dfrac{(-1)^n}{(n+1)(n+2)}$

22. $a_n = \dfrac{(-1)^{n+1}}{(n+1)^2}$

23. $a_n = \dfrac{2^n}{n!}$

24. $a_n = \dfrac{(-1)^n}{(n+1)!}$

25. $a_n = \dfrac{(-2)^{2n-1}}{(n-1)!}$

26. $a_n = \dfrac{e^{-n}}{(n+2)!}$

27. $a_n = -0.1n + 9$

28. $a_n = 0.3n - 0.4$

Find the first four terms and the eighth term of each infinite sequence given by a recursion formula.

29. $a_n = 3a_{n-1} + 2, a_1 = -4$

30. $a_n = 1 - \dfrac{1}{a_{n-1}}, a_1 = 2$

31. $a_n = (a_{n-1})^2 - 3, a_1 = 2$

32. $a_n = (a_{n-1})^2 - 2, a_1 = -2$

33. $a_n = a_{n-1} + 7, a_1 = -15$

34. $a_n = \dfrac{1}{2}a_{n-1}, a_1 = 8$

For Exercises 35–46, write a formula for the nth term of each infinite sequence. Do not use a recursion formula.

35. $2, 4, 6, 8, \ldots$

36. $1, 3, 5, 7, \ldots$

37. $9, 11, 13, 15, \ldots$

38. $14, 16, 18, 20, \ldots$

39. $1, -1, 1, -1, \ldots$

40. $-\dfrac{1}{2}, \dfrac{1}{2}, -\dfrac{1}{2}, \dfrac{1}{2}, \ldots$

41. $1, 8, 27, 64, \ldots$

42. $1, -\dfrac{1}{8}, \dfrac{1}{27}, -\dfrac{1}{64}, \ldots$

43. $e, e^2, e^3, e^4, \ldots$

44. $\pi, 4\pi, 9\pi, 16\pi, \ldots$

45. $1, \dfrac{1}{2}, \dfrac{1}{4}, \dfrac{1}{8}, \ldots$

46. $1, -3, 9, -27, \ldots$

Determine whether each given sequence could be an arithmetic sequence.

47. $2, 3, 4, 5, \ldots$

48. $-7, -4, -2, 0, \ldots$

49. $1, 0.5, 1, 0.5, \ldots$

50. $3, 0, -3, -6, \ldots$

51. $2, 4, 8, 16, \ldots$

52. $1, \dfrac{5}{4}, \dfrac{3}{2}, \dfrac{7}{4}, \ldots$

53. $\dfrac{\pi}{4}, \dfrac{\pi}{2}, \dfrac{3\pi}{4}, \pi, \ldots$

54. $1, 2, 3, 2, 3, \ldots$

Write a formula for the nth term of each arithmetic sequence. Do not use a recursion formula.

55. $1, 6, 11, 16, \ldots$

56. $2, 5, 8, 11, \ldots$

57. $0, 2, 4, 6, \ldots$

58. $-3, 3, 9, 15, \ldots$

59. $5, 1, -3, -7, \ldots$

60. $1, -1, -3, -5, \ldots$

61. $1, 1.1, 1.2, 1.3, \ldots$

62. $2, 2.75, 3.5, 4.25, \ldots$

63. $\dfrac{\pi}{6}, \dfrac{\pi}{3}, \dfrac{\pi}{2}, \dfrac{2\pi}{3}, \ldots$

64. $\dfrac{\pi}{12}, \dfrac{\pi}{6}, \dfrac{\pi}{4}, \dfrac{\pi}{3}, \ldots$

65. $20, 35, 50, 65, \ldots$

66. $70, 60, 50, 40, \ldots$

Find the first four terms and the 10th term of each arithmetic sequence.

67. $a_n = 6 + (n - 1)(-3)$

68. $b_n = -12 + (n - 1)4$

69. $c_n = 1 + (n - 1)(-0.1)$

70. $q_n = 10 - 5n$

71. $w_n = -\dfrac{1}{3}n + 5$

72. $t_n = \dfrac{1}{2}n + \dfrac{1}{2}$

Find the indicated part of each arithmetic sequence.

73. Find the eighth term of the sequence that has a first term of -3 and a common difference of 5.

74. Find the 11th term of the sequence that has a first term of 4 and a common difference of -0.8.

75. Find the 10th term of the sequence whose third term is 6 and whose seventh term is 18.

76. Find the eighth term of the sequence whose second term is 20 and whose fifth term is 10.

77. Find the common difference of the sequence in which the first term is 12 and the 21st term is 96.

78. Find the common difference of the sequence in which the first term is 5 and the 11th term is -10.

79. Find a formula for a_n, given that $a_3 = 10$ and $a_7 = 20$.

80. Find a formula for a_n, given that $a_5 = 30$ and $a_{10} = -5$.

Write a recursion formula for each sequence.

81. $3, 12, 21, 30, \ldots$

82. $30, 25, 20, 15, \ldots$

83. $\dfrac{1}{3}, 1, 3, 9, \ldots$

84. $4, -1, \dfrac{1}{4}, -\dfrac{1}{16}, \ldots$

85. $16, 4, 2, \sqrt{2}, \ldots$

86. $t^2, t^4, t^8, t^{16}, \ldots$

Solve each problem.

87. *Recursive pricing* Each year the price of a new Jeep Cherokee 4 × 4 increases by 6% over the price the previous year. Given that the price this year is $20,000, find the price to the nearest dollar each year for the next five years.

88. *Rising salary* If you get a $1200 raise each year and your income this year is $20,000, what is your income each year for the next five years? Is your salary in five years equal to the price of the Jeep Cherokee 4 × 4 of the previous exercise?

89. *Reading marathon* On November 1 an English teacher had his class read five pages of a long novel. He then told them to increase their daily reading by three pages each day. For example, on November 2 they should read eight pages. Write a formula for the number of pages that they will read on the nth day of November. If they follow the teacher's instructions, then how many pages will they be reading on the last day of November?

90. *Stiff penalty* If a contractor does not complete a multimillion-dollar construction project on time, he must pay a penalty of $500 for the first day that he is late, $700 for the second day, $900 for the third day, and so on. Each day the penalty is $200 larger than the previous day. Write a formula for the penalty on the nth day. What is the penalty for the 10th day?

91. *Nursing home care* The estimated annual cost to reside in a nursing home in 1990 was $29,930 (*Fortune*, 1992 Investor's Guide). If the average increase in the annual cost is $1800 each year, then what will the annual cost be in 2010?

92. *Good planning* Sam's retirement plan gives her a fixed raise of d dollars each year. If her retirement income was $24,500 in her fifth year of retirement and $25,700 in her ninth year, then what was her income her first year? What will her income be in her thirteenth year of retirement?

93. *Approaching e* The sequence $a_n = (1 + 1/n)^n$ is a sequence of numbers that get closer and closer to the number e, the base of the natural logarithm. Use a calculator to find the value of $e - a_{1000}$.

94. *From one to zero* The sequence $a_n = (0.99999)^n$ is a sequence of numbers that get closer and closer to 0. Use a calculator to find the hundredth, thousandth, and millionth terms.

For Writing/Discussion

95. Explain the difference between a function and a sequence.

96. *Cooperative learning* Write your own formula for the *n*th term of a sequence on a piece of paper and list the first five terms. Disclose the terms one at a time to your classmates, giving them the opportunity to guess the formula after each disclosed term.

Graphing Calculator Exercises

A graphing calculator plots only finitely many points. On a graphing calculator that plots 96 points, we can set the display for $1 \leq x \leq 96$. This setting will make the first coordinate of each ordered pair a positive integer. Since a sequence is a function whose domain is the positive integers, set the mode of the calculator so that it plots only 96 dots rather than connecting the dots to form a curve. Graphing a function in this manner shows the first 96 terms of a sequence. The *y*-coordinates of the dots are the terms of the sequence.

1. Graph the function $y = (-1)^x(1/x)$ and use the trace feature to find a_{50} for the sequence $a_n = (-1)^n(1/n)$.

2. Graph the function $y = 100(1.06)^x$ and use the trace feature to find a_{54} for the sequence $a_n = 100(1.06)^n$.

12.2

Series

In this section we continue the study of sequences, but here we concentrate on finding the sum of the terms of a sequence. For example, if an employee starts at $20,000 per year and gets a $1200 raise each year for the next 39 years, then the total pay for 40 years of work is the sum of 40 terms of an arithmetic sequence:

$$20,000 + 21,200 + 22,400 + \cdots + 46,800$$

This sum is an example of a *series*. We could add the 40 terms to find the total pay, but we will soon discover a formula for this sum that involves only the first term, the last term, and the number of terms. (The total investment in an employee over a lifetime of work is often cited as a reason for selecting personnel carefully.)

Summation Notation

As a shorthand way to indicate the sum of the terms of a sequence, we adopt a new notation called **summation notation.** We use the Greek letter Σ (sigma) in summation notation. For example, the sum of the annual salaries for 40 years of work is written as

$$\sum_{n=1}^{40} [20,000 + (n-1)(1200)].$$

Following the letter sigma is the formula for the *n*th term of an arithmetic sequence in which the first term is 20,000 and the common difference is 1200. The numbers below and above the letter sigma indicate that this is an expression for the sum of the first through fortieth terms of this sequence.

As another example, the sum of the squares of the first five positive integers can be written as

$$\sum_{i=1}^{5} i^2.$$

To evaluate this sum, let i take the integral values from 1 through 5 in the expression i^2. Thus

$$\sum_{i=1}^{5} i^2 = 1^2 + 2^2 + 3^2 + 4^2 + 5^2 = 1 + 4 + 9 + 16 + 25 = 55.$$

The letter i in the summation notation is called the **index of summation.** Although we usually use i or n for the index of summation, any letter may be used. For example, the expressions

$$\sum_{n=1}^{5} n^2, \qquad \sum_{j=1}^{5} j^2, \qquad \text{and} \qquad \sum_{i=1}^{5} i^2$$

all have the same value.

In the summation notation, the expression following the letter sigma is the *general term* of a sequence. The numbers below and above sigma indicate which terms of the sequence are to be added.

Example 1   Evaluating summation notation

Find the sum in each case.

a) $\displaystyle\sum_{i=1}^{6} (-1)^i 2^{i-1}$ **b)** $\displaystyle\sum_{n=3}^{7} (2n-1)$ **c)** $\displaystyle\sum_{i=1}^{5} 4$

Solution

a) Evaluate $(-1)^i 2^{i-1}$ for $i = 1$ through 6 and add the resulting terms:

$$\sum_{i=1}^{6} (-1)^i 2^{i-1} = (-1)^1 2^0 + (-1)^2 2^1 + (-1)^3 2^2 + (-1)^4 2^3 + (-1)^5 2^4 + (-1)^6 2^5$$

$$= -1 + 2 - 4 + 8 - 16 + 32 = 21$$

b) Find the third through seventh terms of the sequence whose general term is $2n - 1$ and add the results:

$$\sum_{n=3}^{7} (2n-1) = 5 + 7 + 9 + 11 + 13 = 45$$

c) Every term of this series is 4. The notation $i = 1$ through 5 means that we add the first five 4's from a sequence in which every term is 4:

$$\sum_{i=1}^{5} 4 = 4 + 4 + 4 + 4 + 4 = 20$$

An expression in summation notation or an expression such as $1 + 2 + 3$, in which we have not actually performed the addition, is called an **indicated sum.**

Definition: Series

The indicated sum of the terms of a sequence is called a **series.**

Just as a sequence may be finite or infinite, a series may be finite or infinite. In this section we will discuss only finite series. In Section 12.3 we will discuss one type of infinite series.

To write a series in summation notation, a formula must be found for the nth term of the corresponding sequence.

Example 2   Writing a series in summation notation

Write each series using summation notation.

a) $2 + 4 + 6 + 8 + 10$ **b)** $\dfrac{1}{5} - \dfrac{1}{7} + \dfrac{1}{9} - \dfrac{1}{11} + \dfrac{1}{13}$

Solution

a) The series consists of a sequence of even integers. The nth term for this sequence is $a_n = 2n$. This series consists of five terms of this sequence.

$$2 + 4 + 6 + 8 + 10 = \sum_{i=1}^{5} 2i$$

b) This series has two features of interest: The denominators of the fractions are odd integers and the signs alternate. If we use $a_n = 2n + 1$ as the general term for odd integers, we get $a_1 = 3$ and $a_2 = 5$. We do not always have to use $i = 1$ in the summation notation. In this series it is easier to use $i = 2$ through 6. Use $(-1)^i$ to get the alternating signs:

$$\frac{1}{5} - \frac{1}{7} + \frac{1}{9} - \frac{1}{11} + \frac{1}{13} = \sum_{i=2}^{6} \frac{(-1)^i}{2i + 1} \qquad \blacklozenge$$

Changing the Index of Summation

In Example 2(b) the index of summation i ranged from 2 through 6, but the starting point of the index is arbitrary. The notation

$$\sum_{j=3}^{7} \frac{(-1)^{j-1}}{2j - 1}$$

is another notation for the same sum. The summation notation for a given series can be written so that the index begins at any given number.

Example 3   Adjusting the index of summation

Rewrite the series so that instead of index i, it has index j, where j starts at 1.

$$\sum_{i=2}^{6} \frac{(-1)^i}{2i+1}$$

Solution

In this series, i takes the values 2 through 6. If j starts at 1, then $j = i - 1$, and j takes the values 1 through 5. If $j = i - 1$, then $i = j + 1$. To change the formula for the general term of the series, replace i by $j + 1$:

$$\sum_{i=2}^{6} \frac{(-1)^i}{2i+1} = \sum_{j=1}^{5} \frac{(-1)^{j+1}}{2(j+1)+1} = \sum_{j=1}^{5} \frac{(-1)^{j+1}}{2j+3}$$

Check that these two series have exactly the same five terms.   ◆

The Mean

If you take a sequence of three tests, then your "average" is the sum of the three test scores divided by 3. What is commonly called the "average" is called the *mean* or *arithmetic mean* in mathematics. The mean can be defined using summation notation.

Definition: Mean

The **mean** of the numbers $x_1, x_2, x_3, \ldots, x_n$ is the number $\bar{x}$, given by

$$\bar{x} = \frac{\displaystyle\sum_{i=1}^{n} x_i}{n}.$$

Example 4   Finding the mean of a sequence of numbers

Find the mean of the numbers $-12, 3, 0, 5, -2, 9$.

Solution

To find the mean, divide the total of the six numbers by 6:

$$\bar{x} = \frac{-12 + 3 + 0 + 5 + (-2) + 9}{6} = \frac{1}{2}$$

The mean is $1/2$.   ◆

Arithmetic Series

The indicated sum of an arithmetic sequence is an **arithmetic series.** The sum of a finite arithmetic series can be found without actually adding up all of the terms. Let S represent the sum of the even integers from 2 through 50. We can find S by using the following procedure:

$$S = 2 + 4 + 6 + 8 + \cdots + 48 + 50$$
$$\underline{S = 50 + 48 + 46 + 44 + \cdots + 4 + 2} \qquad \text{\color{blue}{Write terms in reverse order.}}$$
$$2S = 52 + 52 + 52 + 52 + \cdots + 52 + 52 \qquad \text{\color{blue}{Add corresponding terms.}}$$

Since there are 25 numbers in the series of even integers from 2 through 50, the number 52 appears 25 times on the right-hand side of the last equation.

$$2S = 25(52) = 1300$$
$$S = 650$$

So the sum of the even integers from 2 through 50 is 650.

We can use the same idea to develop a formula for the sum of n terms of any arithmetic series. Let $S_n = a_1 + a_2 + a_3 + \cdots + a_n$ be a finite arithmetic series. Since there is a constant difference between the terms, S_n can be written forwards and backwards as follows:

$$S_n = a_1 + (a_1 + d) + (a_1 + 2d) + \cdots + a_n$$
$$\underline{S_n = a_n + (a_n - d) + (a_n - 2d) + \cdots + a_1}$$
$$2S_n = (a_1 + a_n) + (a_1 + a_n) + (a_1 + a_n) + \cdots + (a_1 + a_n) \qquad \text{\color{blue}{Add.}}$$

Now, there are n terms of the type $a_1 + a_n$ on the right-hand side of the last equation, so the right-hand side can be simplified:

$$2S_n = n(a_1 + a_n)$$
$$S_n = \frac{n}{2}(a_1 + a_n)$$

This result is summarized as follows.

Theorem: Sum of an Arithmetic Series

The sum S_n of the first n terms of an arithmetic series with first term a_1 and nth term a_n is given by the formula

$$S_n = \frac{n}{2}(a_1 + a_n).$$

So, to find the sum of any arithmetic series, all we need to know is the first term, the last term, and the number of terms.

Example 5 Finding the sum of an arithmetic series

Find the sum of each arithmetic series.

a) $\displaystyle\sum_{i=1}^{15} (3i - 5)$ b) $36 + 41 + 46 + 51 + \cdots + 91$

Solution

a) To find the sum, we need to know the first term, the last term, and the number of terms. For this series, $a_1 = -2$, $a_{15} = 40$, and $n = 15$. So

$$\sum_{i=1}^{15} (3i - 5) = \frac{15}{2}(-2 + 40) = 285.$$

b) For this series, $a_1 = 36$ and $a_n = 91$, but to find S_n, the number of terms n must be known. We can find n from the formula for the general term of the arithmetic sequence $a_n = a_1 + (n - 1)d$ by using $d = 5$:

$$91 = 36 + (n - 1)5$$
$$55 = (n - 1)5$$
$$11 = n - 1$$
$$12 = n$$

We can now use $n = 12$, $a_1 = 36$, and $a_{12} = 91$ in the formula to get the sum of the 12 terms of this arithmetic series:

$$S_{12} = \frac{12}{2}(36 + 91) = 762 \qquad \blacklozenge$$

In the next example we return to the problem of finding the sum of the 40 annual salaries presented at the beginning of this section.

Example 6 Total salary for 40 years of work

Find the total salary for an employee who is paid \$20,000 for the first year and receives a \$1200 raise each year for the next 39 years. Find the mean of the 40 annual salaries.

Solution

The salaries form an arithmetic sequence whose nth term is $20,000 + (n - 1)1200$. The total of the first 40 terms of the sequence of salaries is given by

$$S_{40} = \sum_{n=1}^{40} [20,000 + (n - 1)1200] = \frac{40}{2}(20,000 + 66,800) = 1,736,000.$$

The total salary for 40 years of work is \$1,736,000. Divide the total salary by 40 to get a mean salary of \$43,400. $\qquad \blacklozenge$

For Thought

True or false? Explain.

1. $\displaystyle\sum_{i=1}^{3} (-2)^i = -6$

2. $\displaystyle\sum_{i=1}^{6} (0 \cdot i + 5) = 30$

3. $\displaystyle\sum_{i=1}^{k} 5i = 5\left(\sum_{i=1}^{k} i\right)$

4. $\displaystyle\sum_{i=1}^{k} (i^2 + 1) = \left(\sum_{i=1}^{k} i^2\right) + k$

5. There are nine terms in the series $\displaystyle\sum_{i=5}^{14} 3i^2$.

6. $\displaystyle\sum_{i=2}^{8} (-1)^i 3i^2 = \sum_{j=1}^{7} (-1)^{j-1} 3(j+1)^2$

7. The series $\displaystyle\sum_{n=1}^{9} (3n-5)$ is an arithmetic series.

8. The sum of the first n counting numbers is $\dfrac{n(n+1)}{2}$.

9. The sum of the even integers from 8 through 68 inclusive is $\dfrac{60}{2}(8 + 68)$.

10. $\displaystyle\sum_{i=1}^{10} i^2 = \dfrac{10}{2}(1 + 100)$

12.2 Exercises

Tape 22   Disk—5.25″: 7   3.5″: 4   Macintosh: 4

Find the sum of each series.

1. $\displaystyle\sum_{i=1}^{5} i^2$

2. $\displaystyle\sum_{j=0}^{4} (j+1)^2$

3. $\displaystyle\sum_{j=1}^{6} (2j-1)$

4. $\displaystyle\sum_{i=1}^{7} (2i+5)$

5. $\displaystyle\sum_{n=2}^{5} 2^{-n}$

6. $\displaystyle\sum_{i=1}^{4} (-1)^n i^2$

7. $\displaystyle\sum_{i=4}^{100} 5i^0$

8. $\displaystyle\sum_{i=7}^{36} (10 + i^0)$

9. $\displaystyle\sum_{i=3}^{47} (-1)^{i+1}$

10. $\displaystyle\sum_{i=1}^{10} 3$

11. $\displaystyle\sum_{j=7}^{44} (-1)^j$

12. $\displaystyle\sum_{i=0}^{5} i(i-1)(i-2)$

Write each series in summation notation. Use the index i and let i begin at 1 in each summation.

13. $1 + 2 + 3 + 4 + 5 + 6$

14. $2 + 4 + 6 + 8 + 10 + 12$

15. $-1 + 3 - 5 + 7 - 9$

16. $3 - 6 + 9 - 12 + 15 - 18$

17. $1 + 4 + 9 + 16 + 25$

18. $1 + 3 + 9 + 27 + 81$

19. $1 - \dfrac{1}{2} + \dfrac{1}{4} - \dfrac{1}{8} + \dfrac{1}{16}$

20. $-1 + \dfrac{1}{2} - \dfrac{1}{3} + \dfrac{1}{4}$

21. $\ln(x_1) + \ln(x_2) + \ln(x_3)$

22. $x^3 + x^4 + x^5 + x^6 + x^7$

23. $a + ar + ar^2 + \cdots + ar^{10}$

24. $b^2 + b^3 + b^4 + \cdots + b^{12}$

Rewrite each series using the new index j as indicated.

25. $\sum_{i=1}^{32} (-1)^i = \sum_{j=0}$

26. $\sum_{i=1}^{10} 2^i = \sum_{j=0}$

27. $\sum_{i=4}^{13} (2i + 1) = \sum_{j=1}$

28. $\sum_{i=7}^{12} (3i - 4) = \sum_{j=1}$

29. $\sum_{x=2}^{10} \frac{10!}{x!(10 - x)!} = \sum_{j=0}$

30. $\sum_{i=2}^{9} \frac{x^i}{i!} = \sum_{j=0}$

31. $\sum_{n=2}^{6} \frac{5^n e^{-5}}{n!} = \sum_{j=5}$

32. $\sum_{i=0}^{5} 3^{2i-1} = \sum_{j=3}$

Write out all of the terms of each series.

33. $\sum_{i=0}^{5} 0.5r^i$

34. $\sum_{i=1}^{6} i^n$

35. $\sum_{j=0}^{4} a^{4-j}b^j$

36. $\sum_{j=0}^{3} (-1)^j x^{3-j} y^j$

37. $\sum_{i=0}^{2} \frac{2}{i!(2 - i)!} a^{2-i}b^i$

38. $\sum_{j=0}^{3} \frac{6}{j!(3 - j)!} a^{3-j}b^j$

Find the mean of each sequence of numbers.

39. 6, 23, 45

40. 33, 42, 78, 19

41. −6, 0, 3, 4, 3, 92

42. 12, 20, 12, 30, 28, 28, 10

43. $\sqrt{2}$, π, 33.6, −19.4, 52

44. $\sqrt{5}$, $-3\sqrt{3}$, $\pi/2$, e, 98.6

Find the sum of each arithmetic series.

45. $1 + 2 + 3 + \cdots + 47$

46. $2 + 4 + 6 + \cdots + 88$

47. $8 + 5 + 2 + (-1) + \cdots + (-16)$

48. $5 + 1 + (-3) + (-7) + \cdots + (-27)$

49. $3 + 7 + 11 + 15 + \cdots + 55$

50. $-6 + 1 + 8 + 15 + \cdots + 50$

51. $\frac{1}{2} + \frac{3}{4} + 1 + \frac{5}{4} + \cdots + 5$

52. $1 + \frac{4}{3} + \frac{5}{3} + 2 + \cdots + \frac{22}{3}$

53. $\sum_{i=1}^{12} (6i - 9)$

54. $\sum_{i=1}^{11} (0.5i + 4)$

55. $\sum_{n=3}^{15} (-0.1n + 1)$

56. $\sum_{i=4}^{20} (-0.3n + 2)$

Solve each problem using the ideas of series.

57. *Total salary* If a graphic artist makes $30,000 his first year and gets a $1000 raise each year, then what will be his total salary for 30 years of work? What is his mean annual salary for 30 years of work?

58. *Assigned reading* An English teacher with a minor in mathematics told her students that if they read $2n + 1$ pages of a long novel on the nth day of October, for each day of October, then they will exactly finish the novel in October. How many pages are there in this novel? What is the mean number of pages read per day?

59. *Annual payments* Wilma deposited $1000 into an account paying 5% compounded annually each January 1 for 10 consecutive years. Given that the first deposit was made January 1, 1980, and the last was made January 1, 1989, write a series in summation notation whose sum is the amount in the account on January 1, 1990.

60. *Compounded quarterly* Duane deposited $100 into an account paying 4% compounded quarterly each January 1 for eight consecutive years. Given that the first deposit was made January 1, 1970, and the last was made January 1, 1977, write a series in summation notation whose sum is the amount in the account on January 1, 1980.

61. *Juan Valdez* A grocer wants to build a "mountain" out of cans of mountain-grown coffee as shown in the figure. The first level is to be a rectangle containing 9 rows of 12 cans in each row. Each level after the first is to contain one less row with 12 cans in each row. Finally, the top level is to contain one row of 12 cans. Write a sequence whose terms are the number of cans at each level. Write the sum of the terms of this sequence in summation notation and find the number of cans in the mountain.

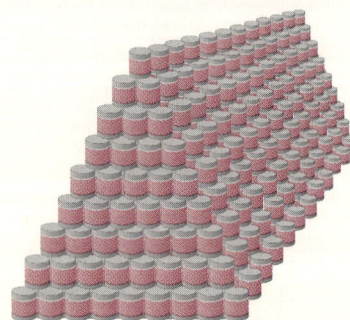

Figure for Exercise 61

62. *Mt. McKinley* Suppose that the grocer in Exercise 61 builds the "mountain" so that each level after the first contains one less row and one less can in each row. Write a sequence whose terms are the number of cans at each level. Write the sum of the terms of this sequence in summation notation and find the number of cans in the mountain.

Find the indicated mean.

63. Find the mean of the 9th through the 60th terms inclusive of the sequence $a_n = 5n + 56$.

64. Find the mean of the 15th through the 55th terms inclusive of the sequence $a_n = 7 - 4n$.

65. Find the mean of the seventh through tenth terms inclusive of the sequence in which $a_1 = -2$ and $a_n = (a_{n-1})^2 - 3$ for $n \geq 2$.

66. Find the mean of the fifth through the eighth terms inclusive of the sequence in which $a_n = (-1/2)^n$.

For Writing/Discussion

67. Explain the difference between a sequence and a series.

68. *Cooperative learning* Write your own formula for the nth term a_n of a sequence on a piece of paper. Find the *partial sums* $S_1 = a_1, S_2 = a_1 + a_2, S_3 = a_1 + a_2 + a_3$, and so on. Disclose the first five partial sums one at a time to your classmates, giving them the opportunity to guess the formula for a_n after each disclosed partial sum.

12.3

Geometric

Sequences

and Series

We saw that in an arithmetic sequence there is a constant difference between consecutive terms. This simple relationship allowed us to find a formula for the sum of n terms of an arithmetic sequence. Another type of sequence that has a simple pattern is a *geometric sequence*. In a geometric sequence consecutive terms have a *constant ratio*. In this section we will make use of the constant ratio to find a formula for the sum of n terms of a geometric sequence. We study arithmetic and geometric sequences in detail because they occur in many applications and because they are two of the very few sequences for which we can find formulas for the sum of n terms.

Geometric Sequences

In the sequence

$$100, 50, 25, 12.5, \ldots$$

each term after the first is half of the term preceding it. This sequence can be written as

$$100, \quad 100\left(\frac{1}{2}\right), \quad 100\left(\frac{1}{2}\right)^2, \quad 100\left(\frac{1}{2}\right)^3, \ldots$$

and its nth term is given by

$$a_n = 100\left(\frac{1}{2}\right)^{n-1}.$$

Any sequence in which each term after the first is a constant multiple of the preceding term is called a geometric sequence.

Definition: Geometric Sequence

A sequence with general term $a_n = ar^{n-1}$ is called a **geometric sequence** with **common ratio r**, where $r \neq 1$ and $r \neq 0$.

According to the definition, every geometric sequence is of the form

$$a, \quad ar, \quad ar^2, \quad ar^3, \quad ar^4, \ldots$$

Notice also that in a geometric sequence the ratio of any term (after the first) and the term preceding it is r. To write a formula for the nth term of a geometric sequence, all we need to know is the first term and the constant ratio.

Example 1   Finding a formula for the nth term

Write a formula for the nth term of each geometric sequence.

a) $0.3, 0.03, 0.003, 0.0003, \ldots$ **b)** $2, -6, 18, -54, \ldots$

Solution

a) Since each term after the first is one-tenth of the term preceding it, we use $r = 0.1$ and $a = 0.3$ in the formula $a_n = ar^{n-1}$:

$$a_n = 0.3(0.1)^{n-1}$$

b) Choose any two consecutive terms and divide the second by the first to obtain the common ratio. So $r = -6/2 = -3$. Since the first term is 2,

$$a_n = 2(-3)^{n-1}.$$

In Section 12.2 we learned some interesting facts about arithmetic sequences. However, this information is useful only if we can determine whether a given sequence is arithmetic. In this section we will learn some facts about geometric sequences. Likewise, we must be able to determine whether a given sequence is geometric even when it is not given in exactly the same form as the definition.

Example 2   Identifying a geometric sequence

Find the first four terms of each sequence and determine whether the sequence is geometric.

a) $b_n = (-2)^{3n}$ **b)** $a_1 = 1.25$ and $a_n = -2a_{n-1}$ for $n \geq 2$ **c)** $c_n = 3n$

Solution

a) Use $n = 1, 2, 3,$ and 4 in the formula $b_n = (-2)^{3n}$ to find the first four terms:

$$-8, \quad 64, \quad -512, \quad 4096, \ldots$$

The ratio of any term and the preceding term can be found from the formula

$$\frac{b_n}{b_{n-1}} = \frac{(-2)^{3n}}{(-2)^{3(n-1)}} = (-2)^3 = -8.$$

Since there is a common ratio of -8, the sequence is geometric.

$$U_n \boxminus -2U_{n-1}$$
$$U_n =$$

Define u_n using the Y= key in sequence mode and set U_nStart = 1.25.

$$U_n(1,4,1)$$
$$\{1.25\ \ -2.5\ \ 5\ \ -1\ldots$$

Evaluate u_n for $n = 1$ through $n = 4$ in increments of 1 to get the first four terms of the geometric sequence in Example 2(b).

b) Use the recursion formula $a_n = -2a_{n-1}$ to obtain each term after the first:

$$a_1 = 1.25$$
$$a_2 = -2a_1 = -2.5$$
$$a_3 = -2a_2 = 5$$
$$a_4 = -2a_3 = -10$$

This recursion formula defines the sequence

$$1.25, -2.5, 5, -10, \ldots$$

The formula $a_n = -2a_{n-1}$ means that each term is a multiple of the term preceding it. The constant ratio is -2 and the sequence is geometric.

c) Use $c_n = 3n$ to find $c_1 = 3$, $c_2 = 6$, $c_3 = 9$, and $c_4 = 12$. Since $6/3 = 2$ and $9/6 = 1.5$, there is no common ratio for consecutive terms. The sequence is not geometric. Because each term is 3 larger than the preceding term, the sequence is arithmetic. ◆

In the next example we see that a geometric sequence occurs in an investment earning compound interest. If we have an initial deposit earning compound interest, then the amounts in the account at the end of each year form a geometric sequence.

Example 3   A geometric sequence in investment

The parents of a newborn decide to start saving early for her college education. On the day of her birth, they invest $3000 at 6% compounded annually. Find the amount of the investment at the end of each of the first four years and find a formula for the amount at the end of the nth year. Find the amount at the end of the 18th year.

Solution

At the end of the first year the amount is $3,000(1.06). At the end of the second year the amount is $3,000(1.06)^2$. Use a calculator to find the amounts at the ends of the first four years.

$$a_1 = 3000(1.06)^1 = \$3180$$
$$a_2 = 3000(1.06)^2 = \$3370.80$$
$$a_3 = 3000(1.06)^3 = \$3573.05$$
$$a_4 = 3000(1.06)^4 = \$3787.43$$

A formula for the nth term of this geometric sequence is $a_n = 3000(1.06)^n$. The amount at the end of the 18th year is

$$a_{18} = 3000(1.06)^{18} = \$8,563.02.$$ ◆

The formula for the general term of a geometric sequence involves a_n, a, n, and r. If we know the value of any three of these quantities, then we can find the value of the fourth.

Example 4   Find the number of terms in a geometric sequence

A certain ball always rebounds 2/3 of the distance from which it falls. If the ball is dropped from a height of 9 feet, and later it is observed rebounding to a height of 64/81 feet, then how many times did it bounce?

Solution

After the first bounce the ball rebounds to $9(2/3) = 6$ feet. After the second bounce the ball rebounds to a height of $9(2/3)^2 = 4$ feet. The first term of this sequence is 6 and the common ratio is 2/3. So after the nth bounce the ball rebounds to a height h_n given by

$$h_n = 6\left(\frac{2}{3}\right)^{n-1}.$$

To find the number of bounces, solve the following equation, which states that the nth bounce rebounds to 64/81 feet.

$$6\left(\frac{2}{3}\right)^{n-1} = \frac{64}{81}$$

$$\left(\frac{2}{3}\right)^{n-1} = \frac{32}{243}$$

$$\left(\frac{2}{3}\right)^{5} = \frac{32}{243}$$

$$n - 1 = 5$$

$$n = 6$$

The ball bounced six times. ◆

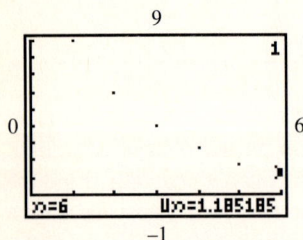

Define $u_n = 2u_{n-1}/3$ with $u_1 = 9$. Then graph the sequence to see the heights from which the ball falls on its first six falls.

Geometric Series

Geometric sequences occur in applications ranging from bouncing balls to investing money. The sum of n terms of a geometric sequence can give the total distance traveled by a bouncing ball or the total value of some periodic deposits earning compound interest. The indicated sum of the terms of a geometric sequence is called a **geometric series.** To find the actual sum of a geometric series, we can use a procedure similar to that used for finding the sum of an arithmetic series.

Consider the geometric sequence $a_n = 2^{n-1}$. Let S_{10} represent the sum of the first ten terms of this geometric sequence:

$$S_{10} = 1 + 2 + 4 + 8 + \cdots + 512$$

If we multiply each side of this equation by -2, the opposite of the common ratio, we get

$$-2S_{10} = -2 - 4 - 8 - 16 - \cdots - 512 - 1024.$$

Adding S_{10} and $-2S_{10}$ eliminates most of the terms:

$$
\begin{array}{lll}
S_{10} = 1 + 2 + 4 + 8 + 16 + \cdots + 512 & & \\
\underline{-2S_{10} = \quad\quad -2 - 4 - 8 - 16 - \cdots - 512 - 1024} & & \\
-S_{10} = 1 & -1024 & \text{\textcolor{blue}{Add.}} \\
-S_{10} = -1023 & & \\
S_{10} = 1023 & &
\end{array}
$$

The "trick" to finding the sum of n terms of a geometric series is to change the signs and shift the terms so that most terms "cancel out" when the two equations are added. This method can also be used to find a general formula for the sum of a geometric series.

Let S_n represent the sum of the first n terms of the geometric sequence $a_n = ar^{n-1}$:

$$S_n = a + ar + ar^2 + \cdots + ar^{n-1}$$

Adding S_n and $-rS_n$ eliminates most of the terms:

$$
\begin{array}{lll}
S_n = a + ar + ar^2 + & \cdots + ar^{n-1} & \\
\underline{-rS_n = \quad\quad - ar - ar^2 - ar^3 - \cdots - ar^{n-1} - ar^n} & & \\
S_n - rS_n = a & - ar^n & \text{\textcolor{blue}{Add.}} \\
(1 - r)S_n = a(1 - r^n) & & \\
S_n = \dfrac{a(1 - r^n)}{1 - r} & \text{\textcolor{blue}{Provided that } r \ne 1.} &
\end{array}
$$

So the sum of a geometric series can be found if we know the first term, the constant ratio, and the number of terms. This result is summarized in the following theorem.

Theorem: Sum of a Finite Geometric Series

If S_n represents the sum of the first n terms of a geometric series with first term a and common ratio r ($r \ne 1$), then

$$S_n = \frac{a(1 - r^n)}{1 - r}.$$

Example 5   Finding the sum of a geometric series

Find the sum of each geometric series.

a) $1 + \dfrac{1}{3} + \dfrac{1}{9} + \cdots + \dfrac{1}{243}$ b) $\displaystyle\sum_{j=0}^{10} 100(1.05)^j$

Solution

a) To find the sum of a finite geometric series, we need the first term a, the ratio r, and the number of terms n. To find n, use $a = 1$, $a_n = 1/243$, and $r = 1/3$ in the formula $a_n = ar^{n-1}$:

$$1\left(\frac{1}{3}\right)^{n-1} = \frac{1}{243}$$

$$n - 1 = 5 \qquad \text{Because } \left(\frac{1}{3}\right)^5 = \frac{1}{243}$$

$$n = 6$$

Now use $n = 6$, $a = 1$, and $r = 1/3$ in the formula $S_n = \dfrac{a(1 - r^n)}{1 - r}$:

$$S_6 = \frac{1\left(1 - \left(\dfrac{1}{3}\right)^6\right)}{1 - \dfrac{1}{3}} = \frac{\dfrac{728}{729}}{\dfrac{2}{3}} = \frac{364}{243}$$

b) First write out some terms of the series:

$$\sum_{j=0}^{10} 100(1.05)^j = 100 + 100(1.05) + 100(1.05)^2 + \cdots + 100(1.05)^{10}$$

In this geometric series, $a = 100$, $r = 1.05$, and $n = 11$:

$$\sum_{j=0}^{10} 100(1.05)^j = S_{11} = \frac{100(1 - (1.05)^{11})}{1 - 1.05} \approx 1420.68$$

One of the most important applications of the sum of a finite geometric series is in projecting the value of an annuity. An **annuity** is a sequence of equal periodic payments. If each payment earns the same rate of compound interest, then the total value of the annuity can be found by using the formula for the sum of a finite geometric series.

Example 6   Finding the value of an annuity

To maintain his customary style of living after retirement, a single man earning $50,000 per year at age 65 must have saved a minimum of $95,700 (*Fortune*, July 26, 1993). If Chad invests $1000 at the beginning of each year for 40 years in an investment paying 6% compounded annually, then what is the value of this annuity at the end of the 40th year?

Solution

The last deposit earns interest for only one year and amounts to 1000(1.06). The second to last deposit earns interest for two years and amounts to $1000(1.06)^2$. This pattern continues down to the first deposit, which earns interest for 40 years and amounts to $1000(1.06)^{40}$. The value of the annuity is the sum of a finite geometric series:

$$S_{40} = 1000(1.06) + 1000(1.06)^2 + \cdots + 1000(1.06)^{40}$$

Since the first term is 1000(1.06), use $a = 1000(1.06)$, $r = 1.06$, and $n = 40$:

$$S_{40} = \frac{1000(1.06)(1 - (1.06)^{40})}{1 - 1.06} = \$164,047.68$$

Infinite Geometric Series

In the geometric series

$$2 + 4 + 8 + 16 + \cdots,$$

in which $r = 2$, the terms get larger and larger. So the sum of the first n terms increases without bound as n increases. In the geometric series

$$\frac{1}{2} + \frac{1}{4} + \frac{1}{8} + \frac{1}{16} + \cdots,$$

in which $r = 1/2$, the terms get smaller and smaller. The sum of n terms of this series is less than 1 no matter how large n is. (To see this, add some terms on your calculator.) We can explain the different behavior of these series by examining the term r^n in the formula

$$S_n = \frac{a(1 - r^n)}{1 - r}.$$

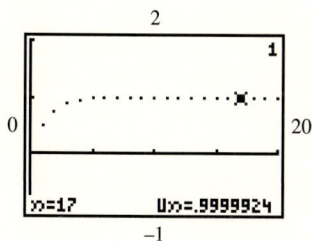

The graph of $u_n = 0.5(1 - 0.5^n)/(1 - 0.5)$ for $n = 1$ to 20 illustrates how u_n approaches 1 as n increases.

If $r = 2$, the values of r^n increase without bound as n gets larger, causing S_n to increase without bound. If $r = 1/2$, the values of $(1/2)^n$ approach 0 as n gets larger. In symbols, $(1/2)^n \to 0$ as $n \to \infty$. If $r^n \to 0$, then $1 - r^n$ is approximately 1. If we replace $1 - r^n$ by 1 in the formula for S_n, we get

$$S_n \approx \frac{a}{1 - r} \qquad \text{for large values of } n.$$

In the above series $r = 1/2$ and $a = 1/2$. So if n is large we have

$$S_n \approx \frac{\dfrac{1}{2}}{1 - \dfrac{1}{2}} = 1.$$

In general, it can be proved that $r^n \to 0$ as $n \to \infty$ provided that $|r| < 1$, and r^n does not get close to 0 as $n \to \infty$ for $|r| \geq 1$. You will better understand these ideas if you use a calculator to find some large powers of r for various values of r. For example,

$$(0.99)^{1000} \approx 0.0000432 \qquad \text{and} \qquad (1.001)^{9000} \approx 8,066.726.$$

So, if $|r| < 1$ and n is large, then S_n is approximately $a/(1 - r)$. Furthermore, by using more terms in the sum we can get a sum that is arbitrarily close to the number $a/(1 - r)$. In this sense we say that the sum of all terms of the infinite geometric series is $a/(1 - r)$.

Theorem: Sum of an Infinite Geometric Series

If $a + ar + ar^2 + \cdots$ is an infinite geometric series with $|r| < 1$, then the sum S of all of the terms is given by

$$S = \frac{a}{1 - r}.$$

We can use the infinity symbol ∞ to indicate the sum of infinitely many terms of an infinite geometric series as follows:

$$a + ar + ar^2 + \cdots = \sum_{i=1}^{\infty} ar^{i-1}$$

Example 7 Finding the sum of an infinite geometric series

Find the sum of each infinite geometric series.

a) $1 + \dfrac{1}{3} + \dfrac{1}{9} + \cdots$ **b)** $\displaystyle\sum_{j=1}^{\infty} 100(-0.99)^j$ **c)** $\displaystyle\sum_{i=0}^{\infty} 3(1.01)^i$

Solution

a) The first term is 1, and the common ratio is 1/3. So the sum of the infinite series is

$$S = \frac{1}{1 - \dfrac{1}{3}} = \frac{3}{2}.$$

b) The first term is -99, and the common ratio is -0.99. So the sum of the infinite geometric series is

$$S = \frac{-99}{1 - (-0.99)} = \frac{-99}{1.99} = -\frac{9900}{199}.$$

c) The first term is 3, and the common ratio is 1.01. Since the absolute value of the ratio is greater than 1, this infinite geometric series has no sum. ◆

In the next example we use the formula for the sum of an infinite geometric series in a physical situation. Since physical processes do not continue infinitely, the formula for the sum of infinitely many terms is used as an approximation for the sum of a large finite number of terms of a geometric series in which $|r| < 1$.

Example 8   Total distance traveled in bungee jumping

A man jumping from a bridge with a bungee cord tied to his legs falls 120 feet before being pulled back upward by the bungee cord. If he always rebounds 1/3 of the distance that he has fallen and then falls 2/3 of the distance of his last rebound, then approximately how far does the man travel before coming to rest?

Solution

In actual practice, the man does not go up and down infinitely on the bungee cord. However, to get an approximate answer, we can model this situation shown in Fig. 12.1 with two infinite geometric sequences, the sequence of falls and the sequence of rises. The man falls 120 feet, then rises 40 feet. He falls 80/3 feet, then rises 80/9 feet. He falls 160/27 feet, then rises 160/81 feet. The total distance the man falls is given by the following series:

$$F = 120 + \frac{80}{3} + \frac{160}{27} + \cdots = \frac{120}{1 - \frac{2}{9}} = \frac{1080}{7} \text{ feet}$$

The total distance the man rises is given by the following series:

$$R = 40 + \frac{80}{9} + \frac{160}{81} + \cdots = \frac{40}{1 - \frac{2}{9}} = \frac{360}{7} \text{ feet}$$

The total distance he travels before coming to rest is the sum of these distances, 1440/7 feet, or approximately 205.7 feet.

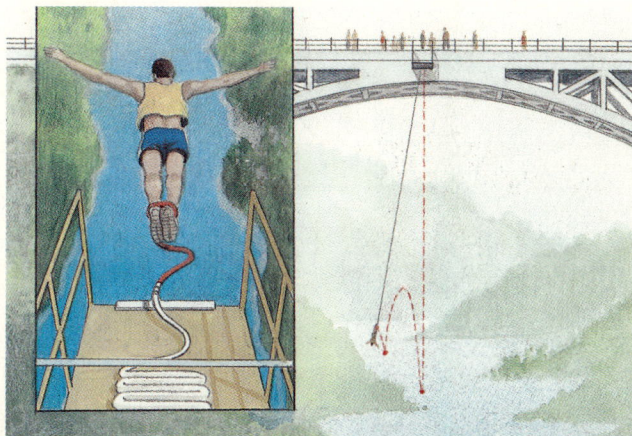

Figure 12.1

Repeating Decimals

Another application of geometric series occurs in rational numbers that are infinite repeating decimals. For example, the fraction 1/3 is a repeating decimal. This repeating decimal can be viewed as the sum of a geometric series:

$$\frac{1}{3} = 0.3333...$$

$$= 0.3 + 0.3 \cdot 10^{-1} + 0.3 \cdot 10^{-2} + 0.3 \cdot 10^{-3} + \cdots$$

The first term is 0.3, and the ratio is 10^{-1}. The formula for the sum of an infinite geometric series can be used to convert the repeating decimal number back into a fraction.

$$0.3333... = \frac{0.3}{1 - 10^{-1}}$$

$$= \frac{0.3}{0.9} = \frac{1}{3}$$

The next example uses this same idea on a decimal number in which some of the digits do not repeat.

Example 9   Converting a repeating decimal to a fraction

Use the formula for the sum of an infinite geometric series to convert the repeating decimal number 1.2417417417... into a fraction.

Solution

Separate the repeating part of the number from the nonrepeating part and convert the repeating part into a fraction, using the formula for the sum of an infinite geometric series:

$$1.2417417417417... = 1.2 + 0.0417417417417...$$

$$= 1.2 + 417 \cdot 10^{-4} + 417 \cdot 10^{-7} + 417 \cdot 10^{-10} + \cdots$$

$$= \frac{12}{10} + \frac{417 \cdot 10^{-4}}{1 - 10^{-3}} \qquad a = 417 \cdot 10^{-4} \text{ and } r = 10^{-3}$$

$$= \frac{12}{10} + \frac{417}{10,000 - 10}$$

$$= \frac{12}{10} + \frac{417}{9990}$$

$$= \frac{12405}{9990}$$

For Thought

True or false? Explain.

1. The sequence 2, 6, 24, . . . is a geometric sequence.
2. The sequence $a_n = 3(2)^{3-n}$ is a geometric sequence.
3. The first term of the geometric sequence $a_n = 5(0.3)^n$ is 1.5.
4. The common ratio in the geometric sequence $a_n = 5^{-n}$ is 5.
5. A geometric series is the indicated sum of a geometric sequence.
6. $3 + 6 + 12 + 24 + \cdots = \dfrac{3}{1-2}$
7. $\displaystyle\sum_{i=1}^{9} 3(0.6)^i = \dfrac{1.8(1-0.6)^9}{1-0.6}$
8. $\displaystyle\sum_{i=0}^{4} 2(10)^i = 22{,}222$
9. $\displaystyle\sum_{i=1}^{\infty} 3(0.1)^i = \dfrac{1}{3}$
10. $\displaystyle\sum_{i=1}^{\infty} \left(\dfrac{1}{2}\right)^i = 1$

12.3 Exercises

Tape 22   Disk—5.25″: 7   3.5″: 4   Macintosh: 4

Write a formula for the nth term of each geometric sequence. Do not use a recursion formula.

1. $\dfrac{1}{6}, \dfrac{1}{3}, \dfrac{2}{3}, \dfrac{4}{3}, \ldots$
2. $\dfrac{1}{6}, 0.5, 1.5, 4.5, \ldots$
3. $0.9, 0.09, 0.009, 0.0009, \ldots$
4. $3, 9, 27, 81, \ldots$
5. $4, -12, 36, -108, \ldots$
6. $5, -1, \dfrac{1}{5}, -\dfrac{1}{25}, \ldots$

Identify each sequence as arithmetic, geometric, or neither.

7. $\dfrac{1}{6}, \dfrac{1}{3}, \dfrac{1}{2}, \dfrac{2}{3}, \ldots$
8. $\dfrac{1}{6}, \dfrac{1}{3}, 1, 4, \ldots$
9. $\dfrac{1}{6}, \dfrac{1}{3}, \dfrac{2}{3}, \dfrac{4}{3}, \ldots$
10. $3, 2, 1, 0, -1, \ldots$
11. $1, 4, 9, 16, 25, \ldots$
12. $5, 1, \dfrac{1}{5}, \dfrac{1}{25}, \ldots$

Find the first four terms of each sequence and identify each sequence as arithmetic, geometric, or neither.

13. $a_n = 2n$
14. $a_n = 2^n$
15. $a_n = n^2$
16. $a_n = n!$
17. $a_n = 2^{-n}$
18. $a_n = n + 2$
19. $b_n = 2^{2n+1}$
20. $d_n = \dfrac{1}{3^{n/2}}$
21. $c_1 = 3, c_n = -3c_{n-1}$ for $n \geq 2$
22. $h_1 = \sqrt{2}, h_n = \sqrt{3}h_{n-1}$ for $n \geq 2$

List the first three terms and the 10th term of the geometric sequence whose nth term is given.

23. $a_n = 3(-2)^{n-1}$
24. $a_n = 2(-0.5)^{n-1}$
25. $a_n = 4(0.1)^{n-1}$
26. $a_n = 0.01(5)^{n-1}$

Find the required part of each geometric sequence.

27. Find the number of terms of a geometric sequence with first term 3, common ratio 1/2, and last term 3/1024.
28. Find the number of terms of a geometric sequence with first term 1/64, common ratio −2, and last term −512.

29. Find the first term of a geometric sequence with sixth term 1/81 and common ratio of 1/3.

30. Find the first term of a geometric sequence with seventh term 1/16 and common ratio 1/2.

31. Find the common ratio for a geometric sequence with first term 2/3 and third term 6.

32. Find the common ratio for a geometric sequence with second term -1 and fifth term -27.

33. Find a formula for the nth term of a geometric sequence with third term -12 and sixth term 96.

34. Find a formula for the nth term of a geometric sequence with second term -40 and fifth term 0.04.

Find the sum of each finite geometric series by using the formula for S_n.

35. $6 + 2 + \dfrac{2}{3} + \dfrac{2}{9} + \dfrac{2}{27}$

36. $2 + 10 + 50 + 250 + 1250$

37. $1.5 - 3 + 6 - 12 + 24 - 48 + 96 - 192$

38. $1 - \dfrac{1}{2} + \dfrac{1}{4} - \dfrac{1}{8} + \dfrac{1}{16} - \dfrac{1}{32} + \dfrac{1}{64}$

39. $\displaystyle\sum_{i=1}^{12} 2(1.05)^{i-1}$

40. $\displaystyle\sum_{i=0}^{30} 300(1.08)^{i}$

Write each geometric series in summation notation.

41. $3 - 1 + \dfrac{1}{3} - \dfrac{1}{9} + \dfrac{1}{27}$

42. $2 + 1 + \dfrac{1}{2} + \dfrac{1}{4} + \dfrac{1}{8} + \dfrac{1}{16}$

43. $0.6 + 0.06 + 0.006 + \cdots$

44. $4 - 1 + \dfrac{1}{4} - \dfrac{1}{16} + \cdots$

45. $-4.5 + 1.5 - 0.5 + \dfrac{1}{6} - \cdots$

46. $a + ab + ab^2 + \cdots + ab^{37}$

Find the sum of each infinite geometric series where possible.

47. $3 - 1 + \dfrac{1}{3} - \dfrac{1}{9} + \dfrac{1}{27} - \cdots$

48. $1 + \dfrac{1}{2} + \dfrac{1}{4} + \dfrac{1}{8} + \dfrac{1}{16} + \cdots$

49. $0.9 + 0.09 + 0.009 + \cdots$

50. $-1 + \dfrac{1}{4} - \dfrac{1}{16} + \dfrac{1}{64} - \cdots$

51. $-9.9 + 3.3 - 1.1 + \cdots$

52. $1.2 - 2.4 + 4.8 - 9.6 + \cdots$

53. $\displaystyle\sum_{i=1}^{\infty} 34(0.01)^{i}$

54. $\displaystyle\sum_{i=0}^{\infty} 300(0.99)^{i}$

55. $\displaystyle\sum_{i=0}^{\infty} 300(-1.06)^{i}$

56. $\displaystyle\sum_{i=0}^{\infty} (0.98)^{i}$

57. $\displaystyle\sum_{i=1}^{\infty} 6(0.1)^{i}$

58. $\displaystyle\sum_{i=0}^{\infty} (0.1)^{i}$

59. $\displaystyle\sum_{i=0}^{\infty} 34(-0.7)^{i}$

60. $\displaystyle\sum_{i=1}^{\infty} 123(0.001)^{i}$

Use the formula for the sum of an infinite geometric series to write each repeating decimal number as a fraction.

61. $0.04444\ldots$

62. $0.0121212\ldots$

63. $8.2545454\ldots$

64. $3.65176176176\ldots$

Use the ideas of geometric series to solve each problem.

65. *Sales goals* A group of college students selling magazine subscriptions during the summer sold one subscription on June 1. The sales manager was encouraged by this and said that their daily goal for each day of June is to double the sales of the previous day. If the students work every day during June and meet this goal, then what is the total number of magazine subscriptions that they will sell during June?

66. *Family tree* Consider yourself, your parents, your grandparents, your great-grandparents, your great-great-grandparents, and so on, back to your grandparents with the word "great" used in front 40 times. What is the total number of people that you are considering?

67. *Compound interest* Given that $4000 is deposited at the beginning of a quarter into an account earning 8% annual interest compounded quarterly, write a formula for the amount in the account at the end of the nth quarter. How much is in the account at the end of 37 quarters?

68. *Compound interest* Given that $8000 is deposited at the beginning of a month into an account earning 6% annual interest compounded monthly, write a formula for the amount in the account at the end of the nth month. How much is in the account at the end of the 56th month?

69. *Value of an annuity* If you deposit $200 on the first of each month for 12 months into an account paying 12% annual interest compounded monthly, then how much is in the account at the end of the 12th month? Note that each deposit earns 1% per month for a different number of months.

70. *Saving for retirement* If you deposit $9000 on the first of each year for 40 years into an account paying 8% compounded annually, then how much is in the account at the end of the 40th year? Use the formula for the sum of a finite geometric series.

71. *Saving for retirement* If $100 is deposited at the end of each month for 30 years in a retirement account earning 9% compounded monthly, then what is the value of this annuity immediately after the last payment?

72. *Down payment* To get a down payment for a house, a couple plan to deposit $800 at the end of each quarter for 36 quarters in an account paying 6% compounded quarterly. What is the value of this annuity immediately after the last deposit?

73. *Bouncing ball* Suppose that a ball always rebounds 2/3 of the distance from which it falls. If this ball is dropped from a height of 9 ft, then approximately how far does it travel before coming to rest?

74. *Saturating the market* A sales manager has set a team goal of $1 million in sales. The team plans to get 1/2 of the goal the first week of the campaign. Every week after the first they will sell only 1/2 as much as the previous week because the market will become saturated. How close will the team be to the sales goal after 15 weeks of selling?

75. *Economic impact* The anticipated payroll of the new General Dynamics plant in Hammond, Louisiana, is $2 million annually. It is estimated that 75% of this money is spent in Hammond by its recipients. The people who receive that money again spend 75% of it in Hammond, and so on. The total of all of this spending is the *total economic impact* of the plant on Hammond. What is the total economic impact of the plant?

76. *Disaster relief* If the federal government provides $300 million in disaster relief to the people of Louisiana after a hurricane, then 80% of that money is spent in Louisiana. If the money is respent over and over at a rate of 80% in Louisiana, then what is the total economic impact of the $300 million on the state?

For Writing/Discussion

77. Can an arithmetic sequence and a geometric sequence have the same first three terms? Explain your answer.

78. *Cooperative learning* Get a "super ball" from a toy store and work in a small group to measure the distance that it rebounds from falls of 8 ft, 6 ft, 4 ft, and 2 ft. Is it reasonable to assume that the rebound distance is a constant percentage of the fall distance? Use an infinite series to find the total distance that your ball travels vertically before coming to rest when it is dropped from a height of 8 ft. Discuss the applicability of the infinite series model to this physical experiment.

Graphing Calculator Exercises

1. Consider the functions $y = r^x$ and $y = x^r$ for $x > 0$. Graph these functions for some values of r with $|r| < 1$ and for some values of r with $|r| \geq 1$. What values do r^x and x^r approach as $x \to \infty$?

2. Consider the function $y = 5(1 - r^x)/(1 - r)$ for $x > 0$. Graph this function for some values of r with $|r| < 1$ and for some values of r with $|r| > 1$. What do these graphs illustrate?

12.4

Counting and

Permutations

In the first three sections of this chapter we studied sequences and series. In this section we will look at sequences of events and count the number of ways that a sequence of events can occur. But in this instance, we actually try to avoid counting in the usual sense because this method of counting would probably be too cumbersome or tedious. In this new context, counting means finding the number of ways in which something can be done without actually listing all of the ways and counting them.

The Fundamental Counting Principle

Let's say that the cafeteria lunch special includes a choice of sandwich and dessert. In the terminology of counting, choosing a sandwich is an **event** and choosing a dessert is another event. Suppose that there are three outcomes, ham, salami, or tuna, for the first event and two outcomes, pie or cake, for the second event. How many different lunches are available using these choices? We can make a diagram showing all of the possibilities as in Fig. 12.2. This diagram is called a **tree diagram.** Considering only the types of sandwich and dessert, the tree diagram shows six different lunches. Of course, 6 can be obtained by multiplying 3 and 2. This example illustrates the fundamental counting principle.

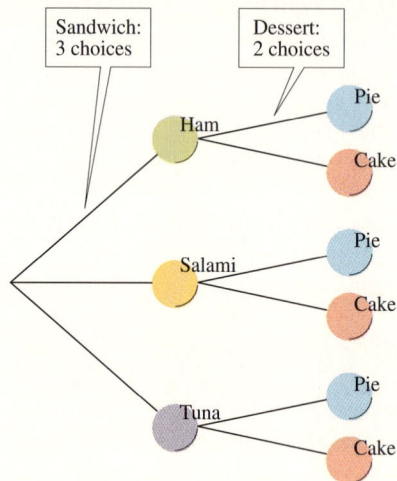

Figure 12.2

**Fundamental Counting
Principle**

If event A has m different outcomes and event B has n different outcomes, then there are mn different ways for events A and B to occur together.

The fundamental counting principle can also be used for more than two events, as is illustrated in the next example.

Example 1   Applying the fundamental counting principle

When ordering a new car, you are given a choice of three engines, two transmissions, six colors, three interior designs, and whether or not to get air conditioning. How many different cars can be ordered considering these choices?

Solution

There are three outcomes to the event of choosing the engine, two outcomes to choosing the transmission, six outcomes to choosing the color, three outcomes to choosing the interior, and two outcomes to choosing the air conditioning (to have it or not). So the number of different cars available is $3 \cdot 2 \cdot 6 \cdot 3 \cdot 2 = 216$. ◆

Example 2   Applying the fundamental counting principle

How many different license plates are possible if each plate consists of three letters followed by a three-digit number? Assume that repetitions in the letters or numbers are allowed and that any of the ten digits may be used.

Solution

Since there are 26 choices for each of the three letters and ten choices for each of the three numbers, by the fundamental counting principle, the number of license plates is $26 \cdot 26 \cdot 26 \cdot 10 \cdot 10 \cdot 10 = 26^3 \cdot 10^3 = 17{,}576{,}000$. ◆

Permutations

In Examples 1 and 2, each choice was independent of the previous choices. But the number of ways in which an event can occur often depends on what has already occurred. For example, in arranging three students in a row, the choice of the student for the second seat depends on which student was placed first. Consider the following six different sequential arrangements (or permutations) of three students, Ann, Bob, and Carol:

Ann, Bob, Carol	Bob, Ann, Carol	Carol, Ann, Bob
Ann, Carol, Bob	Bob, Carol, Ann	Carol, Bob, Ann

These six permutations can also be shown in a tree diagram as we did in Fig. 12.3. Since the students are distinct objects, there can be no repetition of students.

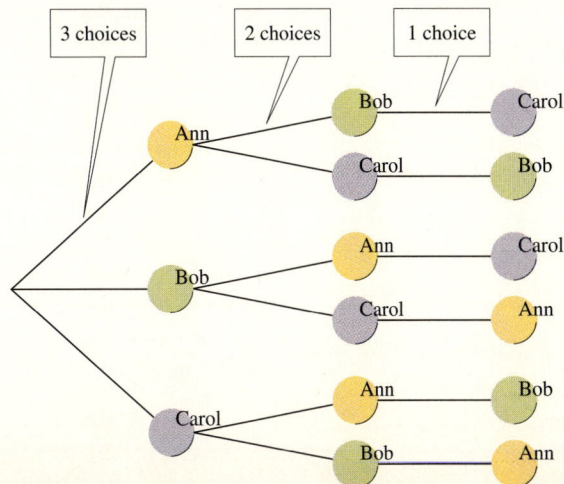

Figure 12.3

An arrangement such as Ann, Ann, Bob is not allowed. A **permutation** is an ordering or arrangement of distinct objects in a sequential manner. There is a first, a second, a third, and so on. High-level diplomats are usually *not* seated so that they are arranged in a sequential order. They are seated at round tables so that no one is first, or second, or third.

Example 3   Finding permutations

A Federal Express driver must make ten deliveries to ten different addresses. In how many ways can she make those deliveries?

Solution

For the event of choosing the first address there are ten outcomes. For the event of choosing the second address, there are nine outcomes (since one has already been chosen). For the third address, there are eight outcomes, and so on. So, according to the fundamental counting principle, the number of permutations of the ten addresses is

$$10 \cdot 9 \cdot 8 \cdot 7 \cdot 6 \cdot 5 \cdot 4 \cdot 3 \cdot 2 \cdot 1 = 10! = 3{,}628{,}800.$$

In general, the number of arrangements of any n distinct objects in a sequential manner is referred to as the number of permutations of n things taken n at a time and the notation $P(n, n)$ is used to represent this number. The phrase "taken n at a time" indicates that all of the n objects are used in the arrangements. (We will soon discuss permutations in which not all of the n objects are used.) Since there are n choices for the first object, $n - 1$ choices for the second object, $n - 2$ choices for the third object, and so on, we have $P(n, n) = n!$.

Theorem: Permutations of *n* Things Taken *n* at a Time

The notation $P(n, n)$ represents the number of permutations of n things taken n at a time, and $P(n, n) = n!$.

Sometimes we are interested in permutations in which not all of the objects are used. For example, if the Federal Express driver wants to deliver three of the ten packages before lunch, then how many ways are there to make those three deliveries? There are ten possibilities for the first delivery, nine for the second, and eight for the third. So the number of ways to make three stops before lunch, or the number of permutations of 10 things taken 3 at a time, is $10 \cdot 9 \cdot 8 = 720$. The notation $P(10, 3)$ is used to represent the number of permutations of 10 things taken 3 at a time. Notice that

$$P(10, 3) = \frac{10!}{7!} = \frac{10 \cdot 9 \cdot 8 \cdot 7 \cdot 6 \cdot 5 \cdot 4 \cdot 3 \cdot 2 \cdot 1}{7 \cdot 6 \cdot 5 \cdot 4 \cdot 3 \cdot 2 \cdot 1} = 10 \cdot 9 \cdot 8 = 720.$$

In general, we have the following theorem.

Theorem: Permutations of *n* Things Taken *r* at a Time

The notation $P(n, r)$ represents the number of permutations of n things taken r at a time, and

$$P(n, r) = \frac{n!}{(n - r)!} \qquad \text{for } 0 \leq r \leq n.$$

Even though n things are usually not taken 0 at a time, 0 is allowed in the formula. Recall that by definition, $0! = 1$. For example, $P(8, 0) = 8!/8! = 1$ and $P(0, 0) = 0!/0! = 1/1 = 1$.

Many calculators have a key that gives the value of $P(n, r)$ when given n and r. The notation nPr is often used on calculators and elsewhere for $P(n, r)$.

```
12 nPr 3
                1320
12!/9!
                1320
```

You can find $P(12, 3)$ by using the permutation function or by using the factorial function.

Example 4   Finding permutations of *n* things taken *r* at a time

The Beau Chene Garden Club has 12 members. They plan to elect a president, a vice-president, and a treasurer. How many different outcomes are possible for the election if each member is eligible for each office and no one can hold two offices?

Solution

The number of ways in which these offices can be filled is precisely the number of permutations of 12 things taken 3 at a time:

$$P(12, 3) = \frac{12!}{(12 - 3)!} = \frac{12!}{9!} = 12 \cdot 11 \cdot 10 = 1320 \qquad \blacklozenge$$

? For Thought

True or false? Explain.

1. If a product code consists of a single letter followed by a two-digit number from 10 through 99, then $89 \cdot 26$ different codes are available.

2. If a fraternity name consists of three Greek letters chosen from the 24 letters in the Greek alphabet with repetitions allowed, then $23 \cdot 22 \cdot 21$ different fraternity names are possible.

3. If an outfit consists of a tie, a shirt, a pair of pants, and a coat, and John has three ties, five shirts, and three coats that all match his only pair of pants, then John has 11 outfits available to wear.

4. The number of ways in which five people can line up to buy tickets is 120.

5. The number of permutations of 10 things taken 2 at a time is 90.

6. The number of different ways to mark the answers to a 20-question multiple-choice test with each question having four choices is $P(20, 4)$.

7. The number of different ways to mark the answers to this sequence of 10 "For Thought" questions is 2^{10}.

8. $\dfrac{1000!}{998!} = 999{,}000$ 9. $P(10, 9) = P(10, 1)$ 10. $P(29, 1) = 1$

12.4 Exercises   ▶ Tape 23   💾 Disk—5.25": 7   3.5": 4   Macintosh: 4

Solve each problem using the fundamental counting principle.

1. *Traveling sales representative* A sales representative can take either of two different routes from Sacramento to Stockton and any one of four different routes from Stockton to San Francisco. How many different routes can she take from Sacramento to San Francisco, going through Stockton?

2. *Optional equipment* A new car can be ordered in any one of nine different colors, with three different engines, two different transmissions, three different body styles, two different interior designs, and four different stereo systems. You must also decide whether to include power windows, power door locks, a six-way adjustable driver's seat, air conditioning, power steering, power brakes, cruise control, and a driver's side air bag. How many different cars are available?

3. *Sleepless night* The ghosts of Christmas Past, Present, and Future plan on visiting Scrooge at 1:00, 2:00, and 3:00 in the morning. All three are available for haunting at all three times. Make a tree diagram showing all of the different orders in which the three ghosts can visit Scrooge. How many different arrangements are possible for this haunting schedule?

4. *Track competition* Juan, Felix, Ronnie, and Ted are in a 100 m race. Make a tree diagram showing all possible orders in which they can finish the race, assuming that there are no ties. In how many ways can they finish the race?

5. *Poker hands* A poker hand consists of five cards drawn from a deck of 52. How many different poker hands are there consisting of an ace, king, queen, jack, and ten?

6. *Drawing cards from a deck* In a certain card game, four cards are drawn from a deck of 52. How many different hands are there containing one heart, one spade, one club, and one diamond?

7. *Have it your way* Wendy's Old Fashioned Hamburgers once advertised that 256 different hamburgers were available at Wendy's. This number was obtained by using the fundamental counting principle and considering whether or not to include each one of several different options on the burger. How many different optional items were used to get this number?

8. *Choosing a pizza* A pizza can be ordered in three sizes with either thick crust or thin crust. You have to decide whether to include each of four different meats on your pizza. You must also decide whether to include green peppers, onions, mushrooms, anchovies, and/or black olives. How many different pizzas can you order?

Evaluate each expression.

9. $P(7, 3)$

10. $P(16, 4)$

11. $P(99, 0)$

12. $P(55, 1)$

13. $\dfrac{P(9, 5)}{5!}$

14. $\dfrac{P(7, 2)}{2!}$

15. $\dfrac{P(11, 3)}{3!}$

16. $\dfrac{P(15, 5)}{5!}$

17. $\dfrac{16!}{4!12!}$

18. $\dfrac{19!}{17!2!}$

19. $\dfrac{88!}{85!3!}$

20. $\dfrac{102!}{100!2!}$

Solve each problem using the idea of permutations.

21. *Parading in order*   A small Mardi Gras parade consists of eight floats and three marching bands. In how many different orders can they line up to parade?

22. *Atonement of Hercules*   The King of Tiryens ordered the Greek hero Hercules to atone for murdering his own family by performing 12 difficult and dangerous tasks. How many different ways are there for Hercules to perform the 12 tasks?

23. *Inspecting restaurants*   A health inspector must visit 3 of 15 restaurants on Monday. In how many ways can she pick a first, second, and third restaurant to visit?

24. *Assigned reading*   In how many ways can an English professor randomly give out one copy each of *War and Peace, The Grapes of Wrath, Moby Dick,* and *Gone with the Wind,* and four copies of *Jurassic Park* to a class of eight students?

25. *Scheduling radio shows*   The program director for a public radio station has 26 half-hour shows available for Sunday evening. How many different schedules are possible for the 6:00 to 10:00 P.M. time period?

26. *Choosing songs*   A disc jockey must choose eight songs from the top 20 to play in the next 30-minute segment of her show. How many different arrangements are possible for this segment?

Solve each counting problem.

27. *Multiple-choice test*   How many different ways are there to mark the answers to a six-question multiple-choice test in which each question has four possible answers?

28. *Choosing a name*   A novelist has decided on four possible first names and three possible last names for the main character in his next book. In how many ways can he name the main character?

29. *Choosing a prize*   A committee has four different VCRs, three different CD players, and six different CDs available. The person chosen as Outstanding Freshman will receive one item from each category. How many different prizes are possible?

30. *Phone extensions*   How many different four-digit extensions are available for a company phone system if the first digit cannot be 0?

31. *Computer passwords*   How many different three-letter computer passwords are available if any letters can be used but repetition of letters is not allowed?

32. *Company cars*   A new Cadillac, a new Dodge, and a used Taurus are to be assigned randomly to three of ten real estate salespersons. In how many ways can the assignment be made?

33. *Phone numbers*   How many different seven-digit phone numbers are available in Jamestown if the first three digits are either 345, 286, or 329?

34. *Phone numbers*   How many different seven-digit phone numbers are there if the first digit may not be zero?

35. *Clark to the rescue*   The archvillain Lex Luther fires a nuclear missile into California's fault line, causing a tremendous earthquake. During the resulting chaos, Superman has to save Lois from a rock slide, rescue Jimmy from a bursting dam, stop a train from derailing, and catch a school bus that is plummeting from the Golden Gate Bridge. How many different ways are there for the Man of Steel to arrange these four rescues?

36. *Bus routes*   A bus picks up passengers from five hotels in downtown Seattle before heading to the Seattle-Tacoma Airport. How many different ways are there to arrange these stops?

37. *Taking a test*   How many ways are there to choose the answers to a test that consists of five true-false questions followed by six multiple-choice questions with four options each?

38. *Granting tenure*   A faculty committee votes on whether or not to grant tenure to each of four candidates. How many different possible outcomes are there to the vote?

39. *Possible words*   Ciara is entering a contest (sponsored by a detergent maker) that requires finding all three-letter words that can be made from the word WASHING. No letter may be used more than once. To make sure that she does not miss any, Ciara plans to write down all possible three-letter words and then look up each one in a dictionary. How many possible three-letter words will be on her list?

40. *Listing permutations*   Make a list of all of the permutations of the letters A, B, C, D, and E taken three at a time. How many permutations should be in your list?

For Writing/Discussion

41. *Listing subsets*   List all of the subsets of each of the sets $\{A\}$, $\{A, B\}$, $\{A, B, C\}$, and $\{A, B, C, D\}$. Find a formula for the number of subsets of a set of n elements.

42. *Number of subsets*   Explain how the fundamental counting principle can be used to find the number of subsets of a set of n elements.

12.5

Combinations, Labeling, and the Binomial Theorem

In Section 12.4 we learned the fundamental counting principle, and we applied it to finding the number of permutations of n objects taken r at a time. In permutations, the r objects are arranged in a sequential manner. In this section, we will find the number of ways to choose r objects from n distinct objects when the order in which the objects are chosen or placed is unimportant.

Combinations of n Things Taken r at a Time

In how many ways can two students be selected to go to the board from a class of four students: Adams, Baird, Campbell, and Dalton? Assuming that the two selected are treated identically, the choice of Adams and Baird is no different from the choice of Baird and Adams. Set notation provides a convenient way of listing all possible choices of two students from the four available students because in set notation $\{A, B\}$ is the same as $\{B, A\}$. We can easily list all subsets or **combinations** of two elements taken from the set $\{A, B, C, D\}$:

$$\{A, B\} \qquad \{A, C\} \qquad \{A, D\} \qquad \{B, C\} \qquad \{B, D\} \qquad \{C, D\}$$

The number of these subsets is the number of combinations of four things taken two at a time, denoted by $C(4, 2)$. Since there are six subsets, $C(4, 2) = 6$.

If we had a first and a second prize to give to two of the four students, then $P(4, 2) = 4 \cdot 3 = 12$ is the number of ways to award the prizes. Since the prizes are different, AB is different from BA, AC is different from CA, and so on. The 12 permutations are listed here:

$$AB \qquad AC \qquad AD \qquad BC \qquad BD \qquad CD$$
$$BA \qquad CA \qquad DA \qquad CB \qquad DB \qquad DC$$

From the list of combinations of four things taken two at a time, we made the list of permutations of four things two at a time by rearranging each combination. So $P(4, 2) = 2 \cdot C(4, 2)$.

In general, we can list all combinations of n things taken r at a time, then rearrange each of those subsets of r things in $r!$ ways to obtain all of the permutations of n things taken r at a time. So $P(n, r) = r! \, C(n, r)$, or

$$C(n, r) = \frac{P(n, r)}{r!}.$$

Since $P(n, r) = n!/(n - r)!$, we have

$$C(n, r) = \frac{n!}{(n - r)! \, r!}.$$

These results are summarized in the following theorem.

Theorem: Combinations of n Things Taken r at a Time

The number of combinations of n things taken r at a time (or the number of subsets of size r from a set of n elements) is given by the formula

$$C(n, r) = \frac{n!}{(n - r)! \, r!} \qquad \text{for } 0 \leq r \leq n.$$

Many calculators can calculate the value of $C(n, r)$ when given the value of n and r. The notations $\binom{n}{r}$ or nCr may be used on your calculator or elsewhere for $C(n, r)$. Note that

$$C(n, n) = \frac{n!}{0! \, n!} = 1 \quad \text{and} \quad C(n, 0) = \frac{n!}{n! \, 0!} = 1.$$

There is only one way to choose all n objects from a group of n objects if the order does not matter, and there is only one way to choose no object ($r = 0$) from a group of n objects.

Example 1   Combinations

To raise money, an alumni association prints lottery tickets with the numbers 1 through 11 printed on each ticket, as shown in Fig. 12.4. To play the lottery, one must circle three numbers on the ticket. In how many ways can three numbers be chosen out of 11 numbers on the ticket?

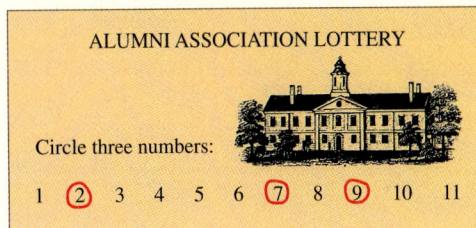

ALUMNI ASSOCIATION LOTTERY

Circle three numbers:

1 ② 3 4 5 6 ⑦ 8 ⑨ 10 11

Figure 12.4

Solution

Choosing three numbers from a list of 11 numbers is the same as choosing a subset of size 3 from a set of 11 elements. So the number of ways to choose the three numbers in playing the lottery is the number of combinations of 11 things taken 3 at a time:

$$C(11, 3) = \frac{11!}{8! \, 3!}$$

$$= \frac{11 \cdot 10 \cdot 9 \cdot 8 \cdot 7 \cdot 6 \cdot 5 \cdot 4 \cdot 3 \cdot 2 \cdot 1}{8 \cdot 7 \cdot 6 \cdot 5 \cdot 4 \cdot 3 \cdot 2 \cdot 1 \cdot 3 \cdot 2 \cdot 1}$$

$$= \frac{11 \cdot 10 \cdot 9}{3 \cdot 2 \cdot 1} \qquad \text{Divide numerator and denominator by 8!.}$$

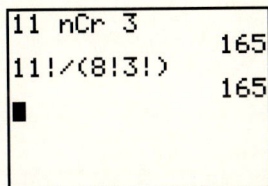

$$= 165$$

```
11 nCr 3
              165
11!/(8!3!)
              165
■
```

You can find $C(11, 3)$ by using the combination function or by using factorial notation.

Note that the combination formula counts the number of subsets of n objects taken r at a time. The objects are not necessarily placed in a subset, but they are all treated alike as in Example 1. In counting the combinations in Example 1, the only important thing is *which* numbers are circled; the order in which the numbers are circled is not important. By contrast, the permutation formula counts the number of ways to select *r* objects from *n*, *in order*. For example, the number of ways to award a first, second, and third prize to three of five people is $P(5, 3) = 60$, because the order matters. The number of ways to give three identical prizes to three of five people is $C(5, 3) = 10$, because the order of the awards doesn't matter.

Do not forget that we also have the fundamental counting principle to count the number of ways in which a sequence of events can occur. In the next example we use both the fundamental counting principle and the combination formula.

Example 2   Combinations and the counting principle

A company employs nine male welders and seven female welders. A committee of three male welders and two female welders is to be chosen to represent the welders in negotiations with management. How many different committees can be chosen?

Solution

First observe that three male welders can be chosen from the nine available in $C(9, 3) = 84$ ways. Next observe that two female welders can be chosen from seven available in $C(7, 2) = 21$ ways. Now use the fundamental counting principle to get $84 \cdot 21 = 1764$ ways to choose the males and then the females.

◆

Labeling

In a **labeling problem,** *n* distinct objects are to be given labels, each object getting exactly one label. For example, each person in a class is "labeled" with a letter grade at the end of the semester. Each student living on campus is "labeled" with the name of the dormitory in which the student resides. In a labeling problem, each distinct object gets one label, but there may be several types of labels and many labels of each type.

Example 3   A labeling problem

Twelve students have volunteered to help with a political campaign. The campaign director needs three telephone solicitors, four door-to-door solicitors, and five envelope stuffers. In how many ways can these jobs (labels) be assigned to these 12 students?

Solution

Since the three telephone solicitors all get the same type of label, the number of ways to select the three students is $C(12, 3)$. The number of ways to select four door-to-door solicitors from the remaining nine students is $C(9, 4)$. The number of ways to select the five envelope stuffers from the remaining five students is $C(5, 5)$. By the fundamental counting principle the number of ways to make all three selections is

$$C(12, 3) \cdot C(9, 4) \cdot C(5, 5) = \frac{12!}{9!\ 3!} \cdot \frac{9!}{5!\ 4!} \cdot \frac{5!}{0!\ 5!}$$

$$= \frac{12!}{3!\ 4!\ 5!}$$

$$= 27{,}720.$$

Note that in Example 3 there were 12 distinct objects to be labeled with three labels of one type, four labels of another type, and five labels of a third type, and the number of ways to assign those labels was found to be $12!/(3!\ 4!\ 5!)$. Instead of using combinations and the fundamental counting principle as in Example 3, we can use the following theorem.

Labeling Theorem

> If each of n distinct objects is to be assigned one label and thère are r_1 labels of the first type, r_2 labels of the second type, . . . , and r_k labels of the kth type, where $r_1 + r_2 + \cdots + r_k = n$, then the number of ways to assign the labels is
>
> $$\frac{n!}{r_1!\ r_2! \cdot \cdots \cdot r_k!}.$$

Example 4   Rearrangements of letters in a word

How many different arrangements are there for the 11 letters in the word MISSISSIPPI?

```
11!/(1!4!4!2!)
          34650
11!/1!/4!/4!/2!
          34650
■
```

To evaluate the expression in Example 4, either enclose the denominator in parentheses or divide by each factorial in the denominator.

Solution

This problem is a labeling problem if we think of the 11 positions for the letters as 11 distinct objects to be labeled. There is one M-label, and there are four S-labels, four I-labels, and two P-labels. So the number of ways to arrange the letters in MISSISSIPPI is

$$\frac{11!}{1!\ 4!\ 4!\ 2!} = 34{,}650.$$

You may think of permutation and combination problems as being very different, but they both are labeling problems in actuality. For example, to find the number of subsets of size 3 from a set of size 5, we are assigning three I-labels and two N-labels (I for "in the subset" and N for "not in the subset") to the five distinct objects of the set. Note that

$$\frac{5!}{3!\,2!} = C(5, 3).$$

To find the number of ways to give a first, second, and third prize to three of ten people, we are assigning one F-label, one S-label, one T-label, and seven N-labels (N for "no prize") to the ten distinct people. Note that

$$\frac{10!}{1!\,1!\,1!\,7!} = P(10, 3).$$

Binomial Expansion

One of the first facts we learn in algebra is

$$(a + b)^2 = a^2 + 2ab + b^2.$$

Our new labeling technique can be used to find the *coefficients* for the square of $a + b$ and any higher power of $a + b$. But before using counting techniques, let's look for patterns in the higher powers of $a + b$ and learn some shortcuts that can be applied in the simple cases.

We can find $(a + b)^3$ by multiplying $(a + b)^2$ by $a + b$. The powers of $a + b$ from the zero power to the fourth power are listed here.

$$(a + b)^0 = 1$$

$$(a + b)^1 = a + b$$

$$(a + b)^2 = a^2 + 2ab + b^2$$

$$(a + b)^3 = a^3 + 3a^2b + 3ab^2 + b^3$$

$$(a + b)^4 = a^4 + 4a^3b + 6a^2b^2 + 4ab^3 + b^4$$

The right-hand sides of these equations are referred to as **binomial expansions.** Each one after the first one is obtained by multiplying the previous expansion by $a + b$. To find the expansion for $(a + b)^5$, multiply the expansion for $(a + b)^4$ by $a + b$:

$$
\begin{array}{r}
a^4 + 4a^3b + 6a^2b^2 + 4ab^3 + b^4 \\
a \quad + b \\
\hline
a^4b + 4a^3b^2 + 6a^2b^3 + 4ab^4 + b^5 \\
a^5 + 4a^4b + 6a^3b^2 + 4a^2b^3 + ab^4 \\
\hline
a^5 + 5a^4b + 10a^3b^2 + 10a^2b^3 + 5ab^4 + b^5
\end{array}
$$

Note that there are six terms in the expansion of $(a + b)^5$ and that the exponents in each term have a sum of 5. The powers of a decrease from left to right, while the powers of b increase. The **binomial coefficients** 1, 5, 10, 10, 5, 1 are obtained by adding the coefficients from consecutive terms in the expansion of $(a + b)^4$ as follows:

$$(a + b)^4 = \mathbf{1}a^4 + \mathbf{4}a^3b + 6a^2b^2 + 4ab^3 + 1b^4$$

$$(a + b)^5 = a^5 + \mathbf{5}a^4b + 10a^3b^2 + 10a^2b^3 + 5ab^4 + b^5$$

It is easy to remember the coefficients for the first few powers of $a + b$ by using **Pascal's triangle.** Pascal's triangle is a triangular array of binomial coefficients:

					1					Coefficient of $(a + b)^0$
				1		1				Coefficients of $(a + b)^1$
			1		2		1			Coefficients of $(a + b)^2$
		1		3		3		1		Coefficients of $(a + b)^3$
	1		4		6		4		1	Coefficients of $(a + b)^4$
1		5		10		10		5		1 Coefficients of $(a + b)^5$

Pascal's triangle

Each row in Pascal's triangle starts and ends with 1 and the coefficients between the 1's are obtained from the previous row by adding consecutive entries.

Example 5   Using Pascal's triangle

Use Pascal's triangle to find each binomial expansion.

a) $(2x - 3y)^4$

b) $(a + b)^6$

Solution

a) Think of $(2x - 3y)^4$ as $(2x + (-3y))^4$ and use the coefficients 1, 4, 6, 4, 1 from Pascal's triangle:

$$(2x - 3y)^4 = 1(2x)^4(-3y)^0 + 4(2x)^3(-3y)^1 + 6(2x)^2(-3y)^2$$
$$+ 4(2x)^1(-3y)^3 + 1(2x)^0(-3y)^4$$
$$= 16x^4 - 96x^3y + 216x^2y^2 - 216xy^3 + 81y^4$$

b) Use the coefficients 1, 5, 10, 10, 5, 1 from Pascal's triangle to obtain the coefficients 1, 6, 15, 20, 15, 6, 1. Use these coefficients with decreasing powers of a and increasing powers of b to get

$$(a + b)^6 = a^6 + 6a^5b + 15a^4b^2 + 20a^3b^3 + 15a^2b^4 + 6ab^5 + b^6.$$

The Binomial Theorem

Pascal's triangle provides an easy way to find a binomial expansion for small powers of a binomial, but it is not practical for large powers. To get the binomial coefficients for larger powers, we count the number of like terms of each type using the idea of labeling. For example, consider

$$(a + b)^3 = (a + b)(a + b)(a + b) = a^3 + 3a^2b + 3ab^2 + b^3.$$

The terms of the product come from all of the different ways there are to select either a or b from each of the three factors and multiply the selections. The coefficient of a^3 is 1 because we get a^3 only from aaa, where a is chosen from each factor. The coefficient of a^2b is 3 because a^2b is obtained from aab, aba, and baa. From the labeling theorem, the number of ways to rearrange the letters aab is $3!/(2! \, 1!) = 3$. So the coefficient of a^2b is the number of ways to label the three factors with two a-labels and one b-label.

In general, the expansion of $(a + b)^n$ contains $n + 1$ terms in which the exponents have a sum of n. The coefficient of $a^{n-r}b^r$ is the number of ways to label n factors with $(n - r)$ of the a-labels and r of the b-labels:

$$\frac{n!}{r!(n - r)!}$$

The *binomial theorem* expresses this result using summation notation.

The Binomial Theorem

If n is a positive integer, then for any real numbers a and b,

$$(a + b)^n = \sum_{r=0}^{n} \binom{n}{r} a^{n-r} b^r, \qquad \text{where} \quad \binom{n}{r} = \frac{n!}{r!(n - r)!}.$$

Example 6   Using the binomial theorem

What is the coefficient of a^7b^2 in the binomial expansion of $(a + b)^9$?

Solution

According to the binomial theorem, the coefficient of a^7b^2 is

$$\binom{9}{2} = \frac{9!}{2! \, 7!} = \frac{9 \cdot 8}{2} = 36.$$

The term $36a^7b^2$ occurs in the expansion of $(a + b)^9$. ◆

Example 7   Using the binomial theorem

Write out the first three terms in the expansion of $(x - 2y)^{10}$.

Solution

Write $(x - 2y)^{10}$ as $(x + (-2y))^{10}$. According to the binomial theorem,

$$(x + (-2y))^{10} = \sum_{r=0}^{10} \binom{10}{r} x^{10-r}(-2y)^r.$$

To find the first three terms, let $r = 0$, 1, and 2:

$$(x - 2y)^{10} = \binom{10}{0} x^{10}(-2y)^0 + \binom{10}{1} x^9(-2y)^1 + \binom{10}{2} x^8(-2y)^2 + \cdots$$

$$= x^{10} - 20x^9 y + 180x^8 y^2 + \cdots$$

In the next example, labeling is used to find the coefficient of a term in a power of a trinomial.

Example 8   Trinomial coefficients

What is the coefficient of $a^3 b^2 c$ in the expansion of $(a + b + c)^6$?

Solution

The terms of the product $(a + b + c)^6$ come from all of the different ways there are to select a, b, or c from each of the six distinct factors and multiply the selections. The number of times that $a^3 b^2 c$ occurs is the same as the number of rearrangements of $aaabbc$, which is a labeling problem. The number of rearrangements of $aaabbc$ is

$$\frac{6!}{3!\, 2!\, 1!} = \frac{6 \cdot 5 \cdot 4 \cdot 3 \cdot 2 \cdot 1}{3 \cdot 2 \cdot 1 \cdot 2 \cdot 1 \cdot 1} = 60.$$

So the term $60a^3 b^2 c$ occurs in the expansion of $(a + b + c)^6$.

Labeling allows us to find the coefficient of any term in the expansion of any polynomial raised to a whole number power.

For Thought

True or false? Explain.

1.   The number of ways to choose three questions to answer out of five questions on an essay test is $C(5, 2)$.

2.   The number of ways to answer a five-question multiple-choice test in which each question has three choices is $P(5, 3)$.

3.   The number of ways to pick a Miss America and the first runner-up from the five finalists is 20.

4.   The binomial expansion for $(x + y)^n$ contains n terms.

5. For any real numbers a and b, $(a + b)^5 = \sum_{i=0}^{5} \binom{5}{i} a^i b^{5-i}$.

6. The sum of the binomial coefficients in $(a + b)^n$ is 2^n.

7. $P(8, 3) < C(8, 3)$ 8. $P(7, 3) = (3!) \cdot C(7, 3)$

9. $P(8, 3) = P(8, 5)$ 10. $C(1, 1) = 1$

12.5 Exercises   ▭ Tape 23   ▭ Disk—5.25″: 8   3.5″: 5   Macintosh: 4

Solve each problem.

1. *Poker hands* How many five-card poker hands are there if you draw five cards from a deck of 52?

2. *Bridge hands* How many 13-card bridge hands are there if you draw 13 cards from a deck of 52?

3. *Playing a lottery* In a certain lottery the player chooses six numbers from the numbers 1 through 49. In how many ways can the six numbers be chosen?

4. *Fantasy five* In a different lottery the player chooses five numbers from the numbers 1 through 39. In how many ways can the five numbers be chosen?

5. *Job candidates* The search committee has narrowed the applicants to five unranked candidates. In how many ways can three be chosen for an in-depth interview?

6. *Spreading the flu* In how many ways can nature select five students out of a class of 20 students to get the flu?

The following problems may involve combinations, permutations, or the fundamental counting principle.

7. *Television schedule* Arnold plans to spend the evening watching television. Each of the three networks runs one-hour shows starting at 7:00, 8:00, and 9:00 P.M. In how many ways can he watch three complete shows from 7:00 to 10:00 P.M.?

8. *Cafeteria meals* For $3.98 you can get a salad, main course, and dessert at the cafeteria. If you have a choice of five different salads, six different main courses, and four different desserts, then how many different meals can you get for $3.98?

9. *Fire code inspections* The fire inspector in Cincinnati must select three night clubs from a list of eight for an inspection of their compliance with the fire code. In how many ways can she select the three night clubs?

10. *Prize-winning pigs* In how many ways can a red ribbon, a blue ribbon, and a green ribbon be awarded to three of six pigs at the county fair?

11. *Choosing a team* From the nine male and six female sales representatives for an insurance company, a team of three men and two women will be selected to attend a national conference on insurance fraud. In how many ways can the team of five be selected?

12. *Poker hands* How many five-card poker hands are there containing three hearts and two spades?

13. *Returning exam papers* In how many different orders can Professor Pereira return 12 exam papers to 12 students?

14. *Saving the best till last* In how many ways can Professor Yang return examination papers to 12 students if she always returns the worst paper first and the best paper last?

15. *Rolling dice* A die is to be rolled twice, and each time the number of dots showing on the top face is to be recorded. If we think of the outcome of two rolls as an ordered pair of numbers, then how many outcomes are there?

16. *Having children* A couple plans to have three children. Considering the sex and order of birth of each of the three children, how many outcomes are possible to this plan?

17. *Marching bands* In how many ways can four marching bands and three floats line up for a parade if a marching band must lead the parade and two bands cannot march next to one another?

18. *Marching bands* In how many ways can four marching bands and four floats line up for a parade if a marching band must lead the parade and two bands cannot march next to one another?

Solve each problem using the idea of labeling.

19. *Rearranging letters* How many permutations are possible using the letters in the word ALABAMA?

20. *Spelling mistakes* How many incorrect spellings of the word FLORIDA are there, using all of the correct letters?

21. *Assigning topics* An instructor in a history class of ten students wants term papers written on World War II, World War I, and the Civil War. If he randomly assigns World War II to five students, World War I to three students, and the Civil War to two students, then in how many ways can these assignments be made?

22. *Parking tickets* Officer O'Reilly is in charge of 12 officers working a sporting event. In how many ways can she choose six to control parking, four to work security, and two to handle emergencies?

23. *Determining chords* How many distinct chords (line segments with endpoints on the circle) are determined by three points lying on a circle? By four points? By five points? By n points?

24. *Determining triangles* How many distinct triangles are determined by six points lying on a circle, where the vertices of each triangle are chosen from the six points?

25. *Assigning volunteers* Ten students volunteered to work in the governor's reelection campaign. Three will be assigned to making phone calls, two will be assigned to stuffing envelopes, and five will be assigned to making signs. In how many ways can the assignments be made?

26. *Assigning vehicles* Three identical Buicks, four identical Fords, and three identical Toyotas are to be assigned to ten traveling salespeople. In how many ways can the assignments be made?

Write the complete binomial expansion for each of the following powers of a binomial.

27. $(a - 2)^3$

28. $(b^2 - 3)^3$

29. $(2a + b^2)^3$

30. $(x + 3)^3$

31. $(x - 2y)^4$

32. $(y - 2)^4$

33. $(x^2 + 1)^4$

34. $(2r + 3t^2)^4$

Write the first three terms of each binomial expansion.

35. $(x + y)^9$

36. $(a - b)^{10}$

37. $(2x - y)^{12}$

38. $(a + 2b)^{11}$

39. $(2s - 0.5t)^8$

40. $(3y^2 + a)^{10}$

41. $(m^2 - 2w^3)^9$

42. $(ab^2 - 5c)^8$

Solve each problem.

43. What is the coefficient of $w^3 y^5$ in the expansion of $(w + y)^8$?

44. What is the coefficient of $a^3 z^9$ in the expansion of $(a + z)^{12}$?

45. What is the coefficient of $a^5 b^8$ in the expansion of $(b - 2a)^{13}$?

46. What is the coefficient of $x^6 y^5$ in the expansion of $(0.5x - y)^{11}$?

47. What is the coefficient of $a^2 b^4 c^6$ in the expansion of $(a + b + c)^{12}$?

48. What is the coefficient of $x^3 y^2 z^3$ in the expansion of $(x - y - 2z)^8$?

49. What is the coefficient of $a^3 b^7$ in the expansion of $(a + b + 2c)^{10}$?

50. What is the coefficient of $w^2 x y^3 z^9$ in the expansion of $(w + x + y + z)^{15}$?

For Writing/Discussion

51. Explain why $C(n, r) = C(n, n - r)$.

52. Is $P(n, r) = P(n, n - r)$? Explain your answer.

53. Explain the difference between permutations and combinations.

54. Prove that $\sum_{i=0}^{n} \binom{n}{i} = 2^n$ for any positive integer n.

55. *Cooperative learning* Make up three counting problems: a combination problem, a permutation problem, and a labeling problem. Give them to a classmate to solve.

12.6

Probability

We often hear statements such as "The probability of rain today is 70%," "I have a 50-50 chance of passing English," or "I have one chance in a million of winning the lottery." It is clear that we have a better chance of getting rained on than we have of winning a lottery, but what exactly is probability? In this section we will study probability and use the counting techniques of Sections 12.4 and 12.5 to make precise statements concerning probabilities.

The Probability of an Event

An **experiment** is any process for which the outcome is uncertain. For example, if we toss a coin, roll a die, draw a poker hand from a deck, or arrange people in a line in a manner that makes the outcome uncertain, then these processes are experiments. A **sample space** is the set of all possible outcomes to an experiment. If each outcome occurs with about the same frequency when the experiment is repeated many times, then the outcomes are called **equally likely.**

The simplest experiments for determining probabilities are ones in which the outcomes are equally likely. For example, if a fair coin is tossed, then the sample space S consists of two equally likely outcomes, heads and tails:

$$S = \{H, T\}$$

An **event** is a subset of a sample space. The subset $E = \{H\}$ is the event of getting heads when the coin is tossed. If $n(S)$ is the number of equally likely outcomes in S, then $n(S) = 2$. If $n(E)$ is the number of equally likely outcomes in E, then $n(E) = 1$. The probability of the event E, $P(E)$, is the ratio of $n(E)$ to $n(S)$:

$$P(E) = \frac{n(E)}{n(S)} = \frac{1}{2}$$

Definition: Probability of an Event

If S is a sample space of equally likely outcomes to an experiment and the event E is a subset of S, then the **probability of E**, $P(E)$, is defined by

$$P(E) = \frac{n(E)}{n(S)}.$$

Since E is a subset of S, $0 \leq n(E) \leq n(S)$. So $P(E)$ is a number between 0 and 1, inclusive. If there are no outcomes in the event E, then $P(E) = 0$ and it is impossible for E to occur. If $n(E) = n(S)$ then $P(E) = 1$ and the event E is certain to occur. For example, if E is the event of getting two heads on a single toss of a coin, then $n(E) = 0$ and $P(E) = 0/2 = 0$. If E is the event of getting fewer than two heads on a single toss of a coin, then for either outcome H or T there are fewer than two heads. So $E = \{H, T\}$, $n(E) = 2$, and $P(E) = 2/2 = 1$.

Probability predicts the future, but not exactly. The probability that heads will occur when a coin is tossed is $1/2$, but there is no way to know whether heads or tails will occur on any future toss. However, the probability $1/2$ indicates that if the coin is tossed many times, then *about* $1/2$ of the tosses will be heads.

Example 1   Rolling a single die

What is the probability of getting a number larger than 4 when a single die is rolled?

Solution

When a die is rolled, the number of dots showing on the upper face of the die is counted. So the sample space of equally likely outcomes is

$$S = \{1, 2, 3, 4, 5, 6\} \quad \text{and} \quad n(S) = 6.$$

Since only 5 and 6 are larger than 4,

$$E = \{5, 6\} \quad \text{and} \quad n(E) = 2.$$

According to the definition of probability,

$$P(E) = \frac{n(E)}{n(S)} = \frac{2}{6} = \frac{1}{3}.$$

Example 2   Tossing a pair of coins

What is the probability of getting at least one head when a pair of coins is tossed?

Solution

Since there are two equally likely outcomes for the first coin and two equally likely outcomes for the second coin, by the fundamental counting principle there are four equally likely outcomes to the experiment of tossing a pair of coins. We can list the outcomes as ordered pairs:

$$S = \{(H, H), (H, T), (T, H), (T, T)\}.$$

The ordered pairs (H, T) and (T, H) must be listed as different outcomes, because the first coordinate is the result from the first coin and the second coordinate is the result from the second coin. Since three of these outcomes result in at least one head,

$$E = \{(H, H), (H, T), (T, H)\}$$

and $n(E) = 3$. So

$$P(E) = \frac{n(E)}{n(S)} = \frac{3}{4}.$$

Example 3   Rolling a pair of dice

What is the probability of getting a sum of 5 when a pair of dice is rolled?

Solution

Since there are six equally likely outcomes for each die, by the fundamental counting principle there are 36 equally likely outcomes to rolling the pair. See Fig. 12.5. We can list the 36 outcomes as ordered pairs as shown on the following page.

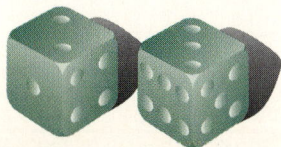

Figure 12.5

$$S = \{(1, 1), (1, 2), (1, 3), (1, 4), (1, 5), (1, 6),$$
$$(2, 1), (2, 2), (2, 3), (2, 4), (2, 5), (2, 6),$$
$$(3, 1), (3, 2), (3, 3), (3, 4), (3, 5), (3, 6),$$
$$(4, 1), (4, 2), (4, 3), (4, 4), (4, 5), (4, 6),$$
$$(5, 1), (5, 2), (5, 3), (5, 4), (5, 5), (5, 6),$$
$$(6, 1), (6, 2), (6, 3), (6, 4), (6, 5), (6, 6)\}$$

The ordered pairs in color in the sample space are the ones in which the sum of the entries is 5. So the phrase "sum of the numbers is 5" describes the event

$$E = \{(4, 1), (3, 2), (2, 3), (1, 4)\},$$

and

$$P(E) = \frac{n(E)}{n(S)} = \frac{4}{36} = \frac{1}{9}.$$

Example 4   Probability of winning a lottery

To play the Alumni Association Lottery, you pick three numbers from the integers from 1 through 11 inclusive. What is the probability that you win the $50 prize by picking the same three numbers as the ones chosen by the Alumni Association?

Solution

The number of ways to select three numbers out of 11 numbers is $C(11, 3) = 165$. Since only one of the 165 equally likely outcomes is the winning combination, the probability of winning is 1/165.

Complementary Events

If the probability of rain today is 60%, then the probability that it does not rain is 40%. Rain and not rain are called complementary events. For complementary events there is no possibility that both occur, and one of them must occur. If A is an event, then A' (read "A prime" or "A complement") represents the complement of the event A.

Definition: Complementary Events

Two events A and A' are called **complementary events** if $A \cap A' = \varnothing$ and $P(A) + P(A') = 1$.

The diagram in Fig. 12.6 illustrates complementary events. The region inside the rectangle represents the sample space S, while the region inside the circle represents the event A. The event A' is the region inside the rectangle, but outside the circle. Note that $A \cup A' = S$ and $A \cap A' = \varnothing$.

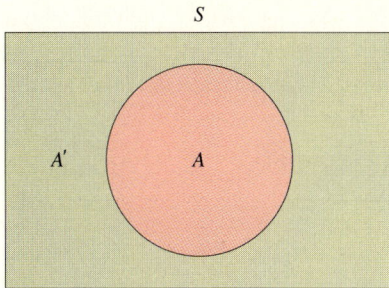

Figure 12.6

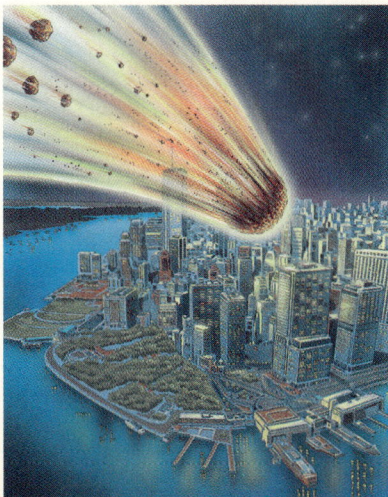

Example 5 The probability of complementary events

What is the probability of getting a sum that is not equal to 5 when rolling a pair of dice?

Solution

From Example 3, the probability that the sum is 5 is 1/9. Getting a sum that is not equal to 5 is the complement of getting a sum that is equal to 5. So

$$P(\text{sum is 5}) + P(\text{sum is not 5}) = 1$$

$$P(\text{sum is not 5}) = 1 - P(\text{sum is 5}).$$

So the probability that the sum is not 5 is $1 - 1/9$, or 8/9. ◆

Scientists estimate that a "killer asteroid," an asteroid with diameter greater than one kilometer, hits the earth on the average of once in a million years. Simply knowing the average number of hits in a million years allows them to find the probability of any number of hits in a million years. The main concern is the probability of no hits. Probabilities of this type involve e, the base of the natural logarithm. In general, the expression $e^{-\alpha}$ gives an estimate of the probability that there are no occurrences of an event in a period of time. Here α is the average number of occurrences in that time period. For the killer asteroid, the average is one hit per one million years, $\alpha = 1$. So the probability that there are no asteroid hits in the next one million years is

$$e^{-1} = 0.3678 \approx 37\%.$$

The probability that there is at least one hit (the complement of no hits) in the next one million years is about 63%, quite high, but a million years is a long time. In the next example, probabilities are calculated for a time period that is more meaningful to humans.

Example 6 The probability of collision with an asteroid

Find the probability that the earth will survive the next 100 years without a hit by a killer asteroid, and find the probability that during the next 100 years the earth will be hit at least once by a killer asteroid. Use the estimate of an average of one hit per one million years.

Solution

To find the required probability, we need the average number of hits in 100 years. Convert the ratio of one hit per one million years into the equivalent ratio of 0.0001 hits per 100 years:

$$\frac{1 \text{ hit}}{1,000,000 \text{ years}} = \frac{0.0001 \text{ hits}}{100 \text{ years}}$$

Since 0.0001 is the average number of hits per 100 years, use $\alpha = 0.0001$ in the expression $e^{-\alpha}$. The probability of no hits in the next 100 years is $e^{-0.0001} \approx$ 0.999900005. The probability of at least one hit (the complement of no hits) in the next 100 years is

$$1 - 0.99990005 = 0.000099995,$$

about 1/10,000. Nothing to worry about. ◆

Odds

If the probability is 4/5 that the Braves will win the World Series and 1/5 that they will lose, then they are four times as likely to win as they are to lose. We say that the *odds* in favor of the Braves winning the World Series are 4 to 1. Notice that odds are *not* probabilities. Odds are ratios of probabilities. Odds are usually written as ratios of whole numbers.

Definition: Odds

> If A is any event, then the **odds in favor of A** are defined as the ratio $P(A)$ to $P(A')$ and the **odds against A** are defined as the ratio of $P(A')$ to $P(A)$.

Example 7   Odds for and against an event

What are the odds in favor of getting a sum of 5 when rolling a pair of dice? What are the odds against a sum of 5?

Solution

From Examples 3 and 5, $P(\text{sum is } 5) = 1/9$ and $P(\text{sum is not } 5) = 8/9$. The odds in favor of getting a sum of 5 are 1/9 to 8/9. Multiply each fraction by 9 to get the odds 1 to 8. The odds against a sum of 5 are 8 to 1. ◆

The odds in favor of A are found from the ratio of $P(A)$ to $P(A')$, but the equation $P(A) + P(A') = 1$ can be used to find $P(A)$ and $P(A')$.

Example 8   Finding probabilities from odds

If the odds in favor of the Rams going to the Super Bowl are 1 to 5, then what is the probability that the Rams will go to the Super Bowl?

Solution

Since 1 to 5 is the ratio of the probability of going to that of not going, the probability of not going is 5 times as large as the probability of going. Let $P(G) = x$ and $P(G') = 5x$. Since

$$P(G) + P(G') = 1,$$

we have

$$x + 5x = 1$$

$$6x = 1$$

$$x = \frac{1}{6}$$

So the probability that the Rams will go to the Super Bowl is 1/6. ◆

We can express the idea found in Example 8 as a theorem relating odds to probabilities.

Theorem: Converting from Odds to Probability

If the odds in favor of event E are a to b, then

$$P(E) = \frac{a}{a + b} \quad \text{and} \quad P(E') = \frac{b}{a + b}.$$

The Addition Rule

One way to find the probability of an event is to list all of the equally likely outcomes to the experiment and count those outcomes that make up the event. The relationship between the probabilities of complementary events allows us to find the probability of the complement of an event by subtraction rather than counting. The addition rule provides another relationship between probabilities of events that can be used to find probabilities without making a list of outcomes and counting.

In rolling a pair of dice, let A be the event that doubles occur and let B be the event that the sum is 4. We have

$$A = \{(1, 1), (2, 2), (3, 3), (4, 4), (5, 5), (6, 6)\}$$

and

$$B = \{(3, 1), (2, 2), (1, 3)\}.$$

The event that doubles *and* a sum of 4 occur is the event $A \cap B$:

$$A \cap B = \{(2, 2)\}$$

Count the ordered pairs to get $P(A) = 6/36$, $P(B) = 3/36$, and $P(A \cap B) = 1/36$. The event that either doubles *or* a sum of 4 occurs is the event $A \cup B$:

$$A \cup B = \{(1, 1), (2, 2), (3, 3), (4, 4), (5, 5), (6, 6), (3, 1), (1, 3)\}$$

Note that $P(A \cup B)$ is 8/36 and

$$\frac{8}{36} = \frac{6}{36} + \frac{3}{36} - \frac{1}{36}.$$

Since $(2, 2)$ is counted in both A and B, $P(A \cap B)$ is subtracted from $P(A) + P(B)$. This example illustrates the addition rule.

The Addition Rule

If A and B are any events in a sample space, then

$$P(A \cup B) = P(A) + P(B) - P(A \cap B).$$

If $P(A \cap B) = 0$, then A and B are called **mutually exclusive** events and

$$P(A \cup B) = P(A) + P(B).$$

Two events are mutually exclusive if it is impossible for both to occur. For example, when a pair of dice is rolled once, getting a sum of 4 and getting a sum of 5 are mutually exclusive events.

The diagram in Fig. 12.7 illustrates the addition rule for two sets A and B whose intersection is not empty. To find $P(A \cup B)$ we add $P(A)$ and $P(B)$, but we must subtract $P(A \cap B)$ because it is counted in both $P(A)$ and $P(B)$. Since $P(A \cap B) = 0$ for mutually exclusive events, the addition rule for mutually exclusive events is a special case of the general addition rule. Note that complementary events are mutually exclusive, but mutually exclusive events are not necessarily complementary.

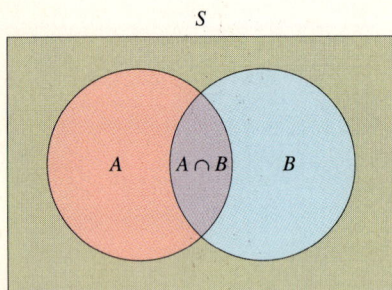

Figure 12.7

Example 9 Using the addition rule

At Hillside Community College, 60% of the students are commuters (C), 50% are female (F), and 30% are female commuters. If a student is selected at random, what is the probability that the student is either a female or a commuter?

Solution

By the addition rule, the probability of selecting either a female or a commuter is

$$
\begin{aligned}
P(F \cup C) &= P(F) + P(C) - P(F \cap C) \\
&= 0.50 + 0.60 - 0.30 \\
&= 0.80
\end{aligned}
$$

Example 10 The addition rule with mutually exclusive events

In rolling a pair of dice, what is the probability that the sum is 12 or at least one die shows a 2?

Solution

Let A be the event that the sum is 12 and B be the event that at least one die shows a 2. Since A occurs on only one of the 36 equally likely outcomes, $(6, 6)$ (see Example 3), $P(A) = 1/36$. B occurs on 11 of the equally likely outcomes, so $P(B) = 11/36$. Since A and B are mutually exclusive, we have

$$P(A \cup B) = P(A) + P(B) = \frac{1}{36} + \frac{11}{36} = \frac{12}{36} = \frac{1}{3}.$$

For Thought

True or false? Explain.

1. If S is a sample space of equally likely outcomes and E is a subset of S, then $P(E) = n(E)$.

2. If an experiment consists of tossing four coins, then the sample space consists of eight equally likely outcomes.

3. If a single coin is tossed twice, then P(at least one tail) = 0.75.

4. If a pair of dice is rolled, then P(at least one 4) = 11/36.

5. If four coins are tossed, then P(at least one head) = 4/16.

6. If two coins are tossed, then the complement of getting exactly two heads is getting exactly two tails.

7. If the probability of getting exactly three tails in a toss of three coins is 1/8, then the probability of getting at least one head is 7/8.

8. If P(snow today) = 0.7, then the odds in favor of snow are 7 to 10.

9. If the odds in favor of an event E are 3 to 4, then $P(E)$ = 3/4.

10. The ratio of 1/5 to 4/5 is equivalent to the ratio of 4 to 1.

12.6 Exercises
Tape 24   Disk—5.25″: 8   3.5″: 5   **Macintosh: 4**

Solve each probability problem.

1. *Business expansion* The board of directors for a major corporation cannot decide whether to build its new assembly plant in Dallas, Memphis, or Chicago. If one of these cities is chosen at random, then what is the probability that
   a) Dallas is chosen?
   b) Memphis is not chosen?
   c) Topeka is chosen?

2. *Scratch and win* A batch of 100,000 scratch-and-win tickets contains 10,000 that are redeemable for a free order of French fries. If you randomly select one of these tickets, then what is the probability that
   a) you win an order of French fries?
   b) you do not win an order of French fries?

3. *Rolling a die* If a single die is rolled, then what is the probability of getting
   a) a number larger than 2?
   b) a number less than or equal to 6?
   c) a number other than 4?
   d) a number larger than 8?
   e) a number smaller than 2?

4. *Tossing a coin* If a single coin is tossed, then what is the probability of getting
   a) heads?
   b) fewer than two tails?
   c) exactly three tails?

5. *Tossing two coins once* If a pair of coins is tossed, then what is the probability of getting
   a) exactly two tails?
   b) at least one head?
   c) exactly two heads?
   d) at most one head?

6. *Tossing one coin twice* If a single coin is tossed twice, then what is the probability of getting
   a) heads followed by tails?
   b) two tails in a row?
   c) heads on the second toss?
   d) exactly one head?

7. *Rolling a pair of dice* If a pair of dice is rolled, then what is the probability of getting
   a) a pair of 3's?
   b) at least one 3?
   c) a sum of 6?
   d) a sum greater than 2?
   e) a sum less than 3?

8. *Rolling a die twice* If a single die is rolled twice, then what is the probability of getting
   a) a 1 followed by a 6?
   b) a sum of 4?
   c) a 5 on the second roll?
   d) no more than two 4's?
   e) an even number followed by an odd number?

9. *Colored marbles* A marble is selected at random from a jar containing three red marbles, four yellow marbles, and six green marbles. What is the probability that
   a) the marble is red?
   b) the marble is not yellow?
   c) the marble is either red or green?
   d) the marble is neither red nor green?

10. *Choosing a chairperson* A committee consists of one Democrat, six Republicans, and seven Independents. If one person is randomly selected from the committee to be the chairperson, then what is the probability that
    a) the person is a Democrat?
    b) the person is either a Democrat or a Republican?
    c) the person is not a Republican?

11. *Numbered marbles* A jar contains nine marbles numbered 1 through 9. Two marbles are randomly selected one at a time without replacement. What is the probability that
    a) 1 is selected first and 9 is selected second?
    b) the sum of the numbers selected is 4?
    c) the sum of the numbers selected is 5?

12. *Foul play* A company consists of a president, a vice-president, and 10 salespeople. If 2 of the 12 people are randomly selected to win a Hawaiian vacation, then what is the probability that none of the salespeople is a winner?

13. *Poker hands* If a five-card poker hand is drawn from a deck of 52, then what is the probability that
    a) the hand contains the ace, king, queen, jack, and 10 of hearts?
    b) the hand contains one 3, one 4, one 5, one 6, and one 7?

14. *Lineup* If four people with different names and different weights randomly line up to buy concert tickets, then what is the probability that
    a) they line up in alphabetical order?
    b) they line up in order of increasing weight?

Solve each problem.

15. *Drive defensively* If the probability of surviving a head-on car accident at 55 mph is 0.001, then what is the probability of not surviving?

16. *Tax time* If the probability of a tax return not being audited by the IRS is 0.91, then what is the probability of a tax return being audited?

17. *Rolling fours* A pair of dice is rolled. What is the probability of
    a) getting a pair of 4's?
    b) not getting a pair of 4's?
    c) getting at least one number that is not a 4?

18. *Tossing triplets* Three coins are tossed. What is the probability of
    a) getting three heads?
    b) not getting three heads?
    c) getting at least one head?

19. *Killer asteroids* Some scientists estimate that the earth takes an average of three hits per one million years by killer asteroids. In this case, what is the probability that there will be no hits in the next one million years? What is the probability that there will be at least one hit in the next million years?

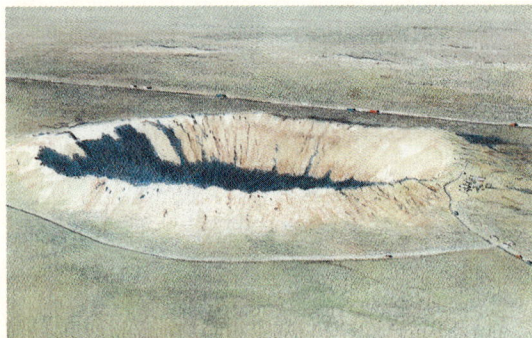

Figure for Exercise 19

20. *Safe at last* Use the estimate of an average of three hits in one million years to find the probability that there will be no hits by a killer asteroid in the next 200 years. Find the probability that there will be at least one hit in the next 200 years.

Solve each problem.

21. *Hurricane Alley* If the probability is 80% that the eye of hurricane Zelda comes ashore within 30 mi of Biloxi, then what are the odds in favor of the eye coming ashore within 30 mi of Biloxi?

22. *On target* If the probability that an arrow hits its target is 7/9, then what are the odds
    a) in favor of the arrow hitting its target?
    b) against the arrow hitting its target?

23. *Stock market rally*   If the probability that the stock market goes up tomorrow is 1/4, then what are the odds
  a) in favor of the stock market going up tomorrow?
  b) against the stock market going up tomorrow?

24. *Read my lips*   If the probability of new taxes this year is 4/5, then what are the odds
  a) in favor of new taxes?
  b) against new taxes?

25. *Morning line*   If the Las Vegas odds makers set the odds at 9 to 1 in favor of the Tigers winning their next game, then
  a) what are the odds against the Tigers winning their next game?
  b) what is the probability that the Tigers win their next game?

26. *Public opinion*   If a pollster says that the odds are 7 to 1 against the reelection of the mayor, then
  a) what are the odds in favor of the reelection of the mayor?
  b) what is the probability that the mayor will be reelected?

27. *Two out of four*   What are the odds in favor of getting exactly two heads in four tosses of a coin?

28. *Rolling once*   What are the odds in favor of getting a 5 in a single roll of a die?

29. *Lucky seven*   What are the odds in favor of getting a sum of 7 when rolling a pair of dice?

30. *Four out of two*   What are the odds in favor of getting at least one 4 when rolling a pair of dice?

31. *Only one winner*   If 2 million lottery tickets are sold and only one of them is the winning ticket, then what are the odds in favor of winning if you hold a single ticket?

32. *Pick six*   What are the odds in favor of winning a lottery in which you must choose six numbers from the numbers 1 through 49?

33. *Five in five*   If the odds in favor of getting five heads in five tosses of a coin are 1 to 31, then what is the probability of getting five heads in five tosses of a coin?

34. *Electing Jones*   If the odds against Jones winning the election are 3 to 5, then what is the probability that Jones will win the election?

Use the addition rule to solve each problem.

35. *Insurance categories*   Among the drivers insured by American Insurance, 64% are women, 38% of the drivers are in a high-risk category, and 24% of the drivers are high-risk women. If a driver is randomly selected from that company, what is the probability that the driver is either high-risk or a woman?

36. *Six or four*   What is the probability of getting either a sum of 6 or at least one 4 in the roll of a pair of dice?

37. *Family of five*   A couple plan to have three children. Assuming that males and females are equally likely, what is the probability that they have either three boys or three girls?

38. *Ten or four*   What is the probability of getting either a sum of 10 or a sum of 4 in the roll of a pair of dice?

39. *Pick a card*   What is the probability of getting either a heart or a king when drawing a single card from a deck of 52 cards?

40. *Any card*   What is the probability of getting either a heart or a diamond when drawing a single card from a deck of 52 cards?

41. *Selecting students*   The accompanying pie chart shows the percentages of freshmen, sophomores, juniors, and seniors at Washington High School. Find the probability that a randomly selected student is
  a) a freshman.
  b) a junior or a senior.
  c) not a sophomore.

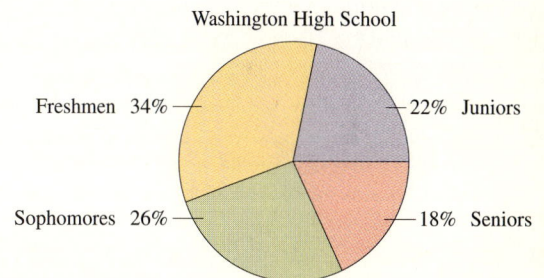

Washington High School

Freshmen 34% · 22% Juniors · Sophomores 26% · 18% Seniors

Figure for Exercise 41

42. *Earth-crossing asteroids*   To reduce the risk of being hit by a killer asteroid, scientists want to locate and track all of the asteroids that are visible through telescopes. Each asteroid would be classified according to whether its diameter D is less than 1 km and whether its orbit crossed the orbit of earth (an earth-crossing orbit). The table on the following page shows a hypothetical classification of 900 asteroids visible to an amateur astronomer. If an amateur astronomer randomly spots an asteroid one night, then what is the probability that
  a) it is an earth-crossing asteroid and its diameter is greater than or equal to 1 km?

b) it is either an earth-crossing asteroid or its diameter is greater than or equal to 1 km?

Asteroid classification	$D < 1$ km	$D \geq 1$ km
Earth-crossing	210	90
Not earth-crossing	340	260

Figure for Exercise 42

For Writing/Discussion

43. Explain the difference between mutually exclusive events and complementary events.

44. Explain the difference between probability and odds.

45. *Cooperative learning* Put three pennies into a can. Shake and toss the pennies 100 times. After each toss have your helper record whether 0, 1, 2, or 3 heads are showing. On what percent of the tosses did 0, 1, 2, and 3 heads occur? What are the theoretical probabilities of obtaining 0, 1, 2, and 3 heads in a toss of three coins? How well does the theoretical model fit the actual coin toss?

Graphing Calculator Exercises

A typical random number generator on a graphing calculator displays a 10-digit randomly selected number between 0 and 1.

1. Generate 100 random numbers to simulate tossing three pennies 100 times. Record 0 heads if the random number x satisfies $0 < x < 0.125$, record 1 head for $0.125 < x < 0.5$, record 2 heads for $0.5 < x < 0.875$, and record 3 heads for $0.875 < x < 1$. On what percent of the simulated tosses did 0, 1, 2, and 3 heads occur? Compare your results with those of Exercise 45.

2. Explain how you could simulate tossing three coins 100 times using a random number generator, knowing only that heads and tails are equally likely on each toss. Use your method and compare the results with those of the previous exercise.

12.7

Mathematical

Induction

A **statement** is a sentence or equation that is either true or false. Statements involving positive integers are common in the study of sequences and series and in other areas of mathematics. For example, the statement

$$1^3 + 2^3 + 3^3 + \cdots + n^3 = \frac{n^2(n + 1)^2}{4}$$

is true for every positive integer n. This statement claims that

$$1^3 = \frac{1^2(1 + 1)^2}{4}, \qquad 1^3 + 2^3 = \frac{2^2(2 + 1)^2}{4}, \qquad 1^3 + 2^3 + 3^3 = \frac{3^2(3 + 1)^2}{4},$$

and so on. You can easily verify that the statement is true in these first three cases, but that does not prove that the statement is true for *every* positive integer n. In this section we will learn how to prove that a statement is true for every positive integer without performing infinitely many verifications.

Statements Involving Positive Integers

Suppose that S_n is used to denote the statement that $2(n + 3)$ is equal to $2n + 6$ for any positive integer n. This statement is written in symbols as follows:

$$S_n: \quad 2(n + 3) = 2n + 6$$

Is S_n true for every positive integer n? We can check that $2(\mathbf{1} + 3) = 2(\mathbf{1}) + 6$, $2(\mathbf{2} + 3) = 2(\mathbf{2}) + 6$, and $2(\mathbf{3} + 3) = 2(\mathbf{3}) + 6$ are all correct, but no matter how many of these equations are checked, it will not prove that S_n is true for every positive integer n. However, we do know that $a(b + c) = ab + ac$ for any real numbers a, b, and c. Since positive integers are real numbers, S_n is true for every positive integer n because of the distributive property of the real numbers.

In Sections 12.2 and 12.3, statements involving positive integers occurred in the study of series. Consider the statement that the sum of the first n positive odd integers is n^2. In symbols,

$$S_n: \quad 1 + 3 + 5 + 7 + \cdots + (2n - 1) = n^2.$$

Note that S_n is a statement about the sum of an arithmetic series. However, we do know that the sum of n terms of an arithmetic series is the sum of the first and last terms multiplied by $n/2$:

$$S = \frac{n}{2}(a_1 + a_n) = \frac{n}{2}(1 + 2n - 1) = n^2$$

So S_n is true for every positive integer n.

The Principle of Mathematical Induction

So far we have proved two statements true for every positive integer n. One statement was proved by using a property of the real numbers, and the other followed from the formula for the sum of an arithmetic series. The techniques used so far will not always work. For this reason, mathematicians have developed another technique, which is called the **principle of mathematical induction** or simply **mathematical induction.**

Suppose you want to prove that you can make a very long (possibly infinite) journey on foot. You could argue that you can take the first step. Next you could argue that after every step you will use your forward momentum to take another step. Therefore, you can make the journey. Note that both parts of the argument are necessary. If every step taken leads to another step, but it is not possible to make the first step, then the journey can't be made. Likewise, if all you can make is the first step, then you will certainly not make the journey. A proof by mathematical induction is like this example of proving you can make a long journey.

The first step in mathematical induction is to prove that a statement S_n is true in the very first case. S_1 must be proved true. The second step is to prove that the truth of S_k implies the truth of S_{k+1} for any positive integer k. Note that S_k is not proved true. We assume that S_k is true and show that this assumption leads to the truth of S_{k+1}. Mathematicians agree that these two steps are necessary and sufficient to prove that S_n is true for all positive integers n, and so the principle of mathematical induction is accepted as an axiom in mathematics.

Principle of Mathematical Induction

Let S_n be a statement for every positive integer n. If

1. S_1 is true, and

2. the truth of S_k implies the truth of S_{k+1} for ever positive integer k, then S_n is true for every positive integer n.

To prove that a statement S_n is true for every positive integer n, the only statements that we work with are S_1, S_k, and S_{k+1}. So in the first example, we will practice simply writing those statements (correctly).

Example 1 Writing S_1, S_k, and S_{k+1}

For the given statement S_n, write the three statements S_1, S_k, and S_{k+1}.

$$S_n: \quad 1^2 + 2^2 + 3^2 + \cdots + n^2 = \frac{n(n + 1)(2n + 1)}{6}$$

Solution

Write S_1 by replacing n by 1 in the statement S_n:

$$1^2 = \frac{1(1 + 1)(2 \cdot 1 + 1)}{6}$$

S_k can be written by replacing n by k in the statement S_n:

$$1^2 + 2^2 + 3^2 + \cdots + k^2 = \frac{k(k + 1)(2k + 1)}{6}$$

S_{k+1} can be written by replacing n by $k + 1$ in the statement S_n:

$$1^2 + 2^2 + 3^2 + \cdots + (k + 1)^2 = \frac{(k + 1)(k + 1 + 1)(2(k + 1) + 1)}{6}$$

$$1^2 + 2^2 + 3^2 + \cdots + (k + 1)^2 = \frac{(k + 1)(k + 2)(2k + 3)}{6}$$

Earlier we proved that the sum of the first n odd integers was n^2 by using the formula for the sum of an arithmetic series. In the next example we prove the statement again, this time using mathematical induction to write a complete proof.

Example 2 Proof by mathematical induction

Use mathematical induction to prove that the statement

$$T_n: \quad 1 + 3 + 5 + \cdots + (2n - 1) = n^2$$

is true for every positive integer n.

Solution

Step 1:   Write T_1 by replacing n by 1 in the given equation.

$$T_1: \quad 1 = 1^2$$

Since $1 = 1^2$ is correct, T_1 is true.

Step 2:   Assume that T_k is true. That is, we assume that the equation

$$1 + 3 + 5 + \cdots + (2k - 1) = k^2$$

is correct. Now we show that the truth of T_k implies that T_{k+1} is true. The next odd integer after $2k - 1$ is $2(k + 1) - 1$. Since T_k is true, we can add $2(k + 1) - 1$ to each side of the equation:

$$1 + 3 + 5 + \cdots + (2k - 1) + (2(k + 1) - 1) = k^2 + 2(k + 1) - 1$$

It is not necessary to write $2k - 1$ on the left-hand side because $2k - 1$ is the odd integer that precedes the odd integer $2(k + 1) - 1$:

$$1 + 3 + 5 + \cdots + 2(k + 1) - 1 = k^2 + 2k + 1 \qquad \text{Simplify.}$$

$$1 + 3 + 5 + \cdots + 2(k + 1) - 1 = (k + 1)^2 \qquad \text{This statement is } T_{k+1}.$$

We have shown that the truth of T_k implies that T_{k+1} is true. So by the principle of mathematical induction, the statement T_n is true for every positive integer n. ◆

Dominoes can be used to illustrate mathematical induction. Imagine an infinite sequence of dominoes, arranged so that when one falls over, it knocks over the one next to it, which knocks over the next one, and so on. If we can topple the first domino, then all dominoes in the infinite sequence will (theoretically) fall over.

In the next example, mathematical induction is used to prove a statement that was presented earlier in this section and would be difficult to prove without it.

Example 3   Proof by mathematical induction

Use mathematical induction to prove that the statement

$$T_n: \quad 1^3 + 2^3 + 3^3 + \cdots + n^3 = \frac{n^2(n + 1)^2}{4}$$

is true for every positive integer n.

Solution

Step 1:   If $n = 1$, then the statement T_1 is

$$1^3 = \frac{1^2(1 + 1)^2}{4}.$$

This equation is correct by arithmetic, so T_1 is true.

Step 2:   If we assume that T_k is true, then

$$1^3 + 2^3 + 3^3 + \cdots + k^3 = \frac{k^2(k + 1)^2}{4}.$$

Add $(k + 1)^3$ to each side of the equation:

$$1^3 + 2^3 + 3^3 + \cdots + k^3 + (k + 1)^3 = \frac{k^2(k + 1)^2}{4} + (k + 1)^3$$

$$1^3 + 2^3 + 3^3 + \cdots + (k + 1)^3 = \frac{k^2(k + 1)^2}{4} + \frac{4(k + 1)^3}{4} \qquad \text{\color{blue}{Find the LCD.}}$$

$$1^3 + 2^3 + 3^3 + \cdots + (k + 1)^3 = \frac{(k + 1)^2(k^2 + 4(k + 1))}{4}$$

$$1^3 + 2^3 + 3^3 + \cdots + (k + 1)^3 = \frac{(k + 1)^2(k + 2)^2}{4}$$

Since the last equation is T_{k+1}, we have shown that the truth of T_k implies the truth of T_{k+1} for every positive integer k. By the principle of mathematical induction, T_n is true for every positive integer n. ◆

Mathematical induction can be applied to the situation of a new college graduate who seeks a lifelong career. The graduate believes he or she can get a first job. The graduate also believes that every job will provide some experience that will guarantee getting a next job. If these two beliefs are really correct, then the graduate will have a lifelong career.

In the next example, we prove a statement involving inequality.

Example 4   Proof by mathematical induction

Use mathematical induction to prove that the statement

$$W_n: \quad 3^n < (n + 2)!$$

is true for every positive integer n.

Solution

Step 1:   If $n = 1$, then the statement W_1 is

$$3^1 < (1 + 2)!$$

This inequality is correct and W_1 is true because $(1 + 2)! = 3! = 6$.

Step 2:   If W_k is assumed to be true for a positive integer k, then

$$3^k < (k + 2)!$$

Multiply each side by 3 to get

$$3^{k+1} < 3 \cdot (k + 2)!$$

Since $3 < k + 3$ for $k \geq 1$, we have

$$3^{k+1} < (k + 3) \cdot (k + 2)!$$
$$3^{k+1} < (k + 3)!$$
$$3^{k+1} < ((k + 1) + 2)!$$

Since the last inequality is W_{k+1}, we have shown that the truth of W_k implies that W_{k+1} is true. By the principle of mathematical induction, W_n is true for all positive integers n. ◆

When writing a mathematical proof we try to convince the reader of the truth of a statement. Most proofs that are included in this text are fairly "mechanical," in that an obvious calculation or simplification gives the desired result. Mathematical induction provides a framework for a certain kind of proof, but within that framework we may still need some human ingenuity (as in the second step of Example 4). How much ingenuity is required in a proof is not necessarily related to the complexity of the statement. The story of Fermat's last theorem provides a classic example of how difficult it can be to prove a simple statement. The French mathematician Pierre de Fermat studied the problem of finding all positive integers that satisfy $a^n + b^n = c^n$. Of course, if $n = 2$ there are solutions such as $3^2 + 4^2 = 5^2$ and $5^2 + 12^2 = 13^2$. In 1637, Fermat stated that there are no solutions with $n \geq 3$. The proof of this simple statement (Fermat's last theorem) stumped mathematicians for over 350 years. Finally, in 1993 Dr. Andrew Wiles of Princeton University announced that he had proved it (*New York Times,* June 24, 1993). Dr. Wiles estimated that the written details of his proof, which he worked on for seven years, will take over 200 pages!

? For Thought

True or false? Explain.

1. The equation $\sum_{i=1}^{n} (4i - 2) = 2n^2$ is true if $n = 1$.

2. The inequality $n^3 < 4n + 15$ is true for $n = 1, 2,$ and 3.

3. For each positive integer n, $\dfrac{n - 1}{n + 1} < 0.9$.

4. Mathematical induction can be used to prove that $3(x + 1) = 3x + 3$ for every real number x.

5. If S_0 is true and the truth of S_{k-1} implies the truth of S_k for every positive integer k, then S_n is true for every nonnegative integer n.

6. If $n = k + 1$, then $\sum_{i=1}^{n} \dfrac{1}{i(i + 1)} = \dfrac{n}{n + 1}$ becomes $\sum_{i=1}^{k} \dfrac{1}{i(i + 1)} = \dfrac{k + 1}{k + 2}$.

7. Mathematical induction can be used to prove that $\sum_{i=1}^{\infty} 2^{-i} = 1$.

8. The statement $n^2 - n > 0$ is true for $n = 1$.

9. The statement $n^2 - n > 0$ is true for $n > 1$.

10. The statement $n^2 - n > 0$ is true for every positive integer n.

12.7 Exercises Tape 24   Disk—5.25": 8   3.5": 5   Macintosh: 4

Determine whether each statement is true for $n = 1, 2,$ and 3.

1. $\sum_{i=1}^{n} (3i - 1) = \dfrac{3n^2 + n}{2}$ **2.** $\sum_{i=1}^{n} \left(\dfrac{1}{2}\right)^i = 1 - 2^{-n}$

3. $\sum_{i=1}^{n} \dfrac{1}{i(i + 1)} = \dfrac{n}{n + 1}$ **4.** $\sum_{i=1}^{n} 3^i = \dfrac{3(3^n - 1)}{2}$

5. $\sum_{i=1}^{n} i^2 = 4n - 3$ **6.** $\sum_{i=1}^{n} 4i = 8n - 4$

7. $n^2 < n^3$ **8.** $(0.5)^{n-1} > 0.5$

For each given statement S_n, write the statements S_1, S_k, and S_{k+1}.

9. S_n: $\sum_{i=1}^{n} 2i = n(n + 1)$ **10.** S_n: $\sum_{i=1}^{n} 5i = \dfrac{5n(n + 1)}{2}$

11. S_n: $2 + 6 + 10 + \cdots + (4n - 2) = 2n^2$

12. S_n: $3 + 8 + 13 + \cdots + (5n - 2) = \dfrac{n(5n + 1)}{2}$

13. S_n: $\sum_{i=1}^{n} 2^i = 2^{n+1} - 2$

14. S_n: $\sum_{i=1}^{n} 5^{i+1} = \dfrac{5^{n+2} - 25}{4}$

15. S_n: $(ab)^n = a^n b^n$ **16.** S_n: $(a + b)^n = a^n + b^n$

17. S_n: If $0 < a < 1$, then $0 < a^n < 1$.

18. S_n: If $a > 1$, then $a^n > 1$.

Use mathematical induction to prove that each statement is true for each positive integer n.

19. $1 + 2 + 3 + \cdots + n = \dfrac{n(n + 1)}{2}$

20. $2 + 4 + 6 + \cdots + 2n = n(n + 1)$

21. $3 + 7 + 11 + \cdots + (4n - 1) = n(2n + 1)$

22. $2 + 7 + 12 + \cdots + (5n - 3) = \dfrac{n(5n - 1)}{2}$

23. $\sum_{i=1}^{n} 2^i = 2^{n+1} - 2$

24. $\sum_{i=1}^{n} 5^{i+1} = \dfrac{5^{n+2} - 25}{4}$

25. $\sum_{i=1}^{n} (3i - 1) = \dfrac{3n^2 + n}{2}$

26. $\sum_{i=1}^{n} \left(\dfrac{1}{2}\right)^i = 1 - 2^{-n}$

27. $1^2 + 2^2 + 3^2 + \cdots + n^2 = \dfrac{n(n + 1)(2n + 1)}{6}$

28. $\dfrac{1}{1 \cdot 2} + \dfrac{1}{2 \cdot 3} + \dfrac{1}{3 \cdot 4} + \cdots + \dfrac{1}{n(n + 1)} = \dfrac{n}{n + 1}$

29. $1 \cdot 3 + 2 \cdot 4 + 3 \cdot 5 + \cdots + n(n + 2) =$
$$\dfrac{n}{6}(n + 1)(2n + 7)$$

30. $\dfrac{1}{1 \cdot 4} + \dfrac{1}{4 \cdot 7} + \dfrac{1}{7 \cdot 10} + \cdots + \dfrac{1}{(3n - 2)(3n + 1)} =$
$$\dfrac{n}{3n + 1}$$

31. $\dfrac{1}{1 \cdot 3} + \dfrac{1}{3 \cdot 5} + \dfrac{1}{5 \cdot 7} + \cdots + \dfrac{1}{(2n - 1)(2n + 1)} =$
$$\dfrac{n}{2n + 1}$$

32. $\dfrac{1}{1 \cdot 2 \cdot 3} + \dfrac{1}{2 \cdot 3 \cdot 4} + \dfrac{1}{3 \cdot 4 \cdot 5} + \cdots +$
$$\dfrac{1}{n(n + 1)(n + 2)} = \dfrac{n(n + 3)}{4(n + 1)(n + 2)}$$

33. If $0 < a < 1$, then $0 < a^n < 1$.

34. If $a > 1$, then $a^n > 1$.

35. $n < 2^n$

36. $2^{n-1} \leq n!$

37. The integer $5^n - 1$ is divisible by 4 for every positive integer n.

38. If x is any real number with $x \neq 1$, then $1 + x + x^2 + x^3 + \cdots + x^n = \dfrac{x^{n+1} - 1}{x - 1}$.

39. If a and b are constants, then $(ab)^n = a^n b^n$.

40. If a and m are constants, then $(a^m)^n = a^{mn}$.

For Writing/Discussion

41. Write a paragraph describing mathematical induction in your own words.

42. Mathematical induction is used to prove that a statement is true for every positive integer. What steps do you think are necessary to prove that a statement is true for every integer?

Highlights

Section 12.1   Sequences

1. A finite sequence is a function whose domain is the set of positive integers less than or equal to some fixed positive integer.

2. An infinite sequence is a function whose domain is the set of all positive integers.

3. The nth term of an arithmetic sequence is $a_n = a_1 + (n - 1)d$, where a_1 is the first term and d is the common difference.

Section 12.2   Series

1. A series is the indicated sum of a sequence.

2. An arithmetic series is the indicated sum of an arithmetic sequence.

3. The sum of the terms a_1 through a_n of an arithmetic series is given by

$$S_n = \frac{n}{2}(a_1 + a_n).$$

Section 12.3   Geometric Sequences and Series

1. The nth term of a geometric sequence is $a_n = ar^{n-1}$, where $r \neq 1$ and $r \neq 0$, and r is the common ratio.

2. A geometric series is an indicated sum of a geometric sequence.

3. The sum of the first n terms of a geometric series is given by $S_n = \dfrac{a(1 - r^n)}{1 - r}$.

4. The sum of all terms of an infinite geometric series in which $|r| < 1$ is given by $S = \dfrac{a}{1 - r}$.

Section 12.4   Counting and Permutations

1. If event A has m outcomes and event B has n outcomes, then there are mn ways for A and B to occur.

2. The number of permutations of n things taken r at a time is denoted by $P(n, r)$ and given by the formula $P(n, r) = \dfrac{n!}{(n - r)!}$ for $0 \leq r \leq n$.

Section 12.5   Combinations, Labeling, and the Binomial Theorem

1. The number of combinations of n things taken r at a time is denoted by $C(n, r)$ or $\dbinom{n}{r}$ and is given by the formula $C(n, r) = \dfrac{n!}{(n - r)! \, r!}$ for $0 \leq r \leq n$.

2. The number of ways to label n objects with r_1 labels of type 1, r_2 labels of type 2, ..., and r_k labels of type k, where $r_1 + r_2 + \cdots + r_k = n$, is

$$\frac{n!}{r_1! \, r_2! \cdot \cdots \cdot r_k!}.$$

3. The binomial theorem says that $(a + b)^n = \sum_{r=0}^{n} \binom{n}{r} a^{n-r}b^r$ for every positive integer n.

Section 12.6 Probability

1. If S is a sample space of equally likely outcomes to an experiment and E is a subset of S, then $P(E) = n(E)/n(S)$.

2. If A and B are any events in a sample space, then
$$P(A \cup B) = P(A) + P(B) - P(A \cap B).$$

3. If A and B are mutually exclusive events, then $P(A \cup B) = P(A) + P(B)$.

4. Two events A and A' are called complementary events if $A \cap A' = \varnothing$ and $P(A) + P(A') = 1$.

5. The odds in favor of event A are defined as the ratio $P(A)$ to $P(A')$, and the odds against A are defined as the ratio $P(A')$ to $P(A)$.

6. If the odds in favor of E are a to b, then $P(E) = a/(a + b)$ and $P(E') = b/(a + b)$.

Section 12.7 Mathematical Induction

1. To prove that the statement S_n is true for every positive integer n, prove that S_1 is true and prove that for any positive integer k, assuming that S_k is true implies that S_{k+1} is true.

Chapter 12 Review Exercises

List all terms of each finite sequence.

1. $a_n = 2^{n-1}$ for $1 \leq n \leq 5$

2. $a_n = 3n - 2$ for $1 \leq n \leq 4$

3. $a_n = \dfrac{(-1)^n}{n!}$ for $1 \leq n \leq 4$

4. $a_n = (n - 2)^2$ for $1 \leq n \leq 6$

List the first three terms of each infinite sequence.

5. $a_n = 3(0.5)^{n-1}$

6. $b_n = \dfrac{1}{n(n + 1)}$

7. $c_n = -3n + 6$

8. $d_n = \dfrac{(-1)^n}{n^2}$

Find the sum of each series.

9. $\sum_{i=1}^{4} (0.5)^i$

10. $\sum_{i=1}^{3} (5i - 1)$

11. $\sum_{i=1}^{50} (4i + 7)$

12. $\sum_{i=1}^{4} 6$

13. $\sum_{i=1}^{\infty} 0.3(0.1)^{n-1}$

14. $\sum_{i=1}^{\infty} 5(-0.8)^n$

15. $\sum_{i=1}^{20} 1000(1.05)^{i-1}$

16. $\sum_{i=1}^{10} 6\left(\dfrac{1}{3}\right)^i$

Write a formula for the nth term of each sequence.

17. $-\dfrac{1}{3}, \dfrac{1}{4}, -\dfrac{1}{5}, \ldots$

18. $20, 17, 14, \ldots$

19. $6, 1, \dfrac{1}{6}, \ldots$

20. $1, -4, 9, -16, \ldots$

Write each series in summation notation. Use the index i and let i begin at 1.

21. $\dfrac{1}{2} - \dfrac{1}{3} + \dfrac{1}{4} - \cdots$

22. $5 + \dfrac{5}{2} + \dfrac{5}{4} + \dfrac{5}{8} + \cdots$

23. $2 + 4 + 6 + \cdots + 28$ **24.** $1 + 4 + 9 + \cdots + n^2$

Solve each problem.

25. *Common ratio* Find the common ratio for a geometric sequence that has a first term of 4 and a seventh term of 256.

26. *Common difference* Find the common difference for an arithmetic sequence that has a first term of 5 and a seventh term of 29.

27. *Quarterly payments* If $100 is deposited in an account paying 9% compounded quarterly, then how much will be in the account at the end of 10 years?

28. *Monthly payments* If $40,000 is deposited in an account paying 6% compounded monthly, then how much will be in the account at the end of nine years?

29. *Annual payments* If $1000 is deposited at the beginning of each year for 10 years in an account paying 6% compounded annually, then what is the total value of the 10 deposits at the end of the 10th year?

30. *Monthly payments* If $50 is deposited at the beginning of each month for 20 years in an account paying 6% compounded monthly, then what is the total value of the 240 deposits at the end of the 20th year?

31. *Long-distance bicycling* Tyrone and Phyllis are trying to bicycle from Houston to Los Angeles. Each day they plan to travel 90% of the distance traveled the previous day. If they traveled 100 miles the first day, then what is the maximum distance that they can travel with this plan? Will they reach Los Angeles?

32. *Bouncing ball* A ball made from a new synthetic rubber will rebound 97% of the distance from which it is dropped. If the ball is dropped from a height of 6 ft, then approximately how far will it travel before coming to rest?

Write the complete binomial expansion for each of the following powers of a binomial.

33. $(a + 2b)^4$

34. $(x - 5)^3$

35. $(2a - b)^5$

36. $(w + 2)^6$

Write the first three terms of each binomial expansion.

37. $(a + b)^{10}$

38. $(x - 2y)^9$

39. $\left(2x + \dfrac{y}{2}\right)^8$

40. $(2a - 3b)^7$

Solve each problem.

41. What is the coefficient of a^4b^9 in the expansion of $(a + b)^{13}$?

42. What is the coefficient of x^8y^7 in the expansion of $(2x - y)^{15}$?

43. What is the coefficient of $w^2x^3y^6$ in the expansion of $(w + 2x + y)^{11}$?

44. What is the coefficient of x^5y^7 in the expansion of $(2w + x + y)^{12}$?

45. How many terms are there in the expansion of $(a + b)^{23}$?

46. Write out the first seven rows of Pascal's triangle.

Solve each counting problem.

47. *Multiple-choice test* How many different ways are there to mark the answers to a nine-question multiple-choice test in which each question has five possible answers?

48. *Scheduling departures* Six airplanes are scheduled to depart at 2:00 on the same runway. In how many ways can they line up for departure?

49. *Three-letter words* John is trying to find all three-letter English words that can be formed without repetition using the letters in the word FLORIDA. He plans to write down all possible three-letter ''words'' and then check each one with a dictionary to see if it is actually a word in the English language. How many possible three-letter ''words'' are there?

50. *Selecting a team* The sales manager for an insurance company must select a team of two agents to give a presentation. If the manager has five male agents and six female agents available, and the team must consist of one man and one woman, then how many different teams are possible?

51. *Placing advertisements* A candidate for city council is going to place one newspaper advertisement, one radio advertisement, and one television advertisement. If there are three newspapers, five radio stations, and four television stations available, then in how many ways can the advertisements be placed?

52. *Signal flags* A ship has nine different flags available. A signal consists of three flags displayed on a vertical pole. How many different signals are possible?

53. *Choosing a vacation* A travel agent offers a vacation in which you can visit any five cities, chosen from Paris, Rome, London, Istanbul, Monte Carlo, Vienna, Madrid, and Berlin. How many different vacations are possible, not counting the order in which the cities are visited?

54. *Counting subsets* How many four-element subsets are there for the set $\{a, b, c, d, e, f, g\}$?

55. *Choosing a committee* A city council consists of five Democrats and three Republicans. In how many ways can four council members be selected by the mayor to go to a convention in San Francisco if the mayor
 a) may choose any four?
 b) must choose four Democrats?
 c) must choose two Democrats and two Republicans?

56. *Full house* In a five-card poker hand, a full house is three cards of one kind and two cards of another. How many full houses are there consisting of three queens and two 10's?

57. *Arranging letters* How many different arrangements are there for the letters in the word KANSAS? In TEXAS?

58. *Marking pickup trucks* Eight Mazda pickups of different colors are on sale through Saturday only. How many ways are there for the dealer to mark two of them $7000, two of them $8000, and four of them $9000?

59. *Possible families* A couple plan to have seven children. How many different families are possible, considering the sex of each child and the order of birth?

60. *Triple feature* The Galaxy Theater has three horror films to show on Saturday night. In how many different ways can the program for a triple feature be arranged?

Solve each probability problem.

61. *Just guessing* If Miriam randomly marks the answers to a ten-question true-false test, then what is the probability that she gets all ten correct? What is the probability that she gets all ten wrong?

62. *Large family* If a couple plan to have six children, then what is the probability that they get six girls?

63. *Jelly beans* There are six red, five green, and two yellow jelly beans in a jar. If a jelly bean is selected at random, then what is the probability that
 a) the selected bean is green?
 b) the selected bean is either yellow or red?
 c) the selected bean is blue?
 d) the selected bean is not purple?

64. *Rolling dice* A pair of dice is rolled. What is the probability that
 a) at least one die shows an even number?
 b) the sum is an even number?
 c) the sum is 4?
 d) the sum is 4 and at least one die shows an even number?
 e) the sum is 4 or at least one die shows an even number?

65. *My three sons* Suppose a couple plan to have three children. What are the odds in favor of getting three boys?

66. *Sum of six* Suppose a pair of dice is rolled. What are the odds in favor of the sum being 6?

67. *Gone fishing* If 90% of the fish in Lake Louise are perch and a single fish is caught at random, then what are the odds in favor of catching a perch?

68. *Future plans* In a survey of high school students, 80% said they planned to attend college. If a student is selected at random from this group, then what are the odds against getting one who plans to attend college?

69. *Enrollment data* At MSU, 60% of the students are enrolled in an English class, 70% are enrolled in a mathematics class, and 40% are enrolled in both. If a student is selected at random, then what is the probability that the student is enrolled in either mathematics or English?

70. *Rolling again* If a pair of dice is rolled, then what is the probability that the sum is either 5 or 6?

Evaluate each expression.

71. $8!$

72. $1!$

73. $\dfrac{5!}{3!}$

74. $\dfrac{7!}{(7-3)!}$

75. $\dfrac{9!}{3!\,3!\,3!}$

76. $\dfrac{10!}{0!\,2!\,3!\,5!}$

77. $C(8, 6)$

78. $\dbinom{12}{3}$

79. $P(8, 4)$

80. $P(4, 4)$

81. $C(8, 1)$

82. $C(12, 0)$

Use mathematical induction to prove that each statement is true for each positive integer n.

83. $3 + 6 + 9 + \cdots + 3n = \dfrac{3}{2}(n^2 + n)$

84. $\displaystyle\sum_{i=1}^{n} 2(3)^{i-1} = 3^n - 1$

Chapter 12 Test

List all terms of each finite sequence.

1. $a_n = 2.3 + (n - 1)(0.5)$, $1 \leq n \leq 4$

2. $c_1 = 20$ and $c_n = \dfrac{1}{2} c_{n-1}$ for $2 \leq n \leq 4$

Write a formula for the nth term of each infinite sequence. Do not use a recursion formula.

3. $0, -1, 4, -9, 16, \ldots$ **4.** $7, 10, 13, 16, \ldots$

5. $\dfrac{1}{3}, -\dfrac{1}{6}, \dfrac{1}{12}, -\dfrac{1}{24}, \ldots$

Find the sum of each series.

6. $\displaystyle\sum_{j=0}^{53} (3j - 5)$ **7.** $\displaystyle\sum_{i=1}^{23} 300(1.05)^i$ **8.** $\displaystyle\sum_{n=1}^{\infty} (0.98)^n$

Solve each problem.

9. Find a formula for the nth term of an arithmetic sequence whose first term is -3 and whose ninth term is 9.

10. How many different nine-letter ''words'' can be made from the nine letters in TENNESSEE?

11. On the first day of April, $300 worth of lottery tickets were sold at the Quick Mart. If sales increased by $10 each day, then what was the mean of the daily sales amounts for April?

12. A customer at an automatic bank teller must enter a four-digit secret number to use the machine. If any of the integers from 0 through 9 can be used for each of the four digits, then how many secret numbers are there? If the machine gives a customer three tries to enter the secret number, then what is the probability that an unauthorized person can guess the secret number and gain access to the account?

13. Desmond and Molly Jones can save $700 per month toward their $120,000 dream house by living with Desmond's parents. If $700 is deposited at the beginning of each month for 300 months into an account paying 6% compounded monthly, then what is the value of this annuity at the end of the 300th month? If the price of a $120,000 house increases 6% each year, then what will it cost at the end of the 25th year? After 25 years of saving, will they have enough to buy the house and move out of his parents' house?

14. Write all of the terms of the binomial expansion for $(a - 2x)^5$.

15. Write the first three terms of the binomial expansion for $(x + y^2)^{24}$.

16. Write the binomial expansion for $(m + y)^{30}$ using summation notation.

17. If a pair of fair dice is rolled, then what is the probability that the sum of the numbers showing is 7? What are the odds in favor of rolling a 7?

18. An employee randomly selects three of the 12 months of the year in which to take a vacation. In how many ways can this selection be made?

19. In a seventh grade class of 12 boys and 10 girls, the teacher randomly selects two boys and two girls to be crossing guards. How many outcomes are there to this process?

20. In a race of eight horses, a bettor randomly selects three horses for the categories of win, place, and show. What is the probability that the bettor gets the horses and the order of finish correct?

21. Use mathematical induction to prove that $\displaystyle\sum_{i=1}^{n} \left(\dfrac{1}{2}\right)^i < 1$ for every positive integer n.

Appendix A
Graphing
Calculator
Lessons

Technology is changing the way that mathematics is taught and learned. For many years computers have been able to do just about any routine procedure in mathematics. In more recent years, the availability of powerful hand-held graphing calculators has incorporated the power of a computer into an inexpensive and portable device.

In this appendix, you will find a graphing calculator lesson corresponding to each chapter of the text. In each lesson, we discuss specific operations and procedures and give examples that will help you to become proficient at using a graphing calculator. In parentheses following each example letter, you will find the section number of the text in which the mathematics relevant to that example is discussed. Exercises that require a graphing calculator are found throughout the exercise sections of this text. Of course, a graphing calculator will be useful for many other exercises as well.

The screens shown in this text are produced with a Texas Instruments TI82 but will look similar on any graphing calculator. Since specific procedures for performing tasks may vary from model to model, you should be sure to learn each new procedure on your calculator as it is discussed. Consult your manual or your instructor if you have difficulties.

Chapter 1

In Chapter 1 we have been reviewing how to evaluate expressions. A graphing calculator is very useful here. With a graphing calculator you can make the expression on the display look nearly the same as it appears on paper, and you can modify it before evaluating it if necessary. After you are satisfied with the expression on the display, press the ENTER or EXE key to evaluate the expression.

Parentheses are used as grouping symbols on a graphing calculator; when no parentheses are used, the calculator follows the standard order of operations. Some grouping symbols used in algebra are not available on all calculators. Many calculators do not use the built-up form of a fraction, and so the fraction bar cannot be used to group the numerator and denominator. In this case, parentheses are used around the numerator and denominator for grouping. Absolute value bars are not usually available as grouping symbols. Instead, the ABS key is used to find the absolute value of a quantity in parentheses.

Example A (Section 1.1)   Using grouping symbols

Evaluate each expression.

(i) $\dfrac{3.4 + 7.8}{5.4 - 5.8}$

(ii) $-3 - 2(8 - 4 \div 2)$

(iii) $|-5| - 2|4 - 6|$

Solution

(i) Enclose both the numerator and the denominator in parentheses. Note that division is indicated by a forward slash on the display.

A-1

```
(3.4+7.8)/(5.4-5
.8)
             -28
```

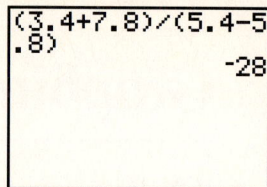

Once the expression is displayed, press the ENTER key to get the value − 28.

(ii) Negative numbers are entered using the negative sign key (-).

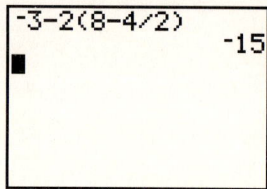

```
-3-2(8-4/2)
             -15
■
```

So $-3 - 2(8 - 4 \div 2) = -15$.

(iii) Note that to find the absolute value of the quantity $4 - 6$ the quantity is put in parentheses, but no parentheses are needed for the absolute value of -5.

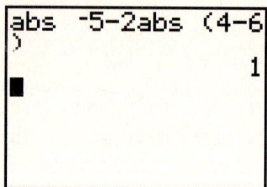

```
abs -5-2abs (4-6
)
              1
■
```

So $|-5| - 2|4 - 6| = 1$.   ◆

Exercises

Evaluate each expression. Round your answers to four decimal places. Answers can be found at the end of this appendix.

1. $\dfrac{7.28 - 6.49}{3.2 - 5.88 + 3.44}$

2. $-4.9 - [9 - 8.4(-3.5 + 4.5 \div 6.6)]$

3. $4|5.6 - 33.4| - 6|-44 + 4 \cdot 3.2|$

The multiplication symbol can be omitted in a product on a graphing calculator. For example, $2*x$, $2x$, and $2(x)$ all indicate the product of 2 and x. When no multiplication symbol is used in an expression like $2x$, some calculators treat the expression as if it were in parentheses.

Example B (Section 1.1)   Omitting the multiplication symbol

Evaluate each expression.

(i) $\dfrac{1}{2}\,\pi$ (ii) $\dfrac{1}{2\pi}$

Solution

(i) To evaluate $\frac{1}{2}\pi$ you can enter $1/2*\pi$ or $(1/2)\pi$.

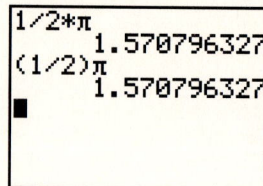

```
1/2*π
        1.570796327
(1/2)π
        1.570796327
■
```

So $\frac{1}{2}\pi \approx 1.570796327$.

(ii) To evaluate $\dfrac{1}{2\pi}$ you can use $1/2\pi$ or $1/(2\pi)$ on a graphing calculator.

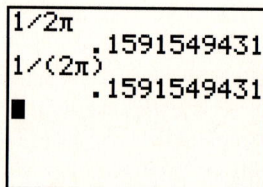

```
1/2π
        .1591549431
1/(2π)
        .1591549431
■
```

So $\dfrac{1}{2\pi} \approx 0.1591549431$. Since π is irrational, these values are approximate. Note that $1/2\pi$ and $1/2*\pi$ do not have the same value on most graphing calculators. Be sure you understand how your calculator evaluates these expressions.   ◆

Exercises

Evaluate each expression. Round your answers to four decimal places.

4. $\dfrac{3}{4\pi} + \dfrac{1}{3\pi}$ 5. $\dfrac{1}{2}\,\pi - \dfrac{1}{4}\,\pi$

For more practice, try Exercises 47–76 in Section 1.1.

Exponential expressions are entered into the graphing calculator using the x^2 key for squaring a number and the $\wedge$ key for other powers.

Example C (Section 1.2)   Expressions with integral exponents

Evaluate $4^2 + 2^{4-6}$ and $\left(\dfrac{10}{9}\right)^{-6}$.

Solution

Parentheses must be used around $4 - 6$ in the first expression and around the fraction in the second expression:

```
4²+2^(4-6)
              16.25
(10/9)^-6
              .531441
■
```

Exercises

Evaluate each expression.

6. $\left(1 + \dfrac{0.08}{365}\right)^{365 \cdot 4}$

7. $\left(\dfrac{2}{3}\right)^{-3} + \left(\dfrac{3}{4}\right)^{5}$

    The EE key or EXP key is used to enter a number in scientific notation. Note that only the power of 10 appears on the display. A number in scientific notation is treated as a single number for the order of operations.

Example D (Section 1.2)   Scientific notation

Evaluate $\dfrac{(1.34 \times 10^4)^3}{(2.4 \times 10^{-5})(3.6 \times 10^3)}$.

Solution

On a graphing calculator, parentheses are not needed around 1.34×10^4, but parentheses are needed to group the product in the denominator.

```
1.34E4^3/(2.4E-5
*3.6E3)
      2.784842593E13
■
```

Rounded to two places, the value of the expression is 2.78×10^{13}.

Exercises

Evaluate each expression.

8. $(2.34 \times 10^5)^3 + (1.8 \times 10^{-4})^{-8}$

9. $\dfrac{9.2 \times 10^{-4}}{84(3.1 \times 10^3) + 5.2 \times 10^6}$

For more practice, try Exercises 23–36 and 89–106 in Section 1.2.

    To evaluate expressions with rational exponents, the exponents can be entered as rational numbers. If a rational number is a terminating decimal, then it could be entered as a decimal number. If you round off a rational number such as $\frac{1}{3}$ to 0.33, you will not get very accurate answers. Due to limitations in the algorithms used by the calculator, calculators will not evaluate certain expressions with negative bases. When this situation occurs, you can use your knowledge of exponents to rewrite the expression using a positive base. Square roots can be evaluated using the radical symbol, but other roots are usually found using rational exponents.

Example E (Section 1.3)   Rational exponents and radicals

Evaluate $(-1.728)^{2/3}$ and $\sqrt{12^2 + 20^2}$.

Solution

Many calculators will not evaluate this expression with a negative base. Since $(-1.728)^{2/3} = 1.728^{2/3}$, we can evaluate $1.728^{2/3}$. If the radical symbol contains an expression other than a single number, then the radicand must be enclosed in parentheses.

```
1.728^(2/3)
                1.44
√(12²+20²)
          23.32380758
■
```

So $(-1.728)^{2/3} = 1.44$. The value of $\sqrt{12^2 + 20^2}$ is approximately 23.323.

Exercises

Evaluate each expression.

10. $\sqrt{34} + \sqrt{92.6 - 5(3.44)}$

11. $\sqrt[3]{433} - \sqrt[5]{589}$

12. $(-8.5)^{4/5} + (-0.89)^{-3/5}$

For more practice, try Exercises 1–10, 37–46, and 111–126 in Section 1.3.

    Probably the most important feature of a graphing calculator is its ability to deal with variables. You can enter a polynomial into a graphing calculator and evaluate it for any value of x that you choose.

Example F (Section 1.4)   Evaluating a polynomial

Let $P(x) = 1.5x^3 - 4.1x^2 + 3.2x + 6$ and $T(x) = 1.1x^4 - 2.3x^2$. Find $P(-2)$ and $T(4.5)$.

Solution

Press the Y= key to define the polynomials as y_1 and y_2.

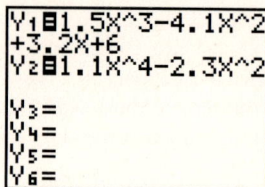

```
Y₁◘1.5X^3-4.1X^2
+3.2X+6
Y₂◘1.1X^4-2.3X^2
Y₃=
Y₄=
Y₅=
Y₆=
```

To evaluate the polynomials, enter $y_1(-2)$ and $y_2(4.5)$:

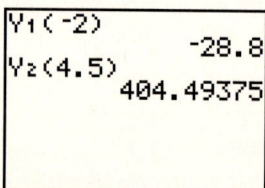

```
Y₁(-2)
              -28.8
Y₂(4.5)
          404.49375
```

So $P(-2) = -28.8$ and $T(4.5) = 404.49375$. If $y_1(-2)$ does not work on your calculator, try using the STO key to store the desired value for x and then find the corresponding value for y_1 or y_2.

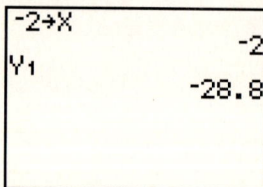

```
-2→X
                -2
Y₁
              -28.8
```

Again we get $P(-2) = -28.8$. You should find $T(4.5)$ by storing 4.5 for x and then evaluating y_2.   ◆

Exercises

Let $P(x) = 4x^3 - 6.5x^2 + 4.8x - 9.2$ and $Q(x) = 2x^3 - 4.5x^2 + 6.4x + 17$. Find the following.

13. $P(3.6)$

14. $Q(-13.2)$

15. $\dfrac{P(5688)}{Q(5688)}$

16. $\dfrac{P(40.1) - P(40)}{0.1}$

For more practice, try Exercises 95–102 in Section 1.4 and Exercises 81–92 in Section 1.6.

Chapter 2

In Chapter 2 we solved many types of equations. Your graphing calculator can help you in solving equations and checking your solutions. A good way to check a possible solution to an equation is with the ANS key. The ANS key contains the value of the last expression calculated on the calculator. You can use the symbol ANS in place of the variable in the original equation to see if the answer satisfies the equation.

Example A (Section 2.1)   Solving an equation and checking

Solve $\sqrt[3]{2}x - \sqrt{5} = 7.4$. Round the solution to three decimal places.

Solution

Solve for x to get the exact solution:

$$\sqrt[3]{2}x - \sqrt{5} = 7.4$$

$$\sqrt[3]{2}x = 7.4 + \sqrt{5}$$

$$x = \frac{7.4 + \sqrt{5}}{\sqrt[3]{2}}$$

Use your calculator to find an approximate solution and to check:

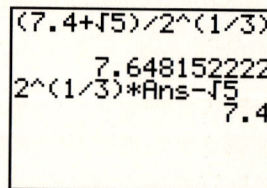

```
(7.4+√5)/2^(1/3)
         7.648152222
2^(1/3)*Ans-√5
                 7.4
```

The solution rounded to three decimal places is 7.648.   ◆

Exercises

Solve each equation and check your answer. Answers can be found at the end of this appendix.

1. $\sqrt[4]{\pi} \cdot x - 9.8 \times 10^{-3} = 8.3215$

2. $\sqrt[3]{2} \cdot y + \dfrac{1}{2\pi} y = 6.9$

For more practice, try Exercises 43–56 and 89–92 in Section 2.1.

   In the next example, we use a graphing calculator to solve a quadratic equation. After finding the value of

$$\frac{-b + \sqrt{b^2 - 4ac}}{2a}$$

the ENTRY key can be used to recall the expression. Then position the cursor on the plus sign and replace it with a minus sign. Press ENTER to evaluate

$$\frac{-b - \sqrt{b^2 - 4ac}}{2a}.$$

Example B (Section 2.4)   Solving a quadratic equation

Solve $2.3x^2 - 6.5x + 3.4 = 0$. Round the solutions to three decimal places.

Solution

Use the quadratic formula to get

$$x = \frac{6.5 \pm \sqrt{(-6.5)^2 - 4(2.3)(3.4)}}{2(2.3)}.$$

Use a graphing calculator to find the first solution and check.

```
(6.5+√((-6.5)²-4
(2.3)(3.4)))/(2*
2.3)
            2.133065007
2.3Ans²-6.5Ans+3
.4
                      0
█
```

Use the ENTRY key twice to recall the original expression for the first solution. Change the plus sign to minus to find the second solution and check:

```
(6.5-√((-6.5)²-4
(2.3)(3.4)))/(2*
2.3)
            .6930219495
2.3Ans²-6.5Ans+3
.4
                      0
```

The solutions are 0.693 and 2.133.   ◆

Exercises

Solve each quadratic equation. Round answers to three decimal places.

3. $4.5x^2 - 5.2x - 8.7 = 0$

4. $-\dfrac{2}{3}w^2 - \dfrac{1}{\sqrt{2}}w + \dfrac{3}{2\pi} = 0$

For more practice, try Exercises 55–58 and 89–92 in Section 2.4.

Even though a graphing calculator is not designed to graph on a number line, we can use the test feature of a graphing calcu-

lator to illustrate the solution to a quadratic or rational inequality. Inequality symbols are found in the TEST menu on the calculator:

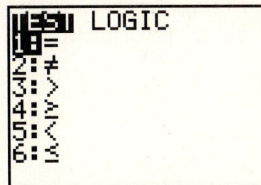

```
TEST LOGIC
1:=
2:≠
3:>
4:≥
5:<
6:≤
```

An inequality on a graphing calculator can be used in a way that is very different from its use in algebra. An inequality such as $x^2 - x > 5$ is treated like an expression that has only two values, 1 or 0. The value of $x^2 - x > 5$ is 1 if it is true for the current value of x, and 0 if it is false for the current value of x.

Inequalities can be used to define other variables. For example, the statement $y_1 = x^2 - x > 5$ means that y_1 has value 1 or 0 depending on the value of x and whether the inequality is satisfied or not. The statement $y_1 = 3(x^2 - x > 5)$ means that y_1 has value 3 or 0 because the inequality in parentheses has value 1 or 0. The next example shows how to use the test feature.

Example C (Section 2.7)   Solving a quadratic inequality

Solve $x^2 - x \le 6$.

Solution

Write the inequality as $x^2 - x - 6 \le 0$ and factor the left side to get

$$(x - 3)(x + 2) \le 0.$$

A sign chart for this inequality yields the solution $[-2, 3]$ as in Example 1(a) of Section 2.7. To support or verify this solution with a calculator, use the Y= key to define y_1 as follows:

```
Y1■X²-X≤6
Y2=
Y3=
Y4=
Y5=
Y6=
Y7=
Y8=
```

Now graph y_1 for $-10 \le x \le 10$ and $-10 \le y \le 10$, the standard viewing window.

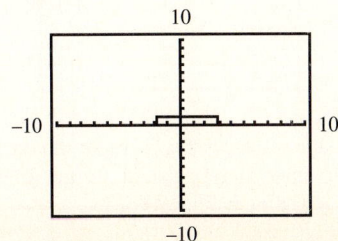

When the inequality is satisfied, the value of y_1 is 1. So a horizontal line appears on the graph above numbers that satisfy the inequality. Since each tick mark on the axis represents one unit, this graph supports the conclusion that the solution is $[-2, 3]$.

◆

Exercises

Solve each inequality with a sign graph and verify your solution with your graphing calculator.

5. $x^2 + 2x - 24 > 0$

6. $-x^2 + x + 30 \geq 0$

For more practice, try Exercises 9–20 in Section 2.7.

Many graphing calculators can generate a table of values for a variable. The numbers in the table can then be used as test points for the inequality.

Example D (Section 2.7)   The test point method

Solve $x^2 - 4x - 6 > 0$.

Solution

First solve $x^2 - 4x - 6 = 0$.

$$x = \frac{4 \pm \sqrt{4^2 - 4(1)(-6)}}{2(1)} = \frac{4 \pm \sqrt{40}}{2}$$

$$= 2 \pm \sqrt{10}$$

Use a calculator to get the approximate solutions

$$2 - \sqrt{10} \approx -1.16$$

and

$$2 + \sqrt{10} \approx 5.16.$$

Define $y_1 = x^2 - 4x - 6$ using the Y= key.

```
Y1■X²-4X-6
Y2=
Y3=
Y4=
Y5=
Y6=
Y7=
Y8=
```

Now make a table of values for y_1 where the range of x values includes the solutions to the equation. In the following table, x ranges from -4 to 8 in increments of 2.

```
  X  | Y1
 -4  | 26
 -2  | 6
  0  | -6
  2  | -10
  4  | -6
  6  | 6
  8  | 26
Y1■X²-4X-6
```

We can use the values of x in the table as test points. If you use a different range or increment in your table, you can scroll through the table to see appropriate test points. You can see that $y_1 > 0$ when $x < -1.16$ or when $x > 5.16$. So the exact solution to the inequality is

$$(-\infty, 2 - \sqrt{10}) \cup (2 + \sqrt{10}, \infty).$$

◆

Exercises

Solve each inequality by using your calculator to generate a table of test points.

7. $2x^2 - 6x + 3 \geq 0$

8. $-0.31z^2 - 5.3z + 12 < 0$

For more practice, try Exercises 45–60 in Section 2.7.

Chapter 3

In Chapter 3 we began studying functions and graphs. Remember that a graph of a function is a representation of the set of all ordered pairs that are in the function. This set of ordered pairs is usually infinite. Since a graphing calculator plots only a finite number of ordered pairs, the graph on your calculator is usually not a very good representation of the graph of a function. On a graphing calculator, even a straight line looks like steps. You can draw a better graph for a straight line using a pencil and ruler. The main benefit of a graphing calculator is that it can quickly and accurately plot nearly 100 points. With the proper interpretation, the points shown on the calculator can be very useful.

Example A (Section 3.1)   Graphing a line

Graph $y = \frac{1}{2}x - 20$.

Solution

First define the function as y_1 using the Y= key.

```
Y1■1/2*X-20
Y2=
Y3=
Y4=
Y5=
Y6=
Y7=
Y8=
```

A "good" graph should show the intercepts $(0, -20)$ and $(40, 0)$. So set the viewing window for $-10 \le x \le 50$ and $-30 \le y \le 20$ with a scale of 10. The scale is the distance between the tick marks on the axes.

```
WINDOW FORMAT
Xmin=-10
Xmax=50
Xscl=10
Ymin=-30
Ymax=20
Yscl=10
```

Now press GRAPH:

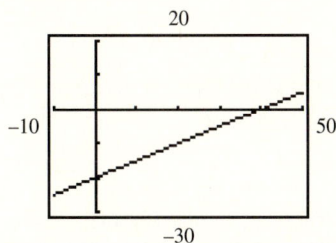

If you look closely at the graph of $y = \frac{1}{2}x - 20$, it looks more like steps than a straight line. Since we know that the graph is supposed to be a straight line, we can view this graph as is or use it as a model for drawing a better graph with a pencil and ruler.

◆

Exercises

Graph each line using a viewing window that shows both the x- and y-intercepts. Answers can be found at the end of this appendix.

1. $y = -400x + 87600$ **2.** $y = 0.0003x + 0.01$

For more practice, try Graphing Calculator Exercises 1–4 in Section 3.1.

The graph of a function can be used to find an approximate solution to an equation. For example, the solution to the equation $\frac{1}{2}x - 20 = 0$ is 40, and 40 is the x-coordinate of the x-intercept for the graph of $y = \frac{1}{2}x - 20$. The TRACE feature on a graphing calculator gives the coordinates of the points that are graphed and the ZOOM feature allows you to zoom in on the x-intercept for greater accuracy.

Example B (Section 3.1)   Solving an equation

Use the TRACE and ZOOM features to find the solution, accurate to two decimal places, to

$$1.33x - 2.44(x - 1.65) = 0.$$

Solution

Define the function

$$y_1 = 1.33x - 2.44(x - 1.65)$$

using the Y= key.

```
Y1■1.33X-2.44(X-
1.65)
Y2=■
Y3=
Y4=
Y5=
Y6=
Y7=
```

Graph the function in the standard viewing window.

Press the TRACE key and move the cursor as close as you can to the x-intercept.

Note that y is positive when x is 3.6 and negative when x is 3.8. So the solution to the equation is between these two values. Press the ZOOM key and then Zoom In to get greater accuracy.

Use the TRACE key on the new graph to see that the solution is between 3.617 and 3.670. Repeat this procedure to see that the solution is between 3.617 and 3.630. Repeat again to see that the solution is between 3.626 and 3.630. So, accurate to two decimal places, the solution is 3.63. ◆

Exercises

For each equation, use TRACE and ZOOM to find an approximate solution accurate to two decimal places.

3. $-3.22x - 4.5(3.61x - 2.97) = 0$

4. $\dfrac{3.55x - 9.4}{4.28} = \dfrac{7.8 - 4.7x}{6.2 - 9.8^2}$

For more practice, try Graphing Calculator Exercises 5–12 in Section 3.1.

In Chapter 2 we saw how to use a graphing calculator to assist us in solving linear and absolute value inequalities. That same technique could be used to solve the inequalities in this chapter. However, in the next example we will see how to use a graph and the TRACE feature to solve a quadratic inequality.

Example C (Section 3.4)   Solving a quadratic inequality

Solve $x^2 - 200 \geq 0$.

Solution

First solve the equation $x^2 - 200 = 0$ to find the x-intercepts.

$$x^2 = 200$$
$$x = \pm\sqrt{200}$$
$$x = \pm 10\sqrt{2}$$

Now graph the function $y_1 = x^2 - 200$ as follows:

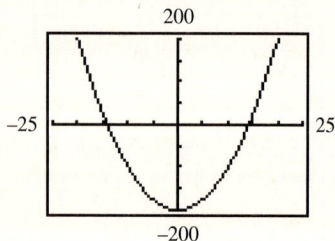

Use the TRACE feature to find the values of x for which y is positive.

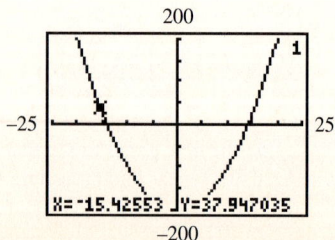

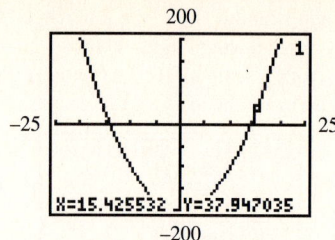

When x is less than $-10\sqrt{2}$ or when x is greater than $10\sqrt{2}$, the y-coordinate is positive. So the solution to the inequality is $(-\infty, -10\sqrt{2}] \cup [10\sqrt{2}, \infty)$. ◆

Exercises

Use the graph of a quadratic function to help you solve each quadratic inequality.

5. $x^2 - 4x + 2 \geq 0$ **6.** $16x - x^2 > -336$

For more practice, try Exercises 77–84 and Graphing Calculator Exercises 1–8 in Section 3.4.

A graphing calculator can be used to help us identify whether two given functions are inverses of each other. We can graph both functions on the same screen and inspect the graphs for symmetry with respect to the line $y = x$. We can also graph the composition of the two functions to see if it is the identity function for all values of x in its domain.

Example D (Section 3.6)   Verifying inverse functions

Let $f(x) = (x - 1)^2 - 3$ for $x \geq 1$ and $g(x) = \sqrt{x + 3} + 1$. Graph $f(x)$, $g(x)$, and $(g \circ f)(x)$. What can you say about f and g?

Solution

Define y_1, y_2, and y_3 as shown here.

Note that we divide the function f by the test condition $(x \geq 1)$. When $x \geq 1$ is satisfied, then $(x \geq 1)$ has value 1 and the function is graphed. When $x \geq 1$ is not satisfied, then $(x \geq 1)$ has value 0 and nothing is graphed because division by 0 is undefined. The graphs of all three functions are shown here in the standard viewing window.

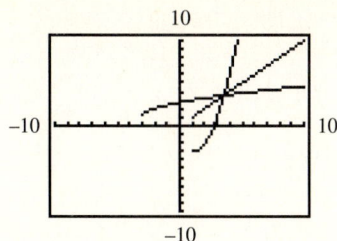

To better see the symmetry of the two functions, draw the graph in the standard viewing window and then use the Square feature (in the ZOOM menu) to make the length of a unit the same on both axes.

The graphs drawn by the calculator appear to be symmetric with respect to the line $y = x$. The composition $g \circ f$ appears to be the identity function when $x \geq 1$. This information leads us to believe that f and g are inverses of each other. To prove that f and g are inverses, we could use an algebraic argument to show that the functions satisfy the conditions given in the theorem in Section 3.6. ◆

Exercises

Use the graphs of $f(x)$, $g(x)$, and $(f \circ g)(x)$, to help you make a conjecture as to whether f and g are inverses of each other.

7. $f(x) = \dfrac{\sqrt[3]{x - 5}}{4}$, $g(x) = 64x^3 + 5$

8. $f(x) = \dfrac{\sqrt[4]{3x - 1}}{2}$, $g(x) = \dfrac{16x^4 + 1}{3}$

For more practice, try Exercises 83–90 and Graphing Calculator Exercises 1–6 in Section 3.6.

Chapter 4

In Chapter 4 we began studying polynomial and rational functions. The graphing calculator can be very useful with the more complicated functions. As always, remember that the graph of a function is usually an infinite set of points and the graphing cal-

culator shows only a finite set of points. So the graph on a graphing calculator can be misleading.

In our first example, we use the TRACE feature to find the intersection of two lines.

Example A (Section 4.1)   Intersection of two lines

Find a viewing window that shows the intersection of the lines $y = \frac{1}{3}x - 20$ and $y = 0.3x - 19$. Use TRACE to find the intersection accurate to the nearest tenth.

Solution

First define the functions using the Y= key.

Start with a viewing window that shows both the x- and y-intercepts.

Use TRACE to move the cursor to the apparent point of intersection of the lines.

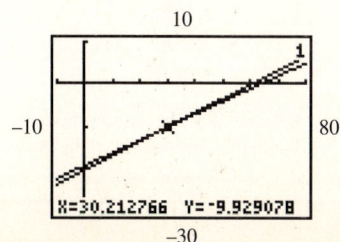

Now examine the coordinates carefully. If $x = 29.25$, the graph of y_2 is above the graph of y_1, and if $x = 30.21$, the graph of y_1 is above the graph of y_2. So the lines cross between 29.25 and 30.21. Zoom In on this area until you get the accuracy desired, and examine the coordinates carefully. We see that the x-coordinate of the intersection is between 29.97 and 30.03, and the y-coordinate is between -9.98 and -10.01. Rounded to the nearest tenth, the intersection is $(30.0, -10.0)$. ◆

Finding an intersection by graphing can be difficult. In some cases, you have to know where the lines intersect to find a viewing window that contains the intersection.

Exercises

For each pair of lines, use TRACE and ZOOM to find the intersection accurate to two decimal places. Answers can be found at the end of this appendix.

1. $y = 2.36x - 34.53$
   $y = -1.83x + 130.45$

2. $4.1a + 5.3b = 16.2$
   $3.1a - 5.2b = 117.4$

For more practice, try the Graphing Calculator Exercises in Section 4.1.

In the next example, we use the graph of a quadratic function to help us factor a quadratic polynomial.

Example B (Section 4.2)   Factoring a quadratic polynomial

Factor $x^2 - 2x - 224$.

Solution

First graph the quadratic function $y_1 = x^2 - 2x - 224$. To see the x-intercepts we need a larger viewing window than the standard one.

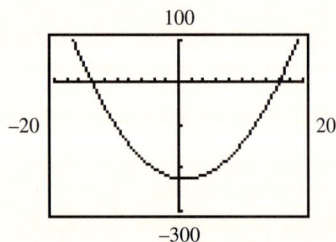

Using TRACE, it appears that the x-intercepts are $(-14, 0)$ and $(16, 0)$. If the polynomial is not prime, then these points are the exact intercepts and $x + 14$ and $x - 16$ are the factors. Since the product of $x + 14$ and $x - 16$ is $x^2 - 2x - 224$, we can be sure

that we have found the correct factors. So
$$x^2 - 2x - 224 = (x + 14)(x - 16).$$ ◆

Some calculators have a feature for graphing functions using only integers for the x-coordinates. In this case, the intercepts and the factors should be easier to find.

Exercises

Use the graph of a quadratic function to help you factor each quadratic polynomial.

3. $x^2 + 8x - 384$

4. $y^2 + 75y + 1386$

For more practice, try the Graphing Calculator Exercises in Section 4.2.

A graphing calculator does not do a very good job of graphing rational functions. However, with the proper interpretation, the graph shown on a calculator can be a big help in drawing a graph of a rational function. Even though the asymptotes are not part of the graph of a rational function, they are essential for drawing an accurate graph.

Example C (Section 4.6)   Graphing a rational function

Graph
$$y = \frac{x^2 + 1}{3x^2 - 1200}$$

and identify the asymptotes.

Solution

If $3x^2 - 1200 = 0$, then $x = \pm 20$. So the vertical asymptotes are the lines $x = \pm 20$. Divide the numerator by the denominator to see that $y = 1/3$ is a horizontal asymptote. If we graph
$$y_1 = \frac{x^2 + 1}{3x^2 - 1200} \text{ and } y_2 = \frac{1}{3}$$

in dot mode, we get the following graph.

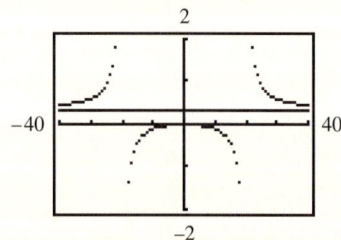

We could draw in the vertical asymptotes and draw the curve through the points shown so that it approaches its asymptotes. In connected mode, we get the following graph.

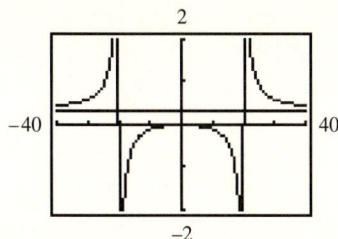

Note that in connected mode the calculator connects a high point on one side of a vertical asymptote with a low point on the other side, making a vertical line that is almost at the correct location of the vertical asymptotes. Again, we could use this graph as a guide for drawing a better graph showing the asymptotes exactly and the curve approaching those asymptotes.   ◆

Exercises

Use a graph to help you identify the horizontal and vertical asymptotes for each rational function.

5. $y = \dfrac{-3x - 2}{25x + 77}$ **6.** $f(x) = \dfrac{20x^3}{x^3 - 0.01x}$

For more practice, try the Graphing Calculator Exercises in Section 4.6.

Chapter 5

In Chapter 5 we are studying exponential and logarithmic functions. A graphing calculator is useful here for graphing functions as well as for evaluating exponential expressions and finding logarithms.

You can graph exponential functions with any base on your calculator, but most calculators have special keys for the bases 10 and e. In the next example, we solve an equation involving a base-e exponential expression.

Example A (Section 5.1)   Solving an exponential equation

Find all solutions to $e^{x-2} = x + 2$. Round the answers to the nearest tenth.

Solution

First define two functions using the Y= key.

The values for x at which the graphs of these functions intersect are the solutions to the equation. Graph the functions in the standard viewing window.

In the standard viewing window, it appears that the functions have two points of intersection. Zoom in on these two points to get the following graphs.

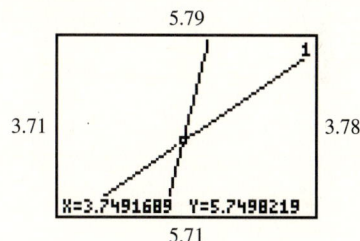

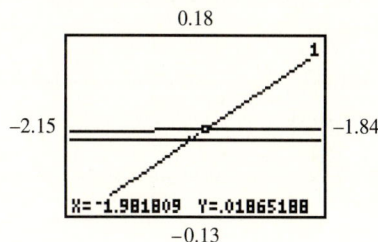

From these graphs we see that the solutions are 3.7 and -2.0, rounded to the nearest tenth. Use your calculator to check these solutions.   ◆

Exercises

Use graphing to find all solutions to each equation. Round answers to one decimal place. Answers can be found at the end of this appendix.

1. $e^{3x-4} = x^2 - 0.1$ **2.** $e^{-x+1} + 1 = -e^{x-5} + 3$

For more practice try the Graphing Calculator Exercises in Section 5.1.

Your graphing calculator has keys labeled LN for the natural logarithm and LOG for the common logarithm. To find the logarithm of a single number, no parentheses are needed, as in log 2.

To find the logarithm of an expression such as $\log(x + 3)$, the parentheses must be used.

Example B (Section 5.2)   Evaluating logarithms

Evaluate each expression. Round answers to three decimal places.

(i) $\ln(6.1)$ (ii) $\log(9)$ (iii) $\ln\left(\dfrac{22}{9}\right)$

Solution

Enter each expression into your calculator as shown.

```
ln 6.1
           1.808288771
log 9
           .9542425094
ln (22/9)
           .893817876
```

So (i) $\ln(6.1) \approx 1.808$, (ii) $\log(9) \approx 0.954$, and (iii) $\ln(22/9) \approx 0.894$.   ◆

Exercises

Evaluate each expression. Round answers to three decimal places.

3. $\ln(5.4) + \ln(3.9)$ **4.** $\log\left(\dfrac{44}{13}\right)$

5. $\log(44) - \log(13)$

For more practice, try Exercises 93–108 in Section 5.2.

Just as we used graphing to solve an exponential equation, we can use graphing to solve a logarithmic equation.

Example C (Section 5.2)   Solving a logarithmic equation

Find the solution to $\log(x - 2) = 4.8$ to the nearest tenth.

Solution

First graph $y = \log(x - 2) - 4.8$. The solution to the equation is the x-coordinate of the x-intercept. We need a wide view to see the x-intercept.

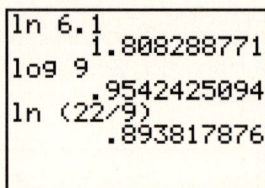

Get the cursor as close as possible to the x-intercept and zoom in until you get the desired accuracy.

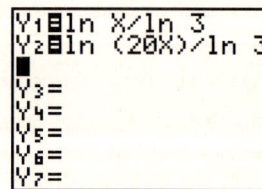

To the nearest tenth, x is 63,097.7.   ◆

Exercises

Use graphing to solve each equation.

6. $\log(x - 3) = 1.446$ **7.** $\ln(x + 3) = x - 5$

For more practice, try the Graphing Calculator Exercises in Section 5.2.

To find logarithms with bases other than 10 or e with a calculator, we use the base-change formula. For example, to find $\log_3(5)$, we calculate $\ln(5)/\ln(3)$ on a calculator. The base-change formula is used also in graphing logarithmic functions with bases other than 10 or e.

Example D (Section 5.3)   Graphing logarithmic functions

Graph the functions $y = \log_3(x)$ and $y = \log_3(20x)$ on the same display and interpret your result.

Solution

Using the base-change formula, we enter the functions as follows:

```
Y₁=ln X/ln 3
Y₂=ln (20X)/ln 3
Y₃=
Y₄=
Y₅=
Y₆=
Y₇=
```

Now graph the functions in the standard viewing window.

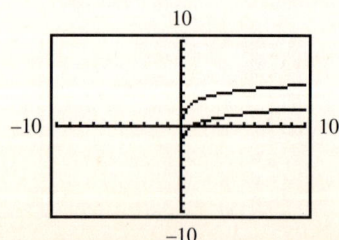

The graph of y_2 appears to be a vertical translation of the graph of y_1. We can verify that this conclusion is correct by writing

$$y_2 = \log_3(20x) = \log_3(20) + \log_3(x).$$

Since $\log_3(20)$ is a constant, y_2 is obtained from y_1 by translating $y = \log_3(x)$ upward a distance of $\log_3(20)$ or about 2.7 units. ◆

Exercises

Graph each pair of functions on the same display and interpret your result.

8. $y = \log_7\left(\dfrac{x}{49}\right)$, $y = \log_7(x) - 2$

9. $y = \log_6(5^x)$, $y = \log_6\left(\dfrac{5^x}{6}\right)$

For more practice, try the Graphing Calculator Exercises in Section 5.3.

Chapter 6

In Chapter 6 we begin our study of trigonometry with the measurement of angles. Angles are measured in degrees or radians. You can set your calculator for degree mode or radian mode by using the MODE key. When evaluating the trigonometric functions, the calculator assumes that numbers are degrees or radians depending on the mode you have selected. If you affix a degree symbol or an r (for radians) to a number, the calculator will treat the number properly, regardless of the mode setting.

Example A (Section 6.1)   Converting angle measurements

Convert 99° to radian measure and $\pi/7$ radians to degree measure. Round answers to four decimal places.

Solution

Press the MODE key and make sure the mode is set to radian.

```
Normal Sci Eng
Float 0123456789
Radian Degree
Func Par Pol Seq
Connected Dot
Sequential Simul
FullScreen Split
```

Display 99° and press ENTER.

```
99°
       1.727875959
```

So 99° is equivalent to 1.7279 radians. Now change the mode to degree, display $\pi/7$, and press ENTER.

```
(π/7)ʳ
       25.71428571
■
```

So $\pi/7$ radians is equivalent to 25.7143°. Note that $\pi/7$ must be in parentheses so that π is divided by 7 before the conversion to degrees. ◆

Exercises

Convert the given radian measure to degrees and the given degree measure to radians. Round your answers to four decimal places. Answers can be found at the end of this appendix.

1. $\dfrac{\pi}{12} + \dfrac{\pi}{6}$ **2.** 476°

For more practice, try Exercises 71–82 in Section 6.1.

A graphing calculator can be used to evaluate the sine and cosine functions. The calculator assumes that the independent variable is an angle in degrees or radians according to what is specified in the mode setting. To override the mode setting, you can affix the degree or radian symbol to the independent variable.

Example B (Section 6.2)   Evaluating sine and cosine

Evaluate each expression.

(i) $\sin\left(\dfrac{3\pi}{2}\right)$ (ii) $\dfrac{1 - \cos^2(33.2°)}{2\sin(33.2°)}$

Solution

To evaluate both expressions without changing mode we use the radian symbol and the degree symbol.

```
sin (3π/2)ʳ
                -1
(1-(cos 33.2°)²)
/(2sin 33.2°)
         .2737816117
```
◆

Exercises

Evaluate each expression. Round your answers to four decimal places.

3. $2 \sin\left(\dfrac{\pi}{12}\right) \cos\left(\dfrac{\pi}{12}\right)$ **4.** $\dfrac{1 - \cos(22.8°)}{1 + \sin(22.8°)}$

For more practice, try Exercises 51–94 and the Graphing Calculator Exercises in Section 6.2.

You can determine the amplitude, period, and phase shift from the equation of a trigonometric function. These characteristics of the function will help you to pick a viewing window that shows a good graph of the function.

Example C (Section 6.3)   Graphing a sine wave

Determine the amplitude, period, and phase shift for $y = 6 \sin(500x)$. Find a viewing window that shows approximately one cycle of the curve.

Solution

From the equation, we see that the amplitude is 6, the period is $2\pi/500$, and the phase shift is zero. Because $2\pi/500 \approx 0.01$, graphing the function for $0 \leq x \leq 0.02$ and $-6 \leq y \leq 6$ will show a little more than one cycle.

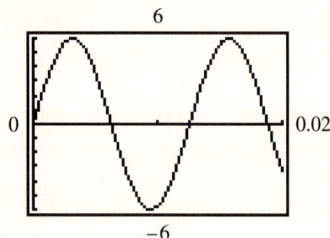

Consider the graph of $y = 6 \sin(500x)$ in the standard viewing window.

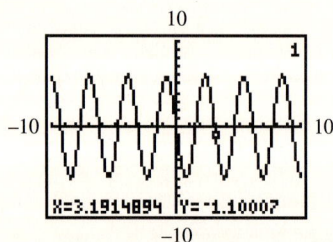

Using TRACE on this graph, we might conclude that the period is about 3.19 or perhaps π. The amplitude looks like it is 6, but the graph appears to go downward rather than upward in its first cycle to the right of the origin. So what went wrong? The ordered pairs that are graphed here satisfy the equation. However, because

of the small period, the curve does a lot of oscillating between these ordered pairs. Because this oscillation does not appear in the standard viewing window, we get a very misleading graph for the function.   ◆

Exercises

Determine the amplitude, period, and phase shift, and find a viewing window that shows approximately one cycle of the function.

5. $y = 12 + 3 \sin(600x - 100)$

6. $y = -4 - 2 \cos(0.02x - 5)$

For more practice, try Exercises 45–68 and the Graphing Calculator Exercises in Section 6.3.

As discussed in Section 6.4, the graphs of the tangent, cotangent, secant, and cosecant functions are periodic and have vertical asymptotes like rational functions. If you graph one of these functions in connected mode, your calculator will show a vertical line almost at the location of the vertical asymptote. Remember that these vertical lines are not really part of the graph of the function. When graphing any trigonometric function you must first determine the period so that a good viewing window can be chosen.

Example D (Section 6.4)   Graphing a tangent function

Determine the period and phase shift for
$$y = 0.03 \tan(30x - 5).$$
Choose a viewing window that shows one cycle of the graph.

Solution

Rewrite the function as
$$y = 0.03 \tan\left(30\left(x - \frac{1}{6}\right)\right)$$
The period of the function is $\pi/30$ and the phase shift is 1/6. So the vertical asymptote of $y = \tan x$ at $\pi/2$ is moved to $x = \pi/60 + 1/6$ or $x = 0.219$. The vertical asymptote at $-\pi/2$ is moved to $-\pi/60 + 1/6$ or $x = 0.114$. So one cycle of the graph should appear for $0.114 \leq x \leq 0.219$. Because of the factor 0.03 preceding the tangent function, we choose $-0.4 \leq y \leq 0.4$.

Exercises

Find the period and phase shift for each function. Find a viewing window that shows one cycle of the graph.

7. $y = 55 \tan(40x - 9)$ **8.** $y = 100 \cot(0.03x + 20)$

9. $y = 0.35 \sec(0.01x - 0.05)$

10. $y = 0.02 \csc(500x + 0.06)$

For more practice, try Exercises 59–90 and the Graphing Calculator Exercises in Section 6.4.

Your calculator has keys for the inverse sine, cosine, and tangent functions. To evaluate the inverse cosecant, secant, or cotangent, you need to use the identities given in Section 6.5.

Example E (Section 6.5)   Evaluating inverse functions

Evaluate each inverse function. Give the answer in radians and round to four decimal places.

(i) $\sin^{-1}(0.673)$ (ii) $\csc^{-1}(2.33)$ (iii) $\cot^{-1}(-4.1)$

Solution

For the inverse sine, no identity is needed. For the other two functions, we use

$$\csc^{-1}(2.33) = \sin^{-1}\left(\frac{1}{2.33}\right)$$

and

$$\cot^{-1}(-4.1) = \tan^{-1}\left(\frac{1}{-4.1}\right) + \pi.$$

So

(i) $\sin^{-1}(0.673) \approx 0.7383$,   (ii) $\csc^{-1}(2.33) \approx 0.4436$,

and

(iii) $\cot^{-1}(-4.1) \approx 2.9024$.   ◆

Exercises

Evaluate each inverse function. Give the answer in degrees, rounded to three decimal places.

11. $\cos^{-1}(-0.815)$ **12.** $\sec^{-1}(55.31)$ **13.** $\cot^{-1}(-12.4)$

For more practice, try Exercises 33–48, 65–84, and the Graphing Calculator Exercises in Section 6.5.

Chapter 7

A trigonometric identity such as

$$\sin^2 x + \cos^2 x = 1$$

is satisfied by all real numbers. A calculator graph of the function

$$y = \sin^2 x + \cos^2 x$$

looks like the horizontal line $y = 1$. So the graph of $y = \sin^2 x + \cos^2 x$ supports our belief that the equation is an identity. Because a graph shows only a finite number of points and even these points are approximations, an identity cannot be proved correct by graphing. However, if the graphs of both sides of an equation fail to coincide for at least one point at which both sides are defined, then the equation is certainly not an identity.

Example A (Section 7.1)   Disproving an identity

Show that the equation

$$\sin^4 x - \cos^4 x = \cos^2 x - \sin^2 x$$

is not an identity.

Solution

Graph $y_1 = \sin^4 x - \cos^4 x$ and $y_2 = \cos^2 x - \sin^2 x$ as shown here.

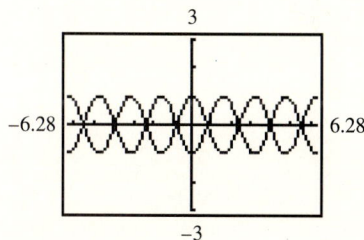

It appears that the functions generally do not coincide. If $x = \pi/2$, we get the following function values:

Because the equation is not satisfied if $x = \pi/2$, it is not an identity.   ◆

Exercises

Show that each equation is not an identity.

1. $2.01 \sin(x) \cos(x) = 1.99 \sin(2x)$

2. $\tan\left(\dfrac{x}{1.999}\right) = \dfrac{\sin(1.001x)}{1 + \cos(x/0.999)}$

For more practice, try Exercises 45–54 in Section 7.1 and Exercises 65–70 in Section 7.2.

If an equation is not an identity, then we are usually interested in solving the equation. By examining the graphs of both sides of the equation, we can see that every point at which the graphs intersect provides a solution to the equation.

Example B (Section 7.6)   Solving a conditional equation

Find all solutions to

$$\sin(3x) = \frac{1}{\sqrt{2}}$$

in the interval $(0, 2\pi)$ and use a graph to verify your answer.

Solution

First solve the equation:

$$\sin(3x) = \frac{1}{\sqrt{2}}$$

$$3x = \frac{\pi}{4} + 2k\pi \quad \text{or} \quad 3x = \frac{3\pi}{4} + 2k\pi$$

$$x = \frac{\pi}{12} + \frac{2k\pi}{3} \quad \text{or} \quad x = \frac{\pi}{4} + \frac{2k\pi}{3}$$

In the interval $(0, 2\pi)$, the solutions are $\pi/12$, $\pi/4$, $3\pi/4$, $11\pi/12$, $17\pi/12$, and $19\pi/12$. Now graph $y_1 = \sin(3x)$ and $y_2 = 1/\sqrt{2}$.

The graphs of y_1 and y_2 have six points of intersection in the interval $(0, 2\pi)$. So the graphs support our solution to the equation. You could also convert the exact solutions to decimal approximations and compare to the points of intersection.   ◆

Exercises   Solving equations

Find the exact solutions to each equation in the interval $[0, \pi]$. Use graphs to verify your answers.

3. $\cos(3x) = -\dfrac{1}{\sqrt{2}}$ **4.** $\tan(8x) = \sin(8x)$

For more practice, try Exercises 41–48, 57–74, and 97–104 in Section 7.6.

Chapter 8

In Chapter 8 we study applications of trigonometry in areas such as vectors, complex numbers, and polar coordinates. In Section 8.6 on polar coordinates, you can use your calculator for converting from polar to rectangular coordinates and vice versa.

Example A (Section 8.6)   Polar-rectangular conversions

Convert

(i) $(8, 120°)$ to rectangular coordinates.
(ii) $(-5, 12)$ to polar coordinates.

Solution

(i) We could perform this conversion in the usual way and simply use the calculator for the computational part. Instead, we will use the built-in conversion feature. Locate the polar-rectangular conversion in the ANGLE menu and display the following information.

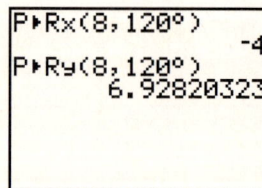

Note that the calculator returns the x- and y-coordinates separately. So the ordered pair $(8, 120°)$ is approximately $(-4, 6.9252)$ in rectangular coordinates.

(ii) To convert $(-5, 12)$ to polar coordinates, use the rectangular-to-polar conversion as follows.

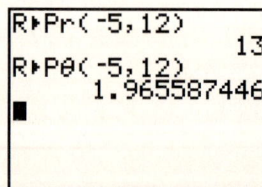

Note that the calculator returns the values of r and θ separately. The value of θ will be in radians or degrees, depending on the mode setting. So $(-5, 12)$ in rectangular coordinates is approximately (13, 1.9656) in polar coordinates with θ in radians, or (13, 112.6199°) with θ in degrees. ◆

Exercises

Perform each conversion using radian measure for the angles.

1. (3.1, − 5.4) to polar coordinates.

2. $\left(6.22, \dfrac{4\pi}{9}\right)$ to rectangular coordinates.

For more practice, try Exercises 17–36 in Section 8.6.

Graphing polar equations is usually difficult when plotting points by hand. With a graphing calculator you can graph polar equations that would not be attempted by hand.

Example B (Section 8.6)   Graphing a polar equation

Graph $r = \theta + 3 \sin(4\theta)$ for $\theta \ge 0$.

Solution

First set the mode of the calculator for graphing polar equations with θ measured in radians.

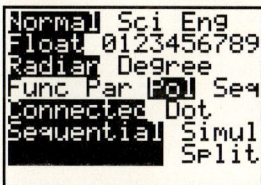

Use the Y= key to enter the function as r_1.

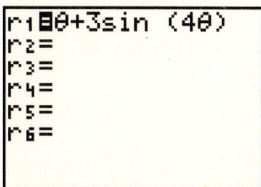

Because the function is not periodic, we let θ range from 0 to 50 with steps of 0.1 and set the viewing window for $-30 \le x \le 30$ and $-30 \le y \le 30$.

Exercises

Graph each polar equation for $\theta \ge 0$.

3. $r = \theta + 10 \sin(4\theta)$ **4.** $r = 10 \tan(9\theta)$

5. $r = \theta \sin(3\theta)$ **6.** $r = \theta \cos(6\theta)$

For more practice, try Exercises 37–56 and the Graphing Calculator Exercises in Section 8.6.

Chapter 9

Although most of the techniques for solving systems of equations and inequalities in this chapter are algebraic, we can increase our understanding of systems by examining their graphs. In Chapter 10 we will study methods that can be more easily performed with a calculator or computer.

The solution to a system of equations consists of all intersection points of the graphs of the equations. For some systems, it is possible to find those points by examining the graphs.

Example A (Section 9.1)   Solving a system by graphing

Solve the system.

$$y = \sqrt{2}x - 3\pi$$
$$y = -\sqrt{3}x + e^2$$

Round answers to the nearest tenth.

Solution

Graph the two lines as shown here:

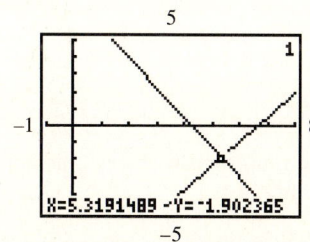

You can find the intersection of the lines by using the TRACE and ZOOM features or the CALC feature. The solution to the system is (5.3, − 1.9). ◆

For some systems it is not easy to find a viewing window that shows the intersection without knowing the intersection to begin with.

Example B (Section 9.1)   Solving a system by graphing

Solve the system.

$$y = 8.987x - 500$$
$$y = 9.001x - 400$$

Solution

Graph the equations as follows:

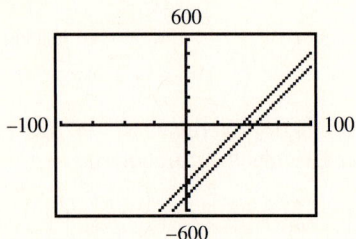

Because the lines are almost parallel, it is difficult to tell where they intersect without solving the system first. So we use substitution:

$$9.001x - 400 = 8.987x - 500$$
$$0.014 = -100$$
$$x \approx -7,142.857$$

Now find y:

$$y = 9.001(-7142.857) - 400$$
$$\approx -64,692.856$$

The solution to the system is

$$(-7,142.857, -64,692.856).$$

Even when you know the solution to the system it might not be easy to find a viewing window that shows the two lines crossing at this point. Try it. ◆

Exercises

Solve each system by graphing, if possible.

1. $y = 6^{2/3}x - 8\pi$
   $y = -3^{3/4}x + e^3$

2. $y = \sqrt{13}x + 452$
   $y = 3.606x - 1$

For more practice, try the Graphing Calculator Exercises in Section 9.1.

In Example 5 of Section 9.2 we solved a system of equations to find the equation of a parabola that passed through three given ordered pairs. This problem can be solved with a graphing calculator using a technique called regression. Using regression, your calculator can find a curve of a specified type that passes as close as possible to a given set of ordered pairs. If there are not too many ordered pairs, then the curve found by the calculator might actually pass through all of them as was the case in Example 5.

Example C (Section 9.2)   Curve fitting

Use regression to find the equation of a line through $(-3, -2)$ and $(2, 1)$.

Solution

Enter the ordered pairs in lists L_1 and L_2 as shown here.

Choose linear regression from the STAT CALC menu:

Press ENTER to perform the calculations.

The equation of the line through the two given points is $y = 0.6x - 0.2$. The value of r, the correlation coefficient, indicates how well the line fits the given ordered pairs. The value $r = 1$ indicates that the line goes exactly through the two given points. If the line has negative slope and goes exactly through the given points, then $r = -1$. ◆

Because any two points determine a line, the calculator will usually be able to find the line using regression. It is not true that any three points determine a parabola, but if you start with three points that lie on a parabola, the calculator can find the parabola using quadratic regression.

Example D (Section 9.2)   Curve fitting

Use quadratic regression to find the equation of the parabola that passes through $(-1, 1)$, $(1, 1)$, and $(2, 7)$.

Solution

Enter the ordered pairs in the lists L_1 and L_2 as shown here:

```
 L1    L2    L3
 -1     1   ▮▮▮▮▮▮
  1     1
  2     7

L3(1)=
```

Choose quadratic regression from the STAT CALC menu.

```
EDIT CALC
1:1-Var Stats
2:2-Var Stats
3:SetUp…
4:Med-Med
5:LinReg(ax+b)
6▮QuadReg
7↓CubicReg
```

Press ENTER to perform the calculations.

```
QuadReg
 y=ax²+bx+c
 a=2
 b=-1E-13
 c=-1
```

The equation of the form $y = ax^2 + bx + c$ that passes through the three given points is $y = 2x^2 - 1$. Check that all three points satisfy this equation. Note that the calculator gave a very small value for b due to round-off errors.   ◆

Exercises

Use regression on your calculator to fit a curve of the given type to the given points.

3. $y = ax + b$; $(2, -3)$, $(5, 9)$

4. $y = ax^2 + bx + c$; $(0, -1)$, $(1, 1)$, $(2, 5)$

5. $y = ax^b$; $(1, 2)$, $(9, 6)$, $(16, 8)$

For more practice, try Exercises 25–32 in Section 9.3.

Chapter 10

The methods presented in this chapter for solving systems of linear equations are commonly used on calculators and computers. Graphing calculators can store matrices and perform the operations that are used in solving systems. Your graphing calculator can also perform row operations, find inverse matrices, and evaluate determinants. Before we solve systems with a calculator, we see how to define matrices and perform operations with them.

Example A (Section 10.2)   Matrix operations

Let $A = \begin{bmatrix} 1 & -2 \\ 3 & 5 \end{bmatrix}$ and $B = \begin{bmatrix} 7 & 6 \\ -8 & 2 \end{bmatrix}$. Find the following matrices.

(i) $A + B$ (ii) AB (iii) A^{-1}

Solution

In the MATRX menu, choose EDIT matrix A.

```
NAMES MATH EDIT
1▮[A]   3×3
2:[B]   2×2
3:[C]   2×2
4:[D]   2×2
5:[E]
```

Change the dimension of matrix A to 2×2 and enter the elements of A.

```
MATRIX[A]  2 ×2▮
[ 1      -2     ]
[ 3       5     ]
```

From the MATRX menu, choose EDIT matrix B and enter the elements of B.

```
MATRIX[B]  2 ×2
[ 7      6      ]
[ -8     2      ]
```

Enter the sum $A + B$ using NAMES in the MATRX menu. Then press ENTER to find the sum. The product AB is found similarly.

```
[A]+[B]
          [[8   4]
           [-5  7]]
[A][B]
          [[23   2 ]
           [-19 28]]
```

The calculator has found that

$$A + B = \begin{bmatrix} 8 & 4 \\ -5 & 7 \end{bmatrix}$$

and

$$AB = \begin{bmatrix} 23 & 2 \\ -19 & 28 \end{bmatrix}.$$

To find A^{-1} use MATRX NAMES to enter A and the x^{-1} key to enter the -1.

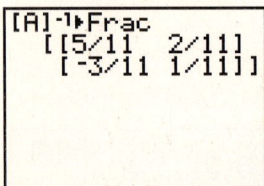

```
[A]-1►Frac
  [[5/11   2/11]
   [-3/11  1/11]]
```

By using the fraction feature, we obtained the exact inverse matrix:

$$A^{-1} = \begin{bmatrix} 5/11 & 2/11 \\ -3/11 & 1/11 \end{bmatrix}. \qquad \blacklozenge$$

Exercises

Let $A = \begin{bmatrix} 4 & 3 \\ -6 & -5 \end{bmatrix}$ and $B = \begin{bmatrix} 1 & 3 \\ 4 & 9 \end{bmatrix}$. Find the following matrices.

1. $A - B$ **2.** $2AB$ **3.** A^{-1}

For more practice, try Exercises 25–40 in Section 10.2, Exercises 27–52 in Section 10.3, and Exercises 19–34 in Section 10.4.

If a system of n linear equations in n unknowns has a unique solution, then that solution can be found using the inverse of the matrix of coefficients. This is probably the simplest method available for use with a calculator.

Example B (Section 10.4) Solving a system with an inverse matrix

Solve the system.

$$x - 2y = 22$$
$$3x + 5y = 44$$

Solution

The matrix of coefficients is the same as matrix A from Example A. Since A is already in the calculator, simply enter the 2×1 matrix $B = \begin{bmatrix} 22 \\ 44 \end{bmatrix}$ using MATRX EDIT:

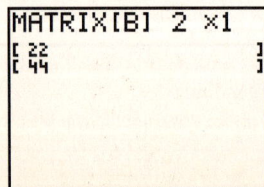

```
MATRIX[B] 2 ×1
[ 22            ]
[ 44            ]
```

Now find the product $A^{-1}B$:

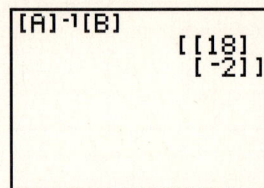

```
[A]-1[B]
            [[18]
             [-2]]
```

Check that $(18, -2)$ satisfies both of the original equations. The solution to the system is $(18, -2)$. $\blacklozenge$

Exercises

Solve each system by using an inverse matrix.

4. $1.2x - 3.4y = -14.62$
    $2.5x + 4.7y = 41.42$

5. $3.2a - 2.4b = 0$
    $5.6a - 3.6b = \dfrac{1}{10}$

For more practice, try Exercises 35–52 and the Graphing Calculator Exercises in Section 10.4.

A calculator can easily find determinants and perform the computations required by Cramer's rule.

Example C (Section 10.5) Solving a system with Cramer's rule

Solve the system.

$$x - 2y = 22$$
$$3x + 5y = 44$$

Solution

First enter the matrices A, B, and C:

$$A = \begin{bmatrix} 1 & -2 \\ 3 & 5 \end{bmatrix} \quad B = \begin{bmatrix} 22 & -2 \\ 44 & 5 \end{bmatrix}$$
$$C = \begin{bmatrix} 1 & 22 \\ 3 & 44 \end{bmatrix}$$

According to Cramer's rule, we get the value of x by dividing the determinant of B by the determinant of A, and the value of y by dividing the determinant of C by the determinant of A. The determinant feature is found in the MATRX MATH menu.

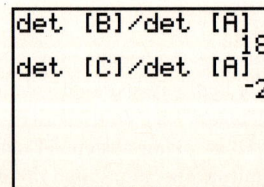

```
det [B]/det [A]
                18
det [C]/det [A]
                -2
```

The solution to the system is $(18, -2)$, as it was in the last example. $\blacklozenge$

Exercises

Use Cramer's rule to solve each system.

6. $\dfrac{1}{2}x - \dfrac{1}{4}y = \dfrac{1}{24}$

$\dfrac{1}{3}x + \dfrac{1}{6}y = \dfrac{5}{36}$

7. $1.8x + 3.9y = 306$

$4.2x - 7.1y = -258$

For more practice, try Exercises 13–32 in Section 10.5 and the Graphing Calculator Exercises in Section 10.6.

If a system has n equations in n unknowns you must input $n + 1$ matrices for Cramer's rule but only two matrices for the inverse matrix method. So it is usually faster to use the inverse matrix method. Since the row operations of the Gauss-Jordan method are rather tedious to perform even with a calculator, it should be saved for the systems on which the other methods fail.

Chapter 11

A parabola opening up or down has an equation of the form $y = ax^2 + bx + c$. This equation is easy to graph because y is written as a function of x. To graph $x = ay^2 + by + c$, a parabola opening left or right, we solve the equation for y and graph two functions.

Example A (Section 11.1)   A parabola opening to the right

Graph the parabola $x = y^2 - 3y + 5$.

Solution

Write the equation as

$$y^2 - 3y + 5 - x = 0.$$

To solve for y, use the quadratic formula with $a = 1$, $b = -3$, and $c = 5 - x$:

$$y = \frac{3 \pm \sqrt{9 - 4(1)(5 - x)}}{2(1)}$$

$$= \frac{3 \pm \sqrt{4x - 11}}{2}$$

Now define y_1 and y_2:

Graph the two functions:

The axis of symmetry is $y = 3/2$. The graphs y_1 and y_2 form the parabola opening to the right.   ◆

Exercises

Graph each parabola.

1. $x = 2y^2 - 6y + 9$

2. $x = -0.5y^2 + 8y - 7$

For more practice, try Exercises 45–56 and the Graphing Calculator Exercises in Section 11.1.

To graph any circle, ellipse, or hyperbola you must solve the equation for y and graph two functions. For hyperbolas you can also graph the asymptotes and see how the curve approaches its asymptotes.

Example B (Section 11.3)   Graphing hyperbolas

Graph $\dfrac{x^2}{32} - \dfrac{y^2}{48} = 1$ and its asymptotes.

Solution

First solve the equation for y:

$$\frac{y^2}{48} = \frac{x^2}{32} - 1$$

$$y^2 = 1.5x^2 - 48$$

$$y = \pm\sqrt{1.5x^2 - 48}$$

The equations of the asymptotes are

$$y = \pm\sqrt{1.5}x.$$

Graph the hyperbola and its asymptotes as shown here:

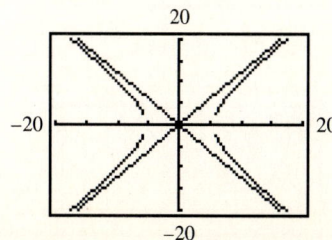

The hyperbola should not have a break at the x-axis as occurs on the calculator graph. This break appears because the calculator did not select the vertex as one of the points to plot. ◆

Exercises

Graph each hyperbola and its asymptotes.

3. $\dfrac{x^2}{88} - \dfrac{y^2}{20} = 1$ **4.** $\dfrac{y^2}{77} - \dfrac{x^2}{15} = 1$

For more practice, try Exercises 1–20 and the Graphing Calculator Exercises in Section 11.3.

Chapter 12

Because sequences are functions, any graphing calculator can evaluate the terms of a sequence or graph a sequence. However, some calculators have special features for handling sequences.

Example A (Section 12.1)   Finding terms of a sequence

Find the first five terms of the sequence

$$a_n = \frac{(-1)^n}{(n-1)!}$$

Solution

First change the mode of the calculator to sequence mode:

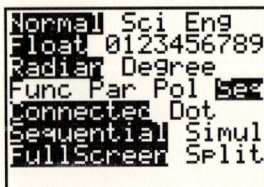

```
Normal Sci Eng
Float 0123456789
Radian Degree
Func Par Pol Seq
Connected Dot
Sequential Simul
FullScreen Split
```

Use the Y= key to define u_n:

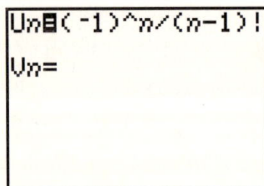

```
Un=(-1)^n/(n-1)!

Vn=
```

Because $a_1 = -1$, we set UnStart = -1 in the WINDOW menu:

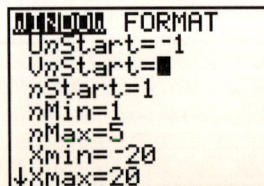

```
WINDOW FORMAT
 UnStart=-1
 VnStart=■
 nStart=1
 nMin=1
 nMax=5
 Xmin=-20
↓Xmax=20
```

Now ask for the values of u_n for $n = 1$ through 5 in increments of 1. Then press ENTER.

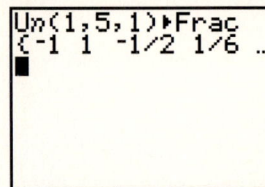

```
Un(1,5,1)▶Frac
{-1 1 -1/2 1/6 …
■
```

Move the cursor to the right to see the fifth term that does not fit on the screen. The first five terms are -1, 1, $-1/2$, 1/6, and $-1/24$. Another way to enter the formula for a sequence is to enclose the formula in quotes and use the STO key to store it as u_n:

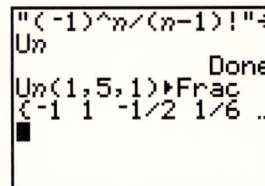

```
"(-1)^n/(n-1)!"→
Un
            Done
Un(1,5,1)▶Frac
{-1 1 -1/2 1/6 …
■
```

The sequence is evaluated in the same manner as before. ◆

Exercises

Find the first five terms of each sequence.

1. $a_n = \dfrac{n!}{2^n}$ **2.** $b_n = \dfrac{(-1)^{n-1}}{(2n-3)^2}$

For more practice, try Exercises 1–28 in Section 12.1.

Sequences can be defined recursively on some calculators.

Example B (Section 12.1)   Using a recursion formula

Find the first five terms of the sequence in which $a_1 = 3$ and $a_n = \dfrac{-a_{n-1}}{3}$ for $n \geq 2$.

Solution

Because $a_1 = 3$, set UnStart = 3 in the WINDOW menu.

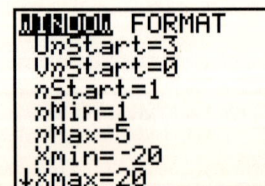

```
WINDOW FORMAT
 UnStart=3
 VnStart=0
 nStart=1
 nMin=1
 nMax=5
 Xmin=-20
↓Xmax=20
```

Use the Y= key to define $u_n = -u_{n-1}/3$.

```
Un8-Un-1/3
Vn=■
```

Now ask for the values of u_n for $n = 1$ through 5 in increments of 1. Then press ENTER.

```
Un(1,5,1)▶Frac
{3 -1 1/3 -1/9 …
■
```

The first five terms of the sequence are 3, − 1, 1/2, − 1/9, and 1/27.   ◆

Exercises

Find the first five terms of each sequence.

3. $a_1 = 6$ and $a_n = \dfrac{-a_{n-1}}{n!}$ for $n \geq 2$

4. $b_1 = \dfrac{1}{8}$ and $b_n = \dfrac{b_{n-1}}{2^{-n}}$

For more practice, try Exercises 29–34 in Section 12.1.

Answers to Appendix Exercises

Chapter 1

1. 1.0395   **2.** − 37.5727   **3.** − 76   **4.** 0.3448

5. 0.7854   **6.** 1.3771   **7.** 3.6123   **8.** 9.07×10^{29}

9. 1.68×10^{-10} **10.** 14.5143   **11.** 3.9842   **12.** 4.4679

13. 110.464   **14.** − 5451.496   **15.** 2.0002

16. 18,732.19

Chapter 2

1. 6.2579   **2.** 4.8623   **3.** 2.083, − 0.928

4. − 1.529, 0.468   **5.** $(-\infty, -6) \cup (4, \infty)$ **6.** $[-5, 6]$

7. $\left(-\infty, \dfrac{3 - \sqrt{3}}{2}\right] \cup \left[\dfrac{3 + \sqrt{3}}{2}, \infty\right)$

8. $(-\infty, -19.121) \cup (2.024, \infty)$

Chapter 3

1.

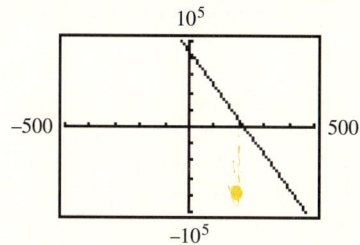

2.

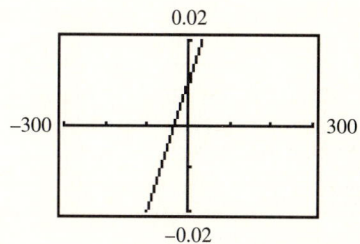

3. 0.69   **4.** 2.71   **5.** $(-\infty, 2 - \sqrt{2}] \cup [2 + \sqrt{2}, \infty)$

6. $(-12, 28)$ **7.** Yes   **8.** Yes

Chapter 4

1. (39.37, 58.39)   **2.** (18.71, − 11.42)   **3.** $(x + 24)(x - 16)$

4. $(y + 33)(y + 42)$ **5.** $x = -\frac{77}{25}, y = -\frac{3}{25}$

6. $x = 0, x = -0.1, x = 0.1, y = 20$

Chapter 5

1. − 0.3, 0.4, 1.7   **2.** 0.3, 5.7   **3.** 3.047   **4.** 0.530

5. 0.530   **6.** 30.925   **7.** 7.234, − 3.000

8. The functions are the same because of the quotient rule for logarithms.

9. The graph of the second function is a vertical translation of the first function, one unit downward.

Chapter 6

1. 45°   **2.** 8.3078   **3.** 0.5   **4.** 0.0563

5. Amplitude 3, period $\pi/300$, phase shift 1/6

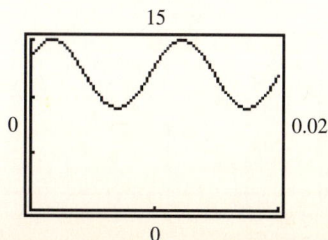

6. Amplitude 2, period 100π, phase shift 250

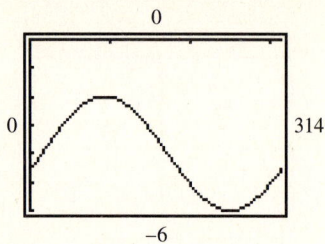

7. Period $\pi/40$, phase shift 9/40

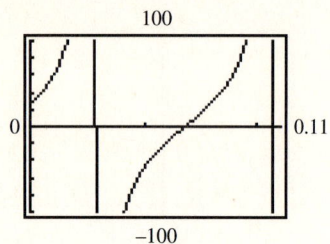

8. Period $100\pi/3$, phase shift 2000/3

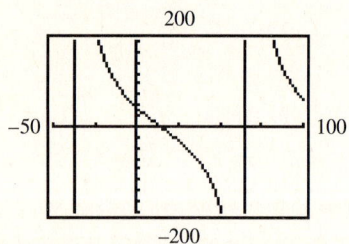

9. Period 200π, phase shift 5

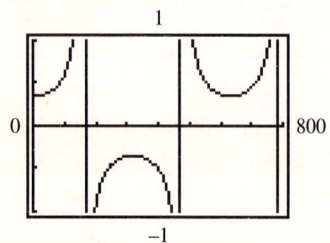

10. Period $\pi/250$, phase shift 1.2×10^{-4}

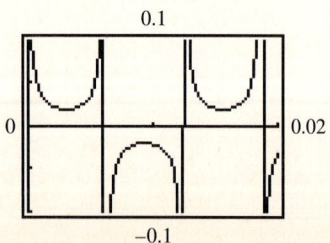

11. $144.587°$ **12.** $88.964°$ **13.** $\sim 4.611°$

Chapter 7

1. The equation is not satisfied if $x = 2$.

2. The equation is not satisfied if $x = 2$.

3. $\dfrac{\pi}{4}, \dfrac{5\pi}{12}, \dfrac{11\pi}{12}$

4. $0, \dfrac{\pi}{8}, \dfrac{\pi}{4}, \dfrac{3\pi}{8}, \dfrac{\pi}{2}, \dfrac{5\pi}{8}, \dfrac{3\pi}{4}, \dfrac{7\pi}{8}, \pi$

Chapter 8

1. $(6.2266, -1.0497)$ **2.** $(1.0801, 6.1255)$

3.

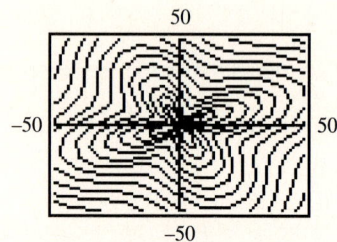

4.

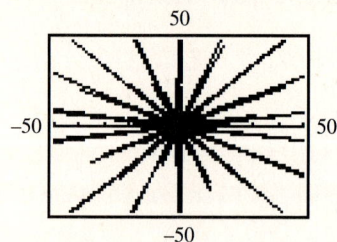

5.

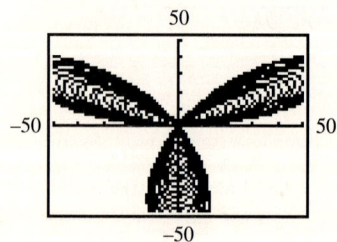

6.

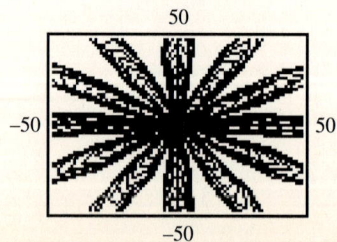

Chapter 9

1. (8.1016, 1.6180)

2. (1009528.037, 3640357.102)

3. $y = 4x - 11$ **4.** $y = x^2 + x - 1$

5. $y = 2\sqrt{x}$

Chapter 10

1. $\begin{bmatrix} 3 & 0 \\ -10 & -14 \end{bmatrix}$ **2.** $\begin{bmatrix} 32 & 78 \\ -52 & -126 \end{bmatrix}$

3. $\begin{bmatrix} 2.5 & 1.5 \\ -3 & -2 \end{bmatrix}$ **4.** (5.1, 6.1)   **5.** (1/8, 1/6)

6. (1/4, 1/3)   **7.** (40, 60)

Chapter 11

1.

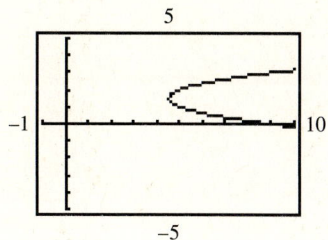

2.

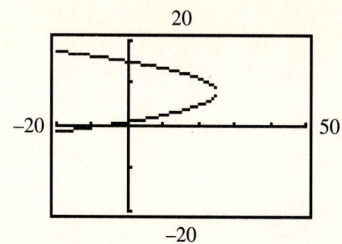

3.

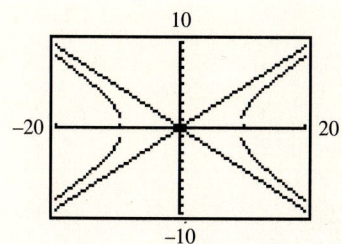

4.

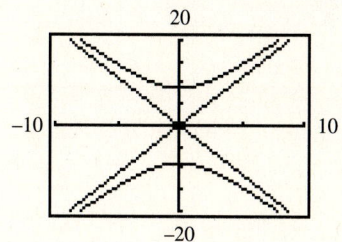

Chapter 12

1. $\dfrac{1}{2}, \dfrac{1}{2}, \dfrac{3}{4}, \dfrac{3}{2}, \dfrac{15}{4}$ **2.** $1, -1, \dfrac{1}{9}, \dfrac{-1}{25}, \dfrac{1}{49}$

3. $6, -3, \dfrac{1}{2}, \dfrac{-1}{48}, \dfrac{1}{5760}$ **4.** $\dfrac{1}{8}, \dfrac{1}{2}, 4, 64, 2048$

Appendix B

Tables

Exponential Functions

x	e^x	e^{-x}	x	e^x	e^{-x}
0.00	1.0000	1.0000	2.5	12.182	0.0821
0.05	1.0513	0.9512	2.6	13.464	0.0743
0.10	1.1052	0.9048	2.7	14.880	0.0672
0.15	1.1618	0.8607	2.8	16.445	0.0608
0.20	1.2214	0.8187	2.9	18.174	0.0550
0.25	1.2840	0.7788	3.0	20.086	0.0498
0.30	1.3499	0.7408	3.1	22.198	0.0450
0.35	1.4191	0.7047	3.2	24.533	0.0408
0.40	1.4918	0.6703	3.3	27.113	0.0369
0.45	1.5683	0.6376	3.4	29.964	0.0334
0.50	1.6487	0.6065	3.5	33.115	0.0302
0.55	1.7333	0.5769	3.6	36.598	0.0273
0.60	1.8221	0.5488	3.7	40.447	0.0247
0.65	1.9155	0.5220	3.8	44.701	0.0224
0.70	2.0138	0.4966	3.9	49.402	0.0202
0.75	2.1170	0.4724	4.0	54.598	0.0183
0.80	2.2255	0.4493	4.1	60.340	0.0166
0.85	2.3396	0.4274	4.2	66.686	0.0150
0.90	2.4596	0.4066	4.3	73.700	0.0136
0.95	2.5857	0.3867	4.4	81.451	0.0123
1.0	2.7183	0.3679	4.5	90.017	0.0111
1.1	3.0042	0.3329	4.6	99.484	0.0101
1.2	3.3201	0.3012	4.7	109.95	0.0091
1.3	3.6693	0.2725	4.8	121.51	0.0082
1.4	4.0552	0.2466	4.9	134.29	0.0074
1.5	4.4817	0.2231	5	148.41	0.0067
1.6	4.9530	0.2019	6	403.43	0.0025
1.7	5.4739	0.1827	7	1096.6	0.0009
1.8	6.0496	0.1653	8	2981.0	0.0003
1.9	6.6859	0.1496	9	8103.1	0.0001
2.0	7.3891	0.1353	10	22026	0.00005
2.1	8.1662	0.1225			
2.2	9.0250	0.1108			
2.3	9.9742	0.1003			
2.4	11.023	0.0907			

Natural Logarithms

x	$\log_e x$	x	$\log_e x$	x	$\log_e x$
0.0	*	4.5	1.5041	9.0	2.1972
0.1	7.6974	4.6	1.5261	9.1	2.2083
0.2	8.3906	4.7	1.5476	9.2	2.2192
0.3	8.7960	4.8	1.5686	9.3	2.2300
0.4	9.0837	4.9	1.5892	9.4	2.2407
0.5	9.3069	5.0	1.6094	9.5	2.2513
0.6	9.4892	5.1	1.6292	9.6	2.2618
0.7	9.6433	5.2	1.6487	9.7	2.2721
0.8	9.7769	5.3	1.6677	9.8	2.2824
0.9	9.8946	5.4	1.6864	9.9	2.2925
1.0	0.0000	5.5	1.7047	10	2.3026
1.1	0.0953	5.6	1.7228	11	2.3979
1.2	0.1823	5.7	1.7405	12	2.4849
1.3	0.2624	5.8	1.7579	13	2.5649
1.4	0.3365	5.9	1.7750	14	2.6391
1.5	0.4055	6.0	1.7918	15	2.7081
1.6	0.4700	6.1	1.8083	16	2.7726
1.7	0.5306	6.2	1.8245	17	2.8332
1.8	0.5878	6.3	1.8405	18	2.8904
1.9	0.6419	6.4	1.8563	19	2.9444
2.0	0.6931	6.5	1.8718	20	2.9957
2.1	0.7419	6.6	1.8871	25	3.2189
2.2	0.7885	6.7	1.9021	30	3.4012
2.3	0.8329	6.8	1.9169	35	3.5553
2.4	0.8755	6.9	1.9315	40	3.6889
2.5	0.9163	7.0	1.9459	45	3.8067
2.6	0.9555	7.1	1.9601	50	3.9120
2.7	0.9933	7.2	1.9741	55	4.0073
2.8	1.0296	7.3	1.9879	60	4.0943
2.9	1.0647	7.4	2.0015	65	4.1744
3.0	1.0986	7.5	2.0149	70	4.2485
3.1	1.1314	7.6	2.0281	75	4.3175
3.2	1.1632	7.7	2.0412	80	4.3820
3.3	1.1939	7.8	2.0541	85	4.4427
3.4	1.2238	7.9	2.0669	90	4.4998
3.5	1.2528	8.0	2.0794	95	4.5539
3.6	1.2809	8.1	2.0919	100	4.6052
3.7	1.3083	8.2	2.1041		
3.8	1.3350	8.3	2.1163		
3.9	1.3610	8.4	2.1282		
4.0	1.3863	8.5	2.1401		
4.1	1.4110	8.6	2.1518		
4.2	1.4351	8.7	2.1633		
4.3	1.4586	8.8	2.1748		
4.4	1.4816	8.9	2.1861		

*Subtract 10 from $\log_e x$ entries for $x < 1.0$.

Common Logarithms

N	0	1	2	3	4	5	6	7	8	9
1.0	.0000	.0043	.0086	.0128	.0170	.0212	.0253	.0294	.0334	.0374
1.1	.0414	.0453	.0492	.0531	.0569	.0607	.0645	.0682	.0719	.0755
1.2	.0792	.0828	.0864	.0899	.0934	.0969	.1004	.1038	.1072	.1106
1.3	.1139	.1173	.1206	.1239	.1271	.1303	.1335	.1367	.1399	.1430
1.4	.1461	.1492	.1523	.1553	.1584	.1614	.1644	.1673	.1703	.1732
1.5	.1761	.1790	.1818	.1847	.1875	.1903	.1931	.1959	.1987	.2014
1.6	.2041	.2068	.2095	.2122	.2148	.2175	.2201	.2227	.2253	.2279
1.7	.2304	.2330	.2355	.2380	.2405	.2430	.2455	.2480	.2504	.2529
1.8	.2553	.2577	.2601	.2625	.2648	.2672	.2695	.2718	.2742	.2765
1.9	.2788	.2810	.2833	.2856	.2878	.2900	.2923	.2945	.2967	.2989
2.0	.3010	.3032	.3054	.3075	.3096	.3118	.3139	.3160	.3181	.3201
2.1	.3222	.3243	.3263	.3284	.3304	.3324	.3345	.3365	.3385	.3404
2.2	.3424	.3444	.3464	.3483	.3502	.3522	.3541	.3560	.3579	.3598
2.3	.3617	.3636	.3655	.3674	.3692	.3711	.3729	.3747	.3766	.3784
2.4	.3802	.3820	.3838	.3856	.3874	.3892	.3909	.3927	.3945	.3962
2.5	.3979	.3997	.4014	.4031	.4048	.4065	.4082	.4099	.4116	.4133
2.6	.4150	.4166	.4183	.4200	.4216	.4232	.4249	.4265	.4281	.4298
2.7	.4314	.4330	.4346	.4362	.4378	.4393	.4409	.4425	.4440	.4456
2.8	.4472	.4487	.4502	.4518	.4533	.4548	.4564	.4579	.4594	.4609
2.9	.4624	.4639	.4654	.4669	.4683	.4698	.4713	.4728	.4742	.4757
3.0	.4771	.4786	.4800	.4814	.4829	.4843	.4857	.4871	.4886	.4900
3.1	.4914	.4928	.4942	.4955	.4969	.4983	.4997	.5011	.5024	.5038
3.2	.5051	.5065	.5079	.5092	.5105	.5119	.5132	.5145	.5159	.5172
3.3	.5185	.5198	.5211	.5224	.5237	.5250	.5263	.5276	.5289	.5302
3.4	.5315	.5328	.5340	.5353	.5366	.5378	.5391	.5403	.5416	.5428
3.5	.5441	.5453	.5465	.5478	.5490	.5502	.5514	.5527	.5539	.5551
3.6	.5563	.5575	.5587	.5599	.5611	.5623	.5635	.5647	.5658	.5670
3.7	.5682	.5694	.5705	.5717	.5729	.5740	.5752	.5763	.5775	.5786
3.8	.5798	.5809	.5821	.5832	.5843	.5855	.5866	.5877	.5888	.5899
3.9	.5911	.5922	.5933	.5944	.5955	.5966	.5977	.5988	.5999	.6010
4.0	.6021	.6031	.6042	.6053	.6064	.6075	.6085	.6096	.6107	.6117
4.1	.6128	.6138	.6149	.6160	.6170	.6180	.6191	.6201	.6212	.6222
4.2	.6232	.6243	.6253	.6263	.6274	.6284	.6294	.6304	.6314	.6325
4.3	.6335	.6345	.6355	.6365	.6375	.6385	.6395	.6405	.6415	.6425
4.4	.6435	.6444	.6454	.6464	.6474	.6484	.6493	.6503	.6513	.6522
4.5	.6532	.6542	.6551	.6561	.6571	.6580	.6590	.6599	.6609	.6618
4.6	.6628	.6637	.6646	.6656	.6665	.6675	.6684	.6693	.6702	.6712
4.7	.6721	.6730	.6739	.6749	.6758	.6767	.6776	.6785	.6794	.6803
4.8	.6812	.6821	.6830	.6839	.6848	.6857	.6866	.6875	.6884	.6893
4.9	.6902	.6911	.6920	.6928	.6937	.6946	.6955	.6964	.6972	.6981
5.0	.6990	.6998	.7007	.7016	.7024	.7033	.7042	.7050	.7059	.7067
5.1	.7076	.7084	.7093	.7101	.7110	.7118	.7126	.7135	.7143	.7152
5.2	.7160	.7168	.7177	.7185	.7193	.7202	.7210	.7218	.7226	.7235
5.3	.7243	.7251	.7259	.7267	.7275	.7284	.7292	.7300	.7308	.7316
5.4	.7324	.7332	.7340	.7348	.7356	.7364	.7372	.7380	.7388	.7396
N		1	2	3	4	5	6	7	8	9

Common Logarithms (continued)

N	0	1	2	3	4	5	6	7	8	9
5.5	.7404	.7412	.7419	.7427	.7435	.7443	.7451	.7459	.7466	.7474
5.6	.7482	.7490	.7497	.7505	.7513	.7520	.7528	.7536	.7543	.7551
5.7	.7559	.7566	.7574	.7582	.7589	.7597	.7604	.7612	.7619	.7627
5.8	.7634	.7642	.7649	.7657	.7664	.7672	.7679	.7686	.7694	.7701
5.9	.7709	.7716	.7723	.7731	.7738	.7745	.7752	.7760	.7767	.7774
6.0	.7782	.7789	.7796	.7803	.7810	.7818	.7825	.7832	.7839	.7846
6.1	.7853	.7860	.7868	.7875	.7882	.7889	.7896	.7903	.7910	.7917
6.2	.7924	.7931	.7938	.7945	.7952	.7959	.7966	.7973	.7980	.7987
6.3	.7993	.8000	.8007	.8014	.8021	.8028	.8035	.8041	.8048	.8055
6.4	.8062	.8069	.8075	.8082	.8089	.8096	.8102	.8109	.8116	.8122
6.5	.8129	.8136	.8142	.8149	.8156	.8162	.8169	.8176	.8182	.8189
6.6	.8195	.8202	.8209	.8215	.8222	.8228	.8235	.8241	.8248	.8254
6.7	.8261	.8267	.8274	.8280	.8287	.8293	.8299	.8306	.8312	.8319
6.8	.8325	.8331	.8338	.8344	.8351	.8357	.8363	.8370	.8376	.8382
6.9	.8388	.8395	.8401	.8407	.8414	.8420	.8426	.8432	.8439	.8445
7.0	.8451	.8457	.8463	.8470	.8476	.8482	.8488	.8494	.8500	.8506
7.1	.8513	.8519	.8525	.8531	.8537	.8543	.8549	.8555	.8561	.8567
7.2	.8573	.8579	.8585	.8591	.8597	.8603	.8609	.8615	.8621	.8627
7.3	.8633	.8639	.8645	.8651	.8657	.8663	.8669	.8675	.8681	.8686
7.4	.8692	.8698	.8704	.8710	.8716	.8722	.8727	.8733	.8739	.8745
7.5	.8751	.8756	.8762	.8768	.8774	.8779	.8785	.8791	.8797	.8802
7.6	.8808	.8814	.8820	.8825	.8831	.8837	.8842	.8848	.8854	.8859
7.7	.8865	.8871	.8876	.8882	.8887	.8893	.8899	.8904	.8910	.8915
7.8	.8921	.8927	.8932	.8938	.8943	.8949	.8954	.8960	.8965	.8971
7.9	.8976	.8982	.8987	.8993	.8998	.9004	.9009	.9015	.9020	.9025
8.0	.9031	.9036	.9042	.9047	.9053	.9058	.9063	.9069	.9074	.9079
8.1	.9085	.9090	.9096	.9101	.9106	.9112	.9117	.9122	.9128	.9133
8.2	.9138	.9143	.9149	.9154	.9159	.9165	.9170	.9175	.9180	.9186
8.3	.9191	.9196	.9201	.9206	.9212	.9217	.9222	.9227	.9232	.9238
8.4	.9243	.9248	.9253	.9258	.9263	.9269	.9274	.9279	.9284	.9289
8.5	.9294	.9299	.9304	.9309	.9315	.9320	.9325	.9330	.9335	.9340
8.6	.9345	.9350	.9355	.9360	.9365	.9370	.9375	.9380	.9385	.9390
8.7	.9395	.9400	.9405	.9410	.9415	.9420	.9425	.9430	.9435	.9440
8.8	.9445	.9450	.9455	.9460	.9465	.9469	.9474	.9479	.9484	.9489
8.9	.9494	.9499	.9504	.9509	.9513	.9518	.9523	.9528	.9533	.9538
9.0	.9542	.9547	.9552	.9557	.9562	.9566	.9571	.9576	.9581	.9586
9.1	.9590	.9595	.9600	.9605	.9609	.9614	.9619	.9624	.9628	.9633
9.2	.9638	.9643	.9647	.9652	.9657	.9661	.9666	.9671	.9675	.9680
9.3	.9685	.9689	.9694	.9699	.9703	.9708	.9713	.9717	.9722	.9727
9.4	.9731	.9736	.9741	.9745	.9750	.9754	.9759	.9763	.9768	.9773
9.5	.9777	.9782	.9786	.9791	.9795	.9800	.9805	.9809	.9814	.9818
9.6	.9823	.9827	.9832	.9836	.9841	.9845	.9850	.9854	.9859	.9863
9.7	.9868	.9872	.9877	.9881	.9886	.9890	.9894	.9899	.9903	.9908
9.8	.9912	.9917	.9921	.9926	.9930	.9934	.9939	.9943	.9948	.9952
9.9	.9956	.9961	.9965	.9969	.9974	.9978	.9983	.9987	.9991	.9996
N	0	1	2	3	4	5	6	7	8	9

Trigonometric Functions

Angle θ									
Degrees	**Radians**	**sin θ**	**csc θ**	**tan θ**	**cot θ**	**sec θ**	**cos θ**		
0° 00′	.0000	.0000	No value	.0000	No value	1.000	1.0000	1.5708	90° 00′
10	.0029	.0029	343.8	.0029	343.8	1.000	1.0000	1.5679	50
20	.0058	.0058	171.9	.0058	171.9	1.000	1.0000	1.5650	40
30	.0087	.0087	114.6	.0087	114.6	1.000	1.0000	1.5621	30
40	.0116	.0116	85.95	.0116	85.94	1.000	.9999	1.5592	20
50	.0145	.0145	68.76	.0145	68.75	1.000	.9999	1.5563	10
1° 00′	.0175	.0175	57.30	.0175	57.29	1.000	.9998	1.5533	89° 00′
10	.0204	.0204	49.11	.0204	49.10	1.000	.9998	1.5504	50
20	.0233	.0233	42.98	.0233	42.96	1.000	.9997	1.5475	40
30	.0262	.0262	38.20	.0262	38.19	1.000	.9997	1.5446	30
40	.0291	.0291	34.38	.0291	34.37	1.000	.9996	1.5417	20
50	.0320	.0320	31.26	.0320	31.24	1.001	.9995	1.5388	10
2° 00′	.0349	.0349	28.65	.0349	28.64	1.001	.9994	1.5359	88° 00′
10	.0378	.0378	26.45	.0378	26.43	1.001	.9993	1.5330	50
20	.0407	.0407	24.56	.0407	24.54	1.001	.9992	1.5301	40
30	.0436	.0436	22.93	.0437	22.90	1.001	.9990	1.5272	30
40	.0465	.0465	21.49	.0466	21.47	1.001	.9989	1.5243	20
50	.0495	.0494	20.23	.0495	20.21	1.001	.9988	1.5213	10
3° 00′	.0524	.0523	19.11	.0524	19.08	1.001	.9986	1.5184	87° 00′
10	.0553	.0552	18.10	.0553	18.07	1.002	.9985	1.5155	50
20	.0582	.0581	17.20	.0582	17.17	1.002	.9983	1.5126	40
30	.0611	.0610	16.38	.0612	16.35	1.002	.9981	1.5097	30
40	.0640	.0640	15.64	.0641	15.60	1.002	.9980	1.5068	20
50	.0669	.0669	14.96	.0670	14.92	1.002	.9978	1.5039	10
4° 00′	.0698	.0698	14.34	.0699	14.30	1.002	.9976	1.5010	86° 00′
10	.0727	.0727	13.76	.0729	13.73	1.003	.9974	1.5981	50
20	.0756	.0756	13.23	.0758	13.20	1.003	.9971	1.5952	40
30	.0785	.0785	12.75	.0787	12.71	1.003	.9969	1.5923	30
40	.0814	.0814	12.29	.0816	12.25	1.003	.9967	1.5893	20
50	.0844	.0843	11.87	.0846	11.83	1.004	.9964	1.5864	10
5° 00′	.0873	.0872	11.47	.0875	11.43	1.004	.9962	1.4835	85° 00′
10	.0902	.0901	11.10	.0904	11.06	1.004	.9959	1.4806	50
20	.0931	.0929	10.76	.0934	10.71	1.004	.9957	1.4777	40
30	.0960	.0958	10.43	.0963	10.39	1.005	.9954	1.4748	30
40	.0989	.0987	10.13	.0992	10.08	1.005	.9951	1.4719	20
50	.1018	.1016	9.839	.1022	9.788	1.005	.9948	1.4690	10
6° 00′	.1047	.1045	9.567	.1051	9.514	1.006	.9945	1.4661	84° 00′
10	.1076	.1074	9.309	.1080	9.255	1.006	.9942	1.4632	50
20	.1105	.1103	9.065	.1110	9.010	1.006	.9939	1.4603	40
30	.1134	.1132	8.834	.1139	8.777	1.006	.9936	1.4573	30
40	.1164	.1161	8.614	.1169	8.556	1.007	.9932	1.4544	20
50	.1193	.1190	8.405	.1198	8.345	1.007	.9929	1.4515	10
7° 00′	.1222	.1219	8.206	.1228	8.144	1.008	.9925	1.4486	83° 00′
10	.1251	.1248	8.016	.1257	7.953	1.008	.9922	1.4457	50
20	.1280	.1276	7.834	.1287	7.770	1.008	.9918	1.4428	40
30	.1309	.1305	7.661	.1317	7.596	1.009	.9914	1.4399	30
40	.1338	.1334	7.496	.1346	7.429	1.009	.9911	1.4370	20
50	.1367	.1363	7.337	.1376	7.269	1.009	.9907	1.4341	10
8° 00′	.1396	.1392	7.185	.1405	7.115	1.010	.9903	1.4312	82° 00′
10	.1425	.1421	7.040	.1435	6.968	1.010	.9899	1.4283	50
20	.1454	.1449	6.900	.1465	6.827	1.011	.9894	1.4254	40
30	.1484	.1478	6.765	.1495	6.691	1.011	.9890	1.4224	30
40	.1513	.1507	6.636	.1524	6.561	1.012	.9886	1.4195	20
50	.1542	.1536	6.512	.1554	6.435	1.012	.9881	1.4166	10
9° 00′	.1571	.1564	6.392	.1584	6.314	1.012	.9877	1.4137	81° 00′
		cos θ	**sec θ**	**cot θ**	**tan θ**	**csc θ**	**sin θ**	**Radians**	**Degrees**
								Angle θ	

Trigonometric Functions (continued)

Angle θ									
Degrees	Radians	sin θ	csc θ	tan θ	cot θ	sec θ	cos θ		
9° 00′	.1571	.1564	6.392	.1584	6.314	1.012	.9877	1.4137	81° 00′
10	.1600	.1593	6.277	.1614	6.197	1.013	.9872	1.4108	50
20	.1629	.1622	6.166	.1644	6.084	1.013	.9868	1.4079	40
30	.1658	.1650	6.059	.1673	5.976	1.014	.9863	1.4050	30
40	.1687	.1679	5.955	.1703	5.871	1.014	.9858	1.4021	20
50	.1716	.1708	5.855	.1733	5.769	1.015	.9853	1.3992	10
10° 00′	.1745	.1736	5.759	.1763	5.671	1.015	.9848	1.3963	80° 00′
10	.1774	.1765	5.665	.1793	5.576	1.016	.9843	1.3934	50
20	.1804	.1794	5.575	.1823	5.485	1.016	.9838	1.3904	40
30	.1833	.1822	5.487	.1853	5.396	1.017	.9833	1.3875	30
40	.1862	.1851	5.403	.1883	5.309	1.018	.9827	1.3846	20
50	.1891	.1880	5.320	.1914	5.226	1.018	.9822	1.3817	10
11° 00′	.1920	.1908	5.241	.1944	5.145	1.019	.9816	1.3788	79° 00′
10	.1949	.1937	5.164	.1974	5.066	1.019	.9811	1.3759	50
20	.1978	.1965	5.089	.2004	4.989	1.020	.9805	1.3730	40
30	.2007	.1994	5.016	.2035	4.915	1.020	.9799	1.3701	30
40	.2036	.2022	4.945	.2065	4.843	1.021	.9793	1.3672	20
50	.2065	.2051	4.876	.2095	4.773	1.022	.9787	1.3643	10
12° 00′	.2094	.2079	4.810	.2126	4.705	1.022	.9781	1.3614	78° 00′
10	.2123	.2108	4.745	.2156	4.638	1.023	.9775	1.3584	50
20	.2153	.2136	4.682	.2186	4.574	1.024	.9769	1.3555	40
30	.2182	.2164	4.620	.2217	4.511	1.024	.9763	1.3526	30
40	.2211	.2193	4.560	.2247	4.449	1.025	.9757	1.3497	20
50	.2240	.2221	4.502	.2278	4.390	1.026	.9750	1.3468	10
13° 00′	.2269	.2250	4.445	.2309	4.331	1.026	.9744	1.3439	77° 00′
10	.2298	.2278	4.390	.2339	4.275	1.027	.9737	1.3410	50
20	.2327	.2306	4.336	.2370	4.219	1.028	.9730	1.3381	40
30	.2356	.2334	4.284	.2401	4.165	1.028	.9724	1.3352	30
40	.2385	.2363	4.232	.2432	4.113	1.029	.9717	1.3323	20
50	.2414	.2391	4.182	.2462	4.061	1.030	.9710	1.3294	10
14° 00′	.2443	.2419	4.134	.2493	4.011	1.031	.9703	1.3265	76° 00′
10	.2473	.2447	4.086	.2524	3.962	1.031	.9696	1.3235	50
20	.2502	.2476	4.039	.2555	3.914	1.032	.9689	1.3206	40
30	.2531	.2504	3.994	.2586	3.867	1.033	.9681	1.3177	30
40	.2560	.2532	3.950	.2617	3.821	1.034	.9674	1.3148	20
50	.2589	.2560	3.906	.2648	3.776	1.034	.9667	1.3119	10
15° 00′	.2618	.2588	3.864	.2679	3.732	1.035	.9659	1.3090	75° 00′
10	.2647	.2616	3.822	.2711	3.689	1.036	.9652	1.3061	50
20	.2676	.2644	3.782	.2742	3.647	1.037	.9644	1.3032	40
30	.2705	.2672	3.742	.2773	3.606	1.038	.9636	1.3003	30
40	.2734	.2700	3.703	.2805	3.566	1.039	.9628	1.2974	20
50	.2763	.2728	3.665	.2836	3.526	1.039	.9621	1.2945	10
16° 00′	.2793	.2756	3.628	.2867	3.487	1.040	.9613	1.2915	74° 00′
10	.2822	.2784	3.592	.2899	3.450	1.041	.9605	1.2886	50
20	.2851	.2812	3.556	.2931	3.412	1.042	.9596	1.2857	40
30	.2880	.2840	3.521	.2962	3.376	1.043	.9588	1.2828	30
40	.2909	.2868	3.487	.2944	3.340	1.044	.9580	1.2799	20
50	.2938	.2896	3.453	.3026	3.305	1.045	.9572	1.2770	10
17° 00′	.2967	.2924	3.420	.3057	3.271	1.046	.9563	1.2741	73° 00′
10	.2996	.2952	3.388	.3089	3.237	1.047	.9555	1.2712	50
20	.3025	.2979	3.357	.3121	3.204	1.048	.9546	1.2683	40
30	.3054	.3007	3.326	.3153	3.172	1.048	.9537	1.2654	30
40	.3083	.3035	3.295	.3185	3.140	1.049	.9528	1.2625	20
50	.3113	.3062	3.265	.3217	3.108	1.050	.9520	1.2595	10
18° 00′	.3142	.3090	3.236	.3249	3.078	1.051	.9511	1.2566	72° 00′
		cos θ	sec θ	cot θ	tan θ	csc θ	sin θ	Radians	Degrees
								Angle θ	

Trigonometric Functions (continued)

Angle θ									
Degrees	Radians	sin θ	csc θ	tan θ	cot θ	sec θ	cos θ		
18° 00′	.3142	.3090	3.236	.3249	3.078	1.051	.9511	1.2566	72° 00′
10	.3171	.3118	3.207	.3281	3.047	1.052	.9502	1.2537	50
20	.3200	.3145	3.179	.3314	3.018	1.053	.9492	1.2508	40
30	.3229	.3173	3.152	.3346	2.989	1.054	.9483	1.2479	30
40	.3258	.3201	3.124	.3378	2.960	1.056	.9474	1.2450	20
50	.3287	.3228	3.098	.3411	2.932	1.057	.9465	1.2421	10
19° 00′	.3316	.3256	3.072	.3443	2.904	1.058	.9455	1.2392	71° 00′
10	.3345	.3283	3.046	.3476	2.877	1.059	.9446	1.2363	50
20	.3374	.3311	3.021	.3508	2.850	1.060	.9436	1.2334	40
30	.3403	.3338	2.996	.3541	2.824	1.061	.9426	1.2305	30
40	.3432	.3365	2.971	.3574	2.798	1.062	.9417	1.2275	20
50	.3462	.3393	2.947	.3607	2.773	1.063	.9407	1.2246	10
20° 00′	.3491	.3420	2.924	.3640	2.747	1.064	.9397	1.2217	70° 00′
10	.3520	.3448	2.901	.3673	2.723	1.065	.9387	1.2188	50
20	.3549	.3475	2.878	.3706	2.699	1.066	.9377	1.2159	40
30	.3578	.3502	2.855	.3739	2.675	1.068	.9367	1.2130	30
40	.3607	.3529	2.833	.3772	2.651	1.069	.9356	1.2101	20
50	.3636	.3557	2.812	.3805	2.628	1.070	.9346	1.2072	10
21° 00′	.3665	.3584	2.790	.3839	2.605	1.071	.9336	1.2043	69° 00′
10	.3694	.3611	2.769	.3872	2.583	1.072	.9325	1.2014	50
20	.3723	.3638	2.749	.3906	2.560	1.074	.9315	1.1985	40
30	.3752	.3665	2.729	.3939	2.539	1.075	.9304	1.1956	30
40	.3782	.3692	2.709	.3973	2.517	1.076	.9293	1.1926	20
50	.3811	.3719	2.689	.4006	2.496	1.077	.9283	1.1897	10
22° 00′	.3840	.3746	2.669	.4040	2.475	1.079	.9272	1.1868	68° 00′
10	.3869	.3773	2.650	.4074	2.455	1.080	.9261	1.1839	50
20	.3898	.3800	2.632	.4108	2.434	1.081	.9250	1.1810	40
30	.3927	.3827	2.613	.4142	2.414	1.082	.9239	1.1781	30
40	.3956	.3854	2.595	.4176	2.394	1.084	.9228	1.1752	20
50	.3985	.3881	2.577	.4210	2.375	1.085	.9216	1.1723	10
23° 00′	.4014	.3907	2.559	.4245	2.356	1.086	.9205	1.1694	67° 00′
10	.4043	.3934	2.542	.4279	2.337	1.088	.9194	1.1665	50
20	.4072	.3961	2.525	.4314	2.318	1.089	.9182	1.1636	40
30	.4102	.3987	2.508	.4348	2.300	1.090	.9171	1.1606	30
40	.4131	.4014	2.491	.4383	2.282	1.092	.9159	1.1577	20
50	.4160	.4041	2.475	.4417	2.264	1.093	.9147	1.1548	10
24° 00′	.4189	.4067	2.459	.4452	2.246	1.095	.9135	1.1519	66° 00′
10	.4218	.4094	2.443	.4487	2.229	1.096	.9124	1.1490	50
20	.4247	.4120	2.427	.4522	2.211	1.097	.9112	1.1461	40
30	.4276	.4147	2.411	.4557	2.194	1.099	.9100	1.1432	30
40	.4305	.4173	2.396	.4592	2.177	1.100	.9088	1.1403	20
50	.4334	.4200	2.381	.4628	2.161	1.102	.9075	1.1374	10
25° 00′	.4363	.4226	2.366	.4663	2.145	1.103	.9063	1.1345	65° 00′
10	.4392	.4253	2.352	.4699	2.128	1.105	.9051	1.1316	50
20	.4422	.4279	2.337	.4734	2.112	1.106	.9038	1.1286	40
30	.4451	.4305	2.323	.4770	2.097	1.108	.9026	1.1257	30
40	.4480	.4331	2.309	.4806	2.081	1.109	.9013	1.1228	20
50	.4509	.4358	2.295	.4841	2.066	1.111	.9001	1.1199	10
26° 00′	.4538	.4384	2.281	.4877	2.050	1.113	.8988	1.1170	64° 00′
10	.4567	.4410	2.268	.4913	2.035	1.114	.8975	1.1141	50
20	.4596	.4436	2.254	.4950	2.020	1.116	.8962	1.1112	40
30	.4625	.4462	2.241	.4986	2.006	1.117	.8949	1.1083	30
40	.4654	.4488	2.228	.5022	1.991	1.119	.8936	1.1054	20
50	.4683	.4514	2.215	.5059	1.977	1.121	.8923	1.1025	10
27° 00′	.4712	.4540	2.203	.5095	1.963	1.122	.8910	1.0996	63° 00′
		cos θ	sec θ	cot θ	tan θ	csc θ	sin θ	Radians	Degrees
								Angle θ	

Trigonometric Functions (continued)

Angle θ Degrees	Angle θ Radians	sin θ	csc θ	tan θ	cot θ	sec θ	cos θ		
27° 00′	.4712	.4540	2.203	.5095	1.963	1.122	.8910	1.0996	63° 00′
10	.4741	.4566	2.190	.5132	1.949	1.124	.8897	1.0966	50
20	.4771	.4592	2.178	.5169	1.935	1.126	.8884	1.0937	40
30	.4800	.4617	2.166	.5206	1.921	1.127	.8870	1.0908	30
40	.4829	.4643	2.154	.5243	1.907	1.129	.8857	1.0879	20
50	.4858	.4669	2.142	.5280	1.894	1.131	.8843	1.0850	10
28° 00′	.4887	.4695	2.130	.5317	1.881	1.133	.8829	1.0821	62° 00′
10	.4916	.4720	2.118	.5354	1.868	1.134	.8816	1.0792	50
20	.4945	.4746	2.107	.5392	1.855	1.136	.8802	1.0763	40
30	.4974	.4772	2.096	.5430	1.842	1.138	.8788	1.0734	30
40	.5003	.4797	2.085	.5467	1.829	1.140	.8774	1.0705	20
50	.5032	.4823	2.074	.5505	1.816	1.142	.8760	1.0676	10
29° 00′	.5061	.4848	2.063	.5543	1.804	1.143	.8746	1.0647	61° 00′
10	.5091	.4874	2.052	.5581	1.792	1.145	.8732	1.0617	50
20	.5120	.4899	2.041	.5619	1.780	1.147	.8718	1.0588	40
30	.5149	.4924	2.031	.5658	1.767	1.149	.8704	1.0559	30
40	.5178	.4950	2.020	.5696	1.756	1.151	.8689	1.0530	20
50	.5207	.4975	2.010	.5735	1.744	1.153	.8675	1.0501	10
30° 00′	.5236	.5000	2.000	.5774	1.732	1.155	.8660	1.0472	60° 00′
10	.5265	.5025	1.990	.5812	1.720	1.157	.8646	1.0443	50
20	.5294	.5050	1.980	.5851	1.709	1.159	.8631	1.0414	40
30	.5323	.5075	1.970	.5890	1.698	1.161	.8616	1.0385	30
40	.5352	.5100	1.961	.5930	1.686	1.163	.8601	1.0356	20
50	.5381	.5125	1.951	.5969	1.675	1.165	.8587	1.0327	10
31° 00′	.5411	.5150	1.942	.6009	1.664	1.167	.8572	1.0297	59° 00′
10	.5440	.5175	1.932	.6048	1.653	1.169	.8557	1.0268	50
20	.5469	.5200	1.923	.6088	1.643	1.171	.8542	1.0239	40
30	.5498	.5225	1.914	.6128	1.632	1.173	.8526	1.0210	30
40	.5527	.5250	1.905	.6168	1.621	1.175	.8511	1.0181	20
50	.5556	.5275	1.896	.6208	1.611	1.177	.8496	1.0152	10
32° 00′	.5585	.5299	1.887	.6249	1.600	1.179	.8480	1.0123	58° 00′
10	.5614	.5324	1.878	.6289	1.590	1.181	.8465	1.0094	50
20	.5643	.5348	1.870	.6330	1.580	1.184	.8450	1.0065	40
30	.5672	.5373	1.861	.6371	1.570	1.186	.8434	1.0036	30
40	.5701	.5398	1.853	.6412	1.560	1.188	.8418	1.0007	20
50	.5730	.5422	1.844	.6453	1.550	1.190	.8403	.9977	10
33° 00′	.5760	.5446	1.836	.6494	1.540	1.192	.8387	.9948	57° 00′
10	.5789	.5471	1.828	.6536	1.530	1.195	.8371	.9919	50
20	.5818	.5495	1.820	.6577	1.520	1.197	.8355	.9890	40
30	.5847	.5519	1.812	.6619	1.511	1.199	.8339	.9861	30
40	.5876	.5544	1.804	.6661	1.501	1.202	.8323	.9832	20
50	.5905	.5568	1.796	.6703	1.492	1.204	.8307	.9803	10
34° 00′	.5934	.5592	1.788	.6745	1.483	1.206	.8290	.9774	56° 00′
10	.5963	.5616	1.781	.6787	1.473	1.209	.8274	.9745	50
20	.5992	.5640	1.773	.6830	1.464	1.211	.8258	.9716	40
30	.6021	.5664	1.766	.6873	1.455	1.213	.8241	.9687	30
40	.6050	.5688	1.758	.6916	1.446	1.216	.8225	.9657	20
50	.6080	.5712	1.751	.6959	1.437	1.218	.8208	.9628	10
35° 00′	.6109	.5736	1.743	.7002	1.428	1.221	.8192	.9599	55° 00′
10	.6138	.5760	1.736	.7046	1.419	1.223	.8175	.9570	50
20	.6167	.5783	1.729	.7089	1.411	1.226	.8158	.9541	40
30	.6196	.5807	1.722	.7133	1.402	1.228	.8141	.9512	30
40	.6225	.5831	1.715	.7177	1.393	1.231	.8124	.9483	20
50	.6254	.5854	1.708	.7221	1.385	1.233	.8107	.9454	10
36° 00′	.6283	.5878	1.701	.7265	1.376	1.236	.8090	.9425	54° 00′
		cos θ	sec θ	cot θ	tan θ	csc θ	sin θ	Radians	Degrees
								Angle θ	

Trigonometric Functions (continued)

Angle θ Degrees	Radians	sin θ	csc θ	tan θ	cot θ	sec θ	cos θ		
36° 00′	.6283	.5878	1.701	.7265	1.376	1.236	.8090	.9425	54° 00′
10	.6312	.5901	1.695	.7310	1.368	1.239	.8073	.9396	50
20	.6341	.5925	1.688	.7355	1.360	1.241	.8056	.9367	40
30	.6370	.5948	1.681	.7400	1.351	1.244	.8039	.9338	30
40	.6400	.5972	1.675	.7445	1.343	1.247	.8021	.9308	20
50	.6429	.5995	1.668	.7490	1.335	1.249	.8004	.9279	10
37° 00′	.6458	.6018	1.662	.7536	1.327	1.252	.7986	.9250	53° 00′
10	.6487	.6041	1.655	.7581	1.319	1.255	.7969	.9221	50
20	.6516	.6065	1.649	.7627	1.311	1.258	.7951	.9192	40
30	.6545	.6088	1.643	.7673	1.303	1.260	.7934	.9163	30
40	.6574	.6111	1.636	.7720	1.295	1.263	.7916	.9134	20
50	.6603	.6134	1.630	.7766	1.288	1.266	.7898	.9105	10
38° 00′	.6632	.6157	1.624	.7813	1.280	1.269	.7880	.9076	52° 00′
10	.6661	.6180	1.618	.7860	1.272	1.272	.7862	.9047	50
20	.6690	.6202	1.612	.7907	1.265	1.275	.7844	.9018	40
30	.6720	.6225	1.606	.7954	1.257	1.278	.7826	.8988	30
40	.6749	.6248	1.601	.8002	1.250	1.281	.7808	.8959	20
50	.6778	.6271	1.595	.8050	1.242	1.284	.7790	.8930	10
39° 00′	.6807	.6293	1.589	.8098	1.235	1.287	.7771	.8901	51° 00′
10	.6836	.6316	1.583	.8146	1.228	1.290	.7753	.8872	50
20	.6865	.6338	1.578	.8195	1.220	1.293	.7735	.8843	40
30	.6894	.6361	1.572	.8243	1.213	1.296	.7716	.8814	30
40	.6923	.6383	1.567	.8292	1.206	1.299	.7698	.8785	20
50	.6952	.6406	1.561	.8342	1.199	1.302	.7679	.8756	10
40° 00′	.6981	.6428	1.556	.8391	1.192	1.305	.7660	.8727	50° 00′
10	.7010	.6450	1.550	.8441	1.185	1.309	.7642	.8698	50
20	.7039	.6472	1.545	.8491	1.178	1.312	.7623	.8668	40
30	.7069	.6494	1.540	.8541	1.171	1.315	.7604	.8639	30
40	.7098	.6517	1.535	.8591	1.164	1.318	.7585	.8610	20
50	.7127	.6539	1.529	.8642	1.157	1.322	.7566	.8581	10
41° 00′	.7156	.6561	1.524	.8693	1.150	1.325	.7547	.8552	49° 00′
10	.7185	.6583	1.519	.8744	1.144	1.328	.7528	.8523	50
20	.7214	.6604	1.514	.8796	1.137	1.332	.7509	.8494	40
30	.7243	.6626	1.509	.8847	1.130	1.335	.7490	.8465	30
40	.7272	.6648	1.504	.8899	1.124	1.339	.7470	.8436	20
50	.7301	.6670	1.499	.8952	1.117	1.342	.7451	.8407	10
42° 00′	.7330	.6691	1.494	.9004	1.111	1.346	.7431	.8378	48° 00′
10	.7359	.6713	1.490	.9057	1.104	1.349	.7412	.8348	50
20	.7389	.6734	1.485	.9110	1.098	1.353	.7392	.8319	40
30	.7418	.6756	1.480	.9163	1.091	1.356	.7373	.8290	30
40	.7447	.6777	1.476	.9217	1.085	1.360	.7353	.8261	20
50	.7476	.6799	1.471	.9271	1.079	1.364	.7333	.8232	10
43° 00′	.7505	.6820	1.466	.9325	1.072	1.367	.7314	.8203	47° 00′
10	.7534	.6841	1.462	.9380	1.066	1.371	.7294	.8174	50
20	.7563	.6862	1.457	.9435	1.060	1.375	.7274	.8145	40
30	.7592	.6884	1.453	.9490	1.054	1.379	.7254	.8116	30
40	.7621	.6905	1.448	.9545	1.048	1.382	.7234	.8087	20
50	.7650	.6926	1.444	.9601	1.042	1.386	.7214	.8058	10
44° 00′	.7679	.6947	1.440	.9657	1.036	1.390	.7193	.8029	46° 00′
10	.7709	.6967	1.435	.9713	1.030	1.394	.7173	.7999	50
20	.7738	.6988	1.431	.9770	1.024	1.398	.7153	.7970	40
30	.7767	.7009	1.427	.9827	1.018	1.402	.7133	.7941	30
40	.7796	.7030	1.423	.9884	1.012	1.406	.7112	.7912	20
50	.7825	.7050	1.418	.9942	1.006	1.410	.7092	.7883	10
45° 00′	.7854	.7071	1.414	1.000	1.000	1.414	.7071	.7854	45° 00′
		cos θ	sec θ	cot θ	tan θ	csc θ	sin θ	Radians	Degrees
								Angle θ	

Answers
to Selected
Exercises

CHAPTER 1

Section 1.1

1. T **3.** T **5.** T **7.** T **9.** T **11.** All
13. $\left\{-\sqrt{2}, \sqrt{3}, \pi, 5.090090009\ldots\right\}$ **15.** $\{0, 1\}$
17. $x + 7$ **19.** $5x + 15$ **21.** $\frac{1}{2}(x+1)$

23. $(-13 + 4) + x$ **25.** 8 **27.** 7/5 **29.** -1 **31.** 5/3
33. 3/2 **35.** $\sqrt{3}$ **37.** $y^2 - x^2$ **39.** 7.2 **41.** $\sqrt{5}$
43. 5.5 **45.** 22 **47.** 72/5 **49.** -38 **51.** 52
53. 490 **55.** 52 **57.** 1 **59.** -3.4 **61.** 143
63. -61 **65.** -14 **67.** 1 **69.** $-14/3$ **71.** 0
73. 12.19 **75.** 35 **77.** $-2x$ **79.** $0.85x$ **81.** $-6xy$
83. $-24wx$ **85.** $-3x - 6$ **87.** $0.97x - 6$ **89.** $2zy - 2$
91. $11x - 17$ **93.** $7x/12$ **95.** $3xy/4$ **97.** $3 - 2x$

99. $3x - y$ **101.** $-\frac{5}{4}y + 3$ **103.** $-\frac{61}{180}x$

105. $-\frac{12}{5}$ **107.** 0 **109.** 0.5198

111. $-1/2, -5/12, -1/3, 0, 1/3, 5/12, 1/2$ **113.** 6 days
115. 3 and 4 **117.** Big gray elephant **121.** Transitive
123. 106

Section 1.2

1. 64 **3.** -16 **5.** 1/8 **7.** 13 **9.** 50 **11.** -5
13. -5 **15.** -35 **17.** -72 **19.** -23 **21.** 2 **23.** 1/81
25. 25 **27.** 11/30 **29.** 24 **31.** 8 **33.** 16 **35.** 5360

37. $-6x^{11}y^{11}$ **39.** 225 **41.** $-4x^6$ **43.** $-\frac{8}{27}x^6$

45. $3x^4$ **47.** $\frac{25}{y^4}$ **49.** $\frac{1}{6y^3}$ **51.** $\frac{n^2}{2}$ **53.** $17a^6$

55. $-\frac{x^2}{3y^6}$ **57.** $\frac{16x^{20}}{81y^8}$ **59.** $\frac{1}{c^6d^2}$ **61.** x^{b+5}

63. $\frac{-125a^{6t}}{b^{9t}}$ **65.** $\frac{-3y^{6v}}{2x^{5w}}$ **67.** $a^{-3s+15}b^{6t-27}$

69. 43,000 **71.** -0.0000356 **73.** 5×10^6
75. -6.72×10^{-5} **77.** 0.000000007 **79.** -2×10^{10}
81. 2×10^{16} **83.** 2×10^{-3} **85.** 4×10^{-29} **87.** 1×10^{23}
89. -9.936×10^{-5} **91.** 1.5732×10^1 **93.** 4.78×10^{-41}
95. 2.7×10^9 **97.** \$5138.34 **99.** 1.577×10^{24} tons
101. 6379 km **103.** 3.3×10^5 **105.** 8.11×10^{16} Btu

Section 1.3

1. -3 **3.** 8 **5.** -4 **7.** 81 **9.** 1/16 **11.** $|x|$
13. a^3 **15.** a^2 **17.** xy^2 **19.** $|a|b^2$ **21.** $x^2y^{1/2}$
23. $6a^{3/2}$ **25.** $3a^{1/6}$ **27.** $a^{7/3}b$ **29.** $a^{2m}b^{4n}$ **31.** a^{5n-3}

33. $\frac{x^2y}{z^3}$ **35.** $\frac{b^{15}}{a^2}$ **37.** 30 **39.** -2 **41.** -2 **43.** $-1/5$

45. 8 **47.** $\sqrt[3]{10^2}$ **49.** $\frac{3}{\sqrt[5]{y^3}}$ **51.** $x^{-1/2}$ **53.** $x^{3/5}$ **55.** $4x$

57. $2y^3$ **59.** $\frac{\sqrt{xy}}{10}$ **61.** $\frac{-2a}{b^5}$ **63.** $2\sqrt{7}$ **65.** $\frac{\sqrt{5}}{5}$

67. $\dfrac{\sqrt{2x}}{4}$ **69.** $2\sqrt[3]{5}$ **71.** $-5x\sqrt[3]{2x}$ **73.** $\dfrac{\sqrt[3]{4}}{2}$ **75.** $2xy\sqrt[4]{2y}$

77. $\dfrac{\sqrt[4]{40x^3}}{2x}$ **79.** $9 - 2\sqrt{6}$ **81.** $2\sqrt{2} + 2\sqrt{5} - 2\sqrt{3}$

83. $-30\sqrt{2}$ **85.** $60a$ **87.** 75 **89.** $\dfrac{3\sqrt{a}}{a^2}$ **91.** $\dfrac{5\sqrt{x}}{x}$

93. $5x\sqrt{5x}$ **95.** $\sqrt[6]{72}$ **97.** $\sqrt[12]{81x^3}$ **99.** $\sqrt[6]{4x^5y^5}$

101. $\sqrt[12]{432a^7b^{11}}$ **103.** $\sqrt[6]{a^5b^4}$ **105.** $\sqrt[6]{7}$ **107.** $\sqrt[15]{6}$

109. $\sqrt[12]{4x^5}$ **111.** 7 **113.** $2\sqrt{6}$ **115.** 6.31 **117.** $2\sqrt{2}$

119. $7\sqrt{2}$ **121.** 5.49 **123.** Yes   **125.** 47.5%

Section 1.4

1. $3, 1$ **3.** Not a polynomial   **5.** $0, 79$ **7.** $5, -0.012$

9. $8x^2 + 3x - 1$ **11.** $-5x^2 + x - 3$

13. $(-5a^2 + 4a)x^3 + 2a^2x - 3$ **15.** $2x - 1$

17. $3x^2 - 3x - 6$ **19.** $-18a^5 + 15a^4 - 6a^3$

21. $3b^3 - 14b^2 + 17b - 6$ **23.** $8x^3 - 1$ **25.** $x^3 + 125$

27. $xz - 4z + 3x - 12$ **29.** $a^3 - b^3$ **31.** $a^2 + 7a - 18$

33. $2y^2 + 15y - 27$ **35.** $4x^2 - 81$ **37.** $4x^2 + 20x + 25$

39. $6x^4 + 22x^2 + 20$ **41.** $5 + 4\sqrt{2}$ **43.** $14 - 11\sqrt{2}$

45. $9 - \sqrt{6}$ **47.** 43 **49.** $8 + 2\sqrt{15}$ **51.** $4 + 4\sqrt{x} + x$

53. $9x^2 + 30x + 25$ **55.** $x^{2n} - 9$ **57.** -23

59. $55 - 6\sqrt{6}$ **61.** $9x^6 - 24x^3 + 16$ **63.** $x - 4\sqrt{x - 4}$

65. $5\sqrt{2} + 2\sqrt{10}$ **67.** $\dfrac{2\sqrt{6} - \sqrt{2}}{11}$ **69.** $-9x^3$

71. $-x + 2$ **73.** $x + 3$ **75.** $a^2 + a + 1$ **77.** $x + 5, 13$

79. $2x - 6, 13$ **81.** $x^2 + x + 1, 0$ **83.** $3x - 5, 7$

85. $ab + b, 0$ **87.** $x + 1 + \dfrac{-1}{x - 1}$ **89.** $2x - 3 + \dfrac{1}{x}$

91. $x + 1 + \dfrac{1}{x}$ **93.** $x + \dfrac{1}{x - 2}$ **95.** 12 **97.** 72

99. 77 **101.** $59/27$ **103.** $x^2 + 2x - 24$

105. $2a^{10} - 3a^5 - 27$ **107.** $-y - 9$ **109.** $6a - 6$

111. $w^2 + 8w + 16$ **113.** $16x^2 - 81$ **115.** $3y^5 - 9xy^2$

117. $2b - 1$ **119.** $x^6 - 64$ **121.** $9w^4 - 12w^2n + 4n^2$

123. 1 **125.** $3 + 2\sqrt{2}$ **127.** $2x^2 + 5x - 3$

129. $x^2 + 5x + 6$ **131.** $4x^2 - 20x + 24$ **133.** $\$1030$

135. $\$80,000$ **137.** $100\ \text{ft}^2$ less

Section 1.5

1. $6x^2(x - 2), -6x^2(-x + 2)$

3. $4a(1 - 2b), -4a(-1 + 2b)$

5. $3x^2(x^4 - 2x^3 + 3), -3x^2(-x^4 + 2x^3 - 3)$

7. $ax(-x^2 + 5x - 6), -ax(x^2 - 5x + 6)$

9. $1(m - n), -1(-m + n)$ **11.** $(x^2 + 5)(x + 2)$

13. $(y^2 - 3)(y - 1)$ **15.** $(d - w)(ay + 1)$

17. $(y^2 - b)(x^2 - a)$ **19.** $(x + 2)(x + 8)$

21. $(x - 6)(x + 2)$ **23.** $(m - 2)(m - 10)$

25. $(t - 7)(t + 12)$ **27.** $(2x + 1)(x - 4)$

29. $(4x + 1)(2x - 3)$ **31.** $(3y + 5)(2y - 1)$

33. $(3b + 2)(4b + 3)$ **35.** $(t - u)(t + u)$ **37.** $(t + 1)^2$

39. $(2w - 1)^2$ **41.** $(y^{2t} - 5)(y^{2t} + 5)$ **43.** $(3zx + 4)^2$

45. $(x^2 + y)^2$ **47.** $(t - u)(t^2 + tu + u^2)$

49. $(a - 2)(a^2 + 2a + 4)$ **51.** $(3y + 2)(9y^2 - 6y + 4)$

53. $(3xy^2 - 2z^3)(9x^2y^4 + 6xy^2z^3 + 4z^6)$ **55.** $(y^3 + 5)^2$

57. $(2a^2b^4 + 1)(2a^2b^4 - 5)$ **59.** $(2a + 7)(2a - 3)$

61. $(b^2 - 2)(b^2 + 1)$ **63.** $-3x(x - 3)(x + 3)$

65. $2t(2t + 3w)(4t^2 - 6tw + 9w^2)$

67. $(a - 2)(a + 2)(a + 1)$ **69.** $(x - 2)^2(x^2 + 2x + 4)$

71. $-2x(6x + 1)(3x - 2)$

73. $(a - 2)(a^2 + 2a + 4)(a + 2)(a^2 - 2a + 4)(a - 1)$

75. $-(3x + 5)(2x - 3)$ **77.** $(a - 1)(a + 1)(a^2 + 1)$

79. $(a + b - 3)(a + b + 3)$ **81.** Yes   **83.** Yes

85. No   **87.** $(x - 1)(x + 2)(x + 3)$

89. $(x - 3)(x^2 + 2x + 2)$

91. $(x + 2)(x + 3)(x - 1)(x + 1)$ **93.** $(x^m - 1)(x^m + 1)$

95. $(x^q - 2)(x^q - 3)$ **97.** $(x^n - 1)(x^{2n} + x^n + 1)$

99. $(x^{w+1} + 2)(x^w + 1)$ **101.** $\left(x - \sqrt{5}\right)\left(x + \sqrt{5}\right)$

103. $\left(a - \sqrt[3]{2}\right)\left(a^2 + a\sqrt[3]{2} + \sqrt[3]{4}\right)$

105. $\left(x - \sqrt[4]{2}\right)\left(x + \sqrt[4]{2}\right)\left(x^2 + \sqrt{2}\right)$

107. $\left(b + \sqrt[6]{2}\right)\left(b^2 - b\sqrt[6]{2} + \sqrt[6]{4}\right)$ **109.** $x^2 + x + 1$

111. $91,204,000\ \text{ft}^3$

Section 1.6

1. $\{x \mid x \neq -2\}$ **3.** $\{x \mid x \neq 4 \text{ and } x \neq -2\}$

5. $\{x \mid x \neq 3 \text{ and } x \neq -3\}$ **7.** All real numbers

9. $\dfrac{3}{x + 2}$ **11.** $-\dfrac{2}{3}$ **13.** $\dfrac{ab^4}{b - a^2}$ **15.** $\dfrac{y^2z}{x^3}$

17. $\dfrac{a^2 + ab + b^2}{a + b}$ **19.** $\dfrac{b + 3}{a - y}$ **21.** $\dfrac{3}{7ab}$ **23.** $\dfrac{42}{a^2}$

25. $\dfrac{a + 3}{3}$ **27.** $\dfrac{2x - 2}{x + y}$ **29.** $x^2y^2 + xy^3$ **31.** $\dfrac{-2}{x + wx}$

33. $\dfrac{16a}{12a^2}$ **35.** $\dfrac{x^2 - 8x + 15}{x^2 - 9}$ **37.** $\dfrac{x^2 + x}{x^2 + 6x + 5}$

39. $\dfrac{t^2 + t}{2t^2 + 4t + 2}$ **41.** $12a^2b^3$ **43.** $6(a + b)$

45. $(x + 2)(x + 3)(x - 3)$ **47.** $\dfrac{9 + x}{6x}$

49. $\dfrac{x + 7}{(x - 1)(x + 1)}$ **51.** $\dfrac{3a + 1}{a}$ **53.** $\dfrac{t^2 - 2}{t + 1}$

55. $\dfrac{2x^2 + 3x - 1}{(x + 1)(x + 2)(x + 3)}$ **57.** $\dfrac{7}{2x - 6}$ **59.** $\dfrac{x^2}{x^3 - y^3}$

61. $\dfrac{-9x - 3}{(2x - 3)(x + 5)(x - 1)}$ **63.** $\dfrac{x^2 + 2x - 1}{x(x^2 - 1)}$

65. $\dfrac{4b^2 - 3ab}{b + 2a}$ **67.** $\dfrac{a^2 - a}{3b^2 + b}$ **69.** $\dfrac{a + 2}{a - 2}$

71. $\dfrac{6t - 3}{2t^2 + t - 4}$ **73.** $\dfrac{1 + x}{1 - x}$ **75.** $\dfrac{a^3b^3 + 1}{ab^3}$

77. $-x^2y - xy^2$ **79.** $\dfrac{m^2n^2}{n^2 - 2mn + m^2}$ **81.** $\dfrac{3}{7}$

83. $\dfrac{3}{506}$ **85.** $\dfrac{1}{5}$ **87.** $\dfrac{1195}{1191}$ **89.** 3.1087   **91.** 3.00001

93. $\dfrac{5}{12}$ **95.** 272.7 mph

Chapter 1 Review Exercises

1. F   **3.** F   **5.** F   **7.** F   **9.** F   **11.** F   **13.** $17x - 12$
15. $\dfrac{3x}{10}$ **17.** $\dfrac{x - 2}{3}$ **19.** $-\dfrac{3}{4}$ **21.** -2 **23.** 4   **25.** 7
27. 28.5   **29.** 625   **31.** 44   **33.** 3/2   **35.** $-16/3$
37. 1/4   **39.** $5x^2$ **41.** 11   **43.** $2s\sqrt{7s}$ **45.** $-10\sqrt[3]{2}$
47. $\dfrac{\sqrt{10a}}{2a}$ **49.** $\dfrac{\sqrt[3]{50}}{5}$ **51.** $8n\sqrt{2n}$ **53.** $3 + \sqrt{3}$ **55.** $\dfrac{\sqrt{3}}{5}$
57. 320,000,000   **59.** -0.000185 **61.** 5.6×10^{-5}
63. -2.34×10^6 **65.** 1.25×10^{20} **67.** 4×10^6
69. $2x^2 + x - 7$ **71.** $-5x^4 + 3x^3 + 3x$
73. $3a^3 - 8a^2 + 9a - 10$ **75.** $b^2 - 6by + 9y^2$
77. $3t^2 - 7t - 6$ **79.** $-5y^3$ **81.** 7   **83.** $23 + 4\sqrt{15}$
85. $2x + 2\sqrt{2x - 1}$

87. $x^2 + 4x - 1$, 1   **89.** $3x + 2$, 4   **91.** $x - 2 + \dfrac{1}{x + 2}$

93. $2 + \dfrac{13}{x - 5}$ **95.** $6x(x - 1)(x + 1)$ **97.** $(3h + 4t)^2$
99. $(t + y)(t^2 - ty + y^2)$ **101.** $(x - 3)(x + 3)^2$
103. $(t - 1)(t^2 + t + 1)(t + 1)(t^2 - t + 1)$
105. $(6x + 5)(3x - 4)$ **107.** $ab(a + 6)(a - 3)$

109. $(x - 1)(x + 1)(2x + y)$ **111.** 2   **113.** $\dfrac{2x + 2}{(x - 2)(x + 4)}$

115. $-\dfrac{1}{2}$ **117.** $\dfrac{bc^{17}}{a^8}$ **119.** $\dfrac{3x + 7}{x^2 - 4}$ **121.** $\dfrac{5x - 21}{30x^2}$

123. $\dfrac{a - 1}{2}$ **125.** $\dfrac{-x^2 + x - 1}{2}$

127. $\dfrac{3a^2 + 8a + 7}{(a + 1)(a - 1)(a + 5)}$ **129.** $\dfrac{7}{2x - 8}$
131. $\dfrac{-3y^2 + 7}{4y^2 - 3}$ **133.** $\dfrac{b^3 - a^2}{ab^2}$ **135.** $\dfrac{q^3 + p^2}{pq^3}$
137. -11 **139.** -9 **141.** 4/11   **143.** 149/91
145. 5.9×10^{26} **147.** 11.12   **149.** 5/6

Chapter 1 Test

1. All   **2.** $\{-1.22, -1, 0, 2, 10/3\}$
3. $\{-\pi, -\sqrt{3}, \sqrt{5}, 6.020020002\ldots\}$ **4.** $\{0, 2\}$ **5.** 13

6. 2   **7.** $-1/9$ **8.** -1 **9.** $6x^5y^7$ **10.** 0
11. $a^2 + 2ab + b^2$ **12.** $-32a^3b^{10}$ **13.** $3\sqrt{3} + 2\sqrt{2}$
14. $\sqrt{3} + 1$ **15.** $\dfrac{\sqrt[3]{2x^2}}{2x^2}$ **16.** $2xy^4\sqrt{3xy}$
17. $3x^3 + 3x^2 - 12x$ **18.** $-5x^2 + 9x - 13$
19. $x^3 + x^2 - 7x - 3$ **20.** $4h^2 + 2h + 1$
21. $x^2 - 6xy - 27y^2$ **22.** $x^2 + 2x + 2$
23. $9x^2 - 48x + 64$ **24.** $4t^8 - 1$
25. $\dfrac{x^2 - 2x + 4}{x}$ **26.** $\dfrac{2x^2 + 7x + 21}{(x + 4)(x - 3)(x - 1)}$
27. $\dfrac{2a^2 - 7}{4a^2 - 9}$ **28.** $\dfrac{6b^2 - 24a^3b^3}{3a + 4a^2b^2}$ **29.** $a(x - 9)(x - 2)$
30. $m(m - 1)(m + 1)(m^2 + 1)$ **31.** $(3x - 1)(x + 5)$
32. $(bx + w)(x - 3)$ **33.** \$379.31   **34.** 5.9×10^{12} mi
35. 176 ft

CHAPTER 2

Section 2.1

1. No   **3.** Yes   **5.** $\{5/3\}$ **7.** $\{-2\}$ **9.** $\{1/2\}$
11. $\{11\}$ **13.** $\{-24\}$ **15.** $\{-6\}$ **17.** $\{\sqrt{2} - 2\}$
19. $\left\{\dfrac{5\sqrt{3} + 3\sqrt{2}}{3}\right\}$ **21.** $\left\{\dfrac{-1}{\pi}\right\}$ **23.** $\{3\sqrt[3]{4}\}$
25. R, identity   **27.** $\varnothing$, inconsistent
29. $\{x \mid x \neq 0\}$, identity   **31.** $\{w \mid w \neq 1\}$, identity
33. $\{x \mid x \neq 0\}$, identity   **35.** $\{9/8\}$, conditional
37. $\{6\}$, conditional   **39.** $\varnothing$, inconsistent
41. $\{-2\}$, conditional   **43.** $\{-19.952\}$ **45.** $\{2.957\}$
47. $\{-2.562\}$ **49.** $\{0.333\}$ **51.** $\{0.199\}$ **53.** $\{0.425\}$
55. $\{-0.380\}$ **57.** $\{200\}$ **59.** $\{-40\}$ **61.** $\{4\}$
63. $\{0\}$ **65.** $\{7\}$ **67.** $\{-10\}$ **69.** $\{5\}$ **71.** $\{18\}$
73. $\{x \mid x \neq 2 \text{ and } x \neq -2\}$, identity
75. $\{x \mid x \neq -3 \text{ and } x \neq 2\}$, identity   **77.** $\{-7/4\}$, conditional
79. $\varnothing$, inconsistent   **81.** $\{13\}$, conditional
83. $\{-45\}$, conditional   **85.** 1   **87.** $-33/2$
89. 498.2 m/min   **91.** 250,000

Section 2.2

1. $r = \dfrac{I}{Pt}$ **3.** $C = \dfrac{5}{9}(F - 32)$ **5.** $r = \dfrac{C}{2\pi}$

7. $y = \dfrac{C - Ax}{B}$ **9.** $R_2 = \dfrac{RR_1R_3}{R_1R_3 - RR_3 - RR_1}$

11. $d = \dfrac{a_n - a_1}{n - 1}$ **13.** 5.4%   **15.** 2.5 hr   **17.** 3.5 in.

19. 12.7%   **21.** 2.0616   **23.** 0.3630   **25.** -6.2710
27. 3.7299   **29.** \$26,600   **31.** \$60,000   **33.** 16 ft, 7 ft
35. 6400 ft^2 **37.** 112.5 mph   **39.** 48 mph
41. \$106,000   **43.** North side 600, South side 900

45. 28.8 hr   **47.** 2 P.M.   **49.** $y = -\dfrac{3}{2}x - 3$

51. $y = \dfrac{3}{2}x - 9$ **53.** $y = \dfrac{3}{4}x + \dfrac{9}{4}$

55. $y = mx - mx_1 + y_1$ **57.** $y = \dfrac{3}{x + 1}$

59. $y = \dfrac{2x + 1}{x - 3}$ **61.** $y = x(1 - \sqrt{2})$ **63.** $y = \dfrac{x^2 + 1}{x}$

65. 64.85 acres   **67.** 225 ft   **69.** 4.15 ft
71. 1.998 hectares   **73.** 34, 35, 36   **75.** \$74,440.65
77. 8/3 liters   **79.** 28 yr   **81.** 4 lb apples, 16 lb apricots
83. 3 dimes, 5 nickels   **85.** 2005

Section 2.3

1. Imaginary, $0 + 6i$ **3.** Imaginary, $\dfrac{1}{3} + \dfrac{1}{3}i$

5. Real, $\sqrt{7} + 0i$ **7.** Real, $\dfrac{\pi}{2} + 0i$ **9.** $7 + 2i$

11. $-2 - 3i$ **13.** $-12 - 18i$ **15.** 26   **17.** 29   **19.** 4
21. $-7 + 24i$ **23.** $1 - 4\sqrt{5}\,i$ **25.** i **27.** -1 **29.** 1
31. $-i$ **33.** 90   **35.** 17/4   **37.** 1   **39.** 12

41. $\dfrac{2}{5} + \dfrac{1}{5}i$ **43.** $\dfrac{3}{2} - \dfrac{3}{2}i$ **45.** $-1 + 3i$ **47.** $3 + 3i$

49. $\dfrac{1}{13} - \dfrac{5}{13}i$ **51.** $-i$ **53.** $-i$ **55.** $-4 + 2i$
57. -6 **59.** -10 **61.** $-1 + i\sqrt{5}$ **63.** $-3 + i\sqrt{11}$
65. $-4 + 8i$ **67.** -6 **69.** $-1 + 2i$ **71.** $-3 - 2i\sqrt{2}$
73. 0   **75.** 0   **77.** $9 + 6i$ **79.** 0   **81.** 21   **83.** Yes
85. No   **87.** Yes   **89.** Yes   **91.** Yes   **93.** $x^2 + 1$

95. $x^2 - 2x + 2$ **97.** $x^2 - 4x + 13$ **99.** $x^2 - x + \dfrac{1}{2}$

101. 4   **103.** $17 - 11i$ **105.** $17 - 11i$ **107.** $2 - 2i$

Section 2.4

1. $\left\{\pm\sqrt{5}\right\}$ **3.** $\left\{\pm i\dfrac{\sqrt{6}}{3}\right\}$ **5.** $\{0, 6\}$ **7.** $\{-2, 3\}$

9. $\{-2 \pm 2i\}$ **11.** $\left\{\dfrac{2}{3} \pm \dfrac{2}{3}i\right\}$ **13.** $\{-4, 5\}$

15. $\{-2, -1\}$ **17.** $\left\{-\dfrac{1}{2}, 3\right\}$ **19.** $\left\{\dfrac{2}{3}, \dfrac{1}{2}\right\}$

21. $\{-7, 6\}$ **23.** $\left\{-\dfrac{9}{2}, 2\right\}$ **25.** $x^2 - 12x + 36$

27. $r^2 + 3r + \dfrac{9}{4}$ **29.** $w^2 + \dfrac{1}{2}w + \dfrac{1}{16}$

31. $\left\{-3 \pm 2\sqrt{2}\right\}$ **33.** $\left\{1 \pm \sqrt{2}\right\}$ **35.** $\left\{\dfrac{-3 \pm \sqrt{13}}{2}\right\}$

37. $\{-1 \pm 2i\}$ **39.** $\left\{-4, \dfrac{3}{2}\right\}$ **41.** $\left\{-\dfrac{1}{3} \pm i\dfrac{\sqrt{2}}{3}\right\}$

43. $\{-4, 1\}$ **45.** $\left\{-\dfrac{1}{2}, 3\right\}$ **47.** $\left\{-\dfrac{1}{3}\right\}$ **49.** $\left\{\pm\dfrac{\sqrt{6}}{2}\right\}$

51. $\{2 \pm i\}$ **53.** $\left\{\dfrac{-1 \pm \sqrt{2}}{3}\right\}$ **55.** $\{-3.24, 0.87\}$

57. $\{0.71 \pm 2.38i\}$ **59.** 0, 1, real   **61.** -4, 2, imaginary

63. 172, 2, real   **65.** $\left\{\dfrac{2 \pm i}{3}\right\}$ **67.** $\left\{\pm i\sqrt[4]{2}\right\}$

69. $\left\{-\dfrac{\sqrt{6}}{6}, \dfrac{\sqrt{6}}{12}\right\}$ **71.** $\{-12, 6\}$ **73.** $\left\{\dfrac{1 \pm \sqrt{5}}{2}\right\}$

75. $\{-6, 8\}$ **77.** $r = \pm\sqrt{\dfrac{A}{\pi}}$ **79.** $x = -k \pm \sqrt{k^2 - 3}$

81. $y = x\left(-1 \pm \dfrac{\sqrt{6}}{2}\right)$ **83.** 5000 or 35,000

85. 2.5 sec   **87.** 340 ft   **89.** $4\sqrt{2}$ and $8\sqrt{2}$

91. $\dfrac{-7 + \sqrt{101}}{2} \approx 1.53$ ft   **93.** 10 ft/hr

95. $\dfrac{9 + \sqrt{53}}{2} \approx 8.14$ days   **97.** 40 lb or 20 lb

99. 4.58 m/sec   **101.** 5.5   **103.** 0.057

Section 2.5

1. $\{\pm 2, -3\}$ **3.** $\left\{-3, \pm\dfrac{\sqrt{2}}{2}\right\}$ **5.** $\left\{0, \dfrac{15 \pm \sqrt{205}}{2}\right\}$

7. $\{0, \pm 2\}$ **9.** $\{\pm 2, \pm 2i\}$ **11.** $\{8\}$ **13.** $\{25\}$

15. $\left\{\dfrac{1}{4}\right\}$ **17.** $\left\{\dfrac{2 + \sqrt{13}}{9}\right\}$ **19.** $\{-4, 6\}$ **21.** $\{9\}$

23. $\{5\}$ **25.** $\{10\}$ **27.** $\left\{\pm 2\sqrt{2}\right\}$ **29.** $\left\{\pm\dfrac{1}{8}\right\}$

31. $\left\{\dfrac{1}{49}\right\}$ **33.** $\left\{\dfrac{5}{4}\right\}$ **35.** $\left\{\pm 3, \pm\sqrt{3}\right\}$

37. $\left\{-\dfrac{17}{2}, \dfrac{13}{2}\right\}$ **39.** $\left\{\dfrac{3}{20}, \dfrac{4}{15}\right\}$ **41.** $\{-2, -1, 5, 6\}$

43. $\{1, 9\}$ **45.** $\{9, 16\}$ **47.** $\{8, 125\}$ **49.** $\left\{-\dfrac{3}{2}, \dfrac{9}{2}\right\}$

51. $\left\{\dfrac{2}{3}\right\}$ **53.** $\left\{\pm\sqrt{7}, \pm 1\right\}$ **55.** $\{0, 8\}$

57. $\{-3, 0, 1, 4\}$ **59.** $\{-2, 4\}$ **61.** $\left\{\dfrac{8}{3}, \dfrac{4}{3}\right\}$ **63.** $\left\{\dfrac{1}{2}\right\}$

65. $\left\{\pm 2, 1 \pm i\sqrt{3}, -1 \pm i\sqrt{3}\right\}$ **67.** $\left\{\sqrt{3}, 2\right\}$ **69.** $\varnothing$

71. $\{-2, \pm 1\}$ **73.** $\{5 \pm 9i\}$ **75.** $\left\{\dfrac{1 \pm 4\sqrt{2}}{3}\right\}$

77. $\left\{\pm 2\sqrt{6}, \pm\sqrt{35}\right\}$ **79.** $\{-3, 2, 3\}$ **81.** $\{-11\}$

83. $\{2\}$ **85.** 279.56 m²   **87.** 23   **89.** $\dfrac{25}{4}$ and $\dfrac{49}{4}$

91. 5 in.　**93.** 1600 ft²　**95.** 462.89 in³　**97.** 1 P.M.
99. 26,874 ft　**101.** 37.5 min, 38.2 min, 1.75 mi or 4.69 mi

Section 2.6

1. $x < 12$　**3.** $[-8, \infty)$　**5.** $(-\infty, \pi/2)$　**7.** $x \geq -7$
9. $(5, \infty)$　　　　　　　**11.** $[2, \infty)$

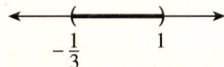　　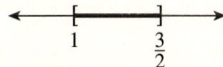

13. $(-\infty, 54)$　　　　　**15.** $(-\infty, 13/3]$

17. $(-\infty, 0]$　　　　　**19.** $[88, \infty)$

21. $(-3, \infty)$　**23.** $\varnothing$　**25.** $(-\infty, 5]$　**27.** $(-3, \infty)$
29. $[3, 7]$　**31.** $[5, 7)$　**33.** No　**35.** Yes　**37.** Yes
39. $(3, 6)$　　　　　　　**41.** $(1/2, \infty)$

43. $(-\infty, -3) \cup (2, \infty)$　**45.** $(-3, \infty)$

47. $(-\infty, \infty)$　　　　　**49.** $\varnothing$

51. $(-1/3, 1)$　　　　　**53.** $[1, 3/2]$

55. $(-\infty, -1) \cup (2, \infty)$　**57.** $\varnothing$

59. $\{4\}$　　　　　　　**61.** $\varnothing$

63. $(-\infty, -1) \cup (5, \infty)$　**65.** $(-\infty, 1) \cup (5, \infty)$

67. $|x - 4| > 1$　**69.** $|x - 6| < 2$　**71.** $|x - 4| > 0$
73. $|x - 2| < 5$　**75.** $|x| \geq 9$　**77.** $|x - 7| \leq 4$
79. $|x - 5| > 2$　**81.** $[2, \infty)$　**83.** $(-\infty, 2)$
85. $(-\infty, -3] \cup [3, \infty)$

87. $(-\infty, 0)$　　　　　**89.** $(2, \infty)$

91. $[\$0, \$7000]$　**93.** $(93, 115)$　**95.** $(86, 102.5)$
97. $|x - 32| > 12$, younger than 20 or older than 44
99. 10 days

Section 2.7

1. $(1.5, \infty)$　　　　　**3.** $(-\infty, -1)$

5. $(5, \infty)$　　　　　**7.** $(-\infty, -3)$

9. $(-1, 3/2)$　　　　　**11.** $(-\infty, -3) \cup (5, \infty)$

13. $(-\infty, -2] \cup [6, \infty)$　**15.** $[-4, 4]$

17. $\{-3\}$　　　　　　　**19.** $(-\infty, 3/2) \cup (3/2, \infty)$

21. 1　**23.** -1　**25.** 12　**27.** -6.3
29. $(-2, 4]$　　　　　　**31.** $(-7, 7)$

33. $(-\infty, 0)$　　　　　**35.** $[-2, 3] \cup (6, \infty)$

37. $(-\infty, -1/2) \cup (3, 5)$　**39.** $(-\infty, -8) \cup (-3, \infty)$

41. $(-2, 3)$　　　　　　**43.** $(-\infty, -2) \cup (4, 5)$

45. $[-1, 3] \cup (5, \infty)$

47. $(-\infty, -5] \cup (-3, 3) \cup [5, \infty)$

49. $(-\infty, -4) \cup (6, \infty)$

51. $(-3/2, 3/2)$

53. $\left(-\infty, 5 - \sqrt{2}\right) \cup \left(5 + \sqrt{2}, \infty\right)$

55. $(-\infty, \infty)$

57. $(1, \infty)$

59. $(3, \infty)$

61. $(-\infty, -2/3) \cup (1, 5)$

63. $(-\infty, 1) \cup (3, 5) \cup (7, \infty)$

65. $[-3, 0] \cup [3, \infty)$

67. $\left(-\infty, \dfrac{3 - \sqrt{409}}{10}\right) \cup \left(\dfrac{3 + \sqrt{409}}{10}, \infty\right)$

69. $[2, 3) \cup [6, \infty)$

71. $(-\infty, -2) \cup (-1/5, 4)$

73. $[1, 2] \cup (3, 4)$

75. $\left(-\infty, \dfrac{5 - \sqrt{17}}{2}\right] \cup (1, 2) \cup \left[\dfrac{5 + \sqrt{17}}{2}, \infty\right)$

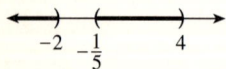

77. $[1/2, \infty)$ **79.** $[-3, 3]$ **81.** $(-\infty, -2) \cup (3, \infty)$
83. $(-\infty, -2) \cup [1, \infty)$ **85.** $(0, 10] \cup [40, 50)$
87. $\sqrt{6}/2 \approx 1.2$ sec **89.** $\left[6 - \sqrt{26}, 6 + \sqrt{26}\right]$
91. $r > 6.38\%$ **93.** 4.788 m/sec **95.** Less than 75%

Chapter 2 Review Exercises

1. $\left\{\dfrac{2}{3}\right\}$ **3.** $\left\{\pm\dfrac{\sqrt{6}}{3}\right\}$ **5.** $\left\{2, -1 \pm i\sqrt{3}\right\}$ **7.** $\left\{\dfrac{32}{15}\right\}$

9. $\{-2\}$ **11.** $\left\{-\dfrac{1}{3}\right\}$ **13.** $\{0\}$ **15.** $\left\{\dfrac{1}{3}, 2\right\}$

17. $\{3 \pm i\}$ **19.** $\left\{2 \pm \sqrt{3}\right\}$ **21.** $\left\{\dfrac{2}{3}, 2\right\}$ **23.** $\{1, 3\}$

25. $\left\{\pm 3i, \pm\sqrt{2}\right\}$ **27.** $\{30\}$ **29.** $\left\{\dfrac{1}{8}, \dfrac{7}{8}\right\}$

31. $\{\pm 2, \pm i\}$
33. $(3, \infty)$ **35.** $(-\infty, 4)$

37. $(-\infty, -14/3)$ **39.** $(-1, 13]$

41. $(1/2, 1)$ **43.** $(-4, \infty)$

45. $(-\infty, 1) \cup (5, \infty)$ **47.** $\{7/2\}$

49. $(-\infty, \infty)$ **51.** $(1/4, 1/2)$ **53.** $[-5, 3]$ **55.** $[3, 5)$
57. $(-\infty, -2) \cup (0, \infty)$ **59.** $(-\infty, 0) \cup (0, 3) \cup (4, \infty)$
61. $\left[1 - \sqrt{5}, 1 + \sqrt{5}\right]$
63. $(-\infty, 1] \cup [2, 3) \cup (4, \infty)$ **65.** $\varnothing$

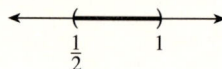

67. $(-\infty, 1) \cup (1, \infty)$ **69.** $y = \dfrac{2}{3}x - 2$

71. $y = \dfrac{1}{x - 3}$ **73.** $y = -\dfrac{a}{b}x + \dfrac{c}{b}$

75. $y = \dfrac{2x}{x + 2}$ **77.** $-1 - i$ **79.** $-9 - 40i$ **81.** 20

83. $-3 - 2i$ **85.** $\dfrac{1}{5} - \dfrac{3}{5}i$ **87.** $-\dfrac{1}{13} + \dfrac{5}{13}i$

89. $3 + i\sqrt{2}$ **91.** $\dfrac{3}{4} - \dfrac{\sqrt{5}}{4}i$ **93.** $-1 - i$ **95.** $\{-7, 9\}$

97. $\varnothing$ **99.** $\{11/4\}$ **101.** $[4/3, \infty)$
103. $(-\infty, -5] \cup [5, \infty)$ **105.** 8, two real
107. 0, one real
109. $\dfrac{19 - \sqrt{209}}{4} \approx 1.14$ in.

111. 136.4 mi **113.** 1600 **115.** 20 mi
117. \$127,659.57 **119.** 2.47×10^9 gal **121.** 20 in.

Chapter 2 Test

1. $\{1\}$ **2.** $\left\{\pm\dfrac{\sqrt{6}}{3}\right\}$ **3.** $\left\{3, \dfrac{-3 \pm 3i\sqrt{3}}{2}\right\}$ **4.** $\{2, 7\}$

5. $(-\infty, -2)$ **6.** $(6, \infty)$

7. $[-1, 2]$ **8.** $(-\infty, 1) \cup (5, \infty)$

9. $7 - 24i$ **10.** $\dfrac{1}{2} - \dfrac{1}{2}i$

11. $-1 + i$ **12.** $-4 + 4i\sqrt{3}$ **13.** $(-2, 4)$
14. $(-\infty, 1/2) \cup (3, \infty)$ **15.** $(-\infty, -3] \cup (-1, 4)$
16. $\left\{3 \pm 3\sqrt{3}\right\}$ **17.** $\{16\}$ **18.** $\{0\}$ **19.** -11

20. 53 **21.** $y = \dfrac{1}{3x + 2}$ **22.** $[5, \infty)$ **23.** $289\ \text{ft}^2$

24. 20 gal

Tying It All Together Chapters 1–2

1. $7x$ **2.** $30x^2$ **3.** $\dfrac{3}{2x}$ **4.** $x^2 + 6x + 9$

5. $6x^2 + x - 2$ **6.** $2x + h$ **7.** $\dfrac{2x}{x^2 - 1}$ **8.** $x^2 + 3x + \dfrac{9}{4}$

9. R **10.** $\left\{0, \dfrac{11}{30}\right\}$ **11.** $(-\infty, 0) \cup (0, \infty)$ **12.** $\{0\}$

13. $\left\{-\dfrac{2}{3}, \dfrac{1}{2}\right\}$ **14.** $\left\{\dfrac{8 \pm \sqrt{89}}{5}\right\}$ **15.** $\{0, 1\}$ **16.** $\{1\}$

17. 0 **18.** -2 **19.** $\dfrac{9}{8}$ **20.** $\dfrac{44}{27}$ **21.** -8 **22.** -2

23. -4 **24.** -2.75

CHAPTER 3

Section 3.1

1. $(4, 1)$, I   **3.** $(1, 0)$, x-axis   **5.** $(5, -1)$, IV
7. $(-4, -2)$, III   **9.** $(-2, 4)$, II
11.   **13.**

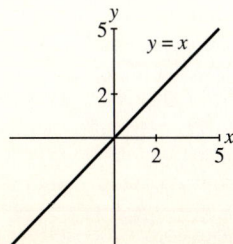

15.

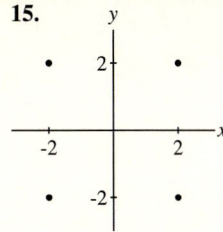

17.

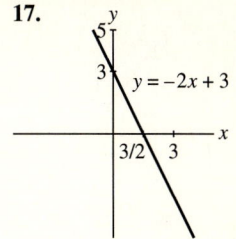

19.

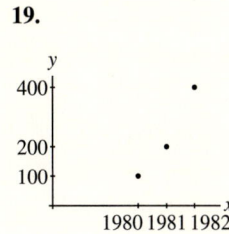

21.

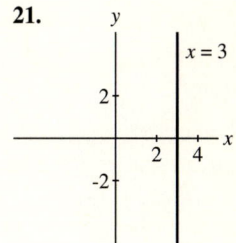

23.

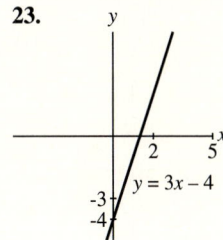

25.

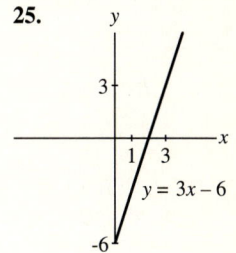

27.

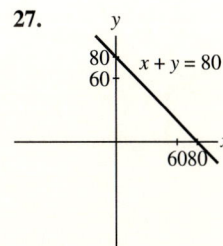

29.

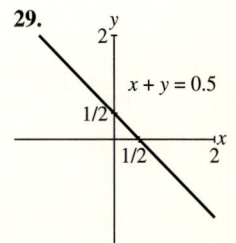

31.

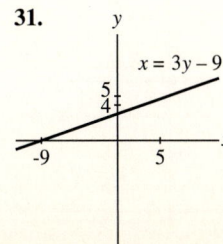

33.

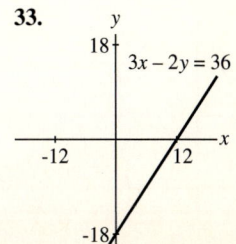

35.

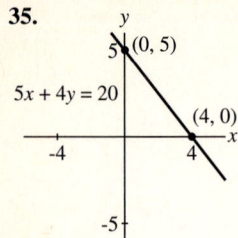

37.

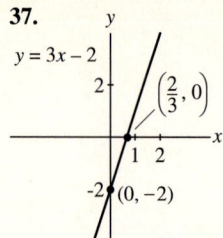

55. 5, (2.5, 5)   **57.** $2\sqrt{2}$, (0, -1)

59. 6, $\left(\dfrac{-2 + 3\sqrt{3}}{2}, \dfrac{5}{2}\right)$ **61.** $\sqrt{5}$, $\left(\dfrac{5\sqrt{2}}{2}, \dfrac{5\sqrt{3}}{2}\right)$

63. $\sqrt{74}$, $(-1.3, 1.3)$ **65.** $|a - b|$, $\left(\dfrac{a + b}{2}, 0\right)$

67. $\dfrac{\sqrt{\pi^2 + 4}}{2}$, $\left(\dfrac{3\pi}{4}, \dfrac{1}{2}\right)$ **69.** No   **71.** Yes

73. $-2 \pm 4\sqrt{2}$ **75.** (5, 7)

85.

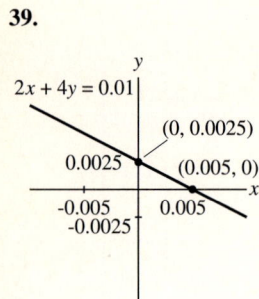

39.

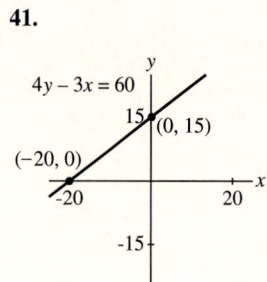

41.

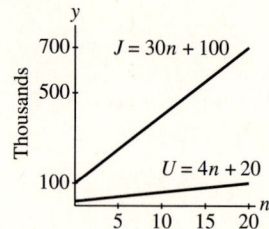

Graphing Calculator Exercises

1.

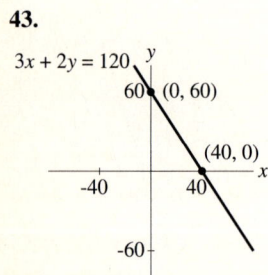

3.

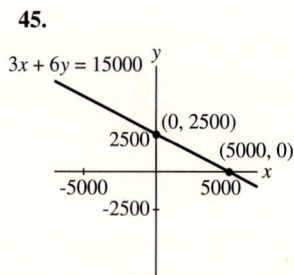

43.

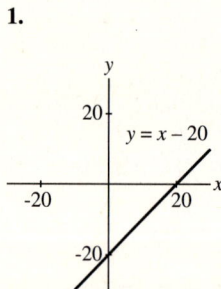

45.

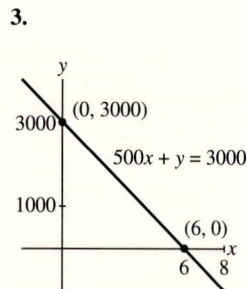

5. $\{-2.83\}$ **7.** $\{558.54\}$ **9.** $\{116,566.67\}$ **11.** $\{4.91\}$

47.

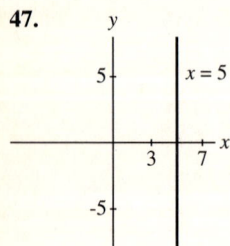

49.

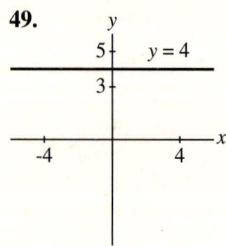

Section 3.2

1. Yes   **3.** No   **5.** No   **7.** Yes   **9.** No   **11.** Yes
13. Yes   **15.** Yes   **17.** Yes   **19.** No   **21.** Yes
23. Yes   **25.** $\{-3, \pi, 5\}$, $\{1, \sqrt{2}, 6\}$
27. Domain $(-\infty, \infty)$, range $[5, \infty)$
29. Domain $(-\infty, 0]$, range $(-\infty, \infty)$
31. Domain $[0, \infty)$, range $[5, \infty)$
33. Domain $(-\infty, \infty)$, range $(-\infty, \infty)$
35. Domain $(0, 3)$, range $(2, 5)$
37. 10   **39.** -6 **41.** 8   **43.** 0.3808   **45.** -2
47. -24 **49.** $3a^2 - a$ **51.** $3a^2 + 17a + 24$
53. $3x^2 + 6xh + 3h^2 - x - h$ **55.** $6xh + 3h^2 - h$
57. $3x^2 + 3x - 2$ **59.** $12x^3 - 10x^2 + 2x$ **61.** 0, 1/3
63. 1, -7 **65.** 114   **67.** 2   **69.** 3   **71.** $6x + 3h$

51.

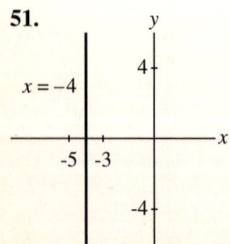

53.

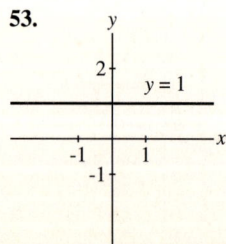

73. $-2x - h + 1$ **75.** $\dfrac{1}{\sqrt{x + h + 2} + \sqrt{x + 2}}$

77. $\dfrac{-1}{x(x + h)}$ **79.** a **81.** $A = s^2$ **83.** $s = \dfrac{d\sqrt{2}}{2}$

85. $P = 4s$ **87.** $A = \dfrac{P^2}{16}$ **89.** $C = 353n$

91. $C = 50 + 35n$ **93.** 0.1125 dollars/card

95. \$16/yd, \$14/yd **97.** -6.2 million hectares/yr

99. $MC(x) = 0.06x + 40.03$, \$46.03

19. $(2, 3/2)$, $5/2$

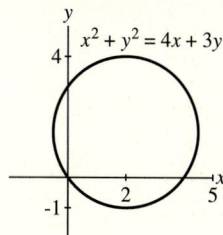
$x^2 + y^2 = 4x + 3y$

21. $(1/4, -1/6)$, $1/6$

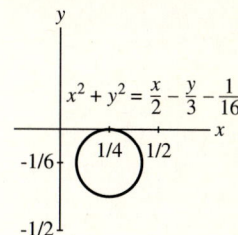
$x^2 + y^2 = \dfrac{x}{2} - \dfrac{y}{3} - \dfrac{1}{16}$

Section 3.3

1. $(0, 0)$, 4

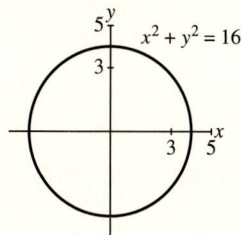
$x^2 + y^2 = 16$

3. $(-6, 0)$, 6

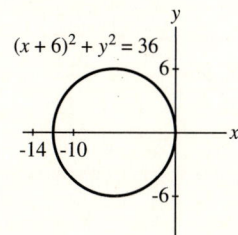
$(x + 6)^2 + y^2 = 36$

23. Domain $[0, \infty)$, range $[0, \infty)$, yes

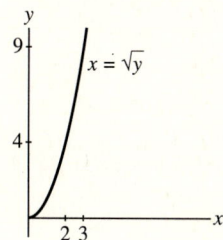
$x = \sqrt{y}$

5. $(2, -2)$ $2\sqrt{2}$

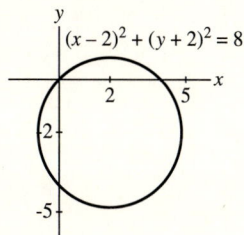
$(x - 2)^2 + (y + 2)^2 = 8$

25. Domain $(-\infty, \infty)$, range $[-1, \infty)$, yes

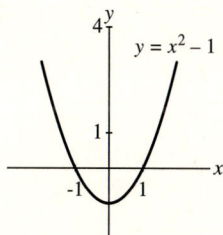
$y = x^2 - 1$

7. $x^2 + y^2 = 7$ **9.** $(x + 2)^2 + (y - 5)^2 = 1/4$

11. $(x - 3)^2 + (y - 5)^2 = 34$ **13.** $(x - 5)^2 + (y + 1)^2 = 32$

15. $(0, -3)$, 3 **17.** $(3, 4)$, 5

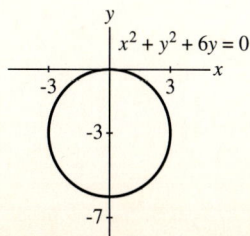
$x^2 + y^2 + 6y = 0$

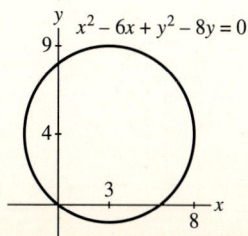
$x^2 - 6x + y^2 - 8y = 0$

27. Domain $(-\infty, \infty)$, range $\{5\}$, yes

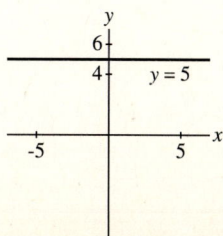
$y = 5$

29. Domain $(-\infty, \infty)$, range $(-\infty, \infty)$, yes

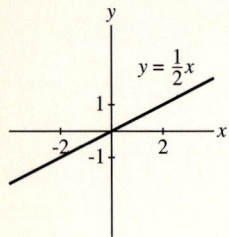

37. Domain $(-\infty, \infty)$, range $[0, \infty)$, yes

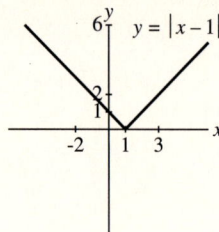

31. Domain $(-\infty, \infty)$, range $(-\infty, \infty)$, yes

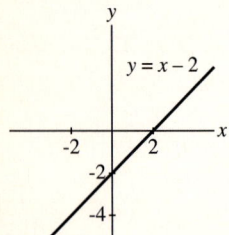

39. Domain $[0, \infty)$, range $(-\infty, \infty)$, no

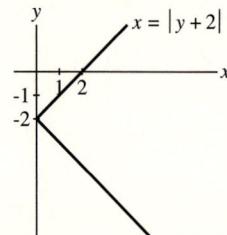

33. Domain $(-\infty, \infty)$, range $[0, \infty)$, yes

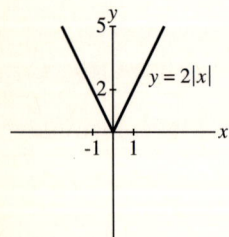

41. Domain $[-1, 1]$, range $[0, 1]$, yes

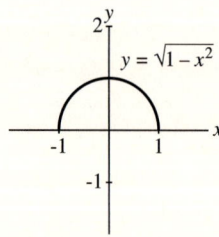

35. Domain $[1, \infty)$, range $(-\infty, \infty)$, no

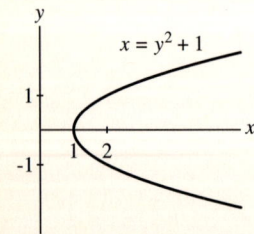

43. Domain $[-1, 1]$, range $[-1, 1]$, no

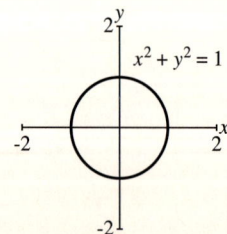

45. Domain $(-\infty, 0]$, range $(-\infty, \infty)$, no

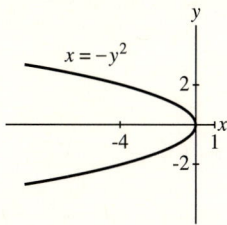

$x = -y^2$

47. Domain $(-\infty, \infty)$, range $(-\infty, 1]$, yes

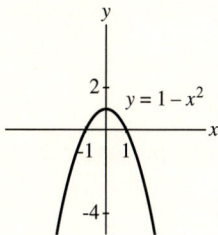

$y = 1 - x^2$

49. Domain $(-\infty, 0) \cup (0, \infty)$, range $(-\infty, 0) \cup (0, \infty)$, yes

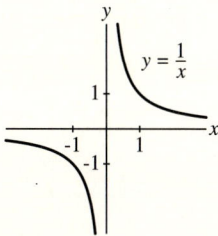

$y = \dfrac{1}{x}$

51. Domain $(-\infty, \infty)$, range $(-\infty, 0]$, yes

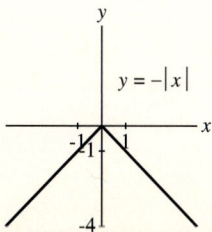

$y = -|x|$

53. No   **55.** Yes   **57.** Yes

59. Domain $(-\infty, \infty)$, range $\{-2, 2\}$

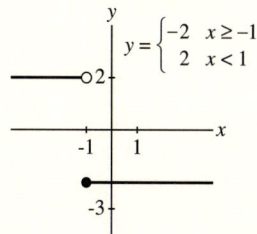

$y = \begin{cases} -2 & x \geq -1 \\ 2 & x < 1 \end{cases}$

61. Domain $(-\infty, \infty)$, range $[0, \infty)$

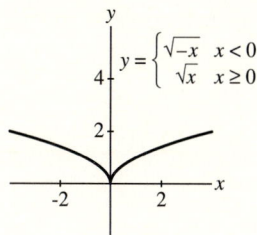

$y = \begin{cases} \sqrt{-x} & x < 0 \\ \sqrt{x} & x \geq 0 \end{cases}$

63. Domain $[-2, \infty)$, range $[0, \infty)$

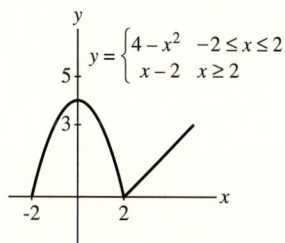

$y = \begin{cases} 4 - x^2 & -2 \leq x \leq 2 \\ x - 2 & x \geq 2 \end{cases}$

65. Domain $(-\infty, \infty)$, range integers

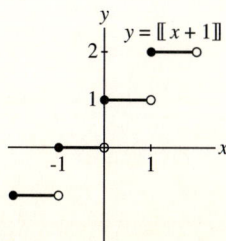

$y = [\![x + 1]\!]$

67. Domain $[0, 4)$, range $\{2, 3, 4, 5\}$

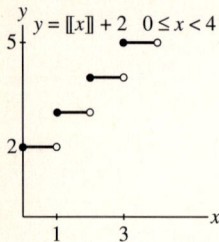

$y = [\![x]\!] + 2 \quad 0 \le x < 4$

69. Domain $(-\infty, \infty)$, range $(-\infty, 4]$, inc on $(-\infty, 0)$, dec on $(0, \infty)$

71. Domain $(-\infty, 2]$, range $(-\infty, 2]$, inc on $(-\infty, -2)$, constant on $(-2, 2)$

73. Domain $[-4, 4]$, range $[0, 4]$, inc on $(-4, 0)$, dec on $(0, 4)$

75. Domain $(-\infty, \infty)$, range $(-\infty, \infty)$, inc on $(-\infty, \infty)$

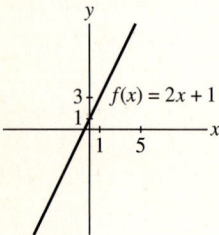

$f(x) = 2x + 1$

77. Domain $(-\infty, \infty)$, range $[0, \infty)$, dec on $(-\infty, 1)$, inc on $(1, \infty)$

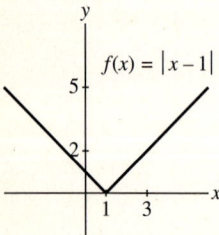

$f(x) = |x - 1|$

79. Domain $(-\infty, 0) \cup (0, \infty)$, range $\{-2, 2\}$, constant on $(-\infty, 0)$ and $(0, \infty)$

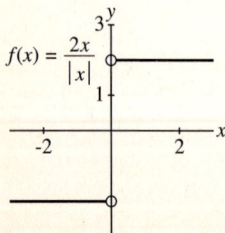

$f(x) = \dfrac{2x}{|x|}$

81. Domain $[-1, 1]$, range $[-1, 0]$, dec on $(-1, 0)$, inc on $(0, 1)$

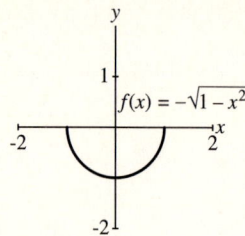

$f(x) = -\sqrt{1 - x^2}$

83. Domain $(-\infty, \infty)$, range $(-\infty, 4) \cup (4, \infty)$, inc on $(-\infty, 3)$ and $(3, \infty)$

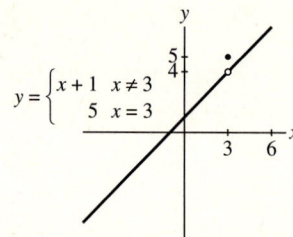

$y = \begin{cases} x + 1 & x \ne 3 \\ 5 & x = 3 \end{cases}$

85. Domain $(-\infty, \infty)$, range $(-\infty, 2]$, inc on $(-\infty, 0)$, dec on $(0, \infty)$

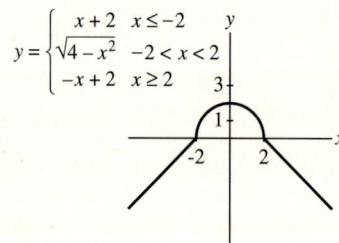

$y = \begin{cases} x + 2 & x \le -2 \\ \sqrt{4 - x^2} & -2 < x < 2 \\ -x + 2 & x \ge 2 \end{cases}$

87. $(x + 1)^2 + (y - 3)^2 = 29$

89. $f(x) = \begin{cases} -4[\![-x]\!] & 0 < x \le 3 \\ 15 & 3 < x \le 8 \end{cases}$

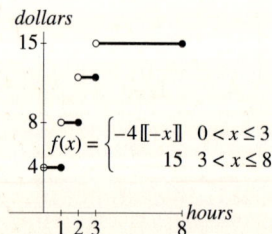

$f(x) = \begin{cases} -4[\![-x]\!] & 0 < x \le 3 \\ 15 & 3 < x \le 8 \end{cases}$

91. $(0, 10^4), (10^4, \infty)$

93. $[5, \infty)$

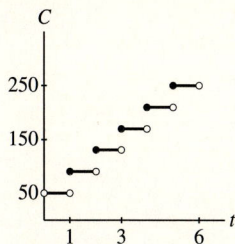

95. $4.75, $6.25, $0.235

Graphing Calculator Exercises

1. $y = -1 \pm \sqrt{5 - (x - 2)^2}$

5. Increasing on $(-\infty, -1)$ and $(1, \infty)$, decreasing on $(-1, 1)$

Section 3.4

1.

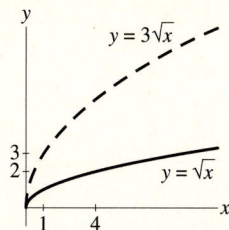

3.

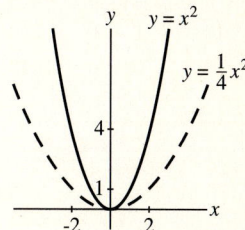

5.

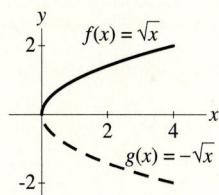

7.

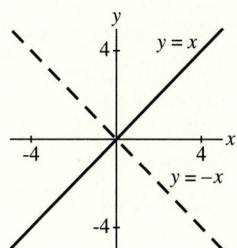

9.

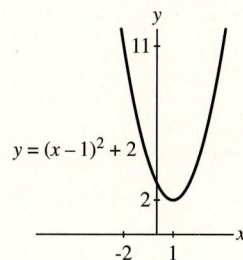

11.

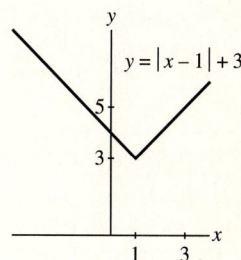

13.

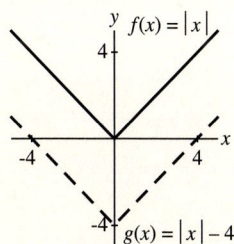

15.

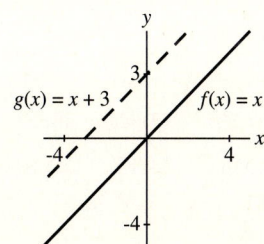

17. (g)   **19.** (b)   **21.** (c)   **23.** (f)

25.

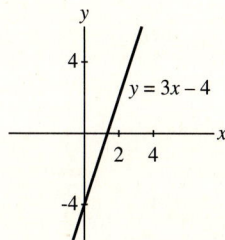

27.

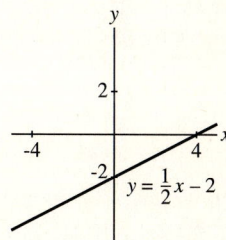

29.

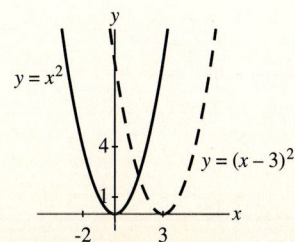

31.

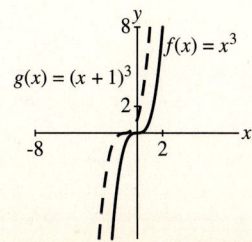

33.

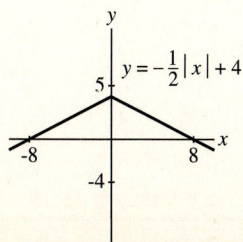

35.

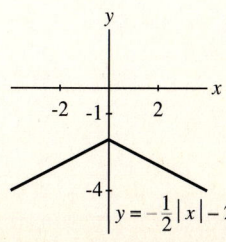

37.

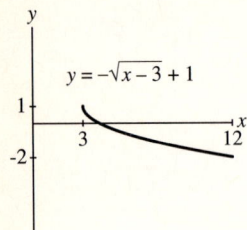

$y = -\sqrt{x-3} + 1$

39.

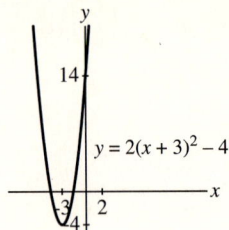

$y = 2(x+3)^2 - 4$

41.

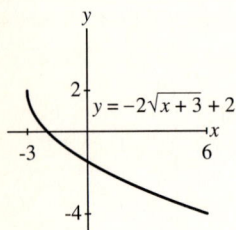

$y = -2\sqrt{x+3} + 2$

43.

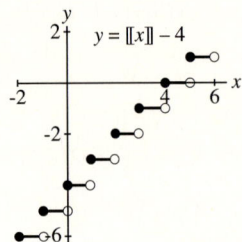

$y = [\![x]\!] - 4$

45. Symmetric about y-axis, even
47. No symmetry, neither
49. Symmetric about $x = -3$, neither
51. No symmetry, neither   **53.** Symmetric about origin, odd
55. No symmetry, neither   **57.** No symmetry, neither
59. Symmetric about $x = 2$, neither
61. Symmetric about y-axis, even
63. Symmetric about y-axis, even   **65.** (e)   **67.** (g)
69. (b)   **71.** (c)   **73.** $(-\infty, -1] \cup [1, \infty)$
75. $(-\infty, -1) \cup (5, \infty)$ **77.** $(-2, 4)$ **79.** $[0, 25]$
81. $\left(-\infty, 2 - \sqrt{3}\right) \cup \left(2 + \sqrt{3}, \infty\right)$ **83.** $(-5, 5)$
85. $N(x) = x + 2000$ **87.** $x \geq 25\%$

Graphing Calculator Exercises

1. $[-1.774, \infty)$ **3.** The function is even.
5. Horizontal translation to the left   **7.** $(-3.360, 1.546)$

Section 3.5

1. 1   **3.** -11 **5.** -8 **7.** $1/12$ **9.** $a^2 - 3$
11. $a^3 - 4a^2 + 3a$ **13.** $y = 6x - 1$
15. $y = x^2 + 6x + 7$ **17.** $y = x$ **19.** $y = x$
21. $\{(-3, 3), (2, 6)\}$ **23.** $\{(-3, 2)\}$
25. $\{(-3, 2), (2, 0)\}$ **27.** $\{(-3, 0), (1, 0), (4, 4)\}$
29. $\{(1, 4)\}$ **31.** $\{(-3, 4), (1, 4)\}$
33. $(f + g)(x) = \sqrt{x} + x - 4, [0, \infty)$

35. $(g \cdot h)(x) = \dfrac{x - 4}{x - 2}, (-\infty, 2) \cup (2, \infty)$

37. $\left(\dfrac{f}{g}\right)(x) = \dfrac{\sqrt{x}}{x - 4}, [0, 4) \cup (4, \infty)$

39. $\left(\dfrac{g}{h}\right)(x) = x^2 - 6x + 8, (-\infty, 2) \cup (2, \infty)$

41. 5   **43.** 5   **45.** 59.8163   **47.** $3x^2 + 2$

49. $9x^2 - 6x + 2$ **51.** $\dfrac{x^2 + 2}{3}$ **53.** $9x - 4$

55. $3x^2 - 2x + 1$ **57.** $(f \circ g)(x) = \sqrt{2x - 1}, \left[\dfrac{1}{2}, \infty\right)$

59. $(h \circ f)(x) = \dfrac{1}{\sqrt{x} - 3}, [0, 9) \cup (9, \infty)$

61. $(h \circ h)(x) = \dfrac{x - 3}{10 - 3x}, (-\infty, 3) \cup \left(3, \dfrac{10}{3}\right) \cup \left(\dfrac{10}{3}, \infty\right)$

63. $(f \circ f)(x) = x^{1/4}, [0, \infty)$
71. $F = g \circ h$ **73.** $H = f \circ g \circ h$ **75.** $N = h \circ g \circ f$
77. $P = g \circ f \circ g$ **79.** $A = d^2/2$ **81.** $(f + g)(x) = 0.55x$

83. $W = \dfrac{(8 + \pi)s^2}{8}$ **85.** $s = \dfrac{d\sqrt{2}}{2}$ **87.** $A = \dfrac{25.6}{\pi d}$ ft^2

89. $P(x) = -20x^2 + 2400x - 4000, -\$1620, \$7500, -\$8880$

Graphing Calculator Exercises

1. $y = -x$, no   **3.** $[-1, \infty), [-7, \infty)$ **5.** $[1, \infty), [0, \infty)$
7. $[-64, \infty), [0, \infty)$

Section 3.6

1. $\{(1, 2), (5, 3)\}, 3, 2$
3. $\{(-3, -3), (5, 0), (-7, 2)\}, 0, 2$ **5.** No
7. Yes, $\{(0, 3), (5, 2), (6, 4), (9, 7)\}$
9. Yes, $\{(1, 1), (2, 2), (4.5, 4.5)\}$
11. Yes, $\{(1, 1), (4, 2), (9, 3), (16, 4)\}$
13. Yes, $\{(x, y) \mid y = x - 2\}$

15. Yes, $\left\{(x, y) \mid y = \dfrac{x - 7}{2}\right\}$ **17.** No   **19.** No

21. Yes   **23.** No   **25.** No   **27.** No   **29.** Yes

31. Yes   **33.** No   **35.** No   **37.** $f^{-1}(x) = \dfrac{x + 7}{3}$

39. $f^{-1}(x) = (x - 2)^2 + 3$ for $x \geq 2$

41. $f^{-1}(x) = -x - 9$ **43.** $f^{-1}(x) = \dfrac{5x + 3}{x - 1}$

45. $f^{-1}(x) = \dfrac{2x - 3}{x - 2}$ **47.** $f^{-1}(x) = -\dfrac{1}{x}$

49. $f^{-1}(x) = (x - 5)^3 + 9$ **51.** $f^{-1}(x) = \sqrt{x} + 2$
53. $(g \circ f)(x) = x$, yes   **55.** $(g \circ f)(x) = |x|$, no
57. $(g \circ f)(x) = x$, yes   **59.** $(g \circ f)(x) = x$, yes

61. $(g \circ f)(x) = x$, yes   **63.** $f^{-1}(x) = \dfrac{x}{5}$

65. $f^{-1}(x) = x + 88$ **67.** $f^{-1}(x) = \dfrac{x + 7}{3}$

69. $f^{-1}(x) = \dfrac{x-4}{-3}$ **71.** $f^{-1}(x) = 2x + 18$

73. $f^{-1}(x) = -x$ **75.** $f^{-1}(x) = (x+9)^3$

77. $f^{-1}(x) = \sqrt[3]{\dfrac{x+7}{2}}$ **79.** No   **81.** Yes

83. $f^{-1}(x) = \dfrac{x-2}{3}$ **85.** $f^{-1}(x) = \sqrt{x+4}$

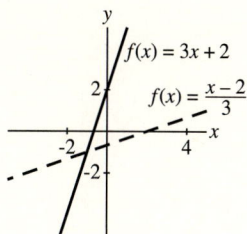

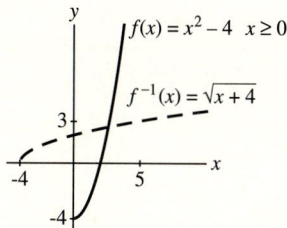

87. $f^{-1}(x) = \sqrt[3]{x}$

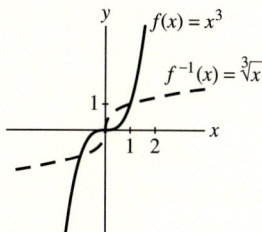

89. $f^{-1}(x) = (x+3)^2$ for $x \geq -3$

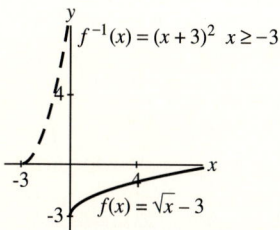

91. $C = 1.08P$, $P = \dfrac{C}{1.08}$

93. $r = \sqrt{5.625 \times 10^{-5} - V/500}$ for $0 \leq V \leq 0.028125$

95. 14.8%, $V = 40{,}000(1 - r)^5$ **99.** $1 + \dfrac{-5}{x+2}$

Graphing Calculator Exercises

5. No

Section 3.7

1. $G = kn$ **3.** $m_1 = \dfrac{k}{m_2}$

5. $C = khr$ **7.** $Y = \dfrac{kx}{\sqrt{z}}$

9. A varies directly as the square of r.
11. The variable a varies jointly as z and w.
13. No variation
15. The variable y is inversely proportional to x.
17. H varies directly as the square root of t and inversely as s.
19. D varies jointly as L and J and inversely as W.

21. $y = \dfrac{5}{9}x$ **23.** $T = \dfrac{-150}{y}$ **25.** $m = 3t^2$

27. $y = \dfrac{1.37x}{\sqrt{z}}$ **29.** $-\dfrac{27}{2}$ **31.** 1   **33.** $\sqrt{6}$ **35.** 7/4

37. Direct, $L_i = 12L_f$ **39.** Inverse, $P = 20/n$
41. Direct, $S_m = 0.6S_k$ **43.** Neither   **45.** Direct, $A = 30W$
47. Inverse, $n = 5/p$ **49.** 12.8 hr   **51.** $50.70
53. $19.84   **55.** 18.125 oz   **57.** 8 ft/yr
59. 1.921×10^{27} kg   **61.** 16.5 ft

Chapter 3 Review Exercises

1. Domain $(-\infty, \infty)$, range $(-\infty, \infty)$, yes

3. Domain $\{0, 1, 2\}$, range $\{-3, -1, 1, 3\}$, no

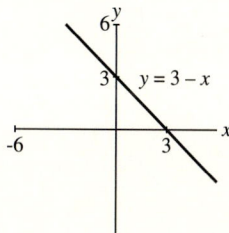

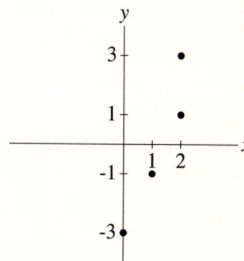

5. Domain $[-0.1, 0.1]$, range $[-0.1, 0.1]$, no

7. Domain $[1, \infty)$, range $(-\infty, \infty)$, no

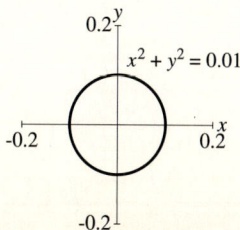

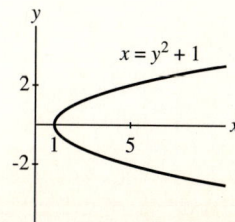

9. Domain $[0, \infty)$, range $[-3, \infty)$, yes

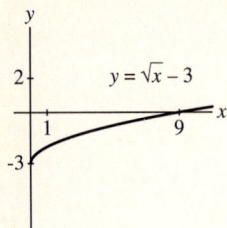

11. 12   **13.** 17   **15.** 17   **17.** 4   **19.** -36 **21.** 12

23. $x^4 + 6x^2 + 12$ **25.** $a^2 + 2a + 4$ **27.** $6 + h$

29. $2x + h$ **31.** x **33.** $\dfrac{x + 7}{2}$

35.

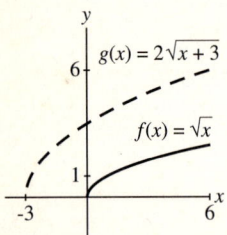

37.

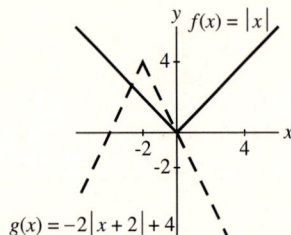

39.

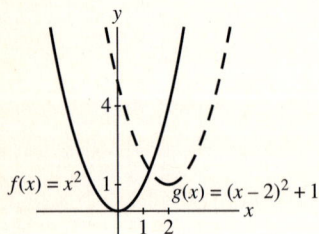

41. $F = f \circ g$ **43.** $H = f \circ h \circ g \circ j$ **45.** $N = h \circ f \circ j$
47. $R = g \circ h \circ j$ **49.** $\sqrt{146}, (-1/2, -1/2)$
51. $\dfrac{\sqrt{73}}{12}, \left(\dfrac{3}{8}, \dfrac{2}{3}\right)$ **53.** -5 **55.** $\dfrac{-1}{2x(x + h)}$

57. Domain $[-10, 10]$,
range $[0, 10]$, inc on
$(-10, 0)$, dec on $(0, 10)$

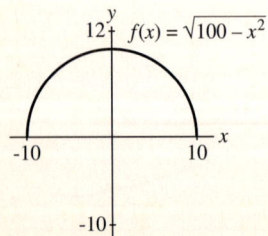

59. Domain $(-\infty, \infty)$, range
$(-\infty, \infty)$, inc on $(-\infty, \infty)$

$$f(x) = \begin{cases} -x^2 & x \le 0 \\ x^2 & x > 0 \end{cases}$$

61. Domain $(-\infty, \infty)$, range $[-2, \infty)$, inc on $(-2, 0)$ and $(2, \infty)$,
dec on $(-\infty, -2)$ and $(0, 2)$

$$f(x) = \begin{cases} -x - 4 & x \le -2 \\ -|x| & -2 < x < 2 \\ x - 4 & x \ge 2 \end{cases}$$

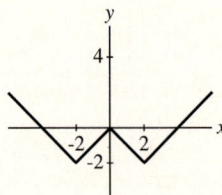

63. $y = |x| - 3, (-\infty, \infty), [-3, \infty)$
65. $y = -2|x| + 4, (-\infty, \infty), (-\infty, 4]$
67. $y = |x + 2| + 1, (-\infty, \infty), [1, \infty)$
69. Symmetric about y-axis
71. Symmetric about origin
73. Neither symmetry
75. Symmetric about y-axis
77. $(-\infty, 2] \cup [4, \infty)$ **79.** $(-2, 2)$ **81.** $\varnothing$
83.   **85.**

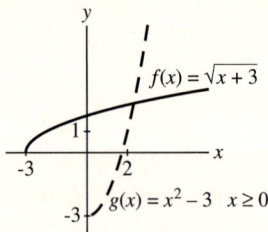

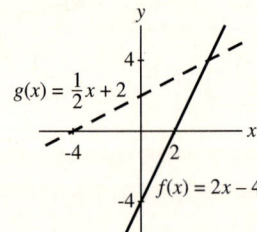

87. Not invertible

89. $f^{-1}(x) = \dfrac{x + 21}{3}, (-\infty, \infty), (-\infty, \infty)$

91. Not invertible
93. $f^{-1}(x) = x^2 + 9$ for $x \ge 0, [0, \infty), [9, \infty)$

95. $f^{-1}(x) = \dfrac{5x + 7}{1 - x}, (-\infty, 1) \cup (1, \infty), (-\infty, -5) \cup (-5, \infty)$

97. $f^{-1}(x) = -\sqrt{x - 1}, [1, \infty), (-\infty, 0]$ **99.** $(3, 0), (0, -9/4)$
101. 36   **103.** $(x + 3)^2 + (y - 5)^2 = 3$
105. $y = -2(x - 1)^2 + 7, (1, 7)$ **107.** $A = 4r^2$

109. 30 mph/sec   **111.** 22.5   **113.** $F = \dfrac{km_1m_2}{d^2}$

Chapter 3 Test

1. No   **2.** Yes   **3.** No   **4.** Yes
5. $\{2, 5\}, \{-3, -4, 7\}$ **6.** $[9, \infty), [0, \infty)$ **7.** $[0, \infty), (-\infty, \infty)$

8.

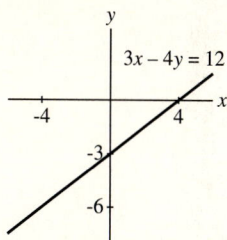

9.

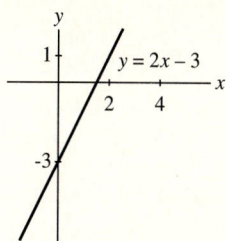

17.

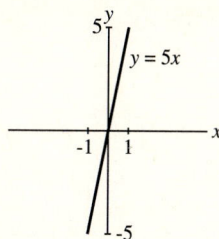

18.

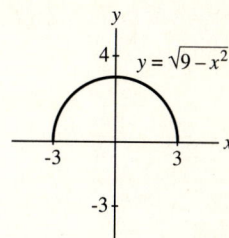

10.

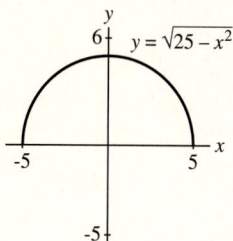

11.

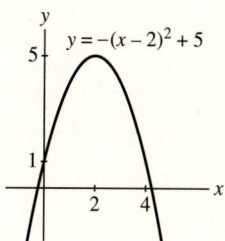

19.

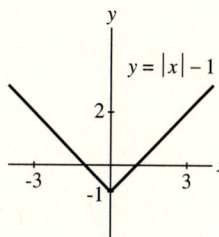

20.

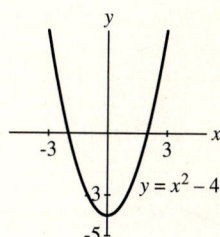

12.

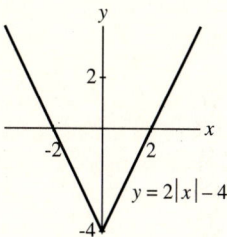

13.

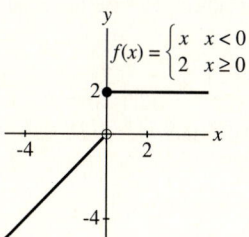

21.

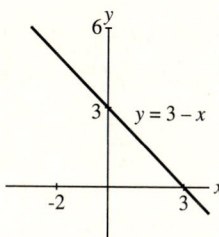

22.

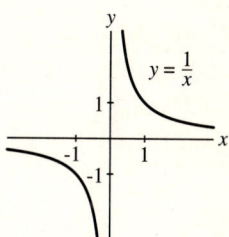

14. 3   **15.** $\sqrt{7}$ **16.** $\dfrac{x+1}{3}$ **17.** 45   **18.** 3

19. $(0, 5.5)$ **20.** $5\sqrt{2}$ **21.** $(x+4)^2 + (y-1)^2 = 7$

22. Dec on $(-\infty, 3)$, inc on $(3, \infty)$

23. Symmetric about y-axis   **24.** $(-2, 4)$

25. $g^{-1}(x) = (x-3)^3 + 2$

26. \$0.125 per envelope   **27.** 12 candlepower   **28.** $V = \dfrac{\sqrt{2}d^3}{4}$

23.

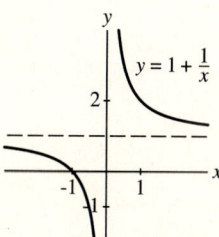

24.

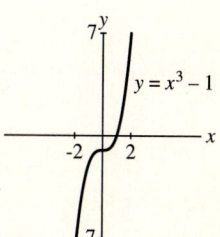

Tying It All Together Chapters 1–3

1. $5x$ **2.** $x^2 - 6x + 9$ **3.** $|x|$ **4.** $x^2 - 4$

5. $x^4 + x^3 - 6x^2$ **6.** 13   **7.** $-2 \pm \sqrt{2}$ **8.** $\dfrac{-3 \pm i\sqrt{6}}{3}$

9. $(-\infty, \infty)$ **10.** $\{0, 3\}$ **11.** $\{-1, 1\}$ **12.** $\{-2, 2\}$

13. $\{2 \pm i\}$ **14.** $\left\{ \dfrac{1 \pm \sqrt{2}}{2} \right\}$ **15.** $\{-3, 0, 3\}$

16. $\{-1, 0, 2\}$

25. $\dfrac{x^2 + 1}{x}$ **26.** $\dfrac{x^2}{x-1}$ **27.** $\dfrac{3x - 10}{x - 4}$ **28.** $\dfrac{-2x - 7}{x + 3}$

CHAPTER 4

Section 4.1

1. $\dfrac{1}{3}$ **3.** -4 **5.** 0 **7.** $-\sqrt{2}$ **9.** $\dfrac{\pi}{8}$ **11.** $\dfrac{3}{2}$

13. $y = \dfrac{3}{5}x - 2, \dfrac{3}{5}, (0, -2)$ **15.** $y = 2x - 5, 2, (0, -5)$

17. $y = \dfrac{1}{2}x + \dfrac{1}{2}, \dfrac{1}{2}, \left(0, \dfrac{1}{2}\right)$

19. $y = -\dfrac{1}{3}x + \dfrac{7}{12}, -\dfrac{1}{3}, \left(0, \dfrac{7}{12}\right)$

21. $y = -\dfrac{1}{3}x + \dfrac{26}{3}, -\dfrac{1}{3}, \left(0, \dfrac{26}{3}\right)$

23. $y = 0.03x - 2.6, 0.03, (0, -2.6)$

25.

27.

29.

31.

33.

35.

37.

39.

41. $y = \dfrac{2}{3}x - 1$ **43.** $y = \dfrac{5}{2}x + \dfrac{3}{2}$ **45.** $y = -2x + 4$

47. $4x - 3y = 12$ **49.** $4x - 5y = -7$ **51.** $x = -4$

53. 0.5 **55.** -1 **57.** 0 **59.** $2x - y = 4$

61. $3x + y = 7$ **63.** $2x + y = -5$ **65.** $y = 5$

67. $3x - 5y = -15$ **69.** $20/3$ **71.** -5 **73.** T   **75.** T

77. F   **79.** T   **81.** $F = \dfrac{9}{5}C + 32, 302°F$

83. $b = 67t - 128,100, \$5766$ **85.** $S = -0.005D + 95$

87. $y = \dfrac{2}{3}x + 6000$ **89.** $P = -2200x + 16,400, \$7600$

91. Japan, never

Graphing Calculator Exercises

1. $(200, 96)$ **3.** $(5.4, 9.2)$ **5.** $(-0.003, -2.00)$
7. $(-1, 1)$ **9.** $y = 2x - 1$

Section 4.2

1. $y = \left(x - \dfrac{3}{2}\right)^2 - \dfrac{9}{4}$

3. $y = 2(x - 3)^2 + 4$

5. $y = -\dfrac{1}{2}(x - 1)^2 + 3$

7. $y = \left(x + \dfrac{3}{2}\right)^2 + \dfrac{1}{4}$

9. $y = -2\left(x - \dfrac{3}{4}\right)^2 + \dfrac{1}{8}$

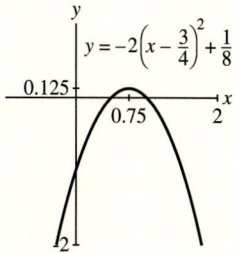

11. $y = -3\left(x - \dfrac{1}{3}\right)^2 + \dfrac{1}{3}$

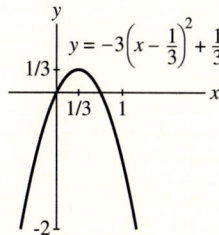

41. Vertex (2, 12), axis $x = 2$

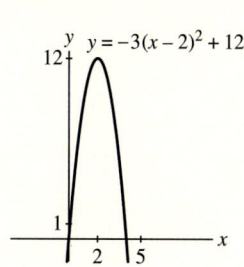

43. Vertex (1, 3), axis $x = 1$

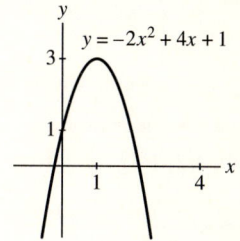

13. (2, −11)   **15.** (4, 1)   **17.** (−1/3, 1/18)
19. Up, (1, −4), $x = 1$, [−4, ∞), min −4, dec on (−∞, 1), inc on (1, ∞)
21. Range [−1, ∞), min value −1, dec (−∞, 1), inc (1, ∞)
23. Range $\left(-\infty, \sqrt{3}\,\right]$, max value $\sqrt{3}$, inc (−∞, 0), dec (0, ∞)
25. Range (−∞, 27/2], max value 27/2, inc (−∞, 3/2), dec (3/2, ∞)
27. Range $\left[-2 - \sqrt{2}, \infty\right)$, min value $-2 - \sqrt{2}$, dec (−∞, −1), inc (−1, ∞)
29. Range [4, ∞), min value 4, dec (−∞, 3), inc (3, ∞)
31. Range (−∞, 9], max value 9, inc (−∞, 1/2), dec (1/2, ∞)

33. Vertex (0, −3), axis $x = 0$

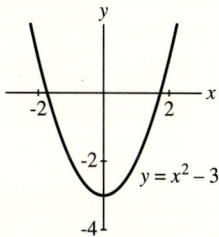

35. Vertex $\left(\dfrac{\pi}{2}, -\dfrac{\pi^2}{4}\right)$, axis $x = \dfrac{\pi}{2}$

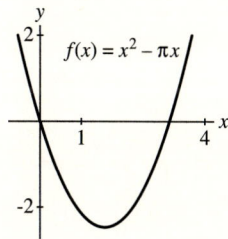

37. Vertex (−3, 0), axis $x = -3$

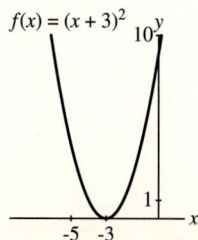

39. Vertex (3, −4), axis $x = 3$

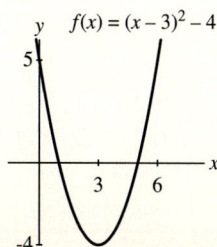

45. (−∞, −1] ∪ [3, ∞)   **47.** (−3, 1)   **49.** [−3, 1]
51. [2, 3]   **53.** $\left(-\infty, 3 - \sqrt{2}\,\right) \cup \left(3 + \sqrt{2}, \infty\right)$
55. (−∞, ∞)   **57.** $x = 0$, (−2, 4)   **59.** $x = 1$, (0, 8)
61. $x = 1$, (3, 16)   **63.** $x = -1/2$, (1, −13)   **65.** −12
67. 3   **69.** $y = 5(x - 1)^2 - 3$ **71.** $y = -\dfrac{1}{2}(x + 3)^2 + 2$
73. 7/2, 7/2   **75.** 25 people, \$625
77. 5 in. wide, 2.5 in. high   **79.** 97.24 mph   **81.** 261 ft
83. 0.9   **85.** 1/2   **87.** $x = 1$ in., $y = \dfrac{8}{3\pi}$ in.
89. (0, 1000), inc on (0, 500), dec on (500, ∞)

Graphing Calculator Exercises

1. $(x - 72)(x + 48)$ **3.** (−0.203, 0.103)
5. $y = x^2 + 4x - 5$

Section 4.3

1. $x - 3$, 1   **3.** $-2x^2 + 6x - 14$, 33   **5.** $s^2 + 2$, 16
7. $x + 6$, 13   **9.** $-x + 7$, −12   **11.** $4x^2 + 2x - 4$, 0
13. $2a^2 - 4a + 6$, 0   **15.** $x^3 + x^2 + x + 1$, −2
17. $x - \dfrac{5}{2}, -\dfrac{1}{4}$ **19.** 0   **21.** −33   **23.** 5   **25.** 55/8
27. 0   **29.** 8   **31.** $(x + 3)(x + 2)(x - 1)$
33. $(x - 4)(x + 3)(x + 5)$ **35.** Yes   **37.** No   **39.** Yes
41. No   **43.** 5/3   **45.** −3, −2   **47.** $\pm i\sqrt{5}$ **49.** $\pm i$, 3
51. −3, 0, 5/2   **53.** $\pm\sqrt{6}$, −1   **55.** 5, $\dfrac{-5 \pm 5i\sqrt{3}}{2}$
57. ± 1, $\pm i$ **59.** ± 1, $\dfrac{-1 \pm i\sqrt{3}}{2}$, $\dfrac{1 \pm i\sqrt{3}}{2}$
61. $\pm(1, 2, 3, 4, 6, 8, 12, 24)$ **63.** $\pm(1, 3, 5, 15)$
65. $\pm\left(1, 3, 5, 15, \dfrac{1}{2}, \dfrac{1}{4}, \dfrac{1}{8}, \dfrac{3}{2}, \dfrac{3}{4}, \dfrac{3}{8}, \dfrac{5}{2}, \dfrac{5}{4}, \dfrac{5}{8}, \dfrac{15}{2}, \dfrac{15}{4}, \dfrac{15}{8}\right)$
67. $\pm\left(1, 2, \dfrac{1}{2}, \dfrac{1}{3}, \dfrac{2}{3}, \dfrac{1}{6}, \dfrac{1}{9}, \dfrac{2}{9}, \dfrac{1}{18}\right)$ **69.** 2, 3, 4
71. −3, $2 \pm i$ **73.** $\dfrac{1}{2}, \dfrac{3}{2}, \dfrac{5}{2}$ **75.** $\dfrac{1}{2}, \dfrac{1 \pm i}{3}$

77. $\pm i, 1, -2$ **79.** $-1, -1, \pm\sqrt{2}$ **81.** $2 + \dfrac{5}{x-2}$

83. $a + \dfrac{5}{a-3}$ **85.** $1 + \dfrac{-3c}{c^2-4}$ **87.** $2 + \dfrac{-7}{2t+1}$

89. $f(x) = -4x^3 + 26x^2 - 42x + 18$, 1.5 in.

Graphing Calculator Exercises

1. 1/4, 1/3, 1/2   **3.** 1/16   **5.** $-6/7, 7/3$

Section 4.4

1. ± 3, 0 with multiplicity 3

3. 0, 1 each with multiplicity 2

5. $\pm\sqrt{5}$ each with multiplicity 2

7. $-4/3, 3/2$ each with multiplicity 2

9. $0, 2 \pm \sqrt{10}$ **11.** $\pm i, -3$ **13.** $x^2 + 9$

15. $x^2 - 2x - 1$ **17.** $x^2 - 6x + 13$

19. $x^3 - 8x^2 + 37x - 50$ **21.** $x^2 - 2x - 15 = 0$

23. $x^2 + 16 = 0$ **25.** $x^2 - 6x + 10 = 0$

27. $x^3 + 2x^2 + x + 2 = 0$ **29.** $x^3 + 3x = 0$

31. $x^3 - 5x^2 + 8x - 6 = 0$ **33.** $x^3 - 6x^2 + 11x - 6 = 0$

35. $x^3 - 5x^2 + 17x - 13 = 0$

37. $24x^3 - 26x^2 + 9x - 1 = 0$

39. $x^4 - 2x^3 + 3x^2 - 2x + 2 = 0$

41. 3 neg; 1 neg, 2 imag   **43.** 1 pos, 2 neg; 1 pos, 2 imag

45. 4 imag   **47.** 4 pos; 2 pos, 2 imag; 4 imag

49. 4 imag and 0   **51.** $-1 < x < 3$ **53.** $-3 < x < 2$

55. $-1 < x < 5$ **57.** $-1 < x < 3$ **59.** $-2, 1, 5$

61. $-3, \dfrac{3 \pm \sqrt{13}}{2}$ **63.** $\pm i, 2, -4$ **65.** $-5, \dfrac{1}{3}, \dfrac{1}{2}$

67. 1, -2 each with multiplicity 2

69. 0, 2 with multiplicity 3   **71.** 0, 1, ± 2, $\pm i\sqrt{3}$

73. 4 sec and 5 sec   **75.** 9 in.   **77.** $7 - 4i$ **79.** $7 - 4i$

81. $14 + 5i$ **87.** $f(x) = -\dfrac{1}{2}x^3 + 3x^2 - \dfrac{11}{2}x + 3$

Graphing Calculator Exercises

1. $-5 < x < 6$, $-5 < x < 6$ **3.** $-6 < x < 6$, $-5 < x < 5$

5. $-1 < x < 23$, $-1 < x < 23$

Section 4.5

1. Neither symmetry; crosses at $(-2, 0)$; does not cross at $(1, 0)$; $y \to \infty$ as $x \to \infty$; $y \to -\infty$ as $x \to -\infty$

3. Symmetric about y-axis; no x-intercepts; $y \to \infty$ as $x \to \infty$; $y \to \infty$ as $x \to -\infty$ **5.** Symmetric about y-axis

7. Symmetric about $x = 3/2$ **9.** Neither symmetry

11. Symmetric about origin   **13.** Symmetric about $x = 5$

15. Symmetric about origin   **17.** Does not cross at $(4, 0)$

19. Crosses at $(1/2, 0)$ **21.** Crosses at $(1/4, 0)$

23. No x-intercepts

25. Does not cross at $(0, 0)$, crosses at $(3, 0)$

27. Crosses at $(1/2, 0)$, does not cross at $(1, 0)$

29. $y \to \infty$ **31.** $y \to -\infty$ **33.** $y \to \infty$ **35.** $y \to \infty$

37. (e)   **39.** (g)   **41.** (b)   **43.** (c)

45.

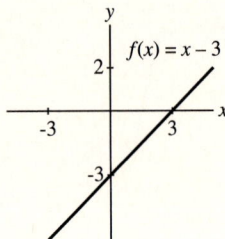
$f(x) = x - 3$

47.

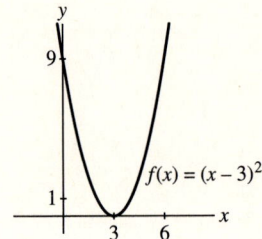
$f(x) = (x - 3)^2$

49.

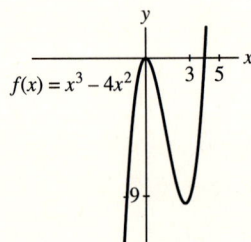
$f(x) = x^3 - 4x^2$

51.

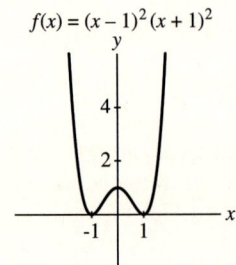
$f(x) = (x - 1)^2 (x + 1)^2$

53.

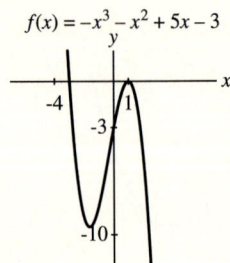
$f(x) = -x^3 - x^2 + 5x - 3$

55.

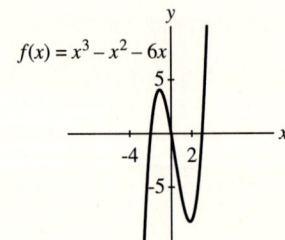
$f(x) = x^3 - x^2 - 6x$

57.

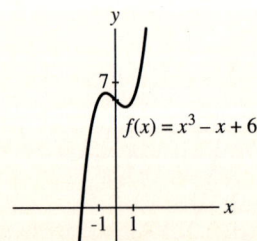
$f(x) = x^3 - x + 6$

59.

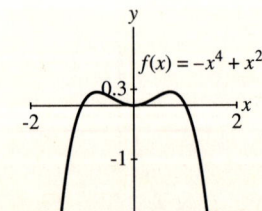
$f(x) = -x^4 + x^2$

61.

$f(x) = x^3 + 3x^2 + 3x + 1$

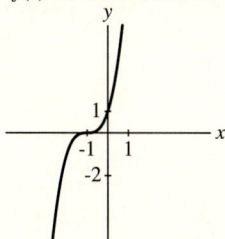

63. Decreases to 0 at 30 stores, then increases.

Graphing Calculator Exercises

1. Loc max value 3.11, loc min value 0.37
3. Loc max value 23.7, loc min value -163.7
5. Loc max value 21.01, loc min value 13.99
7. \$3,400, \$2,600, $x >$ \$2,200

Section 4.6

1. $(-\infty, -2) \cup (-2, \infty)$ **3.** $(-\infty, -2) \cup (-2, 2) \cup (2, \infty)$
5. $(-\infty, 3) \cup (3, \infty)$ **7.** $(-\infty, 0) \cup (0, \infty)$
9. $(-\infty, -1) \cup (-1, 0) \cup (0, 1) \cup (1, \infty)$
11. $(-\infty, -3) \cup (-3, -2) \cup (-2, \infty)$
13. $(-\infty, 2) \cup (2, \infty), y = 0, x = 2$
15. $(-\infty, 0) \cup (0, \infty), y = x, x = 0$ **17.** $x = 2, y = 0$
19. $x = \pm 3, y = 0$ **21.** $x = 1, y = 2$
23. $x = 0, y = x - 2$ **25.** $x = -1, y = 3x - 3$
27. $x = -2, y = -x + 6$

29.

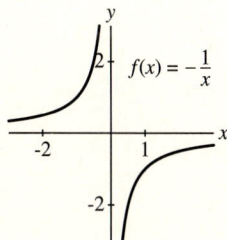

31. $(0, -1/2)$

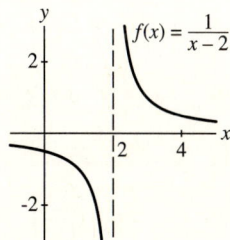

33. $(0, -1/4)$

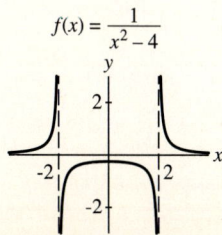

35. $(0, -1)$

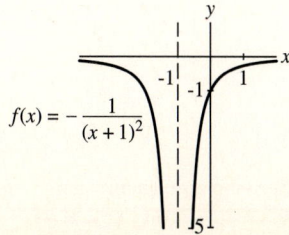

37. $(0, -1), (-1/2, 0)$

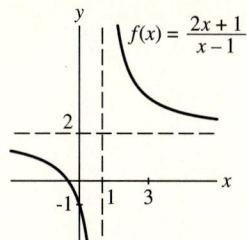

39. $(3, 0), (0, -3/2)$

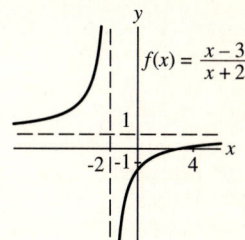

41. $(0, 0)$

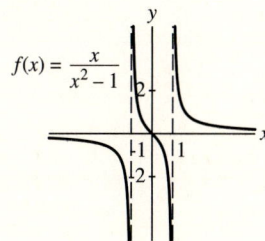

43. $(0, 0)$

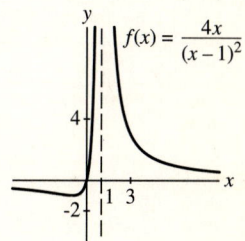

45. $(0, -8/9), (\pm\sqrt{8}, 0)$

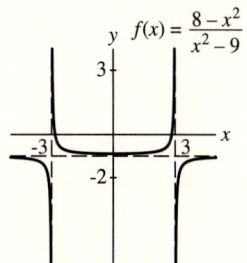

47. $(0, 2), (-2 \pm \sqrt{3}, 0)$

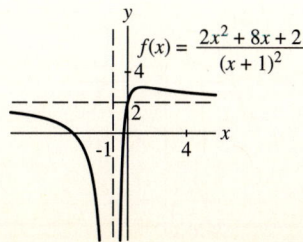

49.

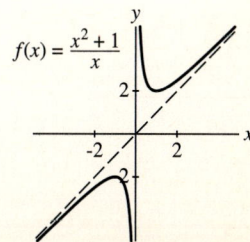

51.

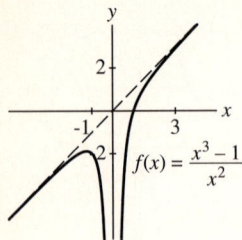

$f(x) = \dfrac{x^3 - 1}{x^2}$

53.

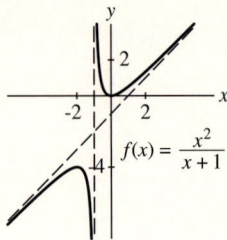

$f(x) = \dfrac{x^2}{x + 1}$

73.

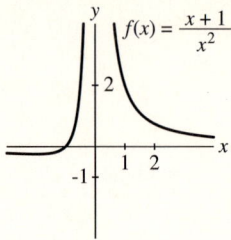

$f(x) = \dfrac{x + 1}{x^2}$

55.

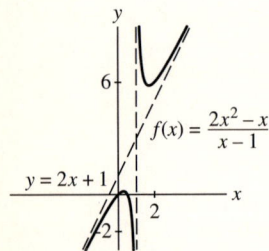

$f(x) = \dfrac{2x^2 - x}{x - 1}$

$y = 2x + 1$

57. (e)   **59.** (a)   **61.** (b)   **63.** (c)

65.

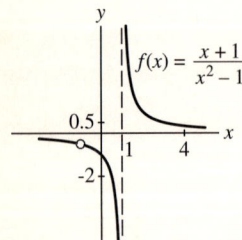

$f(x) = \dfrac{x + 1}{x^2 - 1}$

67.

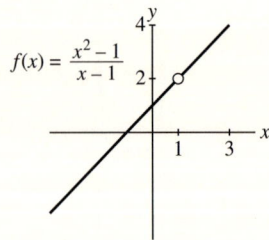

$f(x) = \dfrac{x^2 - 1}{x - 1}$

69.

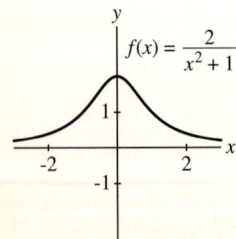

$f(x) = \dfrac{2}{x^2 + 1}$

71.

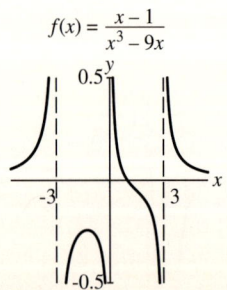

$f(x) = \dfrac{x - 1}{x^3 - 9x}$

75. 1.5   **77.** 0.0   **79.** 3.0   **81.** 0.0

83. $C = \dfrac{100 + x}{x}$, \$2, $C \to \$1$

85. $S = \dfrac{100}{4 - x}$, $S \to \infty$ as $x \to 4$

87. \$400, \$320, $AC(x) \to \$300$

Graphing Calculator Exercises

1. Domain $(-\infty, \infty)$, range $(0, 100]$, $y = 0$
3. Domain $(-\infty, \infty)$, range $(0, 1]$, $y = 0$
5. Domain $(-\infty, \infty)$, range $[-5.55, 0.24]$, $y = 0$

Chapter 4 Review Exercises

1. 3/5   **3.** $y = -\dfrac{3}{2}x - \dfrac{1}{2}$ **5.** $f(x) = 3\left(x - \dfrac{1}{3}\right)^2 + \dfrac{2}{3}$

7. $(1, -3)$, $x = 1$, $\left(\dfrac{2 \pm \sqrt{6}}{2}, 0\right)$, $(0, -1)$

9. $y = -2x^2 + 4x + 6$ **11.** $(-3, 1/2)$ **13.** 1/3

15. $\pm 2\sqrt{2}$ **17.** $\dfrac{1}{2}$, $\dfrac{-1 \pm i\sqrt{3}}{4}$ **19.** $\pm\sqrt{10}$, $\pm i\sqrt{10}$

21. $-\dfrac{1}{2}$ and $\dfrac{1}{2}$ with multiplicity 2   **23.** 0, $-1 \pm \sqrt{7}$

25. 83   **27.** 5   **29.** $\pm\left(1, 2, \dfrac{1}{3}, \dfrac{2}{3}\right)$

31. $\pm\left(1, 3, \dfrac{1}{2}, \dfrac{1}{3}, \dfrac{1}{6}, \dfrac{3}{2}\right)$ **33.** $2x^2 - 5x - 3 = 0$

35. $x^2 - 6x + 13 = 0$ **37.** $x^3 - 4x^2 + 9x - 10 = 0$
39. $x^2 - 4x + 1 = 0$ **41.** 0 with multiplicity 2, 6 imag
43. 1 pos, 2 imag; 3 pos   **45.** 3 neg; 1 neg, 2 imag
47. $-4 < x < 3$ **49.** $-1 < x < 8$ **51.** $-1 < x < 1$

53. 1, 2, 3   **55.** $\dfrac{1}{2}$, $\dfrac{1}{3}$, $\pm i$ **57.** 3, $3 \pm i$ **59.** 2, $1 \pm i\sqrt{2}$

61. 0, $\dfrac{1}{2}$, $1 \pm \sqrt{3}$ **63.** Symmetric about $x = 3/4$

65. Symmetric about y-axis   **67.** Symmetric about origin
69. $(-\infty, -2.5) \cup (-2.5, \infty)$ **71.** $(-\infty, \infty)$

73.

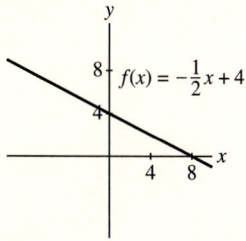

$f(x) = -\frac{1}{2}x + 4$

75.

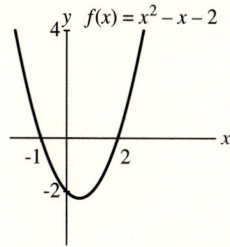

$f(x) = x^2 - x - 2$

89.

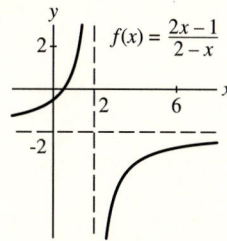

$f(x) = \frac{2x-1}{2-x}$

91. $x^2 - 3x, -15$ **93.** $y = 20x + 14, 90$ **95.** 18 in.

Chapter 4 Test

1. -2 **2.** $\frac{3}{4}$ **3.** $y = \frac{1}{3}x - \frac{14}{3}$ **4.** $y = 3(x - 2)^2 - 11$

5. $(2, -11), x = 2, (0, 1), \left(\frac{6 \pm \sqrt{33}}{3}, 0\right), [-11, \infty)$

6. -11 **7.** $2x^2 - 6x + 14, -37$ **8.** -14

9. $\pm\left(1, 2, 3, 6, \frac{1}{3}, \frac{2}{3}\right)$ **10.** $x^3 + 3x^2 + 16x + 48 = 0$

11. 2 pos, 1 neg; 1 neg, 2 imag   **12.** 256 ft   **13.** ± 3

14. $\pm 2, \pm 2i$ **15.** $1 \pm \sqrt{6}, 2$

16. $\pm i$ each with multiplicity 2

17. 2, 0 with multiplicity 2, $-\frac{3}{2}$ with multiplicity 3

18. $2 \pm i, \frac{1}{2}$

19.

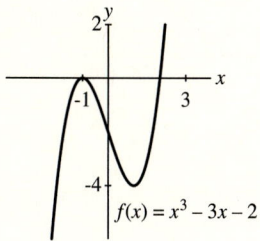

$y = -\frac{1}{2}x + 3$

77.

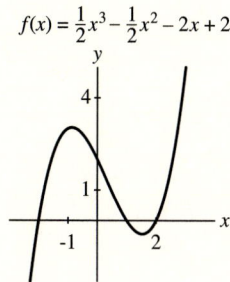

$f(x) = x^3 - 3x - 2$

79.

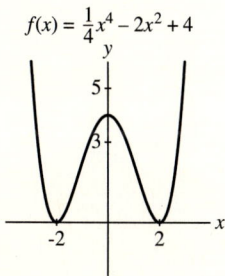

$f(x) = \frac{1}{2}x^3 - \frac{1}{2}x^2 - 2x + 2$

81.

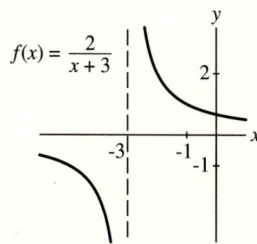

$f(x) = \frac{1}{4}x^4 - 2x^2 + 4$

83.

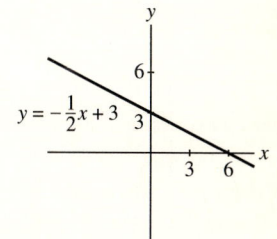

$f(x) = \frac{2}{x+3}$

85.

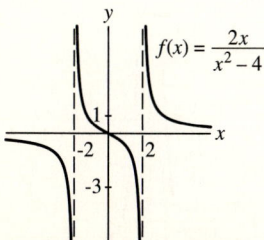

$f(x) = \frac{2x}{x^2 - 4}$

87.

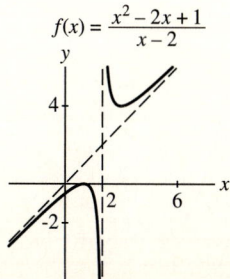

$f(x) = \frac{x^2 - 2x + 1}{x - 2}$

20.

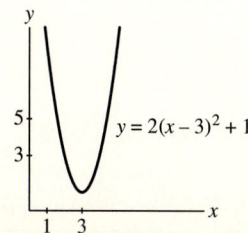

$y = 2(x - 3)^2 + 1$

21.

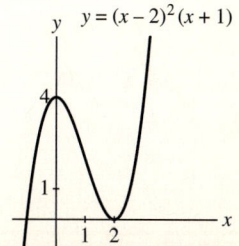

$y = (x - 2)^2(x + 1)$

22.

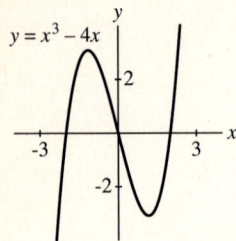
$y = x^3 - 4x$

23.

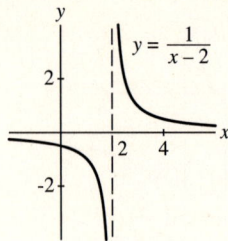
$y = \dfrac{1}{x-2}$

13.

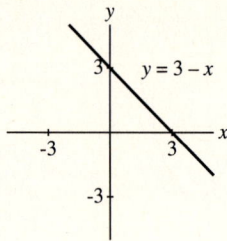
$y = 3 - x$

14.

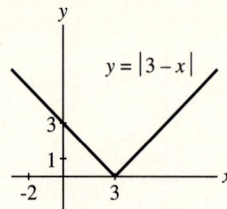
$y = |3 - x|$

24.

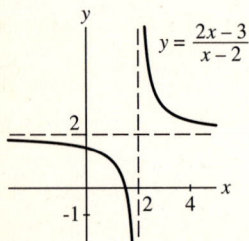
$y = \dfrac{2x-3}{x-2}$

25.

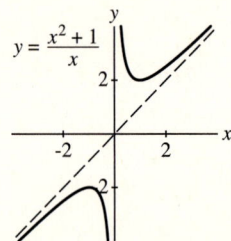
$y = \dfrac{x^2+1}{x}$

15.

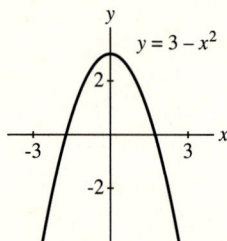
$y = 3 - x^2$

16.

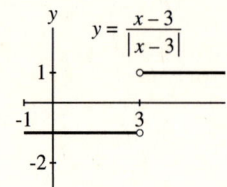
$y = \dfrac{x-3}{|x-3|}$

26.

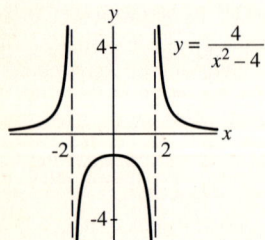
$y = \dfrac{4}{x^2-4}$

17.

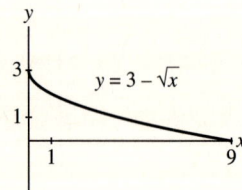
$y = 3 - \sqrt{x}$

18.

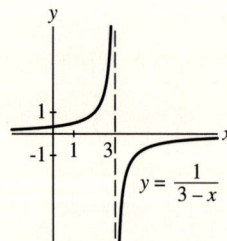
$y = \dfrac{1}{3-x}$

Tying It All Together Chapters 1–4

1. 2   **2.** $-\dfrac{1}{2}, \dfrac{1}{3}$ **3.** $-\dfrac{1}{2}$ **4.** 2   **5.** $1, -\dfrac{13}{6}$

6. $-\dfrac{1}{2}, \dfrac{1}{3}, \pm i$ **7.** 2   **8.** $\pm i\sqrt{3}$ **9.** 0, 4   **10.** 1, −2

11. −14, 13   **12.** 0, 2

19.

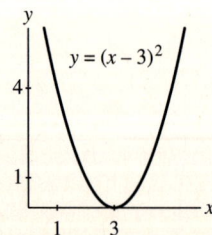
$y = (x-3)^2$

20.

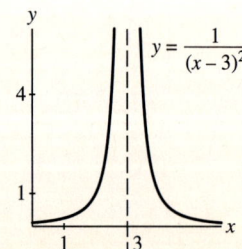
$y = \dfrac{1}{(x-3)^2}$

21.

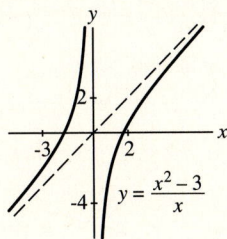

22.

$y = \sqrt{3 - x^2}$

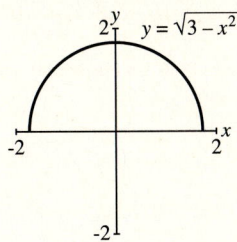

17. Domain $(-\infty, \infty)$, range $(0, \infty)$, dec

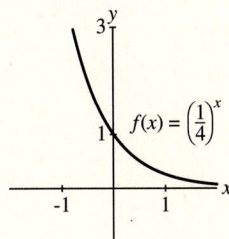

$f(x) = \left(\frac{1}{4}\right)^x$

19. Domain $(-\infty, \infty)$, range $(-3, \infty)$, inc

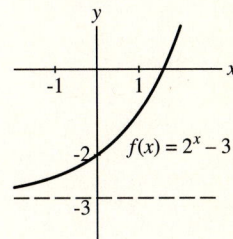

$f(x) = 2^x - 3$

23.

$y = \sqrt{3 - x}$

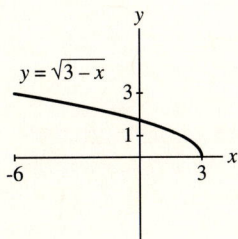

24.

$y = x^3 - 3x^2$

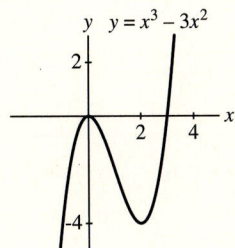

21. Domain $(-\infty, \infty)$, range $(-5, \infty)$, inc

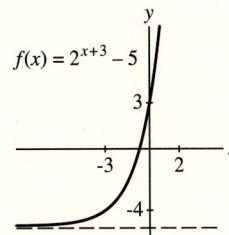

$f(x) = 2^{x+3} - 5$

23. Domain $(-\infty, \infty)$, range $(-\infty, 0)$, inc

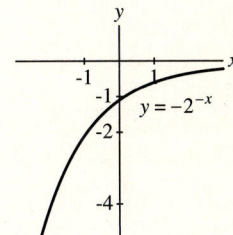

$y = -2^{-x}$

25. 1000   **26.** 1   **27.** $\frac{1}{2}$ **28.** 0.001   **29.** $\frac{1}{9}$ **30.** 3

31. $\frac{1}{9}$ **32.** 8

CHAPTER 5

Section 5.1

1. 9   **3.** 1/9   **5.** 1/2   **7.** 8   **9.** 4   **11.** 2

13. Domain $(-\infty, \infty)$, range $(0, \infty)$, inc

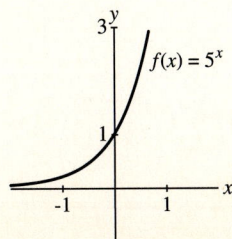

$f(x) = 5^x$

15. Domain $(-\infty, \infty)$, range $(0, \infty)$, dec

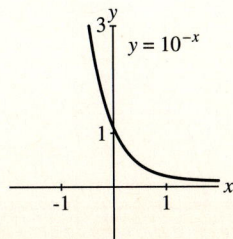

$y = 10^{-x}$

25. Domain $(-\infty, \infty)$, range $(-\infty, 1)$, dec

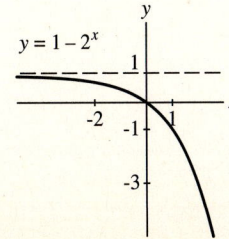

$y = 1 - 2^x$

27. Domain $(-\infty, \infty)$, range $(0, \infty)$, inc

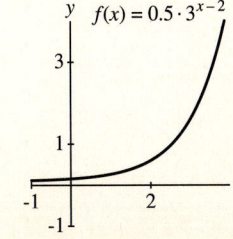

$f(x) = 0.5 \cdot 3^{x-2}$

29. Domain $(-\infty, \infty)$, range $(0, \infty)$, dec

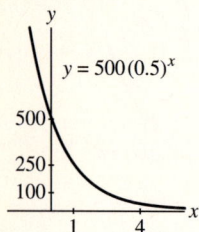

$y = 500(0.5)^x$

31. $\{6\}$ **33.** $\{-1\}$ **35.** $\{3\}$ **37.** $\{-2\}$ **39.** $\{1/3\}$
41. $\{-2\}$ **43.** $\{-3\}$ **45.** $\{-1\}$ **47.** 2   **49.** -1
51. 0   **53.** -3 **55.** 3   **57.** -1 **59.** 1   **61.** -1
63. 9, 1, 1/3, -2 **65.** 1, -2, 5, 1   **67.** -16, -2, $-1/2$, 5
69.   **71.**

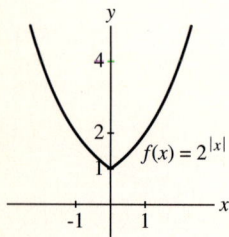

$f(x) = 2^{|x|}$

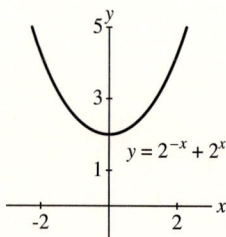

$y = 2^{-x} + 2^x$

73. $7237.97, $2737.97
75. $3251.93, 365 days/yr, 30 days/mo
77. $129,196.51, $129,194.24   **79.** 2,400,000, 3,338,324
81. $P = 10\left(\dfrac{1}{2}\right)^n$ **83.** 6.8×10^{-5} **85.** 6   **87.** 2

Graphing Calculator Exercises

1. $\{1.46\}$ **3.** $\{-2.71\}$ **5.** $\{1.20\}$ **7.** $\{11.55\}$
9. Range $(0, 1]$ **11.** Range $[0, \infty)$

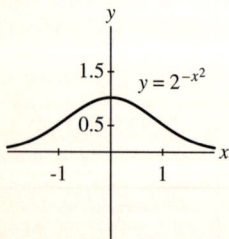

$y = 2^{-x^2}$

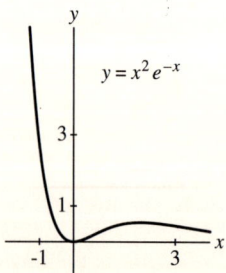

$y = x^2 e^{-x}$

Section 5.2

1. 5   **3.** 1/5   **5.** 1   **7.** -1 **9.** 6   **11.** -4 **13.** 1/4
15. -3 **17.** -1 **19.** 0   **21.** 1   **23.** -5
25. Domain $(0, \infty)$, range $(-\infty, \infty)$ **27.** Domain $(0, \infty)$, range $(-\infty, \infty)$

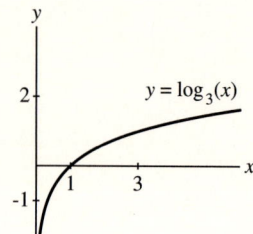

$y = \log_3(x)$

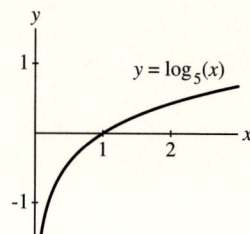

$y = \log_5(x)$

29. Domain $(0, \infty)$, range $(-\infty, \infty)$ **31.** Domain $(0, \infty)$, range $(-\infty, \infty)$

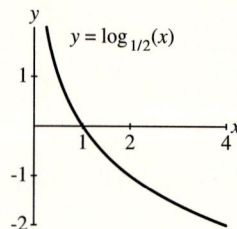

$y = \log_{1/2}(x)$

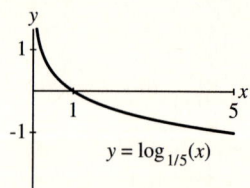

$y = \log_{1/5}(x)$

33. Domain $(1, \infty)$, range $(-\infty, \infty)$ **35.** Domain $(-2, \infty)$, range $(-\infty, \infty)$

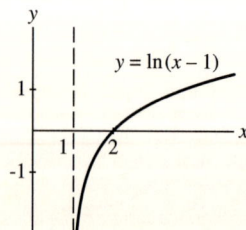

$y = \ln(x - 1)$

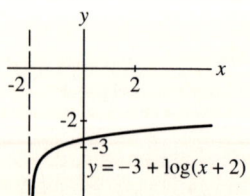

$y = -3 + \log(x + 2)$

37. Domain $(1, \infty)$, range $(-\infty, \infty)$

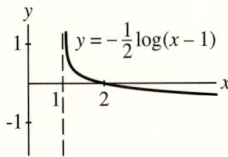

39. Domain $(1, \infty)$, range $(-\infty, \infty)$

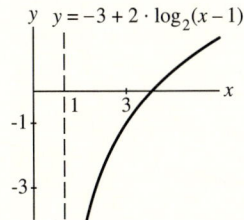

41. $10^y = 30$ **43.** $\log_5(7) = y$ **45.** $2^0 = 1$
47. $\ln(2) = 0.09t$ **49.** $b^3 = N/M$ **51.** $\log_b(y) = 3x$
53. $\sqrt{3}$ **55.** $\log_3(77)$ **57.** 6   **59.** $3\sqrt{2}$
61. $-1 + \log_3(7)$ **63.** 2   **65.** 0.001   **67.** 27   **69.** 1/4
71. 1.3979   **73.** 1.7712   **75.** 3.8451   **77.** -12.3315
79. 4.7381   **81.** $f^{-1}(x) = \log_2(x)$ **83.** $f^{-1}(x) = 7^x$
85. $f^{-1}(x) = (1/8)^x$ **87.** $f^{-1}(x) = e^x + 1$
89. $f^{-1}(x) = \log_3(x) - 2$ **91.** $f^{-1}(x) = 10^{x-3}$
93. 49 yr 125 days   **95.** 23.1%   **97.** 6.9 yr
99. 3.5%   **101.** 9.8 yr   **103.** 1.87%, 6.2 billion
105. 4.1   **107.** 3.7

Graphing Calculator Exercises

1. 26.53   **3.** -0.56

Section 5.3

1. $\log(15)$ **3.** $\log_2(x^2 - x)$ **5.** $\log_4(6)$ **7.** $\ln(x^5)$
9. $2 \cdot \log(5)$ **11.** $-3 \cdot \log(5)$ **13.** $\dfrac{1}{3}\log(5)$
15. $\sqrt{y}$ **17.** $y + 1$ **19.** 999   **21.** $\log(2) + \log(5)$
23. $\log(5) - \log(2)$ **25.** $\log(2) + \dfrac{1}{2}\log(5)$
27. $2 \cdot \log(2) - 2 \cdot \log(5)$ **29.** $\log_3(5) + \log_3(x)$
31. $\log_2(5) - \log_2(2) - \log_2(y)$ **33.** $\log(3) + \dfrac{1}{2}\log(x)$
35. $\log(3) + (x - 1)\log(2)$ **37.** $\dfrac{1}{3}\ln(x) + \dfrac{1}{3}\ln(y) - \dfrac{4}{3}\ln(t)$
39. $\ln(6) + \dfrac{1}{2}\ln(x - 1) - \ln(5) - 3 \cdot \ln(x)$ **41.** $\log_2(5x^3)$
43. $\log_7(x^{-3})$
45. $\log\left(\dfrac{2xy}{z}\right)$ **47.** $\log\left(\dfrac{z\sqrt{x}}{y \cdot \sqrt[3]{w}}\right)$

49. $\log_4(x^{20})$ **51.** 3.1699   **53.** -3.5864 **55.** 11.8957
57. 2.2025   **59.** 1.5850   **61.** 0.3772   **63.** 13.8695
65. 34.3240   **67.** 0.3200   **69.** 0.0479   **71.** 2.0172
73. 1.5928   **75.** T   **77.** F   **79.** T   **81.** F   **83.** F
85. F   **87.** T   **89.** 11 yr 166 days   **91.** 44 quarters
93. 5.7%   **95.** 3.58%   **97.** $\log(I/I_0)$, 3
99. $t = \dfrac{1}{r}\ln(P) - \dfrac{1}{r}\ln(P_0)$
101. $MR(x) = \log\left[\left(\dfrac{x + 2}{x + 1}\right)^{500}\right]$, $MR(x) \to 0$

Graphing Calculator Exercises

1.

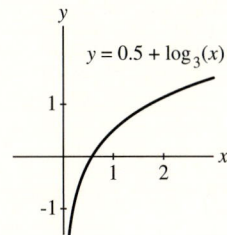

3.

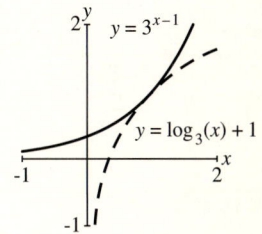

Section 5.4

1. 80   **3.** 3   **5.** 6   **7.** 20   **9.** 2   **11.** $\dfrac{\sqrt{5}}{2}$ **13.** $\dfrac{1}{2}$
15. $\dfrac{2}{1 + \ln(3)}$ **17.** 3.8074   **19.** 3.5502   **21.** -3.0959
23. No solution   **25.** 1.5850   **27.** 0.7677   **29.** -0.2
31. 2.5850   **33.** 1/3   **35.** 1, 100   **37.** 29.4872   **39.** 2
41. $-5/4$ **43.** 19,035 yr ago   **45.** 1507 yr
47. 24,850 yr   **49.** 3,361 million yr   **51.** 34 A.D.
53. 1 hr 11 min, forever   **55.** 5:20 A.M.   **57.** 3.4 yr
59. 19,329 yr 151 days   **61.** 12,300, 13.9 yr
63. 1.32 parsecs   **65.** 10^{-3} watts/m^2 **67.** $6791.91
69. 1.105170833   **71.** 388, 40.5 days

Graphing Calculator Exercises

1. 0.194, 2.70   **3.** -49.73

Chapter 5 Review Exercises

1. 64   **3.** 6   **5.** 0   **7.** 17   **9.** 32   **11.** 3   **13.** 9
15. 3   **17.** -3 **19.** 3   **21.** $\log(x^2 - 3x)$ **23.** $\ln(3x^2y)$
25. $\log(3) + 4 \cdot \log(x)$

27. $\log_3(5) + \dfrac{1}{2}\log_3(x) - 4 \cdot \log_3(y)$

29. $\ln(2) + \ln(5)$ **31.** $\ln(2) + 2 \cdot \ln(5)$ **33.** 10^{10}
35. 3   **37.** -5 **39.** -4 **41.** $2 + \ln(9)$ **43.** -2
45. $100\sqrt{5}$ **47.** 8   **49.** 6   **51.** 3   **53.** $3, -3, \sqrt{3}, 0$
55. (c)   **57.** (b)   **59.** (d)   **61.** (e)
63. Domain $(-\infty, \infty)$, range   **65.** Domain $(-\infty, \infty)$, range
$(0, \infty)$, inc $(0, \infty)$, dec

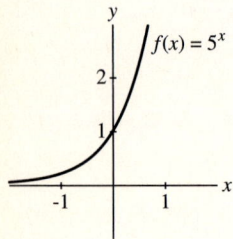
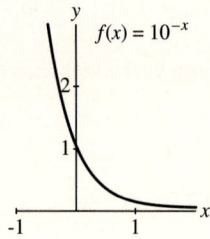

67. Domain $(0, \infty)$, range
$(-\infty, \infty)$, inc

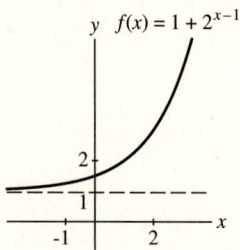

69. Domain $(-3, \infty)$, range
$(-\infty, \infty)$, inc

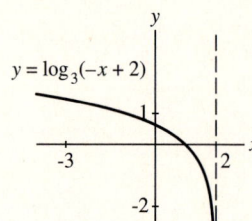

71. Domain $(-\infty, \infty)$, range   **73.** Domain $(-\infty, 2)$, range
$(1, \infty)$, inc $(-\infty, \infty)$, dec

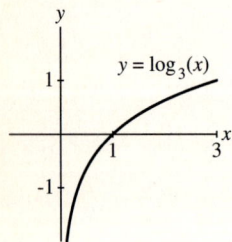
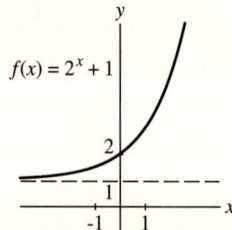

75. $f^{-1}(x) = \log_7(x)$ **77.** $f^{-1}(x) = 5^x$ **79.** 2.0959
81. -4.4243
83. The pH of A is one less than the pH of B.
85. \$122,296.01   **87.** 56 quarters
89. 25 g, 18.15 g, 2166 yr   **91.** 2877 hr

Chapter 5 Test

1. 3   **2.** -2 **3.** 6.47   **4.** $\sqrt{2}$
5. $f^{-1}(x) = e^x$ **6.** $f^{-1}(x) = \log_8(x)$ **7.** $\log(xy^3)$

8. $\ln\left(\dfrac{\sqrt{x-1}}{33}\right)$ **9.** $2 \cdot \log_a(2) + \log_a(7)$

10. $\log_a(7) - \log_a(2)$ **11.** 4   **12.** 18

13. $\dfrac{\ln(5)}{\ln(5) - \ln(3)} \approx 3.1507$ **14.** $1 + 3^{5.46} \approx 403.7931$

15. Domain $(-\infty, \infty)$, range   **16.** Domain $(1, \infty)$, range
$(1, \infty)$, inc $(-\infty, \infty)$, dec

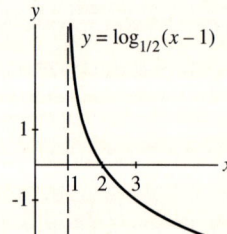
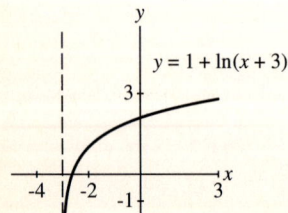

17. $(1,0)$ **18.** \$9750.88, \$9906.06
19. 22.5 watts, 173.3 days, 428.7 days   **20.** 61.5 quarters
21. $p = 0.86$, no

Tying It All Together Chapters 1–5

1. 5, 1   **2.** 5   **3.** 19   **4.** 5   **5.** 19   **6.** 7, −1
7. $2 \pm \sqrt{2}$ **8.** −3   **9.** 9   **10.** $\log_2(3)$ **11.** 4
12. −1, 2, 3
13.

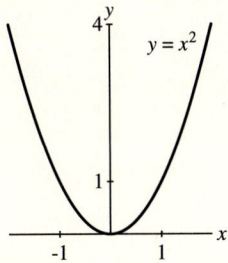

$y = x^2$

14.

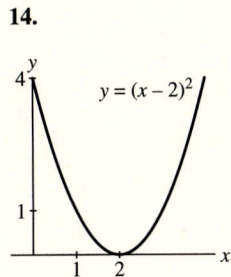

$y = (x - 2)^2$

15.

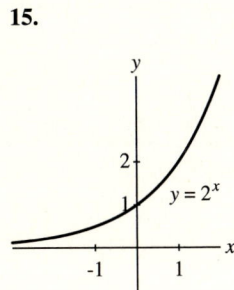

$y = 2^x$

16.

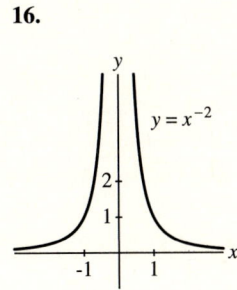

$y = x^{-2}$

17.

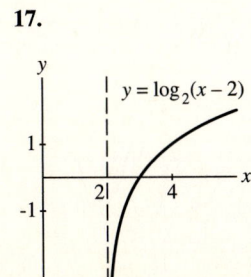

$y = \log_2(x - 2)$

18.

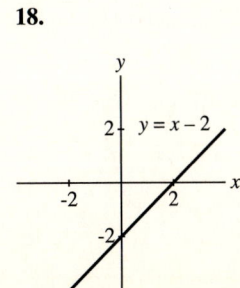

$y = x - 2$

19.

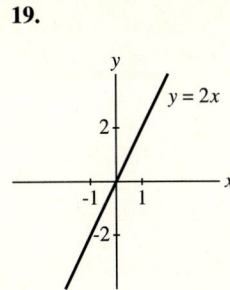

$y = 2x$

20.

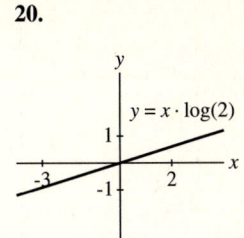

$y = x \cdot \log(2)$

21.

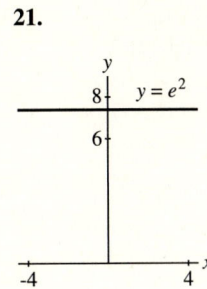

$y = e^2$

22.

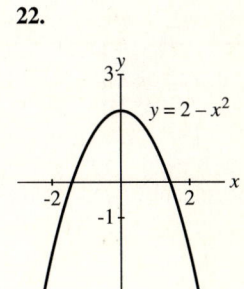

$y = 2 - x^2$

23.

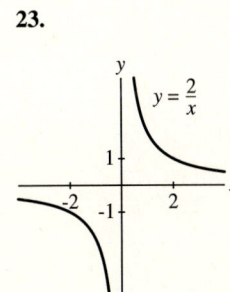

$y = \dfrac{2}{x}$

24.

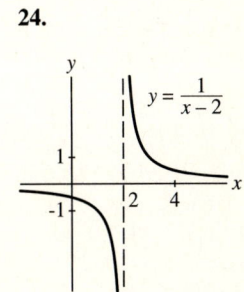

$y = \dfrac{1}{x - 2}$

25. $f^{-1}(x) = 3x$ **26.** $f^{-1}(x) = -\log_3(x)$
27. $f^{-1}(x) = x^2 + 2$ for $x \geq 0$ **28.** $f^{-1}(x) = \sqrt[3]{x - 2} + 5$
29. $f^{-1}(x) = (10^x + 3)^2$ **30.** $f^{-1} = \{(1, 3), (4, 5)\}$
31. $f^{-1}(x) = \dfrac{1}{x - 3} + 5$ **32.** $f^{-1}(x) = (\ln(3 - x))^2$ for $x \leq 2$

CHAPTER 6

Section 6.1

1. 30° = $\pi/6$, 45° = $\pi/4$, 60° = $\pi/3$, 90° = $\pi/2$, 120° = $2\pi/3$, 135° = $3\pi/4$, 150° = $5\pi/6$, 180° = π, 210° = $7\pi/6$, 225° = $5\pi/4$, 240° = $4\pi/3$, 270° = $3\pi/2$, 300° = $5\pi/3$, 315° = $7\pi/4$, 330° = $11\pi/6$, 360° = 2π
3. 420°, 780°, −300°, −660°   **5.** 344°, 704°, −376°, −736°
7. $7\pi/3$, $13\pi/3$, $-5\pi/3$, $-11\pi/3$
9. 5.1, 11.4, −7.5, −13.8   **11.** Yes   **13.** No
15. No   **17.** Yes   **19.** I   **21.** III   **23.** IV   **25.** I
27. III   **29.** IV   **31.** I   **33.** IV   **35.** II   **37.** −120°
39. $9\pi/4$ **41.** $3\pi/2$ **43.** 40°   **45.** 20°   **47.** 340°
49. 180.54°   **51.** 150°27′46″   **53.** 13.2°   **55.** −8.505°
57. 28.0858°   **59.** 75°30′   **61.** −17°19′48″
63. 18°7′23″   **65.** $\pi/10$ **67.** $-3\pi/8$ **69.** $7\pi/2$
71. 75°   **73.** −1080°   **75.** 136.937°   **77.** 0.653
79. −0.241   **81.** 0.936   **83.** π **85.** $\pi/2$ **87.** $\pi/3$
89. $5\pi/3$ **91.** 2.0   **93.** 3π ft   **95.** 209.4 mi   **97.** 1 mi
99. 3.18 km   **101.** 1256.6 in/min   **103.** 166.6 mph
105. 20.5 mph   **107.** 3102 mi   **109.** 29.0°
111. 0.26 mph   **113.** 41,143 km   **115.** 33.5 in²

Section 6.2

1. $(1, 0)$, $\left(\sqrt{2}/2, \sqrt{2}/2\right)$, $(0, 1)$, $\left(-\sqrt{2}/2, \sqrt{2}/2\right)$, $(-1, 0)$, $\left(-\sqrt{2}/2, -\sqrt{2}/2\right)$, $(0, -1)$, $\left(\sqrt{2}/2, -\sqrt{2}/2\right)$
3. 1/2   **5.** 1/2
7. $\sqrt{2}/2$ **9.** $-\sqrt{2}/2$ **11.** −1   **13.** $-\sqrt{2}/2$ **15.** $-\sqrt{3}/2$
17. $\sqrt{3}/2$ **19.** 1/2   **21.** 1/2   **23.** $\sqrt{3}/2$ **25.** 1
27. 30°, $\pi/6$ **29.** 60°, $\pi/3$ **31.** 60°, $\pi/3$ **33.** 30°, $\pi/6$
35. 45°, $\pi/4$ **37.** 45°, $\pi/4$ **39.** $\sqrt{2}/2$ **41.** 1/2
43. $-\sqrt{2}/2$ **45.** $-\sqrt{3}/2$ **47.** $-\sqrt{2}/2$ **49.** −1/2
51. $\sqrt{3}/3$ **53.** −1   **55.** 1   **57.** $2 + \sqrt{3}$ **59.** $\sqrt{2}$
61. $\left(\sqrt{2} - \sqrt{6}\right)/4$ **63.** 1/4   **65.** 1/2   **67.** 0.6820
69. −0.6366   **71.** 0.9999   **73.** 0.4035   **75.** −0.7438
77. 0.7071   **79.** 1.0000   **81.** 0.8948   **83.** −0.2588
85. 0.0314   **87.** 0.8578   **89.** 0.2679   **91.** 1
93. −0.9984   **95.** −12/13   **97.** −4/5   **99.** $2\sqrt{2}/3$
101. 3.53   **103.** 1.708 in., 1.714 in.

Graphing Calculator Exercises

1. 5 terms, 17 terms

Section 6.3

1. 2, 2π, 0   **3.** 1, 2π, $\pi/2$ **5.** 2, 2π, $\pi/3$

7. 1, 0

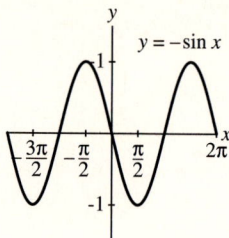
$y = -\sin x$

9. 3, 0

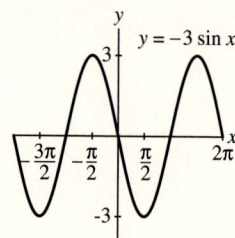
$y = -3\sin x$

11. 1/2, 0

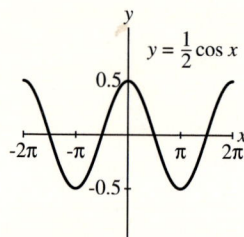
$y = \frac{1}{2}\cos x$

13. 1, π

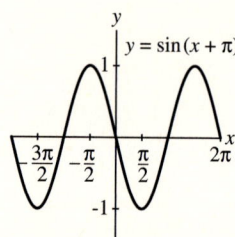
$y = \sin(x + \pi)$

15. 1, $\pi/3$

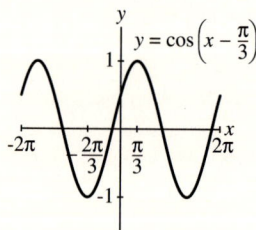
$y = \cos\left(x - \frac{\pi}{3}\right)$

17. 1, 0

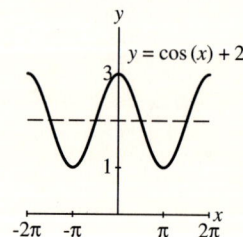
$y = \cos(x) + 2$

19. 1, 0

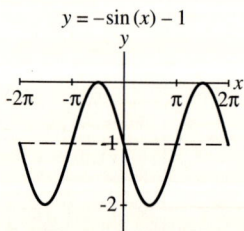
$y = -\sin(x) - 1$

21. 1, $\pi/4$

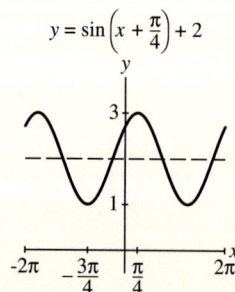
$y = \sin\left(x + \frac{\pi}{4}\right) + 2$

23. 2, $\pi/6$

$$y = 2\cos\left(x + \frac{\pi}{6}\right) + 1$$

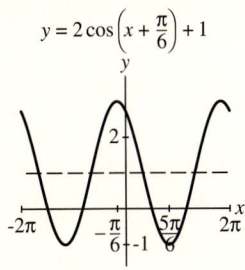

25. 2, $\pi/3$

$$y = -2\sin\left(x - \frac{\pi}{3}\right) + 1$$

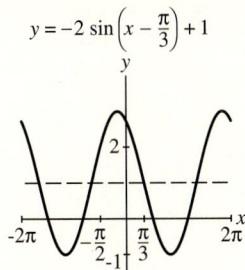

39. 3, $\pi/2$, 0   **41.** 2, π, $\pi/4$
45. $2\pi/3$, 0, $[-1, 1]$

$$y = \sin(3x)$$

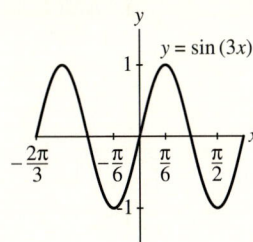

43. 2, 2, 1
47. π, 0, $[-1, 1]$

$$y = -\sin(2x)$$

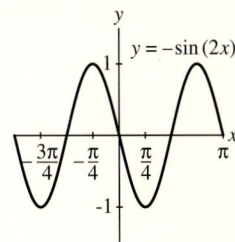

27. 1, 0°

$$y = \sin x$$

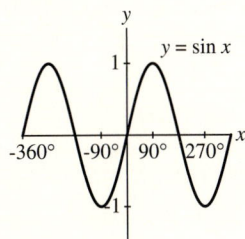

29. 3, 0°

$$y = 3\cos(x) + 1$$

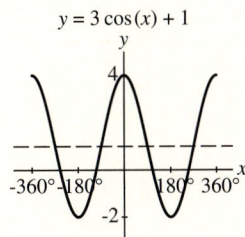

49. $\pi/2$, 0, $[1, 3]$

$$y = \cos(4x) + 2$$

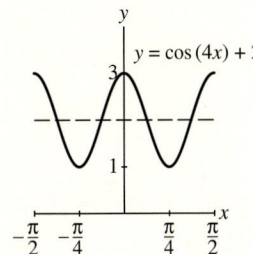

51. 8π, 0, $[1, 3]$

$$y = 2 - \sin\left(\frac{x}{4}\right)$$

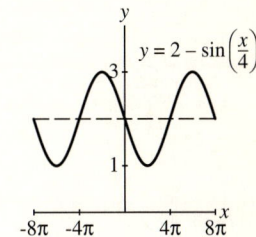

31. 1, 45°

$$y = \sin(x - 45°)$$

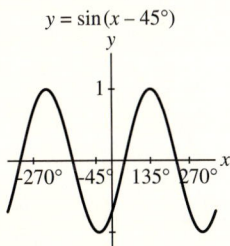

33. 2, 90°

$$y = -2\cos(x + 90°)$$

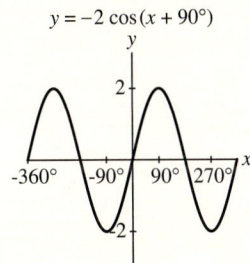

53. 6, 0, $[-1, 1]$

$$y = \sin\left(\frac{\pi}{3}x\right)$$

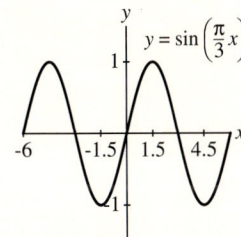

55. π, $\pi/2$, $[-1, 1]$

$$y = \sin\left(2\left[x - \frac{\pi}{2}\right]\right)$$

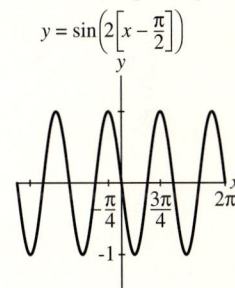

35. 4, 90°

$$y = 4\sin(x - 90°) + 1$$

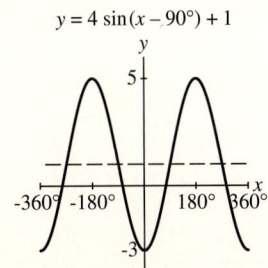

37. 1/2, 30°

$$y = \frac{1}{2}\cos(x - 30°) - 2$$

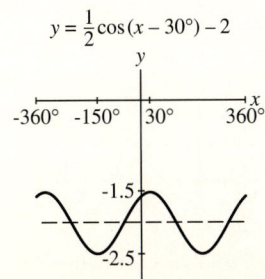

57. 4, 3, $[-1, 1]$

$$y = \sin\left(\frac{\pi}{2}[x + 3]\right)$$

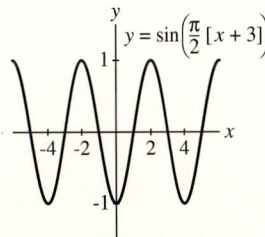

59. π, $\pi/6$, $[-1, 3]$

$$y = 2\cos\left(2\left[x + \frac{\pi}{6}\right]\right) + 1$$

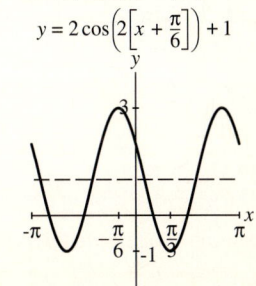

61. $2\pi/3$, $\pi/6$, $[-3/2, -1/2]$

$$y = -\frac{1}{2}\sin\left(3\left[x - \frac{\pi}{6}\right]\right) - 1$$

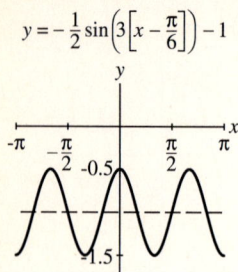

63. $20°$, $0°$, $[-1, 1]$

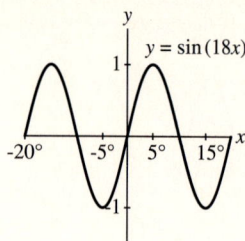

$y = \sin(18x)$

65. $120°$, $30°$, $[-1, 1]$

$y = \cos(3x + 90°)$

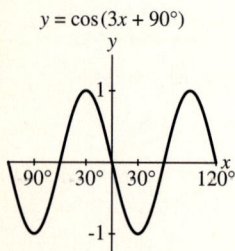

67. $180°$, $60°$, $[-1, 3]$

$y = -2\sin(2[x - 60°]) + 1$

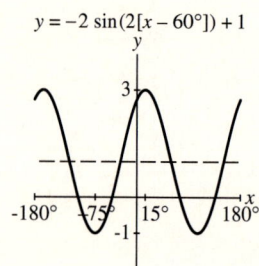

69. $y = 2\sin\left(2\left(x - \frac{\pi}{4}\right)\right)$

71. $y = 3\sin\left(\frac{3}{2}\left(x + \frac{\pi}{3}\right)\right) + 3$

73. 100 cycles/sec

75. 40 cycles/hr

77. $x = 3\sin(2t)$, 3, π

$x = 3\sin(2t)$

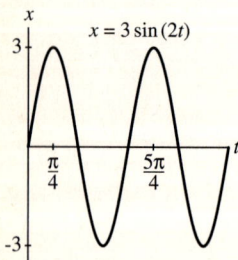

79. 11 yr   **81.** 1300 cc, 500 cc, 30

Graphing Calculator Exercises

1. y_3

Section 6.4

1. $\tan(0) = 0$, $\tan(\pi/4) = 1$, $\tan(\pi/2)$ undefined, $\tan(3\pi/4) = -1$, $\tan(\pi) = 0$, $\tan(5\pi/4) = 1$, $\tan(3\pi/2)$ undefined, $\tan(7\pi/4) = -1$ **3.** $\sqrt{3}$ **5.** -1 **7.** 0
9. $-\sqrt{3}/3$ **11.** $2\sqrt{3}/3$ **13.** Undefined   **15.** Undefined
17. $\sqrt{2}$ **19.** -1 **21.** $\sqrt{3}$ **23.** -2 **25.** $-\sqrt{2}$
27. 48.0785   **29.** -2.8413 **31.** 500.0003   **33.** 1.0353
35. 636.6192   **37.** -1.4318 **39.** 71.6221   **41.** -0.9861
43.   **45.**

$y = \tan(3x)$

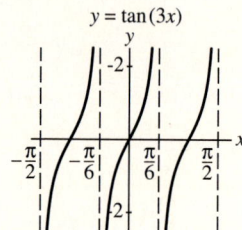

$y = \cot\left(x + \frac{\pi}{4}\right)$

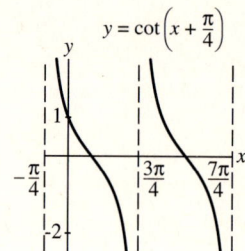

47.   **49.**

$y = \cot\left(\frac{x}{2}\right)$

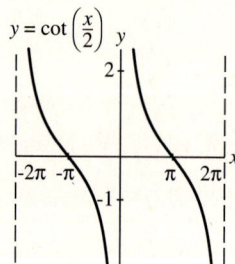

$y = \tan(\pi x)$

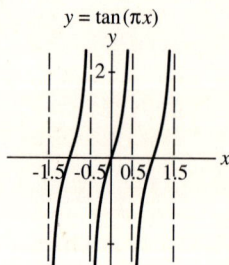

51.   **53.**

$y = -2\tan x$

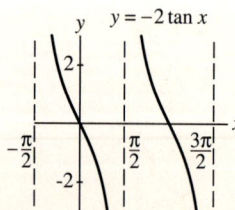

$y = -\cot\left(x + \frac{\pi}{2}\right)$

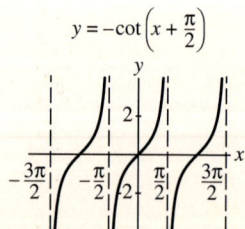

55.

$y = \cot\left(2x - \dfrac{\pi}{2}\right)$

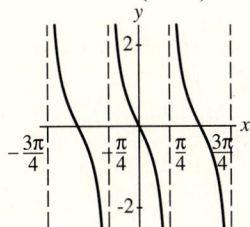

57.

$y = \tan\left(\dfrac{\pi}{2}x - \dfrac{\pi}{2}\right)$

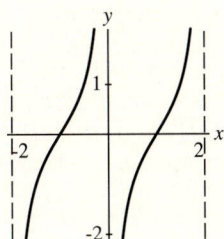

71. $(-\infty, -1] \cup [1, \infty)$

$y = -\csc\left(\dfrac{\pi}{2}x + \dfrac{\pi}{2}\right)$

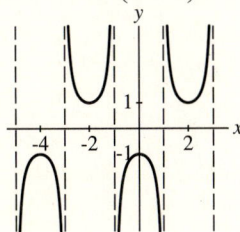

73. $(-\infty, 0] \cup [4, \infty)$

$y = 2 + 2\sec(2x)$

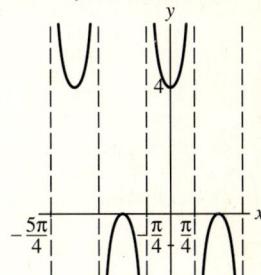

59. $(-\infty, -1] \cup [1, \infty)$

$y = \sec(2x)$

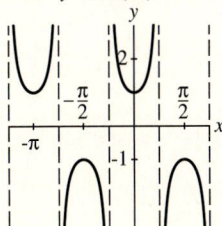

61. $(-\infty, -1] \cup [1, \infty)$

$y = \csc\left(x - \dfrac{\pi}{2}\right)$

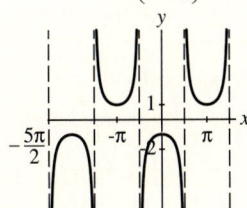

75.

$y = -\cot(3x)$

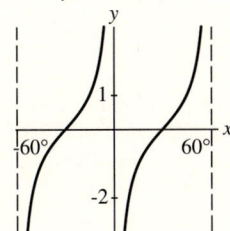

77.

$y = \sec(2x)$

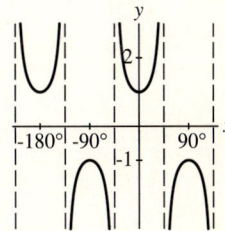

63. $(-\infty, -1] \cup [1, \infty)$

$y = \csc\left(\dfrac{x}{2}\right)$

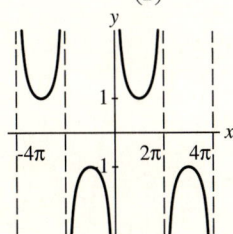

65. $(-\infty, -1] \cup [1, \infty)$

$y = \sec\left(\dfrac{\pi x}{2}\right)$

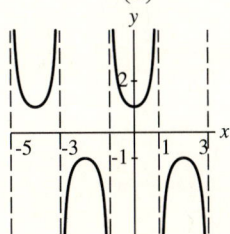

79.

$y = \tan\left(\dfrac{x}{2}\right)$

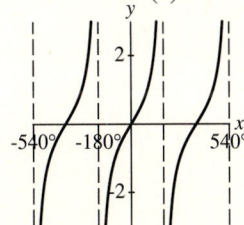

81.

$y = -\csc\left(\dfrac{x}{2}\right)$

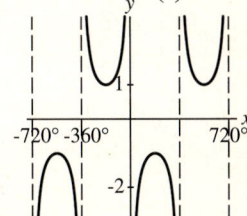

67. $(-\infty, -2] \cup [2, \infty)$

$y = 2\sec(x)$

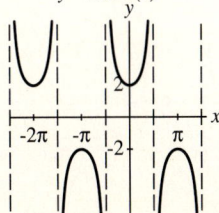

69. $(-\infty, -1] \cup [1, \infty)$

$y = \csc\left(2x - \dfrac{\pi}{2}\right)$

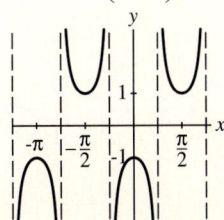

83.

$y = \cot(2x - 180°)$

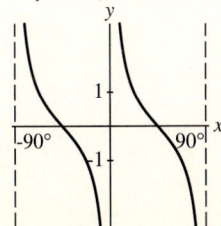

85.

$y = -\sec(x + 45°)$

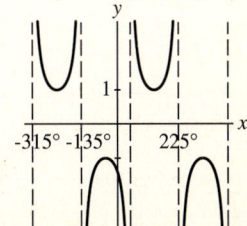

87.

$y = 2 \csc (x - 30°)$

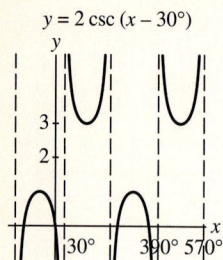

89.

$y = \tan (90x + 90°)$

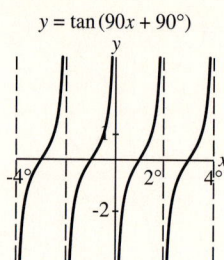

91. $y = x + 1$ **93.** $y = -5.67x + 16.01$

Graphing Calculator Exercises

1. $x = 0$ or $x = \pi/2 + 2k\pi$ for any integer k **3.** No

Section 6.5

1. $-\pi/6$ **3.** $\pi/6$ **5.** $\pi/4$ **7.** $\pi/3$ **9.** $-90°$
11. $135°$ **13.** $30°$ **15.** $180°$ **17.** $-\pi/4$ **19.** $\pi/3$
21. $\pi/4$ **23.** $-\pi/6$ **25.** 0   **27.** $\pi/2$ **29.** $3\pi/4$
31. $2\pi/3$ **33.** 0.60   **35.** 3.02   **37.** -0.14 **39.** 1.87
41. 1.15   **43.** -0.36 **45.** 3.06   **47.** 0.06   **49.** $\sqrt{3}$
51. $-\pi/6$ **53.** $\pi/6$ **55.** $\pi/4$ **57.** 1   **59.** $\pi/2$ **61.** 0
63. $\pi/2$ **65.** 0.8930   **67.** Undefined   **69.** -0.9802
71. -0.4082 **73.** 3.4583   **75.** 1.0183   **77.** 33.8°
79. 86.6°   **81.** 50.3°   **83.** 81.1°

85.

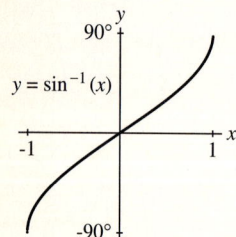

$y = \sin^{-1}(x)$

87.

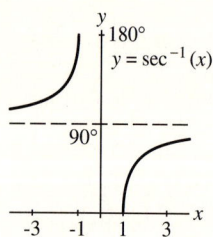

$y = \sec^{-1}(x)$

89.

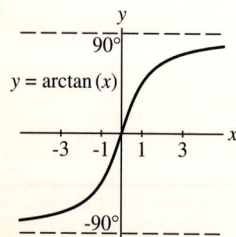

$y = \arctan(x)$

91. $\pi/2$ **93.** 0.2507

Section 6.6

1. 4/5, 3/5, 4/3, 5/4, 5/3, 3/4
3. $3\sqrt{10}/10$, $-\sqrt{10}/10$, -3, $\sqrt{10}/3$, $-\sqrt{10}$, $-1/3$
5. $-\sqrt{3}/3$, $-\sqrt{6}/3$, $\sqrt{2}/2$, $-\sqrt{3}$, $-\sqrt{6}/2$, $\sqrt{2}$
7. $-1/2$, $\sqrt{3}/2$, $-\sqrt{3}/3$, -2, $2\sqrt{3}/3$, $-\sqrt{3}$ **9.** 80.5°
11. 60°   **13.** 1.0   **15.** 0.4
17. $\sqrt{5}/5$, $2\sqrt{5}/5$, 1/2, $2\sqrt{5}/5$, $\sqrt{5}/5$, 2
19. $3\sqrt{34}/34$, $5\sqrt{34}/34$, 3/5, $5\sqrt{34}/34$, $3\sqrt{34}/34$, 5/3
21. 4/5, 3/5, 4/3, 3/5, 4/5, 3/4
23. $c = 10$, $\alpha = 36.9°$, $\beta = 53.1°$
25. $a = 5.73$, $\alpha = 43.7°$, $\beta = 46.3°$
27. $\beta = 74°$, $a = 5.5$, $b = 19.2$
29. $\beta = 50°51'$, $b = 11.1$, $c = 14.3$
31. 25   **33.** 0.831   **35.** 18.8   **37.** -289
39. 50 ft   **41.** 43.2 m   **43.** 25.1 ft   **45.** 22 m, 57.6°
47. No, speed is 11.2 mph   **49.** 153.1 m   **51.** 4.5 km
53. 1.9×10^{13} mi   **55.** 4391 mi

Chapter 6 Review Exercises

1. 28°   **3.** 206°45′33″   **5.** 180°   **7.** 108°   **9.** 300°
11. 270°   **13.** $11\pi/6$ **15.** $-5\pi/3$ **17.** $-\sqrt{2}/2$
19. $\sqrt{3}$ **21.** $-2\sqrt{3}/3$ **23.** 0   **25.** 0   **27.** -1
29. $\sqrt{3}/3$ **31.** $-\sqrt{2}/2$ **33.** -2 **35.** $-\sqrt{3}/3$
37. 5/13, 12/13, 5/12, 13/5, 13/12, 12/5   **39.** 0.6947
41. -0.0923 **43.** 0.1869   **45.** 1.0356   **47.** $-\pi/6$
49. $-\pi/4$ **51.** $\pi/4$ **53.** $\pi/6$ **55.** 90°   **57.** 135°
59. 30°   **61.** 90°   **63.** $c = \sqrt{13}$, $\alpha = 33.7°$, $\beta = 56.3°$
65. $\beta = 68.7°$, $c = 8.8$, $b = 8.2$ **67.** $2\pi/3$, $[-2, 2]$

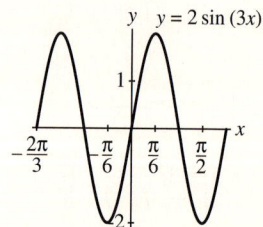

$y = 2 \sin (3x)$

69. $\pi/2$, $(-\infty, \infty)$

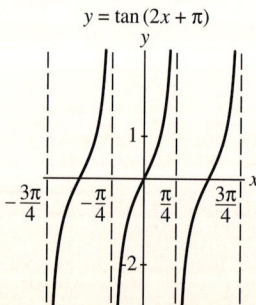

$y = \tan (2x + \pi)$

71. 4π, $(-\infty, -1] \cup [1, \infty)$

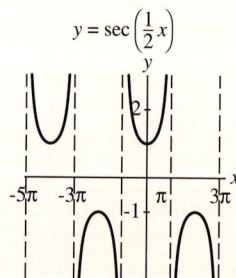

$y = \sec \left(\dfrac{1}{2} x\right)$

73. 180°, $[-1/2, 1/2]$

$y = \frac{1}{2}\cos(2x)$

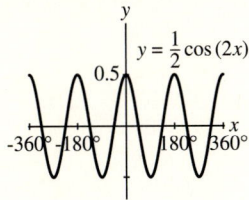

75. 90°, $(-\infty, \infty)$

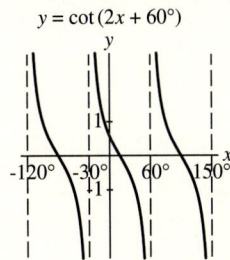

$y = \cot(2x + 60°)$

15. 2, $(-\infty, \infty)$

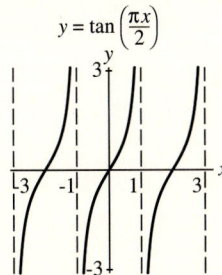

$y = \tan\left(\frac{\pi x}{2}\right)$

16. π, $[-2, 2]$, 2

$y = 2\sin(2x + \pi)$

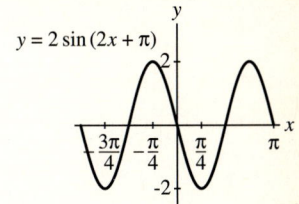

77. 180°, $(-\infty, -1/3] \cup [1/3, \infty)$

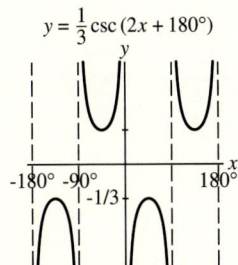

$y = \frac{1}{3}\csc(2x + 180°)$

17. 2π, $(-\infty, -2] \cup [2, \infty)$

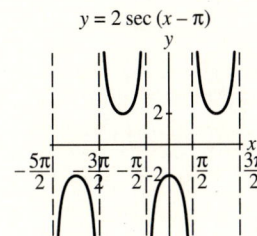

$y = 2\sec(x - \pi)$

18. 2π, $(-\infty, -1] \cup [1, \infty)$

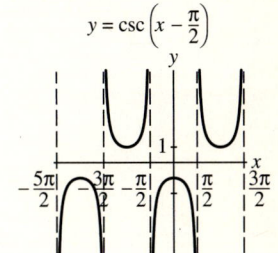

$y = \csc\left(x - \frac{\pi}{2}\right)$

79. $-2\sqrt{6}/5$ **81.** 1.08×10^{-8} sec   **83.** 94.2 ft/hr
85. 6.9813 ft   **87.** No   **89.** 0.33 rad/sec

Chapter 6 Test

1. 1/2   **2.** $-1/2$ **3.** -1 **4.** 2   **5.** -2 **6.** $\sqrt{3}/3$
7. $-\pi/6$ **8.** $2\pi/3$ **9.** $-\pi/4$ **10.** Undefined
11. Undefined   **12.** $2\sqrt{2}/3$
13. $2\pi/3$, $[-3, -1]$, 1   **14.** 2π, $[-1, 1]$, 1

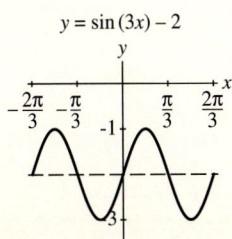

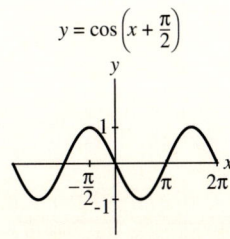

$y = \sin(3x) - 2$

$y = \cos\left(x + \frac{\pi}{2}\right)$

19. $\pi/2$, $(-\infty, \infty)$

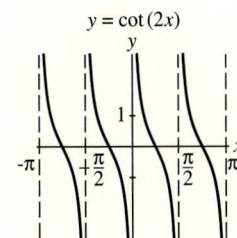

$y = \cot(2x)$

20. 2π, $[-1, 1]$, 1

$y = -\cos\left(x - \frac{\pi}{2}\right)$

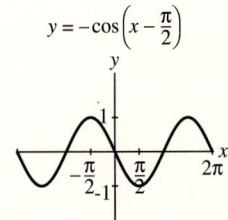

21. 28.85 m   **22.** 134.1°   **23.** $-\sqrt{15}/4$
24. $\sin\alpha = -2/\sqrt{29}$, $\cos\alpha = 5/\sqrt{29}$, $\tan\alpha = -2/5$,
$\csc\alpha = -\sqrt{29}/2$, $\sec\alpha = \sqrt{29}/5$, $\cot\alpha = -5/2$
25. 647.2 rad/min   **26.** 7.97 mph   **27.** 12.2 m   **28.** 1117 ft

Tying It All Together Chapters 1–6

1. $(-\infty, \infty)$, $(2, \infty)$

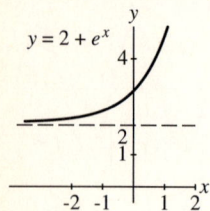

$y = 2 + e^x$

2. $(-\infty, \infty)$, $[2, \infty)$

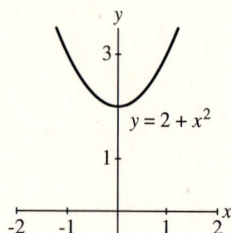

$y = 2 + x^2$

3. $(-\infty, \infty)$, $[1, 3]$

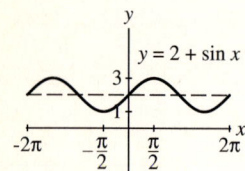

$y = 2 + \sin x$

4. $(0, \infty)$, $(-\infty, \infty)$

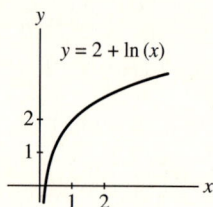

$y = 2 + \ln(x)$

5. $(\pi/4, \infty)$, $(-\infty, \infty)$

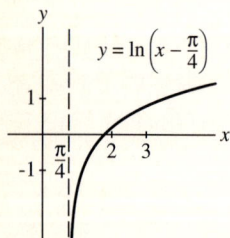

$y = \ln\left(x - \dfrac{\pi}{4}\right)$

6. $(-\infty, \infty)$, $[-1, 1]$

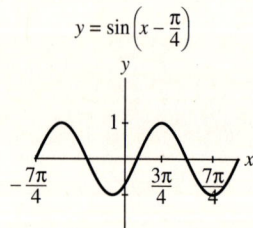

$y = \sin\left(x - \dfrac{\pi}{4}\right)$

7. $(0, \infty)$, $(-\infty, \infty)$

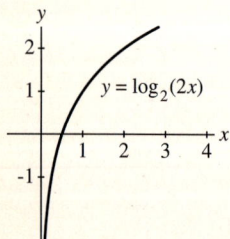

$y = \log_2(2x)$

8. $(-\infty, \infty)$, $[-1, 1]$

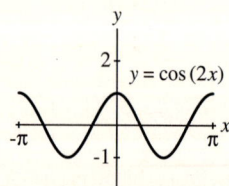

$y = \cos(2x)$

9. Neither   **10.** Odd   **11.** Even   **12.** Even   **13.** Odd
14. Even   **15.** Neither   **16.** Odd   **17.** Increasing
18. Increasing   **19.** Increasing   **20.** Decreasing
21. Decreasing   **22.** Increasing

CHAPTER 7

Section 7.1

1. $\csc x$ **3.** 1   **5.** $\cos^2 \alpha$ **7.** $-\cos^2 \beta$ **9.** $2 \sin \alpha$

11. $\cot x = \pm\sqrt{\csc^2 x - 1}$ **13.** $\sin x = \dfrac{1}{\pm\sqrt{1 + \cot^2 x}}$

15. $\tan x = \dfrac{1}{\pm\sqrt{\csc^2 x - 1}}$ **17.** $\sin \alpha = 1/\sqrt{5}$, $\cos \alpha = 2/\sqrt{5}$,
$\csc \alpha = \sqrt{5}$, $\sec \alpha = \sqrt{5}/2$, $\cot \alpha = 2$ **19.** $\sin \alpha = -\sqrt{22}/5$,
$\tan \alpha = \sqrt{22}/\sqrt{3}$, $\csc \alpha = -5/\sqrt{22}$, $\sec \alpha = -5/\sqrt{3}$,
$\cot \alpha = \sqrt{3}/\sqrt{22}$ **21.** $\sin \alpha = -3/\sqrt{10}$, $\cos \alpha = 1/\sqrt{10}$,
$\tan \alpha = -3$, $\csc \alpha = -\sqrt{10}/3$, $\sec \alpha = \sqrt{10}$ **23.** $\cos x$
25. 0   **27.** 0   **29.** $\cos^2 \alpha$ **31.** $-\cos \beta$ **33.** Odd
35. Neither   **37.** Even   **39.** Even   **41.** Even   **43.** Odd
55. $-\tan^2 x$ **57.** -1 **59.** 1   **61.** $\tan^2 x$ **63.** $\csc x$
65. $\sin^2 x - \cos^2 x$ **67.** $y = -x^2 \sec^2 \theta + x \tan \theta$

Section 7.2

1. D   **3.** A   **5.** B   **7.** H   **9.** G   **11.** $-\cos^2 \alpha$
13. $2 \cos^2 \beta - \cos \beta - 1$ **15.** $\csc^2 x + 2 + \sin^2 x$
17. $4 \sin^2 \theta - 1$ **19.** $9 \sin^2 \theta + 12 \sin \theta + 4$
21. $4 \sin^4 y - 4 + \csc^4 y$ **23.** $(2 \sin \gamma + 1)(\sin \gamma - 3)$
25. $(\tan \alpha - 4)(\tan \alpha - 2)$ **27.** $(2 \sec \beta + 1)^2$
29. $(\tan \alpha - \sec \beta)(\tan \alpha + \sec \beta)$
31. $(\cos \beta)(\sin \beta + 2)(\sin \beta - 1)$ **33.** $(2 \sec^2 x - 1)^2$
35. $(\cos \alpha + 1)(\sin \alpha + 1)$ **65.** Identity
67. Not identity   **69.** Identity

Graphing Calculator Exercises

1. Not identity   **3.** Identity

Section 7.3

1. $7\pi/12$ **3.** $13\pi/12$ **5.** $30° + 45°$ **7.** $120° + 45°$
9. $\dfrac{\sqrt{6} - \sqrt{2}}{4}$ **11.** $\dfrac{\sqrt{6} + \sqrt{2}}{4}$ **13.** $2 + \sqrt{3}$ **15.** $\dfrac{\sqrt{2} - \sqrt{6}}{4}$
17. $\dfrac{-\sqrt{6} - \sqrt{2}}{4}$ **19.** $-2 + \sqrt{3}$ **21.** 1   **23.** $\cos(3\pi/10)$
25. $\tan(5\pi/18)$ **27.** $\tan(13\pi/42)$ **29.** $\sin 49°$
31. $\cos(2\pi/15)$ **33.** $\sin(3k)$ **35.** G   **37.** H   **39.** F
41. A   **43.** $\dfrac{16}{65}$ **45.** $\dfrac{2 - \sqrt{15}}{6}$ **47.** $-\sin \alpha$

49. $-\cos \alpha$ **51.** $-\sin \alpha$ **53.** $\cos \alpha$ **55.** $\dfrac{1 + \tan \alpha}{1 - \tan \alpha}$

57. $\tan \alpha$

Section 7.4

1. 1   **3.** $\sqrt{3}$ **5.** -1 **7.** $\sqrt{3}$ **9.** $\dfrac{\sqrt{2 - \sqrt{3}}}{2}$

11. $2 - \sqrt{3}$ **13.** $\dfrac{\sqrt{2 + \sqrt{2}}}{2}$ **15.** $\dfrac{\sqrt{2 - \sqrt{2}}}{2}$ **17.** Positive

19. Negative   **21.** Negative   **23.** $\sin 26°$ **25.** $\sqrt{2}/2$

27. $\sqrt{3}/6$ **29.** $\sqrt{3}/2$ **31.** $-\sin(7\pi/9)$ **33.** $\tan 6°$

49. $1/\sqrt{5},\ 2/\sqrt{5},\ 1/2,\ \sqrt{5},\ \sqrt{5}/2,\ 2$

51. $-\sqrt{15}/8,\ -7/8,\ \sqrt{15}/7,\ -8/\sqrt{15},\ -8/7,\ 7/\sqrt{15}$

53. $-24/25,\ -7/25,\ 24/7,\ -25/24,\ -25/7,\ 7/24$

55. $\dfrac{5\sqrt{34} - 25}{3}$ **57.** $45°$

Section 7.5

1. $0.5(\cos 4° - \cos 22°)$ **3.** $0.5(\sin 36° - \sin 4°)$

5. $0.5(\cos(\pi/30) + \cos(11\pi/30))$

7. $0.5(\cos 2y^2 + \cos 12y^2)$ **9.** $0.5(\sin(3s) + \sin(s - 2))$

11. $\dfrac{\sqrt{2} - 1}{4}$ **13.** $\dfrac{\sqrt{2} + \sqrt{3}}{4}$ **15.** $2 \cos 10° \sin 2°$

17. $2 \sin 83.5° \sin 3.5°$ **19.** $-2 \cos(4.2) \sin(0.6)$

21. $-2 \sin(4y + 3) \sin(y - 6)$ **23.** $-2 \cos(6.5\alpha) \sin(1.5\alpha)$

25. $-2 \sin(4\pi/15) \sin(\pi/15)$

27. $\dfrac{\sqrt{6}}{2}$ **29.** $\dfrac{\sqrt{2 - \sqrt{2}}}{2}$

31. $\sqrt{2} \sin(x - \pi/4)$ **33.** $\sin(x + 2\pi/3)$ **35.** $\sin(x - \pi/6)$

37. $\sqrt{2},\ 2\pi,\ 3\pi/4$ **39.** $2,\ 2\pi,\ \pi/4$

$y = \sqrt{2} \sin\left(x + \dfrac{3\pi}{4}\right)$

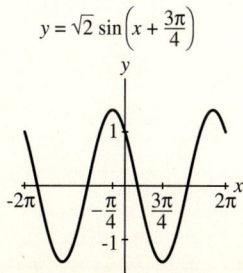

$y = 2 \sin\left(x - \dfrac{\pi}{4}\right)$

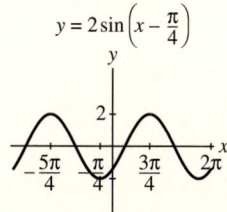

41. $2,\ 2\pi,\ 7\pi/6$

$y = 2 \sin\left(x + \dfrac{7\pi}{6}\right)$

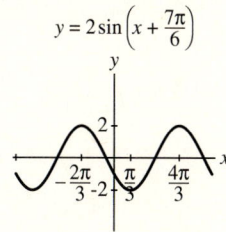

43. $5,\ 0.9$ **45.** $\sqrt{37},\ 3.0$ **47.** $\sqrt{34},\ 4.2$

59. $x = 2 \sin(t + \pi/6),\ 2$

Section 7.6

1. $\left\{ x \,\middle|\, x = \dfrac{\pi}{3} + 2k\pi \text{ or } x = \dfrac{5\pi}{3} + 2k\pi \right\}$

3. $\left\{ x \,\middle|\, x = \dfrac{\pi}{4} + 2k\pi \text{ or } x = \dfrac{3\pi}{4} + 2k\pi \right\}$

5. $\left\{ x \,\middle|\, x = \dfrac{\pi}{4} + k\pi \right\}$

7. $\left\{ x \,\middle|\, x = \dfrac{5\pi}{6} + 2k\pi \text{ or } x = \dfrac{7\pi}{6} + 2k\pi \right\}$

9. $\left\{ x \,\middle|\, x = \dfrac{5\pi}{4} + 2k\pi \text{ or } x = \dfrac{7\pi}{4} + 2k\pi \right\}$

11. $\left\{ x \,\middle|\, x = \dfrac{3\pi}{4} + k\pi \right\}$ **13.** $\{x \,|\, x = 90° + k180°\}$

15. $\{x \,|\, x = 90° + k360°\}$ **17.** $\{x \,|\, x = k180°\}$

19. $\{x \,|\, x = 29.2° + k360° \text{ or } x = 330.8° + k360°\}$

21. $\{x \,|\, x = 345.9° + k360° \text{ or } x = 194.1° + k360°\}$

23. $\{x \,|\, x = 79.5° + k180°\}$

25. $\left\{ x \,\middle|\, x = \dfrac{2\pi}{3} + 4k\pi \text{ or } x = \dfrac{10\pi}{3} + 4k\pi \right\}$

27. $\left\{ x \,\middle|\, x = \dfrac{2k\pi}{3} \right\}$

29. $\left\{ x \,\middle|\, x = \dfrac{\pi}{3} + 4k\pi \text{ or } x = \dfrac{5\pi}{3} + 4k\pi \right\}$

31. $\left\{ x \,\middle|\, x = \dfrac{5\pi}{8} + k\pi \text{ or } x = \dfrac{7\pi}{8} + k\pi \right\}$

33. $\left\{ x \,\middle|\, x = \dfrac{\pi}{6} + \dfrac{k\pi}{2} \right\}$ **35.** $\left\{ x \,\middle|\, x = \dfrac{k\pi}{4} \right\}$

37. $\left\{ x \,\middle|\, x = \dfrac{1}{6} + 2k \text{ or } x = \dfrac{5}{6} + 2k \right\}$

39. $\left\{x \mid x = \dfrac{1}{4} + \dfrac{k}{2}\right\}$

41. $\{240°, 300°\}$ **43.** $\{22.5°, 157.5°, 202.5°, 337.5°\}$
45. $\{45°, 75°, 165°, 195°, 285°, 315°\}$ **47.** $\{60°\}$
49. $\{\alpha \mid \alpha = 6.6° + k120° \text{ or } \alpha = 53.4° + k120°\}$
51. $\{\alpha \mid \alpha = 72.3° + k120° \text{ or } \alpha = 107.7° + k120°\}$
53. $\{\alpha \mid \alpha = 38.6° + k180° \text{ or } \alpha = 141.4° + k180°\}$
55. $\{\alpha \mid \alpha = 668.5° + k720° \text{ or } \alpha = 411.5° + k720°\}$
57. $\{0, 0.3, 2.8, \pi\}$ **59.** $\{\pi, 2\pi/3, 4\pi/3\}$ **61.** $\{11\pi/6\}$
63. $\{1.0, 4.9\}$ **65.** $\{0.7, 2.5, 3.4, 6.0\}$ **67.** $\{0, \pi\}$
69. $\{\pi/2, 3\pi/2\}$ **71.** $\{7\pi/12, 23\pi/12\}$ **73.** $\{7\pi/6, 11\pi/6\}$
75. $\{0°\}$ **77.** $\{26.6°, 206.6°\}$
79. $\{30°, 90°, 150°, 210°, 270°, 330°\}$
81. $\{67.5°, 157.5°, 247.5°, 337.5°\}$ **83.** $\{221.8°, 318.2°\}$
85. $\{30°, 120°, 210°, 300°\}$
87. $\{30°, 45°, 135°, 150°, 210°, 225°, 315°, 330°\}$
89. $\{0°, 60°, 120°, 180°, 240°, 300°\}$ **91.** 10.5
93. 56.4°   **95.** 19.6°

97. $\dfrac{5\pi}{12} + \dfrac{k\pi}{2}$ for k a nonnegative integer

99. $1 + 6k$ and $2 + 6k$ for k a nonnegative integer
101. 44.4° or 45.6°   **103.** 12.5° or 77.5°, 6.3 sec

Graphing Calculator Exercises

1. $\{0.4, 1.9, 2.2, 4.0, 4.4, 5.8\}$ **3.** $\varnothing$

Chapter 7 Review Exercises

1. $\cos^2 \alpha$ **3.** $-\cot^2 x$ **5.** $\sec^2 \alpha$ **7.** $\tan 4s$
9. $-\sin 3\theta$ **11.** $\tan z$ **13.** $\sin \alpha = 12/13$, $\tan \alpha = -12/5$,
$\csc \alpha = 13/12$, $\sec \alpha = -13/5$, $\cot \alpha = -5/12$ **15.** $\sin \alpha =$
$-4/5$, $\cos \alpha = -3/5$, $\tan \alpha = 4/3$, $\csc \alpha = -5/4$, $\sec \alpha =$
$-5/3$, $\cot \alpha = 3/4$ **17.** $\sin \alpha = -24/25$, $\cos \alpha = 7/25$,
$\tan \alpha = -24/7$, $\csc \alpha = -25/24$, $\sec \alpha = 25/7$, $\cot \alpha =$
$-7/24$ **19.** Identity   **21.** Not identity   **23.** Odd

25. Neither   **27.** Even   **43.** $\sqrt{3} - 2$ **45.** $-\dfrac{\sqrt{2 + \sqrt{3}}}{2}$

47. $4\sqrt{2}$, $\pi/4$

$y = 4\sqrt{2} \sin\left(x + \dfrac{\pi}{4}\right)$

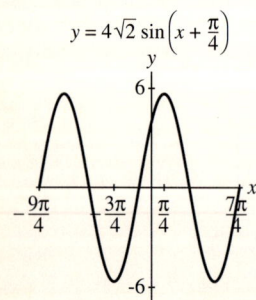

49. $\sqrt{5}$, 2.68

$y = \sqrt{5} \sin (x + 2.68)$

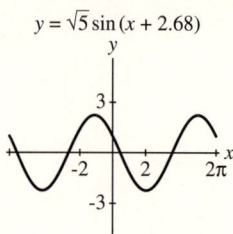

51. $\left\{x \mid x = \dfrac{\pi}{3} + k\pi \text{ or } x = \dfrac{2\pi}{3} + k\pi\right\}$

53. $\left\{x \mid x = \dfrac{\pi}{3} + 2k\pi, \dfrac{2\pi}{3} + 2k\pi, \dfrac{\pi}{6} + 2k\pi, \dfrac{5\pi}{6} + 2k\pi\right\}$

55. $\left\{x \mid x = \dfrac{\pi}{6} + 2k\pi, \dfrac{5\pi}{6} + 2k\pi, \dfrac{\pi}{2} + 2k\pi\right\}$

57. $\left\{x \mid x = \dfrac{2\pi}{3} + 4k\pi \text{ or } x = \dfrac{4\pi}{3} + 4k\pi\right\}$

59. $\left\{x \mid x = \pi + 2k\pi, \dfrac{\pi}{3} + 4k\pi, \dfrac{5\pi}{3} + 4k\pi\right\}$

61. $\left\{x \mid x = \dfrac{\pi}{2} + k\pi\right\}$

63. $\left\{x \mid x = \pi + 2k\pi \text{ or } x = \dfrac{3\pi}{2} + 2k\pi\right\}$

65. $\{45°, 225°\}$ **67.** $\{90°, 180°\}$ **69.** $\varnothing$
71. $\{22.5°, 67.5°, 112.5°, 157.5°, 202.5°, 247.5°, 292.5°, 337.5°\}$
73. $\{0°, 180°\}$ **75.** $\{15°, 75°, 135°, 195°, 255°, 315°\}$

77. $2 \cos 17° \cos 2°$ **79.** $2 \cos \dfrac{\pi}{16} \sin \dfrac{3\pi}{16}$

81. $\sin 24° - \sin 2°$ **83.** $\cos \dfrac{x}{12} + \cos \dfrac{7x}{12}$ **85.** 1.28, 2.85

Chapter 7 Test

1. $2 \cos x$ **2.** $\sin 7t$ **3.** $2 \csc^2 y$ **4.** $\tan(3\pi/10)$

9. $\left\{\theta \mid \theta = \dfrac{3\pi}{2} + 2k\pi\right\}$

10. $\left\{s \mid s = \dfrac{\pi}{9} + \dfrac{2k\pi}{3} \text{ or } s = \dfrac{5\pi}{9} + \dfrac{2k\pi}{3}\right\}$

11. $\left\{t \mid t = \dfrac{\pi}{3} + \dfrac{k\pi}{2}\right\}$

12. $\left\{\theta \mid \theta = \dfrac{\pi}{2} + k\pi \text{ or } \theta = \dfrac{\pi}{6} + 2k\pi \text{ or } \dfrac{5\pi}{6} + 2k\pi\right\}$

13. $\{19.5°, 90°, 160.5°\}$
14. $\{27°, 63°, 99°, 135°, 171°, 207°, 243°, 279°, 315°, 351°\}$

15. 2π, 2, $5\pi/3$

$$y = 2\sin\left(x + \frac{5\pi}{3}\right)$$

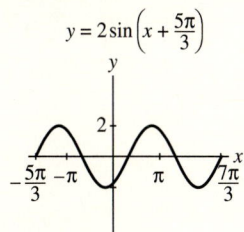

16. $\sin \alpha = 1/2$, $\cos \alpha = -\sqrt{3}/2$, $\tan \alpha = -1/\sqrt{3}$, $\sec \alpha = -2/\sqrt{3}$, $\cot \alpha = -\sqrt{3}$ **17.** Even

18. $-\dfrac{\sqrt{2-\sqrt{3}}}{2}$ **20.** 0.4 sec, 1.4 sec, 2.5 sec, 3.5 sec

Tying It All Together Chapters 1–7

1. Odd   **2.** Even   **3.** Odd   **4.** Even   **5.** Even   **6.** Odd
7. Even   **8.** Even   **9.** Even   **10.** Odd   **11.** Not identity
12. Not identity   **13.** Identity   **14.** Identity
15. Not identity   **16.** Not identity
17. $\beta = 60°$, $b = 4\sqrt{3}$, $c = 8$
18. $\alpha = 60°$, $\beta = 30°$, $c = 2$
19. $\alpha = 17.5°$, $\beta = 72.5°$, $a = 1.6$, $c = 5.2$
20. $\alpha = 36.9°$, $\beta = 53.1°$, $b = 2.7$, $c = 3.3$ **21.** $\dfrac{1}{2} + \dfrac{\sqrt{3}}{2}i$
22. $-1 + \sqrt{3}i$ **23.** i **24.** 1   **25.** $2 + 11i$ **26.** -16

CHAPTER 8

Section 8.1

1. $\gamma = 44°$, $b = 14.4$, $c = 10.5$
3. $\beta = 134.2°$, $a = 5.2$, $c = 13.6$
5. $\beta = 26°$, $a = 14.6$, $b = 35.8$
7. $\alpha = 45.7°$, $b = 587.9$, $c = 160.8$ **9.** None
11. One: $\alpha = 30°$, $\beta = 90°$, $a = 10$
13. One: $\alpha = 26.3°$, $\gamma = 15.6°$, $a = 10.3$
15. Two: $\alpha_1 = 134.9°$, $\gamma_1 = 12.4°$, $c_1 = 11.4$; or $\alpha_2 = 45.1°$, $\gamma_2 = 102.2°$, $c_2 = 51.7$
17. One: $\alpha = 25.4°$, $\beta = 55.0°$, $a = 5.4$ **19.** 9.78
21. 83.4   **23.** 66.3   **25.** 37.7   **27.** $22\sqrt{3} + 6$
29. 28.9 ft, 15.7 ft   **31.** 18.4 mi
33. 97,535 ft² **35.** 277.7 in²

Section 8.2

1. $\alpha = 30.4°$, $\beta = 28.3°$, $c = 5.2$
3. $\alpha = 26.4°$, $\beta = 131.3°$, $\gamma = 22.3°$

5. $\alpha = 163.9°$, $\gamma = 5.6°$, $b = 4.5$
7. $\alpha = 130.3°$, $\beta = 30.2°$, $\gamma = 19.5°$
9. $\beta = 48.3°$, $\gamma = 101.7°$, $a = 6.2$
11. $\alpha = 53.9°$, $\beta = 65.5°$, $\gamma = 60.6°$ **13.** 40.9
15. 13.9   **17.** 40,471.9   **19.** 100   **21.** 20.8   **23.** 20.4
25. 20.6 mi   **27.** 19.2°
29. $\angle A = 78.5°$, $\angle B = 57.1°$, $\angle C = 44.4°$, B and C will not rotate   **31.** 3.8 mi   **33.** 11.76 m   **35.** 15.94 mm

Section 8.3

1. $|\mathbf{v}_x| = 1.9$, $|\mathbf{v}_y| = 4.1$ **3.** $|\mathbf{v}_x| = 7256.4$, $|\mathbf{v}_y| = 3368.3$
5. $|\mathbf{v}_x| = 87.7$, $|\mathbf{v}_y| = 217.0$ **7.** 2, 30°   **9.** 2, 135°
11. 16, 300°   **13.** 5, 0°   **15.** $\sqrt{13}$, 146.3°   **17.** $\sqrt{10}$, 341.6°
19. $\langle 15, -10 \rangle$ **21.** $\langle 18, -22 \rangle$ **23.** $\langle 11, -13 \rangle$
25. $\langle 0, -1 \rangle$ **27.** $\langle 4\sqrt{2}, 4\sqrt{2} \rangle$ **29.** $\langle -237.6, 166.3 \rangle$
31. $\langle 17.5, -4.0 \rangle$ **33.** $2\mathbf{i} + \mathbf{j}$ **35.** $-3\mathbf{i} + \sqrt{2}\mathbf{j}$ **37.** $-9\mathbf{j}$
39. $-7\mathbf{i} - \mathbf{j}$ **41.** $\sqrt{17}$, 76.0°   **43.** $3\sqrt{10}$, 198.4°
45. $\sqrt{29}$, 158.2°   **47.** $\sqrt{65}/2$, 172.9°   **49.** $\sqrt{89}$, 32.0°
51. $2\sqrt{10}$, 108.4°   **53.** $\sqrt{5}$, 243.4°   **55.** $\sqrt{73}$ lb, 69.4°, 20.6°
57. 8.25 newtons, 22.9°, 107.1°   **59.** 5.1 lb, 75.2°
61. 127.0 lb   **63.** 450.3 mph, 260 mph   **65.** 1368.1 lb
67. 108.2°, 451.0 mph   **69.** 198.4°, 6000 ft

Section 8.4

1. $2\sqrt{10}$ **3.** 4   **5.** 8   **7.** 9   **9.** 1   **11.** $3\sqrt{2}$
13. $3\sqrt{2}(\cos 135° + i \sin 135°)$
15. $3(\cos 135° + i \sin 135°)$ **17.** $8(\cos 0° + i \sin 0°)$
19. $\sqrt{3}(\cos 90° + i \sin 90°)$ **21.** $2(\cos 150° + i \sin 150°)$
23. $5(\cos 53.1° + i \sin 53.1°)$
25. $\sqrt{34}(\cos 121.0° + i \sin 121.0°)$
27. $3\sqrt{5}(\cos 296.6° + i \sin 296.6°)$ **29.** $1 + i$
31. $-\dfrac{3}{4} + \dfrac{\sqrt{3}}{4}i$ **33.** $-0.42 - 0.26i$ **35.** $3i$
37. $-i\sqrt{3}$ **39.** $\dfrac{\sqrt{6}}{2} + \dfrac{3\sqrt{2}}{2}i$ **41.** $6i$ **43.** $\dfrac{3\sqrt{2}}{2} + \dfrac{\sqrt{6}}{2}i$
45. $9i$ **47.** $\sqrt{3} + i$ **49.** $0.34 - 0.37i$ **51.** $-40i$, -0.8
53. $8i$, $\dfrac{\sqrt{3}}{4} - \dfrac{1}{4}i$ **55.** $4\sqrt{2}$, $i\sqrt{2}$
57. $-7 - 26i$, $-\dfrac{23}{29} - \dfrac{14}{29}i$ **59.** $-18 + 14i$, $\dfrac{6}{13} + \dfrac{22}{13}i$
61. $-3 + 3i$, $1.5 + 1.5i$ **63.** $-54 + 54i$

Section 8.5

1. $27i$ **3.** $-2 + 2i\sqrt{3}$ **5.** $-\dfrac{1}{2} + \dfrac{\sqrt{3}}{2}i$
7. $-18 + 18i\sqrt{3}$ **9.** $701.5 + 1291.9i$ **11.** $-16 + 16i$
13. $-8 - 8i\sqrt{3}$ **15.** $-3888 + 3888i\sqrt{3}$

17. $-119 - 120i$ **19.** $-7 - 24i$

21. $-44.928 - 31.104i$

23. $2(\cos 45° + i \sin 45°), 2(\cos 225° + i \sin 225°)$

25. $\cos \alpha + i \sin \alpha$ for $\alpha = 30°, 120°, 210°, 300°$

27. $2(\cos \alpha + i \sin \alpha)$ for $\alpha = \dfrac{\pi}{6}, \dfrac{\pi}{2}, \dfrac{5\pi}{6}, \dfrac{7\pi}{6}, \dfrac{3\pi}{2}, \dfrac{11\pi}{6}$

29. $1, -\dfrac{1}{2} + \dfrac{\sqrt{3}}{2}i, -\dfrac{1}{2} - \dfrac{\sqrt{3}}{2}i$

31. $\pm 2, \pm 2i$

33. $\dfrac{\sqrt{2}}{2} \pm \dfrac{\sqrt{2}}{2}i, -\dfrac{\sqrt{2}}{2} \pm \dfrac{\sqrt{2}}{2}i$

35. $\dfrac{\sqrt{3}}{2} + \dfrac{1}{2}i, -\dfrac{\sqrt{3}}{2} + \dfrac{1}{2}i, -i$

37. $1 + i\sqrt{3}, -1 - i\sqrt{3}$

39. $1.272 + 0.786i, -1.272 - 0.786i$

41. $\dfrac{1}{2} + \dfrac{\sqrt{3}}{2}i, -1, \dfrac{1}{2} - \dfrac{\sqrt{3}}{2}i$

43. $\pm 3, \pm 3i$

45. $-1 + i, 1 - i$

47. $0, \pm 2, 1 \pm i\sqrt{3}, -1 \pm i\sqrt{3}$

49. $2^{1/5}(\cos \alpha + i \sin \alpha)$ for $\alpha = 0°, 72°, 144°, 216°, 288°$

51. $10^{1/8}(\cos \alpha + i \sin \alpha)$ for $\alpha = 40.4°, 130.4°, 220.4°, 310.4°$

53. $-\dfrac{1}{4} + \dfrac{1}{4}i$ **55.** $1, -i$

Section 8.6

1. $\left(3\sqrt{2}, \pi/4\right)$ **3.** $(3, \pi/2)$

5–15. Odd

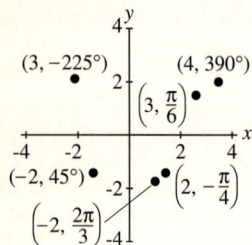

17. $\left(\dfrac{\sqrt{3}}{2}, \dfrac{1}{2}\right)$

19. $(0, 3)$ **21.** $(-1, 1)$ **23.** $\left(-\dfrac{\sqrt{6}}{2}, \dfrac{3\sqrt{2}}{2}\right)$

25. $\left(2\sqrt{3}, 60°\right)$ **27.** $\left(2\sqrt{2}, 135°\right)$ **29.** $(2, 90°)$

31. $\left(3\sqrt{2}, 225°\right)$ **33.** $\left(\sqrt{17}, 75.96°\right)$ **35.** $\left(\sqrt{6}, -54.7°\right)$

37.

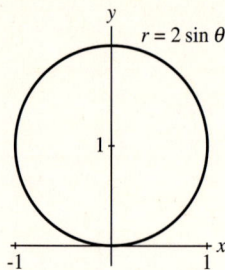

39.

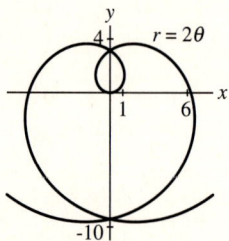

41.

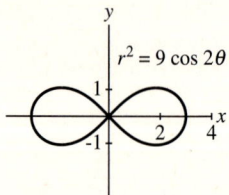

43.

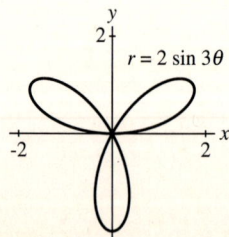

45.

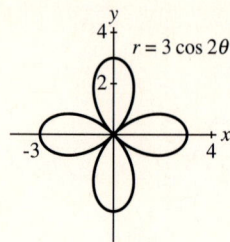

47.

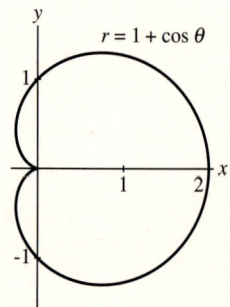

49.

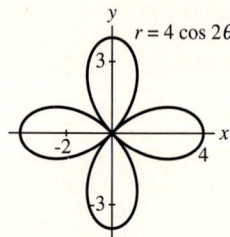

51.

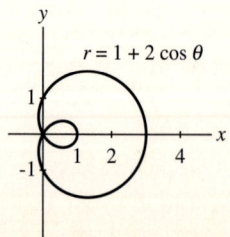

53.

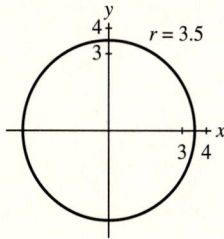

55.

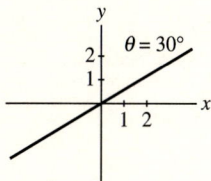

57. $x^2 - 4x + y^2 = 0$ **59.** $y = 3$ **61.** $x = 3$
63. $x^2 + y^2 = 25$ **65.** $y = x$ **67.** $x^2 - 4y = 4$
69. $r \cos \theta = 4$ **71.** $\theta = -\pi/4$ **73.** $r = 4 \tan \theta \sec \theta$
75. $r = 2$ **77.** $r = \dfrac{1}{2 \cos \theta - \sin \theta}$ **79.** $r = 2 \sin \theta$

Chapter 8 Review Exercises

1. $\alpha = 82.7°$, $\beta = 49.3°$, $c = 2.5$
3. $\gamma = 103°$, $a = 4.6$, $b = 18.4$ **5.** No triangle
7. $\alpha = 107.7°$, $\beta = 23.7°$, $\gamma = 48.6°$
9. $\alpha_1 = 110.8°$, $\gamma_1 = 47.2°$, $a_1 = 6.2$; $\alpha_2 = 25.2°$,
$\gamma_2 = 132.8°$, $a_2 = 2.8$ **11.** 92.4 ft²   **13.** 23.0 km²
15. 5.5, 2.4   **17.** 2.0, 2.5   **19.** $\sqrt{13}$, 56.3°
21. 6.0, 237.9°   **23.** $\langle 1, 1 \rangle$ **25.** $\langle -3.0, 8.6 \rangle$
27. $\langle -6, 8 \rangle$ **29.** $\langle 0, -17 \rangle$ **31.** $-4\mathbf{i} + 8\mathbf{j}$
33. $3.6\sqrt{3}\mathbf{i} + 3.6\mathbf{j}$ **35.** $\sqrt{34}$ **37.** $2\sqrt{2}$
39. $5.94(\cos 135° + i \sin 135°)$

41. $7.6(\cos 252.3° + i \sin 252.3°)$ **43.** $-\dfrac{3}{2} + \dfrac{\sqrt{3}}{2}i$

45. $5.4 + 3.5i$ **47.** $-15i, -\dfrac{5}{6}$ **49.** $8 - i, 0.31 + 0.54i$

51. $-4\sqrt{2} + 4i\sqrt{2}$ **53.** $-128 + 128i$

55. $\dfrac{\sqrt{2}}{2} + \dfrac{\sqrt{2}}{2}i, -\dfrac{\sqrt{2}}{2} - \dfrac{\sqrt{2}}{2}i$

57. $2^{1/3}(\cos \alpha + i \sin \alpha)$ for $\alpha = 10°, 130°, 250°$
59. $5^{1/6}(\cos \alpha + i \sin \alpha)$ for $\alpha = 8.9°, 128.9°, 248.9°$
61. $5(\cos \alpha + i \sin \alpha)$ for $\alpha = 22.5°, 112.5°, 202.5°, 292.5°$

63. $\left(2.5, 5\sqrt{3}/2\right)$ **65.** $(-0.3, 1.7)$ **67.** $(4, 4\pi/3)$
69. $\left(\sqrt{13}, -0.98\right)$

71.

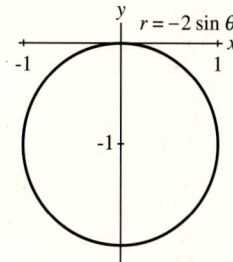

73.

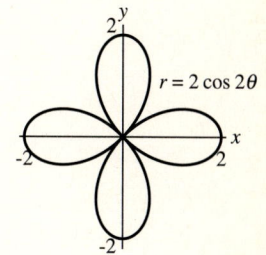

75.

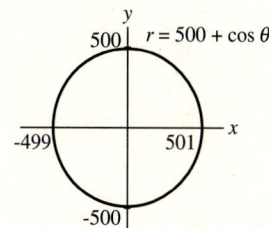

77.

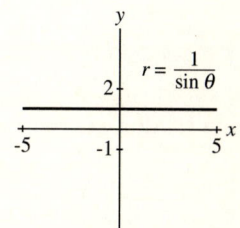

79. $x + y = 1$ **81.** $x^2 + y^2 = 25$ **83.** $r = \dfrac{3}{\sin \theta}$

85. $r = 7$ **87.** 18.4 lb, 11.0°, 19.0°   **89.** Seth, 42,785 ft²
91. \$2639

Chapter 8 Test

1. One: $\beta = 90°$, $\gamma = 60°$, $c = 2\sqrt{3}$ **2.** Two: $\beta_1 = 111.2°$,
$\gamma_1 = 8.8°$, $c_1 = 0.7$; $\beta_2 = 68.9°$, $\gamma_2 = 51.1°$, $c_2 = 3.5$
3. One: $\gamma = 145.6°$, $b = 2.5$, $c = 5.9$
4. One: $\alpha = 33.8°$, $\beta = 129.2°$, $c = 1.5$
5. $\alpha = 28.4°$, $\beta = 93.7°$, $\gamma = 57.9°$ **6.** $2\sqrt{10}$, 108.4°
7. $2\sqrt{5}$, 206.6°   **8.** $3\sqrt{17}$, 76.0°
9. $3\sqrt{2}(\cos 45° + i \sin 45°)$ **10.** $2(\cos 120° + i \sin 120°)$
11. $2\sqrt{5}(\cos 206.6° + i \sin 206.6°)$ **12.** $3\sqrt{2} + 3i\sqrt{2}$

13. $512i$ **14.** $\dfrac{3\sqrt{2}}{4} + \dfrac{3\sqrt{2}}{4}i$ **15.** $\left(\dfrac{5\sqrt{3}}{2}, \dfrac{5}{2}\right)$

16. $\left(-\dfrac{3\sqrt{2}}{2}, \dfrac{3\sqrt{2}}{2}\right)$ **17.** $(-26.4, -19.9)$

18.

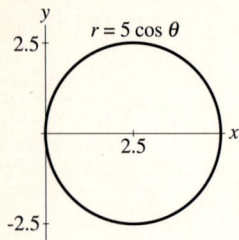
$r = 5 \cos \theta$

19.

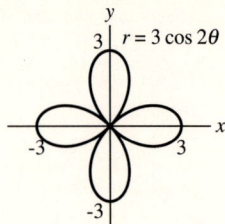
$r = 3 \cos 2\theta$

13.

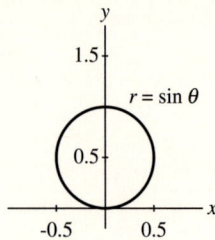
$r = \sin \theta$

14.

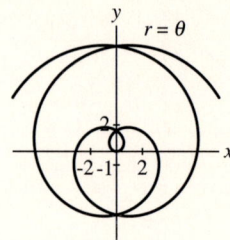
$r = \theta$

20. 12.2 m²
21. $\mathbf{v} = 3.66\mathbf{i} + 2.78\mathbf{j}$
22. $\dfrac{3\sqrt{2}}{2} \pm \dfrac{3\sqrt{2}}{2}i, \ -\dfrac{3\sqrt{2}}{2} \pm \dfrac{3\sqrt{2}}{2}i$
23. $r = -5 \sin \theta$
24. $(x^2 + y^2)^{3/2} = 10xy$ **25.** 47.2°, 239.3 mph

Tying It All Together Chapters 1–8

1. $0, 1, -\dfrac{1}{2} \pm \dfrac{\sqrt{3}}{2}i$ **2.** $-2, 1, 3$

3. $-2, 1, \cos 72° + i \sin 72°, \cos 144° + i \sin 144°,$
$\cos 216° + i \sin 216°, \cos 288° + i \sin 288°$

4. $2^{-1/4} \pm i2^{-1/4}, -2^{-1/4} \pm i2^{-1/4}, 1, -\dfrac{1}{2} \pm \dfrac{\sqrt{3}}{2}i$

5. $\dfrac{\pi}{6} + 2k\pi, \dfrac{5\pi}{6} + 2k\pi, \dfrac{2\pi}{3} + 2k\pi, \dfrac{4\pi}{3} + 2k\pi$

6. $-\dfrac{1}{2}, \dfrac{\pi}{6} + 2k\pi, \dfrac{5\pi}{6} + 2k\pi$ **7.** $k\pi$

8. $\ln\left(\dfrac{\pi}{6} + 2k\pi\right), \ln\left(\dfrac{5\pi}{6} + 2k\pi\right)$ for k a nonnegative
integer
9. 4   **10.** ∅

11.

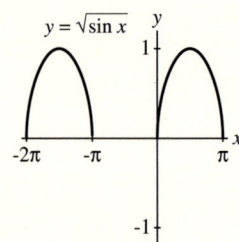
$y = \sin x$

12.

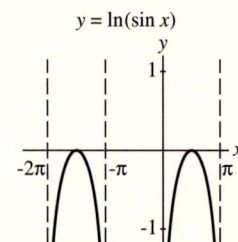
$y = e^x$

15.

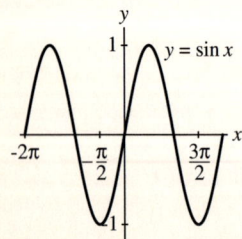
$y = \sqrt{\sin x}$

16.

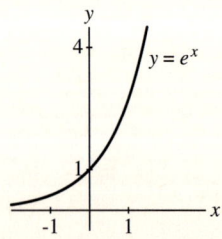
$y = \ln(\sin x)$

17.

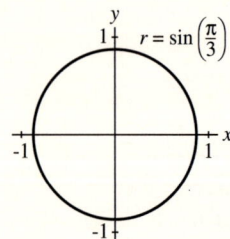
$r = \sin\left(\dfrac{\pi}{3}\right)$

18.

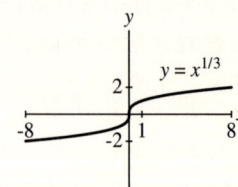
$y = x^{1/3}$

19. 0   **20.** 0   **21.** -1 **22.** 0   **23.** $\dfrac{\pi}{6}$ **24.** $\dfrac{\pi}{3}$

25. $\dfrac{-\pi}{4}$ **26.** $\dfrac{\pi}{4}$ **27.** $\dfrac{2x}{x^2 - 4}$ **28.** $2\sec^2(x)$ **29.** 4

30. $-\log(x^2 - 4)$ **31.** 2   **32.** $\dfrac{1}{4}$ **33.** 2   **34.** 0

35. $\dfrac{2x}{3 - x}$ **36.** 1

CHAPTER 9

Section 9.1

1. $\{(1, 2)\}$ **3.** $\emptyset$ **5.** $\{(3, 2)\}$ **7.** $\{(3, 1)\}$ **9.** $\emptyset$
11. $\{(x, y) | x - 2y = 6\}$ **13.** $\{(-1, -1)\}$, independent
15. $\{(11/5, -6/5)\}$, independent
17. $\{(x, y) | y = 3x + 5\}$, dependent   **19.** $\emptyset$, inconsistent
21. $\{(150, 50)\}$, independent   **23.** $\emptyset$, inconsistent
25. $\{(34, 15)\}$, independent   **27.** $\{(13, 7)\}$, independent
29. $\{(4, -1)\}$, independent   **31.** $\emptyset$, inconsistent
33. $\{(-1, 1)\}$, independent
35. $\{(x, y) | x + 2y = 12\}$, dependent
37. $\{(4, 6)\}$, independent   **39.** $\{(1.5, 3.48)\}$, independent
41. Independent   **43.** Dependent
45. \$10,000 at 10%, \$15,000 at 8%
47. \$6.50 adult, \$4 child
49. 600   **51.** 112 dimes, 54 quarters
53. 5 servings Grape-nuts, 4 servings Rice Krispies

55. Plan A, 12   **57.** $y = -2x + 3$ **59.** $y = -\dfrac{5}{3}x - \dfrac{1}{3}$

Graphing Calculator Exercises

1. $(-1000, -497)$ **3.** $(6.18, -0.54)$

Section 9.2

1. $\{(1, 2, 3)\}$ **3.** $\{(-1, 1, 2)\}$ **5.** $\{(0, -5, 5)\}$ **7.** $\emptyset$
9. $\{(x, x - 1, x + 5) | x \text{ is any real number}\}$
11. $\{(x, y, z) | x + 2y - 3z = 5\}$
13. $\{(2, y, y) | y \text{ is any real number}\}$
15. $\{(x, 5 - x, 3 - x) | x \text{ is any real number}\}$
17. $\{(5, 7, 9)\}$ **19.** $\{(1.2, 1.5, 2.4)\}$
21. $\{(1000, 2000, 6000)\}$ **23.** $\{(15/11, 13/11, -3/11)\}$
25. $y = x^2 - 3$ **27.** $y = -2x^2 + 5x$
29. $y = x^2 + 4x + 4$
31. $E = -0.0009s^2 + 0.0357s - 0.1183$, 19.8 mph
33. \$4000 in stocks, \$7000 in bonds, \$14,000 in mutual fund
35. \$0.60, \$0.80, \$0.50
37. 116 pennies, 48 nickels, 68 dimes
39. No, because the bill is \$6.05.

Section 9.3

1. $\{(2, 4), (3, 9)\}$ **3.** $\{(-1, 0), (3, 2)\}$
5. $\left\{(0, 0), \left(\dfrac{1}{4}, \dfrac{1}{2}\right)\right\}$ **7.** $\{(0, 0), (2, 8), (-2, -8)\}$
9. $\{(0, 0), (2, 2), (-2, 2)\}$ **11.** $\{(1, 0)\}$
13. $\left\{(0, 0), (\sqrt{2}, 2), (-\sqrt{2}, 2)\right\}$
15. $\left\{\left(-2 + \sqrt{3}, -2 - \sqrt{3}\right), \left(-2 - \sqrt{3}, -2 + \sqrt{3}\right)\right\}$
17. $\{(1, 1), (1, -1), (-1, 1), (-1, -1)\}$ **19.** $\{(2, 5)\}$

21. $\{(4, -3), (-1, 2)\}$ **23.** $\{(1, -2), (-1, 2)\}$
25. $\left\{\left(-\dfrac{1}{3}, 2^{2/3}\right)\right\}$ **27.** $\{(2, 1)\}$ **29.** $\{(2, 2)\}$
31. $\{(0, 1)\}$ **33.** 9 m and 12 m
35. $6 - 2\sqrt{3}$ ft, $6\sqrt{3} - 6$ ft, $12 - 4\sqrt{3}$ ft
37. A 9.6 min, B 48 min   **39.** $3 + i$ and $3 - i$
41. 60 ft and 30 ft

Graphing Calculator Exercises

1. $\{(2, 1), (0.3, -1.8)\}$ **3.** $\{(1.9, 0.6), (0.1, -2.0)\}$
5. 0 and 9.66 yr, 29.5 yr

Section 9.4

1. $\dfrac{7x - 5}{(x - 2)(x + 1)}$ **3.** $\dfrac{x^2 - 3x + 5}{(x - 1)(x^2 + 2)}$
5. $\dfrac{3x^3 + x^2 + 8x + 5}{(x^2 + 3)^2}$ **7.** $\dfrac{4x^2 - x - 1}{(x - 1)^3}$
9. $A = 2, B = -2$ **11.** $\dfrac{2}{x + 1} + \dfrac{3}{x - 2}$
13. $\dfrac{1/2}{x + 2} + \dfrac{3/2}{x + 4}$ **15.** $\dfrac{1/3}{x - 3} + \dfrac{-1/3}{x + 3}$
17. $\dfrac{-1}{x} + \dfrac{1}{x - 1}$ **19.** $A = 2, B = 5, C = -1$
21. $\dfrac{-3}{(x + 2)^2} + \dfrac{-2}{x + 2}$ **23.** $\dfrac{2x - 3}{x^2 + 1} + \dfrac{4}{x + 1}$
25. $\dfrac{-8x}{(x^2 + 9)^2} + \dfrac{3x - 1}{x^2 + 9}$ **27.** $\dfrac{5}{x - 2} + \dfrac{-2x + 3}{x^2 + 2x + 4}$
29. $2x + 1 + \dfrac{2}{x - 1} + \dfrac{3}{x + 1}$
31. $\dfrac{3x + 3}{(x^2 + x + 1)^2} + \dfrac{3x - 5}{x^2 + x + 1}$
33. $\dfrac{3x}{x^2 + 4} + \dfrac{1}{x - 2} + \dfrac{-1}{x + 2}$ **35.** $\dfrac{1}{x} + \dfrac{2}{x^2} + \dfrac{3}{x^3} + \dfrac{4}{x - 1}$
37. $\dfrac{3}{x - 2} + \dfrac{-1}{(x - 2)^2} + \dfrac{3}{x - 3}$
39. $\dfrac{4}{x + 5} + \dfrac{2}{x + 2} + \dfrac{3}{x - 3}$
41. $\dfrac{-1}{(x - 1)^3} + \dfrac{2}{(x - 1)^2} + \dfrac{1}{x - 1}$
43. $\dfrac{-b/a}{(ax + b)^2} + \dfrac{1/a}{ax + b}$ **45.** $\dfrac{c/b}{x} + \dfrac{1 - ac/b}{ax + b}$
47. $\dfrac{1/b}{x^2} + \dfrac{-a/b^2}{x} + \dfrac{a^2/b^2}{ax + b}$

Section 9.5

1. (c)   **3.** (d)

5.

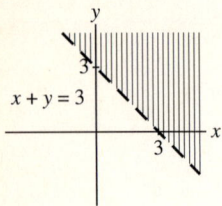

$x + y = 3$

7.

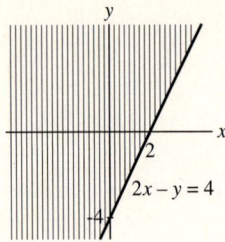

$2x - y = 4$

21.

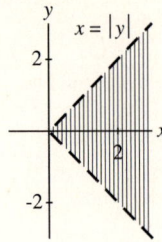

$x = |y|$

23.

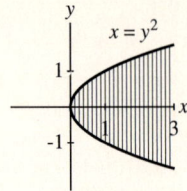

$x = y^2$

9.

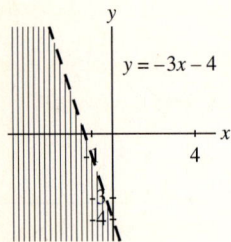

$y = -3x - 4$

11.

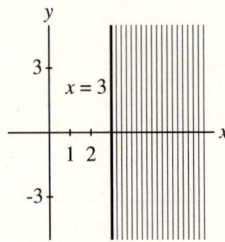

$x = 3$

25.

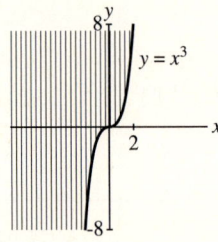

$y = x^3$

27.

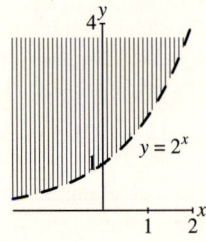

$y = 2^x$

13.

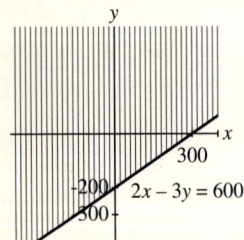

$2x - 3y = 600$

15.

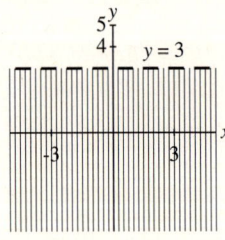

$y = 3$

29.

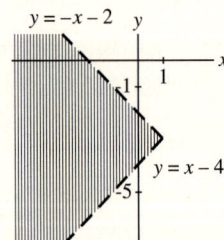

$y = -x - 2$

$y = x - 4$

31.

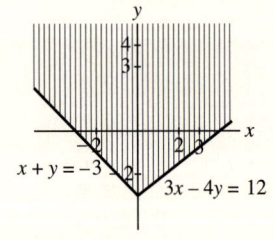

$x + y = -3$

$3x - 4y = 12$

17.

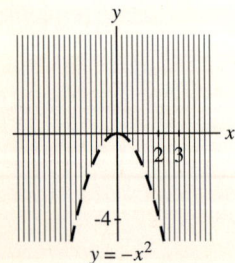

$y = -x^2$

19.

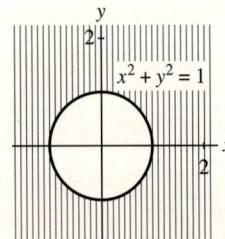

$x^2 + y^2 = 1$

33.

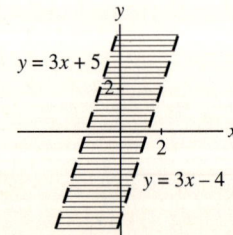

$y = 3x + 5$

$y = 3x - 4$

35. No solution

37.

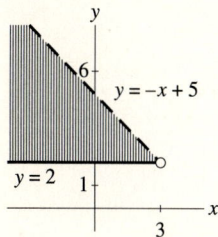

$y = -x + 5$

$y = 2$

39.

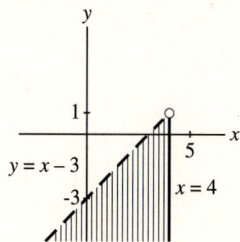

$y = x - 3$

$x = 4$

51.

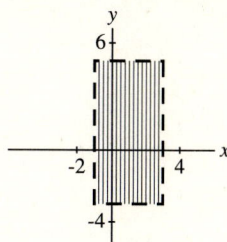

53.

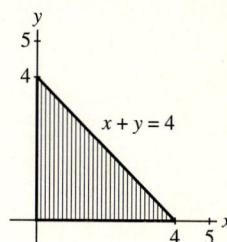

$x + y = 4$

41.

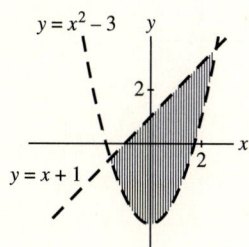

$y = x^2 - 3$

$y = x + 1$

43.

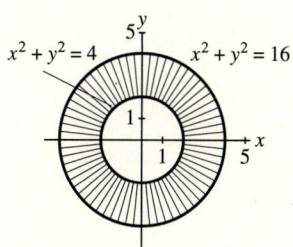

$x^2 + y^2 = 4$ $x^2 + y^2 = 16$

55.

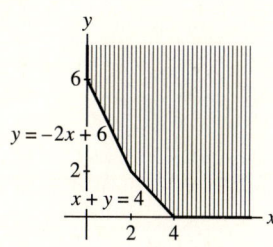

$y = -2x + 6$

$x + y = 4$

57.

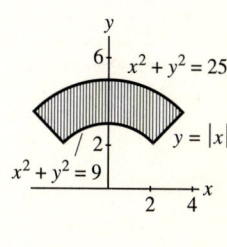

$x^2 + y^2 = 25$

$y = |x|$

$x^2 + y^2 = 9$

45.

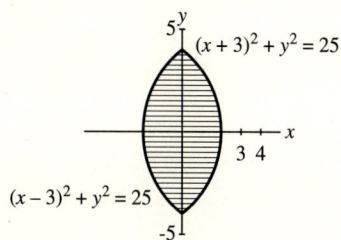

$(x + 3)^2 + y^2 = 25$

$(x - 3)^2 + y^2 = 25$

59.

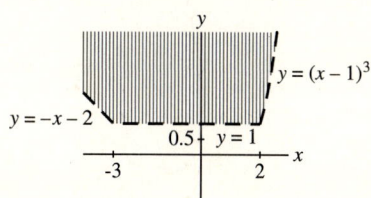

$y = (x - 1)^3$

$y = -x - 2$

$y = 1$

47.

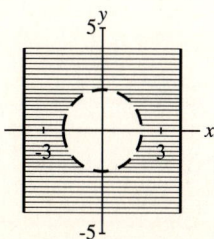

49.

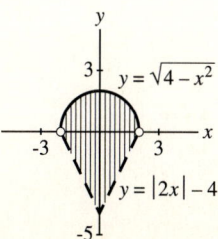

$y = \sqrt{4 - x^2}$

$y = |2x| - 4$

61.

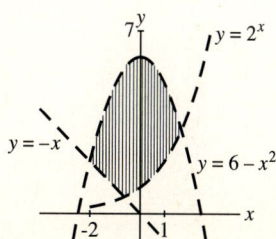

$y = 2^x$

$y = -x$

$y = 6 - x^2$

63. $x \geq 0$, $y \geq 0$, $y \leq -\dfrac{2}{3}x + 5$, $y \leq -3x + 12$

65. $y \geq 0$, $x \geq 0$, $y \geq -\dfrac{1}{2}x + 3$, $y \geq -\dfrac{3}{2}x + 5$

67. $|x| < 2$, $|y| < 2$

69. $x > 0$, $y > 0$, $x^2 + y^2 < 81$

71.

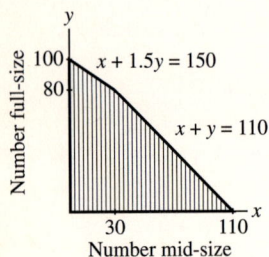

73.

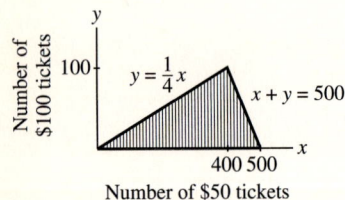

Graphing Calculator Exercises

1. $(-1.17, 1.84)$ **3.** $(-0.65, 0.25)$

Section 9.6

1.

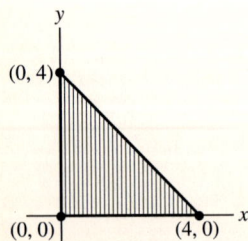

3.

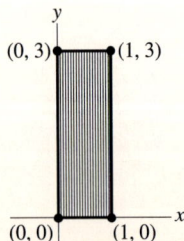

5.

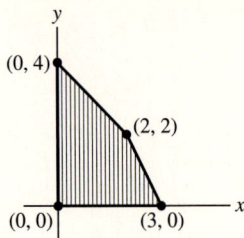

7.

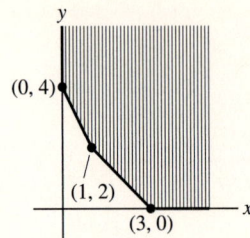

9.

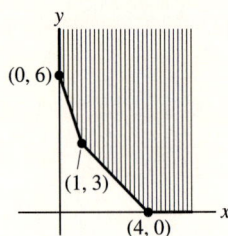

11.

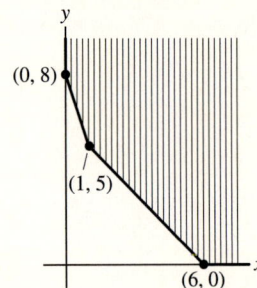

13. 15   **15.** 8   **17.** 30   **19.** 100

21. 8 door harps, 6 mailboxes   **23.** 16 door harps, 0 mailboxes

25. 0 small, 8 large   **27.** 5 small, 3 large

Chapter 9 Review Exercises

1. $\{(3, 5)\}$ **3.** $\{(-1, 3)\}$

5. $\{(-19/2, -19/2)\}$, independent

7. $\{(39/17, 18/17)\}$, independent

9. $\{(x, y)\mid y = -3x + 1\}$, dependent   **11.** $\varnothing$, inconsistent

13. $\{(1, 3, -4)\}$ **15.** $\{(x, 0, 1 - x)\mid x \text{ is any real number}\}$

17. $\varnothing$ **19.** $\left\{\left(\dfrac{-1 + \sqrt{17}}{2}, \pm\sqrt{\dfrac{-1 + \sqrt{17}}{2}}\right)\right\}$

21. $\{(0, 0), (1, 1), (-1, 1)\}$ **23.** $\dfrac{2}{x - 3} + \dfrac{5}{x + 4}$

25. $\dfrac{2x - 1}{x^2 + 4} + \dfrac{5}{x - 3}$

27.

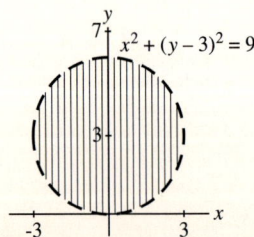

29.

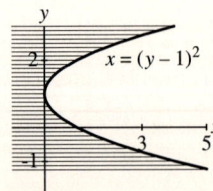

31.

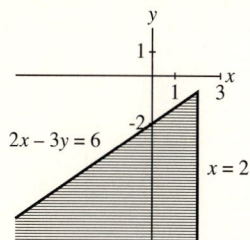

33.

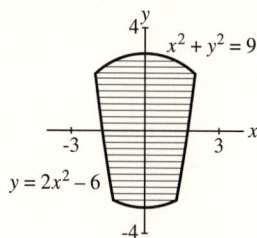

17.

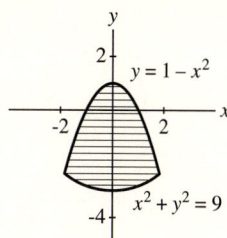

35.

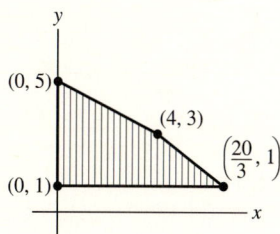

37.

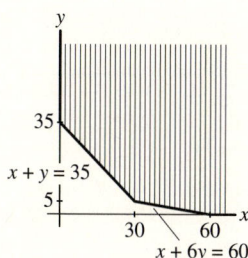

18. 24 males, 28 females   **19.** 5 television, 18 newspaper

Tying It All Together Chapters 1–9

1. $\left\{\dfrac{103}{13}\right\}$ **2.** $\left\{\dfrac{40}{13}, 3\right\}$ **3.** $\left\{-\dfrac{4}{5}\right\}$

4. $\{-1, 4\}$ **5.** $\{-11\}$ **6.** $\left\{\pm\dfrac{2\sqrt{3}}{3}\right\}$ **7.** $\{1\}$

8. $\{1 + \log_2(9)\}$ **9.** $\{1\}$ **10.** $\{\pm 8\}$

11. $\left\{\dfrac{3 \pm \sqrt{33}}{2}\right\}$ **12.** $\left\{\dfrac{6 \pm \sqrt{2}}{2}\right\}$

13. $(-\infty, 3/2)$ **14.** $(-\infty, 1.5) \cup (1.5, \infty)$

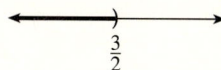

39. $y = -\dfrac{2}{3}x + \dfrac{5}{3}$ **41.** $y = 3x^2 - 4x + 5$

43. 57 tacos, 62 burritos   **45.** $0.95   **47.** 16.8
49. 6 million barrels per day from each source

Chapter 9 Test

1. $\{(-3, 4)\}$ **2.** $\{(25/11, -6/11)\}$ **3.** $\{(4, 6)\}$
4. Inconsistent   **5.** Dependent   **6.** Independent
7. Inconsistent
8. $\{(x, 10 - x, 14 - 3x) \mid x \text{ is any real number}\}$
9. $\{(1, -2, 3)\}$ **10.** $\varnothing$ **11.** $\{(4, 0), (-4, 0)\}$
12. $\left\{\left(2 + \sqrt{2}, -4 - \sqrt{2}\right), \left(2 - \sqrt{2}, -4 + \sqrt{2}\right)\right\}$
13. $\dfrac{3}{x - 4} + \dfrac{-1}{x + 2}$ **14.** $\dfrac{2}{x^2} + \dfrac{1}{x} + \dfrac{3}{x - 1}$

15. $(-\infty, -3] \cup [3, \infty)$ **16.** $[-7, 5]$

15.

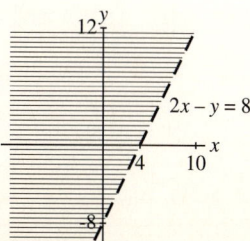

16.

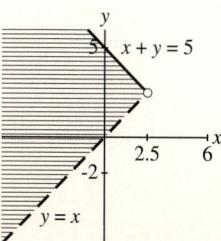

17.

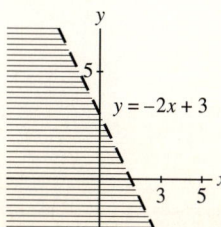

18.

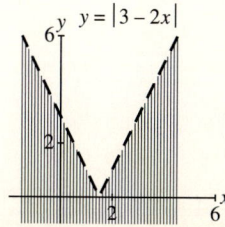

19.

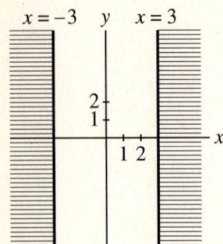

$x = -3$ y $x = 3$

20.

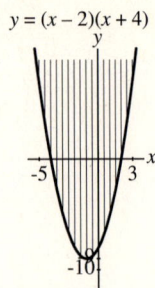

$y = (x-2)(x+4)$

CHAPTER 10

Section 10.1

1. 1×3 **3.** 1×1 **5.** 3×2

7. $\begin{bmatrix} 1 & -2 & | & 4 \\ 3 & 2 & | & -5 \end{bmatrix}$ **9.** $\begin{bmatrix} 1 & -1 & -1 & | & 4 \\ 1 & 3 & -1 & | & 1 \\ 0 & 2 & -5 & | & -6 \end{bmatrix}$

11. $3x + 4y = -2$ **13.** $5x \qquad = 6$
$\quad\; 3x - 5y = 0 \qquad\quad -4x \quad + 2z = -1$
$\qquad\qquad\qquad\qquad\qquad 4x + 4y \quad = 7$

15. Multiply R_1 by 1/2.   **17.** Multiply R_2 by $-1/6$.
19. $\{(2, 3)\}$ **21.** $\{(4, -4)\}$ **23.** $\{(3, 3)\}$ **25.** $\{(0.5, 1)\}$
27. $\varnothing$ **29.** $\{(u, v) | u + 3v = 4\}$ **31.** $\{(3, 2, 1)\}$
33. $\{(1, 1, 1)\}$ **35.** $\{(1, 2, 0)\}$ **37.** $\{(1, 0, -1)\}$
39. $\{(x, y, z) | x - 2y + 3z = 1\}$ **41.** $\varnothing$
43. $\{(x, 5 - 2x, 2 - x) | x \text{ is any real number}\}$

45. $\left\{ \left(x, \dfrac{13 - 5x}{2}, \dfrac{5 - 3x}{2} \right) \middle| x \text{ is any real number} \right\}$

47. $\{(4, 3, 2, 1)\}$
49. 38 hr at Burgers-R-Us, 22 hr at Soap Opera
51. \$20,000 mutual fund, \$11,333.33 treasury bills, \$8666.67 bonds
53. $y = x^3 - 2x + 3$
55. $y = 750 - x$, $z = x - 250$, and $250 \leq x \leq 750$. If $z = 50$, then $x = 300$ and $y = 450$.

Section 10.2

1. $x = 2$, $y = 5$ **3.** $x = 3$, $y = 5/2$, $z = 5/4$ **5.** $\begin{bmatrix} 5 \\ 6 \end{bmatrix}$

7. $\begin{bmatrix} 0.4 & 0.15 \\ 0.7 & 1.1 \end{bmatrix}$ **9.** $\begin{bmatrix} 1/2 & 3/2 \\ 3 & -12 \end{bmatrix}$ **11.** $\begin{bmatrix} 10 & 0 \\ -2 & 16 \end{bmatrix}$

13. Undefined   **15.** $\begin{bmatrix} 0 \\ 0 \\ 7\sqrt{3} \end{bmatrix}$ **17.** $\begin{bmatrix} 13a \\ -b \end{bmatrix}$

19. $\begin{bmatrix} -x & -0.5y \\ -0.7x & 3.5y \end{bmatrix}$ **21.** $\begin{bmatrix} -1 & 13 \\ -9 & 3 \\ 6 & -2 \end{bmatrix}$

23. $\begin{bmatrix} 3x & 2y & -z \\ -6x & 3y & 7z \\ 0 & -7y & -7z \end{bmatrix}$ **25.** $\begin{bmatrix} -5 & -1 \\ 10 & 4 \end{bmatrix}$

27. Undefined   **29.** $\begin{bmatrix} -8 & -5 \\ 12 & -1 \end{bmatrix}$ **31.** $\begin{bmatrix} -8 & -5 \\ 12 & -1 \end{bmatrix}$

33. $\begin{bmatrix} -5 \\ 7 \end{bmatrix}$ **35.** $\begin{bmatrix} -5 \\ 7 \end{bmatrix}$ **37.** $\begin{bmatrix} -7 & 4 \\ -1 & -4 \end{bmatrix}$

39. $\begin{bmatrix} -5 \\ 4 \end{bmatrix}$ **41.** $\{(3, 2)\}$ **43.** $\{(-1, 3)\}$

45. $\{(0.5, 3, 4.5)\}$ **47.** $\begin{bmatrix} \$390 \\ \$160 \\ \$135 \end{bmatrix}$

49. $\begin{bmatrix} 40 & 80 \\ 30 & 90 \\ 80 & 200 \end{bmatrix}, \begin{bmatrix} 60 & 120 \\ 45 & 135 \\ 120 & 300 \end{bmatrix}$ **51.** Yes, yes

53. Yes, yes   **55.** Yes, yes   **57.** $\begin{bmatrix} 0 & 0 \\ 0 & 0 \end{bmatrix}$

Section 10.3

1. 3×5 **3.** 1×1 **5.** 5×5 **7.** 3×3

9. AB is undefined.   **11.** $\begin{bmatrix} 4 & 6 & 8 \\ -6 & -9 & -12 \\ 2 & 3 & 4 \end{bmatrix}$

13. $[2 \quad 5 \quad 9]$ **15.** Undefined   **17.** $\begin{bmatrix} 7 & 8 \\ 5 & 5 \\ 1 & 0 \end{bmatrix}$

19. $\begin{bmatrix} 1 & 1 \\ 14 & 15 \end{bmatrix}$ **21.** Undefined   **23.** $\begin{bmatrix} 0 \\ -2 \\ 1 \end{bmatrix}$

25. Undefined   **27.** $\begin{bmatrix} 2 & 2 \\ 3 & 4 \end{bmatrix}$ **29.** $\begin{bmatrix} 1 & 0 \\ 0 & 1 \end{bmatrix}$

31. $\begin{bmatrix} -0.5 & 4 \\ 9 & 0.7 \end{bmatrix}$ **33.** $[4a \quad -3b]$ **35.** $\begin{bmatrix} -2a & 5a & 3a \\ b & 4b & 6b \end{bmatrix}$

37. $[4 \quad 2 \quad 6]$ **39.** $[17]$

41. $\begin{bmatrix} x^2 & xy \\ xy & y^2 \end{bmatrix}$ **43.** $\begin{bmatrix} 2\sqrt{2} \\ 7\sqrt{2} \end{bmatrix}$ **45.** $\begin{bmatrix} -17/3 & 11 \\ -3 & 6 \end{bmatrix}$

47. Undefined   **49.** $\begin{bmatrix} 2 & 0 & 6 \\ 3 & 5 & 2 \\ 19 & 21 & 16 \end{bmatrix}$

51. $\begin{bmatrix} -0.2 & 1.7 \\ 3 & 3.4 \\ -1.4 & -4.6 \end{bmatrix}$ **53.** $\{(3, 2)\}$ **55.** $\varnothing$

57. $\{(-1, -1, 6)\}$ **59.** $\begin{bmatrix} 2 & 3 \\ 4 & -1 \end{bmatrix} \begin{bmatrix} x \\ y \end{bmatrix} = \begin{bmatrix} 9 \\ 6 \end{bmatrix}$

61. $\begin{bmatrix} 1 & 2 & -1 \\ 3 & -1 & 3 \\ 2 & 1 & -4 \end{bmatrix} \begin{bmatrix} x \\ y \\ z \end{bmatrix} = \begin{bmatrix} 3 \\ 1 \\ 0 \end{bmatrix}$ **63.** $AQ = \begin{bmatrix} \$376,000 \\ \$642,000 \end{bmatrix}$

65. $M^2 = \begin{bmatrix} 0 & 0 & 1 & 2 \\ 1 & 0 & 0 & 1 \\ 1 & 0 & 0 & 0 \\ 0 & 1 & 1 & 0 \end{bmatrix}$ **67.** F   **69.** T   **71.** T

Section 10.4

1. $\begin{bmatrix} -3 & 5 \\ 12 & 6 \end{bmatrix}$ **3.** $\begin{bmatrix} 1 & 0 \\ 0 & 1 \end{bmatrix}$ **5.** $\begin{bmatrix} 1 & 0 \\ 0 & 1 \end{bmatrix}$

7. $\begin{bmatrix} 3 & 5 & 1 \\ 4 & 5 & 7 \\ 4 & 9 & 2 \end{bmatrix}$ **9.** $\begin{bmatrix} 1 & 0 & 0 \\ 0 & 1 & 0 \\ 0 & 0 & 1 \end{bmatrix}$ **11.** $\begin{bmatrix} 1 & 0 & 0 \\ 0 & 1 & 0 \\ 0 & 0 & 1 \end{bmatrix}$

13. Yes   **15.** No   **17.** No   **19.** $\begin{bmatrix} 1 & -2 \\ 0 & 0.5 \end{bmatrix}$

21. $\begin{bmatrix} 3 & -2 \\ -1/3 & 1/3 \end{bmatrix}$ **23.** $\begin{bmatrix} 4 & 3 \\ -3 & -2 \end{bmatrix}$

25. $\begin{bmatrix} -1.5 & -2.5 \\ -0.5 & -0.5 \end{bmatrix}$

27. No inverse   **29.** No inverse

31. $\begin{bmatrix} 0.5 & 0.5 & 0 \\ 0.5 & 0 & -0.5 \\ 0 & -0.5 & 0.5 \end{bmatrix}$

33. $\begin{bmatrix} -3.5 & -1 & 2 \\ 0.5 & 0 & 0 \\ 4.5 & 2 & -3 \end{bmatrix}$

35. $\{(3, -1)\}$ **37.** $\{(2, 1/3)\}$ **39.** $\{(1, -1)\}$
41. $\{(5, 2)\}$ **43.** $\{(1, -1, 3)\}$ **45.** $\{(1, 3, 2)\}$
47. $\{(11.3, -3.9)\}$
49. $\{(-2z - 4, -z - 9, z)|z$ is any real number$\}$

51. $\{(1/2, 1/6, 1/3)\}$ **53.** $\begin{bmatrix} 1/3 & -1/6 & 1/3 \\ -2/3 & 1/3 & 1/3 \\ 2/3 & 1/6 & -1/3 \end{bmatrix}$

55. $\begin{bmatrix} 2.5 & -3 & 1 \\ 1 & -1 & 0 \\ -1.5 & 2 & 0 \end{bmatrix}$ **57.** $\begin{bmatrix} 1 & -2 & 1 & 0 \\ 0 & 1 & -2 & 1 \\ 0 & 0 & 1 & -2 \\ 0 & 0 & 0 & 1 \end{bmatrix}$

59. $\begin{bmatrix} 1 & 7 \\ 3 & 20 \end{bmatrix}, \begin{bmatrix} 2 & 7 \\ 3 & 10 \end{bmatrix}, \begin{bmatrix} 4 & 7 \\ 3 & 5 \end{bmatrix}, \begin{bmatrix} 20 & 7 \\ 3 & 1 \end{bmatrix}, \begin{bmatrix} 10 & 7 \\ 3 & 2 \end{bmatrix}, \begin{bmatrix} 5 & 7 \\ 3 & 4 \end{bmatrix}$

61. $\$0.75$ eggs, $\$1.80$ magazine
63. $\$400$ plywood, $\$150$ insulation   **65.** Good luck

Graphing Calculator Exercises

1. $\{(-165, 97.5, 240)\}$
3. $\{(-1.6842, -9.2632, -9.2632)\}$
5. $\$4.20$ paper, $\$2.75$ pens, $\$0.89$ clips

Section 10.5

1. 2   **3.** 19   **5.** -0.41 **7.** $23/32$ **9.** 0   **11.** 0
13. $\{(-3, 5)\}$ **15.** $\{(11/2, -1/2)\}$ **17.** $\{(12, 6)\}$
19. $\{(500, 400)\}$ **21.** $\{(x, y)|3x + y = 6\}$ **23.** $\varnothing$
25. $\{(-6, -9)\}$ **27.** $\{(\sqrt{2}/2, \sqrt{3})\}$
29. $\{(0, -5), (3, 4), (-3, 4)\}$ **31.** $\{(8, 2), (0, -2)\}$
33. Invertible   **35.** Not invertible
37. 175 boys, 440 girls   **39.** 19
41. 182/3 degrees, 182/3 degrees, 176/3 degrees
43. 2, 3, 6, yes   **47.** 8, no

Graphing Calculator Exercises

1. $\{(146, 237)\}$

Section 10.6

1. 14   **3.** 1   **5.** -23 **7.** -16
9. -33 **11.** -56 **13.** -115 **15.** 14   **17.** -54
19. 0   **21.** -3 **23.** -72 **25.** -137 **27.** -293
29. $\{(1, 2, 3)\}$ **31.** $\{(2, 3, 5)\}$
33. $\{(7/16, -5/16, -13/16)\}$ **35.** $\{(-2/7, 11/7, 5/7)\}$
37. $\{(x, x - 5/3, x - 4/3)|x$ is any real number$\}$ **39.** $\varnothing$
41. Jackie 27, Alisha 11, Rochelle 39   **43.** 30, 50, 100

Graphing Calculator Exercises

1. $\{(16.8, 12.3, 11.2)\}$
3. $\$1.099$ regular, $\$1.219$ plus, $\$1.279$ supreme
5. $\{(1, -2, 3, 2)\}$

Chapter 10 Review Exercises

1. $\begin{bmatrix} 5 & 4 \\ -1 & 6 \end{bmatrix}$ **3.** $\begin{bmatrix} 1 & -13 \\ -5 & 6 \end{bmatrix}$ **5.** $\begin{bmatrix} 3 & 8 \\ -2 & -6 \end{bmatrix}$

7. Undefined   **9.** $\begin{bmatrix} -11 \\ 14 \end{bmatrix}$ **11.** $\begin{bmatrix} 3 & 2 & -1 \\ -12 & -8 & 4 \\ 9 & 6 & -3 \end{bmatrix}$

13. $[1 \quad -1 \quad -1]$ **15.** Undefined   **17.** $\begin{bmatrix} 2 & 1.5 \\ 1 & 1 \end{bmatrix}$

19. $\begin{bmatrix} -1 & 0 & 0 \\ 1 & 1 & 0 \\ -5 & -3 & 1 \end{bmatrix}$ **21.** $\begin{bmatrix} 3 & 4 \\ -1 & -1.5 \end{bmatrix}$ **23.** $\begin{bmatrix} 1 & 0 \\ 0 & 1 \end{bmatrix}$

25. 2   **27.** -1 **29.** $\{(10/3, 17/3)\}$ **31.** $\{(-2, 3)\}$
33. $\{(x, y)|x - 5y = 9\}$ **35.** $\varnothing$ **37.** $\{(1, 2, 3)\}$
39. $\{(-3, 4, 1)\}$

41. $\left\{ \left(\dfrac{3y + 3}{2}, y, \dfrac{1 - y}{2} \right) | y \text{ is any real number} \right\}$ **43.** $\varnothing$

45. $\{(9, -12)\}$ **47.** $\{(2, 4)\}$ **49.** $\{(0, -3)\}$

51. $\{(0.5, 0.5, 0.5)\}$ **53.** $\{(2, -3, 5)\}$

55. 225.56 gal A, 300.74 gal B

57. \$22.88 water, \$55.65 gas, \$111.30 electric

Chapter 10 Test

1. $\{(1, 1/3)\}$ **2.** $\{(3, 2, 1)\}$

3. $\left\{ \left(x, \dfrac{-x - 1}{2}, \dfrac{3x - 1}{2} \right) | x \text{ is any real number} \right\}$

4. $\begin{bmatrix} 3 & -4 \\ -6 & 10 \end{bmatrix}$ **5.** $\begin{bmatrix} 0 & 1 \\ 0 & 2 \end{bmatrix}$ **6.** $\begin{bmatrix} 6 & -9 \\ -20 & 30 \end{bmatrix}$

7. $\begin{bmatrix} -3 \\ 8 \end{bmatrix}$ **8.** Undefined   **9.** $\begin{bmatrix} -2 & 1 & 2 \end{bmatrix}$

10. $\begin{bmatrix} 2 & 0 & -2 \\ 3 & 0 & -3 \\ -1 & 0 & 1 \end{bmatrix}$ **11.** $\begin{bmatrix} 2 & 0.5 \\ 1 & 0.5 \end{bmatrix}$ **12.** $\begin{bmatrix} 7 & -5 & -8 \\ 3 & -2 & -3 \\ 6 & -4 & -7 \end{bmatrix}$

13. 2   **14.** 0   **15.** -1 **16.** $\{(5, 3)\}$ **17.** $\varnothing$

18. $\{(6, 2, 4)\}$ **19.** $\{(-2, -3)\}$ **20.** $\{(15, 6, 13)\}$

21. Bought 46, sold 34   **22.** $y = 0.5x^2 - 4\sqrt{x} + 3$

Tying It All Together Chapters 1–10

1. $\{-1/3\}$ **2.** $\{29/15\}$ **3.** $\{1/3, 2\}$ **4.** $\{-3/2\}$

5. $\{(4, -2)\}$ **6.** $\{(2/5, 1/5)\}$ **7.** $\{(7, -100)\}$

8. $\{(-2, -3)\}$ **9.** $\{(1, 2)\}$ **10.** $\{(2, 1)\}$

11. $\{(-4, -3), (3, 4)\}$ **12.** $\{(-2, 3), (1, 0)\}$

CHAPTER 11

Section 11.1

1. $y = \dfrac{1}{4}x^2$ **3.** $y = -x^2$ **5.** $y = \dfrac{1}{6}(x - 3)^2 + \dfrac{7}{2}$

7. $y = -\dfrac{1}{10}(x - 1)^2 - \dfrac{1}{2}$ **9.** $y = 1.25(x + 2)^2 + 1$

11. $y = (x - 4)^2 - 13$, $(4, -13)$, $(4, -12.75)$, $y = -13.125$

13. $y = 2(x + 3)^2 - 13$, $(-3, -13)$, $(-3, -103/8)$, $y = -105/8$

15. $y = -2(x - 1.5)^2 + 5.5$, $(1.5, 5.5)$, $(1.5, 5.375)$, $y = 5.625$

17. $y = 5(x + 3)^2 - 45$, $(-3, -45)$, $(-3, -44.95)$, $y = -45.05$

19. $y = \dfrac{1}{8}(x - 2)^2 + 4$, $(2, 4)$, $(2, 6)$, $y = 2$

21. $(2, -1)$, $(2, -3/4)$, $y = -5/4$, up

23. $(1, -4)$, $(1, -17/4)$, $y = -15/4$, down

25. $(1, -2)$, $(1, -23/12)$, $y = -25/12$, up

27. $(-3, 13/2)$, $(-3, 6)$, $y = 7$, down

29. $(0, 5)$, $(0, 6)$, $y = 4$, up

31. $(1/2, -9/4)$, $x = 1/2$, $(-1, 0)$, $(2, 0)$, $(0, -2)$

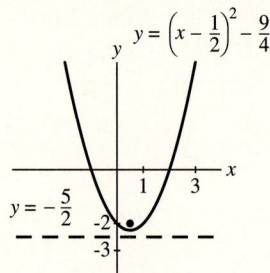

33. $(-1/2, 25/4)$, $x = -1/2$, $(-3, 0)$, $(2, 0)$, $(0, 6)$

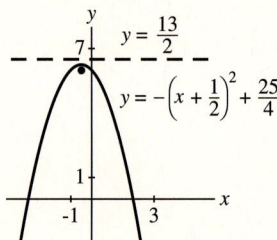

35. $(-2, 2)$, $x = -2$, $(0, 4)$, $(-2, 5/2)$, $y = 3/2$

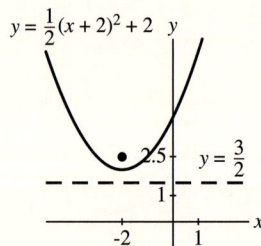

37. $(-4, 2)$, $x = -4$, $\left(-4 \pm 2\sqrt{2}, 0 \right)$, $(0, -2)$, $(-4, 1)$, $y = 3$

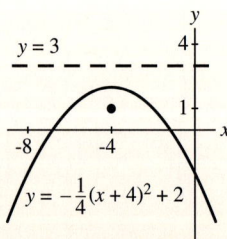

39. $(0, -2)$, $x = 0$, $(\pm 2, 0)$, $(0, -2)$, $(0, -3/2)$, $y = -5/2$

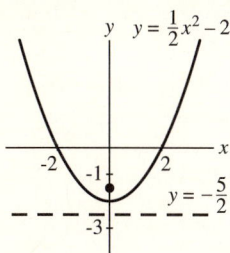

41. $(2, 0)$, $x = 2$, $(2, 0)$, $(0, 4)$, $(2, 1/4)$, $y = -1/4$

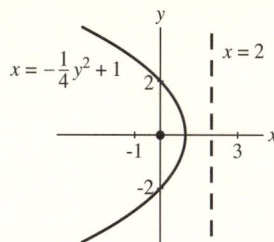

43. $(3/2, -3/4)$, $x = 3/2$, $(0, 0)$, $(3, 0)$, $(0, 0)$, $(3/2, 0)$, $y = -3/2$

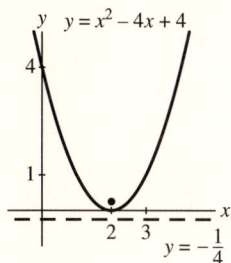

45. $(0, 0)$, $y = 0$, $(0, 0)$, $(0, 0)$, $(-1/4, 0)$, $x = 1/4$

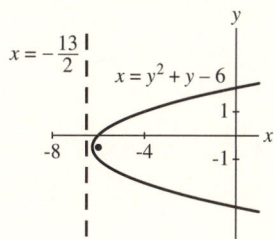

47. $(1, 0)$, $y = 0$, $(1, 0)$, $(0, \pm 2)$, $(0, 0)$, $x = 2$

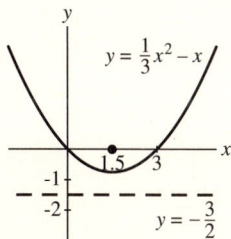

49. $(-25/4, -1/2)$, $y = -1/2$, $(-6, 0)$, $(0, -3)$, $(0, 2)$, $(-6, -1/2)$, $x = -13/2$

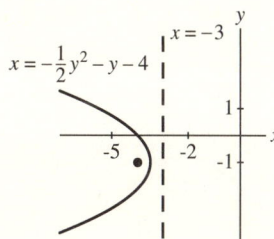

51. $(-7/2, -1)$, $y = -1$, $(-4, 0)$, $(-4, -1)$, $x = -3$

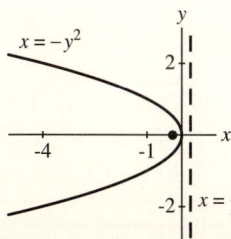

53. $(3, 1)$, $y = 1$, $(5, 0)$, $(25/8, 1)$, $x = 23/8$

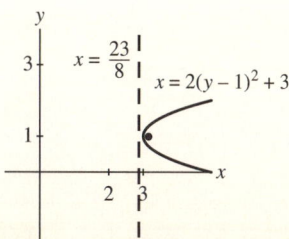

55. $(1, -2)$, $y = -2$, $(-1, 0)$, $(0, -2 \pm \sqrt{2})$, $(1/2, -2)$, $x = 3/2$

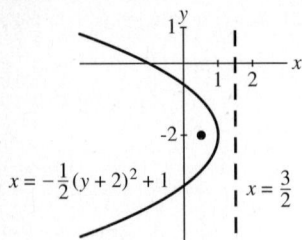

$x = -\frac{1}{2}(y + 2)^2 + 1$ $x = \frac{3}{2}$

57. $y = \frac{1}{4}(x - 1)^2 + 4$ **59.** $x = \frac{1}{8}y^2$

61. $y = \frac{1}{2640}x^2$, 26.8 in.

Section 11.2

1.

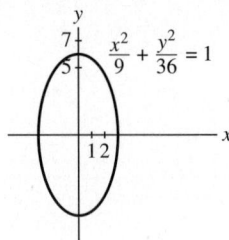

$\frac{x^2}{13} + \frac{y^2}{9} = 1$

3.

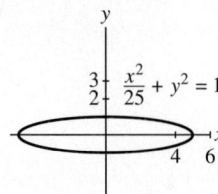

$\frac{x^2}{25} + \frac{y^2}{9} = 1$

5.

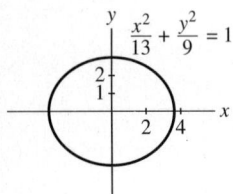

$\frac{x^2}{4} + \frac{y^2}{8} = 1$

7.

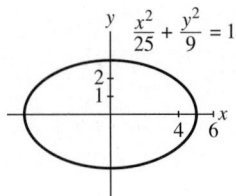

$\frac{x^2}{33} + \frac{y^2}{49} = 1$

9. $\left(\pm 2\sqrt{3}, 0 \right)$

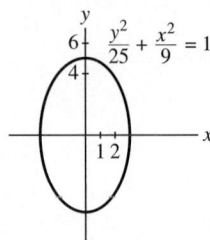

$\frac{x^2}{16} + \frac{y^2}{4} = 1$

11. $\left(0, \pm 3\sqrt{3} \right)$

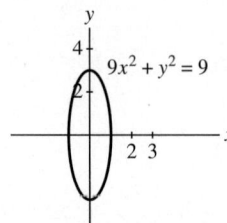

$\frac{x^2}{9} + \frac{y^2}{36} = 1$

13. $\left(\pm 2\sqrt{6}, 0 \right)$

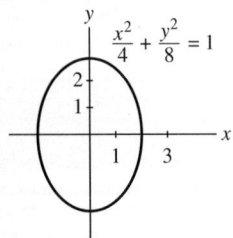

$\frac{x^2}{25} + y^2 = 1$

15. $(0, \pm 4)$

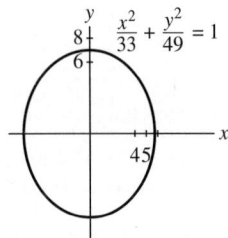

$\frac{y^2}{25} + \frac{x^2}{9} = 1$

17. $\left(0, \pm 2\sqrt{2} \right)$

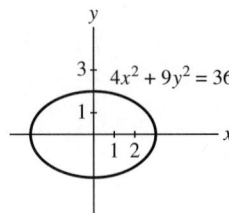

$9x^2 + y^2 = 9$

19. $\left(\pm \sqrt{5}, 0 \right)$

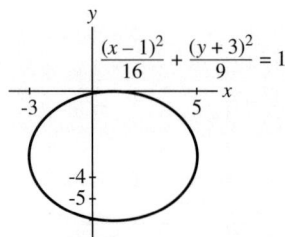

$4x^2 + 9y^2 = 36$

21. $\left(1 \pm \sqrt{7}, -3 \right)$

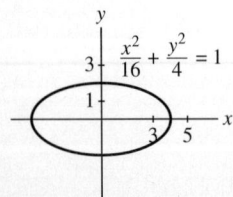

$\frac{(x - 1)^2}{16} + \frac{(y + 3)^2}{9} = 1$

23. $(3, 2)$, $(3, -6)$

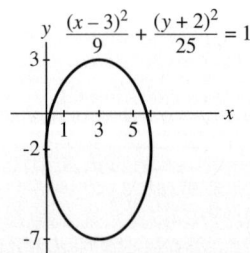

$\frac{(x - 3)^2}{9} + \frac{(y + 2)^2}{25} = 1$

25. $\left(-4 \pm \sqrt{35}, -3 \right)$

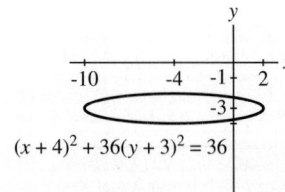

$(x + 4)^2 + 36(y + 3)^2 = 36$

27. $\dfrac{x^2}{16} + \dfrac{y^2}{4} = 1, \left(\pm 2\sqrt{3}, 0\right)$

29. $\dfrac{(x+1)^2}{4} + \dfrac{(y+2)^2}{16} = 1, \left(-1, -2 \pm 2\sqrt{3}\right)$

31. $x^2 + y^2 = 4$ **33.** $x^2 + y^2 = 41$

35. $(x-2)^2 + (y+3)^2 = 20$ **37.** $(x-1)^2 + (y-3)^2 = 5$

39. $(0, 0), 10$ **41.** $(1, 2), 2$

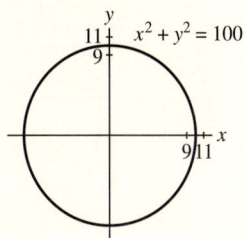
$x^2 + y^2 = 100$

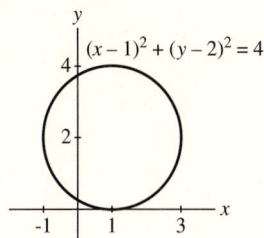
$(x-1)^2 + (y-2)^2 = 4$

43. $(-2, -2), 2\sqrt{2}$

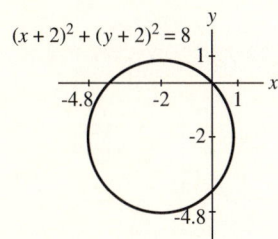
$(x+2)^2 + (y+2)^2 = 8$

45. $(0, -1), 3$ **47.** $(-4, 5), \sqrt{41}$ **49.** $(-2, 0), 3$

51. $(0.5, -0.5), 1$ **53.** $(-1/3, -1/6), 1/2$

55. $(-1, 0), \sqrt{6}/2$ **57.** Circle   **59.** Ellipse

61. Parabola   **63.** Parabola   **65.** Ellipse   **67.** Circle

69. $\dfrac{x^2}{16} + \dfrac{y^2}{12} = 1$ **71.** $(12, 0)$ **73.** $\dfrac{x^2}{100} + \dfrac{y^2}{75} = 1$

75. 5.25×10^9 km   **77.** Parabolic

Section 11.3

1. $\left(\pm\sqrt{13}, 0\right), y = \pm\dfrac{3}{2}x$ **3.** $\left(0, \pm\sqrt{29}\right), y = \pm\dfrac{2}{5}x$

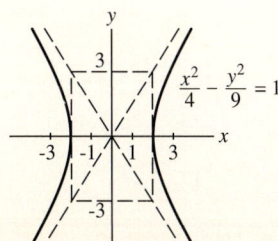
$\dfrac{x^2}{4} - \dfrac{y^2}{9} = 1$

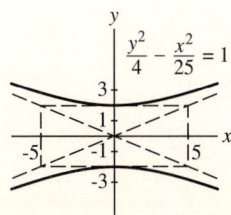
$\dfrac{y^2}{4} - \dfrac{x^2}{25} = 1$

5. $\left(\pm\sqrt{5}, 0\right), y = \pm\dfrac{1}{2}x$ **7.** $\left(\pm\sqrt{10}, 0\right), y = \pm 3x$

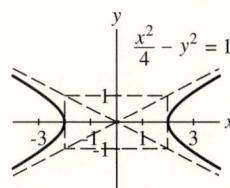
$\dfrac{x^2}{4} - y^2 = 1$

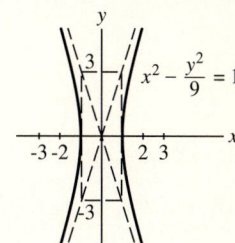
$x^2 - \dfrac{y^2}{9} = 1$

9. $(\pm 5, 0), y = \pm\dfrac{4}{3}x$ **11.** $\left(\pm\sqrt{2}, 0\right), y = \pm x$

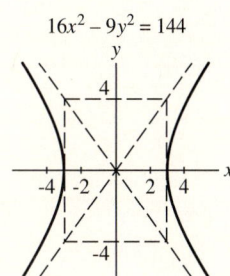
$16x^2 - 9y^2 = 144$

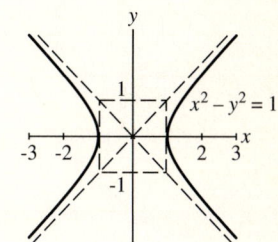
$x^2 - y^2 = 1$

13. $\left(-1 \pm \sqrt{13}, 2\right), y = \dfrac{3}{2}x + \dfrac{7}{2}, y = -\dfrac{3}{2}x + \dfrac{1}{2}$

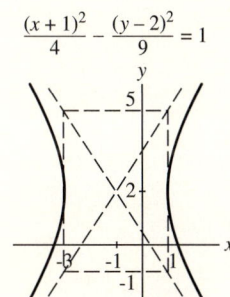
$\dfrac{(x+1)^2}{4} - \dfrac{(y-2)^2}{9} = 1$

15. $\left(-2, 1 \pm \sqrt{5}\right), y = 2x + 5, y = -2x - 3$

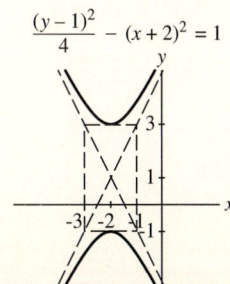
$\dfrac{(y-1)^2}{4} - (x+2)^2 = 1$

17. $(3, 3)$, $(-7, 3)$, $y = \dfrac{3}{4}x + \dfrac{9}{2}$, $y = -\dfrac{3}{4}x + \dfrac{3}{2}$

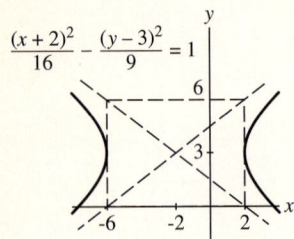

$\dfrac{(x+2)^2}{16} - \dfrac{(y-3)^2}{9} = 1$

19. $\left(3, 3 \pm \sqrt{2}\right)$, $y = x$, $y = -x + 6$

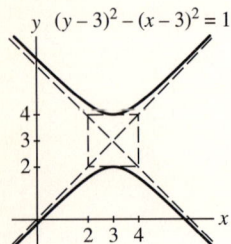

$(y-3)^2 - (x-3)^2 = 1$

21. $\dfrac{x^2}{36} - \dfrac{y^2}{9} = 1$ **23.** $\dfrac{x^2}{9} - \dfrac{y^2}{16} = 1$ **25.** $\dfrac{x^2}{9} - \dfrac{y^2}{25} = 1$

27. $\dfrac{x^2}{9} - \dfrac{y^2}{16} = 1$ **29.** $\dfrac{y^2}{9} - \dfrac{(x-1)^2}{9} = 1$

31. $y^2 - (x-1)^2 = 1$, hyperbola

33. $y = x^2 + 2x$, parabola   **35.** $x^2 + y^2 = 100$, circle

37. $x = -4y^2 + 100$, parabola

39. $(x-1)^2 + (y-2)^2 = \dfrac{1}{2}$, circle

41. $\dfrac{(x+1)^2}{2} + \dfrac{(y+3)^2}{4} = 1$, ellipse   **43.** $\dfrac{y^2}{64} - \dfrac{x^2}{36} = 1$

45. $\left(\dfrac{4\sqrt{14}}{7}, \dfrac{3\sqrt{7}}{7}\right)$

Graphing Calculator Exercises

1. 0.01

Chapter 11 Review Exercises

1. $(-2, -16)$, $x = -2$
$(-2, -63/4)$, $y = -65/4$

3. $(3/2, 9/2)$, $x = 3/2$
$(3/2, 35/8)$, $y = 37/8$

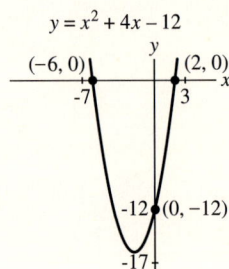

$y = x^2 + 4x - 12$

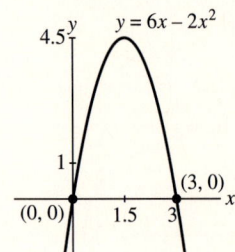

$y = 6x - 2x^2$

5. $(-10, -2)$, $y - -2$
$(-39/4, -2)$, $x = -41/4$

7. $\left(0, \pm 2\sqrt{5}\right)$

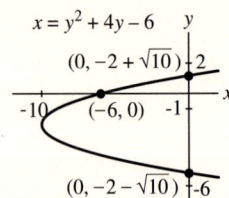

$x = y^2 + 4y - 6$

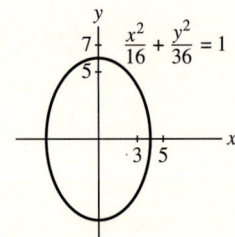

$\dfrac{x^2}{16} + \dfrac{y^2}{36} = 1$

9. $(1, 5)$, $(1, -3)$

11. $\left(1, -3 \pm \sqrt{2}\right)$

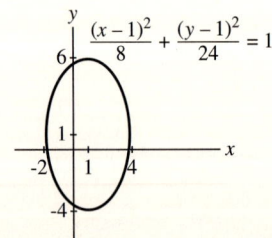

$\dfrac{(x-1)^2}{8} + \dfrac{(y-1)^2}{24} = 1$

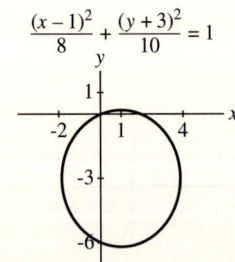

$\dfrac{(x-1)^2}{8} + \dfrac{(y+3)^2}{10} = 1$

13. $(0, 0)$, 9

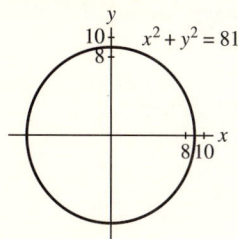
$x^2 + y^2 = 81$

15. $(-1, 0)$, 2

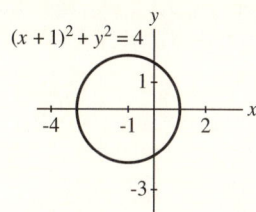
$(x + 1)^2 + y^2 = 4$

29. $\left(2 \pm \sqrt{5}, 4\right)$, $y = \dfrac{1}{2}x + 3$, $y = -\dfrac{1}{2}x + 5$

$$\frac{(x - 2)^2}{4} - (y - 4)^2 = 1$$

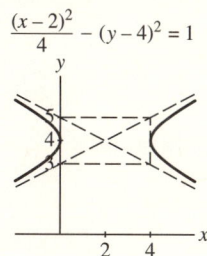

17. $(-5/2, 0)$, $\sqrt{6}$

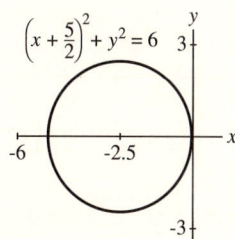
$\left(x + \dfrac{5}{2}\right)^2 + y^2 = 6$

31. Hyperbola   **33.** Ellipse   **35.** Parabola   **37.** Hyperbola
39. $x^2 + y^2 = 4$

41. $y = \dfrac{1}{4}x^2 - 1$

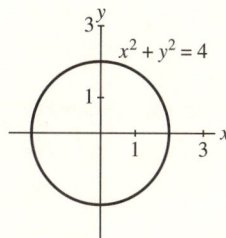
$x^2 + y^2 = 4$

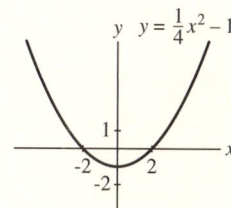
$y = \dfrac{1}{4}x^2 - 1$

19. $x^2 + (y + 4)^2 = 9$ **21.** $(x + 2)^2 + (y + 7)^2 = 6$

23. $(x - 1)^2 + \left(y + \dfrac{1}{3}\right)^2 = \dfrac{3}{4}$ **25.** $(\pm 10, 0)$, $y = \pm \dfrac{3}{4}x$

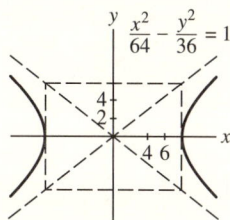
$\dfrac{x^2}{64} - \dfrac{y^2}{36} = 1$

43. $\dfrac{x^2}{4} + y^2 = 1$

45. $y^2 + \dfrac{(x - 2)^2}{4} = 1$

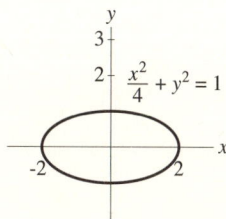
$\dfrac{x^2}{4} + y^2 = 1$

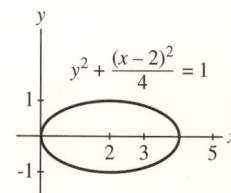
$y^2 + \dfrac{(x - 2)^2}{4} = 1$

27. $\left(4, 2 \pm 4\sqrt{5}\right)$, $y = 2x - 6$, $y = -2x + 10$

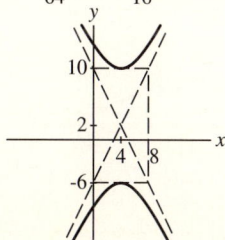
$\dfrac{(y - 2)^2}{64} - \dfrac{(x - 4)^2}{16} = 1$

47. $x = (y - 3)^2 + \dfrac{3}{4}$ **49.** $\dfrac{x^2}{36} + \dfrac{y^2}{20} = 1$

51. $(x - 1)^2 + (y - 3)^2 = 20$ **53.** $\dfrac{x^2}{4} - \dfrac{y^2}{5} = 1$

55. $(x + 2)^2 + (y - 3)^2 = 9$

57. $\dfrac{(x + 2)^2}{9} + (y - 1)^2 = 1$

59. $\dfrac{(y-1)^2}{9} - \dfrac{(x-2)^2}{4} = 1$

61. $\left\{\left(\dfrac{25}{12}, \dfrac{\sqrt{119}}{4}\right)\left(\dfrac{25}{12}, -\dfrac{\sqrt{119}}{4}\right)\left(-\dfrac{25}{12}, \dfrac{\sqrt{119}}{4}\right)\right.$

$\left. \left(-\dfrac{25}{12}, -\dfrac{\sqrt{119}}{4}\right)\right\}$

63. $\dfrac{x^2}{100^2} - \dfrac{y^2}{120^2} = 1$ **65.** $\dfrac{x^2}{34^2} + \dfrac{y^2}{16^2} = 1$, 10.81 ft

Chapter 11 Test

1.

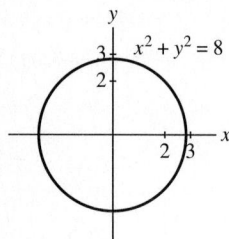

2.

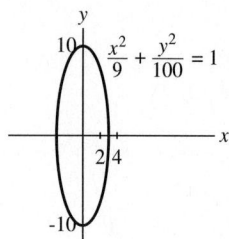

3.

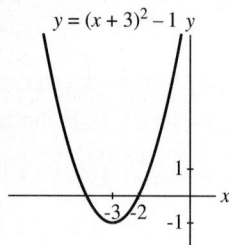

4.

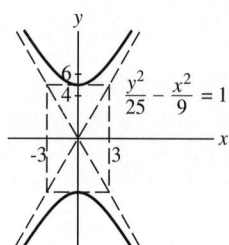

5.

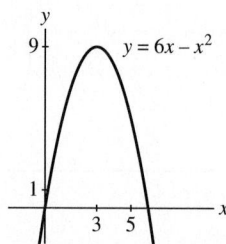

6.

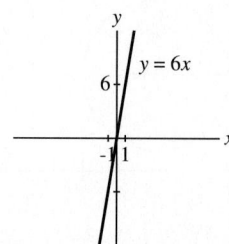

7. Hyperbola   **8.** Parabola   **9.** Circle   **10.** Ellipse

11. $(x+3)^2 + (y-4)^2 = 12$ **12.** $x = \dfrac{1}{8}y^2$

13. $\dfrac{x^2}{4} + \dfrac{y^2}{10} = 1$ **14.** $\dfrac{x^2}{36} - \dfrac{y^2}{28} = 1$

15. $(2, -15/4)$, $y = -17/4$, $(2, -4)$, $x = 2$

16. $(\pm 2\sqrt{3}, 0)$, 8, 4

17. $\left(0, \pm\sqrt{17}\right)$, $(0, \pm 1)$, $y = \pm\dfrac{1}{4}x$, 2, 8

18. $\left(-\dfrac{1}{2}, \dfrac{3}{2}\right), \dfrac{3}{2}$ **19.** 24 cm

Tying It All Together Chapters 1–11

1.

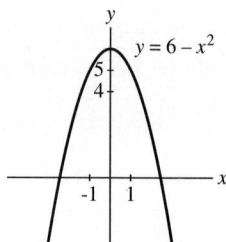

2.

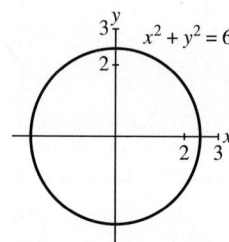

3.

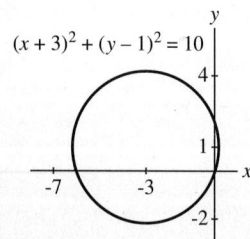

4.

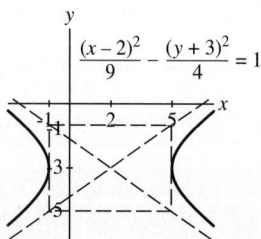

5.

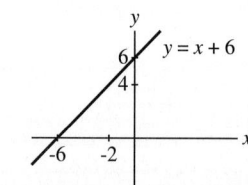

6.

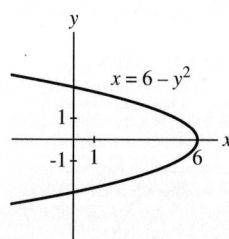

7.

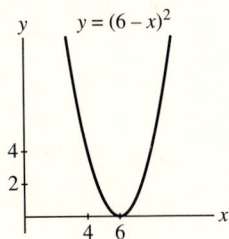

8.

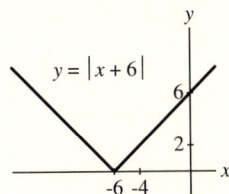

9.

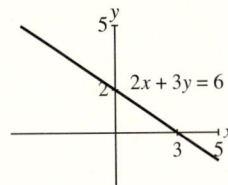

10.

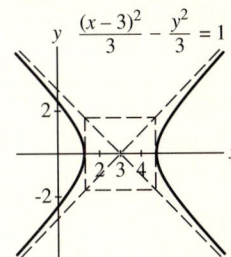

11.

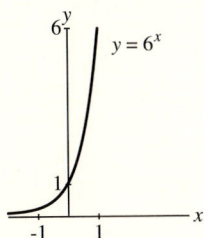

12.

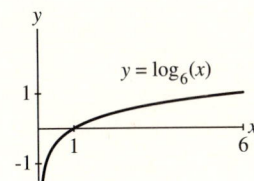

13.

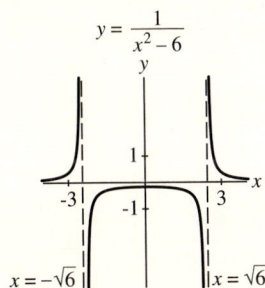

14.

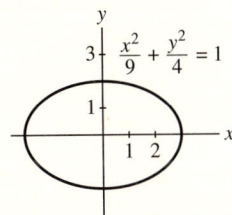

15.

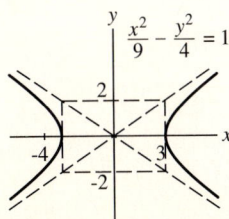

16.

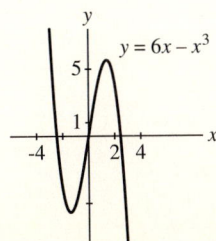

17. $\{-5/3\}$

18. $\{-5/3\}$ **19.** $\{11/15\}$ **20.** $\{0\}$

21. $\{22/3\}$ **22.** $\{50\}$

CHAPTER 12

Section 12.1

1. 1, 4, 9, 16, 25, 36, 49

3. $\dfrac{1}{2}, -\dfrac{1}{3}, \dfrac{1}{4}, -\dfrac{1}{5}, \dfrac{1}{6}, -\dfrac{1}{7}, \dfrac{1}{8}, -\dfrac{1}{9}$

5. $1, -2, 4, -8, 16, -32$ **7.** $2, 1, \dfrac{1}{2}, \dfrac{1}{4}, \dfrac{1}{8}$

9. $-6, -10, -14, -18, -22$ **11.** 5, 5.5, 6, 6.5, 7, 7.5, 8

13. $1, 2, \dfrac{3}{2}, \dfrac{2}{3}, \dfrac{5}{24}$ **15.** 1, 1, 2, 6, 24, 120, 720

17. $8, 5, 2, -1; -19$ **19.** $\dfrac{4}{3}, \dfrac{4}{5}, \dfrac{4}{7}, \dfrac{4}{9}; \dfrac{4}{21}$

21. $-\dfrac{1}{6}, \dfrac{1}{12}, -\dfrac{1}{20}, \dfrac{1}{30}; \dfrac{1}{132}$ **23.** $2, 2, \dfrac{4}{3}, \dfrac{2}{3}; \dfrac{4}{14,175}$

25. $-2, -8, -16, -\dfrac{64}{3}; -\dfrac{4096}{2835}$ **27.** 8.9, 8.8, 8.7, 8.6; 8

29. $-4, -10, -28, -82; -6,562$ **31.** $2, 1, -2, 1; 1$

33. $-15, -8, -1, 6; 34$ **35.** $a_n = 2n$ **37.** $a_n = 2n + 7$

39. $a_n = (-1)^{n+1}$ **41.** $a_n = n^3$ **43.** $a_n = e^n$

45. $a_n = \dfrac{1}{2^{n-1}}$ **47.** Yes   **49.** No   **51.** No   **53.** Yes

55. $a_n = 5n - 4$ **57.** $a_n = 2n - 2$ **59.** $a_n = -4n + 9$

61. $a_n = 0.1n + 0.9$ **63.** $a_n = \dfrac{\pi}{6}n$ **65.** $a_n = 15n + 5$

67. $6, 3, 0, -3; -21$ **69.** 1, 0.9, 0.8, 0.7; 0.1

71. $\dfrac{14}{3}, \dfrac{13}{3}, \dfrac{12}{3}, \dfrac{11}{3}; \dfrac{5}{3}$ **73.** 32   **75.** 27   **77.** 4.2

79. $a_n = 2.5n + 2.5$ **81.** $a_n = a_{n-1} + 9, a_1 = 3$

83. $a_n = 3a_{n-1}, a_1 = 1/3$ **85.** $a_n = \sqrt{a_{n-1}}, a_1 = 16$

87. $21,200, $22,472, $23,820, $25,250, $26,765   **89.** 92

91. $65,930   **93.** 0.00136

Graphing Calculator Exercises

1. 0.02

Section 12.2

1. 55   **3.** 36   **5.** 15/32   **7.** 485   **9.** 1   **11.** 0

13. $\displaystyle\sum_{i=1}^{6} i$ **15.** $\displaystyle\sum_{i=1}^{5} (-1)^i(2i-1)$ **17.** $\displaystyle\sum_{i=1}^{5} i^2$

19. $\displaystyle\sum_{i=1}^{5} \left(-\frac{1}{2}\right)^{i-1}$ **21.** $\displaystyle\sum_{i=1}^{3} \ln(x_i)$ **23.** $\displaystyle\sum_{i=1}^{11} ar^{i-1}$

25. $\displaystyle\sum_{j=0}^{31} (-1)^{j+1}$ **27.** $\displaystyle\sum_{j=1}^{10} (2j+7)$

29. $\displaystyle\sum_{j=0}^{8} \frac{10!}{(j+2)!(8-j)!}$ **31.** $\displaystyle\sum_{j=5}^{9} \frac{5^{j-3}e^{-5}}{(j-3)!}$

33. $0.5 + 0.5r + 0.5r^2 + 0.5r^3 + 0.5r^4 + 0.5r^5$

35. $a^4 + a^3b + a^2b^2 + ab^3 + b^4$ **37.** $a^2 + 2ab + b^2$

39. 74/3   **41.** 16   **43.** 14.151   **45.** 1128   **47.** -36

49. 406   **51.** 52.25   **53.** 360   **55.** 1.3

57. \$1,335,000, \$44,500   **59.** $\displaystyle\sum_{i=1}^{10} 1000(1.05)^i$

61. $\displaystyle\sum_{i=1}^{9} 12i = 540$ **63.** 228.5   **65.** $-1/2$

Section 12.3

1. $a_n = \frac{1}{6}2^{n-1}$ **3.** $a_n = 0.9(0.1)^{n-1}$ **5.** $a_n = 4(-3)^{n-1}$

7. Arithmetic   **9.** Geometric   **11.** Neither

13. 2, 4, 6, 8; arithmetic   **15.** 1, 4, 9, 16; neither

17. $\frac{1}{2}, \frac{1}{4}, \frac{1}{8}, \frac{1}{16}$; geometric

19. 8, 32, 128, 512; geometric

21. 3, -9, 27, -81; geometric   **23.** 3, -6, 12; -1536

25. 4, 0.4, 0.04; 4×10^{-9} **27.** 11   **29.** 3   **31.** ± 3

33. $a_n = -3(-2)^{n-1}$ **35.** 242/27   **37.** -127.5

39. 31.8343

41. $\displaystyle\sum_{n=1}^{5} 3\left(-\frac{1}{3}\right)^{n-1}$ **43.** $\displaystyle\sum_{n=1}^{\infty} 0.6(0.1)^{n-1}$

45. $\displaystyle\sum_{n=1}^{\infty} -4.5\left(-\frac{1}{3}\right)^{n-1}$ **47.** 9/4   **49.** 1   **51.** $-297/40$

53. 34/99   **55.** No sum   **57.** 2/3   **59.** 20

61. 2/45   **63.** 8172/990   **65.** 1,073,741,823

67. $4000(1.02)^n$, \$8322.74   **69.** \$2561.87

71. \$183,074.35   **73.** 45 ft   **75.** \$8,000,000

Section 12.4

1. 8   **3.** 6   **5.** 1024   **7.** 8   **9.** 210   **11.** 1   **13.** 126

15. 165   **17.** 1820   **19.** 109,736   **21.** 39,916,800

23. 2730   **25.** 6.3×10^{10} **27.** 4096   **29.** 72

31. 15,600   **33.** 30,000   **35.** 24   **37.** 131,072   **39.** 210

Section 12.5

1. 2,598,960   **3.** 13,983,816   **5.** 10   **7.** 27   **9.** 56

11. 1260   **13.** 497,001,600   **15.** 36   **17.** 144   **19.** 210

21. 2520   **23.** 3, 6, 10, $C(n, 2)$ **25.** 2520

27. $a^3 - 6a^2 + 12a - 8$ **29.** $8a^3 + 12a^2b^2 + 6ab^4 + b^6$

31. $x^4 - 8x^3y + 24x^2y^2 - 32xy^3 + 16y^4$

33. $x^8 + 4x^6 + 6x^4 + 4x^2 + 1$ **35.** $x^9 + 9x^8y + 36x^7y^2$

37. $4{,}096x^{12} - 24{,}576x^{11}y + 67{,}584x^{10}y^2$

39. $256s^8 - 512s^7t + 448s^6t^2$

41. $m^{18} - 18m^{16}w^3 + 144m^{14}w^6$ **43.** 56   **45.** $-41{,}184$

47. 13,860   **49.** 120

Section 12.6

1. 1/3, 2/3, 0   **3.** 2/3, 1, 5/6, 0, 1/6   **5.** 1/4, 3/4, 1/4, 3/4

7. 1/36, 11/36, 5/36, 35/36, 1/36

9. 3/13, 9/13, 9/13, 4/13   **11.** 1/72, 1/36, 1/18

13. 1/2,598,960, 1024/2,598,960   **15.** 0.999

17. 1/36, 35/36, 35/36   **19.** 0.05, 0.95   **21.** 4 to 1

23. 1 to 3, 3 to 1   **25.** 1 to 9, 9/10   **27.** 3 to 5

29. 1 to 5   **31.** 1 to 1,999,999   **33.** 1/32

35. 78%   **37.** 1/4   **39.** 4/13   **41.** 34%, 40%, 74%

Section 12.7

1. 1, 2, 3   **3.** 1, 2, 3   **5.** 1, 2   **7.** 2, 3

9. S_1: $2(1) = 1(1 + 1)$, S_k: $\displaystyle\sum_{i=1}^{k} 2i = k(k+1)$,

S_{k+1}: $\displaystyle\sum_{i=1}^{k+1} 2i = (k+1)(k+2)$

11. S_1: $2 = 2(1)^2$, S_k: $2 + 6 + \cdots + (4k-2) = 2k^2$,

S_{k+1}: $2 + 6 + \cdots + (4k+2) = 2(k+1)^2$

13. S_1: $2 = 2^2 - 2$, S_k: $\displaystyle\sum_{i=1}^{k} 2^i = 2^{k+1} - 2$,

S_{k+1}: $\displaystyle\sum_{i=1}^{k+1} 2^i = 2^{k+2} - 2$

15. S_1: $(ab)^1 = a^1b^1$, S_k: $(ab)^k = a^kb^k$,

S_{k+1}: $(ab)^{k+1} = a^{k+1}b^{k+1}$

17. S_1: if $0 < a < 1$ then $0 < a^1 < 1$, S_k: if $0 < a < 1$ then

$0 < a^k < 1$, S_{k+1}: if $0 < a < 1$ then $0 < a^{k+1} < 1$

Chapter 12 Review Exercises

1. 1, 2, 4, 8, 16   **3.** $-1, 1/2, -1/6, 1/24$ **5.** 3, 1.5, 0.75

7. $3, 0, -3$ **9.** 0.9375   **11.** 5450   **13.** 1/3

15. 33,065.9541   **17.** $a_n = \dfrac{(-1)^n}{n+2}$ **19.** $a_n = 6\left(\dfrac{1}{6}\right)^{n-1}$

21. $\displaystyle\sum_{i=1}^{\infty} \dfrac{(-1)^{i+1}}{i+1}$ **23.** $\displaystyle\sum_{i=1}^{14} 2i$ **25.** ± 2 **27.** \$243.52

29. \$13,971.64   **31.** 1000 mi, no

33. $a^4 + 8a^3b + 24a^2b^2 + 32ab^3 + 16b^4$

35. $32a^5 - 80a^4b + 80a^3b^2 - 40a^2b^3 + 10ab^4 - b^5$

37. $a^{10} + 10a^9b + 45a^8b^2$ **39.** $256x^8 + 512x^7y + 448x^6y^2$

41. 715   **43.** 36,960   **45.** 24   **47.** 1,953,125   **49.** 210

51. 60   **53.** 56   **55.** 70, 5, 30   **57.** 180, 120   **59.** 128

61. 1/1024, 1/1024   **63.** 5/13, 8/13, 0, 1   **65.** 1 to 7

67. 9 to 1   **69.** 90%   **71.** 40,320   **73.** 20   **75.** 1680

77. 28   **79.** 1680   **81.** 8

Chapter 12 Test

1. 2.3, 2.8, 3.3, 3.8   **2.** 20, 10, 5, 2.5

3. $a_n = (-1)^{n-1}(n-1)^2$ **4.** $a_n = 3n + 4$

5. $a_n = \dfrac{1}{3}\left(-\dfrac{1}{2}\right)^{n-1}$ **6.** 4023   **7.** 13,050.5997

8. 49   **9.** $a_n = 1.5n - 4.5$ **10.** 3780   **11.** \$445

12. 10,000, 3/10,000   **13.** \$487,521.25, \$515,024.49, no

14. $a^5 - 10a^4x + 40a^3x^2 - 80a^2x^3 + 80ax^4 - 32x^5$

15. $x^{24} + 24x^{23}y^2 + 276x^{22}y^4$ **16.** $\displaystyle\sum_{i=0}^{30} \binom{30}{i} m^{30-i}y^i$

17. 1/6, 1 to 5   **18.** 220   **19.** 2970   **20.** 1/336

Index

Trigonometry

Trigonometric Functions

If the angle α
(in standard position)
intersects the unit circle
at (x, y), then

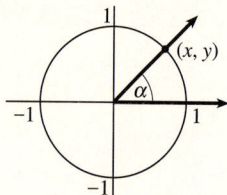

$$\sin \alpha = y \qquad \cos \alpha = x \qquad \tan \alpha = \frac{y}{x}$$

$$\csc \alpha = \frac{1}{y} \qquad \sec \alpha = \frac{1}{x} \qquad \cot \alpha = \frac{x}{y}$$

Trigonometric Ratios

If (x, y) is any point
other than the origin
on the terminal side of α
and $r = \sqrt{x^2 + y^2}$, then

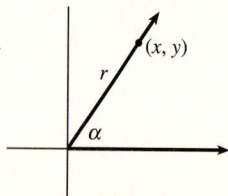

$$\sin \alpha = \frac{y}{r} \qquad \cos \alpha = \frac{x}{r} \qquad \tan \alpha = \frac{y}{x}$$

$$\csc \alpha = \frac{r}{y} \qquad \sec \alpha = \frac{r}{x} \qquad \cot \alpha = \frac{x}{y}$$

Right Triangle Trigonometry

If α is an acute angle
of a right triangle, then

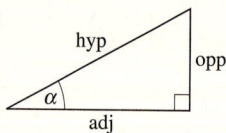

$$\sin \alpha = \frac{\text{opp}}{\text{hyp}} \qquad \cos \alpha = \frac{\text{adj}}{\text{hyp}} \qquad \tan \alpha = \frac{\text{opp}}{\text{adj}}$$

$$\csc \alpha = \frac{\text{hyp}}{\text{opp}} \qquad \sec \alpha = \frac{\text{hyp}}{\text{adj}} \qquad \cot \alpha = \frac{\text{adj}}{\text{opp}}$$

Special Right Triangles

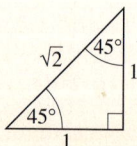

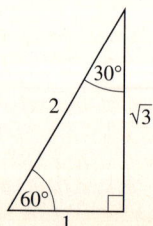

Exact Values of Trigonometric Functions

x degrees	x radians	$\sin x$	$\cos x$	$\tan x$
$0°$	0	0	1	0
$30°$	$\dfrac{\pi}{6}$	$\dfrac{1}{2}$	$\dfrac{\sqrt{3}}{2}$	$\dfrac{\sqrt{3}}{3}$
$45°$	$\dfrac{\pi}{4}$	$\dfrac{\sqrt{2}}{2}$	$\dfrac{\sqrt{2}}{2}$	1
$60°$	$\dfrac{\pi}{3}$	$\dfrac{\sqrt{3}}{2}$	$\dfrac{1}{2}$	$\sqrt{3}$
$90°$	$\dfrac{\pi}{2}$	1	0	—

Basic Identities

$$\tan x = \frac{\sin x}{\cos x} = \frac{1}{\cot x} \qquad\qquad \cot x = \frac{\cos x}{\sin x} = \frac{1}{\tan x}$$

$$\sin x = \frac{1}{\csc x} \qquad\qquad \csc x = \frac{1}{\sin x}$$

$$\cos x = \frac{1}{\sec x} \qquad\qquad \sec x = \frac{1}{\cos x}$$

Pythagorean Identities

$$\sin^2 x + \cos^2 x = 1 \qquad\qquad 1 + \cot^2 x = \csc^2 x$$

$$\tan^2 x + 1 = \sec^2 x$$

Odd Identities

$$\sin(-x) = -\sin(x) \qquad\qquad \csc(-x) = -\csc(x)$$

$$\tan(-x) = -\tan(x) \qquad\qquad \cot(-x) = -\cot(x)$$

Even Identities

$$\cos(-x) = \cos(x) \qquad\qquad \sec(-x) = \sec(x)$$

◈ Trigonometry

Cofunction Identities

$$\sin\left(\frac{\pi}{2} - u\right) = \cos u \qquad \cos\left(\frac{\pi}{2} - u\right) = \sin u$$

$$\tan\left(\frac{\pi}{2} - u\right) = \cot u \qquad \cot\left(\frac{\pi}{2} - u\right) = \tan u$$

$$\sec\left(\frac{\pi}{2} - u\right) = \csc u \qquad \csc\left(\frac{\pi}{2} - u\right) = \sec u$$

Cosine of a Sum or Difference

$$\cos(\alpha + \beta) = \cos \alpha \cos \beta - \sin \alpha \sin \beta$$
$$\cos(\alpha - \beta) = \cos \alpha \cos \beta + \sin \alpha \sin \beta$$

Sine of a Sum or Difference

$$\sin(\alpha + \beta) = \sin \alpha \cos \beta + \cos \alpha \sin \beta$$
$$\sin(\alpha - \beta) = \sin \alpha \cos \beta - \cos \alpha \sin \beta$$

Tangent of a Sum or Difference

$$\tan(\alpha + \beta) = \frac{\tan \alpha + \tan \beta}{1 - \tan \alpha \tan \beta}$$

$$\tan(\alpha - \beta) = \frac{\tan \alpha - \tan \beta}{1 + \tan \alpha \tan \beta}$$

Double-Angle Identities

$$\sin 2x = 2 \sin x \cos x$$
$$\cos 2x = \cos^2 x - \sin^2 x = 2 \cos^2 x - 1 = 1 - 2 \sin^2 x$$

$$\tan 2x = \frac{2 \tan x}{1 - \tan^2 x}$$

Half-Angle Identities

$$\sin \frac{x}{2} = \pm\sqrt{\frac{1 - \cos x}{2}} \qquad \cos \frac{x}{2} = \pm\sqrt{\frac{1 + \cos x}{2}}$$

$$\tan \frac{x}{2} = \pm\sqrt{\frac{1 - \cos x}{1 + \cos x}} = \frac{\sin x}{1 + \cos x} = \frac{1 - \cos x}{\sin x}$$

Product-to-Sum Identities

$$\sin A \cos B = \frac{1}{2}[\sin(A + B) + \sin(A - B)]$$

$$\sin A \sin B = \frac{1}{2}[\cos(A - B) - \cos(A + B)]$$

$$\cos A \sin B = \frac{1}{2}[\sin(A + B) - \sin(A - B)]$$

$$\cos A \cos B = \frac{1}{2}[\cos(A - B) + \cos(A + B)]$$

Sum-to-Product Identities

$$\sin x + \sin y = 2 \sin\left(\frac{x + y}{2}\right)\cos\left(\frac{x - y}{2}\right)$$

$$\sin x - \sin y = 2 \cos\left(\frac{x + y}{2}\right)\sin\left(\frac{x - y}{2}\right)$$

$$\cos x + \cos y = 2 \cos\left(\frac{x + y}{2}\right)\cos\left(\frac{x - y}{2}\right)$$

$$\cos x - \cos y = -2 \sin\left(\frac{x + y}{2}\right)\sin\left(\frac{x - y}{2}\right)$$

Reduction Formula

If α is an angle in standard position whose terminal side contains (a, b), then for any real number x

$$a \sin x + b \cos x = \sqrt{a^2 + b^2}\, \sin(x + \alpha).$$

Oblique Triangle

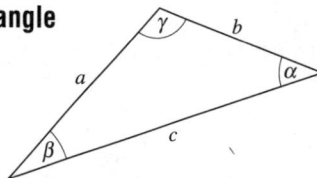

Law of Sines

In any triangle, $\dfrac{\sin \alpha}{a} = \dfrac{\sin \beta}{b} = \dfrac{\sin \gamma}{c}$.

Law of Cosines

$$a^2 = b^2 + c^2 - 2bc \cos \alpha$$
$$b^2 = a^2 + c^2 - 2ac \cos \beta$$
$$c^2 = a^2 + b^2 - 2ab \cos \gamma$$

Algebra

Midpoint Formula

The midpoint of the line segment with endpoints (x_1, y_1) and (x_2, y_2) is

$$\left(\frac{x_1 + x_2}{2}, \frac{y_1 + y_2}{2} \right).$$

Slope Formula

The slope of the line through (x_1, y_1) and (x_2, y_2) is

$$\frac{y_2 - y_1}{x_2 - x_1} \quad \text{(for } x_1 \neq x_2\text{)}.$$

Linear Function

$f(x) = mx + b$ with $m \neq 0$

Graph is a line with slope m.

Quadratic Function

$f(x) = ax^2 + bx + c$ with $a \neq 0$

Graph is a parabola.

Polynomial Function

$f(x) = a_n x^n + a_{n-1} x^{n-1} + \cdots + a_1 x + a_0$ for $a_n \neq 0$
and n a nonnegative integer

Rational Function

$f(x) = \dfrac{p(x)}{q(x)}$, where p and q are polynomial functions
with $q(x) \neq 0$

Exponential and Logarithmic Functions

$f(x) = a^x$ for $a > 0$ and $a \neq 1$

$f(x) = \log_a(x)$ for $a > 0$ and $a \neq 1$

Properties of Logarithms

Base-a logarithm:	$y = \log_a(x) \Leftrightarrow a^y = x$
Natural logarithm:	$y = \ln(x) \Leftrightarrow e^y = x$
Common logarithm:	$y = \log(x) \Leftrightarrow 10^y = x$
One-to-one:	$a^{x_1} = a^{x_2} \Leftrightarrow x_1 = x_2$
	$\log_a(x_1) = \log_a(x_2) \Leftrightarrow x_1 = x_2$

$\log_a(a) = 1 \qquad\qquad \log_a(1) = 0$

$\log_a(a^x) = x \qquad\quad\ a^{\log_a(N)} = N$

$\log_a(MN) = \log_a(M) + \log_a(N)$

$\log_a(M/N) = \log_a(M) - \log_a(N)$

$\log_a(M^x) = x \cdot \log_a(M)$

$\log_a(1/N) = -\log_a(N)$

$$\log_a(M) = \frac{\log_b(M)}{\log_b(a)} = \frac{\ln(M)}{\ln(a)} = \frac{\log(M)}{\log(a)}$$

Compound Interest

$P = $ principal, $t = $ time in years, $r = $ annual interest rate, and $A = $ amount:

$$A = P\left(1 + \frac{r}{n}\right)^{nt} \text{(compounded } n \text{ times/year)}$$

$A = Pe^{rt}$ (compounded continuously)

Variation

Direct: $y = kx$ $\qquad (k \neq 0)$
Inverse: $y = k/x$ $\qquad (k \neq 0)$
Joint: $y = kxz$ $\qquad (k \neq 0)$

Straight Line

Slope-intercept form: $y = mx + b$
Slope: m $\qquad$ y-intercept: $(0, b)$
Point-slope form: $y - y_1 = m(x - x_1)$
Standard form: $Ax + By = C$
Horizontal: $y = k$ $\qquad$ Vertical: $x = k$